STUDENT SOLUTIONS MANUAL

SEVENTH EDITION

CALCULUS EARLY TRANSCENDENTALS

EDWARDS & PENNEY

The University of Georgia, Athens

PEARSON

Prentice
Hall

Upper Saddle River, NJ 07458

Editorial Assistant/Print Supplements Editor: Christine Whitlock
Acquisitions Editor: Adam Jaworski
Vice President & Editorial Director, Mathematics: Christine Hoag
Senior Managing Editor: Linda Behrens
Assistant Managing Editor: Lynn Savino Wendel
Production Editor: Robert Merenoff
Supplement Cover Manager: Paul Gourhan
Supplement Cover Designer: Victoria Colotta
Manufacturing Buyer: Ilene Kahn
Associate Director of Production: Alexis Heydt-Long

© 2008 Pearson Education, Inc.
Pearson Prentice Hall
Pearson Education, Inc.
Upper Saddle River, NJ 07458

All rights reserved. No part of this book may be reproduced in any form or by any means, without permission in writing from the publisher.

Pearson Prentice Hall™ is a trademark of Pearson Education, Inc.

The author and publisher of this book have used their best efforts in preparing this book. These efforts include the development, research, and testing of the theories and programs to determine their effectiveness. The author and publisher make no warranty of any kind, expressed or implied, with regard to these programs or the documentation contained in this book. The author and publisher shall not be liable in any event for incidental or consequential damages in connection with, or arising out of, the furnishing, performance, or use of these programs.

This work is protected by United States copyright laws and is provided solely for teaching courses and assessing student learning. Dissemination or sale of any part of this work (including on the World Wide Web) will destroy the integrity of the work and is not permitted. The work and materials from it should never be made available except by instructors using the accompanying text in their classes. All recipients of this work are expected to abide by these restrictions and to honor the intended pedagogical purposes and the needs of other instructors who rely on these materials.

Printed in the United States of America

10 9 8 7 6 5 4 3

ISBN 13:   978-0-13-614706-0
ISBN 10:   0-13-614706-2

Pearson Education Ltd., *London*
Pearson Education Australia Pty. Ltd., *Sydney*
Pearson Education Singapore, Pte. Ltd.
Pearson Education North Asia Ltd., *Hong Kong*
Pearson Education Canada, Inc., *Toronto*
Pearson Educación de Mexico, S.A. de C.V.
Pearson Education—Japan, *Tokyo*
Pearson Education Malaysia, Pte. Ltd.

Contents

Preface

This manual contains the solutions to the odd-numbered problems in the textbook *Calculus: Early Transcendentals*, 7th Edition (2008) by C. Henry Edwards and David E. Penney.

Many calculus problems can be solved by more than one method, and different methods of solution may yield answers differing in form. We have used the most natural method wherever possible, and only in rare instances have we used instead a "clever" method, and then only when the resulting instructional value justified a departure from more usual practice. With very rare exceptions, we have used only solution methods discussed in the text prior to the statement of any given problem.

We gratefully acknowledge the assistance of both colleagues and student users of earlier editions of this textbook in helping us to correct errors in this edition. For this edition, we especially thank Jill McClain-Wardynski, Kurt Norlin, and Harold Whipple who — under the supervision of Teri Lovelace of Laurel Tech — not only verified the solution of every worked-out example and odd-numbered answer in the book, but made numerous comments and suggestions that greatly improved the presentation of many solutions. For a number of solutions contributed over a period of years, we are indebted to our faculty colleagues Ed Azoff, Brian Boe, Frank Lether, Ted Shifrin, and Robert Varley in the Department of Mathematics at the University of Georgia.

Most of the solutions were checked using various versions of *Maple* or *Mathematica*. Of course, any errors that remain are the responsibility of the authors. We are grateful to all those, especially students, who have written to us, and hope that they will continue to do so. Comments and suggestions can be sent to **h.edwards@mindspring.com**.

C. Henry Edwards
David E. Penney

January 2007

CHAPTER 1

Section 1.1

1.1.1: If $f(x) = \dfrac{1}{x}$, then:

(a) $f(-a) = \dfrac{1}{-a} = -\dfrac{1}{a}$;

(b) $f(a^{-1}) = \dfrac{1}{a^{-1}} = a$;

(c) $f(\sqrt{a}) = \dfrac{1}{\sqrt{a}} = \dfrac{1}{a^{1/2}} = a^{-1/2}$;

(d) $f(a^2) = \dfrac{1}{a^2} = a^{-2}$.

1.1.3: If $f(x) = \dfrac{1}{x^2 + 5}$, then:

(a) $f(-a) = \dfrac{1}{(-a)^2 + 5} = \dfrac{1}{a^2 + 5}$;

(b) $f(a^{-1}) = \dfrac{1}{(a^{-1})^2 + 5} = \dfrac{1}{a^{-2} + 5} = \dfrac{1 \cdot a^2}{a^{-2} \cdot a^2 + 5 \cdot a^2} = \dfrac{a^2}{1 + 5a^2}$;

(c) $f(\sqrt{a}) = \dfrac{1}{\left(\sqrt{a}\right)^2 + 5} = \dfrac{1}{a + 5}$;

(d) $f(a^2) = \dfrac{1}{(a^2)^2 + 5} = \dfrac{1}{a^4 + 5}$.

1.1.5: If $g(x) = 3x + 4$ and $g(a) = 5$, then $3a + 4 = 5$, so $3a = 1$; therefore $a = \frac{1}{3}$.

1.1.7: If $g(x) = \sqrt{x^2 + 16}$ and $g(a) = 5$, then:

$$\sqrt{a^2 + 16} = 5;$$

$$a^2 + 16 = 25;$$

$$a^2 = 9;$$

$$a = 3 \text{ or } a = -3.$$

1.1.9: If $g(x) = \sqrt[3]{x + 25} = (x + 25)^{1/3}$ and $g(a) = 5$, then

$$(a + 25)^{1/3} = 5;$$

$$a + 25 = 5^3 = 125;$$

$$a = 100.$$

1.1.11: If $f(x) = 3x - 2$, then

$$f(a+h) - f(a) = [3(a+h) - 2] - [3a - 2]$$

$$= 3a + 3h - 2 - 3a + 2 = 3h.$$

1.1.13: If $f(x) = x^2$, then

$$f(a+h) - f(a) = (a+h)^2 - a^2$$

$$= a^2 + 2ah + h^2 - a^2 = 2ah + h^2 = h \cdot (2a + h).$$

1.1.15: If $f(x) = \dfrac{1}{x}$, then

$$f(a+h) - f(a) = \frac{1}{a+h} - \frac{1}{a} = \frac{a}{a(a+h)} - \frac{a+h}{a(a+h)}$$

$$= \frac{a - (a+h)}{a(a+h)} = \frac{-h}{a(a+h)}.$$

1.1.17: If $x > 0$ then

$$f(x) = \frac{x}{|x|} = \frac{x}{x} = 1.$$

If $x < 0$ then

$$f(x) = \frac{x}{|x|} = \frac{x}{-x} = -1.$$

We are given $f(0) = 0$, so the range of f is $\{-1, 0, 1\}$. That is, the range of f is the set consisting of the three real numbers -1, 0, and 1.

1.1.19: Given $f(x) = (-1)^{[\![x]\!]}$, we first note that the values of the exponent $[\![x]\!]$ consist of all the integers and no other numbers. So all that matters about the exponent is whether it is an even integer or an odd integer, for if even then $f(x) = 1$ and if odd then $f(x) = -1$. No other values of $f(x)$ are possible, so the range of f is the set consisting of the two numbers -1 and 1.

1.1.21: Given $f(x) = 10 - x^2$, note that for every real number x, x^2 is defined, and for every such real number x^2, $10 - x^2$ is also defined. Therefore the domain of f is the set $\boldsymbol{R}$ of all real numbers.

1.1.23: Given $f(t) = \sqrt{t^2}$, we observe that for every real number t, t^2 is defined and nonnegative, and hence that $\sqrt{t^2}$ is defined as well. Therefore the domain of f is the set $\boldsymbol{R}$ of all real numbers.

1.1.25: Given $f(x) = \sqrt{3x - 5}$, we note that $3x - 5$ is defined for all real numbers x, but that its square root will be defined when and only when $3x - 5$ is nonnegative; that is, when $3x - 5 \geqq 0$, so that $x \geqq \frac{5}{3}$. So the domain of f consists of all those real numbers x in the interval $\left[\frac{5}{3}, +\infty\right)$.

2

1.1.27: Given $f(t) = \sqrt{1 - 2t}$, we observe that $1 - 2t$ is defined for every real number t, but that its square root is defined only when $1 - 2t$ is nonnegative. We solve the inequality $1 - 2t \geq 0$ to find that $f(t)$ is defined exactly when $t \leq \frac{1}{2}$. Hence the domain of f is the interval $\left(-\infty, \frac{1}{2}\right]$.

1.1.29: Given

$$f(x) = \frac{2}{3 - x},$$

we see that $3 - x$ is defined for all real values of x, but that $f(x)$, double its reciprocal, is defined only when $3 - x \neq 0$. So the domain of f consists of those real numbers $x \neq 3$.

1.1.31: Given $f(x) = \sqrt{x^2 + 9}$, observe that for each real number x, $x^2 + 9$ is defined and, moreover, is positive. So its square root is defined for every real number x. Hence the domain of f is the set R of all real numbers.

1.1.33: Given $f(x) = \sqrt{4 - \sqrt{x}}$, note first that we require $x \geq 0$ in order that $\sqrt{x}$ be defined. In addition, we require $4 - \sqrt{x} \geq 0$ so that *its* square root will be defined as well. So we solve [simultaneously] $x \geq 0$ and $\sqrt{x} \leq 4$ to find that $0 \leq x \leq 16$. So the domain of f is the closed interval $[0, 16]$.

1.1.35: Given:

$$g(t) = \frac{t}{|t|}.$$

This fraction will be defined whenever its denominator is nonzero, thus for all real numbers $t \neq 0$. So the domain of g consists of the nonzero real numbers; that is, the union of the two intervals $(-\infty, 0)$ and $(0, +\infty)$.

1.1.37: If a circle has radius r, then its circumference C is given by $C = 2\pi r$ and its area A by $A = \pi r^2$. To express C in terms of A, we first express r in terms of A, then substitute in the formula for C:

$$A = \pi r^2; \qquad r = \sqrt{\frac{A}{\pi}};$$

$$C = 2\pi r = 2\pi \sqrt{\frac{A}{\pi}} = 2\sqrt{\frac{\pi^2 A}{\pi}} = 2\sqrt{\pi A}.$$

Therefore to express C as a function of A, we write

$$C(A) = 2\sqrt{\pi A}, \qquad 0 \leq A < +\infty.$$

It is also permissible simply to write $C(A) = 2\sqrt{\pi A}$ without mentioning the domain, because the "default" domain is correct. In the first displayed equation we do not write $r = \pm\sqrt{A/\pi}$ because we know that r is never negative.

1.1.39: To avoid decimals, we note that a change of 5°C is the same as a change of 9°F, so when the temperature is 10°C it is $32 + 18 = 50°F$; when the temperature is 20°C then it is $32 + 2 \cdot 18 = 68°F$. In general we get the Fahrenheit temperature F by adding 32 to the product of $\frac{1}{10}C$ and 18, where C is the Celsius temperature. That is,

$$F = 32 + \frac{9}{5}C,$$

and therefore $C = \frac{5}{9}(F - 32)$. Answer:

$$C(F) = \frac{5}{9}(F - 32), \quad F > -459.67.$$

1.1.41: Let y denote the height of such a rectangle. The rectangle is inscribed in a circle of diameter 4, so the bottom side x and the left side y are the two legs of a right triangle with hypotenuse 4. Consequently $x^2 + y^2 = 16$, so $y = \sqrt{16 - x^2}$ (not $-\sqrt{16 - x^2}$ because $y \geqq 0$). Because $x \geqq 0$ and $y \geqq 0$, we see that $0 \leqq x \leqq 4$. The rectangle has area $A = xy$, so

$$A(x) = x\sqrt{16 - x^2}, \quad 0 \leqq x \leqq 4.$$

1.1.43: The square base of the box measures x by x centimeters; let y denote its height (in centimeters). Because the volume of the box is 324 cm³, we see that $x^2 y = 324$. The base of the box costs $2x^2$ cents, each of its four sides costs xy cents, and its top costs x^2 cents. So the total cost of the box is

$$C = 2x^2 + 4xy + x^2 = 3x^2 + 4xy. \tag{1}$$

Because $x > 0$ and $y > 0$ (the box has positive volume), but because y can be arbitrarily close to zero (as well as x), we see also that $0 < x < +\infty$. We use the equation $x^2 y = 324$ to eliminate y from Eq. (1) and thereby find that

$$C(x) = 3x^2 + \frac{1296}{x}, \quad 0 < x < +\infty.$$

1.1.45: Let h denote the height of the cylinder. Its radius is r, so its volume is $\pi r^2 h = 1000$. The total surface area of the cylinder is

$$A = 2\pi r^2 + 2\pi rh \quad \text{(look inside the front cover of the book)};$$

$$h = \frac{1000}{\pi r^2}, \quad \text{so}$$

$$A = 2\pi r^2 + 2\pi r \cdot \frac{1000}{\pi r^2} = 2\pi r^2 + \frac{2000}{r}.$$

Now r cannot be negative; r cannot be zero, else $\pi r^2 h \neq 1000$. But r can be arbitrarily small positive as well as arbitrarily large positive (by making h sufficiently close to zero). Answer:

$$A(r) = 2\pi r^2 + \frac{2000}{r}, \quad 0 < r < +\infty.$$

4

1.1.47: The base of the box will be a square measuring $50 - 2x$ in. on each side, so the open-topped box will have that square as its base and four rectangular sides each measuring $50 - 2x$ by x (the height of the box). Clearly $0 \leq x$ and $2x \leq 50$. So the volume of the box will be

$$V(x) = x(50 - 2x)^2, \quad 0 \leq x \leq 25.$$

1.1.49: Recall that the total daily production of the oil field is $p(x) = (20 + x)(200 - 5x)$ if x new wells are drilled (where x is an integer satisfying $0 \leq x \leq 40$). Here is a table of *all* of the values of the production function p:

x	0	1	2	3	4	5	6	7
p	4000	4095	4180	4255	4320	4375	4420	4455

x	8	9	10	11	12	13	14	15
p	4480	4495	4500	4495	4480	4455	4420	4375

x	16	17	18	19	20	21	22	23
p	4320	4255	4180	4095	4000	3895	3780	3655

x	24	25	26	27	28	29	30	31
p	3520	3375	3220	3055	2880	2695	2500	2295

x	32	33	34	35	36	37	38	39
p	2080	1855	1620	1375	1120	855	580	295

and, finally, $p(40) = 0$. Answer: Drill ten new wells.

1.1.51: If x is an integer, then $\mathrm{CEILING}(x) = x$ and $-\mathrm{FLOOR}(-x) = -(-x) = x$. If x is not an integer, then choose the integer n so that $n < x < n + 1$. Then $\mathrm{CEILING}(x) = n + 1$, $-(n + 1) < -x < -n$, and

$$-\mathrm{FLOOR}(-x) = -[-(n + 1)] = n + 1.$$

In both cases we see that $\mathrm{CEILING}(x) = -\mathrm{FLOOR}(-x)$.

1.1.53: By the result of Problem 52, the range of $\mathrm{ROUND}(10x)$ is the set of all integers, so the range of $g(x) = \frac{1}{10}\mathrm{ROUND}(10x)$ is the set of all integral multiple of $\frac{1}{10}$.

1.1.55: Let $\mathrm{ROUND4}(x) = \frac{1}{10000}\mathrm{ROUND}(10000x)$. To verify that $\mathrm{ROUND4}$ has the desired property for [say] positive values of x, write such a number x in the form

$$x = k + \frac{d_1}{10} + \frac{d_2}{100} + \frac{d_3}{1000} + \frac{d_4}{10000} + \frac{d_5}{100{,}000} + r,$$

where k is a nonnegative integer, each d_i is an integer between 0 and 9, and $0 \leqq r < 0.00001$. Application of ROUND4 to x then produces

$$\frac{1}{10000}\text{FLOOR}(10000k + 1000d_1 + 100d_2 + 10d_3 + d_4 + \tfrac{1}{10}(d_5 + 5) + 10000r).$$

Then consideration of the two cases $0 \leqq d_5 \leqq 4$ and $5 \leqq d_5 \leqq 9$ will show that ROUND4 produces the correct four-place rounding of x in both cases.

1.1.57:

x	0.0	0.2	0.4	0.6	0.8	1.0
y	1.0	0.44	−0.04	−0.44	−0.76	−1.0

The sign change occurs between $x = 0.2$ and $x = 0.4$.

x	0.20	0.25	0.30	0.35	0.40
y	0.44	0.3125	0.19	0.0725	−0.04

The sign change occurs between $x = 0.35$ and $x = 0.40$.

x	0.35	0.36	0.37	0.38	0.39	0.40
y	0.0725	0.0496	0.0269	0.0044	−0.0179	−0.04

From this point on, the data for y will be rounded.

x	0.380	0.382	0.384	0.386	0.388	0.390
y	0.0044	−0.0001	−0.0045	−0.0090	−0.0135	−0.0179

Answer (rounded to two places): 0.38. The quadratic formula yields the two roots $\frac{1}{2}\left(3 \pm \sqrt{5}\right)$; the smaller of these is approximately 0.3819660112250105151795.

1.1.59: The sign change intervals are $[1, 2]$, $[1.2, 1.4]$, $[1.20, 1.24]$, $[1.232, 1.240]$, and $[1.2352, 1.2368]$. Answer: $-1 + \sqrt{5} \approx 1.24$.

1.1.61: The sign change intervals are $[0, 1]$, $[0.6, 0.8]$, $[0.68, 0.72]$, $[0.712, 0.720]$, and $[0.7184, 0.7200]$. Answer: $\frac{1}{4}\left(7 - \sqrt{17}\right) \approx 0.72$.

1.1.63: The sign change intervals are $[3, 4]$, $[3.2, 3.4]$, $[3.20, 3.24]$, $[3.208, 3.216]$, and $[3.2080, 3.2096]$. Answer: $\frac{1}{2}\left(11 - \sqrt{21}\right) \approx 3.21$.

1.1.65: The sign change intervals are $[1, 2]$, $[1.6, 1.8]$, $[1.60, 1.64]$, $[1.608, 1.616]$, $[1.6144, 1.6160]$, and $[1.61568, 1.61600]$. Answer: $\frac{1}{6}\left(-23 + \sqrt{1069}\right) \approx 1.62$.

Section 1.2

1.2.1: The slope of L is $m = (3-0)/(2-0) = \frac{3}{2}$, so L has equation

$$y - 0 = \frac{3}{2}(x - 0); \qquad \text{that is,} \qquad 2y = 3x.$$

1.2.3: Because L is horizontal, it has slope zero. Hence an equation of L is

$$y - (-5) = 0 \cdot (x - 3); \qquad \text{that is,} \qquad y = -5.$$

1.2.5: The slope of L is $(3 - (-3))/(5 - 2) = 2$, so an equation of L is

$$y - 3 = 2(x - 5); \qquad \text{that is,} \qquad y = 2x - 7.$$

1.2.7: The slope of L is $\tan(135°) = -1$, so L has equation

$$y - 2 = -1 \cdot (x - 4); \qquad \text{that is,} \qquad x + y = 6.$$

1.2.9: The second line's equation can be written in the form $y = -2x + 10$ to show that it has slope -2. Because L is parallel to the second line, L also has slope -2 and thus equation $y - 5 = -2(x - 1)$.

1.2.11: $x^2 - 4x + 4 + y^2 = 4$: $(x - 2)^2 + (y - 0)^2 = 2^2$. Center $(2, 0)$, radius 2.

1.2.13: $x^2 + 2x + 1 + y^2 + 2y + 1 = 4$: $(x + 1)^2 + (y + 1)^2 = 2^2$. Center $(-1, -1)$, radius 2.

1.2.15: $x^2 + y^2 + x - y = \frac{1}{2}$: $x^2 + x + \frac{1}{4} + y^2 - y + \frac{1}{4} = 1$; $(x + \frac{1}{2})^2 + (y - \frac{1}{2})^2 = 1$. Center: $(-\frac{1}{2}, \frac{1}{2})$, radius 1.

1.2.17: $y = (x - 3)^2$: Opens upward, vertex at $(3, 0)$.

1.2.19: $y - 3 = (x + 1)^2$: Opens upward, vertex at $(-1, 3)$.

1.2.21: $y = 5(x^2 + 4x + 4) + 3 = 5(x + 2)^2 + 3$: Opens upward, vertex at $(-2, 3)$.

1.2.23: $x^2 - 6x + 9 + y^2 + 8y + 16 = 25$: $(x - 3)^2 + (y + 4)^2 = 5^5$. Circle, center $(3, -4)$, radius 5.

1.2.25: $(x + 1)^2 + (y + 3)^2 = -10$: There are no points on the graph.

1.2.27: The graph is the straight line segment connecting the two points $(-1, 7)$ and $(1, -3)$ (including those two points).

1.2.29: The graph is the parabola that opens downward, symmetric around the y-axis, with vertex at $(0, 10)$ and x-intercepts $\pm\sqrt{10}$.

1.2.31: The graph of $y = x^3$ can be visualized by modifying the familiar graph of the parabola with equation $y = x^2$: The former may be obtained by multiplying the y-coordinate of the latter's point (x, x^2) by x. Thus both have flat spots at the origin. For $0 < x < 1$, the graph of $y = x^3$ is below that of $y = x^2$. They cross at $(1, 1)$, and for $x > 1$ the graph of $y = x^2$ is below that of $y = x^3$, with the difference becoming arbitrarily large as x increases without bound. If the graph of $y = x^3$ for $x \geqq 0$ is rotated $180°$ around the point $(0, 0)$, the graph of $y = x^3$ for $x < 0$ is the result.

1.2.33: To graph $y = f(x) = \sqrt{4 - x^2}$, note that $y \geq 0$ and that $y^2 = 4 - x^2$; that is, $x^2 + y^2 = 4$. Hence the graph of f is the *upper half* of the circle with center $(0, 0)$ and radius 2.

1.2.35: To graph $f(x) = \sqrt{x^2 - 9}$, note that there is no graph for $-3 < x < 3$, that $f(\pm 3) = 0$, and that $f(x) > 0$ for $x < -3$ and for $x > 3$. If x is large positive, then $\sqrt{x^2 - 9} \approx \sqrt{x^2} = x$, so the graph of f has x-intercept $(3, 0)$ and rises as x increases, nearly coinciding with the graph of $y = x$ for x large positive. The case $x < -3$ is trickier. In this case, if x is a large negative number, then $f(x) = \sqrt{x^2 - 9} \approx \sqrt{x^2} = -x$ (Note the minus sign!). So for $x \leqq -3$, the graph of f has x-intercept $(-3, 0)$ and, for x large negative, almost coincides with the graph of $y = -x$. Later we will see that the graph of f becomes arbitrarily steep as x gets closer and closer to ± 3.

1.2.37: Note that $f(x)$ is positive and close to zero for x large positive, so that the graph of f is just above the x-axis—and nearly coincides with it—for such x. Similarly, the graph of f is just below the x-axis and nearly coincides with it for x large negative. There is no graph where $x = -2$, but if x is slightly greater than -2 then $f(x)$ is the reciprocal of a very small positive number, so $f(x)$ is large and nearly coincides with the upper half of the vertical line $x = -2$. Similarly, if x is slightly less than -2, then the graph of $f(x)$ is large negative and nearly coincides with the the lower half of the line $x = -3$. The graph of f is decreasing for $x < -2$ and for $x > -2$ and its only intercept is the y-intercept $\left(0, \frac{1}{2}\right)$.

1.2.39: Note that $f(x) > 0$ for all x other than $x = 1$, where f is not defined. If $|x|$ is large, then $f(x)$ is near zero, so the graph of f almost coincides with the x-axis for such x. If x is very close to 1, then $f(x)$ is the reciprocal of a very small positive number, hence $f(x)$ is large positive. So for such x, the graph of $f(x)$ almost coincides with the upper half of the vertical line $x = 1$. The only intercept is $(0, 1)$.

1.2.41: Note that $f(x)$ is undefined when $2x + 3 = 0$; that is, when $x = -\frac{3}{2}$. If x is large positive, then $f(x)$ is positive and close to zero, so the graph of f is slightly above the x-axis and almost coincides with the x-axis. If x is large negative, then $f(x)$ is negative and close to zero, so the graph of f is slightly below the x-axis and almost coincides with the x-axis. If x is slightly greater than $-\frac{3}{2}$ then $f(x)$ is very large positive, so the graph of f almost coincides with the upper half of the vertical line $x = -\frac{3}{2}$. If x is slightly less than $-\frac{3}{2}$ then $f(x)$ is very large negative, so the graph of f almost coincides with the lower half of that vertical line. The graph of f is decreasing for $x < -\frac{3}{2}$ and also decreasing for $x > -\frac{3}{2}$. The only intercept is at $\left(0, \frac{1}{3}\right)$.

1.2.43: Given $y = f(x) = \sqrt{1 - x}$, note that $y \geq 0$ and that $y^2 = 1 - x$; that is, $x = 1 - y^2$. So the graph is the part of the parabola $x = 1 - y^2$ for which $y \geq 0$. This parabola has horizontal axis of symmetry the y-axis, opens to the left (because the coefficient of y^2 is negative), and has vertex $(1, 0)$. Therefore the graph of f is the upper half of this parabola.

1.2.45: Note that $f(x)$ is defined only if $2x + 3 > 0$; that is, if $x > -\frac{3}{2}$. Note also that $f(x) > 0$ for all such x. If x is large positive, then $f(x)$ is positive but near zero, so the graph of f is just above the x-axis and almost coincides with it. If x is very close to $-\frac{3}{2}$ (but larger), then the denominator in $f(x)$ is very tiny positive, so the graph of f almost coincides with the upper half of the vertical line $x = -\frac{3}{2}$ for such x. The graph of f is decreasing for all $x > -\frac{3}{2}$.

1.2.47: Given: $f(x) = |x| + x$. If $x \geq 0$ then $f(x) = x + x = 2x$, so if $x \geq 0$ then the graph of f is the part of the straight line through $(0, 0)$ with slope 2 for which $x \geq 0$. If $x < 0$ then $f(x) = -x + x = 0$, so the rest of the graph of f coincides with the negative x-axis.

1.2.49: Given: $f(x) = |2x + 5|$. The two cases are determined by the point where $2x + 5$ changes sign, which is where $x = -\frac{5}{2}$. If $x \geq -\frac{5}{2}$, then $f(x) = 2x + 5$, so the graph of f consists of the part of the line with slope 2 and y-intercept 5 for which $x \geq -\frac{5}{2}$. If $x < -\frac{5}{2}$, then the graph of f is the part of the straight line $y = -2x - 5$ for which $x < -\frac{5}{2}$.

1.2.51: The graph consists of the horizontal line $y = 0$ for $x < 0$ together with the horizontal line $y = 1$ for $x \geq 0$. As x moves from left to right through the value zero, there is an abrupt and unavoidable "jump" in the value of f from 0 to 1. That is, f is discontinuous at $x = 0$. To see part of the graph of f, enter the *Mathematica* commands

    f[x_] := If[x < 0, 0, 1]

    Plot[f[x], {x, -3.5, 3.5 }, PlotRange -> { { -3.5, 3.5 }, {-1.5, 2.5 }}];

1.2.53: Because the graph of the greatest integer function changes at each integral value of x, the graph of $f(x) = [\![2x]\!]$ changes twice as often—at each integral multiple of $\frac{1}{2}$. So as x moves from left to right through such points, the graph jumps upward one unit. Thus there is a discontinuity at each integral multiple of $\frac{1}{2}$. Because f is constant otherwise, these are the only discontinuities. To see something like the graph of f, enter the *Mathematica* commands

```
f[x_] := Floor[ 2*x ];
Plot[ f[x], {x, −3.5, 3.5 }, PlotRange −> {{ −3.5, 3.5 }, {−4.5, 4.5 }} ];
```

Mathematica will draw vertical lines connecting points that it shouldn't, making the graph look like treads and risers of a staircase, whereas only the treads are on the graph.

1.2.55: Given: $f(x) = [\![x]\!]$. If n is an integer and $n \leq x < n + 1$, then express x as $x = n + (\!(x)\!)$ where $(\!(x)\!) = x - [\![x]\!]$ is the *fractional part* of x. Then $f(x) = n - x = n - [n + (\!(x)\!)] = -(\!(x)\!)$. So $f(x)$ is the negative of the fractional part of x. So as x ranges from n up to (but not including) $n + 1$, $f(x)$ begins at 0 and drops linearly down not quite to -1. That is, on the interval $(n, n + 1)$, the graph of f is the straight line segment connecting the two points $(n, 0)$ and $(n + 1, -1)$ with the first of these points included and the second excluded. There is a discontinuity at each integral value of x.

1.2.57: Because $y = 2x^2 - 6x + 7 = 2(x^2 - 3x + 3.5) = 2(x^2 - 3x + 2.25 + 1.25) = 2(x - 1.5)^2 + 2.5$, the vertex of the parabola is at $(1.5, 2.5)$.

1.2.59: Because $y = 4x^2 - 18x + 22 = 4(x^2 - (4.5)x + 5.5) = 4(x^2 - (4.5)x + 5.0625 + 0.4375) = 4(x - 2.25)^2 + 1.75$, the vertex of the parabola is at $(2.25, 1.75)$.

1.2.61: Because $y = -8x^2 + 36x - 32 = -8(x^2 - (4.5)x + 4) = -8(x^2 - (4.5)x + 5.0625 - 1.0625) = -8(x - 2.25)^2 + 8.5$, the vertex of the parabola is at $(2.25, 8.5)$.

1.2.63: Because $y = -3x^2 - 8x + 3 = -3\left(x^2 + \frac{8}{3}x - 1\right) = -3\left(x^2 + \frac{8}{3}x + \frac{16}{9} - \frac{25}{9}\right) = -3\left(x + \frac{4}{3}\right)^2 + \frac{25}{3}$, the vertex of the parabola is at $\left(-\frac{4}{3}, \frac{25}{3}\right)$.

1.2.65: To find the maximum height $y = -16t^2 + 96t$ of the ball, we find the vertex of the parabola: $y = -16(t^2 - 6t) = -16(t^2 - 6t + 9 - 9) = -16(t - 3)^2 + 144$. The vertex of the parabola is at $(3, 144)$ and therefore the maximum height of the ball is 144 ft.

1.2.67: If two positive numbers x and y have sum 50, then $y = 50 - x$ and $x < 50$ (because $y > 0$). To maximize their product $p(x)$ we find the vertex of the parabola

$$y = p(x) = x(50 - x) = -(x^2 - 50x)$$

$$= -(x^2 - 50x + 625 - 625) = -(x - 25)^2 + 625,$$

which is at $(25, 625)$. Because $0 < 25 < 50$, $x = 25$ is in the domain of the product function $p(x) = x(50 - x)$, and hence the maximum value of the product of x and y is $p(25) = 625$.

1.2.69: The graph looks like the graph of $y = |x|$ because the slope of the left-hand part is -1 and that of the right-hand part is 1; but the vertex is shifted to $(-1, 0)$, so—using the translation principle—the graph in Fig. 1.2.29 must be the graph of $f(x) = |x + 1|$, $-2 \leq x \leq 2$.

10

1.2.71: The graph in Fig. 1.2.31 is much like the graph of the greatest integer function—it takes on only integral values—but the "jumps" occur twice as often, so this must be very like—indeed, it is exactly—the graph of $f(x) = [\![2x]\!]$, $-1 \leq x < 2$.

1.2.73: Clearly $x(t) = 45t$ for the first hour; that is, for $0 \leq t \leq 1$. In the second hour the graph of $x(t)$ must be a straight line (because of constant speed) of slope 75, thus with equation $x(t) = 75t + C$ for some constant C. The constant C is determined by the fact that $45t$ and $75t + C$ must be equal at time $t = 1$, as the automobile cannot suddenly jump from one position to a completely different position in an instant. Hence $45 = 75 + C$, so that $C = -30$. Therefore

$$x(t) = \begin{cases} 45t & \text{if } 0 \leq t \leq 1; \\ 75t - 30 & \text{if } 1 < t \leq 2. \end{cases}$$

To see the graph of $x(t)$, plot in *Mathematica*

$$\text{x}[\text{t\_}] := \text{If}[\text{t} < 1, 45{*}\text{t}, 75{*}\text{t} - 30\,]$$

on the interval $0 \leq t \leq 2$.

1.2.75: The graph must consist of two straight-line segments (because of the constant speeds). The first must have slope 60, so we have $x(t) = 60t$ for $0 \leq t \leq 1$. The second must have slope -30, negative because you're driving in the reverse direction, so $x(t) = -30t + C$ for some constant C if $1 \leq t \leq 3$. The two segments must coincide when $t = 1$, so that $60 = -30 + C$. Thus $C = 90$ and thus a formula for $x(t)$ is

$$x(t) = \begin{cases} 60t & \text{if } 0 \leq t \leq 1, \\ 90 - 30t & \text{if } 1 < t \leq 3. \end{cases}$$

1.2.77: Initially we work in units of pages and cents (to avoid decimals and fractions). The graph of C, as a function of p, must be a straight line segment, and its slope is (by information given)

$$\frac{C(79) - C(34)}{79 - 34} = \frac{305 - 170}{79 - 34} = \frac{135}{45} = 3.$$

Thus $C(p) = 3p + K$ for some constant K. So $3 \cdot 34 + K = 170$, and it follows that $K = 68$. So $C(p) = 3p + 68$, $1 \leq p \leq 100$, if C is to be expressed in cents. If C is to be expressed in dollars, we have

$$C(p) = (0.03)p + 0.68, \quad 1 \leq p \leq 100.$$

The "fixed cost" is incurred regardless of the number of pamphlets printed; it is $0.68. The "marginal cost" of printing each additional page of the pamphlet is the coefficient $0.03 of p.

1.2.79: Suppose that the letter weighs x ounces, $0 < x \leq 16$. If $x \leq 8$, then the cost is simply 8 (dollars). If $8 < x \leq 9$, add $0.80; if $9 < x \leq 10$, add $1.60, and so on. Very roughly, one adds $0.80 if $[\![x - 8]\!] = 1$,

11

$1.60 if $[\![x - 8]\!] = 2$, and so on. But this isn't quite right—we are using the FLOOR function of Section 1.1, whereas we should really be using the CEILING function. By the result of Problem 51 of that section, we see that instead of cost

$$C(x) = 8 + (0.8)[\![x - 8]\!]$$

for $8 < x \leq 16$, we should instead write

$$C(x) = \begin{cases} 8 & \text{if } 0 < x \leq 8, \\ 8 - (0.8)[\![-(x - 8)]\!] & \text{if } 8 < x \leq 16. \end{cases}$$

The graph of the cost function is shown next.

1.2.81: Boyle's law states that under conditions of constant temperature, the product of the pressure p and the volume V of a fixed mass of gas remains constant. If we assume that $pV = c$, a constant, for the given data, we find that the given five data points yield the values $c = 1.68$, 1.68, 1.675, 1.68, and 1.62. The average of these is 1.65 (to two places) and should be a good estimate of the true value of c. Alternatively, you can use a computer algebra program to find c; in *Mathematica*, for example, the command Fit will fit given data points to a sum of constant multiples of functions you specify. We used the commands

data = {{0.25, 6.72}, {1.0, 1.68}, {2.5, 0.67}, {4.0, 0.42}, {6.0, 0.27}};

Fit [data, {1/p}, p]

to find that

$$V(p) = \frac{1.67986}{p}$$

yields the best *least-squares* fit of the given data to a function of the form $V(p) = c/p$. We rounded the numerator to 1.68 to find the estimates $V(0.5) \approx 3.36$ and $V(5) \approx 0.336$ (L). The graph of $V(p)$ together with the given data points is shown next.

12

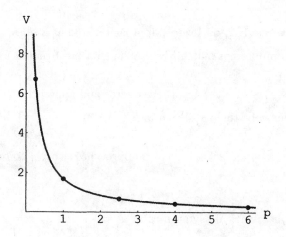

Section 1.3

1.3.1: The domain of f is R, the set of all real numbers; so is the domain of g, but $g(x) = 0$ when $x = 1$ and when $x = -3$. So the domain of $f + g$ and $f \cdot g$ is the set R and the domain of f/g is the set of all real numbers other than 1 and -3. Their formulas are

$$(f + g)(x) = x^2 + 3x - 2,$$

$$(f \cdot g)(x) = (x + 1)(x^2 + 2x - 3) = x^3 + 3x^2 - x - 3, \quad \text{and}$$

$$\left(\frac{f}{g}\right)(x) = \frac{x + 1}{x^2 + 2x - 3}.$$

1.3.3: The domain of f is the interval $[0, +\infty)$ and the domain of g is the interval $[2, +\infty)$. Hence the domain of $f + g$ and $f \cdot g$ is the interval $[2, +\infty)$, but because $g(2) = 0$, the domain of f/g is the open interval $(2, +\infty)$. The formulas for these combinations are

$$(f + g)(x) = \sqrt{x} + \sqrt{x - 2},$$

$$(f \cdot g)(x) = \sqrt{x}\,\sqrt{x - 2} = \sqrt{x^2 - 2x}, \quad \text{and}$$

$$\left(\frac{f}{g}\right)(x) = \frac{\sqrt{x}}{\sqrt{x - 2}} = \sqrt{\frac{x}{x - 2}}.$$

1.3.5: The domain of f is the set R of all real numbers; the domain of g is the open interval $(-2, 2)$. Hence the domain of $f + g$ and $f \cdot g$ is the open interval $(-2, 2)$; because $g(x)$ is never zero, the domain of f/g is the same. Their formulas are

$$(f + g)(x) = \sqrt{x^2 + 1} + \frac{1}{\sqrt{4 - x^2}}, \qquad (f \cdot g)(x) = \frac{\sqrt{x^2 + 1}}{\sqrt{4 - x^2}}, \quad \text{and}$$

$$\left(\frac{f}{g}\right)(x) = \sqrt{x^2 + 1}\,\sqrt{4 - x^2} = \sqrt{4 + 3x^2 - x^4}.$$

13

1.3.7: $f(x) = x^3 - 3x + 1$ has 1, 2, or 3 zeros, approaches $+\infty$ as x does, and approaches $-\infty$ as x does. Because $f(0) \neq 0$, the graph does not match Fig. 1.3.26, so it must match Fig. 1.3.30.

1.3.9: $f(x) = x^4 - 5x^3 + 13x + 1$ has four or fewer zeros and approaches $+\infty$ as x approaches either $+\infty$ or $-\infty$. Hence its graph must be the one shown in Fig. 1.3.31.

1.3.11: $f(x) = 16 + 2x^2 - x^4$ approaches $-\infty$ as x approaches either $+\infty$ or $-\infty$, so its graph must be the one shown in Fig. 1.3.27.

1.3.13: The graph of f has vertical asymptotes at $x = -1$ and at $x = 2$, so its graph must be the one shown in Fig. 1.3.34.

1.3.15: The graph of f has no vertical asymptotes and has maximum value 3 when $x = 0$. Hence its graph must be the one shown in Fig. 1.3.33.

1.3.17: The domain of $f(x) = x\sqrt{x+2}$ is the interval $[-2, +\infty)$, so its graph must be the one shown in Fig. 1.3.38.

1.3.19: The domain of $f(x) = \sqrt{x^2 - 2x}$ consists of those numbers x for which $x^2 - 2x \geq 0$; that is, $x(x - 2) \geq 0$. This occurs when x and $x - 2$ have the same sign and also when either is zero. If $x > 0$ and $x - 2 > 0$, then $x > 2$; if $x < 0$ and $x - 2 < 0$, then $x < 0$. So the domain of f is the union of the two intervals $(-\infty, 0]$ and $[2, +\infty)$. So the graph of f must be the one shown in Fig. 1.3.39.

1.3.21: Good viewing window: $-2.5 \leq x \leq 2.5$. Three zeros, approximately -1.88, 0.35, and 1.53.

1.3.23: Good viewing window: $-3.5 \leq x \leq 2.5$. One zero, approximately -2.10.

1.3.25: Good viewing window: $-1.6 \leq x \leq 2.8$. Three zeros: approximately -1.30, exactly 1, and approximately 2.30.

1.3.27: Good viewing window: $-7.5 \leq x \leq 8.5$. Three zeros: Approximately -5.70, -2.22, and 7.91.

1.3.29: The viewing window $-11 \leq x \leq 8$ shows that there are five zeros, although the two near 2.5 may be only one. The window $1.5 \leq x \leq 3.5$ shows that there are in fact two zeros near 2.5. Approximate values of the five zeros are -10.20, -7.31, 1.98, 3.25, and 7.28.

1.3.31: Every time c increases by 1, the graph is raised 1 unit (in the positive y-direction), but there is no other change.

1.3.33: The graph always passes through $(0, 0)$ and is tangent to the x-axis there. When $c = -5$ there is another zero at $x = 5$. As c increases this zero shifts to the left until it coincides with the one at $x = 0$ when

14

$c = 0$. At this point the "bend" in the graph disappears. As c increases from 1 to 5, the bend reappears to the left of the x-axis and the second zero reappears at $-c$.

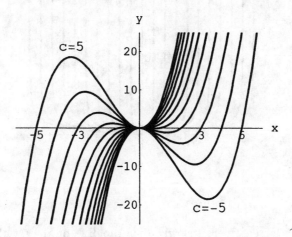

1.3.35: The graph is always symmetric around the origin (and, consequently, always passes through the origin). When $c = -5$ there is another pair of zeros near ± 2.2. As c increases the graph develops positive slope at $x = 0$, two more bends, and two more zeros on either side of the origin. They move outward and, when $c = -2$, they coincide with the outer pair of zeros, which have also been moving toward the origin. They reach the origin when $c = 0$ and thereafter the graph simply becomes steeper and steeper.

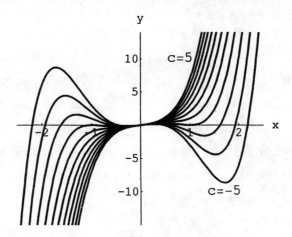

1.3.37: As c increases the graph becomes wider; aside from the horizontal scale, its shape does not seem to change very much.

15

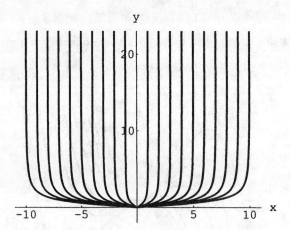

Section 1.4

1.4.1: Because $g(x) = 2^x$ increases—first slowly, then rapidly—on the set of all real numbers, with values in the range $(0, +\infty)$, the given function $f(x) = 2^x - 1$ must increase in the same way, but with values in the range $(-1, +\infty)$. Therefore its graph is the one shown in Fig. 1.4.29.

1.4.3: The graph of $f(x) = 1 + \cos x$ is simply the graph of the ordinary cosine function raised 1 unit—moved upward 1 unit in the positive y-direction. Hence its graph is the one shown in Fig. 1.4.27.

1.4.5: The graph of $g(x) = 2 \cos x$ resembles the graph of the cosine function, but with all values doubled, so that its range is the interval $[-2, 2]$. Add 1 to get $f(x) = 1 + 2 \cos x$ and the range is now the interval $[-1, 3]$. So the graph of f is the one shown in Fig. 1.4.35.

1.4.7: The graph of $g(x) = 2^x$ increases, first slowly, then rapidly, on the set of all real numbers, with range the interval $(0, +\infty)$. So its reciprocal $h(x) = 2^{-x}$ decreases, first rapidly, then slowly, with the same domain and range. Multiply by x to obtain $f(x) = x \cdot 2^{-x}$. The effect of multiplication by x is to change large positive values into large negative values for $x < 0$, to cause $f(0)$ to be zero, and to multiply very small positive values (of 2^{-x}) by somewhat large positive values (of x) for $x > 0$, resulting in values that are still small and positive, even when x is quite large. So the graph of f must increase rapidly through negative values, pass through $(0, 0)$, rise to a maximum, then decrease rapidly through positive values toward zero. Hence the graph of f must be the one shown in Fig. 1.4.31.

1.4.9: The graph of $g(x) = 1 + \cos 6x$ will resemble the graph of the cosine function, but raised 1 unit (so that its range is the interval $[0, 2]$) and with much more "activity" on the x-axis (because of the factor 6). Division by $1 + x^2$ will have little effect until x is no longer close to zero, and then the effect will be to divide values of $g(x)$ by larger and larger positive numbers, so that the cosine oscillations have a much smaller range that $0 \leq x \leq 2$; they will range from 0 to smaller and smaller positive values as $|x|$ increases. So the graph of f is the one shown in Fig. 1.4.34.

16

1.4.11: Given $f(x) = 1 - x^2$ and $g(x) = 2x + 3$,

$$f(g(x)) = 1 - (g(x))^2 = 1 - (2x + 3)^2 = -4x^2 - 12x - 8 \quad \text{and}$$

$$g(f(x)) = 2f(x) + 3 = 2(1 - x^2) + 3 = -2x^2 + 5.$$

1.4.13: If $f(x) = \sqrt{x^2 - 3}$ and $g(x) = x^2 + 3$, then

$$f(g(x)) = \sqrt{(g(x))^2 - 3} = \sqrt{(x^2 + 3)^2 - 3} = \sqrt{x^4 + 6x^2 + 6} \quad \text{and}$$

$$g(f(x)) = (f(x))^2 + 3 = \left(\sqrt{x^2 - 3}\right)^2 + 3 = x^2 - 3 + 3 = x^2.$$

The domain of $f(g)$ is the set R of all real numbers, but the domain of $g(f)$ is the same as the domain of f, the set of all real numbers x such that $x^2 \geq 3$.

1.4.15: If $f(x) = x^3 - 4$ and $g(x) = (x + 4)^{1/3}$, then

$$f(g(x)) = (g(x))^3 - 4 = \left((x + 4)^{1/3}\right)^3 - 4 = x + 4 - 4 = x \quad \text{and}$$

$$g(f(x)) = (f(x) + 4)^{1/3} = \left(x^3 - 4 + 4\right)^{1/3} = \left(x^3\right)^{1/3} = x.$$

The domain of both $f(g)$ and $g(f)$ is the set R of all real numbers, so here is an example of the highly unusual case in which $f(g)$ and $g(f)$ are the same function.

1.4.17: If $f(x) = \sin x$ and $g(x) = x^3$, then

$$f(g(x)) = f\left(x^3\right) = \sin\left(x^3\right) = \sin x^3 \quad \text{and}$$

$$g(f(x)) = g(\sin x) = (\sin x)^3 = \sin^3 x.$$

We note in passing that $\sin x^3$ and $\sin^3 x$ don't mean the same thing!

1.4.19: If $f(x) = 1 + x^2$ and $g(x) = \tan x$, then $f(g(x)) = f(\tan x) = 1 + (\tan x)^2 = 1 + \tan^2 x$ and $g(f(x)) = g(1 + x^2) = \tan(1 + x^2)$.

Note: The answers to Problems 21 through 30 are not unique. We have generally chosen the simplest and most natural answer.

1.4.21: $h(x) = (2 + 3x)^2 = (g(x))^k = f(g(x))$ where $f(x) = x^k$, $k = 2$, and $g(x) = 2 + 3x$.

1.4.23: $h(x) = (2x - x^2)^{1/2} = (g(x))^{1/2} = f(g(x))$ where $f(x) = x^k$, $k = \frac{1}{2}$, and $g(x) = 2x - x^2$.

17

1.4.25: $h(x) = (5 - x^2)^{3/2} = (g(x))^{3/2} = f(g(x))$ where $f(x) = x^k$, $k = \frac{3}{2}$, and $g(x) = 5 - x^2$.

1.4.27: $h(x) = (x + 1)^{-1} = (g(x))^{-1} = f(g(x))$ where $f(x) = x^k$, $k = -1$, and $g(x) = x + 1$.

1.4.29: $h(x) = (x + 10)^{-1/2} = (g(x))^{-1/2} = f(g(x))$ where $f(x) = x^k$, $k = -\frac{1}{2}$, and $g(x) = x + 10$.

1.4.31: Recommended window: $-2 \leqq x \leqq 2$. The graph makes it evident that the equation has exactly one solution (approximately 0.641186).

1.4.33: Recommended window: $-5 \leqq x \leqq 5$. The graph makes it evident that the equation has exactly one solution (approximately 1.42773).

1.4.35: Recommended window: $-8 \leqq x \leqq 8$. The graph makes it evident that the equation has exactly five solutions (approximately -4.08863, -1.83622, 1.37333, 5.65222, and 6.61597).

1.4.37: Recommended window: $0.1 \leqq x \leqq 20$. The graph makes it evident that the equation has exactly three solutions (approximately 1.41841, 5.55211, and 6.86308).

1.4.39: Recommended window: $-11 \leqq x \leqq 11$. The graph makes it evident that the equation has exactly six solutions (approximately -5.92454, -3.24723, 3.04852, 6.75738, 8.59387, and [exactly] 0).

1.4.41: Graphical methods show that the solution of $10 \cdot 2^t = 100$ is slightly less than 3.322. We began with the viewing window $0 \leqq t \leqq 6$ and gradually narrowed it to $3.321 \leqq t \leqq 3.323$.

1.4.43: Graphical methods show that the solution of $(67.4) \cdot (1.026)^t = 134.8$ is approximately 27.0046. We began with the viewing window $20 \leqq t \leqq 30$ and gradually narrowed it to $27.0045 \leqq t \leqq 27.0047$.

1.4.45: Graphical methods show that the solution of $A(t) = 12 \cdot (0.975)^t = 1$ is approximately 98.149. We began with the viewing window $50 \leqq t \leqq 250$ and gradually narrowed it to $98.148 \leqq t \leqq 98.150$.

1.4.47: We plotted $y = \log_{10} x$ and $y = \frac{1}{2}x^{1/5}$ simultaneously. We began with the viewing window $1 \leqq x \leqq 10$ and gradually narrowed it to $4.84890 \leqq x \leqq 4.84892$. Answer: $x \approx 4.84891$.

Chapter 1 Miscellaneous Problems

1.M.1: The domain of $f(x) = \sqrt{x - 4}$ is the set of real numbers x for which $x - 4 \geqq 0$; that is, the interval $[4, +\infty)$.

1.M.3: The domain of f consists of those real numbers for which the denominator is nonzero; that is, the set of all real numbers other than ± 3.

1.M.5: If $x \geqq 0$, then $\sqrt{x}$ exists; there is no obstruction to adding 1 to $\sqrt{x}$ nor to cubing the sum. Hence the domain of f is the set $[0, +\infty)$ of all nonnegative real numbers.

1.M.7: The function $f(x) = \sqrt{2 - 3x}$ is defined whenever the radicand is nonnegative; that is, whenever

$$2 - 3x \geqq 0;$$

$$3x \leqq 2;$$

$$x \leqq \tfrac{2}{3}.$$

Hence the domain of f is the interval $\left(-\infty, \tfrac{2}{3}\right]$.

1.M.9: Regardless of the value of x, it's always possible to subtract 2 from x, to subtract x from 4, and to multiply the results. Hence the domain of f is the set $\boldsymbol{R}$ of all real numbers.

1.M.11: Because $100 \leqq V \leqq 200$ and $p > 0$, it follows that $100p \leqq pV \leqq 200p$. Because $pV = 800$, we see that $100p \leqq 800 \leqq 200p$, so that $p \leqq 8 \leqq 2p$. That is, $p \leqq 8$ and $4 \leqq p$, so that $4 \leqq p \leqq 8$. This is the range of possible values of p.

1.M.13: Because $25 < R < 50$, $25I < IR < 50I$, so that

$$25I < E < 50I;$$

$$25I < 100 < 50I;$$

$$I < 4 < 2I;$$

$$I < 4 \quad \text{and} \quad 2 < I.$$

Therefore the current I lies in the range $2 < I < 4$.

1.M.15: If a cube has edge length x, then its volume is $V = x^3$ and its total surface area is $S = 6x^2$ (because each of its six faces has area x^2). Hence $x = \sqrt{S/6}$, and therefore

$$V(S) = \left(\sqrt{\frac{S}{6}}\right)^3 = \left(\frac{S}{6}\right)^{3/2}, \qquad 0 < S < +\infty.$$

Under certain circumstances it would be both permissible and desirable to let the domain of V be the interval $[0, +\infty)$.

1.M.17: Suppose the given equilateral triangle has sides of length $2x$ and height (perpendicular to its base) of length h. Then the height of the equilateral triangle divides it into two right triangles, each of which has

19

base x, height h, and hypotenuse $2x$. Hence application of the Pythagorean theorem to one of these right triangles gives

$$x^2 + h^2 = (2x)^2, \quad \text{so that} \quad h = x\sqrt{3}.$$

Now the area of the original right triangle is $A = hx$ and its perimeter is $P = 6x$. So

$$A = x^2\sqrt{3} \quad \text{and} \quad x = \frac{P}{6}.$$

Therefore $A(P) = \dfrac{P^2\sqrt{3}}{36}, \quad 0 < P < \infty.$

1.M.19: The slope of L is $\dfrac{13-5}{1-(-3)} = 2$, so an equation of L is

$$y - 5 = 2(x+3); \quad \text{that is,} \quad y = 2x + 11.$$

1.M.21: The point $(0, -5)$ lies on L, so an equation of L is

$$y - (-5) = \frac{1}{2}(x - 0); \quad \text{alternatively,} \quad 2y + 10 = x.$$

1.M.23: The equation $y - 2x = 10$ may be written in the form $y = 2x + 10$, showing that it has slope 2. Hence the perpendicular line L has slope $-\frac{1}{2}$. Therefore an equation of L is

$$y - 7 = -\frac{1}{2}(x - (-3)); \quad \text{that is,} \quad x + 2y = 11.$$

1.M.25: The graph of $y = f(x) = 2 - 2x - x^2$ is a parabola opening downward. The only such graph is shown in Fig. 1.MP.6.

1.M.27: Given: $f(x) = x^4 - 4x^3 + 5$. Because the graph of f has no vertical asymptotes and because $f(x)$ approaches $+\infty$ as x approaches either $+\infty$ or $-\infty$, the graph of f must be the one shown in Fig. 1.MP.4.

1.M.29: Given:

$$f(x) = \frac{5}{x^2 - x + 6} = \frac{20}{4x^2 - 4x + 1 + 23} = \frac{20}{(2x-1)^2 + 23}.$$

The algebra displayed here shows that the denominator in $f(x)$ is never zero, so there are no vertical asymptotes. It also shows that the maximum value of $f(x)$ occurs when the denominator is minimal; that is, when $x = \frac{1}{2}$. Finally, $f(x)$ approaches zero as x approaches either $+\infty$ or $-\infty$. So the graph of $y = f(x)$ must be the one shown in Fig. 1.MP.3.

1.M.31: Given: $f(x) = 2^{-x} - 1$. The graph of $y = 2^x$ is an increasing exponential function, so the graph of $y = 2^{-x}$ is a decreasing exponential function, approaching 0 as x approaches $+\infty$. So the graph of f

approaches -1 as x approaches $+\infty$. Moreover, $f(0) = 0$. Therefore the graph of f is the one shown in Fig. 1.MP.7.

1.M.33: The graph of $y = 3\sin x$ oscillates between its minimum value -3 and its maximum value 3, so the graph of $f(x) = 1 + 3\sin x$ oscillates between -2 and 4. This graph is shown in Fig. 1.MP.8.

1.M.35: The graph of $2x - 5y = 7$ is the straight line with x-intercept $\frac{7}{2}$ and y-intercept $-\frac{7}{5}$.

1.M.37: We complete the square: $x^2 - 2x + 1 + y^2 = 1$, so that $(x-1)^2 + (y-0)^2 = 1^2$. Thus the graph of the given equation is the circle with center $(1, 0)$ and radius 1.

1.M.39: The graph is a parabola opening upward. To find its vertex, we complete the square:

$$y = 2\left(x^2 - 2x - \tfrac{1}{2}\right)$$
$$= 2\left(x^2 - 2x + 1 - \tfrac{3}{2}\right) = 2(x-1)^2 - 3.$$

So the vertex of this parabola is at the point $(1, -3)$.

1.M.41: The graph has a vertical asymptote at $x = -5$ and is shown next.

1.M.43: The graph of f is obtained by shifting the familiar absolute-value graph $y = |x|$ three units to the right, so that the graph of f has its "vertex" (or corner point) at the point $(3, 0)$.

1.M.45: Suppose that a, b, and c are arbitrary real numbers. Then

$$|a + b + c| = |(a + b) + c| \leqq |a + b| + |c| \leqq |a| + |b| + |c|.$$

1.M.47: If $x - 3 > 0$ and $x + 2 > 0$, then $x > 3$ and $x > -2$, so $x > 3$. If $x - 3 < 0$ and $x + 2 < 0$, then $x < 3$ and $x < -2$, so $x < -2$. Answer: $(-\infty, -2) \cup (3, \infty)$.

21

1.M.49: $(x-4)(x+2) > 0$: Either $x > 4$ and $x > -2$ (so that $x > 4$) or $x < 4$ and $x < -2$ (so that $x < -2$). Answer: $(-\infty, -2) \cup (4, +\infty)$.

1.M.51: The viewing window $-3 \leq x \leq 8$ shows a solution near -1 and another near 5. Gradual magnification of the region near -1 shows a solution between -1.1405 and -1.1395. Similarly, the other solution is between 6.1395 and 6.1405. So the solutions are approximately -1.140 and 6.140.

1.M.53: The viewing window $0.5 \leq x \leq 3$ shows one solution near 1.2 and another near 2.3. The method of repeated tabulation with successive intervals $[1.1, 1.3]$, $[1.18, 1.20]$, and $[1.190, 1.192]$ yields the approximation 1.191 to the first solution. The successive intervals $[2.2, 2.4]$, $[2.30, 2.32]$, and $[2.308, 2.310]$ yield the approximation 2.309 to the second solution.

1.M.55: The viewing window $-6 \leq x \leq 2$ shows one solution near -5 and another near 1. The method of repeated tabulation with successive intervals $[-5.1, -4.9]$, $[-5.04, -5.02]$, and $[-5.022, -5.020]$, then with the intervals $[0.8, 1.0]$, $[0.88, 0.90]$, and $[0.896, 0.898]$, yields the two approximations -5.021 and 0.896 to the two solutions.

1.M.57: The viewing window $2 \leq x \leq 3$ shows the low point with x-coordinate near 2.5. The method of repeated tabulation, using the successive intervals $[2.4, 2.6]$, $[2.48, 2.52]$, and $[2.496, 2.504]$, indicates that the low point is very close to $(2.5, 0.75)$.

1.M.59: The viewing window $-0.5 \leq x \leq 4$ shows that the low point has x-coordinate near 1.8. The method of repeated tabulation, with the successive intervals $[1.7, 1.9]$, $[1.72, 1.78]$, and $[1.744, 1.756]$, shows that the low point is very close to $(1.75, -1.25)$.

1.M.61: The viewing window $-5 \leq x \leq 1$ show that the x-coordinate of the low point is close to -2. The method of repeated tabulation shows that the low point is very close to $(-2.0625, 0.96875)$.

1.M.63: The small rectangle has dimensions $10 - 4x$ by $7 - 2x$; $(7)(10) - (10 - 4x)(7 - 2x) = 20$, which leads to the quadratic equation $8x^2 - 48x + 20 = 0$. One solution of this equation is approximately 5.5495, which must be rejected; it is too large. The value of x is the other solution: $x \approx 0.4505$.

1.M.65: The viewing window $-4 \leq x \leq 4$ shows three solutions (and there can be no more).

1.M.67: We plotted $y = \sin x$ and $y = x^3 - 3x + 1$ simultaneously to see where they crossed. The viewing window $-2.2 \leq x \leq 2.2$ shows three solutions, and there can be no more because $|x^3 - 3x + 1| > 1$ if $|x| > 2.2$.

1.M.69: We plotted $y = \cos x$ and $y = \log_{10} x$ simultaneously to see where they crossed. The viewing window $0.1 \leq x \leq 14$ shows three solutions, and there can be no more because $\log_{10} x < -1$ if $0 < x < 0.1$ and $\log_{10} x > 1$ if $x > 14$.

Section 2.1

2.1.1: $f(x) = 0 \cdot x^2 + 0 \cdot x + 5$, so $m(a) = 0 \cdot 2 \cdot a + 0 \equiv 0$. In particular, $m(2) = 0$, so the tangent line has equation $y - 5 = 0 \cdot (x - 0)$; that is, $y \equiv 5$.

2.1.3: Because $f(x) = 1 \cdot x^2 + 0 \cdot x + 0$, the slope-predictor is $m(a) = 2 \cdot 1 \cdot a + 0 = 2a$. Hence the line L tangent to the graph of f at $(2, f(2))$ has slope $m(2) = 4$. So an equation of L is $y - f(2) = 4(x - 2)$; that is, $y = 4x - 4$.

2.1.5: Because $f(x) = 0 \cdot x^2 + 4x - 5$, the slope-predictor for f is $m(a) = 2 \cdot 0 \cdot a + 4 = 4$. So the line tangent to the graph of f at $(2, f(2))$ has slope 4 and therefore equation $y - 3 = 4(x - 2)$; that is, $y = 4x - 5$.

2.1.7: Because $f(x) = 2x^2 - 3x + 4$, the slope-predictor for f is $m(a) = 2 \cdot 2 \cdot a - 3 = 4a - 3$. So the line tangent to the graph of f at $(2, f(2))$ has slope 5 and therefore equation $y - 6 = 5(x - 2)$; that is, $y = 5x - 4$.

2.1.9: Because $f(x) = 2x^2 + 6x$, the slope-predictor for f is $m(a) = 4a + 6$. So the line tangent to the graph of f at $(2, f(2))$ has slope $m(2) = 14$ and therefore equation $y - 20 = 14(x - 2)$; that is, $y = 14x - 8$.

2.1.11: Because $f(x) = -\frac{1}{100}x^2 + 2x$, the slope-predictor for f is $m(a) = -\frac{2}{100}a + 2$. So the line tangent to the graph of f at $(2, f(2))$ has slope $m(2) = -\frac{1}{25} + 2 = \frac{49}{25}$ and therefore equation $y - \frac{99}{25} = \frac{49}{25}(x - 2)$; that is, $25y = 49x + 1$.

2.1.13: Because $f(x) = 4x^2 + 1$, the slope-predictor for f is $m(a) = 8a$. So the line tangent to the graph of f at $(2, f(2))$ has slope $m(2) = 16$ and therefore equation $y - 17 = 16(x - 2)$; that is, $y = 16x - 15$.

2.1.15: If $f(x) = -x^2 + 10$, then the slope-predictor for f is $m(a) = -2a$. A line tangent to the graph of f will be horizontal when $m(a) = 0$, thus when $a = 0$. So the tangent line is horizontal at the point $(0, 10)$ and at no other point of the graph of f.

2.1.17: If $f(x) = x^2 - 2x + 1$, then the slope-predictor for f is $m(a) = 2a - 2$. A line tangent to the graph of f will be horizontal when $m(a) = 0$, thus when $a = 1$. So the tangent line is horizontal at the point $(1, 0)$ and at no other point of the graph of f.

2.1.19: If $f(x) = -\frac{1}{100}x^2 + x$, then the slope-predictor for f is $m(a) = -\frac{1}{50}a + 1$. A line tangent to the graph of f will be horizontal when $m(a) = 0$, thus when $a = 50$. So the tangent line is horizontal at the point $(50, 25)$ and at no other point of the graph of f.

2.1.21: If $f(x) = x^2 - 2x - 15$, then the slope-predictor for f is $m(a) = 2a - 2$. A line tangent to the graph of f will be horizontal when $m(a) = 0$, thus when $a = 1$. So the tangent line is horizontal at the point $(1, -16)$ and at no other point of the graph of f.

23

2.1.23: If $f(x) = -x^2 + 70x$, then the slope-predictor for f is $m(a) = -2a + 70$. A line tangent to the graph of f will be horizontal when $m(a) = 0$, thus when $a = 35$. So the tangent line is horizontal at the point $(35, 1225)$ and at no other point of the graph of f.

2.1.25: If $f(x) = x^2$, then the slope-predictor for f is $m(a) = 2a$. So the line tangent to the graph of f at the point $P(-2, 4)$ has slope $m(-2) = -4$ and the normal line at P has slope $\frac{1}{4}$. Hence an equation for the line tangent to the graph of f at P is $y - 4 = -4(x + 2)$; that is, $y = -4x - 4$. An equation for the line normal to the graph of f at P is $y - 4 = \frac{1}{4}(x + 2)$; that is, $4y = x + 18$.

2.1.27: If $f(x) = 2x^2 + 3x - 5$, then the slope-predictor for f is $m(a) = 4a + 3$. So the line tangent to the graph of f at the point $P(2, 9)$ has slope $m(2) = 11$ and the normal line at P has slope $-\frac{1}{11}$. Hence an equation for the line tangent to the graph of f at P is $y - 9 = 11(x - 2)$; that is, $y = 11x - 13$. An equation for the line normal to the graph of f at P is $y - 9 = -\frac{1}{11}(x - 2)$; that is, $x + 11y = 101$.

2.1.29: If the ball has height $y(t) = -16t^2 + 96t$ (feet) at time t (s), then the slope-predictor for y is $m(a) = -32a + 96$. Assuming that the maximum height of the ball occurs at the point on the graph of y where the tangent line is horizontal, we find that point by solving $m(a) = 0$ and find that $a = 3$. So the highest point on the graph of y is the point $(3, y(3)) = (3, 144)$. Therefore the ball reaches a maximum height of 144 (ft).

2.1.31: If the two positive numbers x and y have sum 50, then $y = 50 - x$, $x > 0$, and $x < 50$ (because $y > 0$). So the product of two such numbers is given by

$$p(x) = x(50 - x), \qquad 0 < x < 50.$$

The graph of $p(x) = -x^2 + 50x$ has a highest point because the graph of $y = p(x)$ is a parabola that opens downward. The slope-predictor for the function p is $m(a) = -2a + 50$. The highest point on the graph of p will occur when the tangent line is horizontal, so that $m(a) = 0$. This leads to $a = 25$, which does lie in the domain of p. Therefore the highest point on the graph of p is $(25, p(25)) = (25, 625)$. Hence the maximum possible value of $p(x)$ is 625. So the maximum possible product of two positive numbers with sum 50 is 625.

2.1.33: Suppose that the "other" line L is tangent to the parabola at the point (a, a^2). The slope-predictor for $y = f(x) = x^2$ is $m(a) = 2a$, so the line L has slope $m(a) = 2a$. (Note that a changes from a variable to a constant in the last sentence. This is dangerous but the notation has forced this situation upon us.) Using the two-point formula for slope, we can compute the slope of L in another way and equate our two results:

$$\frac{a^2 - 0}{a - 3} = 2a;$$

$$a^2 = 2a(a - 3);$$

$$a = 2a - 6; \qquad \text{(because } a \neq 0\text{)};$$

$$a = 6.$$

24

Therefore L has slope $m(6) = 12$. Because L passes through $(3, 0)$, an equation of L is $y - 0 = 12(x - 3)$; that is, $y = 12x - 36$.

2.1.35: Suppose that (a, a^2) is the point on the graph of $y = x^2$ closest to $(3, 0)$. Let L be the line segment from $(3, 0)$ to (a, a^2). Under the plausible assumption that L is normal to the tangent line at (a, a^2), we infer that the slope m of L is $-1/(2a)$ because the slope of the tangent line is $2a$. Because we can also compute m by using the two points known to lie on it, we find that

$$m = -\frac{1}{2a} = \frac{a^2 - 0}{a - 3}.$$

This leads to the equation $0 = 2a^3 + a - 3 = (a - 1)(2a^2 + 2a + 3)$, which has $a = 1$ as its only real solution (note that the discriminant of $2a^2 + 2a + 3$ is negative). Intuitively, it's clear that there *is* a point on the graph nearest $(3, 0)$, so we have found it: That point is $(1, 1)$.

Alternatively, if (x, x^2) is an arbitrary point on the given parabola, then the distance from (x, x^2) to $(3, 0)$ is the square root of $f(x) = (x^2 - 0)^2 + (x - 3)^2 = x^4 + x^2 - 6x + 9$. A positive quantity is minimized when its square is minimized, so we minimize the distance from (x, x^2) to $(3, 0)$ by minimizing $f(x)$. The slope-predictor for f is $m(a) = 4a^3 + 2a - 6 = 2(a - 1)(2a^2 + 2x + 3)$, and (as before) the equation $m(a) = 0$ has only one real solution, $a = 1$. Again appealing to intuition for the existence of a point on the parabola nearest to $(3, 0)$, we see that it can only be the point $(1, 1)$. In Chapter 3 we will see how the existence of the closest point can be established without an appeal to the intuition.

2.1.37: Given: $f(x) = x^3$ and $a = 2$. We computed

$$\frac{f(a + h) - f(a - h)}{2h} \tag{1}$$

for $h = 10^{-1}$, 10^{-2}, $\ldots$, 10^{-10}. The values of the expression in (1) were 12.01, 12.0001, 12.000001, $\ldots$, 12.00000000000000000001. The numerical evidence overwhelmingly suggests that the slope of the tangent line is 12 and thus that it has equation $y = 12x - 16$. The graph of this line and $y = f(x)$ are shown next.

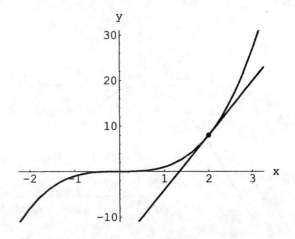

2.1.39: The numerical evidence suggests that the slope of the tangent line is $\frac{1}{2}$, so that its equation is $y = \frac{1}{2}(x+1)$. The graph of the tangent line and the graph of $f(x) = \sqrt{x}$ are shown next.

2.1.41: The numerical evidence suggests that the slope of the tangent line is -1, so that its equation is $y = -x + 2$. The graph of the tangent line and the graph of $f(x) = 1/x$ are shown next.

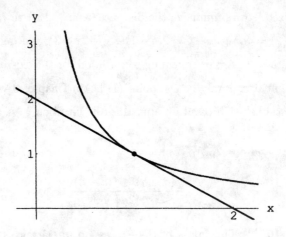

2.1.43: The numerical evidence suggests that the slope of the tangent line is 0, so that its equation is $y = 1$. The graph of the tangent line and the graph of $f(x) = \cos x$ are shown next.

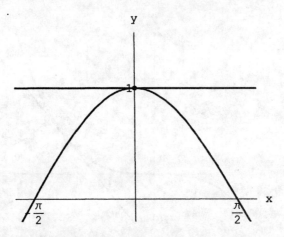

2.1.45: The numerical evidence suggests that the slope of the tangent line is $-1/\sqrt{2}$, so that its equation is

$$y - \frac{\sqrt{2}}{2} = -\frac{\sqrt{2}}{2}\left(x - \frac{\pi}{4}\right).$$

The graph of the tangent line and the graph of $f(x) = \cos x$ are shown next.

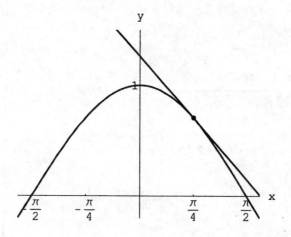

2.1.47: The numerical evidence suggests that the tangent line is horizontal, so that its equation is $y \equiv 5$. The graph of the tangent line and the graph of $f(x) = \sqrt{25 - x^2}$ are shown next.

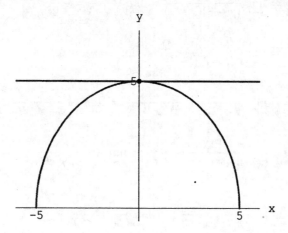

Section 2.2

2.2.1: $\displaystyle\lim_{x \to 3} (3x^2 + 7x - 12) = 3\left(\lim_{x \to 3} x\right)^2 + 7\left(\lim_{x \to 3} x\right) - \lim_{x \to 3} 12 = 3 \cdot 3^2 + 7 \cdot 3 - 12 = 36.$

2.2.3: $\displaystyle\lim_{x \to 1} (x^2 - 1)(x^7 + 7x - 4) = \lim_{x \to 1} (x^2 - 1) \cdot \lim_{x \to 1} (x^7 + 7x - 4) = 0 \cdot 4 = 0.$

2.2.5: $\displaystyle\lim_{x \to 1} \frac{x + 1}{x^2 + x + 1} = \frac{\displaystyle\lim_{x \to 1} (x + 1)}{\displaystyle\lim_{x \to 1} (x^2 + x + 1)} = \frac{2}{3}.$

2.2.7: $\displaystyle \lim_{x \to 3} \frac{(x^2+1)^3}{(x^3-25)^3} = \frac{\displaystyle \lim_{x \to 3} (x^2+1)^3}{\displaystyle \lim_{x \to 3} (x^3-25)^3} = \frac{\left(\displaystyle \lim_{x \to 3} (x^2+1)\right)^3}{\left(\displaystyle \lim_{x \to 3} (x^3-25)\right)^3} = \frac{10^3}{2^3} = \frac{1000}{8} = 125.$

2.2.9: $\displaystyle \lim_{x \to 1} \sqrt{4x+5} = \sqrt{\lim_{x \to 1}(4x+5)} = \sqrt{9} = 3.$

2.2.11: $\displaystyle \lim_{x \to 3} \left(x^2-1\right)^{3/2} = \left(\lim_{x \to 3}\left(x^2-1\right)\right)^{3/2} = 8^{3/2} = 16\sqrt{2}.$

2.2.13: $\displaystyle \lim_{z \to 8} \frac{z^{2/3}}{z - \sqrt{2z}} = \frac{\displaystyle \lim_{z \to 8} z^{2/3}}{\displaystyle \lim_{z \to 8}\left(z - \sqrt{2z}\right)} = \frac{4}{4} = 1.$

2.2.15: $\displaystyle \lim_{w \to 0} \sqrt{(w-2)^4} = \sqrt{\lim_{w \to 0}(w-2)^4} = \sqrt{(-2)^4} = 4.$

2.2.17: $\displaystyle \lim_{x \to -2} \sqrt[3]{\frac{x+2}{(x-2)^2}} = \sqrt[3]{\lim_{x \to -2}\frac{(x+2)}{(x-2)^2}} = 0.$

2.2.19: $\displaystyle \lim_{x \to -1} \frac{x+1}{x^2-x-2} = \lim_{x \to -1} \frac{x+1}{(x+1)(x-2)} = \lim_{x \to -1} \frac{1}{x-2} = -\frac{1}{3}.$

2.2.21: $\displaystyle \lim_{x \to 1} \frac{x^2+x-2}{x^2-4x+3} = \lim_{x \to 1} \frac{(x+2)(x-1)}{(x-3)(x-1)} = \lim_{x \to 1} \frac{x+2}{x-3} = -\frac{3}{2}.$

2.2.23: $\displaystyle \lim_{t \to -3} \frac{t^2+6t+9}{t^2-9} = \lim_{t \to -3} \frac{(t+3)(t+3)}{(t+3)(t-3)} = \lim_{t \to -3} \frac{t+3}{t-3} = 0.$

2.2.25: $\displaystyle \lim_{z \to -2} \frac{(z+2)^2}{z^4-16} = \lim_{z \to -2} \frac{(z+2)(z+2)}{(z+2)(z-2)(z^2+4)} = \lim_{z \to -2} \frac{z+2}{(z-2)(z^2+4)} = 0.$

2.2.27: $\displaystyle \lim_{x \to 1} \frac{x^3-1}{x^4-1} = \lim_{x \to 1} \frac{(x-1)(x^2+x+1)}{(x-1)(x+1)(x^2+1)} = \lim_{x \to 1} \frac{x^2+x+1}{(x+1)(x^2+1)} = \frac{3}{4}.$

2.2.29: $\displaystyle \lim_{x \to 3} \frac{\frac{1}{x} - \frac{1}{3}}{x-3} = \lim_{x \to 3}\left(\frac{3-x}{3x}\right)\left(\frac{1}{x-3}\right) = \lim_{x \to 3} \frac{-1}{3x} = -\frac{1}{9}.$

2.2.31: $\displaystyle \lim_{x \to 4} \frac{x-4}{\sqrt{x}-2} = \lim_{x \to 4} \frac{(\sqrt{x}-2)(\sqrt{x}+2)}{\sqrt{x}-2} = \lim_{x \to 4} \sqrt{x}+2 = 4.$

2.2.33: $\displaystyle \lim_{t \to 0} \frac{\sqrt{t+4}-2}{t} = \lim_{t \to 0}\left(\frac{\sqrt{t+4}-2}{t}\right) \cdot \left(\frac{\sqrt{t+4}+2}{\sqrt{t+4}+2}\right)$

$\displaystyle \qquad\qquad = \lim_{t \to 0} \frac{t+4-4}{t\left(\sqrt{t+4}+2\right)}$

$\displaystyle \qquad\qquad = \lim_{t \to 0} \frac{t}{t\left(\sqrt{t+4}+2\right)} = \lim_{t \to 0} \frac{1}{\sqrt{t+4}+2} = \frac{1}{4}.$

2.2.35: $\displaystyle \lim_{x \to 4} \frac{x^2-16}{2-\sqrt{x}} = \lim_{x \to 4} \frac{(x+4)(\sqrt{x}-2)(\sqrt{x}+2)}{2-\sqrt{x}} = \lim_{x \to 4}\left[-(x+4)(\sqrt{x}+2)\right] = -32.$

2.2.37: $\dfrac{f(x+h)-f(x)}{h} = \dfrac{(x+h)^3 - x^3}{h} = \dfrac{x^3 + 3x^2h + 3xh^2 + h^3 - x^3}{h} = 3x^2 + 3xh + h^2 \to 3x^2$ as $h \to 0$. When $x = 2$, $y = f(2) = x^3 = 8$ and the slope of the tangent line to this curve at $x = 2$ is $3x^2 = 12$, so an equation of this tangent line is $y = 12x - 16$.

2.2.39: $\dfrac{f(x+h)-f(x)}{h} = \dfrac{\dfrac{1}{(x+h)^2} - \dfrac{1}{x^2}}{h} = \dfrac{x^2 - (x+h)^2}{hx^2(x+h)^2} = \dfrac{-2x - h}{x^2(x+h)^2} \to -\dfrac{2}{x^3}$ as $h \to 0$. When $x = 2$, $y = f(2) = \frac{1}{4}$ and the slope of the line tangent to this curve at $x = 2$ is $-\frac{1}{4}$, so an equation of this tangent line is $y - \frac{1}{4} = -\frac{1}{4}(x - 2)$; that is, $y = -\frac{1}{4}(x - 3)$.

2.2.41: $\dfrac{f(x+h)-f(x)}{h} = \dfrac{\left(\dfrac{2}{x+h-1}\right) - \left(\dfrac{2}{x-1}\right)}{h} = \dfrac{2(x - 1 - x - h + 1)}{h(x-1)(x+h-1)} = \dfrac{-2}{(x-1)(x+h-1)}.$

This approaches $\dfrac{-2}{(x-1)^2}$ as h approaches 0. When $x = 2$, $y = f(2) = 2$ and the slope of the line tangent to this curve at $x = 2$ is -2, so an equation of this tangent line is $y - 2 = -2(x - 2)$; alternatively, $y = -2(x - 3)$.

2.2.43:
$$\dfrac{f(x+h)-f(x)}{h} = \dfrac{\left(\dfrac{1}{\sqrt{x+h+2}}\right) - \left(\dfrac{1}{\sqrt{x+2}}\right)}{h}$$
$$= \left(\dfrac{\sqrt{x+2} - \sqrt{x+h+2}}{h\sqrt{x+2}\sqrt{x+h+2}}\right) \cdot \left(\dfrac{\sqrt{x+2} + \sqrt{x+h+2}}{\sqrt{x+2} + \sqrt{x+h+2}}\right)$$
$$= \dfrac{-h}{h\sqrt{x+2}\sqrt{x+h+2}\left(\sqrt{x+2} + \sqrt{x+h+2}\right)} \to \dfrac{-1}{(x+2)\left(2\sqrt{x+2}\right)}$$

as $h \to 0$. When $x = 2$, $y = f(2) = \frac{1}{2}$ and the slope of the line tangent to this curve at $x = 2$ is $-\frac{1}{16}$, so an equation of this tangent line is $y - \frac{1}{2} = -\frac{1}{16}(x - 2)$; that is, $y = -\frac{1}{16}(x - 10)$.

2.2.45:
$$\dfrac{f(x+h)-f(x)}{h} = \dfrac{\sqrt{2(x+h)+5} - \sqrt{2x+5}}{h}$$
$$= \left(\dfrac{\sqrt{2(x+h)+5} - \sqrt{2x+5}}{h}\right) \cdot \left(\dfrac{\sqrt{2(x+h)+5} + \sqrt{2x+5}}{\sqrt{2(x+h)+5} + \sqrt{2x+5}}\right)$$
$$= \dfrac{2}{\sqrt{2(x+h)+5} + \sqrt{2x+5}} \to \dfrac{1}{\sqrt{2x+5}} \text{ as } h \to 0.$$

When $x = 2$, $y = f(2) = 3$ and the slope of the line tangent to this curve at $x = 2$ is $\frac{1}{3}$, so an equation of this tangent line is $y - 3 = \frac{1}{3}(x - 2)$; if you prefer, $y = \frac{1}{3}(x + 7)$.

2.2.47:

x	10^{-2}	10^{-4}	10^{-6}	10^{-8}	10^{-10}
$f(x)$	2.01	2.001	2.	2.	2.

x	-10^{-2}	-10^{-4}	-10^{-6}	-10^{-8}	-10^{-10}
$f(x)$	1.99	1.999	2.	2.	2.

The limit appears to be 2.

2.2.49:

x	10^{-2}	10^{-4}	10^{-6}	10^{-8}	10^{-10}
$f(x)$	0.16662	0.166666	0.166667	0.166667	0.166667

x	-10^{-2}	-10^{-4}	-10^{-6}	-10^{-8}	-10^{-10}
$f(x)$	1.66713	0.166667	0.166667	0.166667	0.166667

The limit appears to be $\frac{1}{6}$.

2.2.51:

x	10^{-2}	10^{-4}	10^{-6}	10^{-8}	10^{-10}
$f(x)$	-0.37128	-0.374963	-0.375	-0.375	-0.375

x	-10^{-2}	-10^{-4}	-10^{-6}	-10^{-8}	-10^{-10}
$f(x)$	-0.378781	-0.375038	-0.375	-0.375	-0.375

The limit appears to be $-\frac{3}{8}$.

2.2.53:

x	10^{-2}	10^{-4}	10^{-6}	10^{-8}	10^{-10}
$f(x)$	0.999983	1.	1.	1.	1.

x	-10^{-2}	-10^{-4}	-10^{-6}	-10^{-8}	-10^{-10}
$f(x)$	0.999983	1.	1.	1.	1.

The limit appears to be 1.

2.2.55:

x	10^{-2}	10^{-4}	10^{-6}	10^{-8}	10^{-10}
$f(x)$	0.166666	0.166667	0.166667	0.166667	0.166667

x	-10^{-2}	-10^{-4}	-10^{-6}	-10^{-8}	-10^{-10}
$f(x)$	0.166666	0.166667	0.166667	0.166667	0.166667

The limit appears to be $\frac{1}{6}$.

2.2.57:

x	2^{-1}	2^{-5}	2^{-10}	2^{-15}	2^{-20}
$f(x)$	2.25	2.67699	2.71696	2.71824	2.71828

x	-2^{-1}	-2^{-5}	-2^{-10}	-2^{-15}	-2^{-20}
$f(x)$	4.	2.76210	2.71961	2.71832	2.71828

Perhaps the limit is the special number $e \approx 2.71828$ that was mentioned in the subsection on natural logarithms in Section 1.4. However, the numerical data here are insufficient to be convincing.

2.2.59: $\lim\limits_{x \to 0} \dfrac{x - \tan x}{x^3} = -\dfrac{1}{3}$.   Answer: -0.3333.

2.2.61: $\sin\left(\dfrac{\pi}{2^{-n}}\right) = \sin\left(2\pi \cdot 2^{(n-1)}\right) = 0$ for every positive integer n. Therefore $\lim\limits_{x \to 0} \sin\left(\dfrac{\pi}{x}\right)$, if it were to exist, would be 0. Notice however that $\sin\left(3^n \cdot \dfrac{\pi}{2}\right)$ alternates between $+1$ and -1 for $n = 1, 2, 3, \ldots$. Therefore $\lim\limits_{x \to 0} \sin\left(\dfrac{\pi}{x}\right)$ does not exist.

2.2.63: The graph of $f(x) = \left(\log_{10}\left(1/|x|\right)\right)^{-1/32}$ is shown next, as well as a table of values of $f(x)$ for x very close to zero. The table was generated by *Mathematica,*, but virtually any computer algebra system will produce similar results.

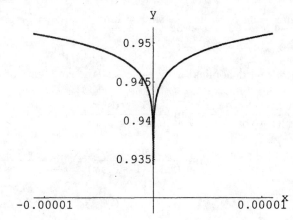

x	$f(x)$	x	$f(x)$	x	$f(x)$	x	$f(x)$
10^{-1}	1.0000	10^{-6}	0.9455	10^{-11}	0.9278	10^{-16}	0.9170
10^{-2}	0.9786	10^{-7}	0.9410	10^{-12}	0.9253	10^{-17}	0.9153
10^{-3}	0.9663	10^{-8}	0.9371	10^{-13}	0.9230	10^{-18}	0.9136
10^{-4}	0.9576	10^{-9}	0.9336	10^{-14}	0.9208	10^{-19}	0.9121
10^{-5}	0.9509	10^{-10}	0.9306	10^{-15}	0.9189	10^{-20}	0.9106

Section 2.3

2.3.1: $\theta \cdot \dfrac{\theta}{\sin \theta} \to 0 \cdot 1 = 0$ as $\theta \to 0$.

2.3.3: Multiply numerator and denominator by $1 + \cos \theta$ (the *conjugate* of the numerator) to obtain

$$\lim_{\theta \to 0} \frac{1 - \cos^2 \theta}{\theta^2 (1 + \cos \theta)} = \lim_{\theta \to 0} \frac{\sin \theta}{\theta} \cdot \frac{\sin \theta}{\theta} \cdot \frac{1}{1 + \cos \theta} = 1 \cdot 1 \cdot \frac{1}{2} = \frac{1}{2}.$$

31

2.3.5: Divide each term in numerator and denominator by t. Then it's clear that the denominator is approaching zero whereas the numerator is not, so the limit does not exist. Because the numerator is positive and the denominator is approaching zero through negative values, the answer $-\infty$ is also correct.

2.3.7: Let $z = 5x$. Then $z \to 0$ as $x \to 0$, and $\dfrac{\sin 5x}{x} = \dfrac{5 \sin z}{z} \to 5 \cdot 1 = 5$.

2.3.9: This limit does not exist because $\sqrt{x}$ is not defined for x near 0 if $x < 0$. But

$$\lim_{x \to 0^+} \frac{\sin x}{\sqrt{x}} = \lim_{x \to 0^+} \left(\sqrt{x}\right) \cdot \left(\frac{\sin x}{x}\right) = 0 \cdot 1 = 0.$$

2.3.11: Let $x = 3z$. Then $x \to 0$ is equivalent to $z \to 0$, and therefore

$$\lim_{x \to 0} \frac{1}{x} \sin \frac{x}{3} = \lim_{z \to 0} \frac{1}{3z} \sin z = \lim_{z \to 0} \frac{1}{3} \cdot \frac{\sin z}{z} = \frac{1}{3} \cdot 1 = \frac{1}{3}.$$

2.3.13: Multiply numerator and denominator by $1 + \cos x$ (the conjugate of the numerator) to obtain

$$\lim_{x \to 0} \frac{(1 - \cos x)(1 + \cos x)}{(\sin x)(1 + \cos x)} = \lim_{x \to 0} \frac{1 - \cos^2 x}{(\sin x)(1 + \cos x)}$$

$$= \lim_{x \to 0} \frac{\sin^2 x}{(\sin x)(1 + \cos x)} = \lim_{x \to 0} \frac{\sin x}{1 + \cos x} = \frac{0}{1 + 1} = 0.$$

2.3.15: Recall that $\sec x = \dfrac{1}{\cos x}$ and $\csc x = \dfrac{1}{\sin x}$. Hence

$$\lim_{x \to 0} x \sec x \csc x = \lim_{x \to 0} \frac{1}{\cos x} \cdot \frac{x}{\sin x} = \frac{1}{1} \cdot 1 = 1.$$

We also used the fact that

$$\lim_{x \to 0} \frac{x}{\sin x} = \lim_{x \to 0} \frac{1}{\dfrac{\sin x}{x}} = \frac{1}{1} = 1.$$

2.3.17: Multiply numerator and denominator by $1 + \cos \theta$ (the conjugate of the numerator) to obtain

$$\lim_{\theta \to 0} \frac{(1 - \cos \theta)(1 + \cos \theta)}{(\theta \sin \theta)(1 + \cos \theta)} = \lim_{\theta \to 0} \frac{\sin^2 \theta}{(\theta \sin \theta)(1 + \cos \theta)} = \lim_{\theta \to 0} \frac{\sin \theta}{\theta(1 + \cos \theta)}$$

$$= \lim_{\theta \to 0} \frac{\sin \theta}{\theta} \cdot \frac{1}{1 + \cos \theta} = 1 \cdot \frac{1}{1 + 1} = \frac{1}{2}.$$

2.3.19: $\displaystyle \lim_{z \to 0} \frac{\tan z}{\sin 2z} = \lim_{z \to 0} \frac{\sin z}{(\cos z)(2 \sin z \cos z)} = \lim_{z \to 0} \frac{1}{2 \cos^2 z} = \frac{1}{2 \cdot 1^2} = \frac{1}{2}.$

2.3.21: $\displaystyle \lim_{x \to 0} x \cot 3x = \lim_{x \to 0} \frac{x \cos 3x}{\sin 3x} = \lim_{x \to 0} \frac{3x}{\sin 3x} \cdot \frac{\cos 3x}{3} = 1 \cdot \frac{1}{3} = \frac{1}{3}$ (see Problem 15, last line).

2.3.23: Let $x = \frac{1}{2}t$. Then $x \to 0$ is equivalent to $t \to 0$, so

$$\lim_{t \to 0} \frac{\sin\left(\dfrac{t}{2}\right)}{\dfrac{t}{2}} = \lim_{x \to 0} \frac{\sin x}{x} = 1.$$

Therefore

$$\lim_{t \to 0} \frac{1}{t^2} \sin^2\left(\frac{t}{2}\right) = \lim_{t \to 0} \frac{1}{4} \cdot \frac{4}{t^2} \sin^2\left(\frac{t}{2}\right) = \lim_{t \to 0} \frac{1}{4} \cdot \left[\frac{\sin\left(\dfrac{t}{2}\right)}{\dfrac{t}{2}}\right]^2 = \frac{1}{4} \cdot 1^2 = \frac{1}{4}.$$

2.3.25: Because $-1 \leqq \cos 10x \leqq 1$ for all x, $-x^2 \leqq x^2 \cos 10x \leqq x^2$ for all x. But both $-x^2$ and x^2 approach zero as $x \to 0$. Therefore $\lim\limits_{x \to 0} x^2 \cos 10x = 0$. The second inequality is illustrated next.

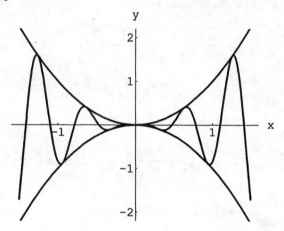

2.3.27: First, $-1 \leqq \cos x \leqq 1$ for all x. Therefore

$$-x^2 \leqq x^2 \cos \frac{1}{\sqrt[3]{x}} \leqq x^2$$

for all $x \neq 0$. Finally, both x^2 and $-x^2$ approach zero as $x \to 0$. Therefore, by the squeeze law,

$$\lim_{x \to 0} x^2 \cos \frac{1}{\sqrt[3]{x}} = 0.$$

2.3.29: $\lim\limits_{x \to 0^+} \left(3 - \sqrt{x}\right) = 3 - \sqrt{\lim\limits_{x \to 0^+} x} = 3 - 0 = 3.$

2.3.31: $\lim\limits_{x \to 1^-} \sqrt{x-1}$ does not exist because if $x < 1$, then $x - 1 < 0$.

2.3.33: Because $x \to 2^+$, $x > 2$, so that $x^2 > 4$. Hence $\sqrt{x^2 - 4}$ is defined for all such x and $\lim\limits_{x \to 2^+} \sqrt{x^2 - 4} = \sqrt{4 - 4} = 0.$

33

2.3.35: Because $x \to 5^-$, $x < 5$ and $x > 0$ for x sufficiently close to 5. Therefore $x(5 - x) > 0$ for such x, so that $\sqrt{x(5 - x)}$ exists for such x. Therefore

$$\lim_{x \to 5^-} \sqrt{x(5 - x)} = \sqrt{\left(\lim_{x \to 5^-} x\right)\left(5 - \lim_{x \to 5^-} x\right)} = \sqrt{5 \cdot 0} = 0.$$

2.3.37: As $x \to 4^+$, $x > 4$, so that both $4x$ and $x - 4$ are positive. Hence the radicand is positive and the square root exists. But the denominator in the radicand is approaching zero through positive values while the numerator is approaching 16. So the fraction is approaching $+\infty$. Therefore

$$\lim_{x \to 4^+} \sqrt{\frac{4x}{x - 4}} = +\infty.$$

It is also correct to say that this limit does not exist.

2.3.39: If $x < 5$, then $x - 5 < 0$, so $\dfrac{x - 5}{|x - 5|} = \dfrac{x - 5}{-(x - 5)} = -1$. Therefore the limit is -1.

2.3.41: If $x > 3$, then $x^2 - 6x + 9 = (x - 3)^2 > 0$ and $x - 3 > 0$, so $\dfrac{\sqrt{x^2 - 6x + 9}}{x - 3} = \dfrac{|x - 3|}{x - 3} = \dfrac{x - 3}{x - 3} \to 1$ as $x \to 3^+$.

2.3.43: If $x > 2$ then $x - 2 > 0$, so $\dfrac{2 - x}{|x - 2|} = \dfrac{2 - x}{x - 2} = -1$. Therefore the limit is also -1.

2.3.45: $\dfrac{1 - x^2}{1 - x} = \dfrac{(1 + x)(1 - x)}{1 - x} = 1 + x$, so the limit is 2.

2.3.47: Recall first that $\sqrt{z^2} = |z|$ for every real number z. Because $x \to 5^+$, $x > 5$, so $5 - x < 0$. Therefore $\sqrt{(5 - x)^2} = |5 - x| = -(5 - x) = x - 5$. Therefore

$$\lim_{x \to 5^+} \frac{\sqrt{(5 - x)^2}}{5 - x} = \lim_{x \to 5^+} \frac{x - 5}{5 - x} = \lim_{x \to 5^+} (-1) = -1.$$

2.3.49: The right-hand and left-hand limits both fail to exist at $a = 1$. The behavior of f near a is best described by observing that

$$\lim_{x \to 1^+} \frac{1}{x - 1} = +\infty \qquad \text{and} \qquad \lim_{x \to 1^-} \frac{1}{x - 1} = -\infty.$$

2.3.51: The right-hand and left-hand limits both fail to exist at $a = -1$. The behavior of f near a is best described by observing that

$$\lim_{x \to -1^+} \frac{x - 1}{x + 1} = -\infty \qquad \text{and} \qquad \lim_{x \to -1^-} \frac{x - 1}{x + 1} = +\infty.$$

34

2.3.53: The right-hand and left-hand limits both fail to exist at $a = -2$. If x is slightly greater than -2, then $1 - x^2$ is close to $1 - 4 = -3$, while $x + 2$ is a positive number close to zero. In this case $f(x)$ is a large negative number. Similarly, if x is slightly less than -2, then $1 - x^2$ is close to -3, while $x + 2$ is a negative number close to zero. In this case $f(x)$ is a large positive number. The behavior of f near -2 is best described by observing that

$$\lim_{x \to -2^+} \frac{1 - x^2}{x + 2} = -\infty \quad \text{and} \quad \lim_{x \to -2^-} \frac{1 - x^2}{x + 2} = +\infty.$$

2.3.55: The left-hand and right-hand limits both fail to exist at $x = 1$. To simplify $f(x)$, observe that

$$f(x) = \frac{|1 - x|}{(1 - x)^2} = \frac{|1 - x|}{|1 - x|^2} = \frac{1}{|1 - x|}.$$

Therefore we can describe the behavior of $f(x)$ near $a = 1$ in this way:

$$\lim_{x \to 1} f(x) = \lim_{x \to 1} \frac{1}{|1 - x|} = +\infty.$$

2.3.57: First simplify $f(x)$: If $x^2 \neq 4$ (that is, if $x \neq \pm 2$), then

$$f(x) = \frac{x - 2}{4 - x^2} = \frac{x - 2}{(2 + x)(2 - x)} = \frac{-1}{2 + x}.$$

So even though $f(2)$ does not exist, there is no real problem with the limit of $f(x)$ as $x \to 2$:

$$\lim_{x \to 2} f(x) = \lim_{x \to 2} \frac{-1}{2 + x} = -\frac{1}{4}.$$

But the left-hand and right-hand limits of $f(x)$ fail to exist at $x = -2$, because

$$\lim_{x \to -2^+} f(x) = \lim_{x \to -2^+} \frac{-1}{2 + x} = -\infty \quad \text{and} \quad \lim_{x \to -2^-} f(x) = \lim_{x \to -2^-} \frac{-1}{2 + x} = +\infty.$$

2.3.59: $\displaystyle\lim_{x \to 2^+} \frac{x^2 - 4}{|x - 2|} = 4$ and $\displaystyle\lim_{x \to 2^-} \frac{x^2 - 4}{|x - 2|} = -4$. The two-sided limit does not exist. The graph is shown next.

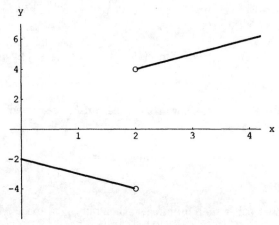

2.3.61: If x is an even integer then $f(x) = 3$, if x is an odd integer then $f(x) = 1$, and $\lim\limits_{x \to a} f(x) = 2$ for *all* real number values of a. The graph of f is shown next.

2.3.63: For any integer n, $\lim\limits_{x \to n^-} f(x) = 10n - 1$ and $\lim\limits_{x \to n^+} f(x) = 10n$. Note: $\lim\limits_{x \to a} f(x)$ exists if and only if $10a$ is not an integer. The graph is shown next.

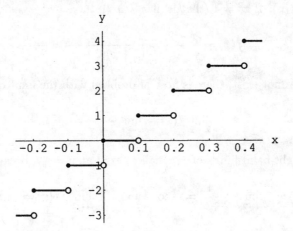

2.3.65: If n is an integer and $n < x < n + 1$, then write $x = n + t$ where $0 < t < 1$. Then $f(x) = n + t - n - \frac{1}{2} = t - \frac{1}{2}$. Moreover, $x \to n^+$ is equivalent to $t \to 0^+$. Therefore

$$\lim_{x \to n^+} f(x) = \lim_{x \to n^+} \left(t - \tfrac{1}{2} \right) = \lim_{t \to 0^+} \left(t - \tfrac{1}{2} \right) = -\frac{1}{2}.$$

Similar reasoning, with $n - 1 < x < n$, shows that if n is an integer, then

$$\lim_{x \to n^-} f(x) = \frac{1}{2}.$$

Finally, if a is a real number other than an integer, then $\lim\limits_{x \to a} f(x)$ exists. The graph of f is next.

36

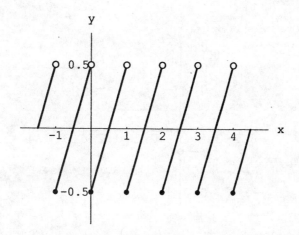

2.3.67: If x is an integer, then $f(x) = x - x = 0$. If x is not an integer, choose the [unique] integer n such that $n < x < n+1$. Then $-(n+1) < -x < -n$, so $f(x) = n - (n+1) = -1$. Therefore

$$\lim_{x \to a} f(x) = \lim_{x \to a} (-1) = -1$$

for every real number a. In particular, for every integer n,

$$\lim_{x \to n^-} f(x) = -1 \quad \text{and} \quad \lim_{x \to n^+} f(x) = -1.$$

The graph of f is shown next.

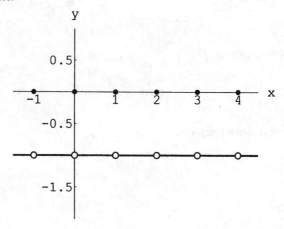

2.3.69: The values of a for which $\lim_{x \to a} g(x)$ exists are those real numbers not integral multiples of $\frac{1}{10}$. If b is an integral multiple of $\frac{1}{10}$, then

$$\lim_{x \to b^-} g(x) = b - \frac{1}{10} \quad \text{and} \quad \lim_{x \to b^+} g(x) = b.$$

The graph of g is shown next.

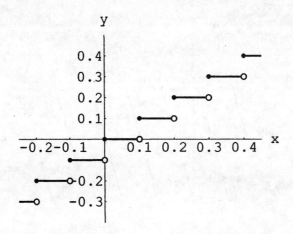

2.3.71: Because $-x^2 \leqq f(x) \leqq x^2$ for all x and because $-x^2 \to 0$ and $x^2 \to 0$ as $x \to 0$, it follows from the squeeze law for limits that $\lim\limits_{x \to 0} f(x) = 0 = f(0)$.

2.3.73: Given: $f(x) = x \cdot [\![1/x]\!]$. Let's first study the right-hand limit of $f(x)$ at $x = 0$. We need consider only values of x in the interval $(0, 1)$, and if $0 < x < 1$ then

$$1 < \frac{1}{x}, \qquad \text{so that} \qquad n \leqq \frac{1}{x} < n + 1$$

for some [unique] positive integer n. Moreover, if so then

$$\frac{1}{n+1} < x \leqq \frac{1}{n}.$$

Therefore $f(x) = x \cdot n$, so that

$$\frac{n}{n+1} < f(x) \leqq \frac{n}{n} = 1. \tag{1}$$

As $x \to 0^+$, $n \to \infty$, so the bounds on $f(x)$ in (1) both approach 1. Therefore

$$\lim_{x \to 0^+} f(x) = 1.$$

A similar (but slightly more delicate) argument shows that $f(x) \to 1$ as $x \to 0^-$ as well. Therefore $\lim\limits_{x \to 0} f(x)$ exists and is equal to 1. The graph of f is next.

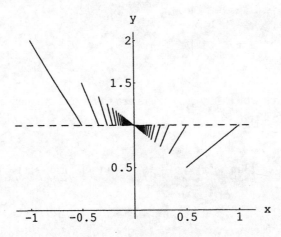

2.3.75: Given $\epsilon > 0$, let $\delta = \epsilon/7$. Suppose that

$$0 < |x - (-3)| < \delta.$$

Then

$$|x + 3| < \frac{\epsilon}{7};$$

$$|7x + 21| < \epsilon;$$

$$|7x - 9 + 30| < \epsilon;$$

$$|(7x - 9) - (-30)| < \epsilon.$$

Therefore, by definition, $\lim_{x \to -3} (7x - 9) = -30$.

2.3.77: Definition: We say that the number L is the **right-hand limit** of the function f at $x = a$ provided that, for every $\epsilon > 0$, there exists $\delta > 0$ such that, if $0 < |x - a| < \delta$ and $x > a$, then $|f(x) - L| < \epsilon$.

To prove that $\lim_{x \to 0^+} \sqrt{x} = 0$, suppose that $\epsilon > 0$ is given. Let $\delta = \epsilon^2$. Suppose that $|x - 0| < \delta$ and that $x > 0$. Then $0 < x < \delta = \epsilon^2$. Hence $\sqrt{x} < \epsilon$, and therefore

$$|\sqrt{x} - 0| < \epsilon.$$

So, by definition, $\lim_{x \to 0^+} \sqrt{x} = 0$.

2.3.79: Suppose that $\epsilon > 0$ is given. Let δ be the minimum of the two numbers 1 and $\epsilon/5$ and suppose that $0 < |x - 2| < \delta$. Then

$$|x - 2| < 1;$$

$$-1 < x - 2 < 1;$$

$$3 < x + 2 < 5;$$

$$|x + 2| < 5.$$

Therefore

$$|x^2 - 4| = |x + 2| \cdot |x - 2| < 5 \cdot \delta \leqq 5 \cdot \frac{\epsilon}{5} = \epsilon.$$

Hence, by definition, $\lim_{x \to 2} x^2 = 4$.

2.3.81: Given $\epsilon > 0$, let δ be the minimum of 1 and $\epsilon/29$. Suppose that $0 < |x - 10| < \delta$. Then

$$0 < |x - 10| < 1;$$

$$-1 < x - 10 < 1;$$

$$-2 < 2x - 20 < 2;$$

$$25 < 2x + 7 < 29;$$

$$|2x + 7| < 29.$$

Thus

$$|(2x^2 - 13x - 25) - 45| = |2x + 7| \cdot |x - 10| < 29 \cdot \delta \leqq 29 \cdot \frac{\epsilon}{29} = \epsilon.$$

Therefore, by definition, $\lim\limits_{x \to 10} (2x^2 - 13x - 25) = 45$.

2.3.83: In Problem 78 we showed that if $a = 0$, then

$$\lim_{x \to a} x^2 = \lim_{x \to 0} x^2 = 0 = 0^2 = a^2,$$

so the result we are to prove here holds when $a = 0$. Next case: Suppose that $a > 0$. Let $\epsilon > 0$ be given. Choose δ to be the minimum of the numbers 1 and $\epsilon/(2a+1)$. Note that $\delta > 0$. Suppose that $0 < |x - a| < \delta$. Then

$$|x - a| < 1;$$

$$-1 < x - a < 1;$$

$$2a - 1 < x + a < 2a + 1;$$

$$|x + a| < 2a + 1.$$

Thus

$$|x^2 - a^2| = |x + a| \cdot |x - a| < (2a + 1) \cdot \frac{\epsilon}{2a + 1} = \epsilon.$$

Therefore, by definition, $\lim\limits_{x \to a} x^2 = a^2$ if $a > 0$.

Final case: $a < 0$. Given $\epsilon > 0$, let

$$\delta = \min\left\{ 1, \frac{\epsilon}{|2a - 1|} \right\}.$$

Note that $\delta > 0$. Suppose that $0 < |x - a| < \delta$. The

$$|x - a| < 1;$$

$$-1 < x - a < 1;$$

$$2a - 1 < x + a < 2a + 1;$$

$$|x + a| < |2a - 1|$$

(because $|2a - 1| > |2a + 1|$ if $a < 0$). It follows that

$$|x^2 - a^2| = |x + a| \cdot |x - a| < |2a - 1| \cdot \frac{\epsilon}{|2a - 1|} = \epsilon.$$

Therefore, by definition, $\lim_{x \to a} x^2 = a^2$ if $a < 0$.

Section 2.4

2.4.1: Suppose that a is a real number. Then

$$\lim_{x \to a} f(x) = \lim_{x \to a} (2x^5 - 7x^2 + 13) = \left(\lim_{x \to a} 2x^5 \right) - \left(\lim_{x \to a} 7x^2 \right) + \left(\lim_{x \to a} 13 \right)$$

$$= \left(\lim_{x \to a} 2 \right) \left(\lim_{x \to a} x \right)^5 - \left(\lim_{x \to a} 7 \right) \left(\lim_{x \to a} x \right)^2 + 13 = 2a^5 - 7a^2 + 13 = f(a).$$

Therefore f is continuous at x for every real number x.

2.4.3: Suppose that a is a real number. Then

$$\lim_{x \to a} g(x) = \lim_{x \to a} \frac{2x - 1}{4x^2 + 1} = \frac{\lim_{x \to a} (2x - 1)}{\lim_{x \to a} (4x^2 + 1)}$$

$$= \frac{\left(\lim_{x \to a} 2 \right) \left(\lim_{x \to a} x \right) - \left(\lim_{x \to a} 1 \right)}{\left(\lim_{x \to a} 4 \right) \left(\lim_{x \to a} x \right)^2 + \left(\lim_{x \to a} 1 \right)} = \frac{2a - 1}{4a^2 + 1} = g(a).$$

Therefore g is continuous at x for every real number x.

2.4.5: Suppose that a is a fixed real number. Then $a^2 + 4a + 5 = (a + 2)^2 + 1 > 0$, so $h(a)$ is defined. Moreover,

$$\lim_{x \to a} h(x) = \lim_{x \to a} \sqrt{x^2 + 4x + 5} = \left(\lim_{x \to a} (x^2 + 4x + 5) \right)^{1/2}$$

$$= \left[\left(\lim_{x \to a} x \right)^2 + \left(\lim_{x \to a} 4 \right) \left(\lim_{x \to a} x \right) + \left(\lim_{x \to a} 5 \right) \right]^{1/2} = \sqrt{a^2 + 4a + 5} = h(a).$$

Therefore, by definition, h is continuous at $x = a$. Because a is arbitrary, h is continuous at x for every real number x.

2.4.7: Suppose that a is a real number. Then $1 + \cos^2 a \neq 0$, so that $f(a)$ is defined. Note also that

$$\lim_{x \to a} \sin x = \sin a \quad \text{and} \quad \lim_{x \to a} \cos x = \cos a$$

because the sine and cosine functions are continuous at every real number (Theorem 1). Moreover,

$$\lim_{x \to a} f(x) = \lim_{x \to a} \frac{1 - \sin x}{1 + \cos^2 x} = \frac{\lim\limits_{x \to a}(1 - \sin x)}{\lim\limits_{x \to a}(1 + \cos^2 x)}$$

$$= \frac{\left(\lim\limits_{x \to a} 1\right) - \left(\lim\limits_{x \to a} \sin x\right)}{\left(\lim\limits_{x \to a} 1\right) + \left(\lim\limits_{x \to a} \cos x\right)^2} = \frac{1 - \sin a}{1 + \cos^2 a} = f(a).$$

Therefore f is continuous at a. Because a is arbitrary, f is continuous at x for every real number x.

2.4.9: If $a > -1$, then $f(a)$ exists because $a \neq -1$. Moreover,

$$\lim_{x \to a} f(x) = \lim_{x \to a} \frac{1}{x + 1} = \frac{\lim\limits_{x \to a} 1}{\left(\lim\limits_{x \to a} x\right) + \left(\lim\limits_{x \to a} 1\right)} = \frac{1}{a + 1} = f(a).$$

Therefore f is continuous on the interval $x > -1$.

2.4.11: Because $-\frac{3}{2} \leqq t \leqq \frac{3}{2}$, $0 \leqq 4t^2 \leqq 9$, so that the radicand in $g(t)$ is never negative. Therefore $g(a)$ is defined for every real number a in the interval $\left[-\frac{3}{2}, \frac{3}{2}\right]$, and

$$\lim_{t \to a} g(t) = \lim_{t \to a} (9 - 4t^2)^{1/2} = \left(\lim_{t \to a} (9 - 4t^2)\right)^{1/2}$$

$$= \left[\left(\lim_{t \to a} 9\right) - 4\left(\lim_{t \to a} t\right)^2\right]^{1/2} = (9 - 4a^2)^{1/2} = g(a).$$

Therefore g is continuous at a for every real number a in $\left[-\frac{3}{2}, \frac{3}{2}\right]$.

2.4.13: If $-\frac{1}{2}\pi < x < \frac{1}{2}\pi$, then $\cos x \neq 0$, so $f(x)$ is defined for all such x. In addition, $\cos x \to \cos a$ as $x \to a$ because the cosine function is continuous everywhere (Theorem 1). Therefore

$$\lim_{x \to a} f(x) = \lim_{x \to a} \frac{x}{\cos x} = \frac{\lim\limits_{x \to a} x}{\lim\limits_{x \to a} \cos x} = \frac{a}{\cos a} = f(a).$$

Therefore f is continuous at x if $-\frac{1}{2}\pi < x < \frac{1}{2}\pi$.

2.4.15: The root law of Section 2.2 implies that $g(x) = \sqrt[3]{x}$ is continuous on the set R of all real numbers. We know that the polynomial $h(x) = 2x$ is continuous on R (Section 2.4, page 88). Hence the sum $f(x) = h(x) + g(x)$ is continuous on R.

42

2.4.17: Because $f(x)$ is the quotient of continuous functions (the numerator and denominator are polynomials, continuous everywhere), f is continuous wherever its denominator is nonzero. Therefore f is continuous on its domain, the set of all real numbers other than -3.

2.4.19: Because $f(x)$ is the quotient of continuous functions (the numerator and denominator are polynomials, continuous everywhere), f is continuous wherever its denominator is nonzero. Therefore f is continuous on its domain, the set of all real numbers.

2.4.21: Note that $f(x)$ is not defined at $x = 5$, so it is not continuous there. Because $f(x) = 1$ for $x > 5$ and $f(x) = -1$ for $x < 5$, f is a polynomial on the interval $(5, +\infty)$ and a [another] polynomial on the interval $(-\infty, 5)$. Therefore f is continuous on its domain, the set of all real numbers other than 5.

2.4.23: Because $f(x)$ is the quotient of continuous functions (the numerator and denominator are polynomials, continuous everywhere), f is continuous wherever its denominator is nonzero. Therefore f is continuous on its domain, the set of all real numbers other than 2.

2.4.25: Let

$$h(x) = \frac{x+1}{x-1}.$$

Because $h(x)$ is the quotient of continuous functions (the numerator and denominator are polynomials, continuous everywhere), h is continuous wherever its denominator is nonzero. Therefore h is continuous on its domain, the set of all real numbers other than 1. Now let $g(x) = \sqrt[3]{x}$. By the root rule of Section 2.2, g is continuous everywhere. Therefore the composition $f(x) = g(h(x))$ is continuous on the set of all real numbers other than 1.

2.4.27: Because $f(x)$ is the quotient of continuous functions (the numerator and denominator are polynomials, continuous everywhere), f is continuous wherever its denominator is nonzero. Therefore f is continuous on its domain, the set of all real numbers other than 0 and 1.

2.4.29: Let $h(x) = 4 - x^2$. Then h is continuous everywhere because $h(x)$ is a polynomial. The root function $g(x) = \sqrt{x}$ is continuous for $x \geqq 0$ by the root rule of Section 2.2. Hence $g(h(x)) = \sqrt{4 - x^2}$ is continuous wherever $x^2 \leqq 4$; that is, on the interval $[-2, 2]$. The quotient

$$f(x) = \frac{x}{\sqrt{4 - x^2}} = \frac{x}{g(h(x))}$$

is continuous wherever the numerator is continuous (that's everywhere) and the denominator is both continuous and nonzero (that's the open interval $(-2, 2)$). Therefore f is continuous on the open interval $(-2, 2)$. That is, f is continuous on its domain.

2.4.31: Because $f(x)$ is the quotient of continuous functions, it is continuous where its denominator is nonzero; that is, if $x \neq 0$. Thus f is continuous on its domain and not continuous at $x = 0$ (because it is undefined there).

2.4.33: Given:

$$f(x) = \frac{1}{\sin 2x}.$$

The numerator in $f(x)$ is a polynomial, thus continuous everywhere. The denominator is the composition of a function continuous on the set of all real numbers (the sine function) with another continuous function (a polynomial), hence is also continuous everywhere. Thus because $f(x)$ is the quotient of continuous functions, it is continuous wherever its denominator is nonzero; that is, its only discontinuities occur when $\sin 2x = 0$. Thus f is continuous at every real number other than an integral multiple of $\pi/2$.

2.4.35: Given: $f(x) = \sin|x|$. The sine function is continuous on the set of all real numbers, as is the absolute value function. Therefore their composition f is continuous on the set R of all real numbers.

2.4.37: The function

$$f(x) = \frac{x}{(x+3)^3}$$

is not continuous when $x = -3$. This discontinuity is not removable because $f(x) \to -\infty$ as $x \to -3^+$, so that the limit of $f(x)$ at $x = -3$ does not exist.

2.4.39: First simplify $f(x)$:

$$f(x) = \frac{x-2}{x^2-4} = \frac{x-2}{(x+2)(x-2)} = \frac{1}{x+2} \quad \text{if} \quad x \neq 2.$$

Now $f(x)$ is not defined at $x = \pm 2$ because $x^2 - 4 = 0$ for such x. The discontinuity at -2 is not removable because $f(x) \to +\infty$ as $x \to -2^+$. But $f(x) \to \frac{1}{4}$ as $x \to 2$, so the discontinuity at $x = 2$ is removable; f can be made continuous at $x = 2$ by defining its value there to be its limit there, $\frac{1}{4}$.

2.4.41: Given:

$$f(x) = \frac{1}{1-|x|}.$$

The function f is not continuous at ± 1 because its denominator is zero if $x = -1$ and if $x = 1$. Because $f(x) \to +\infty$ as $x \to 1^-$ and as $x \to -1^+$ (consider separately the cases $x > 0$ and $x < 0$), these discontinuities are not removable; $f(x)$ has no limit at -1 or at 1.

2.4.43: If $x > 17$, then $x - 17 > 0$, so that

$$f(x) = \frac{x-17}{|x-17|} = \frac{x-17}{x-17} = 1.$$

44

But if $x < 17$, then $x - 17 < 0$, and thus

$$f(x) = \frac{x-17}{|x-17|} = \frac{x-17}{-(x-17)} = -1.$$

Therefore $h(x)$ has no limit as $x \to 17$ because its left-hand and right-hand limits there are unequal. Thus the discontinuity at $x = 17$ is not removable.

2.4.45: Although $f(x)$ is not continuous at $x = 0$ (because it is not defined there), this discontinuity is removable. For it is clear that $f(x) \to 0$ as $x \to 0^+$ and as $x \to 0^-$, so defining $f(0)$ to be 0, the limit of $f(x)$ at $x = 0$, will make f continuous there.

2.4.47: Although $f(x)$ is not continuous at $x = 0$ (because it is not defined there), this discontinuity is removable. For it is clear that $f(x) \to 1$ as $x \to 0^+$ and as $x \to 0^-$, so defining $f(0)$ to be 1, the limit of $f(x)$ at $x = 0$, will make f continuous there.

2.4.49: The given function is clearly continuous for all x *except possibly* for $x = 0$. For continuity at $x = 0$, the left-hand and right-hand limits of $f(x)$ must be the same there. But

$$\lim_{x \to 0^-} f(x) = \lim_{x \to 0^-} (x + c) = c$$

and $f(x) \to 4$ as $x \to 0^+$. So continuity of f at $x = 0$ can occur only if $c = 4$. Moreover, if $c = 4$, then (as we have seen) $f(x) \to 4$ as $x \to 0$ and $f(0) = 4$, so f will be continuous at $x = 0$ if and only if $c = 4$. Answer: $c = 4$.

2.4.51: Note that f is continuous at x if $x \neq 0$, because $f(x)$ is a polynomial for $x < 0$ and for $x > 0$ regardless of the value of c. To be continuous at $x = 0$, it's necessary that the left-hand and right-hand limits exist and are equal there. Now

$$\lim_{x \to 0^-} f(x) = \lim_{x \to 0^-} (c^2 - x^2) = c^2 \qquad \text{and} \qquad \lim_{x \to 0^+} f(x) = \lim_{x \to 0^+} 2(x - c)^2 = 2c^2,$$

and therefore continuity at $x = 0$ will hold if and only if $c^2 = 2c^2$; that is, if $c = 0$. And if so, then $f(0) = \lim_{x \to 0} f(x)$ as well, so f will be continuous at $x = 0$. Answer: $c = 0$.

2.4.53: Let $f(x) = x^2 - 5$. Then f is continuous everywhere because $f(x)$ is a polynomial. So f has the intermediate value property on the interval $[2, 3]$. Also $f(2) = -1 < 0 < 4 = f(3)$, so $f(c) = 0$ for some number c in $[2, 3]$. That is, $c^2 - 5 = 0$. Hence the equation $x^2 - 5 = 0$ has a solution in $[2, 3]$.

2.4.55: Let $f(x) = x^3 - 3x^2 + 1$. Then f is continuous everywhere because $f(x)$ is a polynomial. So f has the intermediate value property on the interval $[0, 1]$. Also $f(0) = 1 > 0 > -1 = f(1)$, so $f(c) = 0$ for some number c in $[0, 1]$. That is, $c^3 - 3c^2 + 1 = 0$. Hence the equation $x^3 - 3x^2 + 1 = 0$ has a solution in $[0, 1]$.

45

2.4.57: Let $f(x) = x^4 + 2x - 1$. Then f is continuous everywhere because $f(x)$ is a polynomial. So f has the intermediate value property on the interval $[0, 1]$. Also $f(0) = -1 < 0 < 2 = f(1)$, so $f(c) = 0$ for some number c in $[0, 1]$. That is, $c^4 + 2c - 1 = 0$. Hence the equation $x^4 + 2x - 1 = 0$ has a solution in $[0, 1]$.

2.4.59: Given: $f(x) = x^3 - 4x + 1$. Values of $f(x)$:

x	-3	-2	-1	0	1	2	3
$f(x)$	-14	1	4	1	-2	1	16

So $f(x_i) = 0$ for x_1 in $(-3, -2)$, x_2 in $(0, 1)$, and x_3 in $(1, 2)$. Because these intervals do not overlap, the equation $f(x) = 0$ has at least three real solutions. Because $f(x)$ is a polynomial of degree 3, that equation also has at most three real solutions. Therefore the equation $x^3 - 4x + 1 = 0$ has exactly three real solutions.

2.4.61: At time t, $[\![t]\!]$ years have elapsed, and at that point your starting salary has been multiplied by 1.06 exactly t times. Thus it is $S(t) = 25 \cdot (1.06)^{[\![t]\!]}$. Of course S is discontinuous exactly when t is an integer between 1 and 5. The graph is next.

2.4.63: The next figure shows the graphs of two such functions f and g, with $[a, b] = [1, 3]$, $p = 2$, and $q = 5$. Because f and g are continuous on $[a, b]$, so is $h = f - g$. Because $p \neq q$, $h(a) = p - q$ and $h(b) = q - p$ have opposite signs, so that 0 is an intermediate value of the continuous function h. Therefore $h(c) = 0$ for some number c in (a, b). That is, $f(c) = g(c)$. This concludes the proof. To construct the figure, we used (the given coefficients are approximate)

$$f(x) = 1.53045 + (0.172739)e^x \quad \text{and} \quad g(x) = 2.27857 + (2.05)x + (1.36429)x^2 - (0.692857)x^3.$$

46

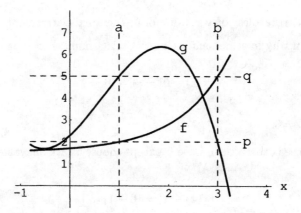

2.4.65: Given $a > 0$, let $f(x) = x^2 - a$. Then f is continuous on $[0, a+1]$ because $f(x)$ is a polynomial. Also $f(a+1) > 0$ because

$$f(a+1) = (a+1)^2 - a = a^2 + a + 1 > 1 > 0.$$

So $f(0) = -a < 0 < f(a+1)$. Therefore, because f has the intermediate value property on the interval $[0, a+1]$, there exists a number r in $(0, a+1)$ such that $f(r) = 0$. That is, $r^2 - a = 0$, so that $r^2 = a$. Therefore a has a square root.

Our proof shows that a has a positive square root. Can you modify it to show that a also has a negative square root? Do you see why we used the interval $[0, a+1]$ rather than the simpler $[0, a]$?

2.4.67: Given the real number a, we need to show that

$$\lim_{x \to a} \cos x = \cos a.$$

Let $h = x - a$, so that $x = a + h$. Then $x \to a$ is equivalent to $h \to 0$; also, $\cos x = \cos(a + h)$. Thus

$$\lim_{x \to a} \cos x = \lim_{h \to 0} \cos(a + h)$$
$$= \lim_{h \to 0} (\cos a \cos h - \sin a \sin h) = (\cos a) \cdot 1 - (\sin a) \cdot 0 = \cos a.$$

Therefore the cosine function is continuous at $x = a$ for every real number a.

2.4.69: Suppose that a is a real number. We appeal to the formal definition of the limit in Section 2.2 (page 74) to show that $f(x)$ has no limit as $x \to a$. Suppose by way of contradiction that $f(x) \to L$ as $x \to a$. Then, for every $\epsilon > 0$, there exists a number $\delta > 0$ such that $|f(x) - L| < \epsilon$ for every number x such that $0 < |x - a| < \delta$. So this statement must hold if $\epsilon = \frac{1}{4}$.

Case 1: $L = 0$. Then there must exist a number $\delta > 0$ such that

$$|f(x) - 0| < \tfrac{1}{4}$$

47

if $0 < |x - a| < \delta$. But, regardless of the value of δ, there exist irrational values of x satisfying this last inequality. (We'll explain why in a moment.) Choose such a number x. Then

$$|f(x) - 0| = |1 - 0| = 1 < \tfrac{1}{4}.$$

This is impossible. So $L \neq 0$.

Case 2: $L = 1$. Proceed exactly as in Case 1, except choose a *rational* value of x such that $0 < |x - a| < \delta$. Then

$$|f(x) - 1| = |0 - 1| = 1 < \tfrac{1}{4}.$$

This, too, is impossible. So $L \neq 1$.

Case 3: L is neither 0 nor 1. Let $\epsilon = \tfrac{1}{3}|L|$. Note that $\epsilon > 0$. Then suppose that there exists $\delta > 0$ such that,

$$\text{if} \qquad 0 < |x - a| < \delta, \qquad \text{then} \qquad |f(x) - L| < \epsilon. \tag{1}$$

Choose a rational number x satisfying the left-hand inequality. Then

$$|f(x) - L| = |0 - L| = |L| = 3\epsilon.$$

It follows from (1) that $3\epsilon < \epsilon$, which is impossible because $\epsilon > 0$.

In summary, L cannot be 0, nor can it be 1, nor can it be any other real number. Therefore $f(x)$ has no limit as $x \to a$. Consequently f is not continuous at $x = a$.

In this proof we relied heavily on the fact that if a is any real number, then we can find both rational and irrational numbers arbitrarily close to a. Rather than providing a formal proof, we illustrate how to do this in the case that

$$a = 1.23456789101112131415\cdots.$$

(It doesn't matter whether a is rational or irrational.) To produce rational numbers arbitrarily close to a, use

$$1.2, \ 1.23, \ 1.234, \ 1.2345, \ 1.23456, \ 1.234567, \ \ldots. \tag{2}$$

The numbers in (2) are all rational because they all have terminating decimal expansions, and the nth number in (2) differs from a by less than 10^{-n}, so there are rational numbers arbitrarily close to a. To get irrational numbers with the same properties, use

$$1.201001000100001000000100\cdots, \ 1.230100100010000100000100\cdots,$$

$$1.234010010001000010000000100\cdots, \ 1.234501001000100001000000100\cdots, \ \ldots.$$

These numbers are irrational because every one of them has a nonrepeating decimal expansion.

2.4.71: Let $g(x) = x - \cos x$. Then $g(x)$ is the sum of continuous functions, thus continuous everywhere, and in particular on the interval $\left[0, \frac{1}{2}\pi\right]$. So g has the intermediate value property there. Also $g(0) = -1 < 0$ and $g(1) = 1 > 0$. Therefore there exists a number c in $\left(0, \frac{1}{2}\pi\right)$ such that $g(c) = 0$. Thus $c - \cos x = 0$, so that $c = \cos c$. Thus the equation $x = \cos x$ has a solution in $\left(0, \frac{1}{2}\pi\right)$. (This solution is approximately 0.7390851.)

2.4.73: Because

$$\lim_{x \to 0^+} 2^{1/x} = +\infty,$$

f is not right continuous at $x = 0$. Because

$$\lim_{x \to 0^-} 2^{1/x} = \lim_{u \to -\infty} 2^u = \lim_{z \to +\infty} \frac{1}{2^z} = 0 = f(0),$$

f is left continuous at $x = 0$.

2.4.75: Because

$$\lim_{x \to 0^+} \frac{1}{1 + 2^{1/x}} = 0 \neq f(0),$$

f is not right continuous at $x = 0$. But

$$\lim_{x \to 0^-} \frac{1}{1 + 2^{1/x}} = \frac{1}{1 + 0} = 1 = f(0),$$

f is left continuous at $x = 0$.

2.4.77: We consider only the discontinuity at $x = a = \pi/2$; the behavior of f is the same near all of its discontinuities (the odd integral multiples of a). Because

$$\lim_{x \to a^+} \frac{1}{1 + 2^{\tan x}} = \frac{1}{1 + 0} = 1 = f(1),$$

the function f is right continuous at $x = a$. But

$$\lim_{x \to a^-} \frac{1}{1 + 2^{\tan x}} = 0 \neq f(0),$$

so f is not left continuous at $x = a$.

Chapter 2 Miscellaneous Problems

2.M.1: $\lim_{x \to 0} (x^2 - 3x + 4) = \left(\lim_{x \to 0} x\right)^2 - 3 \cdot \left(\lim_{x \to 0} x\right) + 4 = 0^2 - 3 \cdot 0 + 4 = 4.$

2.M.3: $\lim_{x \to 2} (4 - x^2)^{10} = \left[4 - \left(\lim_{x \to 2} x\right)^2\right]^{10} = \left(4 - 2^2\right)^{10} = 0^{10} = 0.$

2.M.5: $\displaystyle\lim_{x\to 2}\frac{1+x^2}{1-x^2} = \frac{1+\left(\displaystyle\lim_{x\to 2}x\right)^2}{1-\left(\displaystyle\lim_{x\to 2}x\right)^2} = \frac{1+2^2}{1-2^2} = \frac{1+4}{1-4} = -\frac{5}{3}.$

2.M.7: $\displaystyle\frac{x^2-1}{1-x} = -\frac{(x+1)(x-1)}{x-1} = -(x+1) \to -2$ as $x \to 1.$

2.M.9: $\displaystyle\frac{t^2+6t+9}{9-t^2} = -\frac{(t+3)^2}{(t+3)(t-3)} = -\frac{t+3}{t-3} \to -\frac{-3+3}{-3-3} = 0$ as $t \to -3.$

2.M.11: $\displaystyle\lim_{x\to 3}(x^2-1)^{2/3} = \left[\left(\lim_{x\to 3}x\right)^2 - 1\right]^{2/3} = (3^2-1)^{2/3} = 8^{2/3} = \left(8^{1/3}\right)^2 = 2^2 = 4.$

2.M.13: $\displaystyle\lim_{x\to 3}\left(\frac{5x+1}{x^2-8}\right)^{3/4} = \left[\frac{5\cdot\left(\lim_{x\to 3}x\right)+1}{\left(\lim_{x\to 3}x\right)^2 - 8}\right]^{3/4} = \left(16^{1/4}\right)^3 = 8.$

2.M.15: First multiply numerator and denominator by $\sqrt{x+2}+3$ (the *conjugate* of the numerator) to obtain

$$\lim_{x\to 7}\frac{\sqrt{x+2}-3}{x-7} = \lim_{x\to 7}\frac{x+2-9}{(x-7)\left(\sqrt{x+2}+3\right)} = \lim_{x\to 7}\frac{x-7}{(x-7)\left(\sqrt{x+2}+3\right)}$$

$$= \lim_{x\to 7}\frac{1}{\sqrt{x+2}+3} = \frac{\displaystyle\lim_{x\to 7}1}{\displaystyle\lim_{x\to 7}\left(\sqrt{x+2}+3\right)} = \frac{1}{\left(2+\displaystyle\lim_{x\to 7}x\right)^{1/2}+\displaystyle\lim_{x\to 7}3} = \frac{1}{3+3} = \frac{1}{6}.$$

2.M.17: First simplify:

$$\frac{\dfrac{1}{\sqrt{13+x}}-\dfrac{1}{3}}{x+4} = \frac{1}{x+4}\cdot\frac{3-\sqrt{13+x}}{3\sqrt{13+x}} = \frac{3-\sqrt{13+x}}{3(x+4)\sqrt{13+x}}.$$

Then we multiply both numerator and denominator by $3+\sqrt{13+x}$, the conjugate of the numerator, to obtain

$$\frac{9-(13+x)}{3(x+4)\sqrt{13+x}\left(3+\sqrt{13+x}\right)} = \frac{-(x+4)}{3(x+4)\sqrt{13+x}\left(3+\sqrt{13+x}\right)}$$

$$= -\frac{1}{3\sqrt{13+x}\left(3+\sqrt{13+x}\right)}.$$

Now let $x \to -4$ to obtain the limit $-\dfrac{1}{3\cdot 3\cdot(3+3)} = -\dfrac{1}{54}.$

2.M.19: First, $4-4x+x^2 = (2-x)^2 = (x-2)^2$. Because $x \to 2^+$, $x > 2$, so that $x-2 > 0$. Hence $\sqrt{4-4x+x^2} = \sqrt{(x-2)^2} = |x-2| = x-2$. Therefore

$$\lim_{x\to 2^+}\frac{2-x}{\sqrt{4-4x+x^2}} = \lim_{x\to 2^+}\frac{2-x}{x-2} = \lim_{x\to 2^+}(-1) = -1.$$

2.M.21: As $x \to 4^+$, $x > 4$, so that $x - 4 > 0$. Therefore $|x - 4| = x - 4$, and thus

$$\lim_{x \to 4^+} \frac{x - 4}{|x - 4|} = \lim_{x \to 4^+} \frac{x - 4}{x - 4} = \lim_{x \to 4^+} 1 = 1.$$

2.M.23: As $x \to 2^+$, $x > 2$, so that $4 - x^2 < 0$. Therefore $\sqrt{4 - x^2}$ is undefined for all such x, and consequently $\lim_{x \to 2^+} \sqrt{4 - x^2}$ does not exist.

2.M.25: As $x \to 2$, the denominator $(x - 2)^2$ is approaching zero, while the numerator $x + 2$ is approaching 4. So this limit does not exist. Because the denominator is approaching zero through positive values, it is also correct (and more informative) to write

$$\lim_{x \to 2} \frac{x + 2}{(x - 2)^2} = +\infty.$$

2.M.27: Because $x \to 3^+$, the denominator $x - 3$ is approaching zero, but the numerator x is not. Therefore this limit does not exist. Because the denominator is approaching zero through positive values while the numerator is approaching 3, it is also correct to write

$$\lim_{x \to 3^+} \frac{x}{x - 3} = +\infty.$$

2.M.29: As $x \to 1^-$, the numerator of the fraction is approaching 2, but the denominator is approaching zero. Therefore this limit does not exist. Because the denominator is approaching zero through negative values, it is also correct to write

$$\lim_{x \to 1^-} \frac{x + 1}{(x - 1)^3} = -\infty.$$

2.M.31: Let $u = 3x$. Then $x = \frac{1}{3}u$; also, $x \to 0$ is equivalent to $u \to 0$. Thus

$$\lim_{x \to 0} \frac{\sin 3x}{x} = \lim_{u \to 0} \frac{\sin u}{\frac{1}{3}u} = \lim_{u \to 0} \frac{3 \sin u}{u} = \left(\lim_{u \to 0} 3 \right) \cdot \left(\lim_{u \to 0} \frac{\sin u}{u} \right) = 3 \cdot 1 = 3.$$

2.M.33: The substitution $u = kx$ shows that if $k \neq 0$, then

$$\lim_{x \to 0} \frac{\sin kx}{kx} = 1.$$

It also follows that $\lim_{x \to 0} \frac{kx}{\sin kx} = 1$. Therefore

$$\lim_{x \to 0} \frac{\sin 3x}{\sin 2x} = \lim_{x \to 0} \frac{\sin 3x}{3x} \cdot \frac{2x}{\sin 2x} \cdot \frac{3x}{2x} = \lim_{x \to 0} \frac{\sin 3x}{3x} \cdot \frac{2x}{\sin 2x} \cdot \frac{3}{2} = 1 \cdot 1 \cdot \frac{3}{2} = \frac{3}{2}.$$

51

2.M.35: Let $x = u^2$ where $u > 0$. Then $x \to 0^+$ is equivalent to $u \to 0^+$. Hence

$$\lim_{x \to 0^+} \frac{x}{\sin \sqrt{x}} = \lim_{u \to 0^+} \frac{u^2}{\sin u} = \lim_{u \to 0^+} u \cdot \frac{u}{\sin u} = 0 \cdot 1 = 0.$$

2.M.37: First multiply numerator and denominator by the conjugate $1 + \cos 3x$ of the numerator:

$$\frac{1 - \cos 3x}{2x^2} = \frac{(1 - \cos 3x)(1 + \cos 3x)}{2x^2(1 + \cos 3x)} = \frac{1 - \cos^2 3x}{2x^2(1 + \cos 3x)} = \frac{\sin^2 3x}{2x^2(1 + \cos 3x)}$$

$$= \frac{\sin 3x}{2x} \cdot \frac{\sin 3x}{x} \cdot \frac{1}{1 + \cos 3x} = \frac{\sin 3x}{3x} \cdot \frac{3x}{2x} \cdot \frac{\sin 3x}{3x} \cdot \frac{3x}{x} \cdot \frac{1}{1 + \cos 3x}$$

$$= \frac{\sin 3x}{3x} \cdot \frac{3}{2} \cdot \frac{\sin 3x}{3x} \cdot \frac{3}{1} \cdot \frac{1}{1 + \cos 3x}.$$

Now let $x \to 0$ to obtain the limit $1 \cdot \dfrac{3}{2} \cdot 1 \cdot \dfrac{3}{1} \cdot \dfrac{1}{1+1} = \dfrac{9}{4}$.

2.M.39: Let $u = 2x$; then $x = \frac{1}{2}u$, and $x \to 0$ is then equivalent to $u \to 0$. Also express the secant and tangent functions in terms of the sine and cosine functions. Result:

$$\lim_{x \to 0} \frac{\sec 2x \tan 2x}{x} = \lim_{u \to 0} \frac{\sec u \tan u}{\frac{1}{2}u} = \lim_{u \to 0} \frac{2 \sin u}{u \cos^2 u} = \lim_{u \to 0} \frac{2}{\cos^2 u} \cdot \frac{\sin u}{u} = \frac{2}{1} \cdot 1 = 2.$$

2.M.41: Given $f(x) = 2x^2 + 3$, a slope-predictor for f is $m(x) = 4x$. The slope of the line tangent to the graph of f at $(1, f(1)) = (1, 5)$ is therefore $m(1) = 4$. So an equation of that line is $y - 5 = 4(x - 1)$; that is, $y = 4x + 1$.

2.M.43: Given $f(x) = 3x^2 + 4x - 5$, a slope-predictor for f is $m(x) = 6x + 4$. The slope of the line tangent to the graph of f at $(1, f(1)) = (1, 2)$ is therefore $m(1) = 10$. So an equation of that line is $y - 2 = 10(x-1)$; that is, $y = 10x - 8$.

2.M.45: Given $f(x) = (x - 1)(2x - 1) = 2x^2 - 3x + 1$, a slope-predictor for f is $m(x) = 4x - 3$. The slope of the line tangent to the graph of f at $(1, f(1)) = (1, 0)$ is therefore $m(1) = 1$. So an equation of that line is $y = x - 1$.

2.M.47: If $f(x) = 2x^2 + 3x$, then the slope-predicting function for f is

$$m(x) = \lim_{h \to 0} \frac{f(x+h) - f(x)}{h} = \lim_{h \to 0} \frac{2(x+h)^2 + 3(x+h) - (2x^2 + 3x)}{h}$$

$$= \lim_{h \to 0} \frac{2x^2 + 4xh + 2h^2 + 3x + 3h - 2x^2 - 3x}{h} = \lim_{h \to 0} \frac{4xh + 2h^2 + 3h}{h} = \lim_{h \to 0} \frac{h(4x + 2h + 3)}{h}$$

$$= \lim_{h \to 0} (4x + 2h + 3) = 4x + 3.$$

2.4.49: If $f(x) = \dfrac{1}{3-x}$, then the slope-predicting function for f is

$$m(x) = \lim_{h \to 0} \frac{f(x+h) - f(x)}{h} = \lim_{h \to 0} \frac{\dfrac{1}{3-(x+h)} - \dfrac{1}{3-x}}{h} = \lim_{h \to 0} \frac{1}{h} \cdot \frac{(3-x) - (3-x-h)}{(3-x-h)(3-x)}$$

$$= \lim_{h \to 0} \frac{1}{h} \cdot \frac{3-x-3+x+h}{(3-x-h)(3-x)} = \lim_{h \to 0} \frac{h}{h(3-x-h)(3-x)}$$

$$= \lim_{h \to 0} \frac{1}{(3-x-h)(3-x)} = \frac{1}{(3-x)^2}.$$

2.M.51: If $f(x) = x - \dfrac{1}{x}$, then the slope-predicting function for f is

$$m(x) = \lim_{h \to 0} \frac{f(x+h) - f(x)}{h} = \lim_{h \to 0} \frac{(x+h) - \dfrac{1}{x+h} - \left(x - \dfrac{1}{x}\right)}{h}$$

$$= \lim_{h \to 0} \frac{1}{h} \cdot \left(x + h - \frac{1}{x+h} - x + \frac{1}{x}\right) = \lim_{h \to 0} \frac{1}{h} \cdot \left(h + \frac{1}{x} - \frac{1}{x+h}\right)$$

$$= \lim_{h \to 0} \frac{1}{h} \cdot \left(h + \frac{x+h-x}{x(x+h)}\right) = \lim_{h \to 0} \left(1 + \frac{h}{hx(x+h)}\right)$$

$$= \lim_{h \to 0} \left(1 + \frac{1}{x(x+h)}\right) = 1 + \frac{1}{x^2} = \frac{x^2+1}{x^2}.$$

2.M.53: If $f(x) = \dfrac{x+1}{x-1}$, then the slope-predicting function for f is

$$m(x) = \lim_{h \to 0} \frac{f(x+h) - f(x)}{h} = \lim_{h \to 0} \frac{1}{h} \cdot \left(\frac{x+h+1}{x+h-1} - \frac{x+1}{x-1}\right)$$

$$= \lim_{h \to 0} \frac{1}{h} \cdot \frac{(x+h+1)(x-1) - (x+h-1)(x+1)}{(x+h-1)(x-1)}$$

$$= \lim_{h \to 0} \frac{(x^2 + hx + x - x - h - 1) - (x^2 + hx - x + x + h - 1)}{h(x+h-1)(x-1)}$$

$$= \lim_{h \to 0} \frac{x^2 + hx - h - 1 - x^2 - hx - h + 1}{h(x+h-1)(x-1)} = \lim_{h \to 0} \frac{-2h}{h(x+h-1)(x-1)}$$

$$= \lim_{h \to 0} \frac{-2}{(x+h-1)(x-1)} = -\frac{2}{(x-1)^2}.$$

2.M.55: Following the suggestion, the line tangent to the graph of $y = x^2$ at (a, a^2) has slope $2a$ (because the slope-predicting function for $f(x) = x^2$ is $m(x) = 2x$). But using the two-point formula for slope, this line also has slope

$$\frac{a^2 - 4}{a - 3} = 2a,$$

53

so that $a^2 - 4 = 2a^2 - 6a$; that is, $a^2 - 6a + 4 = 0$. The quadratic formula yields the two solutions $a = 3 \pm \sqrt{5}$, so one of the two lines in question has slope $2(3 + \sqrt{5})$ and the other has slope $2(3 - \sqrt{5})$. Both lines pass through $(3, 4)$, so their equations are

$$y - 4 = 2\left(3 + \sqrt{5}\right)(x - 3) \quad \text{and} \quad y - 4 = 2\left(3 - \sqrt{5}\right)(x - 3).$$

2.M.57: First simplify $f(x)$:

$$f(x) = \frac{1-x}{1-x^2} = \frac{1-x}{(1+x)(1-x)} = \frac{1}{1+x} \tag{1}$$

if $x \neq 1$. Every rational function is continuous wherever it is defined, so f is continuous except at ± 1. The computations in (1) show that $f(x)$ has no limit as $x \to -1$, so f cannot be made continuous at $x = -1$. But the discontinuity at $x = 1$ is removable; if we redefine f at $x = 1$ to be its limit $\frac{1}{2}$ there, then f will be continuous there as well.

2.M.59: First simplify $f(x)$:

$$f(x) = \frac{x^2 + x - 2}{x^2 + 2x - 3} = \frac{(x-1)(x+2)}{(x-1)(x+3)} = \frac{x+2}{x+3} \tag{1}$$

provided that $x \neq 1$. Note that f is a rational function, so f is continuous wherever it is defined: at every number other than 1 and -3. The computations in (1) show that $f(x)$ has no limit at $x = -3$, so it cannot be redefined in such a way to be continuous there. But the discontinuity at $x = 1$ is removable; if we redefine f at $x = 1$ to be its limit $\frac{3}{4}$ there, then f will be continuous everywhere except at $x = -3$.

2.M.61: Let $f(x) = x^5 + x - 1$. Then $f(0) = -1 < 0 < 1 = f(1)$. Because $f(x)$ is a polynomial, it is continuous on $[0, 1]$, so f has the intermediate value property there. Hence there exists a number c in $(0, 1)$ such that $f(c) = 0$. Thus $c^5 + c - 1 = 0$, and so the equation $x^5 + x - 1 = 0$ has a solution. (The value of c is approximately 0.754877666.)

2.M.63: Let $g(x) = x - \cos x$. Then $g(0) = -1 < 0 < \pi/2 = g(\pi/2)$. Because g is continuous, $g(c) = 0$ for some number c in $(0, \pi/2)$. That is, $c - \cos c = 0$, so that $c = \cos c$.

2.M.65: Suppose that L is a straight line through $\left(12, \frac{15}{2}\right)$ that is normal to the graph of $y = x^2$ at the point (a, a^2). The line tangent to the graph of $y = x^2$ at that point has slope $2a$, and the slope of L is then $-1/(2a)$. We can equate this to the slope of L found by using the two-point formula:

$$\frac{a^2 - \frac{15}{2}}{a - 12} = -\frac{1}{2a};$$

$$2a\left(a^2 - \frac{15}{2}\right) = -(a - 12);$$

$$2a^3 - 15a = -a + 12;$$

$$2a^3 - 14a - 12 = 0;$$

$$a^3 - 7a - 6 = 0.$$

54

By inspection, one solution of the last equation is $a = -1$. By the factor theorem of algebra, we know that $a - (-1) = a + 1$ is a factor of the polynomial $a^3 - 7a - 6$, and division of the former into the latter yields

$$a^3 - 7a - 6 = (a+1)(a^2 - a - 6) = (a+1)(a-3)(a+2).$$

So the equation $a^3 - 7a - 6 = 0$ has the three solutions $a = -1$, $a = 3$, and $a = -2$. Therefore there are *three* lines through $\left(12, \frac{15}{2}\right)$ that are normal to the graph of $y = x^2$, and their slopes are $\frac{1}{4}$, $\frac{1}{2}$, and $-\frac{1}{6}$.

Section 3.1

3.1.1: Given $f(x) = 4x - 5$, we have $a = 0$, $b = 4$, and $c = -5$, so $f'(x) = 2ax + b = 4$.

3.1.3: If $h(z) = z(25 - z) = -z^2 + 25z$, then $a = -1$, $b = 25$, and $c = 0$, so $h'(z) = 2az + b = -2z + 25$.

3.1.5: If $y = 2x^2 + 3x - 17$, then $a = 2$, $b = 3$, and $c = -17$, so $\dfrac{dy}{dx} = 2ax + b = 4x + 3$.

3.1.7: If $z = 5u^2 - 3u$, then $a = 5$, $b = -3$, and $c = 0$, so $\dfrac{dz}{du} = 2au + b = 10u - 3$.

3.1.9: If $x = -5y^2 + 17y + 300$, then $a = -5$, $b = 17$, and $c = 300$, so $\dfrac{dx}{dy} = 2ay + b = -10y + 17$.

3.1.11: $f'(x) = \displaystyle\lim_{h \to 0} \frac{f(x+h) - f(x)}{h} = \lim_{h \to 0} \frac{2(x+h) - 1 - (2x - 1)}{h} = \lim_{h \to 0} \frac{2x + 2h - 1 - 2x + 1}{h}$

$= \displaystyle\lim_{h \to 0} \frac{2h}{h} = \lim_{h \to 0} 2 = 2$.

3.1.13: $f'(x) = \displaystyle\lim_{h \to 0} \frac{f(x+h) - f(x)}{h} = \lim_{h \to 0} \frac{(x+h)^2 + 5 - (x^2 + 5)}{h}$

$= \displaystyle\lim_{h \to 0} \frac{x^2 + 2xh + h^2 + 5 - x^2 - 5}{h} = \lim_{h \to 0} \frac{2xh + h^2}{h} = \lim_{h \to 0} (2x + h) = 2x$.

3.1.15: $f'(x) = \displaystyle\lim_{h \to 0} \frac{f(x+h) - f(x)}{h} = \lim_{h \to 0} \frac{\dfrac{1}{2(x+h) + 1} - \dfrac{1}{2x + 1}}{h}$

$= \displaystyle\lim_{h \to 0} \frac{2x + 1 - (2x + 2h + 1)}{h(2x + 2h + 1)(2x + 1)} = \lim_{h \to 0} \frac{2x + 1 - 2x - 2h - 1}{h(2x + 2h + 1)(2x + 1)} = \lim_{h \to 0} \frac{-2h}{h(2x + 2h + 1)(2x + 1)}$

$= \displaystyle\lim_{h \to 0} \frac{-2}{(2x + 2h + 1)(2x + 1)} = \frac{-2}{(2x + 1)^2}$.

3.1.17: $f'(x) = \displaystyle\lim_{h \to 0} \frac{f(x+h) - f(x)}{h} = \lim_{h \to 0} \frac{\sqrt{2(x+h) + 1} - \sqrt{2x + 1}}{h}$

$= \displaystyle\lim_{h \to 0} \frac{\left(\sqrt{2x + 2h + 1} - \sqrt{2x + 1}\right)\left(\sqrt{2x + 2h + 1} + \sqrt{2x + 1}\right)}{h\left(\sqrt{2x + 2h + 1} + \sqrt{2x + 1}\right)}$

$= \displaystyle\lim_{h \to 0} \frac{(2h + 2x + 1) - (2x + 1)}{h\left(\sqrt{2x + 2h + 1} + \sqrt{2x + 1}\right)}$

$= \displaystyle\lim_{h \to 0} \frac{2h}{h\left(\sqrt{2x + 2h + 1} + \sqrt{2x + 1}\right)} = \lim_{h \to 0} \frac{2}{\sqrt{2x + 2h + 1} + \sqrt{2x + 1}}$

$= \dfrac{2}{2\sqrt{2x + 1}} = \dfrac{1}{\sqrt{2x + 1}}$.

3.1.19: $f'(x) = \displaystyle\lim_{h \to 0} \frac{1}{h}\left(f(x+h) - f(x)\right) = \lim_{h \to 0} \frac{1}{h}\left(\frac{x + h}{1 - 2(x+h)} - \frac{x}{1 - 2x}\right)$

$= \displaystyle\lim_{h \to 0} \frac{1}{h} \cdot \frac{(x+h)(1 - 2x) - (1 - 2x - 2h)(x)}{(1 - 2x - 2h)(1 - 2x)} = \lim_{h \to 0} \frac{(x - 2x^2 + h - 2xh) - (x - 2x^2 - 2xh)}{h(1 - 2x - 2h)(1 - 2x)}$

56

$$= \lim_{h \to 0} \frac{x - 2x^2 + h - 2xh - x + 2x^2 + 2xh}{h(1 - 2x - 2h)(1 - 2x)} = \lim_{h \to 0} \frac{h}{h(1 - 2x - 2h)(1 - 2x)}$$

$$= \lim_{h \to 0} \frac{1}{(1 - 2x - 2h)(1 - 2x)} = \frac{1}{(1 - 2x)^2}.$$

3.1.21: The velocity of the particle at time t is $\dfrac{dx}{dt} = v(t) = -32t$, so $v(t) = 0$ when $t = 0$. The position of the particle then is $x(0) = 100$.

3.1.23: The velocity of the particle at time t is $\dfrac{dx}{dt} = v(t) = -32t + 80$, so $v(t) = 0$ when $t = 2.5$. The position of the particle then is $x(2.5) = 99$.

3.1.25: The velocity of the particle at time t is $\dfrac{dx}{dt} = v(t) = -20 - 10t$, so $v(t) = 0$ when $t = -2$. The position of the particle then is $x(-2) = 120$.

3.1.27: The ball reaches its maximum height when its velocity $v(t) = \dfrac{dy}{dt} = -32t + 64$ is zero, and $v(t) = 0$ when $t = 2$. The height of the ball then is $y(2) = 64$ (ft).

3.1.29: The ball reaches its maximum height when its velocity $v(t) = \dfrac{dy}{dt} = -32t + 96$ is zero, and $v(t) = 0$ when $t = 3$. The height of the ball then is $y(3) = 194$ (ft).

3.1.31: Figure 3.1.23 shows a graph first decreasing, then with a horizontal tangent where $x = 1$, then increasing thereafter. So its derivative must be negative for $x < 1$, zero when $x = 1$, and positive for $x > 1$. This matches Fig. 3.1.28(e).

3.1.33: Figure 3.1.25 shows a graph decreasing for $x < -1.5$, increasing for $-1.5 < x < 0$, decreasing for $0 < 1 < 1.5$, and increasing for $1.5 < x$. Hence its derivative is negative for $x < -1.5$, positive for $-1.5 < x < 0$, negative for $0 < x < 1.5$, and positive for $1.5 < x$. Only the graph in Fig. 3.1.28(f) shows these characteristics.

3.1.35: Figure 3.1.27 shows a graph that increases, first slowly, then rapidly. So its derivative must exhibit the same behavior, and thus its graph is the one shown in Fig. 3.1.28(d).

3.1.37: Let r note the radius of the circle. Then $A = \pi r^2$ and $C = 2\pi r$. Thus

$$r = \frac{C}{2\pi}, \quad \text{and so} \quad A(C) = \frac{1}{4\pi}C^2, \quad C > 0.$$

Therefore the rate of change of A with respect to C is

$$A'(C) = \frac{dA}{dC} = \frac{1}{2\pi}C.$$

3.1.39: The velocity of the car (in feet per second) at time t (seconds) is $v(t) = x'(t) = 100 - 10t$. The car comes to a stop when $v(t) = 0$; that is, when $t = 10$. At that time the car has traveled a distance $x(10) = 500$ (ft). So the car skids for 10 seconds and skids a distance of 500 ft.

3.1.41: First, $P(t) = 100 + 30t + 4t^2$. The initial population is 100, so doubling occurs when $P(t) = 200$; that is, when $4t^2 + 30t - 100 = 0$. The quadratic formula yields $t = 2.5$ as the only positive solution of this equation, so the population will take two and one-half months to double. Because $P'(t) = 30 + 8t$, the rate of growth of the population when $P = 200$ will be $P'(2.5) = 50$ (chipmunks per month).

3.1.43: On our graph, the tangent line at the point $(20, 810)$ has slope $m_1 \approx 0.6$ and the tangent line at $(40, 2686)$ has slope $m_2 \approx 0.9$ (roughly). A line of slope 1 on our graph corresponds to a velocity of 125 ft/s (because the line through $(0, 0)$ and $(10, 1250)$ has slope 1), and thus we estimate the velocity of the car at time $t = 20$ to be about $(0.6)(125) = 75$ ft/s, and at time $t = 40$ it is traveling at about $(0.9)(125) = 112.5$ ft/s. The method is crude; another person with his own graph measured slightly smaller values of the two slopes and obtained the two velocities 68 ft/s and 102 ft/s shown in the back of the book. When we used the *Mathematica* function Fit to fit the data to a sixth-degree polynomial, we obtained

$$x(t) \approx 0.0000175721 + (6.500002)x + (1.112083)x^2 + (0.074188)x^3$$

$$- (0.00309375)x^4 + (0.0000481250)x^5 - (0.000000270834)x^6,$$

which yields $x'(20) \approx 74.3083$ and $x'(40) \approx 109.167$. Of course neither method is highly accurate.

3.1.45: With volume V and radius r, the volume of the sphere is $V(r) = \frac{4}{3}\pi r^3$. Then $\dfrac{dV}{dr} = 4\pi r^2$, and this is indeed the surface area of the sphere.

3.1.47: We must compute dV/dt when $t = 30$; $V(r) = \frac{4}{3}\pi r^3$ is the volume of the balloon when its radius is r. We are given $r = \dfrac{60 - t}{12}$, and thus

$$V(t) = \frac{4}{3}\pi\left(\frac{60 - t}{12}\right)^3 = \frac{\pi}{1296}(216000 - 10800t + 180t^2 - t^3).$$

Therefore

$$\frac{dV}{dt} = \frac{\pi}{1296}(-10800 + 360t - 3t^2),$$

and so $V'(t) = -\dfrac{25\pi}{12}$ in.^{3}/s; that is, air is leaking out at approximately 6.545 in.^{3}/s.

3.1.49: Let $V(t)$ denote the volume (in cm^3) of the snowball at time t (in hours) and let $r(t)$ denote its radius then. From the data given in the problem, $r = 12 - t$. The volume of the snowball is

$$V = \frac{4}{3}\pi r^3 = \frac{4}{3}\pi(12 - t)^3 = \frac{4}{3}\pi\left(1728 - 432t + 36t^2 - t^3\right),$$

so its instantaneous rate of change is

$$V'(t) = \frac{4}{3}\pi\left(-432 + 72t - 3t^2\right).$$

Hence its rate of change of volume when $t = 6$ is $V'(6) = -144\pi$ cm^3/h. Its average rate of change of volume from $t = 3$ to $t = 9$ in cm^3/h is

$$\frac{V(9) - V(3)}{9 - 3} = \frac{36\pi - 972\pi}{6} = -156\pi \text{ (cm}^3\text{/h)}.$$

3.1.51: The spaceship hits the ground when $25t^2 - 100t + 100 = 0$, which has solution $t = 2$. The velocity of the spaceship at time t is $y'(t) = 50t - 100$, so the speed of the spaceship at impact is (fortunately) zero.

3.1.53: The average rate of change of the population from January 1, 1990 to January 1, 2000 was

$$\frac{P(10) - P(0)}{10 - 0} = \frac{6}{10} = 0.6 \quad \text{(thousands per year)}.$$

The instantaneous rate of change of the population (in thousands per year, again) at time t was

$$P'(t) = 1 - (0.2)t + (0.018)t^2.$$

Using the quadratic formula to solve the equation $P'(t) = 0.6$, we find two solutions:

$$t = \frac{50 - 10\sqrt{7}}{9} \approx 2.6158318766 \quad \text{and} \quad t = \frac{50 + 10\sqrt{7}}{9} \approx 8.4952792345.$$

These values of t correspond to August 12, 1992 and June 30, 1998, respectively.

3.1.55: The graphs of the function of part (a) is shown next, on the left; the graph of the function of part (b) is on the right.

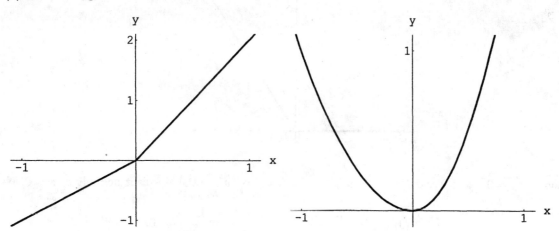

(a) $f'_-(0) = 1$ while $f'_+(0) = 2$. Hence f is not differentiable at $x = 0$. (b) In contrast,

$$f'_-(0) = \lim_{h \to 0^-} \frac{(0 + h)^2 - 2 \cdot 0^2}{h} = 0$$

and

$$f'_+(0) = \lim_{h \to 0^+} \frac{2 \cdot (0+h)^2 - 2 \cdot 0^2}{h} = 0;$$

therefore f is differentiable at $x = 0$ and $f'(0) = 0$.

3.1.57: Clearly f is differentiable except possibly at $x = 3$. Moreover,

$$f'_-(3) = \lim_{h \to 0^-} \frac{11 + 6 \cdot (3+h) - (3+h)^2 - 20}{h}$$

$$= \lim_{h \to 0^-} \frac{11 + 18 + 6h - 9 - 6h - h^2 - 20}{h} = \lim_{h \to 0^-} \frac{-h^2}{h} = 0$$

and

$$f'_+(3) = \lim_{h \to 0^+} \frac{(3+h)^2 - 6 \cdot (3+h) + 29 - 20}{h}$$

$$= \lim_{h \to 0^+} \frac{9 + 6h + h^2 - 18 - 6h + 29 - 20}{h} = \lim_{h \to 0^+} \frac{h^2}{h} = 0.$$

Therefore the function f is also differentiable at $x = 3$; moreover, $f'(3) = 0$.

3.1.59: The graph of $f(x) = x + |x|$ is shown next.

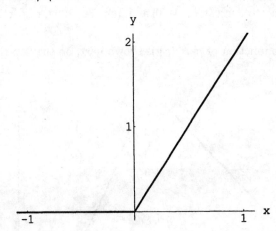

Because $f(x) = 0$ if $x < 0$ and $f(x) = 2x$ if $x > 0$, clearly f is differentiable except possibly at $x = 0$. Next,

$$f'_-(0) = \lim_{h \to 0^-} \frac{0 + h + |0 + h|}{h} = \lim_{h \to 0^-} \frac{h - h}{h} = 0,$$

whereas

$$f'_+(0) = \lim_{h \to 0^+} \frac{0 + h + |0 + h|}{h} = \lim_{h \to 0^+} \frac{2h}{h} = 2 \neq 0.$$

Therefore f is not differentiable at $x = 0$. In summary, $f'(x) = 2$ if $x > 0$ and $f'(x) = 0$ if $x < 0$. For a "single-formula" version of the derivative, consider

$$f'(x) = 1 + \frac{|x|}{x}.$$

Section 3.2

3.2.1: Given: $f(x) = 3x^2 - x + 5$. We apply the rule for differentiating a linear combination and the power rule to obtain

$$f'(x) = 3D_x(x^2) - D_x(x) + D_x(5) = 3 \cdot 2x - 1 + 0 = 6x - 1.$$

3.2.3: Given: $f(x) = (2x + 3)(3x - 2)$. We apply the product rule to obtain

$$f'(x) = (2x + 3)D_x(3x - 2) + (3x - 2)D_x(2x + 3) = (2x + 3) \cdot 3 + (3x - 2) \cdot 2 = 12x + 5.$$

3.2.5: Given: $h(x) = (x + 1)^3$. We rewrite $h(x)$ in the form

$$h(x) = (x + 1)(x + 1)(x + 1)$$

and then apply the extended product rule in Eq. (16) to obtain

$$h'(x) = (x + 1)(x + 1)D_x(x + 1) + (x + 1)(x + 1)D_x(x + 1) + (x + 1)(x + 1)D_x(x + 1)$$

$$= (x + 1)(x + 1)(1) + (x + 1)(x + 1)(1) + (x + 1)(x + 1)(1) = 3(x + 1)^2.$$

Alternatively, we could rewrite $h(x)$ in the form

$$h(x) = x^3 + 3x^2 + 3x + 1$$

and then apply the power rule and the rule for differentiating a linear combination to obtain

$$h'(x) = D_x(x^3) + 3D_x(x^2) + 3D_x(x) + D_x(1) = 3x^2 + 6x + 3.$$

The first method gives the answer in a more useful factored form that makes it easier to determine where $h'(x)$ is positive, where negative, and where zero.

3.2.7: Given: $f(y) = y(2y - 1)(2y + 1)$. We apply the extended product rule in Eq. (16) to obtain

$$f'(y) = (2y - 1)(2y + 1)D_y(y) + y(2y + 1)D_y(2y - 1) + y(2y - 1)D_y(2y + 1)$$

$$= (2y - 1)(2y + 1) \cdot 1 + y(2y + 1) \cdot 2 + y(2y - 1) \cdot 2 = 4y^2 - 1 + 4y^2 + 2y + 4y^2 - 2y = 12y^2 - 1.$$

Alternatively, we could first expand: $f(y) = 4y^3 - y$. Then we could apply the rule for differentiating a linear combination and the power rule to obtain $f'(y) = 4D_y(y^3) - D_y(y) = 12y^2 - 1$.

3.2.9: We apply the rule for differentiating a linear combination and the reciprocal rule (twice) to obtain

$$g'(x) = D_x\left(\frac{1}{x+1}\right) - D_x\left(\frac{1}{x-1}\right)$$

$$= -\frac{D_x(x+1)}{(x+1)^2} + \frac{D_x(x-1)}{(x-1)^2} = -\frac{1}{(x+1)^2} + \frac{1}{(x-1)^2}.$$

Looking ahead to later sections and chapters—in which we will want to find where $g'(x)$ is positive, negative, or zero—it would be good practice to simplify $g'(x)$ to

$$g'(x) = \frac{(x+1)^2 - (x-1)^2}{(x+1)^2(x-1)^2} = \frac{4x}{(x+1)^2(x-1)^2}.$$

3.2.11: First write (or think of) $h(x)$ as

$$h(x) = 3 \cdot \frac{1}{x^2 + x + 1},$$

then apply the rule for differentiating a linear combination and the reciprocal rule to obtain

$$h'(x) = 3 \cdot \left(-\frac{D_x(x^2 + x + 1)}{(x^2 + x + 1)^2}\right) = \frac{-3 \cdot (2x+1)}{(x^2 + x + 1)^2}.$$

Alternatively apply the quotient rule directly to obtain

$$h'(x) = \frac{(x^2 + x + 1)D_x(3) - 3D_x(x^2 + x + 1)}{(x^2 + x + 1)^2} = \frac{-3 \cdot (2x+1)}{(x^2 + x + 1)^2}.$$

3.2.13: Given $g(t) = (t^2 + 1)(t^3 + t^2 + 1)$, apply the product rule, the rule for differentiating a linear combination, and the power rule to obtain

$$g'(t) = (t^2 + 1)D_t(t^3 + t^2 + 1) + (t^3 + t^2 + 1)D_t(t^2 + 1)$$

$$= (t^2 + 1)(3t^2 + 2t + 0) + (t^3 + t^2 + 1)(2t + 0)$$

$$= (3t^4 + 2t^3 + 3t^2 + 2t) + (2t^4 + 2t^3 + 2t) = 5t^4 + 4t^3 + 3t^2 + 4t.$$

Alternatively, first expand: $g(t) = t^5 + t^4 + t^3 + 2t^2 + 1$, then apply the rule for differentiating a linear combination and the power rule.

3.2.15: The easiest way to find $g'(z)$ is first to rewrite $g(z)$:

$$g(z) = \frac{1}{2z} - \frac{1}{3z^2} = \frac{1}{2}z^{-1} - \frac{1}{3}z^{-2}.$$

62

Then apply the linear combination rule and the power rule (for negative integral exponents) to obtain

$$g'(z) = \frac{1}{2}(-1)z^{-2} - \frac{1}{3}(-2)z^{-3} = -\frac{1}{2z^2} + \frac{2}{3z^3} = \frac{4 - 3z}{6z^3}.$$

The last step is advisable should it be necessary to find where $g'(z)$ is positive, where it is negative, and where it is zero.

3.2.17: Apply the extended product rule in Eq. (16) to obtain

$$g'(y) = (3y^2 - 1)(y^2 + 2y + 3)D_y(2y) + (2y)(y^2 + 2y + 3)D_y(3y^2 - 1) + (2y)(3y^2 - 1)D_y(y^2 + 2y + 3)$$

$$= (3y^2 - 1)(y^2 + 2y + 3)(2) + (2y)(y^2 + 2y + 3)(6y) + (2y)(3y^2 - 1)(2y + 2)$$

$$= (6y^4 + 12y^3 + 18y^2 - 2y^2 - 4y - 6) + (12y^4 + 24y^3 + 36y^2) + (12y^4 - 4y^2 + 12y^3 - 4y)$$

$$= 30y^4 + 48y^3 + 48y^2 - 8y - 6.$$

Or if you prefer, first expand $g(y)$, then apply the linear combination rule and the power rule to obtain

$$g(y) = (6y^3 - 2y)(y^2 + 2y + 3) = 6y^5 + 12y^4 + 16y^3 - 4y^2 - 6y, \quad \text{so}$$

$$g'(y) = 30y^4 + 48y^3 + 48y^2 - 8y - 6.$$

3.2.19: Apply the quotient rule to obtain

$$g'(t) = \frac{(t^2 + 2t + 1)D_t(t - 1) - (t - 1)D_t(t^2 + 2t + 1)}{(t^2 + 2t + 1)^2} = \frac{(t^2 + 2t + 1)(1) - (t - 1)(2t + 2)}{[(t + 1)^2]^2}$$

$$= \frac{(t^2 + 2t + 1) - (2t^2 - 2)}{(t + 1)^4} = \frac{3 + 2t - t^2}{(t + 1)^4} = -\frac{(t + 1)(t - 3)}{(t + 1)^4} = \frac{3 - t}{(t + 1)^3}.$$

3.2.21: Apply the reciprocal rule to obtain

$$v'(t) = -\frac{D_t(t^3 - 3t^2 + 3t - 1)}{(t - 1)^6} = -\frac{3t^2 - 6t + 3}{(t - 1)^6} = -\frac{3(t - 1)^2}{(t - 1)^6} = -\frac{3}{(t - 1)^4}.$$

3.2.23: The quotient rule yields

$$g'(x) = \frac{(x^3 + 7x - 5)(3) - (3x)(3x^2 + 7)}{(x^3 + 7x - 5)^2} = \frac{3x^3 + 21x - 15 - 9x^3 - 21x}{(x^3 + 7x - 5)^2} = -\frac{6x^3 + 15}{(x^3 + 7x - 5)^2}.$$

3.2.25: First multiply each term in numerator and denominator by x^4 to obtain

$$g(x) = \frac{x^3 - 2x^2}{2x - 3}.$$

Then apply the quotient rule to obtain

$$g'(x) = \frac{(2x-3)(3x^2-4x)-(x^3-2x^2)(2)}{(2x-3)^2} = \frac{(6x^3-17x^2+12x)-(2x^3-4x^2)}{(2x-3)^2} = \frac{4x^3-13x^2+12x}{(2x-3)^2}.$$

It is usually wise to simplify an expression before differentiating it.

3.2.27: If $y(x) = x^3 - 6x^5 + \frac{3}{2}x^{-4} + 12$, then the linear combination rule and the power rules yield $h'(x) = 3x^2 - 30x^4 - 6x^{-5}$.

3.2.29: Given:

$$y(x) = \frac{5-4x^2+x^5}{x^3} = \frac{5}{x^3} - \frac{4x^2}{x^3} + \frac{x^5}{x^3} = 5x^{-3} - 4x^{-1} + x^2,$$

it follows from the linear combination rule and the power rules that

$$y'(x) = -15x^{-4} + 4x^{-2} + 2x = 2x + \frac{4}{x^2} - \frac{15}{x^4} = \frac{2x^5 + 4x^2 - 15}{x^4}.$$

3.2.31: Because $y(x)$ can be written in the form $y(x) = 3x - \frac{1}{4}x^{-2}$, the linear combination rule and the power rules yield $y'(x) = 3 + \frac{1}{2}x^{-3}$.

3.2.33: If we first combine the two fractions, we will need to use the quotient rule only once:

$$y(x) = \frac{x}{x-1} + \frac{x+1}{3x} = \frac{3x^2+x^2-1}{3x(x-1)} = \frac{4x^2-1}{3x^2-3x},$$

and therefore

$$y'(x) = \frac{(3x^2-3x)(8x)-(4x^2-1)(6x-3)}{(3x^2-3x)^2} = \frac{24x^3-24x^2-24x^3+12x^2+6x-3}{(3x^2-3x)^2} = \frac{-12x^2+6x-3}{(3x^2-3x)^2}.$$

3.2.35: The quotient rule (and other rules, such as the linear combination rule and the power rule) yield

$$y'(x) = \frac{(x^2+9)(3x^2-4)-(x^3-4x+5)(2x)}{(x^2+9)^2}$$

$$= \frac{3x^4+23x^2-36-2x^4+8x^2-10x}{(x^2+9)^2} = \frac{x^4+31x^2-10x-36}{(x^2+9)^2}.$$

3.2.37: First multiply each term in numerator and denominator by $5x^4$ to obtain

$$y(x) = \frac{10x^6}{15x^5-4}.$$

Then apply the quotient rule (among others):

$$y'(x) = \frac{(15x^5-4)(60x^5)-(10x^6)(75x^4)}{(15x^5-4)^2} = \frac{900x^{10}-240x^5-750x^{10}}{(15x^5-4)^2} = \frac{150x^{10}-240x^5}{(15x^5-4)^2} = \frac{30x^5(5x^5-8)}{(15x^5-4)^2}.$$

3.2.39: The quotient rule yields

$$y'(x) = \frac{(x+1)(2x) - (x^2)(1)}{(x+1)^2} = \frac{2x^2 + 2x - x^2}{(x+1)^2} = \frac{x(x+2)}{(x+1)^2}.$$

3.2.41: Given $f(x) = x^3$ and $P(2, 8)$ on its graph, $f'(x) = 3x^2$, so that $f'(2) = 12$ is the slope of the line L tangent to the graph of f at P. So L has equation $y - 8 = 12(x - 2)$; that is, $12x - y = 16$.

3.2.43: Given $f(x) = 1/(x - 1)$ and $P(2, 1)$ on its graph,

$$f'(x) = -\frac{D_x(x-1)}{(x-1)^2} = -\frac{1}{(x-1)^2},$$

so that $f'(2) = -1$ is the slope of the line L tangent to the graph of f at P. So L has equation $y - 1 = -(x - 2)$; that is, $x + y = 3$.

3.2.45: Given $f(x) = x^3 + 3x^2 - 4x - 5$ and $P(1, -5)$ on its graph, $f'(x) = 3x^2 + 6x - 4$, so that $f'(1) = 5$ is the slope of the line L tangent to the graph of f at P. So L has equation $y + 5 = 5(x - 1)$; that is, $5x - y = 10$.

3.2.47: Given $f(x) = 3x^{-2} - 4x^{-3}$ and $P(-1, 7)$ on its graph, $f'(x) = 12x^{-4} - 6x^{-3}$, so that $f'(-1) = 18$ is the slope of the line L tangent to the graph of f at P. So L has equation $y - 7 = 18(x + 1)$; that is, $18x - y = -25$.

3.2.49: Given

$$f(x) = \frac{3x^2}{x^2 + x + 1}$$

and $P(-1, 3)$ on its graph,

$$f'(x) = \frac{(x^2 + x + 1)(6x) - (3x^2)(2x + 1)}{(x^2 + x + 1)^2} = \frac{3x^2 + 6x}{(x^2 + x + 1)^2},$$

so that $f'(-1) = -3$ is the slope of the line L tangent to the graph of f at P. So an equation of the line L is $y - 3 = -3(x + 1)$; that is, $3x + y = 0$.

3.2.51: $V = V_0(1 + \alpha T + \beta T^2 + \gamma T^3)$ where $\alpha \approx -0.06427 \times 10^{-3}$, $\beta \approx 8.5053 \times 10^{-6}$, and $\gamma \approx -6.79 \times 10^{-8}$. Now $dV/dt = V_0(\alpha + 2\beta T + 3\gamma T^2)$; $V = V_0 = 1000$ when $T = 0$. Because $V'(0) = \alpha V_0 < 0$, the water contracts when it is first heated. The rate of change of volume at that point is $V'(0) \approx -0.06427$ cm^3 per °C.

3.2.53: Draw a cross section of the tank through its axis of symmetry. Let r denote the radius of the (circular) water surface when the height of water in the tank is h. Draw a typical radius, label it r, and label the height h. From similar triangles in your figure, deduce that $h/r = 800/160 = 5$, so $r = h/5$. The volume of water in a cone of height h and radius r is $V = \frac{1}{3}\pi r^2 h$, so in this case we have $V = V(h) = \frac{1}{75}\pi h^3$. The rate of change of V with respect to h is $dV/dh = \frac{1}{25}\pi h^2$, and therefore when $h = 600$, we have $V'(600) = 14400\pi$; that is, approximately 45239 cm^3 per cm.

3.2.55: The slope of the tangent line can be computed using dy/dx at $x = a$ and also by using the two points known to lie on the line. We thereby find that

$$3a^2 = \frac{a^3 - 5}{a - 1}.$$

This leads to the equation $(a+1)(2a^2 - 5a + 5) = 0$. The quadratic factor has negative discriminant, so the only real solution of the cubic equation is $a = -1$. The point of tangency is $(-1, -1)$, the slope there is 3, and the equation of the line in question is $y = 3x + 2$.

3.2.57: Suppose that some straight line L is tangent to the graph of $f(x) = x^2$ at the points (a, a^2) and (b, b^2). Our plan is to show that $a = b$, and we may conclude that L cannot be tangent to the graph of f at two *different* points. Because $f'(x) = 2x$ and because (a, a^2) and (b, b^2) both lie on L, the slope of L is equal to both $f'(a)$ and $f'(b)$; that is, $2a = 2b$. Hence $a = b$, so that (a, a^2) and (b, b^2) are the same point. Conclusion: No straight line can be tangent to the graph of $y = x^2$ at two different points.

3.2.59: Given $f(x) = x^n$, we have $f'(x) = nx^{n-1}$. The line tangent to the graph of f at the point $P(x_0, y_0)$ has slope that we compute in two ways and then equate:

$$\frac{y - (x_0)^n}{x - x_0} = n(x_0)^{n-1}.$$

To find the x-intercept of this line, substitute $y = 0$ into this equation and solve for x. It follows that the x-intercept is $x = \dfrac{n-1}{n}x_0$.

3.2.61: $D_x[f(x)]^3 = f'(x)f(x)f(x) + f(x)f'(x)f(x) + f(x)f(x)f'(x) = 3[f(x)]^2 f'(x)$.

3.2.63: Let $u_1(x) = u_2(x) = u_3(x) = \cdots = u_{n-1}(x) = u_n(x) = f(x)$. Then the left-hand side of Eq. (16) is $D_x[(f(x))^n]$ and the right-hand side is

$$f'(x)[f(x)]^{n-1} + f(x)f'(x)[f(x)]^{n-2} + [f(x)]^2 f'(x)[f(x)]^{n-3} + \cdots + [f(x)]^{n-1}f'(x) = n[f(x)]^{n-1} \cdot f'(x).$$

Therefore if n is a positive integer and $f'(x)$ exists, then

$$D_x[(f(x))^n] = n(f(x))^{n-1} \cdot f'(x).$$

3.2.65: Let $f(x) = x^3 - 17x + 35$ and let $n = 17$. Then $g(x) = (f(x))^n$. Hence, by the result in Problem 63,

$$g'(x) = D_x[(f(x))^n] = n(f(x))^{n-1} \cdot f'(x) = 17(x^3 - 17x + 35)^{16} \cdot (3x^2 - 17).$$

3.2.67: If n is a positive integer and

$$f(x) = \frac{x^n}{1+x^2},$$

then

$$f'(x) = \frac{(1+x^2)(nx^{n-1}) - (2x)(x^n)}{(1+x^2)^2} = \frac{nx^{n-1} + nx^{n+1} - 2x^{n+1}}{(1+x^2)^2} = \frac{x^{n-1}[n + (n-2)x^2]}{(1+x^2)^2}. \qquad (1)$$

If $n = 0$, then (by the reciprocal rule)

$$f'(x) = -\frac{2x}{(1+x^2)^2}.$$

If $n = 2$, then by Eq. (1)

$$f'(x) = \frac{2x}{(1+x^2)^2}.$$

In each case there can be but one solution of $f'(x) = 0$, so there is only one horizontal tangent line. If $n = 0$ it is tangent to the graph of f at the point $(0, 1)$; if $n = 2$ it is tangent to the graph of f at the point $(0, 0)$.

3.2.69: If n is a positive integer and $n \geq 3$, $f'(x) = 0$ only when the numerator is zero in Eq. (1) of the solution of Problem 67; that is, when $x^{n-1}(n + [n-2]x^2) = 0$. But this implies that $x = 0$ (because $n \geq 3$) or that $n + [n-2]x^2 = 0$. The latter is impossible because $n > 0$ and $[n-2]x^2 \geq 0$. Therefore the only horizontal tangent to the graph of f is at the point $(0, 0)$.

3.2.71: If

$$f(x) = \frac{x^3}{1+x^2}, \qquad \text{then} \qquad f'(x) = \frac{x^2(3+x^2)}{(1+x^2)^2} = \frac{x^4 + 3x^2}{(1+x^2)^2},$$

by Eq. (1) in the solution of Problem 67. A line tangent to the graph of $y = f'(x)$ will be horizontal when the derivative $f''(x)$ of $f'(x)$ is zero. But

$$D_x\left[f'(x)\right] = f''(x) = \frac{(1+x^2)^2(4x^3 + 6x) - (x^4 + 3x^2)(4x^3 + 4x)}{(1+x^2)^4}$$

$$= \frac{(1+x^2)^2(4x^3 + 6x) - (x^4 + 3x^2)(4x)(x^2 + 1)}{(1+x^2)^4} = \frac{(1+x^2)(4x^3 + 6x) - (x^4 + 3x^2)(4x)}{(1+x^2)^3}$$

$$= \frac{4x^3 + 6x + 4x^5 + 6x^3 - 4x^5 - 12x^3}{(1+x^2)^3} = \frac{6x - 2x^3}{(1+x^2)^3} = \frac{2x(3 - x^2)}{(1+x^2)^3}.$$

So $f''(x) = 0$ when $x = 0$ and when $x = \pm\sqrt{3}$. Therefore there are three points on the graph of $y = f'(x)$ at which the tangent line is horizontal: $(0, 0)$, $\left(-\sqrt{3}, \frac{9}{8}\right)$, and $\left(\sqrt{3}, \frac{9}{8}\right)$.

3.2.73: The graph of $f(x) = |x^3|$ is shown next.

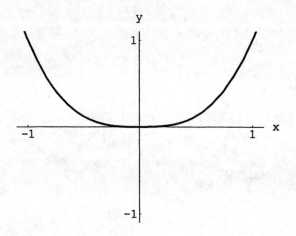

Clearly f is differentiable at x if $x \neq 0$. Moreover,

$$f'_-(0) = \lim_{h \to 0^-} \frac{|(0+h)^3| - 0}{h} = \lim_{h \to 0^-} \frac{-h^3}{h} = 0$$

and $f_+(0) = 0$ by a similar computation. Therefore f is differentiable everywhere.

3.2.75: The graph of

$$f(x) = \begin{cases} 2 + 3x^2 & \text{if } x < 1, \\ 3 + 2x^3 & \text{if } x \geqq 1, \end{cases}$$

is shown next.

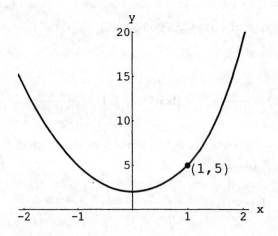

Clearly f is differentiable except possibly at $x = 1$. But

$$f'_-(1) = \lim_{h \to 0^-} \frac{2 + 3(1+h)^2 - 5}{h} = \lim_{h \to 0^-} \frac{6h + 3h^2}{h} = 6$$

and

$$f'_+(1) = \lim_{h \to 0^+} \frac{3 + 2(1+h)^3 - 5}{h} = \lim_{h \to 0^+} \frac{6h + 6h^2 + 2h^3}{h} = 6.$$

Therefore $f'(1)$ exists (and $f'(1) = 6$), and hence $f'(x)$ exists for every real number x.

68

3.2.77: The graph of

$$f(x) = \begin{cases} \dfrac{1}{2-x} & \text{if } x < 1, \\ x & \text{if } x \geq 1 \end{cases}$$

is shown next.

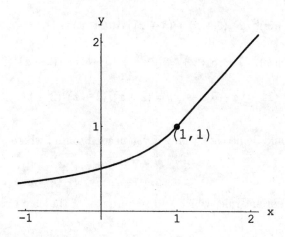

Clearly f is differentiable except possibly at $x = 1$. But

$$f'_-(1) = \lim_{h \to 0^-} \frac{1}{h} \cdot \left(\frac{1}{2 - (1+h)} - 1 \right) = \lim_{h \to 0^+} \frac{1}{h} \cdot \left(\frac{1}{1-h} - 1 \right) = \lim_{h \to 0^-} \frac{h}{h(1-h)} = \lim_{h \to 0^-} \frac{1}{1-h} = 1$$

and

$$f'_+(1) = \lim_{h \to 0^+} \frac{1 + h - 1}{h} = 1.$$

Thus f is differentiable at $x = 1$ as well (and $f'(1) = 1$).

Section 3.3

3.3.1: Given $y = (3x + 4)^5$, the chain rule yields

$$\frac{dy}{dx} = 5 \cdot (3x + 4)^4 \cdot D_x(3x + 4) = 5 \cdot (3x + 4)^4 \cdot 3 = 15(3x + 4)^4.$$

3.3.3: Rewrite the given function in the form $y = (3x - 2)^{-1}$ in order to apply the chain rule. The result is

$$\frac{dy}{dx} = (-1)(3x - 2)^{-2} \cdot D_x(3x - 2) = (-1)(3x - 2)^{-2} \cdot 3 = -3(3x - 2)^{-2} = -\frac{3}{(3x - 2)^2}.$$

3.3.5: Given $y = (x^2 + 3x + 4)^3$, the chain rule yields

$$\frac{dy}{dx} = 3(x^2 + 3x + 4)^2 \cdot D_x(x^2 + 3x + 4) = 3(x^2 + 3x + 4)^2(2x + 3).$$

69

3.3.7: We use the product rule, and in the process of doing so must use the chain rule twice: Given $y = (2-x)^4(3+x)^7$,

$$\frac{dy}{dx} = (2-x)^4 \cdot D_x(3+x)^7 + (3+x)^7 \cdot D_x(2-x)^4$$

$$= (2-x)^4 \cdot 7 \cdot (3+x)^6 \cdot D_x(3+x) + (3+x)^7 \cdot 4(2-x)^3 \cdot D_x(2-x)$$

$$= (2-x)^4 \cdot 7 \cdot (3+x)^6 \cdot 1 + (3+x)^7 \cdot 4(2-x)^3 \cdot (-1) = 7(2-x)^4(3+x)^6 - 4(2-x)^3(3+x)^7$$

$$= (2-x)^3(3+x)^6(14 - 7x - 12 - 4x) = (2-x)^3(3+x)^6(2 - 11x).$$

The last simplifications would be necessary only if you needed to find where $y'(x)$ is positive, where negative, and where zero.

3.3.9: We will use the quotient rule, which will require use of the chain rule to find the derivative of the denominator:

$$\frac{dy}{dx} = \frac{(3x-4)^3 D_x(x+2) - (x+2)D_x(3x-4)^3}{[(3x-4)^3]^2}$$

$$= \frac{(3x-4)^3 \cdot 1 - (x+2) \cdot 3 \cdot (3x-4)^2 \cdot D_x(3x-4)}{(3x-4)^6}$$

$$= \frac{(3x-4)^3 - 3(x+2)(3x-4)^2 \cdot 3}{(3x-4)^6} = \frac{(3x-4) - 9(x+2)}{(3x-4)^4} = -\frac{6x+22}{(3x-4)^4}.$$

3.3.11: Here is a problem in which use of the chain rule contains another use of the chain rule. Given $y = [1 + (1+x)^3]^4$,

$$\frac{dy}{dx} = 4[1 + (1+x)^3]^3 \cdot D_x[1 + (1+x)^3] = 4[1 + (1+x)^3]^3 \cdot [0 + D_x(1+x)^3]$$

$$= 4[1 + (1+x)^3]^3 \cdot 3 \cdot (1+x)^2 \cdot D_x(1+x) = 12[1 + (1+x)^3]^3(1+x)^2.$$

3.3.13: Given: $y = (u+1)^3$ and $u = \dfrac{1}{x^2}$. The chain rule yields

$$\frac{dy}{dx} = \frac{dy}{du} \cdot \frac{du}{dx} = 3(u+1)^2 \cdot \frac{-2}{x^3} = -\frac{6}{x^3}\left(\frac{1}{x^2} + 1\right)^2 = -\frac{6(x^2+1)^2}{x^7}.$$

3.3.15: Given $y = (1 + u^2)^3$ and $u = (4x-1)^2$, the chain rule yields

$$\frac{dy}{dx} = \frac{dy}{du} \cdot \frac{du}{dx} = 6u(1 + u^2)^2 \cdot 8 \cdot (4x-1)$$

$$= 48 \cdot (4x-1)^2(1 + (4x-1)^4)^2(4x-1) = 48(4x-1)^3(1 + (4x-1)^4)^2.$$

70

Without the chain rule, our only way to differentiate $y(x)$ would be first to expand it:

$$y(x) = 8 - 192x + 2688x^2 - 25600x^3 + 181248x^4 - 983040x^5 + 4128768x^6$$

$$- 13369344x^7 + 32636928x^8 - 57671680x^9 + 69206016x^{10} - 50331648x^{11} + 16777216x^{12}.$$

Then we could differentiate $y(x)$ using the linear combination and power rules:

$$y'(x) = -192 + 5376x - 76800x^2 + 724992x^3 - 4915200x^4 + 24772608x^5 - 93585408x^6$$

$$+ 261095424x^7 - 519045120x^8 + 692060160x^9 - 553648128x^{10} + 201326592x^{11}.$$

Fortunately, the chain rule is available—and even if not, we still have *Maple*, *Derive*, *Mathematica*, and MATLAB.

3.3.17: If $y = u(1 - u)^3$ and $u = \dfrac{1}{x^4}$, then the chain rule yields

$$\frac{dy}{dx} = \frac{dy}{du} \cdot \frac{du}{dx} = \left[(1-u)^3 - 3u(1-u)^2 \right] \cdot \left(-4x^{-5} \right) = \left[(1 - x^{-4})^3 - 3x^{-4}(1 - x^{-4})^2 \right] \cdot \left(-4x^{-5} \right),$$

which a very patient person can simplify to

$$\frac{dy}{dx} = \frac{16 - 36x^4 + 24x^8 - 4x^{12}}{x^{17}}.$$

3.3.19: If $y = u^2(u - u^4)^3$ and $u = x^{-2}$, then

$$\frac{dy}{dx} = \frac{dy}{du} \cdot \frac{du}{dx} = \left[2u(u - u^4)^3 + 3u^2(u - u^4)^2(1 - 4u^3) \right] \cdot (-2x^{-3})$$

$$= \left[2x^{-2}(x^{-2} - x^{-8})^3 + 3x^{-4}(x^{-2} - x^{-8})^2(1 - 4x^{-6}) \right] \cdot (-2x^{-3}) = \cdots = \frac{28 - 66x^6 + 48x^{12} - 10x^{18}}{x^{29}}.$$

3.3.21: Let $u(x) = 2x - x^2$ and $n = 3$. Then $f(x) = u^n$, so that

$$f'(x) = nu^{n-1} \cdot \frac{du}{dx} = 3u^2 \cdot (2 - 2x) = 3(2x - x^2)^2(2 - 2x).$$

3.3.23: Let $u(x) = 1 - x^2$ and $n = -4$. Then $f(x) = u^n$, so that

$$f'(x) = nu^{n-1} \cdot \frac{du}{dx} = (-4)u^{-5} \cdot (-2x) = \frac{8x}{(1 - x^2)^5}.$$

3.3.25: Let $u(x) = \dfrac{x + 1}{x - 1}$ and $n = 7$. Then $f(x) = u^n$, and therefore

$$f'(x) = nu^{n-1} \cdot \frac{du}{dx} = 7u^6 \cdot \frac{(x - 1) - (x + 1)}{(x - 1)^2} = 7\left(\frac{x + 1}{x - 1} \right)^6 \cdot \frac{-2}{(x - 1)^2} = -14 \cdot \frac{(x + 1)^6}{(x - 1)^8}.$$

3.3.27: $g'(y) = 1 + 5(2y - 3)^4 \cdot 2 = 1 + 10(2y - 3)^4.$

3.3.29: If $F(s) = (s - s^{-2})^3$, then

$$F'(s) = 3(s - s^{-2})^2(1 + 2s^{-3}) = 3\left(s - \frac{1}{s^2}\right)^2 \cdot \left(1 + \frac{2}{s^3}\right) = 3\left(\frac{s^3 - 1}{s^2}\right)^2 \cdot \frac{s^3 + 2}{s^3}$$

$$= 3 \cdot \frac{(s^3 - 1)^2(s^3 + 2)}{s^7} = 3 \cdot \frac{(s^6 - 2s^3 + 1)(s^3 + 2)}{s^7} = \frac{3(s^9 - 3s^3 + 2)}{s^7}.$$

3.3.31: If $f(u) = (1 + u)^3(1 + u^2)^4$, then

$$f'(u) = 3(1 + u)^2(1 + u^2)^4 + 8u(1 + u)^3(1 + u^2)^3 = (1 + u)^2(1 + u^2)^3(11u^2 + 8u + 3).$$

3.3.33: If $h(v) = \left[v - \left(1 - \dfrac{1}{v}\right)^{-1}\right]^{-2}$, then

$$h'(v) = (-2)\left[v - \left(1 - \frac{1}{v}\right)^{-1}\right]^{-3}\left[1 + \left(1 - \frac{1}{v}\right)^{-2}\left(\frac{1}{v^2}\right)\right] = \frac{2(v - 1)(v^2 - 2v + 2)}{v^3(2 - v)^3}.$$

3.3.35: If $F(z) = (5z^5 - 4z + 3)^{-10}$, then

$$F'(z) = -10(5z^5 - 4z + 3)^{-11}(25z^4 - 4) = \frac{40 - 250z^4}{(5z^5 - 4z + 3)^{11}}.$$

3.3.37: Chain rule: $\dfrac{dy}{dx} = 4(x^3)^3 \cdot 3x^2.$ Power rule: $\dfrac{dy}{dx} = 12x^{11}.$

3.3.39: Chain rule: $\dfrac{dy}{dx} = 2(x^2 - 1) \cdot 2x.$ Without chain rule: $\dfrac{dy}{dx} = 4x^3 - 4x.$

3.3.41: Chain rule: $\dfrac{dy}{dx} = 4(x + 1)^3.$ Without chain rule: $\dfrac{dy}{dx} = 4x^3 + 12x^2 + 12x + 4.$

3.3.43: Chain rule: $\dfrac{dy}{dx} = -2x(x^2 + 1)^{-2}.$ Reciprocal rule: $\dfrac{dy}{dx} = -\dfrac{2x}{(x^2 + 1)^2}.$

3.3.45: If $f(x) = \sin(x^3)$, then $f'(x) = \left[\cos(x^3)\right] \cdot D_x(x^3) = 3x^2 \cos(x^3) = 3x^2 \cos x^3.$

3.3.47: If $g(z) = (\sin 2z)^3$, then

$$g'(z) = 3(\sin 2z)^2 \cdot D_z(\sin 2z) = 3(\sin 2z)^2(\cos 2z) \cdot D_z(2z) = 6\sin^2 2z \cos 2z.$$

3.3.49: The radius of the circular ripple is $r(t) = 2t$ and its area is $a(t) = \pi(2t)^2$; thus $a'(t) = 8\pi t.$ When $r = 10$, $t = 5$, and at that time the rate of change of area with respect to time is $a'(5) = 40\pi$ (in.^{2}/s).

3.3.51: Let A denote the area of the square and x the length of each edge. Then $A = x^2$, so $dA/dx = 2x$. If t denotes time (in seconds), then

$$\frac{dA}{dt} = \frac{dA}{dx} \cdot \frac{dx}{dt} = 2x\frac{dx}{dt}.$$

All that remains is to substitute the given data $x = 10$ and $dx/dt = 2$ to find that the area of the square is increasing at the rate of 40 in.^{2}/s at the time in question.

3.3.53: The volume of the block is $V = x^3$ where x is the length of each edge. So $\frac{dV}{dt} = 3x^2\frac{dx}{dt}$. We are given $dx/dt = -2$, so when $x = 10$ the volume of the block is decreasing at 600 in.^{3}/h.

3.3.55: $G'(t) = f'(h(t)) \cdot h'(t)$. Now $h(1) = 4$, $h'(1) = -6$, and $f'(4) = 3$, so $G'(1) = 3 \cdot (-6) = -18$.

3.3.57: The volume of the balloon is given by $V = \frac{4}{3}\pi r^3$, so

$$\frac{dV}{dt} = \frac{dV}{dr} \cdot \frac{dr}{dt} = 4\pi r^2 \frac{dr}{dt}.$$

Answer: When $r = 10$, $dV/dt = 4\pi \cdot 10^2 \cdot 1 = 400\pi \approx 1256.64$ (cm^3/s).

3.3.59: Given: $\frac{dr}{dt} = -3$. Now $\frac{dV}{dt} = -300\pi = 4\pi r^2 \cdot \left(\frac{dr}{dt}\right)$. So $4\pi r^2 = 100\pi$, and thus $r = 5$ (cm) at the time in question.

3.3.61: Let V denote the volume of the snowball and A its surface area at time t (in hours). Then

$$\frac{dV}{dt} = kA \quad \text{and} \quad A = cV^{2/3}$$

(the latter because A is proportional to r^2, whereas V is proportional to r^3). Therefore

$$\frac{dV}{dt} = \alpha V^{2/3} \quad \text{and thus} \quad \frac{dt}{dV} = \beta V^{-2/3}$$

(α and β are constants). From the last equation we may conclude that $t = \gamma V^{1/3} + \delta$ for some constants γ and δ, so that $V = V(t) = (Pt + Q)^3$ for some constants P and Q. From the information $500 = V(0) = Q^3$ and $250 = V(1) = (P + Q)^3$, we find that $Q = 5\sqrt[3]{4}$ and that $P = -5 \cdot (\sqrt[3]{4} - \sqrt[3]{2})$. Now $V(t) = 0$ when $PT + Q = 0$; it turns out that

$$T = \frac{\sqrt[3]{2}}{\sqrt[3]{2} - 1} \approx 4.8473.$$

Therefore the snowball finishes melting at about 2:50:50 P.M. on the same day.

3.3.63: By the chain rule,

$$\frac{dv}{dx} = \frac{dv}{dw} \cdot \frac{dw}{dx}, \quad \text{and therefore} \quad \frac{du}{dx} = \frac{du}{dv} \cdot \frac{dv}{dx} = \frac{du}{dv} \cdot \frac{dv}{dw} \cdot \frac{dw}{dx}.$$

73

3.3.65: If $h(x) = \sqrt{x+4}$, then

$$h'(x) = \frac{1}{2\sqrt{x+4}} \cdot D_x(x+4) = \frac{1}{2\sqrt{x+4}} \cdot 1 = \frac{1}{2\sqrt{x+4}}.$$

3.3.67: If $h(x) = (x^2+4)^{3/2} = (x^2+4)\sqrt{x^2+4}$, then

$$h'(x) = 2x\sqrt{x^2+4} + (x^2+4) \cdot \frac{1}{2\sqrt{x^2+4}} \cdot 2x = 2x\sqrt{x^2+4} + x\sqrt{x^2+4} = 3x\sqrt{x^2+4}.$$

Section 3.4

3.4.1: Write $f(x) = 4x^{5/2} + 2x^{-1/2}$ to find

$$f'(x) = 10x^{3/2} - x^{-3/2} = 10x^{3/2} - \frac{1}{x^{3/2}} = \frac{10x^3 - 1}{x^{3/2}}.$$

3.4.3: Write $f(x) = (2x+1)^{1/2}$ to find

$$f'(x) = \frac{1}{2}(2x+1)^{-1/2} \cdot 2 = \frac{1}{\sqrt{2x+1}}.$$

3.4.5: Write $f(x) = 6x^{-1/2} - x^{3/2}$ to find that $f'(x) = -3x^{-3/2} - \frac{3}{2}x^{1/2} = -\frac{3(x^2+2)}{2x^{3/2}}.$

3.4.7: $D_x(2x+3)^{3/2} = \frac{3}{2}(2x+3)^{1/2} \cdot 2 = 3\sqrt{2x+3}.$

3.4.9: $D_x(3-2x^2)^{-3/2} = -\frac{3}{2}(3-2x^2)^{-5/2} \cdot (-4x) = \frac{6x}{(3-2x^2)^{5/2}}.$

3.4.11: $D_x(x^3+1)^{1/2} = \frac{1}{2}(x^3+1)^{-1/2} \cdot (3x^2) = \frac{3x^2}{2\sqrt{x^3+1}}.$

3.4.13: $D_x(2x^2+1)^{1/2} = \frac{1}{2}(2x^2+1)^{-1/2} \cdot 4x = \frac{2x}{\sqrt{2x^2+1}}.$

3.4.15: $D_t\left(t^{3/2}\sqrt{2}\right) = \frac{3}{2}t^{1/2}\sqrt{2} = \frac{3\sqrt{t}}{\sqrt{2}}.$

3.4.17: $D_x(2x^2-x+7)^{3/2} = \frac{3}{2}(2x^2-x+7)^{1/2} \cdot (4x-1) = \frac{3}{2}(4x-1)\sqrt{2x^2-x+7}.$

3.4.19: $D_x(x-2x^3)^{-4/3} = -\frac{4}{3}(x-2x^3)^{-7/3} \cdot (1-6x^2) = \frac{4(6x^2-1)}{3(x-2x^3)^{7/3}}.$

74

3.4.21: If $f(x) = x(1 - x^2)^{1/2}$, then (by the product rule and the chain rule, among others)

$$f'(x) = 1 \cdot (1 - x^2)^{1/2} + x \cdot \frac{1}{2}(1 - x^2)^{-1/2} \cdot D_x(1 - x^2)$$

$$= (1 - x^2)^{1/2} + x \cdot \frac{1}{2}(1 - x^2)^{-1/2} \cdot (-2x) = \sqrt{1 - x^2} - \frac{x^2}{\sqrt{1 - x^2}} = \frac{1 - 2x^2}{\sqrt{1 - x^2}}.$$

3.4.23: If $f(t) = \sqrt{\dfrac{t^2 + 1}{t^2 - 1}} = \left(\dfrac{t^2 + 1}{t^2 - 1}\right)^{1/2}$, then

$$f'(t) = \frac{1}{2}\left(\frac{t^2 + 1}{t^2 - 1}\right)^{-1/2} \cdot \frac{(t^2 - 1)(2t) - (t^2 + 1)(2t)}{(t^2 - 1)^2} = \frac{1}{2}\left(\frac{t^2 - 1}{t^2 + 1}\right)^{1/2} \cdot \frac{-4t}{(t^2 - 1)^2} = -\frac{2t}{(t^2 - 1)^{3/2}\sqrt{t^2 + 1}}.$$

3.4.25: $D_x\left(x - \dfrac{1}{x}\right)^3 = 3\left(x - \dfrac{1}{x}\right)^2\left(1 + \dfrac{1}{x^2}\right) = 3\left(\dfrac{x^2 - 1}{x}\right)^2 \cdot \dfrac{x^2 + 1}{x^2} = \dfrac{3(x^2 - 1)^2(x^2 + 1)}{x^4}.$

3.4.27: Write $f(v) = \dfrac{(v + 1)^{1/2}}{v}$. Then

$$f'(v) = \frac{v \cdot \frac{1}{2}(v + 1)^{-1/2} - 1 \cdot (v + 1)^{1/2}}{v^2} = \frac{v \cdot (v + 1)^{-1/2} - 2(v + 1)^{1/2}}{2v^2} = \frac{v - 2(v + 1)}{2v^2(v + 1)^{1/2}} = -\frac{v + 2}{2v^2(v + 1)^{1/2}}.$$

3.4.29: $D_x(1 - x^2)^{1/3} = \dfrac{1}{3}(1 - x^2)^{-2/3} \cdot (-2x) = -\dfrac{2x}{3(1 - x^2)^{2/3}}.$

3.4.31: If $f(x) = x(3 - 4x)^{1/2}$, then (with the aid of the product rule and the chain rule)

$$f'(x) = 1 \cdot (3 - 4x)^{1/2} + x \cdot \frac{1}{2}(3 - 4x)^{-1/2} \cdot (-4) = (3 - 4x)^{1/2} - \frac{2x}{(3 - 4x)^{1/2}} = \frac{3(1 - 2x)}{\sqrt{3 - 4x}}.$$

3.4.33: If $f(x) = (1 - x^2)(2x + 4)^{1/3}$, then the product rule (among others) yields

$$f'(x) = -2x(2x + 4)^{1/3} + \frac{2}{3}(1 - x^2) \cdot (2x + 4)^{-2/3}$$

$$= \frac{-6x(2x + 4) + 2(1 - x^2)}{3(2x + 4)^{2/3}} = \frac{-12x^2 - 24x + 2 - 2x^2}{3(2x + 4)^{2/3}} = \frac{2 - 24x - 14x^2}{3(2x + 4)^{2/3}}.$$

3.4.35: If $g(t) = \left(1 + \dfrac{1}{t}\right)^2(3t^2 + 1)^{1/2}$, then

$$g'(t) = \left(1 + \frac{1}{t}\right)^2 \cdot \frac{1}{2}(3t^2 + 1)^{-1/2}(6t) + 2 \cdot \left(1 + \frac{1}{t}\right)\left(-\frac{1}{t^2}\right)(3t^2 + 1)^{1/2}$$

$$= 3t \cdot \frac{(t + 1)^2}{t^2(3t^2 + 1)^{1/2}} - \frac{2}{t^2} \cdot \frac{t + 1}{t}(3t^2 + 1)^{1/2} = \frac{3t^2(t + 1)^2}{t^3(3t^2 + 1)^{1/2}} - \frac{2(t + 1)(3t^2 + 1)}{t^3(3t^2 + 1)^{1/2}} = \frac{3t^4 - 3t^2 - 2t - 2}{t^3\sqrt{3t^2 + 1}}.$$

75

3.4.37: If $f(x) = \dfrac{2x-1}{(3x+4)^5}$, then

$$f'(x) = \frac{2(3x+4)^5 - (2x-1)\cdot 5(3x+4)^4 \cdot 3}{(3x+4)^{10}} = \frac{2(3x+4) - 15(2x-1)}{(3x+4)^6} = \frac{23-24x}{(3x+4)^6}.$$

3.4.39: If $f(x) = \dfrac{(2x+1)^{1/2}}{(3x+4)^{1/3}}$, then

$$f'(x) = \frac{(3x+4)^{1/3}(2x+1)^{-1/2} - (2x+1)^{1/2}(3x+4)^{-2/3}}{(3x+4)^{2/3}}$$

$$= \frac{(3x+4)-(2x+1)}{(3x+4)^{4/3}(2x+1)^{1/2}} = \frac{x+3}{(3x+4)^{4/3}(2x+1)^{1/2}}.$$

3.4.41: If $h(y) = \dfrac{(1+y)^{1/2} + (1-y)^{1/2}}{y^{5/3}}$, then

$$h'(y) = \frac{y^{5/3}\left[\frac{1}{2}(1+y)^{-1/2} - \frac{1}{2}(1-y)^{-1/2}\right] - \frac{5}{3}y^{2/3}\left[(1+y)^{1/2} + (1-y)^{1/2}\right]}{y^{10/3}}$$

$$= \frac{y\left[\frac{1}{2}(1+y)^{-1/2} - \frac{1}{2}(1-y)^{-1/2}\right] - \frac{5}{3}\left[(1+y)^{1/2} + (1-y)^{1/2}\right]}{y^{8/3}}$$

$$= \frac{y\left[3(1+y)^{-1/2} - 3(1-y)^{-1/2}\right] - 10\left[(1+y)^{1/2} + (1-y)^{1/2}\right]}{6y^{8/3}}$$

$$= \frac{y\left[3(1-y)^{1/2} - 3(1+y)^{1/2}\right] - 10\left[(1+y)(1-y)^{1/2} + (1-y)(1+y)^{1/2}\right]}{6y^{8/3}(1-y)^{1/2}(1+y)^{1/2}}$$

$$= \frac{(7y-10)\sqrt{1+y} - (7y+10)\sqrt{1-y}}{6y^{8/3}\sqrt{1-y}\,\sqrt{1+y}}.$$

3.4.43: If $g(t) = \left[t + (t + t^{1/2})^{1/2}\right]^{1/2}$, then

$$g'(t) = \frac{1}{2}\left[t + (t + t^{1/2})^{1/2}\right]^{-1/2} \cdot \left[1 + \frac{1}{2}(t + t^{1/2})^{-1/2}\left(1 + \frac{1}{2}t^{-1/2}\right)\right].$$

It is possible to write the derivative without negative exponents. The symbolic algebra program *Mathematica* yields

$$g'(t) = -\frac{(t + (t+t^{1/2})^{1/2})^{1/2}\left[1 - 4t^{3/2} - 4t^2 + 3t^{1/2}\left(1 + (t+t^{1/2})^{1/2}\right) + 2t\left(1 + (t+t^{1/2})^{1/2}\right)\right]}{8t(1+t^{1/2})(t^{3/2} - t^{1/2} - 1)}.$$

But the first answer that *Mathematica* gives is

$$g'(t) = \frac{1 + \dfrac{1 + \dfrac{1}{2\sqrt{t}}}{2\sqrt{t+\sqrt{t}}}}{2\sqrt{t + \sqrt{t + \sqrt{t}}}}.$$

76

3.4.45: Because

$$y'(x) = \frac{dy}{dx} = \frac{2}{3x^{1/3}}$$

is never zero, there are no horizontal tangents. Because $y(x)$ is continuous at $x = 0$ and $|y'(x)| \to +\infty$ as $x \to 0$, there is a vertical tangent at $(0, 0)$.

3.4.47: If $g(x) = x^{1/2} - x^{3/2}$, then

$$g'(x) = \frac{1}{2}x^{-1/2} - \frac{3}{2}x^{1/2} = \frac{1}{2\sqrt{x}} - \frac{3\sqrt{x}}{2} = \frac{1 - 3x}{2\sqrt{x}}.$$

Thus there is a horizontal tangent at $\left(\frac{1}{3}, \frac{2}{9}\sqrt{3}\right)$. Also, because g is continuous at $x = 0$ and

$$\lim_{x \to 0^+} |g'(x)| = \lim_{x \to 0^+} \frac{1 - 3x}{2\sqrt{x}} = +\infty,$$

the graph of g has a vertical tangent at $(0, 0)$.

3.4.49: If $y(x) = x(1 - x^2)^{-1/2}$, then

$$y'(x) = \frac{dy}{dx} = (1 - x^2)^{-1/2} - \frac{1}{2}x(1 - x^2)^{-3/2} \cdot (-2x) = \frac{1}{(1 - x^2)^{1/2}} + \frac{x^2}{(1 - x^2)^{3/2}} = \frac{1}{(1 - x^2)^{3/2}}.$$

Thus the graph of $y(x)$ has no horizontal tangents because $y'(x)$ is never zero. The only candidates for vertical tangents are at $x = \pm 1$, but there are none because $y(x)$ is not continuous at either of those two values of x.

3.4.51: Let $f(x) = 2\sqrt{x}$. Then $f'(x) = x^{-1/2}$, so an equation of the required tangent line is $y - f(4) = f'(4)(x - 4)$; that is, $y = \frac{1}{2}(x + 4)$. The graph of f and this tangent line are shown next.

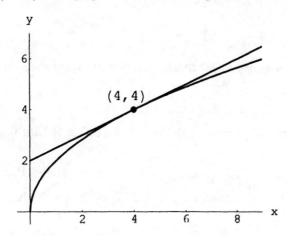

3.4.53: If $f(x) = 3x^{2/3}$, then $f'(x) = 2x^{-1/3}$. Therefore an equation of the required tangent line is $y - f(-1) = f'(-1)(x + 1)$; that is, $y = -2x + 1$. A graph of f and this tangent line are shown next.

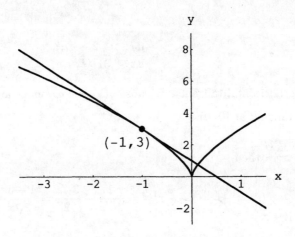

3.4.55: If $f(x) = x(4-x)^{1/2}$, then

$$f'(x) = (4-x)^{1/2} - \frac{1}{2}x(4-x)^{-1/2} = (4-x)^{1/2} - \frac{x}{2(4-x)^{1/2}} = \frac{8-3x}{2\sqrt{4-x}}.$$

So an equation of the required tangent line is $y - f(0) = f'(0)(x-0)$; that is, $y = 2x$. A graph of f and this tangent line are shown next.

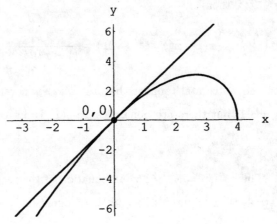

3.4.57: If $x < 0$ then $f'(x) < 0$; as $x \to 0^-$, $f'(x)$ appears to approach $-\infty$. If $x > 0$ then $f'(x) > 0$; as $x \to 0^+$, $f'(x)$ appears to approach $+\infty$. So the graph of f' must be the one shown in Fig. 3.4.13(d).

3.4.59: Note that $f'(x) > 0$ if $x < 0$ whereas $f'(x) < 0$ if $x > 0$. Moreover, as $x \to 0$, $|f'(x)|$ appears to approach $+\infty$. So the graph of f' must be the one shown in Fig. 3.4.13(b).

3.4.61: We see that $f'(x) < 0$ for $-2 < x < -1.4$ (approximately), that $f'(x) > 0$ for $-1.4 < x < 1.4$ (approximately), and that $f'(x) < 0$ for $1.4 < x < 2$. Also $f'(x) = 0$ when $x \approx \pm 1.4$. Therefore the graph of f must be the one shown in Fig. 3.4.13(e).

3.4.63: $L = \dfrac{P^2 g}{4\pi^2}$, so $\dfrac{dL}{dP} = \dfrac{Pg}{2\pi^2}$, and hence $\dfrac{dP}{dL} = \dfrac{2\pi^2}{Pg}$. Given $g = 32$ and $P = 2$, we find the value of the latter to be $\frac{1}{32}\pi^2 \approx 0.308$ (seconds per foot).

78

3.4.65: Whether $y = +\sqrt{1 - x^2}$ or $y = -\sqrt{1 - x^2}$, it follows easily that $dy/dx = -x/y$. The slope of the tangent is -2 when $x = 2y$, so from the equation $x^2 + y^2 = 1$ we see that $x^2 = 4/5$, so that $x = \pm\frac{2}{5}\sqrt{5}$. Because $y = \frac{1}{2}x$, the two points we are to find are $\left(-\frac{2}{5}\sqrt{5}, -\frac{1}{5}\sqrt{5}\right)$ and $\left(\frac{2}{5}\sqrt{5}, \frac{1}{5}\sqrt{5}\right)$.

3.4.67: The line tangent to the parabola $y = x^2$ at the point $Q(a, a^2)$ has slope $2a$, so the normal to the parabola at Q has slope $-1/(2a)$. The normal also passes through $P(18, 0)$, so we can find its slope another way—by using the two-point formula. Thus

$$-\frac{1}{2a} = \frac{a^2 - 0}{a - 18};$$

$$18 - a = 2a^3;$$

$$2a^3 + a - 18 = 0.$$

By inspection, $a = 2$ is a solution of the last equation. Thus $a - 2$ is a factor of the cubic, and division yields

$$2a^3 + a - 18 = (a - 2)(2a^2 + 4a + 9).$$

The quadratic factor has negative discriminant, so $a = 2$ is the only real solution of $2a^3 + a - 18 = 0$. Therefore the normal line has slope $-\frac{1}{4}$ and equation $x + 4y = 18$.

3.4.69: If a line through $P\left(0, \frac{5}{2}\right)$ is normal to $y = x^{2/3}$ at $Q(a, a^{2/3})$, then it has slope $-\frac{3}{2}a^{1/3}$. As in the two previous solutions, we find that

$$\frac{a^{2/3} - \frac{5}{2}}{a} = -\frac{3}{2}a^{1/3},$$

which yields $3a^{4/3} + 2a^{2/3} - 5 = 0$. Put $u = a^{2/3}$; we obtain $3u^2 + 2u - 5 = 0$, so that $(3u + 5)(u - 1) = 0$. Because $u = a^{2/3} > 0$, $u = 1$ is the only solution, so $a = 1$ and $a = -1$ yield the two possibilities for the point Q, and therefore the equations of the two lines are

$$y - \frac{5}{2} = -\frac{3}{2}x \quad \text{and} \quad y - \frac{5}{2} = \frac{3}{2}x.$$

3.4.71: The equation $x^3 = 3x + 8$ is not an *identity* that says the two functions are equal (and would therefore have equal derivatives). It is merely an equation relating *values* of the two functions x^3 and $3x + 8$ that have nothing in particular to do with one another. So it makes no sense to equate their derivatives.

3.4.73: If $f(x) = x^{1/3}$ and $a > 0$, then

$$f'(a) = \lim_{x \to a} \frac{x^{1/3} - a^{1/3}}{x - a} = \lim_{x \to a} \frac{x^{1/3} - a^{1/3}}{(x^{1/3} - a^{1/3})(x^{2/3} + x^{1/3}a^{1/3} + a^{2/3})}$$

$$= \lim_{x \to a} \frac{1}{x^{2/3} + x^{1/3}a^{1/3} + a^{2/3}} = \frac{1}{a^{2/3} + a^{2/3} + a^{2/3}} = \frac{1}{3a^{2/3}}.$$

Therefore $D_x x^{1/3} = \frac{1}{3}x^{-2/3}$ if $x > 0$.

This formula is of course valid for $x < 0$ as well. To show this, observe that the previous argument is valid if $a < 0$, or—if you prefer—you can use the chain rule, laws of exponents, and the preceding result, as follows. Suppose that $x < 0$. Then $-x > 0$; also, $x^{1/3} = -(-x)^{1/3}$. So

$$D_x(x^{1/3}) = D_x\left[-(-x)^{1/3}\right] = -D_x(-x)^{1/3} = -\left[\frac{1}{3}(-x)^{-2/3}\cdot(-1)\right] = \frac{1}{3}(-x)^{2/3} = \frac{1}{3}x^{-2/3}.$$

Therefore $D_x(x^{1/3}) = \frac{1}{3}x^{-2/3}$ if $x \neq 0$.

3.4.75: The preamble to Problems 72 through 75 implies that if q is a positive integer and x and a are positive real numbers, then

$$x - a = (x^{1/q} - a^{1/q})(x^{(q-1)/q} + x^{(q-2)/q}a^{1/q} + x^{(q-3)/q}a^{2/q} + \cdots + x^{1/q}a^{(q-2)/q} + a^{(q-1)/q}).$$

Thus if $f(x) = x^{1/q}$ and $a > 0$, then

$$f'(a) = \lim_{x \to a} \frac{x^{1/q} - a^{1/q}}{x - a}$$

$$= \lim_{x \to a} \frac{x^{1/q} - a^{1/q}}{(x^{1/q} - a^{1/q})(x^{(q-1)/q} + x^{(q-2)/q}a^{1/q} + x^{(q-3)/q}a^{2/q} + \cdots + x^{1/q}a^{(q-2)/q} + a^{(q-1)/q})}$$

$$= \lim_{x \to a} \frac{1}{x^{(q-1)/q} + x^{(q-2)/q}a^{1/q} + x^{(q-3)/q}a^{2/q} + \cdots + a^{(q-1)/q}} \quad (q \text{ terms in the denominator})$$

$$= \frac{1}{a^{(q-1)/q} + a^{(q-1)/q} + a^{(q-1)/q} + \cdots + a^{(q-1)/q}} \quad (\text{still } q \text{ terms in the denominator})$$

$$= \frac{1}{qa^{(q-1)/q}} = \frac{1}{q}a^{-(q-1)/q}.$$

Therefore $D_x(x^{1/q}) = \frac{1}{q}x^{-(q-1)/q}$ if $x > 0$ and q is a positive integer. This result is easy to extend to the case $x < 0$. Therefore if q is a positive integer and $x \neq 0$, then

$$D_x(x^{1/q}) = \frac{1}{q}x^{(1/q)-1}.$$

Section 3.5

3.5.1: Because $f(x) = 1 - x$ is decreasing everywhere, it can attain a maximum only at a left-hand endpoint of its domain and a minimum only at a right-hand endpoint of its domain. Its domain $[-1, 1)$ has no right-hand endpoint, so f has no minimum value. Its maximum value occurs at -1, is $f(-1) = 2$, and is the global maximum value of f on its domain.

3.5.3: Because $f(x) = |x|$ is decreasing for $x < 0$ and increasing for $x > 0$, it can have a maximum only at a left-hand or a right-hand endpoint of its domain $(-1, 1)$. But its domain has no endpoints, so f has no maximum value. It has the global minimum value $f(0) = 0$.

3.5.5: Given: $f(x) = |x - 2|$ on $(1, 4]$. If $x > 2$ then $f(x) = x - 2$, which is increasing for $x > 2$; if $x < 2$ then $f(x) = 2 - x$, which is decreasing for $x < 2$. So f can have a maximum only at an endpoint of its domain; the only endpoint is at $x = 4$, where $f(x)$ has the maximum value $f(4) = 2$. Because $f(x) \to 1$ as $x \to 1^+$, the extremum at $x = 4$ is in fact a global maximum. Finally, $f(2) = 0$ is the global minimum value of f.

3.5.7: Given: $f(x) = x^3 + 1$ on $[-1, 1]$. The only critical point of f occurs where $f'(x) = 3x^2$ is zero; that is, at $x = 0$. But $f(x) < 1 = f(0)$ if $x < 0$ whereas $f(x) > 1 = f(0)$ if $x > 0$, so there is no extremum at $x = 0$. By Theorem 1 (page 142), f must have a global maximum and a global minimum. The only possible locations are at the endpoints of the domain of f, and therefore $f(-1) = 0$ is the global minimum value of f and $f(1) = 2$ is its global maximum value.

3.5.9: If

$$f(x) = \frac{1}{x(1-x)}, \qquad \text{then} \qquad f'(x) = \frac{2x - 1}{x^2(1-x)^2},$$

which does not exist at $x = 0$ or at $x = 1$ and is zero when $x = \frac{1}{2}$. But none of these points lies in the domain $[2, 3]$ of f, so there are no extrema at those three points. By Theorem 1 f must have a global maximum and a global minimum, which therefore must occur at the endpoints of its domain. Because $f(2) = -\frac{1}{2}$ and $f(3) = -\frac{1}{6}$, the former is the global minimum value of f and the latter is its global maximum value.

3.5.11: $f'(x) = 3$ is never zero and always exists. Therefore $f(-2) = -8$ is the global minimum value of f and $f(3) = 7$ is its global maximum value.

3.5.13: $h'(x) = -2x$ always exists and is zero only at $x = 0$, which is not in the domain of h. Therefore $h(1) = 3$ is the global maximum value of h and $h(3) = -5$ is its global minimum value.

3.5.15: $g'(x) = 2(x - 1)$ always exists and is zero only at $x = 1$. Because $g(-1) = 4$, $g(1) = 0$, and $g(4) = 9$, the global minimum value of g is 0 and the global maximum is 9. If $-1 < x < 0$ then $g(x) = (x - 1)^2 < 4$, so the extremum at $x = -1$ is a local maximum.

3.5.17: $f'(x) = 3x^2 - 3 = 3(x + 1)(x - 1)$ always exists and is zero when $x = -1$ and when $x - 1$. Because $f(-2) = -2$, $f(-1) = 2$, $f(1) = -2$, and $f(4) = 52$, the latter is the global maximum value of f and -2 is its global minimum value—note that the minimum occurs at two different points on the graph. Because f is continuous on $[-2, 2]$, it must have a global maximum there, and our work shows that it occurs at $x = -1$.

But because $f(4) = 52 > 2 = f(-1)$, $f(-1) = 2$ is only a local maximum for f on its domain $[-2, 4]$. Summary: Global minimum value -2, local maximum value 2, global maximum value 52.

3.5.19: If

$$h(x) = x + \frac{4}{x}, \qquad \text{then} \qquad h'(x) = 1 - \frac{4}{x^2} = \frac{x^2 - 4}{x^2}.$$

Therefore h is continuous on $[1, 4]$ and $x = 2$ is the only critical point of h in its domain. Because $h(1) = 5$, $h(2) = 4$, and $h(4) = 5$, the global maximum value of h is 5 and its global minimum value is 4.

3.5.21: $f'(x) = -2$ always exists and is never zero, so $f(1) = 1$ is the global minimum value of f and $f(-1) = 5$ is its global maximum value.

3.5.23: $f'(x) = -12 - 18x$ always exists and is zero when $x = -\frac{2}{3}$. Because $f(-1) = 8$, $f\left(-\frac{2}{3}\right) = 9$, and $f(1) = -16$, the global maximum value of f is 9 and its global minimum value is -16. Consideration of the interval $\left[-1, -\frac{2}{3}\right]$ shows that $f(-1) = 8$ is a local minimum of f.

3.5.25: $f'(x) = 3x^2 - 6x - 9 = 3(x + 1)(x - 3)$ always exists and is zero when $x = -1$ and when $x = 3$. Because $f(-2) = 3$, $f(-1) = 10$, $f(3) = -22$, and $f(4) = -15$, the global minimum value of f is -22 and its global maximum is 10. Consideration of the interval $[-2, -1]$ shows that $f(-2) = 3$ is a *local* minimum of f; consideration of the interval $[3, 4]$ shows that $f(4) = -15$ is a *local* maximum of f.

3.5.27: $f'(x) = 15x^4 - 15x^2 = 15x^2(x + 1)(x - 1)$ always exists and is zero at $x = -1$, at $x = 0$, and at $x = 1$. We note that $f(-2) = -56$, $f(-1) = 2$, $f(0) = 0$, $f(1) = -2$, and $f(2) = 56$. So the global minimum value of f is -56 and its global maximum value is 56. Consideration of the interval $[-2, 0]$ shows that $f(-1) = 2$ is a *local* maximum of f on its domain $[-2, 2]$. Similarly, $f(1) = -2$ is a *local* minimum of f there. Suppose that x is near, but not equal, to zero. Then $f(x) = x^3(3x^2 - 5)$ is negative if $x > 0$ and positive if $x < 0$. Therefore there is *no extremum* at $x = 0$.

3.5.29: Given: $f(x) = 5 + |7 - 3x|$ on $[1, 5]$. If $x < \frac{7}{3}$, then $-3x > -7$, so that $7 - 3x > 0$; in this case, $f(x) = 12 - 3x$ and so $f'(x) = -3$. Similarly, if $x > \frac{7}{3}$, then $f(x) = 3x - 2$ and so $f'(x) = 3$. Hence $f'(x)$ is never zero, but it fails to exist at $x = \frac{7}{3}$. Now $f(1) = 9$, $f\left(\frac{7}{3}\right) = 5$, and $f(5) = 13$, so 13 is the global maximum value of f and 5 is its global minimum value. Consideration of the continuous function f on the interval $\left[1, \frac{7}{3}\right]$ shows that $f(1) = 9$ is a *local* maximum of f on its domain.

3.5.31: $f'(x) = 150x^2 - 210x + 72 = 6(5x - 3)(5x - 4)$ always exists and is zero at $x = \frac{3}{5}$ and at $x = \frac{4}{5}$. Now $f(0) = 0$, $f\left(\frac{3}{5}\right) = 16.2$, $f\left(\frac{4}{5}\right) = 16$, and $f(1) = 17$. Hence 17 is the global maximum value of f and 0 is its global minimum value. Consideration of the intervals $\left[0, \frac{4}{5}\right]$ and $\left[\frac{3}{5}, 1\right]$ shows that 16.2 is a *local* maximum value of f on $[0, 1]$ and that 16 is a *local* minimum value of f there.

3.5.33: If

$$f(x) = \frac{x}{x+1}, \qquad \text{then} \qquad f'(x) = \frac{1}{(x+1)^2},$$

so $f'(x)$ exists for all x in the domain $[0, 3]$ of f and is never zero there. Hence $f(0) = 0$ is the global minimum value of f and $f(3) = \frac{3}{4}$ is its global maximum value.

3.5.35: If

$$f(x) = \frac{1-x}{x^2+3}, \qquad \text{then} \qquad f'(x) = \frac{(x+1)(x-3)}{(x^2+3)^2},$$

so $f'(x)$ always exists and is zero when $x = -1$ and when $x = 3$. Now $f(-2) = \frac{3}{7}$, $f(-1) = \frac{1}{2}$, $f(3) = -\frac{1}{6}$, and $f(5) = -\frac{1}{7}$. So the global minimum value of f is $-\frac{1}{6}$ and its global maximum value is $\frac{1}{2}$. Consideration of the interval $[-2, -1]$ shows that $\frac{3}{7}$ is a *local* minimum value of f; consideration of the interval $[3, 5]$ shows that $-\frac{1}{7}$ is a *local* maximum value of f.

3.5.37: Given: $f(x) = x(1-x^2)^{1/2}$ on $[-1, 1]$. First,

$$f'(x) = (1-x^2)^{1/2} + x \cdot \frac{1}{2}(1-x^2)^{-1/2} \cdot (-2x) = (1-x^2)^{1/2} - \frac{x^2}{(1-x^2)^{1/2}} = \frac{1 - 2x^2}{\sqrt{1-x^2}}.$$

Hence $f'(x)$ exists for $-1 < x < 1$ and not otherwise, but we will check the endpoints ± 1 of the domain of f separately. Also $f'(x) = 0$ when $x = \pm\frac{1}{2}\sqrt{2}$. Now $f(-1) = 0$, $f\left(-\frac{1}{2}\sqrt{2}\right) = -\frac{1}{2}$, $f\left(\frac{1}{2}\sqrt{2}\right) = \frac{1}{2}$, and $f(1) = 0$. Therefore the global minimum value of f is $-\frac{1}{2}$ and its global maximum value is $\frac{1}{2}$. Consideration of the interval $\left[-1, -\frac{1}{2}\sqrt{2}\right]$ shows that $f(-1) = 0$ is a *local* maximum value of f on $[-1, 1]$; similarly, $f(1) = 0$ is a *local* minimum value of f there.

3.5.39: Given: $f(x) = x(2-x)^{1/3}$ on $[1, 3]$. Then

$$f'(x) = (2-x)^{1/3} + x \cdot \frac{1}{3}(2-x)^{-2/3} \cdot (-1) = (2-x)^{1/3} - \frac{x}{3(2-x)^{2/3}} = \frac{6-4x}{3(2-x)^{2/3}}.$$

Then $f'(2)$ does not exist and $f'(x) = 0$ when $x = \frac{3}{2}$. Also f is continuous everywhere, and $f(1) = 1$, $f\left(\frac{3}{2}\right) \approx 1.19$, and $f(3) = -3$. Hence the global minimum value of f is -3 and its global maximum value is $f\left(\frac{3}{2}\right) = 3 \cdot 2^{-4/3} \approx 1.190551$. Consideration of the interval $\left[1, \frac{3}{2}\right]$ shows that $f(1) = 1$ is a *local* minimum value of f.

3.5.41: If $A \neq 0$, then $f'(x) \equiv A$ is never zero, but because f is continuous it must have global extrema. Therefore they occur at the endpoints. If $A = 0$, then f is a constant function, and its maximum and minimum value B occurs at every point of the interval, including the two endpoints.

3.5.43: $f'(x) = 0$ if x is not an integer; $f'(x)$ does not exist if x is an integer (we saw in Chapter 2 that $f(x) = [\![x]\!]$ is discontinuous at each integer).

83

3.5.45: If $f(x) = ax^3 + bx^2 + cx + d$ and $a \neq 0$, then $f'(x) = 3ax^2 + 2bx + c$ exists for all x, but the quadratic equation $3ax^2 + 2bx + c = 0$ has two solutions if the discriminant $\Delta = 4b^2 - 12ac$ is positive, one solution if $\Delta = 0$, and no [real] solutions if $\Delta < 0$. Therefore f has either no critical points, exactly one critical point, or exactly two. Examples:

$$f(x) = x^3 + x \qquad \text{has no critical points,}$$

$$f(x) = x^3 \qquad \text{has exactly one critical point, and}$$

$$f(x) = x^3 - 3x \qquad \text{has exactly two critical points.}$$

3.5.47: The derivative is positive on $(-\infty, -1.3)$, negative on $(-1.3, 1.3)$, and positive on $(1.3, +\infty)$. So its graph must be the one in Fig. 3.5.15(c). (Numbers with decimal points are approximations.)

3.5.49: The derivative is positive on $(-\infty, 0.0)$, negative on $(0.0, 2.0)$, and positive on $(2.0, +\infty)$. So its graph must be the one shown in Fig. 3.5.15(d). (Numbers with decimal points are approximations.)

3.5.51: The derivative is negative on $(-\infty, -2.0)$, positive on $(-2.0, 1.0)$, and negative on $(1.0, +\infty)$. Therefore its graph must be the one shown in Fig. 3.5.15(a). (Numbers with decimal points are approximations.)

Note: In Problems 53 through 60, we used *Mathematica* and Newton's method (when necessary), carrying 40 decimal digits throughout all computations. Answers are correct or correctly rounded to the number of digits shown. Your answers may differ in the last (or last few) digits beause of differences in hardware or software. Using a graphing calculator or computer to zoom in on solutions has more limited accuracy when using certain machines.

3.5.53: Global maximum value 28 at the left endpoint $x = -2$, global minimum value approximately 6.828387610996 at the critical point where $x = -1 + \frac{1}{3}\sqrt{30} \approx 0.825741858351$, local maximum value 16 at the right endpoint $x = 2$.

3.5.55: Global maximum value 136 at the left endpoint $x = -3$, global minimum value approximately -8.669500829438 at the critical point $x \approx -0.762212740507$, local maximum value 16 at the right endpont $x = 3$.

3.5.57: Global minimum value -5 at the left endpoint $x = 0$, global maximum value approximately 8.976226903748 at the critical point $x \approx 1.323417756580$, local minimum value 5 at the right endpoint $x = 2$.

3.5.59: Local minimum value -159 at the left endpoint $x = -3$, global maximum value approximately 30.643243080334 at the critical point $x \approx -1.911336401963$, local minimum value approximately

84

−5.767229705222 at the critical point $x \approx -0.460141424682$, local maximum value approximately 21.047667292488 at the critical point $x \approx 0.967947424014$, global minimum value −345 at the right endpoint $x = 3$.

Section 3.6

3.6.1: With $x > 0$, $y > 0$, and $x + y = 50$, we are to maximize the product $P = xy$.

$$P = P(x) = x(50 - x) = 50x - x^2, \qquad 0 < x < 50$$

($x < 50$ because $y > 0$.) The product is not maximal if we let $x = 0$ or $x = 50$, so we adjoin the endpoints to the domain of P; thus the continuous function $P(x) = 50x - x^2$ has a global maximum on the closed interval $[0, 50]$, and the maximum does *not* occur at either endpoint. Because f is differentiable, the maximum must occur at a point where $P'(x) = 0$: $50 - 2x = 0$, and so $x = 25$. Because this is the only critical point of P, it follows that $x = 25$ maximizes $P(x)$. When $x = 25$, $y = 50 - 25 = 25$, so the two positive real numbers with sum 50 and maximum possible product are 25 and 25.

3.6.3: If the coordinates of the "fourth vertex" are (x, y), then $y = 100 - 2x$ and the area of the rectangle is $A = xy$. So we are to maximize

$$A(x) = x(100 - 2x) \qquad 0 \leqq x \leqq 50.$$

By the usual argument the solution occurs where $A'(x) = 0$, thus where $x = 25$, $y = 50$, and the maximum area is 1250.

3.6.5: If x is the length of each edge of the base of the box and y denotes the height of the box, then its volume is given by $V = x^2 y$. Its total surface area is the sum of the area x^2 of its bottom and four times the area xy of each of its vertical sides, so $x^2 + 4xy = 300$. Thus

$$V = V(x) = x^2 \cdot \frac{300 - x^2}{4x} = \frac{300x - x^3}{4}, \qquad 1 \leqq x \leqq 10\sqrt{3}.$$

Hence

$$V'(x) = \frac{300 - 3x^2}{4},$$

so $V'(x)$ always exists and $V'(x) = 0$ when $x = 10$ (we discard the solution $x = -10$; it's not in the domain of V). Then

$$V(1) = \frac{299}{4} = 74.75, \qquad V(10) = 500, \qquad \text{and} \qquad V\left(10\sqrt{3}\right) = 0,$$

so the maximum possible volume of the box is 500 in.[3].

3.6.7: If the two numbers are x and y, then we are to minimize $S = x^2 + y^2$ given $x > 0$, $y > 0$, and $x + y = 48$. So $S(x) = x^2 + (48 - x)^2$, $0 \leq x \leq 48$. Here we adjoin the endpoints to the domain of S to ensure the existence of a maximum, but we must test the values of S at these endpoints because it is not immediately clear that neither $S(0)$ nor $S(48)$ yields the maximum value of S. Now $S'(x) = 2x - 2(48 - x)$; the only interior critical point of S is $x = 24$, and when $x = 24$, $y = 24$ as well. Finally, $S(0) = (48)^2 = 2304 = S(48) > 1152 = S(24)$, so the answer is 1152.

3.6.9: Let x and y be the two numbers. Then $x + y = 10$, $x \geq 0$, and $y \geq 0$. We are to minimize the sum of their cubes,

$$S = x^3 + y^3 : \qquad S(x) = x^3 + (10 - x)^3, \qquad 0 \leq x \leq 10.$$

Now $S'(x) = 3x^2 - 3(10 - x)^2$, so the values of x to be tested are $x = 0$, $x = 5$, and $x = 10$. At the endpoints, $S = 1000$; when $x = 5$, $S = 250$ (the minimum).

3.6.11: As in Fig. 3.6.18, let y denote the length of each of the internal dividers and of the two sides parallel to them; let x denote the length of each of the other two sides. The total length of all the fencing is $2x + 4y = 600$ and the area of the corral is $A = xy$. Hence

$$A = A(y) = \frac{600 - 4y}{2} \cdot y = 300y - 2y^2, \qquad 0 \leq y \leq 150.$$

Now $A'(y) = 0$ only when $y = 75$, and $A(0) = 0 = A(150)$, and therefore the maximum area of the corral is $A(75) = 11250$ yd^2.

3.6.13: If the rectangle has sides x and y, then $x^2 + y^2 = 16^2$ by the Pythagorean theorem. The area of the rectangle is then

$$A(x) = x\sqrt{256 - x^2}, \qquad 0 \leq x \leq 16.$$

A positive quantity is maximized exactly when its square is maximized, so in place of A we maximize

$$f(x) = (A(x))^2 = 256x^2 - x^4.$$

The only solutions of $f'(x) = 0$ in the domain of A are $x = 0$ and $x = 8\sqrt{2}$. But $A(0) = 0 = A(16)$, so $x = 8\sqrt{2}$ yields the maximum value 128 of A.

3.6.15: $V'(T) = -0.06426 + (0.0170086)T - (0.0002037)T^2$. The equation $V'(T) = 0$ is quadratic with the two (approximate) solutions $T \approx 79.532$ and $T \approx 3.967$. The formula for $V(T)$ is valid only in the range $0 \leq T \leq 30$, so we reject the first solution. Finally, $V(0) = 999.87$, $V(30) \approx 1003.763$, and $V(3.967) \approx 999.71$. Thus the volume is minimized when $T \approx 3.967$, and therefore water has its greatest density at about $3.967°$C.

3.6.17: Let x denote the length of each edge of the base and let y denote the height of the box. We are to maximize its volume $V = x^2 y$ given the constraint $2x^2 + 4xy = 600$. Solve the latter for y to write

$$V(x) = 150x - \frac{1}{2}x^3, \qquad 1 \leqq x \leqq 10\sqrt{3}.$$

The solution of $V'(x) = 0$ in the domain of V is $x = 10$. Because $V(10) = 1000 > V(1) = 149.5 > V(10\sqrt{3}) = 0$, this shows that $x = 10$ maximizes V and that the maximum value of V is 1000 cm^3.

3.6.19: Let x be the length of the edge of each of the twelve small squares. Then each of the three cross-shaped pieces will form boxes with base length $1 - 2x$ and height x, so each of the three will have volume $x(1 - 2x)^2$. Both of the two cubical boxes will have edge x and thus volume x^3. So the total volume of all five boxes will be

$$V(x) = 3x(1 - 2x)^2 + 2x^3 = 14x^3 - 12x^2 + 3x, \qquad 0 \leqq x \leqq \frac{1}{2}.$$

Now $V'(x) = 42x^2 - 24x + 3$; $V'(x) = 0$ when $14x^2 - 8x - 1 = 0$. The quadratic formula gives the two solutions $x = \frac{1}{14}\left(4 \pm \sqrt{2}\right)$. These are approximately 0.3867 and 0.1847, and both lie in the domain of V. Finally, $V(0) = 0$, $V(0.1847) \approx 0.2329$, $V(0.3867) \approx 0.1752$, and $V(0.5) = 0.25$. Therefore, to maximize V, one must cut each of the three large squares into four smaller squares of side length $\frac{1}{2}$ each and form the resulting twelve squares into two cubes. At maximum volume there will be only two boxes, not five.

3.6.21: Let x denote the edge length of one square and y that of the other. Then $4x + 4y = 80$, so $y = 20 - x$. The total area of the two squares is $A = x^2 + y^2$, so

$$A = A(x) = x^2 + (20 - x)^2 = 2x^2 - 40x + 400,$$

with domain $(0, 20)$; adjoin the endpoints as usual. Then $A'(x) = 4x - 40$, which always exists and which vanishes when $x = 10$. Now $A(0) = 400 = A(20)$, whereas $A(10) = 200$. So to minimize the total area of the two squares, make two equal squares. To maximize it, make only one square.

3.6.23: Let x be the length of each segment of fence perpendicular to the wall and let y be the length of each segment parallel to the wall.

Case 1: The internal fence is perpendicular to the wall. Then $y = 600 - 3x$ and the enclosure will have area $A(x) = 600x - 3x^2$, $0 \leqq x \leqq 200$. Then $A'(x) = 0$ when $x = 100$; $A(100) = 30000$ (m^2) is the maximum in Case 1.

Case 2: The internal fence is parallel to the wall. Then $y = 300 - x$, and the area of the enclosure is given by $A(x) = 300x - x^2$, $0 \leqq x \leqq 300$. Then $A'(x) = 0$ when $x = 150$; $A(150) = 22500$ (m^2) is the maximum in Case 2.

Answer: The maximum possible area of the enclosure is 30000 m^2. The divider must be perpendicular to the wall and of length 100 m. The side parallel to the wall is to have length 300 m.

3.6.25: Let the dimensions of the box be x by x by y. We are to maximize $V = x^2y$ subject to some conditions on x and y. According to the poster on the wall of the Bogart, Georgia Post Office, the *length* of the box is the larger of x and y, and the *girth* is measured around the box in a plane perpendicular to its length.

Case 1: $x < y$. Then the length is y, the girth is $4x$, and the mailing constraint is $4x + y \leq 100$. It is clear that we take $4x + y = 100$ to maximize V, so that

$$V = V(x) = x^2(100 - 4x) = 100x^2 - 4x^3, \qquad 0 \leq x \leq 25.$$

Then $V'(x) = 4x(50 - 3x)$; $V'(x) = 0$ for $x = 0$ and for $x = 50/3$. But $V(0) = 0$, $V(25) = 0$, and $V(50/3) = 250000/27 \approx 9259$ (in.^{3}). The latter is the maximum in Case 1.

Case 2: $x \geq y$. Then the length is x and the girth is $2x + y$, although you may get some argument from a postal worker who may insist that it's $4x$. So $3x + 2y = 100$, and thus

$$V = V(x) = x^2\left(\frac{100 - 3x}{2}\right) = 50x^2 - \frac{3}{2}x^3, \qquad 0 \leq x \leq 100/3.$$

Then $V'(x) = 100x - \frac{9}{2}x^2$; $V'(x) = 0$ when $x = 0$ and when $x = 200/9$. But $V(0) = 0$, $V(100/3) = 0$, and $V(200/9) = 2000000/243 \approx 8230$ (in.^{3}).

Case 3: You lose the argument in Case 2. Then the box has length x and girth $4x$, so $5x = 100$; thus $x = 20$. To maximize the total volume, no calculus is needed—let $y = x$. Then the box of maximum volume will have volume $20^3 = 8000$ (in.^{3}).

Answer: The maximum is $\dfrac{250000}{27}$ in.^{3}

3.6.27: Suppose that n presses are used, $1 \leq n \leq 8$. The total cost of the poster run would then be

$$C(n) = 5n + (10 + 6n)\left(\frac{50000}{3600n}\right) = 5n + \frac{125}{9}\left(\frac{10}{n} + 6\right)$$

dollars. Temporarily assume that n can take on every real number value between 1 and 8. Then

$$C'(n) = 5 - \frac{125}{9} \cdot \frac{10}{n^2};$$

$C'(n) = 0$ when $n = \frac{5}{3}\sqrt{10} \approx 5.27$ presses. But an integral number of presses must be used, so the actual number that will minimize the cost is either 5 or 6, unless the minimum occurs at one of the two endpoints. The values in question are $C(1) \approx 227.2$, $C(5) \approx 136.1$, $C(6) \approx 136.5$, and $C(8) \approx 140.7$. So to minimize cost and thereby maximize the profit, five presses should be used.

3.6.29: We are to minimize the total cost C over a ten-year period. This cost is the sum of the initial cost and ten times the annual cost:

$$C(x) = 150x + 10\left(\frac{1000}{2 + x}\right), \qquad 0 \leq x \leq 10.$$

Next,

$$C'(x) = 150 - \frac{10000}{(2+x)^2}; \quad C'(x) = 0 \quad \text{when} \quad 150 = \frac{10000}{(2+x)^2},$$

so that $(2 + x)^2 = \frac{200}{3}$. One of the resulting values of x is negative, so we reject it. The other is $x = -2 + \sqrt{200/3} \approx 6.165$ (in.). The problem itself suggests that x must be an integer, so we check $x = 6$ and $x = 7$ along with the endpoints of the domain of C. In dollars, $C(0) = 5000$, $C(6) \approx 2150$, $C(7) \approx 2161$, and $C(10) \approx 2333$. Result: Install six inches of insulation. The annual savings over the situation with no insulation at all then will be one-tenth of $5000 - 2150$, about \$285 per year.

3.6.31: Let x be the number of five-cent fare increases. The resulting revenue will be

$$R(x) = (150 + 5x)(600 - 40x), \qquad -30 \leqq x \leqq 15$$

(the revenue is the product of the price and the number of passengers). Now

$$R(x) = 90000 - 3000x - 200x^2;$$

$$R'(x) = -3000 - 400x; \qquad R'(x) = 0 \quad \text{when} \quad x = -7.5.$$

Because the fare must be an integral number of cents, we check $R(-7) = 1012 = R(-8)$ (dollars). Answer: The fare should be either \$1.10 or \$1.15; this is a reduction of 40 or 35 cents, respectively, and each results in the maximum possible revenue of \$1012 per day.

3.6.33: The following figure shows a cross section of the cone and inscribed cylinder. Let x be the radius of the cylinder and y its height. By similar triangles in the figure,

$$\frac{H}{R} = \frac{y}{R-x}, \quad \text{so} \quad y = \frac{H}{R}(R-x).$$

We are to maximize the volume $V = \pi x^2 y$ of the cylinder, so we write

$$V = V(x) = \pi x^2 \frac{H}{R}(R - x)$$

$$= \pi \frac{H}{R}(Rx^2 - x^3), \qquad 0 \leqq x \leqq R.$$

Because $V(0) = 0 = V(r)$, V is maximized when $V'(x) = 0$; this leads to the equation $2xR = 3x^2$ and thus to the results $x = \frac{2}{3}R$ and $y = \frac{1}{3}H$.

89

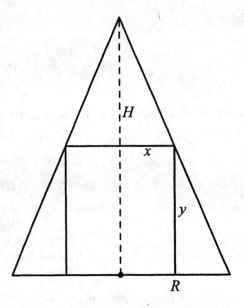

3.6.35: Draw a circle in the plane with center at the origin and with radius R. Inscribe a rectangle with vertical and horizontal sides and let (x, y) be its vertex in the first quadrant. The base of the rectangle has length $2x$ and its height is $2y$, so the perimeter of the rectangle is $P = 4x + 4y$. Also $x^2 + y^2 = R^2$, so

$$P = P(x) = 4x + 4\sqrt{R^2 - x^2}, \qquad 0 \leqq x \leqq R.$$

$$P'(x) = 4 - \frac{4x}{\sqrt{R^2 - x^2}};$$

$$P'(x) = 0 \quad \text{when} \quad 4\sqrt{R^2 - x^2} = 4x;$$

$$R^2 - x^2 = x^2;$$

$$x^2 = \frac{1}{2}R^2.$$

Because $x > 0$, $x = \frac{1}{2}R\sqrt{2}$. The corresponding value of $P(x)$ is $4R\sqrt{2}$, and $P(0) = 4R = P(R)$. So the former value of x maximizes the perimeter P. Because $y^2 = R^2 - x^2$ and because $R^2 - x^2 = x^2$ at maximum, $y = x$ at maximum. Therefore the rectangle of largest perimeter that can be inscribed in a circle is a square.

3.6.37: We are to maximize volume $V = \frac{1}{3}\pi r^2 h$ given $r^2 + h^2 = 100$. The latter relation enables us to write

$$V = V(h) = \frac{1}{3}\pi(100 - h^2)h = \frac{1}{3}\pi(100h - h^3), \qquad 0 \leqq h \leqq 10.$$

Now $V'(h) = \frac{1}{3}\pi(100 - 3h^2)$, so $V'(h) = 0$ when $3h^2 = 100$, thus when $h = \frac{10}{3}\sqrt{3}$. But $V(h) = 0$ at the endpoints of its domain, so the latter value of h maximizes V, and its maximum value is $\frac{2000}{27}\pi\sqrt{3}$.

3.6.39: Let x and y be the two numbers. Then $x \geqq 0$, $y \geqq 0$, and $x + y = 16$. We are to find both the maximum and minimum values of $x^{1/3} + y^{1/3}$. Because $y = 16 - x$, we seek the extrema of

$$f(x) = x^{1/3} + (16 - x)^{1/3}, \qquad 0 \leqq x \leqq 16.$$

Now

90

$$f'(x) = \frac{1}{3}x^{-2/3} - \frac{1}{3}(16 - x)^{-2/3}$$

$$= \frac{1}{3x^{2/3}} - \frac{1}{3(16 - x)^{2/3}};$$

$f'(x) = 0$ when $(16 - x)^{2/3} = x^{2/3}$, so when $16 - x = x$, thus when $x = 8$. Now $f(0) = f(16) = 16^{1/3} \approx 2.52$, so $f(8) = 4$ maximizes f whereas $f(0)$ and $f(16)$ yield its minimum.

3.6.41: If (x, x^2) is a point of the parabola, then its distance from $(0, 1)$ is

$$d(x) = \sqrt{x^2 + (x^2 - 1)^2}\,.$$

So we minimize

$$f(x) = (d(x))^2 = x^4 - x^2 + 1,$$

where the domain of f is the set of all real numbers. But because $f(x)$ is large positive when $|x|$ is large, we will not exclude a minimum if we restrict the domain of f to be an interval of the form $[-a, a]$ where a is a large positive number. On the interval $[-a, a]$, f is continuous and thus has a global minimum, which does not occur at $\pm a$ because $f(\pm a)$ is large positive. Because $f'(x)$ exists for all x, the minimum of f occurs at a point where $f'(x) = 0$:

$$4x^3 - 2x = 0; \qquad 2x(2x^2 - 1) = 0.$$

Hence $x = 0$ or $x = \pm\frac{1}{2}\sqrt{2}$. Now $f(0) = 1$ and $f\left(\pm\frac{1}{2}\sqrt{2}\right) = \frac{3}{4}$. So $x = 0$ yields a local maximum value for $f(x)$, and the minimum possible distance is $\sqrt{0.75} = \frac{1}{2}\sqrt{3}$.

3.6.43: Examine the plank on the right on Fig. 3.6.10. Let its height be $2y$ and its width (in the x-direction) be z. The total area of the four small rectangles in the figure is then $A = 4 \cdot z \cdot 2y = 8yz$. The circle has radius 1, and by Problem 35 the large inscribed square has dimensions $\sqrt{2}$ by $\sqrt{2}$. Thus

$$\left(\frac{1}{2}\sqrt{2} + z\right)^2 + y^2 = 1.$$

This implies that

$$y = \sqrt{\frac{1}{2} - z\sqrt{2} - z^2}\,.$$

Therefore

$$A(z) = 8z\sqrt{\frac{1}{2} - z\sqrt{2} - z^2}\,, \qquad 0 \leqq z \leqq 1 - \frac{1}{2}\sqrt{2}.$$

Now $A(z) = 0$ at each endpoint of its domain and

$$A'(z) = \frac{4\sqrt{2}\left(1 - 3z\sqrt{z} - 4z^2\right)}{\sqrt{1 - 2z\sqrt{z} - 2z^2}}\,.$$

So $A'(z) = 0$ when $z = \frac{1}{8}\left(-3\sqrt{2} \pm \sqrt{34}\right)$; we discard the negative solution, and find that when $A(z)$ is maximized

$$z = \frac{-3\sqrt{2} + \sqrt{34}}{8} \approx 0.198539,$$

$$2y = \frac{\sqrt{7 - \sqrt{17}}}{2} \approx 0.848071, \quad \text{and}$$

$$A(z) = \frac{\sqrt{142 + 34\sqrt{17}}}{2} \approx 0.673500.$$

The four small planks use just under 59% of the wood that remains after the large plank is cut, a very efficient use of what might be scrap lumber.

3.6.45: Set up a coordinate system in which the island is located at $(0, 2)$ and the village at $(6, 0)$, and let $(x, 0)$ be the point at which the boat lands. It is clear that $0 \leq x \leq 6$. The trip involves the land distance $6 - x$ traveled at 20 km/h and the water distance $(4 + x^2)^{1/2}$ traveled at 10 km/h. The total time of the trip is then given by

$$T(x) = \frac{1}{10}\sqrt{4 + x^2} + \frac{1}{20}(6 - x), \qquad 0 \leq x \leq 6.$$

Now

$$T'(x) = \frac{x}{10\sqrt{4 + x^2}} - \frac{1}{20}.$$

Thus $T'(x) = 0$ when $3x^2 = 4$; because $x \geq 0$, we find that $x = \frac{2}{3}\sqrt{3}$. The value of T there is

$$\frac{1}{10}\left(3 + \sqrt{3}\right) \approx 0.473,$$

whereas $T(0) = 0.5$ and $T(6) \approx 0.632$. Therefore the boater should make landfall at $\frac{2}{3}\sqrt{3} \approx 1.155$ km from the point on the shore closest to the island.

3.6.47: The distances involved are $|AP| = |BP| = \sqrt{x^2 + 1}$ and $|CP| = 3 - x$. Therefore we are to minimize

$$f(x) = 2\sqrt{x^2 + 1} + 3 - x, \qquad 0 \leq x \leq 3.$$

Now

$$f'(x) = \frac{2x}{\sqrt{x^2 + 1}} - 1; \quad f'(x) = 0 \quad \text{when} \quad \frac{2x}{\sqrt{x^2 + 1}} = 1.$$

This leads to the equation $3x^2 = 1$, so $x = \frac{1}{3}\sqrt{3}$. Now $f(0) = 5$, $f(3) \approx 6.32$, and at the critical point, $f(x) = 3 + \sqrt{3} \approx 4.732$. Answer: The distribution center should be located at the point $P(\frac{1}{3}\sqrt{3}, 0)$.

3.6.49: We are to minimize total cost

$$C = c_1\sqrt{a^2 + x^2} + c_2\sqrt{(L - x)^2 + b^2}.$$

$$C'(x) = \frac{c_1 x}{\sqrt{a^2 + x^2}} - \frac{c_2(L - x)}{\sqrt{(L - x)^2 + b^2}};$$

$$C'(x) = 0 \text{ when } \frac{c_1 x}{\sqrt{a^2 + x^2}} = \frac{c_2(L - x)}{\sqrt{(L - x)^2 + b^2}}.$$

The result in Part (a) is equivalent to the last equation. For Part (b), assume that $a = b = c_1 = 1$, $c_2 = 2$, and $L = 4$. Then we obtain

$$\frac{x}{\sqrt{1 + x^2}} = \frac{2(4 - x)}{\sqrt{(4 - x)^2 + 1}};$$

$$\frac{x^2}{1 + x^2} = \frac{4(16 - 8x + x^2)}{16 - 8x + x^2 + 1};$$

$$x^2(17 - 8x + x^2) = (4 + 4x^2)(16 - 8x + x^2);$$

$$17x^2 - 8x^3 + x^4 = 64 - 32x + 68x^2 - 32x^3 + 4x^4.$$

Therefore we wish to solve $f(x) = 0$ where

$$f(x) = 3x^4 - 24x^4 + 51x^2 - 32x + 64.$$

Now $f(0) = 64$, $f(1) = 62$, $f(2) = 60$, $f(3) = 22$, and $f(4) = -16$. Because $f(3) > 0 > f(4)$, we interpolate to estimate the zero of $f(x)$ between 3 and 4; it turns out that interpolation gives $x \approx 3.58$. Subsequent interpolation yields the more accurate estimate $x \approx 3.45$. (The equation $f(x) = 0$ has exactly two solutions, $x \approx 3.452462314$ and $x \approx 4.559682567$.)

3.6.51: Let r be the radius of the sphere and x the edge length of the cube. We are to maximize and minimize total volume

$$V = \frac{4}{3}\pi r^3 + x^3 \quad \text{given} \quad 4\pi r^2 + 6x^2 = 1000.$$

The latter equation yields

$$x = \sqrt{\frac{1000 - 4\pi r^2}{6}},$$

so

$$V = V(r) = \frac{4}{3}\pi r^3 + \left(\frac{500 - 2\pi r^2}{3}\right)^{3/2}, \qquad 0 \leqq r \leqq r_1 = 5\sqrt{\frac{10}{\pi}}.$$

Next,

$$V'(r) = 4\pi r^2 - 2\pi r\sqrt{\frac{500 - 2\pi r^2}{3}},$$

and $V'(r) = 0$ when

$$4\pi r^2 = 2\pi r\sqrt{\frac{500 - 2\pi r^2}{3}}.$$

So $r = 0$ or

$$2r = \sqrt{\frac{500 - 2\pi r^2}{3}}.$$

The latter equation leads to

$$r = r_2 = 5\sqrt{\frac{10}{\pi + 6}}.$$

Now $V(0) \approx 2151.66$, $V(r_1) \approx 2973.54$, and $V(r_2) \approx 1743.16$. Therefore, to minimize the sum of the volumes, choose $r = r_2 \approx 5.229$ in. and $x = 2r_2 \approx 10.459$ in. To maximize the sum of their volumes, take $r = r_1 \approx 8.921$ in. and $x = 0$ in.

3.6.53: The graph of $V(x)$ is shown next. The maximum volume seems to occur near the point $(4, V(4)) \approx (4, 95.406)$, so the maximum volume is approximately 95.406 cubic feet.

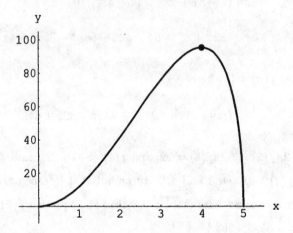

3.6.55: Let V_1 and V_2 be the volume functions of problems 53 and 54, respectively. Then

$$V_1'(x) = \frac{20\sqrt{5}\left(4x - x^2\right)}{3\sqrt{5 - x}},$$

which is zero at $x = 0$ and at $x = 4$, and

$$V_2'(x) = \frac{10\sqrt{5}\left(8x - x^2\right)}{3\sqrt{10 - x}},$$

which is zero at $x = 0$ and at $x = 8$, as expected. Finally, $\dfrac{V_2(8)}{V_1(4)} = 2\sqrt{2}$.

3.6.57: Let x denote the length of each edge of the square base of the box and let y denote its height. Given total surface area A, we have $x^2 + 4xy = A$, and hence

$$y = \frac{A - x^2}{4x}. \tag{1}$$

The volume of the box is $V = x^2 y$, and therefore

$$V(x) = \frac{Ax - x^2}{4}, \quad 0 \le x \le \sqrt{A}.$$

Next,

$$V'(x) = \frac{A - 3x^2}{4}; \qquad V'(x) = 0 \quad \text{when} \quad x = \sqrt{A/3}.$$

Because $V'(x) > 0$ to the left of this critical point and $V'(x) < 0$ to the right, it yields the global maximum value of $V(x)$. By Eq. (1), the corresponding height of the box is $\frac{1}{2}\sqrt{A/3}$. Therefore the open box with square base and maximal volume has height equal to half the length of the edge of its base.

3.6.59: Let r denote the radius of the base of the open cylindrical can and let h denote its height. Its total surface area A then satisfies the equation $\pi r^2 + 2\pi r h = A$, and therefore

$$h = \frac{A - \pi r^2}{2\pi r}. \tag{1}$$

Thus the volume of the can is given by

$$V(r) = \frac{Ar - \pi r^3}{2}, \quad 0 \le r \le \left(\frac{A}{\pi}\right)^{1/2}.$$

Next,

$$V'(r) = \frac{A - 3\pi r^2}{2}; \qquad V'(r) = 0 \quad \text{when} \quad r = \left(\frac{A}{3\pi}\right)^{1/2}.$$

Clearly $V'(r) > 0$ to the left of this critical point and $V'(r) < 0$ to the right, so it determines the global maximum value of $V(r)$. By Eq. (1) the corresponding value of h is the same, so the open cylindrical can of maximum volume has height equal to its radius.

3.6.61: Let

$$f(t) = \frac{1}{1 + t^2}, \quad \text{so that} \quad f'(t) = -\frac{2t}{(1 + t^2)^2}.$$

The line tangent to the graph of $y = f(t)$ at the point $(t, f(t))$ then has x-intercept and y-intercept

$$\frac{1 + 3t^2}{2t} \quad \text{and} \quad \frac{1 + 3t^2}{(1 + t^2)^2},$$

respectively. The area of the triangle bounded by the part of the tangent line in the first quadrant and the coordinate axes is

$$A(t) = \frac{1}{2} \cdot \frac{1 + 3t^2}{2t} \cdot \frac{1 + 3t^2}{(1 + t^2)^2}, \tag{1}$$

and

$$A'(t) = \frac{-9t^6 + 9t^4 + t^2 - 1}{4t^2(1 + t^2)^3}.$$

95

Next, $A'(t) = 0$ when

$$(t-1)(t+1)(3t^2-1)(3t^2+1) = 0,$$

and the only two critical points of A in the interval $[0.5, 2]$ are

$$t_1 = 1 \quad \text{and} \quad t_2 = \frac{\sqrt{3}}{3} \approx 0.57735.$$

Significant values of $A(t)$ are then

$$A(0.5) = 0.98, \quad A(0.57735) \approx 0.97428, \quad A(1) = 1, \quad \text{and} \quad A(2) = 0.845.$$

Therefore $A(t)$ has a local maximum at $t = 0.5$, a local minimum at t_2, its global maximum at $t = 1$, and its global minimum at $t = 2$.

To answer the first question in Problem 61, Eq. (1) makes it clear that $A(t) \to +\infty$ as $t \to 0^+$ and, in addition, that $A(t) \to 0$ as $t \to +\infty$.

Section 3.7

3.7.1: If $f(x) = 3\sin^2 x = 3(\sin x)^2$, then $f'(x) = 6\sin x \cos x$.

3.7.3: If $f(x) = x\cos x$, then $f'(x) = 1 \cdot \cos x + x \cdot (-\sin x) = \cos x - x\sin x$.

3.7.5: If $f(x) = \dfrac{\sin x}{x}$, then $f'(x) = \dfrac{x\cos x - \sin x}{x^2}$.

3.7.7: If $f(x) = \sin x \cos^2 x$, then

$$f'(x) = \cos x \cos^2 x + (\sin x)(2\cos x)(-\sin x) = \cos^3 x - 2\sin^2 x \cos x.$$

3.7.9: If $g(t) = (1 + \sin t)^4$, then $g'(t) = 4(1 + \sin t)^3 \cdot \cos t$.

3.7.11: If $g(t) = \dfrac{1}{\sin t + \cos t}$, then (by the reciprocal rule) $g'(t) = -\dfrac{\cos t - \sin t}{(\sin t + \cos t)^2} = \dfrac{\sin t - \cos t}{(\sin t + \cos t)^2}$.

3.7.13: If $f(x) = 2x\sin x - 3x^2\cos x$, then (by the product rule)

$$f'(x) = 2\sin x + 2x\cos x - 6x\cos x + 3x^2\sin x = 3x^2\sin x - 4x\cos x + 2\sin x.$$

3.7.15: If $f(x) = \cos 2x \sin 3x$, then $f'(x) = -2\sin 2x \sin 3x + 3\cos 2x \cos 3x$.

3.7.17: If $g(t) = t^3\sin^2 2t = t^3(\sin 2t)^2$, then

$$g'(t) = 3t^2(\sin 2t)^2 + t^3 \cdot (2\sin 2t) \cdot (\cos 2t) \cdot 2 = 3t^2\sin^2 2t + 4t^3\sin 2t \cos 2t.$$

3.7.19: If $g(t) = (\cos 3t + \cos 5t)^{5/2}$, then $g'(t) = \frac{5}{2}(\cos 3t + \cos 5t)^{3/2}(-3\sin 3t - 5\sin 5t)$.

3.7.21: If $y = y(x) = \sin^2\sqrt{x} = (\sin x^{1/2})^2$, then

$$\frac{dy}{dx} = 2(\sin x^{1/2})(\cos x^{1/2}) \cdot \frac{1}{2}x^{-1/2} = \frac{\sin\sqrt{x}\cos\sqrt{x}}{\sqrt{x}}.$$

3.7.23: If $y = y(x) = x^2\cos(3x^2 - 1)$, then

$$\frac{dy}{dx} = 2x\cos(3x^2 - 1) - x^2 \cdot 6x \cdot \sin(3x^2 - 1) = 2x\cos(3x^2 - 1) - 6x^3\sin(3x^2 - 1).$$

3.7.25: If $y = y(x) = \sin 2x\cos 3x$, then

$$\frac{dy}{dx} = (\sin 2x) \cdot D_x(\cos 3x) + (\cos 3x) \cdot D_x(\sin 2x) = -3\sin 2x\sin 3x + 2\cos 3x\cos 2x.$$

3.7.27: If $y = y(x) = \dfrac{\cos 3x}{\sin 5x}$, then

$$\frac{dy}{dx} = \frac{(\sin 5x)(-3\sin 3x) - (\cos 3x)(5\cos 5x)}{(\sin 5x)^2} = -\frac{3\sin 3x\sin 5x + 5\cos 3x\cos 5x}{\sin^2 5x}.$$

3.7.29: If $y = y(x) = \sin^2 x^2 = (\sin x^2)^2$, then

$$\frac{dy}{dx} = 2(\sin x^2) \cdot D_x(\sin x^2) = 2(\sin x^2) \cdot (\cos x^2) \cdot D_x(x^2) = 4x\sin x^2\cos x^2.$$

3.7.31: If $y = y(x) = \sin 2\sqrt{x} = \sin(2x^{1/2})$, then

$$\frac{dy}{dx} = \left[\cos(2x^{1/2})\right] \cdot D_x(2x^{1/2}) = x^{-1/2}\cos(2x^{1/2}) = \frac{\cos 2\sqrt{x}}{\sqrt{x}}.$$

3.7.33: If $y = y(x) = x\sin x^2$, then $\dfrac{dy}{dx} = 1 \cdot \sin x^2 + x \cdot (\cos x^2) \cdot 2x = \sin x^2 + 2x^2\cos x^2$.

3.7.35: If $y = y(x) = \sqrt{x}\sin\sqrt{x} = x^{1/2}\sin x^{1/2}$, then

$$\frac{dy}{dx} = \frac{1}{2}x^{-1/2}\sin x^{1/2} + x^{1/2}(\cos x^{1/2}) \cdot \frac{1}{2}x^{-1/2} = \frac{\sin\sqrt{x}}{2\sqrt{x}} + \frac{\cos\sqrt{x}}{2} = \frac{\sin\sqrt{x} + \sqrt{x}\cos\sqrt{x}}{2\sqrt{x}}.$$

3.7.37: If $y = y(x) = \sqrt{x}(x - \cos x)^3 = x^{1/2}(x - \cos x)^3$, then

$$\frac{dy}{dx} = \frac{1}{2}x^{-1/2}(x - \cos x)^3 + 3x^{1/2}(x - \cos x)^2(1 + \sin x)$$

$$= \frac{(x - \cos x)^3}{2\sqrt{x}} + 3\sqrt{x}(x - \cos x)^2(1 + \sin x) = \frac{(x - \cos x)^3 + 6x(x - \cos x)^2(1 + \sin x)}{2\sqrt{x}}.$$

3.7.39: If $y = y(x) = \cos(\sin x^2)$, then $\dfrac{dy}{dx} = \left[-\sin(\sin x^2)\right] \cdot (\cos x^2) \cdot 2x = -2x\left[\sin(\sin x^2)\right]\cos x^2$.

3.7.41: If $x = x(t) = \tan t^7 = \tan(t^7)$, then $\dfrac{dx}{dt} = \left(\sec t^7\right)^2 \cdot D_t(t^7) = 7t^6 \sec^2 t^7$.

3.7.43: If $x = x(t) = (\tan t)^7 = \tan^7 t$, then

$$\frac{dx}{dt} = 7(\tan t)^6 \cdot D_t \tan t = 7(\tan t)^6 \sec^2 t = 7\tan^6 t \sec^2 t.$$

3.7.45: If $x = x(t) = t^7 \tan 5t$, then $\dfrac{dx}{dt} = 7t^6 \tan 5t + 5t^7 \sec^2 5t$.

3.7.47: If $x = x(t) = \sqrt{t}\sec\sqrt{t} = t^{1/2}\sec(t^{1/2})$, then

$$\frac{dx}{dt} = \frac{1}{2}t^{-1/2}\sec(t^{1/2}) + t^{1/2}\left[\sec(t^{1/2})\tan(t^{1/2})\right] \cdot \frac{1}{2}t^{-1/2} = \frac{\sec\sqrt{t} + \sqrt{t}\sec\sqrt{t}\tan\sqrt{t}}{2\sqrt{t}}.$$

3.7.49: If $x = x(t) = \csc\left(\dfrac{1}{t^2}\right)$, then

$$\frac{dx}{dt} = \left[-\csc\left(\frac{1}{t^2}\right)\cot\left(\frac{1}{t^2}\right)\right] \cdot \left(-\frac{2}{t^3}\right) = \frac{2}{t^3}\csc\left(\frac{1}{t^2}\right)\cot\left(\frac{1}{t^2}\right).$$

3.7.51: If $x = x(t) = \dfrac{\sec 5t}{\tan 3t}$, then

$$\frac{dx}{dt} = \frac{5\tan 3t\sec 5t\tan 5t - 3\sec 5t\sec^2 3t}{(\tan 3t)^2} = 5\cot 3t\sec 5t\tan 5t - 3\csc^2 3t\sec 5t.$$

3.7.53: If $x = x(t) = t\sec t\csc t$, then

$$\frac{dx}{dt} = \sec t\csc t + t\sec t\tan t\csc t - t\sec t\csc t\cot t = t\sec^2 t + \sec t\csc t - t\csc^2 t.$$

3.7.55: If $x = x(t) = \sec(\sin t)$, then $\dfrac{dx}{dt} = \left[\sec(\sin t)\tan(\sin t)\right]\cdot\cos t$.

3.7.57: If $x = x(t) = \dfrac{\sin t}{\sec t} = \sin t\cos t$, then $\dfrac{dx}{dt} = \cos^2 t - \sin^2 t = \cos 2t$.

3.7.59: If $x = x(t) = \sqrt{1 + \cot 5t} = (1 + \cot 5t)^{1/2}$, then

$$\frac{dx}{dt} = \frac{1}{2}(1 + \cot 5t)^{-1/2}(-5\csc^2 5t) = -\frac{5\csc^2 5t}{2\sqrt{1 + \cot 5t}}.$$

3.7.61: If $f(x) = x\cos x$, then $f'(x) = -x\sin x + \cos x$, so the slope of the tangent at $x = \pi$ is $f'(\pi) = -\pi\sin\pi + \cos\pi = -1$. Because $f(\pi) = -\pi$, an equation of the tangent line is $y + \pi = -(x - \pi)$; that is, $y = -x$. The graph of f and this tangent line are shown next.

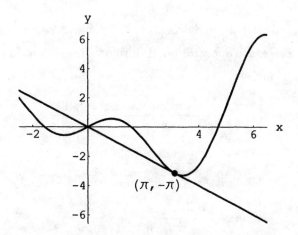

3.7.63: If $f(x) = \dfrac{4}{\pi}\tan\left(\dfrac{\pi x}{4}\right)$, then $f'(x) = \sec^2\left(\dfrac{\pi x}{4}\right)$, so the slope of the tangent at $x = 1$ is $f'(1) = \sec^2\left(\dfrac{\pi}{4}\right) = 2$. Because $f(1) = \dfrac{4}{\pi}$, an equation of the tangent line is $y - \dfrac{4}{\pi} = 2\,(x - 1)$; that is, $y = 2x - 2 + \dfrac{4}{\pi}$. The graph of f and this tangent line are shown next.

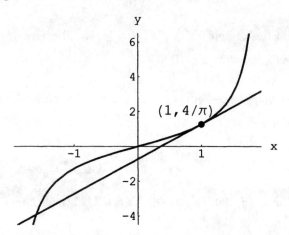

3.7.65: $\dfrac{dy}{dx} = -2\sin 2x$. This derivative is zero at all values of x for which $\sin 2x = 0$; i.e., values of x for which $2x = 0,\ \pm\pi,\ \pm 2\pi,\ \pm 3\pi,\ \dots$. Therefore the tangent line is horizontal at points with x-coordinate an integral multiple of $\frac{1}{2}\pi$. These are points of the form $(n\pi,\ 1)$ for any integer n and $\left(\frac{1}{2}m\pi,\ -1\right)$ for any odd integer m.

3.7.67: If $f(x) = \sin x \cos x$, then $f'(x) = \cos^2 x - \sin^2 x$. This derivative is zero at $x = \frac{1}{4}\pi + n\pi$ and at $x = \frac{3}{4}\pi + n\pi$ for any integer n. The tangent line is horizontal at all points of the form $\left(n\pi + \frac{1}{4}\pi,\ \frac{1}{2}\right)$ and at all points of the form $\left(n\pi + \frac{3}{4}\pi,\ -\frac{1}{2}\right)$ where n is an integer.

3.7.69: Let $f(x) = x - 2\cos x$. Then $f'(x) = 1 + 2\sin x$, so $f'(x) = 1$ when $2\sin x = 0$; that is, when $x = n\pi$ for some integer n. Moreover, if n is an integer then $f(n\pi) = n\pi - 2\cos n\pi$, so $f(n\pi) = n\pi + 2$ if n is even and $f(n\pi) = n\pi - 2$ if n is odd. In particular, $f(0) = 2$ and $f(\pi) = \pi - 2$. So the two lines have equations $y = x + 2$ and $y = x - 2$, respectively.

99

3.7.71: To derive the formulas for the derivatives of the cotangent, secant, and cosecant functions, express each in terms of sines and cosines and apply the quotient rule (or the reciprocal rule) and various trigonometric identities (see Appendix C). Thus

$$D_x \cot x = D_x \frac{\cos x}{\sin x} = \frac{-\sin^2 x - \cos^2 x}{\sin^2 x} = -\frac{1}{\sin^2 x} = -\csc^2 x,$$

$$D_x \sec x = D_x \frac{1}{\cos x} = -\frac{-\sin x}{\cos^2 x} = \frac{1}{\cos x} \cdot \frac{\sin x}{\cos x} = \sec x \tan x, \qquad \text{and}$$

$$D_x \csc x = D_x \frac{1}{\sin x} = -\frac{\cos x}{\sin^2 x} = -\frac{1}{\sin x} \cdot \frac{\cos x}{\sin x} = -\csc x \cot x.$$

3.7.73: Write $R = R(\alpha) = \frac{1}{32} v^2 \sin 2\alpha$. Then

$$R'(\alpha) = \frac{1}{16} v^2 \cos 2\alpha,$$

which is zero when $\alpha = \pi/4$ (we assume $0 \leq \alpha \leq \pi/2$). Because R is zero at the endpoints of its domain, we conclude that $\alpha = \pi/4$ maximizes the range R.

3.7.75: Let h be the altitude of the rocket (in miles) at time t (in seconds) and let α be its angle of elevation then. From the obvious figure, $h = 2 \tan \alpha$, so

$$\frac{dh}{dt} = (2 \sec^2 \alpha) \frac{d\alpha}{dt}.$$

When $\alpha = 5\pi/18$ and $d\alpha/dt = 5\pi/180$, we have $dh/dt \approx 0.4224$ (mi/s; about 1521 mi/h).

3.7.77: Draw a figure in which the observer is located at the origin, the x-axis corresponds to the ground, and the airplane is located at $(x, 20000)$. The observer's line of sight corrects the origin to the point $(x, 20000)$ and makes an angle θ with the ground. Then

$$\tan \theta = \frac{20000}{x},$$

so that $x = 20000 \cot \theta$. Thus

$$\frac{dx}{dt} = (-20000 \csc^2 \theta) \frac{d\theta}{dt}.$$

When $\theta = 60°$, we are given $\frac{d\theta}{dt} = 0.5°/s$; that is, $\frac{d\theta}{dt} = \pi/360$ radians per second when $\theta = \pi/3$. We evaluate dx/dt at this time with these values to obtain

$$\frac{dx}{dt} = (-20000) \frac{1}{\sin^2\left(\frac{\pi}{3}\right)} \cdot \frac{\pi}{360} = -\frac{2000\pi}{27},$$

approximately -232.71 ft/s. Answer: About 158.67 mi/h.

3.7.79: The cross section of the trough is a trapezoid with short base 2, long base $2 + 4\cos\theta$, and height $2\sin\theta$. Thus its cross-sectional area is

$$A(\theta) = \frac{2 + (2 + 4\cos\theta)}{2} \cdot 2\sin\theta$$

$$= 4(\sin\theta + \sin\theta\cos\theta), \qquad 0 \le \theta \le \pi/2$$

(the real upper bound on θ is $2\pi/3$, but the maximum value of A clearly occurs in the interval $[0, \pi/2]$).

$$A'(\theta) = 4(\cos\theta + \cos^2\theta - \sin^2\theta)$$

$$= 4(2\cos^2\theta + \cos\theta - 1)$$

$$= 4(2\cos\theta - 1)(\cos\theta + 1).$$

The only solution of $A'(\theta) = 0$ in the given domain occurs when $\cos\theta = \frac{1}{2}$, so that $\theta = \frac{1}{3}\pi$. It is easy to verify that this value of θ maximizes the function A.

3.7.81: The following figure shows a cross section of the sphere-with-cone through the axis of the cone and a diameter of the sphere. Note that $h = r\tan\theta$ and that

$$\cos\theta = \frac{R}{h - R}.$$

Therefore

$$h = R + R\sec\theta, \qquad \text{and thus} \qquad r = \frac{R + R\sec\theta}{\tan\theta}.$$

Now $V = \frac{1}{3}\pi r^2 h$, so for θ in the interval $(0, \pi/2)$, we have

$$V = V(\theta) = \frac{1}{3}\pi R^3 \cdot \frac{(1 + \sec\theta)^3}{\tan^2\theta}.$$

Therefore

$$V'(\theta) = \frac{\pi R^3}{3\tan^4\theta} \left[3(\tan^2\theta)(1 + \sec\theta)^2 \sec\theta \tan\theta - (1 + \sec\theta)^3(2\tan\theta \sec^2\theta) \right].$$

If $V'(\theta) = 0$ then either $\sec\theta = -1$ (so $\theta = \pi$, which we reject), or $\sec\theta = 0$ (which has no solutions), or $\tan\theta = 0$ (so either $\theta = 0$ or $\theta = \pi$, which we also reject), or (after replacement of $\tan^2\theta$ with $\sec^2\theta - 1$)

$$3(\sec^2\theta - 1) - 2(\sec\theta)(1 + \sec\theta) = 0;$$

$$\sec^2\theta - 2\sec\theta - 3 = 0.$$

It follows that $\sec\theta = 3$ or $\sec\theta = -1$. We reject the latter as before, and find that $\sec\theta = 3$, so $\theta \approx 1.23095$ (radians). The resulting minimum volume of the cone is $\frac{8}{3}\pi R^3$, twice the volume of the sphere!

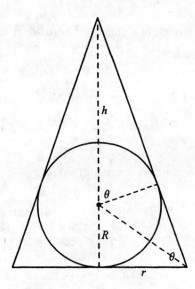

3.7.83: Set up coordinates so the diameter is on the x-axis and the equation of the circle is $x^2 + y^2 = 1$; let (x, y) denote the northwest corner of the trapezoid. The chord from $(1, 0)$ to (x, y) forms a right triangle with hypotenuse 2, side z opposite angle θ, and side w; moreover, $z = 2\sin\theta$ and $w = 2\cos\theta$. It follows that

$$y = w\sin\theta = 2\sin\theta\,\cos\theta \qquad \text{and}$$

$$-x = 1 - w\cos\theta = -\cos 2\theta.$$

Now

$$A = y(1 - x) = (2\sin\theta\cos\theta)(1 - \cos 2\theta) = 4\sin\theta\cos\theta\sin^2\theta,$$

and therefore

$$A = A(\theta) = 4\sin^3\theta\,\cos\theta, \qquad \frac{\pi}{4} \leqq \theta \leqq \frac{\pi}{2}.$$

$$A'(\theta) = 12\sin^2\theta\,\cos^2\theta - 4\sin^4\theta$$

$$= (4\sin^2\theta)(3\cos^2\theta - \sin^2\theta).$$

To solve $A'(\theta) = 0$, we note that $\sin\theta \neq 0$, so we must have $3\cos^2\theta = \sin^2\theta$; that is $\tan^2\theta = 3$. It follows that $\theta = \frac{1}{3}\pi$. The value of A here exceeds its value at the endpoints, so we have found the maximum value of the area—it is $\frac{3}{4}\sqrt{3}$.

3.7.85: The area in question is the area of the sector minus the area of the triangle in Fig. 3.7.19 and turns out to be

$$A = \frac{1}{2}r^2\theta - r^2\cos\frac{\theta}{2}\,\sin\frac{\theta}{2}$$

$$= \frac{1}{2}r^2(\theta - \sin\theta) = \frac{s^2(\theta - \sin\theta)}{2\theta^2}$$

102

because $s = r\theta$. Now

$$\frac{dA}{d\theta} = \frac{s^2(2\sin\theta - \theta\cos\theta - \theta)}{2\theta^3},$$

so $dA/d\theta = 0$ when $\theta(1 + \cos\theta) = 2\sin\theta$. Let $\theta = 2x$; note that $0 < x \leqq \pi$ because $0 < \theta \leqq 2\pi$. So the condition that $dA/d\theta = 0$ becomes

$$x = \frac{\sin\theta}{1 + \cos\theta} = \tan x.$$

But this equation has no solution in the interval $(0, \pi]$. So the only possible maximum of A must occur at an endpoint of its domain, or where x is undefined because the denominator $1 + \cos\theta$ is zero—and this occurs when $\theta = \pi$. Finally,

$$A(2\pi) = \frac{s^2}{4\pi} \quad \text{and} \quad A(\pi) = \frac{s^2}{2\pi},$$

so the maximum area is attained when the arc is a semicircle.

3.7.87: Following the *Suggestion*, we note that if n is a positive integer and

$$h = \frac{2}{(4n + 1)\pi},$$

then

$$\frac{f(h) - f(0)}{h} = \frac{(4n + 1)\pi\sin\frac{1}{2}(4n + 1)\pi}{(4n + 1)\pi} = 1,$$

and that if

$$h = \frac{2}{(4n - 1)\pi},$$

then

$$\frac{f(h) - f(0)}{h} = \frac{(4n - 1)\pi\sin\frac{1}{2}(4n - 1)\pi}{(4n - 1)\pi} = -1.$$

Therefore there are values of h arbitrarily close to zero for which

$$\frac{f(0 + h) - f(0)}{h} = +1$$

and values of h arbitrarily close to zero for which

$$\frac{f(0 + h) - f(0)}{h} = -1.$$

It follows that

$$\lim_{h \to 0} \frac{f(0 + h) - f(0)}{h} \quad \text{does not exist;}$$

that is, $f'(0)$ does not exist, and so f is not differentiable at $x = 0$.

Section 3.8

3.8.1: If $f(x) = e^{2x}$, then $f'(x) = e^{2x} \cdot D_x(2x) = 2e^{2x}$.

3.8.3: If $f(x) = \exp(x^2)$, then $f'(x) = \left[\exp(x^2)\right] \cdot D_x(x^2) = 2x\exp(x^2)$.

3.8.5: If $f(x) = e^{1/x^2}$, then $f'(x) = e^{1/x^2} \cdot D_x(1/x^2) = -\dfrac{2}{x^3}e^{1/x^2}$.

3.8.7: If $g(t) = t\exp(t^{1/2})$, then $g'(t) = \exp(t^{1/2}) + t \cdot \dfrac{1}{2}t^{-1/2}\exp(t^{1/2}) = \dfrac{2+\sqrt{t}}{2}\exp(t^{1/2})$.

3.8.9: If $g(t) = (t^2-1)e^{-t}$, then $g'(t) = 2te^{-t} - (t^2-1)e^{-t} = (1+2t-t^2)e^{-t}$.

3.8.11: If $g(t) = e^{\cos t} = \exp(\cos t)$, then $g'(t) = (-\sin t)\exp(\cos t)$.

3.8.13: If $g(t) = \dfrac{1-e^{-t}}{t}$, then $g'(t) = \dfrac{te^{-t} - (1-e^{-t})}{t^2} = \dfrac{te^{-t}+e^{-t}-1}{t^2}$.

3.8.15: If $f(x) = \dfrac{1-x}{e^x}$, then

$$f'(x) = \frac{(-1)e^x - (1-x)e^x}{(e^x)^2} = \frac{-1-1+x}{e^x} = \frac{x-2}{e^x}.$$

3.8.17: If $f(x) = \exp(e^x)$, then $f'(x) = e^x\exp(e^x)$.

3.8.19: If $f(x) = \sin(2e^x)$, then $f'(x) = 2e^x\cos(2e^x)$.

3.8.21: If $f(x) = \ln(3x-1)$, then $f'(x) = \dfrac{1}{3x-1} \cdot D_x(3x-1) = \dfrac{3}{3x-1}$.

3.8.23: If $f(x) = \ln\left[(1+2x)^{1/2}\right]$, then $f'(x) = \dfrac{\frac{1}{2} \cdot 2(1+2x)^{-1/2}}{(1+2x)^{1/2}} = \dfrac{1}{1+2x}$.

3.8.25: If $f(x) = \ln\left[(x^3-x)^{1/3}\right] = \dfrac{1}{3}\ln(x^3-x)$, then $f'(x) = \dfrac{3x^2-1}{3(x^3-x)}$.

3.8.27: If $f(x) = \cos(\ln x)$, then $f'(x) = -\dfrac{\sin(\ln x)}{x}$.

3.8.29: If $f(x) = \dfrac{1}{\ln x}$, then (by the reciprocal rule) $f'(x) = -\dfrac{1}{x(\ln x)^2}$.

3.8.31: If $f(x) = \ln\left[x(x^2+1)^{1/2}\right]$, then

$$f'(x) = \frac{(x^2+1)^{1/2} + x^2(x^2+1)^{-1/2}}{x(x^2+1)^{1/2}} = \frac{2x^2+1}{x(x^2+1)}.$$

3.8.33: If $f(x) = \ln\cos x$, then $f'(x) = \dfrac{-\sin x}{\cos x} = -\tan x$.

3.8.35: If $f(t) = t^2 \ln(\cos t)$, then $f'(t) = 2t \ln(\cos t) - \dfrac{t^2 \sin t}{\cos t} = t \left[2 \ln(\cos t) - t \tan t \right]$.

3.8.37: If $g(t) = t(\ln t)^2$, then

$$g'(t) = (\ln t)^2 + t \cdot \frac{2 \ln t}{t} = (2 + \ln t) \ln t.$$

3.8.39: Because $f(x) = 3 \ln(2x + 1) + 4 \ln(x^2 - 4)$, we have

$$f'(x) = \frac{6}{2x + 1} + \frac{8x}{x^2 - 4} = \frac{22x^2 + 8x - 24}{(2x + 1)(x^2 - 4)}.$$

3.8.41: If

$$f(x) = \ln \left(\frac{4 - x^2}{9 + x^2} \right)^{1/2} = \frac{1}{2} \ln(4 - x^2) - \frac{1}{2} \ln(9 + x^2),$$

then

$$f'(x) = -\frac{x}{4 - x^2} - \frac{x}{9 + x^2} = \frac{13x}{(x^2 - 4)(x^2 + 9)}.$$

3.8.43: If

$$f(x) = \ln \frac{x + 1}{x - 1} = \ln(x + 1) - \ln(x - 1), \quad \text{then} \quad f'(x) = \frac{1}{x + 1} - \frac{1}{x - 1} = -\frac{2}{(x - 1)(x + 1)}.$$

3.8.45: If

$$g(t) = \ln \frac{t^2}{t^2 + 1} = 2 \ln t - \ln(t^2 + 1), \quad \text{then} \quad g'(t) = \frac{2}{t} - \frac{2t}{t^2 + 1} = \frac{2}{t(t^2 + 1)}.$$

3.8.47: Given: $y = 2^x$. Then

$$\ln y = \ln (2^x) = x \ln 2;$$

$$\frac{1}{y} \cdot \frac{dy}{dx} = \ln 2;$$

$$\frac{dy}{dx} = y(x) \cdot \ln 2 = 2^x \ln 2.$$

3.8.49: Given: $y = x^{\ln x}$. Then

$$\ln y = \ln \left(x^{\ln x} \right) = (\ln x) \cdot (\ln x) = (\ln x)^2;$$

$$\frac{1}{y} \cdot \frac{dy}{dx} = \frac{2 \ln x}{x};$$

$$\frac{dy}{dx} = y(x) \cdot \frac{2 \ln x}{x} = \frac{2 x^{\ln x} \ln x}{x}.$$

105

3.8.51: Given: $y = (\ln x)^{\sqrt{x}}$. Then

$$\ln y = \ln (\ln x)^{\sqrt{x}} = x^{1/2} \ln (\ln x);$$

$$\frac{1}{y} \cdot \frac{dy}{dx} = \frac{1}{2} x^{-1/2} \ln (\ln x) + \frac{x^{1/2}}{x \ln x};$$

$$\frac{dy}{dx} = y(x) \cdot \left[\frac{\ln (\ln x)}{2x^{1/2}} + \frac{1}{x^{1/2} \ln x} \right];$$

$$\frac{dy}{dx} = \frac{2 + (\ln x) \ln (\ln x)}{2x^{1/2} \ln x} \cdot (\ln x)^{\sqrt{x}}.$$

3.8.53: If $y = (1 + x^2)^{3/2}(1 + x^3)^{-4/3}$, then

$$\ln y = \frac{3}{2} \ln(1 + x^2) - \frac{4}{3} \ln(1 + x^3);$$

$$\frac{1}{y} \cdot \frac{dy}{dx} = \frac{3x}{1 + x^2} - \frac{4x^2}{1 + x^3} = \frac{3x - 4x^2 - x^4}{(1 + x^2)(1 + x^3)};$$

$$\frac{dy}{dx} = y(x) \cdot \frac{3x - 4x^2 - x^4}{(1 + x^2)(1 + x^3)} = \frac{3x - 4x^2 - x^4}{(1 + x^2)(1 + x^3)} \cdot \frac{(1 + x^2)^{3/2}}{(1 + x^3)^{4/3}};$$

$$\frac{dy}{dx} = \frac{(3x - 4x^2 - x^4)(1 + x^2)^{1/2}}{(1 + x^3)^{7/3}}.$$

3.8.55: If $y = (x^2 + 1)^{x^2}$, then

$$\ln y = \ln(x^2 + 1)^{x^2} = x^2 \ln(x^2 + 1);$$

$$\frac{1}{y} \cdot \frac{dy}{dx} = \frac{2x^3}{x^2 + 1} + 2x \ln(x^2 + 1);$$

$$\frac{dy}{dx} = y(x) \cdot \left[\frac{2x^3}{x^2 + 1} + 2x \ln(x^2 + 1) \right] = \left[\frac{2x^3}{x^2 + 1} + 2x \ln(x^2 + 1) \right] \cdot (x^2 + 1)^{x^2}.$$

3.8.57: Given: $y = (\sqrt{x})^{\sqrt{x}}$. Then

$$\ln y = \ln (\sqrt{x})^{\sqrt{x}} = x^{1/2} \ln \left(x^{1/2} \right) = \frac{1}{2} x^{1/2} \ln x;$$

$$\frac{1}{y} \cdot \frac{dy}{dx} = \frac{1}{2x^{1/2}} + \frac{\ln x}{4x^{1/2}} = \frac{2 + \ln x}{4\sqrt{x}};$$

$$\frac{dy}{dx} = \frac{(2 + \ln x) (\sqrt{x})^{\sqrt{x}}}{4\sqrt{x}}.$$

3.8.59: If $f(x) = xe^{2x}$, then $f'(x) = e^{2x} + 2xe^{2x}$, so the slope of the graph of $y = f(x)$ at $(1, e^2)$ is $f'(1) = 3e^2$. Hence an equation of the line tangent to the graph at that point is $y - e^2 = 3e^2(x - 1)$; that is, $y = 3e^2 x - 2e^2$.

3.8.61: If $f(x) = x^3 \ln x$, then $f'(x) = x^2 + 3x^2 \ln x$, so the slope of the graph of $y = f(x)$ at the point $(1, 0)$ is $f'(1) = 1$. Hence an equation of the line tangent to the graph at that point is $y - 0 = 1 \cdot (x - 1)$; that is, $y = x - 1$.

3.8.63: If $f(x) = e^{2x}$, then

$$f'(x) = 2e^{2x}, \quad f''(x) = 4e^{2x}, \quad f'''(x) = 8e^{2x}, \quad f^{(4)}(x) = 16e^{2x}, \quad \text{and} \quad f^{(5)}(x) = 32e^{2x}.$$

It appears that $f^{(n)}(x) = 2^n e^{2x}$.

3.8.65: If $f(x) = e^{-x/6} \sin x$, then

$$f'(x) = -\frac{1}{6} e^{-x/6} \sin x + e^{-x/6} \cos x = \frac{6 \cos x - \sin x}{6 e^{x/6}}.$$

Hence the first local maximum point for $x > 0$ occurs when $x = \arctan 6$ and the first local minimum point occurs when $x = \pi + \arctan 6$. The corresponding y-coordinates are, respectively,

$$\frac{6}{e^{(\arctan 6)/6} \sqrt{37}} \quad \text{and} \quad -\frac{6}{e^{(\pi + \arctan 6)/6} \sqrt{37}}.$$

3.8.67: The figure below shows the intersection of the two graphs in the viewing window $1.118 \leq x \leq 1.119$, $3.05 \leq y \leq 3.07$. We see that, accurate to three decimal places, the indicated solution of $e^x = x^{10}$ is $x = 1.118$.

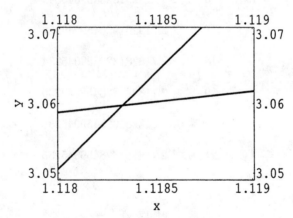

3.8.69: We first let

$$f(k) = \left(1 + \frac{1}{10^k}\right)^{10^k}.$$

Then *Mathematica* yields the following approximations:

k	$f(k)$ (rounded)

107

1	2.593742460100
2	2.704813829422
3	2.716923932236
4	2.718145926825
5	2.718268237174
6	2.718280469319
7	2.718281692545
8	2.718281814868
9	2.718281827100
10	2.718281828323
11	2.718281828445
12	2.718281828458
13	2.718281828459
14	2.718281828459
15	2.718281828459
16	2.718281828459045099
17	2.719291929459045222
18	2.718281828459045234
19	2.718281828459045235
20	2.718281828459045235
21	2.718281828459045235

3.8.71: Solution:

$$\ln y = \ln u + \ln v + \ln w - \ln p - \ln q - \ln r;$$

$$\frac{1}{y} \cdot \frac{dy}{dx} = \frac{1}{u} \cdot \frac{du}{dx} + \frac{1}{v} \cdot \frac{dv}{dx} + \frac{1}{w} \cdot \frac{dw}{dx} - \frac{1}{p} \cdot \frac{dp}{dx} - \frac{1}{q} \cdot \frac{dq}{dx} - \frac{1}{r} \cdot \frac{dr}{dx};$$

$$\frac{dy}{dx} = y \cdot \left(\frac{1}{u} \cdot \frac{du}{dx} + \frac{1}{v} \cdot \frac{dv}{dx} + \frac{1}{w} \cdot \frac{dw}{dx} - \frac{1}{p} \cdot \frac{dp}{dx} - \frac{1}{q} \cdot \frac{dq}{dx} - \frac{1}{r} \cdot \frac{dr}{dx} \right).$$

The solution makes the generalization obvious.

3.8.73: (a): If $f(x) = \log_{10} x$, then the definition of the derivative yields

$$f'(1) = \lim_{h \to 0} \frac{f(1+h) - f(1)}{h} = \lim_{h \to 0} \frac{1}{h} \log_{10}(1+h) = \lim_{h \to 0} \log_{10}(1+h)^{1/h}.$$

(b): When $h = 0.1$ the value of $\log_{10}(1+h)^{1/h}$ is approximately 0.4139. With $h = 0.01$ we get 0.4321, with $h = 0.001$ we get 0.4341, and with $h = \pm 0.0001$ we get 0.4343.

Section 3.9

3.9.1: $2x - 2y\dfrac{dy}{dx} = 0$, so $\dfrac{dy}{dx} = \dfrac{x}{y}$. Also, $y = \pm\sqrt{x^2 - 1}$, so $\dfrac{dy}{dx} = \pm\dfrac{x}{\sqrt{x^2 - 1}} = \dfrac{x}{\pm\sqrt{x^2 - 1}} = \dfrac{x}{y}$.

3.9.3: $32x + 50y\dfrac{dy}{dx} = 0$; $\dfrac{dy}{dx} = -\dfrac{16x}{25y}$. Substituting $y = \pm\frac{1}{5}\sqrt{400 - 16x^2}$ into the derivative, we get $\dfrac{dy}{dx} = \mp\dfrac{16x}{5\sqrt{400 - 16x^2}}$, which is the result obtained by explicit differentiation.

3.9.5: $\frac{1}{2}x^{-1/2} + \frac{1}{2}y^{-1/2}\dfrac{dy}{dx} = 0$: $\quad \dfrac{dy}{dx} = -\sqrt{\dfrac{y}{x}}$.

3.9.7: $\frac{2}{3}x^{-1/3} + \frac{2}{3}y^{-1/3}\dfrac{dy}{dx} = 0$: $\quad \dfrac{dy}{dx} = -\left(\dfrac{x}{y}\right)^{-1/3} = -\left(\dfrac{y}{x}\right)^{1/3}$.

3.9.9: Given: $x^3 - x^2 y = xy^2 + y^3$:

$$3x^2 - x^2\frac{dy}{dx} - 2xy = y^2 + 2xy\frac{dy}{dx} + 3y^2\frac{dy}{dx};$$

$$3x^2 - 2xy - y^2 = (2xy + 3y^2 + x^2)\frac{dy}{dx};$$

$$\frac{dy}{dx} = \frac{3x^2 - 2xy - y^2}{3y^2 + 2xy + x^2}.$$

3.9.11: Given: $x\sin y + y\sin x = 1$:

$$x\cos y\frac{dy}{dx} + \sin y + y\cos x + \sin x\frac{dy}{dx} = 0;$$

$$\frac{dy}{dx} = -\frac{\sin y + y\cos x}{x\cos y + \sin x}.$$

3.9.13: Given: $2x + 3e^y = e^{x+y}$. Differentiation of both sides of this equation (actually, an *identity*) with respect to x yields

$$2 + 3e^y\frac{dy}{dx} = e^{x+y}\left(1 + \frac{dy}{dx}\right), \qquad \text{and so} \qquad \frac{dy}{dx} = \frac{e^{x+y} - 2}{3e^y - e^{x+y}} = \frac{3e^y + 2x - 2}{(3 - e^x)e^y}.$$

3.9.15: $2x + 2y\dfrac{dy}{dx} = 0$: $\quad \dfrac{dy}{dx} = -\dfrac{x}{y}$. At $(3, -4)$ the tangent has slope $\frac{3}{4}$ and thus equation $y + 4 = \frac{3}{4}(x - 3)$.

3.9.17: $x^2\dfrac{dy}{dx} + 2xy = 1$, so $\dfrac{dy}{dx} = \dfrac{1 - 2xy}{x^2}$. At $(2, 1)$ the tangent has slope $-\frac{3}{4}$ and thus equation $3x + 4y = 10$.

3.9.19: $y^2 + 2xy\dfrac{dy}{dx} + 2xy + x^2\dfrac{dy}{dx} = 0$: $\quad \dfrac{dy}{dx} = -\dfrac{2xy + y^2}{2xy + x^2}$. At $(1, -2)$ the slope is zero, so an equation of the tangent there is $y = -2$.

3.9.21: $24x + 24y\dfrac{dy}{dx} = 25y + 25x\dfrac{dy}{dx}$: $\quad \dfrac{dy}{dx} = \dfrac{25y - 24x}{24y - 25x}$. At $(3, 4)$ the tangent line has slope $\frac{4}{3}$ and thus equation $4x = 3y$.

3.9.23: $\dfrac{dy}{dx} = \dfrac{3e^{2x} + 2e^y}{3e^{2x} + e^{x+2y}}$, so the tangent line at $(0, 0)$ has slope $\frac{5}{4}$ and equation $4y = 5x$.

3.9.25: $\frac{2}{3}x^{-1/3} + \frac{2}{3}y^{-1/3}\dfrac{dy}{dx} = 0$: $\quad \dfrac{dy}{dx} = -\dfrac{y^{1/3}}{x^{1/3}}$. At $(8, 1)$ the tangent line has slope $-\frac{1}{2}$ and thus equation $y - 1 = -\frac{1}{2}(x - 8)$; that is, $x + 2y = 10$.

3.9.27: $2\left(x^2 + y^2\right)\left(2x + 2y\dfrac{dy}{dx}\right) = 50x\dfrac{dy}{dx} + 50y$:

$$\dfrac{dy}{dx} = -\dfrac{2x^3 - 25y + 2xy^2}{-25x + 2x^2y + 2y^3}.$$

At $(2, 4)$ the tangent line has slope $\frac{2}{11}$ and thus equation $y - 4 = \frac{2}{11}(x - 2)$; that is, $11y = 2x + 40$.

3.9.29: $3x^2 + 3y^2\dfrac{dy}{dx} = 9x\dfrac{dy}{dx} + 9y$: $\quad \dfrac{dy}{dx} = \dfrac{3y - x^2}{y^2 - 3x}$.

(a): At $(2, 4)$ the tangent line has slope $\frac{4}{5}$ and thus equation $y - 4 = \frac{4}{5}(x - 2)$; that is, $5y = 4x + 12$.

(b): At a point on the curve at which $\dfrac{dy}{dx} = -1$, $3y - x^2 = -y^2 - 3x$ and $x^3 + y^3 = 9xy$. This pair of simultaneous equations has solutions $x = 0$, $y = 0$ and $x = \frac{9}{2}$, $y = \frac{9}{2}$, but the derivative does not exist at the point $(0, 0)$. Therefore the tangent line with slope -1 has equation $y - \frac{9}{2} = -\left(x - \frac{9}{2}\right)$.

3.9.31: Here $\dfrac{dy}{dx} = \dfrac{2 - x}{y - 2}$, so horizontal tangents can occur only if $x = 2$ and $y \neq 2$. When $x = 2$, the original equation yields $y^2 - 4y - 4 = 0$, so that $y = 2 \pm \sqrt{8}$. Thus there are two points at which the tangent line is horizontal: $\left(2, 2 - \sqrt{8}\right)$ and $\left(2, 2 + \sqrt{8}\right)$.

3.9.33: By direct differentiation, $dx/dy = (1 + y)e^y$. By implicit differentiation, $\dfrac{dy}{dx} = \dfrac{1}{(1 + y)e^y}$, and the results are equivalent.

(a): At $(0, 0)$, $dy/dx = 1$, so an equation of the line tangent to the curve at $(0, 0)$ is $y = x$.

(b): At $(e, 1)$, $dy/dx = 1/(2e)$, so an equation of the line tangent to the curve at $(e, 1)$ is $x + e = 2ey$.

3.9.35: From $2(x^2 + y^2)\left(2x + 2y\dfrac{dy}{dx}\right) = 2x - 2y\dfrac{dy}{dx}$ it follows that

$$\frac{dy}{dx} = \frac{x[1 - 2(x^2 + y^2)]}{y[1 + 2(x^2 + y^2)]}.$$

So $dy/dx = 0$ when $x^2 + y^2 = \frac{1}{2}$, but is undefined when $x = 0$, for then $y = 0$ as well. If $x^2 + y^2 = \frac{1}{2}$, then $x^2 - y^2 = \frac{1}{4}$, so that $x^2 = \frac{3}{8}$, and it follows that $y^2 = \frac{1}{8}$. Consequently there are horizontal tangents at all four points where $|x| = \frac{1}{4}\sqrt{6}$ and $|y| = \frac{1}{4}\sqrt{2}$.

Also $dx/dy = 0$ only when $y = 0$, and if so, then $x^4 = x^2$, so that $x = \pm 1$ (dx/dy is undefined when $x = 0$). So there are vertical tangents at the two points $(-1, 0)$ and $(1, 0)$.

3.9.37: Suppose that the pile has height $h = h(t)$ at time t (seconds) and radius $r = r(t)$ then. We are given $h = 2r$ and we know that the volume of the pile at time t is

$$V = V(t) = \frac{\pi}{3}r^2 h = \frac{2}{3}\pi r^3. \quad \text{Now} \quad \frac{dV}{dt} = \frac{dV}{dr} \cdot \frac{dr}{dt}, \quad \text{so} \quad 10 = 2\pi r^2 \frac{dr}{dt}.$$

When $h = 5$, $r = 2.5$; at that time $\dfrac{dr}{dt} = \dfrac{10}{2\pi(2.5)^2} = \dfrac{4}{5\pi} \approx 0.25645$ (ft/s).

3.9.39: We assume that the oil slick forms a solid right circular cylinder of height (thickness) h and radius r. Then its volume is $V = \pi r^2 h$, and we are given $V = 1$ (constant) and $dh/dt = -0.001$. Therefore $0 = \pi r^2 \dfrac{dh}{dt} + 2\pi r h \dfrac{dr}{dt}$. Consequently $2h\dfrac{dr}{dt} = \dfrac{r}{1000}$, and so $\dfrac{dr}{dt} = \dfrac{r}{2000h}$. When $r = 8$, $h = \dfrac{1}{\pi r^2} = \dfrac{1}{64\pi}$. At that time, $\dfrac{dr}{dt} = \dfrac{8 \cdot 64\pi}{2000} = \dfrac{32\pi}{125} \approx 0.80425$ (m/h).

3.9.41: Let x denote the width of the rectangle; then its length is $2x$ and its area is $A = 2x^2$. Thus $\dfrac{dA}{dt} = 4x\dfrac{dx}{dt}$. When $x = 10$ and $dx/dt = 0.5$, we have

$$\left.\frac{dA}{dt}\right|_{x=10} = (4)(10)(0.5) = 20 \ (\text{cm}^2/\text{s}).$$

3.9.43: Let r denote the radius of the balloon and V its volume at time t (in seconds). Then

$$V = \frac{4}{3}\pi r^3, \quad \text{so} \quad \frac{dV}{dt} = 4\pi r^2 \frac{dr}{dt}.$$

We are to find dr/dt when $r = 10$, and we are given the information that $dV/dt = 100\pi$. Therefore

$$100\pi = 4\pi(10)^2 \left.\frac{dr}{dt}\right|_{r=10},$$

and so at the time in question the radius is increasing at the rate of $dr/dt = \frac{1}{4} = 0.25$ (cm/s).

111

3.9.45: Place the person at the origin and the kite in the first quadrant at $(x, 400)$ at time t, where $x = x(t)$ and we are given $dx/dt = 10$. Then the length $L = L(t)$ of the string satisfies the equation $L^2 = x^2 + 160000$, and therefore $2L\dfrac{dL}{dt} = 2x\dfrac{dx}{dt}$. Moreover, when $L = 500$, $x = 300$. So

$$1000\frac{dL}{dt}\bigg|_{L=500} = 600 \cdot 10,$$

which implies that the string is being let out at 6 ft/s.

3.9.47: Locate the observer at the origin and the airplane at $(x, 3)$, with $x > 0$. We are given dx/dt where the units are in miles, hours, and miles per hour. The distance z between the observer and the airplane satisfies the identity $z^2 = x^2 + 9$, and because the airplane is traveling at 8 mi/min, we find that $x = 4$, and therefore that $z = 5$, at the time 30 seconds after the airplane has passed over the observer. Also $2z\dfrac{dz}{dt} = 2x\dfrac{dx}{dt}$, so at the time in question, $10\dfrac{dz}{dt} = 8 \cdot 480$. Therefore the distance between the airplane and the observer is increasing at 384 mi/h at the time in question.

3.9.49: We use $a = 10$ in the formula given in Problem 42. Then

$$V = \frac{1}{3}\pi y^2(30 - y).$$

Hence $(-100)(0.1337) = \dfrac{dV}{dt} = \pi(20y - y^2)\dfrac{dy}{dt}$. Thus $\dfrac{dy}{dt} = -\dfrac{13 \cdot 37}{\pi y(20 - y)}$. Substitution of $y = 7$ and $y = 3$ now yields the two answers:

$$\text{(a):} \quad -\frac{191}{1300\pi} \approx -0.047 \ (\text{ft/min}); \quad \text{(b):} \quad -\frac{1337}{5100\pi} \approx -0.083 \ (\text{ft/min}).$$

3.9.51: Let the positive y-axis represent the wall and the positive x-axis the ground, with the top of the ladder at $(0, y)$ and its lower end at $(x, 0)$ at time t. Given: $dx/dt = 4$, with units in feet, seconds, and feet per second. Also $x^2 + y^2 = 41^2$, and it follows that $y\dfrac{dy}{dt} = -x\dfrac{dx}{dt}$. Finally, when $y = 9$, we have $x = 40$, so at that time $9\dfrac{dy}{dt} = -40 \cdot 4$. Therefore the top of the ladder is moving downward at $\frac{160}{9} \approx 17.78$ ft/s.

3.9.53: Let r be the radius of the cone, h its height. We are given $dh/dt = -3$ and $dr/dt = +2$, with units in centimeters and seconds. The volume of the cone at time t is $V = \frac{1}{3}\pi r^2 h$, so

$$\frac{dV}{dt} = \frac{2}{3}\pi r h \frac{dr}{dt} + \frac{1}{3}\pi r^2 \frac{dh}{dt}.$$

When $r = 4$ and $h = 6$, $\dfrac{dV}{dt} = \dfrac{2}{3} \cdot 24\pi \cdot 2 + \dfrac{1}{3} \cdot 16\pi \cdot (-3) = 16\pi$, so the volume of the cone is increasing at the rate of 16π cm^3/s then.

3.9.55: Locate the radar station at the origin and the rocket at $(4, y)$ in the first quadrant at time t, with y in miles and t in hours. The distance z between the station and the rocket satisfies the equation $y^2 + 16 = z^2$, so $2y\dfrac{dy}{dt} = 2z\dfrac{dz}{dt}$. When $z = 5$, we have $y = 3$, and because $dz/dt = 3600$ it follows that $dy/dt = 6000$ mi/h.

3.9.57: Put the floor on the nonnegative x-axis and the wall on the nonnegative y-axis. Let x denote the distance from the wall to the foot of the ladder (measured along the floor) and let y be the distance from the floor to the top of the ladder (measured along the wall). By the Pythagorean theorem, $x^2 + y^2 = 100$, and we are given $dx/dt = \frac{22}{15}$ (because we will use units of feet and seconds rather than miles and hours). From the Pythagorean relation we find that

$$2x\frac{dx}{dt} + 2y\frac{dy}{dt} = 0,$$

so that $\dfrac{dy}{dt} = -\dfrac{x}{y} \cdot \dfrac{dx}{dt} = -\dfrac{22x}{15y}$.

(a): If $y = 4$, then $x = \sqrt{84} = 2\sqrt{21}$. Hence when the top of the ladder is 4 feet above the ground, it is moving a a rate of

$$\left.\frac{dy}{dt}\right|_{y=4} = -\frac{44\sqrt{21}}{60} = -\frac{11\sqrt{21}}{15} \approx -3.36$$

feet per second, about 2.29 miles per hour downward.

(b): If $y = \frac{1}{12}$ (one inch), then

$$x^2 = 100 - \frac{1}{144} = \frac{14399}{144}, \quad \text{so that} \quad x \approx 9.99965.$$

In this case,

$$\left.\frac{dy}{dt}\right|_{y=1/12} = -\frac{22 \cdot (9.99965)}{15 \cdot \frac{1}{12}} = -\frac{88}{5} \cdot (9.99965) \approx -176$$

feet per second, about 120 miles per hour downward.

(c): If $y = 1$ mm, then $x \approx 10$ (ft), and so

$$\frac{dy}{dt} \approx -\frac{22}{15} \cdot (3048) \approx 4470$$

feet per second, about 3048 miles per hour.

The results in parts (b) and (c) are not plausible. This shows that the assumption that the top of the ladder never leaves the wall is invalid

3.9.59: Locate the military jet at $(x, 0)$ with $x < 0$ and the other aircraft at $(0, y)$ with $y \geq 0$. With units in miles, minutes, and miles per minute, we are given $dx/dt = +12$, $dy/dt = +8$, and when $t = 0$, $x = -208$ and $y = 0$. The distance z between the aircraft satisfies the equation $x^2 + y^2 = z^2$, so

$$\frac{dz}{dt} = \frac{1}{\sqrt{x^2 + y^2}}\left(x\frac{dx}{dt} + y\frac{dy}{dt}\right) = \frac{12x + 8y}{\sqrt{x^2 + y^2}}.$$

The closest approach will occur when $dz/dt = 0$: $y = -3x/2$. Now $x(t) = 12t - 208$ and $y(t) = 8t$. So at closest approach we have

$$8t = y(t) = -\frac{3}{2}x(t) = -\frac{3}{2}(12t - 208).$$

Hence at closest approach, $16t = 624 - 36t$, and thus $t = 12$. At this time, $x = -64$, $y = 96$, and $z = 32\sqrt{13} \approx 115.38$ (mi).

3.9.61: Let x be the radius of the water surface at time t and y the height of the water remaining at time t. If Q is the amount of water remaining in the tank at time t, then (because the water forms a cone) $Q = Q(t) = \frac{1}{3}\pi x^2 y$. But by similar triangles, $\dfrac{x}{y} = \dfrac{3}{5}$, so $x = \dfrac{3y}{5}$. So

$$Q(t) = \frac{1}{3}\pi \frac{9}{25}y^3 = \frac{3}{25}\pi y^3.$$

We are given $dQ/dt = -2$ when $y = 3$. This implies that when $y = 3$, $-2 = \dfrac{dQ}{dt} = \dfrac{9}{25}\pi y^2 \dfrac{dy}{dt}$. So at the time in question,

$$\left.\frac{dy}{dt}\right|_{y=3} = -\frac{50}{81\pi} \approx -0.1965 \text{ (ft/s)}.$$

3.9.63: Let r be the radius of the water surface at time t, h the depth of water in the bucket then. By similar triangles we find that

$$\frac{r-6}{h} = \frac{1}{4}, \text{ so } r = 6 + \frac{h}{4}.$$

The volume of water in the bucket then is

$$\begin{aligned}
V &= \frac{1}{3}\pi h(36 + 6r + r^2) \\
&= \frac{1}{3}\pi\left(36 + 36 + \frac{3}{2}h + 36 + 3h + \frac{1}{16}h^2\right) \\
&= \frac{1}{3}\pi h\left(108 + \frac{9}{2}h + \frac{1}{16}h^2\right).
\end{aligned}$$

Now $\dfrac{dV}{dt} = -10$; we are to find dh/dt when $h = 12$.

$$\frac{dV}{dt} = \frac{1}{3}\pi\left(108 + 9h + \frac{3}{16}h^2\right)\frac{dh}{dt}.$$

Therefore $\left.\dfrac{dh}{dt}\right|_{h=12} = \dfrac{3}{\pi} \cdot \dfrac{-10}{108 + 9 \cdot 12 + \dfrac{3 \cdot 12^2}{16}} = -\dfrac{10}{81\pi} \approx -0.0393$ (in./min).

3.9.65: Set up a coordinate system in which the radar station is at the origin, the plane passes over it at the point $(0, 1)$ (so units on the axes are in miles), and the plane is moving along the graph of the equation $y = x + 1$. Let s be the distance from $(0, 1)$ to the plane and let u be the distance from the radar station to the plane. We are given $du/dt = +7$ mi/min. We may deduce from the law of cosines that $u^2 = s^2 + 1 + s\sqrt{2}$. Let v denote the speed of the plane, so that $v = ds/dt$. Then

$$2u\frac{du}{dt} = 2sv + v\sqrt{2} = v\left(2s + \sqrt{2}\right), \text{ and so } v = \frac{2u}{2s + \sqrt{2}} \cdot \frac{du}{dt}.$$

When $u = 5$, $s^2 + s\sqrt{2} - 24 = 0$. The quadratic formula yields the solution $s = 3\sqrt{2}$, and it follows that $v = 5\sqrt{2}$ mi/min; alternatively, $v \approx 424.26$ mi/h.

3.9.67: Place the pole at the origin in the plane, and let the horizontal strip $0 \leqq y \leqq 30$ represent the road. Suppose that the person is located at $(x, 30)$ with $x > 0$ and is walking to the right, so $dx/dt = +5$. Then the distance from the pole to the person will be $\sqrt{x^2 + 900}$. Let z be the length of the person's shadow. By similar triangles it follows that $2z = \sqrt{x^2 + 900}$, so $4z^2 = x^2 + 900$, and thus $8z\dfrac{dz}{dt} = 2x\dfrac{dx}{dt}$. When $x = 40$, we find that $z = 25$, and therefore that

$$100\frac{dz}{dt}\bigg|_{z=25} = 40 \cdot 5 = 200.$$

Therefore the person's shadow is lengthening at 2 ft/s at the time in question.

Section 3.10

Note: Your answers may differ from ours in the last one or two decimal places because of differences in hardware or in the way the problem was solved. We used *Mathematica* and carried 40 decimal digits throughout all calculations, and our answers are correct or correctly rounded to the number of digits shown here. In most of the first 20 problems the initial guess x_0 was obtained by linear interpolation. Finally, the equals mark is used in this section to mean "equal or approximately equal."

3.10.1: With $f(x) = x^2 - 5$, $a = 2$, $b = 3$, and

$$x_0 = a - \frac{(b-a)f(a)}{f(b) - f(a)} = 2.2,$$

we used the iterative formula

$$x_{n+1} = x_n - \frac{f(x_n)}{f'(x_n)} \qquad \text{for} \quad n \geqq 0.$$

Thus we obtained $x_1 = 2.236363636$, $x_2 = 2.236067997$, and $x_3 = x_4 = 2.236067977$. Answer: 2.2361.

3.10.3: $x_0 = 2.322274882$; we use $f(x) = x^5 - 100$. Then $x_1 = 2.545482651$, $x_2 = 2.512761634$, $x_3 = 2.511887041$, and $x_4 = x_5 = 2.511886432$. Answer: 2.5119.

3.10.5: 0.25, 0.3035714286, 0.3027758133, 0.3027756377. Answer: 0.3028.

3.10.7: $x_0 = -0.5$, $x_1 = -0.8108695652$, $x_2 = -0.7449619516$, $x_3 = -0.7402438226$, $x_4 = -0.7402217826 = x_5$. Answer: 0.7402.

3.10.9: With $f(x) = x - \cos x$, $f'(x) = 1 + \sin x$, and calculator set in *radian* mode, we obtain $x_0 = 0.5854549279$, $x_1 = 0.7451929664$, $x_2 = 0.7390933178$, $x_3 = 0.7390851332$, and $x_4 = x_3$. Answer: 0.7391.

3.10.11: With $f(x) = 4x - \sin x - 4$ and calculator in *radian* mode, we get the following results: $x_0 = 1.213996400$, $x_1 = 1.236193029$, $x_2 = 1.236129989 = x_3$. Answer: 1.2361.

3.10.13: With $x_0 = 2.188405797$ and the iterative formula

$$x - \frac{x^4(x+1) - 100}{x^3(5x+4)} \longrightarrow x,$$

we obtain $x_1 = 2.360000254$, $x_2 = 2.339638357$, $x_3 = 2.339301099$, and $x_4 = 2.339301008 = x_5$. Answer: 2.3393.

3.10.15: The nearest discontinuities of $f(x) = x - \tan x$ are at $\pi/2$ and at $3\pi/2$, approximately 1.571 and 4.712. Therefore the function $f(x) = x - \tan x$ has the intermediate value property on the interval $[2, 3]$. Results: $x_0 = 2.060818495$, $x_1 = 2.027969226$, $x_2 = 2.028752991$, and $x_3 = 2.028757838 = x_4$. Answer: 2.0288.

3.10.17: $x - e^{-x} = 0$; $[0, 1]$: $x_0 = 0.5$, $x_1 \approx 0.5663$, $x_2 \approx 0.5671$, $x_3 \approx 0.5671$.

3.10.19: $e^x + x - 2 = 0$; $[0, 1]$: $x_0 = 0.5$, $x_1 \approx 0.4439$, $x_2 \approx 0.4429$, $x_3 \approx 0.4429$.

3.10.21: Let $f(x) = x^3 - a$. Then the iteration of Newton's method in Eq. (6) takes the form

$$x_{n+1} = x_n - \frac{f(x_n)}{f'(x_n)} = x_n - \frac{(x_n)^3 - a}{3(x_n)^2} = \frac{2(x_n)^3 + a}{3(x_n)^2} = \frac{1}{3}\left(2x_n + \frac{a}{(x_n)^2}\right).$$

Because $1 < \sqrt[3]{2} < 2$, we begin with $x_0 = 1.5$ and apply this formula with $a = 2$ to obtain $x_1 = 1.296296296$, $x_2 = 1.260932225$, $x_3 = 1.259921861$, and $x_4 = 1.259921050 = x_5$. Answer: 1.25992.

3.10.23: We get $x_0 = 0.5$, $x_1 = 0.4387912809$, $x_2 = 0.4526329217$, $x_3 = 0.4496493762$, $\ldots$, $x_{14} = 0.4501836113 = x_{15}$. The method of repeated substitution tends to converge much more slowly than Newton's method, has the advantage of not requiring that you compute a derivative or even that the functions involved are differentiable, and has the disadvantage of more frequent failure than Newton's method when both are applicable (see Problems 24 and 25).

3.10.25: Beginning with $x_0 = 0.5$, the first formula yields $x_0 = 0.5$, $x_1 = -1$, $x_2 = 2$, $x_3 = 2.75$, $x_4 = 2.867768595$, $\ldots$, $x_{12} = 2.879385242 = x_{13}$. Wrong root! At least the method converged. If your calculator or computer balks at computing the cube root of a negative number, then you can rewrite the second formula in Problem 25 in the form

$$x = \mathrm{Sgn}(3x^2 - 1) \cdot |3x^2 - 1|^{1/3}.$$

The results, again with $x_0 = 0.5$, are $x_1 = -0.629960525$, $x_2 = 0.575444686$, $x_3 = -0.187485243$, $x_4 = -0.963535808$, $\ldots$, $x_{25} = 2.877296053$, $x_{26} = 2.877933902$, $\ldots$, and $x_{62} = 2.879385240 = x_{63}$.

Not only is convergence extremely slow, the method of repeated substitution again leads to the wrong root. Finally, the given equation can also be written in the form

$$x = \frac{1}{\sqrt{3 - x}},$$

and in this case, again with $x_0 = 0.5$, we obtain $x_1 = 0.632455532$, $x_2 = 0.649906570$, $x_3 = 0.652315106$, $x_4 = 0.652649632$, ..., and $x_{12} = 0.652703645 = x_{13}$.

3.10.27: Let $f(x) = x^5 + x - 1$. Then $f(x)$ is a polynomial, thus is continuous everywhere, and thus has the intermediate value property on every interval. Also $f(0) = -1$ and $f(1) = 1$, so $f(x)$ must assume the intermediate value 0 somewhere in the interval $[0, 1]$. Thus the equation $f(x) = 0$ has *at least* one solution. Next, $f'(x) = 5x^4 + 1$ is positive for all x, so f is an increasing function. Because f is continuous, its graph can therefore cross the x-axis at most once, and so the equation $f(x) = 0$ has *at most* one solution. Thus it has exactly one solution. Incidentally, Newton's method yields the approximate solution 0.75487766624669276. To four places, 0.7549.

3.10.29: Let $f(x) = x - 2 \sin x$. The graph of f on $[-2, 2]$ shows that there are exactly three solutions, the largest of which is approximately $x_0 = 1.9$. With Newton's method we obtain $x_1 = 1.895505940$, and $x_2 = 1.895494267 = x_3$. Because $f(-x) = -f(x)$, the other two solutions are 0 and -1.895494267. Answer: ± 1.8955 and 0.

3.10.31: Let $f(x) = x^7 - 3x^3 + 1$. Then $f(x)$ is a polynomial, so f is continuous on every interval of real numbers, including the intervals $[-2, -1]$, $[0, 1]$, and $[1, 2]$. Also $f(-2) = -103 < 0 < 3 = f(-1)$, $f(0) = 1 > 0 > -1 = f(1)$, and $f(1) = -1 < 0 < 105 = f(2)$. Therefore the equation $f(x) = 0$ has one solution in $(-2, -1)$, another in $(0, 1)$, and a third in $(1, 2)$. (It has no other real solutions.) The graph of f shows that the first solution is near -1.4, the second is near 0.7, and the third is near 1.2. Then Newton's method yields

n	First x_n	Second x_n	Third x_n
1	-1.362661201	0.714876604	1.275651936
2	-1.357920265	0.714714327	1.258289744
3	-1.357849569	0.714714308	1.256999591
4	-1.357849553	0.714714308	1.256992779
5	-1.357849553	0.714714308	1.256992779

Answers: -1.3578, 0.7147, and 1.2570.

3.10.33: There is only one solution of $x^3 = \cos x$ for the following reasons: $x^3 < -1 \leqq \cos x$ if $x < -1$, $x^3 < 0 < \cos x$ if $-1 < x < 0$, x^3 is increasing on $[0, 1]$ whereas $\cos x$ is decreasing there (and their graphs

117

cross in this interval as a consequence of the intermediate value property of continuous functions), and $x^3 > 1 \geqq \cos x$ for $x > 1$. The graph of $f(x) = x^3 - \cos x$ crosses the x-axis near $x_0 = 0.9$, and Newton's method yields $x_1 = 0.866579799$, $x_2 = 0.865475218$, and $x_3 = 0.865474033 = x_4$. Answer: Approximately 0.8654740331016145.

3.10.35: With $x_0 = 3.5$, we obtain the sequence $x_1 = 3.451450588$, $x_2 = 3.452461938$, and finally $x_3 = 3.452462314 = x_4$. Answer: Approximately 3.452462314057969.

3.10.37: If the plane cuts the sphere at distance x from its center, then the smaller spherical segment has height $h = a - x = 1 - x$ and the larger has height $h = a + x = 1 + x$. So the smaller has volume

$$V_1 = \frac{1}{3}\pi h^2 (3a - h) = \frac{1}{3}\pi(1 - x)^2 (2 + x)$$

and the larger has volume

$$V_2 = \frac{1}{3}\pi h^2 (3a - h) = \frac{1}{3}\pi(1 + x)^2 (2 - x) = 2V_1.$$

These equations leads to

$$(1 + x)^2 (2 - x) = 2(1 - x)^2 (2 + x);$$

$$(x^2 + 2x + 1)(x - 2) + 2(x^2 - 2x + 1)(x + 2) = 0;$$

$$x^3 - 3x - 2 + 2x^3 - 6x + 4 = 0;$$

$$3x^3 - 9x + 2 = 0.$$

The last of these equations has three solutions, one near -1.83 (out of range), one near 1.61 (also out of range), and one near $x_0 = 0.2$. Newton's method yields $x_1 = 0.225925926$, $x_2 = 0.226073709$, and $x_3 = 0.226073714 = x_4$. Answer: 0.2261.

3.10.39: We iterate using the formula

$$x - \frac{x + \tan x}{1 + \sec^2 x} \longrightarrow x.$$

Here is a sequence of simple *Mathematica* commands to find approximations to the four least positive solutions of the given equation, together with the results. (The command **list=g[list]** was executed repeatedly,

118

but deleted from the output to save space.)

list={2.0, 5.0, 8.0, 11.0}

f[x_]:=x+Tan[x]

g[x_]:=N[x−f[x]/f′[x], 10]

list=g[list]

2.027314579, 4.879393859, 7.975116372, 11.00421012

2.028754298, 4.907699753, 7.978566616, 11.01202429

2.028757838, 4.913038110, 7.978665635, 11.02548807

2.028757838, 4.913180344, 7.978665712, 11.04550306

2.028757838, 4.913180439, 7.978665712, 11.06778114

2.028757838, 4.913180439, 7.978665712, 11.08205766

2.028757838, 4.913130439, 7.978665712, 11.08540507

2.028757838, 4.913180439, 7.978665712, 11.08553821

2.028757838, 4.913180439, 7.978665712, 11.08553841

Answer: 2.029 and 4.913.

3.10.41: Similar triangles show that

$$\frac{x}{u+v} = \frac{5}{v} \quad \text{and} \quad \frac{y}{u+v} = \frac{5}{u},$$

so that

$$x = 5 \cdot \frac{u+v}{v} = 5(1+t) \quad \text{and} \quad y = 5 \cdot \frac{u+v}{u} = 5\left(1+\frac{1}{t}\right).$$

Next, $w^2 + y^2 = 400$ and $w^2 + x^2 = 225$, so that:

$$400 - y^2 = 225 - x^2;$$

$$175 + x^2 = y^2;$$

$$175 + 25(1+t)^2 = 25\left(1+\frac{1}{t}\right)^2;$$

$$175t^2 + 25t^2(1+t)^2 = 25(1+t)^2;$$

$$7t^2 + t^4 + 2t^3 + t^2 = t^2 + 2t + 1;$$

$$t^4 + 2t^3 + 7t^2 - 2t - 1 = 0.$$

The graph of $f(t) = t^4 + 2t^3 + 7t^2 - 2t - 1$ shows a solution of $f(t) = 0$ near $x_0 = 0.5$. Newton's method yields $x_1 = 0.491071429$, $x_2 = 0.490936940$, and $x_3 = 0.490936909 = x_4$. It now follows that $x = 7.454684547$, that $y = 15.184608052$, that $w = 13.016438772$, that $u = 4.286063469$, and that $v = 8.730375303$. Answers: $t = 0.4909$ and $w = 13.0164$.

3.10.43: Let $f(\theta) = (100 + \theta)\cos\theta - 100$. The iterative formula of Newton's method is

$$\theta_{i+1} = \theta_i - \frac{f(\theta_i)}{f'(\theta_i)} \tag{1}$$

where, of course, $f'(\theta) = \cos\theta - (100 + \theta)\sin\theta$. Beginning with $\theta_0 = 1$, iteration of the formula in (1) yields

$$0.4620438212, \quad 0.2325523723, \quad 0.1211226155, \quad 0.0659741863,$$

$$0.0388772442, \quad 0.0261688780, \quad 0.0211747166, \quad 0.0200587600,$$

$$0.0199968594, \quad 0.0199966678, \quad 0.0199966678, \quad 0.0199966678.$$

We take the last value of θ_i to be sufficiently accurate. The corresponding radius of the asteriod is thus approximately $1000/\theta_{12} \approx 50008.3319$ ft, about 9.47 mi.

Chapter 3 Miscellaneous Problems

3.M.1: If $y = y(x) = x^2 + 3x^{-2}$, then $\dfrac{dy}{dx} = 2x - 6x^{-3} = 2x - \dfrac{6}{x^3}$.

3.M.3: If $y = y(x) = \sqrt{x} + \dfrac{1}{\sqrt[3]{x}} = x^{1/2} + x^{-1/3}$, then

$$\frac{dy}{dx} = \frac{1}{2}x^{-1/2} - \frac{1}{3}x^{-4/3} = \frac{1}{2x^{1/2}} - \frac{1}{3x^{4/3}} = \frac{3x^{5/6} - 2}{6x^{4/3}}.$$

3.M.5: Given $y = y(x) = (x - 1)^7(3x + 2)^9$, the product rule and the chain rule yield

$$\frac{dy}{dx} = 7(x - 1)^6(3x + 2)^9 + 27(x - 1)^7(3x + 2)^8 = (x - 1)^6(3x + 2)^8(48x - 13).$$

3.M.7: If $y = y(x) = \left(3x - \dfrac{1}{2x^2}\right)^4 = \left(3x - \frac{1}{2}x^{-2}\right)^4$, then

$$\frac{dy}{dx} = 4\left(3x - \tfrac{1}{2}x^{-2}\right)^3 \cdot (3 + x^{-3}) = 4\left(3x - \frac{1}{2x^2}\right)^3 \cdot \left(3 + \frac{1}{x^3}\right).$$

3.M.9: Given $xy = 9$, implicit differentiation with respect to x yields

$$x\frac{dy}{dx} + y = 0, \qquad \text{so that} \qquad \frac{dy}{dx} = -\frac{y}{x}.$$

Alternatively, $y = y(x) = \dfrac{9}{x}$, so that $\dfrac{dy}{dx} = -\dfrac{9}{x^2}$.

3.M.11: Given $y = y(x) = \dfrac{1}{\sqrt{(x^3 - x)^3}} = (x^3 - x)^{-3/2}$,

$$\frac{dy}{dx} = -\frac{3}{2}(x^3 - x)^{-5/2}(3x^2 - 1) = -\frac{3(3x^2 - 1)}{2(x^3 - x)^{5/2}}.$$

3.M.13: $\dfrac{dy}{dx} = \dfrac{dy}{du} \cdot \dfrac{du}{dx} = \dfrac{-2u}{(1 + u^2)^2} \cdot \dfrac{-2x}{(1 + x^2)^2}$. Now $1 + u^2 = 1 + \dfrac{1}{(1 + x^2)^2} = \dfrac{x^4 + 2x^2 + 2}{(1 + x^2)^2}$.

So $\dfrac{dy}{du} = \dfrac{-2u}{(1 + u^2)^2} = \dfrac{-2}{1 + x^2} \cdot \dfrac{(1 + x^2)^4}{(x^4 + 2x^2 + 2)^2} = \dfrac{-2(1 + x^2)^3}{(x^4 + 2x^2 + 2)^2}$.

Therefore $\dfrac{dy}{dx} = \dfrac{-2(1 + x^2)^3}{(x^4 + 2x^2 + 2)^2} \cdot \dfrac{-2x}{(1 + x^2)^2} = \dfrac{4x(1 + x^2)}{(x^4 + 2x^2 + 2)^2}$.

3.M.15: Given $y = y(x) = \left(x^{1/2} + 2^{1/3}x^{1/3}\right)^{7/3}$,

$$\frac{dy}{dx} = \frac{7}{3}\left(x^{1/2} + 2^{1/3}x^{1/3}\right)^{4/3} \cdot \left(\frac{1}{2}x^{-1/2} + \frac{2^{1/3}}{3}x^{-2/3}\right).$$

3.M.17: If $y = \dfrac{u + 1}{u - 1}$ and $u = (x + 1)^{1/2}$, then

$$\frac{dy}{dx} = \frac{dy}{du} \cdot \frac{du}{dx} = \frac{(u - 1) - (u + 1)}{(u - 1)^2} \cdot \frac{1}{2}(x + 1)^{-1/2} = -\frac{2}{(u - 1)^2} \cdot \frac{1}{2\sqrt{x + 1}} = -\frac{1}{\left(\sqrt{x + 1} - 1\right)^2 \sqrt{x + 1}}.$$

3.M.19: Given $x^2y^2 = x + y$, we differentiate (both sides) implicitly with respect to x and obtain

$$2xy^2 + 2x^2y\frac{dy}{dx} = 1 + \frac{dy}{dx};$$

$$2x^2y\frac{dy}{dx} - \frac{dy}{dx} = 1 - 2xy^2;$$

$$\frac{dy}{dx} = \frac{1 - 2xy^2}{2x^2y - 1};$$

$$\frac{dy}{dx} = \frac{y}{x} \cdot \frac{x - 2x^2y^2}{2x^2y^2 - y};$$

$$\frac{dy}{dx} = \frac{y}{x} \cdot \frac{x - 2(x + y)}{2(x + y) - y};$$

$$\frac{dy}{dx} = \frac{y}{x} \cdot \frac{-x - 2y}{2x + y} = -\frac{(x + 2y)y}{(2x + y)x}.$$

121

Of course you may stop with the third line, but to find horizontal and vertical lines tangent to the graph of $x^2 y^2 = x + y$, the last line is probably the most convenient.

3.M.21: Given $y = y(x) = \sqrt{x + \sqrt{2x + \sqrt{3x}}} = \left(x + \left[2x + (3x)^{1/2}\right]^{1/2}\right)^{1/2}$,

$$\frac{dy}{dx} = \frac{1}{2}\left(x + \left[2x + (3x)^{1/2}\right]^{1/2}\right)^{-1/2} \cdot \left(1 + \frac{1}{2}\left[2x + (3x)^{1/2}\right]^{-1/2} \cdot \left[2 + \frac{3}{2}(3x)^{-1/2}\right]\right).$$

The symbolic algebra program *Mathematica* writes this answer without exponents as follows:

$$\frac{dy}{dx} = \frac{1 + \dfrac{2 + \dfrac{\sqrt{3}}{2\sqrt{x}}}{2\sqrt{2x + \sqrt{3x}}}}{2\sqrt{x + \sqrt{2x + \sqrt{3x}}}}.$$

3.M.23: Given $x^{1/3} + y^{1/3} = 4$, differentiate both sides with respect to x:

$$\frac{1}{3}x^{-2/3} + \frac{1}{3}y^{-2/3}\frac{dy}{dx} = 0, \qquad \text{so} \qquad \frac{dy}{dx} = -\left(\frac{y}{x}\right)^{2/3}.$$

3.M.25: Given $y = (1 + 2u)^3$ where $u = (1 + x)^{-3}$:

$$\frac{dy}{dx} = \frac{dy}{du} \cdot \frac{du}{dx} = 6(1 + 2u)^2 \cdot (-3)(1 + x)^{-4} = -\frac{18(1 + 2u)^2}{(1 + x)^4} = -\frac{18(1 + 2(1 + x)^{-3})^2}{(1 + x)^4}$$

$$= -\frac{18(1 + x)^6(1 + 2(1 + x)^{-3})^2}{(1 + x)^{10}} = -\frac{18((1 + x)^3 + 2)^2}{(1 + x)^{10}} = -18 \cdot \frac{(x^3 + 3x^2 + 3x + 3)^2}{(x + 1)^{10}}.$$

3.M.27: Given $y = y(x) = \left(\dfrac{\sin^2 x}{1 + \cos x}\right)^{1/2}$,

$$\frac{dy}{dx} = \frac{1}{2}\left(\frac{\sin^2 x}{1 + \cos x}\right)^{-1/2} \cdot \frac{(1 + \cos x)(2\sin x \cos x) + \sin^3 x}{(1 + \cos x)^2}$$

$$= \left(\frac{1 + \cos x}{\sin^2 x}\right)^{1/2} \cdot \frac{2\sin x \cos x + 2\sin x \cos^2 x + \sin^3 x}{2(1 + \cos x)^2}.$$

3.M.29: Given: $y = y(x) = \dfrac{\cos 2x}{\sqrt{\sin 3x}} = (\cos 2x)(\sin 3x)^{-1/2}$,

$$\frac{dy}{dx} = (-2\sin 2x)(\sin 3x)^{-1/2} + (\cos 2x)\left(-\frac{1}{2}(\sin 3x)^{-3/2}\right)(3\cos 3x)$$

$$= -\frac{2\sin 2x}{\sqrt{\sin 3x}} - \frac{3\cos 2x \cos 3x}{2(\sin 3x)^{3/2}} = -\frac{4\sin 2x \sin 3x + 3\cos 2x \cos 3x}{2(\sin 3x)^{3/2}}.$$

3.M.31: $\dfrac{dy}{dx} = (e^x) \cdot \cos x + e^x \cdot (-\sin x) = e^x(\cos x - \sin x)$.

3.M.33: $\dfrac{dy}{dx} = -\dfrac{3e^x}{(2+3e^x)^{5/2}\left[1+(2+3e^x)^{-3/2}\right]^{1/3}}$.

3.M.35: $\dfrac{dy}{dx} = -\dfrac{\cos^2\left([1+\ln x]^{1/3}\right)\sin\left([1+\ln x]^{1/3}\right)}{x[1+\ln x]^{2/3}}$.

3.M.37: If $f(x) = \sin^2(e^{-x}) = \left[\sin(e^{-x})\right]^2$, then $f'(x) = -2e^{-x}\sin(e^{-x})\cos(e^{-x})$.

3.M.39: If $f(x) = e^x\cos 2x$, then $f'(x) = e^x\cos 2x - 2e^x\sin 2x$.

3.M.41: If $g(t) = \ln(te^{t^2}) = (\ln t) + t^2\ln e = (\ln t) + t^2$, then

$$g'(t) = \frac{1}{t} + 2t = \frac{1+2t^2}{t}.$$

3.M.43: If $g(t) = \sin(e^t)\cos(e^{-t})$, then

$$g'(t) = e^t\cos(e^t)\cos(e^{-t}) + e^{-t}\sin(e^t)\sin(e^{-t}).$$

3.M.45: If $g(t) = \dfrac{1+e^t}{1-e^t}$, then $g'(t) = \dfrac{(1-e^t)e^t + (1+e^t)e^t}{(1-e^t)^2} = \dfrac{2e^t}{(1-e^t)^2}$.

3.M.47: Given $\sin(e^{xy}) = x$, we apply D_x to both sides and find that

$$[\cos(e^{xy})]\cdot e^{xy}\cdot\left(y + x\frac{dy}{dx}\right) = 1; \qquad xe^{xy}[\cos(e^{xy})]\cdot\frac{dy}{dx} = 1 - ye^{xy}\cos(e^{xy});$$

$$\frac{dy}{dx} = \frac{1 - ye^{xy}\cos(e^{xy})}{xe^{xy}\cos(e^{xy})}.$$

3.M.49: Given $x = ye^y$, we apply D_x to both sides and obtain

$$1 = e^y\frac{dy}{dx} + ye^y\frac{dy}{dx}; \qquad \frac{dy}{dx} = \frac{1}{e^y + ye^y} = \frac{y}{ye^y + y^2e^y} = \frac{y}{x+xy}.$$

We used the fact that $ye^y = x$ in the simplification in the last step.

Here is an alternative approach to finding dy/dx. Beginning with $x = ye^y$, we differentiate with respect to y and find that

$$\frac{dx}{dy} = e^y + ye^y, \qquad \text{so that} \qquad \frac{dy}{dx} = \frac{1}{e^y + ye^y}.$$

123

3.M.51: Given $x \ln y = x + y$, we apply D_x to both sides and find that

$$\ln y + \frac{x}{y} \cdot \frac{dy}{dx} = 1 + \frac{dy}{dx};$$

$$\left(\frac{x}{y} - 1 \right) \cdot \frac{dy}{dx} = 1 - \ln y;$$

$$\frac{x - y}{y} \cdot \frac{dy}{dx} = 1 - \ln y;$$

$$\frac{dy}{dx} = \frac{y(1 - \ln y)}{x - y}.$$

3.M.53: Given: $y = (3 - x^2)^{1/2}(x^4 + 1)^{-1/4}$. Thus

$$\ln y = \frac{1}{2} \ln(3 - x^2) - \frac{1}{4} \ln(x^4 + 1),$$

and therefore

$$\frac{1}{y} \cdot \frac{dy}{dx} = -\frac{x}{3 - x^2} - \frac{x^3}{x^4 + 1} = \frac{x(3x^2 + 1)}{(x^2 - 3)(x^4 + 1)}.$$

Thus

$$\frac{dy}{dx} = y(x) \cdot \frac{x(3x^2 + 1)}{(x^2 - 3)(x^4 + 1)} = -\frac{x(3x^2 + 1)(3 - x^2)^{1/2}}{(3 - x^2)(x^4 + 1)(x^4 + 1)^{1/4}} = -\frac{x(3x^2 + 1)}{(3 - x^2)^{1/2}(x^4 + 1)^{5/4}}.$$

3.M.55: If $y = (x + 1)^{1/2}(x + 2)^{1/3}(x + 3)^{1/4}$, then

$$\ln y = \frac{1}{2} \ln(x + 1) + \frac{1}{3} \ln(x + 2) + \frac{1}{4} \ln(x + 3);$$

$$\frac{1}{y} \cdot \frac{dy}{dx} = \frac{1}{2(x + 1)} + \frac{1}{3(x + 2)} + \frac{1}{4(x + 3)};$$

$$\frac{dy}{dx} = y(x) \cdot \frac{13x^2 + 55x + 54}{12(x + 1)(x + 2)(x + 3)} = \frac{13x^2 + 55x + 54}{12(x + 1)^{1/2}(x + 2)^{2/3}(x + 3)^{3/4}}.$$

3.M.57: Given: $y = (\ln x)^{\ln x}$, $x > 1$. Then

$$\ln y = (\ln x) \ln(\ln x);$$

$$\frac{1}{y} \cdot \frac{dy}{dx} = \frac{1}{x} \ln(\ln x) + \frac{\ln x}{x \ln x} = \frac{1 + \ln(\ln x)}{x};$$

$$\frac{dy}{dx} = \frac{1 + \ln(\ln x)}{x} \cdot (\ln x)^{\ln x}.$$

3.M.59: $1 = (2 \cos 2y)\dfrac{dy}{dx}$, so $\dfrac{dy}{dx} = \dfrac{1}{2 \cos 2y}$. Because $\dfrac{dy}{dx}$ is undefined at $(1, \pi/4)$, there may well be a vertical tangent at that point. And indeed there is: $\dfrac{dx}{dy} = 0$ at $(1, \pi/4)$. So an equation of the tangent line is $x = 1$.

124

3.M.61: $\dfrac{dy}{dx} = \dfrac{2x+1}{3y^2}$; at $(0,0)$, $\dfrac{dx}{dy} = 0$, so the tangent line is vertical. Its equation is $x = 0$.

3.M.63: Let r be the radius of the sandpile, h its height, each a function of time t. We know that $2r = h$, so the volume of the sandpile at time t is

$$V = \frac{1}{3}\pi r^2 h = \frac{2}{3}\pi r^3.$$

$$\text{So} \quad 25\pi = \frac{dV}{dt} = 2\pi r^2 \frac{dr}{dt};$$

substitution of $r = 5$ yields the answer: $dr/dt = \frac{1}{2}$ (ft/min) when $r = 5$ (ft).

3.M.65: $x \cot 3x = \dfrac{1}{3} \cdot \dfrac{3x}{\sin 3x} \to \dfrac{1}{3} \cdot 1 = \dfrac{1}{3}$ as $x \to 0$.

3.M.67: $x^2 \csc 2x \cot 2x = \dfrac{1}{4} \cdot \dfrac{2x}{\sin 2x} \cdot \dfrac{2x}{\sin 2x} \cdot \cos 2x \to \dfrac{1}{4} \cdot 1 \cdot 1 \cdot 1 = \dfrac{1}{4}$ as $x \to 0$.

3.M.69: $-1 \le \sin u \le 1$ for all u. So

$$-\sqrt{x} \le \sqrt{x} \sin \frac{1}{x} \le \sqrt{x}$$

for all $x > 0$. But $\sqrt{x} \to 0$ as $x \to 0^+$, so the limit is zero.

3.M.71: $h(x) = (x^2 + 25)^{-1/2} = f(g(x))$ where $f(x) = x^{-1/2}$ and $g(x) = x^2 + 25$. Therefore $h'(x) = f'(g(x)) \cdot g'(x) = -\frac{1}{2}(x^2 + 25)^{-3/2} \cdot 2x$.

3.M.73: One solution: $h(x) = (x-1)^{5/3} = f(g(x))$ where $f(x) = x^{5/3}$ and $g(x) = x - 1$. Therefore $h'(x) = f'(g(x)) \cdot g'(x) = \frac{5}{3}(x-1)^{2/3} \cdot 1 = \frac{5}{3}(x-1)^{2/3}$. You might alternatively choose $f(x) = x^{1/3}$ and $g(x) = (x-1)^5$.

3.M.75: $h(x) = \cos(x^2 + 1) = f(g(x))$ where $f(x) = \cos x$ and $g(x) = x^2 + 1$. Therefore $h'(x) = f'(g(x)) \cdot g'(x) = -2x \sin(x^2 + 1)$.

3.M.77: Of course r denotes the radius of the sphere. First, $\dfrac{dV}{dA} \cdot \dfrac{dA}{dr} = \dfrac{dV}{dr}$. Now $V = \dfrac{4}{3}\pi r^3$ and $A = 4\pi r^2$, so $\dfrac{dV}{dA} \cdot 8\pi r = 4\pi r^2$, and therefore $\dfrac{dV}{dA} = \dfrac{r}{2} = \dfrac{1}{4}\sqrt{\dfrac{A}{\pi}}$.

3.M.79: Let $y = y(t)$ denote the altitude of the rocket at time t; let $u = u(t)$ denote the angle of elevation of the observer's line of sight at time t. Then $\tan u = y/3$, so that $y = 3\tan u$ and, therefore,

$$\frac{dy}{dt} = (3\sec^2 u) \frac{du}{dt}.$$

When $u = 60°$, we take $du/dt = \dfrac{\pi}{3}$ and find that the speed of the rocket is

$$\left.\frac{dy}{dt}\right|_{u=60°} = \frac{3}{\cos^2(\pi/3)} \cdot \frac{\pi}{30} = \frac{2}{5}\pi \approx 1.2566 \text{ (mi/s)},$$

about 4524 mi/h, or about 6635 ft/s.

3.M.81: Let the circle be the one with equation $x^2 + y^2 = R^2$ and let the base of the triangle lie on the x-axis; denote the opposite vertex of the triangle by (x, y). The area of the triangle $A = Ry$ is clearly maximal when y is maximal; that is, when $y = R$. To solve this problem using calculus, let θ be the angle of the triangle at $(-R, 0)$. Because the triangle has a right angle at (x, y), its two short sides are $2R \cos \theta$ and $2R \sin \theta$, so its area is

$$A(\theta) = 2R^2 \sin \theta \cos \theta = R^2 \sin 2\theta, \qquad 0 \leq \theta \leq \frac{\pi}{2}.$$

Then $A'(\theta) = 2R^2 \cos 2\theta$; $A'(\theta) = 0$ when $\cos 2\theta = 0$; because θ lies in the first quadrant, $\theta = \frac{1}{4}\pi$. Finally, $A(0) = 0 = A(\pi/2)$, but $A(\pi/4) = R^2 > 0$. Hence the maximum possible area of such a triangle is R^2.

3.M.83: Let one sphere have radius r; the other, s. We seek the extrema of $A = 4\pi(r^2 + s^2)$ given $\frac{4}{3}\pi(r^3 + s^3) = V$, a constant. We illustrate here the **method of auxiliary variables:**

$$\frac{dA}{dr} = 4\pi \left(2r + 2s \frac{ds}{dr} \right);$$

the condition $dA/dr = 0$ yields $ds/dr = -r/s$. But we also know that $\frac{4}{3}\pi(r^3 + s^3) = V$; differentiation of both sides of this *identity* with respect to r yields

$$\frac{4}{3}\pi \left(3r^2 + 3s^2 \frac{ds}{dr} \right) = 0, \quad \text{and so}$$

$$3r^2 + 3s^2 \left(-\frac{r}{s} \right) = 0;$$

$$r^2 - rs = 0.$$

Therefore $r = 0$ or $r = s$. Also, ds/dr is undefined when $s = 0$. So we test these three critical points. If $r = 0$ or if $s = 0$, there is only one sphere, with radius $(3V/4\pi)^{1/3}$ and surface area $(36\pi V^2)^{1/3}$. If $r = s$, then there are two spheres of equal size, both with radius $\frac{1}{2}(3V/\pi)^{1/3}$ and surface area $(72\pi V^2)^{1/3}$. Therefore, for maximum surface area, make two equal spheres. For minimum surface area, make only one sphere.

3.M.85: Let r be the radius of the cone; let its height be $h = R + y$ where $0 \leq y \leq R$. (Actually, $-R \leq y \leq R$, but the cone will have maximal volume if $y \geq 0$.) A central vertical cross section of the figure (*draw it!*) shows a right triangle from which we read the relation $y^2 = R^2 - r^2$. We are to maximize $V = \frac{1}{3}\pi r^2 h$, so we write

$$V = V(r) = \frac{1}{3}\pi \left[r^2 \left(R + \sqrt{R^2 - r^2} \right) \right], \qquad 0 \leq r \leq R.$$

The condition $V'(r) = 0$ leads to the equation $r \left(2R^2 - 3r^2 + 2R\sqrt{R^2 - r^2} \right) = 0$, which has the two solutions $r = 0$ and $r = \frac{2}{3}R\sqrt{2}$. Now $V(0) = 0$, $V(R) = \frac{1}{3}\pi R^3$ (which is one-fourth the volume of the sphere), and $V\left(\frac{2}{3}R\sqrt{2}\right) = \frac{32}{81}\pi R^3$ (which is 8/27 of the volume of the sphere). Answer: The maximum volume is $\frac{32}{81}\pi R^3$.

3.M.87: First, $R'(x) = kM - 2kx$; because $k \neq 0$, $R'(x) = 0$ when $x = M/2$. Moreover, because $R(0) = 0 = R(M)$ and $R(M/2) > 0$, the latter is the maximum value of $R(x)$. Therefore the incidence of the disease is the highest when half the susceptible individuals are infected.

3.M.89: Let x be the width of the base of the box, so that the base has length $2x$; let y be the height of the box. Then the volume of the box is $V = 2x^2 y$, and for its total surface area to be 54 ft^2, we require $2x^2 + 6xy = 54$. Therefore the volume of the box is given by

$$V = V(x) = 2x^2 \left(\frac{27 - x^2}{3x} \right) = \frac{2}{3}(27x - x^3), \qquad 0 < x \leqq 3\sqrt{3}.$$

Now $V'(x) = 0$ when $x^2 = 9$, so that $x = 3$. Also $V(x) \to 0$ as $x \to 0^+$ and $V(3\sqrt{3}) = 0$, so $V(3) = 36$ (ft^3) is the maximum possible volume of the box.

3.M.91: Let (x, y) be the coordinates of the vertex of the trapezoid lying properly in the first quadrant and let θ be the angle that the radius of the circle to (x, y) makes with the x-axis. The bases of the trapezoid have lengths 4 and $4\cos\theta$ and its altitude is $2\sin\theta$, so its area is

$$A(\theta) = \frac{1}{2}(4 + 4\cos\theta)(2\sin\theta) = 4(1 + \cos\theta)\sin\theta, \quad 0 \leq \theta \leq \frac{\pi}{2}.$$

Now

$$A'(\theta) = 4(\cos\theta + \cos^2\theta - \sin^2\theta)$$

$$= 4(2\cos^2\theta + \cos\theta - 1)$$

$$= 4(2\cos\theta - 1)(\cos\theta + 1).$$

The only zero of A' in its domain occurs at $\theta = \pi/3$. At the endpoints, we have $A(0) = 0$ and $A(\pi/2) = 4$. But $A(\pi/3) = 3\sqrt{3} \approx 5.196$, so the latter is the maximum possible area of such a trapezoid.

C03S0M.093: If $Ax + By + C = 0$ is an equation of a straight line L, then not both A and B can be zero.

Case 1: $A = 0$ and $B \neq 0$. Then L has equation $y = -C/B$ and thus is a horizontal line. So the shortest segment from $P(x_0, y_0)$ to Q on L is a vertical segment that therefore meets L in the point $Q(x_0, -C/B)$. Therefore, because $A = 0$, the distance from P to Q is

$$\left| y_0 + \frac{C}{B} \right| = \frac{|By_0 + C|}{|B|} = \frac{|Ax_0 + By_0 + C|}{\sqrt{A^2 + B^2}}.$$

Case 2: $A \neq 0$ and $B = 0$. Then L has equation $x = -C/A$ and thus is a vertical line. So the shortest segment from $P(x_0, y_0)$ to Q on L is a horizontal segment that therefore meets L in the point $Q(-C/A, y_0)$. Therefore, because $B = 0$, the distance from P to Q is

$$\left| x_0 + \frac{C}{A} \right| = \frac{|Ax_0 + C|}{|A|} = \frac{|Ax_0 + By_0 + C|}{\sqrt{A^2 + B^2}}.$$

Case 3: $A \neq 0$ and $B \neq 0$. Then L is neither horizontal nor vertical, and the segment joining $P(x_0, y_0)$ to the nearest point $Q(u, v)$ on L is also neither horizontal nor vertical. The equation of L may be written in the form

$$y = -\frac{A}{B} - \frac{C}{B},$$

so L has slope $-A/B$. Thus the slope of PQ is B/A (by the result in Problem 70), and therefore PQ lies on the line K with equation

$$y - y_0 = \frac{B}{A}(x - x_0).$$

Consequently $A(v - y_0) = B(u - x_0)$. But $Q(u, v)$ also lies on L, and so $Au + Bv = -C$. Thus we have the simultaneous equations

$$Au + Bv = -C;$$

$$Bu - Av = Bx_0 - Ay_0.$$

These equations may be solved for

$$u = \frac{-AC + B^2 x_0 - ABy_0}{A^2 + B^2} \quad \text{and} \quad v = \frac{-BC - ABx_0 + A^2 y_0}{A^2 + B^2},$$

and it follows that

$$u - x_0 = \frac{A(-C - Ax_0 - By_0)}{A^2 + B^2} \quad \text{and} \quad v - y_0 = \frac{B(-C - Ax_0 - By_0)}{A^2 + B^2}.$$

Therefore

$$(u - x_0)^2 + (v - y_0)^2 = \frac{A^2(-C - Ax_0 - By_0)^2}{(A^2 + B^2)^2} + \frac{B^2(-C - Ax_0 - By_0)^2}{(A^2 + B^2)^2}$$

$$= \frac{(A^2 + B^2)(-C - Ax_0 - By_0)^2}{(A^2 + B^2)^2} = \frac{(Ax_0 + By_0 + C)^2}{A^2 + B^2}.$$

The square root of this expression then gives the distance from P to Q as

$$\frac{|Ax_0 + By_0 + C|}{\sqrt{A^2 + B^2}},$$

and the proof is complete.

3.M.95: As the following diagram suggests, we are to minimize the sum of the lengths of the two diagonals. Fermat's principle of least time may be used here, so we know that the angles at which the roads meet the shore are equal, and thus so are the tangents of those angles: $\frac{x}{1} = \frac{6 - x}{2}$. It follows that the pier should

128

be built two miles from the point on the shore nearest the first town. To be sure that we have found a minimum, consider the function that gives the total length of the two diagonals:

$$f(x) = \sqrt{x^2 + 1} + \sqrt{(6-x)^2 + 4}, \qquad 0 \leqq x \leqq 6.$$

(The domain certainly contains the global minimum value of f.) Moreover, $f(0) = 1 + \sqrt{40} \approx 7.32$, $f(6) = 2 + \sqrt{37} \approx 8.08$, and $f(2) = \sqrt{5} + \sqrt{20} \approx 6.71$. This establishes that $x = 2$ yields the global minimum of $f(x)$.

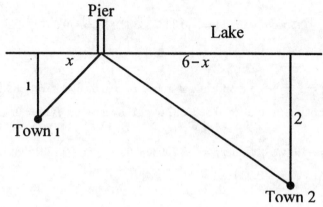

3.M.97: Denote the initial velocity of the arrow by v. First, we have

$$\frac{dy}{dx} = m - \frac{32x}{v^2}(m^2 + 1);$$

$dy/dx = 0$ when $mv^2 = 32x(m^2+1)$, so that $x = \dfrac{mv^2}{32(m^2+1)}$. Substitution of this value of x in the formula given for y in the problem yields the maximum height

$$y_{\text{max}} = \frac{m^2 v^2}{64(m^2 + 1)}.$$

For part (b), we set $y = 0$ and solve for x to obtain the range

$$R = \frac{mv^2}{16(m^2 + 1)}.$$

Now R is a continuous function of the slope m of the arrow's path at time $t = 0$, with domain $0 \leqq m < +\infty$. Because $R(m) = 0$ and $R(m) \to 0$ as $m \to +\infty$, the function R has a global maximum; because R is differentiable, this maximum occurs at a point where $R'(m) = 0$. But

$$\frac{dR}{dm} = \frac{v^2}{16} \cdot \frac{(m^2 + 1) - 2m^2}{(m^2 + 1)^2},$$

so $dR/dm = 0$ when $m = 1$ and only then. So the maximum range occurs when $\tan \alpha = 1$; that is, when $\alpha = \frac{1}{4}\pi$.

3.M.99: With initial guess $x_0 = 2.5$ (the midpoint of the given interval $[2, 3]$), the iteration

$$x_{n+1} = x_n - \frac{f(x_n)}{f'(x_n)} = x_n - \frac{(x_n)^2 - 7}{2x_n}$$

of Newton's method yields $x_1 = 2.65$, $x_2 = 2.645754717$, and $x_3 = 2.645751311$. Answer: 2.6458.

3.M.101: With $x_0 = 2.5$, we obtain $x_1 = 2.384$, $x_2 = 2.371572245$, $x_3 = 2.371440624$, and $x_4 = 2.371440610$. Answer: 2.3714.

3.M.103: With $x_0 = -0.5$ we obtain $x_1 = -0.333333333$, $x_2 = -0.347222222$, $x_3 = -0.347296353$, and $x_4 = -0.347296355$. Answer: -0.3473.

3.M.105: With $f(x) = e^{-x} - \sin x$ and initial guess $x_0 = 0.6$, five iterations of the formula of Newton's method yields the approximate solution 0.588532744 of the equation $f(x) = 0$.

3.M.107: With $x_0 = -1.0$ we obtain $x_1 = -0.750363868$, $x_2 = -0.739112891$, $x_3 = -0.739085133$, and $x_4 = -0.739085133$. Answer: -0.7391.

3.M.109: With $x_0 = -1.5$, we obtain $x_1 = -1.244861806$, $x_2 = -1.236139793$, $x_3 = -1.236129989$, and $x_4 = -1.236129989$. Answer: -1.2361.

3.M.111: The volume of a spherical segment of height h is

$$V = \frac{1}{3}\pi h^2(3r - h)$$

if the sphere has radius r. If ρ is the density of water and the ball sinks to the depth h, then the weight of the water that the ball displaces is equal to the total weight of the ball, so

$$\frac{1}{3}\pi\rho h^2(3r - h) = \frac{4}{32}\pi\rho r^3.$$

Because $r = 2$, this leads to the equation $p(h) = 3h^3 - 18h^2 + 32 = 0$. This equation has at most three [real] solutions because $p(h)$ is a polynomial of degree 3, and it turns out to have exactly three solutions because $p(-2) = -64$, $p(-1) = 11$, $p(2) = -16$, and $p(6) = 32$. Newton's method yields the three approximate solutions $h = -1.215825766$, $h = 1.547852572$, and $h = 5.667973193$. Only one is plausible, so the answer is that the ball sinks to a depth of approximately 1.54785 ft, about 39% of the way up a diameter.

3.M.113: Let $f(x) = x^5 - 3x^3 + x^2 - 23x + 19$. Then $f(-3) = -65$, $f(0) = 19$, $f(1) = -5$, and $f(3) = 121$. So there are at least three, and at most five, real solutions. Newton's method produces three real solutions, specifically $r_1 = -2.722493355$, $r_2 = 0.8012614801$, and $r_3 = 2.309976541$. If one divides the polynomial $f(x)$ by $(x - r_1)(x - r_2)(x - r_3)$, one obtains the quotient polynomial $x^2 + (0.38874466)x + 3.770552031$, which has no real roots—the quadratic formula yields the two complex roots $-0.194372333 \pm (1.932038153)i$. Consequently we have found all three real solutions.

3.M.115: The number of summands on the right is variable, and we have no formula for finding its derivative. One thing is certain: Its derivative is *not* $2x^2$.

3.M.117: We factor:

$$z^{2/3} - x^{2/3} = (z^{1/3})^2 - (x^{1/3})^2 = (z^{1/3} - x^{1/3})(z^{1/3} + x^{1/3}) \qquad \text{and}$$

$$z - x = (z^{1/3})^3 - (x^{1/3})^3 = (z^{1/3} - x^{1/3})(z^{2/3} + z^{1/3}x^{1/3} + x^{2/3}).$$

Therefore

$$\frac{z^{2/3} - x^{2/3}}{z - x} = \frac{z^{1/3} + x^{1/3}}{z^{2/3} + z^{1/3}x^{1/3} + z^{2/3}} \ \to\ \frac{2x^{1/3}}{3x^{2/3}} = \frac{2}{3}x^{-1/3} \quad \text{as} \quad x \to x.$$

3.M.119: The balloon has volume $V = \frac{4}{3}\pi r^3$ and surface area $A = 4\pi r^2$ where r is its radius and V, A, and r are all functions of time t. We are given $dV/dt = +10$, and we are to find dA/dt when $r = 5$.

$$\frac{dV}{dt} = 4\pi r^2 \frac{dr}{dt}, \quad \text{so} \quad 10 = 4\pi \cdot 25 \cdot \frac{dr}{dt}.$$

$$\text{Thus} \quad \frac{dr}{dt} = \frac{10}{100\pi} = \frac{1}{10\pi}.$$

$$\text{Also} \quad \frac{dA}{dt} = 8\pi r \frac{dr}{dt}, \quad \text{and therefore}$$

$$\left.\frac{dA}{dt}\right|_{r=5} = 8\pi \cdot 5 \cdot \frac{1}{10\pi} = 4.$$

Answer: At 4 in.^{2}/s.

3.M.121: Let Q be the amount of water in the cone at time t, r the radius of its upper surface, and h its height. From similar triangles we find that $h = 2r$, so

$$Q = \frac{1}{3}\pi r^2 h = \frac{2}{3}\pi r^3 = \frac{1}{12}\pi h^3.$$

Now

$$-50 = \frac{dQ}{dt} = \frac{1}{4}\pi h^2 \frac{dh}{dt}, \quad \text{so} \quad \frac{dh}{dt} = -\frac{200}{36\pi}.$$

Therefore $\dfrac{dh}{dt} = -\dfrac{50}{9\pi} \approx -1.7684$ (ft/min).

3.M.123: $\dfrac{dV}{dt} = 3\sqrt{y}\,\dfrac{dy}{dt} = -3\sqrt{y}$, so $\dfrac{dy}{dt} = -1$. Answer: At 1 in./min—a constant rate. The tank is a *clock!*

3.M.125: The straight line through $P(x_0, y_0)$ and $Q(a, a^2)$ has slope $\dfrac{a^2 - y_0}{a - x_0} = 2a$, a consequence of the two-point formula for slope and the fact that the line is tangent to the parabola at Q. Hence $a^2 - 2ax_0 + y_0 = 0$. Think of this as a quadratic equation in the unknown a. It has two real solutions when the discriminant is positive: $(x_0)^2 - y_0 > 0$, and this establishes the conclusion in part (b). There are no real solutions when $(x_0)^2 - y_0 < 0$, and this establishes the conclusion in part (c). What if $(x_0)^2 - y_0 = 0$?

Section 4.2

4.2.1: $y = y(x) = 3x^2 - 4x^{-2}$: $\dfrac{dy}{dx} = 6x + 8x^{-3}$, so $dy = \left(6x + \dfrac{8}{x^3}\right) dx$.

4.2.3: $y = y(x) = x - (4 - x^3)^{1/2}$: $\dfrac{dy}{dx} = 1 - \dfrac{1}{2}(4 - x^3)^{-1/2} \cdot (-3x^2)$, so

$$dy = \left(1 + \frac{3x^2}{2\sqrt{4 - x^3}}\right) dx = \frac{3x^2 + 2\sqrt{4 - x^3}}{2\sqrt{4 - x^3}} \ dx.$$

4.2.5: $y = y(x) = 3x^2(x - 3)^{3/2}$, so

$$dy = \left[6x(x - 3)^{3/2} + \frac{9}{2}x^2(x - 3)^{1/2}\right] dx = \frac{3}{2}(7x^2 - 12x)\sqrt{x - 3} \ dx.$$

4.2.7: $y = y(x) = x(x^2 + 25)^{1/4}$, so

$$dy = (x^2 + 25)^{1/4} + \frac{1}{4}x(x^2 + 25)^{-3/4} \cdot 2x \ dx = \frac{3x^2 + 50}{2(x^2 + 25)^{3/4}} \ dx.$$

4.2.9: $y = y(x) = \cos\sqrt{x}$, so $dy = -\dfrac{\sin\sqrt{x}}{2\sqrt{x}} \ dx$.

4.2.11: $y = y(x) = \sin 2x \cos 2x$, so $dy = (2\cos^2 2x - 2\sin^2 2x) \ dx$.

4.2.13: $y = y(x) = \dfrac{\sin 2x}{3x}$, so $dy = \dfrac{2x\cos 2x - \sin 2x}{3x^2} \ dx$.

4.2.15: $y = y(x) = \dfrac{1}{1 - x\sin x}$, so $dy = \dfrac{x\cos x + \sin x}{(1 - x\sin x)^2} \ dx$.

4.2.17: $f'(x) = \dfrac{1}{(1 - x)^2}$, so $f'(0) = 1$. Therefore

$$f(x) = \frac{1}{1 - x} \approx f(0) + f'(0)(x - 0) = 1 + 1 \cdot x = 1 + x.$$

4.2.19: $f'(x) = 2(1 + x)$, so $f'(0) = 2$. Therefore $f(x) = (1 + x)^2 \approx f(0) + f'(0)(x - 0) = 1 + 2x$.

4.2.21: $f'(x) = -3\sqrt{1 - 2x}$, so $f'(0) = -3$; $f(x) = (1 - 2x)^{3/2} \approx f(0) + f'(0)(x - 0) = 1 - 3x$.

4.2.23: If $f(x) = \sin x$, then $f'(x) = \cos x$, so that $f'(0) = 1$. Therefore

$$f(x) = \sin x \approx f(0) + f'(0)(x - 0) = 0 + 1 \cdot x = x.$$

4.2.25: Choose $f(x) = x^{1/3}$ and $a = 27$. Then $f'(x) = \dfrac{1}{3x^{2/3}}$, so that $f'(a) = \dfrac{1}{27}$. So the linear approximation to $f(x)$ near $a = 27$ is $L(x) = 2 + \dfrac{1}{27}x$. Hence

$$\sqrt[3]{25} = f(25) \approx L(25) = \frac{79}{27} \approx 2.9259.$$

A calculator reports that $f(25)$ is actually closer to 2.9240, but the linear approximation is fairly accurate, with an error of only about -0.0019.

4.2.27: Choose $f(x) = x^{1/4}$ and $a = 16$. Then $f'(x) = \dfrac{1}{4x^{3/4}}$, so that $f'(a) = \dfrac{1}{32}$. So the linear approximation to $f(x)$ near $a = 16$ is $L(x) = \dfrac{3}{2} + \dfrac{1}{32}x$. Hence

$$\sqrt[4]{15} = f(15) \approx L(15) = \frac{63}{32} = 1.96875.$$

A calculator reports that $f(15)$ is actually closer to 1.96799.

4.2.29: Choose $f(x) = x^{-2/3}$ and $a = 64$. Then $f'(x) = -\dfrac{2}{3x^{5/3}}$, so that $f'(a) = -\dfrac{1}{1536}$. So the linear approximation to $f(x)$ near $a = 64$ is $L(x) = \dfrac{5}{48} - \dfrac{1}{1536}x$. Hence

$$65^{-2/3} = f(65) \approx L(65) = \frac{95}{1536} \approx 0.06185.$$

A calculator reports that $f(65)$ is actually closer to 0.06186.

4.2.31: Choose $f(x) = \cos x$ and $a = \dfrac{45}{180}\pi = \dfrac{1}{4}\pi$. Then $f'(x) = -\sin x$, so that $f'(a) = -\dfrac{1}{2}\sqrt{2}$. So the linear approximation to $f(x)$ near a is $L(x) = \dfrac{1}{2}\sqrt{2}\left(\dfrac{1}{4}\pi + 1\right) - \dfrac{1}{2}x\sqrt{2}$. Hence

$$\cos 43° = f\left(\frac{43}{180}\pi\right) \approx L\left(\frac{43}{180}\pi\right) = \frac{\pi + 90}{90\sqrt{2}} \approx 0.7318.$$

A calculator reports that $\cos 43°$ is actually closer to 0.7314.

4.2.33: Choose $f(x) = e^x$ and $a = 0$. Then $f'(x) = e^x$, so that $f'(a) = 1$. So the linear approximation to $f(x)$ near a is $L(x) = x + 1$. Thus $e^{0.1} = f(0.1) \approx L(0.1) = 1.1$. A calculator reports that $e^{0.1} \approx 1.105171$.

4.2.35: Given $x^2 + y^2 = 1$, we compute the differential of both sides and obtain

$$2x\,dx + 2y\,dy = 0;$$

$$y\,dy = -x\,dx;$$

$$\frac{dy}{dx} = -\frac{x}{y}.$$

134

4.2.37: Given $x^3 + y^3 = 3xy$, we compute the differential of each side and obtain

$$3x^2 \, dx + 3y^2 \, dy = 3y \, dx + 3x \, dy;$$

$$(y^2 - x) \, dy = (y - x^2) \, dx;$$

$$\frac{dy}{dx} = \frac{y - x^2}{y^2 - x}.$$

4.2.39: If $f(x) = (1 + x)^k$, then $f'(x) = k(1 + x)^{k-1}$, and so $f'(0) = k$. Hence the linear approximation to $f(x)$ near zero is $L(x) = 1 + kx$.

4.2.41: If the square has edge length x and area A, then $A = x^2$. Therefore $dA = 2x \, dx$, and so $\Delta A \approx 2x \, \Delta x$. With $x = 10$ and $\Delta x = -0.2$, we obtain $\Delta A \approx 2 \cdot 10 \cdot (-0.2) = -4$. So the area of the square decreases by 4 in.^{2}.

4.2.43: A [right circular] cylinder of base radius r and height h has volume $V = \pi r^2 h$, and hence $dV = \pi r^2 \, dh + 2\pi rh \, dr$. Therefore $\Delta V \approx \pi r^2 \, \Delta h + 2\pi rh \, \Delta r$. With $r = h = 15$ and $\Delta r = \Delta h = -0.3$ we find that $\Delta V \approx (225\pi)(-0.3) + (450\pi)(-0.3) = \frac{405}{2}\pi$, so the volume of the cylinder decreases by approximately 636.17 cm^3.

4.2.45: Because $\theta = 45°$, the range R of the shell is a function of its initial velocity v alone, and $R = \frac{1}{32}v^2$. Hence $dR = \frac{1}{16}v \, dv$. With $v = 80$ and $dv = 1$, we find that $dR = \frac{1}{16} \cdot 80 = 5$, so the range is increased by approximately 5 ft.

4.2.47: Technically, if $W = RI^2$, then $dW = I^2 \, dR + 2IR \, dI$. But in this problem, R remains constant, so that $dR = 0$ and hence $dW = 2IR \, dI$. We take $R = 10$, $I = 3$, and $dI = 0.1$, and find that $dW = 6$. So the wattage increases by approximately 6 watts.

4.2.49: Let V be the volume of the ball and let r be its radius, so that $V = \frac{4}{3}\pi r^3$. Then the calculated value of the volume is $V_{\text{calc}} = \frac{4}{3}(1000\pi) \approx 4188.7902$ in.^{3}, whereas $\Delta V \approx 4\pi(10)^2\frac{1}{16} = 25\pi \approx 78.5398$ in.^{3} (the true value of ΔV is approximately 79.0317).

4.2.51: With surface area S and radius r, we have $S = 2\pi r^2$ (*half* the surface area of a sphere of radius r), so that $dS = 4\pi r \, dr \approx 4\pi(100)(0.01) = 4\pi$. That is, $\Delta S \approx 12.57$ square meters.

4.2.53: We plotted $f(x) = x^2$ and its linear approximation $L(x) = 1 + 2(x - 1)$ on the interval $[0.5, 1.5]$, and it was clear that the interval $I = (0.58, 1.42)$ would be an adequate answer to this problem. We then used Newton's method to find a "better" interval, which turns out to be $I = (0.5528, 1.4472)$.

4.2.55: We plotted $f(x) = 1/x$ and its linear approximation $L(x) = \frac{1}{2} + \frac{1}{4}(2 - x)$ on the interval $[1.73, 2.32]$, and it was clear that the interval was a little too large to be a correct answer. We used Newton's method to find more accurate endpoints, and came up with the answer $I = (1.7365, 2.3035)$. Of course, any open subinterval of this interval that contains $a = 2$ is also a correct answer.

4.2.57: We plotted $f(x) = \sin x$ and its linear approximation $L(x) = x$ on the interval $[-2.8, 2.8]$ and it was clear that the interval $(-0.5, 0.5)$ would suffice. We then used Newton's method to find "better" endpoints and found that $(-0.6746, 0.6745)$ would suffice.

4.2.59: We plotted $f(x) = \sin x$ and its linear approximation

$$L(x) = \frac{\sqrt{2}}{2} \left(1 - \frac{\pi}{4} + x \right)$$

on the interval $[0.5, 1.1]$, and it was clear that the interval $(0.6, 0.95)$ would be adequate. We then used Newton's method to "improve" the endpoints and found that the interval $(0.5364, 1.0151)$ would suffice.

Section 4.3

4.3.1: $f'(x) = -2x$; f is increasing on $(-\infty, 0)$ and decreasing on $(0, +\infty)$. Matching graph: (c).

4.3.3: $f'(x) = 2x + 4$; f is increasing on $(-2, +\infty)$, decreasing on $(-\infty, -2)$. Matching graph: (f).

4.3.5: $f'(x) = x^2 - x - 2 = (x + 1)(x - 2)$; $f'(x) = 0$ when $x = -1$ and when $x = 2$; f is increasing on $(-\infty, -1)$ and on $(2, +\infty)$, decreasing on $(-1, 2)$. Matching graph: (d).

4.3.7: $f(x) = 2x^2 + C$; $5 = f(0) = C$: $f(x) = 2x^2 + 5$.

4.3.9: $f(x) = -\dfrac{1}{x} + C$; $1 = f(1) = C - 1$: $f(x) = -\dfrac{1}{x} + 2$.

4.3.11: $f'(x) \equiv 3 > 0$ for all x, so f is increasing for all x.

4.3.13: $f'(x) = -4x$, so f is increasing on $(-\infty, 0)$ and decreasing on $(0, +\infty)$. The graph of $y = f(x)$ is shown next.

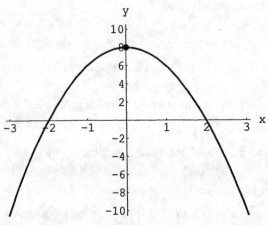

136

4.3.15: $f'(x) = 6 - 4x$. Therefore f is increasing for $x < \frac{3}{2}$ and decreasing for $x > \frac{3}{2}$. The graph of $y = f(x)$ is shown next.

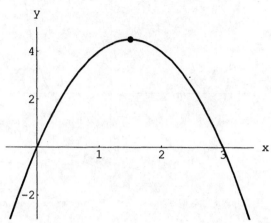

4.3.17: $f'(x) = 4x^3 - 4x = 4x(x+1)(x-1)$. The intervals on which $f'(x)$ cannot change sign are $x < -1$, $-1 < x < 0$, $0 < x < 1$, and $1 < x$. Because $f'(-2) = -24$, $f'(-0.5) = 1.5$, $f'(0.5) = -1.5$, and $f'(2) = 24$, we may conclude that f is increasing if $-1 < x < 0$ or if $x > 1$, decreasing for $x < -1$ and for $0 < x < 1$. The graph of $y = f(x)$ is shown next.

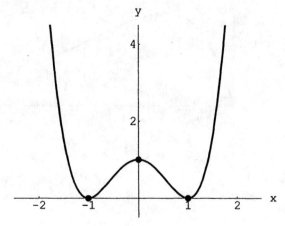

4.3.19: $f'(x) = 12x^3 + 12x^2 - 24x = 12x(x+2)(x-1)$, so the only points where $f'(x)$ can change sign are -2, 0, and 1. If $0 < x < 1$ then $12x > 0$, $x + 2 > 0$, and $x - 1 < 0$, and therefore $f'(x) < 0$ if $0 < x < 1$. Therefore f is decreasing on the interval $(0, 1)$. A similar analysis shows that f is also decreasing on $(-\infty, -2)$ and is increasing on $(-2, 0)$ and on $(1, +\infty)$. The graph of $y = f(x)$ is shown next.

137

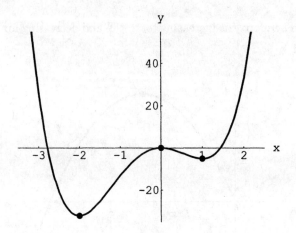

4.3.21: After simplifications,

$$f'(x) = \frac{2 - x}{2e^{x/2}}.$$

Hence $f(x)$ can change sign only at $x = 2$. Therefore f is increasing if $x < 2$ and is decreasing if $x > 2$. The graph of $y = f(x)$ is next.

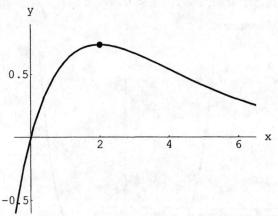

4.3.23: Because

$$f'(x) = -\frac{(x - 1)(x - 3)}{e^x},$$

$f(x)$ can change sign only at $x = 1$ and at $x = 3$. Also $f(0) < 0$, $f(2) < 0$, and $f(4) < 0$, so f is decreasing on $(-\infty, 1)$, increasing on $(1, 3)$, and decreasing on $(3, +\infty)$. The graph of $y = f(x)$ is next.

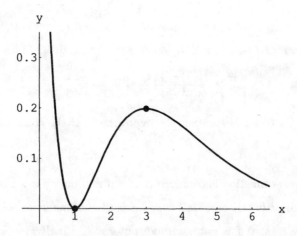

4.3.25: $f(0) = 0$, $f(2) = 0$, f is continuous for $0 \leq x \leq 2$, and $f'(x) = 2x - 2$ exists for $0 < x < 2$. To find the numbers c satisfying the conclusion of Rolle's theorem, we solve $f'(c) = 0$ to find that $c = 1$ is the only such number.

4.3.27: Given: $f(x) = 2 \sin x \cos x$ on the interval $I = [0, \pi]$. Then

$$f'(x) = 2 \cos^2 x - 2 \sin^2 x = 2 \cos 2x,$$

and it is clear that all the hypotheses of Rolle's theorem are satisfied by f on the interval I. Also $f'(x) = 0$ when $2x = \pi/2$ and when $2x = 3\pi/2$, so the numbers whose existence is guaranteed by Rolle's theorem are $c = \pi/4$ and $c = 3\pi/4$.

4.3.29: On the interval $(-1, 0)$, $f'(x) = 1$; on the interval $(0, 1)$, we have $f'(x) = -1$. Because f is not differentiable at $x = 0$, it does not satisfy the hypotheses of Rolle's theorem, so there is no guarantee that the equation $f'(x) = 0$ has a solution—and, indeed, it has no solution in $(-1, 1)$.

4.3.31: If $f(x) = xe^x$, then $f(1) = e \neq 0$, so f does not satisfy the hypotheses of Rolle's theorem on $[0, 1]$. Because $f'(x) = (x+1)e^x$ is zero only when $x = -1$, f also does not satisfy the conclusion of Rolle's theorem there.

4.3.33: Here, $f'(x) = 6x + 6$; $f(-2) = -5$ and $f(1) = 4$. So we are to solve the equation

$$6(c + 1) = \frac{4 - (-5)}{1 - (-2)} = 3. \qquad \text{It follows that} \quad c = -\frac{1}{2}.$$

4.3.35: First, $f'(x) = \frac{2}{3}(x - 1)^{-1/3}$ is defined on $(1, 2)$; moreover, f is continuous for $1 \leq x \leq 2$ (the only "problem point" is $x = 1$, but the limit of f there is equal to its value there). To find c, we solve

$$f'(c) = \frac{2}{3(c - 1)^{1/3}} = \frac{f(2) - f(1)}{2 - 1} = \frac{1 - 0}{1} = 1.$$

This leads to the equation $(c - 1)^{1/3} = \frac{2}{3}$, and thereby $c = \frac{35}{27}$. Note that c does lie in the interval $(1, 2)$.

139

4.3.37: First, $f(x) = |x - 2|$ is not differentiable at $x = 2$, so does not satisfy the hypotheses of the mean value theorem on the given interval $1 \leqq x \leqq 4$. Wherever $f'(x)$ is defined, its value is 1 or -1, but

$$\frac{f(4) - f(1)}{4 - 1} = \frac{2 - 1}{3} = \frac{1}{3}$$

is never a value of $f'(x)$. So f satisfies neither the hypotheses nor the conclusion of the mean value theorem on the interval $1 \leqq x \leqq 4$.

4.3.39: The greatest integer function is continuous at x if and only if x is not an integer. Consequently all the hypotheses of the mean value theorem fail here: f is discontinuous at -1, 0, and 1, and also $f'(0)$ does not exist because (for one reason) f is not continuous at $x = 0$. Finally, the average slope of the graph of f is 1, but $f'(x) = 0$ wherever it is defined. Thus the conclusion of the mean value theorem also fails to hold.

4.3.41: Let $f(x) = x^5 + 2x - 3$. Then $f'(x) = 5x^4 + 2$, so $f'(x) > 0$ for all x. This implies that f is an increasing continuous function, and therefore $f(x)$ can have at most one zero in any interval. To show that f has at least one zero in the interval $0 \leqq x \leqq 1$, it is sufficient to notice that $f(1) = 0$. Therefore the equation $f(x) = 0$ has exactly one solution in $[0, 1]$.

4.3.43: Let $f(x) = -3 + x \ln x$. Note that $\frac{1}{2} < \ln 2 < 1$ because $\ln 2 \approx 0.693147$. Hence $1 < 2 \ln 2 < 2$, so $f(2) < 0$. Also $\ln 4 = 2 \ln 2$, so $1 < \ln 4 < 2$, and hence $4 < 4 \ln 4 < 8$. Therefore $f(4) > 0$. Because f is differentiable on $[2, 4]$, it is continuous there. Hence, by the intermediate value property of continuous functions, $f(c) = 0$ for some number c in $[2, 4]$. Thus $c \ln c = 3$, and therefore the equation $x \ln x = 3$ as at *least* one solution in $[2, 4]$.

Next, $f'(x) = 1 + \ln x$ is positive on $[2, 4]$ (use inequalities similar to those in the previous paragraph if this is not perfectly clear), and therefore f is increasing on that interval. Hence the equation $f(x) = 0$ has at *most* one solution in $[2, 4]$. In conclusion, the equation $x \ln x = 3$ has exactly one solution in $[2, 4]$. (By Newton's method, the solution is approximately 2.85739078.)

4.3.45: The car traveled 35 miles in 18 minutes, which is an average speed of $\frac{250}{3} \approx 83.33$ miles per hour. By the mean value theorem, the car must have been traveling over 83 miles per hour at some time between 3:00 P.M. and 3:18 P.M.

4.3.47: Let $f(t)$ be the distance that the first car has traveled from point A on its way to point B at time t, with t measured in hours and with $t = 0$ corresponding to 9:00 A.M. (so that $t = 1$ corresponds to 10:00 A.M.). Let $g(t)$ be the corresponding function for the second car. Let $h(t) = f(t) - g(t)$. We make the very plausible assumption that the functions f and g are differentiable on $(0, 1)$ and continuous on $[0, 1]$, so h has the same properties. In addition, $h(0) = f(0) - g(0) = 0$ and $h(1) = 0$ as well. By Rolle's theorem, $h'(c) = 0$ for some c in $(0, 1)$. But this implies that $f'(c) = g'(c)$. That is, the velocity of the first car is exactly the same as that of the second car at time $t = c$.

140

4.3.49: Because

$$f'(x) = \frac{3}{2}(1+x)^{1/2} - \frac{3}{2} = \frac{3}{2}\left(\sqrt{1+x} - 1\right),$$

it is clear that $f'(x) > 0$ for $x > 0$. Also $f(0) = 0$; it follows that $f(x) > 0$ for all $x > 0$. That is,

$$(1+x)^{3/2} > 1 + \frac{3}{2}x \quad \text{for} \quad x > 0.$$

4.3.51: Proof: Suppose that $f'(x)$ is a polynomial of degree $n-1$ on the interval $I = [a, b]$. Then $f'(x)$ has the form

$$f'(x) = a_{n-1}x^{n-1} + a_{n-2}x^{n-2} + \ldots + a_2x^2 + a_1x + a_0$$

where $a_{n-1} \neq 0$. Note that $f'(x)$ is the derivative of the function

$$g(x) = \frac{1}{n}a_{n-1}x^n + \frac{1}{n-1}a_{n-2}x^{n-1} + \ldots + \frac{1}{3}a_2x^3 + \frac{1}{2}a_1x^2 + a_0x.$$

By Corollary 2, $f(x)$ and $g(x)$ can differ only by a constant, and this is sufficient to establish that $f(x)$ must also be a polynomial, and one of degree n because the coefficient of x^n in $f(x)$ is the same as the coefficient of x^n in $g(x)$, and that coefficient is nonzero.

4.3.53: First note that $f'(x) = \frac{1}{2}x^{-1/2}$, and that the hypotheses of the mean value theorem are all satisfied for the given function f on the given interval $[100, 101]$. Thus there does exist a number c between 100 and 101 such that

$$\frac{1}{2c^{1/2}} = \frac{f(101) - f(100)}{101 - 100} = \sqrt{101} - \sqrt{100}.$$

Therefore $1/(2\sqrt{c}) = \sqrt{101} - 10$, and thus we have shown that $\sqrt{101} = 10 + 1/(2\sqrt{c})$ for some number c in $(100, 101)$.

Proof for part (b): If $0 \leq \sqrt{c} \leq 10$, then $0 \leq c \leq 100$; because $c > 100$, we see that $0 \leq \sqrt{c} \leq 10$ is impossible. If $10.5 \leq \sqrt{c}$ then $110.25 \leq c$, which is also impossible because $c < 110$. So we may conclude that $10 < \sqrt{c} < 10.5$. Finally,

$$10 < \sqrt{c} < 10.5 \quad \text{implies that} \quad 20 < 2\sqrt{c} < 21.$$

Consequently

$$\frac{1}{21} < \frac{1}{2c^{1/2}} < \frac{1}{20}, \quad \text{so} \quad 10 + \frac{1}{21} < \sqrt{101} < 10 + \frac{1}{20}.$$

The decimal expansion of $1/21$ begins $0.047619047619\ldots$, and therefore $10.0476 < \sqrt{101} < 10.05$.

4.3.55: Let $f(x) = (\tan x)^2$ and let $g(x) = (\sec x)^2$. Then

$$f'(x) = 2(\tan x)(\sec^2 x) \quad \text{and} \quad g'(x) = 2(\sec x)(\sec x \tan x) = f'(x) \quad \text{on} \quad (-\pi/2, \pi/2).$$

141

Therefore there exists a constant C such that $f(x) = g(x) + C$ for all x in $(-\pi/2, \pi/2)$. Finally, $f(0) = 0$ and $g(0) = 1$, so $C = f(0) - g(0) = -1$.

4.3.57: The average slope of the graph of f on the given interval $[-1, 2]$ is

$$\frac{f(2) - f(-1)}{2 - (-1)} = \frac{5 - (-1)}{3} = 2$$

and f satisfies the hypotheses of the mean value theorem there. Therefore $f'(c) = 2$ for some number c, $-1 < c < 2$. This implies that the tangent line to the graph of f at the point $(c, f(c))$ has slope 2 and is therefore parallel to the line with equation $y = 2x$ because the latter line also has slope 2.

4.3.59: Use the definition of the derivative:

$$g'(0) = \lim_{h \to 0} \frac{g(0 + h) - g(0)}{h}$$

$$= \lim_{h \to 0} \left[\frac{1}{2} + \frac{1}{h} h^2 \sin\left(\frac{1}{h}\right) \right]$$

$$= \lim_{h \to 0} \left[\frac{1}{2} + h \sin\left(\frac{1}{h}\right) \right]$$

$$= \frac{1}{2} + 0 \quad \text{(by the squeeze law)}$$

$$= \frac{1}{2} > 0.$$

If $x \neq 0$ then

$$g'(x) = \frac{1}{2} + 2x \sin\left(\frac{1}{x}\right) - \cos\left(\frac{1}{x}\right).$$

Because $\cos(1/x)$ oscillates between $+1$ and -1 near $x = 0$ and $2x \sin(1/x)$ is near zero for x close to zero, it follows that every interval about $x = 0$ contains subintervals on which $g'(x) > 0$ and subintervals on which $g'(x) < 0$. They are not clearly visible near $x = 0$ in the graph of $y = g(x)$ (shown next) because they are very short intervals.

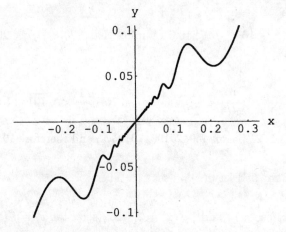

142

4.3.61: Let $h(x) = 1 - \frac{1}{2}x^2 - \cos x$. Then $h'(x) = -x + \sin x$. By Example 8, $\sin x < x$ for all $x > 0$, so $h'(x) < 0$ for all $x > 0$. If $x > 0$, then $\dfrac{h(x) - 0}{x - 0} = h'(c)$ for some $c > 0$, so $h(x) < 0$ for all $x > 0$; that is, $\cos x > 1 - \frac{1}{2}x^2$ for all $x > 0$.

4.3.63: (a): Let $K(x) = 1 - \frac{1}{2}x^2 + \frac{1}{24}x^4 - \cos x$. Then

$$K'(x) = -x + \frac{1}{6}x^3 + \sin x = \sin x - \left(x - \frac{1}{6}x^3\right).$$

By Problem 62, part (a), $K'(x) > 0$ for all $x > 0$. So if $x > 0$, $\dfrac{K(x) - 0}{x - 0} = K'(c)$ for some $c > 0$. Therefore $K(x) > 0$ for all $x > 0$. That is,

$$\cos x < 1 - \frac{1}{2}x^2 + \frac{1}{24}x^4 \quad \text{for all} \quad x > 0.$$

(b): By Problem 61 and part (a),

$$1 - \frac{1}{2}x^2 < \cos x < 1 - \frac{1}{2}x^2 + \frac{1}{24}x^4$$

for all $x > 0$. In particular,

$$1 - \frac{1}{2}\left(\frac{\pi}{18}\right)^2 < \cos\frac{\pi}{18} < 1 - \frac{1}{2}\left(\frac{\pi}{18}\right)^2 + \frac{1}{24}\left(\frac{\pi}{18}\right)^4;$$

hence $0.984769 < \cos 10° < 0.984808$. So $\cos 10° \approx 0.985$.

Section 4.4

4.4.1: $f'(x) = 2x - 4$; $x = 2$ is the only critical point. Because $f'(x) > 0$ for $x > 2$ and $f'(x) < 0$ for $x < 2$, it follows that $f(2) = 1$ is the global minimum value of $f(x)$. The graph of $y = f(x)$ is shown next.

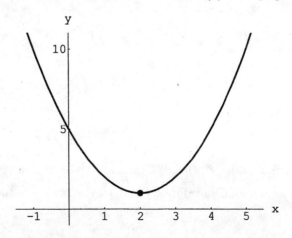

4.4.3: $f'(x) = 3x^2 - 6x = 3x(x - 2)$, so $x = 0$ and $x = 2$ are the only critical points. If $x < 0$ or if $x > 2$ then $f'(x)$ is positive, but $f'(x) < 0$ for $0 < x < 2$. So $f(0) = 5$ is a local maximum and $f(2) = 1$ is a local minimum. The graph of $y = f(x)$ is shown next.

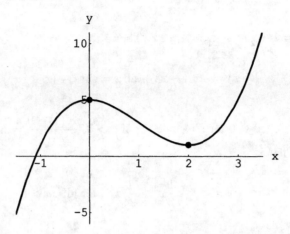

4.4.5: $f'(x) = 3x^2 - 6x + 3 = 3(x - 1)^2$, so $x = 1$ is the only critical point of f. $f'(x) > 0$ if $x \neq 1$, so the graph of f is increasing for *all* x; so f has no extrema of any sort. The graph of $y = f(x)$ is shown next.

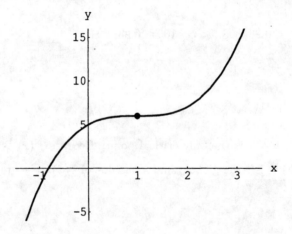

4.4.7: $f'(x) = -6(x - 5)(x + 2)$; $f'(x) < 0$ if $x < -2$ and if $x > 5$, but $f'(x) > 0$ for $-2 < x < 5$. Hence $f(-2) = -58$ is a local minimum value of f and $f(5) = 285$ is a local maximum value. The graph of $y = f(x)$ is shown next.

144

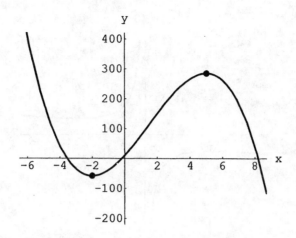

4.4.9: $f'(x) = 4x(x-1)(x+1)$; $f'(x) < 0$ for $x < -1$ and on the interval $(0, 1)$, whereas $f'(x) > 0$ for $x > 1$ and on the interval $(-1, 0)$. Consequently, $f(-1) = -1 = f(1)$ is the global minimum value of $f(x)$ and $f(0) = 0$ is a local maximum value. Note that the [unique] global minimum *value* occurs at two different *points* on the graph of f, which is shown next.

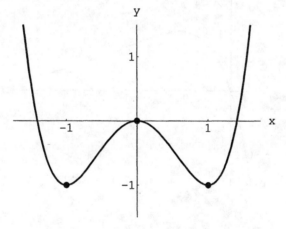

4.4.11: $f'(x) = 1 - 9x^{-2}$, so the critical points occur where $x = -3$ and $x = 3$ (horizontal tangents); note that f is not defined at $x = 0$. If $x^2 > 9$ then $f'(x) > 0$, so f is increasing if $x > 3$ and if $x < -3$. If $x^2 < 9$ then $f'(x) < 0$, so f is decreasing on $(-3, 0)$ and on $(0, 3)$. Therefore $f(-3) = -6$ is a local maximum value and $f(3) = 6$ is a local minimum value for $f(x)$. The graph of $y = f(x)$ is next.

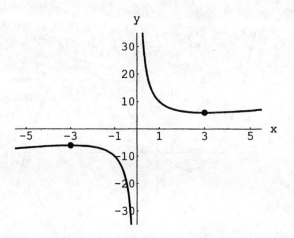

4.4.13: If $f(x) = xe^{-2x}$, then

$$f'(x) = e^{-2x} - 2xe^{-2x} = \frac{1 - 2x}{e^{2x}}.$$

The only critical point occurs when $x = \frac{1}{2}$. The graph of f is increasing to the left of this point and decreasing to the right, so f has a global maximum there. The graph of $y = f(x)$ is next.

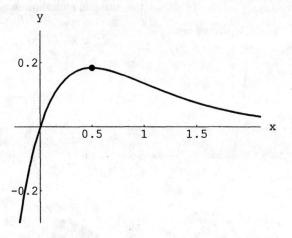

4.4.15: If $f(x) = (x + 4)^2 e^{-x/5}$, then

$$f'(x) = \frac{1}{5}(24 + 2x - x^2)e^{-x/5} = \frac{(6 - x)(4 + x)}{5e^{x/5}}.$$

Therefore the only critical points occur where $x = -4$ and $x = 6$. Because $f'(-5) \approx -5.98022 < 0$, $f'(0) = 4.8 > 0$, and $f'(7) \approx -0.54251 < 0$, the graph of f is decreasing if $x < -4$ and if $x > 6$, increasing on $(-4, 6)$. Because $f(x) \geqq 0$ for all x and $f(x) = 0$ only if $x = -4$, there is a global minimum at $(-4, 0)$. Because $f(x) \to +\infty$ as $x \to -\infty$, the maximum where $x = 6$ is local but not global. The graph of $y = f(x)$ is next.

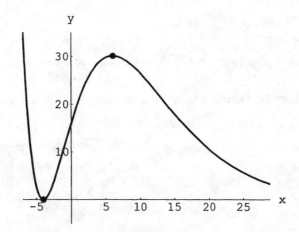

4.4.17: $f'(x) = 2\sin x \cos x$; $f'(x) = 0$ when x is any integral multiple of $\pi/2$. In $(0, 3)$, $f'(x) = 0$ when $x = \pi/2$. Because $f'(x) > 0$ if $0 < x < \pi/2$ and $f'(x) < 0$ if $\pi/2 < x < 3$, $f(x)$ has the global maximum value $f(\pi/2) = 1$.

4.4.19: $f'(x) = 3\sin^2 x \cos x$; $f'(x) = 0$ when $x = -\pi/2$, 0, or $\pi/2$. f is decreasing on $(-3, -\pi/2)$ and on $(\pi/2, 3)$, but increasing on $(-\pi/2, \pi/2)$. So f has a global minimum at $(-\pi/2, -1)$ and a global maximum at $(\pi/2, 1)$; there is no extremum at the critical point $(0, 0)$.

4.4.21: $f'(x) = x\sin x$; $f'(x) = 0$ at $-\pi$, 0, and π. $f'(x) > 0$ on $(-\pi, \pi)$, $f'(x) < 0$ on $(-5, -\pi)$ and on $(\pi, 5)$. So f has a global maximum at (π, π) and a global minimum at $(-\pi, -\pi)$. Note that the critical point $(0, 0)$ is not an extremum.

4.4.23: If

$$f(x) = \frac{\ln x}{x^2}, \qquad \text{then} \qquad f'(x) = \frac{1 - 2\ln x}{x^3}.$$

So $f'(x) = 0$ when $\ln x = \frac{1}{2}$; that is, $x = e^{1/2}$. Because $f'(x) < 0$ if $0 < x < e^{1/2}$ and $f'(x) > 0$ when $x > e^{1/2}$, the critical point at $x = e^{1/2}$ is a global maximum. The graph of $y = f(x)$ is next.

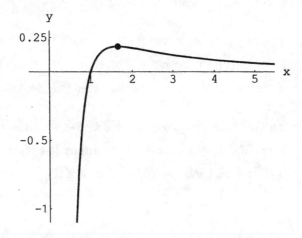

4.4.25: If $f(x) = e^x \sin x$, then $f'(x) = e^x(\cos x + \sin x)$. Hence $f'(x) = 0$ in $(-3, 3)$ when $\tan x = -1$, thus at $x = a = -\pi/4$ and at $x = b = 3\pi/4$. Because $f'(x) < 0$ if $-3 < x < a$ and if $b < x < 3$, the graph

147

of f is decreasing on these intervals; it is increasing on (a, b). Inspection of the values of $f(x)$ near -3 and near 3 shows that $f(x)$ has a global minimum value at $x = a$ and a global maximum value at $x = b$.

4.4.27: Let x be the smaller of the two numbers; then the other is $x+20$ and their product is $f(x) = x^2+20x$. Consequently $f'(x) = 2x + 20$, so $x = -10$ is the only critical point of f. The graph of f is decreasing for $x < -10$ and increasing for $x > -10$. Therefore $(-10, -100)$ is the lowest point on the graph of f. Answer: The two numbers are -10 and 10.

4.4.29: Let us minimize

$$g(x) = (x - 3)^2 + (3 - 2x - 2)^2 = (x - 3)^2 + (1 - 2x)^2,$$

the square of the distance from (x, y) on the line $2x + y = 3$ to the point $(3, 2)$. We have

$$g'(x) = 2(x - 3) - 4(1 - 2x) = 10x - 10,$$

so $x = 1$ is the only critical point of $g(x)$. If $x > 1$ then $g'(x) > 0$, but $g'(x) < 0$ for $x < 1$. Thus $x = 1$ minimizes $g(x)$, and so the point on the line $2x+y = 3$ closest to the point $(3, 2)$ is $(1, 1)$. As an independent check, note that the slope of the line segment joining $(3, 2)$ and $(1, 1)$ is $\frac{1}{2}$, whereas the slope of the line $2x + y = 3$ is -2, so the segment and the line are perpendicular; see Miscellaneous Problem 70 of Chapter 3.

4.4.31: Base of box: x wide, $2x$ long. Height: y. Then the box has volume $2x^2y = 972$, so $y = 486x^{-2}$. Its total surface area is $A = 2x^2 + 6xy$, so we minimize

$$A = A(x) = 2x^2 + \frac{2916}{x}, \qquad x > 0.$$

Now

$$A'(x) = 4x - \frac{2916}{x^2},$$

so the only critical point of $A(x)$ occurs when $4x^3 = 2916$; that is, when $x = 9$. It is easy to verify that $A'(x) < 0$ for $0 < x < 9$ and that $A'(x) > 0$ for $x > 9$. Therefore $A(9)$ is the global minimum value of $A(x)$. Answer: The dimensions of the box are 9 inches wide, 18 inches long, and 6 inches high.

4.4.33: Let r denote the radius of the pot and h its height. We are given the constraint $\pi r^2 h = 250$, so $h = 250/(\pi r^2)$. Now the bottom of the pot has area πr^2, and thus costs $4\pi r^2$ cents. The curved side of the pot has area $2\pi rh$, and thus costs $4\pi rh$ cents. So the total cost of the pot is

$$C = 4\pi r^2 + 4\pi rh; \qquad \text{thus} \qquad C = C(r) = 4\pi r^2 + \frac{1000}{r}, \qquad r > 0.$$

Now

$$C'(r) = 8\pi r - \frac{1000}{r^2};$$

$C'(r) = 0$ when $8\pi r^3 = 1000$, so that $r = 5/\sqrt[3]{\pi}$. It is clear that this is the only (positive) value of r at which $C'(r)$ can change sign, and that $C'(r) < 0$ for r positive and near zero, but $C'(r) > 0$ for r large positive. Therefore we have found the value of r that minimizes $C(r)$. The corresponding value of h is $10/\sqrt[3]{\pi}$, so the pot of minimal cost has height equal to its diameter, each approximately 6.828 centimeters.

4.4.35: If the sides of the rectangle are x and y, then $xy = 100$, so that $y = \dfrac{100}{x}$. Therefore the perimeter of the rectangle is

$$P = P(x) = 2x + \frac{200}{x}, \qquad x > 0.$$

Then

$$P'(x) = 2 - \frac{200}{x^2};$$

$P'(x) = 0$ when $x = 10$ (-10 is not in the domain of $P(x)$). Now $P'(x) < 0$ on $(0, 10)$ and $P'(x) > 0$ for $x > 10$, and so $x = 10$ minimizes $P(x)$. A little thought about the behavior of $P(x)$ for x near zero and for x large makes it clear that we have found the global minimum value for P: $P(10) = 40$. When $x = 10$, also $y = 10$, so the rectangle of minimal perimeter is indeed a square.

4.4.37: Let the square base of the box have edge length x and let its height be y, so that its total volume is $x^2 y = 62.5$ and the surface area of this box-without-top will be $A = x^2 + 4xy$. So

$$A = A(x) = x^2 + \frac{250}{x}, \qquad x > 0.$$

Now

$$A'(x) = 2x - \frac{250}{x^2},$$

so $A'(x) = 0$ when $x^3 = 125$: $x = 5$. In this case, $y = 2.5$. Also $A'(x) < 0$ if $0 < x < 5$ and $A'(x) > 0$ if $x > 5$, so we have found the global minimum for $A(x)$. Answer: Square base of edge length 5 inches, height 2.5 inches.

4.4.39: Let x denote the radius and y the height of the cylinder (in inches). Then its cost (in cents) is $C = 8\pi x^2 + 4\pi xy$, and we also have the constraint $\pi x^2 y = 100$. So

$$C = C(x) = 8\pi x^2 + \frac{400}{x}, \qquad x > 0.$$

Now $dC/dx = 16\pi x - 400/(x^2)$; $dC/dx = 0$ when $x = (25/\pi)^{1/3}$ (about 1.9965 inches) and consequently, when $y = (1600/\pi)^{1/3}$ (about 7.9859 inches). Because $C'(x) < 0$ if $x^3 < 25/\pi$ and $C'(x) > 0$ if $x^3 > 25/\pi$, we have indeed found the dimensions that minimize the total cost of the can. For simplicity, note that $y = 4x$ at minimum: The height of the can is twice its diameter.

4.4.41: Let $(x, y) = (x, x^2)$ denote an arbitrary point on the curve. The square of its distance from $(0, 2)$ is then

$$f(x) = x^2 + (x^2 - 2)^2.$$

Now $f'(x) = 2x(2x^2 - 3)$, and therefore $f'(x) = 0$ when $x = 0$, when $x = -\sqrt{3/2}$, and when $x = +\sqrt{3/2}$. Now $f'(x) < 0$ if $x < -\sqrt{3/2}$ and if $0 < x < \sqrt{3/2}$; $f'(x) > 0$ if $-\sqrt{3/2} < x < 0$ and if $x > \sqrt{3/2}$. Therefore $x = 0$ yields a local maximum for f; the other two zeros of $f'(x)$ yield its global minimum. Answer: There are exactly two points on the curve that are nearest $(0, 2)$; they are $(+\sqrt{3/2}, 3/2)$ and $(-\sqrt{3/2}, 3/2)$.

4.4.43: If the dimensions of the rectangle are x by y, and the line segment bisects the side of length x, then the square of the length of the segment is

$$f(x) = \left(\frac{x}{2}\right)^2 + y^2 = \frac{x^2}{4} + \frac{4096}{x^2}, \qquad x > 0,$$

because $y = 64/x$. Now

$$f'(x) = \frac{x}{2} - \frac{8192}{x^3}.$$

When $f'(x) = 0$, we must have $x = +8\sqrt{2}$, so that $y = 4\sqrt{2}$. We have found the minimum of f because if $0 < x < 8\sqrt{2}$ then $f'(x) < 0$, and $f'(x) > 0$ if $x > 8\sqrt{2}$. The minimum length satisfies $L^2 = f(8\sqrt{2})$, so that $L = 8$ centimeters.

4.4.45: If the end of the rod projects the distance y into the narrower hall, then we have the proportion $y/2 = 4/x$ by similar triangles. So $y = 8/x$. The square of the length of the rod is then

$$f(x) = (x + 2)^2 + \left(4 + \frac{8}{x}\right)^2, \qquad x > 0.$$

It follows that

$$f'(x) = 4 + 2x - \frac{64}{x^2} - \frac{128}{x^3},$$

and that $f'(x) = 0$ when $(x + 2)(x^3 - 32) = 0$. The only admissible solution is $x = \sqrt[3]{32}$, which indeed minimizes $f(x)$ by the usual argument ($f(x)$ is very large positive if x is either large positive or positive and very close to zero). The minimum length is

$$L = \left(20 + 12\sqrt[3]{4} + 12\sqrt[3]{16}\right)^{1/2} \approx 8.323876 \text{ (meters)}.$$

4.4.47: If the pyramid has base edge length x and altitude y, then its volume is $V = \frac{1}{3}x^2 y$. From Fig. 4.4.30 we see also that

$$\frac{2y}{x} = \tan\theta \quad \text{and} \quad \frac{a}{y - a} = \cos\theta$$

150

where θ is the angle that each side of the pyramid makes with its base. It follows, successively, that

$$\left(\frac{a}{y-a}\right)^2 = \cos^2\theta;$$

$$\sin^2\theta = 1 - \cos^2\theta = \frac{(y-a)^2 - a^2}{(y-a)^2}$$

$$= \frac{y^2 - 2ay}{(y-a)^2} = \frac{y(y-2a)}{(y-a)^2}.$$

$$\sin\theta = \frac{(y(y-2a))^{1/2}}{y-a}.$$

$$y = \frac{x\sin\theta}{2\cos\theta} = \left(\frac{x}{2}\right)\left(\frac{(y(y-2a))^{1/2}}{y-a}\right)\left(\frac{y-a}{a}\right);$$

$$2y = \frac{x}{a}\sqrt{y(y-2a)};$$

$$x^2 = \frac{4a^2y^2}{y(y-2a)}.$$

Therefore

$$V = \frac{1}{3}x^2y = V(y) = \frac{4a^2y^2}{3(y-2a)}, \qquad y > 2a.$$

Now

$$\frac{dV}{dy} = \frac{24a^2y(y-2a) - 12a^2y^2}{9(y-2a)^2}.$$

The condition $dV/dy = 0$ then implies that $2(y-2a) = y$, and thus that $y = 4a$. Consequently the minimum volume of the pyramid is

$$V(4a) = \frac{(4a^2)(16a^2)}{(3)(2a)} = \frac{32}{3}a^3.$$

The ratio of the volume of the smallest pyramid to that of the sphere is then

$$\frac{32/3}{4\pi/3} = \frac{32}{4\pi} = \frac{8}{\pi}.$$

4.4.49: Let z be the length of the segment from the top of the tent to the midpoint of one side of its base. Then $x^2 + y^2 = z^2$. The total surface area of the tent is

$$A = 4x^2 + (4)(\tfrac{1}{2})(2x)(z) = 4x^2 + 4xz = 4x^2 + 4x(x^2 + y^2)^{1/2}.$$

Because the [fixed] volume V of the tent is given by

$$V = \frac{1}{3}(4x^2)(y) = \frac{4}{3}x^2y,$$

we have $y = 3V/(4x^2)$, so

$$A = A(x) = 4x^2 + \frac{1}{x}(16x^6 + 9V^2)^{1/2}.$$

151

After simplifications, the condition $dA/dx = 0$ takes the form

$$8x\left(16x^6 + 9V^2\right)^{1/2} - \frac{1}{x^2}\left(16x^6 + 9V^2\right) + 48x^4 = 0,$$

which has solution $x = 2^{-7/6}\sqrt[3]{3V}$. Because this is the only positive solution of the equation, and because it is clear that neither large values of x nor values of x near zero will yield small values of the surface area, this is the desired value of x.

4.4.51: Let x denote the length of each edge of the square base of the box and let y denote the height of the box. Then $x^2 y = V$ where V is the fixed volume of the box. The surface total area of this closed box is

$$A = 2x^2 + 4xy, \quad \text{and hence} \quad A(x) = 2x^2 + \frac{4V}{x}, \quad 0 < x < +\infty.$$

Then

$$A'(x) = 4x - \frac{4V}{x^2} = \frac{4(x^3 - V)}{x^2},$$

so $A'(x) = 0$ when $x = V^{1/3}$. This is the only critical point of A, and $A'(x) < 0$ if x is near zero while $A'(x) > 0$ if x is large positive. Thus the global minimum value of the surface area occurs at this critical point. And if $x = V^{1/3}$, then the height of the box is

$$y = \frac{V}{x^2} = \frac{V}{V^{2/3}} = V^{1/3} = x,$$

and hence the closed box with square base, fixed volume, and minimal surface area is a cube.

4.4.53: Let r denote the radius of the base (and top) of the closed cylindrical can and let h denote its height. Then its fixed volume is $V = \pi r^2 h$ and its total surface area is $A = 2\pi r^2 + 2\pi rh$. Hence

$$A(r) = 2\pi r^2 + 2\pi r \cdot \frac{V}{\pi r^2} = 2\pi r^2 + \frac{2V}{r}, \quad 0 < r < +\infty.$$

Then

$$A'(r) = 4\pi r - \frac{2V}{r^2} = \frac{4\pi r^3 - 2V}{r^2};$$

$A'(r) = 0$ when $r = (V/2\pi)^{1/3}$. This critical point minimizes total surface area $A(r)$ because $A'(r) < 0$ when r is small positive and $A'(r) > 0$ when r is large positive. And at this critical point we have

$$h = \frac{V}{\pi r^2} = \frac{V}{\pi(V/2\pi)^{2/3}} = \frac{2\cdot(V/2\pi)}{(V/2\pi)^{2/3}} = 2(V/2\pi)^{1/3} = 2r.$$

Therefore the closed cylindrical can with fixed volume and minimal total surface area has height equal to the diameter of its base.

4.4.55: Finding the exact solution of this problem is quite challenging. In the spirit of mathematical modeling we accept the very good approximation that—if the thickness of the material of the can is small in comparison with its other dimensions—the total volume of material used to make the can may be approximated sufficiently accurately by multiplying the area of the bottom by its thickness, the area of the curved side by its thickness, the area of the top by its thickness, then adding these three products. Thus let r denote the radius of the inside of the cylindrical can, let h denote the height of the inside, and let t denote the thickness of its bottom and curved side; $3t$ will be the thickness of its top. The total (inner) volume of the can is the fixed number $V = \pi r^2 h$. The amount of material to make the can will (approximately, but accurately)

$$M = \pi r^2 t + 2\pi rht + 3\pi r^2 t = 4\pi r^2 t + 2\pi rht, \tag{1}$$

so that

$$M(r) = 4\pi r^2 t + 2\pi rt \cdot \frac{V}{\pi r^2} = \left(4\pi r^2 + \frac{2V}{r}\right) \cdot t, \quad 0 < r < L,$$

where L is some rather large positive number that we don't actually need to evaluate. (You can find L from Eq. (1) by setting $h = 0$ there and solving for r in terms of M and t.) Next,

$$M'(r) = \left(8\pi r - \frac{2V}{r^2}\right) \cdot t = \frac{2t(4\pi r^3 - V)}{r^2};$$

$M'(r) = 0$ when $r = (V/4\pi)^{1/3}$. This critical point yields the global minimum value of $M(r)$ because $M'(r) < 0$ when r is small positive and $M'(r) > 0$ when $(V/2\pi)^{1/3} < r < L$. (You need to verify that, under reasonable assumptions about the relative sizes of the linear measurements, $L > (V/4\pi)^{1/3}$.) And at this critical point, we have

$$h = \frac{V}{\pi r^2} = \frac{4 \cdot (V/4\pi)}{(V/4\pi)^{2/3}} = 4 \cdot (V/4\pi)^{1/3} = 4r.$$

Therefore the pop-top soft drink can of fixed internal volume V, with thickness as described in the problem, and using the minimal total volume of material for its top, bottom, and curved side, will have height (approximately) twice the diameter of its base.

In support of this conclusion, the smallest commonly available pop-top can of a popular blend of eight vegetable juices has height about 9 cm and base diameter about 5 cm. Most of the 12-oz pop-top soft drink cans we measured had height about 12.5 cm and base diameter about 6.5 cm.

Section 4.5

4.5.1: $f(x) \to +\infty$ as $x \to +\infty$, $f(x) \to -\infty$ as $x \to -\infty$. Matching graph: 4.5.13(c).

4.5.3: $f(x) \to -\infty$ as $x \to +\infty$, $f(x) \to +\infty$ as $x \to -\infty$. Matching graph: 4.5.13(d).

4.5.5: $y'(x) = 4x - 10$, so the only critical point is $\left(\frac{5}{2}, -\frac{39}{2}\right)$, the lowest point on the graph of y because $y'(x) < 0$ if $x < \frac{5}{2}$ and $y'(x) > 0$ if $x > \frac{5}{2}$.

4.5.7: $y'(x) = 12x^2 - 6x - 90$ is zero when $x = -\frac{5}{2}$ and when $x = 3$. The graph of y is increasing if $x < -\frac{5}{2}$, decreasing if $-\frac{5}{2} < x < 3$, and increasing if $x > 3$. Consequently there is a local maximum at $(-2.5, 166.75)$ and a local minimum at $(3, -166)$.

4.5.9: $y'(x) = 12x^3 + 12x^2 - 72x = 12(x - 2)x(x + 3)$, so there are critical points at $P(-3, -149)$, at $Q(0, 40)$, and $R(2, -24)$. The graph is decreasing to the left of P and between Q and R; it is increasing otherwise.

4.5.11: $y'(x) = 15x^4 - 300x^2 + 960 = 15(x + 4)(x + 2)(x - 2)(x - 4)$, so there are critical points at $P(-4, -512)$, $Q(-2, -1216)$, $R(2, 1216)$, and $S(4, 512)$. The graph is increasing to the left of P, between Q and R, and to the right of S; it is decreasing otherwise.

4.5.13: $y'(x) = 21x^6 - 420x^4 + 1344x^2 = 21(x + 4)(x + 2)x^2(x - 2)(x - 4)$, so there are critical points at $P(-4, 8192)$, $Q(-2, -1280)$, $R(0, 0)$, $S(2, 1280)$, and $T(4, -8192)$. The graph is increasing to the left of P, between Q and S (there is no extremum at R), and to the right of T; it is decreasing otherwise.

4.5.15: $f'(x) = 6x - 6$, so there is a critical point at $(1, 2)$. Because $f'(x) < 0$ if $x < 1$ and $f'(x) > 0$ if $x > 1$, there is a global minimum at the critical point. The graph of $y = f(x)$ is shown next.

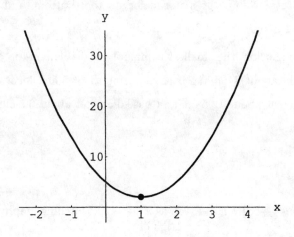

4.5.17: $f'(x) = 3(x^2 - 4)$. There is a local maximum at $(-2, 16)$ and a local minimum at $(2, -16)$; neither is global. The graph of $y = f(x)$ is shown next.

154

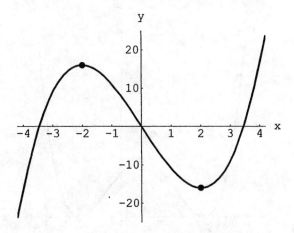

4.5.19: $f'(x) = 3x^2 - 12x + 9 = 3(x-1)(x-3)$, so f is increasing for $x < 1$ and for $x > 3$, decreasing for $1 < x < 3$. It has a local maximum at $(1, 4)$ and a local minimum at $(3, 0)$. Its graph is shown next.

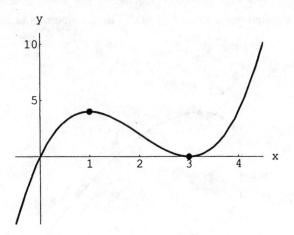

4.5.21: $f'(x) = 3(x^2 + 2x + 3) = 3[(x+1)^2 + 2]$ is positive for all x, so the graph of f is increasing on the set of all real numbers; there are no extrema. The graph of f is shown next.

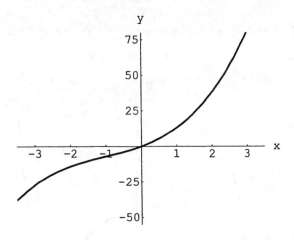

4.5.23: $f'(x) = 2(x-1)(x+2)(2x+1)$; there are global minima at $(-2, 0)$ and $(1, 0)$ and a local maximum at $\left(-\frac{1}{2}, \frac{81}{16}\right)$. The minimum value 0 is global because [clearly] $f(x) = (x-1)^2(x+2)^2 \geqq 0$ for all x. The maximum value is local because $f(x) \to +\infty$ as $x \to \pm\infty$. The graph of $y = f(x)$ is next.

155

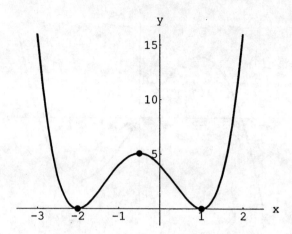

4.5.25: $f'(x) = \dfrac{3(1-x)}{2\sqrt{x}}$, so $f'(x) > 0$ if $0 < x < 1$ and $f'(x) < 0$ if $x > 1$. Therefore there is a local minimum at $(0,\,0)$ and a global maximum at $(1,\,2)$. The minimum is only local because $f(x) \to -\infty$ as $x \to +\infty$. The graph of $y = f(x)$ is shown next.

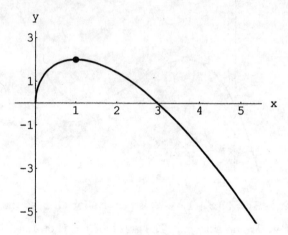

4.5.27: $f'(x) = 15x^2(x-1)(x+1)$, so f is increasing for $x < -1$ and for $x > 1$, decreasing for $-1 < x < 1$. Hence there is a local maximum at $(-1,\,2)$ and a local minimum at $(1,\,-2)$. The critical point at $(0,\,0)$ is not an extremum. The graph of $y = f(x)$ is shown next.

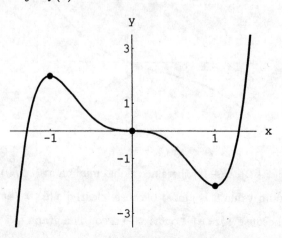

156

4.5.29: $f'(x) = 4x(x-2)(x+2)$, so the graph of f is decreasing for $x < -2$ and for $0 < x < 2$; it is increasing if $-2 < x < 0$ and if $x > 2$. Therefore the global minimum value -9 of $f(x)$ occurs at $x = \pm 2$ and the extremum at $(0, 7)$ is a maximum, but not global because $f(x) \to +\infty$ as $x \to \pm\infty$. The graph of $y = f(x)$ is shown next.

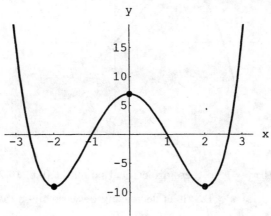

4.5.31: Because $f'(x) = 4x - 3$, there is a critical point at $\left(\frac{3}{4}, -\frac{81}{8}\right) = (0.75, -10.125)$. The graph of f is decreasing to the left of this point and increasing to its right, so there is a global minimum at this critical point and no other extrema. The graph of $y = f(x)$ is shown next.

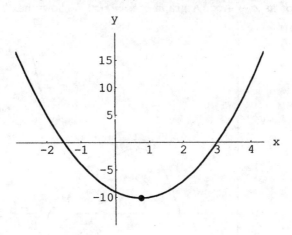

4.5.33: $f'(x) = 6(x-1)(x+2)$, so the graph of f is increasing for $x < -2$ and for $x > 1$, decreasing if $-2 < x < 1$. Hence there is a local maximum at $(-2, 20)$ and a local minimum at $(1, -7)$. The first is not a global maximum because $f(10) = 2180 > 20$; the second is not a global minimum because $f(-10) = -1580 < -7$. The graph of $y = f(x)$ is shown next.

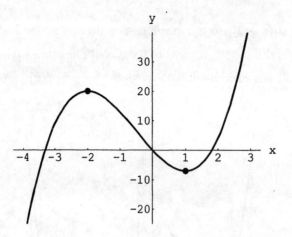

4.5.35: $f'(x) = 6(5x - 3)(5x - 4)$, so there are critical points at $(0.6, 16.2)$ and $(0.8, 16)$. The graph of f is increasing for $x < 0.6$ and for $x > 0.8$; it is decreasing between these two points. Hence there is a local maximum at the first and a local minimum at the second. Neither is global because

$$\lim_{x \to \infty} f(x) = \lim_{x \to \infty} x^3 \left(50 - \frac{105}{x} + \frac{72}{x^2} \right) = +\infty$$

and, similarly, $f(x) \to -\infty$ as $x \to -\infty$. A graph of $y = f(x)$ is shown next.

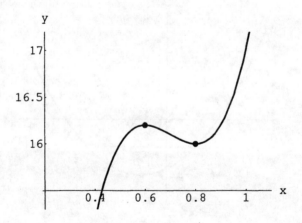

4.5.37: $f'(x) = 12x(x-2)(x+1)$, so f is decreasing for $x < -1$ and for $0 < x < 2$, increasing for $-1 < x < 0$ and for $x > 2$. There is a local minimum at $(-1, 3)$, a local maximum at $(0, 8)$, and a global minimum at $(2, -24)$. The latter is global rather than local because $f(x) \to +\infty$ as $x \to \pm\infty$. The graph of f is shown next.

158

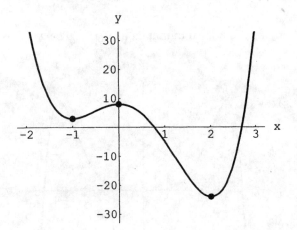

4.5.39: $f'(x) = 15x^2(x-2)(x+2)$, so f is increasing if $|x| > 2$ and decreasing if $|x| < 2$. There is a local maximum at $(-2, 64)$ and a local minimum at $(2, -64)$; the critical point at $(0, 0)$ is not an extremum. The graph of $y = f(x)$ appears next.

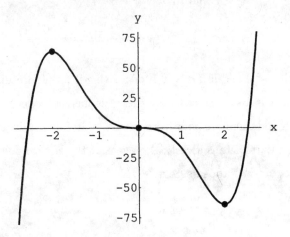

4.5.41: $f'(x) = 6(x^2 + x + 1)$ is positive for all x, so the graph of f is increasing everywhere, with no critical points and thus no extrema. The only intercept is $(0, 0)$. The graph of f is shown next.

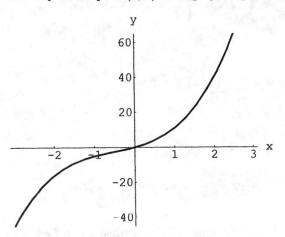

4.5.43: $f'(x) = 32x^3 - 8x^7 = -8x^3(x^2 + 2)(x^2 - 2)$, so the graph of f is increasing if $x < -\sqrt{2}$ and if $0 < x < \sqrt{2}$ but decreasing if $-\sqrt{2} < x < 0$ and if $x > \sqrt{2}$. The global maximum value of $f(x)$ is

$16 = f(\sqrt{2}) = f(-\sqrt{2})$ and there is a local minimum at $(0, 0)$. The graph of f is shown next.

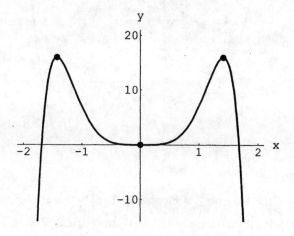

4.5.45: Given $f(x) = x^{1/3}(4 - x)$, we find that

$$f'(x) = \frac{1}{3}x^{-2/3}(4 - x) - x^{1/3} = \frac{4(1 - x)}{3x^{2/3}}.$$

Therefore f is increasing if $x < 0$ and if $0 < x < 1$, decreasing if $x > 1$. Because f is continuous everywhere, including the point $x = 0$, it is also correct to say that f is increasing on $(-\infty, 1)$. The point $(1, 3)$ is the highest point on the graph and there are no other extrema. Careful examination of the behavior of $f(x)$ and $f'(x)$ for x near zero shows that there is a vertical tangent at the critical point $(0, 0)$. The point $(4, 0)$ is an x-intercept. The graph of $y = f(x)$ appears next.

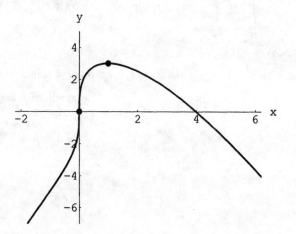

4.5.47: Given $f(x) = x(x - 1)^{2/3}$, we find that

$$f'(x) = (x - 1)^{2/3} + \frac{2}{3}x(x - 1)^{-1/3} = \frac{5x - 3}{3(x - 1)^{1/3}}.$$

Thus $f'(x) = 0$ when $x = \frac{3}{5}$ and $f'(x)$ does not exist when $x = 1$. So f is increasing for $x < \frac{3}{5}$ and for $x > 1$, decreasing if $\frac{3}{5} < x < 1$. Thus there is a local maximum at $\left(\frac{3}{5}, 0.3257\right)$ (y-coordinate approximate) and a local minimum at $(1, 0)$. Examination of $f(x)$ and $f'(x)$ for x near 1 shows that there is a vertical tangent

160

at $(1, 0)$. There are no global extrema because $f(x) \to +\infty$ as $x \to +\infty$ and $f(x) \to -\infty$ as $x \to -\infty$. The graph of f is shown next.

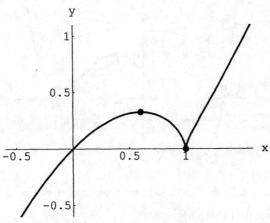

4.5.49: The graph of $f(x) = 2x^3 + 3x^2 - 36x - 3$ is shown next.

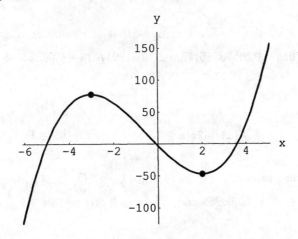

4.5.51: The graph of $f(x) = -2x^3 - 3x^2 + 36x + 15$ is shown next.

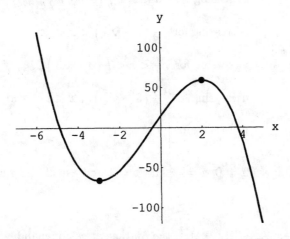

4.5.53: The graph of $f(x) = 3x^4 - 8x^3 - 30x^2 + 72x + 45$ is next.

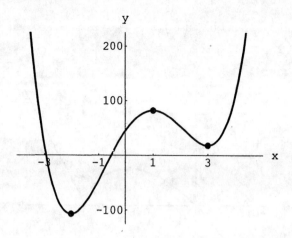

4.5.55: Let $f(x) = x^3 - 3x + 3$. For part (a), we find that if we let $x = -2.1038034027$, then $f(x) \approx 7.58 \times 10^{-9}$. For part (b), we find that

$$f(x) \approx (x + 2.1038034027)(x^2 - (2.1038034027)x + 1.4259887573).$$

And in part (c), we find by the quadratic formula that the complex conjugate roots of $f(x) = 0$ are approximately $1.0519017014 + 0.5652358517i$ and $1.0519017014 - 0.5652358517i$.

4.5.57: If $f(x) = [x(x-1)(2x-1)]^2$, then

$$f'(x) = 2x(x-1)(2x-1)(6x^2 - 6x + 1),$$

so the critical points of the graph of f will be $(0, 0)$, $\left(\frac{1}{2}, 0\right)$, $(1, 0)$, $\left(\frac{1}{6}\left[3 - \sqrt{3}\right], 0.009259259\right)$ (ordinate approximate), and $\left(\frac{1}{6}\left[3 + \sqrt{3}\right], 0.009259259\right)$. The graph of f will be

$$\text{decreasing for} \quad x < 0,$$
$$\text{increasing for} \quad 0 < x < \frac{1}{6}\left(3 - \sqrt{3}\right),$$
$$\text{decreasing for} \quad \frac{1}{6}\left(3 - \sqrt{3}\right) < x < \frac{1}{2},$$
$$\text{increasing for} \quad \frac{1}{2} < x < \frac{1}{6}\left(3 + \sqrt{3}\right),$$
$$\text{decreasing for} \quad \frac{1}{6}\left(3 + \sqrt{3}\right) < x < 1, \quad \text{and}$$
$$\text{increasing for} \quad 1 < x.$$

There will be global minima at $x = 0$, $x = \frac{1}{2}$, and $x = 1$ and [equal] local maxima at $x = \frac{1}{6}\left(3 - \sqrt{3}\right)$ and $x = \frac{1}{6}\left(3 + \sqrt{3}\right)$.

4.5.59: Given $f(x) = \left[\frac{1}{6}x(9x - 5)(x - 1)\right]^4$, the *Mathematica* command

```
Plot[ f[x], { x, -1, 2 }, PlotPoints → 97,
```

162

```
PlotRange → {{ −0.6, 1.6 }, { −0.1, 1.1 }} ];
```

generated the graph below. As predicted, the graph appears to have a "flat spot" on the interval $[0, 1]$.

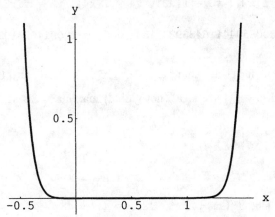

Then we modified the `Plot` command to restrict the range of y-values to the interval $[−0.00001, 0.00011]$:

```
Plot[ f[x], { x, −1, 2 }, PlotPoints → 197,

        PlotRange → {{ −0.2, 1.2 }, { −0.00001, 0.00011 }} ];
```

Then we used *Mathematica* to identify the extrema:

```
soln = Solve[ f'[x] == 0, x ]
```

$$\{\{ x \to 0 \}, \{ x \to 0 \}, \{ x \to 0 \}, \{ x \to \frac{5}{9} \}, \{ x \to \frac{5}{9} \},$$

$$\{ x \to \frac{5}{9} \}, \{ x \to 1 \}, \{ x \to 1 \}, \{ x \to 1 \},$$

$$\{ x \to \frac{14 - \text{Sqrt}[61]}{27} \}, \{ x \to \frac{14 + \text{Sqrt}[61]}{27} \}\}$$

```
{ x1 = soln[[1,1,2]], x2 = soln[[4,1,2]], x3 = soln[[7,1,2]],

x4 = N[soln[[10,1,2]], 20], x5 = N[soln[[11,1,2]], 20] }
```

163

$$\{ \ 0, \ \frac{5}{9}, \ 1, \ 0.22925001200345724466, \ 0.8077870250335797924 \ \}$$

$$\{ \ \text{y1 = f[x1], y2 = f[x2], y3 = f[x3], y4 = f[x4], y5 = f[x5]} \ \}$$

$$\{ \ 0, \ 0, \ 0, \ 0.00005594411164359303138, \ 0.0000119091402810978625 \ \}$$

The second graph makes it clear that $(x_1, 0)$, $(x_2, 0)$, and $(x_3, 0)$ are local (indeed, global) minima, while (x_4, y_4) and (x_5, y_5) are local (not global) maxima.

Section 4.6

4.6.1: $f'(x) = 8x^3 - 9x^2 + 6$, $f''(x) = 24x^2 - 18x$, $f'''(x) = 48x - 18$.

4.6.3: $f'(x) = -8(2x - 1)^{-3}$, $f''(x) = 48(2x - 1)^{-4}$, $f'''(x) = -384(2x - 1)^{-5}$.

4.6.5: $g'(t) = 4(3t - 2)^{1/3}$, $g''(t) = 4(3t - 2)^{-2/3}$, $g'''(t) = -8(3t - 2)^{-5/3}$.

4.6.7: $h'(y) = (y + 1)^{-2}$, $h''(y) = -2(y + 1)^{-3}$, $h'''(y) = 6(y + 1)^{-4}$.

4.6.9: $g'(t) = t(1 + 2\ln t)$, $g''(t) = 3 + 2\ln t$, $g'''(t) = \dfrac{2}{t}$.

4.6.11: $f'(x) = 3\cos 3x$, $f''(x) = -9\sin 3x$, $f'''(x) = -27\cos 3x$.

4.6.13: $f'(x) = \cos^2 x - \sin^2 x$, $f''(x) = -4\sin x \cos x$, $f'''(x) = 4\sin^2 x - 4\cos^2 x$.

4.6.15: $f'(x) = \dfrac{x\cos x - \sin x}{x^2}$, $f''(x) = \dfrac{(2 - x^2)\sin x - 2x\cos x}{x^3}$,

$f'''(x) = \dfrac{(6 - x^2)x\cos x + (3x^2 - 6)\sin x}{x^4}$.

4.6.17: Given: $x^2 + xy + y^2 = 3$.

$$2x + y + x\frac{dy}{dx} + 2y\frac{dy}{dx} = 0, \qquad \text{so} \qquad y'(x) = -\frac{2x + y}{x + 2y}.$$

$$y''(x) = -\frac{2(1 + y'(x) + [y'(x)]^2)}{x + 2y} = -\frac{6(x^2 + xy + y^2)}{x + 2y} = -\frac{18}{(x + 2y)^3}.$$

4.6.19: Given: $y^3 + x^2 + x = 5$.

$$3y^2\frac{dy}{dx} + 2x + 1 = 0, \qquad \text{so} \qquad y'(x) = -\frac{2x + 1}{3y^2}.$$

$$y''(x) = -\frac{2(1 + 3y(x)[y'(x)]^2)}{3[y(x)]^2} = -\frac{2[(2x + 1)^2 + 3y^3]}{9y^5}.$$

164

4.6.21: Given: $\sin y = xy$.

$$(\cos y)\frac{dy}{dx} = y + x\frac{dy}{dx}, \qquad \text{so} \qquad y'(x) = -\frac{y}{x - \cos y}.$$

$$y''(x) = -\frac{[x - \cos y(x)]y'(x) - y(x)[1 + y'(x)\sin y]}{(x - \cos y)^2} = -\frac{(y\sin y + 2\cos y - 2x)y}{(x - \cos y)^3}.$$

4.6.23: $f'(x) = 3x^2 - 6x - 45 = 3(x + 3)(x - 5)$, so there are critical points at $(-3, 81)$ and $(5, -175)$. $f''(x) = 6(x - 1)$, so the inflection point is located at $(1, -47)$.

4.6.25: $f'(x) = 12x^2 - 12x - 189 = 3(2x + 7)(2x - 9)$, so there are critical points at $(-3.5, 553.5)$ and $(4.5, -470.5)$. $f''(x) = 24x - 12$, so the inflection point is located at $(0.5, 41.5)$.

4.6.27: $f'(x) = 4x^3 - 108x = 4x(x^2 - 27)$, so there are critical points at $(0, 237)$, $(-3\sqrt{3}, -492)$, and $(3\sqrt{3}, -492)$. Next, $f''(x) = 12x^2 - 108 = 12(x - 3)(x + 3)$, so the inflection points are at $(-3, -168)$ and $(3, -168)$.

$$x - \sqrt{27}$$
$$x + \sqrt{27}$$

4.6.29: $f'(x) = 15x^4 - 80x^3 = 5x^3(3x - 16)$, so there are critical points at $(0, 1000)$ and $\left(\frac{16}{3}, -\frac{181144}{81}\right)$ (approximately $(5.333333, -2236.345679)$). Now $f''(x) = 60x^3 - 240x^2 = 60x^2(x - 4)$, so the *possible* inflection points are $(0, 1000)$ and $(4, -1048)$. However, only the latter is an actual inflection point.

4.6.31: $f'(x) = 2x - 4$ and $f''(x) \equiv 2$, so there is a critical point at $(2, -1)$ and no inflection points. Because $f''(2) = 2 > 0$, there is a local minimum at the critical point. The first derivative test shows that it is in fact a global minimum.

4.6.33: $f'(x) = 3(x + 1)(x - 1)$ and $f''(x) = 6x$. $f(-1) = 3$ and $f''(-1) = -6$, so the critical point at $(-1, 3)$ is a local maximum. Similarly, the critical point at $(1, -1)$ is a local minimum. The point $(0, 1)$ is an inflection point because $f''(x) < 0$ for $x < 0$ and $f''(x) > 0$ for $x > 0$. The extrema are not global because $f(x) \to +\infty$ as $x \to +\infty$ and $f(x) \to -\infty$ as $x \to -\infty$.

4.6.35: If $f(x) = xe^{-x}$, then $f'(x) = (1 - x)e^{-x}$ and $f''(x) = (x - 2)e^{-x}$. Hence f is increasing for $x < 1$ and decreasing for $x > 1$; its graph is concave downward for $x < 2$ and concave upward for $x > 2$. Hence there is a global maximum at $(1, e^{-1})$ and an inflection point at $(2, 2e^{-2})$. The graph of $y = f(x)$ is next.

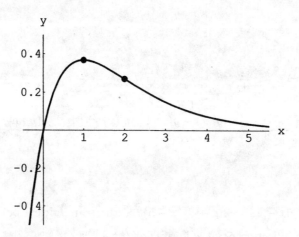

4.6.37: $f'(x) = 5x^4 + 2$ and $f''(x) = 20x^3$, so there are no critical points ($f'(x) > 0$ for all x) and $(0, 0)$ is the only possible inflection point. And it is an inflection point because $f''(x)$ changes sign at $x = 0$.

4.6.39: $f'(x) = 2x(x-1)(2x-1)$ and $f''(x) = 2(6x^2 - 6x + 1)$. So the critical points are $(0, 0)$, $\left(\frac{1}{2}, \frac{1}{16}\right)$, and $(1, 0)$. The second derivative test indicates that the first and third are local minima and that the second is a local maximum. The only possible inflection points are $\left(\frac{1}{6}\left(3 - \sqrt{3}\right), \frac{1}{36}\right)$ and $\left(\frac{1}{6}\left(3 + \sqrt{3}\right), \frac{1}{36}\right)$. Why are they both inflection points? The graph of $f''(x) = 2(6x^2 - 6x + 1)$ is a parabola opening upward and with its vertex below the x-axis (because $f''(x) = 0$ has two real solutions). The possible inflection points are located at the zeros of $f''(x)$, so it should now be clear that $f''(x)$ changes sign at each of the two possible inflection points. Because $f(x)$ is never negative, the two local minima are actually global, but the local maximum is not (because $f(x) \to +\infty$ as $x \to \pm\infty$).

4.6.41: $f'(x) = \cos x$ and $f''(x) = -\sin x$. $f'(x) = 0$ when $x = \pi/2$ and when $x = 3\pi/2$; $f''(\pi/2) = -1 < 0$ and $f''(3\pi/2) = 1 > 0$, so the first of these critical points is a local maximum and the second is a local minimum. Because $|\sin x| \leq 1$ for all x, these extrema are in fact global. $f''(x) = 0$ when $x = \pi$ and clearly changes sign there, so there is an inflection point at $(\pi, 0)$.

4.6.43: $f'(x) = \sec^2 x \geq 1$ for $-\pi/2 < x < \pi/2$, so there are no extrema. $f''(x) = 2\sec^2 x \tan x$, so there is a possible inflection point at $(0, 0)$. Because $\tan x$ changes sign at $x = 0$ (and $2\sec^2 x$ does not), $(0, 0)$ is indeed an inflection point.

4.6.45: $f'(x) = -2\sin x \cos x$ and $f''(x) = 2\sin^2 x - 2\cos^2 x$. Hence $(0, 1)$, $(\pi/2, 0)$, and $(\pi, 1)$ are critical points. By the second derivative test the first and third are local maxima and the second is a local minimum. (Because $0 \leq \cos^2 x \leq 1$ for all x, these extrema are all global.) There are possible inflection points at $(-\pi/4, 1/2)$, $(\pi/4, 1/2)$, $(3\pi/4, 1/2)$, and $(5\pi/4, 1/2)$. A close examination of $f''(x)$ reveals that it changes sign at all four of these points, so each is an inflection point.

4.6.47: If $f(x) = 10(x - 1)e^{-2x}$, then

$$f'(x) = 10(3 - 2x)e^{-2x} \qquad \text{and} \qquad f''(x) = 40(x - 2)e^{-2x}.$$

166

Thus the graph of f is increasing for $0 < x < \frac{3}{2}$ and decreasing for $x > \frac{3}{2}$. It is concave downward for $0 < x < 2$ and concave upward if $x > 2$. Hence there is a global maximum at $(1.5, 5e^{-3}) \approx (1.5, 0.248935)$ and an inflection point at $(2, 10e^{-4}) \approx (2, 0.183156)$. The graph of f is next.

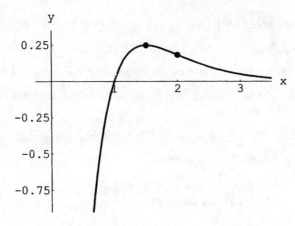

4.6.49: If $f(x) = (x^2 - 2x - 1)e^{-x}$, then

$$f'(x) = -(x^2 - 4x + 1)e^{-x} \quad \text{and} \quad f''(x) = (x-1)(x-5)e^{-x}.$$

Thus the graph of f is decreasing for $x < 2 - \sqrt{3}$ and for $x > 2 + \sqrt{3}$, increasing otherwise; it is concave upward for $x < 1$ and for $x > 5$ and concave downward otherwise. Thus there is a global minimum near $\left(2 - \sqrt{3}, -1.119960\right)$, a local maximum near $\left(2 + \sqrt{3}, 0.130831\right)$, and inflection points at $(1, -2e^{-1})$ and $(5, 14e^{-5})$. The graph of f is next.

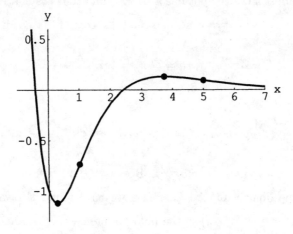

4.6.51: We are to minimize the product of two numbers whose difference is 20; thus if x is the smaller, we are to minimize $f(x) = x(x + 20)$. Now $f'(x) = 20 + 2x$, so $f'(x) = 0$ when $x = -10$. But $f''(x) \equiv 2$ is positive when $x = -10$, so there is a local minimum at $(-10, -100)$. Because the graph of f is a parabola opening upward, this local minimum is in fact the global minimum. Answer: The two numbers are -10 and 10.

4.6.53: Let us minimize

$$g(x) = (x-3)^2 + (3 - 2x - 2)^2 = (x-3)^2 + (1 - 2x)^2,$$

the square of the distance from the point (x, y) on the line $2x + y = 3$ to the point $(3, 2)$. We have $g'(x) = 2(x-3) - 4(1 - 2x) = 10x - 10$; $g''(x) \equiv 10$. So $x = 1$ is the only critical point of g. Because $g''(x)$ is always positive, the graph of g is concave upward on the set $\mathbb{R}$ of all real numbers, and therefore $(1, g(1)) = (1, 1)$ yields the global minimum for g. So the point on the given line closest to $(3, 2)$ is $(1, 1)$.

4.6.55: Base of box: x wide, $2x$ long. Height: y. Then the box has volume $2x^2 y = 972$, so $y = 486x^{-2}$. Its total surface area is $A = 2x^2 + 6xy$, so we minimize

$$A = A(x) = 2x^2 + \frac{2916}{x}, \qquad x > 0.$$

Now

$$A'(x) = 4x - \frac{2916}{x^2} \qquad \text{and} \qquad A''(x) = 4 + \frac{5832}{x^3}.$$

The only critical point of A occurs when $x = 9$, and $A''(x)$ is always positive. So the graph of $y = A(x)$ is concave upward for all $x > 0$; consequently, $(9, A(9))$ is the lowest point on the graph of A. Answer: The dimensions of the box are 9 inches wide, 18 inches long, and 6 inches high.

4.6.57: Let r denote the radius of the pot and h its height. We are given the constraint $\pi r^2 h = 250$, so $h = 250/(\pi r^2)$. Now the bottom of the pot has area πr^2, and thus costs $4\pi r^2$ cents. The curved side of the pot has area $2\pi rh$, and thus costs $4\pi rh$ cents. So the total cost of the pot is

$$C = 4\pi r^2 + 4\pi rh = C(r) = 4\pi r^2 + \frac{1000}{r}, \qquad r > 0.$$

Now

$$C'(r) = 8\pi r - \frac{1000}{r^2} \qquad \text{and} \qquad C''(r) = 8\pi + \frac{2000}{r^3}.$$

$C'(r) = 0$ when $8\pi r^3 = 1000$, so that $r = 5/\sqrt[3]{\pi}$. Because $C''(r) > 0$ for all $r > 0$, the graph of $y = C(r)$ is concave upward on the domain of C. Therefore we have found the value of r that minimizes $C(r)$. The corresponding value of h is $10/\sqrt[3]{\pi}$, so the pot of minimal cost has height equal to its diameter, each approximately 6.828 centimeters.

4.6.59: Let the square base of the box have edge length x and let its height be y, so that its total volume is $x^2 y = 62.5$ and the surface area of this box-without-top will be $A = x^2 + 4xy$. So

$$A = A(x) = x^2 + \frac{250}{x}, \qquad x > 0.$$

168

Now

$$A'(x) = 2x - \frac{250}{x^2} \qquad \text{and} \qquad A''(x) = 2 + \frac{500}{x^3}.$$

The only critical point occurs when $x = 5$, and $A''(x) > 0$ for all x in the domain of A, so $x = 5$ yields the global minimum for A. Answer: Square base of edge length $x = 5$ inches, height $y = 2.5$ inches.

4.6.61: Let x denote the radius and y the height of the cylinder (in inches). Then its cost (in cents) is $C = 8\pi x^2 + 4\pi xy$, and we also have the constraint $\pi x^2 y = 100$. So

$$C = C(x) = 8\pi x^2 + \frac{400}{x}, \qquad x > 0.$$

Now

$$C'(x) = 16\pi x - \frac{400}{x^2} \qquad \text{and} \qquad C''(x) = 16\pi + \frac{800}{x^3}.$$

The only critical point in the domain of C is $x = \sqrt[3]{25/\pi}$ (about 1.9965 inches) and, consequently, when $y = \sqrt[3]{1600/\pi}$ (about 7.9859 inches). Because $C''(x) > 0$ for all x in the domain of C, we have indeed found the dimensions that minimize the cost of the can. For simplicity, note that $y = 4x$ at the minimum: The height of the can is twice its diameter.

4.6.63: Given: $f(x) = 2x^3 - 3x^2 - 12x + 3$. We have

$$f(x) = 6(x - 2)(x + 1) \qquad \text{and} \qquad f''(x) = 12x - 6.$$

Hence $(-1, 10)$ and $(2, -17)$ are critical points and $(0.5, -3.5)$ is a possible inflection point. Because $f''(x) > 0$ if $x > 0.5$ and $f''(x) < 0$ if $x < 0.5$, the possible inflection point is an actual inflection point, there is a local maximum at $(-1, 10)$, and a local minimum at $(2, -17)$. The extrema are not global because $f(x) \to +\infty$ as $x \to +\infty$ and $f(x) \to -\infty$ as $x \to -\infty$. The graph of f is next.

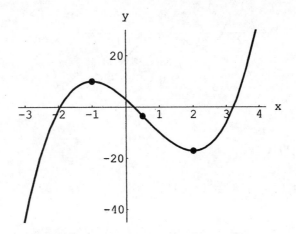

4.6.65: If $f(x) = 6 + 8x^2 - x^4$, then $f'(x) = -4x(x + 2)(x - 2)$ and $f''(x) = 16 - 12x^2$. So f is increasing for $x < -2$ and for $0 < x < 2$, decreasing otherwise; its graph is concave upward on $\left(-\frac{2}{3}\sqrt{3}, \frac{2}{3}\sqrt{3}\right)$ and concave downward otherwise. Therefore the global maximum value of f is $f(-2) = f(2) = 22$ and there is a local minimum at $f(0) = 6$. There are inflection points at $\left(-\frac{2}{3}\sqrt{3}, \frac{134}{9}\right)$ and at $\left(\frac{2}{3}\sqrt{3}, \frac{134}{9}\right)$. The graph of f is next.

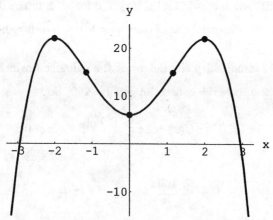

4.6.67: If $f(x) = 3x^4 - 4x^3 - 12x^2 - 1$, then

$$f'(x) = 12x^3 - 12x^2 - 24x = 12x(x-2)(x+1) \quad \text{and} \quad f''(x) = 36x^2 - 24x - 24 = 12(3x^2 - 2x - 2).$$

So the graph of f is decreasing for $x < -1$ and for $0 < x < 2$ and increasing otherwise; it is concave upward for $x < \frac{1}{3}\left(1 - \sqrt{7}\right)$ and for $x > \frac{1}{3}\left(1 + \sqrt{7}\right)$ and concave downward otherwise. So there is a local minimum at $(-1, -6)$, a local maximum at $(0, -1)$, and a global minimum at $(2, -33)$. There are inflection points at $\left(\frac{1}{3}\left(1 - \sqrt{7}\right), \frac{1}{27}\left(-311 + 80\sqrt{7}\right)\right)$ and at $\left(\frac{1}{3}\left(1 + \sqrt{7}\right), \frac{1}{27}\left(-311 - 80\sqrt{7}\right)\right)$. The graph of $y = f(x)$ is next.

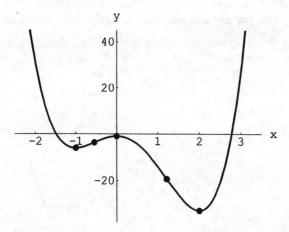

4.6.69: If $f(x) = x^3(x-1)^4$, then

$$f'(x) = 3x^2(x-1)^4 + 4x^3(x-1)^3 = x^2(x-1)^3(7x-3) \qquad \text{and}$$

$$f''(x) = 6x(x-1)^4 + 24x^2(x-1)^3 + 12x^3(x-1)^2 = 6x(x-1)^2(7x^2 - 6x + 1).$$

Hence the graph of f is increasing for $x < \frac{3}{7}$ and for $x > 1$, decreasing otherwise; concave upward for $0 < x < \frac{1}{7}\left(3 - \sqrt{2}\right)$ and for $x > \frac{1}{7}\left(3 + \sqrt{2}\right)$. So there is a local maximum at $\left(\frac{3}{7}, \frac{6912}{823543}\right)$ and a local minimum at $(1, 0)$. Also there are inflection points at $(0, 0)$ and at the two points with x-coordinates $\frac{1}{7}(3 \pm \sqrt{2})$. The graph of $y = f(x)$ is shown next. Note the scale on the y-axis.

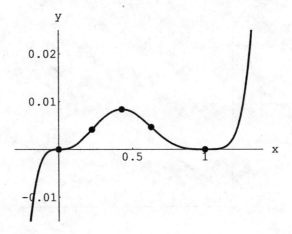

4.6.71: If $f(x) = 1 + x^{1/3}$ then

$$f'(x) = \frac{1}{3}x^{-2/3} = \frac{1}{3x^{2/3}} \quad \text{and} \quad f''(x) = -\frac{2}{9}x^{-5/3} = -\frac{2}{9x^{5/3}}.$$

Therefore $f'(x) > 0$ for all $x \neq 0$; because f is continuous even at $x = 0$, the graph of f is increasing for all x, but $(0, 1)$ is a critical point. Because $f''(x)$ has the sign of $-x$, the graph of f is concave upward for $x < 0$ and concave downward for $x > 0$. Thus there is an inflection point at $(0, 1)$. Careful examination of the first derivative shows also that there is a vertical tangent at $(0, 1)$. The graph is next.

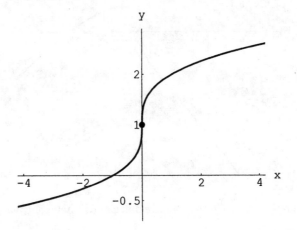

4.6.73: Given $f(x) = (x + 3)\sqrt{x}$,

$$f'(x) = \frac{3(x + 1)}{2\sqrt{x}} \quad \text{and} \quad f''(x) = \frac{3(x - 1)}{4x\sqrt{x}}.$$

Note that $f(x)$ has domain $x \geq 0$. Hence $f'(x) > 0$ for all $x > 0$; in fact, because f is continuous (from the right) at $x = 0$, f is increasing on $[0, +\infty)$. It now follows that $(1, 4)$ is an inflection point, but the change in the concavity of the graph there — from bending downward on the left to upward on the right —

171

is visually almost imperceptible. The point $(0, 0)$ is, of course, the location of the global minimum of f and is of particular interest because $f'(x) \to +\infty$ as $x \to 0^+$. The graph of $y = f(x)$ is shown next.

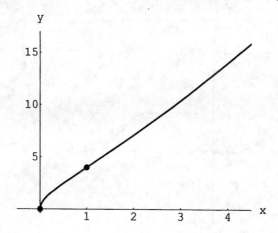

4.6.75: Given $f(x) = (4 - x)x^{1/3}$, we have

$$f'(x) = \frac{4(1-x)}{3x^{2/3}} \qquad \text{and} \qquad f''(x) = -\frac{4(x+2)}{9x^{5/3}}.$$

There is a global maximum at $(1, 3)$, a vertical tangent, dual intercept, and inflection point at $(0, 0)$, an x-intercept at $(4, 0)$, and an inflection point at $\left(-2, -6\sqrt[3]{2}\right)$. The graph of f is next.

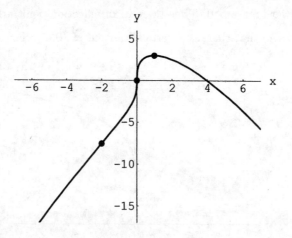

4.6.77: Figure 4.6.34 shows a graph that is concave downward, then concave upward, so its second derivative is negative, then zero, then positive. This matches the graph in (c).

4.6.79: Figure 4.6.36 shows a graph that is concave upward, then downward, then upward again, so the second derivative will be positive, then negative, then positive again. This matches the graph in (b).

4.6.81: Figure 4.6.38 shows a graph that is concave upward, then almost straight, then strongly concave downward, so the second derivative must be positive, then close to zero, then large negative. This matches the graph in (d).

4.6.83: (a): Proof: The result holds when $n = 1$. Suppose that it holds for $n = k$ where $k \geqq 1$. Then $f^{(k)}(x) = k!$ if $f(x) = x^k$. Now if $g(x) = x^{k+1}$, then $g(x) = xf(x)$. So by the product rule,

$$g'(x) = xf'(x) + f(x) = x\,(kx^{k-1}) + x^k = (k+1)x^k.$$

Thus

$$g^{(k+1)}(x) = (k+1)D_x^k(x^k) = (k+1)f^{(k)}(x) = (k+1)(k!) = (k+1)!.$$

That is, whenever the result holds for $n = k$, it follows for $n = k + 1$. Therefore, by induction, it holds for all integers $n \geqq 1$.

(b): Because the nth derivative of x^n is constant, any higher order derivative of x^n is zero. The result now follows immediately.

4.6.85: $\dfrac{dz}{dx} = \dfrac{dz}{dy} \cdot \dfrac{dy}{dx}$. So $\dfrac{d^2z}{dx^2} = \dfrac{dz}{dy} \cdot \dfrac{d^2y}{dx^2} + \dfrac{dy}{dx} \cdot \dfrac{d^2z}{dy^2} \cdot \dfrac{dy}{dx}$.

4.6.87: If $f(x) = ax^3 + bx^2 + cx + d$ with $a \neq 0$, then both $f'(x)$ and $f''(x)$ exist for all x and $f''(x) = 6ax + 2b$. The latter is zero when and only when $x = -b/(3a)$, and this is the abscissa of an inflection point because $f''(x)$ changes sign at $x = -b/(3a)$. Therefore the graph of a cubic polynomial has exactly one inflection point.

4.6.89: First, $p = p(V) = \dfrac{RT}{V - b} - \dfrac{a}{V^2}$, so

$$p'(V) = \frac{2a}{V^3} - \frac{RT}{(V - b)^2} \quad \text{and} \quad p''(V) = \frac{2RT}{(V - b)^3} - \frac{6a}{V^4}.$$

From now on, use the constant values $p = 72.8$, $V = 128.1$, and $T = 304$; we already have $n = 1$. Then

$$p = \frac{RT}{V - b} - \frac{a}{V^2}, \quad \frac{2a}{V^3} = \frac{RT}{(V - b)^2}, \quad \text{and} \quad \frac{3a}{V^4} = \frac{RT}{(V - b)^3}.$$

The last two equations yield

$$\frac{RTV^3}{a(V - b)^2} = 2 = \frac{3(V - b)}{V},$$

and thus $b = \frac{1}{3}V$ and $V - b = \frac{2}{3}V$. Next,

$$a = \frac{V^3RT}{2(V - b)^2} = \frac{V^3RT}{2(2V/3)^2} = \frac{9V^3RT}{8V^2} = \frac{9}{8}VRT.$$

Finally, $\dfrac{RT}{V - b} = p + \dfrac{a}{V^2}$, so

$$R = \frac{2V}{3T}\left(p + \frac{a}{V^2}\right) = \frac{2V}{3T}\left(p + \frac{9RT}{8V}\right) = \frac{2Vp}{3T} + \frac{3R}{4}.$$

Therefore $R = \dfrac{8Vp}{3T}$. We substitute this into the earlier formula for a, in order to determine that $a = \frac{9}{8}VRT = 3V^2p$. In summary, and using the values given in the problem, we find that

$$b = \frac{1}{3}V = 42.7, \quad a = 3V^2p \approx 3{,}583{,}859, \quad \text{and} \quad R = \frac{8Vp}{3T} \approx 81.8.$$

4.6.91: If $f(x) = 1000x^3 - 3051x^2 + 3102x + 1050$, then

$$f'(x) = 3000x^2 - 6102x + 3102 \quad \text{and} \quad f''(x) = 6000x - 6102.$$

So the graph of f has horizontal tangents at the two points $(1, 2101)$ and $(1.034, 2100.980348)$ (coordinates exact) and there is a possible inflection point at $(1.017, 2100.990174)$ (coordinates exact). Indeed, the usual tests show that the first of these is a local maximum, the second is a local minimum, and the third is an inflection point. The *Mathematica* command

```
Plot[ f[x], { x, 0.96, 1.07 } ];
```

produces a graph that shows all three points clearly; it's next.

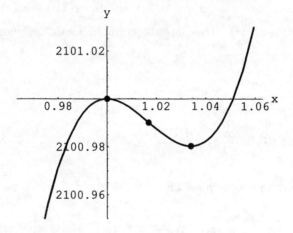

Section 4.7

4.7.1: It is almost always a good idea, when evaluating the limit of a *rational* function of x as $x \to \pm\infty$, to divide each term in numerator and denominator by the highest power of x that appears there. This technique occasionally succeeds with more complicated functions. Here we obtain

$$\lim_{x \to \infty} \frac{x}{x+1} = \lim_{x \to \infty} \frac{\dfrac{x}{x}}{\dfrac{x}{x} + \dfrac{1}{x}} = \lim_{x \to \infty} \frac{1}{1 + \dfrac{1}{x}} = \frac{1}{1+0} = 1.$$

4.7.3: $\dfrac{x^2 + x - 2}{x - 1} = \dfrac{(x+2)(x-1)}{x-1} = x + 2$ for $x \neq 1$, so $\dfrac{x^2 + x - 2}{x - 1} \to 1 + 2 = 3$ as $x \to 1$.

4.7.5: $\dfrac{2x^2 - 1}{x^2 - 3x} = \dfrac{2 - (1/x^2)}{1 - (3/x)} \to \dfrac{2 - 0}{1 - 0} = 2$ as $x \to +\infty$.

4.7.7: The numerator is equal to the denominator for all $x \neq -1$, so the limit is 1. If you'd prefer to write the solution more symbolically, try this:

$$\lim_{x \to -1} \frac{x^2 + 2x + 1}{(x+1)^2} = \lim_{x \to -1} \frac{x^2 + 2x + 1}{x^2 + 2x + 1} = \lim_{x \to -1} 1 = 1.$$

4.7.9: Factor the numerator: $x - 4 = (\sqrt{x} + 2)(\sqrt{x} - 2)$. Thus the fraction is equal to $\sqrt{x} + 2$ if $x \neq 4$. Therefore as $x \to 4$, the fraction approaches $\sqrt{4} + 2 = 4$. Alternatively, for a more symbolic solution, write

$$\lim_{x \to 4} \frac{x - 4}{\sqrt{x} - 2} = \lim_{x \to 4} \frac{(\sqrt{x} + 2)(\sqrt{x} - 2)}{\sqrt{x} - 2} = \lim_{x \to 4} (\sqrt{x} + 2) = \sqrt{4} + 2 = 2 + 2 = 4.$$

For another approach, use the conjugate (as discussed in Chapter 2):

$$\lim_{x \to 4} \frac{x - 4}{\sqrt{x} - 2} = \lim_{x \to 4} \frac{(x - 4)(\sqrt{x} + 2)}{(\sqrt{x} - 2)(\sqrt{x} + 2)} = \lim_{x \to 4} \frac{(x - 4)(\sqrt{x} + 2)}{x - 4} = \lim_{x \to 4} (\sqrt{x} + 2) = \sqrt{4} + 2 = 4.$$

4.7.11: Divide each term in numerator and denominator by x to obtain

$$\frac{\left(\dfrac{8}{x}\right) - \left(\dfrac{1}{\sqrt[3]{x^2}}\right)}{\left(\dfrac{2}{x}\right) + 1} \to \frac{0 - 0}{0 + 1} = 0 \quad \text{as} \quad x \to -\infty.$$

4.7.13: Ignoring the radical for a moment, we divide each term in numerator and denominator by x^2, the higest power of x that appears in either. (Remember that this technique is effective with limits as $x \to \pm\infty$ but is not likely to be productive in other cases.) Result:

$$\lim_{x \to +\infty} \sqrt{\frac{4x^2 - x}{x^2 + 9}} = \lim_{x \to +\infty} \sqrt{\frac{4 - (1/x)}{1 + (9/x^2)}} = \sqrt{\frac{4 - 0}{1 + 0}} = \sqrt{4} = 2.$$

4.7.15: As $x \to -\infty$, $x^2 + 2x = x(x + 2)$ is the product of two very large negative numbers, so $x^2 + 2x \to +\infty$. Therefore

$$\lim_{x \to -\infty} \sqrt{x^2 + 2x} = +\infty$$

as well. If the large *positive* number $-x$ is added to $\sqrt{x^2 + 2x}$, then the resulting sum also approaches $+\infty$. But arguments such as this are sometimes misleading (this subject will be taken up in detail in Sections 4.8 and 4.9), so it is more reliable to reason analytically, as follows:

$$\lim_{x \to -\infty} \left(\sqrt{x^2 + 2x} - x \right) = \lim_{x \to -\infty} \frac{\left(\sqrt{x^2 + 2x} - x\right)\left(\sqrt{x^2 + 2x} + x\right)}{\sqrt{x^2 + 2x} + x} = \lim_{x \to -\infty} \frac{x^2 + 2x - x^2}{\sqrt{x^2 + 2x} + x}$$

$$= \lim_{x \to -\infty} \frac{2x}{\sqrt{x^2 + 2x} + x} = \lim_{x \to -\infty} \frac{\dfrac{2x}{x}}{\dfrac{\sqrt{x^2 + 2x} + x}{x}} = \lim_{x \to -\infty} \frac{2}{\dfrac{x}{x} - \sqrt{\dfrac{x^2 + 2x}{x^2}}} \quad \text{(see Note 1)}$$

$$= \lim_{x \to -\infty} \frac{2}{1 - \sqrt{1 + \dfrac{2}{x}}} = +\infty$$

175

because, if x is large negative, then $1 + \dfrac{2}{x}$ is slightly smaller than 1, so that

$$1 - \sqrt{1 + \frac{2}{x}}$$

is a very small positive number, approaching zero through positive values as $x \to -\infty$.

Note 1: The minus sign is necessary because $x < 0$, and therefore $\sqrt{x^2} = -x$, not x. It's important not to miss this detail because the other sign will give the incorrect limit 1.

Note 2: There are so many dangers associated with minus signs and negative numbers in this problem that it would probably be better to let $u = -x$ and recast the problem in the form

$$\lim_{x \to -\infty} \left(\sqrt{x^2 + 2x} - x \right) = \lim_{u \to +\infty} \left(\sqrt{u^2 - 2u} + u \right),$$

then multiply numerator and denominator by the conjugate of the numerator as in the previous calculation.

4.7.17: Matches Fig. 4.7.20(g), because $f(x) \to +\infty$ as $x \to 1^+$, $f(x) \to -\infty$ as $x \to 1^-$, and $f(x) \to 0$ as $x \to \pm\infty$.

4.7.19: Matches Fig. 4.7.20(a), because $f(x) \to +\infty$ as $x \to 1$ and $f(x) \to 0$ as $x \to \pm\infty$.

4.7.21: Matches Fig. 4.7.20(f), because $f(x) \to +\infty$ as $x \to 1^+$, $f(x) \to -\infty$ as $x \to 1^-$, $f(x) \to +\infty$ as $x \to -1^-$, $f(x) \to -\infty$ as $x \to -1^+$, and $f(x) \to 0$ as $x \to \pm\infty$.

4.7.23: Matches Fig. 4.7.20(j), because $f(x) \to +\infty$ as $x \to 1^+$, $f(x) \to -\infty$ as $x \to 1^-$, $f(x) \to -\infty$ as $x \to -1^-$, $f(x) \to +\infty$ as $x \to -1^+$, and $f(x) \to 0$ as $x \to \pm\infty$.

4.7.25: Matches Fig. 4.7.20(l), because $f(x) \to +\infty$ as $x \to 1^+$, $f(x) \to -\infty$ as $x \to 1^-$, and $f(x) \to 1$ as $x \to \pm\infty$.

4.7.27: Matches Fig. 4.7.20(k), because $f(x) \to +\infty$ as $x \to 1^+$, $f(x) \to -\infty$ as $x \to 1^-$, and because

$$f(x) = \frac{x^2}{x - 1} = x + 1 + \frac{1}{x - 1} \qquad \text{if } x \neq 1,$$

the graph of f has the slant asymptote with equation $y = x + 1$.

4.7.29: Given $f(x) = \dfrac{2}{x - 3}$, we find that

$$f'(x) = -\frac{2}{(x - 3)^2} \qquad \text{and} \qquad f''(x) = \frac{4}{(x - 3)^3}.$$

So there are no extrema or inflection points, the only intercept is $\left(0, -\frac{2}{3}\right)$, and $f(x) \to +\infty$ as $x \to 3^+$ whereas $f(x) \to -\infty$ as $x \to 3^-$. So the line $x = 3$ is a vertical asymptote. Also $f(x) \to 0$ as $x \to \pm\infty$, so the line $y = 0$ is a [two-way] horizontal asymptote. A *Mathematica*-generated graph of $y = f(x)$ is next.

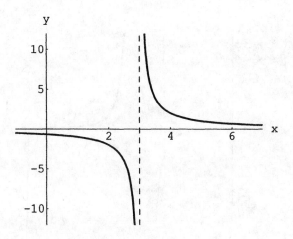

4.7.31: Given $f(x) = \dfrac{3}{(x+2)^2}$, we find that

$$f'(x) = -\frac{6}{(x+2)^3} \quad \text{and} \quad f''(x) = \frac{18}{(x+2)^4}.$$

So there are no extrema or inflection points, the only intercept is $\left(0, \frac{3}{4}\right)$, the line $x = -2$ is a vertical asymptote, and the x-axis is a horizontal asymptote. A *Mathematica*-generated graph of $y = f(x)$ is shown next.

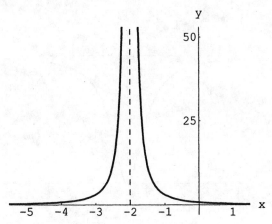

4.7.33: If $f(x) = \dfrac{1}{(2x-3)^3}$, it follows that

$$f'(x) = -\frac{6}{(2x-3)^4} \quad \text{and} \quad f''(x) = \frac{48}{(2x-3)^5}.$$

So there are no extrema or inflection points, the only intercept is $\left(0, -\frac{1}{27}\right)$, the line $x = \frac{3}{2}$ is a vertical asymptote, and the x-axis is a horizontal asymptote. A *Mathematica*-generated graph of $y = f(x)$ is next.

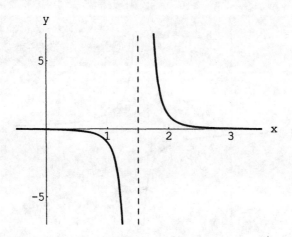

4.7.35: Given: $f(x) = \dfrac{x^2}{x^2+1}$. Then

$$f'(x) = \frac{2x}{(x^2+1)^2} \qquad \text{and} \qquad f''(x) = \frac{2(1-3x^2)}{(x^2+1)^3}.$$

Thus the only intercept is $(0, 0)$, where there is a global minimum, there are inflection points at $\left(\pm\frac{1}{3}\sqrt{3}, \frac{1}{4}\right)$, and the line $y = 1$ is a horizontal asymptote. A *Mathematica*-generated graph of $y = f(x)$ is shown next (without the asymptote).

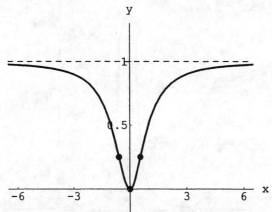

4.7.37: If $f(x) = \dfrac{1}{x^2-9}$, then

$$f'(x) = -\frac{2x}{(x^2-9)^2} \qquad \text{and} \qquad f''(x) = \frac{6(x^2+3)}{(x^2-9)^3}.$$

So there is a local maximum at $\left(0, -\frac{1}{9}\right)$, which is also the only intercept; there are vertical asymptotes at $x = -3$ and at $x = 3$, and the x-axis is a horizontal asymptote. A *Mathematica*-generated graph of $y = f(x)$ is shown next

178

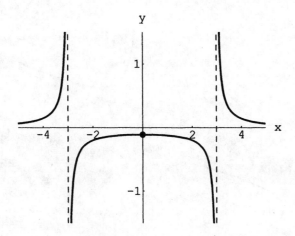

4.7.39: Given $f(x) = \dfrac{1}{x^2 + x - 6} = \dfrac{1}{(x-2)(x+3)}$, we find that

$$f'(x) = -\frac{2x+1}{(x^2+x-6)^2} \quad \text{and} \quad f''(x) = \frac{2(x^2+3x+7)}{(x^2+x-6)^3}.$$

Thus there is a local maximum at $\left(-\frac{1}{2}, -\frac{4}{25}\right)$ and the only intercept is at $\left(0, -\frac{1}{6}\right)$. The lines $x = -3$ and $x = 2$ are vertical asymptotes and the x-axis is a horizontal asymptote. A *Mathematica*-generated graph of $y = f(x)$ is shown next.

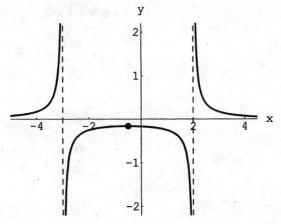

4.7.41: Given: $f(x) = x + \dfrac{1}{x} = \dfrac{x^2 + 1}{x}$, we find that

$$f'(x) = \frac{x^2 - 1}{x^2} \quad \text{and that} \quad f''(x) = \frac{2}{x^3}.$$

Hence there is a local maximum at $(-1, -2)$, a local minimum at $(1, 2)$, and no inflection points or intercepts. The y-axis is a vertical asymptote and the line $y = x$ is a slant asymptote. A *Mathematica*-generated graph of $y = f(x)$ is next.

179

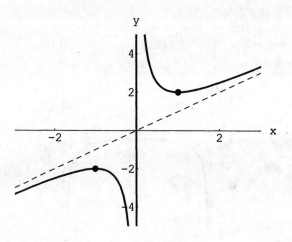

4.7.43: If $f(x) = \dfrac{x^2}{x-1} = x + 1 + \dfrac{1}{x-1}$, then

$$f'(x) = \frac{x(x-2)}{(x-1)^2} \quad \text{and} \quad f''(x) = \frac{2}{(x-1)^3}.$$

So $(0, 0)$ is a local maximum and the only intercept, there is a local minimum at $(2, 4)$, the line $x = 1$ is a vertical asymptote, and the line $y = x + 1$ is a slant asymptote. A *Mathematica*-generated graph of $y = f(x)$ is shown next.

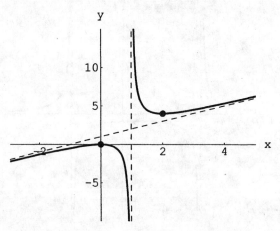

4.7.45: If $f(x) = \dfrac{1}{(x-1)^2}$, then

$$f'(x) = -\frac{2}{(x-1)^3} \quad \text{and} \quad f''(x) = \frac{6}{(x-1)^4}.$$

Hence there are no extrema or inflection points, the only intercept is $(0, 1)$, the line $x = 1$ is a vertical asymptote, and the x-axis is a horizontal asymptote. A *Mathematica*-generated graph of $y = f(x)$ is next.

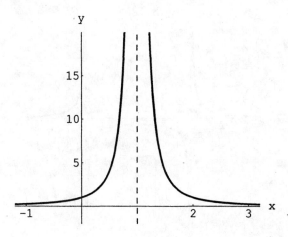

4.7.47: If

$$f(x) = \frac{e^x}{1 + e^x}, \quad \text{then} \quad f'(x) = \frac{e^x}{(1 + e^x)^2} \quad \text{and} \quad f''(x) = \frac{e^x(1 - e^x)}{(1 + e^x)^3}.$$

Therefore the graph of f is increasing everywhere, concave upward for $x < 0$, and concave downward for $x > 0$. So there are no extrema, an inflection point at $\left(0, \frac{1}{2}\right)$, and two horizontal asymptotes: the line $y = 1$ to the right, the line $y = 0$ to the left. The graph of f is next.

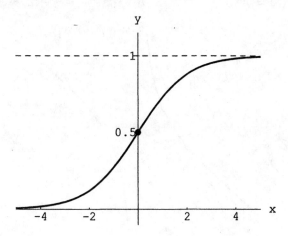

4.7.49: If $f(x) = \dfrac{1}{x^2 - x - 2} = \dfrac{1}{(x - 2)(x + 1)}$, then

$$f'(x) = \frac{1 - 2x}{(x^2 - x - 2)^2} \quad \text{and} \quad f''(x) = \frac{6(x^2 - x + 1)}{(x^2 - x - 2)^3}.$$

Therefore $\left(\frac{1}{2}, -\frac{4}{9}\right)$ is a local maximum and the only extremum, $\left(0, -\frac{1}{2}\right)$ is the only intercept, the lines $x = -1$ and $x = 2$ are vertical asymptotes, and the x-axis is a horizonal asymptote. A *Mathematica*-generated graph of $y = f(x)$ is next.

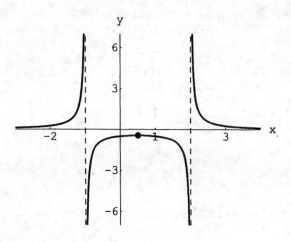

4.7.51: Given: $f(x) = \dfrac{x^2 - 4}{x} = x - \dfrac{4}{x}$, we find that

$$f'(x) = \frac{x^2 + 4}{x^2} \quad \text{and that} \quad f''(x) = -\frac{8}{x^3}.$$

Therefore $(-2, 0)$ and $(2, 0)$ are the only intercepts, there are no inflection points or extrema, the y-axis is a vertical asymptote, and the line $y = x$ is a slant asymptote (not shown in the figure). A *Mathematica*-generated graph of $y = f(x)$ is next.

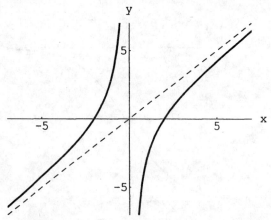

4.7.53: If $f(x) = \dfrac{x^3 - 4}{x^2} = x - \dfrac{4}{x^2}$, then

$$f'(x) = \frac{x^3 + 8}{x^3} \quad \text{and} \quad f''(x) = -\frac{24}{x^4}.$$

Thus $\left(\sqrt[3]{4}, 0\right)$ is the only intercept, there is a local maximum at $(-2, -3)$ and no other extrema, and there are no inflection points. The y-axis is a vertical asymptote and the line $y = x$ is a slant asymptote. A *Mathematica*-generated graph of $y = f(x)$ is next.

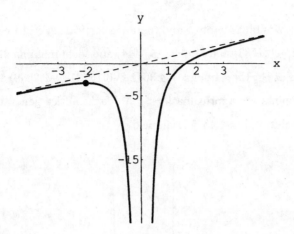

4.7.55: The x-axis is a horizontal asymptote, and there are vertical asymptotes at $x = 0$ and $x = 2$. There are local minima at $(-1.9095, -0.3132)$ and $(1.3907, 3.2649)$ and a local maximum at $(4.5188, 0.1630)$ (all coordinates approximate, of course), and inflection points at $(-2.8119, -0.2768)$ and $(6.0630, 0.1449)$. A *Mathematica*-generated graph of $y = f(x)$ is shown next, with the local minimum point at $x = 1.3907$ marked.

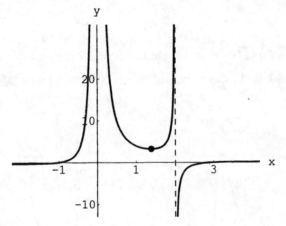

A "close-up" of the graph for $-7 \leqq x < 0$ is next, on the left, and another for $2 < x \leqq 10$ is on the right; these two graphs show the remaining four critical and inflection points of f.

183

4.7.57: The x-axis is a horizontal asymptote and there are vertical asymptotes at $x = 0$ and $x = 2$. There are local minima at $(-2.8173, -0.1783)$ and $(1.4695, 5.5444)$ and local maxima at $(-1, 0)$ and $(4.3478, 0.1998)$. There are inflection points at the three points $(-4.3611, -0.1576)$, $(-1.2569, -0.0434)$, and $(5.7008, 0.1769)$. (Numbers with decimal points are approximations.) A *Mathematica*-generated graph of $y = f(x)$ is next, and the local minimum point at $x = 1.4695$ is marked.

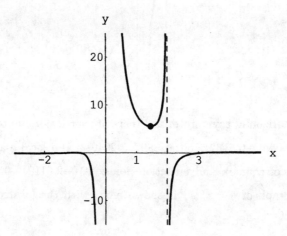

A "close-up" of the graph for $-7 \leq x < 0$ is shown next, on the left; the graph for $2 < x \leq 10$ is on the right; these two graphs exhibit the other six critical and inflection points of f.

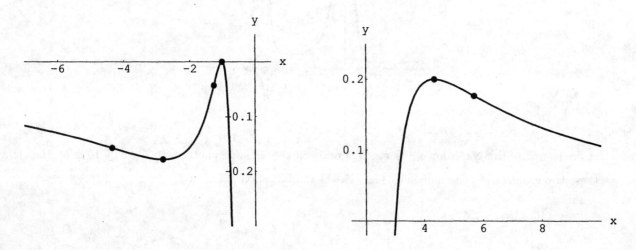

4.7.59: The horizontal line $y = 0$ is an asymptote, as are the vertical lines $x = 0$ and $x = 2$. There are local minima at $(-2.6643, -0.2160)$, $(1.2471, 14.1117)$, and $(3, 0)$; there are local maxima at $(-1, 0)$ and $(5.4172, 0.1296)$. There are inflection points at $(-4.0562, -0.1900)$, $(-1.2469, -0.0538)$, $(3.3264, 0.0308)$, and $(7.4969, 0.1147)$. (Numbers with decimal points are approximations.) A *Mathematica*-generated graph of $y = f(x)$ is shown next, and the local minimum point at $x = 1.2471$ is marked.

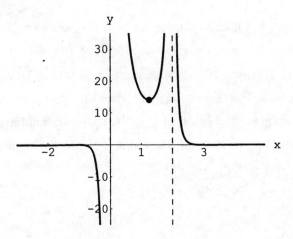

The graph for $-7 \leqq x < 0$ is next, on the left; the graph for $2 < x \leqq 12$ is on the right, and these graphs show the other eight critical and inflection points of the function f.

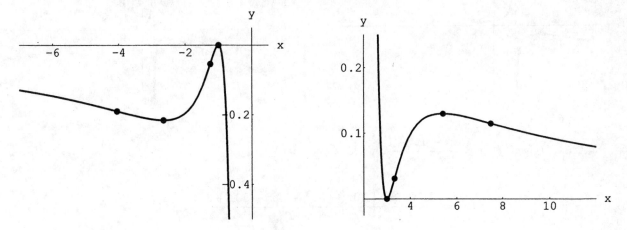

4.7.61: The x-axis is a horizontal asymptote; there are vertical asymptotes at $x = -0.5321$, $x = 0.6527$, and $x = 2.8794$. There is a local minimum at $(0, 0)$ and a local maximum at $(\sqrt[3]{2}, -0.9008)$. There are no inflection points (Numbers with decimal points are approximations.) A *Mathematics*-generated graph of $y = f(x)$ is next, and it exhibits the two local extreme points.

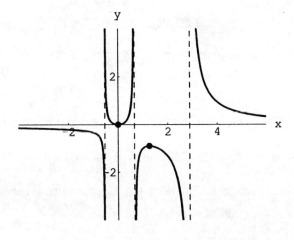

4.7.63: The line $y = x + 3$ is a slant asymptote in both the positive and negative directions; thus there is no horizontal asymptote. There is a vertical asymptote at $x = -1.1038$. There are local maxima at $(-2.3562, -1.8292)$ and $(2.3761, 18.5247)$, local minima at $(0.8212, 0.6146)$ and $(5.0827, 11.0886)$. There are inflection points at $(1.9433, 11.3790)$ and $(2.7040, 16.8013)$. (Numbers with decimal points are approximations.) A *Mathematica*-generated graph of $y = f(x)$ is next, exhibiting the four local extrema and two inflection points on the left; on the right the graph is shown on a wider scale, together with its slant asymptote.

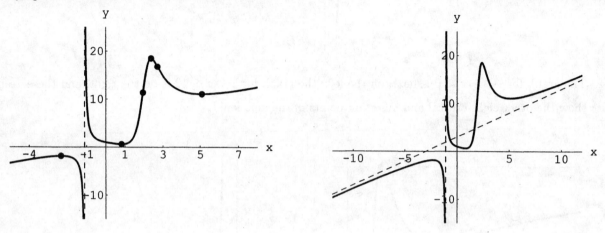

4.7.65: Given $f(x) = \dfrac{x^5 - 4x^2 + 1}{2x^4 - 3x + 2}$, we first find that

$$f'(x) = \frac{2x^8 + 4x^5 + 10x^4 - 8x^3 + 12x^2 - 16x + 3}{(2x^4 - 3x + 2)^2} \quad \text{and}$$

$$f''(x) = -\frac{2(30x^8 + 24x^7 - 40x^6 + 90x^5 - 102x^4 - 28x^3 + 24x^2 + 7)}{(2x^4 - 3x + 2)^3}.$$

The line $2y = x$ is a slant asymptote in both the positive and negative directions; thus there is no horizontal asymptote. There also are no vertical asymptotes. There is

a local maximum at $(0.2200976580, 0.6000775882)$,

a local minimum at $(0.8221567934, -2.9690453671)$,

and inflection points at

$(-2.2416918017, -1.2782199626)$, $(-0.5946286318, -0.1211409770)$,

$(0.6700908810, -1.6820255735)$, and $(0.96490314661, -2.2501145861)$.

(Numbers with decimal points are approximations.) A *Mathematica*-generated graph of $y = f(x)$ showing these two local extrema and four inflection points is next.

186

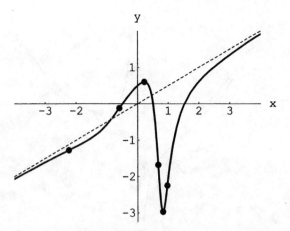

4.7.67: The line $2y = x$ is a slant asymptote in both the positive and negative directions; thus there is no horizontal asymptote. There is a vertical asymptote at $x = -1.7277$. There are local maxima at $(-3.1594, -2.3665)$ and $(1.3381, 1.7792)$, local minima at $(-0.5379, -0.3591)$ and $(1.8786, 1.4388)$. There are inflection points at $(0, 0)$, $(0.5324, 0.4805)$, $(1.1607, 1.4294)$, and $(1.4627, 1.6727)$. (Numbers with decimal points are approximations.) A *Mathematica*-generated graph of $y = f(x)$ is next on the left; it shows the two asmptotes and the local maximum point on the left-hand branch. The figure on the right shows the right-hand branch of the graph with its three local extrema and four inflection points.

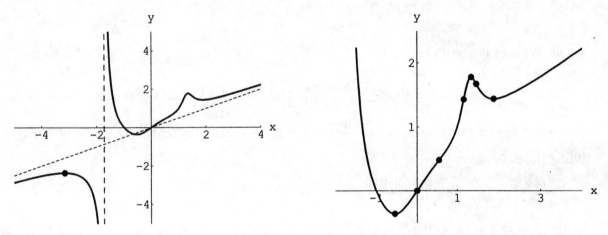

4.7.69: Sketch the parabola $y = x^2$, but modify it by changing its behavior near $x = 0$: Let $y \to +\infty$ as $x \to 0^+$ and let $y \to -\infty$ as $x \to 0^-$. Using calculus, we compute

$$f'(x) = \frac{2(x^3 - 1)}{x^2} \qquad \text{and} \qquad f''(x) = \frac{2(x^3 + 2)}{x^3}.$$

It follows that the graph of f is decreasing for $0 < x < 1$ and for $x < 0$, increasing for $x > 1$. It is concave upward for $x < -\sqrt[3]{2}$ and also for $x > 0$, concave downward for $-\sqrt[3]{2} < x < 0$. The only intercept is at $\left(-\sqrt[3]{2}, 0\right)$; this is also the only inflection point. There is a local minimum at $(1, 3)$. The y-axis is a vertical asymptote. A *Mathematica*-generated graph of $y = f(x)$ is shown next, together with its asymptotic parabola (dashed).

187

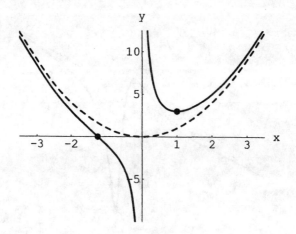

Section 4.8

4.8.1: You don't need l'Hôpital's rule to evaluate this limit, but you may use it:

$$\lim_{x \to 1} \frac{x-1}{x^2-1} = \lim_{x \to 1} \frac{1}{2x} = \frac{1}{2}.$$

4.8.3: You don't need l'Hôpital's rule to evaluate this limit, but you may use it (twice):

$$\lim_{x \to \infty} \frac{2x^2-1}{5x^2+3x} = \lim_{x \to \infty} \frac{4x}{10x+3} = \lim_{x \to \infty} \frac{4}{10} = \frac{2}{5}.$$

4.8.5: Without l'Hôpital's rule:

$$\lim_{x \to 0} \frac{\sin x^2}{x} = \lim_{x \to 0} x \cdot \frac{\sin x^2}{x^2} = 0 \cdot 1 = 0.$$

(We used Theorem 1 of Section 2.3, $\lim_{x \to 0} \dfrac{\sin x}{x} = 1$, and the product law for limits.)

With l'Hôpital's rule:

$$\lim_{x \to 0} \frac{\sin x^2}{x} = \lim_{x \to 0} \frac{2x \cos x^2}{1} = 2 \cdot 0 \cdot 1 = 0.$$

4.8.7: You may *not* use l'Hôpital's rule! The numerator is approaching zero but the denominator is not. Hence use the quotient law for limits (Section 2.2):

$$\lim_{x \to 1} \frac{x-1}{\sin x} = \frac{\lim_{x \to 1} (x-1)}{\lim_{x \to 1} \sin x} = \frac{0}{\sin 1} = 0.$$

Note that illegal use of l'Hôpital's rule in this problem will result in the incorrect value $\sec 1 \approx 1.8508157177$ for the limit.

4.8.9: Without l'Hôpital's rule we might need to resort to the Taylor series methods of Section 10.9 to evaluate this limit. But l'Hôpital's rule may be applied (twice):

$$\lim_{x \to 0} \frac{e^x - x - 1}{x^2} = \lim_{x \to 0} \frac{e^x - 1}{2x} = \lim_{x \to 0} \frac{e^x}{2} = \frac{1}{2}.$$

188

4.8.11: The numerator and denominator are both approaching zero, so l'Hôpital's rule may be applied:

$$\lim_{u \to 0} \frac{u \tan u}{1 - \cos u} = \lim_{u \to 0} \frac{\tan u + u \sec^2 u}{\sin u}$$

$$= \lim_{u \to 0} \left(\sec u + \frac{u}{\sin u} \cdot \sec^2 u \right) = 1 + 1 \cdot 1^2 = 2.$$

Note that we used the sum law for limits (Section 2.2) and the fact that

$$\lim_{u \to 0} \frac{u}{\sin u} = 1,$$

a consequence of Theorem 1 of Section 2.3 and the quotient law for limits.

4.8.13: $\lim\limits_{x \to \infty} \dfrac{\ln x}{x^{1/10}} = \lim\limits_{x \to \infty} \dfrac{1}{x \cdot \frac{1}{10} x^{-9/10}} = \lim\limits_{x \to \infty} \dfrac{10}{x^{1/10}} = 0.$

4.8.15: $\lim\limits_{x \to 10} \dfrac{\ln(x - 9)}{x - 10} = \lim\limits_{x \to 10} \dfrac{1}{1 \cdot (x - 9)} = 1.$

4.8.17: Always verify that the hypotheses of l'Hôpital's rule are satisfied.

$$\lim_{x \to 0} \frac{e^x + e^{-x} - 2}{x \sin x} = \lim_{x \to 0} \frac{e^x - e^{-x}}{x \cos x + \sin x} = \lim_{x \to 0} \frac{e^x + e^{-x}}{2 \cos x - x \sin x} = \frac{1 + 1}{2 - 0} = 1.$$

4.8.19: Methods of Section 2.3 may be used, or l'Hôpital's rule yields

$$\lim_{x \to 0} \frac{\sin 3x}{\tan 5x} = \lim_{x \to 0} \frac{3 \cos 3x}{5 \sec^2 5x} = \frac{3 \cdot 1}{5 \cdot 1} = \frac{3}{5}.$$

4.8.21: The factoring techniques of Section 2.2 work well here, or l'Hôpital's rule yields

$$\lim_{x \to 1} \frac{x^3 - 1}{x^2 - 1} = \lim_{x \to 1} \frac{3x^2}{2x} = \frac{3}{2}.$$

4.8.23: Both numerator and denominator approach $+\infty$ as x does, so we may attempt to find the limit with l'Hôpital's rule:

$$\lim_{x \to \infty} \frac{x + \sin x}{3x + \cos x} = \lim_{x \to \infty} \frac{1 + \cos x}{3 - \sin x},$$

but the latter limit does not exist (and so the equals mark in the previous equation is invalid). The reason: As $x \to +\infty$, x runs infinitely many times through numbers of the form $n\pi$ where n is a positive even integer. At such real numbers the value of

$$f(x) = \frac{1 + \cos x}{3 - \sin x}$$

is $\frac{2}{3}$. But x also runs infinitely often through numbers of the form $n\pi$ where n is a positive odd integer. At these real numbers the value of $f(x)$ is 0. Because $f(x)$ takes on these two distinct values infinitely often as $x \to +\infty$, $f(x)$ has no limit as $x \to +\infty$.

189

This does not imply that the limit given in Problem 22 does not exist. (Read Theorem 1 carefully.) In fact, the limit does exist, and we have here the rare phenomenon of failure of l'Hôpital's rule. Other techniques must be used to solve this problem. Perhaps the simplest is this:

$$\lim_{x \to \infty} \frac{x + \sin x}{3x + \cos x} = \lim_{x \to \infty} \frac{1 + \dfrac{\sin x}{x}}{3 - \dfrac{\cos x}{x}} = \frac{1 + 0}{3 - 0} = \frac{1}{3}.$$

4.8.25: $\displaystyle \lim_{x \to 0} \frac{2^x - 1}{3^x - 1} = \lim_{x \to 0} \frac{2^x \ln 2}{3^x \ln 3} = \frac{\ln 2}{\ln 3} \approx 0.6309297536.$

4.7.27: You can work this problem without l'Hôpital's rule, but if you intend to use it you should probably proceed as follows:

$$\lim_{x \to \infty} \frac{\sqrt{x^2 - 1}}{\sqrt{4x^2 - x}} = \lim_{x \to \infty} \left(\frac{x^2 - 1}{4x^2 - x} \right)^{1/2} = \left(\lim_{x \to \infty} \frac{x^2 - 1}{4x^2 - x} \right)^{1/2}$$

$$= \left(\lim_{x \to \infty} \frac{2x}{8x - 1} \right)^{1/2} = \left(\lim_{x \to \infty} \frac{2}{8} \right)^{1/2} = \left(\frac{1}{4} \right)^{1/2} = \frac{1}{2}.$$

4.8.29: $\displaystyle \lim_{x \to 0} \frac{\ln(1 + x)}{x} = \lim_{x \to 0} \frac{1}{1 \cdot (1 + x)} = 1.$

4.8.31: Three applications of l'Hôpital's rule yield

$$\lim_{x \to 0} \frac{2e^x - x^2 - 2x - 2}{x^3} = \lim_{x \to 0} \frac{2e^x - 2x - 2}{3x^2} = \lim_{x \to 0} \frac{2e^x - 2}{6x} = \lim_{x \to 0} \frac{2e^x}{6} = \frac{1}{3}.$$

4.8.33: $\displaystyle \lim_{x \to 0} \frac{2 - e^x - e^{-x}}{2x^2} = \lim_{x \to 0} \frac{e^{-x} - e^x}{4x} = \lim_{x \to 0} \frac{-e^{-x} - e^x}{4} = -\frac{1}{2}.$

4.8.35: $\displaystyle \lim_{x \to \pi/2} \frac{2x - \pi}{\tan 2x} = \lim_{x \to \pi/2} \frac{2}{2\sec^2 2x} = \frac{1}{\sec^2 \pi} = 1.$

4.8.37: $\displaystyle \lim_{x \to 2} \frac{x - 2\cos \pi x}{x^2 - 4} = \lim_{x \to 2} \frac{1 + 2\pi \sin \pi x}{2x} = \frac{1 + 0}{4} = \frac{1}{4}.$

4.8.39: We first simplify (using laws of logarithms), *then* apply l'Hôpital's rule:

$$\lim_{x \to 0^+} \frac{\ln(2x)^{1/2}}{\ln(3x)^{1/3}} = \lim_{x \to 0^+} \frac{\frac{1}{2}(\ln 2 + \ln x)}{\frac{1}{3}(\ln 3 + \ln x)} = \frac{3}{2} \left(\lim_{x \to 0^+} \frac{\dfrac{1}{x}}{\dfrac{1}{x}} \right) = \frac{3}{2}.$$

4.8.41: Two applications of l'Hôpital's rule yield

$$\lim_{x \to 0} \frac{\exp(x^3) - 1}{x - \sin x} = \lim_{x \to 0} \frac{3x^2 \exp(x^3)}{1 - \cos x} = \lim_{x \to 0} \frac{(6x + 9x^4) \exp(x^3)}{\sin x}$$

$$= \lim_{x \to 0} \frac{(6 + 9x^3) \exp(x^3)}{\dfrac{\sin x}{x}} = \frac{(6 + 0) \cdot 1}{1} = 6.$$

190

Alternatively, three applications of l'Hôpital's rule yield

$$\lim_{x\to0}\frac{\exp(x^3)-1}{x-\sin x}=\lim_{x\to0}\frac{3x^2\exp(x^3)}{1-\cos x}=\lim_{x\to0}\frac{(6x+9x^4)\exp(x^3)}{\sin x}$$

$$=\lim_{x\to0}\frac{(6+54x^3+27x^6)\exp(x^3)}{\cos x}=\frac{(6+54\cdot0+27\cdot0)\cdot1}{1}=6.$$

4.8.43: $\displaystyle\lim_{x\to0}\frac{(1+4x)^{1/3}-1}{x}=\lim_{x\to0}\frac{\frac{1}{3}(1+4x)^{-2/3}\cdot4}{1}=\lim_{x\to0}\frac{4}{3(1+4x)^{2/3}}=\frac{4}{3}.$

4.8.45: If you want to use the conjugate technique to find this limit, you need to know that the conjugate of $a^{1/3}-b^{1/3}$ is $a^{2/3}+a^{1/3}b^{1/3}+b^{2/3}$, and the algebra becomes rather long. Here l'Hôpital's rule is probably the easy way:

$$\lim_{x\to0}\frac{(1+x)^{1/3}-(1-x)^{1/3}}{x}=\lim_{x\to0}\frac{\frac{1}{3}(1+x)^{-2/3}+\frac{1}{3}(1-x)^{-2/3}}{1}$$

$$=\lim_{x\to0}\left[\frac{1}{3(1+x)^{2/3}}+\frac{1}{3(1-x)^{2/3}}\right]=\frac{1}{3}+\frac{1}{3}=\frac{2}{3}.$$

4.8.47: $\displaystyle\lim_{x\to0}\frac{\ln(1+x^2)}{e^x-\cos x}=\lim_{x\to0}\frac{2x}{(1+x^2)(e^x+\sin x)}=\frac{0}{(1+0)(1+0)}=0.$

4.8.49: If $f(x)=\dfrac{\sin^2 x}{x}$, then

$$\lim_{x\to0}f(x)=\lim_{x\to0}\frac{2\sin x\cos x}{1}=2\cdot0\cdot1=0.$$

The graph of $y=f(x)$ is next.

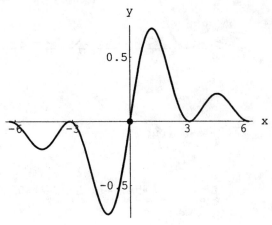

4.8.51: Here we have

$$\lim_{x\to\pi}\frac{\sin x}{x-\pi}=\lim_{x\to\pi}\frac{\cos x}{1}=\cos\pi=-1.$$

The graph is next.

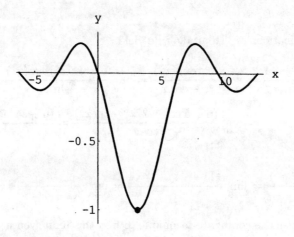

4.8.53: Two applications of l'Hôpital's rule yield

$$\lim_{x \to 0} \frac{1 - \cos x}{x^2} = \lim_{x \to 0} \frac{\sin x}{2x} = \lim_{x \to 0} \frac{\cos x}{2} = \frac{1}{2}.$$

The graph is next.

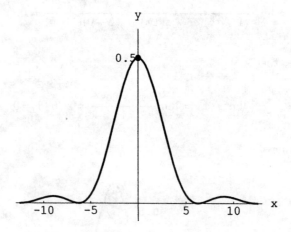

4.8.55: As $x \to -\infty$, $f(x) = xe^{-x} \to -\infty$, but

$$\lim_{x \to \infty} f(x) = \lim_{x \to \infty} \frac{x}{e^x} = \lim_{x \to \infty} \frac{1}{e^x} = 0.$$

Also $f'(x) = (1 - x)e^{-x}$ and $f''(x) = (x - 2)e^{-x}$. It follows that the graph of f is increasing for $x < 1$, decreasing for $x > 1$, concave downward for $x < 2$, and concave upward for $x > 2$. The positive x-axis is a horizontal asymptote and the only intercept is $(0, 0)$. The graph of $y = f(x)$ is shown next.

4.8.57: If $f(x) = x \exp\left(-x^{1/2}\right)$, then

$$\lim_{x \to \infty} f(x) = \lim_{x \to \infty} \frac{x}{\exp\left(x^{1/2}\right)} = \lim_{x \to \infty} \frac{2x^{1/2}}{\exp\left(x^{1/2}\right)} = \lim_{x \to \infty} \frac{2x^{1/2}}{x^{1/2}\exp\left(x^{1/2}\right)} = \lim_{x \to \infty} \frac{2}{\exp\left(x^{1/2}\right)} = 0.$$

Thus the positive x-axis is a horizontal asymptote. Next,

$$f'(x) = \frac{2 - x^{1/2}}{2\exp\left(x^{1/2}\right)} \qquad \text{and} \qquad f''(x) = \frac{x^{1/2} - 3}{4x^{1/2}\exp\left(x^{1/2}\right)}.$$

Hence the graph of f is increasing for $0 < x < 4$ and decreasing for $x > 4$; it is concave downward if $0 < x < 9$ and concave upward if $x > 9$. The only intercept is $(0, 0)$. The graph of $y = f(x)$ is shown next.

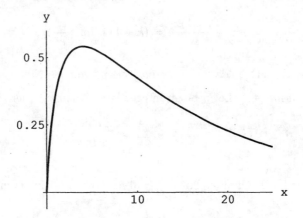

4.8.59: Given: $f(x) = \dfrac{\ln x}{x}$. So

$$\lim_{x \to 0^+} f(x) = \lim_{x \to 0^+} \frac{\ln x}{x} = -\infty$$

because the numerator is approaching $-\infty$ and the denominator is approaching 0 through positive values. Next, using l'Hôpital's rule,

$$\lim_{x \to \infty} f(x) = \lim_{x \to \infty} \frac{\ln x}{x} = \lim_{x \to \infty} \frac{1}{1 \cdot x} = 0,$$

so the positive y-axis is a horizontal asymptote and the negative y-axis is a vertical asymptote. Moreover,

$$f'(x) = \frac{1 - \ln x}{x^2} \qquad \text{and} \qquad f''(x) = \frac{-3 + 2\ln x}{x^3},$$

193

and thus the graph of f is increasing if $0 < x < e$, decreasing if $x > e$, concave downward for $0 < x < e^{3/2}$, and concave upward if $x > e^{3/2}$. The inflection point where $x = e^{3/2}$ is not visible because the curvature of the graph is very small for $x > 3$. The graph of $y = f(x)$ is next.

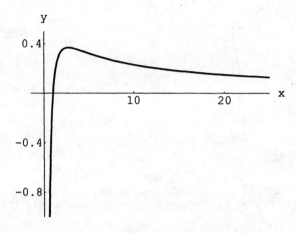

4.8.61: The computation in the solution of Problem 55 establishes that

$$\lim_{x \to \infty} \frac{x^n}{e^x} = 0 \tag{1}$$

in the case $n = 1$. Suppose that Eq. (1) holds for $n = k$, a positive integer. Then (by l'Hôpital's rule)

$$\lim_{x \to \infty} \frac{x^{k+1}}{e^x} = \lim_{x \to \infty} \frac{(k+1)x^k}{e^x} = (k+1)\left(\lim_{x \to \infty} \frac{x^k}{e^x}\right) = (k+1) \cdot 0 = 0.$$

Therefore, by induction, Eq. (1) holds for every positive integer n. Now suppose that $k > 0$ (a positive number not necessarily an integer). Let n be an integer larger than k. Then, for x large positive, we have

$$0 < \frac{x^k}{e^x} < \frac{x^n}{e^x}.$$

Therefore, by the squeeze law for limits, $\displaystyle\lim_{x \to \infty} \frac{x^k}{e^x} = 0$ for every positive number k.

4.8.63: Given: $f(x) = x^n e^{-x}$ where n is a positive integer larger than 1. Then

$$\lim_{x \to \infty} f(x) = \lim_{x \to \infty} \frac{x^n}{e^x} = 0$$

by the result in Problem 61. So the positive x-axis is a horizontal asymptote. Next,

$$f'(x) = \frac{(n-x)x^{n-1}}{e^x} \qquad \text{and} \qquad f''(x) = \frac{(x^2 - 2nx + n^2 - n)x^{n-2}}{e^x}.$$

Therefore $f'(x) = 0$ at the two points $(0, 0)$ and $(n, n^n e^{-n})$. We consider only the part of the graph for which $x > 0$, and the graph of f is increasing for $0 < x < n$ and decreasing if $x > n$, so there is a local maximum at $x = n$. Next, $f''(x) = 0$ when $x = a = n - \sqrt{n}$ and when $x = b = n + \sqrt{n}$. It is easy to

194

establish that $f''(x) > 0$ if $0 < x < a$ and if $x > b$, but that $f''(x) < 0$ if $a < x < b$. (Use the fact that the graph of $g(x) = x^2 - 2nx + n^2 - n$ is a parabola opening upward.) Therefore the graph of f has two inflection points for $x > 0$.

4.8.65: The substitution $y = \dfrac{1}{x}$ yields

$$\lim_{x \to 0^+} x^k \ln x = \lim_{y \to \infty} \frac{-\ln y}{y^k} = -\left(\lim_{y \to \infty} \frac{1}{ky^k} \right) = 0.$$

4.8.67: In the following computations we take derivatives with respect to h in the first step. By l'Hôpital's rule,

$$\lim_{h \to 0} \frac{f(x+h) - f(x-h)}{2h} = \lim_{h \to 0} \frac{f'(x+h) + f'(x-h)}{2} = \frac{2f'(x)}{2} = f'(x).$$

The continuity of $f'(x)$ is needed for two reasons: It implies that f is also continuous, so the first numerator approaches zero as $h \to 0$; moreover, continuity of $f'(x)$ is needed to ensure that $f'(x+h)$ and $f'(x-h)$ both approach $f'(x)$ as $h \to 0$.

4.8.69: If

$$f(x) = \frac{(2x - x^4)^{1/2} - x^{1/3}}{1 - x^{4/3}},$$

then both the numerator $n(x) = (2x - x^4)^{1/2} - x^{1/3}$ and the denominator $d(x) = 1 - x^{4/3}$ approach zero as $x \to 1$, and both are differentiable, so l'Hôpital's rule may be applied. After simplifications we find that

$$n'(x) = \frac{3x^{2/3} - 6x^{11/3} - (2x - x^4)^{1/2}}{3x^{2/3}(2x - x^4)^{1/2}} \quad \text{and} \quad d'(x) = -\frac{4x^{1/3}}{3}.$$

Therefore

$$\lim_{x \to 1} f(x) = \lim_{x \to 1} \frac{n'(x)}{d'(x)} = \lim_{x \to 1} \frac{-3x^{2/3} + 6x^{11/3} + (2x - x^4)^{1/2}}{4x(2x - x^4)^{1/2}} = \frac{-3 + 6 + 1}{4 \cdot 1} = \frac{4}{4} = 1.$$

4.8.71: $\displaystyle\lim_{x \to \infty} \left(\frac{x}{e}\right)^x \geqq \lim_{x \to \infty} \left(\frac{e^2}{e}\right)^x = \lim_{x \to \infty} e^x = +\infty.$

4.8.73: If $f(x) = x^n e^{-x}$ (with n a fixed positive integer), then $f'(x) = (n - x)x^{n-1}e^{-x}$. Because $f(x) \geqq 0$ for $x \geqq 0$, $f(0) = 0$, and $f(x) \to 0$ as $x \to +\infty$, $f(x)$ must have a maximum value, and the critical point where $x = n$ is the sole candidate. Hence the global maximum value of $f(x)$ is $f(n) = n^n e^{-n}$.

Next, $f(n-1) = (n-1)^n e^{-(n-1)} < n^n e^{-n}$, so

$$\left(\frac{n-1}{n}\right)^n < \frac{e^{n-1}}{e^n} = \frac{1}{e}.$$

Therefore

$$e < \left(\frac{n}{n-1}\right)^n = \left(\frac{n-1}{n}\right)^{-n} = \left(1 - \frac{1}{n}\right)^{-n}.$$

Also $f(n+1) = (n+1)^n e^{-(n+1)} < \dfrac{n^n}{e^n}$. Therefore, by similar computations,

$$\left(1 + \frac{1}{n}\right)^n < e.$$

When we substitute $n = 10^6$ (using a computer algebra program, of course) we find that

$$2.7182804690 < e < 2.7182831877$$

(round *down* on the left, *up* on the right). Thus, to five places, $e = 2.71828$.

Section 4.9

4.9.1: We use Theorem 1 of Section 2.3 and the quotient and product laws for limits in Section 2.2:

$$\lim_{x \to 0} x \cot x = \lim_{x \to 0} \frac{x}{\sin x} \cdot \cos x = 1 \cdot 1 = 1.$$

4.9.3: Both the "numerator" $\ln \dfrac{7x+8}{4x+8}$ and the denominator x approach zero as $x \to 0$, so l'Hôpital's rule may be applied. The trick is to use a law of logarithms to make the "numerator" easier to differentiate.

$$\lim_{x \to 0} \frac{1}{x} \cdot \ln \frac{7x+8}{4x+8} = \lim_{x \to 0} \frac{1}{x} \cdot [\ln(7x+8) - \ln(4x+8)]$$

$$= \lim_{x \to 0} \frac{1}{1} \left(\frac{7}{7x+8} - \frac{4}{4x+8} \right) = \lim_{x \to 0} \frac{24}{(7x+8)(4x+8)} = \frac{24}{64} = \frac{3}{8}.$$

4.9.5: $\displaystyle \lim_{x \to 0} x^2 \csc^2 x = \lim_{x \to 0} \left(\frac{x}{\sin x} \right)^2 = 1^2 = 1.$

4.9.7: $\displaystyle \lim_{x \to \infty} x \left(e^{1/x} - 1 \right) = \lim_{x \to \infty} \frac{e^{1/x} - 1}{x^{-1}} = \lim_{x \to \infty} \frac{-(x^{-2} e^{1/x})}{-(x^{-2})} = \lim_{x \to \infty} e^{1/x} = 1.$

4.9.9: $\displaystyle \lim_{x \to 0^+} x \ln x = \lim_{x \to 0^+} \frac{\ln x}{x^{-1}} = \lim_{x \to 0^+} \frac{1}{-x \cdot x^{-2}} = \lim_{x \to 0^+} (-x) = 0.$

4.9.11: $\displaystyle \lim_{x \to \pi} (x - \pi) \csc x = \lim_{x \to \pi} \frac{x - \pi}{\sin x} = \lim_{x \to \pi} \frac{1}{\cos x} = -1.$

4.9.13: First combine terms to form a single fraction, apply l'Hôpital's rule once, then multiply each term in numerator and denominator by $2x^{1/2}$. Result:

$$\lim_{x \to 0^+} \left(\frac{1}{\sqrt{x}} - \frac{1}{\sin x} \right) = \lim_{x \to 0^+} \frac{(\cos x) - \frac{1}{2} x^{-1/2}}{x^{1/2} \cos x + \frac{1}{2} x^{-1/2} \sin x}$$

$$= \lim_{x \to 0^+} \frac{(2x^{1/2} \cos x) - 1}{2x \cos x + \sin x} = -\infty.$$

The last limit follows because the numerator is approaching -1 as $x \to 0^+$, while the denominator is approaching zero through positive values.

4.9.15: First combine terms to form a single fraction, then apply l'Hôpital's rule:

$$\lim_{x \to 1^+} \left(\frac{x}{x^2 + x - 2} - \frac{1}{x - 1} \right) = \lim_{x \to 1^+} \frac{2}{(1 - x)(2 + x)} = -\infty.$$

Note that l'Hôpital's rule is not actually required.

4.9.17: In this solution we first combine the two terms into a single fraction, apply l'Hôpital's rule a first time, make algebraic simplifications, then apply the rule a second time.

$$\lim_{x \to 0} \left(\frac{1}{x} - \frac{1}{\ln(1 + x)} \right) = \lim_{x \to 0} \frac{\ln(1 + x) - x}{x \ln(1 + x)}$$

$$= \lim_{x \to 0} \frac{\dfrac{1}{1 + x} - 1}{\dfrac{x}{1 + x} + \ln(1 + x)} = \lim_{x \to 0} \frac{1 - (1 + x)}{x + (1 + x) \ln(1 + x)}$$

$$= \lim_{x \to 0} \frac{-x}{x + (1 + x) \ln(1 + x)} = \lim_{x \to 0} \frac{-1}{1 + 1 + \ln(1 + x)} = -\frac{1}{2}$$

4.9.19: The conjugate of $a^{1/3} - b^{1/3}$ is $a^{2/3} + a^{1/3} b^{1/3} + b^{2/3}$ because

$$(a^{1/3} - b^{1/3})(a^{2/3} + a^{1/3} b^{1/3} + b^{2/3}) = a - b.$$

Therefore we multiply "numerator" and denominator (which is 1) by the conjugate of the numerator. The result:

$$\lim_{x \to \infty} \left[(x^3 + 2x + 5)^{1/3} - x \right]$$

$$= \lim_{x \to \infty} \frac{\left[(x^3 + 2x + 5)^{1/3} - x \right] \cdot \left[(x^3 + 2x + 5)^{2/3} + x(x^3 + 2x + 5)^{1/3} + x^2 \right]}{(x^3 + 2x + 5)^{2/3} + x(x^3 + 2x + 5)^{1/3} + x^2}$$

$$= \lim_{x \to \infty} \frac{x^3 + 2x + 5 - x^3}{(x^3 + 2x + 5)^{2/3} + x(x^3 + 2x + 5)^{1/3} + x^2}$$

$$= \lim_{x \to \infty} \frac{2x + 5}{(x^3 + 2x + 5)^{2/3} + x(x^3 + 2x + 5)^{1/3} + x^2}$$

$$= \lim_{x \to \infty} \frac{\dfrac{2}{x} + \dfrac{5}{x^2}}{\left(1 + \dfrac{2}{x} + \dfrac{5}{x^2} \right)^{2/3} + \left(1 + \dfrac{2}{x} + \dfrac{5}{x^2} \right)^{1/3} + 1} = \frac{0}{1 + 1 + 1} = 0.$$

There was no need—certainly, no temptation—to use l'Hôpital's rule.

4.9.21: Apply the natural logarithm function:

$$\ln\left(\lim_{x\to0^+} x^{\sin x}\right) = \lim_{x\to0^+} \ln\left(x^{\sin x}\right) = \lim_{x\to0^+} (\sin x)(\ln x)$$

$$= \lim_{z\to0^+} \frac{\ln x}{\csc x} = \lim_{x\to0^+} \frac{1}{-x\csc x \cot x} = \lim_{x\to0^+} \frac{\tan x}{-\dfrac{x}{\sin x}} = \frac{0}{-1} = 0.$$

Therefore $\lim_{x\to0^+} x^{\sin x} = e^0 = 1.$

4.9.23: Apply the natural logarithm function:

$$\ln\left(\lim_{x\to\infty} (\ln x)^{1/x}\right) = \lim_{x\to\infty} \ln(\ln x)^{1/x} = \lim_{x\to\infty} \frac{\ln(\ln x)}{x} = \lim_{x\to\infty} \frac{1}{x\ln x} = 0.$$

Therefore $\lim_{x\to\infty} (\ln x)^{1/x} = e^0 = 1.$

4.9.25: Apply the natural logarithm function:

$$\ln\left(\lim_{x\to0}\left(\frac{\sin x}{x}\right)^{1/x^2}\right) = \lim_{x\to0} \ln\left(\frac{\sin x}{x}\right)^{1/x^2} = \lim_{x\to0} \frac{1}{x^2}\ln\frac{\sin x}{x}$$

$$= \lim_{x\to0} \frac{\ln(\sin x)-\ln x}{x^2} = \lim_{x\to0} \frac{\dfrac{\cos x}{\sin x}-\dfrac{1}{x}}{2x} = \lim_{x\to0} \frac{x\cos x - \sin x}{2x^2\sin x}$$

$$= \lim_{x\to0} \frac{-x\sin x}{2x^2\cos x + 4x\sin x} = \lim_{x\to0} \frac{-\dfrac{\sin x}{x}}{2\cos x + \dfrac{4\sin x}{x}} = -\frac{1}{2\cdot1 + 4\cdot1} = -\frac{1}{6}.$$

Therefore $\lim_{x\to0}\left(\frac{\sin x}{x}\right)^{1/x^2} = e^{-1/6} \approx 0.846481724890614.$

This limit is particularly resistant to numerical approximation by a conventional (hand-held) calculator. To simulate the behavior of such a calculator, we used *Mathematica* and defined

$$g[x\_] := (N[(Sin[x])/x, 14])\wedge(N[1/(x*x), 14]) \tag{1}$$

thus asking the computer to carry 14 digits (and no more) in its computations of values of

$$f(x) = \left(\frac{\sin x}{x}\right)^{1/x^2}. \tag{2}$$

Here are the results.

x	$g(x)$
0.1	0.846435

198

0.01	0.846481
0.001	0.846482
0.0001	0.846482
0.00001	0.846482
0.000001	0.846501
0.0000001	0.837247
0.00000001	1.0
0.000000001	1.0
0.0000000001	1.0
0.00000000001	$1.848091019 \times 10^{-482164}$
0.000000000001	1.0
0.0000000000001	1.0
0.00000000000001	1.0
0.000000000000001	1.0
0.0000000000000001	1.0

Of course, *Mathematica* is capable of essentially arbitrarily great accuracy (if you have the time and the memory). When we replaced the parameter 14 in Eq. (1) with 200 and asked for results to 20 places, here's what we found.

x	$g(x)$
0.1	0.8464346695553817290
0.01	0.8464812546201339178
0.001	0.8464817201879375391
0.0001	0.8464817248435873115
0.00001	0.8464817248901438064
0.000001	0.8464817248906093714
0.0000001	0.8464817248906140270
0.00000001	0.8464817248906140736
0.000000001	0.8464817248906140740

and all such values of x through 10^{-20} gave the same answer as that in the last row, which agrees with $e^{-1/6}$ to the number of digits shown. We also tried direct evaluation of $f(1/10^n)$, but exhausted time and memory when $n = 4$. Results with *Derive* with precision set to `Exact: 60 Digits` were similar to those in the second table here. The typical hand-held calculator we tried generally gave results similar to those in the first table, except jumping to the value 1 somewhere between $x = 10^{-4}$ and $x = 10^{-6}$. We seemed to get the very best results with *Maple*: With $f(x)$ defined as in Eq. (1), the command

```
evalf(f(1.0*10∧(−25)),100);
```

agreed with the exact value of the limit with 50 digits correct to the right of the decimal point, and this program's approximations began to break down—and only slightly—at $x = 10^{-35}$. If this discussion provokes your interest in numerical mathematics and its computer implementation, it is worth the space.

4.9.27: Apply the natural logarithm function:

$$\ln\left(\lim_{x\to\infty}\left(\cos\frac{1}{x^2}\right)^{(x^4)}\right) = \lim_{x\to\infty}\ln\left(\cos\frac{1}{x^2}\right)^{(x^4)} = \lim_{x\to\infty} x^4\ln\left(\cos\frac{1}{x^2}\right)$$

$$= \lim_{x\to\infty}\frac{\ln\left(\cos\frac{1}{x^2}\right)}{x^{-4}} = \lim_{x\to\infty}\frac{2\tan\frac{1}{x^2}}{-4x^{-5}\cdot x^3} = \lim_{x\to\infty}\frac{\tan\frac{1}{x^2}}{-2x^{-2}}$$

$$= \lim_{x\to\infty}\frac{-\frac{2}{x^3}\sec^2\frac{1}{x^2}}{4x^{-3}} = \lim_{x\to\infty}\frac{-\sec^2\frac{1}{x^2}}{2} = -\frac{1}{2}.$$

Therefore $\lim_{x\to\infty}\left(\cos\frac{1}{x^2}\right)^{(x^4)} = e^{-1/2} \approx 0.6065306597$.

4.9.29: Apply the natural logarithm function:

$$\ln\left(\lim_{x\to 0^+}(x+\sin x)^x\right) = \lim_{x\to 0^+}\ln(x+\sin x)^x = \lim_{x\to 0^+} x\ln(x+\sin x)$$

$$= \lim_{x\to 0^+}\frac{\ln(x+\sin x)}{x^{-1}} = \lim_{x\to 0^+}\frac{\frac{1+\cos x}{x+\sin x}}{-(x^{-2})} = \lim_{x\to 0^+}\frac{-x^2(1+\cos x)}{x+\sin x}$$

$$= \lim_{x\to 0^+}\frac{x^2\sin x - 2x(1+\cos x)}{1+\cos x} = \frac{0\cdot 0 - 2\cdot 0\cdot 2}{1+1} = 0.$$

Therefore $\lim_{x\to 0^+}(x+\sin x)^x = e^0 = 1$.

4.9.31: Apply the natural logarithm function:

$$\ln\left(\lim_{x\to 1} x^{1/(1-x)}\right) = \lim_{x\to 1}\ln x^{1/(1-x)} = \lim_{x\to 1}\frac{\ln x}{1-x} = \lim_{x\to 1}\frac{1}{-x} = -1.$$

Therefore $\lim_{x\to 1} x^{1/(1-x)} = e^{-1} \approx 0.3678795512$.

4.9.33: First combine the two terms to form a single fraction, then apply l'Hôpital's rule, and finally simplify:

$$\lim_{x \to 2^+} \left(\frac{1}{(x^2-4)^{1/2}} - \frac{1}{x-2} \right) = \lim_{x \to 2^+} \frac{x - 2 - (x^2-4)^{1/2}}{(x^2-4)^{1/2}(x-2)} = \lim_{x \to 2^+} \frac{1 - x(x^2-4)^{-1/2}}{x(x^2-4)^{-1/2}(x-2) + (x^2-4)^{1/2}}$$

$$= \lim_{x \to 2^+} \frac{(x^2-4)^{1/2} - x}{x(x-2) + (x^2-4)} = -\infty$$

because, in the last limit, the numerator is approaching -2 while the denominator is approaching zero through *positive* values.

4.9.35: Given: $f(x) = x^{1/x}$ for $x > 0$. We plotted the graph of $y = f(x)$ on the interval $10^{-6} \leqq x \leqq 1$ and obtained strong evidence that $f(x)$ approaches zero as $x \to 0^+$. We also plotted $y = f(x)$ on the interval $100 \leqq x \leqq 1000$ and obtained some evidence that $f(x) \to 1$ as $x \to +\infty$. Then we verified these limits with l'Hôpital's rule as follows:

$$\ln \left(\lim_{x \to \infty} x^{1/x} \right) = \lim_{x \to \infty} \frac{1}{x} \ln x = \lim_{x \to \infty} \frac{1}{x} = 0,$$

so that $\lim_{x \to \infty} x^{1/x} = e^0 = 1$.

But $\lim_{x \to 0^+} x^{1/x}$ is not indeterminate, because the exponent is approaching $+\infty$; this limit is clearly zero.

The graph that follows this solution indicates that the global maximum value of $f(x)$ occurs close to 2.71828 (surely no coincidence). We found that

$$f'(x) = \frac{x^{1/x}(1 - \ln x)}{x^2},$$

and it follows that the maximum value of $f(x)$ is $f(e) = e^{1/e} \approx 1.4446678610$.

4.9.37: Given: $f(x) = (x^2)^{1/x}$ for $x > 0$. We plotted $y = f(x)$ for $0 < x \leqq 1$ and obtained strong evidence that $f(x) \to 0$ as $x \to 0^+$. We plotted $f(x)$ for $0 < x \leqq 10000$ and obtained suggestive (but inconclusive)

201

evidence that $f(x) \to 1$ as $x \to +\infty$. (See the two graphs below.) The first limit is clear because x^2 is approaching zero while the exponent $1/x$ is approaching $+\infty$. For the second limit, we found

$$\ln\left(\lim_{x\to\infty} f(x)\right) = \lim_{x\to\infty} \frac{2\ln x}{x} = \lim_{x\to\infty} \frac{2}{x} = 0,$$

so that $\lim_{x\to\infty} f(x) = e^0 = 1$.

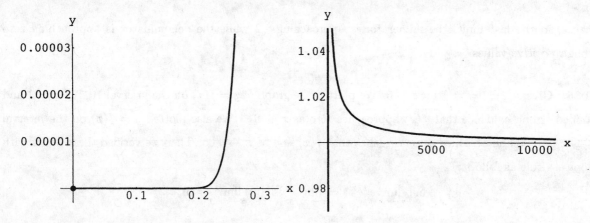

Next we plotted $f(x)$ for $2.71827 \leqq x \leqq 2.71829$ (by the "method of successive zooms") and saw a clear maximum near where $x = 2.71828$. (See the graph below.) We found that

$$f'(x) = \frac{(2 - 2\ln x)(x^2)^{1/x}}{x^2},$$

and it follows that the maximum value of $f(x)$ is $f(e) = e^{2/e} \approx 2.0870652286$.

4.9.39: We graphed $f(x) = (1 + x^2)^{1/x}$ for $0 < x \leqq 1$; the graph provides strong evidence that $f(x) \to 1$ as $x \to 0^+$. Then we graph $y = f(x)$ for $0 < x \leqq 10000$; the graph suggests (but does not prove) that $f(x) \to 1$ as $x \to +\infty$. These two graphs are shown next.

202

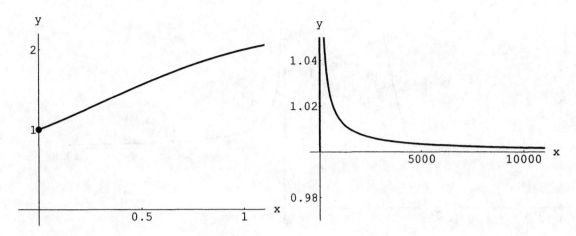

To be sure about these limits, we computed

$$\ln\left(\lim_{x \to 0^+} f(x)\right) = \lim_{x \to 0^+} \frac{\ln(1 + x^2)}{x} = \lim_{x \to 0^+} \frac{2x}{1 + x^2} = 0,$$

and therefore $f(x) \to e^0 = 1$ as $x \to 0^+$. Next,

$$\ln\left(\lim_{x \to \infty} f(x)\right) = \lim_{x \to \infty} \frac{\ln(1 + x^2)}{x} = \lim_{x \to \infty} \frac{2x}{1 + x^2} = 0,$$

and therefore $f(x) \to e^0 = 1$ as $x \to +\infty$.

Then we used the "method of successive zooms" and thereby found that the graph of $y = f(x)$ for $1.9802 \leqq x \leqq 1.9804$ shows a maximum near where $x = 1.9803$. (See the graph below.) Then we found that

$$f'(x) = \frac{(1 + x^2)^{1/x} \left[2x^2 - (1 + x^2)\ln(1 + x^2)\right]}{x^2(1 + x^2)}$$

but could not solve the transcendental equation $2x^2 = (1 + x^2)\ln(1 + x^2)$ exactly. So we used Newton's method to solve $f'(x) = 0$, and our conclusion is that the global maximum value of $f(x)$ is approximately $2.2361202715 \approx f(1.9802913004)$.

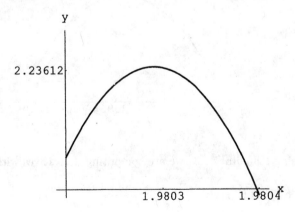

4.9.41: Given: $f(x) = (x + \sin x)^{1/x}$. The graph of $y = f(x)$ for $0 < x \leqq 1$ provides strong evidence that $f(x) \to 0$ as $x \to 0^+$. The graph of $y = f(x)$ for $0 < x \leqq 10000$ provides fairly good evidence that $f(x) \to 1$ as $x \to +\infty$. These graphs are shown next.

203

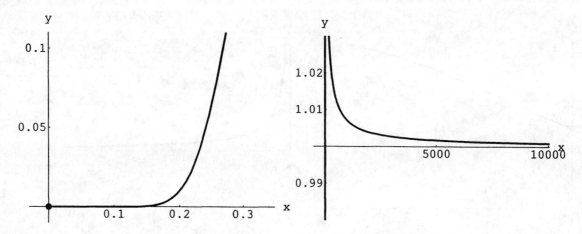

Next, we verified these limits as follows:

$$\ln\left(\lim_{x\to\infty}(x+\sin x)^{1/x}\right) = \lim_{x\to\infty}\frac{\ln(x+\sin x)}{x} = \lim_{x\to\infty}\frac{1+\cos x}{x+\sin x} = 0,$$

and therefore $\lim_{x\to\infty}(x+\sin x)^{1/x} = e^0 = 1$.

But $(x+\sin x)^{1/x}$ is not indeterminate as $x\to 0^+$ because if x is very small and positive, then $x+\sin x$ is positive and near zero while $1/x$ is very large positive. Therefore $\lim_{x\to 0^+}(x+\sin x)^{1/x} = 0$.

Then a plot of $y = f(x)$ for $0.5 \leq x \leq 2$ revealed a global maximum near where $x = 1.2$. A plot of f for $1.2095 \leq x \leq 1.2097$ (by the "method of repeated zooms") showed the maximum near the midpoint of that interval. That graph is shown next. The equation $f'(x) = 0$ appeared to be impossible to solve exactly, so we used Newton's method to find that the maximum of $f(x)$ is very close to $(1.2095994645, 1.8793598343)$.

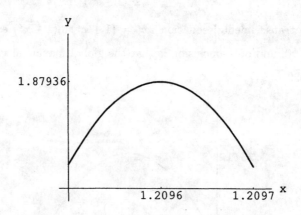

4.9.43: Note that in using l'Hôpital's rule we are computing derivatives with respect to h.

$$\ln\left(\lim_{h\to 0}(1+hx)^{1/h}\right) = \lim_{h\to 0}\frac{\ln(1+hx)}{h} = \lim_{h\to 0}\frac{x}{1+hx} = x,$$

and therefore $\lim_{h\to 0}(1+hx)^{1/h} = e^x$.

4.9.45: We let $f(x) = x^{\tan x}$ and applied Newton's method to the equation $f'(x) = 0$ with initial guess $x_0 = 0.45$. Results: $x_1 = 0.4088273642$, $x_2 = 0.4099763617$, $x_3 = x_4 = 0.4099776300$; $f(x_4) = 0.6787405265$.

204

4.9.47: Replace b with x to remind us that it's the only variable in this problem; note also that $0 < x < a$. The surface area of the ellipsoid is then

$$A(x) = 2\pi ax \left[\frac{x}{a} + \frac{a}{(a^2 - x^2)^{1/2}} \arcsin \frac{(a^2 - a^2)^{1/2}}{a} \right]$$

$$= 2\pi x^2 + 2\pi a^2 \cdot \frac{x \arcsin \dfrac{(a^2 - x^2)^{1/2}}{a}}{(a^2 - x^2)^{1/2}}.$$

Therefore

$$\lim_{x \to a^-} A(x) = 2\pi a^2 + 2\pi a^2 \left(\lim_{x \to a^-} \frac{x}{a} \cdot \frac{\arcsin \dfrac{(a^2 - x^2)^{1/2}}{a}}{\dfrac{(a^2 - x^2)^{1/2}}{a}} \right).$$

Let $u = \dfrac{(a^2 - x^2)^{1/2}}{a}$. Then $u \to 0^+$ as $x \to a^-$. Hence

$$\lim_{x \to a^-} A(x) = 2\pi a^2 + 2\pi a^2 \left(\lim_{x \to a^-} \frac{x}{a} \right) \cdot \left(\lim_{u \to 0^+} \frac{\arcsin u}{u} \right)$$

$$= 2\pi a^2 + 2\pi a^2 \left(\lim_{u \to 0^+} \frac{1}{\sqrt{1 - u^2}} \right) = 4\pi a^2.$$

4.9.49: Given: $f(x) = |\ln x|^{1/x}$ for $x > 0$. The graph of $y = f(x)$ for $0.2 \leq x \leq 0.3$ shows $f(x)$ taking on values in excess of 10^{28}, so it seems quite likely that $f(x) \to +\infty$ as $x \to 0^+$. Indeed, this is the case, because as $x \to 0^+$, we see that $|\ln x| \to +\infty$ and also the exponent $1/x$ is increasing without bound.

Below we show the graph of $y = f(x)$ for $0.4 \leq x \leq 1$ (on the left) and for $0.5 \leq x \leq 1.5$ (on the right). The first indicates a likely inflection point near where $x = 0.8$ and the second shows a clear global minimum at $(1, 0)$. To find the inflection point, we redefined $f(x) = (-\ln x)^{1/x}$ and computed

$$f''(x) = \frac{f(x)}{x^4 (\ln x)^2} \left(1 - x - 3x \ln x - 2(\ln x) \ln(-\ln x) + 2x(\ln x)^2 \ln(-\ln x) + (\ln x)^2 (\ln(-\ln x))^2 \right)$$

(assisted by *Mathematica*, of course). We let

$$g(x) = 1 - x - 3x \ln x - 2(\ln x) \ln(-\ln x) + 2x(\ln x)^2 \ln(-\ln x) + (\ln x)^2 (\ln(-\ln x))^2$$

and applied Newton's method to solve the equation $g(x) = 0$. With initial guess $x_0 = 0.8$, six iterations yielded over 20 digits of accuracy, and the inflection point shown in the figure is located close to $(0.8358706352, 0.1279267691)$.

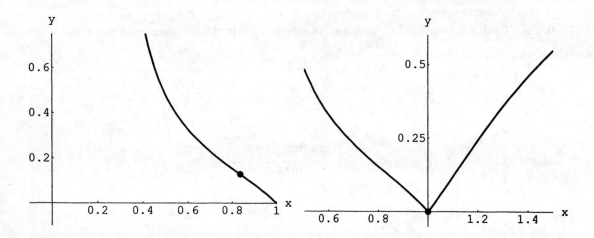

It is clear that $f(x) > 0$ if $0 < x < 1$ and if $x > 1$, so the graph of f has a global minimum at $(1, 0)$. Still using $f(x) = (-\ln x)^{1/x}$, we computed

$$f'(x) = \frac{f(x)(1 - (\ln x)(\ln(-\ln x)))}{x^2 \ln x},$$

and thereby found that

$$\lim_{x \to 1^-} f'(x) = -1.$$

Then we redefined $f(x) = (\ln x)^{1/x}$ and computed

$$f'(x) = \frac{f(x)}{x^2 \ln x} \left(1 - (\ln x)\ln(\ln x)\right).$$

Then we found that

$$\lim_{x \to 1^+} f'(x) = 1.$$

Thus the shape of the graph of f near $(1, 0)$ is quite similar to the shape of $y = |x|$ near $(0, 0)$.

Next we plotted the graph of $y = f(x)$ for $1 \leqq x \leqq 2$ (shown next, on the left) and for $4 \leqq x \leqq 10$ (next, on the right). The first of these indicates a possible inflection point near where $x = 1.2$ and the second shows a clear local maximum near where $x = 5.8$. To locate the inflection point more accurately, we computed

$$f''(x) = \frac{f(x)}{x^4 (\ln x)^2} \left(1 - x - 3x \ln x - 2(\ln x)\ln(\ln x) + 2x(\ln x)^2 \ln(\ln x) + (\ln x)^2 (\ln(\ln x))^2\right)$$

and applied Newton's method to the solution of $g(x) = 0$, where

$$g(x) = 1 - x - 3x \ln x - 2(\ln x)\ln(\ln x) + 2x(\ln x)^2 \ln(\ln x) + (\ln x)^2 (\ln(\ln x))^2.$$

206

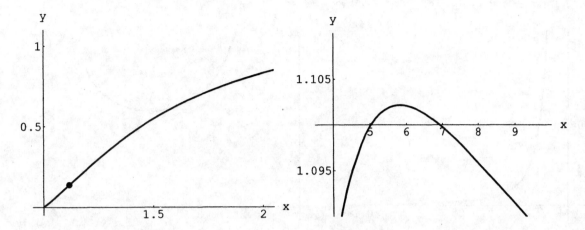

Beginning with the initial guess $x_0 = 1.2$, seven iterations yielded more than 20 digits of accuracy; the inflection point is located close to $(1.1163905964, 0.1385765415)$.

To find the local maximum, we applied Newton's method to the equation $g(x) = 0$, where

$$g(x) = 1 - (\ln x) \ln(\ln x).$$

Beginning with the initial guess $x_0 = 5.8$, six iterations yielded more than 20 digits of accuracy; the local maximum is very close to $(5.8312001357, 1.1021470392)$.

Finally, we plotted $y = f(x)$ for $10 \leqq x \leqq 1000$ (shown after this solution). The change in concavity indicates that there must be yet another inflection point near where $x = 9$. Newton's method again yielded its approximate coordinates as $(8.9280076968, 1.0917274397)$. The last graph also suggests that $f(x) \to 1$ as $x \to +\infty$. This is indeed the case;

$$\ln\left(\lim_{x \to \infty} f(x)\right) = \lim_{x \to \infty} \frac{\ln(\ln x)}{x} = \lim_{x \to \infty} \frac{1}{x \ln x} = 0,$$

and therefore $f(x) \to e^0 = 1$ as $x \to +\infty$.

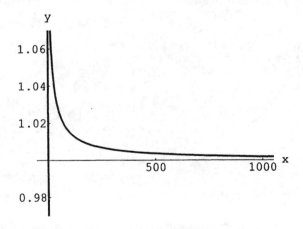

4.9.51: Given: $f(x) = |\ln x|^{|\ln x|}$ for $x > 0$. The graph of $y = f(x)$ for $0.01 \leqq x \leqq 0.1$ (on the left below) indicates that $f(x) \to +\infty$ as $x \to 0^+$, and this is clear. The graph for $0.25 \leqq x \leqq 1$ (on the right) indicates a local minimum near where $x = 0.7$.

We rewrote $f(x)$ in the form $f(x) = (-\ln x)^{-\ln x}$ for $0 < x < 1$ and found that

$$f'(x) = -\frac{(-\ln x)^{-\ln x}(1 + \ln(-\ln x))}{x}. \tag{1}$$

Then it is easy to solve $f'(x) = 0$ for $x = e^{-1/e}$. So the graph of $y = f(x)$ has a local minimum at $(e^{-1/e}, e^{-1/e})$. Both the abscissa and the ordinate are approximately 0.6922006276. Next, the graph on the left below shows a cusp at the point $(1, 1)$; there is a local maximum at that point, and $|f'(x)| \to +\infty$ as $x \to 1$. To see why, rewrite $f(x) = (\ln x)^{\ln x}$ for $x > 1$. Then

$$f'(x) = \frac{(\ln x)^{\ln x}(1 + \ln(\ln x))}{x}, \tag{2}$$

and it is clear that $f'(x) \to -\infty$ as $x \to 1^+$. You can also use Eq. (1) to show that $f'(x) \to +\infty$ as $x \to 1^-$. This is not apparent from the graph of $y = f(x)$ for $0.9 \le x \le 1.1$, shown next (on the left).

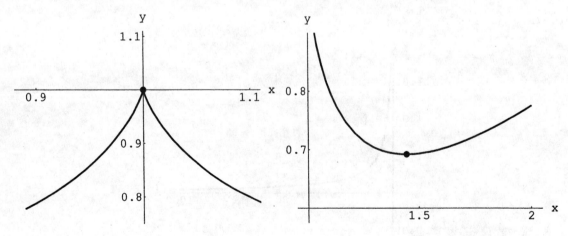

Then we plotted $y = f(x)$ for $1 \le x \le 2$ and found another local minimum near where $x = 1.4$. (See the graph above on the right.) It is easy to solve $f'(x) = 0$ (use Eq. (2)), and you'll find that the coordinates of this point are $(e^{1/e}, e^{-1/e})$, so the two local minima are actually global minima. Finally, it's clear that $f(x) \to +\infty$ as $x \to +\infty$.

208

4.9.53: The figure on the left below shows the graph of $y = f(x)$ on the interval $[-1, 1]$. The figure on the right shows the graph of $y = f(x)$ on the interval $[-0.00001, 0.00001]$. It is clear from the latter figure that $e \approx 2.71828$ to five places. When the removable discontinuity at $x = 0$ is removed in such a way to make f continuous there, the y-intercept will be e because

$$\ln\left[\lim_{x \to 0}\left(1 + \frac{1}{x}\right)^x\right] = \lim_{x \to 0} x \ln\left(\frac{x+1}{x}\right) = \lim_{x \to 0}\frac{\ln(x+1) - \ln x}{\frac{1}{x}} = \lim_{x \to 0}\left(\frac{-x^2}{x+1} + \frac{x^2}{x}\right)$$

$$= \lim_{x \to 0}\frac{-x^3 + x^2 + x^2}{x(x+1)} = \lim_{x \to 0}\frac{x^2}{x(x+1)} = \lim_{x \to 0}\frac{x}{x+1} = 1,$$

and therefore $\displaystyle\lim_{x \to 0}\left(1 + \frac{1}{x}\right)^x = e^1 = e.$

Chapter 4 Miscellaneous Problems

4.M.1: $dy = \frac{3}{2}(4x - x^2)^{1/2}(4 - 2x)\, dx.$

4.M.3: $dy = -\dfrac{2}{(x-1)^2}\, dx.$

4.M.5: $dy = \left(2x\cos\sqrt{x} - \frac{1}{2}x^{3/2}\sin\sqrt{x}\right) dx.$

4.M.7: Let $f(x) = x^{1/2}$; $f'(x) = \frac{1}{2}x^{-1/2}$. Then

$$\sqrt{6401} = f(6400 + 1) \approx f(6400) + 1 \cdot f'(6400)$$

$$= 80 + \frac{1}{160} = \frac{12801}{160} = 80.00625.$$

(A calculator reports that $\sqrt{6401} \approx 80.00624976$.)

4.M.9: Let $f(x) = x^{10}$; then $f'(x) = 10x^9$. Choose $x = 2$ and $\Delta x = 0.0003$. Then

$$(2.0003)^{10} = f(x + \Delta x) \approx f(x) + f'(x)\,\Delta x$$

$$= 2^{10} + 10 \cdot 2^9 \cdot (0.0003)$$

$$= 1024 + (5120)(0.0003) = 1024 + 1.536 = 1025.536.$$

A calculator reports that $(2.0003)^{10} \approx 1025.537$.

4.M.11: Let $f(x) = x^{1/3}$; then $f'(x) = \frac{1}{3}x^{-2/3}$. Choose $x = 1000$ and $\Delta x = 5$. Then

$$\sqrt[3]{1005} = f(x + \Delta x) \approx f(x) + f'(x)\,\Delta x$$

$$= (1000)^{1/3} + \frac{1}{3(1000)^{2/3}} \cdot 5 = 10 + \frac{5}{3 \cdot 100} = \frac{601}{60} \approx 10.0167.$$

A calculator reports that $\sqrt[3]{1005} \approx 10.0166$.

4.M.13: Let $f(x) = x^{3/2}$; then $f'(x) = \frac{3}{2}x^{1/2}$. Let $x = 25$ and let $\Delta x = 1$. Then

$$26^{3/2} = f(x + \Delta x) \approx f(x) + f'(x)\,\Delta x$$

$$= (25)^{3/2} + \frac{3}{2} \cdot (25)^{1/2} \cdot 1 = 125 + 7.5 = 132.5.$$

4.M.15: Let $f(x) = x^{1/4}$; then $f'(x) = \frac{1}{4}x^{-3/4}$. Let $x = 16$ and $\Delta x = 1$. Then

$$\sqrt[4]{17} = (17)^{1/4} = f(x + \Delta x) \approx f(x) + f'(x)\,\Delta x$$

$$= (16)^{1/4} + \frac{1}{4 \cdot (16)^{3/4}} \cdot 1 = 2 + \frac{1}{4 \cdot 8} = \frac{65}{32} = 2.03125.$$

4.M.17: The volume V of a cube of edge s is given by $V(s) = s^3$. So $dV = 3s^2\,ds$, and thus with $s = 5$ and $\Delta s = 0.1$ we obtain $\Delta V \approx 3(5)^2(0.1) = 7.5$ (cubic inches).

4.M.19: The volume V of a sphere of radius r is given by $V(r) = \frac{4}{3}\pi r^3$. Hence $dV = 4\pi r^2\,dr$, so with $r = 5$ and $\Delta r = \frac{1}{10}$ we obtain

$$\Delta V \approx 4\pi \cdot 25 \cdot \frac{1}{10} = 10\pi \quad (\text{cm}^3).$$

4.M.21: If

$$T = 2\pi\sqrt{\frac{L}{32}} = 2\pi\left(\frac{L}{32}\right)^{1/2}, \quad \text{then} \quad dT = \pi\left(\frac{L}{32}\right)^{-1/2} \cdot \frac{1}{32}\,dL = \frac{\pi}{32}\left(\frac{32}{L}\right)^{1/2} dL.$$

Therefore if $L = 2$ and $\Delta L = \frac{1}{12}$, we obtain

$$\Delta T \approx dT = \frac{\pi}{32}\left(\frac{32}{2}\right)^{1/2} \cdot \frac{1}{12} = \frac{\pi}{32} \cdot \frac{4}{12} = \frac{\pi}{96} \approx 0.0327 \quad (\text{seconds}).$$

4.M.23: First, $f'(x) = 1 + \dfrac{1}{x^2}$, so $f'(x)$ exists for $1 < x < 3$ and f is continuous for $1 \leqq x \leqq 3$. So we are to solve

$$\frac{f(3) - f(1)}{3 - 1} = f'(c);$$

that is,

$$\frac{3 - \frac{1}{3} - 1 + 1}{2} = 1 + \frac{1}{c^2}.$$

After simplifications we find that $c^2 = 3$. Therefore, because $1 < c < 3$, $c = +\sqrt{3}$.

4.M.25: Every polynomial is continuous and differentiable everywhere, so all hypotheses of the mean value theorem are satisfied. Then

$$\frac{f(2) - f(-1)}{2 - (-1)} = \frac{8 + 1}{3} = 3 = f'(c) = 3c^2,$$

so $c^2 = 1$. But -1 does not lie in the interval $(-1, 2)$, so the number whose existence is guaranteed by the mean value theorem is $c = 1$.

4.M.27: Given: $f(x) = \frac{11}{5}x^5$ on the interval $[-1, 2]$. Because $f(x)$ is a polynomial, it is continuous and differentiable everywhere, so the hypotheses of the mean value theorem are satisfied on the interval $[-1, 2]$. Moreover,

$$\frac{f(2) - f(-1)}{2 - (-1)} = \frac{\frac{11}{5} \cdot 32 + \frac{11}{5}}{3} = \frac{11 \cdot 33}{5 \cdot 3} = \frac{121}{5} = f'(c) = 11c^4.$$

It follows that $c^4 = \frac{11}{5}$, so that $c = \pm \left(\frac{11}{5}\right)^{1/4}$. Only the positive root lies in the interval $[-1, 2]$, so the number whose existence is guaranteed by the mean value theorem is $c = \left(\frac{11}{5}\right)^{1/4}$.

4.M.29: $f'(x) = 2x - 6$, so $f'(x) = 0$ when $x = 3$; $f''(x) \equiv 2$ is always positive, so there are no inflection points and there is a global minimum at $(3, -5)$. The y-intercept is $(0, 4)$ and the x-intercepts are $\left(3 \pm \sqrt{5}, 0\right)$. The graph of f is shown next.

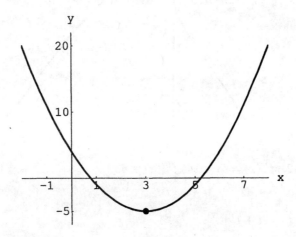

4.M.31: $f(x) = (3x^4 - 5x^2 + 60)x$ and $f'(x) = 15(x^4 - x^2 + 4)$, so $f'(x) > 0$ for all x and hence $(0, 0)$ is the only intercept. $f''(x) = 30x(2x^2 - 1)$, so there are inflection points at $\left(-\frac{1}{2}\sqrt{2}, -\frac{233}{8}\sqrt{2}\right)$, $(0, 0)$, and $\left(\frac{1}{2}\sqrt{2}, \frac{233}{8}\sqrt{2}\right)$. The graph is actually concave upward between the first and second of these inflection points and concave downward between the second and third, but so slightly that this is not visible on the graph of f that is shown next.

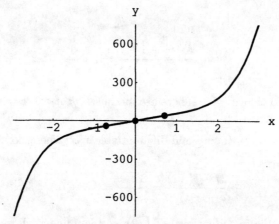

4.M.33: Given $f(x) = (1 - x)x^{1/3}$, we find that

$$f'(x) = \frac{1 - 4x}{3x^{2/3}} \quad \text{and} \quad f''(x) = -\frac{2(2x + 1)}{9x^{5/3}}.$$

So there are intercepts at $(0, 0)$ and $(1, 0)$ and a critical point at $\left(\frac{1}{4}, \frac{3}{8}\sqrt[3]{2}\right)$. There is an inflection point at $\left(-\frac{1}{2}, -\frac{3}{4}\sqrt[3]{4}\right)$; the graph of f is actually concave upward between this point and $(0, 0)$ and concave downward to its left, but the latter is not visible in the scale of the accompanying figure below. The origin is also an inflection point and there is a vertical tangent there too.

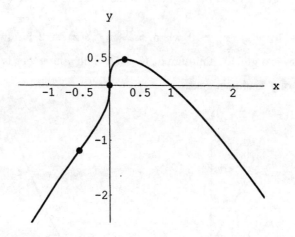

4.M.35: $f'(x) = 3x^2 - 2$, $f''(x) = 6x$, and $f'''(x) \equiv 6$.

4.M.37: Given $g(t) = \frac{1}{t} - \frac{1}{2t + 1}$,

212

$$g'(t) = \frac{2}{(2t+1)^2} - \frac{1}{t^2}, \qquad g''(t) = \frac{2}{t^3} - \frac{8}{(2t+1)^3}, \qquad \text{and} \qquad g'''(t) = \frac{48}{(2t+1)^4} - \frac{6}{t^4}.$$

4.M.39: $f'(t) = 3t^{1/2} - 4t^{1/3}$, $f''(t) = \frac{3}{2}t^{-1/2} - \frac{4}{3}t^{-2/3}$, and $f'''(t) = \frac{8}{9}t^{-5/3} - \frac{3}{4}t^{-3/2}$.

4.M.41: $h'(t) = -\dfrac{4}{(t-2)^2}$, $h''(t) = \dfrac{8}{(t-2)^3}$, and $h'''(t) = -\dfrac{24}{(t-2)^4}$.

4.M.43: $g'(x) = -\dfrac{4}{3(5-4x)^{2/3}}$, $g''(x) = -\dfrac{32}{9(5-4x)^{5/3}}$, and $g'''(x) = -\dfrac{640}{27(5-4x)^{8/3}}$.

4.M.45: $x^{-2/3} + y^{-2/3}\dfrac{dy}{dx} = 0$, so $\dfrac{dy}{dx} = -(y/x)^{2/3}$.

$$\frac{d^2y}{dx^2} = -\frac{2}{3}(y/x)^{-1/3} \cdot \frac{x(dy/dx) - y}{x^2} = -\frac{2}{3} \cdot \frac{x(dy/dx) - y}{x^{5/3}y^{1/3}}$$

$$= \frac{2}{3} \cdot \frac{y + x^{1/3}y^{2/3}}{x^{5/3}y^{1/3}} = \frac{2}{3}y^{2/3} \cdot \frac{x^{1/3} + y^{1/3}}{x^{5/3}y^{1/3}} = \frac{2}{3}(y/x^5)^{1/3}.$$

4.M.47: Given $y^5 - 4y + 1 = x^{1/2}$, we differentiate both sides of this equation (actually, an *identity*) with respect to x and obtain

$$5y^4 \frac{dy}{dx} - 4\frac{dy}{dx} = \frac{1}{2x^{1/2}}, \qquad \text{so that} \qquad \frac{dy}{dx} = y'(x) = \frac{1}{2(5y^4 - 4)\sqrt{x}}. \qquad (1)$$

Another differentiation yields

$$\frac{d^2y}{dx^2} = -\frac{1 + 80x^{3/2}y^3[y'(x)]^3}{4(5y^4 - 4)x^{3/2}},$$

then substitution of $y'(x)$ from Eq. (1) yields

$$y''(x) = \frac{40y^4 - 25y^8 - 20x^{1/3}y^3 - 16}{4x^{3/2}(5y^4 - 4)^3}.$$

4.M.49: Given $x^2 + y^2 = 5xy + 5$, we differentiate both sides with respect to x to obtain

$$2x + 2y\frac{dy}{dx} = 5y + 5x\frac{dy}{dx}, \qquad \text{so that} \qquad \frac{dy}{dx} = y'(x) = \frac{2x - 5y}{5x - 2y}. \qquad (1)$$

Another differentiation yields

$$y''(x) = \frac{2[(y'(x))^2 - 5y'(x) + 1]}{5x - 2y},$$

then substitution of $y'(x)$ from Eq. (1) yields

$$y''(x) = -\frac{42(x^2 - 5xy + y^2)}{(5x - 2y)^3} = -\frac{210}{(5x - 2y)^3}.$$

In the last step we used the original equation in which y is defined implicitly as a function of x.

4.M.51: Given: $y^3 - y = x^2y$:

$$\frac{dy}{dx} = y'(x) = -\frac{2xy}{x^2 + 1 - 3y^2}. \tag{1}$$

Then

$$y''(x) = -\frac{2[y + 2xy'(x) - 3y(y'(x))^2]}{x^2 + 1 - 3y^2}. \tag{2}$$

Substitution of $y'(x)$ from Eq. (1) in the right-hand side of Eq. (2) then yields

$$y''(x) = \frac{2y\left[3x^4 - 9y^4 + 6(x^2 + 1)y^2 + 2x^2 - 1\right]}{(x^2 + 1 - 3y^2)^3}.$$

4.M.53: $f'(x) = 4x^3 - 32$, so there is a critical point at $(2, -48)$. $f''(x) = 12x^2$, but there is no inflection point at $(0, 0)$ because the graph of f is concave upward for all x. But $(0, 0)$ is a dual intercept, and there is an x-intercept at $\left(\sqrt[3]{32}, 0\right)$. The graph of $y = f(x)$ is shown next.

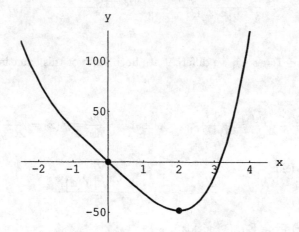

4.M.55: $f'(x) = 2x^3(3x^2 - 4)$ and $f''(x) = 6x^2(5x^2 - 4)$. There are global minima at $\left(-\frac{2}{3}\sqrt{3}, -\frac{32}{27}\right)$ and $\left(\frac{2}{3}\sqrt{3}, -\frac{32}{27}\right)$ and a local maximum at $(0, 0)$, which is also a dual intercept. There are inflection points at $\left(-\frac{2}{5}\sqrt{5}, -\frac{95}{125}\right)$ and $\left(\frac{2}{5}\sqrt{5}, -\frac{95}{125}\right)$ but not at $(0, 0)$. There are x-intercepts at $\left(-\sqrt{2}, 0\right)$ and $\left(\sqrt{2}, 0\right)$. The graph of $y = f(x)$ is shown next.

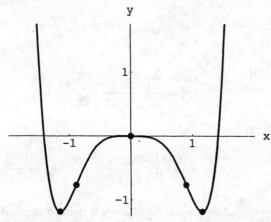

214

4.M.57: If $f(x) = x(4-x)^{1/3}$, then

$$f'(x) = \frac{4(3-x)}{3(4-x)^{2/3}} \quad \text{and} \quad f''(x) = \frac{4(x-6)}{9(4-x)^{5/3}}.$$

So there is a global maximum at $(3, 3)$ intercepts at $(0, 0)$ and $(4, 0)$, a vertical tangent and inflection point at the latter, and an inflection point at $\left(6, -6\sqrt[3]{2}\right)$. The graph is next.

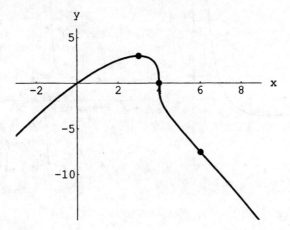

C4.M.59: If $f(x) = \dfrac{x^2+1}{x^2-4} = 1 + \dfrac{5}{x^2-4}$, then

$$f'(x) = -\frac{10x}{(x^2-4)^2} \quad \text{and} \quad f''(x) = \frac{10(3x^2+4)}{(x^2-4)^3}.$$

Thus there is a local maximum at $\left(0, -\frac{1}{4}\right)$, no other extrema, no inflection points, and no intercepts. The lines $x = -2$ and $x = 2$ are vertical asymptotes and $y = 1$ is a horizontal asymptote; the graph is next.

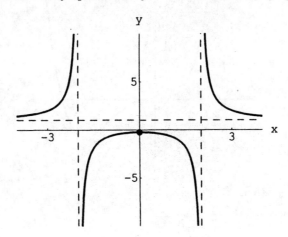

4.M.61: If $f(x) = \dfrac{2x^2}{(x-2)(x+1)}$, then

$$f'(x) = -\frac{2x(x+4)}{(x-2)^2(x+1)^2} \quad \text{and} \quad f''(x) = \frac{4(x^3+6x^2+4)}{(x^2-x-2)^3}.$$

215

Thus $(0, 0)$ is a local maximum and the only intercept; there is a local minimum at $\left(-4, \frac{16}{9}\right)$. There is an inflection point with the approximate coordinates $(-6.107243, 1.801610)$. The lines $x = -1$ and $x = 2$ are vertical asymptotes and the line $y = 2$ is a horizontal asymptote. The left-hand branch of $y = f(x)$ for $x < -2$ is shown next on the left; the figure on the right shows a wider view of the whole graph.

4.M.63: Here we have $f(x) = x^3(3x - 4)$, $f'(x) = 12x^2(x - 1)$, and $f''(x) = 12x(3x - 2)$. Hence there are intercepts at $(0, 0)$ and $\left(\frac{4}{3}, 0\right)$; the graph is increasing for $x > 1$ and decreasing for $x < 1$; it is concave upward for $x > \frac{2}{3}$ and for $x < 0$, concave downward on the interval $\left(0, \frac{2}{3}\right)$. Consequently there is a global minimum at $(1, -1)$ and inflection points at $(0, 0)$ and $\left(\frac{2}{3}, -\frac{16}{27}\right)$. There are no asymptotes, no other extrema, and $f(x) \to +\infty$ as $x \to +\infty$ and as $x \to -\infty$. The graph of f is shown next.

4.M.65: If $f(x) = \dfrac{x^2}{x^2 - 1}$, then

$$f'(x) = -\frac{2x}{(x^2 - 1)^2} \quad \text{and} \quad f''(x) = \frac{2(3x^2 + 1)}{(x^2 - 1)^3}.$$

Thus $(0, 0)$ is the only intercept and is a local maximum; there are no inflection points, the lines $x = -1$ and $x = 1$ are vertical asymptotes, and the line $y = 1$ is a horizontal asymptote. The graph of $y = f(x)$ is next.

216

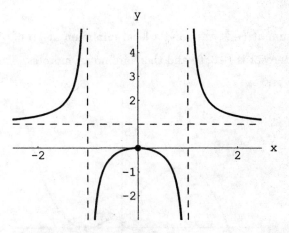

4.M.67: If $f(x) = -10 + 6x^2 - x^3$, then $f'(x) = 3x(4-x)$ and $f''(x) = 6(2-x)$. It follows that there is a local minimum at $(0, -10)$, a local maximum at $(4, 22)$, and an inflection point at $(2, 6)$. The intercepts are approximately $(-1.180140, 0)$, $(1.488872, 0)$, $(5.691268, 0)$, and [exactly] $(0, -10)$. The graph of f is next.

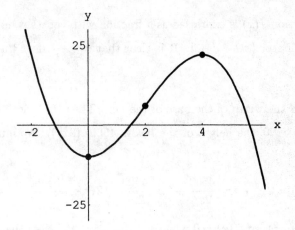

4.M.69: If $f(x) = x^3 - 3x$, then $f'(x) = 3(x+1)(x-1)$ and $f''(x) = 6x$. So there is a local maximum at $(-1, 2)$, a local minimum at $(1, -2)$, and an inflection point at $(0, 0)$. There are also intercepts at $\left(-\sqrt{3}, 0\right)$ and $\left(\sqrt{3}, 0\right)$ but no asymptotes. The graph is next.

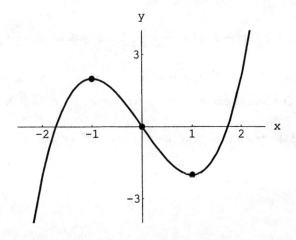

4.M.71: If $f(x) = x^3 + x^2 - 5x + 3 = (x-1)^2(x+3)$, then $f'(x) = (3x+5)(x-1)$ and $f''(x) = 6x + 2$.

So there is a local maximum at $\left(-\frac{5}{3}, \frac{256}{27}\right)$ and a local minimum at $(1, 0)$. There is an inflection point at $\left(-\frac{1}{3}, \frac{128}{27}\right)$. A second x-intercept is $(-3, 0)$ and there are no asymptotes. The graph of $y = f(x)$ is next.

4.M.73: The given function $f(x)$ is expressed as a fraction with constant numerator, so we maximize $f(x)$ by minimizing its denominator $(x + 1)^2 + 1$. It is clear that $x = -1$ does the trick, so the maximum value of $f(x)$ is $f(-1) = 1$.

4.M.75: Let x represent the width of the base of the box. Then the length of the base is $2x$ and, because the volume of the box is 4500, the height of the box is $4500/\left(2x^2\right)$. We minimize the surface area of the box, which is given by

$$f(x) = 2x^2 + 4x \cdot \frac{4500}{2x^2} + 2x \cdot \frac{4500}{2x^2} = 2x^2 + \frac{13500}{x}, \quad 0 < x < \infty.$$

Now $f'(x) = 4x - (13500/x^2)$, so $f'(x) = 0$ when $x = \sqrt[3]{3375} = 15$. Note that $f''(15) > 0$, so surface area is minimized when $x = 15$. The box of minimal surface area is 15 cm wide, 30 cm long, and 10 cm high.

4.M.77: Let x represent the width of the base of the box. Then the box has base of length $2x$ and height $200/x^2$. We minimize the cost C of the box, where

$$C = C(x) = 7 \cdot 2x^2 + 5 \cdot \left(6x \cdot \frac{200}{x^2}\right) + 5 \cdot 2x^2 = 24x^2 + \frac{6000}{x}, \quad 0 < x < \infty.$$

Now $C'(x) = 48x - (6000/x^2)$, so $C'(x) = 0$ when $x = \sqrt[3]{125} = 5$. Because $C''(5) > 0$, the cost C is minimized when $x = 5$. The box of minimal cost is 5 in. wide, 10 in. long, and 8 in. high.

4.M.79: If the speed of the truck is v, then the trip time is $T = 1000/v$. So the resulting cost is

$$C(v) = \frac{10000}{v} + (1000)\left(1 + (0.0003)v^{3/2}\right),$$

so that

$$\frac{C(v)}{1000} = \frac{10}{v} + 1 + (0.0003)v^{3/2}.$$

218

Thus

$$\frac{C'(v)}{1000} = -\frac{10}{v^2} + \frac{3}{2}(0.0003)\sqrt{v}.$$

Then $C'(v) = 0$ when $v = (200,000/9)^{2/5} \approx 54.79$ mi/h. This clearly minimizes the cost, because $C''(v) > 0$ for *all* $v > 0$.

4.M.81: First, given $y^2 = x(x-1)(x-2)$, we differentiate implicitly and find that

$$2y\frac{dy}{dx} = 3x^2 - 6x + 2, \qquad \text{so} \qquad \frac{dy}{dx} = \frac{3x^2 - 6x + 2}{2y}.$$

The only zero of dy/dx in the domain is $1 - \frac{1}{3}\sqrt{3}$, so there are two horizontal tangent lines (the y-coordinates are approximately ± 0.6204). Moreover, $dx/dy = 0$ when $y = 0$; that is, when $x = 0$, when $x = 1$, and when $x = 2$. So there are three vertical tangent lines. After lengthy simplifications, one can show that

$$\frac{d^2y}{dx^2} = \frac{3x^4 - 12x^3 + 12x^2 - 4}{4y^3}.$$

The only zero of $y''(x)$ in the domain is about 2.4679, and there the graph has the two values $y \approx \pm 1.3019$. These are the two inflection points. The graph of the given equation is shown next.

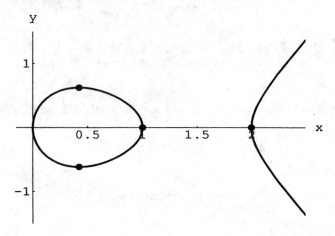

4.M.83: Let x represent the length of each of the dividers. Then the field measures x by $1800/x$ ft. We minimize the total length of fencing, given by:

$$f(x) = 4x + \frac{3600}{x}, \qquad 0 < x < \infty.$$

Now $f'(x) = 4 - 3600x^{-2}$, which is zero only when $x = 30$. Verification: $f'(x) > 0$ if $x > 30$ and $f'(x) < 0$ if $x < 30$, so $f(x)$ is minimized when $x = 30$. The minimum length of fencing required for this field is $f(30) = 240$ ft.

4.M.85: Let x represent the length of each of the dividers. Then the field measures x by A/x ft. We minimize the total length of fencing, given by:

$$f(x) = (n+2)x + \frac{2A}{x}, \quad 0 < x < \infty.$$

Now $f'(x) = n + 2 - 2Ax^{-2}$, which is zero only when $x = \sqrt{2A/(n+2)}$. Verification: $f'(x) > 0$ if $x > \sqrt{2A/(n+2)}$ and $f'(x) < 0$ if $x < \sqrt{2A/(n+2)}$, so $f(x)$ is minimized when $x = \sqrt{2A/(n+2)}$. The minimum length of fencing required for this field is

$$f\left(\sqrt{\frac{2A}{n+2}}\right) = (n+2)\sqrt{\frac{2A}{n+2}} + \frac{2A\sqrt{n+2}}{\sqrt{2A}}$$

$$= \sqrt{2A(n+2)} + \sqrt{2A(n+2)} = 2\sqrt{2A(n+2)} \quad \text{(ft)}.$$

4.M.87: Let L be the line segment with endpoints at $(0, c)$ and $(b, 0)$ on the coordinate axes and suppose that L is tangent to the graph of $y = 1/x^2$ at $(x, 1/x^2)$. We compute the slope of L in several ways: as the value of dy/dx at the point of tangency, as the slope of the line segment between $(x, 1/x^2)$ and $(b, 0)$, and as the slope of the line segment between $(x, 1/x^2)$ and $(0, c)$:

$$-\frac{2}{x^3} = \frac{\frac{1}{x^2} - 0}{x - b}, \quad \text{so} \quad x = -2(x - b), \quad \text{hence} \quad b = \frac{3}{2}x.$$

$$-\frac{2}{x^3} = \frac{\frac{1}{x^2} - c}{x - 0}; \quad -2x = x^3\left(\frac{1}{x^2} - 3\right), \quad cx^2 = 3, \quad \text{hence} \quad c = \frac{3}{x^2}.$$

Now the length of the base of the right triangle is b and its height is c, so its area is given by

$$A(x) = \frac{1}{2} \cdot \frac{3x}{2} \cdot \frac{3}{x^2} = \frac{9}{4x}, \quad 0 < x < +\infty.$$

Clearly A is a strictly decreasing function of x, so it has neither a maximum nor a minimum—not even any local extrema.

4.M.89: Let x be the length of the shorter sides of the base and let y be the height of the box. Then its volume is $3x^2y$, so that $y = 96/x^2$. The total surface area of the box is $6x^2 + 8xy$, thus is given as a function of x by

$$A(x) = 6x^2 + 8x \cdot \frac{96}{x^2} = 6x^2 + \frac{768}{x}, \quad 0 < x < +\infty.$$

So $A'(x) = 12x - 768x^{-2}$; $A'(x) = 0$ when $x^3 = 64$, so that $x = 4$. Because $A''(x) > 0$ for all $x > 0$, we have found the value of x that yields the global minimum value of A, which is $A(4) = 288$ (in.^{2}). The fact that 288 is also the numerical value of the volume is a mere coincidence.

4.M.91: Let x be the length of the shorter sides of the base and let y be the height of the box. Then its volume is $5x^2y$, so that $y = 45/x^2$. The total surface area of the box is $10x^2 + 12xy$, thus is given as a function of x by

$$A(x) = 10x^2 + 12x \cdot \frac{45}{x^2} = 10x^2 + \frac{540}{x}, \quad 0 < x < +\infty.$$

So $A'(x) = 20x - 540x^{-2}$; $A'(x) = 0$ when $x^3 = 27$, so that $x = 3$. Because $A''(x) > 0$ for all $x > 0$, we have found the value of x that yields the global minimum value of A, which is $A(3) = 270$ (cm^2).

4.M.93: First,

$$m = \lim_{x \to \pm\infty} \frac{f(x)}{x} = \lim_{x \to \pm\infty} \frac{(1-x)^{2/3}}{x^{2/3}}$$

$$= \lim_{x \to \pm\infty} \left(\frac{\frac{1}{x} - 1}{1} \right)^{2/3} = +1.$$

Then

$$b = \lim_{x \to \infty} [f(x) - mx] = \lim_{x \to \infty} \left(x^{1/3}(1-x)^{2/3} - x \right)$$

$$= \lim_{x \to \infty} \frac{\left(x^{1/3}(1-x)^{2/3} - x \right)\left(x^{2/3}(1-x)^{4/3} + x^{4/3}(1-x)^{2/3} + x^2 \right)}{x^{2/3}(1-x)^{4/3} + x^{4/3}(1-x)^{2/3} + x^2}$$

$$= \lim_{x \to \infty} \frac{x(1-x)^2 - x^3}{x^{2/3}(1-x)^{4/3} + x^{4/3}(1-x)^{2/3} + x^2}$$

$$= \lim_{x \to \infty} \frac{x - 2x^2}{x^{2/3}(1-x)^{4/3} + x^{4/3}(1-x)^{2/3} + x^2}$$

$$= \lim_{x \to \infty} \frac{\frac{1}{x} - 2}{\left(\frac{1-x}{x} \right)^{4/3} + \left(\frac{1-x}{x} \right)^{2/3} + 1}$$

$$= \frac{0 - 2}{1 + 1 + 1} = -\frac{2}{3}.$$

The limit is the same as $x \to -\infty$. So the graph of $f(x) = x^{1/3}(1-x)^{2/3}$ has the oblique asymptote $y = x - \frac{2}{3}$.

4.M.95: By l'Hôpital's rule,

$$\lim_{x \to 2} \frac{x - 2}{x^2 - 4} = \lim_{x \to 2} \frac{1}{2x} = \frac{1}{4}.$$

Without l'Hôpital's rule,

$$\lim_{x \to 2} \frac{x - 2}{x^2 - 4} = \lim_{x \to 2} \frac{x - 2}{(x + 2)(x - 2)} = \lim_{x \to 2} \frac{1}{x + 2} = \frac{1}{4}.$$

4.M.97: By l'Hôpital's rule, $\displaystyle \lim_{x \to \pi} \frac{1 + \cos x}{(x - \pi)^2} = \lim_{x \to \pi} \frac{-\sin x}{2(x - \pi)} = \lim_{x \to \pi} \frac{-\cos x}{2} = \frac{1}{2}.$

221

4.M.99: By l'Hôpital's rule (applied three times),

$$\lim_{t\to 0}\frac{\tan t-\sin t}{t^3}=\lim_{t\to 0}\frac{\sec^2 t-\cos t}{3t^2}$$

$$=\lim_{t\to 0}\frac{2\sec^2 t\tan t+\sin t}{6t}=\lim_{t\to 0}\frac{4\sec^2 t\tan^2 t+2\sec^4 t+\cos t}{6}=\frac{1}{2}.$$

4.M.101: By l'Hôpital's rule,

$$\lim_{x\to 0}(\cot x)\ln(1+x)=\lim_{x\to 0}\frac{\ln(1+x)}{\tan x}=\lim_{x\to 0}\frac{1}{(1+x)\sec^2 x}=\frac{1}{1\cdot 1}=1.$$

4.M.103: After combining the two fractions, we apply l'Hôpital's rule once, then use a little algebra:

$$\lim_{x\to 0}\left(\frac{1}{x^2}-\frac{1}{1-\cos x}\right)=\lim_{x\to 0}\frac{1-\cos x-x^2}{x^2(1-\cos x)}=\lim_{x\to 0}\frac{(\sin x)-2x}{2x(1-\cos x)+x^2\sin x}=\lim_{x\to 0}\frac{\dfrac{\sin x}{x}-2}{2(1-\cos x)+x\sin x}.$$

Now if x is close to (but not equal to) zero, $(\sin x)/x\approx 1$, so the numerator in the last limit is near -1. Moreover, for such x, $\cos x<1$ and x and $\sin x$ have the same sign, so the denominator in the last limit is close to zero *and positive*. Therefore the limit is $-\infty$.

4.M.105: Here it is easier not to use l'Hôpital's rule:

$$\lim_{x\to\infty}\left(\sqrt{x^2-x-1}-\sqrt{x}\right)=\lim_{x\to\infty}\frac{x^2-x-1-x}{\sqrt{x^2-x-1}+\sqrt{x}}$$

$$=\lim_{x\to\infty}\frac{x^2-2x-1}{\sqrt{x^2-x-1}+\sqrt{x}}=\lim_{x\to\infty}\frac{x-2-\dfrac{1}{x}}{\sqrt{1-\dfrac{1}{x}-\dfrac{1}{x^2}}-\sqrt{\dfrac{1}{x}}}=+\infty.$$

4.M.107: First we need an auxiliary result:

$$\lim_{x\to\infty}2xe^{-2x}=\lim_{x\to\infty}\frac{2x}{e^{2x}}=\lim_{x\to\infty}\frac{2}{2e^{2x}}=0.$$

Then

$$\ln\left(\lim_{x\to\infty}(e^{2x}-2x)^{1/x}\right)=\lim_{x\to\infty}\ln(e^{2x}-2x)^{1/x}=\lim_{x\to\infty}\frac{\ln(e^{2x}-2x)}{x}$$

$$=\lim_{x\to\infty}\frac{2e^{2x}-2}{e^{2x}-2x}=\lim_{x\to\infty}\frac{2-2e^{-2x}}{1-2xe^{-2x}}=\frac{2-0}{1-0}=2.$$

Therefore $\lim\limits_{x\to\infty}(e^{2x}-2x)^{1/x}=e^2$.

4.M.109: This is one of the more challenging problems in the book, and we almost regret publication of this solution. First let $u = 1/x$. Then

$$L = \lim_{x \to \infty} x \cdot \left[\left(1 + \frac{1}{x} \right)^x - e \right] = \lim_{u \to 0^+} \frac{(1+u)^{1/u} - e}{u}.$$

Apply l'Hôpital's rule once:

$$L = \lim_{u \to 0^+} (1+u)^{1/u} \left(\frac{u - (1+u)\ln(1+u)}{u^2(1+u)} \right).$$

Now apply the product rule for limits!

$$L = e \cdot \left(\lim_{u \to 0^+} \frac{u - (1+u)\ln(1+u)}{u^2(1+u)} \right).$$

Finally apply l'Hôpital's rule twice:

$$L = e \cdot \left(\lim_{u \to 0^+} \frac{1 - 1 - \ln(1+u)}{2u + 3u^2} \right) = e \cdot \left(\lim_{u \to 0^+} \frac{-1}{(1+u)(2+6u)} \right) = e \cdot \frac{-1}{1 \cdot 2} = -\frac{e}{2}.$$

Section 5.2

5.2.1: $\int (3x^2 + 2x + 1)\,dx = x^3 + x^2 + x + C.$

5.2.3: $\int (1 - 2x^2 + 3x^3)\,dx = \frac{3}{4}x^4 - \frac{2}{3}x^3 + x + C.$

5.2.5: $\int (3x^{-3} + 2x^{3/2} - 1)\,dx = -\frac{3}{2}x^{-2} + \frac{4}{5}x^{5/2} - x + C.$

5.2.7: $\int \left(\frac{3}{2}t^{1/2} + 7\right) dt = t^{3/2} + 7t + C.$

5.2.9: $\int (x^{2/3} + 4x^{-5/4})\,dx = \frac{3}{5}x^{5/3} - 16x^{-1/4} + C.$

5.2.11: $\int (4x^3 - 4x + 6)\,dx = x^4 - 2x^2 + 6x + C.$

5.2.13: $\int 7e^{x/7}\,dx = 49e^{x/7} + C.$

5.2.15: $\int (x+1)^4\,dx = \frac{1}{5}(x+1)^5 + C.$ Note that many computer algebra systems give the answer

$$C + x + 2x^2 + 2x^3 + x^4 + \frac{x^5}{5}.$$

5.2.17: $\int \frac{1}{(x-10)^7}\,dx = \int (x-10)^{-7}\,dx = -\frac{1}{6}(x-10)^{-6} + C = -\frac{1}{6(x-10)^6} + C.$

5.2.19: $\int \sqrt{x}\,(1-x)^2\,dx = \int (x^{1/2} - 2x^{3/2} + x^{5/2})\,dx = \frac{2}{3}x^{3/2} - \frac{4}{5}x^{5/2} + \frac{2}{7}x^{7/2} + C.$

5.2.21: $\int \frac{2x^4 - 3x^3 + 5}{7x^2}\,dx = \int \left(\frac{2}{7}x^2 - \frac{3}{7}x + \frac{5}{7}x^{-2}\right) dx = \frac{2}{21}x^3 - \frac{3}{14}x^2 - \frac{5}{7}x^{-1} + C.$

5.2.23: $\int (9t+11)^5\,dt = \frac{1}{54}(9t+11)^6 + C.$ Mathematica gives the answer

$$C + 161051t + \frac{658845t^2}{2} + 359370t^3 + \frac{441045t^4}{2} + 72171t^5 + \frac{19683t^6}{2}.$$

5.2.25: $\int (e^{2x} + e^{-2x})\,dx = \frac{1}{2}e^{2x} - \frac{1}{2}e^{-2x} + C.$

5.2.27: $\int (5\cos 10x - 10\sin 5x)\,dx = \frac{1}{2}\sin 10x + 2\cos 5x + C.$

5.2.29: $\int (3\cos \pi t + \cos 3\pi t)\,dt = \frac{3}{\pi}\sin \pi t + \frac{1}{3\pi}\sin 3\pi t + C.$

224

5.2.31: $D_x\left(\frac{1}{2}\sin^2 x + C_1\right) = \sin x \cos x = D_x\left(-\frac{1}{2}\cos^2 x + C_2\right)$. Because

$$\frac{1}{2}\sin^2 x + C_1 = -\frac{1}{2}\cos^2 x + C_2, \quad \text{it follows that} \quad C_2 - C_1 = \frac{1}{2}\sin^2 x + \frac{1}{2}\cos^2 x = \frac{1}{2}.$$

5.2.33: $\displaystyle\int \sin^2 x \, dx = \int \left(\frac{1}{2} - \frac{1}{2}\cos 2x\right) dx = \frac{1}{2}x - \frac{1}{4}\sin 2x + C$ and

$\displaystyle\int \cos^2 x \, dx = \int \left(\frac{1}{2} + \frac{1}{2}\cos 2x\right) dx = \frac{1}{2}x + \frac{1}{4}\sin 2x + C.$

5.2.35: $y(x) = x^2 + x + C$; $y(0) = 3$, so $y(x) = x^2 + x + 3$.

5.2.37: $y(x) = \frac{2}{3}x^{3/2} + C$ and $y(4) = 0$, so $y(x) = \frac{2}{3}x^{3/2} - \frac{16}{3}$.

5.2.39: $y(x) = 2\sqrt{x+2} + C$ and $y(2) = -1$, so $y(x) = 2\sqrt{x+2} - 5$.

5.2.41: $y(x) = \frac{3}{4}x^4 - 2x^{-1} + C$; $y(1) = 1$, so $y(x) = \frac{3}{4}x^4 - 2x^{-1} + \frac{9}{4}$.

5.2.43: $y(x) = \frac{1}{4}(x-1)^4 + C$; $y(0) = 2 = \frac{1}{4} + C$, so $C = \frac{7}{4}$.

5.2.45: $y(x) = 3e^{2x} + C$; $y(0) = 10$, so $C = 7$.

5.2.47: $v(t) = 6t^2 - 4t + C$; $v(0) = -10$, so $v(t) = 6t^2 - 4t - 10$. Next, $x(t) = 2t^3 - 2t^2 - 10t + K$; $x(0) = 0$, so $K = 0$.

5.2.49: $v(t) = \frac{2}{3}t^3 + C$; $v(0) = 3$, so $v(t) = \frac{2}{3}t^3 + 3$. Next, $x(t) = \frac{1}{6}t^4 + 3t + K$; $x(0) = -7$, so $K = -7$.

5.2.51: $v(t) = C - \cos t$; $v(0) = 0$, so $v(t) = 1 - \cos t$. Next, $x(t) = t - \sin t + K$; $x(0) = 0$, so $K = 0$.

5.2.53: Note that $v(t) = 5$ for $0 \leq t \leq 5$ and that $v(t) = 10 - t$ for $5 \leq t \leq 10$. Hence $x(t) = 5t + C_1$ for $0 \leq t \leq 5$ and $x(t) = 10t - \frac{1}{2}t^2 + C_2$ for $5 \leq t \leq 10$. Also $C_1 = 0$ because $x(0) = 0$ and continuity of $x(t)$ requires that $5t + C_1$ and $10t - \frac{1}{2}t^2 + C_2$ agree when $t = 5$. This implies that $C_2 = -\frac{25}{2}$. The graph of x is next.

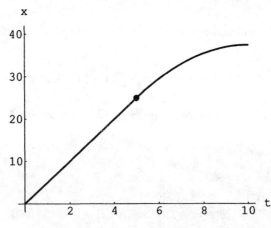

5.2.55: First, $v(t) = t$ if $0 \leq t \leq 5$ and $v(t) = 10 - t$ if $5 \leq t \leq 10$. Hence $x(t) = \frac{1}{2}t^2 + C_1$ if $0 \leq t \leq 5$ and $x(t) = 10t - \frac{1}{2}t^2 + C_2$ if $5 \leq t \leq 10$. Finally, $C_1 = 0$ because $x(0) = 0$ and continuity of $x(t)$ requires that $\frac{1}{2}t^2 + C_1 = 10t - \frac{1}{2}t^2 + C_2$ when $t = 5$, so that $C_2 = -25$. The graph of $x(t)$ is next.

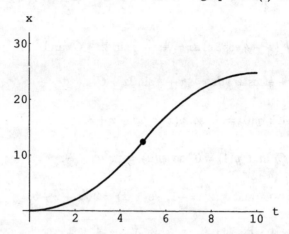

In the solutions for Problems 57–78, unless otherwise indicated, we will take the upward direction to be the positive direction, $s = s(t)$ for position (in feet) at time t (in seconds) with $s = 0$ corresponding to ground level, and $v(t)$ velocity at time t in feet per second, $a = a(t)$ acceleration at time t in feet per second per second. The initial position will be denoted by s_0 and the initial velocity by v_0.

5.2.57: Here, $a = -32$, $v(t) = -32t + 96$, $s(t) = -16t^2 + 96t$. The maximum height is reached when $v = 0$, thus when $t = 3$. The maximum height is therefore $s(3) = 144$. The ball remains aloft until $s(t) = 0$ for $t > 0$; $t = 6$. So it remains aloft for six seconds.

5.2.59: Here it is more convenient to take the downward direction as the positive direction. Thus

$$a(t) = +32, \quad v(t) = +32t \quad \text{(because } v_0 = 0), \quad \text{and} \quad s(t) = 16t^2 \quad \text{(because } s_0 = 0).$$

The stone hits the bottom when $t = 3$, and at that time we have $s = s(3) = 144$. Answer: The well is 144 feet deep.

5.2.61: Here, $v(t) = -32t + 48$ and $s(t) = -16t^2 + 48t + 160$. The ball strikes the ground at that value of $t > 0$ for which $s(t) = 0$:

$$0 = s(t) = -16(t - 5)(t + 2), \quad \text{so} \quad t = 5.$$

Therefore the ball remains aloft for 5 seconds. Its velocity at impact is $v(5) = -112$ (ft/s), so the ball strikes the ground at a speed of 112 ft/s.

5.2.63: One solution: Take $s_0 = 960$, $v_0 = 0$. Then

$$v(t) = -32t \quad \text{and} \quad s(t) = -16t^2 + 960.$$

The ball hits the street for that value of $t > 0$ for which $s(t) = 0$—that is, when $t = 2\sqrt{15}$. It therefore takes the ball approximately 7.746 seconds to reach the street. Its velocity then is $v(2\sqrt{15}) = -64\sqrt{15}$ (ft/s)—approximately 247.87 ft/s (downward), almost exactly 169 miles per hour.

5.2.65: With $v(t) = -32t + v_0$ and $s(t) = -16t^2 + v_0 t$, we have the maximum altitude $s = 225$ occurring when $v(t) = 0$; that is, when $t = v_0/32$. So

$$225 = s(v_0/32) = (-16)(v_0/32)^2 + (v_0)^2/32 = (v_0)^2/64.$$

It follows that $v_0 = +120$. So the initial velocity of the ball was 120 ft/s.

5.2.67: In this problem, $v(t) = -32t + v_0 = -32t$ and $s(t) = -16t^2 + s_0 = -16t^2 + 400$. The ball reaches the ground when $s = 0$, thus when $16t^2 = 400$: $t = +5$. Therefore the impact velocity is $v(5) = (-32)(5) = -160$ (ft/s).

5.2.69: For this problem, we take $s(t) = -16t^2 + 160t$ and $v(t) = -32t + 160$. Because $s = 0$ when $t = 0$ and when $t = 10$, the time aloft is 10 seconds. The velocity is zero at maximum altitude, and that occurs when $32t = 160$: $t = 5$. So the maximum altitude is $s(5) = 400$ (ft).

5.2.71: Because $v_0 = -40$, $v(t) = -32t - 40$. Thus $s(t) = -16t^2 - 40t + 555$. Now $s(t) = 0$ when $t = \frac{1}{4}\left(-5 + 2\sqrt{145}\right) \approx 4.77$ (s). The speed at impact is $|v(t)|$ for that value of t; that is, $16\sqrt{145} \approx 192.6655$ (ft/s), over 131 miles per hour.

5.2.73: Bomb equations: $a = -32$, $v = -32t$, $s_B = s = -16t^2 + 800$. Here we have $t = 0$ at the time the bomb is released. Projectile equations: $a = -32$, $v = -32(t - 2) + v_0$, and

$$s_P = s = -16(t - 2)^2 + v_0(t - 2), \quad t \geqq 2.$$

We require $s_B = s_P = 400$ at the same time. The equation $s_B = 400$ leads to $t = 5$, and for $s_P(5) = 400$, we must have $v_0 = 544/3 \approx 181.33$ (ft/s).

5.2.75: The deceleration $a = k > 0$ is unknown at first. But the velocity of the car is $v(t) = -kt + 88$, and so the distance it travels after the brakes are applied at time $t = 0$ is

$$x(t) = -\frac{1}{2}kt^2 + 88t.$$

But $x = 176$ when $v = 0$, so the stopping time t_1 is $88/k$ because that is the time at which $v = 0$. Therefore

$$176 = -\frac{1}{2}k\left(\frac{88}{k}\right)^2 + (88)\left(\frac{88}{k}\right) = \frac{3872}{k}.$$

It follows that $k = 22$ (ft/s^2), about $0.69g$.

5.2.77: (a): With the usual coordinate system, the ball has velocity $v(t) = -32t + v_0$ (ft/s) at time t (seconds) and altitude $y(t) = -16t^2 + v_0 t$ (ft). We require $y(T) = 144$ when $v(T) = 0$: $v_0 = 32T$, so

$$144 = -16T^2 + 32T^2 = 16T^2,$$

and thus $T = 3$ and $v_0 = 96$. Answer: 96 ft/s.

(b): Now $v(t) = -\frac{26}{5}t + 96$, $y(t) = -\frac{13}{5}t^2 + 96t$. Maximum height occurs when $v(t) = 0$: $t = \frac{240}{13}$. The maximum height is

$$y\left(\frac{240}{13}\right) = -\frac{13}{5} \cdot \frac{240^2}{13^2} + 96 \cdot \frac{240}{13} = \frac{11520}{13} \approx 886 \text{ ft.}$$

5.2.79: Let a denote the deceleration constant of the car when braking. In the police experiment, we have the distance the car travels from $x_0 = 0$ at time t to be

$$x(t) = -\frac{1}{2}at^2 + 25 \cdot \frac{22}{15}t$$

(the factor $\frac{22}{15}$ converts 25 miles per hour to feet per second). When we solve simultaneously the equations $x(t) = 45$ and $x'(t) = 0$, we find that $a = \frac{1210}{81} \approx 14.938$. When we use this value of a and substitute data from the accident, we find the position function of the car to be

$$x(t) = -\frac{1}{2} \cdot \frac{1210}{81}t^2 + v_0 t$$

where v_0 is its initial velocity. Now when we solve simultaneously $x(t) = 210$ and $x'(t) = 0$, we find that $v_0 = \frac{110}{9}\sqrt{42} \approx 79.209$ feet per second, almost exactly 54 miles per hour.

Section 5.3

5.3.1: $\displaystyle\sum_{i=1}^{5} 3^i = 3^1 + 3^2 + 3^3 + 3^4 + 3^5 = 3 + 9 + 27 + 81 + 243.$

5.3.3: $\displaystyle\sum_{j=1}^{5} \frac{1}{j+1} = \frac{1}{2} + \frac{1}{3} + \frac{1}{4} + \frac{1}{5} + \frac{1}{6}.$

5.3.5: $\displaystyle\sum_{k=1}^{6} \frac{1}{k^2} = 1 + \frac{1}{4} + \frac{1}{9} + \frac{1}{16} + \frac{1}{25} + \frac{1}{36}.$

5.3.7: $\displaystyle\sum_{n=1}^{5} x^n = x + x^2 + x^3 + x^4 + x^5.$

5.3.9: $\displaystyle 1 + 4 + 9 + 16 + 25 = \sum_{n=1}^{5} n^2.$

5.3.11: $1 + \dfrac{1}{2} + \dfrac{1}{3} + \dfrac{1}{4} + \dfrac{1}{5} = \displaystyle\sum_{k=1}^{5} \dfrac{1}{k}.$

5.3.13: $\dfrac{1}{2} + \dfrac{1}{4} + \dfrac{1}{8} + \dfrac{1}{16} + \dfrac{1}{32} + \dfrac{1}{64} = \displaystyle\sum_{m=1}^{6} \dfrac{1}{2^m}.$

5.3.15: $\dfrac{2}{3} + \dfrac{4}{9} + \dfrac{8}{27} + \dfrac{16}{81} + \dfrac{32}{243} = \displaystyle\sum_{n=1}^{5} \left(\dfrac{2}{3}\right)^n.$

5.3.17: $x + \dfrac{x^2}{2} + \dfrac{x^3}{3} + \cdots + \dfrac{x^{10}}{10} = \displaystyle\sum_{n=1}^{10} \dfrac{1}{n} x^n.$

5.3.19: Using Eqs. (3), (4), (6), and (7), we find that

$$\sum_{i=1}^{10}(4i - 3) = 4 \cdot \sum_{i=1}^{10} i - 3 \cdot \sum_{i=1}^{10} 1 = 4 \cdot \frac{10 \cdot 11}{2} - 3 \cdot 10 = 190.$$

5.3.21: We use Eqs. (3), (4), (6), and (8) to obtain

$$\sum_{i=1}^{10}(3i^2 + 1) = 3 \cdot \sum_{i=1}^{10} i^2 + \sum_{i=1}^{10} 1 = 3 \cdot \frac{10 \cdot 11 \cdot 21}{6} + 10 = 1165.$$

5.3.23: First expand, then use Eqs. (3), (4), (6), (7), and (8):

$$\sum_{r=1}^{8}(r - 1)(r + 2) = \sum_{r=1}^{8}(r^2 + r - 2) = \sum_{r=1}^{8} r^2 + \sum_{r=1}^{8} r - 2 \cdot \sum_{r=1}^{8} 1 = \frac{8 \cdot 9 \cdot 17}{6} + \frac{8 \cdot 9}{2} - 2 \cdot 8 = 224.$$

5.3.25: $\displaystyle\sum_{i=1}^{6} \left(i^3 - i^2\right) = \frac{6^2 \cdot 7^2}{4} - \frac{6 \cdot 7 \cdot 13}{6} = 3^2 \cdot 7^2 - 7 \cdot 13 = 441 - 91 = 350.$

5.3.27: $\displaystyle\sum_{i=1}^{100} i^2 = \frac{100 \cdot 101 \cdot 201}{6} = 50 \cdot 101 \cdot 67 = 338350.$

5.3.29: $\displaystyle\lim_{n\to\infty} \frac{1^2 + 2^2 + \cdots + n^2}{n^3} = \lim_{n\to\infty} \frac{n(n+1)(2n+1)}{6n^3} = \lim_{n\to\infty} \left(\frac{1}{3} + \frac{1}{2n} + \frac{1}{6n^2}\right) = \frac{1}{3}.$

5.3.31: $\displaystyle\sum_{i=1}^{n}(2i - 1) = 2 \cdot \frac{n(n+1)}{2} - n = n^2.$

5.3.33: $\underline{A}_5 = \displaystyle\sum_{i=1}^{5} \frac{i-1}{5} \cdot \frac{1}{5} = \frac{2}{5}, \qquad \overline{A}_5 = \displaystyle\sum_{i=1}^{5} \frac{i}{5} \cdot \frac{1}{5} = \frac{3}{5}.$

5.3.35: $\underline{A}_6 = \displaystyle\sum_{i=1}^{6} \left[2\left(\frac{i-1}{2}\right) + 3\right] \cdot \frac{3}{6} = \frac{33}{2}, \qquad \overline{A}_6 = \displaystyle\sum_{i=1}^{6} \left[2 \cdot \frac{i}{2} + 3\right] \frac{3}{6} = \frac{39}{2}.$

5.3.37: $\underline{A}_5 = \sum_{i=1}^{5} \left(\frac{i-1}{5}\right)^2 \cdot \frac{1}{5} = \frac{6}{25}, \qquad \overline{A}_5 = \sum_{i=1}^{5} \left(\frac{i}{5}\right)^2 \cdot \frac{1}{5} = \frac{11}{25}.$

5.3.39: $\underline{A}_5 = \sum_{i=1}^{5} \left[9 - \left(\frac{3i}{5}\right)^2\right] \frac{3}{5} = \frac{378}{25}, \qquad \overline{A}_5 = \sum_{i=1}^{5} \left[9 - \left(\frac{3i-3}{5}\right)^2\right] \cdot \frac{3}{5} = \frac{513}{25}.$

5.3.41: $\underline{A}_{10} = \sum_{i=1}^{10} \left(\frac{i-1}{10}\right)^3 \cdot \frac{1}{10} = \frac{81}{400}, \qquad \overline{A}_{10} = \sum_{i=1}^{10} \left(\frac{i}{10}\right)^3 \cdot \frac{1}{10} = \frac{121}{400}.$

5.3.43: When we add the two given equations, we obtain

$$2 \cdot \sum_{i=1}^{n} i = (n+1) + (n+1) + \cdots + (n+1) \qquad (n \text{ terms})$$

$$= n(n+1), \qquad \text{and therefore} \qquad \sum_{i=1}^{n} i = \frac{n(n+1)}{2}.$$

5.3.45: $\sum_{i=1}^{n} \frac{i}{n^2} = \frac{n(n+1)}{2n^2} \to \frac{1}{2}$ as $n \to \infty$.

5.3.47: $\sum_{i=1}^{n} \left(\frac{3i}{n}\right)^3 \cdot \frac{3}{n} = \frac{81n^2(n+1)^2}{4n^4} \to \frac{81}{4}$ as $n \to \infty$.

5.3.49: $\sum_{i=1}^{n} \left(5 - \frac{3i}{n}\right) \cdot \left(\frac{1}{n}\right) = 5n \cdot \frac{1}{n} - \frac{3n(n+1)}{2n^2} \to \frac{7}{2}$ as $n \to \infty$.

5.3.51: $\sum_{i=1}^{n} f(x_i) \, \Delta x = \sum_{i=1}^{n} \frac{h}{b} \cdot \frac{bi}{n} \cdot \frac{b}{n} = \frac{bh}{n^2} \cdot \frac{n(n+1)}{2} \to \frac{1}{2}bh$ as $n \to \infty$.

5.3.53: Using the formulas derived in the solution of Problem 52, we have

$$\lim_{n \to \infty} \frac{A_n}{C_n} = \lim_{n \to \infty} \frac{r}{2} \cos \frac{\pi}{n} = \frac{r}{2}.$$

Thus $A = \frac{1}{2}rC$. Hence if $A = \pi r^2$, then $C = 2\pi r$.

Section 5.4

5.4.1: $\lim_{n \to \infty} \sum_{i=1}^{n} (2x_i - 1) \, \Delta x = \int_{1}^{3} (2x - 1) \, dx.$

5.4.3: $\lim_{n \to \infty} \sum_{i=1}^{n} (x_i^2 + 4) \, \Delta x = \int_{0}^{10} (x^2 + 4) \, dx.$

5.4.5: $\lim_{n \to \infty} \sum_{i=1}^{n} \sqrt{m_i} \, \Delta x = \int_{4}^{9} \sqrt{x} \, dx.$

5.4.7: $\displaystyle \lim_{n \to \infty} \sum_{i=1}^{n} \frac{1}{\sqrt{1+m_i}} \, \Delta x = \int_3^8 \frac{1}{\sqrt{1+x}} \, dx.$

5.4.9: $\displaystyle \lim_{n \to \infty} \sum_{i=1}^{n} (\sin 2\pi m_i) \, \Delta x = \int_0^{1/2} \sin 2\pi x \, dx.$

5.4.11: $\displaystyle \sum_{i=1}^{n} f(x_i^\star) \, \Delta x = \sum_{i=1}^{5} \left(\frac{i}{5}\right)^2 \cdot \left(\frac{1}{5}\right) = \frac{1}{125} \cdot \frac{5}{6} \cdot 6 \cdot 11 = \frac{11}{25}.$

5.4.13: $\displaystyle \sum_{i=1}^{n} f(x_i^\star) \, \Delta x = \sum_{i=1}^{5} \frac{1}{1+i} = \frac{29}{20}.$

5.4.15: $\displaystyle \sum_{i=1}^{n} f(x_i^\star) \, \Delta x = \sum_{i=1}^{6} \left[2\left(1 + \frac{i}{2}\right) + 1\right] \cdot \frac{1}{2} = \frac{39}{2}.$

5.4.17: $\displaystyle \sum_{i=1}^{n} f(x_i^\star) \, \Delta x = \sum_{i=1}^{5} \left[\left(1 + \frac{3i}{5}\right)^3 - 3\left(1 + \frac{3i}{5}\right)\right] \cdot \frac{3}{5} = \frac{294}{5}.$

5.4.19: $\displaystyle \sum_{i=1}^{n} f(x_i^\star) \, \Delta x = \sum_{i=1}^{6} \left(\cos \frac{i\pi}{6}\right) \cdot \frac{\pi}{6} = -\frac{\pi}{6}.$

5.4.21: $\displaystyle \sum_{i=1}^{n} f(x_i^\star) \, \Delta x = \sum_{i=1}^{5} \left(\frac{i-1}{5}\right)^2 \cdot \left(\frac{1}{5}\right) = \frac{6}{25}.$

5.4.23: $\displaystyle \sum_{i=1}^{n} f(x_i^\star) \, \Delta x = \sum_{i=1}^{5} \frac{1}{1 + (i-1)} = \frac{137}{60}.$

5.4.25: $\displaystyle \sum_{i=1}^{n} f(x_i^\star) \, \Delta x = \sum_{i=1}^{6} \left[2\left(1 + \frac{i-1}{2}\right) + 1\right] \cdot \frac{1}{2} = \frac{33}{2}.$

5.4.27: $\displaystyle \sum_{i=1}^{n} f(x_i^\star) \, \Delta x = \sum_{i=1}^{5} \left[\left(1 + \frac{3(i-1)}{5}\right)^3 - 3\left(1 + \frac{3(i-1)}{5}\right)\right] \cdot \frac{3}{5} = \frac{132}{5}.$

5.4.29: $\displaystyle \sum_{i=1}^{n} f(x_i^\star) \, \Delta x = \sum_{i=1}^{6} \left(\cos \frac{(i-1)\pi}{6}\right) \cdot \frac{\pi}{6} = \frac{\pi}{6}.$

5.4.31: $\displaystyle \sum_{i=1}^{n} f(x_i^\star) \, \Delta x = \sum_{i=1}^{5} \left(\frac{2i-1}{10}\right)^2 \cdot \left(\frac{1}{5}\right) = \frac{33}{100}.$

5.4.33: $\displaystyle \sum_{i=1}^{n} f(x_i^\star) \, \Delta x = \sum_{i=1}^{5} \frac{2}{1 + 2i} = \frac{6086}{3465}.$

5.4.35: $\displaystyle \sum_{i=1}^{n} f(x_i^\star) \, \Delta x = \sum_{i=1}^{6} \left[2\left(1 + \frac{2i-1}{4}\right) + 1\right] \cdot \frac{1}{2} = 18.$

5.4.37: $\displaystyle\sum_{i=1}^{n} f(x_i^\star)\,\Delta x = \sum_{i=1}^{5}\left[\left(1+\frac{6i-3}{10}\right)^3 - 3\left(1+\frac{6i-3}{10}\right)\right]\cdot\frac{3}{5} = \frac{1623}{40}.$

5.4.39: $\displaystyle\sum_{i=1}^{n} f(x_i^\star)\,\Delta x = \sum_{i=1}^{6}\left(\cos\frac{(2i-1)\pi}{12}\right)\cdot\frac{\pi}{6} = 0.$

5.4.41: $\displaystyle\sum_{i=1}^{n} f(x_i^\star)\,\Delta x = \sum_{i=1}^{5}\frac{5}{5i+2} = \frac{259775}{141372} \approx 1.837527940469.$

5.4.43: $\displaystyle\sum_{i=1}^{n}\left(\frac{2i}{n}\right)^2\cdot\frac{2}{n} = \frac{8n(n+1)(2n+1)}{6n^3} \to \frac{8}{3}$ as $n\to\infty.$

5.4.45: $\displaystyle\sum_{i=1}^{n}\left(2\cdot\frac{3i}{n}+1\right)\cdot\frac{3}{n} = \frac{18n(n+1)}{2n^2} + n\cdot\frac{3}{n} = 12 + \frac{9}{n} \to 12$ as $n\to\infty.$

5.4.47: $\displaystyle\sum_{i=1}^{n}\left[3\cdot\left(\frac{3i}{n}\right)^2+1\right]\cdot\frac{3}{n} = \frac{81n(n+1)(2n+1)}{6n^3} + n\cdot\frac{3}{n} \to 27 + 3 = 30$ as $n\to\infty.$

5.4.49: Choose $x_i^\star = x_i = \dfrac{bi}{n}$ and $\Delta x = \dfrac{b}{n}.$ Then

$$\int_0^b x^2\,dx = \lim_{n\to\infty}\sum_{i=1}^{n}\left(\frac{bi}{n}\right)^2\cdot\frac{b}{n} = \lim_{n\to\infty}\frac{n(n+1)(2n+1)}{6n^3}b^3 = \frac{1}{3}b^3.$$

5.4.51: If $x_i^\star = \frac{1}{2}(x_{i-1}+x_i)$ for each i, then $x_i^\star$ is the midpoint of each subinterval of the partition, and hence $\{x_i^\star\}$ is a selection for the partition. Moreover, $\Delta x_i = x_i - x_{i-1}$ for each i. So

$$\sum_{i=1}^{n} x_i^\star\,\Delta x_i = \sum_{i=1}^{n}\tfrac{1}{2}(x_i + x_{i-1})(x_i - x_{i-1})$$

$$= \tfrac{1}{2}\sum_{i=1}^{n}\left(x_i^2 - x_{i-1}^2\right)$$

$$= \tfrac{1}{2}\left(x_1^2 - x_0^2 + x_2^2 - x_1^2 + x_3^2 - x_2^2 + \cdots + x_n^2 - x_{n-1}^2\right)$$

$$= \tfrac{1}{2}\left(x_n^2 - x_0^2\right) = \tfrac{1}{2}(b^2 - a^2).$$

Therefore $\displaystyle\int_a^b x\,dx = \lim_{n\to\infty}\sum_{i=1}^{m} x_i^\star\,\Delta x_i = \tfrac{1}{2}b^2 - \tfrac{1}{2}a^2.$

5.4.53: Suppose that $a < b$. Let $\mathcal{P} = \{x_0, x_1, x_2, \ldots, x_n\}$ be a partition of $[a, b]$ and let $\{x_i^\star\}$ be a selection for $\mathcal{P}$. Then

$$\int_a^b c\,dx = \lim_{n\to\infty}\sum_{i=1}^{n} c\,\Delta x_i$$

$$= \lim_{n\to\infty} c\cdot(x_1 - x_0 + x_2 - x_1 + x_3 - x_2 + \cdots + x_n - x_{n-1})$$

$$= \lim_{n\to\infty} c(x_n - x_0) = c(b - a).$$

232

The proof is similar in the case $b < a$.

5.4.55: Whatever partition $\mathcal{P}$ is given, a selection $\{x_i^\star\}$ with all $x_i^\star$ irrational can be made because irrational numbers can be found in any interval of the form $[x_{i-1}, x_i]$ with $x_{i-1} < x_i$. (An explanation of why this is possible is given after the solution of Problem 2.4.69 of this manual.) For such a selection, we have

$$\sum_{i=1}^n f\left(x_i^\star\right) \Delta x_i = \sum_{i=1}^n 1 \cdot \Delta x_i = 1$$

regardless of the choice of $\mathcal{P}$ or n. Similarly, by choosing $x_i^\star$ rational for all i, we get

$$\sum_{i=1}^n f\left(x_i^\star\right) \Delta x_i = \sum_{i=1}^n 0 \cdot \Delta x_i = 0$$

regardless of the choice of $\mathcal{P}$ or n. Therefore the limit of Riemann sums for f on $[0, 1]$ does not exist, and therefore

$$\int_0^1 f(x)\, dx \quad \text{does not exist.}$$

5.4.57: Choose $x_i^\star = x_i = \dfrac{3i}{n}$ and $\Delta x = \dfrac{3}{n}$. Then

$$\sum_{i=1}^n \exp\left(x_i^\star\right) \Delta x = \sum_{i=1}^n e^{-3i/n} \cdot \frac{3}{n} = \frac{3}{n}\left(e^{-3/n} + e^{-6/n} + e^{-9/n} + \cdots + e^{-3n/n}\right)$$

$$= \frac{3e^{-3/n}}{n}\left(1 + e^{-3/n} + e^{-6/n} + \cdots + e^{-3(n-1)/n}\right)$$

$$= \frac{3r}{n}\left(1 + r + r^2 + \cdots + r^{n-1}\right) \qquad (\text{where } r = e^{-3/n})$$

$$= \frac{3r}{n} \cdot \frac{r^n - 1}{r - 1} = \frac{3e^{-3/n}}{n} \cdot \frac{e^{-3} - 1}{e^{-3/n} - 1} = \frac{3(e^{-3} - 1)}{n(1 - e^{3/n})} = \frac{3(1 - e^{-3})}{n(e^{3/n} - 1)}.$$

Now by l'Hôpital's rule,

$$\lim_{n \to \infty} n \cdot (e^{3/n} - 1) = \lim_{n \to \infty} \frac{e^{3/n} - 1}{\dfrac{1}{n}} = \lim_{n \to \infty} \frac{-\dfrac{3}{n^2} e^{3/n}}{-\dfrac{1}{n^2}} = \lim_{n \to \infty} 3e^{3/n} = 3.$$

Therefore

$$\int_0^3 e^{-x}\, dx = \lim_{n \to \infty} \sum_{i=1}^n \exp\left(x_i^\star\right) \Delta x = \lim_{n \to \infty} \frac{3(1 - e^{-3})}{n(e^{3/n} - 1)} = \frac{3(1 - e^{-3})}{3} = 1 - e^{-3}.$$

5.4.59: Let $x_i^\star = x_i = \dfrac{k\pi}{n}$ and $\Delta x = \dfrac{\pi}{n}$. Then

$$\int_0^\pi \sin x\, dx = \lim_{n \to \infty} \sum_{k=1}^n \sin\left(x_i^\star\right) \Delta x = \lim_{n \to \infty} \frac{\pi}{n} \sum_{k=1}^n \sin\left(\frac{k\pi}{n}\right) = \lim_{n \to \infty} \frac{\pi}{n} \cot\left(\frac{\pi}{2n}\right),$$

which is of the indeterminate form $0 \cdot \infty$. Hence (no need for l'Hôpital's rule)

$$\int_0^\pi \sin x\, dx = \lim_{n\to\infty} \frac{2 \cdot \dfrac{\pi}{2n} \cdot \cos \dfrac{\pi}{2n}}{\sin \dfrac{\pi}{2n}} = \lim_{n\to\infty} \left(2\cos\frac{\pi}{2n}\right) \cdot \frac{\dfrac{\pi}{2n}}{\sin \dfrac{\pi}{2n}} = 2 \cdot 1 \cdot 1 = 2.$$

5.4.61: Let $h = b - a$, $x_i^\star = x_i = a + \dfrac{ih}{n}$, and $\Delta x = \dfrac{h}{n}$. Then

$$\sum_{i=1}^n \sin(x_i^\star)\,\Delta x = \frac{h}{n}\sum_{i=1}^n \sin\left(a + \frac{ih}{n}\right).$$

According to *Mathematica*,

$$\sum_{i=1}^n \sin\left(a + \frac{ih}{n}\right) = \csc\frac{h}{2n} \sin\frac{h}{2}\sin\left(\frac{1}{2}\left[2a + h + \frac{h}{n}\right]\right) = \csc\frac{b-a}{2n}\sin\frac{b-a}{2}\sin\left(\frac{1}{2}\left[b + a + \frac{b-a}{n}\right]\right).$$

But then,

$$\frac{b-a}{n}\sum_{i=1}^n \sin\left(a + i \cdot \frac{b-a}{n}\right) = \frac{b-a}{n}\csc\frac{b-a}{2n}\sin\frac{b-a}{2}\sin\left(\frac{1}{2}\left[b + a + \frac{b-a}{n}\right]\right)$$

$$= \frac{\dfrac{2(b-a)}{2n}}{\sin\dfrac{b-a}{2n}} \cdot \left(\sin\frac{b-a}{2}\right) \cdot \sin\left(\frac{1}{2}\left[b + a + \frac{b-a}{n}\right]\right) \to 2 \cdot \left(\sin\frac{b-a}{2}\right) \cdot \left(\sin\frac{b+a}{2}\right)$$

as $n \to +\infty$. But using one of the trigonometric identities that immediately precede Problems 59 through 62 in Section 7.4, we find that

$$2 \cdot \left(\sin\frac{b-a}{2}\right) \cdot \left(\sin\frac{b+a}{2}\right) = \cos\left(\frac{b+a}{2} - \frac{b-a}{2}\right) - \cos\left(\frac{b+a}{2} + \frac{b-a}{2}\right) = \cos a - \cos b.$$

Therefore

$$\int_a^b \sin x\, dx = \cos a - \cos b.$$

Section 5.5

5.5.1: $\displaystyle\int_0^1 \left(3x^2 + 2\sqrt{x} + 3\sqrt[3]{x}\right) dx = \left[x^3 + \frac{4}{3}x^{3/2} + \frac{9}{4}x^{4/3}\right]_0^1 = \frac{55}{12}.$

5.5.3: $\displaystyle\int_0^1 x^3(1+x)^2\, dx = \left[\frac{1}{4}x^4 + \frac{2}{5}x^5 + \frac{1}{6}x^6\right]_0^1 = \frac{49}{60}.$

5.5.5: $\displaystyle\int_0^1 \left(x^4 - x^3\right) dx = \left[\frac{1}{5}x^5 - \frac{1}{4}x^4\right]_0^1 = -\frac{1}{20}.$

234

5.5.7: $\displaystyle\int_{-1}^{0} (x+1)^3 \, dx = \left[\tfrac{1}{4}(x+1)^4 \right]_{-1}^{0} = \tfrac{1}{4}.$

5.5.9: $\displaystyle\int_{0}^{4} \sqrt{x} \, dx = \left[\tfrac{2}{3}x^{3/2} \right]_{0}^{4} = \tfrac{16}{3}.$

5.5.11: $\displaystyle\int_{-1}^{2} (3x^2 + 2x + 4) \, dx = \left[x^3 + x^2 + 4x \right]_{-1}^{2} = 20 - (-4) = 24.$

5.5.13: $\displaystyle\int_{-1}^{1} x^{99} \, dx = \left[\tfrac{1}{100}x^{100} \right]_{-1}^{1} = 0.$

5.5.15: $\displaystyle\int_{1}^{3} (x-1)^5 \, dx = \left[\tfrac{1}{6}(x-1)^6 \right]_{1}^{3} = \tfrac{32}{3}.$

5.5.17: $\displaystyle\int_{-1}^{0} (2x+1)^3 \, dx = \int_{-1}^{0} (8x^3 + 12x^2 + 6x + 1) \, dx = \left[2x^4 + 4x^3 + 3x^2 + x \right]_{-1}^{0} = 0 - 0 = 0.$

5.5.19: $\displaystyle\int_{1}^{8} x^{2/3} \, dx = \left[\tfrac{3}{5}x^{5/3} \right]_{1}^{8} = \tfrac{93}{5}.$

5.5.21: $\displaystyle\int_{-1}^{1} (e^x - e^{-x}) \, dx = \left[e^x + e^{-x} \right]_{-1}^{1} = \left(e + \tfrac{1}{e} \right) - \left(\tfrac{1}{e} + e \right) = 0.$

5.5.23: $\displaystyle\int_{0}^{2} \sqrt{e^{3t}} \, dt = \int_{0}^{2} e^{3t/2} \, dt = \left[\tfrac{2}{3}e^{3t/2} \right]_{0}^{2} = \tfrac{2}{3}e^3 - \tfrac{2}{3} = \tfrac{2}{3}(e^3 - 1) \approx 12.723691282.$

5.5.25: $\displaystyle\int_{1}^{2} \frac{1}{t} \, dt = \left[\ln t \right]_{1}^{2} = \ln 2 - 0 = \ln 2 \approx 0.69314718055999453094.$

5.5.27: $\displaystyle\int_{0}^{1} (e^x - 1)^2 \, dx = \int_{0}^{1} (e^{2x} - 2e^x + 1) \, dx = \left[\tfrac{1}{2}e^{2x} - 2e^x + x \right]_{0}^{1} = \tfrac{1}{2}e^2 - 2e + 1 + \tfrac{3}{2}$

$= \dfrac{e^2 - 4e + 5}{2} = \tfrac{1}{2}e^2 - 2e + \tfrac{5}{2} \approx 0.7579643925.$

5.5.29: $\displaystyle\int_{0}^{\pi/4} \sin x \cos x \, dx = \left[\tfrac{1}{2}(\sin x)^2 \right]_{0}^{\pi/4} = \tfrac{1}{4}.$

5.5.31: $\displaystyle\int_{0}^{\pi} \sin 5x \, dx = \left[-\tfrac{1}{5}\cos x \right]_{0}^{\pi} = \tfrac{2}{5}.$

5.5.33: $\displaystyle\int_{0}^{\pi/2} \cos 3x \cdot dx = \left[\tfrac{1}{3}\sin 3x \right]_{0}^{\pi/2} = -\tfrac{1}{3}.$

5.5.35: $\displaystyle\int_{0}^{2} \cos \frac{\pi x}{4} \, dx = \left[\frac{4}{\pi}\sin \frac{\pi x}{4} \right]_{0}^{2} = \frac{4}{\pi}.$

5.5.37: Choose $x_i = i/n$, $\Delta x = 1/n$, $x_0 = 0$, and $x_n = 1$. Then the limit in question is the limit of a Riemann sum for the function $f(x) = 2x - 1$ on the interval $0 \leqq x \leqq 1$, and its value is therefore

$$\int_0^1 (2x - 1)\, dx = \left[x^2 - x \right]_0^1 = 1 - 1 = 0.$$

5.5.39: Choose $x_i = i/n$, $\Delta x = 1/n$, $x_0 = 0$, and $x_n = 1$. Then the limit in question is the limit of a Riemann sum for the function $f(x) = x$ on the interval $0 \leq x \leq 1$, and therefore

$$\lim_{n \to \infty} \frac{1 + 2 + 3 + \cdots + n}{n^2} = \lim_{n \to \infty} \sum_{i=1}^{n} \frac{i}{n^2} = \int_0^1 x\, dx = \frac{1}{2}.$$

5.5.41: Choose $x_i = i/n$, $\Delta x = 1/n$, $x_0 = 0$, and $x_n = 1$. Then the given limit is the limit of a Riemann sum for the function $f(x) = \sqrt{x}$ on the interval $0 \leq x \leq 1$, and therefore

$$\lim_{n \to \infty} \frac{\sqrt{1} + \sqrt{2} + \sqrt{3} + \cdots + \sqrt{n}}{n\sqrt{n}} = \lim_{n \to \infty} \sum_{i=1}^{n} \frac{\sqrt{i}}{n\sqrt{n}} = \int_0^1 \sqrt{x}\, dx = \frac{2}{3}.$$

5.5.43: The graph is shown next. The region represented by the integral consists of two triangles above the x-axis, one with base 3 and height 3, the other with base 1 and height 1. So the value of the integral is the total area $\frac{9}{2} + \frac{1}{2} = 5$.

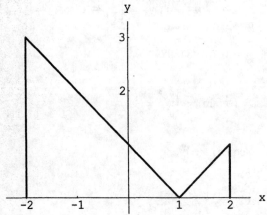

5.5.45: The graph is next. The region represented by the integral consists of two triangles, one above the x-axis with base 2 and height 2, the other below the x-axis with base 3 and height 3, so the total value of the integral is $2 - \frac{9}{2} = -\frac{5}{2}$.

236

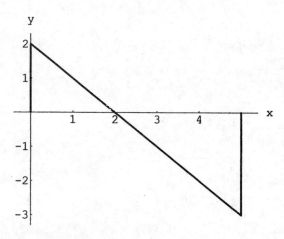

5.5.47: If $y = \sqrt{25 - x^2}$ for $0 \leq x \leq 5$, then

$$x^2 + y^2 = 25, \quad 0 \leq x \leq 5, \quad \text{and} \quad 0 \leq y \leq 5.$$

Therefore the region represented by the integral consists of the quarter of the circle $x^2 + y^2 = 25$ that lies in the first quadrant. This circle has radius 5, and therefore the value of the integral is $\frac{25}{4}\pi$. The region is shown next.

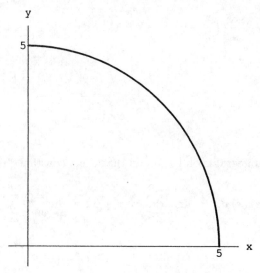

5.5.49: $0 \leq x^2 \leq x$ if $0 \leq x \leq 1$. Hence $1 \leq 1 + x^2 \leq 1 + x$ for such x. Therefore

$$1 \leq \sqrt{1 + x^2} \leq \sqrt{1 + x}$$

if $0 \leq x \leq 1$. Hence, by the comparison property, the inequality in Problem 49 follows.

5.5.51: $x^2 \leq x$ and $x \leq \sqrt{x}$ if $0 \leq x \leq 1$. So $1 + x^2 \leq 1 + \sqrt{x}$ for such x. Therefore

$$\frac{1}{1 + \sqrt{x}} \leq \frac{1}{1 + x^2}$$

if $0 \leq x \leq 1$. The inequality in Problem 51 now follows from the comparison property for definite integrals.

5.5.53: $\sin t \le 1$ for all t. Therefore $\displaystyle\int_0^2 \sin\left(\sqrt{x}\right)\,dx \le \int_0^2 1\,dx = 2$.

5.5.55: If $0 \le x \le 1$, then

$$1 \le 1 + x \le 2;$$

$$\frac{1}{2} \le \frac{1}{1+x} \le 1;$$

$$\frac{1}{2} \cdot (1-0) \le \int_0^1 \frac{1}{1+x}\,dx \le 1 \cdot (1-0).$$

So the value of the integral lies between 0.5 and 1.0.

5.5.57: If $0 \le x \le \frac{1}{6}\pi$, then

$$\frac{\sqrt{3}}{2} \le \cos x \le 1;$$

$$\frac{3}{4} \le \cos^2 x \le 1;$$

$$\frac{3}{4} \cdot \left(\frac{\pi}{6} - 0\right) \le \int_0^{\pi/6} \cos^2 x\,dx \le 1 \cdot \left(\frac{\pi}{6} - 0\right).$$

Therefore

$$\frac{\pi}{8} \le \int_0^{\pi/6} \cos^2 x\,dx \le \frac{\pi}{6}.$$

5.5.59: Suppose that f is integrable on $[a, b]$ and that c is a constant. Then

$$\int_a^b cf(x)\,dx = \lim_{\Delta x \to 0} \sum_{i=1}^n cf(x_i^\star)\,\Delta x = \lim_{\Delta x \to 0} c \cdot \sum_{i=1}^n f(x_i^\star)\,\Delta x$$

$$= c \cdot \left(\lim_{\Delta x \to 0} \sum_{i=1}^n f(x_i^\star)\,\Delta x\right) = c\int_a^b f(x)\,dx.$$

5.5.61: Suppose that f is integrable on $[a, b]$ and that $f(x) \le M$ for all x in $[a, b]$. Let $g(x) \equiv M$. Then by the first comparison property,

$$\int_a^b f(x)\,dx \le \int_a^b g(x)\,dx = \Big[Mx\Big]_a^b = M(b - a).$$

The proof of the other inequality is similar.

5.5.63: $1000 + \displaystyle\int_0^{30} V'(t)\,dt = 1000 + \Big[(0.4)t^2 - 40t\Big]_0^{30} = 160$ (gallons). Alternatively, the tank contains

$$V(t) = (0.4)t^2 - 40t + 1000$$

gallons at time $t \geq 0$, so at time $t = 30$ it contains $V(30) = 160$ gallons.

5.5.65: In Fig. 5.5.11 of the text we see that

$$\frac{12 - 4x}{9} \leq \frac{1}{x} \leq \frac{3 - x}{2}.$$

Therefore

$$\frac{2}{3} \leq \int_1^2 \frac{1}{x} \, dx \leq \frac{3}{4}.$$

Another way to put it would be to write

$$\int_1^2 \frac{1}{x} \, dx = 0.708333 \pm 0.041667.$$

For a really sophisticated answer, you could point out that, from the figure, it appears that the low estimate of the integral is about twice as accurate as (has half the error of) the high estimate. So

$$\int_1^2 \frac{1}{x} \, dx \approx \frac{2}{3} \cdot \frac{2}{3} + \frac{1}{3} \cdot \frac{3}{4} \approx 0.6944.$$

Section 5.6

5.6.1: $\dfrac{1}{2 - 0} \displaystyle\int_0^2 x^4 \, dx = \dfrac{1}{2} \cdot \left[\dfrac{1}{5} x^5 \right]_0^2 = \dfrac{16}{5}.$

5.6.3: $\dfrac{1}{2 - 0} \displaystyle\int_0^2 3x^2 (x^3 + 1)^{1/2} \, dx = \dfrac{1}{2} \cdot \left[\dfrac{2}{3} (x^3 + 1)^{3/2} \right]_0^2 = \dfrac{26}{3}.$

5.6.5: $\dfrac{1}{4 - (-4)} \displaystyle\int_{-4}^4 8x \, dx = \dfrac{1}{4} \cdot \left[4x^2 \right]_{-4}^4 = 0.$

5.6.7: $\dfrac{1}{5 - 0} \displaystyle\int_0^5 x^3 \, dx = \dfrac{1}{5} \cdot \left[\dfrac{1}{4} x^4 \right]_0^5 = \dfrac{125}{4}.$

5.6.9: $\dfrac{1}{3 - 0} \displaystyle\int_0^3 \sqrt{x + 1} \, dx = \dfrac{1}{3} \cdot \left[\dfrac{2}{3} (x + 1)^{3/2} \right]_0^3 = \dfrac{14}{9}.$

5.6.11: $\dfrac{1}{\pi - 0} \displaystyle\int_0^\pi \sin 2x \, dx = \dfrac{1}{\pi} \cdot \left[-\dfrac{1}{2} \cos 2x \right]_0^\pi = 0.$

5.6.13: $\displaystyle\int_{-1}^3 1 \, dx = \left[x \right]_{-1}^3 = 3 - (-1) = 4.$

5.6.15: $\displaystyle\int_1^4 \dfrac{1}{\sqrt{9x^3}} \, dx = \int_1^4 \dfrac{1}{3} x^{-3/2} \, dx = \left[-\dfrac{2}{3} x^{-1/2} \right]_1^4 = -\dfrac{2}{3} \cdot \left(\dfrac{1}{2} - 1 \right) = \dfrac{1}{3}.$

5.6.17: $\displaystyle\int_1^3 \frac{3t-5}{t^4}\,dt = \int_1^3 (3t^{-3} - 5t^{-4})\,dt = \left[\frac{5}{3t^3} - \frac{3}{2t^2}\right]_1^3 = -\frac{22}{81}.$

5.6.19: $\displaystyle\int_0^\pi \sin x \cos x\,dx = \left[\frac{1}{2}\sin^2 x\right]_0^\pi = 0.$

5.6.21: $\displaystyle\int_1^2 \left(t - \frac{1}{2}t^{-1}\right)^2 dt = \int_1^2 \left(t^2 - 1 + \frac{1}{4}t^{-2}\right)dt = \left[\frac{1}{3}t^3 - t - \frac{1}{4}t^{-1}\right]_1^2 = \frac{35}{24}.$

5.6.23: $\displaystyle\int_0^1 \frac{e^{2x}-1}{e^x}\,dx = \int_0^1 (e^x - e^{-x})\,dx = \left[e^x + e^{-x}\right]_0^1 = \left(e + \frac{1}{e}\right) - (1+1) = \frac{(e-1)^2}{e} \approx 1.086161.$

5.6.25: Because $x^2 - 1 \geqq 0$ if $|x| \geqq 1$ and $x^2 - 1 < 0$ if $|x| < 1$, we split the integral into three:

$$\int_{-2}^2 |x^2 - 1|\,dx = \int_{-2}^{-1}(x^2 - 1)\,dx + \int_{-1}^1 (1 - x^2)\,dx + \int_1^2 (x^2 - 1)\,dx = \frac{4}{3} + \frac{4}{3} + \frac{4}{3} = 4.$$

5.6.27: $\displaystyle\int_4^8 \frac{1}{x}\,dx = \left[\ln x\right]_4^8 = \ln 8 - \ln 4 = \ln\frac{8}{4} = \ln 2 \approx 0.6931471805599453094.$

5.6.29: $\displaystyle\int_{-1}^0 (1 - x^4)\,dx + \int_0^1 (1 - x^3)\,dx = \left[x - \frac{1}{5}x^5\right]_{-1}^0 + \left[x - \frac{1}{4}x^4\right]_0^1 = 1 - \frac{1}{5} + 1 - \frac{1}{4} = \frac{31}{20}.$

5.6.31: $\displaystyle\int_{-3}^0 (x^3 - 9x)\,dx - \int_0^3 (x^3 - 9x)\,dx = \left[\frac{1}{4}x^4 - \frac{9}{2}x^2\right]_{-3}^0 - \left[\frac{1}{4}x^4 - \frac{9}{2}x^2\right]_0^3 = \frac{81}{4} + \frac{81}{4} = \frac{81}{2}.$

5.6.33: Height: $s(t) = 400 - 16t^2$. Velocity: $v(t) = -32t$. Time T of impact occurs when $T^2 = 25$, so that $T = 5$. Average height:

$$\frac{1}{5}\int_0^5 s(t)\,dt = \frac{1}{5}\left[400t - \frac{16}{3}t^3\right]_0^5 = \frac{800}{3} \approx 266.666667 \quad (\text{ft}).$$

Average velocity:

$$\frac{1}{5}\int_0^5 v(t)\,dt = \frac{1}{5}\left[-16t^2\right]_0^5 = -80 \quad (\text{ft/s}).$$

5.6.35: Clearly the tank empties itself in the time interval $[0, 10]$. So the average amount of water in the tank during that time interval is

$$\frac{1}{10}\int_0^{10} V(t)\,dt = \frac{1}{10}\left[\frac{50}{3}t^3 - 500t^2 + 5000t\right]_0^{10} = \frac{5000}{3} \approx 1666.666667 \quad (\text{L}).$$

As an independent check,

$$\frac{V(0) + V(2) + V(4) + V(6) + V(8) + V(10)}{6} = \frac{5000 + 3200 + 1800 + 800 + 200 + 0}{6} \approx 1833.333.$$

5.6.37: The average temperature of the rod is

$$\frac{1}{10}\int_0^{10} T(x)\,dx = \frac{1}{10}\left[20x^2 - \frac{4}{3}x^3\right]_0^{10} = \frac{200}{3} \approx 66.666667.$$

As an independent check,

$$\frac{T(0) + T(2) + T(4) + T(6) + T(8) + T(10)}{6} = \frac{0 + 64 + 96 + 96 + 64 + 0}{6} \approx 53.33.$$

5.6.39: Similar triangles yield $y/r = 2/1$, so $r = y/2$. So the area of the cross section at y is $A(y) = \pi(y/2)^2$, and thus the average area of such a cross section is

$$\frac{1}{2} \int_0^2 A(y)\ dt = \frac{1}{2} \left[\frac{\pi}{12} y^3 \right]_0^2 = \frac{\pi}{3}.$$

5.6.41: First, $A(x) = 3(9 - x^2)$ for $-3 \leq x \leq 3$. Hence the average value of A on that interval is

$$\frac{1}{6} \int_{-3}^3 A(x)\ dx = \frac{1}{6} \left[27x - x^3 \right]_{-3}^3 = 18.$$

Next, $A(x) = 18$ has the two solutions $x = \pm\sqrt{3}$ in the interval $[-3, 3]$, so there are two triangles having the same area as the average. One is shown next.

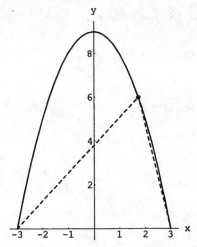

5.6.43: The area function is $A(x) = 2x\sqrt{16 - x^2}$ for $0 \leq x \leq 4$, so the average value of A is

$$\frac{1}{4} \int_0^4 A(x)\ dx = \frac{1}{4} \left[-\frac{2}{3}(16 - x^2)^{3/2} \right]_0^4 = \frac{32}{3}.$$

The equation $A(x) = \frac{32}{3}$ has the two solutions $x = 2\sqrt{\frac{2}{3}\left(3 \pm \sqrt{5}\right)}$, so there are two rectangles having the same area as the average. One is shown next.

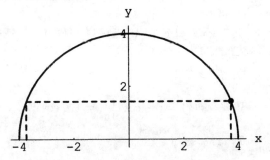

5.6.45: $f'(x) = (x^2 + 1)^{17}$.

5.6.47: $h'(z) = (z - 1)^{1/3}$.

5.6.49: Because

$$f(x) = -\int_{10}^{x} (e^t - e^{-t})\, dt,$$

part (1) of the fundamental theorem of calculus (Section 5.6) implies that

$$f'(x) = -(e^x - e^{-x}) = e^{-x} - e^x.$$

To use part (2) of the fundamental theorem of calculus, first compute

$$f(x) = \int_{x}^{10} (e^t - e^{-t})\, dt = -e^x - e^{-x} + e^{10} - e^{-10},$$

and differentiation now shows that $f'(x) = e^{-x} - e^x$.

5.6.51: $G(x) = \int_{0}^{x} f(t)\, dt$, so $G'(x) = f(x) = \sqrt{x + 4}$.

5.6.53: $G(x) = \int_{1}^{x} f(t)\, dt$, so $G'(x) = f(x) = \sqrt{x^3 + 1}$.

5.6.55: Let $u(x) = 3x$. Then

$$f(x) = \int_{2}^{3x} \sin t^2\, dt = g(u) = \int_{2}^{u} \sin t^2\, dt.$$

Therefore

$$f'(x) = D_x\, g(u) = g'(u) \cdot u'(x) = 3 \sin u^2 = 3 \sin 9x^2.$$

5.6.57: Let $u(x) = x^2$. Then

$$f(x) = \int_{0}^{x^2} \sin t\, dt = g(u) = \int_{0}^{u} \sin t\, dt.$$

Therefore

$$f'(x) = D_x\, g(u) = g'(u) \cdot u'(x) = 2x \sin u = 2x \sin x^2.$$

For an independent verification, note that

$$f(x) = \Big[-\cos t \Big]_{0}^{x^2} = 1 - \cos x^2,$$

and therefore that $f'(x) = 2x \sin x^2$.

5.6.59: Let $u(x) = x^2 + 1$. Then

$$f(x) = \int_1^{x^2+1} \frac{1}{t}\, dt = g(u) = \int_1^u \frac{1}{t}\, dt.$$

Therefore

$$f'(x) = D_x\, g(u) = g'(u) \cdot u'(x) = 2x \cdot \frac{1}{u} = 2x \cdot \frac{1}{x^2+1}.$$

5.6.61: $y(x) = \int_1^x \frac{1}{t}\, dt.$

5.6.63: $y(x) = 10 + \int_5^x \sqrt{1+t^2}\, dt.$

5.6.65: The fundamental theorem does not apply because the integrand is not continuous on $[-1, 1]$. We will see how to handle integrals such as the one in this problem in Section 9.8. One thing is certain: Its value is *not* -2.

5.6.67: If $0 \leqq x \leqq 2$, then

$$g(x) = \int_0^x f(t)\, dt = \int_0^x 2t\, dt = x^2.$$

Thus $g(0) = 0$ and $g(2) = 4$. If $2 \leqq x \leqq 6$, then continuity of g at $x = 2$ implies that

$$g(x) = g(2) + \int_2^x f(t)\, dt = 4 + \int_2^x (8 - 2t)\, dt = 4 + \left[8t - t^2\right]_2^x = 4 + 8x - x^2 - 16 + 4 = 8x - x^2 - 8.$$

Therefore $g(4) = 8$ and $g(6) = 4$. If $6 \leq x \leq 8$, then continuity of g at $x = 6$ implies that

$$g(x) = g(6) + \int_6^x f(t)\, dt = 4 + \int_6^x (-4)\, dt = 4 - 4x + 24 = 28 - 4x.$$

Thus $g(8) = -4$. Finally, if $8 \leqq x \leqq 10$, then

$$g(x) = g(8) + \int_8^x (2t - 20)\, dt = -4 + \left[t^2 - 20t\right]_8^x = -4 + x^2 - 20x - 64 + 160 = x^2 - 20x + 92,$$

and therefore $g(10) = -8$. Next, $g(x)$ is increasing where $f(x) > 0$ and decreasing where $f(x) < 0$, so g is increasing on $(0, 4)$ and decreasing on $(4, 10)$. The global maximum of g will therefore occur at $(4, 8)$ and its global minimum at $(10, -8)$. The graph of $y = g(x)$ is next.

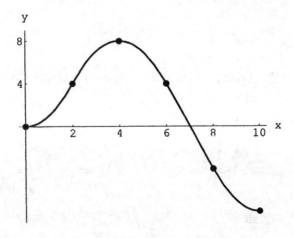

5.6.69: The local extrema of $g(x)$ occur only where $g'(x) = f(x) = 0$, thus at π, 2π, and 3π, and at the endpoints 0 and 4π of the domain of g. Figure 5.6.19 shows us that g is increasing on $(0, \pi)$ and $(2\pi, 3\pi)$, decreasing on $(\pi, 2\pi)$ and $(3\pi, 4\pi)$, so there are local minima at $(0, 0)$, $(2\pi, -2\pi)$, and $(4\pi, -4\pi)$; there are local maxima at (π, π) and $(3\pi, 3\pi)$. The global maximum occurs at $(3\pi, 3\pi)$ and the global minimum occurs at $(4\pi, -4\pi)$. To find the inflection points, we used Newton's method to solve the equation $f'(x) = 0$ and found the x-coordinates of the inflection points—it's clear from the graph that there are four of them—to be approximately 2.028758, 4.913180, 7.978666, and 11.085538. (The values of $f(x)$ at these four points are approximately 1.819706, -4.814470, 7.916727, and -11.040708.) The graph of $y = g(x)$ is shown next.

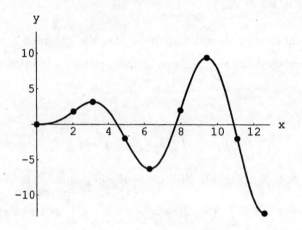

Section 5.7

5.7.1: Let $u = 3x - 5$. Then $du = 3\,dx$, so $dx = \frac{1}{3}\,du$. Thus

$$\int (3x - 5)^{17}\,dx = \int \frac{1}{3}u^{17}\,du = \frac{1}{54}u^{18} + C = \frac{1}{54}(3x - 5)^{18} + C.$$

5.7.3: The given substitution yields $\int \frac{1}{2}u^{1/2}\,du = \frac{1}{3}u^{3/2} + C = \frac{1}{3}\left(x^2 + 9\right)^{3/2} + C.$

5.7.5: The given substitution yields $\int \frac{1}{5}\sin u\,du = -\frac{1}{5}\cos u + C = -\frac{1}{5}\cos 5x + C.$

5.7.7: The given substitution yields $\int \frac{1}{4}\sin u\,du = -\frac{1}{4}\cos u + C = -\frac{1}{4}\cos\left(2x^2\right) + C.$

5.7.9: The given substitution yields $\int u^5\,du = \frac{1}{6}u^6 + C = \frac{1}{6}(1 - \cos x)^6 + C.$

5.7.11: If necessary, let $u = x + 1$. In any case, $\int (x + 1)^6\,dx = \frac{1}{7}(x + 1)^7 + C.$

5.7.13: If necessary, let $u = 4 - 3x$. In any case, $\int (4 - 3x)^7\,dx = -\frac{1}{24}(4 - 3x)^8 + C.$

5.7.15: If necessary, let $u = 7x + 5$. In any case, $\displaystyle\int \frac{1}{\sqrt{7x+5}}\, dx = \frac{2}{7}(7x+5)^{1/2} + C$.

5.7.17: If necessary, let $u = \pi x + 1$. In any case, $\displaystyle\int \sin(\pi x + 1)\, dx == -\frac{1}{\pi}\cos(\pi x + 1) + C$.

5.7.19: If necessary, let $u = 2\theta$. In any case, $\displaystyle\int \sec 2\theta \tan 2\theta\, d\theta = \frac{1}{2}\sec 2\theta + C$.

5.7.21: $\displaystyle\int e^{1-2x}\, dx = -\frac{1}{2}e^{1-2x} + C$.

5.7.23: $\displaystyle\int x^2 \exp\left(3x^3 - 1\right)\, dx = \frac{1}{9}\exp\left(3x^3 - 1\right) + C$.

If you prefer integration by substitution, let $u = 3x^3 - 1$.

5.7.25: $\displaystyle\int \frac{1}{2x-1}\, dx = \frac{1}{2}\ln|2x - 1| + C$.

5.7.27: $\displaystyle\int \frac{1}{x}(\ln x)^2\, dx = \frac{1}{3}(\ln x)^3 + C$.

5.7.29: $\displaystyle\int \frac{x + e^{2x}}{x^2 + e^{2x}}\, dx = \frac{1}{2}\ln\left(x^2 + e^{2x}\right) + C$. For integration by substitution, let $u = x^2 + e^{2x}$.

5.7.31: If necessary, let $u = x^2 - 1$, so that $du = 2x\, dx$; that is, $x\, dx = \frac{1}{2}\, du$. Then

$$\int x\sqrt{x^2 - 1}\, dx = \int \frac{1}{2}u^{1/2}\, du = \frac{1}{3}u^{3/2} + C = \frac{1}{3}(x^2 - 1)^{3/2} + C.$$

Comment: An alternative and frequently faster method for an integral like this runs as follows. Given

$$\int x\sqrt{x^2 - 1}\, dx,$$

it should be evident that the antiderivative involves $(x^2 - 1)^{3/2}$. Because

$$D_x\left(x^2 - 1\right)^{3/2} = \frac{3}{2}(x^2 - 1)^{1/2} \cdot 2x = 3x\sqrt{x^2 - 1},$$

the fact that if c is constant then $D_x\, cf(x) = cf'(x)$ now allows us to modify the initial guess $(x^2 - 1)^{3/2}$ for the antiderivative by multiplying it by $\frac{1}{3}$. Therefore

$$\int x\sqrt{x^2 - 1}\, dx = \frac{1}{3}(x^2 - 1)^{3/2} + C.$$

All that's necessary is to remember the constant of integration. Note that this technique is self-checking, requires less time and space, and is immune to the dangers of various illegal substitutions. The computation of the "correction factor" (in this case, $\frac{1}{3}$) can usually be done mentally.

5.7.33: If necessary, let $u = 2 - 3x^2$. In any case, $\displaystyle\int x(2 - 3x^2)^{1/2}\, dx = -\frac{1}{9}(2 - 3x^2)^{3/2} + C$.

5.7.35: If necessary, let $u = x^4 + 1$. In any case, $\displaystyle\int x^3(x^4 + 1)^{1/2}\,dx = \frac{1}{6}(x^4 + 1)^{3/2} + C$.

5.7.37: If necessary, let $u = 2x^3$. In any case, $\displaystyle\int x^2 \cos(2x^3)\,dx = \frac{1}{6}\sin(2x^3) + C$.

5.7.39: If necessary, let $u = -x^2$ or let $u = x^2$. In the latter case $du = 2x\,dx$, so that $x\,dx = \frac{1}{2}\,du$. In any case,

$$\int x \exp(-x^2)\,dx = -\frac{1}{2}\exp(-x^2) + C.$$

5.7.41: If necessary, let $u = \cos x$. In any case, $\displaystyle\int \cos^3 x \sin x\,dx = -\frac{1}{4}\cos^4 x + C$.

5.7.43: If necessary, let $u = \tan\theta$. In any case, $\displaystyle\int \tan^3\theta \sec^2\theta\,d\theta = \frac{1}{4}\tan^4\theta + C$.

5.7.45: If necessary, let $u = \sqrt{x}$. In any case, $\displaystyle\int x^{-1/2}\cos\left(x^{1/2}\right)\,dx = 2\sin\left(x^{1/2}\right) + C$.

5.7.47: If necessary, let $u = x^2 + 2x + 1$. This yields

$$\int (x^2 + 2x + 1)^4(x + 1)\,dx = \int \frac{1}{2}u^4\,du = \frac{1}{10}u^5 + C = \frac{1}{10}(x^2 + 2x + 1)^5 + C.$$

Alternatively,

$$\int (x^2 + 2x + 1)^4(x + 1)\,dx = \int (x + 1)^9\,dx = \frac{1}{10}(x + 1)^{10} + C.$$

5.7.49: If necessary, let $u = x^2 + 4x + 3$, so that $du = (2x + 4)\,dx$. Thus $(x + 2)\,dx = \frac{1}{2}\,du$. In any case,

$$\int \frac{x + 2}{x^2 + 4x + 3}\,dx = \frac{1}{2}\ln(x^2 + 4x + 3) + C,$$

although, because the quadratic can be negative, a better answer is $\frac{1}{2}\ln|x^2 + 4x + 3| + C$.

5.7.51: $\displaystyle\int_1^2 \frac{dt}{(t + 1)^3} = \left[-\frac{1}{2(t + 1)^2}\right]_1^2 = -\frac{1}{18} - \left(-\frac{1}{8}\right) = \frac{5}{72}$.

5.2.53: $\displaystyle\int_0^4 x\sqrt{x^2 + 9}\,dx = \left[\frac{1}{3}(x^2 + 9)^{3/2}\right]_0^4 = \frac{125}{3} - 9 = \frac{98}{3}$.

5.7.55: Given: $\displaystyle J = \int_0^8 t\sqrt{t + 1}\,dt$. Let $u = t + 1$. Then $du = dt$, and so

$$J = \int_{t=0}^8 (u - 1)u^{1/2}\,du = \int_{t=0}^8 \left(u^{3/2} - u^{1/2}\right)\,du = \left[\frac{2}{5}u^{5/2} - \frac{2}{3}u^{3/2}\right]_{t=0}^8$$

$$= \left[\frac{2}{5}(t + 1)^{5/2} - \frac{2}{3}(t + 1)^{3/2}\right]_0^8 = \frac{396}{5} - \left(-\frac{4}{15}\right) = \frac{1192}{15}.$$

Alternatively,

$$J = \int_{u=1}^{9} (u-1)u^{1/2}\,du = \int_{1}^{9} \left(u^{3/2} - u^{1/2} \right) du = \left[\frac{2}{5} u^{5/2} - \frac{2}{3} u^{3/2} \right]_{1}^{9} = \frac{396}{5} - \left(-\frac{4}{15} \right) = \frac{1192}{15}.$$

Thus you have at least two options:

1.   Make the substitution for u in place of t, find the antiderivative, then express the antiderivative in terms of t before substituting the original limits of integration, or

2.   Make the substitution for u in place ot t, find the antiderivative, then substitute the new limits of integration in terms of u.

Whichever option you use, be sure to use the notation correctly (as shown here).

5.7.57: $\displaystyle\int_{0}^{\pi/6} \sin 2x \cos^3 2x\,dx = \left[-\frac{1}{8} \cos^4 2x \right]_{0}^{\pi/6} = -\frac{1}{128} - \left(-\frac{1}{8} \right) = \frac{15}{128}.$

5.7.59: $\displaystyle\int_{0}^{\pi/2} (1 + 3\sin\theta)^{3/2} \cos\theta\,d\theta = \left[\frac{2}{15} (1 + 3\sin\theta)^{5/2} \right]_{0}^{\pi/2} = \frac{64}{15} - \frac{2}{15} = \frac{62}{15}.$

5.7.61: Given: $K = \displaystyle\int_{0}^{\pi/2} \exp(\sin x) \cos x\,dx.$ Let $u = \sin x.$ Then $du = \cos x\,dx.$ Hence

$$K = \int_{0}^{1} e^{u}\,du = \left[e^{u} \right]_{0}^{1} = e - 1 \approx 1.71828182845904523546.$$

5.7.63: Let $u = \dfrac{1}{x}.$ Then $du = -\dfrac{1}{x^2}\,dx,$ and therefore

$$\int_{1}^{2} \frac{e^{-1/x}}{x^2}\,dx = -\int_{1}^{1/2} e^{-u}\,du = \left[e^{-u} \right]_{1}^{1/2} = e^{-1/2} - e^{-1} = \frac{\sqrt{e} - 1}{e} \approx 0.2386512185.$$

5.7.65: $\displaystyle\int \sin^2 x\,dx = \int \frac{1 - \cos 2x}{2}\,dx = \frac{1}{2}x - \frac{1}{4}\sin 2x + C = \frac{1}{2}x - \frac{1}{2}\sin x \cos x + C.$

5.7.67: $\displaystyle\int_{0}^{\pi} \sin^2 3t\,dt = \int_{0}^{\pi} \frac{1 - \cos 6t}{2}\,dt = \left[\frac{t}{2} - \frac{1}{12}\sin 6t \right]_{0}^{\pi} = \frac{\pi}{2} - 0 = \frac{\pi}{2}.$

5.7.69: $\displaystyle\int \tan^2 x\,dx = \int \left(\sec^2 x - 1 \right) dx = -x + \tan x + C.$

5.7.71: $\displaystyle\int \sin^3 x\,dx = \int \left(\sin x - \cos^2 x \sin x \right) dx = -\cos x + \frac{1}{3}\cos^3 x + C.$

5.7.73: If you solve the equation

$$\frac{1}{2}\sin^2\theta + C_1 = -\frac{1}{2}\cos^2\theta + C_2$$

247

for $C_2 - C_1$, you will find its value to be $\frac{1}{2}$. That is, $\cos^2\theta + \sin^2\theta = 1$. Two functions with the same derivative differ by a constant; in this case $\sin^2\theta - (-\cos^2\theta) = 1$. The graphs of $f(x) = \frac{1}{2}\sin^2 x$ and $g(x) = -\frac{1}{2}\cos^2 x$ are shown next.

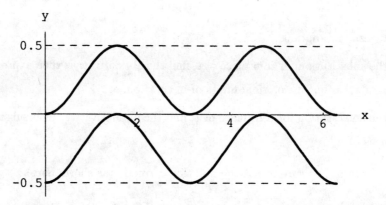

5.7.75: First, if $f(x) = \dfrac{x}{1-x}$, then

$$f'(x) = \frac{(1-x)\cdot 1 - x\cdot(-1)}{(1-x)^2} = \frac{1}{(1-x)^2}.$$

Next, if $u = 1 - x$ then $x = 1 - u$ and $dx = -du$, so

$$\int \frac{dx}{(1-x)^2} = \int \frac{-du}{u^2} = \frac{1}{u} + C_2 = \frac{1}{1-x} + C_2.$$

If $g(x) = \dfrac{1}{1-x}$, then

$$g(x) - f(x) = \frac{1}{1-x} - \frac{x}{1-x} = \frac{1-x}{1-x} \equiv 1.$$

As expected, because $g(x)$ and $f(x)$ have the same derivative on $(1, +\infty)$, they differ by a constant there. (This also holds on the interval $(-\infty, 1)$.) The graphs of $y = f(x)$ and $y = g(x)$ are shown next.

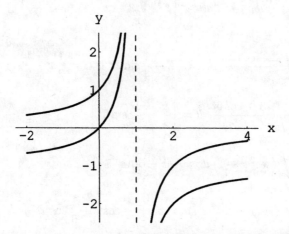

5.7.77: Suppose that f is continuous and odd. The substitution $u = -x$ requires $dx = -du$, so that

$$\int_{-a}^{0} f(x)\,dx = \int_{a}^{0} -f(-u)\,du = \int_{0}^{a} f(-u)\,du = -\int_{0}^{a} f(u)\,du = -\int_{0}^{a} f(x)\,dx.$$

248

(The last equality follows because in the next-to-last integral, u is merely a "dummy" variable of integration, and can simply be replaced with x.) Therefore

$$\int_{-a}^{a} f(x)\, dx = \int_{-a}^{0} f(x)\, dx + \int_{0}^{a} f(x)\, dx = -\int_{0}^{a} f(x)\, dx + \int_{0}^{a} f(x)\, dx = 0.$$

5.7.79: Because the tangent function is continuous and odd on $[-1, 1]$, it follows from the result in Problem 77 that

$$\int_{-1}^{1} \tan x\, dx = 0.$$

Next, $f(x) = x^{1/3}$ is continuous and odd on $[-1, 1]$ while $g(x) = (1 + x^2)^7$ is continuous and even there, so (Exercise!) their quotient is odd. Thus

$$\int_{-1}^{1} \frac{x^{1/3}}{(1 + x^2)^7}\, dx = 0.$$

Finally, $h(x) = x^{17}$ is continuous and odd on $[-1, 1]$ while $j(x) = \cos x$ is continuous and even there, so (Exercise!) their product is odd. Thus

$$\int_{-1}^{1} x^{17} \cos x\, dx = 0.$$

5.7.81: Given

$$I = \int_{a}^{b} f(x + k)\, dx,$$

let $u = x + k$. Then $x = u - k$, $dx = du$, $u = a + k$ when $x = a$, and $u = b + k$ when $x = b$. Hence

$$I = \int_{a+k}^{b+k} f(u)\, du = \int_{a+k}^{b+k} f(x)\, dx$$

(because it doesn't matter whether the variable of integration is called u or x).

5.7.83: (a) $D_u[e^u(u - 1)] = e^u(u - 1) + e^u \cdot 1 = ue^u$. (b) Let $u = x^{1/2}$. Then $x = u^2$, $dx = 2u\, du$, $u = 0$ when $x = 0$, and $u = 1$ when $x = 1$. Therefore

$$\int_{0}^{1} \exp\left(\sqrt{x}\,\right) dx = \int_{0}^{1} 2ue^u\, du = \left[2e^u(u - 1)\right]_{0}^{1} = 0 - (-2e^0) = 2.$$

Section 5.8

5.8.1: To find the limits of integration, we solve $25 - x^2 = 9$ for $x = \pm 4$. Hence the area is

$$\int_{-4}^{4} (25 - x^2 - 9)\, dx = \left[16x - \frac{1}{3}x^3\right]_{-4}^{4} = \frac{128}{3} - \left(-\frac{128}{3}\right) = \frac{256}{3}.$$

249

5.8.3: To find the limits of integration, we solve $x^2 - 3x = 0$ for $x = 0$, $x = 3$. Hence the area is

$$\int_0^3 (3x - x^2)\, dx = \left[\frac{3}{2}x^2 - \frac{1}{3}x^3\right]_0^3 = \frac{9}{2} - 0 = \frac{9}{2}.$$

5.8.5: To find the limits of integration, we solve $12 - 2x^2 = x^2$ for $x = \pm 2$. Therefore the area is

$$\int_{-2}^2 (12 - 3x^2)\, dx = \left[12x - x^3\right]_{-2}^2 = 16 - (-16) = 32.$$

5.8.7: To find the limits of integration, we solve $4 - x^2 = 3x^2 - 12$ for $x = \pm 2$. So the area is

$$\int_{-2}^2 (16 - 4x^2)\, dx = \left[16x - \frac{4}{3}x^3\right]_{-2}^2 = \frac{64}{3} - \left(-\frac{64}{3}\right) = \frac{128}{3}.$$

5.8.9: To find the limits of integration, we solve $x^2 - 3x = 6$ for $x = a = \frac{1}{2}\left(3 - \sqrt{33}\right)$ and $x = b = \frac{1}{2}\left(3 + \sqrt{33}\right)$. So the area is

$$\int_a^b (6 - x^2 + 3x)\, dx = \left[6x + \frac{3}{2}x^2 - \frac{1}{3}x^3\right]_a^b = \frac{45 + 11\sqrt{33}}{4} - \frac{45 - 11\sqrt{33}}{4} = \frac{11\sqrt{33}}{2}.$$

5.8.11: $\displaystyle\int_0^1 (x - x^3)\, dx = \left[\frac{1}{2}x^2 - \frac{1}{4}x^4\right]_0^1 = \frac{1}{4}.$

5.8.13: $\displaystyle\int_0^1 (x^3 - x^4)\, dx = \left[\frac{1}{4}x^4 - \frac{1}{5}x^5\right]_0^1 = \frac{1}{20}.$

5.8.15: $\displaystyle\int_0^2 \frac{1}{x+1}\, dx = \left[\ln(x+1)\right]_0^2 = \ln 3 - \ln 1 = \ln 3 \approx 1.0986122887.$

5.8.17: To find the limits of integration, we first solve $y^2 = 4$ for $y = \pm 2$. So the area of the region R is

$$\int_{-2}^2 (4 - y^2)\, dy = \left[4y - \frac{1}{3}y^3\right]_{-2}^2 = \frac{16}{3} - \left(-\frac{16}{3}\right) = \frac{32}{3}.$$

5.8.19: To find the limits of integration, we first solve $8 - y^2 = y^2 - 8$ for $y = a = -2\sqrt{2}$ and $y = b = 2\sqrt{2}$. So the area of R is

$$\int_a^b (16 - 2y^2)\, dy = \left[16y - \frac{2}{3}y^3\right]_a^b = \frac{64\sqrt{2}}{3} - \left(-\frac{64\sqrt{2}}{3}\right) = \frac{128\sqrt{2}}{3}.$$

5.8.21: The area of the region—shown next—is

$$A = \int_0^2 (2x - x^2)\, dx = \left[x^2 - \frac{1}{3}x^3\right]_0^2 = \frac{4}{3}.$$

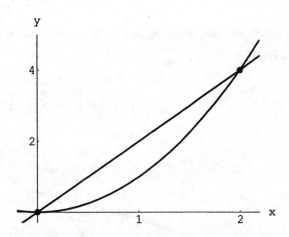

5.8.23: We find the limits of integration by solving $y^2 = 25$ for $y = \pm 5$. The area of the region—shown next—is therefore

$$\int_{-5}^{5} \left(25 - y^2\right) dy = \left[25y - \frac{1}{3}y^3\right]_{-5}^{5} = \frac{250}{3} - \left(-\frac{250}{3}\right) = \frac{500}{3}.$$

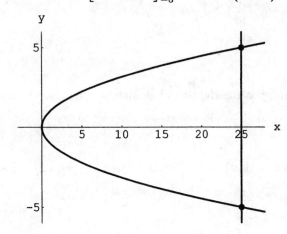

5.8.25: We find the limits of integration by solving $x^2 = 2x + 3$ for $x = -1$, $x = 3$. Thus the area of the region—shown next—is

$$A = \int_{-1}^{3} \left(2x + 3 - x^2\right) dx = \left[3x + x^2 - \frac{1}{3}x^3\right]_{-1}^{3} = 9 - \left(-\frac{5}{3}\right) = \frac{32}{3}.$$

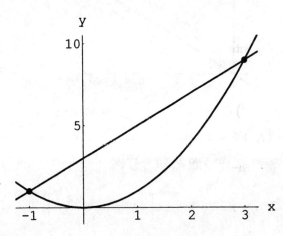

5.8.27: We first find the limits of integration by solving $y^2 = y + 6$ for $y = -2$, $y = 3$. The area of the region—shown next—is therefore

$$A = \int_{-2}^{3} \left(y + 6 - y^2\right)\, dy = \left[6y + \frac{1}{2}y^2 - \frac{1}{3}y^3\right]_{-2}^{3} = \frac{27}{2} - \left(-\frac{22}{3}\right) = \frac{125}{6}.$$

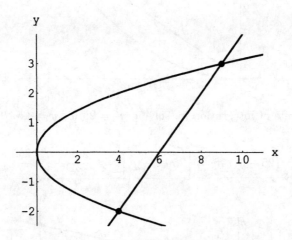

5.8.29: The two graphs meet at the right-hand endpoint of the given interval, where $x = \pi/4$. Therefore the area of the region they bound—shown next—is

$$A = \int_{0}^{\pi/4} \left(\cos x - \sin x\right)\, dx = \left[\sin x + \cos x\right]_{0}^{\pi/4} = \sqrt{2} - 1 \approx 0.414213562373.$$

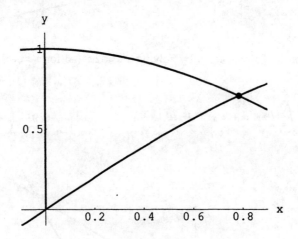

5.8.31: Solution of $4y^2 + 12y + 5 = 0$ yields the limits of integration $y = a = -\frac{5}{2}$ and $y = b = -\frac{1}{2}$. Hence the area of the region bounded by the two given curves—shown next—is

$$A = \int_{a}^{b} \left(-12y - 5 - 4y^2\right)\, dy = \left[-5y - 6y^2 - \frac{4}{3}y^3\right]_{a}^{b} = \frac{7}{6} - \left(-\frac{25}{6}\right) = \frac{16}{3}.$$

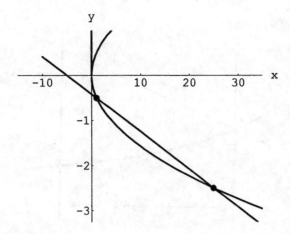

5.8.33: Solution of $3y^2 = 12y - y^2 - 5$ yields the limits of integration $y = a = \frac{1}{2}$ and $y = b = \frac{5}{2}$. Hence the area of the region bounded by the given curves—shown next—is

$$A = \int_a^b (-5 + 12y - 4y^2)\, dy = \left[-5y + 6y^2 - \frac{4}{3}y^3 \right]_a^b = \frac{25}{6} - \left(-\frac{7}{6} \right) = \frac{16}{3}.$$

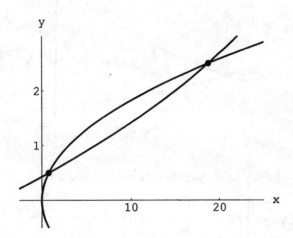

5.8.35: The only solution of $x + 1 = 1/(x+1)$ in the given interval $I = [0, 1]$ is $x = 0$. The area between the two curves over I is thus

$$A = \int_0^1 \left(x + 1 - \frac{1}{x+1} \right)\, dx = \left[x + \frac{1}{2}x^2 - \ln(x+1) \right]_0^1 = \frac{3}{2} - \ln 2 \approx 0.80685281944005469.$$

The region bounded by the two curves over the interval I is shown next.

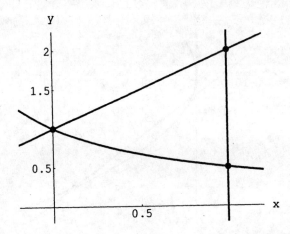

5.8.37: It's clear that the first two curves cross only at (0, 1). Thus the area of the region bounded by all three—shown next—is

$$A = \int_0^1 (e^x - e^{-x})\, dx = \left[e^x + e^{-x} \right]_0^1 = e + \frac{1}{e} - 2 = \frac{(e-1)^2}{e} \approx 1.0861612696.$$

5.8.39: The following figure makes it clear that the area bounded by the three curves is

$$A = \int_0^1 x \exp(-x^2)\, dx = \left[-\frac{1}{2} \exp(-x^2) \right]_0^1 = -\frac{1}{2e} + \frac{1}{2} = \frac{e-1}{2e} \approx 0.3160602794.$$

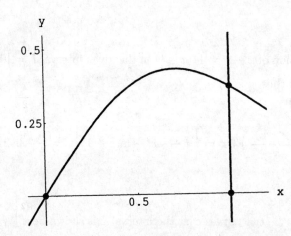

5.8.41: The curves meet where $x = -1$ and where $x = 1$, so the area of the region they bound—shown next—is

$$\int_{-1}^{1} (1 - x^3 - x^2 + x)\,dx = \left[x + \frac{1}{2}x^2 - \frac{1}{3}x^3 - \frac{1}{4}x^4 \right]_{-1}^{1} = \frac{11}{12} - \left(-\frac{5}{12} \right) = \frac{4}{3}.$$

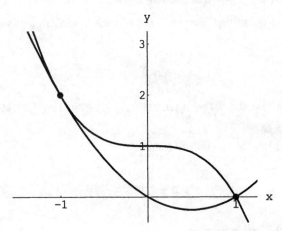

5.8.43: We solve $x^2 = x^3 - 2x$ to find $x = -1$, $x = 0$, and $x = 2$. So the two given curves meet at $(-1, 1)$, $(0, 0)$, and $(2, 4)$, as shown in the following figure. The area of the region on the left is

$$A_1 = \int_{-1}^{0} (x^3 - 2x - x^2)\,dx = \left[\frac{1}{4}x^4 - \frac{1}{3}x^3 - x^2 \right]_{-1}^{0} = 0 - \left(-\frac{5}{12} \right) = \frac{5}{12}.$$

The area of the region on the right is

$$A_2 = \int_{0}^{2} (2x + x^2 - x^3)\,dx = \left[x^2 + \frac{1}{3}x^3 - \frac{1}{4}x^4 \right]_{0}^{2} = \frac{8}{3}.$$

Therefore the total area bounded by the two regions is $A_1 + A_2 = \dfrac{37}{12}$.

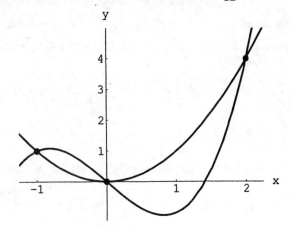

5.8.45: The first integral is

$$I_1 = \int_{-3}^{3} 4x(9 - x^2)^{1/2}\,dx = \left[-\frac{4}{3}(9 - x^2)^{3/2} \right]_{-3}^{3} = 0.$$

The second is

$$I_2 = \int_{-3}^{3} 5\sqrt{9 - x^2} \ dx = 5 \int_{-3}^{3} \sqrt{9 - x^2} \ dx.$$

Because the graph of $y = \sqrt{9 - x^2}$ is a semicircle of radius 3, centered at the origin, and lying in the first and second quadrants, I_2 is thus five times the area of such a semicircle, so that

$$I_2 = 5 \cdot \frac{1}{2} \cdot \pi \cdot 3^2 = \frac{45}{2}\pi,$$

and because $I_1 = 0$, this is also the value of the integral given in Problem 45.

5.8.47: We solve the equation of the ellipse for

$$y = \frac{b}{a}(a^2 - x^2)^{1/2},$$

and hence the area of the ellipse is given by

$$A = 4 \int_0^a \frac{b}{a}(a^2 - x^2)^{1/2} \ dx = \frac{4b}{a} \int_0^a \sqrt{a^2 - x^2} \ dx. \tag{1}$$

The last integral in Eq. (1) is the area of the quarter-circle in the first quadrant with center at the origin and radius a, and therefore

$$A = \frac{4b}{a} \cdot \frac{1}{4}\pi a^2 = \pi ab.$$

5.8.49: By solving the equations of the line and the parabola simultaneously, we find that $A = (-1, 1)$ and $B = (2, 4)$. The slope of the tangent line at C is the same as the slope 1 of the line through A and B, and it follows that C has x-coordinate $\frac{1}{2}$ and thus y-coordinate $\frac{1}{4}$. It now follows that the distance from A to B is $AB = 3\sqrt{2}$, the distance from B to C is $BC = \frac{3}{4}\sqrt{29}$, and the distance from A to C is $AC = \frac{3}{4}\sqrt{5}$. Heron's formula then allows us to find the area of triangle ABC; it is the square root of the product of

$$\frac{3}{8}\left(4\sqrt{2} + \sqrt{5} + \sqrt{29}\right), \qquad -3\sqrt{2} + \frac{3}{8}\left(4\sqrt{2} + \sqrt{5} + \sqrt{29}\right),$$

$$-\frac{3}{4}\sqrt{5} + \frac{3}{8}\left(4\sqrt{2} + \sqrt{5} + \sqrt{29}\right), \qquad \text{and} \qquad -\frac{3}{4}\sqrt{29} + \frac{3}{8}\left(4\sqrt{2} + \sqrt{5} + \sqrt{29}\right).$$

The product can be simplified to $\frac{729}{64}$, so the area of triangle ABC is $\frac{27}{8}$. The area of the parabolic segment is

$$\int_{-1}^{2} (x + 2 - x^2) \ dx = \frac{9}{2} = \frac{4}{3} \cdot \frac{27}{8},$$

exactly as Archimedes proved in more general form over 2000 years ago. *Mathematica* did the arithmetic for us in this problem. If you prefer to do it by hand, show that the line through C perpendicular to the tangent line there has equation $y = \frac{3}{4} - x$. Show that this line meets the line through AB at the point

$\left(-\frac{5}{8}, \frac{11}{8}\right)$. Show that the perpendicular from C to that line has length $h = \frac{9}{8}\sqrt{2}$. Show that AB has length $3\sqrt{2}$. Then triangle ABC has base AB and height h, so its area is

$$\frac{1}{2} \cdot \left(\frac{9}{8}\sqrt{2}\right) \cdot 3\sqrt{2} = \frac{27}{8}.$$

5.8.51: The graph of the cubic $y = 2x^3 - 2x^2 - 12x$ meets the x-axis at $x = -2$, $x = 0$, and $x = 3$. The graph is above the x-axis for $-2 < x < 0$ and below it for $0 < x < 3$, so the graph of the cubic and the x-axis form two bounded plane regions. The area of the one on the left is

$$A_1 = \int_{-2}^{0} (2x^3 - 2x^2 - 12x)\, dx = \left[\frac{1}{2}x^4 - \frac{2}{3}x^3 - 6x^2\right]_{-2}^{0} = \frac{32}{3}$$

and the area of the one on the right is

$$A_2 = \int_{0}^{3} (12x + 2x^2 - 2x^3)\, dx = \left[6x^2 + \frac{2}{3}x^3 - \frac{1}{2}x^4\right]_{0}^{3} = \frac{63}{2}.$$

Therefore the total area required in Problem 51 is $A_1 + A_2 = \dfrac{253}{6}$.

5.8.53: Given $y^2 = x(5 - x)^2$, the loop lies above and below the interval $[0, 5]$, so by symmetry (around the x-axis) its area is

$$2\int_{0}^{5} (5 - x)x^{1/2}\, dx = 2\left[\frac{2}{15}(25x^{3/2} - 3x^{5/2})\right]_{0}^{5} = 2 \cdot \frac{20}{3}\sqrt{5} = \frac{40}{3}\sqrt{5}.$$

5.8.55: We applied Newton's method to the equation $x^2 - \cos x = 0$ to find that the two curves $y = x^2$ and $y = \cos x$ meet at the two points $(-0.824132, 0.679194)$ and $(0.824132, 0.679194)$ (numbers involving decimals are approximations). Let $a = -0.824132$ and $b = 0.824132$. The region between the two curves (shown next) has area

$$\int_{a}^{b} (\cos x - x^2)\, dx = \left[\sin x - \frac{1}{3}x^3\right]_{a}^{b} \approx 1.09475.$$

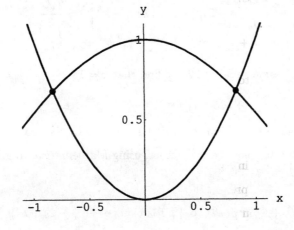

257

5.8.57: We used Newton's method to solve $f(x) = 0$ where

$$f(x) = \frac{1}{1 + x^2} - x^2 + 1,$$

and found the points at which the two curves intersect to be $(-1.189207, 0.414214)$ and $(1.189207, 0.414214)$ (numbers involving decimals are approximations). With $b = 1.189207$ and $a = -b$, the area bounded by the two curves (shown next) is

$$A = \int_a^b \left(\frac{1}{1 + x^2} - x^2 + 1 \right) dx = \left[\arctan(x) - \frac{1}{3} x^3 + x \right]_a^b \approx 3.00044$$

(the antiderivative was computed with the aid of Formula 17 of the endpapers).

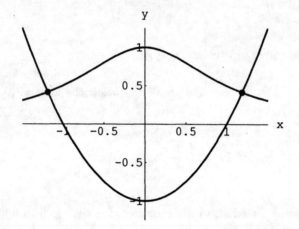

5.8.59: The curves $y = x^2$ and $y = k - x^2$ meet at the two points where $x = a = -(k/2)^{1/2}$ and $x = b = (k/2)^{1/2}$. So the area between the two curves is

$$\int_a^b (k - 2x^2) \, dx = \left[kx - \frac{2}{3} x^3 \right]_a^b = \frac{2\sqrt{2}}{3} k^{3/2}.$$

When we set the last expression equal to 72, we find that $k = 18$.

5.8.61: We are interested in the region (or regions) bounded by the graphs of the two functions $f(x) = x$ and $g(x) = x(x - 4)^2$. First we plot the two curves, using *Mathematica*, to guide our future computations.

```
Plot[ { f[x], g[x] }, { x, -1, 6 }, PlotRange → { -2, 11 } ];
```

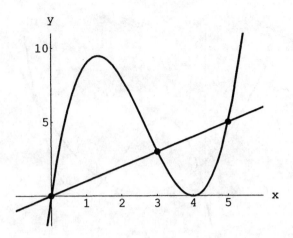

It is easy to see that the curves cross where $x = 0$, $x = 3$, and $x = 5$. Note that $g(x) \geqq f(x)$ if $0 \leqq x \leqq 3$ and that $f(x) \geqq g(x)$ if $3 \leqq x \leqq 5$. Hence we compute the values of two integrals and add the results to get the total area of the (bounded) regions bounded by the two curves.

```
a1 = Integrate[ g[x] - f[x], { x, 0, 3 } ]
```

$$\frac{63}{4}$$

```
a2 = Integrate[ g[x] - f[x], { x, 3, 5 } ]
```

$$\frac{16}{3}$$

```
a1 + a2
```

$$\frac{253}{12}$$

```
N[ %, 20 ]
```

```
21.083333333333333333
```

5.8.63: We are given $f(x) = (x - 2)^2$ and $g(x) = x(x - 4)^2$. We begin by plotting the graphs of both functions.

```
Plot[ { f[x], g[x] }, { x, -2, 6 }, PlotRange -> { -4, 20 } ];
```

259

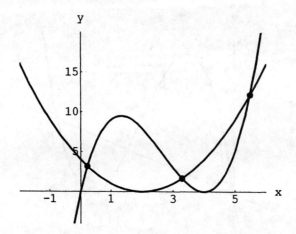

The `Solve` command in *Mathematica* returns exact solutions all of which include the number $i = \sqrt{-1}$, but their numerical values are numbers such as $5.48929 + 10^{-51}i$, so all solutions are pure real, as indicated in the preceding graph. We entered the numerical approximations to the roots:

```
r1 = 0.22154288174161126812;  r2 = 3.28916854644830996908;

r3 = 5.48928857181007876279;
```

Then we computed two integrals and added the results to find the total area bounded by the two curves.

```
a1 = Integrate[ g[x] - f[x], { x, r1, r2 } ]

    17.96479813911499075801


a2 = Integrate[ f[x] - g[x], { x, r2, r3 } ]

    7.39746346656350500268


a1 + a2

    25.36226160567849576069
```

5.8.65: Given $f(x) = 10 \ln x$ and $g(x) = (x-5)^2$, we first plotted the graphs of both functions:

```
Plot[ { f[x], g[x] }, { x, 0.1, 10.5 } ];
```

260

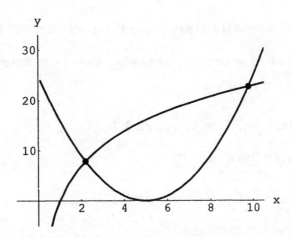

We then applied Newton's method, carrying 50 digits initially, to approximate the solutions of $f(x) = g(x)$. The initial guesses $x_0 = 2.1$ and $x_0 = 9.8$ produced (after seven iterations) the following solutions, accurate to the number of digits shown:

r1 = 2.19561040829474667642305282648227; r2 = 9.77472510546165546062557839624217;

and a single integral yields a very good approximation to the area bounded by the two curves:

Integrate[f[x] − g[x], { x, r1, r2 }]

86.14890547673214185465304844927136 5936

5.8.67: Given $f(x) = e^{-x/2}$ and $g(x) = x^4 - 6x^2 - 2x + 4$, we first plotted the graphs of f and g:

Plot[{ f[x], g[x] }, { x, −2.5, 2.7 }];

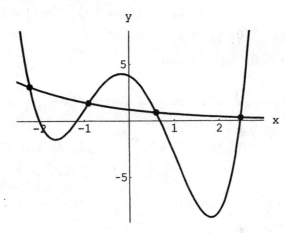

Clearly the graphs cross at four points, near where $x = -2.2, -0.9, 0.6,$ and 2.5. Using these estimates as initial values, seven iterations of Newton's method (carrying 60 digits initially) yields these four solutions of $f(x) = g(x)$, accurate to the number of digits shown here:

r1 = −2.21286965649560596689486746473117; r2 = −0.904163806104821356267362 26590897;

r3 = 0.60360685508638066445123518977576; r4 = 2.49090865808642591763321220641991;

Then three integrations, followed by the sum of the results, gives a very accurate approximation to the total area bounded by the graphs of f and g:

```
a1 = Integrate[ f[x] - g[x], { x, r1, r2 } ]
```

3.2920140909512522761730952582957

```
a2 = Integrate[ g[x] - f[x], { x, r2, r3 } ]
```

3.0387718742311950555162025363337

```
a3 = Integrate[ f[x] - g[x], { x, r3, r4 } ]
```

10.502231444117474603531902088120

```
a1 + a2 + a3
```

16.833017409299921935221199882749

Section 5.9

5.9.1: With $\Delta x = 1$, $f(x) = x$, $n = 4$, and $x_i = i \cdot \Delta x$, we have

$$T_n = \frac{\Delta x}{2} \cdot \left(f(x_0) + f(x_n) + 2 \cdot \sum_{i=1}^{n-1} f(x_i) \right) = 8,$$

which is also the true value of the given integral.

5.9.3: With $\Delta x = 0.5$, $f(x) = \sqrt{x}$, $n = 5$, and $x_i = i \cdot \Delta x$, we have

$$T_n = \frac{\Delta x}{2} \cdot \left(f(x_0) + f(x_n) + 2 \cdot \sum_{i=1}^{n-1} f(x_i) \right) \approx 0.6497385976 \approx 0.65.$$

The exact value of the integral is $\frac{2}{3} \approx 0.666667$.

5.9.5: With $\Delta x = \pi/6$, $f(x) = \cos x$, $n = 3$, and $x_i = i \cdot \Delta x$, we have

$$T_n = \frac{\Delta x}{2} \cdot \left(f(x_0) + f(x_n) + 2 \cdot \sum_{i=1}^{n-1} f(x_i) \right) = \frac{\pi}{12} \left(2 + \sqrt{3} \right) \approx 0.9770486167 \approx 0.98.$$

The exact value of the integral is 1.

262

5.9.7: With $\Delta x = 1$, $f(x) = x$, $n = 4$, and $m_i = \left(i - \frac{1}{2}\right) \cdot \Delta x$, we have

$$M_n = (\Delta x) \cdot \sum_{i=1}^{n} f(m_i) = 8.$$

The exact value of the integral is also 8.

5.9.9: With $\Delta x = 0.2$, $f(x) = \sqrt{x}$, $n = 5$, and $m_i = \left(i - \frac{1}{2}\right) \cdot \Delta x$, we have

$$M_n = (\Delta x) \cdot \sum_{i=1}^{n} f(m_i) \approx 0.6712800859 \approx 0.67.$$

The exact value of the integral is $\frac{2}{3} \approx 0.666667$.

5.9.11: With $\Delta x = \pi/6$, $f(x) = \cos x$, $n = 3$, and $m_i = \left(i - \frac{1}{2}\right) \cdot \Delta x$, we have

$$M_n = (\Delta x) \cdot \sum_{i=1}^{n} f(m_i) = \frac{\pi}{12}\left(\sqrt{2} + \sqrt{6}\right) \approx 1.01.$$

The exact value of the integral is 1.

5.9.13: With $\Delta x = 0.5$, $n = 4$, $f(x) = x^2$, and $x_i = 1 + i \cdot \Delta x$, we have

$$T_n = \frac{\Delta x}{2} \cdot \left(f(x_0) + f(x_n) + 2 \cdot \sum_{i=1}^{n-1} f(x_i)\right) = \frac{35}{4} = 8.75$$

and

$$S_n = \frac{\Delta x}{3} \cdot [f(x_0) + 4f(x_1) + 2f(x_2) + \cdots + 4f(x_{n-1}) + f(x_n)] = \frac{26}{3} \approx 8.67.$$

The true value of the integral is also $\frac{26}{3}$ (see Problem 29).

5.9.15: With $a = 0$, $b = 2$, $n = 4$, $\Delta x = 0.5$, $f(x) = e^{-x}$, and $x_i = i \cdot \Delta x$, we obtain

$$T_n = \frac{\Delta x}{2} \cdot \left[f(x_0) + f(x_n) + 2 \cdot \sum_{i=1}^{n-1} f(x_i)\right] \approx 0.8826039513$$

and

$$S_n = \frac{\Delta x}{3} \cdot [f(x_0) + 4f(x_1) + 2f(x_2) + 4f(x_3) + f(x_4)] \approx 0.8649562408.$$

The true value of the integral is $\int_0^2 e^{-x}\, dx = \left[-e^{-x}\right]_0^2 = 1 - \frac{1}{e^2} \approx 0.8646647168$.

5.9.17: With $\Delta x = \frac{1}{3}$, $n = 6$, $f(x) = \sqrt{1 + x^3}$, and $x_i = i \cdot \Delta x$, we have

$$T_n = \frac{\Delta x}{2} \cdot \left(f(x_0) + f(x_n) + 2 \cdot \sum_{i=1}^{n-1} f(x_i)\right) \approx 3.2598849023 \approx 3.26$$

263

and

$$S_n = \frac{\Delta x}{3} \cdot [f(x_0) + 4f(x_1) + 2f(x_2) + \cdots + 4f(x_{n-1}) + f(x_n)] \approx 3.2410894400 \approx 3.24.$$

The true value of the integral is approximately 3.24131. The antiderivative of f is known to be a nonelementary function.

5.9.19: With $a = 1$, $b = 5$, $n = 8$, $\Delta x = 0.5$, $f(x) = (1 + \ln x)^{1/3}$, and $x_i = 1 + i \cdot \Delta x$, we obtained

$$T_n = \frac{\Delta x}{2} \cdot \left(f(x_0) + f(x_n) + 2 \cdot \sum_{i=1}^{n-1} f(x_i) \right) \approx 5.0139700316$$

and

$$S_n = \frac{\Delta x}{3} \cdot [f(x_0) + 4f(x_1) + 2f(x_2) + \cdots + 4f(x_{n-1}) + f(x_n)] \approx 5.0196759933.$$

The true value of the integral is approximately 5.02005.

5.9.21: With $\Delta x = 0.25$ and $n = 6$, we have

$$T_n = \frac{\Delta x}{2} [3.43 + 2 \cdot (2.17 + 0.38 + 1.87 + 2.65 + 2.31) + 1.97] = 3.02$$

and

$$S_n = \frac{\Delta x}{3} [3.43 + 4 \cdot (2.17 + 1.87 + 2.31) + 2 \cdot (0.38 + 2.65) + 1.97] \approx 3.07167.$$

5.9.23: We read the following data from Fig. 5.9.16:

0	1	2	3	4	5	6	7	8	9	10
250	300	320	327	318	288	250	205	158	110	80

With $\Delta x = 1$ and $n = 10$, we have

$$T_n = \frac{\Delta x}{2} [250 + 2 \cdot (300 + 320 + 327 + 318 + 288 + 250 + 205 + 158 + 110) + 80] = 2441$$

and

$$S_n = \frac{\Delta x}{3} [250 + 4 \cdot (300 + 327 + 288 + 205 + 110) + 2 \cdot (320 + 318 + 250 + 158) + 80] = \frac{7342}{3} \approx 2447.33.$$

5.9.25: With $\Delta x = 50$ and $n = 12$, we find

$$T_n = \frac{\Delta x}{2} [0 + 2 \cdot (165 + 192 + 146 + 63 + 42 + 84 + 155 + 224 + 270 + 267 + 215) + 0] = 91150.$$

This result is in square feet. Divide by 9 to convert to square yards, then divide by 4840 to convert to acres; the result is approximately 2.093 acres. Next,

$$S_n = \frac{\Delta x}{3} [0 + 4 \cdot (165 + 146 + 42 + 155 + 270 + 215) + 2 \cdot (192 + 63 + 84 + 224 + 267) + 0] = \frac{281600}{3}.$$

As before, divide by 9 and by 4840 to obtain the estimate 2.155 acres.

5.9.27: We have $f''(x) = \dfrac{2}{x^3}$, so the constant in the error estimate is $K_2 = 2$. With $a = 1$, $b = 2$, and n subintervals, we require

$$\frac{K_2(b-a)^3}{12n^2} < 0.0005,$$

which implies that $n^2 > 333.33$ and thus that $n > 18.2$; $n = 19$ will suffice. (In fact, with only $n = 13$ subintervals, the trapezoidal estimate of 0.6935167303 differs from the true value of $\ln 2$ by less than 0.00037.)

5.9.29: If $p(x) = ax^3 + bx^2 + cx + d$ is a polynomial of degree 3 or smaller, then $p^{(4)}(x) \equiv 0$, so the constant K_4 in Simpson's error estimate is zero, which implies that the error will also be zero no matter what the value of the even positive integer n may be.

5.9.31: With the usual meanings of the notation, we have

$$M_n + T_n = (\Delta x) \cdot [f(m_1) + f(m_2) + f(m_3) + \cdots + f(m_n)]$$

$$+ \frac{\Delta x}{2} \cdot [f(x_0) + 2f(x_1) + 2f(x_2) + \cdots + 2f(x_{n-1}) + f(x_n)]$$

$$= \frac{\Delta x}{2} \cdot [f(x_0) + 2f(m_1) + 2f(x_1) + 2f(m_2) + 2f(x_2) + \cdots + 2f(x_{n-1}) + 2f(m_n) + f(x_n)] = 2T_{2n}.$$

The result in Problem 31 follows immediately.

5.9.33: Suppose first that $f(x) > 0$ and $f''(x) > 0$ for $a \leqq x \leqq b$. Then the graph of f is concave upward on $[a, b]$. Now examine Fig. 5.9.11, where it is shown that the midpoint approximation is the same as the tangent approximation. The tangent line will lie under the graph of f because the graph is concave upward; the chord connecting $(x_{i-1}, f(x_{i-1}))$ and $(x_i, f(x_i))$ will lie over the graph. Hence every term in the midpoint sum will underestimate the area under the graph of f and every term in the trapezoidal sum will overestimate it. Thus

$$M_n < \int_a^b f(x)\, dx < T_n$$

no matter what the choice of n. If $f''(x) < 0$ on $[a, b]$, then Figs. 5.9.11 and 5.9.12 show that the inequalities will be reversed.

5.9.35: We estimate the integral $\displaystyle\int_{90000}^{100000} \frac{1}{\ln x}\, dx$ first with the midpoint rule:

$$\frac{10000}{\ln(95000)} \approx 872.476;$$

then with the trapezoidal rule:

$$(5000)\left(\frac{1}{\ln(90000)} + \frac{1}{\ln(100000)}\right) \approx 872.600.$$

The first is an underestimate and the second is an overestimate because the graph of $y = 1/(\ln x)$ is concave upward for $x > 0$. The true value of the integral, incidentally, is approximately 872.5174045.

Chapter 5 Miscellaneous Problems

5.M.1: $\displaystyle\int (5x^{-3} - 2x^{-2} + x^2)\, dx = -\frac{5}{2}x^{-2} + 2x^{-1} + \frac{1}{3}x^3 + C.$

5.M.3: $\displaystyle\int (1 - 3x)^9\, dx = -\frac{1}{30}(1 - 3x)^{10} + C.$

5.M.5: $\displaystyle\int (9 + 4x)^{1/3}\, dx = \frac{3}{16}(9 + 4x)^{4/3} + C.$

5.M.7: $\displaystyle\int x^3(1 + x^4)^5\, dx = \frac{1}{24}(1 + x^4)^6 + C.$

5.M.9: $\displaystyle\int x(1 - x^2)^{1/3}\, dx = -\frac{3}{8}(1 - x^2)^{4/3} + C.$

5.M.11: $\displaystyle\int (7 \cos 5x - 5 \sin 7x)\, dx = \frac{1}{35}(25 \cos 7x + 49 \sin 5x) + C.$

5.M.13: If $u = x^4$, then $du = 4x^3\, dx$, so that $x^3\, dx = \frac{1}{4}\, du$. Hence

$$\int x^3(1 + x^4)^{1/2} dx = \int \frac{1}{4}(1 + u)^{1/2}\, du = \frac{1}{4} \cdot \frac{2}{3}(1 + u)^{3/2} + C = \frac{1}{6}(1 + x^4)^{3/2} + C.$$

5.M.15: If $u = 1 + x^{1/2}$ then $du = \frac{1}{2}x^{-1/2}\, dx$, so that $x^{-1/2}\, dx = 2\, du$. Therefore

$$\int \frac{x^{-1/2}}{\left(1 + x^{1/2}\right)^2}\, dx = \int \frac{2}{u^2}\, du = \int 2u^{-2}\, du = -2u^{-1} + C = -\frac{2}{1 + \sqrt{x}} + C.$$

5.M.17: If $u = 4x^3$ then $du = 12x^2\, dx$, so that $x^2\, dx = \frac{1}{12}\, du$. Hence

$$\int x^2 \cos 4x^3\, dx = \int \frac{1}{12} \cos u\, du = \frac{1}{12} \sin u + C = \frac{1}{12} \sin 4x^3 + C.$$

5.M.19: If $u = x^2 + 1$, then $du = 2x\, dx$, so that $x\, dx = \frac{1}{2}\, du$. Thus

$$\int x(x^2 + 1)^{14}\, dx = \int \frac{1}{2}u^{14}\, du = \frac{1}{30}u^{15} + C = \frac{1}{30}(x^2 + 1)^{15} + C.$$

An unpleasant alternative is to expand $(x^2 + 1)^{14}$ using the binomial formula, multiply by x, then integrate the resulting polynomial to obtain

$$\frac{1}{30}x^{30} + \frac{1}{2}x^{28} + \frac{7}{2}x^{26} + \frac{91}{6}x^{24} + \frac{91}{2}x^{22} + \frac{1001}{10}x^{20} + \frac{1001}{6}x^{18}$$

$$+ \frac{429}{2}x^{14} + \frac{1001}{6}x^{12} + \frac{1001}{10}x^{10} + \frac{91}{2}x^8 + \frac{91}{6}x^6 + \frac{7}{2}x^4 + \frac{1}{2}x^2 + C.$$

5.M.21: If $u = 4 - x$ then $x = 4 - u$ and $dx = -du$. Hence

$$\int x(4-x)^{1/2}\,dx = \int (u-4)u^{1/2}\,du = \int (u^{3/2} - 4u^{1/2})\,du$$

$$= \frac{2}{5}u^{5/2} - \frac{8}{3}u^{3/2} + C = \frac{2}{5}(4-x)^{5/2} - \frac{8}{3}(4-x)^{3/2} + C.$$

5.M.23: If $u = x^4$, then $du = 4x^3\,dx$, so that $2x^3\,dx = \frac{1}{2}\,du$. Therefore

$$\int 2x^3(1+x^4)^{-1/2}\,dx = \int \frac{1}{2}(1+u)^{-1/2}\,du = (1+u)^{1/2} + C = \sqrt{1+x^4} + C.$$

5.M.25: $y(x) = \int_0^x (3t^2 + 2t)\,dt + 5 = \left[t^3 + t^2\right]_0^x + 5 = x^3 + x^2 + 5.$

5.M.27: If $\dfrac{dy}{dx} = (2x+1)^5$, then $y(x) = \dfrac{1}{12}(2x+1)^6 + C$. Then

$$2 = y(0) = \frac{1}{12} + C \quad\text{implies that}\quad C = \frac{23}{12},$$

and therefore $y(x) = \dfrac{1}{12}(2x+1)^6 + \dfrac{23}{12}.$

5.M.29: $y(x) = \int_1^x t^{-1/3}\,dt + 1 = \left[\dfrac{3}{2}t^{2/3}\right]_1^x + 1 = \dfrac{3}{2}x^{2/3} - \dfrac{3}{2} + 1 = \dfrac{3x^{2/3} - 1}{2}.$

5.M.31: First convert 90 mi/h to 132 ft/s (just multiply by $\frac{22}{15}$). Let $x(t)$ denote the distance (in feet) the automobile travels after its brakes are first applied at time $t = 0$ (s). Then $x(t) = -11t^2 + 132t$, so the automobile first comes to a stop when $v(t) = x'(t) = -22t + 132 = 0$; that is, when $t = 6$. Therefore the total distance it travels while braking will be $x(6) = 396$ (ft).

5.M.33: Let v_0 denote the initial velocity of the automobile and let $x(t)$ denote the distance it has skidded since its brakes were applied at time $t = 0$ (units are in feet and seconds). Then $x(t) = -20t^2 + v_0 t$. Let T denote the time at which the automobile first comes to a stop. Then

$$x(T) = 180 \qquad\text{and}\qquad x'(T) = 0.$$

That is,

$$-20T^2 + v_0T = 180 \quad \text{and} \quad -40T + v_0 = 0.$$

The second of these equations implies that $v_0 = 40T$, and substitution of this datum in the first of these equations yields $-20T^2 + 40T^2 = 180$, so that $T^2 = 9$. Thus $T = 3$, and therefore $v_0 = 120$. Hence the initial velocity of the automobile was 120 feet per second, slightly less than 82 miles per hour (slightly less than 132 kilometers per hour).

5.M.35: Let q denote the acceleration of gravity on the planet Zorg. First consider the ball dropped from a height of 20 feet. Then its altitude at time t will be $y(t) = -\frac{1}{2}qt^2 + 20$ (because its initial velocity is $v_0 = 0$). Then the information that $x(2) = 0$ yields the information that $q = 10$. Now suppose that the ball is dropped from an initial height of 200 feet. Then its altitude at time t will be $y(t) = -5t^2 + 200$, so the ball will reach then ground when $5t^2 = 200$; that is, when $t = 2\sqrt{10}$. Its velocity during its descent will be $v(t) = y'(t) = -10t$, so its impact velocity will be $v(2\sqrt{10}) = -20\sqrt{10}$ feet per second. Thus its impact speed will be $|v(2\sqrt{10})| = 20\sqrt{10} \approx 63.25$ feet per second.

5.M.37: First we need to find the deceleration constant a of the car. Let $x(t)$ denote the distance (in feet) the car has skidded at time t (in seconds) if its brakes are applied at time $t = 0$. Then $x(t) = -\frac{1}{2}at^2 + 44t$. (We converted 30 miles per hour to 44 feet per second.) Thus in the first skid, if the car first comes to a stop at time T, then both

$$x(T) = 44 \quad \text{and} \quad x'(T) = 0.$$

Because the velocity of the car is $v(t) = x'(t) = -at + 44$, we must solve simultaneously the equations

$$-\frac{1}{2}aT^2 + 44T = 44 \quad \text{and} \quad -aT + 44 = 0.$$

The second of these equations yields $T = \dfrac{44}{a}$, then substitution in the first equation yields

$$-\frac{1}{2} \cdot \frac{44^2}{a} + \frac{44^2}{a} = 44;$$

$$-\frac{22}{a} + \frac{44}{a} = 1;$$

$$\frac{22}{a} = 1; \quad a = 22.$$

Now suppose that the initial velocity of the car is 60 miles per hour; that is, 88 feet per second. With the same meaning of $x(t)$ as before, we now have $x(t) = -11t^2 + 88t$. Thus $v(t) = x'(t) = -22t + 88$, so the car comes to a stop when $t = 4$. The distance it now skids will be $x(t) = 176$ feet. The point of the problem is that doubling the initial speed of the car *quadruples* the stopping distance.

5.M.39: $\displaystyle\sum_{i=1}^{100} 17 = 17 \cdot \sum_{i=1}^{100} 1 = 17 \cdot 100 = 1700.$

5.M.41: $\displaystyle\sum_{n=1}^{10} (3n-2)^2 = 9 \cdot \sum_{n=1}^{10} n^2 - 12 \cdot \sum_{n=1}^{10} n + 4 \cdot \sum_{n=1}^{10} 1 = 9 \cdot \frac{10 \cdot 11 \cdot 21}{6} - 12 \cdot \frac{10 \cdot 11}{2} + 4 \cdot 10 = 2845.$

5.M.43: On $[1, 2]$, we have $\displaystyle\lim_{n \to \infty} \sum_{i=1}^{n} \frac{\Delta x}{\sqrt{x_i^\star}} = \int_1^2 \frac{1}{\sqrt{x}}\, dx = \left[2\sqrt{x} \right]_1^2 = 2\sqrt{2} - 2.$

5.M.45: On $[0, 1]$, we have $\displaystyle\lim_{n \to \infty} \sum_{i=1}^{n} 2\pi x_i^\star \sqrt{1 + (x_i^\star)^2}\, \Delta x = \int_0^1 2\pi x \sqrt{1 + x^2}\, dx$

$$= \left[\frac{2\pi}{3}(1 + x^2)^{3/2} \right]_0^1 = \frac{2\pi}{3}\left(2\sqrt{2} - 1 \right).$$

5.M.47: If $f(x) \equiv c$ (a constant), then for every partition of $[a, b]$ and every selection for each such partition, we have $f(x_i^\star) = c$. Therefore

$$\sum_{i=1}^{n} f(x_i^\star)\, \Delta x = \sum_{i=1}^{n} c \cdot \frac{b-a}{n} = c(b-a) \cdot \sum_{i=1}^{n} \frac{1}{n} = c(b-a) \cdot n \cdot \frac{1}{n} = c(b-a).$$

Then, because every Riemann sum is equal to $c(b-a)$, this is also the limit of those Riemann sums. Therefore, by definition,

$$\int_a^b f(x)\, dx = c(b-a).$$

5.M.49: Given: f continuous on $[a, b]$ and $f(x) > 0$ there. Let m be the global minimum value of f on $[a, b]$; then $m > 0$. Hence, by the second comparison property (Section 5.5),

$$\int_a^b f(x)\, dx \geqq m(b-a) > 0.$$

5.M.51: $\displaystyle\int \left[(2x)^{1/2} - (3x^3)^{-1/2} \right] dx = \int \left(x^{1/2}\sqrt{2} - \frac{\sqrt{3}}{3} x^{-3/2} \right) dx = \frac{2\sqrt{2}}{3} x^{3/2} + \frac{2\sqrt{3}}{3x^{1/2}} + C.$

5.M.53: $\displaystyle\int \frac{4 - x^3}{2x^2}\, dx = \int \left(2x^{-2} - \frac{1}{2}x \right) dx = -2x^{-1} - \frac{1}{4}x^2 + C.$

5.M.55: $\displaystyle\int x^{1/2} \cos x^{3/2}\, dx = \frac{2}{3} \sin x^{3/2} + C.$

5.M.57: $\displaystyle\int \frac{1}{t^2} \sin \frac{1}{t}\, dt = \cos \frac{1}{t} + C.$

5.M.59: $\displaystyle\int \frac{u^{1/3}}{(1 + u^{4/3})^3}\, du = -\frac{3}{8}(1 + u^{4/3})^{-2} + C.$

5.M.61: $\displaystyle\int_1^4 \frac{(1+t^{1/2})^2}{t^{1/2}}\, dt = \int_1^4 (t^{1/2} + 2 + t^{-1/2})\, dt = \left[\frac{2}{3}t^{3/2} + 2t + 2t^{1/2}\right]_1^4 = \frac{52}{3} - \frac{14}{3} = \frac{38}{3}.$

5.M.63: Let $u = \dfrac{1}{x}$, so that $x = \dfrac{1}{u}$ and $dx = -\dfrac{1}{u^2}\, du$. Then

$$I = \int \frac{(4x^2-1)^{1/2}}{x^4}\, dx = \int -\frac{u^4}{u^2}\left(\frac{4}{u^2}-1\right)^{1/2} du = -\int u^2 \left(\frac{4}{u^2}-1\right)^{1/2} du.$$

Next move one copy of u from outside the square root to inside, where it becomes u^2, and we see

$$I = -\int u(4-u^2)^{1/2}\, du = \frac{1}{3}(4-u^2)^{3/2} + C = \frac{1}{3}\left(4 - \frac{1}{x^2}\right)^{3/2} + C.$$

To make this answer more appealing, multiply numerator and denominator by x^3; when the x^3 in the numerator is moved into the 3/2-power, it becomes x^2 and the final version of the answer is

$$I = \frac{(4x^2-1)^{3/2}}{3x^3} + C.$$

5.M.65: The area is $\displaystyle\int_0^1 (x^4 - x^5)\, dx = \left[\frac{1}{5}x^5 - \frac{1}{6}x^6\right]_0^1 = \frac{1}{5} - \frac{1}{6} = \frac{1}{30}.$

5.M.67: Solve $x^4 = 2 - x^2$ to find that the two curves cross where $x = \pm 1$. Hence the area between them is

$$\int_{-1}^1 (2 - x^2 - x^4)\, dx = \left[2x - \frac{1}{3}x^3 - \frac{1}{5}x^5\right]_{-1}^1 = \frac{22}{15} - \left(-\frac{22}{15}\right) = \frac{44}{15}.$$

5.M.69: Solve $(x-2)^2 = 10 - 5x$ to find that the two curves cross where $x = -3$ and where $x = 2$. The area between them is

$$\int_{-3}^2 (10 - 5x - (x-2)^2)\, dx = \left[6x - \frac{1}{2}x^2 - \frac{1}{3}x^3\right]_{-3}^2 = \frac{22}{3} - \left(-\frac{27}{2}\right) = \frac{125}{6}.$$

5.M.71: If $y = \sqrt{2x - x^2}$, then $y \geq 0$ and $x^2 - 2x + y^2 = 0$, so that $(x-1)^2 + y^2 = 1$. Therefore the graph of the integrand is the top half of a circle of radius 1 centered at $(1, 0)$, and so the value of the integral is the area of that semicircle:

$$\int_0^2 \sqrt{2x - x^2}\, dx = \frac{1}{2} \cdot \pi \cdot 1^2 = \frac{\pi}{2}.$$

5.M.73: If

$$x^2 = 1 + \int_1^x \sqrt{1 + [f(t)]^2}\, dt,$$

then differentiation of both sides of this *identity* with respect to x (using Part 1 of the fundamental theorem of calculus, Section 5.6) yields

$$2x = \sqrt{1 + [f(x)]^2},$$

so that $4x^2 = 1 + [f(x)]^2$. Therefore if $x > 1$, one solution is $f(x) = \sqrt{4x^2 - 1}$.

5.M.75: Because $f(x) = \sqrt{1 + x^2}$ is increasing on $[0, 1]$, the left-endpoint approximation will be an underestimate of the integral and the right-endpoint approximation will be an overestimate. With $n = 5$, $\Delta x = \frac{1}{5}$, and $x_i = i \cdot \Delta x$, we find the left-endpoint approximation to be

$$\sum_{i=1}^{5} f(x_{i-1}) \, \Delta x \approx 1.10873$$

and the right-endpoint approximation is

$$\sum_{i=1}^{5} f(x_i) \, \Delta x \approx 1.19157.$$

The average of the two approximations is 1.15015 and half their difference is 0.04142, and therefore

$$\int_0^1 \sqrt{1 + x^2} \, dx = 1.15015 \pm 0.04143.$$

The true value of this integral is approximately 1.1477935747. When we used $n = 4$ subintervals the error in the approximation was larger than 0.05.

5.M.77: $M_5 \approx 0.2866736772$ and $T_5 \approx 0.2897075147$. Because the graph of the integrand is concave upward on the interval $[1, 2]$,

$$M_5 < \int_1^2 \frac{1}{x + x^2} \, dx < T_5$$

for the reasons given in the solution of Problem 33 of Section 5.9.

5.M.79: Suppose that $0 < a < b$, that n is a positive integer, that $\mathcal{P} = \{x_0, x_1, x_2, \ldots, x_n\}$ is a partition of $[a, b]$, and that $\Delta x_i = x_i - x_{i-1}$ for $1 \le i \le n$. Let $x_i^\star = \sqrt{x_{i-1} x_i}$ for $1 \le i \le n$. Then $S = \{x_1^\star, x_2^\star, \ldots, x_n^\star\}$ is a selection for $\mathcal{P}$ because

$$x_{i-1} = \sqrt{(x_{i-1})^2} < \sqrt{x_{i-1} x_i} < \sqrt{(x_i)^2} = x_i$$

for $1 \le i \le n$. Next,

$$\sum_{i=1}^{n} \frac{1}{(x_i^\star)^2} \, \Delta x_i = \sum_{i=1}^{n} \frac{x_i - x_{i-1}}{x_{i-1} x_i} = \sum_{i=1}^{n} \left(\frac{1}{x_{i-1}} - \frac{1}{x_i} \right)$$

$$= \left(\frac{1}{x_0} - \frac{1}{x_1} \right) + \left(\frac{1}{x_1} - \frac{1}{x_2} \right) + \left(\frac{1}{x_2} - \frac{1}{x_3} \right) + \cdots + \left(\frac{1}{x_{n-2}} - \frac{1}{x_{n-1}} \right) + \left(\frac{1}{x_{n-1}} - \frac{1}{x_n} \right)$$

$$= \frac{1}{x_0} - \frac{1}{x_n} = \frac{1}{a} - \frac{1}{b}.$$

Therefore

$$\sum_{i=1}^{n} \frac{1}{(x_i^\star)^2} \, \Delta x_i \to \frac{1}{a} - \frac{1}{b}$$

as $|\mathcal{P}| \to 0$. But

$$\sum_{i=1}^{n} \frac{1}{(x_i^\star)^2} \, \Delta x_i \quad \text{is a Riemann sum for} \quad \int_a^b \frac{1}{x^2} \, dx.$$

Because f is continuous on $[a, b]$, all such Riemann sums converge to the same limit, which must therefore be the same as the particular limit just computed. Therefore

$$\int_a^b \frac{1}{x^2} \, dx = \frac{1}{a} - \frac{1}{b}.$$

Section 6.1

6.1.1: With $a = 0$ and $b = 1$, $\displaystyle\lim_{n\to\infty} \sum_{i=1}^{n} 2x_i^\star \, \Delta x = \int_0^1 2x \, dx = \left[x^2 \right]_0^1 = 1.$

6.1.3: With $a = 0$ and $b = 1$, $\displaystyle\lim_{n\to\infty} \sum_{i=1}^{n} (\sin \pi x_i^\star) \, \Delta x = \int_0^1 \sin \pi x \, dx = \left[-\frac{1}{\pi} \cos \pi x \right]_0^1 = \frac{2}{\pi}.$

6.1.5: With $a = 0$ and $b = 4$,

$$\lim_{n\to\infty} \sum_{i=1}^{n} x_i^\star \sqrt{(x_i^\star)^2 + 9} \, \Delta x = \int_0^4 x \sqrt{x^2 + 9} \, dx = \left[\frac{1}{3}(x^2+9)^{3/2} \right]_0^4 = \frac{125}{3} - 9 = \frac{98}{3}.$$

6.1.7: The limit is $\displaystyle\int_0^1 e^{-x} \, dx = \left[-e^{-x} \right]_0^1 = e^0 - e^{-1} = 1 - \frac{1}{e}.$

6.1.9: The limit is $\displaystyle\int_0^6 \frac{x}{x^2+9} \, dx = \left[\frac{1}{2} \ln(x^2+9) \right]_0^6 = \frac{1}{2} \ln 45 - \frac{1}{2} \ln 9 = \frac{1}{2} \ln 5.$

6.1.11: With $a = 1$ and $b = 4$, $\displaystyle\lim_{n\to\infty} \sum_{i=1}^{n} 2\pi x_i^\star f(x_i^\star) \, \Delta x = \int_1^4 2\pi x f(x) \, dx.$ Compare this with Eq. (2) in Section 6.3.

6.1.13: With $a = 0$ and $b = 10$, $\displaystyle\lim_{n\to\infty} \sum_{i=1}^{n} \sqrt{1 + [f(x_i^\star)]^2} \, \Delta x = \int_0^{10} \sqrt{1 + [f(x)]^2} \, dx.$

6.1.15: $M = \displaystyle\int_0^{100} \frac{1}{5} x \, dx = \left[\frac{1}{10} x^2 \right]_0^{100} = 1000 - 0 = 1000$ (grams).

6.1.17: $M = \displaystyle\int_0^{10} x(10 - x) \, dx = \left[5x^2 - \frac{1}{3} x^3 \right]_0^{10} = \frac{500}{3} - 0 = \frac{500}{3}$ (grams).

6.1.19: The net distance is

$$\int_0^{10} (-32) \, dt = \left[-32t \right]_0^{10} = -320$$

and the total distance is 320.

6.1.21: The net distance is

$$\int_0^{10} (4t - 25) \, dt = \left[2t^2 - 25t \right]_0^{10} = 200 - 250 = -50.$$

Because $v(t) = 4t - 25 \leqq 0$ for $0 \leqq t \leqq 6.25$ and $v(t) \geqq 0$ for $6.25 \leqq t \leqq 10$, the total distance is

$$-\int_0^{6.25} v(t) \, dt + \int_{6.25}^{10} v(t) \, dt = \frac{625}{8} + \frac{225}{8} = \frac{425}{4} = 106.25.$$

6.1.23: The net distance is

$$\int_{-2}^{3} 4t^3 \, dt = \left[t^4 \right]_{-2}^{3} = 81 - 16 = 65.$$

Because $v(t) \leq 0$ for $-2 \leq t \leq 0$, the total distance is

$$-\int_{-2}^{0} 4t^3 \, dt + \int_{0}^{3} 4t^3 \, dt = 16 + 81 = 97.$$

6.1.25: Because $v(t) = \sin 2t \geq 0$ for $0 \leq t \leq \pi/2$, the net distance and the total distance are both equal to

$$\int_{0}^{\pi/2} \sin 2t \, dt = \left[-\frac{1}{2} \cos 2t \right]_{0}^{\pi/2} = \frac{1}{2} - \left(-\frac{1}{2} \right) = 1.$$

6.1.27: The net distance is

$$\int_{-1}^{1} \cos \pi t \, dt = \left[\frac{1}{\pi} \sin \pi t \right]_{-1}^{1} = 0 - 0 = 0.$$

Because $v(t) = \cos \pi t \leq 0$ for $-1 \leq t \leq -0.5$ and for $0.5 \leq t \leq 1$, the total distance is

$$-\int_{-1}^{-0.5} \cos \pi t \, dt + \int_{-0.5}^{0.5} \cos \pi t \, dt - \int_{0.5}^{1} \cos \pi t \, dt = \frac{1}{\pi} + \frac{2}{\pi} + \frac{1}{\pi} = \frac{4}{\pi}.$$

6.1.29: The net distance is

$$\int_{0}^{10} (t^2 - 9t + 14) \, dt = \left[\frac{1}{3} t^3 - \frac{9}{2} t^2 + 14t \right]_{0}^{10} = \frac{70}{3} \approx 23.333333,$$

but because $v(t) = t^2 - 9t + 14 \leq 0$ for $2 \leq t \leq 7$, the total distance is

$$\int_{0}^{2} v(t) \, dt - \int_{2}^{7} v(t) \, dt + \int_{7}^{10} v(t) \, dt = \frac{38}{3} + \frac{125}{6} + \frac{63}{2} = 65.$$

6.1.31: If $v(t) = t^3 - 7t + 4$ for $0 \leq t \leq 3$, then $v(t) < 0$ for $\alpha = 0.602705 < t < \beta = 2.29240$ (numbers with decimal points are approximations), so the net distance is

$$\int_{0}^{3} (t^3 - 7t + 4) \, dt = \left[\frac{1}{4} t^4 - \frac{7}{2} t^2 + 4t \right]_{0}^{3} = \frac{3}{4}$$

but the total distance is approximately

$$\int_{0}^{\alpha} (t^3 - 7t + 4) \, dt - \int_{\alpha}^{\beta} (t^3 - 7t + 4) \, dt + \int_{\beta}^{3} (t^3 - 7t + 4) \, dt \approx 1.17242 + 3.49165 + 3.06923 = 7.73330.$$

6.1.33: Here, $v(t) = t \sin t - \cos t$ is negative for $0 \leq t < \alpha = 0.860333589019$ (numbers with decimals are approximations), so the net distance is

$$\int_{0}^{\pi} (t \sin t - \cos t) \, dt = \left[-t \cos t \right]_{0}^{\pi} = \pi$$

274

but the total distance is

$$-\int_0^\alpha (t\sin t - \cos t)\, dt + \int_\alpha^\pi (t\sin t - \cos t)\, dt \approx 0.561096338191 + 3.702688991781 = 4.263785329972.$$

6.1.35: If n is a large positive integer, $\Delta x = r/n$, $x_i = i \cdot \Delta x$, and $x_i^\star$ is chosen in the interval $[x_{i-1},\, x_i]$ for $1 \leq i \leq n$, then the area of the annular ring between x_{i-1} and x_i is approximately $2\pi x_i^\star \, \Delta x$ and its average density is approximately $\rho\,(x_i^\star)$, so a good approximation to the total mass of the disk is

$$\sum_{i=1}^n 2\pi x_i^\star \rho\,(x_i^\star)\, \Delta x.$$

But this is a Riemann sum for

$$\int_0^r 2\pi x \rho(x)\, dx,$$

and therefore such sums approach this integral as $\Delta x \to 0$ and $n \to +\infty$ because $2\pi x \rho(x)$ is (we presume) continuous for $0 \leqq x \leqq r$. But such Riemann sums also approach the total mass M of the disk and this establishes the equation in Problem 35.

6.1.37: The mass is $M = \displaystyle\int_0^5 2\pi x (25 - x^2)\, dx = \left[-\frac{1}{2}\pi(x^4 - 50x^2) \right]_0^5 = \frac{625\pi}{2} \approx 981.747704.$

6.1.39: The amount of water that flows into the tank from time $t = 10$ to time $t = 20$ is

$$\int_{10}^{20} (100 - 3t)\, dt = \left[100t - \frac{3}{2}t^2 \right]_{10}^{20} = 1400 - 850 = 550 \quad \text{(gallons)}.$$

6.1.41: Answer:

$$375000 + \int_0^{20} \left[(1000(16 + t) - 1000\left(5 + \frac{1}{2}t\right) \right]\, dt = 375000 + \left[11000t + 250t^2 \right]_0^{20} = 695000.$$

6.1.43: We solved $r(0) = 0.1$ and $r(182.5) = 0.3$ for $a = 0.2$ and $b = 0.1$. Then we found that

$$\int_0^{365} r(t)\, dt = \left[\frac{73}{4\pi} \cdot \left(\frac{4\pi t}{365} - \sin\frac{2\pi t}{365} \right) \right]_0^{365} = 73$$

inches per year, average annual rainfall.

6.1.45: If $f(x) = x^{1/3}$ on $[0, 1]$, n is a positive integer, $\Delta x = 1/n$, and $x_i = x_i^\star = i \cdot \Delta x$, then

$$\lim_{n\to\infty} \sum_{i=1}^n \frac{i^{1/3}}{n^{4/3}} = \lim_{n\to\infty} \sum_{i=1}^n f\,(x_i^\star)\, \Delta x = \int_0^1 x^{1/3}\, dx = \left[\frac{3}{4}x^{4/3} \right]_0^1 = \frac{3}{4}.$$

6.1.47: Let n be a large positive integer, let $\Delta x = 1/n$, and let $x_i = i \cdot \Delta x$. Then $\mathcal{P} = \{x_0,\, x_1,\, x_2,\, \ldots,\, x_n\}$ is a regular partition of $[0, 1]$. Let $x_i^\star$ be the midpoint of the interval $[x_{i-1},\, x_i]$. Then the weight of the spherical shell with inner radius x_{i-1} and outer radius x_i will be approximately

$$100\,(1 + x_i^\star) \cdot 4\pi\,(x_i^\star)^2\, \Delta x$$

(the approximate volume of the shell multiplied by its approximate average density). Therefore the total weight of the ball will be approximately

$$\sum_{i=1}^{n} 100\left(1 + x_i^\star\right) \cdot 4\pi\left(x_i^\star\right)^2 \ \Delta x.$$

This is a Riemann sum for $f(x) = 100(1 + x) \cdot 4\pi x^2$ on $[0, 1]$, and such sums approach the total weight W of the ball as $n \to +\infty$. Therefore

$$W = \int_0^1 100(1 + x) \cdot 4\pi x^2 \ dx = \left[\frac{400}{3}\pi x^3 + 100\pi x^4\right]_0^1 = \frac{700}{3}\pi \approx 733.038286 \ \text{(pounds)}.$$

6.1.49: Because the pressure P is inversely proportional to the fourth power of the radius r, the values $r = 1.00,\ 0.95,\ 0.90,\ 0.85,\ 0.80$, and 0.75 yield the values $r^{-4} = 1.00,\ 1.22774,\ 1.52416,\ 1.91569,\ 2.44141$, and 3.16049. We subtract 1 from each of the latter and then multiply by 100 to obtain *percentage increase* in pressure (and multiply each value of r by 100 to convert to percentages). Result:

100	0.000
95	22.774
90	52.416
85	91.569
80	144.141
75	216.049

6.1.51: Given: $A = 4.5$ (mg). Simpson's rule applied to the concentration data

$$c = 0,\quad 2.32,\quad 9.80,\quad 10.80,\quad 7.61,\quad 4.38,\quad 2.21,\quad 1.06,\quad 0.47,\quad 0.18,\quad 0.0$$

measured at the times $t = 0, 1, 2, \ldots, 10$ (seconds) yields

$$\int_0^{10} c(t) \ dt \approx S_{10} = \frac{1}{3}\left[1 \cdot 0 + 4 \cdot (2.32) + 2 \cdot (9.8) + 4 \cdot (10.8) + 2 \cdot (7.61) + 4 \cdot (4.38)\right.$$

$$\left. + 2 \cdot (2.21) + 4 \cdot (1.06) + 2 \cdot (0.47) + 4 \cdot (0.18) + 1 \cdot 1\right] = \frac{1919}{50} = 38.38.$$

Hence—multiplying by 60 to convert second to minutes—the cardiac output is approximately

$$F \approx \frac{60A}{38.38} \approx 7.0349$$

L/min (liters per minute).

The *Mathematica* code is straightforward, although you must remember that an array's initial subscript is 1 rather than zero (unless you decree it otherwise):

276

```
c = {0, 232/100, 98/10, 108/10, 761/100, 438/100, 221/100, 106/100, 47/100, 18/100, 0};

c[[1]] + c[[11]] + 4*Sum[c[[i]], {i, 2, 10, 2}] + 2*Sum[c[[i]], {i, 3, 9, 2}]
```

$$\frac{5757}{50}$$

```
N[ 5757/(3*50), 12 ]
```

    38.3800000000

```
N[ (60*45/10)/(3838/100), 12 ]
```

    7.03491401772

Section 6.2

6.2.1: The volume is $V = \int_0^1 \pi x^4 \, dx = \left[\frac{1}{5}\pi x^5\right]_0^1 = \frac{\pi}{5}$.

6.2.3: The volume is $V = \int_0^4 \pi y \, dy = \left[\frac{1}{2}\pi y^2\right]_0^4 = 8\pi$.

6.2.5: The volume is $V = \int_0^1 \pi(1-x)^2 \, dx = -\frac{\pi}{3}\left[(1-x)^3\right]_0^1 = \frac{\pi}{3}$.

6.2.7: Rotation of the given figure around the x-axis produces annular rings of inner radius $y = x^2$ and outer radius $y = \sqrt{x}$ (because $x^2 \leq \sqrt{x}$ if $0 \leq x \leq 1$). Hence the volume of the solid is

$$V = \int_0^1 \pi\left[(\sqrt{x})^2 - x^4\right] dx = \pi\left[\frac{1}{2}x^2 - \frac{1}{5}x^5\right]_0^1 = \pi\left(\frac{1}{2} - \frac{1}{5}\right) = \frac{3}{10}\pi.$$

6.2.9: The volume is $\int_1^5 \frac{\pi}{x}\, dx = \left[\pi \ln x\right]_1^5 = \pi \ln 5 \approx 5.05619832$.

6.2.11: Volume: $V = \int_{-1}^1 \pi(1-x^2)^2 \, dx = \pi\left[x - \frac{2}{3}x^3 + \frac{1}{5}x^5\right]_{-1}^1 = \frac{16\pi}{15} \approx 3.351032$.

6.2.13: A horizontal cross section "at" y has radius $x = \sqrt{1-y}$, so the volume of the solid is

$$\int_0^1 \pi(1-y) \, dy = \pi\left[y - \frac{1}{2}y^2\right]_0^1 = \frac{\pi}{2}.$$

6.2.15: The region between the two curves extends from $y = 2$ to $y = 6$ and the radius of a horizontal cross section "at" y is $x = \sqrt{6-y}$. Therefore the volume of the solid is

$$V = \int_2^6 \pi(6-y) \, dy = \pi\left[6y - \frac{1}{2}y^2\right]_2^6 = 18\pi - 10\pi = 8\pi.$$

6.2.17: When the region bounded by the given curves is rotated around the horizontal line $y = -1$, it generates annular regions with outer radius $x - x^3 + 1$ and inner radius 1. Hence the volume of the solid thereby generated is

$$V = \int_0^1 \pi \left[(x - x^3 + 1)^2 - 1^2 \right] dx = \pi \int_0^1 (2x + x^2 - 2x^3 - 2x^4 + x^6) \, dx$$

$$= \pi \left[x^2 + \frac{1}{3} x^3 - \frac{1}{2} x^4 - \frac{2}{5} x^5 + \frac{1}{7} x^7 \right]_0^1 = \frac{121\pi}{210} - 0 = \frac{121\pi}{210} \approx 1.8101557671.$$

6.2.19: When the region bounded by the given curves is rotated around the y-axis, it generates circular regions; the one "at" y has radius $x = \sqrt{y}$, so the volume generated is

$$V = \int_0^4 \pi y \, dy = \pi \left[\frac{1}{2} y^2 \right]_0^4 = 8\pi.$$

6.2.21: When the region bounded by the given curves is rotated around the horizontal line $y = -2$, it generates annular regions with outer radius $\sqrt{x} - (-2)$ and inner radius $x^2 - (-2)$. Hence the volume of the solid thereby generated is

$$V = \int_0^1 \pi \left[\left(\sqrt{x} + 2 \right)^2 - (2 + x^2)^2 \right] dx = \pi \int_0^1 (4x^{1/2} + x - 4x^2 - x^4) \, dx$$

$$= \pi \left[\frac{8}{3} x^{3/2} + \frac{1}{2} x^2 - \frac{4}{3} x^3 - \frac{1}{5} x^5 \right]_0^1 = \frac{49\pi}{30} \approx 5.1312680009.$$

6.2.23: When the region bounded by the given curves is rotated around the vertical line $x = 3$, it generates annular regions with outer radius $3 - y^2$ and inner radius $3 - \sqrt{y}$. Hence the volume of the solid thereby generated is

$$V = \int_0^1 \pi \left[(3 - y^2)^2 - (3 - \sqrt{y})^2 \right] dy = \pi \int_0^1 (6y^{1/2} - y - 6y^2 + y^4) \, dy$$

$$= \pi \left[4y^{3/2} - \frac{1}{2} y^2 - 2y^3 + \frac{1}{5} y^5 \right]_0^1 = \frac{17\pi}{10} \approx 5.340708.$$

6.2.25: The volume generated by rotation of R around the x-axis is

$$V = \int_0^\pi \pi \sin^2 x \, dx = \pi \int_0^\pi \frac{1 - \cos 2x}{2} \, dx = \pi \left[\frac{1}{2} x - \frac{1}{4} \sin 2x \right]_0^\pi = \frac{\pi^2}{2} \approx 4.9348022005.$$

6.2.27: The curves $y = \cos x$ and $y = \sin x$ cross at $\pi/4$ and the former is above the latter for $0 \leqq x < \pi/4$. So the region between them, when rotated around the x-axis, generates annular regions with outer radius $\cos x$ and inner radius $\sin x$. Therefore the volume of the solid generated will be

$$V = \int_0^{\pi/4} \pi (\cos^2 x - \sin^2 x) \, dx = \pi \int_0^{\pi/4} \cos 2x \, dx = \pi \left[\frac{1}{2} \sin 2x \right]_0^{\pi/4} = \frac{\pi}{2}.$$

278

6.2.29: The volume is

$$V = \int_0^{\pi/4} \pi \tan^2 x \, dx = \pi \int_0^{\pi/4} \frac{\sin^2 x}{\cos^2 x} \, dx = \pi \int_0^{\pi/4} \frac{1 - \cos^2 x}{\cos^2 x} \, dx$$

$$= \pi \int_0^{\pi/4} (\sec^2 x - 1) \, dx = \pi \left[(\tan x) - x \right]_0^{\pi/4} = \frac{\pi(4-\pi)}{4} \approx 0.6741915533.$$

6.2.31: The two curves cross near $a = -0.532089$, $b = 0.652704$, and $c = 2.87939$. When the region between the curves for $a \le x \le b$ is rotated around the x-axis, the solid it generates has approximate volume

$$V_1 = \int_a^b \pi \left[(x^3 + 1)^2 - (3x^2)^2 \right] \, dx \approx 1.68838 - (-1.39004) = 2.99832.$$

When the region between the curves for $b \le x \le c$ is rotated around the x-axis, the solid it generates has approximate volume

$$V_2 = \int_b^c \pi \left[(3x^2)^2 - (x^3 + 1)^2 \right] \, dx \approx 265.753 - (-1.68838) \approx 267.442.$$

The total volume generated is thus $V_1 + V_2 \approx 270.440.$

6.2.33: The two curves cross near $a = -0.8244962453$ and $b = 0.8244962453$. When the region between them is rotated around the x-axis, the volume of the solid it generates is

$$V \approx \int_a^b \pi \left[(\cos^2 x) - x^4 \right] \, dx = \frac{\pi}{20} \left[10x - 4x^5 + 5\sin 2x \right]_a^b \approx 1.83871 - (-1.83871) \approx 3.67743.$$

6.2.35: The two curves cross at the two points where $x = 6$, and the one on the right meets the x-axis at $(3, 0)$. When the region they bound is rotated around the x-axis, the volume of the solid it generates is

$$V = \int_0^6 \pi x \, dx - \int_3^6 2\pi(x-3) \, dx = \pi \left[\frac{1}{2}x^2 \right]_0^6 - \pi \left[x^2 - 6x \right]_3^6 = (18\pi - 0) - (0 + 9\pi) = 9\pi.$$

6.2.37: The right half of the ellipse is the graph of the function

$$g(y) = \frac{a}{b}\sqrt{b^2 - y^2}, \quad -b \le y \le b.$$

Hence when the region bounded on the right by the graph of $x = g(y)$ and on the left by the y-axis is rotated around the y-axis, the volume of the ellipsoid thereby swept out will be

$$V = \int_{-b}^b \pi \left[g(y) \right]^2 \, dy = \pi \left[\frac{a^2 y (3b^2 - y^2)}{3b^2} \right]_{-a}^a = \frac{4}{3}\pi a^2 b.$$

6.2.39: Locate the base of the observatory in the xy-plane with the center at the origin and the diameter AB on the x-axis. Then the boundary of the base has equation $x^2 + y^2 = a^2$. A typical vertical cross-section of the observatory has as *its* base a chord of that circle perpendicular to the x-axis at [say] x, so that the length of this chord is $2\sqrt{a^2 - x^2}$. The square of this length gives the area of that vertical cross section, and it follows that the volume of the observatory is

$$V = \int_{-a}^{a} 4(a^2 - x^2)\, dx = \left[\frac{4}{3}(3a^2 x - x^3)\right]_{-a}^{a} = \frac{8}{3}a^3 - \left(-\frac{8}{3}a^2\right) = \frac{16}{3}a^3.$$

6.2.41: Locate the base of the solid in the xy-plane with the center at the origin and the diameter AB on the x-axis. Then the boundary of the base has equation $x^2 + y^2 = a^2$. A typical vertical cross-section of the base has as *its* base a chord of that circle perpendicular to the x-axis at [say] x, so that the length of this chord is $2\sqrt{a^2 - x^2}$. This chord is one of the three equal sides of the vertical cross section—which is an equilateral triangle—and it follows that this triangle has

$$\text{Base:} \quad b = 2\sqrt{a^2 - x^2} \quad \text{and height:} \quad h = \frac{\sqrt{3}}{2}b = \sqrt{3a^2 - 3x^2}.$$

So the area of this triangle is $\frac{1}{2}bh = \left(\sqrt{3}\right)(a^2 - x^2)$, and therefore the volume of the solid is

$$\int_{-a}^{a} \left(\sqrt{3}\right)(a^2 - x^2)\, dx = \left[\frac{\sqrt{3}}{3}(3a^2 x - x^3)\right]_{-a}^{a} = \frac{4\sqrt{3}}{3}a^3.$$

6.2.43: The volume of the paraboloid is

$$V_p = \int_{0}^{h} 2\pi px\, dx = \left[\pi p x^2\right]_{0}^{h} = \pi p h^2.$$

If r is the radius of the cylinder, then the equation $y^2 = 2px$ yields $r^2 = 2ph$, so the volume of the cylinder is $V_c = \pi r^2 h = 2\pi p h^2 = 2V_p$.

6.2.45: Consider a cross section of the pyramid parallel to its base and at distance x from the vertex of the pyramid. Similar triangles show that the lengths of the edges of this triangular cross section are proportional to x. If the edges have lengths p, q, and r, then Heron's formula tells us that the area of this triangular cross section is

$$\frac{1}{4}\sqrt{(p+q+r)(p+q-r)(p-q+r)(q+r-p)}\,,$$

and thus the area $g(x)$ of this triangular cross section is proportional to x^2; that is, $g(x) = kx^2$ for some constant k. But $g(h) = A$, which tells us that $kh^2 = A$ and thus that $k = A/(h^2)$. So the total volume of the pyramid will be

$$V = \int_{0}^{h} g(x)\, dx = \int_{0}^{h} \frac{A}{h^2} \cdot x^2\, dx = \left[\frac{Ax^3}{3h^2}\right]_{0}^{h} = \frac{Ah^3}{3h^2} = \frac{1}{3}Ah.$$

6.2.47: Set up a coordinate system in which one cylinder has axis the x-axis and the other has axis the y-axis. Introduce a z-axis perpendicular to the xy-plane and passing through $(0, 0)$. We will find the volume of the eighth of the intersection that lies in the *first octant*, where x, y, and z are nonnegative, then multiply by 8 to find the answer.

A cross section of the eighth perpendicular to the z-axis (thus parallel to the xy-plane) is a square; if this cross section meets the z-axis at z, then one of its edges lies in the yz-plane and reaches from the z-axis (where $y = 0$) to the side of the cylinder symmetric around the x-axis. That cylinder has equation the same as the equation as the circle in which it meets the yz-plane: $y^2 + z^2 = a^2$. Hence the edge of the square under consideration has length $y = \sqrt{a^2 - z^2}$. So the area of the square cross section "at" z is $a^2 - z^2$. So the volume of the eighth of the solid in the first octant is

$$\int_0^a (a^2 - z^2)\, dz = \left[a^2 z - \frac{1}{3} z^3 \right]_0^a = a^3 - \frac{1}{3} a^3 = \frac{2}{3} a^3.$$

Therefore the total volume of the intersection of the two cylinders is $\dfrac{16}{3} a^3$.

As an independent check, it's fairly easy to see that the sphere of radius a centered at the origin is enclosed in the intersection and occupies most of the volume of the intersection; the ratio of the volume of the intersection to the volume of that sphere is

$$\frac{\frac{16}{3} a^3}{\frac{4}{3} \pi a^3} = \frac{4}{\pi} \approx 1.273240,$$

a very plausible result.

6.2.49: The cross section of the torus perpendicular to the y-axis "at" y is an annular ring with outer radius $x = b + \sqrt{a^2 - y^2}$ and inner radius $x = b - \sqrt{a^2 - y^2}$, a consequence of the fact that the circular disk that generates the torus has equation $(x - b)^2 + y^2 = a^2$. So the cross section has area

$$\pi \left[\left(b + \sqrt{a^2 - y^2} \right)^2 - \left(b - \sqrt{a^2 - y^2} \right)^2 \right] = 4\pi b \sqrt{a^2 - y^2}.$$

Therefore the volume of the torus is

$$V = 4\pi b \int_{-a}^a \sqrt{a^2 - y^2}\, dy = 4\pi b \cdot \frac{1}{2} \pi a^2$$

because the integral is the area of a semicircle of radius a centered at the origin. Therefore $V = 2\pi^2 a^2 b$.

6.2.51: First, r is the value of $y = R - kx^2$ when $x = \frac{1}{2} h$, so $r = R - \frac{1}{4} k h^2 = R - \delta$ where $4\delta = kh^2$. Next, the volume of the barrel is

$$V = 2\int_0^{h/2} \pi(R - kx^2)^2 \, dx = 2\pi \int_0^{h/2} \left(R^2 - 2Rkx^2 + k^2x^4\right) \, dx$$

$$= 2\pi \left[R^2 x - \frac{2}{3}Rkx^3 + \frac{1}{5}k^2x^5 \right]_0^{h/2} = 2\pi \left(\frac{1}{2}R^2h - \frac{1}{12}Rkh^3 + \frac{1}{160}k^2h^5 \right)$$

$$= \pi h \left(R^2 - \frac{1}{6}Rkh^2 + \frac{1}{80}k^2h^4 \right) = \pi h \left(R^2 - \frac{1}{6}R \cdot 4\delta + \frac{1}{80} \cdot 16\delta^2 \right)$$

$$= \pi h \left(R^2 - \frac{2}{3}R\delta + \frac{1}{5}\delta^2 \right) = \frac{\pi h}{3} \cdot \left(3R^2 - 2R\delta + \frac{3}{5}\delta^2 \right)$$

$$= \frac{\pi h}{3} \cdot \left(2R^2 + R^2 - 2R\delta + \delta^2 - \frac{2}{5}\delta^2 \right) = \frac{\pi h}{3} \cdot \left(2R^2 + (R-\delta)^2 - \frac{2}{5}\delta^2 \right)$$

$$= \frac{\pi h}{3} \cdot \left(2R^2 + r^2 - \frac{2}{5}\delta^2 \right).$$

6.2.53: The factors 27 and 3.3 in the following computations convert cubic feet to cubic yards and cubic yards to dollars. The trapezoidal approximation gives

$$T_6 = \frac{(10)(3.3)}{(2)(27)} \cdot (1513 + 2 \cdot 882 + 2 \cdot 381 + 2 \cdot 265 + 2 \cdot 151 + 2 \cdot 50 + 0) \approx 3037.83$$

and Simpson's approximation gives

$$S_6 = \frac{(10)(3.3)}{(3)(27)} \cdot (1513 + 4 \cdot 882 + 2 \cdot 381 + 4 \cdot 265 + 2 \cdot 151 + 4 \cdot 50 + 0) \approx 3000.56.$$

To the nearest hundred dollars, each answer rounds to $3000.

6.2.55: First "finish" the frustum; that is, complete the cone of which it is a frustum. We measure all distances from the vertex of the completed cone and perpendicular to the bases of the frustum. Let H be the height of the cone and suppose that $H - h \le y \le H$, so that a cross section of the cone at distance y from its vertex is a circular cross section of the frustum. The area $A(y)$ of such a cross section is proportional to the square of its radius, which is proportional to y^2, so that $A(y) = ky^2$ where k is a positive proportionality constant. By similar triangles,

$$\frac{H-h}{r} = \frac{H}{R}, \quad \text{and so} \quad H = \frac{hR}{R-r}.$$

Also, $A(H) = kH^2 = \pi R^2$, and it follows that

$$k = \frac{\pi R^2}{H^2}, \quad \text{so that} \quad A(y) = \frac{\pi R^2 y^2}{H^2}.$$

Therefore the volume of the frustum is

$$V = \int_{H-h}^{H} \frac{\pi R^2 y^2}{H^2} \, dy = \frac{\pi R^2}{3H^2} \cdot \left(H^3 - (H-h)^3 \right).$$

Next note that $A(H - h) = \pi r^2 = \dfrac{\pi R^2}{H^2}(H - h)^2$. Therefore

$$V = \frac{\pi R^2}{3H^2}(3H^2 h - 3Hh^2 + h^3) = \frac{\pi R^2}{3h^2 R^2}(R - r)^2\left(\frac{3h^3 R^2}{(R - r)^2} - \frac{3h^3 R}{R - r} + h^3\right)$$

$$= \frac{\pi}{3h^2}\left(3h^3 R^2 - 3h^3 R(R - r) + h^3(R - r)^2\right) = \frac{\pi h}{3}\left(3R^2 - 3R(R - r) + (R - r)^2\right)$$

$$= \frac{\pi h}{3}(3R^2 - 3R^2 + 3rR + R^2 - 2Rr + r^2) = \frac{\pi h}{3}(R^2 + Rr + r^2).$$

6.2.57: The solid of intersection of the two spheres can be generated by rotating around the x-axis the region R common to the two circles shown in the following figure.

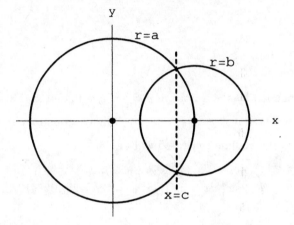

The circle on the left, of radius a and centered at the origin, has equation $x^2 + y^2 = a^2$. The circle on the right, of radius $b < a$ and centered at $(a, 0)$, has equation $(x - a)^2 + y^2 = b^2$. The x-coordinate of their two points of intersection can be found by solving

$$x^2 - a^2 = (x - a)^2 - b^2 \quad \text{for} \quad x = \frac{2a^2 - b^2}{2a} = c.$$

The part of R above the x-axis is comprised of two smaller regions, one to the left of the vertical line $x = c$ and one to its right. The width of the region R_1 on the left—measured along the x-axis—is

$$h_1 = c - (a - b) = \frac{2a^2 - b^2}{2a} - a + b = \frac{b(2a - b)}{2a}.$$

The width of the region R_2 on the right is

$$h_2 = a - c = \frac{b^2}{2a}.$$

One formula will tell us the volume of the solid generated by rotation of either R_1 or R_2 around the x-axis. Suppose that C is the circle of radius r centered at the origin and that $0 < h < r$. Let us find the volume generated by rotation of the region above the x-axis, within the circle, and to the right of the vertical line $x = r - h$ around the x-axis. It is

$$V(r, h) = \int_{r-h}^{r} \pi(r^2 - x^2)\, dx = \pi\left[r^2 x - \frac{1}{3}x^3\right]_{r-h}^{r} = \cdots = \frac{\pi h^2}{3}(3r - h)$$

(we worked this problem by hand—it is not difficult, merely tedious—but checked our results with *Mathematica*. Hence the volume of intersection of the two original spheres is

$$V(b, h_1) + V(a, h_2) = \cdots = \frac{\pi b^3}{12a}(8a - 3b)$$

(also by hand, tedious but not difficult, and checked with *Mathematica*).

Section 6.3

6.3.1: The region R, bounded by the graphs of $y = x^2$, $y = 0$, and $x = 2$, is shown next. If we rotate R around the y-axis, then the vertical strip in R "at" x will move around a circle of radius x and the height of the strip will be $y = x^2$, so the volume generated will be

$$\int_0^2 2\pi x^3\, dx = 2\pi \left[\frac{1}{4}x^4\right]_0^2 = 8\pi \approx 25.1327412287.$$

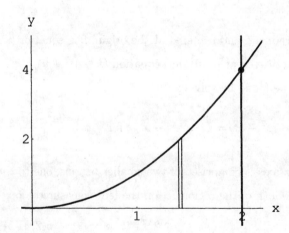

6.3.3: To obtain all the cylindrical shells, we need only let x range from 0 to 5, so the volume of the solid is

$$V = \int_0^5 2\pi x(25 - x^2)\, dx = \pi\left[25x^2 - \frac{1}{2}x^4\right]_0^5 = \frac{625\pi}{2} \approx 981.7477042468.$$

6.3.5: The volume is $\displaystyle\int_0^2 2\pi x(8 - 2x^2)\, dx = 2\pi\left[4x^2 - \frac{1}{2}x^4\right]_0^2 = 16\pi \approx 50.2654824574.$

6.3.7: A horizontal strip of the region R (shown next) "at" y stretches from $x = y$ to $x = 3 - 2y$ and thus has length $3 - 3y$. Hence the volume swept out by rotation of R around the x-axis is

$$V = \int_0^1 2\pi y(3 - 3y) \, dy = \pi \left[3y^2 - 2y^3 \right]_0^1 = \pi \approx 3.1415926536.$$

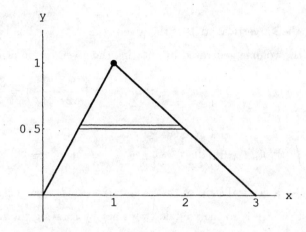

6.3.9: A horizontal strip of the region R "at" y $(0 \le y \le 2)$ stretches from $y^2/4$ to $x = (y/2)^{1/2}$ and is rotated around a circle of radius y, so the volume swept out when R is rotated around the x-axis is

$$V = \int_0^4 2\pi y \left[\left(\frac{y}{2} \right)^{1/2} - \frac{y^2}{4} \right] dy = 2\pi \int_0^4 \left(\frac{\sqrt{2}}{2} y^{3/2} - \frac{1}{4} y^3 \right) dy$$

$$= 2\pi \left[\frac{\sqrt{2}}{5} y^{5/2} - \frac{1}{16} y^4 \right]_0^2 = \frac{6\pi}{5} \approx 3.7699111843.$$

6.3.11: The graph of $y = 4x - x^3$ crosses the x-axis at $x = -2$, $x = 0$, and $x = 2$. It is below the x-axis for $-2 < x < 0$ and above it for $0 < x < 2$. A vertical strip "at" x for $0 \le x \le 2$ has height $4x - x^3$ and moves around a circle of radius x, whereas a vertical strip "at" x for $-2 \le x \le 0$ has height $-(4x - x^3)$ and moves around a circle of radius $-x$. Therefore the total volume swept out when the given region is rotated around the y-axis is

$$V = \int_0^2 2\pi x(4x - x^3) \, dx + \int_{-2}^0 (-2\pi x)(x^3 - 4x) \, dx = \int_{-2}^2 2\pi x(4x - x^3) \, dx$$

$$= 2\pi \left[\frac{4}{3} x^3 - \frac{1}{5} x^5 \right]_{-2}^2 = \frac{256\pi}{15} \approx 53.6165146213.$$

6.3.13: The volume is $\displaystyle \int_0^1 2\pi x(x - x^3) \, dx = 2\pi \left[\frac{1}{3} x^3 - \frac{1}{5} x^5 \right]_0^1 = \frac{4\pi}{15} \approx 0.8377580410.$

6.3.15: A vertical strip "at" x of the given region has height $y = x - x^3$ and moves around a circle of radius $2 - x$, so the volume the region sweeps out is

$$V = \int_0^1 2\pi(2 - x)(x - x^3) \, dx = 2\pi \left[x^2 - \frac{1}{3} x^3 - \frac{1}{2} x^4 + \frac{1}{5} x^5 \right]_0^1 = \frac{11\pi}{15} \approx 2.3038346126.$$

6.3.17: If $0 \leq x \leq 2$, then a vertical strip of the given region "at" x has height x^3 and moves around a circle of radius $3 - x$, so the volume swept out is

$$V = \int_0^2 2\pi(3 - x)x^3 \, dx = 2\pi\left[\frac{3}{4}x^4 - \frac{1}{5}x^5\right]_0^2 = \frac{56\pi}{5} \approx 35.1858377202.$$

6.3.19: If $-1 \leq x \leq 1$, then a vertical strip of the given region "at" x has height x^2 and moves around a circle of radius $2 - x$, so the volume generated by rotating the given region around the vertical line $x = 2$ is

$$V = \int_{-1}^1 2\pi(2 - x)x^2 \, dx = 2\pi\left[\frac{2}{3}x^3 - \frac{1}{4}x^4\right]_{-1}^1 = 2\pi\left(\frac{5}{12} + \frac{11}{12}\right) = \frac{8\pi}{3} \approx 8.3775804096.$$

6.3.21: The volume is $\int_0^1 2\pi y(y^{1/2} - y) \, dy = 2\pi\left[\frac{2}{5}y^{5/2} - \frac{1}{3}y^3\right]_0^1 = \frac{2\pi}{15} \approx 0.4188790205.$

6.3.23: If $0 \leq x \leq 1$, then a vertical strip of the given region "at" x has height $x - x^2$ and moves around a circle of radius $x - (-1) = x + 1$, so the volume swept out by rotating the region around the vertical line $x = -1$ is

$$V = \int_0^1 2\pi(x + 1)(x - x^2) \, dx = 2\pi\int_0^1 (x - x^3) \, dx = 2\pi\left[\frac{1}{2}x^2 - \frac{1}{4}x^4\right]_0^1 = \frac{\pi}{2} \approx 1.5707963268.$$

6.3.25: The volume is

$$V = \int_0^1 2\pi x \exp(-x^2) \, dx = \left[-\pi \exp(-x^2)\right]_0^1 = -\frac{\pi}{e} - (-\pi) = \frac{\pi(e - 1)}{e} \approx 1.9858653038.$$

6.3.27: If $0 \leq x \leq \sqrt{\pi}$, then a vertical strip of the given region "at" x has height $2\sin x^2$ and is rotated around a circle of radius x. Hence the volume thereby generated is

$$V = \int_0^{\sqrt{\pi}} 4\pi x \sin x^2 \, dx = \left[-2\pi \cos x^2\right]_0^{\sqrt{\pi}} = 2\pi - (-2\pi) = 4\pi \approx 12.5663706144.$$

6.3.29: The curves cross to the right of the y-axis at the points $x = a = 0.17248$ and $x = b = 1.89195$ (numbers with decimals are approximations). The quadratic is above the cubic if $a < x < b$, so the volume of the region generated by rotation of R around the y-axis is

$$V = \int_a^b 2\pi x(6x - x^2 - x^3 - 1) \, dx = 2\pi\left[-\frac{1}{5}x^5 - \frac{1}{4}x^4 + 2x^3 - \frac{1}{2}x^2\right]_a^b \approx 23.2990983139.$$

6.3.31: We used Newton's method to find that the two curves cross at $x = a = -0.8241323123$ and $x = b = -a$. But because the region R is symmetric around the y-axis, the interval of integration must be $[0, b]$. Also $\cos x \geq x^2$ on this interval, so the volume generated by rotation of R around the y-axis is

$$\int_0^b 2\pi x(\cos x - x^2) \, dx = 2\pi\left[\cos x + x \sin x - \frac{1}{4}x^4\right]_0^b \approx 1.0602688478.$$

6.3.33: We used Newton's method to find that the two curves cross where $x = a = 0.1870725959$ and $x = b = 1.5758806791$. Because $\cos x \geq 3x^2 - 6x + 2$ on $[a, b]$, the volume generated by rotation of R around the y-axis is

$$V = \int_a^b 2\pi x(\cos x - 3x^2 + 6x - 2)\, dx = 2\pi \left[\cos x + x\sin x - x^2 + 2x^3 - \frac{3}{4}x^4\right]_a^b \approx 8.1334538068.$$

6.3.35: The slant side of the cone is the graph of

$$f(x) = h - \frac{hx}{r} \quad \text{for} \quad 0 \leq x \leq r.$$

Therefore the volume of a cone of radius r and height h is

$$V = \int_0^r 2\pi x f(x)\, dx = 2\pi \left[\frac{1}{2}hx^2 - \frac{hx^3}{3r}\right]_0^r = \frac{1}{3}\pi r^2 h.$$

6.3.37: The top half of the ellipse is the graph of

$$y = f(x) = \frac{b}{a}(a^2 - x^2)^{1/2}, \quad -a \leq x \leq a.$$

The height of a vertical cross section of the ellipse "at" the number x is therefore $2f(x)$, so the volume of the ellipsoid will be

$$V = \int_0^a 4\pi x f(x)\, dx = \frac{4\pi b}{a}\int_0^a x(a^2 - x^2)^{1/2}\, dx = \frac{4\pi b}{a}\left[-\frac{1}{3}(a^2 - x^2)^{3/2}\right]_0^a = \frac{4\pi b}{3a}\cdot a^3 = \frac{4}{3}\pi a^2 b.$$

Note the lower limit of integration: zero, not $-a$.

6.3.39: The torus is generated by rotating the circular disk D in the xy-plane around the y-axis; the boundary of D has equation $(x - b)^2 + y^2 = a^2$ where $0 < a \leq b$. Thus D has its center at $(b, 0)$ on the positive x-axis. If $b - a \leq x \leq b + a$, then a vertical slice through D at x has height $2\sqrt{a^2 - (x - b)^2}$, and so the volume of the torus is

$$V = \int_{b-a}^{b+a} 2\pi x \cdot 2\sqrt{a^2 - (x - b)^2}\, dx.$$

The substitution $u = x - b$, with $x = u + b$ and $dx = du$, transforms this integral into

$$V = \int_{-a}^a 2\pi(u + b) \cdot 2\sqrt{a^2 - u^2}\, du = 4\pi \int_{-a}^a \left[u(a^2 - u^2)^{1/2} + b(a^2 - u^2)^{1/2}\right] du$$

$$= 4\pi \left[-\frac{1}{3}(a^2 - u^2)^{3/2}\right]_{-a}^a + 4\pi b \int_{-a}^a \sqrt{a^2 - u^2}\, du = 0 + 4\pi b \cdot \frac{1}{2}\pi a^2 = 2\pi^2 a^2 b$$

because the last integral represents the area of a semicircle of radius a.

6.3.41: If $-a \leqq x \leqq a$, then the vertical cross section through the disk "at" x has length $2\sqrt{a^2 - x^2}$ and moves through a circle of radius $a - x$, so the volume of the so-called *pinched torus* the disk generates is

$$V = \int_{-a}^{a} 2\pi(a - x) \cdot 2(a^2 - x^2)^{1/2} \, dx$$

$$= 4\pi a \int_{-a}^{a} (a^2 - x^2)^{1/2} \, dx - 4\pi \left[-\frac{1}{3}(a^2 - x^2)^{3/2} \right]_{-a}^{a} = 4\pi a \cdot \frac{1}{2}\pi a^2 + 0 = 2\pi^2 a^3.$$

The value of the last integral is $\frac{1}{2}\pi a^2$ because it represents the area of a semicircle of radius a.

6.3.43: The next figure shows the central cross section of the sphere-with-hole; the radius of the hole is a, its height is h, and the radius of the sphere is b.

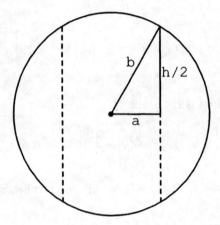

The figure shows that $\frac{1}{2}h = \sqrt{b^2 - a^2}$, and substitution in the volume formula $V = \frac{4}{3}\pi(b^2 - a^2)^{3/2}$ of Example 2 yields

$$V = \frac{4}{3}\pi \cdot \left(\frac{h}{2}\right)^3 = \frac{\pi}{6}h^3,$$

quite independent of the values of a or b.

6.3.45: (a) The method of cross sections yields volume

$$V = \int_0^5 \pi x(5 - x)^2 \, dx = \frac{\pi}{12}\left[3x^4 - 40x^3 + 150x^2 \right]_0^5 = \frac{625\pi}{12} \approx 163.6246173745.$$

(b) The method of cylindrical shells yields volume

$$V = \int_0^5 4\pi x^{3/2}(5 - x) \, dx = 2\pi \left[4x^{5/2} - \frac{4}{7}x^{7/2} \right]_0^5 = \frac{400\pi\sqrt{5}}{7} \approx 401.4179846309.$$

(c) The method of cylindrical shells yields volume

$$V = \int_0^5 4\pi x^{1/2}(5 - x)^2 \, dx = 4\pi \left[\frac{50}{3}x^{3/2} - 4x^{5/2} + \frac{2}{7}x^{7/2} \right]_0^5 = \frac{1600\pi\sqrt{5}}{21} \approx 535.2239795079.$$

6.3.47: Given

$$f(x) = 1 + \frac{x^2}{5} - \frac{x^4}{500} \quad \text{and} \quad g(x) = \frac{x^4}{10000},$$

the curves cross where $f(x) = g(x)$. The *Mathematica* command

```
Solve[ f[x] == g[x], x ]
```

returns two complex conjugate solutions and two real solutions, $x = \pm 10$. Hence the volume of the solid obtained by rotating the region between the two curves around the y-axis can be computed in this way (we include extra steps for the reader's benefit):

```
Integrate[ 2*Pi*x*(f[x] - g[x]), x ]
```

$$\pi x^2 + \frac{\pi x^4}{10} - \frac{7\pi x^6}{10000}$$

```
(% /.   x → 10) − (% /.   x → 0)
```

400π

```
N[%, 20]
```

1256.6370614359172954

To find the volume of water the birdbath will hold when full, we need to find the highest points on the graph of $y = f(x)$.

```
Solve[ D[ f[x], x] == 0, x ]
```

$$\{\{\, x \to 0 \,\}, \{\, x \to -5\sqrt{2} \,\}, \{\, x \to 5\sqrt{2} \,\}\}$$

```
f[ 5*Sqrt[2] ]
```

6

Thus the amount of water the birdbath will hold can be found as follows:

```
Integrate[ 2*Pi*x*(6 - f[x]), x ]
```

$$5\pi x^2 - \frac{\pi x^4}{10} + \frac{\pi x^6}{1500}$$

```
(% /.   x → 5*Sqrt[2]) − (% /.   x → 0)
```

$$\frac{250\pi}{3}$$

289

```
N[%, 20]
```

261.79938779914943654

Section 6.4

Note: In problems 1–20, we will also provide the exact answer (when the antiderivative is elementary) and an approximation to the exact answer (in every case). This information is not required of students working these problems, but it provides an opportunity for extra practice for them in techniques of integration (after they complete Chapter 7) and in numerical integration (which most of them have already completed). The approximations are accurate to the number of decimal places given.

6.4.1: The length is $\int_0^1 \sqrt{1 + 4x^2} \, dx = \frac{1}{4} \left(2\sqrt{5} + \sinh^{-1} 2 \right) \approx 1.4789428575$.

6.4.3: The length is $\int_0^2 \left[1 + 36x^2(x-1)^2 \right]^{1/2} dx \approx 6.6617$.

6.4.5: The length is $\int_0^{100} (1 + 4x^2)^{1/2} \, dx = 50\sqrt{40001} + \frac{1}{4} \sinh^{-1} 200 \approx 10001.6228669180$.

6.4.7: The length is $\int_{-1}^2 (1 + 16y^6)^{1/2} \, dy \approx 18.2471$.

6.4.9: The arc length is $\int_1^2 \left(1 + \frac{1}{x^2} \right)^{1/2} dx$

$$= \sqrt{5} - \sqrt{2} + \ln\left(1 + \sqrt{2}\right) + \ln 2 - \ln\left(1 + \sqrt{5}\right) \approx 1.2220161771.$$

6.4.11: The surface area is $\int_0^4 2\pi x^2 (1 + 4x^2)^{1/2} \, dx = \frac{\pi}{32} \left(1032\sqrt{65} - \sinh^{-1} 8 \right) \approx 816.5660537285$.

6.4.13: The surface area is $\int_0^1 2\pi(x - x^2)(4x^2 - 4x + 2)^{1/2} \, dx$

$$= \frac{\pi}{16} \left(\sqrt{2} + 5\sinh^{-1} 1 \right) \approx 1.1429666793.$$

6.4.15: The surface area is $\int_0^1 2\pi(2 - x)(1 + 4x^2)^{1/2} \, dx$

$$= \frac{\pi}{6} \left(7\sqrt{5} + 1 + 6\sinh^{-1} 2 \right) \approx 13.2545305651.$$

6.4.17: Given $f(x) = \ln(x^2 - 1)$,

$$1 + \left[f'(x) \right]^2 = 1 + \frac{4x^2}{(x^2 - 1)^2} = \frac{(x^2 + 1)^2}{(x^2 - 1)^2}.$$

290

Hence the surface area of revolution around the y-axis is

$$A = \int_2^3 2\pi x \cdot \frac{x^2+1}{x^2-1}\, dx = \pi(5 + 6\ln 2 - 2\ln 3) \approx 21.8706952193.$$

6.4.19: If $f(x) = \ln(x+1)$, then

$$1 + [f'(x)]^2 = 1 + \frac{1}{(x+1)^2} = \frac{x^2+2x+2}{(x+1)^2}.$$

Therefore the surface area of revolution around the line $x = -1$ is

$$A = \int_0^1 2\pi(x+1) \cdot \frac{\sqrt{x^2+2x+2}}{x+1}\, dx = \pi\left(2\sqrt{5} - \sqrt{2} + \sinh^{-1} 2 - \sinh^{-1} 1\right) \approx 11.3731443412.$$

6.4.21: The length is $\displaystyle\int_0^2 (1 + 2x^2)\, dx = \left[x + \frac{2}{3}x^3\right]_0^2 = \frac{22}{3}.$

6.4.23: First let $f(x) = \dfrac{1}{6}x^3 + \dfrac{1}{2x}$. Then

$$1 + [f'(x)]^2 = 1 + \left(\frac{1}{2}x^2 - \frac{1}{2}x^{-2}\right)^2 = 1 + \frac{1}{4}x^4 - \frac{1}{2} + \frac{1}{4}x^{-4} = \frac{1}{4}x^4 + \frac{1}{2} + \frac{1}{4}x^{-4} = \left(\frac{1}{2}x^2 + \frac{1}{2}x^{-2}\right)^2.$$

Therefore the length is

$$\frac{1}{2}\int_1^3 (x^2 + x^{-2})\, dx = \frac{1}{2}\left[\frac{1}{3}x^3 - \frac{1}{x}\right]_1^3 = \frac{13}{3} - \left(-\frac{1}{3}\right) = \frac{14}{3}.$$

6.4.25: First solve for $y = f(x) = \dfrac{2x^6+1}{8x^2} = \dfrac{1}{4}x^4 + \dfrac{1}{8}x^{-2}$. Then

$$1 + [f'(x)]^2 = 1 + \left(x^3 - \frac{1}{4}x^{-3}\right)^2 = 1 + x^6 - \frac{1}{2} + \frac{1}{16}x^{-6} = x^6 + \frac{1}{2} + \frac{1}{16}x^{-6} = \left(x^3 + \frac{1}{4}x^{-3}\right)^2.$$

Therefore the length is

$$\int_1^2 \left(x^3 + \frac{1}{4}x^{-3}\right) dx = \left[\frac{1}{4}x^4 - \frac{1}{8}x^{-2}\right]_1^2 = \frac{127}{32} - \frac{1}{8} = \frac{123}{32} = 3.84375.$$

6.4.27: Given $f(x) = \frac{1}{2}(e^x + e^{-x})$,

$$1 + [f'(x)]^2 = 1 + \frac{1}{4}(e^x - e^{-x})^2 = \left[\frac{1}{2}(e^x + e^{-x})\right]^2.$$

Therefore the length of the given arc is

$$L = \int_0^1 \frac{1}{2}(e^x + e^{-x})\, dx = \left[\frac{1}{2}(e^x - e^{-x})\right]_0^1 = \frac{e^2-1}{2e} \approx 1.1752011936.$$

6.4.29: Because

$$1 + [f'(x)]^2 = 1 + \left(\frac{1}{2}x^{-1/2}\right)^2 = \frac{4x+1}{4x},$$

the surface area is

$$\int_0^1 \frac{1}{2}\pi x^{1/2} \left(\frac{4x+1}{x}\right)^{1/2} dx = \int_0^1 \pi(4x+1)^{1/2} dx = \left[\frac{1}{6}\pi(4x+1)^{3/2}\right]_0^1 = \frac{5\sqrt{5}-1}{6}\pi \approx 5.3304135003.$$

6.4.31: First,

$$1 + [f'(x)]^2 = 1 + \left(x^4 - \frac{1}{4x^4}\right)^2 = x^8 + \frac{1}{2} + \frac{1}{16x^8} = \left(x^4 + \frac{1}{4x^4}\right)^2.$$

Therefore the surface area of revolution is

$$\int_1^2 2\pi x \left(x^4 + \frac{1}{4x^4}\right) dx = \pi \int_1^2 \left(2x^5 + \frac{1}{2}x^{-3}\right) dx$$

$$= \pi \left[\frac{1}{3}x^6 - \frac{1}{4}x^{-2}\right]_1^2 = \frac{339}{16}\pi \approx 66.5624943479.$$

6.4.33: Let $f(x) = (3x)^{1/3}$. Then

$$\sqrt{1 + [f'(x)]^2} = \left(1 + \frac{1}{(3x)^{4/3}}\right)^{1/2}.$$

Therefore the surface area of revolution is

$$\int_0^9 2\pi x \left(1 + \frac{1}{(3x)^{4/3}}\right)^{1/2} dx = \left[\frac{\pi}{27x}\left(9x^2 + (3x)^{2/3}\right)^{3/2}\right]_0^9 = \frac{82\sqrt{82}-1}{9}\pi \approx 258.8468426921.$$

6.4.35: Given $f(x) = x^2 - \frac{1}{8}\ln x$, $1 \leqq x \leqq 2$,

$$1 + [f'(x)]^2 = 1 + \left(2x - \frac{1}{8x}\right)^2 = \frac{(16x^2+1)^2}{64x^2}.$$

Therefore the surface area of revolution around the y-axis is

$$A = \int_1^2 2\pi x \cdot \frac{16x^2+1}{8x} dx = \pi \left[\frac{1}{4}x + \frac{4}{3}x^3\right]_1^2 = \frac{115\pi}{12} \approx 30.1069296.$$

6.4.37: We take $f(x) = \sqrt{1 + \cos^2 x}$, $\Delta x = \pi/6$, $x_i = i \cdot \Delta x$, and compute

$$S_6 = \frac{\Delta x}{3}\left[f(x_0) + 4f(x_1) + 2f(x_2) + 4f(x_3) + 2f(x_4) + 4f(x_5) + f(x_6)\right]$$

$$= \frac{\pi}{18}\left(4 + 2\sqrt{2} + 2\sqrt{5} + 4\sqrt{7}\right) \approx 3.8194031934.$$

6.4.39: The line segment from $(r_1, 0)$ to (r_2, h) is part of the line with equation

$$y = f(x) = \frac{h(x - r_1)}{r_2 - r_1},$$

and

$$\sqrt{1 + [f'(x)]^2} = \sqrt{1 + \frac{h^2}{(r_2 - r_1)^2}}.$$

Therefore the area of the conical frustum is

$$\int_{r_1}^{r_2} 2\pi x \sqrt{1 + [f'(x)]^2} \; dx = \pi \left[x^2 \left(\frac{(r_2 - r_1)^2 + h^2}{(r_2 - r_1)^2} \right)^{1/2} \right]_{r_1}^{r_2} = \pi(r_1 + r_2)\sqrt{h^2 + (r_2 - r_1)^2} \, .$$

With $\bar{r} = (r_1 + r_2)/2$ the "average radius" of the frustum and $L = \sqrt{h^2 + (r_2 - r_1)^2}$ its slant height, this yields the result in Eq. (6) of the text.

6.4.41: Let $f(x) = (1 - x^{2/3})^{3/2}$ for $0 \leq x \leq 1$. Then the graph of f is the part of the astroid that lies in the first quadrant in Fig. 6.4.16. Next,

$$\sqrt{1 + [f'(x)]^2} = \frac{1}{x^{1/3}},$$

so it would appear that the total length of the astroid is

$$4 \int_0^1 \frac{1}{x^{1/3}} \; dx.$$

But this integral is not a Riemann integral—the integrand approaches $+\infty$ as $x \to 0^+$. (It is an *improper integral*, the topic of Section 9.8.) But we can avoid the difficulty at $x = 0$ by integrating from the *midpoint* of the graph of f to $x = 1$. That midpoint occurs where $y = x$, so that $2x^{2/3} = 1$, and thus $x = a = \frac{1}{4}\sqrt{2}$. Therefore the total length of the astroid is

$$L = 8 \int_a^1 \frac{1}{x^{1/3}} \; dx = 8 \left[\frac{3}{2} x^{2/3} \right]_a^1 = 8 \cdot \left(\frac{3}{2} - \frac{3}{4} \right) = 6.$$

6.4.43: We will solve this problem by first rotating Fig. 6.4.18 through an angle of 90°. Let $f(x) = (r^2 - x^2)^{1/2}$. Think of the sphere as generated by rotation of the graph of f around the x-axis for $-r \leq x \leq r$. Suppose that the spherical zone Z of height h is the part of the sphere between $x = a$ and $x = a + h$, where

$$-r \leq a \leq a + h \leq r.$$

Also,

$$2\pi f(x)\sqrt{1 + [f'(x)]^2} = 2\pi(r^2 - x^2)^{1/2} \cdot \frac{r}{(r^2 - x^2)^{1/2}} = 2\pi r.$$

Therefore the area of the spherical zone Z is

$$A = \int_a^{a+h} 2\pi r \, dx = \left[2\pi r x \right]_a^{a+h} = 2\pi r(a + h) - 2\pi r a = 2\pi r h.$$

As noted in the statement of the problem, the area of Z depends only on the radius r of the sphere and the height (width) h of the zone and not on the location of the two planes (at a and $a + h$) that determine the zone. You can also "test" the answer by substituting the value $2r$ for h.

6.4.45: The right-hand endpoint of the cable is located at the point (S, H), so that $H = kS^2$. It follows that

$$y(x) = \frac{H}{S^2} x^2, \quad \text{so that} \quad \frac{dy}{dx} = \frac{2H}{S^2} x.$$

Therefore the total length of the cable is

$$L = \int_{-S}^{S} \sqrt{1 + \left(\frac{dy}{dx} \right)^2} \, dx = 2 \int_0^S \sqrt{1 + \frac{4H^2}{S^4} x^2} \, dx.$$

Section 6.5

6.5.1: The work is $W = \int_a^b F(x) \, dx = \int_{-2}^1 10 \, dx = \left[10x \right]_{-2}^1 = 30.$

6.5.3: The work is $W = \int_a^b F(x) \, dx = \int_1^{10} 10x^{-2} \, dx = \left[-\frac{10}{x} \right]_1^{10} = -1 - (-10) = 9.$

6.5.5: The work is $W = \int_a^b F(x) \, dx = \int_{-1}^1 \sin \pi x \, dx = \left[-\frac{1}{\pi} \cos \pi x \right]_{-1}^1 = \frac{1}{\pi} - \frac{1}{\pi} = 0.$

6.5.7: The spring constant is 30 lb/ft, so the force function for this spring is $F(x) = 30x$. Hence the work done is

$$W = \int_0^1 30x \, dx = \left[15x^2 \right]_0^1 = 15 \quad \text{(ft·lb)}.$$

6.5.9: With $k = 16 \times 10^9$, we find that the work done is

$$W = \int_{5000}^{6000} \frac{k}{x^2} \, dx = \left[-\frac{k}{x} \right]_{5000}^{6000} = \frac{k}{30000} = \frac{16}{3} \times 10^5 \quad \text{(mi·lb)}.$$

We multiply by 5280 to convert the answer to 2.816×10^9 ft·lb.

6.5.11: By similar triangles, if the height of water in the tank is y and the radius of the circular water surface is r, then $r = \frac{1}{2}(10 - y)$. Hence the area of the water surface is $A(y) = \frac{1}{4}\pi(10 - y)^2$. With water density $\rho = 62.4$ (lb/ft^3), the work done to fill the tank is

$$W = \int_0^{10} \rho y A(y) \, dy = \left[\frac{13}{10} \pi (3y^4 - 80y^3 + 600y^2) \right]_0^{10} = 13000\pi \approx 40840.704497 \quad \text{(ft·lb)}.$$

6.5.13: The work is $\int_0^5 50(y+10) \cdot 5\pi y \, dy = \pi \left[\frac{250}{3} y^3 + 1250 y^2 \right]_0^5 = \frac{125000\pi}{3} \approx 130899.69$ (ft·lb).

6.5.15: $W = \int_0^{10} \rho(15-y) \cdot 25\pi \, dy = \pi \left[23400y - 780y^2 \right]_0^{10} = 156000\pi \approx 490088.454$ (ft·lb).

6.5.17: With water density $\rho = 62.4$ lb/ft^3, the work will be

$$\int_{-10}^{10} \rho(50+y) \cdot \pi(100-y^2) \, dy = \frac{26}{5}\pi \left[60000y + 600y^2 - 200y^3 - 3y^4 \right]_{-10}^{10} = 4160000\pi \approx 13069025 \quad \text{(ft·lb)}.$$

6.5.19: Let $y = 0$ at the surface of the water in the well, so the top of the well is at $y = 100$. The weight of water in the bucket is $100 - \frac{1}{4}y$ when the bucket is at level y, so the total work done in lifting the water to the top of the well is

$$W = \int_0^{100} \left(100 - \frac{y}{4} \right) dy = \left[100y - \frac{1}{8}y^2 \right]_0^{100} = 8750 \quad \text{(ft·lb)}.$$

6.5.21: The weight of the rope and the water will be $w(y) = 100 + (100-y)/4$ when the bucket is y feet above the water surface. So the work to lift the rope and the water to the top of the well will be

$$W = \int w(y) \, dy = \left[125y - \frac{1}{8}y^2 \right]_0^{100} = 11250 \quad \text{(ft·lb)}.$$

6.5.23: Given: $pV^{1.4} = c$. When $p_1 = 200$, $V_1 = 50$. Hence

$$c = 200 \cdot 50^{7/5} = 200 \cdot 50^{7/5}, \quad \text{so} \quad p = \frac{c}{V^{7/5}} = 200 \cdot \left(\frac{50}{V} \right)^{7/5}.$$

Therefore the work done by the engine in each cycle is

$$W = \int_{50}^{500} 200 \cdot \left(\frac{50}{V} \right)^{7/5} dV = \left[200 \cdot 50^{7/5} \cdot \left(-\frac{5}{2} V^{-2/5} \right) \right]_{50}^{500} = 2500(10 - 10^{3/5}) \approx 15047.320736.$$

The answer is in "inch-pounds" (in.·lb); divide by 12 to convert the answer into $W \approx 1253.94339468$ foot-pounds.

6.5.25: $W = \int_0^1 60\pi(1-y)\sqrt{y} \, dy = 16\pi \approx 50.265482$ (ft·lb).

6.5.27: It is convenient to set up a coordinate system in which the center of the tank is at the origin, the x-axis horizontal, and the y-axis vertical. A horizontal cross section at y is circular with radius x satisfying $x^2 + y^2 = 144$, so the work to fill the tank is

$$W = \int_{-12}^{12} (50\pi)(y+12)\left(144-y^2\right) dy = \pi \left[86400y + 3600y^2 - 200y^3 - \frac{25}{2}y^4 \right]_{-12}^{12}$$

$$= 950400\pi - (-432000\pi) = 1382400\pi \approx 4.342938 \times 10^6 \quad \text{(ft·lb)}.$$

6.5.29: Let the string begin its journey stretched out straight along the x-axis from $x = 0$ to $x = 500\sqrt{2}$. Imagine it reaching its final position by simply pivoting at the origin up to a 45° angle while remaining straight. A small segment of the string initially at location x and of length dx is lifted from $y = 0$ to the final height $y = x/\sqrt{2}$, so the total work done in lifting the string is

$$W = \int_0^{500\sqrt{2}} \frac{x}{16\sqrt{2}} \, dx = \left[\frac{1}{32\sqrt{2}} x^2 \right]_0^{500\sqrt{2}} = \frac{15625\sqrt{2}}{2} \quad \text{(ft·oz)}.$$

Divide by 16 to convert the answer into ft·lb. Answer: Approximately 690.533966 ft·lb. To check the answer without using calculus, note that the string is lifted an average distance of 250 feet. Multiply this by the weight of the string in pounds to obtain the answer in ft·lb.

6.5.31: Set up a coordinate system with the y-axis vertical and the x-axis coinciding with the bottom of one end of the trough. A horizontal section of the trough at y is $2 - y$ feet below the water surface, so the total force on the end of the trough is given by

$$F = \int_0^2 (2)(2 - y)\rho \, dy = \rho \left[4y - y^2 \right]_0^2 = 4\rho = 249.6 \quad \text{(lb)}.$$

6.5.33: Set up a coordinate system in which one end of the trough lies in the [vertical] xy-plane with its base on the x-axis and bisected by the y-axis. Thus the trapezoidal end of the trough has vertices at the points $(1,0)$, $(-1,0)$, $(2,3)$, and $(-2,3)$. Because the width of a horizontal section at height y is $2x = \frac{2}{3}(y+3)$, the total force on the end of the trough is

$$F = \int_0^3 \rho \frac{2}{3}(y+3)(3-y) \, dy = \rho \left[6y - \frac{2}{9}y^3 \right]_0^3 = 12\rho = 748.8 \quad \text{(lb)}.$$

6.5.35: Let $\rho = 62.4$ lb/ft^3 and set up a coordinate system in which the y-axis is vertical and $y = 0$ is the location of the bottom of the square gate. Then the pressure at a horizontal cross section of the plate at location y will be $(15 - y)\rho$ and the area of the cross section will be $5 \, dy$, so the total force on the gate will be

$$F = \int_0^5 5\rho(15 - y) \, dy = \rho \left[75y - \frac{5}{2}y^2 \right]_0^5 = 19500 \quad \text{(lb)}.$$

6.5.37: Let ρ denote the density of water, as usual. Place the origin at the low vertex of the triangular gate with the y-axis vertical. Then a horizontal cross section of the gate at y has width $\frac{8}{5}y$ and depth $15 - y$, so the total force on the gate will be

$$F = \int_0^5 \frac{8}{5}\rho y(15 - y) \, dy = \rho \left[12y^2 - \frac{8}{15}y^3 \right]_0^5 = 14560 \quad \text{(lb)}.$$

6.5.39: Set up a coordinate system in which the bottom of the *vertical* face of the dam lies on the x-axis and the y-axis is vertical, with the origin at the center of the bottom of the vertical face of the dam. The vertical face occupies the interval $0 \leq y \leq 100$; form a regular partition of this interval and suppose that $[y_{i-1}, y_i]$ is one of the subintervals in this partition. Horizontal lines perpendicular to the vertical face through the endpoints of this interval determine a strip on the slanted face of length 200 (units are in feet) and (by similar triangles) width

$$\frac{\sqrt{100^2 + 30^2}}{100}(y_i - y_{i-1}) = \frac{\sqrt{100^2 + 30^2}}{100} \Delta y.$$

If y_i^* lies in this interval, then the strip on the slanted face opposite $[y_{i-1}, y_i]$ has approximate depth $100 - y_i^*$. So the total force of water on this strip—acting normal to the slanted face—is

$$200\rho(100 - y_i^*) \cdot \frac{\sqrt{100^2 + 30^2}}{100} \Delta y$$

where $\rho \doteq 62.4$ lb/ft^3 is the density of the water. Therefore the total force the water exerts on the slanted face of the dam—normal to that face—is

$$F = \int_0^{100} 200\rho(100 - y) \cdot \frac{\sqrt{100^2 + 30^2}}{100} \, dy$$

$$= \rho(109)^{1/2} \left[2000y - 10y^2 \right]_0^{100} = 100000\rho\sqrt{109} \approx 6.514721 \times 10^7 \quad \text{(lb)}.$$

The horizontal component of this force can be found by multiplying the total force by $10/\sqrt{109}$, and is thus 6.24×10^7 (lb).

Section 6.6

6.6.1: By symmetry, the centroid is located at $(2, 3)$.

6.6.3: By symmetry, the centroid is at $(1, 1)$.

6.6.5: The area of the triangular region is $A = 4$, and

$$M_y = \int_0^4 x\left(2 - \frac{1}{2}x\right) dx = \left[x^2 - \frac{1}{6}x^3 \right]_0^4 = 16 - \frac{32}{3} = \frac{16}{3}.$$

Hence $\overline{x} = \dfrac{4}{3}$. Next,

$$M_x = \int_0^4 \frac{1}{2} \cdot \left(\frac{4 - x}{2}\right)^2 dx = \frac{1}{8} \int_0^4 (16 - 8x + x^2) \, dx = \frac{1}{8} \left[16x - 4x^2 + \frac{1}{3}x^3 \right]_0^4 = \frac{8}{3},$$

and therefore $\overline{y} = \dfrac{2}{3}$.

6.6.7: The area of the region is

$$A = \int_0^2 x^2 \, dx = \left[\frac{1}{3}x^3 \right]_0^2 = \frac{8}{3}.$$

Next,

$$M_y = \int_0^2 x^3 \, dx = \left[\frac{1}{4}x^4 \right]_0^2 = 4 \quad \text{and}$$

$$M_x = \frac{1}{2} \int_0^2 x^4 \, dx = \left[\frac{1}{10}x^5 \right]_0^2 = \frac{16}{5}.$$

Therefore $(\bar{x}, \bar{y}) = \left(\frac{3}{2}, \frac{6}{5} \right)$.

6.6.9: The area of the region is

$$A = 2 \int_0^2 (4 - x^2) \, dx = 2 \left[4x - \frac{1}{3}x^3 \right]_0^2 = 2 \cdot \left(8 - \frac{8}{3} \right) = \frac{32}{3}.$$

Now $\bar{x} = 0$ by symmetry, but

$$M_x = -\frac{1}{2} \int_{-2}^2 (4 - x^2)^2 \, dx = -\int_0^2 (x^2 - 4)^2 \, dx = -\int_0^2 (x^4 - 8x^2 + 16) \, dx$$

$$= -\left[\frac{1}{5}x^5 - \frac{8}{3}x^3 + 16x \right]_0^2 = -\left(\frac{32}{5} - \frac{64}{3} + 32 \right) = -\frac{256}{15}.$$

Therefore $\bar{y} = -\frac{256}{15} \cdot \frac{3}{32} = -\frac{8}{5}$.

6.6.11: The area of the region is

$$A = 2 \int_0^2 (4 - x^2) \, dx = \frac{32}{3}, \quad \text{and}$$

$$M_x = \int_{-2}^2 \frac{1}{2}(4 - x^2)^2 \, dx = \frac{256}{15}.$$

Therefore $\bar{y} = \frac{256 \cdot 3}{15 \cdot 32} = \frac{8}{5}$; $\bar{x} = 0$ by symmetry.

6.6.13: The area of the region is $A = 1$. Moreover,

$$M_y = \int_0^1 3x^3 \, dx = \frac{3}{4} \quad \text{and} \quad M_x = \int_0^1 \frac{9}{2}x^4 \, dx = \frac{9}{10}.$$

Therefore $(\bar{x}, \bar{y}) = \left(\frac{3}{4}, \frac{9}{10} \right)$.

6.6.15: The parabola and the line meet at the two points $P(-3, -3)$ and $Q(2, 2)$. Hence

$$A = \int_{-3}^2 (6 - x - x^2) \, dx = \frac{125}{6},$$

298

$$M_y = \int_{-3}^{2} (6x - x^2 - x^3)\, dx = -\frac{125}{12}, \quad \text{and}$$

$$M_x = \int_{-3}^{2} \left[\frac{1}{2}(6 - x^2)^2 - \frac{1}{2}x^2 \right] dx = \frac{125}{3}.$$

Therefore the centroid is located at the point $\left(-\frac{1}{2}, 2 \right)$.

6.6.17: The region has area $A = \frac{1}{12}$. Next,

$$M_y = \int_{0}^{1} (x^3 - x^4)\, dx = \frac{1}{20} \quad \text{and}$$

$$M_x = \frac{1}{2} \int_{0}^{1} (x^4 - x^6)\, dx = \frac{1}{35}.$$

Thus $(\bar{x}, \bar{y}) = \left(\frac{3}{5}, \frac{12}{35} \right)$.

6.6.19: First note that $\bar{y} = \bar{x}$ by symmetry. Next,

$$M_y = \int_{0}^{r} x(r^2 - x^2)^{1/2}\, dx = \left[-\frac{1}{3}(r^2 - x^2)^{3/2} \right]_{0}^{r} = \frac{1}{3}r^3$$

and the area of the quarter circle is $\frac{1}{4}\pi r^2$. Therefore the centroid is at

$$(\bar{x}, \bar{y}) = \left(\frac{4r}{3\pi}, \frac{4r}{3\pi} \right).$$

6.6.21: By symmetry, $\bar{y} = \bar{x}$. Because $y = (r^2 - x^2)^{1/2}$,

$$1 + \left(\frac{dy}{dx} \right)^2 = \frac{r^2}{r^2 - x^2}.$$

So

$$M_y = \int_{0}^{r} \frac{rx}{(r^2 - x^2)^{1/2}}\, dx = \left[-r(r^2 - x^2)^{1/2} \right]_{0}^{r} = r^2.$$

The length of the quarter circle is $\frac{1}{2}\pi r$, so $\bar{x} = \frac{2r}{\pi}$.

6.6.23: First,

$$M_y = \int_{0}^{r} x \left(h - \frac{hx}{r} \right) dx = \frac{1}{6}hr^2 \quad \text{and} \quad A = \frac{1}{2}rh.$$

Therefore $\bar{x} = r/3$. By interchanging the roles of x and y, we find that $\bar{y} = \bar{x}$.

Next, the midpoint of the hypotenuse is $(r/2, h/2)$ and its slope is $-h/r$. The line L from $(0, 0)$ to the midpoint has equation

$$y = \frac{h}{r}x.$$

If $x = r/3$, then $y = h/3$, so $(r/3, h/3)$ lies on the line L. The distance from $(0, 0)$ to $(r/3, h/3)$ is

$$D_1 = \frac{1}{3}(r^2 + h^2)^{1/2};$$

the distance from $(0, 0)$ to $(r/2, h/2)$ is

$$D_2 = \frac{1}{2}(r^2 + h^2)^{1/2}.$$

Therefore

$$\frac{D_1}{D_2} = \frac{2}{3},$$

and this concludes the proof.

6.6.25: $A = \left(2 \cdot \frac{\pi r}{2}\right)\sqrt{r^2 + h^2} = \pi r(r^2 + h^2)^{1/2} = \pi r L.$

6.6.27: The radius of revolution is $\frac{1}{2}(r_1 + r_2)$, so the lateral area is

$$A = 2\pi \cdot \left(\frac{r_1 + r_2}{2}\right)\left[(r_2 - r_1)^2 + h^2\right]^{1/2} = \pi(r_1 + r_2)L.$$

6.6.29: The semicircular region has centroid (x, y) where $x = 0$ (by symmetry) and $y = b + \frac{4a}{3\pi}$ (by earlier work). So, for the semicircular region,

$$M_x = \left(b + \frac{4a}{3\pi}\right) \cdot \left(\frac{1}{2}\pi a^2\right).$$

For the rectangle, we have

$$M_x = \frac{b}{2} \cdot 2ab = ab^2.$$

The sum of these two moments is the moment of the entire region:

$$M_x = \frac{2}{3}a^3 + \frac{\pi}{2}a^2 b + ab^2.$$

When we divide this moment by the area $2ab + \frac{1}{2}\pi a^2$ of the entire region, we find that

$$\overline{y} = \frac{4a^2 + 3\pi ab + 6b^2}{12b + 3\pi a}.$$

Of course $\overline{x} = 0$ by symmetry.

For Part (b), we use the fact that the radius of the circle of rotation is $\overline{y}$, so the volume generated by rotation around the x-axis is

$$V = 2\pi \overline{y} A = 2\pi \overline{y}\left(2ab + \frac{1}{2}\pi a^2\right)$$

$$= \pi\bar{y}(4ab + \pi a^2) = \pi a\bar{y}(4b + \pi a) = \frac{1}{3}\pi a\bar{y}(12b + 3\pi a);$$

that is,

$$V = \frac{1}{3}\pi a(4a^2 + 3\pi ab + 6b^2).$$

6.6.31: Because $y = \sqrt{1 - x^2}$, as in Example 6 of this section we have

$$ds = \frac{1}{\sqrt{1 - x^2}}\, dx.$$

Hence the moment with respect to the y-axis is

$$M_x = 2\int_0^{\sin\alpha} \frac{y}{\sqrt{1 - x^2}}\, dx = \left[2x\right]_0^{\sin\alpha} = 2\sin\alpha.$$

Clearly the arc length is $s = 2\alpha$, and therefore the y-coordinate of the centroid is

$$\bar{y} = \frac{\sin\alpha}{\alpha}.$$

6.6.33: Let $f(x) = x$ and $g(x) = x^2$. The area of the region bounded by the graphs of f and g is

$$A = \int_0^1 (x - x^2)\, dx = \frac{1}{2} - \frac{1}{3} = \frac{1}{6}.$$

The moments with respect to the coordinate axes are

$$M_y = \int_0^1 (x^2 - x^3)\, dx = \frac{1}{3} - \frac{1}{4} = \frac{1}{12} \quad \text{and}$$

$$M_x = \int_0^1 \frac{1}{2}(x^2 - x^4)\, dx = \frac{1}{2}\cdot\left(\frac{1}{3} - \frac{1}{5}\right) = \frac{1}{15}.$$

Therefore the centroid is located at the point

$$(\bar{x}, \bar{y}) = C\left(\frac{1}{2}, \frac{2}{5}\right).$$

The axis L of rotation is the line $y = x$; the line through the centroid perpendicular to L has equation

$$y = \frac{9}{10} - x$$

and this perpendicular meets L at the point $P\left(\dfrac{9}{20}, \dfrac{9}{20}\right)$. The distance from P to C is

$$d = \sqrt{\left(\frac{1}{2} - \frac{9}{20}\right)^2 + \left(\frac{2}{5} - \frac{9}{20}\right)^2} = \frac{\sqrt{2}}{20}.$$

Because d is the radius of the circle through which C is rotated, the volume generated is (by the first theorem of Pappus)

$$V = 2\pi d A = 2\pi \cdot \frac{\sqrt{2}}{20} \cdot \frac{1}{6} = \frac{\pi\sqrt{2}}{60} \approx 0.07404804897.$$

Section 6.7

6.7.1: If $f(x) = 10^x$, then $f'(x) = 10^x \ln 10$ by Eq. (28).

6.7.3: If $f(x) = \dfrac{3^x}{4^x} = \left(\dfrac{3}{4}\right)^x$, then $f'(x) = \left(\dfrac{3}{4}\right)^x \ln \dfrac{3}{4}$ by Eq. (28).

6.7.5: If $f(x) = 7^{\cos x}$, then $f'(x) = -\left(7^{\cos x} \ln 7\right) \cdot \sin x$ by Eq. (29).

6.7.7: If $f(x) = 2^{x\sqrt{x}} = 2^{x^{3/2}}$, then $f'(x) = \dfrac{3}{2} x^{1/2} \left(2^{x^{3/2}} \ln 2\right)$.

6.7.9: If $f(x) = 2^{\ln x}$, then $f'(x) = \dfrac{1}{x}\left(2^{\ln x} \ln 2\right)$.

6.7. 11: If $f(x) = 17^x$, then $f'(x) = 17^x \ln 17$.

6.7.13: If $f(x) = 10^{1/x}$, then $f'(x) = -\dfrac{1}{x^2}\left(10^{1/x} \ln 10\right)$.

6.7.15: If $f(x) = 2^{2^x} = 2^{(2^x)}$, then $f'(x) = 2^{2^x} \cdot 2^x \cdot (\ln 2)^2$.

6.7.17: If $f(x) = \log_3 \sqrt{x^2 + 4} = \dfrac{1}{2}\log_3(x^2 + 4)$, then by Eq. (40) we find that

$$f'(x) = \frac{1}{2} \cdot \frac{1}{(x^2 + 4)\ln 3} \cdot 2x = \frac{x}{(x^2 + 4)\ln 3}.$$

6.7.19: If $f(x) = \log_3(2^x) = x \log_3 2$, then $f'(x) = \log_3 2 = \dfrac{\ln 2}{\ln 3}$ (by Eq. (40)).

6.7.21: If $f(x) = \log_2(\log_3 x)$, then by Eq. (35)

$$f(x) = (\log_2 e)\ln(\log_3 x) = (\log_2 e)\ln\left[(\log_3 e)\ln x\right] = (\log_2 e)\left[\ln(\log_3 e) + \ln(\ln x)\right].$$

Therefore

$$f'(x) = (\log_2 e)\left(\frac{1}{x \ln x}\right) = \frac{1}{(x \ln x)\ln 2}.$$

6.7.23: If $f(x) = \exp(\log_{10} x)$, then $f'(x) = \dfrac{\exp(\log_{10} x)}{x \ln 10}$. *Mathematica* reports that

$$f'(x) = \frac{x^{-1+(1/\ln 10)}}{\ln 10},$$

but a moment's work shows that the two answers are the same.

6.7.25: $\displaystyle\int 3^{2x}\,dx = \frac{3^{2x}}{2\ln 3} + C$ by Eq. (30).

6.7.27: $\displaystyle\int \frac{2^{\sqrt{x}}}{\sqrt{x}}\,dx = \frac{2 \cdot 2^{\sqrt{x}}}{\ln 2} + C.$

Comment: The easiest way to find an antiderivative such as this is to make an "educated guess" as to the form of the answer, differentiate it, and then modify the guess so it becomes correct. Here, for example, we guess $2^{\sqrt{x}}$ for the antiderivative. Then

$$D_x\left(2^{\sqrt{x}}\right) = \left(2^{\sqrt{x}}\ln 2\right) \cdot D_x\left(x^{1/2}\right) = \left(2^{\sqrt{x}}\ln 2\right) \cdot \frac{1}{2\sqrt{x}} = \frac{\ln 2}{2} \cdot \frac{2^{\sqrt{x}}}{\sqrt{x}}.$$

Thus we should multiply $2^{\sqrt{x}}$ by $\dfrac{2}{\ln 2}$ to correct it.

Of course this technique will succeed only if the correction consists of multiplication by a *constant*. If something else is needed, make a better guess or try integration by substitution.

6.7.29: Given: $\displaystyle\int x^2 \cdot 7^{x^3+1}\,dx$. Let $u = x^3 + 1$. Then $du = 3x^2\,dx$, so that $x^2\,dx = \dfrac{1}{3}\,du$. Thus

$$\int x^2 \cdot 7^{x^3+1}\,dx = \int \frac{1}{3}7^u\,du = \frac{7^u}{3\ln 7} + C = \frac{7^{x^3+1}}{3\ln 7} + C.$$

6.7.31: $\displaystyle\int \frac{\log_2 x}{x}\,dx = \int \frac{(\log_2 e)\ln x}{x}\,dx = \frac{1}{\ln 2}\int \frac{\ln x}{x}\,dx = \frac{1}{2\ln 2}(\ln x)^2 + C.$

6.7.33: Taking logarithms transforms the equation $R = kW^m$ into $\ln R = \ln k + m\ln W$, an equation linear in the two unknown coefficients $\ln k$ and m. We put the data given in Fig. 6.7.13 into the array

```
datapoints = { {25,131}, {67,103}, {127,88}, {175,81}, {240,75}, {975,53} },
```

then entered the *Mathematica* command

```
logdatapoints = N[Log[datapoints], 10]
```

to obtain the logarithms of the values of W and R to ten significant figures. We then set up the graph

```
pts = ListPlot[ logdatapoints ];
```

to see if the data points lay on a straight line. They very nearly did. Next we used *Mathematica*'s Fit command to find the coefficients of the equation of the straight line that best fit the data points (by minimizing the sum of the squares of the deviations of the data points from the straight line—see Miscellaneous Problem 51 of Chapter 13). The command

```
Fit[ logdatapoints, {1, x}, x ]
```

finds the best-fitting linear combination $a \cdot 1 + b \cdot x$ to the given data. The result was

```
5.67299196 - 0.24730011 x
```

which told us that $\ln k \approx 5.67299$, so that $k \approx 290.903$, and that $m \approx -0.2473$. Thus we obtained the formula $R = (290.903) \cdot W^{-0.2473}$. This formula predicts the following values for R:

		predicted	experimental
	W	R	R
	25	131.231	131
	67	102.839	103
	127	87.7966	88
	175	81.1045	81
	240	75.0105	75
	975	53.0356	53

The agreement between the predicted and experimental results is quite good. The graph of the logarithms of the data points and the line $y = 5.67299 - (0.2473)x$ is shown next.

6.7.35: If $f(x) = x \cdot 2^{-x}$, then $f'(x) = (1 - x \ln 2) \cdot 2^{-x}$, so $f'(x) = 0$ when $x = a = 1/\ln 2$. Because $f'(x) > 0$ if $x < a$ and $f'(x) < 0$ if $x > a$, we have found the highest point on the graph of f; it is

$$\left(\frac{1}{\ln 2}, \frac{1}{2^{1/(\ln 2)} \ln 2} \right) \approx (1.4426950409, 0.5307378454).$$

6.7.37: Let $f(x) = 2^{-x}$ and $g(x) = (x - 1)^2$. We saw in the solution of Problem 36 that the graphs meet where $x = 0$ and where $x = a \approx 1.5786206361$ (obtained by applying Newton's method to the equation

$f(x) - g(x) = 0$ with initial estimate $x_0 = 1.6$). There we saw also that $f(x) > g(x)$ if $0 < x < a$. So the method of parallel cross sections gives the volume of revolution around the x-axis as

$$V = \int_0^a \pi \left[2^{-2x} - (x-1)^4 \right] \, dx = \pi \left[2x^2 - x - 2x^3 + x^4 - \frac{1}{5}x^5 - \frac{2^{-2x}}{2\ln 2} \right]_0^a \approx 1.343088216395.$$

6.7.39: By definition of z, x, and y, respectively, we have

$$a^z = c, \quad a^x = b, \quad \text{and} \quad b^y = c.$$

Therefore $a^{xy} = b^y = c = a^z$. Because $a > 0$ and $a \neq 1$, it now follows that $z = xy$.

6.7.41: Beginning with the equation $x^y = 2$, we first write $y \cdot \ln x = \ln 2$, then differentiate implicitly with respect to x to obtain

$$\frac{y}{x} + \frac{dy}{dx} \ln x = 0.$$

Thus

$$\frac{dy}{dx} = -\frac{y}{x\ln x} = -\frac{\ln 2}{x(\ln x)^2}.$$

Section 6.8

6.8.1: $\arcsin\left(\dfrac{1}{2}\right) = \dfrac{\pi}{6}$ because $\sin\left(\dfrac{\pi}{6}\right) = \dfrac{1}{2}$ and $-\dfrac{\pi}{2} \leqq \dfrac{\pi}{6} \leqq \dfrac{\pi}{2}$. Similarly,

$$\arcsin\left(-\frac{1}{2}\right) = -\frac{\pi}{6}, \qquad \arcsin\left(\frac{1}{2}\sqrt{2}\right) = \frac{\pi}{4}, \qquad \text{and} \qquad \arcsin\left(-\frac{1}{2}\sqrt{3}\right) = -\frac{\pi}{3}.$$

6.8.3: $\arctan(0) = 0$ because $\tan(0) = 0$ and $-\dfrac{\pi}{2} < 0 < \dfrac{\pi}{2}$. Similarly,

$$\arctan(1) = \frac{\pi}{4}, \qquad \arctan(-1) = -\frac{\pi}{4}, \qquad \text{and} \qquad \arctan\left(\sqrt{3}\right) = \frac{\pi}{3}.$$

6.8.5: If $f(x) = \sin^{-1}\left(x^{100}\right)$, then $f'(x) = \dfrac{100x^{99}}{\sqrt{1 - x^{200}}}$.

6.8.7: If $f(x) = \sec^{-1}(\ln x)$, then $f'(x) = \dfrac{1}{x\,|\ln x|\sqrt{(\ln x)^2 - 1}}$.

6.8.9: If $f(x) = \arcsin(\tan x)$, then $f'(x) = \dfrac{\sec^2 x}{\sqrt{1 - \tan^2 x}}$.

6.8.11: If $f(x) = \sin^{-1} e^x$, then $f'(x) = \dfrac{e^x}{\sqrt{1 - e^{2x}}}$.

6.8.13: If $f(x) = \cos^{-1} x + \sec^{-1}\left(\dfrac{1}{x}\right)$ and $0 < x < 1$, then

$$f'(x) = -\frac{1}{\sqrt{1-x^2}} - \frac{1}{x^2 \cdot \dfrac{1}{x} \cdot \sqrt{\left(\dfrac{1}{x}\right)^2 - 1}}$$

$$= -\frac{1}{\sqrt{1-x^2}} - \frac{1}{x\sqrt{x^{-2}-1}} = -\frac{2}{\sqrt{1-x^2}}.$$

But if $-1 < x < 0$, then

$$f'(x) = -\frac{1}{\sqrt{1-x^2}} - \frac{1}{x^2 \cdot \left|\dfrac{1}{x}\right| \cdot \sqrt{x^{-2}-1}} = -\frac{1}{\sqrt{1-x^2}} + \frac{1}{x^2 \cdot \dfrac{1}{x} \cdot \sqrt{x^{-2}-1}}$$

$$= -\frac{1}{\sqrt{1-x^2}} - \frac{1}{(-x)\sqrt{x^{-2}-1}} = -\frac{2}{\sqrt{1-x^2}}.$$

In the last line in the second derivation, we needed to replace $x < 0$ with $-x > 0$ in order to move it under the radical.

6.8.15: If $f(x) = \csc^{-1} x^2$, then

$$f'(x) = -\frac{2x}{|x^2|\sqrt{x^4-1}} = -\frac{2x}{x^2\sqrt{x^4-1}} = -\frac{2}{x\sqrt{x^4-1}}.$$

6.8.17: If $f(x) = \dfrac{1}{\arctan x} = (\arctan x)^{-1}$, then $f'(x) = -\dfrac{1}{(1+x^2)(\arctan x)^2}$.

6.8.19: If $f(x) = \tan^{-1}(\ln x)$, then

$$f'(x) = \frac{1}{1+(\ln x)^2} \cdot \frac{1}{x} = \frac{1}{x\left[1+(\ln x)^2\right]}.$$

6.8.21: If $f(x) = \tan^{-1} e^x + \cot^{-1} e^{-x}$, then

$$f'(x) = \frac{e^x}{1+e^{2x}} - \frac{-e^{-x}}{1+e^{-2x}} = \frac{2e^x}{1+e^{2x}} = \frac{2}{e^x + e^{-x}}.$$

The last expression here is $\operatorname{sech} x = 1/\cosh x$, the *hyperbolic secant* of x (Section 6.9). Thus we have accidentally discovered an antiderivative of the hyperbolic secant function.

6.8.23: If $f(x) = \sin(\arctan x)$, then

$$f'(x) = \frac{\cos(\arctan x)}{1+x^2}.$$

But this problem has a twist. A reference right triangle with acute angle $\theta = \arctan x$, adjacent side 1, opposite side x, and hypotenuse $\sqrt{1+x^2}$ shows that

$$\sin(\arctan x) = \frac{x}{\sqrt{1+x^2}}.$$

Therefore, by the quotient rule,

$$f'(x) = \frac{(1+x^2)^{1/2} - x^2(1+x^2)^{-1/2}}{1+x^2} = \frac{1+x^2-x^2}{(1+x^2)^{3/2}} = \frac{1}{(1+x^2)^{3/2}}.$$

The second version of the derivative is potentially more useful than the first version.

6.8.25: If $f(x) = \dfrac{\arctan x}{(1+x^2)^2}$, then

$$f'(x) = \frac{(1+x^2)^2 \cdot \dfrac{1}{1+x^2} - 4x(1+x^2)\arctan x}{(1+x^2)^4} = \frac{1 - 4x\arctan x}{(1+x^2)^3}.$$

6.8.27: Given: $\tan^{-1} x + \tan^{-1} y = \dfrac{\pi}{2}$:

$$\frac{1}{1+x^2} + \frac{1}{1+y^2} \cdot \frac{dy}{dx} = 0, \quad \text{so} \quad \frac{dy}{dx} = -\frac{1+y^2}{1+x^2}.$$

So the slope of the line tangent to the graph at $P(1, 1)$ is -1, and therefore an equation of that line is $y - 1 = -(x - 1)$; that is, $y = 2 - x$.

6.8.29: Given: $(\sin^{-1} x)(\sin^{-1} y) = \dfrac{\pi^2}{16}$:

$$\frac{\sin^{-1} y}{\sqrt{1-x^2}} + \frac{\sin^{-1} x}{\sqrt{1-y^2}} \cdot \frac{dy}{dx} = 0, \quad \text{so} \quad \frac{dy}{dx} = -\frac{(1-y^2)^{1/2}\sin^{-1} y}{(1-x^2)^{1/2}\sin^{-1} x}.$$

So the slope of the line tangent to the graph at $P\left(\frac{1}{2}\sqrt{2}, \frac{1}{2}\sqrt{2}\right)$ is -1, and therefore an equation of that line is

$$y - \frac{1}{2}\sqrt{2} = -\left(x - \frac{1}{2}\sqrt{2}\right); \quad \text{that is,} \quad y = -x + \sqrt{2}.$$

6.8.31: $\displaystyle\int_0^1 \frac{1}{1+x^2}\, dx = \left[\arctan x\right]_0^1 = \frac{\pi}{4} - 0 = \frac{\pi}{4}.$

6.8.33: $\displaystyle\int_{\sqrt{2}}^2 \frac{1}{x\sqrt{x^2-1}}\, dx = \left[\operatorname{arcsec} x\right]_{\sqrt{2}}^2 = \frac{\pi}{3} - \frac{\pi}{4} = \frac{\pi}{12} \approx 0.261799387799.$

In comparison, *Mathematica* yields the result

$$\int_{\sqrt{2}}^2 \frac{1}{x\sqrt{x^2-1}}\, dx = \left[-\arctan\left(\frac{1}{\sqrt{x^2-1}}\right)\right]_{\sqrt{2}}^2 = -\frac{\pi}{6} + \frac{\pi}{4} = \frac{\pi}{12}.$$

6.8.35: Let $x = 3u$. Then $dx = 3\, du$, and as x ranges from 0 to 3, u ranges from 0 to 1. Therefore

$$\int_0^3 \frac{1}{9+x^2}\, dx = \int_0^1 \frac{3}{9+9u^2}\, du = \frac{1}{3}\int_0^1 \frac{1}{1+u^2}\, du = \frac{1}{3}\left[\arctan u\right]_0^1 = \frac{\pi}{12} - 0 = \frac{\pi}{12}.$$

307

Alternatively, the antiderivative can be expressed as a function of x before evaluation:

$$\int_0^3 \frac{1}{9+x^2}\,dx = \int_0^1 \frac{3}{9+9u^2}\,du = \frac{1}{3}\int_0^1 \frac{1}{1+u^2}\,du = \frac{1}{3}\Big[\arctan u\Big]_0^1$$

$$= \frac{1}{3}\left[\arctan\left(\frac{x}{3}\right)\right]_0^3 = \frac{\pi}{12} - 0 = \frac{\pi}{12} \approx 0.261799387799.$$

Note that in the latter case the original x-limits of integration must be restored before evaluation of the antiderivative.

6.8.37: Let $u = 2x$, so that $4x^2 = u^2$ and $dx = \frac{1}{2}\,du$. Then

$$\int \frac{1}{\sqrt{1-4x^2}}\,dx = \frac{1}{2}\int \frac{1}{\sqrt{1-u^2}}\,du = \frac{1}{2}\arcsin u + C = \frac{1}{2}\arcsin 2x + C.$$

6.8.39: Let $x = 5u$. Then $dx = 5\,du$ and $x^2 - 25 = 25u^2 - 25 = 25(u^2 - 1)$. Thus

$$\int \frac{1}{x\sqrt{x^2-25}}\,dx = \int \frac{5}{5\cdot 5u\sqrt{u^2-1}}\,du = \frac{1}{5}\operatorname{arcsec}|u| + C = \frac{1}{5}\operatorname{arcsec}\frac{|x|}{5} + C.$$

For comparison, *Mathematica* reports that

$$\int \frac{1}{x\sqrt{x^2-25}}\,dx = C - \frac{1}{5}\arctan\left(\frac{5}{\sqrt{x^2-25}}\right).$$

6.8.41: $\displaystyle\int \frac{e^x}{1+e^{2x}}\,dx = \int \frac{e^x}{1+(e^x)^2}\,dx = \arctan(e^x) + C.$

If you prefer integration by substitution, use $u = e^x$, so that $du = e^x\,dx$. Then

$$\int \frac{e^x}{1+e^{2x}}\,dx = \int \frac{1}{1+u^2}\,du = \arctan u + C = \arctan(e^x) + C.$$

6.8.43: Let $u = \frac{1}{5}x^3$. Then $5u = x^3$, $3x^2\,dx = 5\,du$, and $x^6 - 25 = 25(u^2 - 1)$. So

$$\int \frac{1}{x\sqrt{x^6-25}}\,dx = \int \frac{3x^2}{3x^3(x^6-25)^{1/2}}\,dx = \int \frac{5}{3\cdot 5u\,[25(u^2-1)]^{1/2}}\,du$$

$$= \frac{1}{15}\int \frac{1}{u(u^2-1)^{1/2}}\,du = \frac{1}{15}\operatorname{arcsec}|u| + C = \frac{1}{15}\operatorname{arcsec}\frac{|x^3|}{5} + C.$$

Mathematica gives the answer in the form $C + \dfrac{1}{15}\arctan\left(\dfrac{1}{5}\sqrt{x^6-25}\right)$.

6.8.45: The radicand is

$$x(1-x) = x - x^2 = -(x^2 - x) = -\frac{1}{4}(4x^2 - 4x)$$

$$= -\frac{1}{4}(4x^2 - 4x + 1) + \frac{1}{4} = \frac{1}{4}\left[1 - (2x-1)^2\right] = \frac{1}{4}(1 - u^2)$$

308

if we let $u = 2x - 1$. If so, $du = 2\,dx$, and then

$$\int \frac{1}{\sqrt{x(1-x)}}\,dx = \frac{1}{2}\int \frac{2}{\sqrt{1-u^2}}\,du = \arcsin u + C = \arcsin(2x - 1) + C.$$

The more "obvious" substitution $x = u^2$, so that $u = x^{1/2}$ and $dx = 2u\,du$, leads to

$$\int \frac{1}{\sqrt{x(1-x)}}\,dx = \int \frac{2u}{\sqrt{u^2(1-u^2)}}\,du = 2\int \frac{u}{|u|\sqrt{1-u^2}}\,du$$

$$= 2\int \frac{1}{\sqrt{1-u^2}}\,du = 2\arcsin u + C = 2\arcsin\sqrt{x} + C.$$

Replacement of $|u|$ with u here is permitted because $u = \sqrt{x} > 0$. Test your skill at trigonometry by showing that $f(x) = \arcsin(2x - 1)$ and $g(x) = 2\arcsin\sqrt{x}$ differ by a constant (if $0 \le x \le 1$).

6.8.47: Let $u = x^{50}$, so that $du = 50x^{49}\,dx$. Then

$$\int \frac{x^{49}}{1 + x^{100}}\,dx = \frac{1}{50}\int \frac{1}{1 + u^2}\,du = \frac{1}{50}\arctan u + C = \frac{1}{50}\arctan\left(x^{50}\right) + C.$$

6.8.49: $\displaystyle\int \frac{1}{x\left[1 + (\ln x)^2\right]}\,dx = \arctan(\ln x) + C.$ (Use $u = \ln x$ if necessary.)

6.8.51: $\displaystyle\int_0^1 \frac{1}{1 + (2x - 1)^2}\,dx = \left[\frac{1}{2}\arctan(2x - 1)\right]_0^1 = \frac{\pi}{8} - \left(-\frac{\pi}{8}\right) = \frac{\pi}{4} \approx 0.7853981634.$

If integration by substitution is preferred, let $u = 2x - 1$, $du = 2\,dx$, and do not forget to change the limits of integration to $u = -1$ and $u = 1$.

6.8.53: Let $u = \ln x$ (if necessary) to find that

$$\int_1^e \frac{1}{x\sqrt{1 - (\ln x)^2}}\,dx = \left[\arcsin(\ln x)\right]_1^e = \frac{\pi}{2} - 0 = \frac{\pi}{2}.$$

6.8.55: If $u = x^{1/2}$, then $du = \frac{1}{2}x^{-1/2}\,dx.$. Moreover, $u = 1$ when $x = 1$ and $u = \sqrt{3}$ when $x = 3$. Therefore

$$\int_1^3 \frac{1}{2x^{1/2}(1+x)}\,dx = \int_1^3 \frac{\frac{1}{2}x^{-1/2}}{1 + (x^{1/2})^2}\,dx = \int_1^{\sqrt{3}} \frac{1}{1 + u^2}\,du$$

$$= \left[\arctan u\right]_1^{\sqrt{3}} = \left[\arctan\left(\sqrt{x}\right)\right]_1^3 = \frac{\pi}{3} - \frac{\pi}{4} = \frac{\pi}{12} \approx 0.2617993878.$$

6.8.57: Suppose that $u < -1$ and let $x = -u$. Then $x > 0$, so

$$y = \operatorname{arcsec}|u| = \operatorname{arcsec} x,$$

and then the chain rule yields

$$\frac{dy}{du} = \frac{dy}{dx} \cdot \frac{dx}{du} = \frac{1}{|x|\sqrt{x^2-1}} \cdot (-1) = \frac{-1}{x\sqrt{(-x)^2-1}} = \frac{1}{(-x)\sqrt{(-x)^2-1}} = \frac{1}{u\sqrt{u^2-1}}.$$

6.8.59: If $a > 0$ and $u = ax$, then $du = a\,dx$ and $a^2 + u^2 = a^2 + a^2x^2 = a^2(1+x^2)$. So

$$\int \frac{1}{a^2+u^2}\,du = \int \frac{a}{a^2(1+x^2)}\,dx = \frac{1}{a}\arctan x + C = \frac{1}{a}\arctan\left(\frac{u}{a}\right) + C.$$

6.8.61: If $x > 1$, then

$$f'(x) = \frac{1}{x^2\sqrt{1-\dfrac{1}{x^2}}} = \frac{1}{x\sqrt{x^2-\dfrac{x^2}{x^2}}} = \frac{1}{|x|\sqrt{x^2-1}}.$$

If $x < -1$, then

$$f'(x) = \frac{1}{x^2\sqrt{1-\dfrac{1}{x^2}}} = \frac{1}{(-x)^2\sqrt{1-\dfrac{1}{x^2}}} = \frac{1}{(-x)\sqrt{(-x)^2-\dfrac{(-x)^2}{x^2}}} = \frac{1}{|x|\sqrt{x^2-1}}.$$

6.8.63: Let g denote the "alternative secant function," so that $y = g(x)$ if and only if $\sec y = x$ and either $0 \leqq y < \pi/2$ or $\pi \leqq y < 3\pi/2$. We differentiate implicitly the *identity* $\sec y = x$ on both the intervals $0 < y < \pi/2$ and $\pi < y < 3\pi/2$ and find that

$$(\sec y \tan y)\frac{dy}{dx} = 1, \quad \text{so that} \quad g'(x) = \frac{1}{\sec y \tan y} = \pm\frac{1}{x\sqrt{\sec^2 y - 1}} = \pm\frac{1}{x\sqrt{x^2-1}}.$$

Then Fig. 6.8.13 shows that $g'(x) < 0$ if $x < -1$ and that $g'(x) > 0$ if $x > 1$. Therefore the choice of the plus sign in the previous equation is correct in both cases:

$$g'(x) = \frac{1}{x\sqrt{x^2-1}}.$$

6.8.65: See the following figure for the meanings of the variables.

We are required to maximize the angle θ, and from the figure and the data given in the problem we may express θ as a function of the distance x of the billboard from the motorist:

$$\theta = \theta(x) = \arctan\frac{16}{x} - \arctan\frac{4}{x}, \quad 0 < x < +\infty.$$

After simplifications we find that

$$\frac{d\theta}{dx} = \frac{4}{x^2+16} - \frac{16}{x^2+256}.$$

The solution of $\theta'(x) = 0$ certainly maximizes θ because θ is near zero if x is near zero and if x is large positive. We solve $\theta'(x) = 0$:

$$4(x^2+16) = x^2 + 256; \qquad 3x^2 = 192;$$

$$x^2 = 64; \qquad\qquad x = 8.$$

Answer: The billboard should be placed so that it will be 8 meters (horizontal distance) from the eyes of passing motorists. Many alert students have pointed out that such a billboard wouldn't be visible long enough to be effective. This illustrates that once you have used mathematics to solve a problem, you must *interpret* the results!

6.8.67: If $f(x) = (a^2 - x^2)^{1/2}$, then

$$1 + [f'(x)]^2 = 1 + \frac{x^2}{a^2 - x^2} = \frac{a^2}{a^2 - x^2},$$

so the circumference of a circle of radius a is

$$C = 8\int_0^{a/\sqrt{2}} \frac{a}{\sqrt{a^2-x^2}}\,dx = 8\cdot\left[a\arcsin\left(\frac{x}{a}\right)\right]_0^{a/\sqrt{2}} = 8\left(\frac{\pi a}{4} - 0\right) = 2\pi a.$$

The integration was carried out by using the result in Problem 58.

6.8.69: Let A_a denote the area under the graph of $y(x)$ for $0 \leq x \leq a$. Therefore

$$A_a = \int_0^a \frac{1}{1+x^2}\,dx = \left[\arctan x\right]_0^a = \arctan a.$$

Then $\displaystyle\lim_{a\to\infty} A_a = \frac{\pi}{2}$ by Eq. (2) and Fig. 6.8.4.

6.8.71: For $x > 1$: $f(x) = \text{arcsec}\, x + A$:

$$1 = f(2) - \frac{\pi}{3} + 1, \quad \text{so} \quad A = 1 - \frac{\pi}{3}.$$

For $x < -1$: $f(x) = -\text{arcsec}\, x + B$;

$$1 = f(-2) = -\frac{2\pi}{3} + B, \quad \text{so} \quad B = 1 + \frac{2\pi}{3}.$$

311

Therefore $f(x) = \text{arcsec } x + 1 - \dfrac{\pi}{3}$ if $x > 1$, $f(x) = -\text{arcsec } x + 1 + \dfrac{2\pi}{3}$ if $x < -1$. The graph of $y = f(x)$ is next.

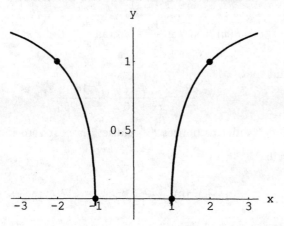

6.8.73: If

$$f(x) = \arctan(x^2 - 1)^{1/2} \quad \text{for} \quad x > 1,$$

then

$$f'(x) = \frac{1}{1 + (x^2 - 1)} \cdot \frac{1}{2}(x^2 - 1)^{-1/2} \cdot 2x = \frac{x}{x^2(x^2 - 1)^{1/2}} = \frac{1}{x\sqrt{x^2 - 1}}.$$

Therefore $\text{arcsec } x = C + \arctan(x^2 - 1)^{1/2}$ for some constant C for all $x > 1$. Now substitute $x = \sqrt{2}$ to show that $C = 0$.

If

$$g(x) = \pi - \arctan(x^2 - 1)^{1/2} \quad \text{for} \quad x < -1,$$

then (using the earlier result)

$$g'(x) = -\frac{1}{x\sqrt{x^2 - 1}}.$$

Therefore

$$\text{arcsec } x = C + \pi - \arctan\left(\sqrt{x^2 - 1}\right)$$

for some constant C for all $x < -1$. Now substitute $x = -\sqrt{2}$ to show that $C = 0$.

6.8.75: The graph of f on $[0, 10]$ (following this solution) indicates a global maximum near $x = 2.7$. We used Newton's method to show that its location is close to $(2.6892200292, 0.9283427321)$.

312

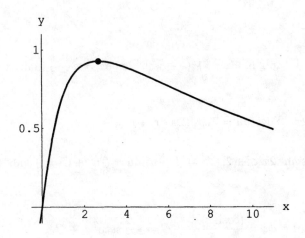

Section 6.9

6.9.1: If $f(x) = \cosh(3x - 2)$, then $f'(x) = 3\sinh(3x - 2)$.

6.9.3: If $f(x) = x^2 \tanh\left(\dfrac{1}{x}\right)$, then

$$f'(x) = 2x \tanh\left(\frac{1}{x}\right) + \left[x^2 \operatorname{sech}^2\left(\frac{1}{x}\right)\right] \cdot \left[-\frac{1}{x^2}\right] = 2x \tanh\left(\frac{1}{x}\right) - \operatorname{sech}^2\left(\frac{1}{x}\right).$$

6.9.5: If $f(x) = \coth^3 4x = (\coth 4x)^3$, then

$$f'(x) = 4 \cdot (3 \coth 4x)^2 \left[-(\operatorname{csch} 4x)^2\right] = -12 \coth^2 4x \ \operatorname{csch}^2 4x.$$

6.9.7: If $f(x) = e^{\operatorname{csch} x}$, then $f'(x) = e^{\operatorname{csch} x} \cdot D_x \operatorname{csch} x = -e^{\operatorname{csch} x} \operatorname{csch} x \coth x$.

6.9.9: If $f(x) = \sin(\sinh x)$, then

$$f'(x) = [\cos(\sinh x)] \cdot D_x(\sinh x) = (\cosh x) \cos(\sinh x).$$

6.9.11: If $f(x) = \sinh x^4 = \sinh(x^4)$, then $f'(x) = 4x^3 \cosh x^4$.

6.9.13: If $f(x) = \dfrac{1}{x + \tanh x}$, then (by the reciprocal rule)

$$f'(x) = -\frac{1 + \operatorname{sech}^2 x}{(x + \tanh x)^2} = \frac{(\tanh^2 x) - 2}{(x + \tanh x)^2}.$$

6.9.15: If necessary, use the substitution $u = x^2$, $du = 2x\,dx$ to show that

$$\int x \sinh x^2 \, dx = \frac{1}{2} \cosh x^2 + C.$$

313

6.9.17: By Eq. (5) we have

$$\int \tanh^2 3x \, dx = \int \left(1 - \operatorname{sech}^2 3x\right) dx = x - \frac{1}{3} \tanh 3x + C.$$

6.9. 19: Let $u = \sinh 2x$. Then $du = 2 \cosh 2x \, dx$, so

$$\int \sinh^2 2x \cosh 2x \, dx = \int \frac{1}{2} u^2 \, du = \frac{1}{6} u^3 + C = \frac{1}{6} \sinh^3 2x + C.$$

6.9.21: $\displaystyle \int (\cosh x)^{-3} \sinh x \, dx = \frac{(\cosh x)^{-2}}{-2} + C = -\frac{1}{2} \operatorname{sech}^2 x + C.$

6.9.23: Let $u = \coth x$. Then $du = - \operatorname{csch}^2 x \, dx$. So

$$\int \coth x \operatorname{csch}^2 x \, dx = -\int u \, du = -\frac{1}{2} u^2 + C = -\frac{1}{2} \coth^2 x + C = -\frac{1}{2} \operatorname{csch}^2 x + C_1.$$

6.9.25: If necessary, let $u = 1 + \cosh x$ or let $u = \cosh x$. But simply "by inspection,"

$$\int \frac{\sinh x}{1 + \cosh x} \, dx = \ln \left(1 + \cosh x\right) + C.$$

6.9.27: One solution:

$$\int \frac{1}{(e^x + e^{-x})^2} \, dx = \frac{1}{4} \int \left(\frac{2}{e^x + e^{-x}}\right)^2 dx = \frac{1}{4} \int \operatorname{sech}^2 x \, dx = \frac{1}{4} \tanh x + C.$$

Another solution: Let $u = e^x$, so that $du = e^x \, dx$; thus $dx = e^{-x} \, du = \dfrac{1}{u} \, du$. Then

$$\int \frac{1}{(e^x + e^{-x})^2} \, dx = \int \frac{u^{-1}}{(u + u^{-1})^2} \, du = \int \frac{u^{-1}}{u^2 + 2 + u^{-2}} \, dx$$

$$= \int \frac{u}{u^4 + 2u^2 + 1} \, dx = \int u(u^2 + 1)^{-2} \, du = -\frac{1}{2} (u^2 + 1)^{-1} + C$$

$$= -\frac{1}{2} \cdot \frac{1}{e^{2x} + 1} + C = -\frac{1}{2} \cdot \frac{e^{-x}}{e^x + e^{-x}} + C = -\frac{1}{4e^x} \cdot \frac{2}{e^x + e^{-x}} + C = -\frac{\operatorname{sech} x}{4e^x} + C.$$

The two answers appear quite different, but they are both correct (although they differ only by the constant $\frac{1}{4}$).

6.9.29: $f'(x) = \dfrac{1}{\sqrt{1 + 4x^2}} \cdot D_x (2x) = \dfrac{2}{\sqrt{1 + 4x^2}}.$

6.9.31: $f'(x) = \dfrac{1}{1 - (\sqrt{x})^2} \cdot D_x \left(x^{1/2}\right) = \dfrac{1}{2(1 - x)\sqrt{x}}.$

314

6.9.33: If $f(x) = \text{sech}^{-1}\left(\dfrac{1}{x}\right)$, then

$$f'(x) = -\frac{1}{\dfrac{1}{x}\sqrt{1 - \dfrac{1}{x^2}}} \cdot D_x\left(\frac{1}{x}\right)$$

$$= -\frac{x}{\sqrt{1 - \dfrac{1}{x^2}}} \cdot \left(-\frac{1}{x^2}\right) = \frac{x}{x^2\sqrt{1 - \dfrac{1}{x^2}}} = \frac{x}{|x|\sqrt{x^2 - 1}}.$$

(You need to write x^2 in the form $|x|^2$ in order to move one copy of $|x|$ underneath the radical, where it becomes x^2. Compare this result with Eq. (29).)

6.9.35: If $f(x) = \left(\sinh^{-1} x\right)^{3/2}$, then

$$f'(x) = \frac{3}{2}\left(\sinh^{-1} x\right)^{1/2} \cdot D_x\left(\sinh^{-1} x\right) = \frac{3\left(\sinh^{-1} x\right)^{1/2}}{2(1 + x^2)^{1/2}}.$$

6.9.37: If $f(x) = \ln\left(\tanh^{-1} x\right)$, then

$$f'(x) = \frac{1}{\tanh^{-1} x} \cdot D_x\left(\tanh^{-1} x\right) = \frac{1}{(1 - x^2)\tanh^{-1} x}.$$

6.9.39: Let $x = 3u$. Then $dx = 3\,du$, so

$$\int \frac{1}{\sqrt{x^2 + 9}}\,dx = \int \frac{3}{\sqrt{9u^2 + 9}}\,du = \int \frac{1}{\sqrt{u^2 + 1}}\,du = \text{arcsinh } u + C = \text{arcsinh } \frac{x}{3} + C.$$

6.9.41: Let $x = 2u$. Then $dx = 2\,du$, so

$$I = \int_{1/2}^{1} \frac{1}{4 - x^2}\,dx = \int_{x=1/2}^{1} \frac{2}{4 - 4u^2}\,du = \frac{1}{2}\int_{x=1/2}^{1} \frac{1}{1 - u^2}\,du$$

$$= \frac{1}{2}\left[\tanh^{-1} u\right]_{x=1/2}^{1} = \frac{1}{2}\left[\tanh^{-1}\left(\frac{x}{2}\right)\right]_{1/2}^{1} = \frac{1}{2}\left[\tanh^{-1}\left(\frac{1}{2}\right) - \tanh^{-1}\left(\frac{1}{4}\right)\right].$$

Now use Eq. (36) to transform the answer into a more familiar form:

$$I = \frac{1}{4}\left(\ln\frac{1 + \frac{1}{2}}{1 - \frac{1}{2}} - \ln\frac{1 + \frac{1}{4}}{1 - \frac{1}{4}}\right) = \frac{1}{4}\left(\ln 3 - \ln\frac{5}{3}\right) = \frac{1}{4}\ln\frac{9}{5} \approx 0.1469466662.$$

6.9.43: Let $x = \dfrac{2}{3}u$. Then $\sqrt{4 - 9x^2} = \sqrt{4 - 4u^2}$ and $dx = \dfrac{2}{3}\,du$. Therefore

$$\int \frac{1}{x\sqrt{4 - 9x^2}}\,dx = \frac{2}{3}\int \frac{1}{\frac{4}{3}u\sqrt{1 - u^2}}\,du$$

$$= \frac{1}{2}\int \frac{1}{u\sqrt{1 - u^2}}\,du = -\text{sech}^{-1}|u| + C = -\frac{1}{2}\text{sech}^{-1}\left|\frac{3}{2}x\right| + C.$$

315

6.9.45: Let $u = e^x$: $du = e^x \, dx$. Hence

$$\int \frac{e^x}{\sqrt{e^{2x}+1}} \, dx = \int \frac{1}{\sqrt{u^2+1}} \, du = \sinh^{-1} u + C = \sinh^{-1}(e^x) + C.$$

6.9.47: Let $u = e^x$: $x = \ln u$ and $ds = \frac{1}{u} \, du$. Thus

$$\int \frac{1}{\sqrt{1-e^{2x}}} \, dx = \int \frac{1}{u\sqrt{1-u^2}} \, du = -\operatorname{sech}^{-1}|u| + C = -\operatorname{sech}^{-1}(e^x) + C.$$

6.9.49: $\sinh x \cosh y + \cosh x \sinh y - \sinh(x+y)$

$$= \frac{e^x - e^{-x}}{2} \cdot \frac{e^y + e^{-y}}{2} + \frac{e^x + e^{-x}}{2} \cdot \frac{e^y - e^{-y}}{2} - \frac{e^{x+y} - e^{-x-y}}{2}$$

$$= \frac{1}{4} \left(e^{x+y} + e^{x-y} - e^{-x+y} - e^{-x-y} + e^{x+y} - e^{x-y} + e^{-x+y} - e^{-x-y} - 2e^{x+y} + 2e^{-x-y} \right) = 0.$$

Therefore $\sinh(x+y) = \sinh x \cosh y + \cosh x \sinh y$.

6.9.51: Substitute $y = x$ in the identity

$$\cosh(x+y) = \cosh x \cosh y + \sinh x \sinh y$$

to prove that $\cosh 2x = \cosh^2 x + \sinh^2 x$. Then, with the aid of Eq. (4), we find that

$$\cosh 2x = 2\cosh^2 x - 1, \qquad \text{so that} \qquad \cosh^2 x = \frac{1}{2}(1 + \cosh 2x).$$

6.9.53: The length is

$$L = \int_0^a \sqrt{1 + \sinh^2 x} \, dx = \int_0^a \cosh x \, dx = \left[\sinh x \right]_0^a = \sinh a.$$

6.9.55: Beginning with the equation

$$A(\theta) = \frac{1}{2} \cosh\theta \, \sinh\theta - \int_1^{\cosh\theta} (x^2 - 1)^{1/2} \, dx,$$

we take the derivative of each side with respect to θ (using the fundamental theorem of calculus on the right-hand side). The result is

$$A'(\theta) = \frac{1}{2} \cosh^2\theta + \frac{1}{2} \sinh^2\theta - \left[(\cosh^2\theta - 1)^{1/2} \sinh\theta \right]$$

$$= \frac{1}{2} \cosh^2\theta + \frac{1}{2} \sinh^2\theta - \sinh^2\theta = \frac{1}{2}(\cosh^2\theta - \sinh^2\theta) = \frac{1}{2}.$$

Therefore $A(\theta) = \frac{1}{2}\theta + C$ for some constant C. Evaluation of both sides of this equation when $\theta = 0$ yields the information that $C = 0$, and therefore

$$A(\theta) = \frac{1}{2}\theta.$$

6.9.57: Let $y = \sinh^{-1} 1$. Then

$$1 = \sinh y = \frac{e^y - e^{-y}}{2}; \qquad e^y - e^{-y} = 2;$$

$$e^{2y} - 2e^y - 1 = 0; \qquad u^2 - 2u - 1 = 0 \quad \text{where} \quad u = e^y;$$

$$u = \frac{2 \pm \sqrt{4 + 4}}{2} = 1 \pm \sqrt{2}.$$

But $u = e^y > 0$, so $u = 1 + \sqrt{2}$. Hence

$$\sinh^{-1} 1 = y = \ln u = \ln\left(1 + \sqrt{2}\right) \approx 0.8813735870.$$

6.9.59: Let $y = \sinh^{-1} x$ and remember that $\cosh y > 0$ for all y. Hence

$$\sinh y = x; \qquad (\cosh y)\,\frac{dy}{dx} = 1;$$

$$\frac{dy}{dx} = \frac{1}{\cosh y} = \frac{1}{\sqrt{\cosh^2 y}} = \frac{1}{\sqrt{1 + \sinh^2 y}} = \frac{1}{\sqrt{1 + x^2}}.$$

6.9.61: Let

$$f(x) = \tanh^{-1} x \qquad \text{and} \qquad g(x) = \frac{1}{2}\ln\left(\frac{1 + x}{1 - x}\right)$$

for $-1 < x < 1$. Equation (36) states that $f(x) = g(x)$. To prove this, note that

$$f'(x) = \frac{1}{1 - x^2}$$

and $g(x) = \frac{1}{2}\ln(1 + x) - \frac{1}{2}\ln(1 - x)$, so that

$$g'(x) = \frac{1}{2}\left(\frac{1}{1 + x} + \frac{1}{1 - x}\right) = \frac{1 - x + 1 + x}{2(1 - x^2)} = \frac{1}{1 - x^2},$$

and therefore $f'(x) = g'(x)$. Hence $f(x) = g(x) + C$ for some constant C and for all x in $(-1, 1)$. To evaluate C, note that

$$0 = f(x) = \tanh^{-1} 0 = g(0) + C = \frac{1}{2}\ln 1 + C = 0 + C = C.$$

Therefore $f(x) = g(x)$ if $-1 < x < 1$.

6.9.63: Let $u = e^y$. Then

$$x = \coth y = \frac{e^y + e^{-y}}{e^y - e^{-y}} = \frac{e^{2y} + 1}{e^{2y} - 1} = \frac{u^2 + 1}{u^2 - 1}.$$

Therefore

$$u^2 + 1 = xu^2 - x; \qquad (x - 1)u^2 = x + 1; \qquad u^2 = \frac{x + 1}{x - 1}.$$

317

Therefore $y = \ln u = \frac{1}{2}\ln u^2 = \frac{1}{2}\ln\dfrac{x+1}{x-1}$ for all x such that $|x| > 1$.

6.9.65: Let

$$f(x) = \operatorname{csch}^{-1} x \quad \text{and} \quad g(x) = \ln\left(\frac{1}{x} + \frac{\sqrt{1+x^2}}{|x|}\right)$$

for $x \neq 0$. Then

$$f'(x) = -\frac{1}{|x|\sqrt{1+x^2}}.$$

If $x > 0$, then $g(x) = \ln\left(\dfrac{1 + \sqrt{1+x^2}}{x}\right)$. Thus

$$g'(x) = \frac{x}{1 + (1+x^2)^{1/2}} \cdot \frac{x \cdot \frac{1}{2}(1+x^2)^{-1/2} \cdot 2x - 1 - (1+x^2)^{1/2}}{x^2}$$

$$= \frac{1}{1 + (1+x^2)^{1/2}} \cdot \frac{x^2(1+x^2)^{-1/2} - 1 - (1+x^2)^{1/2}}{x}$$

$$= \frac{1}{1 + (1+x^2)^{1/2}} \cdot \frac{x^2 - (1+x^2)^{1/2} - 1 - x^2}{x(1+x^2)^{1/2}}$$

$$= -\frac{1}{x(1+x^2)^{1/2}} = -\frac{1}{|x|\sqrt{1+x^2}}.$$

But if $x < 0$, then $g(x) = \ln\left(\dfrac{1 - \sqrt{1+x^2}}{x}\right)$, and hence

$$g'(x) = \frac{x}{1 - (1+x^2)^{1/2}} \cdot \frac{-x \cdot \frac{1}{2}(1+x^2)^{-1/2} \cdot 2x - 1 + (1+x^2)^{1/2}}{x^2}$$

$$= \frac{1}{1 - (1+x^2)^{1/2}} \cdot \frac{-x^2(1+x^2)^{-1/2} - 1 + (1+x^2)^{1/2}}{x}$$

$$= \frac{1}{1 - (1+x^2)^{1/2}} \cdot \frac{-x^2 - (1+x^2)^{1/2} + 1 + x^2}{x(1+x^2)^{1/2}}$$

$$= \frac{1}{x(1+x^2)^{1/2}} = -\frac{1}{|x|\sqrt{1+x^2}}.$$

Therefore there exist constants C_1 and C_2 such that

$$f(x) = g(x) + C_1 \quad \text{if} \quad x > 0 \quad \text{and} \quad f(x) = g(x) + C_2 \quad \text{if} \quad x < 0.$$

Let us now express $u = \operatorname{csch}^{-1} 1$ in a more useful way.

$$\operatorname{csch} u = 1; \qquad \frac{2}{e^u - e^{-u}} = 1;$$

$$e^u - e^{-u} = 2; \qquad e^{2u} - 1 = 2e^u;$$

$$e^{2u} - 2e^u - 1 = 0; \qquad e^u = \frac{2 \pm \sqrt{4+4}}{2} = 1 \pm \sqrt{2}.$$

318

Therefore $e^u = 1 + \sqrt{2}$, and so $f(1) = \operatorname{csch}^{-1} 1 = u = \ln\left(1 + \sqrt{2}\right)$. But

$$g(1) = \ln\left(\frac{1}{1} + \frac{\sqrt{2}}{1}\right) = \ln\left(1 + \sqrt{2}\right).$$

Therefore $C_1 = 0$, and so

$$\operatorname{csch}^{-1} x = \ln\left(\frac{1}{x} + \frac{\sqrt{1 + x^2}}{|x|}\right)$$

if $x > 0$. This argument may be repeated for $u = \operatorname{csch}^{-1}(-1)$ with few changes other than minus signs here and there, and it follows that $C_2 = 0$ as well. This establishes Eq. (39).

6.9.67: Given: $f(x) = e^{-2x} \tanh x$. First,

$$\lim_{x \to \infty} f(x) = \lim_{x \to \infty} \frac{e^x - e^{-x}}{e^{2x}(e^x + e^{-x})} = \lim_{x \to \infty} \frac{1 - e^{-2x}}{e^{2x} + 1} = 0.$$

Next, a plot of $y = f(x)$ for $0 \leq x \leq 1$ reveals a local maximum near where $x = 0.45$. We applied Newton's method to solve $f'(x) = 0$ numerically and after seven iterations found that the maximum is close to the point $(0.4406867935, 0.1715728753)$. Indeed,

$$f'(x) = -\frac{2(e^{4x} - 2e^{2x} - 1)}{e^{2x}(e^{2x} + 1)^2},$$

so $f'(x) = 0$ when $e^{4x} - 2e^{2x} - 1 = 0$; that is, when

$$e^{2x} = \frac{2 \pm \sqrt{4 + 4}}{2} = 1 \pm \sqrt{2}, \qquad \text{so that} \qquad x = \frac{1}{2}\ln\left(1 + \sqrt{2}\right) \approx 0.4406867935.$$

6.9.69: Given: $y(x) = y_0 + \dfrac{1}{k}(-1 + \cosh kx)$. Then

$$\frac{dy}{dx} = \sinh kx \qquad \text{and} \qquad \frac{d^2 y}{dx^2} = k \cosh kx.$$

So

$$k\sqrt{1 + [y'(x)]^2} = k\sqrt{1 + \sinh^2 kx} = k\sqrt{\cosh^2 kx} = k\cosh kx.$$

Therefore $\dfrac{d^2 y}{dx^2} = k\sqrt{1 + \left(\dfrac{dy}{dx}\right)^2}$. Moreover,

$$y(0) = y_0 + \frac{1}{k}(-1 + \cosh 0) = y_0 + \frac{0}{k} = y_0$$

and

$$y'(0) = \sinh(k \cdot 0) = \sinh(0) = 0.$$

Chapter 6 Miscellaneous Problems

6.M.1: The net distance is

$$\int_0^3 v(t)\,dt = \left[\frac{1}{3}t^3 - \frac{1}{2}t^2 - 2t\right]_0^3 = -\frac{3}{2} - 0 = -\frac{3}{2}.$$

Because $v(t) < 0$ for $0 < t < 2$, the total distance is

$$-\int_0^2 v(t)\,dt + \int_2^3 v(t)\,dt = \left(\frac{10}{3} - 0\right) + \left(-\frac{3}{2} + \frac{10}{3}\right) = \frac{31}{6} \approx 5.166667.$$

6.M.3: Because $v(t) < 0$ for $0 < t < \frac{1}{2}$ but $v(t) > 0$ for $\frac{1}{2} < t < \frac{3}{2}$, the net distance is

$$\int_0^{3/2} v(t)\,dt = \left[-\cos\left(\frac{1}{2}\pi(2t-1)\right)\right]_0^{3/2} = 1 - 0 = 1$$

and the total distance is

$$-\int_0^{1/2} v(t)\,dt + \int_{1/2}^{3/2} v(t)\,dt = (1-0) + (1-(-1)) = 3.$$

6.M.5: The volume is $\displaystyle\int_1^4 x^{1/2}\,dx = \left[\frac{2}{3}x^{3/2}\right]_1^4 = \frac{16}{3} - \frac{2}{3} = \frac{14}{3} \approx 4.666667.$

6.M.7: The volume is $\displaystyle\int_0^1 \pi(x^2 - x^4)\,dx = \pi\left[\frac{1}{3}x^3 - \frac{1}{5}x^5\right]_0^1 = \frac{2\pi}{15} \approx 0.4188790205.$

6.M.9: Between time $t = 0$ and time $t = 12$, the rainfall in inches is

$$\int_0^{12} \frac{1}{12}(t+6)\,dt = \left[\frac{1}{2}t + \frac{1}{24}t^2\right]_0^{12} = 12 - 0 = 12.$$

6.M.11: The region R of Problem 10 is bounded above by the graph of $f(x) = 2x - x^2$ and below by the graph of $g(x) = x^3$ for $0 \le x \le 1$. A cross section of the solid S of Problem 11 perpendicular to the x-axis at x is an annular region with outer radius $f(x)$ and inner radius $g(x)$, thus of cross-sectional area $A(x) = \pi[f(x)]^2 - \pi[g(x)]^2$. Therefore the volume of S is

$$V = \int_0^1 A(x)\,dx = \pi\left[\frac{4}{3}x^3 - x^4 + \frac{1}{5}x^5 - \frac{1}{7}x^7\right]_0^1 = \frac{41}{105}\pi - 0 \approx 1.2267171314.$$

6.M.13: Each cross section perpendicular to the x-axis has area $A(x) = \frac{1}{16}\pi$, so the total mass of the helix is

$$m = \int_0^{20} (8.5)\cdot A(x)\,dx = \left[\frac{17\pi}{32}x\right]_0^{20} = \frac{85\pi}{8} \approx 33.3794219444 \quad \text{(grams)}.$$

6.M.15: Let z denote the distance from P to the origin. A horizontal cross-section of the elliptical cone "at" z (thus at distance z from P) is an ellipse with major axis and minor axis each proportional to z. So the area $A(z)$ of this cross section is proportional to z^2: $A(z) = kz^2$ where k is a positive constant. But

$$A(h) = kh^2 = \pi ab$$

by the result of Problem 47 of Section 5.8, and hence $k = \pi ab/h^2$. Therefore $A(z) = \pi ab z^2/h^2$. So the volume of the elliptical cone is

$$V = \int_0^h \frac{\pi ab}{h^2} z^2\, dz = \left[\frac{\pi ab}{3h^2} z^3\right]_0^h = \frac{1}{3}\pi abh,$$

one-third the product of the area of the base and the height of the elliptical cone.

6.M.17: Because $(a + h, r)$ lies on the hyperbola,

$$\frac{(a + h)^2}{a^2} - \frac{r^2}{b^2} = 1.$$

It follows that

$$r^2 = \frac{b^2(2ah + h^2)}{a^2}. \tag{1}$$

Moreover, the equation of the hyperbola may be written in the form

$$y^2 = \frac{b^2}{a^2}(x^2 - a^2).$$

Therefore the "segment of the hyperboloid" has volume

$$V = \int_a^{a+h} \pi y^2\, dx = \frac{\pi b^2}{a^2} \int_a^{a+h} (x^2 - a^2)\, dx = \frac{\pi b^2}{a^2}\left[\frac{1}{3}x^3 - a^2 x\right]_a^{a+h}$$

$$= \frac{\pi b^2}{3a^2}\left[x^3 - 3a^2 x\right]_a^{a+h} = \frac{\pi b^2}{3a^2}\left(a^3 + 3a^2 h + 3ah^2 + h^3 - 3a^3 - 3a^2 h - a^3 + 3a^3\right)$$

$$= \frac{1}{3}\pi \frac{b^2}{a^2}h^2(3a + h).$$

But by Eq. (1), $b^2 = \dfrac{a^2 r^2}{2ah + h^2}$. So $V = \dfrac{1}{3}\pi \dfrac{h^2}{a^2}(3a + h)\dfrac{a^2 r^2}{h(2a + h)} = \dfrac{1}{3}\pi r^2 h\, \dfrac{3a + h}{2a + h}$.

6.M.19: $V = \displaystyle\int_1^t \pi\left(f(x)\right)^2\, dx = \dfrac{\pi}{6}\left[(1 + 3t)^2 - 16\right]$. Thus

$$\pi\left(f(x)\right)^2 = \frac{\pi}{6}\left[(2)(1 + 3x)(3)\right] = \pi(1 + 3x).$$

Therefore $f(x) = \sqrt{1 + 3x}$.

6.M.21: The graphs of $f(x) = \sin\left(\frac{1}{2}\pi x\right)$ and $g(x) = x$ cross at $(0,0)$ and $(1,1)$, and $g(x) < f(x)$ if $0 < x < 1$. When the region they bound is rotated around the y-axis, the method of cylindrical shells yields the volume of the solid thus generated to be

$$V = \int_0^1 2\pi x \left[f(x) - g(x) \right] dx = 2\pi \int_0^1 \left[\left(x \sin \frac{\pi x}{2} \right) - x^2 \right] dx.$$

Now let $u = \dfrac{\pi x}{2}$, so that $x = \dfrac{2u}{\pi}$. This substitution yields

$$V = 2\pi \int_0^{\pi/2} \left(\frac{2}{\pi} u \sin u - \frac{4}{\pi^2} u^2 \right) \cdot \frac{2}{\pi} \, du = \int_0^{\pi/2} \left(\frac{8}{\pi} u \sin u - \frac{16}{\pi^2} u^2 \right) du$$

$$= \left[\frac{8}{\pi} (\sin u - u \cos u) - \frac{16}{3\pi^2} u^3 \right]_0^{\pi/2} = \frac{8}{\pi} - \frac{16}{3\pi^2} \cdot \frac{\pi^3}{8} = \frac{8}{\pi} - \frac{2\pi}{3} \approx 0.4520839871.$$

6.M.23: $\dfrac{dy}{dx} = \frac{1}{2} x^{1/2} - \frac{1}{2} x^{-1/2}$, so

$$1 + \left(\frac{dy}{dx} \right)^2 = \left(\frac{1}{2} x^{1/2} + \frac{1}{2} x^{-1/2} \right)^2. \tag{1}$$

So the length of the curve is

$$L = \int_1^4 \left(\frac{1}{2} x^{1/2} + \frac{1}{2} x^{-1/2} \right) dx = \left[x^{1/2} + \frac{1}{3} x^{3/2} \right]_1^4 = \frac{14}{3} - \frac{4}{3} = \frac{10}{3}.$$

6.M.25: Let $x = f(y) = \frac{3}{8}(y^{4/3} - 2y^{2/3})$. Then

$$1 + [f'(y)]^2 = 1 + \frac{9}{64} \left(\frac{4}{3} y^{1/3} - \frac{4}{3} y^{-1/3} \right)^2 = \frac{1}{4} y^{2/3} + \frac{1}{2} + \frac{1}{4} y^{-2/3} = \frac{(1 + y^{2/3})^2}{4y^{2/3}},$$

and therefore $ds = \frac{1}{2}(y^{1/3} + y^{-1/3}) \, dy$. Hence the length of the graph of g from $y = 1$ to $y = 8$ is

$$L = \int_1^8 1 \, ds = \int_1^8 \frac{1}{2}(y^{1/3} + y^{-1/3}) \, dy = \left[\frac{3}{8} y^{4/3} + \frac{3}{4} y^{2/3} \right]_1^8 = 9 - \frac{9}{8} = \frac{63}{8} = 7.875.$$

6.M.27: Let $f(x) = \frac{1}{3} x^{3/2} - x^{1/2}$, $1 \leqq x \leqq 4$. Then

$$1 + [f'(x)]^2 = 1 + \left(\frac{1}{2} x^{1/2} - \frac{1}{2} x^{-1/2} \right)^2 = \frac{1}{4} x + \frac{1}{2} + \frac{1}{4} x^{-1} = \left(\frac{1}{2} x^{1/2} + \frac{1}{2} x^{-1/2} \right)^2.$$

Therefore $ds = \frac{1}{2} \left(x^{1/2} + x^{-1/2} \right) dx$. Therefore the area of the surface generated when the graph of f is rotated around the vertical line $x = 1$ is

$$A = \int_1^4 2\pi(x - 1) \, ds = \pi \int_1^4 (x^{3/2} - x^{-1/2}) \, dx = \pi \left[\frac{2}{5} x^{5/2} - 2x^{1/2} \right]_1^4$$

$$= \frac{44\pi}{5} - \left(-\frac{8\pi}{5} \right) = \frac{52\pi}{5} \approx 32.6725635973.$$

6.M.29: This is merely a matter of substituting $2r$ for h in the area formula $A = 2\pi rh$ derived in Problem 28. Thus the area of a sphere of radius r is $A = 2\pi r \cdot 2r = 4\pi r^2$.

6.M.31: Denote the spring constant by K. The information given in the problem yields

$$\int_2^5 K(x - L)\,dx = 5\int_2^3 K(x - L)\,dx;$$

$$\int_2^5 (x - L)\,dx = 5\int_2^3 (x - L)\,dx;$$

$$\left[\frac{1}{2}(x-L)^2\right]_2^5 = 5\left[\frac{1}{2}(x-L)^2\right]_2^3;$$

$$(5-L)^2 - (2-L)^2 = 5(3-L)^2 - 5(2-L)^2;$$

$$25 - 10L + L^2 - 4 + 4L - L^2 = 45 - 30L + 5L^2 - 20 + 20L - 5L^2;$$

$$4L = 4.$$

Therefore the natural length of the spring is $L = 1$ (ft).

6.M.33: Set up a coordinate system in which the center of the tank is at the origin and the y-axis is vertical. A horizontal cross section of the oil at positive y ($-R \leq y \leq R$) is circular with radius $x = \sqrt{R^2 - y^2}$, so its area is $\pi(R^2 - y^2)$. Hence the work to pump the oil to its final position $y = 3R$ is

$$W = \int_{-R}^R (3R - y)\pi\rho(R^2 - y^2)\,dy = \pi\rho\int_{-R}^R (y^3 - 3Ry^2 - R^2y + 3R^3)\,dy$$

$$= \pi\rho\left[\frac{1}{4}y^4 - Ry^3 - \frac{1}{2}R^2y^2 + 3R^3y\right]_{-R}^R = \pi\rho\left(\frac{7}{4}R^4 + \frac{9}{4}R^4\right) = 4\pi\rho R^4.$$

6.M.35: Set up a coordinate system in which the center of the earth is at the origin and the hole extends upward along the vertical y-axis, with its top where $y = R$, the radius of the earth in feet. A 1-pound weight at position y ($0 \leq y \leq R$) weighs y/R pounds, so the total work to lift the weight from $y = 0$ to $y = R$ is

$$W = \int_0^R \frac{y}{R}\,dy = \left[\frac{y^2}{2R}\right]_0^R = \frac{R}{2} = \frac{3960 \cdot 5280}{2} = 10454400 \quad \text{(ft·lb)}.$$

The assumption of constant density of the earth is required to draw the conclusion that the gravitational force is proportional to the distance from the center of the earth.

6.M.37: If the coordinate system is chosen with the origin at the midpoint of the bottom of the dam and with the x-axis horizontal, then the equation of the slanted edge of the dam is $y = 2x - 200$ (with units in feet). Therefore the width of the dam at level y is $2x = y + 200$. Let $\rho = 62.4$ be the density of water in pounds per cubic foot. Then the total force on the dam is

$$F = \int_0^{100} \rho(100 - y)(y + 200)\, dy = \rho\left[20000y - 50y^2 - \frac{1}{3}y^3\right]_0^{100} = \frac{3500000\rho}{3} = 72800000 \quad \text{(lb)}.$$

6.M.39: The volume of the solid is

$$V = \int_0^c 2\pi\left(y + \frac{1}{c}\right)\frac{2}{c}\sqrt{y}\, dy = \frac{4\pi}{c}\cdot\left[\frac{2}{5}y^{5/2} + \frac{2}{3c}y^{3/2}\right]_0^c = 8\pi\left(\frac{1}{5}c^{3/2} + \frac{1}{3}c^{-1/2}\right).$$

It is clear that there is no maximum volume, because $V \to +\infty$ as $c \to 0^+$. But $V \to +\infty$ as $c \to +\infty$ as well, so there is a minimum volume; $V'(c) = 0$ when $c = \frac{1}{3}\sqrt{5}$, so this value of c minimizes V.

6.M.41: Here,

$$L = \int_1^2 \frac{1}{2}(y^3 + y^{-3})\, dy = \frac{33}{16},$$

$$M_y = \int_1^2 \frac{1}{2}\cdot\left(\frac{1}{8}y^4 + \frac{1}{4}y^{-2}\right)\cdot(y^3 + y^{-3})\, dy = \frac{1179}{512}, \quad \text{and}$$

$$M_x = \int_1^2 \frac{1}{2}(y^4 + y^{-2})\, dy = \frac{67}{20}.$$

Therefore

$$\overline{x} = \frac{393}{352} \approx 1.116477 \quad \text{and} \quad \overline{y} = \frac{268}{165} \approx 1.624242.$$

6.M.43: To begin with,

$$1 + \left(\frac{dx}{dy}\right)^2 = \left(\frac{1}{2}y^{1/3} + \frac{1}{2}y^{-1/3}\right)^2;$$

it follows that the length of the curve is $L = \frac{63}{8}$. Next,

$$M_y = \int_1^8 \frac{3}{8}\cdot\frac{1}{2}\cdot(y^{4/3} - 2y^{2/3})\cdot(y^{1/3} + y^{-1/3})\, dy = \frac{999}{128} = 7.8046875 \quad \text{and}$$

$$M_x = \int_1^8 \frac{1}{2}(y^{4/3} + y^{2/3})\, dy = \frac{1278}{35} \approx 36.51428571.$$

Therefore

$$\overline{x} = \frac{111}{112} \approx 0.991071427 \quad \text{and} \quad \overline{y} = \frac{1136}{245} \approx 4.636734694.$$

6.M.45: The curves meet at $(2, 1)$, at $(0, 0)$, and at $(2, -1)$. It follows that $\overline{y} = 0$ by symmetry and that we may compute $\overline{x}$ by using only the upper half of the figure. In that case we have

$$A = \int_0^1 (y^2 + 1 - 2y^4)\, dy = \frac{14}{15} \quad \text{and}$$

$$M_y = \int_0^1 \frac{1}{2}\left[(y^2 + 1)^2 - (2y^4)^2\right]\, dy = \frac{32}{45}.$$

Therefore $\bar{x} = \dfrac{16}{21} \approx 0.7619047619$.

6.M.47: $2\pi\bar{y} \cdot \dfrac{\pi ab}{2} = \dfrac{4}{3}\pi ab^2$, and it follows that $\bar{y} = \dfrac{4b}{3\pi}$.

6.M.49: (a) The area A of the triangle T can be computed in several ways; we chose the most direct which, elementary, is easy to do by hand. Let O denote the vertex of the triangle at $(0, 0)$, $C = C(c, 0)$, $A = A(a, 0)$, $B = B(a, b)$, and $D = D(c, d)$. Then A is the area of triangle OCD *plus* the area of trapezoid $CABD$ *minus* the area of triangle OAB:

$$A = \frac{cd}{2} + \frac{(a - c)(b + d)}{2} - \frac{ab}{2}$$

$$= \frac{cd}{2} + \frac{ab}{2} + \frac{ad}{2} - \frac{bc}{2} - \frac{cd}{2} - \frac{ab}{2} = \frac{ad - bc}{2}.$$

(b) In Problem 46 we saw that the centroid of a triangle lies on the intersection of its medians. From plane geometry we also know that the point of intersection is two-thirds of the way from any vertex to the midpoint of the opposite side. The midpoint of L has y-coordinate $(b + d)/2$, and hence

$$\bar{y} = \frac{2}{3} \cdot \frac{b + d}{2} = \frac{b + d}{3}.$$

(c) $V = 2\pi\bar{y}A = 2\pi \cdot \dfrac{b + d}{3} \cdot \dfrac{ad - bc}{2} = \dfrac{1}{3}\pi(b + d)(ad - bc).$

(d) $\dfrac{1}{2}pw = A = \dfrac{ad - bc}{2}$, so $p = \dfrac{ad - bc}{w}$.

(e) $S = 2\pi \cdot \dfrac{b + d}{2} \cdot w = \pi w(b + d).$

(f) $V = 2\pi\bar{y}A = 2\pi \cdot \dfrac{b + d}{3} \cdot \dfrac{1}{2}pw = \pi pw \cdot \dfrac{b + d}{3} = \dfrac{1}{3}pS.$

6.M.51: A *Mathematica* solution: First let $f(x) = x^m$ and $g(x) = x^n$ where m and n are positive integers and $n > m$.

```
a = Integrate[ f[x] - g[x], { x, 0, 1 } ]
```

$$\frac{1}{m + 1} - \frac{1}{n + 1}$$

```
area = a /.  m → 1
```

$$\frac{1}{2} - \frac{1}{n + 1}$$

Formulas (11) and (12) in the text give the moments

```
my = Integrate[ x*(f[x] - g[x]), { x, 0, 1 } ];
```

```
mx = Integrate[ (1/2)*((f[x])^2 - (g[x])^2), { x, 0, 1 } ];
```

```
My = my /.  m → 1
```
$$\frac{n-1}{3(n+2)}$$

```
Mx = mx /.  m → 1
```
$$\frac{\frac{1}{3} - \frac{1}{2n+1}}{2}$$

Hence the centroid has coordinates

```
{ xc, yc } = { My/area, Mx/area } // Simplify
```
$$\{ \frac{(n+1)}{3(n+2)}, \frac{2(n+1)}{3(2n+1)} \}$$

```
Limit[ { xc, yc }, n → Infinity ]
```
$$\{ \frac{2}{3}, \frac{1}{3} \}$$

Obviously this is the centroid of the triangle with vertices $(0, 0)$, $(1, 0)$, and $(1, 1)$—which the area of the region bounded by the graphs of f and g "exhausts" as $n \to +\infty$.

6.M.53: Let $u = 1 - 2x$. Then $dx = -\frac{1}{2}\, du$, and so

$$\int \frac{dx}{1-2x} = -\frac{1}{2}\int \frac{1}{u}\, du = -\frac{1}{2}\ln|u| + C = -\frac{1}{2}\ln|1-2x| + C.$$

6.M.55: Let $u = 1 + 6x - x^2$. Then $du = (6-2x)\, dx$, so that $(3-x)\, dx = \frac{1}{2}\, du$. Thus

$$\int \frac{3-x}{1+6x-x^2}\, dx = \frac{1}{2}\int \frac{1}{u}\, du = \frac{1}{2}\ln|u| + C = \frac{1}{2}\ln|1+6x-x^2| + C.$$

6.M.57: Let $u = 2 + \cos x$. Then $du = -\sin x\, dx$, and thus

$$\int \frac{\sin x}{2+\cos x}\, dx = -\int \frac{1}{u}\, du = -(\ln u) + C = -\ln(2+\cos x) + C.$$

6.M.59: Let $u = 10^{\sqrt{x}}$. Then

$$du = (10^{\sqrt{x}}\ln 10)\cdot \frac{1}{2}x^{-1/2}\, dx = \frac{10^{\sqrt{x}}}{\sqrt{x}}\cdot \frac{\ln 10}{2}\, dx.$$

Therefore

$$\int \frac{10^{\sqrt{x}}}{\sqrt{x}}\, dx = \int \frac{2}{\ln 10}\, du = \frac{2u}{\ln 10} + C = \frac{2\cdot 10^{\sqrt{x}}}{\ln 10} + C.$$

6.M.61: Let $u = 1 + e^x$. Then $du = e^x \, dx$, and thus

$$\int e^x (1 + e^x)^{1/2} \, dx = \int u^{1/2} \, du = \frac{2}{3} u^{3/2} + C = \frac{2}{3}(1 + e^x)^{3/2} + C.$$

6.M.63: $\displaystyle \int 2^x \cdot 3^x \, dx = \int 6^x \, dx = \frac{6^x}{\ln 6} + C.$

6.M.65: The revenue realized upon selling after t months will be

$$f(t) = B \cdot \left(2 + \frac{t}{12}\right) \cdot 2^{-t/12}, \quad \text{for which} \quad f'(t) = B \cdot \frac{12 - 24 \ln 2 - t \ln 2}{144 \cdot 2^{t/12}}.$$

Thus $f'(t) = 0$ when

$$t = \frac{12 - 24 \ln 2}{\ln 2} \approx -6.68765951.$$

But this value of t is negative, and in addition $f'(t) < 0$ for all larger values of t. Thus the revenue is a decreasing function of t for all $t \geq 0$. Therefore the grain should be sold immediately.

6.M.67: If lots composed of x pooled samples are tested, there will be $1000/x$ lots, so there will be $1000/x$ tests. In addition, if a lot tests positive, there will be x additional tests. The probability of a lot testing positive is $1 - (0.99)^x$, so the expected number of lots that require additional tests will be the product of the number of lots and the probability $1 - (0.99)^x$ that a lot tests positive. Hence the total number of tests to be expected will be

$$f(x) = \frac{1000}{x} + \frac{1000}{x}[1 - (0.99)^x] \cdot x = \frac{1000}{x} + 1000 - 1000(0.99)^x$$

if $x \geq 2$. Next,

$$f'(x) = -\frac{1000}{x^2} - 1000(0.99)^x \ln(0.99);$$

$$f'(x) = 0 \quad \text{when} \quad \frac{1}{x^2} = (0.99)^x \ln(100/99);$$

$$x^2 = \frac{(0.99)^{-x}}{\ln(100/99)};$$

$$x = \frac{(0.99)^{-x/2}}{\sqrt{\ln(100/99)}}.$$

The form of the last equation is exactly what we need to implement the method of repeated substitution (see Problems 23 through 25 of Section 3.10). We substitute our first "guess" $x_0 = 10$ into the right-hand side of the last equation, thus obtaining a "better" (we hope) value x_1 and continue this process until the digits in these successive approximations stabilize. Results: $x_1 = 10.488992$, $x_2 = 10.514798$, $x_3 = 10.516161$, $x_4 = 10.516233$, and $x_5 = 10.516237 = x_6$. The method is not as fast as Newton's method but the formula is simpler. (The graph of $y = f(x)$ follows this solution to convince you that we have actually found the minimum value of f.) We must use an integral number of samples, so we find that $f(10) \approx 195.618$ and $f(11) = 195.571$, so there should be 90 lots of 11 samples each and one lot of 10 for the most economical results. Alternatively, it might be simpler to use 10 samples in every lot; the extra cost would be only about 24 cents. The total cost of the batch method will be about \$978, significantly less than the \$5000 cost of testing each sample individually.

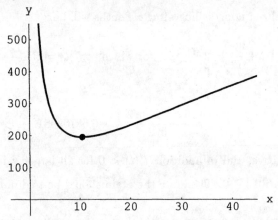

6.M.69: If $f(x) = \sin^{-1} 3x$, then $f'(x) = \dfrac{1}{\sqrt{1-(3x)^2}} \cdot D_x(3x) = \dfrac{3}{\sqrt{1-9x^2}}$.

6.M.71: If $g(t) = \sec^{-1} t^2$, then

$$g'(t) = \frac{1}{|t^2|\sqrt{(t^2)^2 - 1}} \cdot D_t\left(t^2\right) = \frac{2t}{t^2\sqrt{t^4-1}} = \frac{2}{t\sqrt{t^4-1}}.$$

6.M.73: If $f(x) = \sin^{-1}(\cos x)$, then

$$f'(x) = \frac{1}{\sqrt{1-(\cos x)^2}} \cdot D_x(\cos x) = -\frac{\sin x}{\sqrt{1-\cos^2 x}} = -\frac{\sin x}{|\sin x|}.$$

6.M.75: If $g(t) = \cosh^{-1} 10t$, then $g'(t) = \dfrac{1}{\sqrt{(10t)^2-1}} \cdot D_t(10t) = \dfrac{10}{\sqrt{100t^2-1}}$, $t > \frac{1}{10}$.

6.M.77: If $f(x) = \sin^{-1}\left(\dfrac{1}{x^2}\right)$, then

$$f'(x) = \frac{1}{\sqrt{1-\left(\dfrac{1}{x^2}\right)^2}} \cdot D_x\left(\frac{1}{x^2}\right) = \frac{1}{\sqrt{1-\dfrac{1}{x^4}}} \cdot \frac{-2}{x^3} = -\frac{2}{x\sqrt{x^4-1}}.$$

6.M.79: If $f(x) = \arcsin\sqrt{x}$, then

$$f'(x) = \frac{1}{\sqrt{1-(\sqrt{x})^2}} \cdot D_x\left(x^{1/2}\right) = \frac{1}{\sqrt{1-x}} \cdot \frac{1}{2}x^{-1/2} = \frac{1}{2\sqrt{x-x^2}}.$$

328

6.M.81: If $f(x) = \tan^{-1}(x^2 + 1)$, then

$$f'(x) = \frac{1}{1 + (x^2 + 1)^2} \cdot D_x(x^2 + 1) = \frac{2x}{x^4 + 2x^2 + 2}.$$

6.M.83: If $f(x) = e^x \sinh e^x$, then $f'(x) = e^{2x} \cosh e^x + e^x \sinh e^x$.

6.M.85: If $f(x) = \tanh^2 3x + \operatorname{sech}^2 3x$, then $f(x) \equiv 1$ by Eq. (5) in Section 6.9. Therefore $f'(x) \equiv 0$. Alternatively, $f'(x) = 6 \tanh 3x \operatorname{sech}^2 3x - 6 \operatorname{sech} 3x \operatorname{sech} 3x \tanh 3x \equiv 0$.

6.M.87: If $f(x) = \cosh^{-1} \sqrt{x^2 + 1}$, then

$$f'(x) = \frac{1}{\sqrt{\left(\sqrt{x^2+1}\right)^2 - 1}} \cdot D_x(x^2 + 1)^{1/2} = \frac{1}{\sqrt{x^2}} \cdot \frac{1}{2}(x^2 + 1)^{-1/2} \cdot 2x = \frac{x}{|x|\sqrt{x^2+1}}.$$

6.M.89: Let $u = 2x$. Then $du = 2\,dx$, so

$$\int \frac{1}{\sqrt{1 - 4x^2}}\,dx = \frac{1}{2}\int \frac{1}{\sqrt{1 - u^2}}\,du = \frac{1}{2}\sin^{-1}u + C = \frac{1}{2}\sin^{-1}2x + C.$$

6.M.91: Let $x = 2u$. Then $dx = 2\,du$ and $\sqrt{4 - x^2} = \sqrt{4 - 4u^2} = 2\sqrt{1 - u^2}$. Therefore

$$\int \frac{1}{\sqrt{4 - x^2}}\,dx = \int \frac{2}{2\sqrt{1 - u^2}}\,du = \arcsin u + C = \arcsin\left(\frac{x}{2}\right) + C.$$

6.M.93: Let $u = e^x$. Then $du = e^x\,dx$ and $\sqrt{1 - e^{2x}} = \sqrt{1 - u^2}$. Hence

$$\int \frac{e^x}{\sqrt{1 - e^{2x}}}\,dx = \int \frac{1}{\sqrt{1 - u^2}}\,du = \arcsin u + C = \arcsin e^x + C.$$

6.M.95: Let $u = \frac{2}{3}x$, so that $x = \frac{3}{2}u$. Then $dx = \frac{3}{2}\,du$ and $\sqrt{9 - 4x^2} = \sqrt{9 - 9u^2} = 3\sqrt{1 - u^2}$. So

$$\int \frac{1}{\sqrt{9 - 4x^2}}\,dx = \frac{3}{2} \cdot \frac{1}{3}\int \frac{1}{\sqrt{1 - u^2}}\,du = \frac{1}{2}\arcsin u + C = \frac{1}{2}\arcsin\left(\frac{2x}{3}\right) + C.$$

6.M.97: The integrand resembles the derivative of the inverse tangent of *something*, so we let $u = x^3$. Then $du = 3x^2\,dx$, so that $x^2\,dx = \frac{1}{3}du$. Then

$$\int \frac{x^2}{1 + x^6}\,dx = \frac{1}{3}\int \frac{1}{1 + u^2}\,du = \frac{1}{3}\arctan u + C = \frac{1}{3}\arctan x^3 + C.$$

6.M.99: Let $u = 2x$. Then $du = 2\,dx$ and $\sqrt{4x^2 - 1} = \sqrt{u^2 - 1}$. Thus

$$\int \frac{1}{x\sqrt{4x^2 - 1}}\,dx = \frac{1}{2}\int \frac{1}{\frac{1}{2}u\sqrt{u^2 - 1}}\,du = \int \frac{1}{u\sqrt{u^2 - 1}}\,du = \operatorname{arcsec}|u| + C = \operatorname{arcsec}|2x| + C.$$

Mathematica returns the equivalent alternative answer

$$\int \frac{1}{x\sqrt{4x^2-1}}\,dx = -\arctan\left(\frac{1}{\sqrt{4x^2-1}}\right)+C,$$

which in turn is equal to $\arctan\left(\sqrt{4x^2-1}\right)+C.$

6.M.101: The integrand is slightly evocative of the derivative of the inverse secant function, so we try the substitution $u = e^x$, so that $\sqrt{e^{2x}-1} = \sqrt{u^2-1}$. Then $du = e^x\,dx$, so

$$\int \frac{1}{\sqrt{e^{2x}-1}}\,dx = \int \frac{e^x}{e^x\sqrt{e^{2x}-1}}\,dx = \int \frac{1}{u\sqrt{u^2-1}}\,du = \operatorname{arcsec}|u| + C = \operatorname{arcsec}(e^x)+C.$$

Mathematica returns the equivalent answer $\arctan\left(\sqrt{e^{2x}-1}\right)+C.$

6.M.103: Let $u = \sqrt{x}$ if necessary, but by inspection an antiderivative is $f(x) = k\cosh x^{1/2}$ for some constant k. Because

$$f'(x) = \left(k\sinh x^{1/2}\right)\cdot D_x\left(x^{1/2}\right) = \frac{k\sinh x^{1/2}}{2x^{1/2}},$$

it follows that $k = 2$ and therefore that

$$\int \frac{\sinh\sqrt{x}}{\sqrt{x}}\,dx = 2\cosh\sqrt{x}+C.$$

6.M.105: Let $u = \arctan x$ if necessary, but evidently

$$\int \frac{\arctan x}{1+x^2}\,dx = \frac{1}{2}(\arctan x)^2+C.$$

6.M.107: Let $u = \frac{2}{3}x$, so that $x = \frac{3}{2}u$. Then $dx = \frac{3}{2}\,du$ and $\sqrt{4x^2+9} = \sqrt{9u^2+9} = 3\sqrt{u^2+1}$. Therefore

$$\int \frac{1}{\sqrt{4x^2+9}}\,dx = \frac{3}{2}\cdot\frac{1}{3}\int \frac{1}{\sqrt{u^2+1}}\,du = \frac{1}{2}\sinh^{-1}u + C = \frac{1}{2}\sinh^{-1}\left(\frac{2x}{3}\right)+C.$$

6.M.109: The volume is

$$V = \int_0^{1/\sqrt{2}} \frac{2\pi x}{\sqrt{1-x^4}}\,dx.$$

Let $u = x^2$. Then $du = 2x\,dx$, and hence

$$\int \frac{2\pi x}{\sqrt{1-x^4}}\,dx = \pi\int \frac{1}{\sqrt{1-u^2}}\,du = \pi\arcsin u + C = \pi\arcsin x^2 + C.$$

Therefore

$$V = \left[\pi\arcsin x^2\right]_0^{1/\sqrt{2}} = \frac{\pi^2}{6} = \zeta(2) \approx 1.644934066848226436472415.$$

See the Index in the textbook for references to further information on the Riemann zeta function $\zeta(z)$.

6.M.111: We use some results in Section 6.9. By Eqs. (36) and (37),

$$\tanh^{-1}\left(\frac{1}{x}\right) = \frac{1}{2}\ln\frac{1+\frac{1}{x}}{1-\frac{1}{x}} = \frac{1}{2}\ln\frac{x+1}{x-1} = \coth^{-1}x$$

by Eq. (37) (provided that $|x| > 1$). By Eq. (35), if $0 < x \leq 1$ then

$$\cosh^{-1}\left(\frac{1}{x}\right) = \ln\left(\frac{1}{x} + \sqrt{\frac{1}{x^2}-1}\right) = \ln\left(\frac{1}{x} + \frac{\sqrt{1-x^2}}{\sqrt{x^2}}\right) = \ln\left(\frac{1+\sqrt{1-x^2}}{x}\right) = \operatorname{sech}^{-1}x$$

by Eq. (38). Note that $\sqrt{x^2} = x$ because $x > 0$.

6.M.113: The graphs of $y = \cos x$ and $y = \operatorname{sech} x$ are shown following this solution, graphed for $0 \leq x \leq 6$. One wonders if there is a solution of $\cos x = \operatorname{sech} x$ in the interval $(0, \pi/2)$. We graphed $y = f(x) = \operatorname{sech} x - \cos x$ for $0 \leq x \leq 1$ and it was clear that $f(x) > 0$ if $x > 0.25$. We graphed $y = f(x)$ for $0 \leq x \leq 0.25$ and it was clear that $f(x) > 0$ if $x > 0.06$. We repeated this process several times and could see that $f(x) > 0$ if $x > 0.0002$. Instability in the hardware or software made further progress along these lines impossible. Methods of Section 10.8 can be used to show the desired inequality, but Ted Shifrin provided the following elegant argument.

First, $\operatorname{sech} x \leq 1 \leq \sec x$ if x is in $I = [0, \pi/2)$. Moreover, $\operatorname{sech} x = \sec x$ only for $x = 0$ in that interval; otherwise, $\operatorname{sech} x < \sec x$ for x in $J = (0, \pi/2)$. So $\operatorname{sech}^2 x < \sec^2 x$ if x is in J.

But $\tanh x = \tan x$ if $x = 0$. Therefore $\tanh x < \tan x$ for x in J because $D_x \tanh x < D_x \tan x$ for such x. That is,

$$\frac{\sinh x}{\cosh x} < \frac{\sin x}{\cos x}$$

if x is in J. All the expressions involved here are positive, so

$$\sinh x \cos x < \cosh x \sin x$$

if x is in J. That is,

$$-\cosh x \sin x + \sinh x \cos x < 0$$

for x in J. But $\cosh x \cos x = 1$ if $x = 0$, and we have now shown that $D_x(\cosh x \cos x) < 0$ if x is in J. Therefore

$$\cosh x \cos x < 1$$

for x in J. In other words, $\operatorname{sech} x \geq \cos x$ for x in I and $\operatorname{sech} x > \cos x$ for x in J.

331

Because $\cos x \leqq 0 < \operatorname{sech} x$ for $\pi/2 < x < 3\pi/2$ and because $\cos x$ is increasing, while $\operatorname{sech} x$ is decreasing, for $3\pi/2 \leqq x < 2\pi$, it follows that the least positive solution of $f(x) = 0$ is slightly larger than $3\pi/2$, about 4.7, exactly as the figure suggests.

We then applied Newton's method to the solution of $f(x) = 0$ with $x_0 = 4.7$, with the following results: $x_1 \approx 4.7300338216$, $x_2 \approx 4.7300407449$, $x_3 \approx 4.7300407449$. Thus x_3 is a good approximation to the least positive solution of $\cos x \cosh x = 1$.

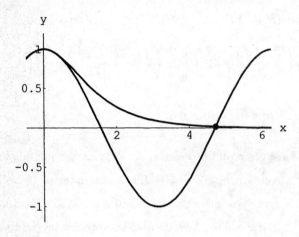

6.M.115: Given $f(x) = x^{1/2}$, let $F(x) = f(x) - \ln x$. Then

$$F'(x) = \frac{1}{2x^{1/2}} - \frac{1}{x} = \frac{x^{1/2} - 2}{2x},$$

so $F'(x) = 0$ when $x = 4$. Clearly F is decreasing on $(0, 4)$ and increasing on $(0, +\infty)$, so the global minimum value of $F(x)$ is

$$F(4) = 2 - \ln 4 = 2 - 2\ln 2 > 2 - 2 \cdot 1$$

because $\ln 2 < 1$. Therefore $f(x) > \ln x$ for all $x > 0$.

For part (b), we need to solve $x^{1/3} - \ln x = 0$. The iteration of Newton's method takes the form

$$x \longleftarrow x - \frac{x^{1/3} - \ln x}{\frac{1}{3}x^{-2/3} - \frac{1}{x}}.$$

Beginning with $x_0 = 100$, we get $x_5 \approx 93.354461$.

For part (c), suppose that $j(x) = x^{1/p}$ is tangent to the graph of $g(x) = \ln x$ at the point $(q, \ln q)$. Then $q^{1/p} = \ln q$ and $j'(q) = g'(q)$. Hence

$$\frac{1}{p}q^{(1/p)-1} = \frac{1}{q}; \qquad q^{1/p} = p;$$

$$p = \ln q = \ln p^p = p\ln p; \qquad \ln p = 1, \quad \text{so} \quad p = e.$$

Section 7.2

7.2.1: Let $u = 2 - 3x$. Then $du = -3\,dx$, and so

$$\int (2 - 3x)^4\,dx = -\frac{1}{3}\int (2-3x)^4(-3)\,dx = -\frac{1}{3}\int u^4\,du = -\frac{1}{3}\cdot\frac{1}{5}u^5 + C = -\frac{1}{15}(2-3x)^5 + C.$$

7.2.3: Let $u = 2x^3 - 4$. Then $du = 6x^2\,dx$, so that

$$\int x^2(2x^3-4)^{1/2}\,dx = \frac{1}{6}\int (2x^3-4)^{1/2}\cdot 6x^2\,dx = \frac{1}{6}\int u^{1/2}\,du = \frac{1}{6}\cdot\frac{2}{3}u^{3/2} + C = \frac{1}{9}(2x^3-4)^{3/2} + C.$$

7.2.5: Let $u = 2x^2 + 3$. Then $du = 4x\,dx$, and so

$$\int 2x(2x^2+3)^{-1/3}\,dx = \frac{1}{2}\int (2x^2+3)^{-1/3}\cdot 4x\,dx = \frac{1}{2}\int u^{-1/3}\,du = \frac{1}{2}\cdot\frac{3}{2}u^{2/3} + C = \frac{3}{4}(2x^2+3)^{2/3} + C.$$

7.2.7: Let $u = y^{1/2}$, so that $du = \frac{1}{2}y^{-1/2}\,dy$. Then

$$\int y^{-1/2}\left(\cot y^{1/2}\right)\left(\csc y^{1/2}\right)\,dy = 2\int \cot u\,\csc u\,du = -2\csc u + C = -2\csc\sqrt{y} + C.$$

7.2.9: Let $u = 1 + \sin\theta$. Then $du = \cos\theta\,d\theta$, and therefore

$$\int (1+\sin\theta)^5\cos\theta\,d\theta = \int u^5\,du = \frac{1}{6}u^6 + C = \frac{1}{6}(1+\sin\theta)^6 + C.$$

7.2.11: Let $u = -\cot x$. Then $du = \csc^2 x\,dx$. So

$$\int e^{-\cot x}\csc^2 x\,dx = \int e^u\,du = e^u + C = e^{-\cot x} + C = \exp\left(-\cot x\right) + C.$$

7.2.13: Let $u = \ln t$. Then $du = \frac{1}{t}\,dt$, so

$$\int \frac{(\ln t)^{10}}{t}\,dt = \int u^{10}\,du = \frac{1}{11}u^{11} + C = \frac{1}{11}(\ln t)^{11} + C.$$

7.2.15: Let $u = 3t$, so that $du = 3\,dt$. Then

$$\int \frac{1}{\sqrt{1-9t^2}}\,dt = \frac{1}{3}\int \frac{1}{\sqrt{1-9t^2}}\cdot 3\,dt = \frac{1}{3}\int \frac{1}{\sqrt{1-u^2}}\,du = \frac{1}{3}\arcsin u + C = \frac{1}{3}\arcsin(3t) + C.$$

7.2.17: Let $u = e^{2x}$. Then $du = 2e^{2x}\,dx$. Therefore

$$\int \frac{e^{2x}}{1+e^{4x}}\,dx = \frac{1}{2}\int \frac{2e^{2x}}{1+(e^{2x})^2}\,dx = \frac{1}{2}\int \frac{1}{1+u^2}\,du = \frac{1}{2}\arctan u + C = \frac{1}{2}\arctan\left(e^{2x}\right) + C.$$

7.2.19: Let $u = x^2$, so that $du = 2x\,dx$, and so

$$\int \frac{3x}{\sqrt{1-x^4}}\,dx = \frac{3}{2}\int \frac{2x}{\sqrt{1-x^4}}\,dx = \frac{3}{2}\int \frac{1}{\sqrt{1-u^2}}\,du = \frac{3}{2}\arcsin u + C = \frac{3}{2}\arcsin\left(x^2\right) + C.$$

7.2.21: Let $u = \tan 3x$. Then $du = 3\sec^2 3x\,dx$. Hence

$$\int (\tan 3x)^4 \sec^2 3x\,dx = \frac{1}{3}\int (\tan 3x)^4(3\sec^2 3x)\,dx = \frac{1}{3}\int u^4\,du = \frac{1}{15}u^5 + C = \frac{1}{15}\tan^5 3x + C.$$

7.2.23: Let $u = \sin\theta$. Then $du = \cos\theta\,d\theta$. Thus

$$\int \frac{\cos\theta}{1+\sin^2\theta}\,d\theta = \int \frac{1}{1+u^2}\,du = \arctan u + C = \arctan(\sin\theta) + C.$$

7.2.25: Let $u = 1 + x^{1/2}$. Then $du = \frac{1}{2}x^{-1/2}\,dx$, and so

$$\int (1+x^{1/2})^4 \cdot x^{-1/2}\,dx = 2\int (1+x^{1/2})^4 \cdot \frac{1}{2}x^{-1/2}\,dx = 2\int u^4\,du = \frac{2}{5}u^5 + C = \frac{2}{5}\left(1+\sqrt{x}\right)^5 + C.$$

7.2.27: Let $u = \arctan t$. Then $du = \dfrac{1}{1+t^2}\,dt$, and thus

$$\int \frac{1}{(1+t^2)\arctan t}\,dt = \int \frac{1}{u}\,du = \ln|u| + C = \ln|\arctan t| + C.$$

7.2.29: Let $u = e^x$. Then $du = e^x\,dx$. The first equality that follows is motivated by the similarity of the integrand to part of the derivative of the inverse secant function:

$$\int \frac{1}{\sqrt{e^{2x}-1}}\,dx = \int \frac{e^x}{|e^x|\sqrt{(e^x)^2 - 1}}\,dx = \int \frac{1}{|u|\sqrt{u^2-1}}\,du$$

$$= \operatorname{arcsec} u + C = \operatorname{arcsec}(e^x) + C = \arctan\left(\sqrt{e^{2x}-1}\right) + C.$$

To obtain the last equality, and thus the answer in the form given by *Mathematica*, draw a right triangle and label an acute angle θ. Let the hypotenuse have length e^x and the side adjacent θ have length 1. Then $\sec\theta = e^x$, so that $\theta = \operatorname{arcsec}(e^x)$. By the Pythagorean theorem, the side opposite θ has length $\sqrt{e^{2x}-1}$, and it follows that $\theta = \arctan\left(\sqrt{e^{2x}-1}\right)$.

7.2.31: Let $u = x - 2$; then $x = u + 2$ and $dx = du$. Thus

$$\int x^2(x-2)^{1/2}\,dx = \int (u+2)^2 u^{1/2}\,du = \int \left(u^{5/2} + 4u^{3/2} + 4u^{1/2}\right)\,du$$

$$= \frac{2}{7}u^{7/2} + \frac{8}{5}u^{5/2} + \frac{8}{3}u^{3/2} + C = \frac{2}{7}(x-2)^{7/2} + \frac{8}{5}(x-2)^{5/2} + \frac{8}{3}(x-2)^{3/2} + C.$$

The answer is quite acceptable in this form, but simplifications are possible; for example, before replacing u with $x - 2$ in the last line, we could proceed as follows:

$$\cdots = \frac{2u^{3/2}}{105}\left(15u^2 + 84u + 140\right) + C = \frac{2}{105}(x-2)^{3/2}\left[15(x^2 - 4x + 4) + 84(x-2) + 140\right] + C$$

$$= \frac{2}{105}(x-2)^{3/2}\left(15x^2 - 60x + 60 + 84x - 168 + 140\right) + C = \frac{2}{105}(x-2)^{3/2}(15x^2 + 24x + 32) + C.$$

7.2.33: Let $u = 2x + 3$, so that $x = \frac{1}{2}(u-3)$ and $ds = \frac{1}{2}\,du$. Then

$$\int x(2x+3)^{-1/2}\,dx = \frac{1}{2}\int x(2x+3)^{-1/2}\cdot 2\,dx = \frac{1}{2}\int \frac{1}{2}(u-3)u^{-1/2}\,du$$

$$= \frac{1}{4}\int\left(u^{1/2} - 3u^{-1/2}\right)du = \frac{1}{4}\left(\frac{2}{3}u^{3/2} - 6u^{1/2}\right) + C = \frac{1}{6}(2x+3)^{3/2} - \frac{3}{2}(2x+3)^{1/2} + C.$$

If you need further simplifications, proceed as follows before replacing u with $2x+3$:

$$\cdots = \frac{1}{6}u^{1/2}(u-9) + C = \frac{1}{6}(2x+3)^{1/2}(2x-6) + C = \frac{1}{3}(x-3)\sqrt{2x+3} + C.$$

7.2.35: Let $u = x + 1$, so that $x = u - 1$ and $dx = du$. Then

$$\int x(x+1)^{-1/3}\,dx = \int (u-1)u^{-1/3}\,du = \int\left(u^{2/3} - u^{-1/3}\right)du$$

$$= \frac{3}{5}u^{5/3} - \frac{3}{2}u^{2/3} + C = \frac{3}{5}(x+1)^{5/3} - \frac{3}{2}(x+1)^{2/3} + C.$$

If additional simplifications are required, proceed as follows before replacing u with $x+1$ in the last line:

$$\cdots = 3u^{2/3}\left(\frac{1}{5}u - \frac{1}{2}\right) + C = \frac{3}{10}u^{2/3}(2u-5) + C = \frac{3}{10}(x+1)^{2/3}(2x-3) + C.$$

7.2.37: Let $u = 3x$. Then $du = 3\,dx$, and thus

$$I = \int \frac{1}{100 - 9x^2}\,dx = \frac{1}{3}\int \frac{1}{10^2 - u^2}\,du = \frac{1}{3}\left(\frac{1}{20}\ln\left|\frac{u+10}{u-10}\right|\right) + C = \frac{1}{60}\ln\left|\frac{3x+10}{3x-10}\right| + C.$$

Mathematica and *Maple* return the answer

$$I = C - \frac{1}{60}\ln(-10 + 3x) + \frac{1}{60}\ln(10 + 3x).$$

The only difference is that both omit the absolute value symbols. The antiderivative produced by *Derive* is almost the same:

$$I = -\frac{1}{60}\ln\frac{3x-10}{3x+10} + C.$$

7.2.39: Let $u = 3x$, so that $du = 3\,dx$. Then

$$J = \int (4 + 9x^2)^{1/2}\,dx = \frac{1}{3}\int (4 + u^2)^{1/2}\,du$$

$$= \frac{1}{3}\left[\frac{1}{2}u(4 + u^2)^{1/2} + 2\ln\left|u + (4 + u^2)^{1/2}\right|\right] = \frac{1}{2}x(4 + 9x^2)^{1/2} + \frac{2}{3}\ln\left(3x + (4 + 9x^2)^{1/2}\right) + C.$$

Mathematica and *Maple* return instead

$$J = C + \frac{1}{2}x\sqrt{4 + 9x^2} + \frac{2}{3}\operatorname{arcsinh}\left(\frac{3x}{2}\right).$$

But Eq. (34) in Section 6.9 makes it clear that the two answers agree. Here's how: Eq. (34) tells us that

$$\sinh^{-1} x = \ln\left(x + \sqrt{x^2 + 1}\right)$$

for all x. Thus

$$\sinh^{-1}\left(\frac{3x}{2}\right) = \ln\left(\frac{3x}{2} + \sqrt{\left(\frac{3x}{2}\right)^2 + 1}\right) = \ln\left(\frac{3x}{2} + \sqrt{\frac{9x^2 + 4}{4}}\right)$$

$$= \ln\left(\frac{3x + \sqrt{9x^2 + 4}}{2}\right) = \ln\left(3x + \sqrt{4 + 9x^2}\right) - \ln 2.$$

Therefore the two answers differ by a constant, as expected. *Derive* yields the antiderivative in the first form given here.

7.2.41: Let $u = 4x$; then $du = 4\,dx$ and we obtain

$$I = \int \frac{x^2}{\sqrt{16x^2 + 9}}\,dx = \frac{1}{4}\int \frac{\frac{1}{16}u^2}{\sqrt{u^2 + 9}}\,du = \frac{1}{64}\int \frac{u^2}{\sqrt{u^2 + 9}}\,du$$

$$= \frac{1}{64}\left(\frac{1}{2}u\sqrt{u^2 + 9} - \frac{9}{2}\ln\left|u + \sqrt{u^2 + 9}\right|\right) + C$$

$$= \frac{1}{32}x\sqrt{16x^2 + 9} - \frac{9}{128}\ln\left(4x + \sqrt{16x^2 + 9}\right) + C.$$

Mathematica returns the equivalent answer

$$I = C + \frac{1}{32}x\sqrt{9 + 16x^2} - \frac{9}{128}\operatorname{arcsinh}\left(\frac{4x}{3}\right).$$

See the solution of Problem 39 for an explanation of why the two antiderivatives are equivalent (that is, they differ by a constant).

336

7.2.43: Let $u = 4x$, so that $du = 4\,dx$. Thus

$$\int x^2 \sqrt{25 - 16x^2}\,dx = \frac{1}{4}\int \frac{1}{16}u^2\sqrt{25 - u^2}\,du$$

$$= \frac{1}{64}\left[\frac{1}{8}u(2u^2 - 25)\sqrt{25 - u^2} + \frac{625}{8}\arcsin\frac{u}{5}\right] + C$$

$$= \frac{1}{128}x(32x^2 - 25)\sqrt{25 - 16x^2} + \frac{625}{512}\arcsin\left(\frac{4x}{5}\right) + C.$$

Mathematica and *Maple* return an almost identical answer.

7.2.45: Let $u = e^x$. Then $du = e^x\,dx$, and thus

$$K = \int e^x\sqrt{9 + e^{2x}}\,dx = \int \sqrt{9 + u^2}\,du = \frac{1}{2}u\sqrt{9 + u^2} + \frac{9}{2}\ln\left|u + \sqrt{9 + u^2}\right| + C$$

$$= \frac{1}{2}e^x\sqrt{9 + e^{2x}} + \frac{9}{2}\ln\left(e^x + \sqrt{9 + e^{2x}}\right) + C.$$

Mathematica returns the equivalent answer

$$K = C + \frac{1}{2}e^x\sqrt{9 + e^{2x}} + \frac{9}{2}\operatorname{arcsinh}\left(\frac{e^x}{3}\right).$$

See the solution of Problem 39 for the reason that the two answers differ only by a constant.

7.2.47: Let $u = x^2$, so that $du = 2x\,dx$. Then

$$J = \int \frac{(x^4 - 1)^{1/2}}{x}\,dx = \frac{1}{2}\int \frac{(x^4 - 1)^{1/2}}{x^2}\cdot 2x\,dx = \frac{1}{2}\int \frac{(u^2 - 1)^{1/2}}{u}\,du$$

$$= \frac{1}{2}\left(\sqrt{u^2 - 1} - \operatorname{arcsec} u\right) + C = \frac{1}{2}\sqrt{x^4 - 1} - \frac{1}{2}\operatorname{arcsec}\left(x^2\right) + C.$$

Mathematica and *Maple* return the antiderivative in the form

$$J = C + \frac{1}{2}\sqrt{-1 + x^4} + \frac{1}{2}\arctan\left(\frac{1}{\sqrt{-1 + x^4}}\right).$$

Derive returns the antiderivative in the form

$$J = \frac{1}{2}\sqrt{x^4 - 1} - \frac{1}{2}\arctan\left(\sqrt{x^4 - 1}\right)$$

(like most computer algebra programs, *Derive* omits the constant of integration). Because

$$\arctan\left(\frac{1}{x}\right) = \pi - \arctan x$$

if $x \neq 0$, the last two answers differ by a constant. See the solution to Problem 29 for an explanation of why our answer is the same as the one given by *Derive*.

337

7.2.49: Let $u = \ln x$. Then $du = \dfrac{1}{x}\,dx$, and hence

$$I = \int \frac{(\ln x)^2}{x}\sqrt{1+(\ln x)^2}\,dx = \int u^2\sqrt{1+u^2}\,du = \frac{1}{8}u(2u^2+1)\sqrt{1+u^2} - \frac{1}{8}\ln\left|u+\sqrt{1+u^2}\right| + C$$

$$= \frac{1}{8}(\ln x)\left[2(\ln x)^2+1\right]\sqrt{1+(\ln x)^2} - \frac{1}{8}\ln\left(\ln x + \sqrt{1+(\ln x)^2}\right) + C.$$

Maple, *Mathematica*, and *Derive* all yield

$$I = C - \frac{1}{8}\operatorname{arcsinh}(\ln x) + \sqrt{1+(\ln x)^2}\left[\frac{1}{8}\ln x + \frac{1}{4}(\ln x)^3\right],$$

which differs from our first answer only by a constant (see the explanation in the solution of Problem 39).

7.2.51: The substitution is illegal: $x = \sqrt{u} \geqq 0$, but $x < 0$ for many x in $[-1, 1]$. To use this substitution correctly, let

$$x = u^{1/2}, \quad dx = \frac{1}{2u^{1/2}}\,du, \quad 0 \leqq x \leqq 1.$$

Then

$$\int_0^1 x^2\,dx = \int_0^1 \frac{1}{2}u^{1/2}\,du = \frac{1}{2}\left[\frac{2}{3}u^{3/2}\right]_0^1 = \frac{1}{3}.$$

Then let

$$x = -u^{1/2}, \quad dx = -\frac{1}{2u^{1/2}}\,du, \quad -1 \leqq x \leqq 0.$$

Then

$$\int_{-1}^0 x^2\,dx = \int_1^0 -\frac{1}{2}u^{1/2}\,du = \frac{1}{2}\left[\frac{2}{3}u^{3/2}\right]_0^1 = \frac{1}{3}.$$

And, finally,

$$\int_{-1}^1 x^2\,dx = \int_{-1}^0 x^2\,dx + \int_0^1 x^2\,dx = \frac{1}{3} + \frac{1}{3} = \frac{2}{3}.$$

7.2.53: The substitution $u = x - 1$, $x = u + 1$, $dx = du$ yields

$$\int \frac{1}{\sqrt{2x-x^2}}\,dx = \int \frac{1}{\sqrt{1-(x-1)^2}}\,dx = \int \frac{1}{\sqrt{1-u^2}}\,du = \arcsin u + C = \arcsin(x-1) + C.$$

7.2.55: First note that

$$D_x\left(\frac{1}{2}\tan^{-1} x^2\right) = \frac{x}{1+x^4}$$

for all x and that

$$D_x\left(-\frac{1}{2}\tan^{-1} x^{-2}\right) = -\frac{1}{2}\cdot\frac{1}{1+x^{-4}}\cdot(-2x^{-3}) = \frac{x^4\cdot x^{-3}}{x^4+1} = \frac{x}{1+x^4}$$

provided that $x \neq 0$. Therefore

$$\frac{1}{2}\tan^{-1} x^2 + \frac{1}{2}\tan^{-1} x^{-2} = C_1,$$

a constant, for all $x > 0$ and

$$\frac{1}{2}\tan^{-1}x^2 + \frac{1}{2}\tan^{-1}x^{-2} = C_2,$$

a constant, for all $x < 0$. Moreover, substitution of $x = 1$ yields

$$C_1 = \frac{1}{2}\tan^{-1}1 + \frac{1}{2}\tan^{-1}1 = \frac{\pi}{4}$$

and substitution of $x = -1$ yields

$$C_2 = \frac{1}{2}\tan^{-1}(-1)^2 + \frac{1}{2}\tan^{-1}(-1)^2 = \frac{\pi}{4}.$$

Therefore

$$\frac{1}{2}\tan^{-1}x^2 = \frac{\pi}{4} - \frac{1}{2}\tan^{-1}x^{-2}$$

for all $x \neq 0$.

7.2.57: Here is a *Mathematica* solution:

```
G = (x/2)*Sqrt[x∧2 + 1] + (1/2)*Log[x + Sqrt[x∧2 + 1]];

D[G,x]
```

$$\frac{x^2}{2\sqrt{1+x^2}} + \frac{\sqrt{1+x^2}}{2} + \frac{1 + \dfrac{x}{\sqrt{1+x^2}}}{2\left(x + \sqrt{1+x^2}\right)}$$

```
% // Together
```

$$\sqrt{1+x^2}$$

```
H = (1/8)*((x + Sqrt[x∧2 + 1])∧2 + 4*Log[x + Sqrt[x∧2 + 1]]

    - (x + Sqrt[x∧2 + 1])∧(−2))
```

$$\frac{1}{8}\left[-\left(x + \sqrt{1+x^2}\right)^{-2} + \left(x + \sqrt{1+x^2}\right)^2 + 4\ln\left(x + \sqrt{1+x^2}\right)\right]$$

```
DH = D[H,x]
```

$$\frac{1}{8}\left[\frac{2\left(1 + \dfrac{x}{\sqrt{1+x^2}}\right)}{\left(x + \sqrt{1+x^2}\right)^3} + \frac{4\left(1 + \dfrac{x}{\sqrt{1+x^2}}\right)}{x + \sqrt{1+x^2}} + 2\left(1 + \dfrac{x}{\sqrt{1+x^2}}\right)\left(x + \sqrt{1+x^2}\right)\right]$$

```
DH // Together
```

$$\frac{1 + 3x^2 + 2x^4 + 2x\sqrt{1+x^2} + 2x^3\sqrt{1+x^2}}{\left(x + \sqrt{1+x^2}\right)^2\sqrt{1+x^2}}$$

```
DH // FullSimplify
```

$$\sqrt{1+x^2}$$

The only thing we might add is that because $G(0) = H(0)$, it now follows that $G(x) = H(x)$ for all x.

Section 7.3

7.3.1: Let $u = x$ and $dv = e^{2x}\,dx$: $du = dx$ and choose $v = \frac{1}{2}e^{2x}$. Then

$$\int xe^{2x}\,dx = \frac{1}{2}xe^{2x} - \int \frac{1}{2}e^{2x}\,dx = \frac{1}{2}xe^{2x} - \frac{1}{4}e^{2x} + C.$$

7.3.3: Let $u = t$ and $dv = \sin t\,dt$: $du = dt$ and choose $v = -\cos t$. Then

$$\int t\sin t\,dt = -t\cos t + \int \cos t\,dt = -t\cos t + \sin t + C.$$

7.3.5: Let $u = x$ and $dv = \cos 3x\,dx$: $du = dx$ and choose $v = \frac{1}{3}\sin 3x$. Then

$$\int x\cos 3x\,dx = \frac{1}{3}x\sin 3x - \frac{1}{3}\int \sin 3x\,dx = \frac{1}{3}x\sin 3x + \frac{1}{9}\cos 3x + C.$$

7.3.7: Let $u = \ln x$ and $dv = x^3\,dx$: $du = \frac{1}{x}\,dx$ and choose $v = \frac{1}{4}x^4$. Then

$$\int x^3 \ln x\,dx = \frac{1}{4}x^4 \ln x - \frac{1}{4}\int x^3\,dx = \frac{1}{4}x^4 \ln x - \frac{1}{16}x^4 + C.$$

7.3.9: Let $u = \arctan x$ and $dv = dx$: $du = \dfrac{1}{1+x^2}\,dx$ and choose $v = x$. Then

$$\int \arctan x\,dx = x\arctan x - \int \frac{x}{1+x^2}\,dx = x\arctan x - \frac{1}{2}\ln(1+x^2) + C.$$

7.3.11: Let $u = \ln y$ and $dv = y^{1/2}\,dy$: $du = \dfrac{1}{y}\,dy$ and choose $v = \frac{2}{3}y^{3/2}$. Then

$$\int y^{1/2}\ln y\,dy = \frac{2}{3}y^{3/2}\ln y - \frac{2}{3}\int y^{1/2}\,dy = \frac{2}{3}y^{3/2}\ln y - \frac{4}{9}y^{3/2} + C.$$

7.3.13: Let $u = (\ln t)^2$ and $dv = dt$: $du = \dfrac{2\ln t}{t}\,dt$ and choose $v = t$. Then

$$\int (\ln t)^2\,dt = t(\ln t)^2 - 2\int \ln t\,dt.$$

Next let $u = \ln t$ and $dv = dt$: $du = \dfrac{1}{t}\, dt$ and choose $v = t$. Thus

$$\int (\ln t)^2\, dt = t(\ln t)^2 - 2\left(t \ln t - \int 1\, dt\right) = t(\ln t)^2 - 2t \ln t + 2t + C.$$

7.3.15: Let $u = x$ and $dv = (x+3)^{1/2}\, dx$: $du = dx$ and choose $v = \dfrac{2}{3}(x+3)^{3/2}$. Then

$$\int x(x+3)^{1/2}\, dx = \frac{2}{3}x(x+3)^{3/2} - \frac{2}{3}\int (x+3)^{3/2}\, dx = \frac{2}{3}x(x+3)^{3/2} - \frac{4}{15}(x+3)^{5/3} + C$$

$$= (x+3)^{3/2}\left(\frac{2}{3}x - \frac{4}{15}x - \frac{4}{5}\right) + C = (x+3)^{3/2}\left(\frac{6x-12}{15}\right) + C$$

$$= \frac{2}{5}(x-2)(x+3)^{3/2} + C = \frac{2}{5}(x^2 + x - 6)\sqrt{x+3} + C.$$

7.3.17: Let $u = x^3$ and $dv = x^2(x^3+1)^{1/2}\, dx$: $du = 3x^2\, dx$ and choose $v = \dfrac{2}{9}(x^3+1)^{3/2}$. Then

$$\int x^5(x^3+1)^{1/2}\, dx = \frac{2}{9}x^3(x^3+1)^{3/2} - \frac{2}{3}\int x^2(x^3+1)^{3/2}\, dx = \frac{2}{9}x^3(x^3+1)^{3/2} - \frac{4}{45}(x^3+1)^{5/2} + C$$

$$= \frac{1}{45}(x^3+1)^{3/2}\left[10x^3 - 4(x^3+1)\right] + C = \frac{1}{45}(x^3+1)^{3/2}(6x^3 - 4) + C$$

$$= \frac{2}{45}(x^3+1)^{3/2}(3x^3 - 2) + C = \frac{2}{45}(x^3+1)^{1/2}(3x^6 + x^3 - 2) + C.$$

7.3.19: Let $u = \csc\theta$ and $dv = \csc^2\theta\, d\theta$: $du = -\csc\theta \cot\theta$ and choose $v = -\cot\theta$. Then

$$\int \csc^3\theta\, d\theta = -\csc\theta \cot\theta - \int \csc\theta \cot^2\theta\, d\theta$$

$$= -\csc\theta \cot\theta - \int (\csc\theta)(\csc^2\theta - 1)\, d\theta = -\csc\theta \cot\theta - \int \csc^3\theta\, d\theta + \int \csc\theta\, d\theta;$$

$$2\int \csc^3\theta\, d\theta = -\csc\theta \cot\theta + \ln|\csc\theta - \cot\theta| + 2C;$$

$$\int \csc^3\theta\, d\theta = -\frac{1}{2}\csc\theta \cot\theta + \frac{1}{2}\ln|\csc\theta - \cot\theta| + C.$$

Mathematica returns the antiderivative in the form

$$C - \frac{1}{2}\cot\theta \csc\theta - \frac{1}{2}\ln\left(\cos\frac{\theta}{2}\right) + \frac{1}{2}\ln\left(\sin\frac{\theta}{2}\right),$$

whereas *Maple* yields an answer that is essentially the same as the one we obtained "by hand."

7.3.21: Let $u = \arctan x$ and $dv = x^2\,dx$: $du = \dfrac{1}{1+x^2}\,dx$ and choose $v = \dfrac{1}{3}x^3$. Then

$$\int x^2 \arctan x\,dx = \frac{1}{3}x^3 \arctan x - \frac{1}{3}\int \frac{x^3}{x^2+1}\,dx$$

$$= \frac{1}{3}x^3 \arctan x - \frac{1}{3}\int \left(x - \frac{x}{x^2+1}\right) + C = \frac{1}{3}x^2 \arctan x - \frac{1}{6}x^2 + \frac{1}{6}\ln(x^2+1) + C.$$

7.3.23: Let $u = \operatorname{arcsec}(x^{1/2})$ and $dv = dx$: $du = \dfrac{1}{2x(x-1)^{1/2}}\,dx$ and choose $v = x$. Then

$$\int \operatorname{arcsec}(x^{1/2})\,dx = x\operatorname{arcsec}(x^{1/2}) - \frac{1}{2}\int (x-1)^{-1/2}\,dx = x\operatorname{arcsec}(x^{1/2}) - (x-1)^{1/2} + C.$$

7.3.25: Let $u = \arctan(x^{1/2})$ and $dv = dx$: $du = \dfrac{1}{2(x+1)x^{1/2}}\,dx$ and cleverly choose $v = x + 1$. Then

$$\int \arctan(x^{1/2})\,dx = (x+1)\arctan(x^{1/2}) - \frac{1}{2}\int x^{-1/2}\,dx = (x+1)\arctan(x^{1/2}) - x^{1/2} + C.$$

7.3.27: Let $u = x$ and $dv = \csc^2 x\,dx$: $du = dx$; choose $v = -\cot x$. Then

$$\int x\csc^2 x\,dx = -x\cot x + \int \frac{\cos x}{\sin x}\,dx = -x\cot x + \ln|\sin x| + C.$$

7.3.29: Let $u = x^2$ and $dv = x\cos x^2\,dx$: $du = 2x\,dx$ and choose $v = \dfrac{1}{2}\sin x^2$. Then

$$\int x^3 \cos x^2\,dx = \frac{1}{2}x^2 \sin x^2 - \int x\sin x^2\,dx = \frac{1}{2}x^2 \sin x^2 + \frac{1}{2}\cos x^2 + C.$$

7.3.31: Let $u = \ln x$ and $dv = x^{-3/2}\,dx$: $du = \dfrac{1}{x}\,dx$ and choose $v = -2x^{-1/2}$. Then

$$\int \frac{\ln x}{x^{3/2}}\,dx = -\frac{2\ln x}{x^{1/2}} + 2\int x^{-3/2}\,dx = -\frac{2\ln x}{x^{1/2}} - \frac{4}{x^{1/2}} + C.$$

7.3.33: Let $u = x$ and $dv = \cosh x\,dx$: $du = dx$ and choose $v = \sinh x$. Then

$$\int x\cosh x\,dx = x\sinh x - \int \cosh x\,dx = x\sinh x - \cosh x + C.$$

7.3.35: Let $t = x^2$. Then $dt = 2x\,dx$, so $\frac{1}{2}t\,dt = x^3\,dx$. This substitution transforms the given integral into

$$I = \tfrac{1}{2}\int t\sin t\,dt.$$

Then integrate by parts: Let $u = t$, $dv = \sin t \, dt$. Thus $du = dt$ and $v = -\cos t$, and hence

$$2I = -t \cos t + \int \cos t \, dt = -t \cos t + \sin t + C.$$

Therefore

$$\int x^3 \sin x^2 \, dx = \tfrac{1}{2} \left(-x^2 \cos x^2 + \sin x^2 \right) + C.$$

7.3.37: Let $t = \sqrt{x}$, so that $x = t^2$ and $dx = 2t \, dt$. Thus

$$I = \int \exp\left(-\sqrt{x}\right) dx = \int 2t \exp(-t) \, dt.$$

Now let $u = 2t$ and $dv = \exp(-t) \, dt$. Then $du = 2 \, dt$ and $v = -\exp(-t)$. Hence

$$I = -2t \exp(-t) + \int 2 \exp(-t) \, dt = -2t \exp(-t) - 2 \exp(-t) + C.$$

Therefore

$$I = -2\sqrt{x} \exp\left(-\sqrt{x}\right) - 2 \exp\left(-\sqrt{x}\right) + C.$$

7.3.39: The volume is

$$V = \int_0^{\pi/2} 2\pi x \cos x \, dx = 2\pi \int_0^{\pi/2} x \cos x \, dx.$$

Let $u = x$ and $dv = \cos x \, dx$: $du = dx$ and choose $v = \sin x$. Then

$$\int x \cos x \, dx = x \sin x - \int \sin x \, dx = x \sin x + \cos x + C.$$

Therefore

$$V = 2\pi \left[x \sin x + \cos x \right]_0^{\pi/2} = 2\pi \left(\frac{\pi}{2} - 1 \right) = \pi^2 - 2\pi \approx 3.5864190939.$$

7.3.41: The volume is

$$V = \int_1^e 2\pi x \ln x \, dx = 2\pi \int_1^e x \ln x \, dx.$$

Let $u = \ln x$ and $dv = x \, dx$: $du = \dfrac{1}{x} \, dx$ and choose $v = \dfrac{1}{2} x^2$. Then

$$\int x \ln x \, dx - \frac{1}{2} x^2 \ln x - \int \frac{1}{2} x \, dx = \frac{1}{2} x^2 \ln x - \frac{1}{4} x^2 + C.$$

Therefore

$$V = 2\pi \left[\frac{1}{2} x^2 \ln x - \frac{1}{4} x^2 \right]_1^e = 2\pi \left(\frac{1}{2} e^2 - \frac{1}{4} e^2 + \frac{1}{4} \right) = \frac{\pi}{2} (e^2 + 1) \approx 13.1774985055.$$

7.3.43: The curves intersect at the point (a, b) in the first quadrant for which $a \approx 0.824132312$. The volume is

$$V = \int_0^a 2\pi x \left[(\cos x) - x^2 \right] \, dx = 2\pi \int_0^a \left(x \cos x - x^3 \right) \, dx.$$

To find the antiderivative of $x \cos x$, let $u = x$ and $dv = \cos x \, dx$. Then $du = dx$; choose $v = \sin x$. Thus

$$\int x \cos x \, dx = x \sin x - \int \sin x \, dx = x \sin x + \cos x + C.$$

Therefore

$$V = 2\pi \left[x \sin x + \cos x - \frac{1}{4} x^4 \right]_0^a \approx 1.06027.$$

7.3.45: The curves intersect where $x = 0$ and where $x = a \approx 2.501048238$. The volume is

$$V = \int_0^a 2\pi x \left[2x - x^2 + \ln(x+1) \right] \, dx = 2\pi \int_0^a \left[2x^2 - x^3 + x \ln(x+1) \right] \, dx.$$

Let $u = \ln(x+1)$ and $dv = x \, dx$. Then $du = \dfrac{1}{x+1} \, dx$; choose $v = \dfrac{1}{2} x^2 - \dfrac{1}{2}$. Then

$$\int x \ln(x+1) \, dx = \frac{x^2-1}{2} \ln(x+1) - \frac{1}{2} \int \frac{x^2-1}{x+1} \, dx = \frac{x^2-1}{2} \ln(x+1) - \frac{1}{4} x^2 + \frac{1}{2} x + C.$$

Therefore

$$V = 2\pi \left[\frac{2}{3} x^3 - \frac{1}{4} x^4 + \frac{1}{2} \left(x^2 - 1 \right) \ln(x+1) - \frac{1}{4} x^2 + \frac{1}{2} x \right]_0^a \approx 22.7894.$$

7.3.47: First choose $u = xe^x$ and $dv = \cos x \, dx$. This yields

$$I = \int xe^x \cos x \, dx = xe^x \sin x - \int (x+1) e^x \sin x \, dx.$$

Now choose $u = (x+1) e^x$ and $dv = \sin x \, dx$;

$$I = xe^x \sin x + (x+1) e^x \cos x - \int (x+2) e^x \cos x \, dx$$

$$= xe^x \sin x + (x+1) e^x \cos x - 2 \int e^x \cos x \, dx - I.$$

Thus

$$2I = xe^x \sin x + (x+1) e^x \cos x - 2 \int e^x \cos x \, dx. \qquad (1)$$

Compute the right-hand integral by parts separately: Let $u = e^x$ and $dv = \cos x \, dx$. Then $du = e^x \, dx$; choose $v = \sin x$. Thus

$$J = \int e^x \cos x \, dx = e^x \sin x - \int e^x \sin x \, dx.$$

344

Now let $u = e^x$ and $dv = \sin x \, dx$. So $du = e^x \, dx$; choose $v - \cos x$. Thus

$$J = e^x \sin x - \left(-e^x \cos x + \int e^x \cos x \, dx \right)$$

$$= e^x \sin x + e^x \cos x - J.$$

Thus $J = \frac{1}{2}(\sin x + \cos x)e^x + C$. Substitute this result in Eq. (1), then solve for I:

$$I = \frac{1}{2}xe^x \cos x + \frac{1}{2}(x-1)\, e^x \sin x + C.$$

7.3.49: Let $u = x^n$ and $dv = e^x \, dx$: $du = nx^{n-1} \, dx$ and choose $v = e^x$. Then

$$\int x^n e^x \, dx = x^n e^x - n \int x^{n-1} e^x \, dx, \qquad n \geq 1.$$

7.3.51: Let $u = (\ln x)^n$ and $dv = dx$: $du = \dfrac{n(\ln x)^{n-1}}{x} \, dx$; choose $v = x$. Then

$$\int (\ln x)^n \, dx = x(\ln x)^n - n \int (\ln x)^{n-1} \, dx, \qquad n \geq 1.$$

7.3.53: Let $u = (\sin x)^{n-1}$ and $dv = \sin x \, dx$. Then $du = (n-1)(\sin x)^{n-2} \cos x \, dx$; choose $v = -\cos x$.
Then

$$I_n = \int (\sin x)^n \, dx = -(\sin x)^{n-1} \cos x + (n-1) \int (\sin x)^{n-2} \cos^2 x \, dx$$

$$= -(\sin x)^{n-1} \cos x + (n-1) \int (\sin x)^{n-2} \, dx - (n-1) \int (\sin x)^n \, dx;$$

$$nI_n = -(\sin x)^{n-1} \cos x + (n-1)I_{n-2};$$

$$I_n = -\frac{1}{n}(\sin x)^{n-1} \cos x + \frac{n-1}{n} I_{n-2}, \qquad n \geq 2.$$

7.3.55: The formula in Problem 49 yields

$$\int_0^1 x^3 e^x \, dx = \left[x^3 e^x \right]_0^1 - 3 \int_0^1 x^2 e^x \, dx = e - 3 \left(\left[x^2 e^x \right]_0^1 - 2 \int_0^1 xe^x \, dx \right)$$

$$= e - 3e + 6 \left(\left[xe^x \right]_0^1 - \int_0^1 e^x \, dx \right) = -2e + 6e - 6 \left[e^x \right]_0^1$$

$$= 4e - 6e + 6 = 6 - 2e \approx 0.5634363431.$$

7.3.57: $\displaystyle \int (\ln x)^3 \, dx = x(\ln x)^3 - 3 \int x(\ln x)^2 \, dx - 2 \int x \ln x \, dx - \int 1 \, dx$. Therefore

$$\int_1^e (\ln x)^3 \, dx = \left[x(\ln x)^3 - 3x(\ln x)^2 + 6x(\ln x) - 6x \right]_1^e = e - 3e + 6e - 6e + 6 = 6 - 2e \approx 0.5634363431.$$

7.3.59: Part (a): Let $u = x + 10$: $x = u - 10$, $dx = du$. Thus

$$\int \ln(x+10)\,dx = \int \ln u\,du = u \ln u - u + C_1$$

$$= (x+10)\ln(x+10) - (x+10) + C_1 = (x+10)\ln(x+10) - x + C.$$

Part (b): Let $u = \ln(x+10)$ and $dv = dx$: $du = \dfrac{1}{x+10}\,dx$; choose $v = x$. Then

$$\int \ln(x+10)\,dx = x\ln(x+10) - \int \frac{x}{x+10}\,dx = x\ln(x+10) - \int \left(1 - \frac{10}{x+10}\right)dx$$

$$= x\ln(x+10) - x + 10\ln(x+10) + C = (x+10)\ln(x+10) - x + C.$$

Part (c): Let $u = \ln(x+10)$ and $dv = dx$: $du = \dfrac{1}{x+10}\,dx$; choose $v = x + 10$. Then

$$\int \ln(x+10)\,dx = (x+10)\ln(x+10) - \int 1\,dx = (x+10)\ln(x+10) - x + C.$$

7.3.61: Part (a):

$$J_0 = \int_0^1 e^{-x}\,dx = \left[\,-e^{-x}\,\right]_0^1 = 1 - \frac{1}{e}.$$

If $n \geq 1$, then let $u = x^n$ and $dv = e^{-x}\,dx$. Then $du = nx^{n-1}\,dx$; choose $v = -e^{-x}$. Thus

$$J_n = \left[\,-x^n e^{-x}\,\right]_0^1 + n\int_0^1 x^{n-1}e^{-x}\,dx = nJ_{n-1} - \frac{1}{e}.$$

Part (b): If $n = 1$, then

$$n! - \frac{n!}{e}\sum_{k=0}^{n}\frac{1}{k!} = 1 - \frac{1}{e}\left(\frac{1}{0!} + \frac{1}{1!}\right) = 1 - \frac{2}{e}$$

and

$$J_1 = 1 \cdot J_0 - \frac{1}{e} = 1 - \frac{2}{e}.$$

Therefore the formula in part (b) holds if $n = 1$. Assume that

$$J_m = m! - \frac{m!}{e}\sum_{k=0}^{m}\frac{1}{k!}$$

for some integer $m \geq 1$. Then

$$J_{m+1} = (m+1)J_m - \frac{1}{e} = m!(m+1) - \frac{m!(m+1)}{e}\sum_{k=0}^{m}\frac{1}{k!} - \frac{1}{e}$$

$$= (m+1)! - \left[\frac{(m+1)!}{e}\sum_{k=0}^{m}\frac{1}{k!} + \frac{(m+1)!}{(m+1)!\,e}\right]$$

$$= (m+1)! - \frac{(m+1)!}{e}\left[\frac{1}{(m+1)!} + \sum_{k=0}^{m}\frac{1}{k!}\right] = (m+1)! - \frac{(m+1)!}{e}\sum_{k=0}^{m+1}\frac{1}{k!}.$$

Therefore, by induction,

$$J_n = n! - \frac{n!}{e} \sum_{k=0}^{n} \frac{1}{k!}$$

for every integer $n \geqq 1$.

Part (c): The next figure will aid in understanding the following proof.

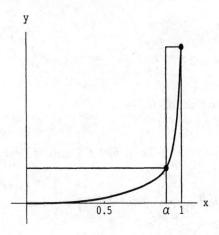

The curve represents the graph of $y = x^n e^{-x}$ on $[0, 1]$. (It really isn't; it's the graph of $y = \frac{1}{10} x^3 + \frac{10}{13} x^{30} e^{-x}$.)
Given the positive integer k, choose the real number α, $0 < \alpha < 1$, so close to 1 that

$$\frac{1 - \alpha}{e} < \frac{1}{2k}.$$

Because $\alpha^n \to 0$ as $n \to \infty$, choose the positive integer N so large that

$$\alpha^{N+1} < \frac{1}{2k}.$$

Then

$$x^N e^{-x} \leqq \alpha^N \quad \text{if} \quad 0 \leqq x \leqq \alpha \quad \text{and} \quad x^N e^{-x} \leqq \frac{1}{e} \quad \text{if} \quad \alpha \leqq x \leqq 1. \tag{1}$$

The area of the short wide rectangle in the figure is

$$\alpha \cdot \alpha^N e^{-\alpha} < \alpha \cdot \alpha^N = \alpha^{N+1}$$

and the area of the tall narrow rectangle there is $(1 - \alpha)/e$. The inequalities in (1) show that the graph of $y = x^N e^{-x}$ is enclosed in the two rectangles, and hence

$$\int_0^1 x^N e^{-x} \, dx \leqq \alpha^{N+1} + \frac{1 - \alpha}{e} < \frac{1}{2k} + \frac{1}{2k} = \frac{1}{k}.$$

Moreover, if $n \geqq N$ and $0 \leqq x \leqq 1$, then $x^n e^{-x} \leqq x^N e^{-x}$. Therefore, for every positive integer k, there exists a positive integer N such that

$$0 \leqq \int_0^1 x^n e^{-x} \, dx < \frac{1}{k}$$

347

if $n \geq N$. Let $k \to \infty$. By the squeeze law for limits,

$$\lim_{n \to \infty} \int_0^1 x^n e^{-x} \, dx = 0.$$

Therefore $J_n \to 0$ as $n \to +\infty$. ◀

Part (d): By part (c),

$$\lim_{n \to \infty} \frac{eJ_n}{n!} = 0 = \lim_{n \to \infty} \left(e - \sum_{k=0}^n \frac{1}{k!} \right).$$

Therefore $e = \lim_{n \to \infty} \sum_{k=0}^n \frac{1}{k!}$. (See Eq. (20) in Section 10.4.)

7.3.63: The expansion of $(k \ln x - 2x^3 + 3x^2 + b)^4$ is a sum of 35 terms, including terms as formidable to antidifferentiate as

$$-32kx^9 \ln x, \qquad 54k^2 x^4 (\ln x)^2, \qquad \text{and} \qquad k^4 (\ln x)^4,$$

as well as several polynomial terms. The reduction formula of Problem 62 handles the three shown here as follows:

$$\int x^9 \ln x \, dx = \frac{x^{10}}{10} \ln x - \frac{1}{10} \int x^9 \, dx = \frac{x^{10}}{10} \ln x - \frac{1}{100} x^{10} + C,$$

$$\int x^4 (\ln x)^2 \, dx = \frac{x^5}{5} (\ln x)^2 - \frac{2}{5} \int x^4 \ln x \, dx$$

$$= \frac{x^5}{5} (\ln x)^2 - \frac{2}{5} \left[\frac{x^5}{5} \ln x - \frac{1}{5} \int x^4 \, dx \right] = \frac{x^5}{5} (\ln x)^2 - \frac{2}{5} \left[\frac{x^5}{5} \ln x - \frac{x^5}{25} \right] + C$$

$$= \frac{x^5}{5} (\ln x)^2 - \frac{2x^5}{25} \ln x + \frac{2x^5}{125} + C,$$

and

$$\int (\ln x)^4 \, dx = x(\ln x)^4 - 4 \int (\ln x)^3 \, dx$$

$$= x(\ln x)^4 - 4 \left[x(\ln x)^3 - 3 \int (\ln x)^2 \, dx \right]$$

$$= x(\ln x)^4 - -4x(\ln x)^3 + 12 \left[x(\ln x)^2 - 2 \int \ln x \, dx \right]$$

$$= x(\ln x)^4 - 4x(\ln x)^3 + 12x(\ln x)^2 - 24 \left[x \ln x - \int 1 \, dx \right]$$

$$= x(\ln x)^4 - 4x(\ln x)^3 + 12x(\ln x)^2 - 24x \ln x + 24x + C.$$

A very patient person can in this way discover that the engineer's antiderivative is

$$(b^4 - 4b^3k + 12b^2k^2 - 24bk^3 + 24k^4)x + \frac{4}{9}(9b^3 - 9b^2k + 6bk^2 - 2k^3)x^3 + \frac{1}{16}(-32b^3 + 24b^2k - 12bk^2$$

$$+ 3k^3)x^4 + \frac{54}{125}(25b^2 - 10bk + 2k^2)x^5 - \frac{2}{3}(18b^2 - 6bk + k^2)x^6 + \frac{12}{343}(441b + 98b^2 - 63k - 28bk + 4k^2)x^7$$

$$- \frac{27}{8}(8b - k)x^8 + \frac{1}{9}(81 + 144b - 16k)x^9 - \frac{4}{25}(135 + 20b - 2k)x^{10} + \frac{216}{11}x^{11} - 8x^{12} + \frac{16}{13}x^{13}$$

$$- \frac{1}{14700}kx(-58800b^3 + 176400b^2k - 352800bk^2 + 352800k^3 - 176400b^2x^2 + 117600bkx^2 - 39200k^2x^2$$

$$+ 82200b^2x^3 - 44100bkx^3 + 11025k^2x^3 - 317520bx^4 + 63504kx^4 + 352800bx^5 - 58800kx^5 - 226800x^6$$

$$- 100800bx^6 + 14400kx^6 + 396900x^7 - 235200x^8 + 47040x^9)\ln x + \frac{1}{70}k^2x(420b^2 - 840bk + 840k^2 + 840bx^2$$

$$- 280kx^2 - 420bx^3 + 105kx^3 + 756x^4 - 840x^5 + 240x^6)(\ln x)^2 - 2k^3x(-2b + 2k - 2x^2 + x^3)(\ln x)^3$$

$$+ k^4x(\ln x)^4 + C.$$

7.3.65: Volume: $V = \displaystyle\int_0^\pi \pi \left(\tfrac{1}{2}x^2 \sin x\right)^2 dx = \frac{\pi}{4}\int_0^\pi x^4 \sin^2 x \, dx = \frac{\pi}{8}\int_0^\pi x^4(1 - \cos 2x)\, dx.$

Let $u = 2x$: $x = \tfrac{1}{2}u$, $dx = \tfrac{1}{2}\,du.$

$$V = \frac{\pi}{8}\int_0^{2\pi} \frac{u^4}{16}(1 - \cos u) \cdot \tfrac{1}{2}\,du$$

$$= \frac{\pi}{256}\int_0^{2\pi} (u^4 - u^4 \cos u)\, du$$

$$= \frac{\pi}{256}\left(\left[\tfrac{1}{5}u^5\right]_0^{2\pi} - \int_0^{2\pi} u^4 \cos u \, du\right)$$

$$= \frac{\pi}{256}\left(\tfrac{32}{5}\pi^5 - \left[u^4 \sin u - 4\int u^3 \sin u \, du\right]_0^{2\pi}\right)$$

$$= \frac{\pi^6}{40} - \frac{\pi}{256}\left[u^4 \sin u - 4\left(-u^3 \cos u + 3\left\{u^2 \sin u - 2\left[-u \cos u + \sin u\right]\right\}\right)\right]_0^{2\pi}$$

$$= \frac{\pi^6}{40} - \frac{\pi}{256}\left[4(2\pi)^3 - 24(2\pi)\right]$$

$$= \frac{\pi^6}{40} - \frac{\pi^4}{8} + \frac{3\pi^2}{16} = \frac{\pi^2}{80}\left(2\pi^4 - 10\pi^2 + 15\right).$$

7.3.67: A *Mathematica* solution:

```
f = x∧2;    g = 2∧x;

R = Plot[ { f, g }, { x, 1.5, 4.5 },
PlotStyle → { RGBColor[0,0,1], RGBColor[1,0,0] } ];
```

The different colors enable us to more easily distinguish the graphs. Area:

```
A = Integrate[ f - g, { x, 2, 4 } ]
```

$$-4 \cdot \frac{-9\ln 2 + 6(\ln 2)^2 + 12\ln 8 - 16(\ln 2)(\ln 8)}{3(\ln 2)(\ln 8)}$$

```
A = A /. { Log[8] -> 3*Log[2] }
```

$$\frac{-4\left[27\ln 2 - 42(\ln 2)^2\right]}{9(\ln 2)^2}$$

(*Mathematica* writes `Log[x]` where we write $\ln x$.)

```
A = A // Simplify
```

$$\frac{-36 + 56\ln 2}{\ln 8}$$

```
A = A /. { Log[8] -> 3*Log[2] }
```

$$\frac{-36 + 56\ln 2}{3\ln 2}$$

Next we find the x-coordinate of the centroid.

```
xc = (1/A)*Integrate[ x*(f - g), { x, 2, 4 } ]
```

$$\frac{12\left[3 + 2\ln 2 + 15(\ln 2)^2 - 4\ln 16\right]}{(-36 + 56\ln 2)(\ln 2)}$$

```
xc = xc /. { Log[16] -> 4*Log[2] }
```

$$\frac{12\left[3 - 14\ln 2 + 15(\ln 2)^2\right]}{(-36 + 56\ln 2)(\ln 2)}$$

Finally, we find the y-coordinate of the centroid.

```
yc = (1/A)*(1/2)*Integrate[ f^2 - g^2, { x, 2, 4 } ]
```

$$24 \cdot \frac{25\ln 2 - 10(\ln 2)(\ln 4) - 40\ln(1024) + 64(\ln 2)(\ln 1024)}{5(-36 + 56\ln 2)(\ln 1024)}$$

```
yc = yc /. { Log[4] -> 2*Log[2], Log[1024] -> 10*Log[2] }
```

$$\frac{12\left[-375\ln 2 + 620(\ln 2)^2\right]}{(-36 + 56\ln 2)(25\ln 2)}$$

```
yc = yc // Simplify
```

$$\frac{225 - 372\ln 2}{45 - 70\ln 2}$$

Thus the centroid of the region has approximate coordinates (3.0904707864762604, 9.3317974433586819).

Section 7.4

7.4.1: $\int \sin^2 x \, dx = \int \frac{1-\cos 2x}{2} \, dx = \frac{1}{2}x - \frac{1}{4}\sin 2x + C = \frac{1}{2}(x - \sin x \, \cos x) + C.$

7.4.3: $\int \sec^2 \frac{x}{2} \, dx = 2 \tan \frac{x}{2} + C.$

7.4.5: $\int \tan 3x \, dx = \int \frac{\sin 3x}{\cos 3x} \, dx = -\frac{1}{3}\ln|\cos 3x| + C = \frac{1}{3}\ln|\sec 3x| + C.$

7.4.7: $\int \sec 3x \, dx = \frac{1}{3}\ln|\sec 3x + \tan 3x| + C.$

7.4.9: $\int \frac{1}{\csc^2 x} \, dx = \int \sin^2 x \, dx = \frac{1}{2}(x - \sin x \, \cos x) + C$ (by Problem 1).

7.4.11: $\int \sin^3 x \, dx = \int (1 - \cos^2 x) \sin x \, dx = \int (\sin x - \cos^2 x \, \sin x) \, dx = \frac{1}{3}\cos^3 x - \cos x + C.$

7.4.13: $\int \sin^2 \theta \, \cos^3 \theta \, d\theta = \int (\sin^2 \theta)(1 - \sin^2 \theta) \cos \theta \, d\theta = \int (\sin^2 \theta \, \cos \theta - \sin^4 \theta \, \cos \theta) \, d\theta$

$= \frac{1}{3}\sin^3 \theta - \frac{1}{5}\sin^5 \theta + C.$

7.4.15: $\int \cos^5 x \, dx = \int (1 - \sin^2 x)^2 \cos x \, dx = \int (\sin^4 x \, \cos x - 2\sin^2 x \, \cos x + \cos x) \, dx$

$= \frac{1}{5}\sin^5 x - \frac{2}{3}\sin^3 x + \sin x + C.$

7.4.17: $\int (\sin^3 x)(\cos x)^{-1/2} \, dx = \int (1 - \cos^2 x)(\cos x)^{-1/2} \sin x \, dx$

$= \int \left[(\cos x)^{-1/2} \sin x - (\cos x)^{3/2} \sin x \right] dx = \frac{2}{5}(\cos x)^{5/2} - 2(\cos x)^{1/2} + C.$

7.4.19: $\int \sin^5 2z \, \cos^2 2z \, dz = \int (1 - \cos^2 2z)^2 \cos^2 2z \, \sin 2z \, dz$

$= \int (\cos^6 2z \, \sin 2z - 2\cos^4 2z \, \sin 2z + \cos^2 2z \, \sin 2z) \, dz = -\frac{1}{14}\cos^7 2z + \frac{1}{5}\cos^5 2z - \frac{1}{6}\cos^3 2z + C.$

The computer algebra program *Derive* returns the answer

$$-\frac{1}{14}\sin^4 2z \, \cos^3 2z - \frac{2}{35}\sin^2 2z \, \cos^3 2z - \frac{4}{105}\cos^3 2z.$$

7.4.21: $\int \frac{\sin^3 4x}{\cos^2 4x} \, dx = \int \frac{(1 - \cos^2 4x)\sin 4x}{\cos^2 4x} \, dx$

$= \int \left[(\cos 4x)^{-2} \sin 4x - \sin 4x \right] dx = \frac{1}{4}(\cos 4x)^{-1} + \frac{1}{4}\cos 4x + C = \frac{1}{4}(\sec 4x + \cos 4x) + C.$

7.4.23: $\int \sec^4 t \, dt = \int (\sec^2 t)(1 + \tan^2 t) \, dt = \int (\sec^2 t + \sec^2 t \, \tan^2 t) \, dt = \tan t + \frac{1}{3}\tan^3 t + C.$

351

7.4.25: $\displaystyle\int \cot^3 2x \, dx = \int (\csc^2 2x - 1) \cot 2x \, dx = \int \left[(\csc 2x)(\csc 2x \cot 2x) - \frac{\cos 2x}{\sin 2x} \right] dx$

$\displaystyle = -\frac{1}{4} \csc^2 2x - \frac{1}{2} \ln |\sin 2x| + C.$

7.4.27: $\displaystyle\int \tan^5 2x \sec^2 2x \, dx = \frac{1}{12} \tan^6 2x + C.$ Alternatively,

$\displaystyle\int \tan^5 2x \sec^2 2x \, dx = \int (\sec^2 2x - 1)^2 \sec^2 2x \tan 2x \, dx = \int (\sec^6 2x - 2\sec^4 2x + \sec^2 2x) \tan 2x \, dx$

$\displaystyle = \int (\sec^5 2x - 2\sec^3 2x + \sec 2x)(\sec 2x \tan 2x) \, dx = \frac{1}{12} \sec^6 2x - \frac{1}{4} \sec^4 2x + \frac{1}{4} \sec^2 2x + C.$

7.4.29: $\displaystyle\int \csc^6 2t \, dt = \int (1 + \cot^2 2t)^2 \csc^2 2t \, dt$

$\displaystyle = \int (\cot^4 2t \csc^2 2t + 2\cot^2 2t \csc^2 2t + \csc^2 2t) \, dt = -\frac{1}{10} \cot^5 2t - \frac{1}{3} \cot^3 2t - \frac{1}{2} \cot 2t + C.$

7.4.31: $\displaystyle\int \frac{\tan^3 \theta}{\sec^4 \theta} \, d\theta = \int \frac{\sin^3 \theta \cos^4 \theta}{\cos^3 \theta} \, d\theta = \int \sin^3 \theta \cos \theta \, d\theta = \frac{1}{4} \sin^4 \theta + C.$ Alternatively,

$\displaystyle\int \frac{\tan^3 \theta}{\sec^4 \theta} \, d\theta = \int \frac{(\sec^2 \theta - 1) \tan \theta}{\sec^4 \theta} \, d\theta = \int \frac{\sec^2 \theta - 1}{\sec^5 \theta} \sec \theta \tan \theta \, d\theta$

$\displaystyle = \int \left[(\sec \theta)^{-3} \sec \theta \tan \theta - (\sec \theta)^{-5} \sec \theta \tan \theta \right] d\theta$

$\displaystyle = -\frac{1}{2} (\sec \theta)^{-2} + \frac{1}{4} (\sec \theta)^{-4} + C = \frac{1}{4} \cos^4 \theta - \frac{1}{2} \cos^2 \theta + C.$

7.4.33: $\displaystyle\int \frac{\tan^3 t}{\sqrt{\sec t}} \, dt = \int (\sec t)^{-1/2} (\sec^2 t - 1) \tan t \, dt = \int \left[(\sec t)^{3/2} \tan t - (\sec t)^{-1/2} \tan t \right] dt$

$\displaystyle = \int \left[(\sec t)^{1/2} (\sec t \tan t) - (\sec t)^{-3/2} (\sec t \tan t) \right] dt = \frac{2}{3} (\sec t)^{3/2} + 2(\sec t)^{-1/2} + C$

$\displaystyle = \frac{2}{3} (\sec t)^{3/2} + 2(\cos t)^{1/2} + C.$

7.4.35: $\displaystyle\int \frac{\cot \theta}{\csc^3 \theta} \, d\theta = \int \frac{\cos \theta \sin^3 \theta}{\sin \theta} \, d\theta = \int \sin^2 \theta \cos \theta \, d\theta = \frac{1}{3} \sin^3 \theta + C.$

7.4.37: $\displaystyle\int \cos^3 5t \, dt = \int (1 - \sin^2 5t) \cos 5t \, dt = \frac{1}{5} \sin 5t - \frac{1}{15} \sin^3 5t + C.$

7.4.39: $\displaystyle\int \cot^4 3t \, dt = \int (\csc^2 3t - 1) \cot^2 3t \, dt = \int \left[\cot^2 3t \csc^2 3t - (\csc^2 3t - 1) \right] dt$

$\displaystyle = -\frac{1}{9} \cot^3 3t + \frac{1}{3} \cot 3t + t + C.$

352

7.4.41: $\displaystyle\int \sin^5 2t \, (\cos 2t)^{3/2} \, dt = \int (1 - \cos^2 2t)^2 (\cos 2t)^{3/2} \sin 2t \, dt$

$$= \int \left[(\cos 2t)^{3/2} - 2(\cos 2t)^{7/2} + (\cos 2t)^{11/2} \right] \sin 2t \, dt$$

$$= -\frac{1}{5}(\cos 2t)^{5/2} + \frac{2}{9}(\cos 2t)^{9/2} - \frac{1}{13}(\cos 2t)^{13/2} + C.$$

7.4.43: $\displaystyle\int \frac{\tan x + \sin x}{\sec x} \, dx = \int (\sin x + \sin x \cos x) \, dx = \frac{1}{2}\sin^2 x - \cos x + C.$

7.4.45: The area is

$$A = \int_0^\pi \sin^3 x \, dx = \int_0^\pi (1 - \cos^2 x) \sin x \, dx = \left[-\cos x + \frac{1}{3}\cos^3 x \right]_0^\pi = 1 - \frac{1}{3} + 1 - \frac{1}{3} = \frac{4}{3}.$$

7.4.47: The area is

$$A = \int_{\pi/4}^\pi (\sin^2 x - \sin x \cos x) \, dx = \int_{\pi/4}^\pi \left(\frac{1 - \cos 2x}{2} - \sin x \cos x \right) dx$$

$$= \left[\frac{1}{2}x - \frac{1}{4}\sin 2x - \frac{1}{2}\sin^2 x \right]_{\pi/4}^\pi = \frac{\pi}{2} - \frac{\pi}{8} + \frac{1}{4} + \frac{1}{2}\cdot\frac{1}{2} = \frac{3\pi + 4}{8}.$$

7.4.49: The following graph makes it appear that the value of the integral is zero.

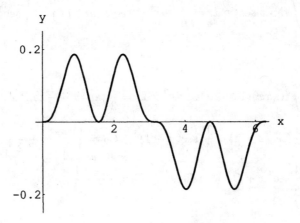

Sure enough,

$$\int_0^{2\pi} \sin^3 x \cos^2 x \, dx = \int_0^{2\pi} (1 - \cos^2 x) \cos^2 x \sin x \, dx = \left[-\frac{1}{3}\cos^3 x + \frac{1}{5}\cos^5 x \right]_0^{2\pi} = 0.$$

7.4.51: The volume is

$$V = \int_0^\pi \pi \sin^4 x \, dx = 2\pi \int_0^{\pi/2} \sin^4 x \, dx = 2\pi \cdot \frac{\pi}{2} \cdot \frac{1}{2} \cdot \frac{3}{4} = \frac{3\pi^2}{8}$$

(we used the result in Problem 58 of Section 7.3 to evaluate the definite integral).

7.4.53: The volume is

$$V = 2\pi \int_0^{\pi/3} (4 - \sec^2 x)\, dx = 2\pi \left[4x - \tan x \right]_0^{\pi/3} = \frac{2\pi}{3} \left(4\pi - 3\sqrt{3} \right) \approx 15.4361488842.$$

7.4.55: Part (a): The area is

$$A = \int_0^{\pi/4} (\sec^2 x - \tan^2 x)\, dx = \int_0^{\pi/4} 1\, dx = \frac{\pi}{4}.$$

Part (b): The volume is

$$V = \pi \int_0^{\pi/4} (\sec^4 x - \tan^4 x)\, dx = \pi \int_0^{\pi/4} (\sec^2 x + \tan^2 x)(\sec^2 x - \tan^2 x)\, dx$$

$$= \pi \int_0^{\pi/4} (\sec^2 x + \tan^2 x)\, dx = \pi \int_0^{\pi/4} (2\sec^2 x - 1)\, dx$$

$$= \pi \left[2\tan x - x \right]_0^{\pi/4} = \frac{\pi}{4}(8 - \pi) \approx 3.8157842069.$$

7.4.57: First way:

$$\int \tan x \sec^4 x\, dx = \int (\sec^3 x)(\sec x \tan x)\, dx = \frac{1}{4}\sec^4 x + C_1.$$

Second way:

$$\int \tan x \sec^4 x\, dx = \int (\tan x)(1 + \tan^2 x) \sec^2 x\, dx = \frac{1}{2}\tan^2 x + \frac{1}{4}\tan^4 x + C_2$$

$$= \frac{1}{2}(\sec^2 x - 1) + \frac{1}{4}(\sec^2 x - 1)^2 + C_2 = \frac{1}{2}\sec^2 x - \frac{1}{2} + \frac{1}{4}\sec^4 x - \frac{1}{2}\sec^2 x + \frac{1}{4} + C_2$$

$$= \frac{1}{4}\sec^4 x + C_1 \qquad \text{where} \qquad C_1 = C_2 - \frac{1}{4}.$$

7.4.59: First, $\sin 3x \cos 5x = \frac{1}{2}\left[\sin(3x - 5x) + \sin(3x + 5x)\right] = \frac{1}{2}\sin 8x - \frac{1}{2}\sin 2x$. Thus

$$\int \sin 3x \cos 5x\, dx = \frac{1}{4}\cos 2x - \frac{1}{16}\cos 8x + C.$$

Because *Mathematica* gives identical answers in this and the next two problems, it appears that it uses the same formulas for integrating such products.

7.4.61: First, $\cos x \cos 4x = \frac{1}{2}\left[\cos(x - 4x) + \cos(x + 4x)\right] = \frac{1}{2}\cos 3x + \frac{1}{2}\cos 5x$. Thus

354

$$\int \cos x \cos 4x \, dx = \frac{1}{6} \sin 3x + \frac{1}{10} \sin 5x + C.$$

7.4.63: $\int \sec x \csc x \, dx = \int \dfrac{\sec^2 x}{\tan x} \, dx = \ln |\tan x| + C.$

7.4.65: We use the substitution $x = \frac{1}{2}\pi - u$, $dx = -du$. Because (from Problem 64)

$$\int \csc x \, dx = \ln \left| \tan \frac{x}{2} \right| + C,$$

we have

$$-\int \csc \left(\frac{1}{2}\pi - u \right) du = \ln \left| \tan \left(\frac{\pi}{4} - \frac{u}{2} \right) \right| + C.$$

But

$$\sin \left(\frac{1}{2}\pi - u \right) = \cos u \qquad \text{and so} \qquad \csc \left(\frac{1}{2}\pi - u \right) = \sec u.$$

Therefore

$$-\int \sec u \, du = \ln \left| \tan \left(\frac{\pi}{4} - \frac{u}{2} \right) \right| C;$$

$$\int \sec u \, du = \ln \left| \tan \left(\frac{\pi}{4} - \frac{u}{2} \right) \right|^{-1} + C;$$

$$\int \sec x \, dx = \ln \left| \cot \left(\frac{\pi}{4} - \frac{x}{2} \right) \right| + C.$$

7.4.67: The reduction formula in Eq. (12) tells us that if n is an integer and $n \geq 2$, then

$$\int \tan^n x \, dx = \frac{(\tan x)^{n-1}}{n - 1} - \int (\tan x)^{n-2} dx.$$

Hence

$$\int \tan^4 x \, dx = \frac{1}{3} \tan^3 x - \int \tan^2 x$$

$$= \frac{1}{3} \tan^3 x - \tan x + \int 1 \, dx = \frac{1}{3} \tan^3 x - \tan x + x + C.$$

Mathematica gives the antiderivative in the form

$$\frac{1}{12} (\sec^3 x)(9x \cos x + 3x \cos 3x - 4 \sin 3x).$$

Now

$$\cos 3x = \cos 2x \cos x - \sin 2x \sin x = \cos^3 x - \sin^2 x \cos x - 2 \sin^2 x \cos x$$

$$= \cos^3 x - 3(1 - \cos^2 x) \cos x = \cos^3 x + 3 \cos^3 x - 3 \cos x = 4 \cos^3 x - 3 \cos x$$

355

and

$$\sin 3x = \sin 2x \cos x + \cos 2x \sin x = 2 \sin x \cos^2 x + \cos^2 x \sin x - \sin^3 x$$

$$= 3 \sin x \cos^2 x - (1 - \cos^2 x) \sin x = 4 \sin x \cos^2 x - \sin x.$$

Therefore

$$\frac{1}{12} (\sec^3 x)(9x \cos x + 3x \cos 3x - 4 \sin 3x)$$

$$= \frac{1}{12} (\sec^3 x)(9x \cos x + 12x \cos^3 x - 9x \cos x - 16 \sin x \cos^2 x + 4 \sin x)$$

$$= \frac{1}{12} (12x - 16 \tan x + 4 \sec^2 x \tan x) = \frac{1}{12} (12x - 16 \tan x + 4(1 + \tan^2 x) \tan x)$$

$$= \frac{1}{12} (12x - 16 \tan x + 4 \tan x + 4 \tan^3 x) = x - \tan x + \frac{1}{3} \tan^3 x.$$

The two antiderivatives are exactly the same.

Section 7.5

7.5.1: $\dfrac{x^2}{x+1} = x - 1 + \dfrac{1}{x+1}$, so $\displaystyle\int \frac{x^2}{x+1}\, dx = \frac{1}{2}x^2 - x + \ln|x+1| + C.$

7.5.3: Given:

$$\frac{1}{x^2 - 3x} = \frac{1}{x(x-3)} = \frac{A}{x} + \frac{B}{x-3},$$

so $Ax - 3A + Bx = 1$, and thus $A + B = 0$ and $-3A = 1$. So

$$\frac{1}{x^2 - 3x} = \frac{-\frac{1}{3}}{x} + \frac{\frac{1}{3}}{x-3},$$

and therefore

$$\int \frac{1}{x^2 - 3x}\, dx = \frac{1}{3} \left(\ln|x-3| - \ln|x| \right) + C = \frac{1}{3} \ln \left| \frac{x-3}{x} \right| + C.$$

7.5.5: $x^2 + x - 6 = (x-2)(x+3)$, so

$$\frac{1}{x^2 + x - 6} = \frac{A}{x-2} + \frac{B}{x+3}.$$

Therefore $Ax + 3A + Bx - 2B = 1$, so that $A + B = 0$ and $3A - 2B = 1$. Thus

$$\int \frac{1}{x^2 + x - 6}\, dx = \int \left(\frac{\frac{1}{5}}{x-2} - \frac{\frac{1}{5}}{x+3} \right) dx = \frac{1}{5} \left(\ln|x-2| - \ln|x+3| \right) + C.$$

7.5.7: $\dfrac{1}{x^3 + 4x} = \dfrac{A}{x} + \dfrac{Bx + C}{x^2 + 4}$ leads to $Ax^2 + 4A + Bx^2 + Cx = 1$. Thus

$$A + B = 0, \quad C + 0, \quad \text{and} \quad 4A = 1.$$

It follows that $A = \frac{1}{4}$, $B = -\frac{1}{4}$, and $C = 0$. Hence

$$\int \frac{1}{x^3 + 4x} \, dx = \int \left(\frac{\frac{1}{4}}{x} - \frac{\frac{1}{4}x}{x^2 + 4} \right) dx = \frac{1}{4} \ln |x| - \frac{1}{8} \ln(x^2 + 4) + C.$$

7.5.9: Division of denominator into numerator leads to

$$\frac{x^4}{x^2 + 4} = x^2 - 4 + \frac{16}{x^2 + 4},$$

and therefore

$$\int \frac{x^4}{x^2 + 4} \, dx = \frac{1}{3} x^3 - 4x + 8 \arctan \left(\frac{x}{2} \right) + C.$$

7.5.11: Division of denominator into numerator yields

$$\int \frac{x - 1}{x + 1} \, dx = \int \left(1 - \frac{2}{x + 1} \right) dx = x - 2 \ln |x + 1| + C.$$

7.5.13: Division of denominator into numerator yields

$$\int \frac{x^2 + 2x}{(x + 1)^2} \, dx = \int \left(1 - \frac{1}{(x + 1)^2} \right) dx = x + \frac{1}{x + 1} + C.$$

7.5.15: $\dfrac{1}{x^2 - 4} = \dfrac{A}{x - 2} + \dfrac{B}{x + 2}$, so that $Ax + 2A + Bx - 2B = 1$. So

$$\int \frac{1}{x^2 - 4} \, dx = \int \left(\frac{\frac{1}{4}}{x - 2} - \frac{\frac{1}{4}}{x + 2} \right) dx = \frac{1}{4} \ln |x - 2| - \frac{1}{4} \ln |x + 2| + C = \frac{1}{4} \ln \left| \frac{x - 2}{x + 2} \right| + C.$$

7.5.17: $\dfrac{x + 10}{2x^2 + 5x - 3} = \dfrac{A}{x + 3} + \dfrac{B}{2x - 1}$ yields $2Ax - A + Bx + 3B = x + 10$. Thus

$$\int \frac{x + 10}{2x^2 + 5x - 3} \, dx = \int \left(\frac{3}{2x - 1} - \frac{1}{x + 3} \right) dx = \frac{3}{2} \ln |2x - 1| - \ln |x + 3| + C.$$

7.5.19: $\dfrac{x^2 + 1}{x^3 + 2x^2 + x} = \dfrac{A}{x} + \dfrac{B}{x + 1} + \dfrac{C}{(x + 1)^2}$ yields $A(x + 1)^2 + Bx(x + 1) + Cx = x^2 + 1$. So

$$A + B = 1, \quad 2A + B + C = 0, \quad \text{and} \quad A = 1.$$

Hence $B = 0$ and $C = -2$. Therefore

$$\int \frac{x^2 + 1}{x^3 + 2x^2 + x} \, dx = \int \left(\frac{1}{x} - \frac{2}{(x + 1)^2} \right) dx = \frac{2}{x + 1} + \ln |x| + C.$$

7.5.21: $\dfrac{4x^3 - 7x}{x^4 - 5x^2 + 4} = \dfrac{A}{x-2} + \dfrac{B}{x-1} + \dfrac{C}{x+1} + \dfrac{D}{x+2}$ yields

$$A(x^3 + 2x^2 - x - 2) + B(x^3 + x^2 - 4x - 4) + C(x^3 - x^2 - 4x + 4) + D(x^3 - 2x^2 - x + 2) = 4x^3 - 7x.$$

Thus

$$A + B + C + D = 4,$$

$$2A + B - C - 2D = 0,$$

$$-A - 4B - 4C - D = -7,$$

$$-2A - 4B + 4C + 2D = 0.$$

This system of equations has the solution $A = \frac{3}{2}$, $B = \frac{1}{2}$, $C = \frac{1}{2}$, $D = \frac{3}{2}$. Therefore

$$\int \frac{4x^3 - 7x}{x^4 - 5x^2 + 4} = \frac{1}{2} \int \left(\frac{3}{x-2} + \frac{1}{x-1} + \frac{1}{x+1} + \frac{3}{x+2} \right) dx$$

$$= \frac{1}{2} \left(3\ln|x-2| + \ln|x-1| + \ln|x+1| + 3\ln|x+2| \right) + C.$$

7.5.23: $\dfrac{x^2}{(x+2)^3} = \dfrac{A}{x+2} + \dfrac{B}{(x+2)^2} + \dfrac{C}{(x+2)^3}$ yields

$$A(x^2 + 4x + 4) + B(x+2) + C = x^2,$$

so that $A = 1$, $4A + B = 0$, and $4A + 2B + C = 0$. It follows that $B = -4$ and $C = 4$. Hence

$$\int \frac{x^2}{(x+2)^3}\, dx = \int \left(\frac{1}{x+2} - \frac{4}{(x+2)^2} + \frac{4}{(x+2)^3} \right) dx = \frac{4}{x+2} - \frac{2}{(x+2)^2} + \ln|x+2| + C.$$

7.5.25: $\dfrac{1}{x^3 + x} = \dfrac{A}{x} + \dfrac{Bx + C}{x^2 + 1}$, so $Ax^2 + A + Bx^2 + Cx = 1$. Thus $A + B = 0$, $C = 0$, and $A = 1$.
Therefore

$$\int \frac{1}{x^3 + x}\, dx = \int \left(\frac{1}{x} - \frac{x}{x^2 + 1} \right) dx = \ln|x| - \frac{1}{2}\ln(x^2 + 1) + C = \frac{1}{2}\ln\left(\frac{x^2}{x^2 + 1} \right) + C.$$

7.7.27: $\dfrac{x+4}{x^3 + 4x} = \dfrac{A}{x} + \dfrac{Bx + C}{x^2 + 4}$ leads to $Ax^2 + 4A + Bx^2 + Cx = x + 4$. So

$$A + B = 0, \quad C = 1, \quad \text{and} \quad 4A = 4. \qquad \text{So} \quad A = 1, \quad B = -1.$$

Thus

$$\int \frac{x+4}{x^3 + 4x}\, dx = \int \left(\frac{1}{x} - \frac{x}{x^2 + 4} + \frac{1}{x^2 + 4} \right) dx = \ln|x| - \frac{1}{2}\ln(x^2 + 4) + \frac{1}{2}\arctan\left(\frac{x}{2} \right) + C.$$

7.5.29: $\dfrac{x}{(x+1)(x^2+1)} = \dfrac{A}{x+1} + \dfrac{Bx+C}{x^2+1}$ yields $Ax^2 + A + Bx^2 + Bx + Cx + C = x$. Thus

$$A + B = 0, \quad B + C = 1, \quad \text{and} \quad A + C = 0.$$

It follows that $A = -\frac{1}{2}$, $B = \frac{1}{2}$, and $C = \frac{1}{2}$. Therefore

$$\int \frac{x}{(x+1)(x^2+1)} \, dx = \frac{1}{2} \int \left(-\frac{1}{x+1} + \frac{x+1}{x^2+1} \right) dx = -\frac{1}{2} \ln|x+1| + \frac{1}{4} \ln(x^2+1) + \frac{1}{2} \arctan x + C.$$

7.5.31: $\dfrac{x^2 - 10}{2x^4 + 9x^2 + 4} = \dfrac{A}{x^2 + 4} + \dfrac{B}{2x^2 + 1}$ implies that $2Ax^2 + A + Bx^2 + 4B = x^2 - 10$, and thus

$$2A + B = 1 \quad \text{and} \quad A + 4B = -10, \quad \text{so that} \quad A = 2 \quad \text{and} \quad B = -3.$$

Therefore

$$\int \frac{x^2 - 10}{2x^4 + 9x^2 + 4} \, dx = \int \left(\frac{2}{x^2 + 4} - \frac{3}{2x^2 + 1} \right) dx = \arctan \left(\frac{x}{2} \right) - \frac{3\sqrt{2}}{2} \arctan \left(x\sqrt{2} \right) + C.$$

A substitution to integrate the second fraction is $u = x\sqrt{2}$.

7.5.33: $\dfrac{x^3 + x^2 + 2x + 3}{x^4 + 5x^2 + 6} = \dfrac{Ax + B}{x^2 + 2} + \dfrac{Cx + D}{x^2 + 3}$, and so

$$Ax^3 + 3Ax + Bx^2 + 3B + Cx^3 + 2Cx + Dx^2 + 2D = x^3 + x^2 + 2x + 3.$$

Therefore

$$A + C = 1, \qquad B + D = 1,$$

$$3A + 2C = 2, \qquad 3B + 2D = 3.$$

It follows that $A = 0$, $B = 1$, $C = 1$, and $D = 0$. Hence

$$\int \frac{x^3 + x^2 + 2x + 3}{x^4 + 5x^2 + 6} \, dx = \int \left(\frac{1}{x^2 + 2} + \frac{x}{x^2 + 3} \right) dx = \frac{\sqrt{2}}{2} \arctan \left(\frac{x\sqrt{2}}{2} \right) + \frac{1}{2} \ln(x^2 + 3) + C.$$

7.5.35: Expand the denominator to $x^4 - 2x^3 + 2x^2 - 2x + 1$ and then divide it into the numerator to find that

$$\frac{x^4 + 3x^2 - 4x + 5}{(x^2 + 1)(x - 1)^2} = 1 + \frac{2x^3 + x^2 - 2x + 4}{(x^2 + 1)(x - 1)^2}.$$

Then

$$\frac{2x^3 + x^2 - 2x + 4}{(x^2 + 1)(x - 1)^2} = \frac{A}{x - 1} + \frac{B}{(x - 1)^2} + \frac{Cx + D}{x^2 + 1}$$

leads to

$$A(x^3 - x^2 + x - 1) + B(x^2 + 1) + C(x^3 - 2x^2 + x) + D(x^2 - 2x + 1) = 2x^3 + x^2 - 2x + 4.$$

Therefore

$$A + C = 2, \qquad -A + B - 2C + D = 1,$$
$$A + C - 2D = -2, \qquad -A + B + D = 4;$$

the solution is $A = \frac{1}{2}$, $B = \frac{5}{2}$, $C = \frac{3}{2}$, $D = 2$. Therefore

$$\int \frac{x^4 + 3x^2 - 4x + 5}{(x^2 + 1)(x - 1)^2}\, dx = \int \left(1 + \frac{\frac{1}{2}}{x - 1} + \frac{\frac{5}{2}}{(x - 1)^2} + \frac{\frac{3}{2}x}{x^2 + 1} + \frac{2}{x^2 + 1} \right) dx$$

$$= x + \frac{1}{2} \ln|x - 1| - \frac{5}{2(x - 1)} + \frac{3}{4} \ln(x^2 + 1) + 2 \arctan x + C.$$

7.5.37: Let $x = e^{2t}$; then $dx = 2e^{2t}\, dt$. Thus

$$\int \frac{e^{4t}}{(e^{2t} - 1)^3}\, dt = \frac{1}{2} \int \frac{x}{(x - 1)^3}\, dx.$$

Then $\dfrac{x}{(x - 1)^3} = \dfrac{A}{x - 1} + \dfrac{B}{(x - 1)^2} + \dfrac{C}{(x - 1)^3}$ leads to $A(x^2 - 2x + 1) + B(x - 1) + C = x$, and so

$$A = 0, \quad -2A + B = 1, \quad \text{and} \quad A - B + C = 0.$$

Therefore $B = C = 1$. Hence

$$\int \frac{x}{(x - 1)^3}\, dx = \int \left(\frac{1}{(x - 1)^2} + \frac{1}{(x - 1)^3} \right) dx = -\frac{1}{x - 1} - \frac{1}{2(x - 1)^2} + C.$$

Finally,

$$\int \frac{e^{4t}}{(e^{2t} - 1)^3}\, dt = -\frac{1}{2(x - 1)} - \frac{1}{4(x - 1)^2} + C = -\frac{1}{2(e^{2t} - 1)} - \frac{1}{4(e^{2t} - 1)^2} + C.$$

7.5.39: Let $u = \ln t$; then $du = \dfrac{1}{t}\, dt$. Therefore

$$J = \int \frac{1 + \ln t}{t(3 + 2\ln t)^2}\, dt = \int \frac{u + 1}{(2u + 3)^2}\, du = \frac{1}{2} \int \left(\frac{1}{2u + 3} - \frac{1}{(2u + 3)^2} \right) du$$

$$= \frac{1}{4} \ln|2u + 3| + \frac{1}{4} \cdot \frac{1}{2u + 3} + C = \frac{1}{4} \ln|3 + 2\ln t| + \frac{1}{4(3 + 2\ln t)} + C.$$

7.5.41: $\dfrac{x - 9}{x^2 - 3x} = \dfrac{3}{x} - \dfrac{2}{x - 3}$. So

360

$$\int_1^2 \frac{x-9}{x^2-3x}\, dx = \Big[3\ln|x| - 2\ln|x-3|\Big]_1^2 = 3\ln 2 - (-2\ln 2) = 5\ln 2 \approx 3.4657359028.$$

7.5.43: $\dfrac{3x-15-2x^2}{x^3-9x} = \dfrac{1}{3}\left(\dfrac{5}{x} - \dfrac{7}{x+3} - \dfrac{4}{x-3}\right)$. Hence

$$\int_0^2 \frac{3x-15-2x^2}{x^3-9x}\, dx = \frac{1}{3}\Big[5\ln|x| - 7\ln|x+3| - 4\ln|x-3|\Big]_0^2$$

$$= \frac{1}{3}\left(5\ln 2 - 7\ln 5 + 4\ln 2 + 7\ln 4\right) = \frac{1}{3}\left(23\ln 2 - 7\ln 5\right) \approx 1.5587732553.$$

7.5.45: $x \cdot \dfrac{x-9}{x^2-3x} = 1 - \dfrac{6}{x-3}$. Hence the volume is

$$V = 2\pi \int_1^2 \left(1 - \frac{6}{x-3}\right) dx = 2\pi\Big[x - 6\ln|x-3|\Big]_1^2 = 2\pi(1 + 6\ln 2) \approx 32.4142183908.$$

7.5.47: $x \cdot \dfrac{3x-15-2x^2}{x^3-9x} = -2 - \dfrac{4}{x-3} + \dfrac{7}{x+3}$. So the volume is

$$V = 2\pi \int_1^2 \left(-2 - \frac{4}{x-3} + \frac{7}{x+3}\right) dx$$

$$= 2\pi\Big[-2x - 4\ln|x-3| + 7\ln|x+3|\Big]_1^2 = 2\pi(7\ln 5 - 2 - 10\ln 2) \approx 14.66868411.$$

7.5.49: $\left(\dfrac{x-9}{x^2-3x}\right)^2 = \dfrac{4}{x} + \dfrac{9}{x^2} - \dfrac{4}{x-3} + \dfrac{4}{(x-3)^2}$. So the volume is

$$V = \pi \int_1^2 \left(\frac{x-9}{x^2-3x}\right)^2 dx = \pi\Big[4\ln|x| - \frac{9}{x} - 4\ln|x-3| - \frac{4}{x-3}\Big]_1^2 = \frac{\pi}{2}(13 + 16\ln 2) \approx 37.8410409708.$$

7.5.51: The volume is $V = \displaystyle\int_0^1 \pi y^2\, dx$. Now $y^2 = \dfrac{1-x}{1+x}x^2 = -x^2 + 2x - 2 + \dfrac{2}{x+1}$, and so

$$V = \pi \int \left(-x^2 + 2x - 2 + \frac{2}{x+1}\right) dx$$

$$= \pi\Big[-\frac{1}{3}x^3 + x^2 - 2x + 2\ln|x+1|\Big]_0^1 = \frac{\pi}{3}(-4 + 6\ln 2) \approx 0.1663819758.$$

7.5.53: $f(x) = \dfrac{A}{x-7} + \dfrac{B}{x-5} + \dfrac{C}{x} + \dfrac{D}{x^2} = \dfrac{93}{x-7} + \dfrac{49}{x-5} - \dfrac{44}{x} + \dfrac{280}{x^2}$. Thus

$$\int f(x)\,dx = 93\ln|x-7| + 49\ln|x-5| - 44\ln|x| - \frac{280}{x} + C,$$

both by *Mathematica* and by hand (except that *Mathematica* omits the absolute value symbols).

7.5.55: $f(x) = \dfrac{A}{x-4} + \dfrac{B}{(x-4)^2} + \dfrac{C}{x-3} + \dfrac{D}{x+5} + \dfrac{E}{(x+5)^2}$

$$= -\frac{\frac{104}{3}}{x-4} + \frac{48}{(x-4)^2} + \frac{\frac{567}{16}}{x-3} - \frac{\frac{37}{48}}{x+5} - \frac{\frac{39}{2}}{(x+5)^2}.\quad \text{Therefore}$$

$$\int f(x)\,dx = -\frac{104}{3}\ln|x-4| - \frac{48}{x-4} + \frac{567}{16}\ln|x-3| - \frac{37}{48}\ln|x+5| + \frac{39}{2(x+5)} + C,$$

both by hand and using *Mathematica* (except that *Mathematica* omits the absolute value symbols).

7.5.57: *Mathematica* yields the partial fraction decomposition

$$\frac{4}{2x-1} + \frac{6}{3x-1} + \frac{3}{(3x-1)^2} + \frac{2x}{x^2+25} + \frac{5}{x^2+25}$$

and the antiderivative

$$-\frac{1}{3x-1} - \arctan\frac{5}{x} + 2\ln(1-3x) + 2\ln(1-2x) + \ln(25+x^2).$$

By hand, we get

$$-\frac{1}{3x-1} + \arctan\frac{x}{5} + 2\ln|3x-1| + 2\ln|2x-1| + \ln(x^2+25) + C.$$

Mathematica normally omits the absolute value symbols in logarithmic integrals as well as the constant of integration; of course,

$$-\arctan\frac{5}{x} = \arctan\frac{x}{5} + C,$$

so the two answers are basically the same.

7.5.59: *Maple* and *Mathematica* both yield

$$\int \frac{ax^2 + bx + c}{x^2(x-1)}\,dx = \frac{c}{x} + (a+b+c)\ln(x-1) - (b+c)\ln x + C.$$

The term including $\ln x$ drops out if we let $c = -b$, and then the term including $\ln(x-1)$ drops out if $a = 0$. Thus to obtain a rational antiderivative, let $a = 0$, $b \neq 0$ (but otherwise arbitrary), and $c = -b$.

7.5.61: According to *Mathematica*,

$$\int \frac{ax^2 + bx + c}{x^3(x-4)^4}\,dx$$

362

$$= -\frac{16a + 4b + c}{192(x-4)^3} + \frac{16a + 8b + 3c}{512(x-4)^2} - \frac{8a + 6b + 3c}{512(x-4)} - \frac{c}{512x^2} - \frac{b+c}{256x}$$

$$- \frac{(8a + 8b + 5c)\ln(x-4)}{2048} + \frac{(8a + 8b + 5c)\ln x}{2048} + C.$$

The logarithmic terms drop out if $8a + 8b + 5c = 0$. Hence choose a and b not both zero (but otherwise arbitrary) and let $c = -(8a + 8b)/5$. For example, if $a = b = 5$ and $c = -16$, then the antiderivative is

$$-\frac{7}{16(x-4)^3} + \frac{9}{64(x-4)^2} - \frac{11}{256(x-4)} + \frac{1}{32x^2} + \frac{11}{256x} + C.$$

Section 7.6

7.6.1: Let $x = 4\sin\theta$. Then

$$\sqrt{16 - x^2} = \sqrt{16 - 16\sin^2\theta} = \sqrt{16\cos^2\theta} = 4\cos\theta \qquad \text{and} \qquad dx = 4\cos\theta \, d\theta.$$

Therefore

$$\int \frac{1}{\sqrt{16 - x^2}} \, dx = \int \frac{1}{4\cos\theta} \cdot 4\cos\theta \, d\theta = \int 1 \, d\theta = \theta + C = \arcsin\left(\frac{x}{4}\right) + C.$$

7.6.3: Let $x = 2\sin u$. Then $dx = 2\cos u \, du$ and

$$x^2\sqrt{4 - x^2} = (4\sin^2 u)\sqrt{4 - 4\sin^2 u} = 8\sin^2 u \cos u.$$

Thus

$$I = \int \frac{1}{x^2\sqrt{4 - x^2}} \, dx = \int \frac{2\cos u}{8\sin^2 u \cos u} \, du = \frac{1}{4} \int \csc^2 u \, du = -\frac{1}{4}\cot u + C.$$

The reference triangle with acute angle u, opposite side x, and hypotenuse 2 has adjacent side of length $\sqrt{4 - x^2}$, and thus

$$I = -\frac{1}{4} \cdot \frac{\sqrt{4 - x^2}}{x} + C = -\frac{\sqrt{4 - x^2}}{4x} + C.$$

7.6.5: Let $x = 4\sin u$. Then $dx = 4\cos u \, du$ and $\dfrac{x^2}{\sqrt{16 - x^2}} = \dfrac{16\sin^2 u}{4\cos u}$. Therefore

$$K = \int \frac{x^2}{\sqrt{16 - x^2}} \, dx = \int 16\sin^2 u \, du = 8\int (1 - \cos 2u) \, du$$

$$= 8\left(u - \frac{1}{2}\sin 2u\right) + C = 8(u - \sin u \cos u) + C.$$

A reference triangle with acute angle u, opposite side x, and hypotenuse 4 has adjacent side $\sqrt{16-x^2}$, and therefore

$$K = 8\left[\arcsin\left(\frac{x}{4}\right) - \frac{x}{4}\cdot\frac{\sqrt{16-x^2}}{4}\right] + C = 8\arcsin\left(\frac{x}{4}\right) - \frac{x\sqrt{16-x^2}}{2} + C.$$

7.6.7: Let $x = \frac{3}{4}\sin u$: $dx = \frac{3}{4}\cos u\,du$, $(9-16x^2)^{3/2} = (9-9\sin^2 u)^{3/2} = 27\cos^3 u$. Hence

$$J = \int \frac{1}{(9-16x^2)^{3/2}}\,dx = \frac{3}{4}\int \frac{\cos u}{27\cos^3 u}\,du = \frac{1}{36}\int \sec^2 u\,du = \frac{1}{36}\tan u + C.$$

The reference triangle with acute angle u, opposite side $4x$, and hypotenuse 3 has adjacent side $\sqrt{9-16x^2}$, and hence

$$J = \frac{1}{36}\cdot\frac{4x}{\sqrt{9-16x^2}} + C = \frac{x}{9\sqrt{9-16x^2}} + C.$$

7.6.9: Let $x = \sec\theta$: $\sqrt{x^2-1} = \tan\theta$, $dx = \sec\theta\tan\theta\,d\theta$. Thus

$$I = \int \frac{\sqrt{x^2-1}}{x^2}\,dx = \int \frac{\tan\theta}{\sec^2\theta}\sec\theta\tan\theta\,d\theta = \int \frac{\tan^2\theta}{\sec\theta}\,d\theta$$

$$= \int \frac{\sec^2\theta - 1}{\sec\theta}\,d\theta = \int (\sec\theta - \cos\theta)\,d\theta = \ln|\sec\theta + \tan\theta| - \sin\theta + C.$$

The reference triangle with acute angle θ, adjacent side x, and hypotenuse x has opposite side $\sqrt{x^2-1}$. Therefore

$$I = \ln\left|x + \sqrt{x^2-1}\right| - \frac{\sqrt{x^2-1}}{x} + C.$$

7.6.11: Let $x = \frac{3}{2}\tan u$: $dx = \frac{3}{2}\sec^2 u\,du$, $9 + 4x^2 + 9 + 9\tan^2 u = 9\sec^2 u$. Thus

$$K = \int x^3\sqrt{9+4x^2}\,dx = \int \left(\frac{27}{8}\tan^3 u\right)(3\sec u)\left(\frac{3}{2}\sec^2 u\right)du$$

$$= \frac{243}{16}\int \sec^3 u\tan^3 u\,du = \frac{243}{16}\int (\sec^3 u)(\sec^2 u - 1)\tan u\,du$$

$$= \frac{243}{16}\int (\sec^4 u - \sec^2 u)\sec u\tan u\,du = \frac{243}{16}\left(\frac{1}{5}\sec^5 u - \frac{1}{3}\sec^3 u\right) + C.$$

The reference triangle with acute angle u, opposite side $2x$, and adjacent side 3 has hypotenuse $\sqrt{9+4x^2}$. Hence

$$K = \frac{243}{16}\left[\frac{1}{5}\cdot\frac{(9+4x^2)^{5/2}}{243} - \frac{1}{3}\cdot\frac{(9+4x^2)^{3/2}}{27}\right] + C = \frac{1}{80}(9+4x^2)^{5/2} - \frac{3}{16}(9+4x^2)^{3/2} + C$$

$$= \frac{1}{80}\left[(9+4x^2)^{5/2} - 15(9+4x^2)^{3/2}\right] + C = \frac{\sqrt{9+4x^2}}{80}\left[(9+4x^2)^2 - 15(9+4x^2)\right] + C$$

$$= \frac{\sqrt{9+4x^2}}{80}\left(16x^4 + 72x^2 + 81 - 60x^2 - 135\right) + C = \frac{\sqrt{9+4x^2}}{80}(16x^4 + 12x^2 - 54) + C$$

$$= \frac{1}{40}(8x^4 + 6x^2 - 27)\sqrt{9+4x^2} + C.$$

Such extensive algebraic simplifications are not normally necessary.

7.6.13: Let $x = \dfrac{1}{2}\sin\theta$: $dx = \dfrac{1}{2}\cos\theta\,d\theta$, $1 - 4x^2 = 1 - \sin^2\theta = \cos^2\theta$. So

$$I = \int \frac{\sqrt{1 - 4x^2}}{x}\,dx = \int \frac{2\cos\theta}{\sin\theta}\cdot\frac{1}{2}\cos\theta\,d\theta = \int \frac{\cos^2\theta}{\sin\theta}\,d\theta = \int \frac{1 - \sin^2\theta}{\sin\theta}\,d\theta$$

$$= \int (\csc\theta - \sin\theta)\,d\theta = \ln|\csc\theta - \cot\theta| + \cos\theta + C.$$

The reference triangle with acute angle θ, opposite side $2x$, and hypotenuse 1 has adjacent side $\sqrt{1 - 4x^2}$. Therefore

$$I = \ln\left|\frac{1 - \sqrt{1 - 4x^2}}{2x}\right| + \sqrt{1 - 4x^2} + C = \ln\left|\frac{1 - 1 + 4x^2}{2x\left(1 + \sqrt{1 - 4x^2}\right)}\right| + \sqrt{1 - 4x^2} + C$$

$$= \ln(4x^2) - \ln|2x| - \ln\left(1 + \sqrt{1 - 4x^2}\right) + \sqrt{1 - 4x^2} + C$$

$$= \ln 4 + 2\ln|x| - \ln 2 - \ln|x| - \ln\left(1 + \sqrt{1 - 4x^2}\right) + \sqrt{1 - 4x^2} + C$$

$$= \ln|x| + \ln 2 - \ln\left(1 + \sqrt{1 - 4x^2}\right) + \sqrt{1 - 4x^2} + C$$

$$= \ln|x| - \ln\left(1 + \sqrt{1 - 4x^2}\right) + \sqrt{1 - 4x^2} + C_1$$

where $C_1 = C - \ln 2$.

7.6.15: Let $x = \dfrac{3}{2}\tan u$: $9 + 4x^2 = 9 + 9\tan^2 u = 9\sec^2 u$, $ds = \dfrac{3}{2}\sec^2 u\,du$. Thus

$$J = \int \frac{1}{\sqrt{9 + 4x^2}}\,dx = \int \frac{1}{3\sec u}\cdot\frac{3}{2}\sec^2 u\,du = \frac{1}{2}\int \sec u\,du = \frac{1}{2}\ln|\sec u + \tan u| + C.$$

The reference triangle with acute angle u, opposite side $3x$, and adjacent side 3 has hypotenuse $\sqrt{9 + 4x^2}$. Therefore

$$J = \frac{1}{2}\ln\left|\frac{\sqrt{9 + 4x^2}}{3} + \frac{2x}{3}\right| + C = \frac{1}{2}\ln\left(2x + \sqrt{9 + 4x^2}\right) + C_1$$

where $C_1 = C - \dfrac{1}{2}\ln 3$.

7.6.17: Let $x = 5\sin\theta$: $dx = 5\cos\theta\,d\theta$, $25 - x^2 = 25\cos^2\theta$. Thus

$$I = \int \frac{x^2}{\sqrt{25 - x^2}}\,dx = \int \frac{25\sin^2\theta}{5\cos\theta}\cdot 5\cos\theta\,d\theta = \frac{25}{2}\int (1 - \cos 2\theta)\,d\theta = \frac{25}{2}(\theta - \sin\theta\cos\theta) + C.$$

The reference triangle with acute angle θ, opposite side x, and hypotenuse 5 has adjacent side $\sqrt{25 - x^2}$. Therefore

$$I = \frac{25}{2}\left[\arcsin\left(\frac{x}{5}\right) - \frac{x\sqrt{25 - x^2}}{25}\right] + C = \frac{25}{2}\arcsin\left(\frac{x}{5}\right) - \frac{x\sqrt{25 - x^2}}{2} + C.$$

365

7.6.19: Let $x = \tan\theta$: $1 + x^2 = \sec^2\theta$, $ds = \sec^2\theta\, d\theta$. Thus

$$K = \int \frac{x^2}{\sqrt{1+x^2}}\, dx = \int \frac{\tan^2\theta}{\sec\theta} \sec^2\theta\, d\theta = \int \sec\theta \tan^2\theta\, d\theta = \int (\sec^3\theta - \sec\theta)\, d\theta.$$

For the antiderivatives, refer to Formulas 14 and 28 of the endpapers of the text or use the reduction formula in Example 6 of Section 7.3. Thus we obtain

$$K = \frac{1}{2}\sec\theta \tan\theta - \frac{1}{2}\ln|\sec\theta + \tan\theta| + C.$$

A reference triangle with acute angle θ, opposite side x, and adjacent side 1 has hypotenuse $\sqrt{1+x^2}$. Therefore

$$K = \frac{1}{2}\left[x\sqrt{1+x^2} - \ln\left(x + \sqrt{1+x^2}\right)\right] + C.$$

7.6.21: Let $x = \frac{2}{3}\tan u$: $dx = \frac{2}{3}\sec^2 u\, du$, $4 + 9x^2 = 4 + 4\tan^2 u = 4\sec^2 u$. Therefore

$$J = \int \frac{x^2}{\sqrt{4+9x^2}}\, dx = \int \frac{\frac{4}{9}\tan^2 u}{2\sec u} \cdot \frac{2}{3}\sec^2 u\, du = \frac{4}{27}\int \sec u \tan^2 u\, du = \frac{4}{27}\int (\sec^3 u - \sec u)\, du$$

$$= \frac{4}{27}\left(\frac{1}{2}\sec u \tan u - \frac{1}{2}\ln|\sec u + \tan u|\right) + C = \frac{2}{27}(\sec u \tan u - \ln|\sec u + \tan u|) + C.$$

A reference triangle with acute angle u, opposite side $3x$, and adjacent side 2 yields hypotenuse of length $\sqrt{4+9x^2}$. Therefore

$$J = \frac{2}{27}\left[\frac{3x\sqrt{4+9x^2}}{4} - \ln\left(\frac{3x + \sqrt{4+9x^2}}{2}\right)\right] + C$$

$$= \frac{1}{18}x\sqrt{4+9x^2} - \frac{2}{27}\ln\left(3x + \sqrt{4+9x^2}\right) + C_1.$$

7.6.23: Let $x = \tan u$: $dx = \sec^2 u\, du$, $1 = x^2 = \sec^2 u$. Hence

$$I = \int \frac{1}{(1+x^2)^{3/2}}\, dx = \int \frac{1}{\sec^3 u}\sec^2 u\, du = \int \cos u\, du = \sin u + C.$$

A reference triangle with acute angle u, opposite side z, and adjacent side 1 has hypotenuse $\sqrt{1+x^2}$. Therefore

$$I = \frac{x}{\sqrt{1+x^2}} + C.$$

7.6.25: Let $x = 2\sin\theta$: $4 - x^2 = 4 - 4\sin^2\theta = 4\cos^2\theta$, $dx = 2\cos\theta\, d\theta$. Therefore

$$K = \int \frac{1}{(4-x^2)^3}\, dx = \int \frac{2\cos\theta}{64\cos^6\theta}\, d\theta = \frac{1}{32}\int \sec^5\theta\, d\theta.$$

From Example 6 in Section 7.3, we know that if n is an integer and $n \geq 2$, then

$$\int \sec^n x\, dx = \frac{(\sec x)^{n-2} \tan x}{n-1} + \frac{n-2}{n-1} \int (\sec x)^{n-2}\, dx.$$

Hence, beginning with $n = 5$, we see that

$$K = \frac{1}{32}\left[\frac{1}{4}\sec^3\theta\tan\theta + \frac{3}{4}\int\sec^3\theta\, d\theta\right]$$

$$= \frac{1}{32}\left[\frac{1}{4}\sec^3\theta\tan\theta + \frac{3}{4}\left(\frac{1}{2}\sec\theta\tan\theta + \frac{1}{2}\int\sec\theta\, d\theta\right)\right]$$

$$= \frac{1}{32}\left(\frac{1}{4}\sec^3\theta\tan\theta + \frac{3}{8}\sec\theta\tan\theta + \frac{3}{8}\ln|\sec\theta + \tan\theta|\right) + C$$

$$= \frac{1}{128}\sec^3\theta\tan\theta + \frac{3}{256}\sec\theta\tan\theta + \frac{3}{256}\ln|\sec\theta + \tan\theta| + C.$$

A reference triangle with acute angle θ, opposite side x, and hypotenuse 2 has adjacent side $\sqrt{4-x^2}$. So

$$K = \frac{1}{512}\left[4\cdot\frac{8x}{(4-x^2)^2} + 6\cdot\frac{2x}{4-x^2} + 6\ln\left|\frac{x+2}{\sqrt{4-x^2}}\right|\right] + C$$

$$= \frac{1}{512}\left[\frac{32x}{(4-x^2)^2} + \frac{12x}{4-x^2} + 3\ln\left|\frac{(x+2)^2}{4-x^2}\right|\right] + C = \frac{1}{512}\left[\frac{32x}{(4-x^2)^2} + \frac{12x}{4-x^2} + 3\ln\left|\frac{2+x}{2-x}\right|\right] + C.$$

7.6.27: Let $x = \frac{3}{4}\tan\theta$: $ds = \frac{3}{4}\sec^2\theta\, d\theta$, $9 + 16x^2 = 9 + 9\tan^2 x^3 = 9\sec^2\theta$. Hence

$$I = \int\sqrt{9+16x^2}\, dx = \int(3\sec\theta)\cdot\frac{3}{4}\sec^2\theta\, d\theta = \frac{9}{4}\int\sec^3\theta\, d\theta = \frac{9}{8}(\sec\theta\tan\theta + \ln|\sec\theta + \tan\theta|) + C.$$

A reference triangle with acute angle θ, opposite side $4x$, and adjacent side 3 has hypotenuse $\sqrt{9+16x^2}$. Thus

$$I = \frac{9}{8}\left[\frac{4x\sqrt{9+16x^2}}{9} + \ln\left(\frac{4x\sqrt{9+16x^2}}{3}\right)\right] + C = \frac{1}{2}x\sqrt{9+16x^2} + \frac{9}{8}\ln\left(4x + \sqrt{9+16x^2}\right) + C_1.$$

7.6.29: Let $x = 5\sec\theta$: $dx = 5\sec\theta\tan\theta\, d\theta$, $x^2 - 25 = 25\sec^2\theta - 25 = 25\tan^2\theta$. So

$$J = \int\frac{\sqrt{x^2-25}}{x}\, dx = \int\frac{5\tan\theta}{5\sec\theta}\cdot 5\sec\theta\tan\theta\, d\theta$$

$$= 5\int\tan^2\theta\, d\theta = 5\int(\sec^2\theta - 1)\, d\theta = 5(\tan\theta - \theta) + C.$$

A reference triangle with acute angle θ, adjacent side 5, and hypotenuse x has opposite side $\sqrt{x^2-25}$. Therefore

$$J = 5\left[\frac{\sqrt{x^2-25}}{5} - \operatorname{arcsec}\left(\frac{x}{5}\right)\right] + C = \sqrt{x^2-25} - 5\operatorname{arcsec}\left(\frac{x}{5}\right) + C$$

$$= \sqrt{x^2-25} + 5\arctan\left(\frac{5}{\sqrt{x^2-25}}\right) + C = \sqrt{x^2-25} - 5\arctan\left(\frac{\sqrt{x^2-25}}{5}\right) + C.$$

7.6.31: Let $x = \sec\theta$: $x^2 - 1 = \sec^2\theta - 1$, $= \sec\theta\tan\theta\,d\theta$. Then

$$I = \int x^2\sqrt{x^2 - 1}\,dx = \int (\sec^2\theta)(\tan\theta)(\sec\theta\tan\theta)\,d\theta = \int \sec^3\theta\tan^2\theta\,d\theta = \int (\sec^5\theta - \sec^3\theta)\,d\theta.$$

Use the result in Example 6 of Section 7.3 (if you haven't memorized it by now!): If n is an integer and $n \geq 2$, then

$$\int \sec^n x\,dx = \frac{(\sec x)^{n-2}\tan x}{n-1} + \frac{n-2}{n-1}\int (\sec x)^{n-2}\,dx.$$

Thus

$$I = \int (\sec^5\theta - \sec^3\theta)\,d\theta = \frac{1}{4}\sec^3\theta\tan\theta + \frac{3}{4}\int \sec^3\theta\,d\theta - \int \sec^3\theta\,d\theta$$

$$= \frac{1}{4}\sec^3\theta\tan\theta - \frac{1}{4}\left(\frac{1}{2}\sec\theta\tan\theta + \frac{1}{2}\ln|\sec\theta + \tan\theta|\right) + C$$

$$= \frac{1}{4}\sec^3\theta\tan\theta - \frac{1}{8}\sec\theta\tan\theta - \frac{1}{8}\ln|\sec\theta + \tan\theta| + C.$$

A reference triangle with acute angle θ, adjacent side 1, and hypotenuse x has opposite side $\sqrt{x^2 - 1}$, and therefore

$$I = \frac{1}{4}x^3\sqrt{x^2 - 1} - \frac{1}{8}x\sqrt{x^2 - 1} - \frac{1}{8}\ln\left|x + \sqrt{x^2 - 1}\right| + C.$$

Additional algebraic simplifications of the answer are possible but not normally required (except possibly to reconcile the answer with that of a computer algebra system such as *Mathematica*, *Maple*, or *Derive*).

7.6.33: Let $x = \frac{1}{2}\sec\theta$: $dx = \frac{1}{2}\sec\theta\tan\theta\,d\theta$, $(4x^2 - 1)^{3/2} = (\sec^2\theta - 1)^{3/2} = \tan^3\theta$. Thus

$$J = \int \frac{1}{(4x^2 - 1)^{3/2}}\,dx = \int \frac{1}{\tan^3\theta}\cdot\frac{1}{2}\sec\theta\tan\theta\,d\theta$$

$$= \frac{1}{2}\int \frac{\sec\theta}{\tan^2\theta}\,d\theta = \frac{1}{2}\int \frac{\cos\theta}{\sin^2\theta}\,d\theta = -\frac{1}{2\sin\theta} + C = -\frac{1}{2}\csc\theta + C.$$

A reference triangle with acute angle θ, adjacent side 1, and hypotenuse $2x$ has opposite side $\sqrt{4x^2 - 1}$. Therefore

$$J = -\frac{1}{2}\cdot\frac{2x}{\sqrt{4x^2 - 1}} + C = -\frac{x}{\sqrt{4x^2 - 1}} + C.$$

7.6.35: Let $x = \left(\sqrt{5}\right)\sec u$: $x^2 - 5 = 5\sec^2 u - 5 = 5\tan^2 u$, $dx = \left(\sqrt{5}\right)\sec u\tan u\,du$. Therefore

$$I = \int \frac{\sqrt{x^2 - 5}}{x^2}\,dx = \int \frac{\left(\sqrt{5}\right)\tan u}{5\sec^2 u}\cdot\left(\sqrt{5}\right)\sec u\tan u\,du = \int \frac{\tan^2 u}{\sec u}\,du$$

$$= \int \frac{\sin^2 u}{\cos u}\,du = \int \frac{1 - \cos^2 u}{\cos u}\,du = \int (\sec u - \cos u)\,du = \ln|\sec u + \tan u| - \sin u + C.$$

A reference triangle with acute angle u, adjacent side $\sqrt{5}$, and hypotenuse x has opposite side $\sqrt{x^2 - 5}$. Thus

$$I = \ln\left|\frac{x + \sqrt{x^2 - 5}}{\sqrt{5}}\right| - \frac{\sqrt{x^2 - 5}}{x} + C = \ln\left|x + \sqrt{x^2 - 5}\right| - \frac{\sqrt{x^2 - 5}}{x} + C_1.$$

7.6.37: Let $x = 5\sinh\theta$. Then $25 + x^2 = 25 + 25\sinh^2\theta = 25\cosh^2\theta$ and $dx = 5\cosh\theta\,d\theta$. So

$$\int \frac{1}{\sqrt{25 + x^2}}\,dx = \int \frac{5\cosh\theta}{5\cosh\theta}\,d\theta = \theta + C = \sinh^{-1}\left(\frac{x}{5}\right) + C.$$

7.6.39: Let $x = 2\cosh\theta$. Then $x^2 - 4 = 4\cosh^2\theta - 4 = 4\sinh^2\theta$ and $dx = 2\sinh\theta\,d\theta$. So—at one point using Eq. (5) of Section 6.9—

$$\int \frac{\sqrt{x^2 - 4}}{x^2}\,dx = \int \frac{2\sinh\theta}{4\cosh^2\theta} \cdot 2\sinh\theta\,d\theta = \int \tanh^2\theta\,d\theta = \int \left(1 - \operatorname{sech}^2\theta\right)\,d\theta$$

$$= \theta - \tanh\theta + C = \theta - \frac{\sinh\theta}{\cosh\theta} + C = \cosh^{-1}\left(\frac{x}{2}\right) - \frac{\sqrt{x^2 - 4}}{x} + C.$$

7.6.41: We will use Eqs. (9), (12), and (10) of Section 6.9 . Let $x = \sinh\theta$. Then $1 + x^2 = \cosh^2\theta$ and $dx = \cosh\theta\,d\theta$. Hence

$$\int x^2\sqrt{1 + x^2}\,dx = \int \sinh^2\theta\,\cosh^2\theta\,d\theta = \frac{1}{4}\int (2\sinh\theta\,\cosh\theta)^2\,d\theta = \frac{1}{4}\int (\sinh 2\theta)^2\,d\theta$$

$$= \frac{1}{8}\int (\cosh 4\theta - 1)\,d\theta = \frac{1}{8}\left(\frac{1}{4}\sinh 4\theta - \theta\right) + C = \frac{1}{8}\left(\frac{1}{2}\sinh 2\theta\,\cosh 2\theta - \theta\right) + C$$

$$= \frac{1}{8}\left[(\sinh\theta\,\cosh\theta)(\cosh^2\theta + \sinh^2\theta) - \theta\right] + C$$

$$= \frac{1}{8}\left(\sinh^3\theta\,\cosh\theta + \sinh\theta\,\cosh^3\theta - \theta\right) + C$$

$$= \frac{1}{8}x^3\sqrt{1 + x^2} + \frac{1}{8}x(1 + x^2)^{3/2} - \frac{1}{8}\sinh^{-1}x + C$$

$$= \frac{1}{8}x^3\sqrt{1 + x^2} + \frac{1}{8}x(1 + x^2)\sqrt{1 + x^2} - \frac{1}{8}\sinh^{-1}x + C$$

$$= \frac{\sqrt{1 + x^2}}{8}(x^3 + x^3 + x) - \frac{1}{8}\sinh^{-1}x + C$$

$$= \frac{1}{8}\left[x(2x^2 + 1)\sqrt{1 + x^2} - \sinh^{-1}x\right] + C.$$

7.6.43: The area of triangle OAC in Fig. 7.6.8 is

$$\frac{1}{2}xy = \frac{1}{2}x\sqrt{a^2 - x^2}.$$

369

The area of the region ABC is (by Example 2)

$$\int_x^a \sqrt{a^2 - u^2}\, du = \left[\frac{u}{2}\sqrt{a^2 - u^2} + \frac{a^2}{2}\arcsin\left(\frac{u}{a}\right)\right]_x^a = \frac{a^2}{2}\cdot \arcsin(1) - \frac{x}{2}\sqrt{a^2 - x^2} - \frac{a^2}{2}\arcsin\left(\frac{x}{a}\right).$$

The area A of sector OBC is therefore their sum:

$$A = \frac{a^2}{2}\cdot\frac{\pi}{2} - \frac{a^2}{2}\arcsin\left(\frac{x}{a}\right).$$

But $x = a\cos\theta$, so

$$A = \frac{\pi a^2}{4} - \frac{a^2}{2}\arcsin\left(\cos\theta\right) = \frac{\pi a^2}{4} - \frac{a^2}{2}\left(\frac{\pi}{2} - \theta\right) = \frac{\pi a^2}{4} - \frac{\pi a^2}{4} + \frac{1}{2}a^2\theta = \frac{1}{2}a^2\theta.$$

7.6.45: Given $y = x^2$, we have $ds = \sqrt{1 + (dy/dx)^2}\, dx = \sqrt{1 + 4x^2}\, dx$. So the surface area of revolution around the x-axis is

$$A = \int_0^1 2\pi x^2 \sqrt{1 + 4x^2}\, dx.$$

Let $x = \frac{1}{2}\tan\theta$: $1 + 4x^2 = \sec^2\theta$, $dx = \frac{1}{2}\sec^2\theta\, d\theta$. So

$$A = \int_{x=0}^1 2\pi\left(\frac{1}{4}\tan^2\theta\right)(\sec\theta)\left(\frac{1}{2}\sec^2\theta\right)d\theta = \frac{\pi}{4}\int_{x=0}^1 \sec^3\theta \tan^2\theta\, d\theta.$$

Now

$$\int \sec^3\theta \tan^2\theta\, d\theta = \int \left(\sec^5\theta - \sec^4\theta\right)d\theta$$

$$= \frac{1}{4}\sec^3\theta\tan\theta + \frac{3}{8}\sec\theta\tan\theta + \frac{3}{8}\ln|\sec\theta + \tan\theta| - \frac{1}{2}\sec\theta\tan\theta - \frac{1}{2}\ln|\sec\theta + \tan\theta| + C$$

$$= \frac{1}{4}\sec^3\theta\tan\theta - \frac{1}{8}\sec\theta\tan\theta - \frac{1}{8}\ln|\sec\theta + \tan\theta| + C.$$

A reference triangle with acute angle θ, opposite side $2x$, and adjacent side 1 has hypotenuse $\sqrt{1 + 4x^2}$. Therefore

$$S = \frac{\pi}{4}\left[\frac{1}{2}x(1 + 4x^2)^{3/2} - \frac{1}{4}x\sqrt{1 + 4x^2} - \frac{1}{8}\ln\left(2x + \sqrt{1 + 4x^2}\right)\right]_0^1$$

$$= \frac{\pi}{4}\left[\frac{1}{2}\cdot 5\sqrt{5} - \frac{1}{4}\sqrt{5} - \frac{1}{8}\ln\left(2 + \sqrt{5}\right)\right] = \frac{\pi}{32}\left[18\sqrt{5} - \ln\left(2 + \sqrt{5}\right)\right] \approx 3.8097297049.$$

7.6.47: Given $y = \ln x$, it follows that the arc length element is $ds = \frac{1}{x}\sqrt{x^2 + 1}\, dx$, so the arc length in question is

$$L = \int_1^2 \frac{1}{x}\sqrt{x^2 + 1}\, dx.$$

370

The substitution $x = \sinh u$ can be made to work, but we prefer to use $x = \tan u$. This results in the definite integral

$$L = \int_{x=1}^{2} (\csc u + \sec u \tan u)\, du = \left[\ln|\csc u - \cot u| + \sec u\right]_{x=1}^{2}$$

$$= \left[\ln\left|\frac{-1 + \sqrt{1 + x^2}}{x}\right| + \sqrt{1 + x^2}\right]_{1}^{2} = \ln\left(\frac{-1 + \sqrt{5}}{2}\right) - \ln\left(-1 + \sqrt{2}\right) + \sqrt{5} - \sqrt{2}$$

$$= \ln\left(\frac{1}{\sqrt{2} - 1}\right) - \ln\left(\frac{2}{\sqrt{5} - 1}\right) + \sqrt{5} - \sqrt{2} = \ln\left(\sqrt{2} + 1\right) - \ln\left(\frac{\sqrt{5} + 1}{2}\right) + \sqrt{5} - \sqrt{2}$$

$$= \ln\left(\sqrt{2} + 1\right) - \ln\left(\sqrt{5} + 1\right) + \ln 2 + \sqrt{5} - \sqrt{2} \approx 1.222016177.$$

7.6.49: First solve for $y = [a^2 - (x - b)^2]^{1/2}$. Then

$$\frac{dy}{dx} = \frac{-(x - b)}{[a^2 - (x - b)^2]^{1/2}}, \quad \text{and so} \quad 1 + \left(\frac{dy}{dx}\right)^2 = \frac{a^2}{a^2 - (x - b)^2}.$$

Therefore the surface area of the torus is

$$S = 2\int_{b-a}^{b+a} 2\pi x\, ds = 4\pi a \int_{b-a}^{b+a} \frac{x}{\sqrt{a^2 - (x - b)^2}}\, dx.$$

The substitution we want should produce

$$a^2 - (x - b)^2 = a^2 - a^2 \sin^2\theta = a^2 \cos^2\theta,$$

so we choose $x = b + a\sin\theta$. Then $dx = a\cos\theta\, d\theta$,

$$\sin\theta = \frac{x - b}{a}, \quad \text{and} \quad \theta = \arcsin\left(\frac{x - b}{a}\right).$$

Before proceeding, note that when $x = b + a$, we have

$$a\cos\theta = [a^2 - (x - b)^2]^{1/2} = 0 \quad \text{and} \quad \theta = \arcsin(1) = \frac{\pi}{2},$$

and when $x = b - a$,

$$a\cos\theta = [a^2 - (x - b)^2]^{1/2} = 0 \quad \text{and} \quad \theta = \arcsin(-1) = -\frac{\pi}{2}.$$

Consequently

$$S = 4\pi a \int_{x=b-a}^{b+a} \frac{b + a\sin\theta}{a\cos\theta} \cdot a\cos\theta\, d\theta$$

$$= 4\pi a \left[b\theta - a\cos\theta\right]_{x=b-a}^{b+a} = 4\pi a \left[b \cdot \frac{\pi}{2} - b \cdot \left(-\frac{\pi}{2}\right) - 0 + 0\right] = 4\pi^2 ab.$$

7.6.51: $A = 4\pi \int_0^{\pi/2} (\sin x)\sqrt{1 + \cos^2 x} \, dx$. With $u = \cos x$ and $du = -\sin x \, dx$, we obtain

$$A = 4\pi \int_0^1 \sqrt{1 + u^2} \, du.$$

To find the antiderivative, we let $u = \sinh z$, $du = \cosh z \, dz$. Then we obtain

$$A = 4\pi \int_{u=0}^{u=1} \cosh^2 z \, dz = \left[2\pi(z + \sinh z \cosh z) \right]_{u=0}^1$$

$$= 2\pi \left[\sinh^{-1} u + u\sqrt{1 + u^2} \right]_0^1 = 2\pi \left(\sinh^{-1}(1) + \sqrt{2} \right)$$

$$= 2\pi \left[\sqrt{2} + \ln\left(1 + \sqrt{2}\right) \right] \approx 14.4236.$$

7.6.53: This is the case in which $0 < a < b$. First we solve the equation of the ellipse for

$$y = \frac{b}{a}(a^2 - x^2)^{1/2}. \tag{1}$$

Thus

$$\frac{dy}{dx} = \frac{b}{a} \cdot \frac{1}{2}(a^2 - x^2)^{1/2} \cdot (-2x) = \frac{-bx}{a(a^2 - x^2)^{1/2}},$$

so that

$$1 + \left(\frac{dy}{dx}\right)^{1/2} = 1 + \frac{b^2 x^2}{a^2(a^2 - x^2)} = \frac{a^4 - a^2 x^2 + b^2 x^2}{a^2(a^2 - x^2)}.$$

Equation (1) also gives the radius of the circle of revolution "at" x, so the surface area of revolution around the x-axis is

$$A = 2\int_0^a 2\pi \cdot \frac{b}{a}(a^2 - x^2)^{1/2} \cdot \frac{\sqrt{a^4 + (b^2 - a^2)x^2}}{a\sqrt{a^2 - x^2}} \, dx = \frac{4\pi b}{a^2} \int_0^a \sqrt{a^4 + (b^2 - a^2)x^2} \, dx.$$

Let

$$x = \frac{a^2 \tan u}{\sqrt{b^2 - a^2}}, \quad \text{so that} \quad dx = \frac{a^2 \sec^2 u}{\sqrt{b^2 - a^2}} \, du.$$

Then

$$a^4 + (b^2 - a^2)x^2 = a^2 + (b^2 - a^2) \cdot \frac{a^4 \tan^2 u}{b^2 - a^2} = a^4(1 + \tan^2 u) = a^2 \sec^2 u. \tag{2}$$

Thus

$$A = \frac{4\pi b}{a^2} \int_{x=0}^a (a^2 \sec^2 u) \cdot \frac{a^2 \sec^2 u}{\sqrt{b^2 - a^2}} \, du = \frac{4\pi a^2 b}{\sqrt{b^2 - a^2}} \int_{x=0}^a \sec^3 u \, du$$

$$= \frac{4\pi a^2 b}{\sqrt{b^2 - a^2}} \left[\frac{1}{2}\left(\sec u \tan u + \ln|\sec u + \tan u| \right) \right]_{x=0}^a.$$

Now $\tan u = \dfrac{x\sqrt{b^2 - a^2}}{a^2}$, and by Eq. (2),

$$\sec u = \frac{\sqrt{a^4 + (b^2 - a^2)x^2}}{a^2}.$$

So

$$A = \frac{2\pi a^2 b}{\sqrt{b^2 - a^2}}\left[\frac{x\sqrt{b^2 - a^2}}{a^2}\cdot\frac{\sqrt{a^4 + (b^2 - a^2)x^2}}{a^2} + \ln\left|\frac{\sqrt{a^4 + (b^2 - a^2)x^2}}{a^2} + \frac{x\sqrt{b^2 - a^2}}{a^2}\right|\;\right]_{x=0}^{a}$$

$$= \frac{2\pi a^2 b}{\sqrt{b^2 - a^2}}\left[\frac{\sqrt{b^2 - a^2}}{a}\cdot\frac{\sqrt{a^2 b^2}}{a^2} + \ln\left(\frac{\sqrt{a^2 b^2}}{a^2} + \frac{\sqrt{b^2 - a^2}}{a}\right)\right].$$

Let $c = \sqrt{b^2 - a^2}$. Then

$$A = \frac{2\pi a^2 b}{c}\left[\frac{c}{a}\cdot\frac{b}{a} + \ln\left(\frac{b}{a} + \frac{c}{a}\right)\right] = 2\pi ab\left[\frac{b}{a} + \frac{a}{c}\ln\left(\frac{b+c}{a}\right)\right].$$

Now let $b \to a^+$. Then

$$\frac{b+c}{a} \approx 1 + \frac{c}{a}, \quad\text{so that}\quad \ln\left(\frac{b+c}{a}\right) \approx \ln\left(1 + \frac{c}{a}\right) \approx \frac{c}{a}.$$

So

$$\frac{b}{a} \to 1 \quad\text{and}\quad \frac{a}{c}\ln\left(\frac{b+c}{a}\right) \to 1$$

as $b \to a^+$. Therefore as $b \to a^+$, $A \to 2\pi ab(1 + 1) = 4\pi ab$.

7.6.55: Given: $y = -1 + 2(x-1)^{1/2}$. Then

$$\frac{dy}{dx} = (x-1)^{-1/2}, \quad\text{so}\quad 1 + \left(\frac{dy}{dx}\right)^2 = 1 + \frac{1}{x-1} = \frac{x}{x-1}.$$

Therefore the cost is

$$C = \int_2^5 \sqrt{x}\left(\frac{x}{x-1}\right)^{1/2}\,dx = \int_2^5 \frac{x}{\sqrt{x-1}}\,dx.$$

Let $x - 1 = u^2$. Then $x = 1 + u^2$ and $dx = 2u\,du$. Hence

$$C = \int_{x=2}^5 \frac{1+u^2}{u}\cdot 2u\,du = 2\int_{x=2}^5 (1+u^2)\,du = 2\left[u + \frac{1}{3}u^3\right]_{x=2}^5$$

$$= 2\left[\sqrt{x-1} + \frac{1}{3}(x-1)^{1/3}\right]_2^5 = 2\left(2 - 1 + \frac{8}{3} - \frac{1}{3}\right) = \frac{20}{3} \quad\text{(million dollars)}.$$

Section 7.7

7.7.1: $\displaystyle\int \frac{1}{x^2+4x+5}\,dx = \int \frac{1}{(x+2)^2+1}\,dx = \arctan(x+2)+C.$

7.7.3: $\displaystyle\int \frac{5-3x}{x^2+4x+5}\,dx = -\frac{3}{2}\int \frac{2x+4-\frac{22}{3}}{x^2+4x+5}\,dx = -\frac{3}{2}\ln(x^2+4x+5)+11\int \frac{1}{x^2+4x+5}\,dx$

$\displaystyle = -\frac{3}{2}\ln(x^2+4x+5)+11\arctan(x+2)+C \quad \text{(see Problem 1)}.$

7.7.5: $3-2x-x^2 = -(x^2+2x-3) = -(x^2+2x+1-4) = 4-(x+1)^2 - 4 - 4\sin^2\theta = 4\cos^2\theta$ if we (and we do) let $x+1 = 2\sin\theta$. Then $x = -1+2\sin\theta$, $dx = 2\cos\theta\,d\theta$, and $\sin\theta = \frac{1}{2}(x+1)$. So

$$\int \frac{1}{\sqrt{3-2x-x^2}}\,dx = \int \frac{1}{2\cos\theta}\cdot 2\cos\theta\,d\theta = \int 1\,d\theta = \theta + C = \arcsin\left(\frac{x+1}{2}\right)+C.$$

7.7.7: $3-2x-x^2 = -(x^2+2x-3) = -(x^2+2x+1-4) = 4-(x+1)^2 - 4 - 4\sin^2\theta = 4\cos^2\theta$ if we (and we do) let $x+1 = 2\sin\theta$. Then $x = -1+2\sin\theta$, $dx = 2\cos\theta\,d\theta$, and $\sin\theta = \frac{1}{2}(x+1)$. So

$$\int x\sqrt{3-2x-x^2}\,dx = \int (-1+2\sin\theta)\cdot(2\cos\theta)\cdot(2\cos\theta)\,d\theta = 4\int (2\cos^2\theta\sin\theta - \cos^2\theta)\,d\theta$$

$$= 4\int \left(2\cos^2\theta\sin\theta - \frac{1+\cos 2\theta}{2}\right)d\theta = 4\left(-\frac{2}{3}\cos^3\theta - \frac{1}{2}\theta - \frac{1}{2}\sin\theta\cos\theta\right)+C$$

$$= -\frac{1}{3}(2\cos\theta)^3 - 2\theta - 2\sin\theta\cos\theta + C$$

$$= -\frac{1}{3}(3-2x-x^2)^{3/2} - 2\arcsin\left(\frac{x+1}{2}\right) - \frac{x+1}{2}\sqrt{3-2x-x^2}+C$$

$$= -2\arcsin\left(\frac{x+1}{2}\right) - \left[\frac{1}{3}(3-2x-x^2) + \frac{1}{2}(x+1)\right]\sqrt{3-2x-x^2}+C$$

$$= -2\arcsin\left(\frac{x+1}{2}\right) - \frac{1}{6}\left[2(3-2x-x^2) + 3(x+1)\right]\sqrt{3-2x-x^2}+C$$

$$= -2\arcsin\left(\frac{x+1}{2}\right) - \frac{1}{6}(-2x^2 - 4x + 6 + 3x + x)\sqrt{3-2x-x^2}+C$$

$$= -2\arcsin\left(\frac{x+1}{2}\right) + \frac{1}{6}(2x^2 + x - 9)\sqrt{3-2x-x^2}+C.$$

7.7.9: $4x^2+4x-3 = 4x^2+4x+1-4 = (2x+1)^2 - 4 = 4\sec^2\theta - 4 = 4\tan^2\theta$ if we (and we do) let $2\sec\theta = 2x+1$. Thus $x = \sec\theta - \frac{1}{2}$, $\sec\theta = x+\frac{1}{2}$, and $dx = \sec\theta\tan\theta\,d\theta$. Thus

$$I = \int \frac{3x+2}{4x^2+4x-3}\,dx = \int \frac{3\sec\theta+\frac{1}{2}}{4\tan^2\theta}\cdot\sec\theta\tan\theta\,d\theta = \frac{1}{8}\int \frac{6\sec\theta+1}{\tan\theta}\cdot\sec\theta\,d\theta$$

$$= \frac{1}{8}\int \left(6\cdot\frac{\sec^2\theta}{\tan\theta} + \frac{\sec\theta}{\tan\theta}\right)d\theta = \frac{1}{8}\int \left(6\cdot\frac{\sec^2\theta}{\tan\theta} + \csc\theta\right)d\theta$$

$$= \frac{1}{8}(6\ln|\tan\theta| + \ln|\csc\theta - \cot\theta|)+C = \frac{3}{4}\ln|\tan\theta| + \frac{1}{8}\ln|\csc\theta - \cot\theta|+C.$$

374

A reference triangle with acute angle θ, adjacent side 2, and hypotenuse $2x + 1$ has opposite side of length $\sqrt{4x^2 + 4x - 2}$. Therefore

$$I = \frac{3}{4} \ln \left| \frac{\sqrt{4x^2 + 4x - 3}}{2} \right| + \frac{1}{8} \ln \left| \frac{2x + 1 - 2}{\sqrt{4x^2 + 4x - 3}} \right| + C$$

$$= \frac{3}{4} \ln \sqrt{4x^2 + 4x - 3} + \frac{1}{8} \ln |2x - 1| - \frac{1}{8} \ln \sqrt{4x^2 + 4x - 3} + C_1$$

$$= \frac{5}{16} \ln |4x^2 + 4x - 3| + \frac{1}{8} \ln |2x - 1| + C_1 = \frac{7}{16} \ln |2x - 1| + \frac{5}{16} \ln |2x + 3| + C_1.$$

7.7.11: $x^2 + 4x + 13 = x^2 + 4x + 4 + 9 = (x + 2)^2 + 9 = 9 + 9 \tan^2 \theta = 9 \sec^2 \theta$ if we (and we do) let $3 \tan \theta = x + 2$; that is, $x = -2 + 3 \tan \theta$, $dx = 3 \sec^2 \theta \, d\theta$, and $\tan \theta = \frac{1}{3}(x + 2)$. Then

$$\int \frac{1}{x^2 + 4x + 13} \, dx = \int \frac{1}{9 \sec^2 \theta} \cdot 3 \sec^2 \theta \, d\theta = \int \frac{1}{3} \, d\theta = \frac{1}{3} \theta + C = \frac{1}{3} \arctan \left(\frac{x + 2}{3} \right) + C.$$

7.7.13: $3 + 2x - x^2 = -(x^2 - 2x - 3) = -(x^2 - 2x + 1 - 4) = 4 - (x - 1)^2 = 4 - 4 \sin^2 \theta = 4 \cos^2 \theta$ if $2 \sin \theta = x - 1$; that is, $x = 1 + 2 \sin \theta$, $dx = 2 \cos \theta \, d\theta$, and $\sin \theta = \frac{1}{2}(x - 1)$. Therefore

$$K = \int \frac{1}{3 + 2x - x^2} \, dx = \int \frac{1}{4 \cos^2 \theta} \cdot 2 \cos \theta \, d\theta = \frac{1}{2} \int \sec \theta \, d\theta = \frac{1}{2} \ln |\sec \theta + \tan \theta| + C.$$

A reference triangle with acute angle θ, opposite side $x - 1$, and hypotenuse 2 has adjacent side of length $\sqrt{3 + 2x - x^2}$. Thus

$$K = \frac{1}{2} \left| \frac{x + 1}{\sqrt{3 + 2x - x^2}} \right| + C = \frac{1}{4} \ln \left| \frac{(x + 1)^2}{(x + 1)(x - 3)} \right| + C = \frac{1}{4} \ln \left| \frac{x + 1}{x - 3} \right| + C.$$

7.7.15: $x^2 + 2x + 2 = (x + 1)^2 + 1 = 1 + \tan^2 \theta = \sec^2 \theta$ if $x + 1 = \tan \theta$; that is, $x = -1 + \tan \theta$, $dx = \sec^2 \theta \, d\theta$, and $\tan \theta = x + 1$. Therefore

$$\int \frac{2x - 5}{x^2 + 2x + 2} \, dx = \int \frac{-2 + 2 \tan \theta - 5}{\sec^2 \theta} \cdot \sec^2 \theta \, d\theta = 2 \ln |\sec \theta| - 7\theta + C$$

$$= 2 \ln \sqrt{x^2 + 2x + 2} - 7 \arctan(x + 1) + C = \ln(x^2 + 2x + 2) - 7 \arctan(x + 1) + C.$$

7.7.17: $5 + 12x - 9x^2 = -(9x^2 - 12x - 5) = -(9x^2 - 12x + 4 - 9) = 9 - (3x - 2)^2 = 9 - 9 \sin^2 u = 9 \cos^2 u$ if $3x - 2 = 3 \sin u$. So let $x = \frac{2}{3} + \sin u$, so that $dx = \cos u \, du$ and $\sin u = \frac{1}{3}(3x - 2)$. Then

$$\int \frac{x}{\sqrt{5 + 12x - 9x^2}} \, dx = \int \frac{\frac{2}{3} + \sin u}{3 \cos u} \cdot \cos u \, du = \int \left(\frac{2}{3} + \frac{1}{3} \sin u \right) du$$

$$= \frac{2}{9} u - \frac{1}{3} \cos u + C = \frac{2}{9} \arcsin \left(\frac{3x - 2}{3} \right) - \frac{1}{9} \sqrt{5 + 12x - 9x^2} + C.$$

7.7.19: $9 + 16x - 4x^2 = -(4x^2 - 16x - 9) = 25 - (2x - 4)^2 = 25 - 25\sin^2 u = 25\cos^2 u$ if $2x - 4 = 5\sin u$. So let $x = \frac{1}{2}(4 + 5\sin u)$. Then $dx = \frac{5}{2}\cos u\,du$ and $2\sin u = \frac{1}{5}(2x - 4)$. Thus

$$\int (7 - 2x)\sqrt{9 + 16x - 4x^2}\,dx = \int (7 - 4 - 5\sin u)(5\cos u)\cdot\frac{5}{2}\cos u\,du = \frac{1}{2}\int (3 - 5\sin u)(25\cos^2 u)\,du$$

$$= \frac{1}{2}\int (75\cos^2 u - 125\sin u\cos^2 u)\,du = \frac{25}{2}\int\left(3\cdot\frac{1 + \cos 2u}{2} - 5\sin u\cos^2 u\right)du$$

$$= \frac{25}{4}\int (3 + 3\cos 2u - 10\sin u\cos^2 u)\,du = \frac{25}{4}\left(3u + 3\sin u\cos u + \frac{10}{3}\cos^3 u\right) + C$$

$$= \frac{25}{4}\left[3\arcsin\left(\frac{2x - 4}{5}\right) + 3\cdot\frac{2x - 4}{5}\cdot\frac{\sqrt{9 + 16x - 4x^2}}{5} + \frac{10}{3}\cdot\frac{(9 + 16x - 4x^2)^{3/2}}{125}\right] + C$$

$$= \frac{75}{4}\arcsin\left(\frac{2x - 4}{5}\right) + \frac{3(x - 2)}{2}\sqrt{9 + 16x - 4x^2} + \frac{1}{6}(9 + 16x - 4x^2)\sqrt{9 + 16x - 4x^2} + C$$

$$= \frac{75}{4}\arcsin\left(\frac{2x - 4}{5}\right) + \left[\frac{9}{6}(x - 2) + \frac{1}{6}(9 + 16x - 4x^2)\right]\sqrt{9 + 16x - 4x^2} + C$$

$$= \frac{75}{4}\arcsin\left(\frac{2x - 4}{5}\right) + \frac{1}{6}\left(-4x^2 + 25x - 9\right)\sqrt{9 + 16x - 4x^2} + C$$

$$= \frac{75}{4}\arcsin\left(\frac{2x - 4}{5}\right) - \frac{4x^2 - 25x + 9}{6}\sqrt{9 + 16x - 4x^2} + C.$$

7.7.21: $6x - x^2 = -(x^2 - 6x) = 9 - (x^2 - 6x + 9) = 9 - (x - 3)^2 = 9 - 9\sin^2\theta = 9\cos^2\theta$ if $3\sin\theta = x - 3$, so we let $x = 3 + 3\sin\theta$. Then $dx = 3\cos\theta\,d\theta$, $\sin\theta = \frac{1}{3}(x - 3)$, and

$$I = \int\frac{x + 4}{(6x - x^2)^{3/2}}\,dx = \int\frac{7 + 3\sin\theta}{27\cos^3\theta}\cdot 3\cos\theta\,d\theta = \frac{1}{9}\int\frac{7 + 3\sin\theta}{\cos^2\theta}\,d\theta$$

$$= \frac{1}{9}\int (7\sec^2\theta + 3\sec\theta\tan\theta)\,d\theta = \frac{7}{9}\tan\theta + \frac{1}{3}\sec\theta + C$$

$$= \frac{7}{9}\cdot\frac{x - 3}{\sqrt{6x - x^2}} + \frac{1}{\sqrt{6x - x^2}} + C = \frac{7x - 12}{9\sqrt{6x - x^2}} + C.$$

To obtain the last line, we used a reference triangle with acute angle θ, opposite side $x - 3$, and hypotenuse 3, which therefore has adjacent side of length $\sqrt{6x - x^2}$.

7.7.23: $4x^2 + 12x + 13 = 4x^2 + 12x + 9 + 4 = (2x + 3)^2 + 4 = 4 + 4\tan^2 u = 4\sec^2 u$ if $2\tan u = 2x + 3$. Hence we let $x = \frac{1}{2}(-3 + 2\tan u)$, so that $dx = \sec^2 u\,du$ and $\tan u = \frac{1}{2}(2x + 3)$. Then

$$K = \int\frac{2x + 3}{(4x^2 + 12x + 13)^2}\,dx = \int\frac{2\tan u\sec^2 u}{16\sec^4 u}\,du$$

$$= \frac{1}{8}\int\tan u\cos^2 u\,du = \frac{1}{8}\int\sin u\cos u\,du = \frac{1}{16}\sin^2 u + C.$$

376

A reference triangle with acute angle u, opposite side $2x + 3$, and adjacent side 2 has hypotenuse of length $\sqrt{4x^2 + 12x + 13}$. Therefore

$$K = \frac{1}{16} \cdot \frac{(2x+3)^2}{4x^2 + 12x + 13} + C = \frac{1}{16} \cdot \frac{4x^2 + 12x + 13 - 4}{4x^2 + 12x + 13} + C$$

$$= \frac{1}{16}\left(1 - \frac{4}{4x^2 + 12x + 13}\right) + C = -\frac{1}{4(4x^2 + 12x + 13)} + C_1.$$

7.7.25: $x^2 + x + 1 = x^2 + x + \frac{1}{4} + \frac{3}{4} = \left(x + \frac{1}{2}\right)^2 + \frac{3}{4} = \frac{3}{4}\left[\frac{4}{3}\left(x + \frac{1}{2}\right)^2 + 1\right] = \frac{3}{4}\left(1 + \tan^2 u\right) = \frac{3}{4}\sec^2 u$
provided that

$$\tan u = \frac{2}{\sqrt{3}}\left(x + \frac{1}{2}\right).$$

Therefore we let

$$x = \frac{-1 + \sqrt{3}\,\tan u}{2} : \qquad dx = \frac{\sqrt{3}}{2}\sec^2 u\,du, \quad \tan u = \frac{\sqrt{3}}{2}\sec^2 u\,du.$$

Thus

$$J = \int \frac{3x - 1}{x^2 + x + 1}\,dx = \int \frac{\frac{1}{2}\left(-3 + 3\sqrt{3}\,\tan u\right) - 1}{\frac{3}{4}\sec^2 u} \cdot \frac{\sqrt{3}}{2}\sec^2 u\,du$$

$$= \frac{4}{3} \cdot \frac{\sqrt{3}}{2} \cdot \frac{1}{2}\int\left(-5 + 3\sqrt{3}\,\tan u\right)du = \frac{\sqrt{3}}{3}\left(-5u + 3\sqrt{3}\,\ln|\sec u|\right) + C.$$

A reference triangle with acute angle u, opposite side $2x + 1$, and adjacent side $\sqrt{3}$ has hypotenuse of length $2\sqrt{x^2 + x + 1}$. Therefore

$$J = \frac{\sqrt{3}}{3}\left[-5\arctan\left(\frac{\sqrt{3}}{3}[2x+1]\right) + 3\sqrt{3}\,\ln\left(\frac{2\sqrt{x^2 + x + 1}}{\sqrt{3}}\right)\right] + C$$

$$= -\frac{5\sqrt{3}}{3}\arctan\left(\frac{\sqrt{3}}{3}[2x+1]\right) + \frac{3}{2}\ln(x^2 + x + 1) + C_1.$$

7.7.27: The method of partial fractions yields

$$\frac{1}{(x^2 - 4)^2} = \frac{A}{x - 2} + \frac{B}{(x - 2)^2} + \frac{C}{x + 2} + \frac{D}{(x + 2)^2} = \frac{1}{32}\left[-\frac{1}{x - 2} + \frac{2}{(x - 2)^2} + \frac{1}{x + 2} + \frac{2}{(x + 2)^2}\right].$$

Therefore

$$I = \int \frac{1}{(x^2 - 4)^2}\,dx = \frac{1}{32}\ln\left|\frac{x + 2}{x - 2}\right| - \frac{x}{8(x^2 - 4)} + C.$$

Alternatively, if we let $x = 2\sec u$, then $x^2 - 4 = 4\sec^2 u - 4 = 4\tan^2 u$ and $dx = 2\sec u\tan u\,du$. Then

$$I = \int \frac{2\sec u\tan u}{16\tan^4 u}\,du = \frac{1}{8}\int \frac{\sec u}{\tan^3 u}\,du = \frac{1}{8}\int \frac{\cos^2 u}{\sin^3 u}\,du = \frac{1}{8}\int \frac{1 - \sin^2 u}{\sin^3 u}\,du = \frac{1}{8}\int\left(\csc^3 u - \csc u\right)du.$$

Then Formulas 15 and 29 from the endpapers of the text yield

$$I = \left(-\frac{1}{2} \csc u \cot u - \frac{1}{2} \ln|\csc u - \cot u| \right) + C = -\frac{1}{16} \left(\csc u \cot u + \ln|\csc u - \cot u| \right) + C.$$

A reference triangle with acute angle u, adjacent side 2, and hypotenuse side x has opposite side of length $\sqrt{x^2 - 4}$. Therefore

$$I = -\frac{1}{16} \left(\frac{2x}{x^2 - 4} + \ln \left| \frac{x - 2}{\sqrt{x^2 - 4}} \right| \right) + C$$

$$= -\frac{x}{8(x^2 - 4)} + \frac{1}{32} \ln \left| \frac{x^2 - 4}{(x - 2)^2} \right| + C = \frac{1}{32} \ln \left| \frac{x + 2}{x - 2} \right| - \frac{x}{8(x^2 - 4)} + C.$$

7.7.29: The partial fractions decomposition

$$\frac{x^2 + 1}{x(x^2 + x + 1)} = \frac{A}{x} + \frac{Bx + C}{x^2 + x + 1}$$

yields the equation $Ax^2 + Ax + A + Bx^2 + Cx = x^2 + 1$, so that

$$A + B = 1, \qquad A + C = 0, \qquad \text{and} \qquad A = 1.$$

It follows that $B = 0$ and $C = 1$. Hence

$$\frac{x^2 + 1}{x(x^2 + x + 1)} = \frac{1}{x} - \frac{1}{x^2 + x + 1}.$$

Now $x^2 + x + 1 = x^2 + x + \frac{1}{4} + \frac{3}{4} = \left(x + \frac{1}{2} \right)^2 + \frac{3}{4} = \frac{3}{4} \tan^2 u + \frac{3}{4} = \frac{3}{4} \sec^2 u$ if $\frac{1}{2}\sqrt{3} \tan u = x + \frac{1}{2}$, so we let

$$x = \frac{1}{2} \left(-1 + \sqrt{3} \tan u \right). \qquad \text{Then} \quad dx = \frac{1}{2}\sqrt{3} \sec^2 u \, du \quad \text{and} \quad \tan u = \frac{1}{3}\sqrt{3} \, (2x + 1).$$

Thus

$$\int \frac{1}{x^2 + x + 1} \, dx = \int \frac{\frac{1}{2}\sqrt{3} \sec^2 u}{\frac{3}{4} \sec^2 u} \, du = \frac{2u\sqrt{3}}{3} + C = \frac{2\sqrt{3}}{3} \arctan \left(\frac{\sqrt{3}}{3} [2x + 1] \right) + C.$$

Therefore

$$\int \frac{x^2 + 1}{x(x^2 + x + 1)} \, dx = \int \left(\frac{1}{x} - \frac{1}{x^2 + x + 1} \right) dx = \ln|x| - \frac{2\sqrt{3}}{3} \arctan \left(\frac{\sqrt{3}}{3} [2x + 1] \right) + C.$$

7.7.31: Because $x^4 - 2x^2 + 1 = (x^2 - 1)^2$, we let $x = \sec\theta$. Then $(x^2 - 1)^2 = (\sec^2\theta - 1)^2 = \tan^4\theta$ and $dx = \sec\theta \tan\theta \, d\theta$. Thus

$$K = \int \frac{2x^2 + 3}{x^4 - 2x^2 + 1} \, dx = \int \frac{3 + 2\sec^2\theta}{\tan^4\theta} \cdot \sec\theta \tan\theta \, d\theta = \int \left(\frac{3\sec\theta}{\tan^3\theta} + \frac{2\sec^3\theta}{\tan^3\theta} \right) d\theta$$

$$= \int \left(\frac{3\cos^2\theta}{\sin^3\theta} + \frac{2}{\sin^3\theta} \right) d\theta = \int \left(\frac{3(1 - \sin^2\theta)}{\sin^3\theta} + 2\csc^3\theta \right) d\theta = \int \left(5\csc^3\theta - 3\csc\theta \right) d\theta.$$

378

Formulas 15 and 29 of the endpapers of the text now yield

$$K = 5\left(-\frac{1}{2}\csc\theta\cot\theta + \frac{1}{2}\ln|\csc\theta - \cot\theta|\right) - 3\ln|\csc\theta - \cot\theta| + C$$

$$= -\frac{5}{2}\csc\theta\cot\theta - \frac{1}{2}\ln|\csc\theta - \cot\theta| + C.$$

A reference triangle with acute angle θ, adjacent side 1, and hypotenuse x has opposite side of length $\sqrt{x^2 - 1}$. Therefore

$$K = -\frac{5}{2}\cdot\frac{x}{x^2 - 1} - \frac{1}{2}\ln\left|\frac{x-1}{\sqrt{x^2-1}}\right| + C$$

$$= -\frac{5x}{2(x^2-1)} + \frac{1}{4}\ln\left|\frac{x^2-1}{(x-1)^2}\right| + C = \frac{1}{4}\ln\left|\frac{x+1}{x-1}\right| - \frac{5x}{2(x^2-1)} + C.$$

7.7.33: $x^2 + 2x + 5 = (x+1)^2 + 4 = 4\tan^2 u + 4 = 4\sec^2 u$ if $2\tan u = x + 1$, so we let $x = -1 + \tan u$. Then $dx = 2\sec^2 u\,du$ and $\tan u = \frac{1}{2}(x+1)$. Therefore

$$I = \int\frac{3x+1}{(x^2+2x+5)^2}\,dx = \int\frac{-3+6\tan u+1}{16\sec^4 u}\cdot 2\sec^2 u\,du = \frac{1}{8}\int(-2 + 6\tan u)\cos^2 u\,du$$

$$= \frac{1}{8}\int(6\sin u\cos u - 1 - \cos 2u)\,du = \frac{1}{8}(3\sin^2 u - u - \sin u\cos u) + C.$$

A reference triangle with acute angle u, opposite side $x + 1$, and adjacent side 2 has hypotenuse of length $\sqrt{x^2 + 2x + 5}$. Therefore

$$I = \frac{1}{8}\left[3\cdot\frac{(x+1)^2}{x^2+2x+5} - \arctan\left(\frac{x+1}{2}\right) - \frac{2(x+1)}{x^2+2x+5}\right] + C = \frac{3x^2+4x+1}{8(x^2+2x+5)} - \frac{1}{8}\arctan\left(\frac{x+1}{2}\right) + C.$$

7.7.35: The substitution $u = a\tan\theta$ entails $a^2 + u^2 = a^2\sec^2\theta$ and $du = a\sec^2\theta\,d\theta$. Thereby we find that

$$\int\frac{1}{(a^2+u^2)^n}\,du = \int\frac{a\sec^2\theta}{(a^2\sec^2\theta)^n}\,d\theta = \int\frac{a\sec^2\theta}{a^{2n}(\sec\theta)^{2n}}\,d\theta = \frac{1}{a^{2n-1}}\int(\cos\theta)^{2n-2}\,d\theta.$$

7.7.37: $x^2 - 2x + 5 = (x-1)^2 + 4 = 4 + 4\tan^2 u = 4\sec^2 u$ if $2\tan u = x - 1$, therefore we let $x = 1 + 2\tan u$. Then $dx = 2\sec^2 u\,du$ and $\tan u = \frac{1}{2}(x-1)$. Thus the area is

$$A = \int_0^5\frac{1}{x^2-2x+5}\,dx = \int_{x=0}^5\frac{2\sec^2 u}{4\sec^2 u}\,du = \int_{x=0}^5\frac{1}{2}\,du = \left[\frac{1}{2}u\right]_{x=0}^5 = \left[\frac{1}{2}\arctan\left(\frac{x-1}{2}\right)\right]_0^5$$

$$= \frac{1}{2}\left[\arctan 2 - \arctan\left(-\frac{1}{2}\right)\right] = \frac{1}{2}\left[\arctan 2 + \arctan\left(\frac{1}{2}\right)\right] = \frac{1}{2}\left(\arctan 2 + \frac{\pi}{2} - \arctan 2\right) = \frac{\pi}{4}.$$

7.7.39: $x^2 - 2x + 5 = (x-1)^2 + 4 = 4 + 4\tan^2 u = 4\sec^2 u$ if $2\tan u = x - 1$, therefore we let $x = 1 + 2\tan u$. Then $dx = 2\sec^2 u \, du$ and $\tan u = \frac{1}{2}(x-1)$. Thus the volume of revolution around the x-axis is

$$V = \pi \int_0^5 \frac{1}{(x^2 - 2x + 5)^2} \, dx = \pi \int_{x=0}^5 \frac{2\sec^2 u}{16\sec^4 u} \, du = \frac{\pi}{8} \int_{x=0}^5 \frac{1 + \cos 2u}{2} \, du = \frac{\pi}{16} \Big[u + \sin u \cos u \Big]_{x=0}^5.$$

A reference triangle with acute angle u, opposite side $x - 1$, and adjacent side 2 has hypotenuse of length $\sqrt{x^2 - 2x + 5}$. Therefore

$$V = \frac{\pi}{16} \left[\arctan\left(\frac{x-1}{2}\right) + \frac{2(x-1)}{x^2 - 2x + 5} \right]_0^5 = \frac{\pi}{16} \left[\arctan 2 - \arctan\left(-\frac{1}{2}\right) + \frac{8}{20} + \frac{2}{5} \right]$$

$$= \frac{\pi}{16} \left[\arctan 2 + \arctan\left(\frac{1}{2}\right) + \frac{4}{5} \right] = \frac{\pi}{16} \left(\frac{\pi}{2} + \frac{4}{5} \right) = \frac{5\pi^2 + 8\pi}{160} \approx 0.4655047702.$$

7.7.41: $4x^2 - 20x + 29 = (2x - 5)^2 + 4 = 4\tan^2 u + 4 = 4\sec^2 u$ if $2\tan u = 2x - 5$. Hence we let $x = \frac{5}{2} + \tan u$. Then $\tan u = \frac{1}{2}(2x - 5)$ and $dx = \sec^2 u \, du$. Therefore the volume generated by rotating the given region around the y-axis is

$$V = \int_1^4 \frac{2\pi x}{4x^2 - 20x + 29} \, dx = \pi \int_{x=1}^4 \frac{\frac{5}{2} + \tan u}{4\sec^2 u} \cdot \sec^2 u \, du = \frac{\pi}{2} \left[\frac{5}{2} u + \ln|\sec u| \right]_{x=1}^4.$$

A reference triangle with acute angle u, opposite side $2x - 5$, and adjacent side 2 has hypotenuse of length $\sqrt{4x^2 - 20x + 29}$. Therefore

$$V = \frac{\pi}{2} \left[\frac{5}{2} \arctan\left(\frac{2x-5}{2}\right) + \ln\left(\frac{\sqrt{4x^2 - 20x + 29}}{2}\right) \right]_1^4$$

$$= \frac{\pi}{2} \left[\frac{5}{2} \arctan\left(\frac{3}{2}\right) - \frac{5}{2} \arctan\left(-\frac{3}{2}\right) + \frac{1}{4} \ln(13) - \frac{1}{4} \ln(13) \right] = \frac{5\pi}{2} \arctan\left(\frac{3}{2}\right) \approx 7.7188438524.$$

7.7.43: Given $(4x + 4)^2 + (4y - 19)^2 = 377$, implicit differentiation yields

$$8(4x + 4) + 8(4y - 19) = 0, \qquad \text{so that} \qquad \frac{dy}{dx} = -\frac{4x + 4}{4y - 19}.$$

Thus

$$1 + \left(\frac{dy}{dx}\right)^2 = 1 + \frac{(4x + 4)^2}{(4y - 19)^2} = \frac{(4x + 4)^2 + (4y - 19)^2}{377 - (4x + 4)^2} = \frac{377}{377 - (4x + 4)^2}.$$

Therefore the length of the road is

$$L = \int_0^3 \frac{\sqrt{377}}{\sqrt{377 - (4x + 4)^2}} \, dx.$$

Now $377 - (4x + 4)^2 = 377 - 16(x + 1)^2 = 377 - 377\sin^2 u = 377\cos^2 u$ if $377\sin^2 u = 16(x + 1)^2$. So we let $x = -1 + \frac{1}{4}\left(\sqrt{377}\sin u\right)$. Then $dx = \frac{1}{4}\sqrt{377}\cos u \, du$, and this substitution yields

$$L = \int_{x=0}^3 \frac{\sqrt{377}}{\sqrt{377}\cos u} \cdot \frac{\sqrt{377}}{4}\cos u \, du = \int_{x=0}^3 \frac{\sqrt{377}}{4} \, du = \left[\frac{\sqrt{377}}{4} u \right]_{x=0}^3$$

$$= \left[\frac{\sqrt{377}}{4} \arcsin\left(\frac{4x + 4}{\sqrt{377}}\right) \right]_0^3 = \frac{\sqrt{377}}{4} \left[\left(\arcsin\frac{16}{\sqrt{377}} \right) - \arcsin\left(\frac{4}{\sqrt{377}}\right) \right].$$

It is easy to show that $\arcsin x - \arcsin y = \arcsin\left(x\sqrt{1-y^2} - y\sqrt{1-x^2}\right)$. Therefore (after a bit of arithmetic) you can also show that

$$L = \frac{\sqrt{377}}{4}\arcsin\left(\frac{260}{377}\right) \approx 3.6940487219 \quad \text{(miles)}.$$

For an alternative approach, the equation of the road may be put into the form

$$(x+1)^2 + \left(y - \tfrac{19}{4}\right)^2 = \frac{377}{16},$$

so the given curve joining $A(0, 0)$ with $B(3, 2)$ is an arc of a circle having center at $C\left(-1, \tfrac{19}{4}\right)$ and radius $\frac{1}{4}\sqrt{377}$. The straight line segment AB has length $\sqrt{13}$, so the law of cosines may be used to find the angle θ between the two radii CA and CB:

$$13 = \frac{377}{16} + \frac{377}{16} - 2\cdot\frac{377}{16}\cos\theta.$$

Hence the length of the circular arc AB is simply the product of the radius of the circle with the angle θ (in radians):

$$\frac{\sqrt{377}}{4}\arccos\left(\frac{273}{377}\right) \approx 3.6940487219 \quad \text{(miles)}.$$

7.7.45: The equation

$$\frac{3x+2}{(x-1)(x^2+2x+2)} = \frac{A}{x-1} + \frac{Bx+C}{x^2+2x+2}$$

leads to $A(x^2+2x+2) + B(x^2-x) + C(x-1) = 3x+12$, and thereby to

$$A + B = 0, \qquad 2A - B + C = 3, \qquad 2A - C = 2.$$

Thus

$$\frac{3x+2}{(x-1)(x^2+2x+2)} = \frac{1}{x-1} - \frac{x}{x^2+2x+2}.$$

Now $x^2 + 2x + 2 = (x+1)^2 + 1 = 1 + \tan^2\theta = \sec^2\theta$ provided that $x + 1 = \tan\theta$. Therefore we let $x = -1 + \tan\theta$, and thus $dx = \sec^2\theta\,d\theta$ and $\theta = \arctan(x+1)$. Hence

$$J = \int \frac{x}{(x+1)^2+1}\,dx = \int \frac{-1+\tan\theta}{\sec^2\theta}\cdot\sec^2\theta\,d\theta = -\theta + \ln|\sec\theta| + C.$$

A reference triangle with acute angle θ, opposite side $x+1$, and adjacent side 1 has hypotenuse of length $\sqrt{x^2+2x+2}$, and therefore

$$J = -\arctan(x+1) + \ln\sqrt{x^2+2x+2} + C = \frac{1}{2}\ln(x^2+2x+2) - \arctan(x+1) + C.$$

In conclusion,

$$\int \frac{3x+2}{x^3+x^2-2}\,dx = \ln|x-1| - \frac{1}{2}(x^2+2x+2) + \arctan(x+1) + C.$$

7.7.47: Division of denominator into numerator reveals that

$$\frac{x^4 + 2x^2}{x^3 - 1} = x + \frac{2x^2 + x}{(x-1)(x^2 + x + 1)}. \tag{1}$$

The partial fraction decomposition of the last term in Eq. (1) has the form

$$\frac{2x^2 + x}{(x-1)(x^2 + x + 1)} = \frac{A}{x - 1} + \frac{Bx + C}{x^2 + x + 1},$$

and it follows that $A(x^2 + x + 1) + B(x^2 - x) + C(x - 1) = 2x^2 + x$, and thus

$$A + B = 2, \quad A - B + C = 1, \quad \text{and} \quad A - C = 0.$$

Therefore $A = B = C = 1$, and so

$$\frac{2x^2 + x}{x^3 - 1} = \frac{1}{x - 1} + \frac{x + 1}{x^2 + x + 1}.$$

Now write

$$\frac{x + 1}{x^2 + x + 1} = \frac{1}{2} \cdot \frac{2x + 2}{x^2 + x + 1} = \frac{1}{2}\left(\frac{2x + 1}{x^2 + x + 1} + \frac{1}{x^2 + x + 1} \right).$$

Then

$$K = \int \frac{1}{x^2 + x + 1}\, dx = \int \frac{1}{\left(x + \frac{1}{2}\right)^2 + r^2}\, dx$$

where $r = \frac{1}{2}\sqrt{3}$. Let $u = x + \frac{1}{2}$. Then use integral formula 17 from the endpapers of the text:

$$K = \int \frac{1}{u^2 + r^2}\, du = \frac{1}{r} \arctan\left(\frac{u}{r} \right) + C = \frac{2\sqrt{3}}{3} \arctan\left(\frac{2x + 1}{\sqrt{3}} \right) + C.$$

Finally put all this work together to obtain

$$J = \frac{1}{2}x^2 + \ln|x - 1| + \frac{1}{2}\ln(x^2 + x + 1) + \frac{\sqrt{3}}{3}\arctan\left(\frac{2x + 1}{\sqrt{3}} \right) + C.$$

7.7.49: The trial factorization

$$x^4 + x^2 + 1 = (x^2 + ax + 1)(x^2 + bx + 1) = x^4 + (a + b)x^3 + (ab + 2)x^2 + (a + b)x + 1$$

yields $a + b = 0$ and $ab + 2 = 1$. Hence $b = -a$ and $ab = -1$, so that $a = \pm 1$ and $b = \mp 1$. Either way,

$$x^4 + x^2 + 1 = (x^2 + x + 1)(x^2 - x + 1).$$

The partial fractions decomposition

$$\frac{2x^3 + 3x}{x^4 + x^2 + 1} = \frac{Ax + B}{x^2 + x + 1} + \frac{Cx + D}{x^2 - x + 1}$$

382

then yields

$$A(x^3 - x^2 + x) + B(x^2 - x + 1) + C(x^3 + x^2 + x) + D(x^2 + x + 1) = 2x^3 + 3x,$$

and thus

$$A + C = 2, \qquad -A + B + C + D = 0,$$

$$A - B + C + D = 3, \qquad\qquad B + D = 0.$$

These equations are easily solved for $A = C = 1$, $B = -\frac{1}{2}$, and $D = \frac{1}{2}$. Therefore

$$\frac{2x^3 + 3x}{x^4 + x^2 + 1} = \frac{1}{2}\left(\frac{2x - 1}{x^2 + x + 1} + \frac{2x + 1}{x^2 - x + 1}\right).$$

Next, $x^2 + x + 1 = x^2 + x + \frac{1}{4} + \frac{3}{4} = \left(x + \frac{1}{2}\right)^2 + \frac{3}{4} = \frac{3}{4}\tan^2 u + \frac{32}{4} = \frac{3}{4}\sec^2 u$ if $x + \frac{1}{2} = \frac{1}{2}\sqrt{3}\,\tan u$. Therefore we let

$$x = -\frac{-1 + \sqrt{3}\,\tan u}{2}, \quad \text{and so} \quad dx = \frac{\sqrt{3}}{2}\sec^2 u\,du \quad \text{and} \quad \tan u = \frac{2x + 1}{\sqrt{3}}.$$

Thus

$$J_1 = \int \frac{2x - 1}{x^2 + x + 1}\,dx = \int \frac{-1 + \sqrt{3}\,\tan u - 1}{\frac{3}{4}\sec^2 u} \cdot \frac{\sqrt{3}}{2}\sec^2 u\,du$$

$$= \frac{2\sqrt{3}}{3}\int(-2 + \sqrt{3}\,\tan u)\,du = -\frac{4\sqrt{3}}{3}u + 2\ln|\sec u| + C.$$

A reference triangle with acute angle u, opposite side $2x + 1$, and adjacent side 3 has hypotenuse of length $2\sqrt{x^2 + x + 1}$. Therefore

$$J_1 = -\frac{4\sqrt{3}}{3}\arctan\left(\frac{2x + 1}{\sqrt{3}}\right) + 2\ln\left|\frac{2\sqrt{x^2 + x + 1}}{3}\right| + C = -\frac{4\sqrt{3}}{3}\arctan\left(\frac{2x + 1}{\sqrt{3}}\right) + \ln(x^2 + x + 1) + C_1.$$

Similarly, $x^2 - x + 1 = x^2 - x + \frac{1}{4} + \frac{3}{4} = \left(x - \frac{1}{2}\right)^2 + \frac{3}{4} = \frac{3}{4}\tan^2 v + \frac{3}{4} = \frac{3}{4}\sec^2 v$ if $x - \frac{1}{2} = \frac{1}{2}\sqrt{3}\,\tan v$. Therefore we let

$$x = \frac{1 + \sqrt{3}\,\tan v}{2}, \quad \text{so that} \quad dx = \frac{\sqrt{3}}{2}\sec^2 v\,dv \quad \text{and} \quad \tan v = \frac{2x - 1}{\sqrt{3}}.$$

Hence

$$J_2 = \int \frac{2x + 1}{x^2 - x + 1}\,dx = \int \frac{1 + \sqrt{3}\,\tan v + 1}{\frac{3}{4}\sec^2 v} \cdot \frac{\sqrt{3}}{2}\sec^2 v\,dv$$

$$= \frac{2\sqrt{3}}{3}\int(2 + \sqrt{3}\,\tan v)\,dv = \frac{2\sqrt{3}}{3}\left(2v + \sqrt{3}\,\ln|\sec v|\right) + C = \frac{4\sqrt{3}}{3}v + 2\ln|\sec v| + C.$$

A reference triangle with acute angle v, opposite side $2x-1$, and adjacent side $\sqrt{3}$ has hypotenuse of length $2\sqrt{x^2 - x + 1}$. Therefore

$$J_2 = \frac{4\sqrt{3}}{3} \arctan\left(\frac{2x-1}{\sqrt{3}}\right) + \ln(x^2 - x + 1) + C_2.$$

Thus

$$\int \frac{2x^3 + 3x}{x^4 + x^2 + 1}\, dx = \frac{1}{2}\left(J_1 + J_2\right) + C$$

$$= \frac{2\sqrt{3}}{3}\arctan\left(\frac{2x-1}{\sqrt{3}}\right) + \frac{1}{2}\ln(x^2 - x + 1) - \frac{2\sqrt{3}}{3}\arctan\left(\frac{2x+1}{\sqrt{3}}\right) + \frac{1}{2}\ln(x^2 + x + 1) + C.$$

Using the result in Problem 64 of Section 6.8, the antiderivative can be further simplified to

$$\frac{1}{2}\ln(x^4 + x^2 + 1) - \frac{2\sqrt{3}}{3}\arctan\left(\frac{\sqrt{3}}{2x^2 + 1}\right) + C.$$

7.7.51: *Mathematica* gives the partial fraction decomposition

$$\frac{7x^4 + 28x^3 + 50x^2 + 67x + 23}{(x-1)(x^2 + 2x + 2)^2} = \frac{A}{x-1} + \frac{Bx + C}{x^2 + 2x + 2} + \frac{Dx + E}{(x^2 + 2x + 2)^2} = \frac{7}{x-1} - \frac{6x - 5}{(x^2 + 2x + 2)^2}.$$

Next, $x^2 + 2x + 2 = (x+1)^2 + 1 = 1 + \tan^2 u = \sec^2 u$ if $x + 1 = \tan u$. Hence we let

$$x = -1 + \tan u, \quad \text{so that} \quad dx = \sec^2 u\, du.$$

Then

$$K = \int \frac{6x - 5}{(x^2 + 2x + 2)^2}\, dx = \int \frac{-6 + 6\tan u - 5}{\sec^4 u}\cdot \sec^2 u\, du$$

$$= \int (-11\cos^2 u + 6\sin u \cos u)\, du = -\frac{11}{2}(u + \sin u \cos u) + 3\sin^2 u + C.$$

A reference triangle with acute angle u, opposite side $x + 1$, and adjacent side 1 has hypotenuse of length $\sqrt{x^2 + 2x + 2}$. Therefore

$$K = -\frac{11}{2}\arctan(x+1) - \frac{11}{2}\cdot\frac{x+1}{x^2 + 2x + 2} + 3\cdot\frac{(x+1)^2}{x^2 + 2x + 2} + C.$$

Therefore

$$\int \frac{7x^4 + 28x^3 + 50x^2 + 67x + 23}{(x-1)(x^2 + 2x + 2)^2}\, dx = 7\ln|x - 1| + \frac{11}{2}\arctan(x+1) + \frac{11(x+1)}{2(x^2 + 2x + 2)} - \frac{3(x+1)^2}{x^2 + 2x + 2} + C.$$

Mathematica returns the antiderivative in the form

$$7\ln|x - 1| + \frac{11}{2}\arctan(x+1) + \frac{11x + 17}{2(x^2 + 2x + 2)}.$$

384

Ignoring C, the difference between the two answers is

$$\frac{11(x+1)}{2(x^2+2x+2)} - \frac{3(x^2+2x+1)}{x^2+2x+2} - \frac{11x+17}{2(x^2+2x+2)} = -3,$$

a constant.

7.7.53: Given:

$$h(x) = \frac{32x^5 + 16x^4 + 19x^3 - 98x^2 - 107x - 15}{(x^2 - 2x - 15)(4x^2 + 4x + 5)^2}.$$

First we factor: $x^2 - 2x - 15 = (x-5)(x+3)$. Then *Mathematica* yields the partial fraction decomposition

$$h(x) = \frac{A}{x-5} + \frac{B}{x+3} + \frac{Cx+D}{4x^2+4x+5} + \frac{Ex+F}{(4x^2+4x+5)^2} = \frac{7}{8(x-5)} + \frac{9}{8(x+3)} + \frac{3x-4}{(4x^2+4x+5)^2}.$$

Then $4x^2 + 4x + 5 = (2x+1)^2 + 4 = 4\tan^2 u + 4 = 4\sec^2 u$ if $2\tan u = 2x+1$, so we let

$$x = \frac{-1 + 2\tan u}{2}. \quad \text{Then} \quad dx = \sec^2 u \, du \quad \text{and} \quad \tan u = \frac{2x+1}{2}.$$

Hence

$$K = \int \frac{3x-4}{(4x^2+4x+5)^2} \, dx = \int \frac{-\frac{3}{2} + 3\tan u - 4}{16\sec^4 u} \cdot \sec^2 u \, du = \frac{1}{32} \int \frac{-11 + 6\tan u}{\sec^2 u} \, du$$

$$= \frac{1}{32} \int \left[-\frac{11}{2}(1 + \cos 2u) + 6\sin u \cos u \right] du = \frac{1}{64} \left(-11u - 11\sin u \cos u + 6\sin^2 u \right) + C.$$

A reference triangle with acute angle u, opposite side $2x+1$, and adjacent side 2 has hypotenuse of length $\sqrt{4x^2 + 4x + 5}$. Therefore

$$K = \frac{1}{64} \left[-11 \arctan\left(\frac{2x+1}{2}\right) - 11 \cdot \frac{2(2x+1)}{4x^2+4x+5} + \frac{6(2x+1)^2}{4x^2+4x+5} \right] + C$$

$$= -\frac{11}{64} \arctan\left(\frac{2x+1}{2}\right) - \frac{11}{32} \cdot \frac{2x+1}{4x^2+4x+5} + \frac{3}{32} \cdot \frac{(2x+1)^2}{4x^2+4x+5} + C.$$

Therefore

$$\int h(x) \, dx = \frac{7}{8} \ln|x-5| + \frac{9}{8} \ln|x+3| - \frac{11}{64} \arctan\left(\frac{2x+1}{2}\right) - \frac{11}{32} \cdot \frac{2x+1}{4x^2+4x+5} + \frac{3}{32} \cdot \frac{(2x+1)^2}{4x^2+4x+5} + C.$$

Mathematica returns for the antiderivative

$$\frac{7}{8} \ln|x-5| + \frac{9}{8} \ln|x+3| - \frac{11}{64} \arctan\left(\frac{2x+1}{2}\right) - \frac{22x+23}{32(4x^2+4x+5)}.$$

Ignoring C, the two answers differ by

$$\frac{22x+23}{32(4x^2+4x+5)} - \frac{22x+11}{32(4x^2+4x+5)} + \frac{3(4x^2+4x+1)}{32(4x^2+4x+5)} = \frac{12x^2+12x+15}{32(4x^2+4x+5)} = \frac{3}{32},$$

385

a constant.

7.7.55: *Mathematica* reports that the antiderivative is

$$\frac{2b - 5a + (b - 2a)x}{2(x^2 + 4x + 5)} + \frac{(b - 2a)\arctan(x + 2)}{2} + C.$$

The inverse tangent term will disappear if $b = 2a$. Hence choose $a \neq 0$ and $b = 2a$; then the antiderivative will becom

$$\frac{2b - 5a}{2(x^2 + 4x + 5)} + C = -\frac{a}{2(x^2 + 4x + 5)} + C.$$

7.7.57: *Mathematica* reports that the antiderivative is

$$2(b + c - 4a)\arctan(x + 1) + 2(11a - 4b + c)\arctan(x + 2)$$

$$-\frac{2a - 3b + 2c}{10} \cdot \left[\ln(x^2 + 2x + 2) - \ln(x^2 + 4x + 5)\right] + C.$$

In order for the inverse tangent and logarithmic terms to drop out, we impose the conditions

$$-4a + b + c = 0;$$

$$11a - 4b + c = 0;$$

$$2a - 3b + 2c = 0.$$

Mathematica reports that the only solution is $a = b = c = 0$. Thus there are *no* nonzero coefficients that yield a rational function as the antiderivative.

Section 7.8

7.8.1: The integral converges because

$$\int_2^\infty x^{-3/2}\, dx = \lim_{z \to \infty}\left[-2x^{-1/2}\right]_2^z = \frac{2}{\sqrt{2}} - \lim_{z \to \infty}\frac{2}{\sqrt{z}} = \sqrt{2}.$$

Computer algebra programs generally have little difficulty with improper integrals. The *Maple* command

```
int(x∧(−3/2), x=2..infinity);
```

and the *Mathematica* command

```
Integrate[ x∧(−3/2), { x, 2, Infinity } ];
```

386

both immediately produce the response $\sqrt{2}$. In *Derive* (which is menu-driven), the commands

> Author   [Enter]
>
> x∧(−3/2)   [Enter]
>
> Calculus   [Enter]
>
> Integrate   [Enter]
>
> Lower Limit   2   Upper Limit   inf   [Enter]

produce the display

$$\int_2^\infty x^{-3/2}\, dx;$$

then the command Evaluate produces the immediate result $\sqrt{2}$.

7.8.3: Divergent: $\displaystyle\int_0^4 x^{-3/2}\, dx = \lim_{z\to 0+}\left[2x^{-1/2}\right]_z^4 = -\frac{2}{\sqrt{4}} + \lim_{z\to 0+}\frac{2}{\sqrt{z}} = +\infty.$

7.8.5: Divergent: $\displaystyle\int_1^\infty \frac{1}{x+1}\, dx = \lim_{z\to\infty}\left[\ln(x+1)\right]_1^z = -\ln 2 + \lim_{z\to\infty}\ln(z+1) = +\infty.$

7.8.7: Convergent: $\displaystyle\int_5^\infty (x-1)^{-3/2}\, dx = \lim_{z\to\infty}\left[-2(x-1)^{-1/2}\right]_5^\infty = \frac{2}{\sqrt{4}} - \lim_{z\to\infty}\frac{2}{\sqrt{z-1}} = 1.$

7.8.9: Divergent: $\displaystyle\int_0^9 (9-x)^{-3/2}\, dx = \lim_{z\to 9-}\left[2(9-x)^{-1/2}\right]_0^z = -\frac{2}{\sqrt{9}} + \lim_{z\to 9-}\frac{2}{\sqrt{9-z}} = +\infty.$

7.8.11: Convergent: $\displaystyle\int_{-\infty}^{-2} \frac{1}{(x+1)^3}\, dx = \lim_{z\to -\infty}\left[-\frac{1}{2(x+1)^2}\right]_z^{-2} = -\frac{1}{2} + \lim_{z\to -\infty}\frac{1}{2(z+1)^2} = -\frac{1}{2}.$

7.8.13: Convergent: $\displaystyle\int_{-1}^8 x^{-1/3}\, dx = \lim_{z\to 8-}\left[\frac{3}{2}x^{2/3}\right]_{-1}^z = -\frac{3}{2} + \lim_{z\to 8-}\left(\frac{3}{2}z^{2/3}\right) = -\frac{3}{2} + 6 = \frac{9}{2}.$

7.8.15: Divergent: $\displaystyle\int_2^\infty (x-1)^{-1/3}\, dx = \lim_{z\to\infty}\left[\frac{3}{2}(x-1)^{2/3}\right]_2^z = +\infty.$

7.8.17: Divergent:

$$\int_{-\infty}^\infty \frac{x}{x^2+4}\, dx = \int_{-\infty}^0 \frac{x}{x^2+4}\, dx + \int_0^\infty \frac{x}{x^2+4}\, dx = \lim_{z\to -\infty}\left[\frac{1}{2}\ln(x^2+4)\right]_z^0 + \lim_{w\to\infty}\left[\frac{1}{2}\ln(x^2+4)\right]_0^w,$$

and neither of the last two limits exists.

7.8.19: Convergent: $\displaystyle\int_0^1 \frac{\exp(\sqrt{x})}{\sqrt{x}}\, dx = \lim_{z\to 0+}\left[2\exp(\sqrt{x})\right]_z^1 = 2e - 2 = 2(e-1).$

7.8.21: Convergent: $\displaystyle\int_0^\infty xe^{-3x}\,dx = \lim_{z\to\infty}\left[-\frac{3x+1}{9}e^{-3x}\right]_0^z = \frac{1}{9}.$ To find the antiderivative:

$$\text{Let}\quad u = x \quad\text{and}\quad dv = e^{-3x}\,dx.$$

$$\text{Then}\quad du = dx \quad\text{and}\quad v = -\frac{1}{3}e^{-3x}.\quad\text{Hence}$$

$$\int xe^{-3x}\,dx = -\frac{1}{3}xe^{-3x} + \int \frac{1}{3}e^{-3x}\,dx = -\frac{1}{3}xe^{-3x} - \frac{1}{9}e^{-3x} + C = -\frac{3x+1}{9}e^{-3x} + C.$$

7.8.23: Convergent: $\displaystyle\int_0^\infty x\exp\left(-x^2\right)\,dx = \lim_{z\to\infty}\left[-\frac{1}{2}\exp\left(-x^2\right)\right]_0^z = 0 - \left(-\frac{1}{2}\right) = \frac{1}{2}.$

7.8.25: Convergent: $\displaystyle\int_0^\infty \frac{1}{1+x^2}\,dx = \lim_{z\to\infty}\left[\arctan x\right]_0^z = \lim_{z\to\infty}\arctan z = \frac{\pi}{2}.$

7.8.27: $\displaystyle\int_0^{2n\pi}\cos x\,dx = 0$ if n is a positive integer, but

$$\int_0^{2n\pi+(\pi/2)}\cos x\,dx = \left[\sin x\right]_0^{2n\pi+(\pi/2)} = 1.$$

Therefore $\displaystyle\lim_{z\to\infty}\int_0^z\cos x\,dx$ does not exist; this improper integral diverges.

7.8.29: Divergent: $\displaystyle\int_1^\infty \frac{\ln x}{x}\,dx = \lim_{z\to\infty}\left[\frac{1}{2}(\ln x)^2\right]_1^z = \lim_{z\to\infty}\frac{1}{2}(\ln z)^2 = +\infty.$

(But the divergence is relatively slow: $\displaystyle\int_1^{10^9}\frac{\ln x}{x}\,dx \approx 214.7268734744.$)

7.8.31: Convergent: $\displaystyle\int_2^\infty \frac{1}{x(\ln x)^2}\,dx = \lim_{z\to\infty}\left[-\frac{1}{\ln x}\right]_2^z = \frac{1}{\ln 2}.$

7.8.33: Convergent: $\displaystyle\int_0^{\pi/2} \frac{\cos x}{\sqrt{\sin x}}\,dx = \lim_{z\to0^+}\left[2\sqrt{\sin x}\right]_z^{\pi/2} = 2 - 0 = 2.$

7.8.35: First note that—by l'Hôpital's rule—

$$\lim_{z\to0^+} z\ln z = \lim_{z\to0^+}\frac{\ln z}{\dfrac{1}{z}} = \lim_{z\to0^+}\frac{\dfrac{1}{z}}{-\dfrac{1}{z^2}} = \lim_{z\to0^+}(-z) = 0.$$

Therefore the given improper integral converges because

$$\int_0^1 \ln x\,dx = \lim_{z\to0^+}\left[-x+x\ln x\right]_z^1 = -1 + \lim_{z\to0^+}(z - z\ln z) = -1 + 0 - 0 = -1.$$

See Example 1 of Section 7.3 for the computation of the antiderivative.

7.8.37: Let $u = \ln x$ and $dv = \dfrac{1}{x^2}\,dx$. Then $du = \dfrac{1}{x}\,dx$; choose $v = -\dfrac{1}{x}$. Hence

$$\int \frac{\ln x}{x^2}\,dx = -\frac{1}{x}\ln x + \int \frac{1}{x^2}\,dx = -\frac{1}{x}\ln x - \frac{1}{x} + C.$$

Therefore the given improper integral diverges because

$$\int_0^1 \frac{\ln x}{x^2}\,dx = \lim_{z \to 0^+} \left[-\frac{1}{x}(1 + \ln x) \right]_z^1 = -1 + \left(\lim_{z \to 0^+} \frac{1 + \ln z}{z} \right) = -\infty.$$

(If you use l'Hôpital's rule on the last limit, you'll get the *wrong answer!*)

7.8.39: The first improper integral diverges because

$$\int_0^1 \frac{1}{x + x^2}\,dx = \int_0^1 \left(\frac{1}{x} - \frac{1}{x+1} \right) dx = \lim_{z \to 0^+} \left[\ln \left| \frac{x}{x+1} \right| \right]_z^1 = -\ln 2 - \left(\lim_{z \to 0^+} \ln \frac{z}{z+1} \right) = +\infty.$$

The second integral converges because

$$\int_1^\infty \frac{1}{x + x^2}\,dx = -\ln \frac{1}{2} + \left(\lim_{z \to \infty} \ln \frac{z}{z+1} \right) = 0 + \ln 2 = \ln 2.$$

7.8.41: Let $x = u^2$. Then $dx = 2u\,du$, and thus

$$\int \frac{1}{x^{1/2} + x^{3/2}}\,dx = \int \frac{2u}{x + x^2}\,du = \int \frac{2}{1 + u^2}\,du = 2 \arctan u + C = 2 \arctan \sqrt{x} + C.$$

Hence both improper integrals converge, because

$$\int_0^1 \frac{1}{x^{1/2} + x^{3/2}}\,dx = \lim_{z \to 0^+} \left[2 \arctan \sqrt{x} \right]_z^1 = 2 \cdot \frac{\pi}{4} - 2 \cdot 0 = \frac{\pi}{2}$$

and

$$\int_1^\infty \frac{1}{x^{1/2} + x^{3/2}}\,dx = \lim_{z \to \infty} \left[2 \arctan \sqrt{x} \right]_1^z = 2 \cdot \frac{\pi}{2} - 2 \cdot \frac{\pi}{4} = \frac{\pi}{2}.$$

7.8.43: If $k = 1$, then

$$\int_0^1 \frac{1}{x^k}\,dx = \lim_{z \to 0^+} \left[\ln x \right]_z^1 = +\infty.$$

If $k \neq 1$, then

$$\int_0^1 x^{-k}\,dx = \lim_{x \to 0^+} \left[\frac{x^{1-k}}{1-k} \right]_z^1 = \frac{1}{1-k} - \left(\lim_{z \to 0^+} \frac{z^{1-k}}{1-k} \right) = \begin{cases} \dfrac{1}{1-k} & \text{if } k < 1, \\[2mm] +\infty & \text{if } k > 1. \end{cases}$$

Therefore the given improper integral converges precisely when $k < 1$.

7.8.45: If $k = -1$, then

$$\int_0^1 x^k \ln x \, dx = \int_0^1 \frac{\ln x}{x} \, dx = \lim_{z \to 0^+} \left[\frac{1}{2} (\ln x)^2 \right]_z^1 = - \left[\lim_{z \to 0^+} \frac{1}{2} (\ln z)^2 \right] = -\infty.$$

If $k \neq -1$, then use integration by parts:

$$\text{Let} \quad u = \ln x \qquad \text{and} \quad dv = x^k \, dx;$$

$$\text{then} \quad du = \frac{1}{x} \, dx \qquad \text{and} \quad v = \frac{x^{k+1}}{k+1}.$$

Hence

$$\int x^k \ln x \, dx = \frac{x^{k+1} \ln x}{k+1} - \int \frac{x^k}{k+1} \, dx$$

$$= \frac{x^{k+1} \ln x}{k+1} - \frac{x^{k+1}}{(k+1)^2} + C = \frac{x^{k+1}}{(k+1)^2} \left[(k+1)(\ln x) - 1 \right] + C.$$

Therefore if $k \neq -1$, then

$$I = \int_0^1 x^k \ln x \, dx = \lim_{z \to 0^+} \left[\frac{x^{k+1}}{(k+1)^2} \left\{ (k+1)(\ln x) - 1 \right\} \right]_z^1$$

$$= -\frac{1}{(k+1)^2} - \lim_{z \to 0^+} \left[\frac{z^{k+1}}{(k+1)^2} \left\{ (k+1)(\ln z) - 1 \right\} \right].$$

Suppose first that $k < -1$. Write $k = -1 - \epsilon$ where $\epsilon > 0$. Then

$$I = \int_0^1 \frac{\ln x}{x^{1+\epsilon}} \, dx.$$

But $0 < x \leq 1$, so $x^{1+\epsilon} = x \cdot x^\epsilon \leq x$. Therefore

$$\frac{\ln x}{x^{1+\epsilon}} \geq \frac{\ln x}{x}.$$

Therefore I diverges to $-\infty$ if $k \leq -1$.

Now suppose that $k > -1$. Recall that

$$I = -\frac{1}{(k+1)^2} - \lim_{z \to 0^+} \left[\frac{z^{k+1}}{(k+1)^2} \left\{ (k+1)(\ln z) - 1 \right\} \right].$$

Now

$$\lim_{z \to 0^+} z^{k+1} \ln z = \lim_{z \to 0^+} z^\epsilon \ln z \qquad \text{(where } \epsilon > 0\text{)}$$

$$= \lim_{z \to 0^+} \frac{\ln z}{z^{-\epsilon}} = \lim_{z \to 0^+} \frac{1}{-\epsilon z^{-\epsilon-1} z} = -\frac{1}{\epsilon} \left(\lim_{z \to 0^+} \frac{1}{z^{-\epsilon}} \right) = - \left(\lim_{z \to 0^+} \frac{z^\epsilon}{\epsilon} \right) = 0.$$

Also,

$$\lim_{z \to 0^+} z^{k+1} = \lim_{z \to 0^+} z^\epsilon = 0.$$

Therefore $I = -\dfrac{1}{(k+1)^2}$ if $k > -1$.

Consequently $\displaystyle\int_0^1 x^k \ln x \, dx$ converges exactly when $k > -1$, and its value for such k is $-\dfrac{1}{(k+1)^2}$.

7.8.47: Given: $\Gamma(t) = \displaystyle\int_0^\infty x^{t-1} e^{-x} \, dx$ for $t > 0$. Thus

$$\Gamma(t+1) = \int_0^\infty x^t e^{-x} \, dx.$$

Let $u = x^t$ and $dv = e^{-x} \, dx$. Then $du = tx^{t-1} \, dx$ and $v = -e^{-x}$. Thus

$$\Gamma(t+1) = \left[-(x^t e^{-x}) \right]_{x=0}^{\infty} + t \int_0^\infty x^{t-1} e^{-x} \, dx = -\left(\lim_{z \to \infty} \frac{z^t}{e^z} \right) + 0 + t\Gamma(t) = t\Gamma(t).$$

To evaluate the last limit, we used the result in Problem 61 of Section 4.8.

7.8.49: The area is

$$A = \int_1^\infty \frac{1}{x} \, dx = \lim_{z \to \infty} \left[\ln x \right]_1^z = \lim_{z \to \infty} \ln z = +\infty.$$

7.8.51: Because $\dfrac{dy}{dx} = -\dfrac{1}{x^2}$, we have arc length element

$$ds = \left(1 + \frac{1}{x^4} \right)^{1/2} dx = \frac{\sqrt{x^4 + 1}}{x^2} \, dx.$$

Therefore the area of the surface of Gabriel's horn satisfies the inequality

$$S = \int_1^\infty 2\pi \cdot \frac{1}{x} \cdot \frac{\sqrt{x^4 + 1}}{x^2} \, dx = 2\pi \left(\lim_{z \to \infty} \int_1^z \frac{\sqrt{x^4 + 1}}{x^3} \, dx \right)$$

$$\geqq 2\pi \left(\lim_{z \to \infty} \int_1^z \frac{\sqrt{x^4}}{x^3} \, dx \right) = 2\pi \left(\lim_{z \to \infty} \left[\ln x \right]_1^z \right) = +\infty.$$

7.8.53: We will use the definition

$$\Gamma(t) = \int_{x-0}^\infty x^{t-1} e^{-x} \, dx$$

for $t > 0$ and the result of Problem 48 to the effect that $\Gamma(n+1) = n!$ if n is a positive integer. For fixed nonnegative integers m and n, let

$$J(m, n) = \int_0^1 x^m (\ln x)^n \, dx.$$

391

Then

$$J(m, 0) = \int_0^1 x^m \, dx = \frac{1}{m+1} = \frac{0!(-1)^0}{(m+1)^1} = \frac{n!(-1)^n}{(m+1)^{n+1}}$$

where $n = 0$. Therefore

$$J(m, n) = \frac{n!(-1)^n}{(m+1)^{n+1}}$$

if $n = 0$ and $m \geq 0$. Assume that

$$J(m, k) = \frac{k!(-1)^k}{(m+1)^{k+1}}$$

for some integer $k \geq 0$ and all integers $m \geq 0$. Then

$$J(m, k+1) = \int_0^1 x^m (\ln x)^{k+1} \, dx.$$

Let $u = (\ln x)^{k+1}$ and $dv = x^m \, dx$. Then

$$du = \frac{(k+1)(\ln x)^k}{x} \, dx \quad \text{and} \quad v = \frac{x^{m+1}}{m+1}.$$

Therefore

$$J(m, k+1) = \left[\frac{(\ln x)^{k+1} x^{m+1}}{m+1} \right]_0^1 - \frac{k+1}{m+1} \int_0^1 x^m (\ln x)^k \, dx.$$

The value of the evaluation bracket is zero because

$$\lim_{x \to 0^+} \frac{(\ln x)^{k+1}}{\frac{1}{x^{m+1}}} = \left(\lim_{x \to 0^+} \frac{\ln x}{x^{(m+1)/(k+1)}} \right)^{k+1} = \left(\lim_{x \to 0^+} \frac{k+1}{(m+1)x^{(m+1)/(k+1)}} \right)^{k+1} = 0^{k+1} = 0.$$

Therefore

$$J(m, k+1) = -\frac{k+1}{m+1} \cdot J(m, k) = -\frac{k+1}{m+1} \cdot \frac{k!(-1)^k}{(m+1)^{k+1}} = \frac{(-1)^{k+1}(k+1)!}{(m+1)^{k+2}}.$$

Therefore, by induction,

$$\int_0^1 x^m (\ln x)^n \, dx = \frac{n!(-1)^n}{(m+1)^{n+1}}$$

for all positive integers m and n.

7.8.55: We assume that $a > 0$. A short segment of the rod "at" position $x \geq 0$ and of length dx has mass $\delta \, dx$, and thereby exerts on m the force

$$\frac{Gm\delta}{(x+a)^2} \, dx.$$

Therefore the total force exerted by the rod on m is

$$F = \int_0^\infty \frac{Gm\delta}{(x+a)^2} \, dx = \left[-\frac{Gm\delta}{x+a} \right]_0^\infty = \frac{Gm\delta}{a}.$$

What if $a = 0$? Then the total force is

$$F = \int_0^\infty \frac{Gm\delta}{x^2}\, dx = \left[-\frac{Gm\delta}{x} \right]_0^\infty = \lim_{z \to 0^+} \frac{Gm\delta}{z} = +\infty.$$

You will obtain the same result if $a < 0$.

7.8.57: In the integral

$$\Gamma\left(\frac{1}{2}\right) = \int_0^\infty x^{-1/2} e^{-x}\, dx$$

we substitute $x = u^2$, so that $dx = 2u\, du$. Then

$$\Gamma\left(\frac{1}{2}\right) = \int_0^\infty \frac{1}{u} \cdot e^{-u^2} \cdot 2u\, du = 2\int_0^\infty e^{-u^2}\, du = 2\int_0^\infty e^{-x^2}\, dx.$$

7.8.59: The volume of revolution around the y-axis is

$$V = \int_0^\infty 2\pi x \exp\left(-x^2\right) dx = \pi\left[-\exp\left(-x^2\right) \right]_0^\infty = \pi.$$

7.8.61: Part (a): Suppose that $k > 1$ and let

$$I = \int_0^\infty x^k \exp\left(-x^2\right) dx.$$

Let $u = x^{k-1}$ and $dv = x\exp\left(-x^2\right) dx$. Then

$$du = (k-1)x^{k-2}\, dx \quad \text{and} \quad v = -\frac{1}{2}\exp\left(-x^2\right).$$

Therefore

$$I = \left[-\frac{x^{k-1}}{2}\exp\left(-x^2\right) \right]_0^\infty + \frac{k-1}{2}\int_0^\infty x^{k-2}\exp\left(-x^2\right) dx.$$

The evaluation bracket is zero by Problem 62 of Section 4.8. This concludes the proof in part (a).

Part (b): Now suppose that n is a positive integer. If $n = 1$, then

$$\int_0^\infty x^{n-1}\exp\left(-x^2\right) dx = \int_0^\infty \exp\left(-x^2\right) dx = \frac{1}{2}\Gamma\left(\frac{1}{2}\right)$$

by Eq. (9) of the text. Assume that for some integer $k \geqq 1$,

$$\int_0^\infty x^{k-1}\exp\left(-x^2\right) dx - \frac{1}{2}\Gamma\left(\frac{k}{2}\right)$$

and *in addition* that

$$\int_0^\infty x^{m-1}\exp\left(-x^2\right) dx = \frac{1}{2}\Gamma\left(\frac{m}{2}\right)$$

393

for every integer m such that $1 \leqq m \leqq k$. Then

$$\int_0^\infty x^k \exp\left(-x^2\right) dx = \frac{k-1}{2} \int_0^\infty x^{k-2} \exp\left(-x^2\right) dx$$

$$= \frac{k-1}{2} \cdot \frac{1}{2} \cdot \Gamma\left(\frac{k-1}{2}\right) = \frac{1}{2}\Gamma\left(1 + \frac{k-1}{2}\right) = \frac{1}{2}\Gamma\left(\frac{k+1}{2}\right).$$

Therefore, by induction,

$$\int_0^\infty x^{n-1} \exp\left(-x^2\right) dx = \frac{1}{2}\Gamma\left(\frac{n}{2}\right)$$

for every positive integer n.

7.8.63: Substitute $t = \dfrac{x}{\sqrt{2}}$ in the given interval. This routinely gives Eq. (10).

7.8.65: We defined

$$I_b = \int_0^b x^5 e^{-x} \, dx$$

and asked *Mathematica* to evaluate I_b for increasing large positive values of b. The results:

b	I_b (approximately)
10	111.949684454516
20	119.991370939137
30	119.999997291182
40	119.999999999505
50	119.999999999999933
60	119.99999999999999999258

It seems very likely that

$$k = \lim_{b \to \infty} \frac{1}{60} I_b = 2.$$

Indeed, because

$$\int x^5 e^{-x} \, dx = -(x^5 + 5x^4 + 20x^3 + 60x^2 + 120x + 120)e^{-x} + C,$$

it follows that

$$I_b = 120 - \frac{b^5 + 5b^4 + 20b^3 + 60b^2 + 120b + 12}{e^b} \to 120$$

as $b \to +\infty$, and this proves that $k = 2$.

7.8.67: We defined

$$k_b = \frac{\pi}{\left(\sqrt{2}\right) \cdot \int_0^b \frac{1}{x^2+2}\, dx}$$

and asked *Mathematica* to evaluate k_b for various large increasing values of b. Here are the results:

b	k_b (approximately)
10	2.1964469705919963
20	2.0941114933683164
40	2.0460327426191207
80	2.0227515961720504
160	2.0113173416565956
320	2.0056428162376127
640	2.0028174473301223
1280	2.0014077338319199
2560	2.0007036195035935
5120	2.0003517479044710
10240	2.0001758584911264

There is good evidence that $k = 2$. In fact, because

$$I_b = \int_0^b \frac{1}{x^2+2}\, dx = \left[\frac{1}{\sqrt{2}} \arctan\left(\frac{x}{\sqrt{2}}\right)\right]_0^b = \frac{1}{\sqrt{2}} \arctan\left(\frac{b}{\sqrt{2}}\right),$$

it follows that

$$\lim_{b \to \infty} \frac{\sqrt{2}}{\pi} I_b = \frac{\sqrt{2}}{\pi} \cdot \frac{1}{\sqrt{2}} \cdot \frac{\pi}{2} = \frac{1}{2},$$

and therefore $k = 2$ exactly.

7.8.69: We defined

$$k(b) = \frac{\pi}{e \cdot \int_0^b \exp\left(-x^2\right) \cos 2x\, dx}$$

and asked *Mathematica* to evaluate $k(b)$ for some large positive values of b. The results:

$$k(10) = k(100) \approx 3.5449077018110320545963349666 8229$$

395

to the number of decimal places shown. We conclude that there is no such integer k.

7.8.71: In the notation of this section, we have $\mu = 100$ and $\sigma = 15$. For part (a), we let $a = 10/\sigma$. Then

$$P = \frac{1}{\sqrt{2\pi}} \int_{-a}^{a} \exp\left(-\tfrac{1}{2}x^2\right) \, dx = \operatorname{erf}\left(\frac{\sqrt{2}}{3}\right) \approx 0.4950149249.$$

Thus just under 50% of students have IQs between 90 and 110. For part (b), we let $a = 25/\sigma$. Then

$$P = \frac{1}{\sqrt{2\pi}} \int_{a}^{\infty} \exp\left(-\tfrac{1}{2}x^2\right) \, dx = \frac{1}{2}\left[1 - \operatorname{erf}\left(\frac{5}{3\sqrt{2}}\right)\right] \approx 0.0477903523.$$

Thus just under 5% of students have IQs of 125 or higher.

7.8.73: Here we take $p = q = 1/2$ and $N = 900$; we have $\mu = Np = 450$ and $\sigma = \sqrt{Npq} = 15$. So we let $a = 25/\sigma = 5/3$ and compute

$$P = \frac{1}{\sqrt{2\pi}} \int_{-a}^{a} \exp\left(-\tfrac{1}{2}x^2\right) \, dx = \operatorname{erf}\left(\frac{5}{3\sqrt{2}}\right) \approx 0.9044192954.$$

Thus there is over a 90% probability of 425 to 475 heads. For part (b), we let $a = 50/\sigma = 10/3$ and compute

$$P = \frac{1}{\sqrt{2\pi}} \int_{a}^{\infty} \exp\left(-\tfrac{1}{2}x^2\right) \, dx = \frac{1}{2}\left[1 - \operatorname{erf}\left(\frac{5\sqrt{2}}{3}\right)\right] \approx 0.0004290603.$$

Thus there is only a probability of 0.04%—les than one chance in 2000—of 500 or more heads.

7.8.75: Here we have $p = q = 1/2$ and $N = 50$. Then $\mu = Np = 25$ and $\sigma = \sqrt{Npq} = 5/\sqrt{2}$. Part (a): We let $a = 5/\sigma = \sqrt{2}$ and compute

$$P = \frac{1}{\sqrt{2\pi}} \int_{a}^{\infty} \exp\left(-\tfrac{1}{2}x^2\right) \, dx = \frac{1}{2}\left[1 - \operatorname{erf}(1)\right] \approx 0.0786496035.$$

Thus there is slightly more than a 1 in 13 chance of passing by pure guessing. Part (b): We let $a = 10/\sigma$ and compute

$$P = \frac{1}{\sqrt{2\pi}} \int_{a}^{\infty} \exp\left(-\tfrac{1}{2}x^2\right) \, dx = \frac{1}{2}\left[1 - \operatorname{erf}(2)\right] \approx 0.0023388675.$$

Thus there is less than 1 chance in 425 of making a C by pure guessing.

7.8.77: Let $p = 0.55$, $q = 0.45$, and $N = 750$. In the notation of Section 7.8, we have $\mu = Np = 412.5$ and $\sigma = \sqrt{Npq} = \frac{3}{4}\sqrt{330} \approx 13.62443$. Now 59% of the $N = 750$ voters amounts to 442.5, so we let

$$a = \frac{442.5 - \mu}{\sigma} = \frac{4}{33}\sqrt{330} \approx 2.20183$$

and evaluate

$$P = \frac{1}{\sqrt{2\pi}} \int_{-a}^{a} \exp\left(-\tfrac{1}{2}x^2\right) \, dx = \operatorname{erf}\left(\frac{4}{33}\sqrt{165}\right) \approx 0.9723295720.$$

Thus there is a 97.23% probability that between 51% and 59% will say that they are Democratic voters, and thus that between 41% and 49% will say that they are Republican voters.

Chapter 7 Miscellaneous Problems

7.M.1: The substitution $x = u^2$, $dx = 2u\,du$ yields

$$\int \frac{1}{(1+x)\sqrt{x}}\,dx = \int \frac{2u}{(1+u^2)u}\,du = 2\arctan\sqrt{x} + C.$$

7.M.3: $\displaystyle\int \sin x \sec x\,dx = \int \frac{\sin x}{\cos x}\,dx = -\ln|\cos x| + C = \ln|\sec x| + C.$

7.M.5: $\displaystyle\int \frac{\tan\theta}{\cos^2\theta}\,d\theta = \int (\cos\theta)^{-3}\sin\theta\,d\theta = \frac{1}{2}(\cos\theta)^{-2} + C = \frac{1}{2}\sec^2\theta + C.$

7.M.7: Let $u = x$ and $dv = \tan^2 x\,dx = (\sec^2 x - 1)\,dx$. Then $du = dx$ and $v = -x + \tan x$. Thus

$$\int x\tan^2 x\,dx = x\tan x - x^2 + \int (x - \tan x)\,dx$$

$$= x\tan x - x^2 + \frac{1}{2}x^2 + \ln|\cos x| + C = x\tan x + \ln|\cos x| - \frac{1}{2}x^2 + C.$$

7.M.9: Let $u = x^3$ and $dv = x^2(2 - x^3)^{1/2}\,dx$. Then $du = 3x^2\,dx$ and $v = -\frac{2}{9}(2 - x^3)^{3/2}$. So

$$\int x^5(2 - x^3)^{1/2}\,dx = -\frac{2}{9}x^3(2 - x^3)^{3/2} + \frac{2}{3}\int x^2(2 - x^3)^{3/2}\,dx = -\frac{2}{9}x^3(2 - x^3)^{3/2} - \frac{4}{45}(2 - x^3)^{5/2} + C.$$

7.M.11: Let $x = 5\tan u$. Then $25 + x^2 = 25 + 25\tan^2 u = 25\sec^2 u$ and $dx = 5\sec^2 u\,du$. Thus

$$K = \int \frac{x^2}{\sqrt{25 + x^2}}\,dx = \int \frac{125\tan^2 u \sec^2 u}{5\sec u}\,du$$

$$= 25\int (\sec^3 u - \sec u)\,du = 25\left(\frac{1}{2}\sec u \tan u - \frac{1}{2}\ln|\sec u + \tan u|\right) + C.$$

The antiderivative is a consequence of integral formulas 14 and 28 of the endpapers of the text. Next, a reference triangle with acute angle u, opposite side x, and adjacent side 5 has hypotenuse of length $\sqrt{25 - x^2}$. Therefore

$$K = \frac{25}{2}\left[\frac{x\sqrt{25 + x^2}}{25} - \ln\left(\frac{x + \sqrt{25 + x^2}}{5}\right)\right] + C = \frac{1}{2}x\sqrt{25 + x^2} - \frac{25}{2}\ln\left(x + \sqrt{25 + x^2}\right) + C_1.$$

When we simplify an answer by allowing a constant such as $\frac{25}{2}\ln 5$ to be absorbed by the constant C of integration, we will generally indicate this by replacing C with C_1, as in this solution.

7.M.13: Complete the square:

$$x^2 - x + 1 = x^2 - x + \frac{1}{4} + \frac{3}{4} = \left(x - \frac{1}{2}\right)^2 + \frac{3}{4} = \frac{3}{4} + \frac{3}{4}\tan^2 u = \frac{3}{4}\sec^2 u$$

if $x - \frac{1}{2} = \frac{\sqrt{3}}{2}\tan u$. So we let

$$x = \frac{1 + \sqrt{3}\,\tan u}{2}; \quad dx = \frac{\sqrt{3}}{2}\sec^2 u\,du \quad \text{and} \quad \tan u = \frac{2x - 1}{\sqrt{3}}.$$

Therefore

$$J = \int \frac{1}{x^2 - x + 1}\,dx = \frac{\sqrt{3}}{2}\int \frac{\sec^2 u}{\frac{3}{4}\sec^2 u}\,du = \frac{2\sqrt{3}}{3}u + C.$$

. A reference triangle with acute angle u, opposite side $2x - 1$, and adjacent side $\sqrt{3}$ has hypotenuse of length $2\sqrt{x^2 - x + 1}$. Therefore

$$J = \frac{2\sqrt{3}}{3}\arctan\left(\frac{\sqrt{3}}{3}[2x - 1]\right) + C.$$

7.M.15: Given: $\displaystyle\int \frac{5x + 31}{3x^2 - 4x + 11}\,dx$. First complete the square in the denominator:

$$3x^2 - 4x + 11 = \frac{1}{3}(9x^2 - 12x + 33) = \frac{1}{3}\left([3x - 2]^2 + 29\right) = \frac{1}{3}\left(29\tan^2\theta + 29\right) = \frac{29}{3}\sec^2\theta$$

if $3x - 2 = \sqrt{29}\,\tan\theta$. So let

$$x = \frac{2 + \sqrt{29}\,\tan\theta}{3}; \quad \text{then} \quad dx = \frac{\sqrt{29}}{3}\sec^2\theta \quad \text{and} \quad \tan\theta = \frac{3x - 2}{\sqrt{29}}.$$

Then

$$I = \int \frac{\frac{10}{3} + \frac{5}{3}\sqrt{29}\,\tan\theta + \frac{93}{3}}{\frac{29}{3}\sec^2\theta} \cdot \frac{\sqrt{29}}{3}\sec^2\theta\,d\theta$$

$$= \frac{\sqrt{29}}{87}\int \left(103 + 5\sqrt{29}\,\tan\theta\right)d\theta = \frac{\sqrt{29}}{87}\left(103\theta + 5\sqrt{29}\,\ln|\sec\theta|\right) + C.$$

A reference triangle with acute angle θ, opposite side $3x - 2$, and adjacent side $\sqrt{29}$ has hypotenuse of length $\sqrt{9x^2 - 12x + 33}$. Therefore

$$I = \frac{\sqrt{29}}{87} \cdot 103 \cdot \arctan\left(\frac{3x - 2}{\sqrt{29}}\right) + \frac{5 \cdot 29}{87}\ln\left|\frac{\sqrt{3}\,\sqrt{3x^2 - 4x + 11}}{\sqrt{29}}\right| + C$$

$$= \frac{\sqrt{29}}{87} \cdot 103 \cdot \arctan\left(\frac{3x - 2}{\sqrt{29}}\right) + \frac{5}{3}\ln\left(\sqrt{3x^2 - 4x + 11}\right) + C_1$$

$$= \frac{103\sqrt{29}}{87}\arctan\left(\frac{3x - 2}{\sqrt{29}}\right) + \frac{5}{6}\ln(3x^2 - 4x + 11) + C_1.$$

7.M.17: $\int (x^4 + x^7)^{1/2}\, dx = \int x^2 (1 + x^3)^{1/2}\, dx = \dfrac{2}{9}(1 + x^3)^{3/2} + C.$

7.M.19: We use integral formula 16 of the endpapers and the substitution $u = \sin x$, $du = \cos x\, dx$:

$$\int \frac{\cos x}{\sqrt{4 - \sin^2 x}}\, dx = \int \frac{1}{\sqrt{4 - u^2}}\, du = \arcsin\left(\frac{u}{2}\right) + C = \arcsin\left(\frac{\sin x}{2}\right) + C.$$

7.M.21: Let $u = \ln(\cos x)$. Then $du = -\dfrac{\sin x}{\cos x}\, dx = -\tan x\, dx$, and thus

$$\int \frac{\tan x}{\ln(\cos x)}\, dx = -\int \frac{1}{u}\, du = -\ln|u| + C = -\ln|\ln(\cos x)| + C.$$

7.M.23: Let $u = \ln(1 + x)$, $dv = dx$. Then $du = \dfrac{1}{1 + x}\, dx$, and we let $v = 1 + x$. Then

$$\int \ln(1 + x)\, dx = (1 + x)\ln(1 + x) - \int 1\, dx = (1 + x)\ln(1 + x) - x + C.$$

The choice $v = x$ will produce an answer that appears different: $x\ln(1 + x) - x + \ln(1 + x) + C.$

7.M.25: Let $x = 3\tan u$: $dx = 3\sec^2 u\, du$ and $\sqrt{x^2 + 9} = \sqrt{9\tan^2 u + 9} = 3\sec u$. Thus

$$K = \int \sqrt{x^2 + 9}\, dx = \int 9\sec^3 u\, du = \frac{9}{2}\left(\sec u \tan u + \ln|\sec u + \tan u|\right) + C.$$

A reference triangle with acute angle u, opposite side x, and adjacent side 3 has hypotenuse of length $\sqrt{x^2 + 9}$. Thus

$$K = \frac{9}{2}\left[\frac{x\sqrt{x^2 + 9}}{9} + \ln\left(\frac{x + \sqrt{x^2 + 9}}{3}\right)\right] + C = \frac{1}{2}x\sqrt{x^2 + 9} + \frac{9}{2}\ln\left(x + \sqrt{x^2 + 9}\right) + C_1.$$

A hyperbolic substitution will yield the antiderivative in the form $K = \dfrac{1}{2}x\sqrt{x^2 + 9} + \dfrac{9}{2}\sinh^{-1}\left(\dfrac{x}{3}\right) + C.$

7.M.27: Note that $2x - x^2 = -(x^2 - 2x) = 1 - (x - 1)^2 = 1 - \sin^2 u = \cos^2 u$ if $x - 1 = \sin u$. Hence we let $x = 1 + \sin u$, so that $dx = \cos u\, du$. Then

$$I = \int \sqrt{2x - x^2}\, dx = \int \cos^2 u\, du = \frac{1}{2}\int (1 + \cos 2u)\, du = \frac{1}{2}(u + \sin u \cos u) + C.$$

A reference triangle with acute angle u, opposite side $x - 1$, and hypotenuse 1 has adjacent side of length $\sqrt{2x - x^2}$. Therefore

$$I = \frac{1}{2}\arcsin(x - 1) + \frac{1}{2}(x - 1)\sqrt{2x - x^2} + C.$$

Mathematica returns an answer that differs only in that $\arcsin(x - 1)$ is replaced with $-\arcsin(1 - x)$.

7.M.29: First divide denominator into numerator to obtain

$$\frac{x^4}{x^2-2} = x^2 + 2 + \frac{4}{x^2-2}.$$

Then the partial fractions decomposition of the last term yields

$$\frac{4}{x^2-2} = \frac{A}{x+\sqrt{2}} + \frac{B}{x-\sqrt{2}}$$

and thus the equation $A\left(x-\sqrt{2}\right) + B\left(x+\sqrt{2}\right) = 4$. Consequently

$$A+B=0 \quad \text{and} \quad -A\sqrt{2} + B\sqrt{2} = 4,$$

and it follows that $A = -\sqrt{2}$ and $B = \sqrt{2}$. Therefore

$$\int \frac{x^4}{x^2-2}\,dx = \frac{1}{3}x^3 + 2x - \sqrt{2}\,\ln\left|x+\sqrt{2}\right| + \sqrt{2}\,\ln\left|x-\sqrt{2}\right| + C = \frac{1}{3}x^3 + 2x + \sqrt{2}\,\ln\left|\frac{x-\sqrt{2}}{x+\sqrt{2}}\right| + C.$$

Mathematica and *Maple* V version apparently prefer hyperbolic substitutions to the method of partial fractions; they both yield instead the equivalent answer

$$2x + \frac{x^3}{3} - 2\sqrt{2}\,\tanh^{-1}\left(\frac{x}{\sqrt{2}}\right)$$

(remember that most computer algebra programs omit the "$+\,C$").

7.M.31: First, $x^2 + 2x + 2 = 1 + (x+1)^2 = 2 + \tan^2 u = \sec^2 u$ if $\tan u = x+1$. Hence we let $x = -1 + \tan u$, so that $dx = \sec^2 u\,du$ and

$$J = \int \frac{x}{(x^2+2x+2)^2} = \int \frac{-1+\tan u}{\sec^4 u}\cdot \sec^2 u\,du = \int\left(-\cos^2 u + \sin u\,\cos u\right)du$$

$$= \int\left(\sin u\,\cos u - \frac{1+\cos 2u}{2}\right)du = \frac{1}{2}\sin^2 u - \frac{1}{2}u - \frac{1}{2}\sin u\,\cos u + C.$$

Then a reference triangle with acute angle u, opposite side $x+1$, and adjacent side 1 has hypotenuse of length $\sqrt{x^2+x+2}$. Therefore

$$J = \frac{1}{2}\cdot\frac{(x+1)^2}{x^2+2x+2} - \frac{1}{2}\arctan(x+1) - \frac{1}{2}\cdot\frac{x+1}{x^2+2x+2} + C = \frac{x^2+x}{2(x^2+2x+2)} - \frac{1}{2}\arctan(x+1) + C.$$

Mathematica, *Derive*, and *Maple* all yield instead the equivalent result

$$J = -\frac{x+2}{2(x^2+x+2)} - \frac{1}{2}\arctan(x+1) + C.$$

7.M.33: We use the identity $\cos^2 \theta = \dfrac{1 + \cos 2\theta}{2}$:

$$\int \frac{1}{1 + \cos 2\theta}\, d\theta = \frac{1}{2} \int \frac{2}{1 + \cos 2\theta}\, d\theta = \frac{1}{2} \int \frac{1}{\cos^2 \theta}\, d\theta = \frac{1}{2} \int \sec^2 \theta\, d\theta = \frac{1}{2} \tan \theta + C.$$

7.M.35: $\displaystyle \int \sec^3 x \tan^3 x\, dx = \int (\sec^2 x)(\sec^2 x - 1) \sec x \tan x\, dx = \frac{1}{5} \sec^5 x - \frac{1}{3} \sec^3 x + C.$

7.M.37: It's almost always wise to develop a reduction formula for problems of this sort. Suppose that n is a positive integer. Let

$$I_n = \int x(\ln x)^n\, dx.$$

Integration by parts: Let $u = (\ln x)^n$ and $dv = x\, dx$. Then

$$du = \frac{n(\ln x)^{n-1}}{x}\, dx \quad \text{and} \quad v = \frac{1}{2} x^2.$$

Therefore

$$I_n = \int x(\ln x)^n\, dx = \frac{1}{2} x^2 (\ln x)^n - \frac{n}{2} \int x(\ln x)^{n-1}\, dx.$$

And thus

$$\begin{aligned}
I_3 = \int x(\ln x)^3\, dx &= \frac{1}{2} x^2 (\ln x)^3 - \frac{3}{2} \int x(\ln x)^2\, dx \\[2mm]
&= \frac{1}{2} x^2 (\ln x)^3 - \frac{3}{2} \left[\frac{1}{2} x^2 (\ln x)^2 - \int x(\ln x)\, dx \right] \\[2mm]
&= \frac{1}{2} x^2 (\ln x)^3 - \frac{3}{4} x^2 (\ln x)^2 + \frac{3}{2} \left[\frac{1}{2} x^2 \ln x - \frac{1}{2} \int x\, dx \right] \\[2mm]
&= \frac{1}{2} x^2 (\ln x)^3 - \frac{3}{4} x^2 (\ln x)^2 + \frac{3}{4} x^2 \ln x - \frac{3}{8} x^2 + C \\[2mm]
&= \frac{x^2}{8} \left[4(\ln x)^3 - 6(\ln x)^2 + 6 \ln x - 3 \right] + C.
\end{aligned}$$

7.M.39: Let $u = e^x$. Then $du = e^x\, dx$ and $x = \ln u$. Therefore

$$K = \int e^x \sqrt{1 + e^{2x}}\, dx = \int \sqrt{1 + u^2}\, du.$$

Now let $u = \tan \theta$. Then $du = \sec^2 \theta\, d\theta$ and $1 + u^2 = \sec^2 \theta$. So

$$K = \int \sec^3 \theta\, d\theta = \frac{1}{2} \Big(\sec \theta \tan \theta + \ln |\sec \theta + \tan \theta| \Big) + C.$$

Then a reference triangle with acute angle θ, opposite side u, and adjacent side 1 has hypotenuse of length $\sqrt{1 + u^2}$. Hence

$$K = \frac{1}{2} \left[u\sqrt{1 + u^2} + \ln \left(u + \sqrt{1 + u^2} \right) \right] + C = \frac{1}{2} e^x \sqrt{1 + e^{2x}} + \frac{1}{2} \ln \left(e^x + \sqrt{1 + e^{2x}} \right) + C.$$

Mathematica, with its well-known penchant for using hyperbolic functions, returns the equivalent

$$K = \frac{1}{2} e^x \sqrt{1 + e^{2x}} + \frac{1}{2} \sinh^{-1}(e^x) + C.$$

Maple yields the same antiderivative, except that *Maple* writes $(e^x)^2$ instead of e^{2x}.

7.M.41: Let $x = 3 \sec u$. Then $\sqrt{x^2 - 9} = \sqrt{9 \sec^2 u - 9} = \sqrt{9 \tan^2 u} = 3 \tan u$; moreover, $dx = 3 \sec u \tan u \, du$. A reference triangle for this substitution has acute angle u, hypotenuse x, and adjacent side 3, thus opposite side of length $\sqrt{x^2 - 9}$. Therefore

$$\int \frac{1}{x^3 \sqrt{x^2 - 9}} \, dx = \int \frac{3 \sec u \tan u}{(27 \sec^3 u)(3 \tan u)} \, du = \frac{1}{27} \int \frac{1 + \cos 2u}{2} = \frac{1}{54}(u + \sin u \cos u) + C$$

$$= \frac{1}{54} \left[\operatorname{arcsec}\left(\frac{x}{3}\right) + \frac{3\sqrt{x^2 - 9}}{x^2} \right] + C = \frac{1}{54} \operatorname{arcsec}\left(\frac{x}{3}\right) + \frac{\sqrt{x^2 - 9}}{18x^2} + C.$$

There is a technical point that we have glossed over too many times not to mention. In our substitution, $\sqrt{9 \tan^2 u} = 3 \tan u$ is true only if $\tan u \geq 0$. Nevertheless, our antiderivative is correct for all x such that $|x| > 3$, including values of x for which $\tan u < 0$. We confess to a pragmatic approach to such problems. If the substitution is "legal" only for certain values of the variables, use it anyway. Find the antiderivative, then verify its validity for *all* meaningful values of the variable by differentiation. Almost always, you will find that an antiderivative valid for an interval of values of the variables is valid for all meaningful values of the variables.

    Maple returns for the antiderivative the equivalent

$$\int \frac{1}{x^3 \sqrt{x^2 - 9}} \, dx = \frac{\sqrt{-9 + x^2}}{18x^2} - \frac{1}{54} \arctan\left(\frac{3}{\sqrt{-9 + x^2}}\right) + C,$$

as does *Mathematica*.

7.M.43: The partial fractions decomposition of the integrand has the form

$$\frac{4x^2 + x + 1}{4x^3 + x} = \frac{A}{x} + \frac{Bx + C}{4x^2 + 1},$$

which leads to $A(4x^2 + 1) + Bx^2 + Cx = 4x^2 + x + 1$, and thereby to the simultaneous equations

$$4A + B = 4, \quad C = 1, \quad A = 1$$

which are easily solved for $A = 1$, $B = 0$, and $C = 1$. Therefore

$$\int \frac{4x^2 + x + 1}{4x^3 + x} \, dx = \int \left(\frac{1}{x} + \frac{1}{4x^2 + 1}\right) dx = \ln|x| + \frac{1}{2} \arctan(2x) + C.$$

The easiest way to find the antiderivative of the second fraction is by judicious guesswork.

7.M.45: $\int \tan^2 x \sec x \, dx = \int (\sec^3 x - \sec x) \, dx = \frac{1}{2} \left(\sec x \tan x - \ln|\sec x + \tan x| \right) + C.$

7.M.47: The partial fractions decomposition

$$\frac{x^4 + 2x + 2}{x^5 + x^4} = \frac{A}{x} + \frac{B}{x^2} + \frac{C}{x^3} + \frac{D}{x^4} + \frac{E}{x+1}$$

yields the equation $A(x^4 + x^3) + B(x^3 + x^2) + C(x^2 + x) + D(x+1) + Ex^4 = x^4 + 2x + 2$, and thereby the simultaneous equations

$$A + E = 1,$$

$$A + B = 0,$$

$$B + C = 0,$$

$$C + D = 2,$$

$$D = 2.$$

These equations are easy to solve "from the bottom up," and you'll find that $D = 2$, $C = 0$, $B = 0$, $A = 0$, and $E = 1$. Therefore

$$\int \frac{x^4 + 2x + 2}{x^5 + x^4} \, dx = \int \left(\frac{2}{x^4} + \frac{1}{x+1} \right) dx = \ln|x+1| - \frac{2}{3x^3} + C.$$

7.M.49: The partial fractions decomposition of the integrand is

$$\frac{3x^5 - x^4 + 2x^3 - 12x^2 - 2x + 1}{(x^3 - 1)^2} = \frac{A}{x-1} + \frac{B}{(x-1)^2} + \frac{Cx+D}{x^2+x+1} + \frac{Ex+F}{(x^2+x+1)^2},$$

and thus

$$A(x-1)(x^4 + 2x^3 + 3x^2 + 2x + 1) + B(x^4 + 2x^3 + 3x^2 + 2x + 1)$$

$$+ (Cx+D)(x^4 - x^3 - x + 1) + (Ex+F)(x^2 - 2x + 1) = 3x^5 - x^4 + 2x^3 - 12x^2 - 2x + 1.$$

Thus we obtain the following simultaneous equations:

$$
\begin{array}{rcrcrcrcrcrcr}
A & & & +C & & & & & & = & 3, \\
A & + & B & - & C & + & D & & & = & -1, \\
A & + & 2B & & & - & D & + & E & = & 2, \\
-A & + & 3B & - & C & & & - & 2E & + & F & = & -12, \\
-A & + & 2B & + & C & - & D & + & E & - & 2F & = & -2, \\
-A & + & B & & & + & D & & & + & F & = & 1.
\end{array}
$$

403

It follows that $A = 1$, $B = -1$, $C = 2$, $D = 1$, $E = 4$, and $F = 2$. Hence

$$\frac{3x^5 - x^4 + 2x^3 - 12x^2 - 2x + 1}{(x^3 - 1)^2} = \frac{1}{x - 1} - \frac{1}{(x - 1)^2} + \frac{2x + 1}{x^2 + x + 1} + \frac{2(2x + 1)}{(x^2 + x + 1)^2},$$

and therefore the required antiderivative is

$$\ln|x - 1| + \frac{1}{x - 1} + \ln(x^2 + x + 1) - \frac{2}{x^2 + x + 1} + C.$$

7.M.51: Another problem in which it is probably wise to develop a reduction formula. If n is a positive integer, let

$$I_n = \int (\ln x)^n \, dx.$$

Then use integration by parts with $u = (\ln x)^n$ and $dv = dx$. Then

$$du = \frac{n(\ln x)^{n-1}}{x} \, dx; \quad \text{choose} \quad v = x.$$

Hence

$$I_n = x(\ln x)^n - n \int (\ln x)^{n-1} \, dx = x(\ln x)^n - nI_{n-1}.$$

So

$$\int (\ln x)^6 \, dx = x(\ln x)^6 - 6 \int (\ln x)^5 \, dx$$

$$\doteq x(\ln x)^6 - 6 \left[x(\ln x)^5 - 5 \int (\ln x)^4 \, dx \right]$$

$$= x(\ln x)^6 - 6x(\ln x)^5 + 30 \left[x(\ln x)^4 - 4 \int (\ln x)^3 \, dx \right]$$

$$= x(\ln x)^6 - 6x(\ln x)^5 + 30x(\ln x)^4 - 120 \left[x(\ln x)^3 - 3 \int (\ln x)^2 \, dx \right]$$

$$= x(\ln x)^6 - 6x(\ln x)^5 + 30x(\ln x)^4 - 120x(\ln x)^3 + 360 \left[x(\ln x)^2 - 2 \int (\ln x) \, dx \right]$$

$$= x(\ln x)^6 - 6x(\ln x)^5 + 30x(\ln x)^4 - 120x(\ln x)^3 + 360x(\ln x)^2 - 720 \left[x \ln x - \int 1 \, dx \right]$$

$$= x(\ln x)^6 - 6x(\ln x)^5 + 30x(\ln x)^4 - 120x(\ln x)^3 + 360x(\ln x)^2 - 720x \ln x + 720x + C.$$

7.M.53: $\int \frac{(\arcsin x)^2}{\sqrt{1 - x^2}} \, dx = \frac{1}{3}(\arcsin x)^3 + C.$

7.M.55: Here we have

$$\int \tan^3 z \, dz = \int (\sec^2 z - 1) \tan z \, dz = \int (\sec^2 z \tan z - \tan z) \, dz$$

$$= \frac{1}{2} \tan^2 z + \ln|\cos z| + C = \frac{1}{2} \sec^2 z + \ln|\cos z| + C_1.$$

404

7.M.57: Let $u = \exp(x^2)$. Then $du = 2x \exp(x^2)\, dx$ and $\exp(2x^2) = \left(e^{x^2}\right)^2 = u^2$. Hence

$$\int \frac{x \exp(x^2)}{1 + \exp(2x^2)}\, dx = \frac{1}{2}\int \frac{1}{1 + u^2}\, du = \frac{1}{2}\arctan u + C = \frac{1}{2}\arctan\left(\exp(x^2)\right) + C.$$

7.M.59: Let $u = x^2$ and $dv = x \exp(-x^2)\, dx$. Then $du = 2x\, dx$; choose $v = -\frac{1}{2}\exp(-x^2)$. Then

$$\int x^3 \exp(-x^2)\, dx = -\frac{1}{2}x^2 \exp(-x^2) + \int x \exp(-x^2)\, dx$$

$$= -\frac{1}{2}x^2 \exp(-x^2) - \frac{1}{2}\exp(-x^2) + C = -\frac{x^2 + 1}{2}\exp(-x^2) + C.$$

7.M.61: Integrate by parts with $u = \arcsin x$ and $dv = \frac{1}{x^2}\, dx$, so that

$$du = \frac{1}{\sqrt{1 - x^2}}\, dx; \quad \text{choose} \quad v = -\frac{1}{x}.$$

Then

$$K = \int \frac{\arcsin x}{x^2}\, dx = -\frac{1}{x}\arcsin x + \int \frac{1}{x\sqrt{1 - x^2}}\, dx = -\frac{1}{x}\arcsin x - \operatorname{sech}^{-1}|x| + C.$$

If you prefer to avoid hyperbolic functions, substitute $x = \sin\theta$ in the last integral. With $1 - x^2 = \cos^2\theta$ and $dx = \cos\theta\, d\theta$, you'll get

$$K = -\frac{1}{x}\arcsin x + \int \frac{\cos\theta}{\sin\theta \cos\theta}\, d\theta = -\frac{1}{x}\arcsin x + \ln|\csc\theta - \cot\theta| + C.$$

Then a reference triangle with acute angle θ, opposite side x, and hypotenuse 1 has adjacent side of length $\sqrt{1 - x^2}$. Therefore

$$K = -\frac{1}{x}\arcsin x + \ln\left|\frac{1 - \sqrt{1 - x^2}}{x}\right| + C = -\frac{1}{x}\arcsin x + \ln\left(1 - \sqrt{1 - x^2}\right) - \ln|x| + C.$$

7.M.63: Let $x = \sin u$. Then $1 - x^2 = 1 - \sin^2 u = \cos^2 u$ and $dx = \cos u\, du$. A reference triangle for this substitution has acute angle u, opposite side x, and hypotenuse 1, thus adjacent side of length $\sqrt{1 - x^2}$. Therefore

$$\int x^2\sqrt{1 - x^2}\, dx = \int \sin^2 u \cos^2 u\, du = \frac{1}{4}\int (2\sin u \cos u)^2\, du = \frac{1}{4}\int \sin^2 2u\, du = \frac{1}{8}\int (1 - \cos 4u)\, du$$

$$= \frac{1}{8}\left(u - \frac{1}{4}\sin 4u\right) + C = \frac{1}{8}u - \frac{1}{16}\sin 2u \cos 2u + C$$

$$= \frac{1}{8}u - \frac{1}{8}(\sin u \cos u)(\cos^2 u - \sin^2 u) + C = \frac{1}{8}\left(u - \sin u \cos^3 u + \sin^3 u \cos u\right) + C$$

$$= \frac{1}{8}\left[\arcsin x - x(1 - x^2)^{3/2} + x^3(1 - x^2)^{1/2}\right] + C$$

$$= \frac{1}{8}x(2x^2 - 1)\sqrt{1 - x^2} + \frac{1}{8}\arcsin x + C.$$

7.M.65: The partial fractions decomposition of the integrand has the form

$$\frac{x-2}{4x^2+4x+1} = \frac{A}{2x+1} + \frac{B}{(2x+1)^2},$$

and therefore $A(2x+1) + B = x - 2$. It follows that $A = \frac{1}{2}$ and $B = -\frac{5}{2}$, and thus

$$\int \frac{x-2}{(2x+1)^2}\, dx = \frac{1}{2}\int \left(\frac{1}{2x+1} - \frac{5}{(2x+1)^2} \right) dx = \frac{1}{4}\ln|2x+1| + \frac{5}{4(2x+1)} + C.$$

7.M.67: $\displaystyle\int \frac{e^{2x}}{e^{2x}-1}\, dx = \frac{1}{2}\ln|e^{2x}-1| + C.$

7.M.69: The partial fractions decomposition of the integrand has the form

$$\frac{2x^3+3x^2+4}{(x+1)^4} = \frac{A}{x+1} + \frac{B}{(x+1)^2} + \frac{C}{(x+1)^3} + \frac{D}{(x+1)^4}.$$

It follows that $A(x^3+3x^2+3x+1) + B(x^2+2x+1) + C(x+1) + D = 2x^3 + 3x^2 + 4$, and thus

$$A = 2,$$

$$3A + B = 3,$$

$$3A + 2B + C = 0,$$

$$A + B + C + D = 4.$$

The triangular form of this system of equations makes it easy to solve for $A = 2$, $B = -3$, $C = 0$, and $D = 5$. Therefore

$$\int \frac{2x^3+3x^2+4}{(x+1)^4}\, dx = 2\ln|x+1| + \frac{3}{x+1} - \frac{5}{3(x+1)^3} + C.$$

7.M.71: The partial fractions decomposition of the integrand has the form

$$\frac{x^3+x^2+2x+1}{x^4+2x^2+1} = \frac{Ax+B}{x^2+1} + \frac{Cx+D}{(x^2+1)^2},$$

so that $A(x^3+x) + B(x^2+1) + Cx + D = x^3 + x^2 + 2x + 1$. It follows that $A = 1$, $B = 1$, $C = 1$, and $D = 0$. Therefore

$$\int \frac{x^3+x^2+2x+1}{x^4+2x^2+1}\, dx = \int \left(\frac{x}{x^2+1} + \frac{1}{x^2+1} + \frac{x}{(x^2+1)^2} \right) dx = \frac{1}{2}\ln(x^2+1) + \arctan x - \frac{1}{2(x^2+1)} + C.$$

7.M.73: Use integration by parts with $u = x^3$ and $dv = c^2(x^3-1)^{1/2}\, dx$. Then

$$du = 3x^2\, dx; \quad \text{choose} \quad v = \frac{2}{9}(x^3-1)^{3/2}.$$

Then

$$K = \int x^5 \sqrt{x^3 - 1} \; dx = \frac{2}{9} x^3 (x^3 - 1)^{3/2} - \int \frac{2}{3} x^2 (x^3 - 1)^{3/2} \; dx$$

$$= \frac{2}{9} x^3 (x^3 - 1)^{3/2} - \frac{4}{45} (x^3 - 1)^{5/2} + C = \frac{1}{45} \left[10 x^3 (x^3 - 1)^{3/2} - 4(x^3 - 1)^{5/2} \right] + C$$

$$= \frac{2(x^3 - 1)^{1/2}}{45} \left[5 x^3 (x^3 - 1) - 2(x^3 - 1)^2 \right] + C = \frac{2}{45} (3 x^6 - x^3 - 2) \sqrt{x^3 - 1} + C.$$

7.M.75: $\displaystyle\int \frac{\sqrt{1 + \sin x}}{\sec x} \; dx = \int (1 + \sin x)^{1/2} \cos x \; dx = \frac{2}{3} (1 + \sin x)^{3/2} + C.$

7.M.77: $\displaystyle\int \frac{\sin x}{\sin 2x} \; dx = \int \frac{\sin x}{2 \sin x \cos x} \; dx = \frac{1}{2} \int \sec x \; dx = \frac{1}{2} \ln |\sec x + \tan x| + C.$

7.M.79: We multiply numerator and denominator by $\sqrt{1 - \sin t}$ and thus obtain

$$\int \sqrt{1 + \sin t} \; dt = \int \frac{\sqrt{1 - \sin^2 t}}{\sqrt{1 - \sin t}} \; dt = \int \frac{\sqrt{\cos^2 t}}{\sqrt{1 - \sin t}} \; dt = \int (1 - \sin t)^{-1/2} |\cos t| \; dt.$$

Therefore

$$\int \sqrt{1 + \sin t} \; dt = \begin{cases} -2\sqrt{1 - \sin t} + C & \text{if } \cos t \geqq 0, \\[2mm] 2\sqrt{1 - \sin t} + C & \text{if } \cos t \leqq 0. \end{cases}$$

Mathematica returns (in effect)

$$\int \sqrt{1 + \sin t} \; dt = \frac{2\left(\sin \dfrac{t}{2} - \cos \dfrac{t}{2}\right)}{\sin \dfrac{t}{2} + \cos \dfrac{t}{2}} \cdot \sqrt{1 + \sin t} + C$$

and *Maple* yields

$$\int \sqrt{1 + \sin t} \; dt = \frac{2(-1 + \sin t)\sqrt{1 + \sin t}}{\cos t} + C.$$

7.M.81: Integrate by parts with $u = \ln(x^2 + x + 1)$ and $dv = dx$. Then

$$du = \frac{2x + 1}{x^2 + x + 1} \; dx; \quad \text{choose} \quad v = x.$$

Then

$$J = \int \ln(x^2 + x + 1) \; dx = x \ln(x^2 + x + 1) - \int \frac{2x^2 + x}{x^2 + x + 1} \; dx = x \ln(x^2 + x + 1) - \int \left(2 - \frac{x + 2}{x^2 + x + 1} \right) dx.$$

Now

$$x^2 + x + 1 = \left(x + \frac{1}{2} \right)^2 + \frac{3}{4} = \frac{3}{4} \tan^2 u + \frac{3}{4} = \frac{3}{4} \sec^2 u$$

if $x + \frac{1}{2} - \frac{\sqrt{3}}{2} \tan u$, so we let

$$x = \frac{-1 + \sqrt{3}\,\tan u}{2}, \quad dx = \frac{\sqrt{3}}{2}\sec^2 u\,du, \quad \tan u = \frac{2x+1}{\sqrt{3}}.$$

A reference triangle for this substitution has acute angle u, opposite side $2x + 1$, and adjacent side $\sqrt{3}$, so its hypotenuse has length $2\sqrt{x^2 + x + 1}$. Thus

$$\int \frac{x+2}{x^2+x+1}\,dx = \int \frac{\frac{1}{2}\left(3 + \sqrt{3}\,\tan u\right)}{\frac{3}{4}\sec^2 u} \cdot \frac{\sqrt{3}}{2}\sec^2 u\,du = \frac{\sqrt{3}}{3}\int \left(3 + \sqrt{3}\,\tan u\right)\,du$$

$$= \frac{\sqrt{3}}{3}\left(3u + \sqrt{3}\,\ln|\sec u|\right) + C = \frac{\sqrt{3}}{3}\left[3\arctan\left(\frac{2x+1}{\sqrt{3}}\right) + \sqrt{3}\,\ln\left(\frac{2\sqrt{x^2+x+1}}{\sqrt{3}}\right)\right] + C.$$

Therefore

$$J = x\ln(x^2 + x + 1) - 2x + \sqrt{3}\,\arctan\left(\frac{2x+1}{\sqrt{3}}\right) + \frac{1}{2}\ln(x^2 + x + 1) + C.$$

7.M.83: Integrate by parts with

$$u = \arctan x \quad \text{and} \quad dv = \frac{1}{x^2}\,dx.$$

Then

$$du = \frac{1}{1+x^2}\,dx; \quad \text{choose} \quad v = -\frac{1}{x}.$$

Thus

$$\int \frac{\arctan x}{x^2}\,dx = -\frac{1}{x}\arctan x + \int \frac{1}{x(x^2+1)}\,dx$$

$$= -\frac{1}{x}\arctan x + \int \left(\frac{1}{x} - \frac{x}{x^2+1}\right)\,dx = -\frac{1}{x}\arctan x + \ln|x| - \frac{1}{2}\ln(x^2+1) + C.$$

7.M.85: We use the idea of the method of partial fractions but avoid the algebra as follows:

$$\frac{x^3}{(x^2+1)^2} = \frac{x^3 + x - x}{(x^2+1)^2} = \frac{x(x^2+1)}{(x^2+1)^2} - \frac{x}{(x^2+1)^2} = \frac{x}{x^2+1} - \frac{x}{(x^2+1)^2}.$$

Therefore

$$\int \frac{x^3}{(x^2+1)^2}\,dx = \int \left(\frac{x}{x^2+1} - \frac{x}{(x^2+1)^2}\right)\,dx = \frac{1}{2}\ln(x^2+1) + \frac{1}{2(x^2+1)} + C.$$

7.M.87: First, $x^2 + 4 = 4\tan^2 u + 4 = 4\sec^2 u$ if $x = 2\tan u$, so that $dx = 2\sec^2 u\,du$ and $\tan u = \frac{1}{2}x$. A reference triangle for this substitution has acute angle u, opposite side x, and adjacent side 2, thus its hypotenuse has length $\sqrt{x^2 + 4}$. Therefore

$$\int \frac{3x+2}{(x^2+4)^{3/2}}\,dx = \int \frac{2 + 6\tan u}{8\sec^3 u} \cdot 2\sec^2 u\,du = \frac{1}{2}\int (1 + 3\tan u)\cos u\,du = \frac{1}{2}\int (\cos u + 3\sin u)\,du$$

$$= \frac{1}{2}(\sin u - 3\cos u) + C = \frac{1}{2}\left(\frac{x}{\sqrt{x^2+4}} - \frac{6}{\sqrt{x^2+4}}\right) + C = \frac{x-6}{2\sqrt{x^2+4}} + C.$$

7.M.89: $\int \dfrac{\sqrt{1 + \sin^2 x}}{\sec x \, \csc x} \, dx = \int (1 + \sin^2 x)^{1/2} (\sin x \, \cos x) \, dx = \dfrac{1}{3} (1 + \sin^2 x)^{3/2} + C.$

7.M.91: Integration by parts is indicated, but there are several choices. We found that $u = \sin x$ and $dv = x e^x \, dx$ was a bad choice, leading to the correct but complicated antiderivative

$$I = \frac{1}{2}(x - 1) e^x \sin x - \frac{1}{2}(x - 2) e^x \cos x + \frac{1}{2} e^x \sin x - \frac{1}{2} e^x \cos x + C.$$

A better choice is $u = x$, $dv = e^x \sin x \, dx$. Even so, we need a preliminary computation to find v. We integrate by parts with $p = e^x$ and $dq = \sin x \, dx$. Then $dp = e^x \, dx$ and we may choose $q = -\cos x$. Then

$$v = \int e^x \sin x \, dx = -e^x \cos x + \int e^x \cos x \, dx. \tag{1}$$

We integrate by parts a second time, with $p = e^x$ and $dq = \cos x \, dx$. Then with $dp = e^x \, dx$ and $q = \sin x$ we find that

$$v = -e^x \cos x + e^x \sin x - \int e^x \sin x \, dx = e^x \sin x - e^x \cos x - v,$$

and therefore we may choose $v = \dfrac{1}{2} e^x (\sin x - \cos x)$. Also $du = dx$, and thus

$$I = \int x e^x \sin x \, dx = u \cdot v - \int v \, du = \frac{1}{2} x e^x (\sin x - \cos x) - \frac{1}{2} \int e^x \sin x \, dx + \frac{1}{2} \int e^x \cos x \, dx.$$

Now by Eq. (1),

$$\int e^x \cos x \, dx = e^x \cos x + v = e^x \cos x + \frac{1}{2} e^x (\sin x - \cos x) + C = \frac{1}{2} e^x (\sin x + \cos x) + C.$$

Therefore

$$I = \frac{1}{2} x e^x (\sin x - \cos x) - \frac{1}{2} \cdot \frac{1}{2} e^x (\sin x - \cos x) + \frac{1}{2} \cdot \frac{1}{2} e^x (\sin x + \cos x) + C$$

$$= \frac{1}{2} x e^x (\sin x - \cos x) + \frac{1}{2} e^x \cos x + C = \frac{1}{2} e^x (x \sin x - x \cos x + \cos x) + C.$$

Maple yields essentially the same answer:

$$I = \frac{1}{2} \big[x e^x \sin x - (x - 1) e^x \cos x \big] + C.$$

7.M.93: First integrate by parts with $u = \arctan x$ and $dv = (x - 1)^{-3} \, dx$. Then the new integrand will be a rational function of x, to which the method of partial fractions can be applied if necessary. We have

$$du = \frac{1}{1 + x^2} \, dx \quad \text{and we choose} \quad v = -\frac{1}{2}(x - 1)^{-2}.$$

Then

$$J = \int \frac{\arctan x}{(x-1)^3}\, dx = -\frac{\arctan x}{2(x-1)^2} + \frac{1}{2} \int \frac{1}{(x-1)^2(x^2+1)}\, dx.$$

The partial fractions decomposition

$$\frac{1}{(x-1)^2(x^2+1)} = \frac{A}{x-1} + \frac{B}{(x-1)^2} + \frac{Cx+D}{x^2+1}$$

leads to the equation $A(x^3 - x^2 + x - 1) + B(x^2 + 1) + C(x^3 + 2x^2 + x) + D(x^2 - 2x + 1) = 1$, and thus to the simultaneous equations

$$A + C = 0, \qquad -A + B - 2C + D = 0,$$

$$A + C - 2D = 0, \qquad -A + B + D = 1.$$

Then $C = -A$ and $D = 0$, so that $A + B = 0$ and $-A + B = 1$. Hence $B = \frac{1}{2}$, $A = -\frac{1}{2}$, and $C = \frac{1}{2}$. Therefore

$$\frac{1}{2} \cdot \frac{1}{(x-1)^2(x^2+1)} = \frac{1}{4}\left(-\frac{1}{x-1} + \frac{1}{(x-1)^2} + \frac{x}{x^2+1} \right).$$

So, finally,

$$J = -\frac{\arctan x}{2(x-1)^2} + \frac{1}{4}\left(\frac{1}{2}\ln(x^2+1) - \frac{1}{x-1} - \ln|x-1| \right) + C.$$

7.M.95: First,

$$3 + 6x - 9x^2 = -(9x^2 - 6x - 3) = -(9x^2 - 6x + 1 - 4) = 4 - (3x-1)^2 = 4 - 4\sin^2 u = 4\cos^2 u$$

if $3x - 1 = 2\sin u$, so we let

$$x = \frac{1 + 2\sin u}{3}; \quad dx = \frac{2}{3}\cos u\, du \quad \text{and} \quad \sin u = \frac{3x-1}{2}.$$

A reference triangle for this substitution has acute angle u, opposite side $3x - 1$, and hypotenuse 2, so its adjacent (to u) side has length $\sqrt{3 + 6x - 9x^2}$. Therefore

$$\int \frac{2x+3}{\sqrt{3+6x-9x^2}}\, dx = \int \frac{\frac{2}{3}\left(1 + 2\sin u + \frac{9}{2}\right)}{2\cos u} \cdot \frac{2}{3}\cos u\, du = \frac{2}{9}\int \left(\frac{11}{2} + 2\sin u \right) du$$

$$= \frac{11}{9}u - \frac{4}{9}\cos u + C = \frac{11}{9}\arcsin\left(\frac{3x-1}{2} \right) - \frac{2}{9}\sqrt{3+6x-9x^2} + C.$$

7.M.97: The method of partial fractions can be avoided with the substitution $u = x - 1$, so that $x = u + 1$ and $dx = du$. Then

$$\int \frac{x^4}{(x-1)^2}\, dx = \int \frac{(u+1)^4}{u^2}\, du = \int \frac{u^4 + 4u^3 + 6u^2 + 4u + 1}{u^2}\, du$$

$$= \int \left(u^2 + 4u + 6 + \frac{4}{u} + \frac{1}{u^2} \right) du = \frac{1}{3}u^3 + 2u^2 + 6u + 4\ln|u| - \frac{1}{u} + C$$

$$= \frac{1}{3}(x-1)^3 + 2(x-1)^2 + 6(x-1) + 4\ln|x-1| - \frac{1}{x-1} + C$$

$$= \frac{1}{3}x^3 + x^2 + 3x - \frac{1}{x-1} + 4\ln|x-1| + C_1$$

where $C_1 = C + \dfrac{13}{3}$.

If you used the method of partial fractions, you should have found that

$$\frac{x^4}{(x-1)^2} = x^2 + 2x + 3 + \frac{4}{x-1} + \frac{1}{(x-1)^2}.$$

7.M.99: Integrate by parts with $u = \operatorname{arcsec}(\sqrt{x}\,)$ and $dv = dx$. Then

$$du = \frac{1}{\sqrt{x}\,\sqrt{x-1}} \cdot \frac{1}{2\sqrt{x}}\ dx = \frac{1}{2x\sqrt{x-1}}\ dx; \quad \text{choose}\quad v = x.$$

Then

$$\int \operatorname{arcsec}(\sqrt{x}\,)\ dx = x\operatorname{arcsec}(\sqrt{x}\,) - \int \frac{1}{2}(x-1)^{-1/2}\ dx = x\operatorname{arcsec}(\sqrt{x}\,) - \sqrt{x-1} + C.$$

7.M.101: If $y = \cosh x$, then

$$1 + \left(\frac{dy}{dx}\right)^2 = 1 + \sinh^2 x = \cosh^2 x,$$

so that $ds = \cosh x\ dx$. We will also use Eq. (11) of Section 6.8 to find that the surface area of revolution is

$$A = \int_0^1 2\pi \cosh^2 x\ dx = \pi \int_0^1 (1 + \cosh 2x)\ dx = \pi\left[x + \frac{1}{2}\sinh 2x\right]_0^1$$

$$= \pi\left(1 + \frac{e^2 - e^{-2}}{4}\right) = \frac{\pi}{4}\left(4 + e^2 - e^{-2}\right) \approx 8.83865166003373.$$

7.M.103: Given $y = e^{-x}$,

$$\frac{dy}{dx} = -e^{-x}, \quad \text{so that}\quad ds = \sqrt{1 + e^{-2x}}\ dx.$$

Hence the surface area of revolution is

$$A_t = \int_0^t 2\pi e^{-x}\sqrt{1 + e^{-2x}}\ dx.$$

Let $u = e^{-x}$. Then $du = -e^{-x}\ dx$, and so

$$A_t = \int_{x=0}^t -2\pi\sqrt{1 + u^2}\ du.$$

Next, let $u = \tan\theta$. Then $du = \sec^2\theta\ d\theta$ and $\sqrt{1 + u^2} = \sec\theta$. Therefore

$$A_t = -2\pi \int_{x=0}^t \sec^3\theta\ d\theta = -\pi\left[\sec\theta\tan\theta + \ln|\sec\theta + \tan\theta|\right]_{x=0}^t$$

$$= -\pi\left[u\sqrt{1 + u^2} + \ln\left(u + \sqrt{1 + u^2}\,\right)\right]_{x=0}^t = -\pi\left[e^{-x}\sqrt{1 + e^{-2x}} + \ln\left(e^{-x} + \sqrt{1 + e^{-2x}}\,\right)\right]_0^t$$

$$= \pi\left[\sqrt{2} + \ln\left(1 + \sqrt{2}\,\right) - e^{-t}\sqrt{1 + e^{-2t}} - \ln\left(e^{-t} + \sqrt{1 + e^{-2t}}\,\right)\right].$$

411

Clearly

$$\lim_{t \to \infty} A_t = \pi \left[\sqrt{2} + \ln\left(1 + \sqrt{2}\right) \right] \approx 7.211799724207.$$

7.M.105: Given $y = (x^2 - 1)^{1/2}$, we have

$$\frac{dy}{dx} = \frac{x}{\sqrt{x^2 - 1}},$$

so the arc length element is

$$ds = \sqrt{1 + \frac{x^2}{x^2 - 1}} \, dx = \left(\frac{2x^2 - 1}{x^2 - 1}\right)^{1/2} dx.$$

We will use integral formula 44 from the endpapers to help us find that the surface area of revolution is

$$A = \int_1^2 2\pi (x^2 - 1)^{1/2} \left(\frac{2x^2 - 1}{x^2 - 1}\right)^{1/2} dx = 2\pi \int_1^2 \sqrt{2x^2 - 1} \, dx = 2\pi\sqrt{2} \int_1^2 \left(x^2 - \frac{1}{2}\right)^{1/2} dx$$

$$= 2\pi\sqrt{2} \left[\frac{x}{2}\left(x^2 - \frac{1}{2}\right)^{1/2} - \frac{1}{4}\ln\left(x + \left(x^2 - \frac{1}{2}\right)^{1/2}\right) \right]_1^2 \qquad \text{(using integral formula 44)}$$

$$= 2\pi\sqrt{2} \left[\sqrt{\frac{7}{2}} - \frac{1}{4}\ln\left(2 + \sqrt{\frac{7}{2}}\right) - \frac{1}{2}\sqrt{\frac{1}{2}} + \frac{1}{4}\ln\left(1 + \sqrt{\frac{1}{2}}\right) \right] \approx 11.663528688558.$$

7.M.107: Let $u = (\sin x)^{m-1}$ and $dv = (\cos x)^n \sin x \, dx$ (m and n are integers with $n \geqq 0$ and $m \geqq 2$). Then

$$du = (m-1)(\sin x)^{m-2}\cos x \, dx \quad \text{and} \quad v = -\frac{1}{n+1}(\cos x)^{n+1}.$$

Therefore

$$I = \int (\sin x)^m (\cos x)^n \, dx = -\frac{1}{n+1}(\sin x)^{m-1}(\cos x)^{n+1} + \frac{m-1}{n+1}\int (\sin x)^{m-2}(\cos x)^{n+2} \, dx$$

$$= -\frac{1}{n+1}(\sin x)^{m-1}(\cos x)^{n+1} + \frac{m-1}{n+1}\int (\sin x)^{m-2}(\cos x)^n (1 - \sin^2 x) \, dx$$

$$= -\frac{1}{n+1}(\sin x)^{m-1}(\cos x)^{n+1} + \frac{m-1}{n+1}\int (\sin x)^{m-2}(\cos x)^n \, dx - \frac{m-1}{n+1}I.$$

Thus

$$\left(\frac{m-1}{n+1} + 1\right)I = -\frac{1}{n+1}(\sin x)^{m-1}(\cos x)^{n+1} + \frac{m-1}{n+1}\int (\sin x)^{m-2}(\cos x)^n \, dx;$$

$$\frac{m+n}{n+1}I = -\frac{1}{n+1}(\sin x)^{m-1}(\cos x)^{n+1} + \frac{m-1}{n+1}\int (\sin x)^{m-2}(\cos x)^n \, dx;$$

and, finally,

$$I = -\frac{n+1}{m+n} \cdot \frac{1}{n+1}(\sin x)^{m-1}(\cos x)^{n+1} + \frac{n+1}{m+n} \cdot \frac{m-1}{n+1}\int (\sin x)^{m-2}(\cos x)^n \, dx$$

$$= -\frac{1}{m+n}(\sin x)^{m-1}(\cos x)^{n+1} + \frac{m-1}{m+n}\int (\sin x)^{m-2}(\cos x)^n \, dx.$$

7.M.109: We need a result in Problem 58 of Section 7.3: If n is a positive integer, then

$$\int_0^{\pi/2} (\sin x)^{2n} \, dx = \frac{\pi}{2} \cdot \frac{1}{2} \cdot \frac{3}{4} \cdot \frac{5}{6} \cdots \frac{2n-1}{2n}.$$

The area in question is

$$A = 2 \int_0^2 x^{5/2} \sqrt{2-x} \, dx.$$

Let $x = 2\sin^2 \theta$. Then $\sqrt{2-x} = \sqrt{2} \, \cos\theta$ and $dx = 4\sin\theta \cos\theta \, d\theta$. Therefore

$$A = 2 \int_0^{\pi/2} (2^{5/2} \sin^5 \theta) \left(\sqrt{2} \, \cos\theta\right) (4\sin\theta \cos\theta) \, d\theta = 64 \int_0^{\pi/2} \sin^6 \theta \cos^2 \theta \, d\theta$$

$$= 64 \int_0^{\pi/2} \left[\sin^6 \theta - \sin^8 \theta\right] \, d\theta = 64 \left[\left(\frac{\pi}{2} \cdot \frac{1}{2} \cdot \frac{3}{4} \cdot \frac{5}{6}\right) - \left(\frac{\pi}{2} \cdot \frac{1}{2} \cdot \frac{3}{4} \cdot \frac{5}{6} \cdot \frac{7}{8}\right)\right]$$

$$= 64 \cdot \frac{\pi}{2} \cdot \frac{1}{2} \cdot \frac{3}{4} \cdot \frac{5}{6} \cdot \frac{1}{8} = \frac{5\pi}{4} \approx 3.926990816987241548078304.$$

7.M.111: First,

$$\int_0^1 t^4 (1-t)^4 \, dt = \int_0^1 (t^8 - 4t^7 + 6t^6 - 4t^5 + t^4) \, dt$$

$$= \frac{1}{9} - \frac{1}{2} + \frac{6}{7} - \frac{2}{3} + \frac{1}{5} = \frac{70 - 315 + 540 - 420 + 126}{630} = \frac{1}{630}.$$

But $\dfrac{1}{2} \leqq \dfrac{1}{t^2+1} \leqq 1$ if $0 \leqq t \leqq 1$. Therefore

$$\frac{1}{1260} < \int_0^1 \frac{t^4(1-t)^4}{t^2+1} \, dt < \frac{1}{630};$$

$$\frac{1}{1260} < \frac{22}{7} - \pi < \frac{1}{630};$$

$$-\frac{1}{630} < \pi - \frac{22}{7} < -\frac{1}{1260};$$

$$\frac{22}{7} - \frac{1}{630} < \pi < \frac{22}{7} - \frac{1}{1260}.$$

7.M.113: Given $y = \dfrac{4}{3} x^{3/4}$,

$$\frac{dy}{dx} = x^{-1/4}, \quad \text{so} \quad ds = \sqrt{1 + x^{-1/2}} \, dx.$$

Therefore the length of the given curve is

$$L = \int_1^4 \sqrt{1 + x^{-1/2}} \, dx.$$

413

Now $1 + x^{-1/2} = 1 + \tan^2 u = \sec^2 u$ if $x^{-1/2} = \tan^2 u$; that is, if $x^{1/2} = \cot^2 u$. So we let $x = \cot^4 u$; then $dx = -4 \cot^3 u \csc^2 u \, du$, and this substitution yields

$$L = \int_{x=1}^{4} (-4 \sec u)(\cot^3 u \csc^3 u) \, du = -4 \int_{x=1}^{4} \frac{\cos^3 u}{\sin^5 u \cos u} \, du$$

$$= 4 \int_{x=1}^{4} \frac{\sin^2 u - 1}{\sin^5 u} \, du = 4 \int_{x=1}^{4} (\csc^3 u - \csc^5 u) \, du.$$

Now we need to pause to develop a reduction formula. Suppose that n is an integer and $n \geq 3$. Let

$$J_n = \int \csc^n x \, dx.$$

Now integrate by parts: Let $u = (\csc x)^{n-2}$ and $dv = \csc^2 x \, dx$. Then

$$du = -(n-2)(\csc x)^{n-3}(\csc x \cot x) \, dx; \quad \text{choose } v = -\cot x.$$

Then

$$J_n = -(\csc x)^{n-2} \cot x - (n-2) \int (\csc x)^{n-2} \cot^2 x \, dx$$

$$= -(\csc x)^{n-2} \cot x - (n-2) \int (\csc x)^{n-2}(\csc^2 x - 1) \, dx$$

$$= -(\csc x)^{n-2} \cot x - (n-2)J_n + (n-2) \int (\csc x)^{n-2} \, dx.$$

Thus

$$(n-1)J_n = -(\csc x)^{n-2} \cot x + (n-2) \int (\csc x)^{n-2} \, dx,$$

and, finally,

$$J_n = \int \csc^n x \, dx = -\frac{1}{n-1}(\csc x)^{n-2} \cot x + \frac{n-2}{n-1} \int (\csc x)^{n-2} \, dx.$$

With the aid of this reduction formula, we find that

$$\int \csc^3 x \, dx = -\frac{1}{2} \csc x \cot x + \frac{1}{2} \int \csc x \, dx = -\frac{1}{2} \csc x \cot x + \frac{1}{2} \ln|\csc x - \cot x| + C$$

and

$$\int \csc^5 x \, dx = -\frac{1}{4} \csc^3 x \cot x + \frac{3}{4} \int \csc^3 x \, dx$$

$$= -\frac{1}{4} \csc^3 x \cot x - \frac{3}{8} \csc x \cot x + \frac{3}{8} \ln|\csc x - \cot x| + C.$$

Therefore

$$L = 4 \left[\frac{1}{4} \csc^3 u \cot u - \frac{1}{8} \csc u \cot u + \frac{1}{8} \ln|\csc u - \cot u| \right]_{x=1}^{4}$$

414

$$= \left[x^{1/4} \left(1 + \sqrt{x} \right)^{3/2} - \frac{1}{2} x^{1/4} \left(1 + \sqrt{x} \right)^{1/2} + \frac{1}{2} \ln \left(\sqrt{1 + \sqrt{x}} - x^{1/4} \right) \right]_1^4$$

$$= 3\sqrt{6} - \frac{1}{2}\sqrt{6} + \frac{1}{2} \ln \left(\sqrt{3} - \sqrt{2} \right) - 2\sqrt{2} + \frac{1}{2}\sqrt{2} - \frac{1}{2} \ln \left(\sqrt{2} - 1 \right)$$

$$= \frac{5}{2}\sqrt{6} - \frac{3}{2}\sqrt{2} + \frac{1}{2} \ln \left(\frac{\sqrt{3} - \sqrt{2}}{\sqrt{2} - 1} \right) = \frac{5}{2}\sqrt{6} - \frac{3}{2}\sqrt{2} + \frac{1}{2} \ln \left[\left(\sqrt{3} - \sqrt{2} \right) \left(\sqrt{2} + 1 \right) \right]$$

$$= \frac{1}{2} \left[5\sqrt{6} - 3\sqrt{2} + \ln \left(\sqrt{6} + \sqrt{3} - 2 - \sqrt{2} \right) \right] \approx 3.869982889518.$$

7.M.115: Let $u = e^x$: $du = e^x \, dx = u \, dx$, so $dx = \frac{1}{u} \, du$. Hence

$$\int \frac{1}{1 + e^x + e^{-x}} \, dx = \int \frac{1}{u(1 + u + u^{-1})} \, du = \int \frac{1}{u^2 + u + 1} \, du = \int \frac{1}{\left(u + \frac{1}{2} \right)^2 + \frac{3}{4}} \, du$$

$$= \frac{4}{3} \int \frac{1}{\left(\frac{2u+1}{\sqrt{3}} \right)^2 + 1} \, du = \frac{2\sqrt{3}}{3} \arctan \left(\frac{2u+1}{\sqrt{3}} \right) + C = \frac{2\sqrt{3}}{3} \arctan \left(\frac{2e^x + 1}{\sqrt{3}} \right) + C.$$

This substitution will always succeed in integrals of this ilk because you'll always obtain a rational function of u after making the substitution.

7.M.117: $\int \frac{1}{1 + e^x} \, dx = \int \frac{e^{-x}}{e^{-x} + 1} \, dx = - \ln \left(1 + e^{-x} \right) + C.$

Mathematica returns the antiderivative $x - \ln \left(1 + e^x \right)$.

7.M.119: Let $u = \tan \theta$, so that $du = \sec^2 \theta \, d\theta$, $\theta = \arctan u$, $1 + u^2 = \sec^2 \theta$, and

$$d\theta = \frac{1}{1 + u^2} \, du.$$

Thus

$$H = \int \sqrt{\tan \theta} \, d\theta = \int \frac{u^{1/2}}{1 + u^2} \, du.$$

Now let $u = x^2$, so that $x = u^{1/2}$ and $du = 2x \, dx$. Then

$$H = \int \frac{2x^2}{1 + x^4} \, dx.$$

The partial fractions decomposition of the last integrand has the form

$$\frac{2x^2}{x^4 + 1} = \frac{Ax + B}{x^2 - x\sqrt{2} + 1} + \frac{Cx + D}{x^2 + x\sqrt{2} + 1},$$

and we find that

$$A(x^3 + x^2\sqrt{2} + x) + B(x^2 + x\sqrt{2} + 1) + C(x^3 - x\sqrt{2} + x) + D(x^2 - x\sqrt{2} + 1) = 2x^2.$$

415

Thus we obtain the simultaneous equations

$$A + C = 0, \qquad A\sqrt{2} + B - C\sqrt{2} + D = 2,$$

$$A + B\sqrt{2} + C - D\sqrt{2} = 0, \qquad B + D = 0.$$

Then it's easy to solve for $B = D = 0$, $A = \frac{1}{2}\sqrt{2}$, and $C = -\frac{1}{2}\sqrt{2}$. Therefore

$$\frac{2x^2}{1 + x^4} = \frac{\sqrt{2}}{2}\left(\frac{x}{x^2 - x\sqrt{2} + 1} - \frac{x}{x^2 + x\sqrt{2} + 1}\right).$$

Now let $r = \frac{1}{2}\sqrt{2}$. Then

$$\frac{x}{x^2 - 2rx + 1} = \frac{1}{2} \cdot \frac{2x - 2r}{x^2 - 2rx + 1} + \frac{r}{x^2 - 2rx + 1}.$$

It's easy to antidifferentiate the second fraction; you'll obtain $\frac{1}{2}\ln(x^2 - 2rx + 1) + C$. For the last fraction, a trigonometric substitution is one technique that will succeed:

$$x^2 - 2rx + 1 = x^2 - 2rx + r^2 + 1 - r^2 = (x - r)^2 + \frac{1}{2} = \frac{1}{2}\tan^2\omega + \frac{1}{2} = \frac{1}{2}\sec^2\omega$$

if $x - r = r\tan\omega$. So we let

$$x = r + r\tan\omega, \quad dx = r\sec^2\omega\, d\omega, \quad \tan\omega = \frac{x - r}{r} = 2rx - 1 = x\sqrt{2} - 1.$$

Then

$$\int \frac{r}{x^2 - 2rx + 1}\, dx = r \int \frac{r}{\frac{1}{2}\sec^2\omega} \cdot \sec^2\omega\, d\omega = 2r^2\omega + C = \omega + C = \arctan\left(x\sqrt{2} - 1\right) + C.$$

The case of $x^2 + 2rx + 1$ is handled similarly—only a few sign changes—and the result is that

$$H = \frac{\sqrt{2}}{4}\left[2\arctan\left(x\sqrt{2} - 1\right) + 2\arctan\left(x\sqrt{2} + 1\right) + \ln\left(x^2 - x\sqrt{2} + 1\right) - \ln\left(x^2 + x\sqrt{2} + 1\right)\right] + C.$$

Therefore, because $x = \sqrt{u} = \sqrt{\tan\theta}$, we finally obtain

$$H = \int \sqrt{\tan\theta}\, d\theta$$

$$= \frac{\sqrt{2}}{4}\left[2\arctan\left(-1 + \sqrt{2\tan\theta}\right) + 2\arctan\left(1 + \sqrt{2\tan\theta}\right)\right.$$

$$\left. + \ln\left(\tan\theta - \sqrt{2\tan\theta} + 1\right) - \ln\left(\tan\theta + \sqrt{2\tan\theta} + 1\right)\right] + C.$$

7.M.121: Let $u^2 = 3x - 2$. Then

$$x = \frac{u^2 + 2}{3}, \quad dx = \frac{2}{3}u\, du, \quad \text{and} \quad u = (3x - 2)^{1/2}.$$

Hence

$$\int x^3\sqrt{3x-2}\,dx = \int \frac{1}{27}(u^2+2)^3 \cdot u \cdot \frac{2}{3}u\,du = \frac{2}{81}\int (u^8+6u^6+12u^4+8u^2)\,du$$

$$= \frac{2}{81}\left(\frac{1}{9}u^9 + \frac{6}{7}u^7 + \frac{12}{5}u^5 + \frac{8}{3}u^3\right) + C$$

$$= \frac{2}{729}(3x-2)^{9/2} + \frac{4}{189}(3x-2)^{7/2} + \frac{8}{135}(3x-2)^{5/2} + \frac{16}{243}(3x-2)^{3/2} + C$$

$$= \frac{1}{25515}\left[70(3x-2)^{9/2} + 540(3x-2)^{7/2} + 1512(3x-2)^{5/2} + 1680(3x-2)^{3/2}\right] + C$$

$$= \frac{2\sqrt{3x-2}}{25515}\left[35(3x-2)^4 + 270(3x-2)^3 + 756(3x-2)^2 + 840(3x-2)\right] + C$$

$$= \frac{2(3x-2)^{3/2}}{25515}(945x^3 + 540x^2 + 288x + 128) + C.$$

Mathematica obtains the equivalent

$$\int x^3\sqrt{3x-2}\,dx = \frac{\sqrt{3x-2}}{25515}(5670x^4 - 540x^3 - 432x^2 - 384x - 512) + C,$$

and *Maple* gives the antiderivative in the form shown in the third line of the display here.

7.M.123: Let $u^3 = x^2 - 1$. Then $3u^2\,du = 2x\,dx$, $x\,dx = \frac{3}{2}u^2\,du$, $x^2 = u^3 + 1$, and $(x^2-1)^{4/3} = u^4$. Thus

$$\int \frac{x^3}{(x^2-1)^{4/3}}\,dx = \int \frac{u^3+1}{u^4}\cdot\frac{3}{2}u^2\,du = \frac{3}{2}\int\left(u+\frac{1}{u^2}\right)du = \frac{3}{2}\left(\frac{1}{2}u^2 - \frac{1}{u}\right) + C = \frac{3}{4}\left(u^2 - \frac{2}{u}\right) + C$$

$$= \frac{3}{4}\cdot\frac{u^3-2}{u} + C = \frac{3(u^3-2)}{4u} + C = \frac{3(x^2-1-2)}{4(x^2-1)^{1/3}} + C = \frac{3(x^2-3)}{4(x^2-1)^{1/3}} + C.$$

7.M.125: Let $u^2 = x^3 + 1$. Then $u = (x^3+1)^{1/2}$, $x^3 = u^2 - 1$, $3x^2\,dx = 2u\,du$, and $x^2\,dx = \frac{2}{3}u\,du$. Therefore

$$\int \frac{x^5}{\sqrt{x^3+1}}\,dx = \int \frac{x^3}{(x^3+1)^{1/2}}\cdot x^2\,dx = \int \frac{u^2-1}{u}\cdot\frac{2}{3}u\,du = \frac{2}{3}\int (u^2-1)\,du = \frac{2}{3}\left(\frac{1}{3}u^3 - u\right) + C$$

$$= \frac{2}{9}(u^3-3u) + C = \frac{2u}{9}(u^2-3) + C = \frac{2}{9}(x^2-2)\sqrt{x^3+1} + C.$$

7.M.127: Let

$$u^2 = \frac{1+x}{1-x}. \quad\text{Then}\quad x = \frac{u^2-1}{u^2+1} \quad\text{and}\quad dx = \frac{2u(u^2+1) - 2u(u^2-1)}{(u^2+1)^2}\,du = \frac{4u}{(u^2+1)^2}\,du.$$

Therefore

$$J = \int \left(\frac{1+x}{1-x}\right)^{1/2}dx = \int \frac{4u^2}{(u^2+1)^2}\,du.$$

417

The method of partial fractions requires us to integrate $4(u^2+1)^{-2}$ with a trigonometric substitution, so we might as well go directly to the trigonometry. Let $u = \tan\theta$. Then $du = \sec^2\theta\,d\theta$. A reference triangle for this substitution has acute angle θ, opposite side u, and adjacent side 1, and therefore has hypotenuse of length $\sqrt{1+u^2}$. Therefore

$$J = \int \frac{4\tan^2\theta}{\sec^4\theta}\cdot\sec^2\theta\,d\theta = \int 4\sin^2\theta\,d\theta = \int 2(1-\cos 2\theta)\,d\theta = 2(\theta - \sin\theta\cos\theta) + C$$

$$= 2\left(\arctan u - \frac{u}{u^2+1}\right) + C = 2\left[\arctan\left(\frac{1+x}{1-x}\right)^{1/2} - \frac{\left(\frac{1+x}{1-x}\right)^{1/2}}{\frac{1+x}{1-x}+1}\right] + C$$

$$= 2\left[\arctan\left(\frac{1+x}{1-x}\right)^{1/2} - \frac{1}{2}(1-x)\left(\frac{1+x}{1-x}\right)^{1/2}\right] + C = 2\arctan\left(\frac{1+x}{1-x}\right)^{1/2} - \sqrt{1-x^2} + C.$$

7.M.129: Let $u^3 = x+1$. Then $u = (x+1)^{1/3}$, $x = u^3 - 1$, and $dx = 3u^2\,du$. Thus

$$K = \int \frac{(x+1)^{1/3}}{x}\,dx = \int \frac{u}{u^3-1}\cdot 3u^2\,du = 3\int \frac{u^3}{u^3-1}\,du = 3\int\left(1 + \frac{1}{u^3-1}\right)du.$$

The partial fractions decomposition of the last fraction has the form

$$\frac{1}{u^3-1} = \frac{A}{u-1} + \frac{Bu+C}{u^2+u+1}.$$

Thus we find that $A(u^2+u+1) + b(u^2-u) + C(u-1) = 1$, and thus we obtain the simultaneous equations

$$A+B = 0,\quad A-B+C = 0,\quad\text{and}\quad A-C = 1.$$

It follows that $A = \frac{1}{3}$, $B = -\frac{1}{3}$, and $C = -\frac{2}{3}$. Thus

$$\frac{1}{u^3-1} = \frac{1}{3}\left(\frac{1}{u-1} - \frac{u+2}{u^2+u+1}\right).$$

Now

$$u^2+u+1 = \left(u+\frac{1}{2}\right)^2 + \frac{3}{4} = \frac{3}{4}\tan^2\theta + \frac{3}{4} = \frac{3}{4}\sec^2\theta \quad\text{if}\quad \frac{\sqrt{3}}{2}\tan\theta = u + \frac{1}{2}.$$

Therefore we let

$$u = \frac{-1 + \sqrt{3}\,\tan\theta}{2};\quad\text{thus}\quad \tan\theta = \frac{2u+1}{\sqrt{3}},\quad du = \frac{\sqrt{3}}{2}\sec^2\theta\,d\theta.$$

Note that a reference triangle for this substitution has acute angle θ, opposite side $2u+1$, and adjacent side $\sqrt{3}$, thus hypotenuse of length $2\sqrt{u^2+u+1}$. Therefore

$$\int \frac{u+2}{u^2+u+1}\,du = \frac{1}{2}\int \frac{3 + \sqrt{3}\,\tan\theta}{\frac{3}{4}\sec^2\theta}\cdot\frac{\sqrt{3}}{2}\sec^2\theta\,d\theta = \frac{\sqrt{3}}{3}\int\left(3 + \sqrt{3}\,\tan\theta\right)d\theta$$

$$= \frac{\sqrt{3}}{3}\left(3\theta + \sqrt{3}\,\ln|\sec\theta|\right) + C = \frac{\sqrt{3}}{3}\left[3\arctan\left(\frac{2u+1}{\sqrt{3}}\right) + \sqrt{3}\,\ln\left(\sqrt{u^2+u+1}\right)\right] + C_1$$

$$= \sqrt{3}\,\arctan\left(\frac{2u+1}{\sqrt{3}}\right) + \frac{1}{2}\ln(u^2+u+1) + C_1.$$

418

Therefore

$$K = 3 \int \left(1 + \frac{1}{u^3 - 1} \right) du = \int \left(3 + \frac{1}{u - 1} - \frac{u + 2}{u^2 + u + 1} \right) du$$

$$= 3u + \ln|u - 1| - \sqrt{3} \arctan\left(\frac{2u + 1}{\sqrt{3}} \right) - \frac{1}{2} \ln(u^2 + u + 1) + C$$

$$= 3(x + 1)^{1/3} + \ln\left| (x + 1)^{1/3} - 1 \right|$$

$$- \sqrt{3} \arctan\left(\frac{2(x + 1)^{1/3} + 1}{\sqrt{3}} \right) - \frac{1}{2} \ln\left((x + 1)^{2/3} + (x + 1)^{1/3} + 1 \right) + C.$$

Mathematica returns

$$K = 3(x + 1)^{1/3} - \frac{3(1 + x^{-1})^{2/3}}{2(1 + x)^{2/3}} \cdot \text{Hypergeometric2F1}\left[\frac{2}{3}, \frac{2}{3}, \frac{5}{3}, -\frac{1}{x} \right]$$

(the generalized hypergeometric function is discussed more fully in some of the solutions in Chapter 10). *Derive* and *Maple* yield, with minor algebraic changes, the answer we obtained "by hand."

7.M.131: The substitution $u^2 = 1 + e^x$ entails $e^{2x} = u^2 - 1$, $2x = \ln(u^2 - 1)$,

$$x = \frac{1}{2} \ln(u^2 - 1), \quad \text{and} \quad dx = \frac{1}{2} \cdot \frac{2u}{u^2 - 1} du = \frac{u}{u^2 - 1} du.$$

Therefore

$$\int \sqrt{1 + e^{2x}} \, dx = \int \frac{u^2}{u^2 - 1} \, du = \int \left(1 + \frac{1}{u^2 - 1} \right) du = \int \left(1 + \frac{\frac{1}{2}}{x - 1} - \frac{\frac{1}{2}}{u + 1} \right) du$$

$$= \frac{1}{2} \int \left(2 + \frac{1}{u - 1} - \frac{1}{u + 1} \right) du = \frac{1}{2} \left(2u + \ln\left| \frac{u - 1}{u + 1} \right| \right) + C = \sqrt{1 + e^{2x}} + \frac{1}{2} \ln\left| \frac{-1 + \sqrt{1 + e^{2x}}}{1 + \sqrt{1 + e^{2x}}} \right| + C.$$

7.M.133: The area is

$$A = 2 \int_0^1 x\sqrt{1 - x} \, dx.$$

Let $u = 1 - x$. Then $x = 1 - u$ and $dx = -du$. Hence

$$A = -2 \int_{u=1}^0 (1 - u)u^{1/2} \, du = 2 \int_0^1 (u^{1/2} - u^{3/2}) \, du = 2 \left[\frac{2}{3} u^{3/2} - \frac{2}{5} u^{5/2} \right]_0^1 = 2 \left(\frac{2}{3} - \frac{2}{5} \right) = \frac{8}{15}.$$

7.M.135: Using the recommended substitution, we find that

$$\int \frac{1}{1 + \cos\theta} \, d\theta = \int \frac{1}{1 + \frac{1 - u^2}{1 + u^2}} \cdot \frac{2}{1 + u^2} \, du = \int \frac{2}{1 + u^2 + 1 - u^2} \, du$$

$$= u + C = \tan\frac{\theta}{2} + C = \frac{1 - \cos\theta}{\sin\theta} + C = \frac{\sin\theta}{1 + \cos\theta} + C.$$

419

7.M.137: The recommended substitution yields

$$\int \frac{1}{1+\sin\theta}\,d\theta = \int \frac{1}{1+\dfrac{2u}{1+u^2}} \cdot \frac{2}{1+u^2}\,du = \int \frac{2}{1+u^2+2u}\,du = \int 2(u+1)^{-2}\,du = -\frac{2}{u+1}+C$$

$$= -\frac{2}{1+\tan\dfrac{\theta}{2}}+C = -\frac{2\sin\theta}{1+\sin\theta-\cos\theta}+C = -\frac{2+2\cos\theta}{1+\sin\theta+\cos\theta}+C.$$

7.M.139: The substitution $u = \tan\dfrac{\theta}{2}$ yields

$$I = \int \frac{1}{\sin\theta+\cos\theta}\,d\theta = \int \frac{1}{\dfrac{2u}{1+u^2}+\dfrac{1-u^2}{1+u^2}} \cdot \frac{2}{1+u^2}\,du = \int \frac{2}{2u+1-u^2}\,du.$$

Now

$$-u^2 + 2u + 1 = -(u^2-2u-1) = -(u^2-2u+1-2) = 2-(u-1)^2 = 2-2\sin^2 w = 2\cos^2 w$$

if $u-1 = \sqrt{2}\,\sin w$, so we let $u = 1+\sqrt{2}\,\sin w$, so that

$$du = \sqrt{2}\,\cos w\,dw \quad \text{and} \quad \sin w = \frac{u-1}{\sqrt{2}}.$$

Thus

$$I = \int \frac{2}{2u+1-u^2}\,du = \int \frac{2}{2\cos^2 w} \cdot \sqrt{2}\,\cos w\,dw = \sqrt{2}\,\ln|\sec w + \tan w|+C.$$

A reference triangle for the trigonometric substitution has acute angle w, opposite side $u-1$, and hypotenuse $\sqrt{2}$. Therefore its adjacent side has length $\sqrt{2u+1-u^2}$, and thus

$$I = \sqrt{2}\,\ln\left|\frac{\sqrt{2}+u-1}{\sqrt{2u+1-u^2}}\right|+C$$

$$= \sqrt{2}\left[\ln\left(-1+\sqrt{2}+\frac{1-\cos\theta}{\sin\theta}\right) - \frac{1}{2}\ln\left(1+\frac{2-2\cos\theta}{\sin\theta}-\frac{(1-\cos\theta)^2}{\sin^2\theta}\right)\right]+C.$$

7.M.141: The substitution $u = \tan\dfrac{\theta}{2}$ yields

$$K = \int \frac{\sin\theta}{2+\cos\theta}\,d\theta = \int \frac{2u}{2(1+u^2)+(1-u^2)} \cdot \frac{2}{1+u^2}\,du = \int \frac{4u}{(u^2+3)(u^2+1)}\,du.$$

Next, the partial fractions decomposition

$$\frac{4u}{(u^2+1)(u^2+3)} = \frac{Au+B}{u^2+1}+\frac{Cu+D}{u^2+3}$$

leads to the equation $A(u^3 + 3u) + B(u^2 + 3) + C(u^3 + 3) + D(u^2 + 1) = 4u$, and thus to the system

$$A + C = 0, \qquad B + D = 0,$$

$$3A + C = 4, \qquad 3B + D = 0$$

with solution $A = 2$, $C = -2$, $B = D = 0$. Therefore

$$K = \int \left(\frac{2u}{u^2 + 1} - \frac{2u}{u^2 + 3} \right) du = \ln(u^2 + 1) - \ln(u^2 + 3) + C = \ln \left(\frac{u^2 + 1}{u^2 + 3} \right) + C$$

$$= \ln \left(\frac{(1 - \cos\theta)^2 + \sin^2\theta}{(1 - \cos\theta)^2 + 3\sin^2\theta} \right) + C = \ln \left(\frac{1 - 2\cos\theta + 1}{1 - 2\cos\theta + 1 + 2\sin^2\theta} \right) + C$$

$$= \ln \left(\frac{2 - 2\cos\theta}{2 + 2\sin^2\theta - 2\cos\theta} \right) + C = \ln \left(\frac{1 - \cos\theta}{1 + \sin^2\theta - \cos\theta} \right) + C$$

$$= \ln \left(\frac{1 - \cos\theta}{2 - \cos\theta - \cos^2\theta} \right) + C = \ln \left(\frac{1 - \cos\theta}{(1 - \cos\theta)(2 + \cos\theta)} \right) + C = -\ln(2 + \cos\theta) + C.$$

7.M.143: The substitution $u = \tan\dfrac{\theta}{2}$ yields

$$\int \sec\theta \, d\theta = \int \frac{1}{\cos\theta} \, d\theta = \int \frac{1 + u^2}{1 - u^2} \cdot \frac{2}{1 + u^2} \, du = \int \frac{2}{1 - u^2} \, du = \int \left(\frac{1}{1 + u} + \frac{1}{1 - u} \right) du$$

$$= \ln \left| \frac{1 + u}{1 - u} \right| + C = \ln \left| \frac{1 + \tan\dfrac{\theta}{2}}{1 - \tan\dfrac{\theta}{2}} \right| + C = \ln \left| \frac{1 + \left(\dfrac{1 - \cos\theta}{1 + \cos\theta} \right)^{1/2}}{1 - \left(\dfrac{1 - \cos\theta}{1 + \cos\theta} \right)^{1/2}} \right| + C$$

$$= \ln \left| \frac{\sqrt{1 + \cos\theta} + \sqrt{1 - \cos\theta}}{\sqrt{1 + \cos\theta} - \sqrt{1 - \cos\theta}} \right| + C = \ln \left| \frac{1 + \cos\theta + 2\sqrt{1 - \cos^2\theta} + 1 - \cos\theta}{1 + \cos\theta - (1 - \cos\theta)} \right| + C$$

$$= \ln \left| \frac{2 + 2\sin\theta}{2\cos\theta} \right| + C = \ln |\sec\theta + \tan\theta| + C.$$

Section 8.1

8.1.1: Separate the variables:

$$\frac{dy}{dx} = 2y; \qquad \frac{1}{y}\, dy = 2\, dx;$$

$$\ln y = C + 2x; \qquad y(x) = Ae^{2x}.$$

$$3 = y(1) = Ae^2 : \qquad A = 3e^{-2}.$$

Therefore $y(x) = 3\exp(2x - 2) = 3e^{2x-2}$.

8.1.3: Given: $\dfrac{dy}{dx} = 2y^2,\ y(7) = 3$:

$$-\frac{1}{y^2}\, dy = -2\, dx; \qquad \frac{1}{y} = C - 2x;$$

$$y(x) = \frac{1}{C - 2x}. \qquad 3 = y(7) = \frac{1}{C - 14} :$$

$$C = \frac{43}{3}. \qquad y(x) = \frac{1}{\frac{43}{3} - 2x} = \frac{3}{43 - 6x}.$$

8.1.5: Given: $\dfrac{dy}{dx} = 2y^{1/2},\ y(0) = 9$:

$$y^{-1/2}\, dy = 2\, dx; \qquad 2y^{1/2} = C + 2x;$$

$$y^{1/2} = A + x; \qquad y(x) = (A + x)^2.$$

$$y(0) = 9 : \qquad A = 3.$$

Therefore $y(x) = (x + 3)^2$.

8.1.7: Given: $\dfrac{dy}{dx} = 1 + y,\ y(0) = 5$:

$$\frac{1}{1 + y}\, dy = 1\, dx; \qquad \ln(1 + y) = x + C;$$

$$1 + y = Ae^x; \qquad y(x) = Ae^x - 1.$$

$$5 = y(0) = A - 1 : \qquad A = 6.$$

Therefore $y(x) = 6e^x - 1$.

8.1.9: Given: $\dfrac{dy}{dx} = e^{-y},\ y(0) = 2$:

$$e^y \, dy = 1 \, dx; \qquad e^y = x + C;$$

$$y(x) = \ln(x + C). \qquad 2 = y(0) = \ln C :$$

$$C = e^2. \qquad y(x) = \ln(x + e^2).$$

8.1.11: If the slope of $y = g(x)$ at the point (x, y) is the sum of x and y, then we expect $y = g(x)$ to be a solution of the differential equation

$$\frac{dy}{dx} = x + y. \tag{1}$$

Some solutions of this differential equation with initial conditions $y(0) = -1$, -0.5, 0, 0.5, and 1 are shown next. The figure is constructed with the same scale on the x- and y-axes, so you can confirm with a ruler that the solution curves agree with the differential equation in (1).

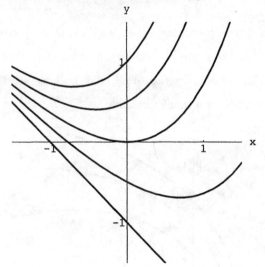

8.1.13: If every straight line normal to the graph of $y = g(x)$ passes through the point $(0, 1)$, then we expect that $y = g(x)$ will be a solution of the differential equation

$$\frac{dy}{dx} = \frac{x}{1 - y}. \tag{1}$$

The general solution of this equation is implicitly defined by $x^2 + (y - 1)^2 = C$ where $C > 0$. Some solution curves for $C = 0.16$, $C = 0.49$, and $C = 1$ are shown next. Note that two functions are solutions for each value of C. Because the figure was constructed with the same scale on the x- and y-axes, you can confirm with a ruler that the solution curves agree with the differential equation in (1).

423

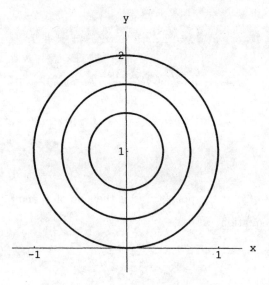

8.1.15: If, for each (x, y) on the graph of $y = g(x)$, the line tangent to the graph at that point passes through $(-y, x)$, then we expect for $y = g(x)$ to satisfy the differential equation

$$\frac{dy}{dx} = \frac{y - x}{y + x}. \tag{1}$$

The substitution of the new dependent variable $u = y/x$ leads to the general solution of this equation—see the discussion of *Homogeneous Equations* in Section 1.6 of Edwards and Penney: *Differential Equations: Computing and Modeling* (3rd edition, Prentice Hall, 2004). Some solution curves are shown next; note than none appears to be the graph of a single function.

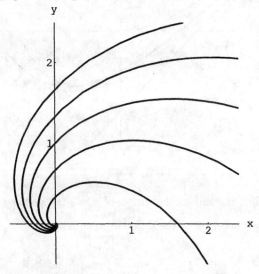

8.1.17: $\dfrac{dv}{dt} = -kv^2$ (where k is a positive constant).

8.1.19: $\dfrac{dN}{dt} = k(P - N)$ (where k is a positive constant).

8.1.21: The principal (in dollars) at time t (in years) is $A(t) = 1000e^{(0.08)t}$. Therefore $A'(t) = 80e^{(0.08)t}$. So the answers to the two questions in Problem 21 are these: $A'(5) = \$119.35$ and $A'(20) = \$396.24$.

424

8.1.23: Let $P(t)$ be the number of bacteria present at time t (in hours), with initial number $P_0 = P(0)$. Then $P(t) = P_0 e^{kt}$ where k is a constant. We are given $P(10) = 6P_0$, and therefore

$$P_0 e^{10k} = 6P_0; \qquad 10k = \ln 6; \qquad k = \frac{1}{10} \ln 6.$$

If T is the doubling time, then

$$P(T) = 2P_0 = P_0 \exp\left(\frac{T}{10} \ln 6\right);$$

$$\frac{T}{10} \ln 6 = \ln 2;$$

$$T = \frac{10 \ln 2}{\ln 6} \approx 3.8685280723.$$

Thus the doubling time is approximately 3 h 52 min.

8.1.25: We take $t = 0$ (in years) corresponding to the time of formation of the relic and let t denote its age. Then the amount of ^{14}C it contains at time t is $N(t) = N_0 e^{-kt}$, where $k \approx 0.0001216$. Therefore

$$\left(5.0 \times 10^{10}\right) e^{-kt} = 4.6 \times 10^{10}, \qquad \text{so} \qquad t \approx \frac{0.0833816}{k} \approx 686.$$

We conclude that the relic is between 650 and 700 years old and is probably not authentic (although it may be much older if it recently has been contaminated with "modern" carbon).

8.1.27: After t years the fine will grow to $A(t) = 30e^{it}$ (cents) where $i = 0.05$. Thus the answer is $30 \exp(0.05 \cdot 100) = 30e^5 = 4452$ cents; that is, $44.52.

8.1.29: Let $S(t)$ represent the sales t weeks after advertising is discontinued and let $S(0) = S_0$. Then for some constant λ,

$$\frac{dS}{dt} = -\lambda S, \quad \text{so} \quad S(t) = S_0 e^{-\lambda t}.$$

Because $S(1) = (0.95)S_0 = S_0 e^{-\lambda}$, $\lambda = \ln \frac{20}{19}$. Therefore at time $t = T$, when sales have declined to 75% of the initial rate,

$$S(T) = \tfrac{3}{4} S_0 = S_0 e^{-\lambda T} : \quad e^{\lambda T} = \tfrac{4}{3}; \quad T = \frac{\ln(\frac{4}{3})}{\lambda} = \frac{\ln(\frac{4}{3})}{\ln(\frac{20}{19})} \approx 5.608.$$

So the company plans to resume advertising about 5.6 weeks (about 39 days) after cessation of advertising.

8.1.31: Let Q denote the amount of radioactive cobalt remaining at time t (in years), with the occurrence of the accident set at time $t = 0$. Then

$$Q = Q_0 e^{-(t \ln 2)/(5.27)}.$$

If T is the number of years until the level of radioactivity has dropped to a hundredth of its initial value, then

$$\frac{1}{100} = e^{-(T \ln 2)/(5.27)}, \qquad \text{so that} \qquad T = (5.27)\frac{\ln 100}{\ln 2},$$

approximately 35 years.

8.1.33: Of course we set up coordinates so that $t = 0$ corresponds to 12 noon. (a) $P(t) = 49e^{kt}$, and

$$294 = 49e^k : \qquad e^k = 6; \qquad k = \ln 6.$$

Therefore $P(t) = 49\exp(t\ln 6) = 49\exp(\ln 6^t) = 49 \cdot 6^t$. (b) At 1:40 P.M. we have $t = \frac{5}{3}$, so the number of bacteria present at that time is

$$P\left(\frac{5}{3}\right) = 49 \cdot 6^{5/3} \approx 971.$$

(c) If $P(t) = 20000$, then

$$49 \cdot 6^T = 20000; \qquad 6^T = \frac{20000}{49};$$

$$T\ln 6 = \ln\frac{20000}{49}; \qquad T = \frac{1}{\ln 6} \cdot \ln\frac{20000}{49} \approx 3.355175377985.$$

Answer: The clock time then will be approximately 3:21:19 P.M.

8.1.35: (a) $A(t) = 15e^{-kt}$; $10 = A(5) = 15e^{-5k}$, so

$$\frac{3}{2} = e^{5k} : \qquad k = \frac{1}{5}\ln\frac{3}{2}.$$

Therefore

$$A(t) = 15\exp\left(-\frac{t}{5}\ln\frac{3}{2}\right) = 15 \cdot \left(\frac{3}{2}\right)^{-t/5} = 15 \cdot \left(\frac{2}{3}\right)^{t/5}.$$

(b) After 8 months we have

$$A(8) = 15 \cdot \left(\frac{2}{3}\right)^{8/5} \approx 7.8405261683.$$

(c) $A(T) = 1$ when

$$\left(\frac{2}{3}\right)^{T/5} = \frac{1}{15}; \qquad \left(\frac{2}{3}\right)^T = \left(\frac{1}{15}\right)^5;$$

$$T\ln\frac{2}{3} = 5\ln\frac{1}{15}; \qquad T = 5 \cdot \frac{\ln\left(\frac{1}{15}\right)}{\ln\left(\frac{2}{3}\right)} = \frac{5\ln 15}{\ln(3/2)}.$$

Answer: It will be safe to return after approximately 33.394368 months.

8.1.37: If $L(t)$ denotes the number of Native American language families at time t (in years), then $L(t) = e^{kt}$ where k is a positive constant. To find k, we use the given data

$$\frac{3}{2} = L(6000) = e^{6600k}: \qquad k = \frac{1}{6000} \ln \frac{3}{2}.$$

If $t = T$ corresponds to the present time, then $150 = L(T) = e^{kT}$, so that $kT = \ln 150$. Therefore

$$T = \frac{1}{k} \ln 150 = \frac{6000 \ln 150}{\ln(3/2)} \approx 74146.483047.$$

This analysis suggests that the ancestors of today's Native Americans first arrived in the western hemisphere about 74000 years ago.

8.1.39: Let $y(t)$ denote the height of the water in the tank (in feet) at time t (in hours). Then $y(0) = 9$ and $y(1) = 4$. By Eq. (29) of Section 8.1, we have

$$\frac{dy}{dt} = -ky^{1/2},$$

and it follows that $y^{-1/2}\, dy = -k\, dt$, so that $2y^{1/2} = C - kt$ where C is a constant. The condition $y(0) = 9$ now yields $C = 6$, so that $y^{1/2} = 3 - \frac{1}{2}kt$. The condition $y(1) = 4$ then yields $2 = 3 - \frac{1}{2}k$, so that $k = 2$. Hence $y(t) = (3 - t)^2$ for $0 \leq t \leq 3$. Therefore $y(t) = 0$ when $t = 3$, and thus it takes three hours total for the tank to drain completely.

8.1.41: We will use Eq. (28) of Section 8.1, which allows us to work in units of feet and hours (instead of seconds) because the conversion of the gravitational acceleration g into such units is taken care of by the proportionality constant c there. Let $y(t)$ denote the depth of the water in the tank at time t. If $r(t)$ is the radius of the circular surface of the water then, using similar triangles, we have

$$\frac{r(t)}{y(t)} = \frac{5}{16}, \qquad \text{so that} \qquad r(t) = \frac{5}{16}y(t).$$

Hence the volume of water in the tank at time t will be

$$V(t) = \frac{1}{3}\pi \cdot \frac{25}{144}\left[y(t)\right]^3,$$

and therefore (with the aid of Eq. (28))

$$\frac{dV}{dt} = \frac{25\pi}{144}y^2\frac{dy}{dt} = -cy^{1/2};$$

$$y^{3/2}\, dy = -\frac{144c}{25\pi}\, dt = -k\, dt \qquad \text{(where } k \text{ is constant)};$$

$$\frac{2}{5}y^{5/2} = C - kt.$$

427

Therefore $y(t) = (A - Bt)^{2/5}$ for some constants A and B. Because $16 = y(0) = A^{2/5}$, we see that $A = 1024$, so that $y(t) = (1024 - Bt)^{2/5}$. Moreover, $9 = y(1) = (1024 - B)^{2/5}$, and it follows that $B = 781$. Therefore

$$y(t) = (1024 - 781t)^{2/5}.$$

Consequently the tank will be empty when

$$t = \frac{1024}{781} \approx 1.31114 \quad \text{(h)};$$

that is, in a little less than 1 h 19 min.

8.1.43: Let $y(t)$ be the height of water in the tank (in feet) at time t (in hours), with $t = 0$ corresponding to 12 noon. Then $y(0) = 12$ and $y(1) = 6$. When the height of water in the tank is h, then—by the method of cross sections—the volume of water in the tank will be

$$V = \int_0^h \pi y^{3/2} \, dy = \left[\frac{2}{5}\pi y^{5/2}\right]_0^h = \frac{2}{5}\pi h^{5/2}.$$

Therefore at time t, the volume of water in the tank will be

$$V(t) = \frac{2}{5}\pi \left[y(t)\right]^{5/2}.$$

We will use Eq. (28) of Section 8.1 because the constant of proportionality there will allow us to use hours and feet for our units rather than seconds and feet. Thus we find that

$$\frac{dV}{dt} = \pi \left[y(t)\right]^{3/2} \cdot \frac{dy}{dt} = -cy^{1/2};$$

$$\pi y \, dy = -c \, dt;$$

$$\frac{\pi}{2}y^2 = A - ct;$$

$$y(t) = (B - Ct)^{1/2}.$$

The condition $12 = y(0) = \sqrt{B}$ now yields $B = 144$ (not -144; look at the last equation in the display), so that $y(t) = (144 - Ct)^{1/2}$. Next, $6 = y(1) = (144 - C)^{1/2}$, so that $C = 108$. Therefore $y(t) = \sqrt{144 - 108t}$. The tank will be empty when $y(t) = 0$, which will occur when $t = \frac{4}{3}$ (h). Answer: The tank will be empty at 1:20 P.M.

8.1.45: Set up a coordinate system in which one end of the tank lies in the xy-plane with its lowest point at the origin, thus bounded by the circle with equation $x^2 + (y-3)^2 = 9$. A horizontal cross section of the tank "at" location y ($0 \leq y \leq 6$) has width $2\sqrt{6y - y^2}$ and length 5, so if the depth of xylene in the tank is h, then its volume is

$$V = \int_0^h 10(6y - y^2)^{1/2} \, dy.$$

428

Thus if the depth of xylene in the tank is y, then its volume is given by

$$V(y) = \int_0^y 10(6u - u^2)^{1/2} \, du.$$

Hence by the chain rule and the fundamental theorem of calculus,

$$\frac{dV}{dt} = \frac{dV}{dy} \cdot \frac{dy}{dt} = 10(6y - y^2)^{1/2} \frac{dy}{dt}.$$

We will use Eq. (27) of Section 8.1,

$$\frac{dV}{dt} = -a\sqrt{2gy},$$

in which we take $g = 32$ ft/s^2; a is the area of the hole in the bottom of the tank, so—converting inches into feet—we have $a = \pi/144$ (ft^2). As usual, $y = y(t)$ is the depth of water in the tank at time t (distances will be measured in feet and time in seconds). Thus

$$10(6y - y^2)^{1/2} \frac{dy}{dt} = -a\sqrt{2gy} = -8ay^{1/2} = -\frac{\pi}{18} y^{1/2};$$

$$(6 - y)^{1/2} \, dy = -\frac{\pi}{180} \, dt;$$

$$\frac{2}{3}(6 - y)^{3/2} = C + \frac{\pi t}{180}.$$

Now $y(0) = 3$, so $C = \frac{2}{3}(3^{3/2}) = 2\sqrt{3}$. Therefore

$$\frac{2}{3}(6 - y)^{3/2} = 2\sqrt{3} + \frac{\pi t}{180}.$$

Hence $y = 0$ when

$$t = \frac{180}{\pi} \left(4\sqrt{6} - 2\sqrt{3} \right) \approx 362.9033$$

(seconds); that is, just a little less than 6 min 3 s.

8.1.47: Let $h = f^{-1}$, let a be the area of the hole, and let $y(t)$ be the depth of water in the tank at time t. We use $c = 1$, $g = 32$, and $\dfrac{dy}{dt} = -\dfrac{1}{10800}$ (feet per second). Then by Eq. (26) of Section 8.1,

$$-\frac{1}{10800} A(y) = -a\sqrt{64y}$$

where $A(y) = \pi \left[h(y) \right]^2$. Therefore

$$[h(y)]^2 = \frac{86400}{\pi} a y^{1/2},$$

and thus

$$[h(y)]^4 = \frac{7464960000}{\pi^2} a^2 y.$$

429

Finally, because $y = f(x)$ and $x = h(y)$, we have

$$f(x) = \frac{\pi^2}{(86400a)^2} x^4.$$

Now $f(1) = 4$, so $86400a = \pi/2$; it follows that

$$a = \frac{\pi}{172800} = \pi r^2$$

where r is the radius of the hole. Therefore

$$r = \frac{1}{240\sqrt{3}} \quad \text{(feet).} \quad \text{That is,} \quad r \approx 0.02887 \quad \text{(in.)}$$

Section 8.2

8.2.1: The following sequence of *Mathematica* commands will generate the slope field and the solution curves through the given points. Begin with the differential equation

$$\frac{dy}{dx} = f(x, y)$$

where

```
f[x_, y_] := -y - Sin[x]
```

Then set up the viewing window $a \leqq x \leqq b,\ c \leqq y \leqq d$:

```
a = -3;  b = 3;  c = -3;  d = 3;
```

The unit vectors that comprise the components of the short line segments tangent to the solution curves—those that form the slope field—are these:

```
u[x_, y_] := 1/Sqrt[1 + (f[x,y])^2]

v[x_, y_] := f[x,y]/Sqrt[1 + (f[x,y])^2]
```

The next commands construct the slope field.

```
Needs["Graphics`PlotField`"]

dfield = PlotVectorField[ { u[x,y], v[x,y] }, { x, a, b }, { y, c, d },
    HeadWidth -> 0, HeadLength -> 0, PlotPoints -> 19,
    PlotRange -> {{ a, b, }, { c, d }}, Axes -> True,
```

430

```
Frame → True, AspectRatio → 1 ];
```

To set up the first initial point and solution curve:

```
x0 = −2.5;  y0 = 2.0;

point1 = Graphics[ { PointSize[0.025], Point[ { x0, y0 } ] } ];

soln = NDSolve[ { y'[x] == f[x, y[x]], y[x0] == y0 }, y[x], { x, a, b } ];

soln[[1, 1, 2]];

curve1 = Plot[ soln[[1, 1, 2]], { x, a, b },

    PlotStyle → { Thickness[0.0065] } ];
```

The last option may be omitted; it's used to thicken the solution curve to make it more visible. Repeat the last sequence of commands with the remaining initial points; for example,

```
x7 = −2.5;  y7 = 1.0;

point7 = Graphics[ { PointSize[0.025], Point[ { x7, y7 } ] } ];

soln = NDSolve[ { y'[x] == f[x, y[x]], y[x0] == y0 }, y[x], { x, a, b } ];

soln[[1, 1, 2]];

curve7 = Plot[ soln[[1, 1, 2]], { x, a, b },

    PlotStyle → { Thickness[0.0065] } ];
```

The final version of the figure can be generated by the *Mathematica* command

```
Show[ dfield, point1, curve1, point2, curve2, point3, curve3, point4, curve4,

point5, curve5, point6, curve6, point7, curve7, point8, curve8, point9, curve9,

    point10, curve10, point11, curve11, point12, curve12, point13, curve13 ];
```

The resulting figure is next.

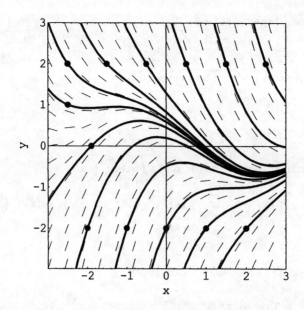

8.2.3: Follow the template given in the solution of Problem 1, with the obvious changes to $f(x, y)$ and the initial points (x_i, y_i) and the viewing window. The resulting figure is next.

8.2.5: Follow the template given in the solution of Problem 1, with the obvious changes to $f(x, y)$ and the initial points (x_i, y_i) and the viewing window. The resulting figure is next.

8.2.7: Follow the template given in the solution of Problem 1, with the obvious changes to $f(x, y)$ and the initial points (x_i, y_i) and the viewing window. The resulting figure is next.

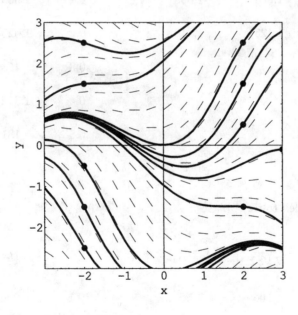

8.2.9: Follow the template given in the solution of Problem 1, with the obvious changes to $f(x, y)$ and the initial points (x_i, y_i) and the viewing window. The resulting figure is next.

8.2.11: Results, rounded to three places:

x	y ($h = 0.25$)	y (true)	x	y ($h = 0.1$)	y (true)
0.25	1.500	1.558	0.1	1.800	1.810
0.50	1.125	1.213	0.2	1.620	1.637
			0.3	1.458	1.482
			0.4	1.312	1.341
			0.5	1.181	1.213

8.2.13: Results, rounded to three places:

x	y ($h = 0.25$)	y (true)	x	y ($h = 0.1$)	y (true)
0.25	1.500	1.568	0.1	1.200	1.210
0.50	2.125	2.297	0.2	1.420	1.443
			0.3	1.662	1.700
			0.4	1.928	1.984
			0.5	2.221	2.297

8.2.15: Results, rounded to three places:

x	$y\ (h = 0.25)$	y (true)	x	$y\ (h = 0.1)$	y (true)
0.25	1.000	0.966	0.1	1.000	0.995
0.50	0.938	0.851	0.2	0.990	0.979
			0.3	0.969	0.950
			0.4	0.936	0.908
			0.5	0.889	0.851

8.2.17: Results, rounded to three places:

x	$y\ (h = 0.25)$	y (true)	x	$y\ (h = 0.1)$	y (true)
0.25	3.000	2.953	0.1	3.000	2.997
0.50	2.859	2.647	0.2	2.991	2.976
			0.3	2.955	2.920
			0.4	2.875	2.814
			0.5	2.737	2.647

8.2.19: Results, rounded to three places:

x	$y\ (h = 0.25)$	y (true)	x	$y\ (h = 0.1)$	y (true)
0.25	1.125	1.134	0.1	1.050	1.051
0.50	1.267	1.287	0.2	1.103	1.105
			0.3	1.158	1.162
			0.4	1.216	1.223
			0.5	1.278	1.287

8.2.21: We follow the template given in the solution of Problem 1 for using *Mathematica* to generate both the slope field and the desired solution curve. The result is shown at the end of this solution. The exact solution of the initial value problem

$$\frac{dy}{dx} = x + y, \qquad y(0) = 0$$

435

is $y(x) = e^x - x - 1$, and $y(-4) = e^{-4} + 3 \approx 3.018316$, in agreement with the following figure.

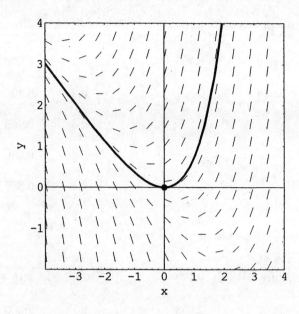

8.2.23: We followed the template given in the solution of Problem 1. Thus *Mathematica* generated the figure shown below. In the solution of Problem 27 we find that $y(2) \approx 1.0044$ (although the last digit is in question).

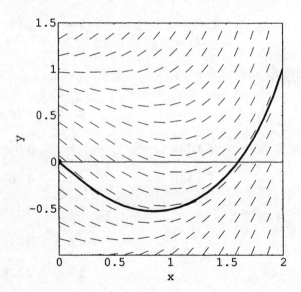

8.2.25: We constructed the slope field and solution curve as in the solution of Problem 1. The result is shown after this solution. The exact solution of the given initial value problem is

$$v(t) = 20 \left[1 - \exp\left(-\frac{8t}{5} \right) \right],$$

so—in accord with the figure—the limiting velocity is 20 ft/s. Landing with this velocity is about the same as landing after jumping off a wall 6.25 feet high, so the landing is perfectly subvivable (but be sure to relax and bend your knees). A "strategically located haystack" would nevertheless be welcome. Your velocity will be 95% of your limiting velocity when

$$t = \frac{5}{8} \ln 20 \approx 1.872333$$

seconds.

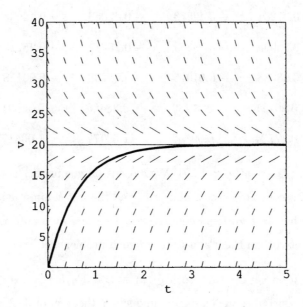

8.2.27: With $f(x, y) = x^2 + y^2 - 1$, we began with the initial values $x = 0$, $y = 0$, $h = 0.1$, $k = 2$, and $n = 0$ and executed the *Mathematica* command

```
While[ n < 10*k, { n = n + 1, y = y + h*f[x,y], x = x + h,

    If[IntegerQ[n/k], Print[ { x, y } ]] } ]
```

We repeated with $h = 0.01$ and $k = 20$, again with $h = 0.001$ and $k = 200$, again with $h = 0.0001$ and

$k = 2000$, and once more with $h = 0.00001$ and $k = 20000$. The results are shown next.

	$h = 0.1$	$h = 0.01$	$h = 0.001$	$h = 0.0001$	$h = 0.00001$
x	y	y	y	y	y
0.2	-0.19800	-0.19513	-0.19479	-0.19475	-0.19475
0.4	-0.37267	-0.36119	-0.35999	-0.35987	-0.35985
0.6	-0.49817	-0.47591	-0.47367	-0.47345	-0.47343
0.8	-0.55948	-0.52741	-0.52423	-0.52391	-0.52388
1.0	-0.55135	-0.51131	-0.50734	-0.50694	-0.50690
1.2	-0.47281	-0.42533	-0.42056	-0.42008	-0.42003
1.4	-0.32094	-0.26359	-0.25772	-0.25713	-0.25707
1.6	-0.08538	-0.01026	-0.00238	-0.00158	-0.00150
1.8	0.26121	0.37239	0.38466	0.38591	0.38603
2.0	0.77724	0.99768	1.00172	1.00417	1.00442

It appears that, to three places, $y(2) \approx 1.004$. We did not pursue further accuracy because the next-to-last column required over 12 seconds to execute and print and the last column required over 121 seconds. Linear extrapolation suggests that an additional column would require over an hour; we prefer to use numerical methods more sophisticated than Euler's method for such problems.

8.2.29: Direct substitution verifies that both $y_1(x) \equiv 1$ and $y_2(x) = \cos x$ satisfy the given initial value problem. This does not contradict the existence-uniqueness theorem of Section 8.2 because

$$\frac{\partial}{\partial y}\left(-\sqrt{1 - y^2}\right) = \frac{y}{\sqrt{1 - y^2}}$$

is not continuous at the point $(0, 1)$, so the theorem does not guarantee uniqueness of a solution passing through that point.

8.2.31: If $a \geqq 0$ and $y(x) = (x - a)^3$, then

$$\frac{dy}{dx} = 3(x - a)^2 = 3\left[(x - a)^3\right]^{2/3} = 3\left[y(x)\right]^{2/3}.$$

Therefore $y(x) = (x - a)^{2/3}$ is a solution of the given differential equation on $[0, +\infty)$. Consequently, if $a \geqq 0$, then

$$y(x) = \begin{cases} x^3 & \text{if } x \leqq 0, \\ 0 & \text{if } 0 \leqq x \leqq a, \\ (x - a)^3 & \text{if } a \leqq x \end{cases}$$

438

is a solution of the given initial value problem. (It is easy to verify that $y(x)$ is differentiable for all x.) This does not contradict the existence-uniqueness theorem because that theorem guarantees existence and uniqueness only on some open interval containing $x = -1$; here, uniqueness fails only at the "distant" point $x = 0$.

Section 8.3

8.3.1: Given: $\dfrac{dy}{dx} = 2x\sqrt{y}$. Then

$$y^{-1/2}\, dy = 2x\, dx;$$

$$2y^{1/2} = x^2 + C;$$

$$y^{1/2} = \frac{x^2 + C}{2};$$

$$y(x) = \left(\frac{x^2 + C}{2}\right)^2.$$

8.3.3: Given: $\dfrac{dy}{dx} = x^2 y^2$. Then

$$y^{-2}\, dy = x^2\, dx;$$

$$y^{-1} = C - \frac{1}{3}x^3;$$

$$y(x) = \frac{1}{C - \frac{1}{3}x^3} = \frac{3}{K - x^3}$$

where $K = 3C$ is a constant.

8.3.5: Given: $\dfrac{dy}{dx} = 2x(y-1)^{1/2}$. Then

$$(y-1)^{-1/2}\, dy = 2x\, dx;$$

$$2(y-1)^{1/2} = x^2 + C;$$

$$(y-1)^{1/2} = \frac{x^2 + C}{2};$$

$$y - 1 = \left(\frac{x^2 + C}{2}\right)^2;$$

$$y(x) = 1 + \left(\frac{x^2 + C}{2}\right)^2.$$

8.3.7: Given: $\dfrac{dy}{dx} = \dfrac{1 + \sqrt{x}}{1 + \sqrt{y}}$. Then

$$(1 + \sqrt{y})\, dy = (1 + \sqrt{x})\, dx; \qquad y + \frac{2}{3}y^{3/2} = x + \frac{2}{3}x^{3/2} + C.$$

It is possible to solve explicitly for $y(x)$. To see the explicit form, enter the *Mathematica* command

```
DSolve[ y'[x] == (1 + Sqrt[x])/(1 + Sqrt[y[x]]), y[x], x ]
```

and be prepared for about 32 lines of output.

8.3.9: Given: $\dfrac{dy}{dx} = \dfrac{x^2 + 1}{x^2(3y^2 + 1)}$. Then

$$(3y^2 + 1)\, dy = \left(1 + \frac{1}{x^2}\right) dx; \qquad y^3 + y = x - \frac{1}{x} + C.$$

It is possible to solve explicitly for $y(x)$. Enter the *Mathematica* command

```
DSolve[ y'[x] == (x^2 + 1)/(x^2*(3*(y[x])^2 + 1)), y[x], x ]
```

to see the result.

8.3.11: Given: $\dfrac{dy}{dx} = y^2$, $y(0) = 1$. Then

$$y^{-2}\, dy = dx; \qquad y^{-1} = C - x; \qquad y(x) = \frac{1}{C - x}.$$

Then the initial condition yields

$$1 = y(0) = \frac{1}{C}, \quad \text{and thus} \quad y(x) = \frac{1}{1 - x}.$$

8.3.13: Given: $\dfrac{dy}{dx} = \dfrac{1}{4y^3}$, $y(0) = 1$. Then

$$4y^3\, dy = dx; \qquad y^4 = x + C; \qquad 1^4 = [y(0)]^4 = 0 + C;$$

$$C = 1; \qquad [y(x)]^4 = x + 1; \qquad y(x) = (x + 1)^{1/4}.$$

We take the positive root in the last step because $y(0) > 0$.

8.3.15: Given: $\dfrac{dy}{dx} = \sqrt{xy^3}$, $y(0) = 4$. Then

$$y^{-3/2}\, dy = x^{1/2}\, dx; \qquad 3y^{-3/2}\, dy = 3x^{1/2}\, dx; \qquad 6y^{-1/2} = C_1 - 2x^{3/2};$$

$$y^{-1/2} = C_2 - \frac{1}{3}x^{3/2}; \qquad y^{1/2} = \frac{1}{C_2 - \frac{1}{3}x^{3/2}}; \qquad y^{1/2} = \frac{3}{C - x^{3/2}};$$

$$2 = [y(0)]^{1/2} = \frac{3}{C}; \qquad C = \frac{3}{2}; \qquad y(x) = \frac{9}{\left(\frac{3}{2} - x^{3/2}\right)^2};$$

$$y(x) = \frac{36}{\left(3 - 2x^{3/2}\right)^2}.$$

8.3.17: Given: $\dfrac{dy}{dx} = -\dfrac{x}{y}$, $y(12) = -5$. Then

$$y\,dy = -x\,dx; \qquad\qquad 2y\,dy = -2x\,dx; \qquad y^2 = C - x^2;$$

$$25 = [y(12)]^2 = C - 144; \qquad C = 169; \qquad\qquad y^2 = 169 - x^2;$$

$$y(x) = -\sqrt{169 - x^2}.$$

We took the negative root in the last step because $y(12) < 0$.

8.3.19: Given: $\dfrac{dy}{dx} = 3x^2 y^2 - y^2$, $y(0) = 1$. Then

$$y^{-2}\,dy = (3x^2 - 1)\,dx; \qquad y^{-1} = x - x^3 + C; \qquad y = \dfrac{1}{x - x^3 + C};$$

$$1 = y(0) = \dfrac{1}{C}; \qquad\qquad y(x) = \dfrac{1}{x - x^3 + 1}.$$

8.3.21: Given: $\dfrac{dy}{dx} = y + 1$, $y(0) = 1$.

$$\int \dfrac{dy}{y+1} = \int 1\,dx; \qquad \ln(y+1) = x + C;$$

$$y + 1 = e^{x+C} = Ae^x; \qquad y(x) = Ae^x - 1;$$

$$1 = y(0) = A - 1; \qquad A = 2.$$

Answer: $y(x) = 2e^x - 1$.

8.3.23: Given: $\dfrac{dy}{dx} = 2y - 3$, $y(0) = 2$.

$$\int \dfrac{2\,dy}{2y-3} = \int 2\,dx; \qquad \ln(2y-3) = 2x + C; \qquad 2y - 3 = e^{2x+C} = Ae^{2x};$$

$$y(x) = \dfrac{Ae^{2x} + 3}{2}; \qquad 2 = y(0) = \dfrac{A+3}{2}; \qquad y(x) = \dfrac{e^{2x} + 3}{2}.$$

8.3.25: Given: $\dfrac{dx}{dt} = 2(x - 1)$, $x(0) = 0$.

$$\int \dfrac{dx}{x-1} = \int 2\,dt; \qquad \ln(x-1) = 2t + C; \qquad x - 1 = e^{2t+C} = Ae^{2t};$$

$$x(t) = 1 + Ae^{2t}; \qquad 0 = x(0) = 1 + A; \qquad x(t) = 1 - e^{2t}.$$

8.3.27: Given: $\dfrac{dx}{dt} = 5(x+2)$, $x(0) = 25$.

$$\int \frac{dx}{x+2} = \int 5\,dt; \qquad \ln(x+2) = 5t + C; \qquad x+2 = e^{5t+C} = Ae^{5t};$$

$$x(t) = Ae^{5t} - 2; \qquad 25 = x(0) = A - 2; \qquad x(t) = 27e^{5t} - 2.$$

8.3.29: Given: $\dfrac{dv}{dt} = 10(10 - v)$, $v(0) = 0$.

$$\int \frac{dv}{v-10} = \int (-10)dt; \qquad \ln(v-10) = C - 10t; \qquad v - 10 = e^{C-10t} = Ae^{-10t};$$

$$v(t) = 10 + Ae^{-10t}; \qquad 0 = v(0) = 10 + A; \qquad v(t) = 10\left(1 - e^{-10t}\right).$$

8.3.31: Let the population at time t (in years) be $Q(t)$; $t = 0$ corresponds to the year 1990. From the data given in the problem, we know that $\dfrac{dQ}{dt} = (0.04)Q + 50000$; $Q(0) = 1,500,000$.

$$25\,\frac{dQ}{dt} = Q + 1,250,000;$$

$$\frac{1}{Q + 1,250,000} \cdot \frac{dQ}{dt} = \frac{1}{25};$$

$$\ln(Q + 1,250,000) = (0.04)t + C;$$

$$Q(t) + 1,250,000 = Ke^{t/25}.$$

Now from the condition $Q(0) = 1,500,000$ it follows that $1,500,000 + 1,250,000 = K$, so

$$Q(t) + 1,250,000 = 2,750,000e^{t/25}.$$

In the year 2010, we have $Q(20) = -1,250,000 + 2,750,000e^{0.8} \approx 4,870,238$, so the population in the year 2010 will be approximately 4.87 million people.

8.3.33: One effective way to derive a differential equation is to estimate the changes that take place in the dependent variable over a short interval $[t,\ t + \Delta t]$ where t is the independent variable. In this problem t is measured in months, and the change in the principal balance from time t to time $t + \Delta t$ is

$$P(t + \Delta t) - P(t) \approx rP(t)\,\Delta t - c\,\Delta t.$$

The reason is that the interest added to the principal balance is $rP(t)\,\Delta t$ and the monthly payment decreases the principal by $c\,\Delta t$. Thus

$$\frac{P(t + \Delta t) - P(t)}{\Delta t} \approx rP(t) - c. \tag{1}$$

The errors in this approximation will approach zero as $\Delta t \to 0$, and when we evaluate the limits of both sides of the approximation in (1) we obtain

$$\frac{dP}{dt} = rP - c, \qquad P(0) = P_0.$$

8.3.35: Let $P = P(t)$ denote the number of people who have heard the rumor after t days. Then

$$\frac{dP}{dt} = k(100000 - P); \qquad \int \frac{dP}{P - 100000} = \int (-k)\, dt;$$

$$\ln(P - 100000) = C - kt; \qquad P - 100000 = Ae^{-kt};$$

$$P(t) = 100000 - Ae^{-kt}.$$

We assume that $P(0) = 0$, so that $A = 100000$ and thus $P(t) = 100000\left(1 - e^{-kt}\right)$. Next, $P(7) = 10000$, so

$$100000\left(1 - e^{-7k}\right) = 10000; \qquad 1 - e^{-7k} = \frac{1}{10};$$

$$e^{-7k} = \frac{9}{10}; \qquad e^{7k} = \frac{10}{9}; \qquad k = \frac{1}{7}\ln\frac{10}{9}.$$

Half the population of the city will have heard the rumor when $P(T) = 50000$, so that

$$100000\left(1 - e^{-kT}\right) = 50000; \qquad 1 - e^{-kT} = \frac{1}{2}; \qquad e^{kT} = 2; \qquad T = \frac{\ln 2}{k} \approx 46.05169435.$$

Therefore half the population will have heard the rumor 46 days after it begins.

8.3.37: Assuming that you begin with nothing, the value of the account $P(t)$ (in thousands of dollars, at time t in years) satisfies $P(0) = P_0 = 0$. If I is your yearly investment, which we assume is made continuously (well approximated by equal monthly deposits), then

$$\frac{dP}{dt} = \frac{1}{10}P(t) + I; \qquad 10\frac{dP}{dt} = P + 10I;$$

$$\frac{10}{P + 10I}\, dP = dt; \qquad \frac{1}{P + 10I}\, dP = \frac{1}{10}\, dt;$$

$$\ln(P + 10I) = \frac{1}{10}t + C; \qquad P + 10I = A\exp\left(\frac{t}{10}\right);$$

$$P(t) = -10I + A\exp\left(\frac{t}{10}\right). \qquad 0 = P(0) = A - 10I:$$

$$A = 10I. \qquad P(t) = 10I\left[-1 + \exp\left(\frac{t}{10}\right)\right].$$

Thus the account will grow to a value of 5,000,000 in 30 years when

$$5000 = 10I\left[-1 + \exp\left(\frac{30}{10}\right)\right]: \qquad I = \frac{500}{e^3 - 1} \approx 26.19785.$$

443

Hence your monthly investment should be $I/12 \approx 2.18315$; that is, approximately \$2183.15 per month. It is of interest to note that your total investment will be \$654,946.21 and that the accrued interest will be \$4,345,053.79. You should now recompute the "real" answer to this problem under the assumption that your interest income will be subject to federal, state, and local taxes. Don't forget to compute the total tax you expect to pay over the 30 years of investing.

8.3.39: Given: $\dfrac{dN}{dt} = k(10000 - N)$, with time t measured in months.

$$-\frac{dN}{10000 - N} = -k\,dt;$$

$$\ln(10000 - N) = C_1 - kt;$$

$$10000 - N = Ce^{-kt}.$$

On January 1, $t = 0$ and $N = 1000$. On April 1, $t = 3$ and $N = 2000$. On October 1, $t = 9$; we want to determine the value of N then.

$$9000 = Ce^0 = C, \quad \text{so} \quad N(t) = 10000 - 9000e^{-kt}.$$

$$2000 = 10000 - 9000e^{-3k}, \quad \text{so} \quad 8 = 9e^{-3k}.$$

Therefore $k = \frac{1}{3}\ln\left(\frac{9}{8}\right)$. So $N(9) = 1000 - 9000e^{-9k} = 1000\left(10 - 9e^{-3\ln(9/8)}\right) \approx 3679$.

8.3.41: Let $t = 0$ when it began to snow, with $t = t_0$ at 7:00 A.M. Let $x(t)$ denote the distance traveled by the snowplow along the road, so that $x(t_0) = 0$. If $y = ct$ is the depth of the snow at time t, w is the width of the road, and $v = x'(t)$ is the velocity of the snowplow, then "plowing at a constant rate" means that the product wyv is constant. Hence $x(t)$ satisfies the differential equation

$$k\frac{dx}{dt} = \frac{1}{t}$$

where k is a positive constant. The solution for which $x(t_0) = 0$ satisfies the equation

$$t = t_0 e^{kx}.$$

We are given $x = 2$ when $t = t_0 + 1$ and $x = 4$ when $t = t_0 + 3$, and it follows that

$$t_0 + 1 = t_0 e^{2k} \quad \text{and} \quad t_0 + 3 = t_0 e^{4k}.$$

Elimination of t_0 yields the equation

$$e^{4k} - 3e^{2k} + 2 = 0; \quad \text{that is,} \quad \left(e^{2k} - 1\right)\left(e^{2k} - 2\right) = 0.$$

Thus it follows (because $k > 0$) that $e^{2k} = 2$. Hence $t_0 + 1 = 2t_0$, and so $t_0 = 1$. Therefore it began to snow at 6:00 A.M.

8.3.43: Substitution of $v = dy/dx$ in the differential equation for $y = y(x)$ yields

$$a\frac{dv}{dx} = \sqrt{1 + v^2},$$

and separation of variables then yields

$$\frac{1}{\sqrt{1 + v^2}}\, dv = \frac{1}{a}\, dx;$$

$$\sinh^{-1} v = \frac{x}{a} + C_1;$$

$$\frac{dy}{dx} = \sinh\left(\frac{x}{a} + C_1\right).$$

Because $y'(0) = 0$, it follows that $C_1 = 0$, and therefore

$$\frac{dy}{dx} = \sinh\left(\frac{x}{a}\right);$$

$$y(x) = a\cosh\left(\frac{x}{a}\right) + C.$$

Of course the (vertical) position of the x-axis may be adjusted so that $C = 0$, and the units in which T and ρ are measured may be adjusted so that $a = 1$. In essence, then, the shape of the hanging cable is the graph of $y = \cosh x$.

Section 8.4

8.4.1: $\rho(x) = \exp\left(\displaystyle\int 1\, dx\right) = e^x$:

$$e^x\frac{dy}{dx} + e^x y = 2e^x; \qquad e^x y(x) = 2e^x + C;$$

$$y(x) = 2 + Ce^{-x}. \qquad 0 = y(0) = C + 2:$$

$$y(x) = 2(1 - e^{-x}).$$

8.4.3: $\rho(x) = \exp\left(\displaystyle\int 3\, dx\right) = e^{3x}$:

$$e^{3x}\frac{dy}{dx} + 3e^{3x}y(x) = 2x; \qquad e^{3x}y(x) = x^2 + C; \qquad y(x) = e^{-3x}(x^2 + C).$$

445

8.4.5: Given: $\dfrac{dy}{dx} + \dfrac{2}{x}y = 3$, $y(1) = 5$:

$$\rho(x) = \exp\left(\int \frac{2}{x}\,dx\right) = \exp(2\ln x) = x^2.$$

Therefore

$$x^2 \frac{dy}{dx} + 2xy = 3x^2; \qquad x^2 y(x) = x^3 + C; \qquad y(x) = x + \frac{C}{x^2}.$$

$$5 = y(1) = 1 + C: \qquad\qquad C = 4. \qquad\qquad y(x) = x + \frac{4}{x^2}.$$

8.4.7: Given: $\dfrac{dy}{dx} + \dfrac{1}{2x}y = 5x^{-1/2}$.

$$\rho(x) = \exp\left(\int \frac{1}{2x}\,dx\right) = \exp\left(\frac{1}{2}\ln x\right) = x^{1/2}.$$

Therefore

$$x^{1/2}\frac{dy}{dx} + \frac{1}{2}x^{-1/2}y = 5; \qquad x^{1/2}y(x) = 5x + C; \qquad y(x) = 5x^{1/2} + Cx^{-1/2}.$$

8.4.9: Given $\dfrac{dy}{dx} - \dfrac{1}{x}y = 1$, an integrating factor is

$$\rho(x) = \exp\left(\int -\frac{1}{x}\,dx\right) = \exp(-\ln x) = \frac{1}{x}.$$

Thus

$$\frac{1}{x}\cdot\frac{dy}{dx} - \frac{1}{x^2}\cdot y(x) = \frac{1}{x}; \qquad \frac{1}{x}\cdot y(x) = C + \ln x; \qquad y(x) = Cx + x\ln x.$$

$$7 = y(1) = C: \qquad\qquad y(x) = 7x + x\ln x.$$

8.4.11: Given: $x\dfrac{dy}{dx} + (1 - 3x)y = 0$, $y(1) = 0$, write the equation in the form

$$\frac{dy}{dx} + \left(\frac{1}{x} - 3\right)y = 0.$$

Then an integrating factor is

$$\rho(x) = \exp\left(\int \left[\frac{1}{x} - 3\right] dx\right) = xe^{-3x}.$$

Therefore

$$xe^{-3x}\frac{dy}{dx} + \left(e^{-3x} - 3xe^{-3x}\right)\cdot y = 0; \qquad xe^{-3x}y(x) = C; \qquad y(x) = \frac{C}{x}e^{3x}.$$

$$0 = y(0) = 0: \qquad\qquad y(x) \equiv 0.$$

446

8.4.13: $\rho(x) = \exp\left(\displaystyle\int 1\,dx\right) = e^x$:

$$e^x \frac{dy}{dx} + e^x y(x) = e^{2x}; \qquad e^x y(x) = C + \frac{1}{2}e^{2x}; \qquad y(x) = Ce^{-x} + \frac{1}{2}e^x.$$

$$1 = y(0) = C + \frac{1}{2}: \qquad C = \frac{1}{2}. \qquad y(x) = \frac{1}{2}\left(e^x + e^{-x}\right) = \cosh x.$$

8.4.15: An integrating factor is $\rho(x) = \exp\left(\displaystyle\int 2x\,dx\right) = \exp(x^2)$. Hence

$$\exp(x^2)\frac{dy}{dx} + 2x\exp(x^2)y(x) = x\exp(x^2); \qquad y(x)\exp(x^2) = C + \frac{1}{2}\exp(x^2);$$

$$y(x) = \frac{1}{2} + C\exp(-x^2). \qquad -2 = y(0) = \frac{1}{2} + C:$$

$$C = -\frac{5}{2}. \qquad y(x) = \frac{1 - 5\exp(-x^2)}{2}.$$

8.4.17: $\dfrac{dy}{dx} + \dfrac{1}{1+x}y = \dfrac{\cos x}{1+x}$, so an integrating factor is

$$\rho(x) = \exp\left(\int \frac{1}{1+x}\,dx\right) = 1 + x.$$

Therefore

$$(1+x)\frac{dy}{dx} + y = \cos x; \qquad (1+x)y(x) = C + \sin x; \qquad y(x) = \frac{C + \sin x}{1 + x}.$$

$$1 = y(0) = C: \qquad y(x) = \frac{1 + \sin x}{1 + x}.$$

8.4.19: An integrating factor is

$$\rho(x) = \exp\left(\int \cot x\,dx\right) = \exp(\ln \sin x) = \sin x.$$

Hence

$$(\sin x)\frac{dy}{dx} + (\cos x)y(x) = \sin x \cos x; \qquad y(x)\sin x = C + \frac{1}{2}\sin^2 x; \qquad y(x) = C\csc x + \frac{1}{2}\sin x.$$

Mathematica yields

```
DSolve[ y'[x] + y[x] Cot[x] == Cos[x], y[x], x]
```

$$y(x) = C_1 \csc x - \frac{1}{2}\cos x \cot x.$$

8.4.21: An integrating factor for the equation $\dfrac{dy}{dx} - 2xy = 1$ is

$$\rho(x) = \exp\left(\int -2x\,dx\right) = \exp\left(-x^2\right),$$

which yields

$$\exp\left(-x^2\right)\frac{dy}{dx} - 2x\exp\left(-x^2\right)y(x) = \exp\left(-x^2\right);$$

$$\exp\left(-x^2\right)y(x) = \int_0^x \exp\left(-t^2\right)\,dt + C = \frac{\sqrt{\pi}}{2}\operatorname{erf}(x) + C;$$

$$y(x) = \left[\exp\left(x^2\right)\right]\left[\frac{\sqrt{\pi}}{2}\operatorname{erf}(x) + C\right].$$

8.4.23: Let $A(t)$ denote the amount of salt (in kilograms) in the tank at time t (in seconds) and measure volume in liters. Then

$$\frac{dA}{dt} = -\frac{1}{200}A(t), \quad A(0) = 100,$$

and the solution of this familiar initial value problem is

$$A(t) = 100\exp\left(-\frac{t}{200}\right).$$

Hence $A(t) = 10$ when $t = 200\ln 10 \approx 460.517$ (s), about 7 min 40.517 s.

8.4.25: Substitute of $V = 1640$ km^3 and $r = 410$ km^3/y in the last equation in the solution of Example 5 yields

$$t = \frac{V}{r}\ln 4 = 4\ln 4 \approx 5.5452$$

(years).

8.4.27: If $V(t)$ is the volume of brine (in gallons) in the tank at time t (in minutes), then it's easy to see that $V(t) = 100 + 2t$. Let $x(t)$ denote the number of pounds of salt in the tank at time t. Then

$$\frac{dx}{dt} = 5 - \frac{3x}{100 + 2t}, \quad x(0) = 50.$$

An integrating factor is

$$\rho(t) = \exp\left(\int \frac{3}{100 + 2t}\,dt\right) = \exp\left(\tfrac{3}{2}\ln(100 + 2t)\right) = (100 + 2t)^{3/2},$$

and thereby the differential equation takes the form

$$(100 + 2t)^{3/2}\frac{dx}{dt} + 3(100 + 2t)^{1/2}x(t) = 5(100 + 2t)^{3/2};$$

$$(100 + 2t)^{3/2}x(t) = (100 + 2t)^{5/2} + C;$$

$$x(t) = 100 + 2t + \frac{C}{(100 + 2t)^{3/2}}.$$

Then the initial condition $x(0) = 50$ yields $C = -50000$, and therefore

$$x(t) = 100 + 2t - \frac{50000}{(100 + 2t)^{3/2}}.$$

The tank is full when $t = 150$, and at that time the tank will contain

$$x(150) = \frac{1575}{4} = 393.75$$

pounds of salt.

8.4.29: Part (a):

$$A(t + \Delta t) \approx A(t) + (0.12)(30e^{t/20}) \Delta t + (0.06)A(t);$$

$$\frac{A(t + \Delta t) - A(t)}{\Delta t} \approx (3.6)e^{t/20} + (0.06)A(t);$$

$$\frac{dA}{dt} - (0.06)A(t) = (3.6)e^{t/20} = (3.6)e^{(0.05)t}, \quad A(0) = 0.$$

An integrating factor for this linear equation is

$$\rho(t) = \exp\left(\int (-0.06) \, dt\right) = e^{(-0.06)t}.$$

Thus

$$e^{(-0.06)t}\frac{dA}{dt} - (0.06)e^{(-0.06)t}A(t) = (3.6)e^{(-0.01)t};$$

$$e^{(-0.06)t}A(t) = C = 360e^{(-0.01)t};$$

$$A(t) = Ce^{(0.06)t} - 360e^{(0.05)t}.$$

Now $0 = A(0) = C - 360$, so

$$A(t) = 360\left[e^{(0.06)t} - e^{(0.05)t}\right].$$

Part (b): $A(40) = 360\left(e^{12/5} - e^2\right) \approx 1308.28330$. Units are in thousand of dollars, so this amounts to $1,308,283.30 (less taxes).

8.4.31: Let $v(t)$ denote the velocity of the sports car (in kilometers per hour) at time t (in seconds). We thereby obtain the linear initial value problem

$$\frac{dv}{dt} = k(250 - v), \quad v(0) = 0.$$

This equation is easy to solve by the method of separation of variables, but we choose to solve it as a linear differential equation:

$$\frac{dv}{dt} + kv = 250k.$$

An integrating factor is

$$\rho(t) = \exp\left(\int k\,dt\right) = e^{kt},$$

and thus

$$e^{kt}\frac{dv}{dt} + ke^{kt}v = 250ke^{kt};$$

$$e^{kt}v(t) = 250e^{kt} + C;$$

$$v(t) = 250 + Ce^{-kt}.$$

The initial condition yields $C = -250$, so that

$$v(t) = 250\left(1 - e^{-kt}\right).$$

We are also given $v(10) = 100$, and thus

$$100 = 250\left(1 - e^{-10k}\right); \quad \text{thus} \quad k = \frac{1}{10}\ln\frac{5}{3}.$$

To find when $v(t) = 200$, we solve

$$200 = 250\left(1 - e^{-kt}\right); \qquad \frac{4}{5} = 1 - e^{-kt};$$

$$e^{-kt} = \frac{1}{5}; \qquad t = \frac{\ln 5}{k} = \frac{10\ln 5}{\ln(5/3)}.$$

Answer: About 31.506601 seconds.

8.4.33: We use the formulas given in the statement of Problem 32:

$$v(t) = v_0 e^{-kt} \quad \text{and} \quad x(t) = x_0 + \frac{v_0}{k}\left(1 - e^{-kt}\right).$$

The initial condition $v_0 = 40$ and the information $v(10) = 20$ yields

$$20 = v(10) = 40e^{-10k}; \qquad e^{10k} = 2; \qquad k = \frac{1}{10}\ln 2.$$

In the solution of Problem 32 we saw that the total distance traveled by the motorboat is

$$\left[\lim_{t\to\infty} x(t)\right] - x_0 = \frac{v_0}{k},$$

and in this case we have

$$\frac{v_0}{k} = \frac{400}{\ln 2} \approx 577.078 \quad \text{(ft)}.$$

8.4.35: The formulas in Problem 34 for velocity and position (distance traveled) are

$$v(t) = \frac{v_0}{kv_0 t + 1} \quad \text{and} \quad x(t) = x_0 + \frac{1}{k}\ln(kv_0 t + 1).$$

From the data given in Problem 33, we have $v_0 = 40$ and $v(10) = 20$. Thus

$$20 = v(10) = \frac{40}{1 + 400k}; \qquad 20 + 8000k = 40;$$

$$8000k = 20; \qquad k = \frac{1}{400}.$$

Therefore

$$v(t) = \frac{40}{1 + (t/10)} = \frac{400}{t + 10}.$$

To find the total distance traveled, we may assume that $x_0 = 0$. Then

$$x(t) = 400\ln\left(1 + \frac{t}{10}\right), \qquad \text{so that}$$

$$x(60) = 400\ln 7 \approx 778.364 \quad \text{(ft)}.$$

8.4.37: With the usual notation of this section, we have

$$\frac{dv}{dt} = 10 - \frac{1}{10}v; \qquad v(0) = v_0 = 0, \quad x(0) = x_0 = 0.$$

An integration factor for this linear equation is $\rho(t) = e^{t/10}$, and thus (in the usual way)

$$e^{t/10}v(t) = 100e^{t/10} + C; \qquad v(t) = 100 + Ce^{-t/10};$$

$$0 = v(0) = 100 + C : \qquad v(t) = 100\left(1 - e^{-t/10}\right).$$

Part (a): It is clear that the limiting velocity of the car is 100 ft/s. Part (b):

$$x(t) = C_1 + 100t + 1000e^{-t/10}; \qquad 0 = x(0) = C_1 + 1000;$$

$$x(t) = 100t - 1000\left(1 - e^{-t/10}\right).$$

The car reaches 90% of its limiting velocity at that time t for which

$$v(t) = 90: \qquad 1 - e^{-t/10} = \frac{9}{10}; \qquad e^{-t/10} = \frac{1}{10}.$$

451

Therefore $t = 10 \ln 10 \approx 23.025851$ (s). The distance the car travels from rest until that time is

$$x(10 \ln 10) = -900 + 1000 \ln 10 \approx 1402.585093 \quad \text{(ft)}.$$

8.4.39: We are to solve the initial value problem

$$\frac{dv}{dt} = 5 - \frac{1}{10}v, \quad v(0) = 0.$$

The usual integrating factor $\rho(t) = e^{t/10}$ yields the solution $v(t) = 50(1 - e^{-t/10})$, and it is clear that $v(t) \to 50$ as $t \to +\infty$.

8.4.41: Here we have $y_0 = 0$, $v_0 = 49$, and $v_\tau = -g/\rho = -245$. Thus the velocity and position (altitude) functions are

$$v(t) = 294e^{-t/25} - 245 \quad \text{and}$$

$$y(t) = 7350 - 245t - 7350e^{-t/25}.$$

If the maximum height occurs at time t_m, then we solve $v(t_m) = 0$ and find that

$$t_m = 25 \ln \frac{294}{245} \approx 4.5580389198,$$

and hence the maximum height is $y(t_m) \approx 108.2804646370$ (meters). Impact occurs when $y(t) = 0$; that is, when

$$7350 - 245t - 7350e^{-t/25} = 0.$$

A few iterations of Newton's method with initial "guess" $t_0 = 10$ yields the solution $t = t_i \approx 9.4109499312$. The impact speed will be

$$|v(t_i)| \approx 43.2273093261 \quad \text{(m/s)}.$$

8.4.43: Beginning with

$$\frac{dv}{dt} = -g - kv, \quad v(0) = v_0,$$

the integrating factor $\rho(t) = e^{kt}$ yields

$$e^{kt}\frac{dv}{dt}ke^{kt}v(t) = -ge^{kt}; \qquad e^{kt}v(t) = C - \frac{g}{k}e^{kt};$$

$$v(t) = Ce^{-kt} - \frac{g}{k}. \qquad v_0 = v(0) = C - \frac{g}{k}:$$

$$C = v_0 + \frac{g}{k}.$$

Therefore

$$v(t) = \left(v_0 + \frac{g}{k}\right)e^{-kt} - \frac{g}{k} = v_0 e^{-kt} + \frac{g}{k}(e^{-kt} - 1).$$

Finally, the limiting velocity of the projectile is

$$v_\tau = \lim_{t \to \infty} v(t) = -\frac{g}{k}.$$

Section 8.5

8.5.1: First note that $\dfrac{1}{x(1-x)} = \dfrac{1}{x} + \dfrac{1}{1-x}$. Then

$$\int \frac{1}{x(1-x)}\,dx = \int 1\,dt; \qquad \ln\frac{x}{1-x} = t + C_1;$$

$$\frac{x}{1-x} = Ce^t; \qquad \frac{2}{1-2} = \frac{x(0)}{1-x(0)} = C = -2;$$

$$\frac{x}{1-x} = -2e^t; \qquad x = 2(x-1)e^t;$$

$$(1 - 2e^t)x = -2e^t; \qquad x(t) = \frac{2e^t}{2e^t - 1} = \frac{2}{2 - e^{-t}}.$$

8.5.3: First note that $\dfrac{1}{1-x^2} = \dfrac{1}{2}\left(\dfrac{1}{1-x} + \dfrac{1}{1+x}\right)$. Hence

$$\int \frac{1}{1-x^2}\,dx = \int 1\,dt; \qquad \int\left(\frac{1}{1+x} + \frac{1}{1-x}\right)dx = 2t + C_1;$$

$$\ln\frac{1+x}{1-.x} = 2t + C_1; \qquad \frac{1+x}{1-x} = Ce^{2t};$$

$$\frac{1+3}{1-3} = C = -2; \qquad (1+x) = -2e^{2t}(1-x);$$

$$(1 - 2e^{2t})x = -(1 + 2e^{2t}); \qquad x(t) = \frac{2e^{2t} + 1}{2e^{2t} - 1}.$$

8.5.5: First note that

$$\frac{1}{x(5-x)} = \frac{1}{5}\left(\frac{1}{x} + \frac{1}{5-x}\right).$$

Therefore

$$\left(\frac{1}{x} + \frac{1}{5-x}\right) dx = 15\, dt; \qquad \ln\left|\frac{x}{5-x}\right| = 15t + C;$$

$$\left|\frac{x}{5-x}\right| \quad Ae^{15t} \qquad \text{(where } A = e^C > 0);$$

$$x(0) = 8, \quad \text{so} \quad A = \frac{8}{3}. \qquad \text{Also } x > 0 \text{ and } 5 - x < 0.$$

Hence

$$\frac{x}{x-5} = \frac{8}{3}e^{15t}; \qquad 3x = 8xe^{15t} - 40e^{15t};$$

$$x(t) = -\frac{40e^{15t}}{3 - 8e^{15t}}; \qquad x(t) = \frac{40}{8 - 3e^{-15t}}.$$

8.5.7: We are to solve

$$\left(\frac{1}{x} + \frac{1}{7-x}\right) dx = 28\, dt, \qquad x(0) = 11.$$

$$\left|\frac{x}{7-x}\right| = 28t + C; \qquad \left|\frac{x}{7-x}\right| = Ae^{28t} \quad (A = e^C > 0).$$

$$x(0) = 11: \qquad A = \frac{11}{4}, \quad x > 0, \quad 7 - x < 0.$$

$$4x = 11(x-7)e^{28t}; \qquad 4x - 11xe^{28t} = -77e^{28t};$$

$$x(t) = \frac{77e^{28t}}{11e^{28t} - 4}; \qquad x(t) = \frac{77}{11 - 4e^{-28t}}.$$

8.5.9: Given:

$$\frac{dP}{dt} = kP^{1/2}; \qquad P(0) = 100, \quad P'(0) = 20.$$

Separation of variables yields

$$P^{-1/2}\, dP = k\, dt; \qquad 2P^{1/2} = C + kt.$$

$$20 = 2(P_0)^{1/2} = C: \qquad P(t) = \left(10 + \tfrac{1}{2}kt\right)^2.$$

$$20 = P'(0) = k \cdot 10: \quad k = 2; \qquad P(t) = (10 + t)^2.$$

Therefore in one year there will be $P(12) = 22^2 = 484$ rabbits.

8.5.11: If $\beta = aP^{-1/2}$ and $\delta = bP^{-1/2}$, then

$$\frac{dP}{dt} = (a - b)P^{-1/2} \cdot P = kP^{1/2}$$

where $k = a - b$. In part (b) we will use the information that $P_0 = P(0) = 100$ and $P(6) = 169$. For Part (a):

$$P^{-1/2} \, dP = k \, dt; \qquad 2P^{1/2} = C + kt \qquad (C > 0);$$

$$P(t) = \left(\tfrac{1}{2}kt + \tfrac{1}{2}C\right)^2. \qquad P_0 = \left(\tfrac{1}{2}C\right)^2: \quad \tfrac{1}{2}C = \sqrt{P_0} \quad \text{(because } C > 0).$$

$$\text{Therefore} \quad P(t) = \left(\tfrac{1}{2}kt + \sqrt{P_0}\right)^2.$$

Part (b): Here we have

$$P(t) = \left(\tfrac{1}{2}kt + 10\right)^2.$$

Thus

$$196 = P(6) = (3k + 10)^2; \qquad 3k + 10 = \pm 13;$$

$$k = 1 \quad \text{(because } k > 0). \qquad P(t) = \left(10 + \tfrac{1}{2}t\right)^2.$$

Thus after one year there will be $P(12) = 16^2 = 256$ fish in the lake.

8.5.13: The birth rate is $\beta = aP(t)$ and the death rate is $\delta = bP(t)$ where $a > b > 0$. Thus

$$\frac{dP}{dt} = [aP(t) - bP(t)] \cdot P(t) = kP^2$$

where $k = a - b > 0$. As usual, let $P_0 = P(0)$. Then

$$-\frac{1}{P^2} \, dP = -k \, dt; \qquad \frac{1}{P} = C - kt; \qquad P(t) = \frac{1}{C - kt}.$$

Part (a):

$$P_0 = P(0) = \frac{1}{C}, \quad \text{so} \quad P(t) = \frac{P_0}{1 - kP_0 t}.$$

Part (b): With $P_0 = 6$, $P(10) = 9$, and t measured in months:

$$P(t) = \frac{6}{1 - 6kt}; \qquad 9 = P(10) = \frac{6}{1 - 60k};$$

$$k = \frac{1}{180}; \qquad P(t) = \frac{180}{30 - t}.$$

Because $P(t) \to +\infty$ as $t \to 30^-$, "doomsday" occurs when $t = 30$ months.

8.5.15: Measure P in millions and t in years, with $t = 0$ corresponding to the year 1940. Given: $P(0) = 100$, $P'(0) = 1$, and

$$\frac{dP}{dt} = kP(200 - P) \qquad (k \text{ constant}). \tag{1}$$

Note that $\dfrac{1}{P(200 - P)} = \dfrac{1}{200}\left(\dfrac{1}{P} + \dfrac{1}{200 - P}\right)$. Thus

$$\int \frac{1}{P(200 - P)}\, dP = \int k\, dt; \qquad \int \left(\frac{1}{P} + \frac{1}{200 - P}\right) dP = 200kt + C_1;$$

$$\ln \frac{P}{200 - P} = 200kt + C_1; \qquad \frac{P}{200 - P} = Ce^{200kt}$$

$$\frac{100}{100} = Ce^0 = C; \qquad \frac{P}{200 - P} = e^{200kt}.$$

By Eq. (1), $1 = P'(0) = k \cdot 100(200 - 100) = 10000k$, so $k = 1/10000$. Therefore

$$\frac{P}{200 - P} = e^{t/50}; \qquad \frac{200 - P}{P} = e^{-t/50};$$

$$\frac{200}{P} = 1 + e^{-t/50}; \qquad P(t) = \frac{200}{1 + e^{-t/50}}.$$

Therefore the population in the year 2000 (corresponding to $t = 60$) will be

$$P(60) = \frac{200}{1 + e^{-6/5}} \approx 153.7 \quad \text{(million)}.$$

8.5.17: Given:

$$\frac{dx}{dt} = \frac{4}{5}x - \frac{1}{250}x^2 = \frac{200x - x^2}{250}; \quad x(0) = 50.$$

First note that $\dfrac{1}{200x - x^2} = \dfrac{1}{x(200 - x)} = \dfrac{1}{200}\left(\dfrac{1}{x} + \dfrac{1}{200 - x}\right)$. Then

$$\frac{1}{200}\int \left(\frac{1}{x} + \frac{1}{200 - x}\right) dx = \int \frac{1}{250}\, dt; \qquad \ln \frac{x}{200 - x} = \frac{4}{5}t + C_1;$$

$$\frac{x}{200 - x} = Ce^{4t/5}; \qquad \frac{1}{3} = Ce^0 = C;$$

$$\frac{x}{200 - x} = \frac{1}{3}e^{4t/5}; \qquad x = \frac{200}{3}e^{4t/5} - \frac{1}{3}e^{4t/5}x;$$

$$\left(1 + \frac{1}{3}e^{4t/5}\right)x = \frac{200}{3}e^{4t/5}; \qquad x(t) = \frac{200e^{4t/5}}{3 + e^{4t/5}} = \frac{200}{1 + 3e^{-4t/5}}.$$

Part (a): We need to solve $x(T) = 100$:

$$100 = \frac{200}{1 + 3e^{-4T/5}}; \qquad 1 + 3e^{-4T/5} = 2;$$

$$e^{-4T/5} = \frac{1}{3}; \qquad \frac{4}{5}T = \ln 3.$$

456

Thus $T = \dfrac{5}{4} \ln 3 \approx 1.373$ (seconds).

Part (b):   As $t \to +\infty$, $x(t) \to 200$. So there is no "maximum" amount of salt that will dissolve, but for all practical purposes, the maximum is 200 g. (The amount that dissolves becomes arbitrarily close to, but remains always less than, 200 g.)

8.5.19:   We are given an animal population $P(t)$ at time t (in years) such that

$$\frac{dP}{dt} = kP^2 - \frac{1}{100}P; \qquad P(0) = 200, \quad P'(0) = 2$$

where k is a constant. Substitution of the numerical data in the differential equation yields

$$2 = P'(0) = 40000k - 2, \quad \text{so that} \quad k = \frac{1}{10000}.$$

Thus

$$\frac{dP}{dt} = \frac{P^2 - 100}{10000} = \frac{P(P - 100)}{10000}.$$

Because $\dfrac{1}{P(P - 100)} = \dfrac{1}{100}\left(\dfrac{1}{P - 100} - \dfrac{1}{P} \right)$,   we have

$$\frac{1}{100} \int \left(\frac{1}{P - 100} - \frac{1}{P} \right) dP = \int \frac{1}{10000}\, dt.$$

Therefore

$$\ln \frac{P - 100}{P} = C_1 + \frac{1}{100}t; \qquad \frac{P - 100}{P} = Ce^{t/100};$$

$$\frac{200 - 100}{200} = C = \frac{1}{2}; \qquad 1 - \frac{100}{P} = \frac{1}{2}e^{t/100};$$

$$\frac{100}{P} = 1 - \frac{1}{2}e^{t/100}; \qquad P(t) = \frac{100}{1 - \frac{1}{2}e^{t/100}} = \frac{200}{2 - e^{t/100}}.$$

Part (a):   We need to solve $P(T) = 1000$:

$$1000 = P(t) = \frac{200}{2 - e^{T/100}}; \qquad 2 - e^{T/100} = \frac{1}{5};$$

$$e^{T/100} = \frac{9}{5}; \qquad T = 100 \ln \frac{9}{5}.$$

Answer: In approximately 58.779 years.   Part (b): Doomsday will occur when the denominator in $P(t)$ is zero; that is, when $e^{t/100} = 2$, so that $t = 100 \ln 2$.   Answer: In approximately 69.315 years.

8.5.21: If we write $P' = bP(a/b - P)$ we see that $M = a/b$. Hence

$$\frac{B_0 P_0}{D_0} = \frac{(aP_0)P_0}{bP_0^2} = \frac{a}{b} = M.$$

Note also (for Problems 22 and 23) that $a = B_0/P_0$ and $b = D_0 P_0^2 = k$.

8.5.23: The relations in Problem 21 give $k = 1/2400$ and $M = 180$. The solution is $P(t) = 43200/(240 - 60e^{-3t/80})$. We find that $P = 1.05M$ after about 44.22 months.

8.5.25: The relations in Problem 24 give $k = 1/1000$ and $M = 90$. The solution is $P(t) = 9000/(100 - 10e^{9t/10})$. We find that $P = 10M$ after about 24.41 months.

8.5.27: We work in thousands of persons, so $M = 100$ for the total fixed population. We substitute $M = 100$, $P'(0) = 1$, and $P_0 = 50$ in the logistic equation, and thereby obtain

$$1 = k(50)(100 - 50), \quad \text{so} \quad k = 0.0004.$$

If t denotes the number of days until 80 thousand people have heard the rumor, then Eq. (7) gives

$$80 = \frac{50 \times 100}{50 + (100 - 50)e^{-0.04t}},$$

so that t is approximately 34.66. Thus the rumor will have spread to 80% of the population in a little less than 35 days.

8.5.29: The solution of the initial value problem given in the statement of Problem 29 is

$$P(t) = \frac{1}{\dfrac{1489}{313500} + \dfrac{341881}{1358500}\exp\left(-\dfrac{627t}{20000}\right)} \approx \frac{1}{0.0047496013 + (0.0276636323)e^{-(0.03135)t}}. \tag{1}$$

Part (a): The year 1930 corresponds to $t = 140$, for which the equation in (1) predicts $P(140) \approx 127.008$ (million). Part (b):

$$\lim_{t \to \infty} P(t) = \frac{313500}{1489} \approx 210.544 \quad \text{(million)}.$$

Part (c): The following table gives the year, the population predicted by Eq. (1), and the actual U.S. population (in millions) from the 1992 *World Almanac and Book of Facts* (New York: Pharos Books, 1991, pp. 74–75). The population data are rounded.

Year	Predicted population	Actual population
1790	3.900	3.930
1800	5.300	5.308
1810	7.185	7.240
1820	9.708	9.638
1830	13.061	12.861
1840	17.471	17.063
1850	23.193	23.192
1860	30.499	31.443
1870	39.616	38.558
1880	50.690	50.189
1890	63.707	62.980
1900	78.427	76.212
1910	94.362	92.228
1920	110.819	106.022
1930	127.008	123.203
1940	142.191	132.165
1950	155.803	151.326
1960	167.525	179.323
1970	177.272	203.302
1980	185.146	226.542
1990	191.358	248.710
2000	196.169	*281.422

*Note: This datum is from `www.census.gov/main/www/cen2000/html`, where the U.S. population on April 1, 2000 is given as 281,421,906.

8.5.31: We begin with

$$\frac{dP}{dt} = kP(M - P), \qquad P(0) = P_0. \tag{6}$$

Thus

$$\frac{1}{P(M - P)} \, dP = k \, dt; \qquad \frac{1}{M} \left(\frac{1}{P} + \frac{1}{M - P} \right) dP = k \, dt;$$

$$\ln \left| \frac{P}{M - P} \right| = kMt + C; \qquad \left| \frac{P}{M - P} \right| = Ae^{kMt} \qquad (A = e^C > 0);$$

$$\frac{P}{M - P} = Be^{kMt} \qquad (B = \pm A).$$

For later use, we note at this point that $B = \dfrac{P_0}{M - P_0}$. Next,

$$P = MBe^{kMt} - PBE^{kMt};$$

$$P(t) = \frac{MBe^{kMt}}{1 + Be^{kMt}} = \frac{MB}{e^{-kMt} + B} = \frac{\dfrac{MP_0}{M - P_0}}{e^{-kMt} + \dfrac{P_0}{M - P_0}}.$$

Therefore

$$P(t) = \frac{MP_0}{(M - P_0)e^{-kMt} + P_0}. \tag{7}$$

8.5.33: We begin with

$$\frac{dP}{dt} = kP(M - P) \tag{3}$$

and differentiate both sides with respect to t (using the chain rule on the right-hand side). Thus

$$\frac{d^2P}{dP^2} = \left\{ \frac{d}{dP} [kP(M - P)] \right\} \cdot \frac{dP}{dt} = k(M - P - P) \cdot kP(M - P)$$

$$= k^2 P(M - P)(M - 2P) = 2k^2 P(P - M)\left(P - \tfrac{1}{2}M\right).$$

The conclusions stated in Problem 33 are now clear.

Section 8.6

8.6.1: Characteristic equation: $r^2 - 7r + 10 = (r - 2)(r - 5) = 0$: $r_1 = 2$, $r_2 = 5$;

$$y(x) = c_1 e^{2x} + c_2 e^{5x}.$$

460

8.6.3: C. E.: $4r^2 - 4r - 3 = (2r+1)(2r-3) = 0$: $r_1 = -\frac{1}{2}$, $r_2 = \frac{3}{2}$; $y(x) = c_1 e^{-x/2} + c_2 e^{3x/2}$.

8.6.5: C. E.: $r^2 + 4r + 1 = 0$:

$$r = \frac{-4 \pm \sqrt{16-4}}{2} = -2 \pm \sqrt{3};$$

$$y(x) = c_1 \exp\left(\left[-2 + \sqrt{3}\right]x\right) + c_2 \exp\left(\left[-2 - \sqrt{3}\right]x\right).$$

8.6.7: C. E.: $4r^2 + 12r + 9 = 0$; $(2r+3)^2 = 0$; $r_1 = r_2 = -\frac{3}{2}$. Hence

$$y(x) = (c_1 + c_2 x)e^{-3x/2}.$$

8.6.9: C. E.: $25r^2 - 20r + 4 = 0$; $(5r-2)^2 = 0$; $r_1 = r_2 = \frac{2}{5}$. Therefore

$$y(x) = (c_1 + c_2 x)e^{2x/5}.$$

8.6.11: Characteristic equation: $r^2 + 6r + 13 = 0$:

$$r = \frac{-6 \pm \sqrt{36-52}}{2} = -3 \pm 2i.$$

Therefore $y(x) = e^{-3x}\left(c_1 \cos 2x + c_2 \sin 2x\right)$.

8.6.13: Characteristic equation: $9r^2 + 6r + 226 = 0$:

$$r = \frac{-6 \pm \sqrt{36-8136}}{18} = \frac{-6 \pm 90i}{18} = -\frac{1}{3} \pm 5i;$$

$$y(x) = e^{-x/3}(c_1 \cos 5x + c_2 \sin 5x).$$

8.6.15: Characteristic equation: $2r^2 - 11r + 12 = 0$:

$$r = \frac{11 \pm \sqrt{121-96}}{4} = \frac{11 \pm 5}{4}.$$

Thus the general solution is $y(x) = c_1 e^{3x/2} + c_2 e^{4x}$. Also

$$y'(x) = \frac{3}{2}c_1 e^{3x/2} + 4c_2 e^{4x};$$

$$5 = y(0) = c_1 + c_2;$$

$$15 = y'(0) = \frac{3}{2}c_1 + 4c_2.$$

Therefore $c_1 = 2$, $c_2 = 3$, and $y(x) = 2e^{3x/2} + 3e^{4x}$.

461

8.6.17: The roots of the characteristic equation are $r_1 = 7$ and $r_2 = 11$; the solution of the given initial value problem is $y(x) = 9e^{7x} - 5e^{11x}$.

8.6.19: The roots of the characteristic equation are $r_1 = r_2 = -11$; the solution of the given initial value problem is $y(x) = (2 - 3x)e^{-11x}$.

8.6.21: The roots of the characteristic equation are $r_1 = 5i$ and $r_2 = -5i$; the solution of the given initial value problem is $y(x) = 7\cos 5x + 2\sin 5x$.

8.6.23: The roots of the characteristic equation are $r_1 = -2 + 4i$ and $r_2 = -2 - 4i$; the solution of the given initial value problem is $y(x) = e^{-2x}(9\cos 4x + 7\sin 4x)$.

8.6.25: The roots of the characteristic equation are $r_1 = -\frac{1}{2} + 5i$ and $r_2 = -\frac{1}{2} - 5i$; the solution of the given initial value problem is $y(x) = e^{-x/2}(10\cos 5x + 6\sin 5x)$.

8.6.27: The roots of the characteristic equation are $r_1 = 0$ and $r_2 = -10$:

$$r(r + 10) = 0; \qquad r^2 + 10r = 0; \qquad y'' + 10y' = 0.$$

8.6.29: The roots of the characteristic equation are $r_1 = r_2 = -10$:

$$(r + 10)^2 = 0; \qquad r^2 + 20r + 100 = 0; \qquad y'' + 20y' + 100y = 0.$$

8.6.31: The roots of the characteristic equation are $r_1 = r_2 = 0$:

$$(r - 0)(r - 0) = 0; \qquad r^2 = 0; \qquad y'' = 0.$$

8.6.33: The roots of the characteristic equation are $r_1 = -5 + \frac{1}{5}i$ and $r_2 = -5 - \frac{1}{5}i$:

$$\left(r + 5 - \frac{1}{5}i\right)\left(r + 5 + \frac{1}{5}i\right) = 0; \qquad r^2 + 10r + \frac{626}{25} = 0; \qquad 25y'' + 250y' + 626y = 0.$$

8.6.35: The characteristic equation is $r^2 + 25 = 0$, and hence the general solution is

$$y(x) = c_1 \cos 5x + c_2 \sin 5x.$$

Part (a): The condition $y(0) = 0$ yields $c_1 = 0$, and hence $y(x) = c_2 \sin 5x$. But the second condition $y(\pi) = 0$ is satisfied for every choice of the constant c_2. Moreover, if so, then

$$y'' + 25y = -25c_2 \sin x + 25c_2 \sin x \equiv 0$$

462

for every choice of the constant c_2, and therefore the given *boundary value problem* has infinitely many solutions, one for every choice of the constant c_2. Part (b): The condition $y(0) = 0$ yields $c_1 = 0$, and hence $y(x) = c_2 \sin 5x$. But the second condition

$$0 = y(3) = c_2 \sin 15 \approx (0.6502878401)c_2$$

is satisfied only if $c_2 = 0$. Therefore the given *boundary value problem* has at most the trivial solution $y(x) \equiv 0$ (and substitution verifies that, indeed, this is a solution). Hence the problem has no nontrivial solutions. The point of this problem is to draw a very sharp distinction between second-order initial value problems, with initial conditions $y(a) = b_0$, $y'(a) = b_1$ given at the *same* abscissa, and second-order *boundary value problems*, which typically have values of y (and/or y') imposed at two different values of the abscissa. In particular, the vital existence-uniqueness theorem stated in Section 8.6 does not hold for such boundary value problems.

Section 8.7

8.7.1: Given: $2x'' + 50x = 0$; $x(0) = 4$, $x'(0) = 15$. The general solution and its derivative are

$$x(t) = A \cos 5t + B \sin 5t \qquad \text{and}$$

$$x'(t) = -5A \sin 5t + 5B \cos 5t.$$

The initial conditions yield $A = 4$ and $B = 3$. In the notation is Section 8.7, we have

$$C = \sqrt{16 + 9} = 5, \quad \cos\alpha = \frac{4}{5}, \quad \text{and} \quad \sin\alpha = \frac{3}{5}.$$

Therefore $x(t) = 5 \cos\left(5t - \tan^{-1}\frac{3}{4}\right) \approx 5 \cos(5t - 0.64350111)$.

8.7.3: Given: $4x'' + 36x = 0$; $x(0) = -5$, $x'(0) = -36$. Then the general solution and its derivative are

$$x(t) = A \cos 3t + B \sin 3t \qquad \text{and}$$

$$x'(t) = -3A \sin 3t + 4B \cos 3t.$$

Then the initial conditions yields $A = -5$ and $B = -12$, so the solution is

$$x(t) = -5 \cos 3t - 12 \sin 3t.$$

In the notation of Section 8.7 we have

$$C = \sqrt{25 + 144} = 13, \quad \cos\alpha = -\frac{5}{13}, \quad \text{and} \quad \sin\alpha = -\frac{12}{13}.$$

Thus the phase angle α lies in the third quadrant; $\alpha = \pi + \tan^{-1}\left(\frac{12}{5}\right)$. Hence

$$x(t) = 13\cos\left(3t - \pi - \tan^{-1}\tfrac{12}{5}\right) \approx 13\cos(3t - 4.31759786).$$

8.7.5: Given: $x'' + 6x' + 8x = 0$. The characteristic equation is

$$r^2 + 6r + 8 = (r+2)(r+4) = 0$$

with roots -2 and -4. In the notation of Section 8.7, $c^2 = 9 > 8 = 4km$, so the motion is overdamped. The general solution is $x(t) = Ae^{-2t} + Be^{-4t}$, and the initial conditions yield $A = 4$ and $B = -2$. Hence the solution is

$$x(t) = 4e^{-2t} - 2e^{-4t}.$$

8.7.7: The characteristic equation is

$$r^2 + 8r + 16 = (r+4)^2 = 0,$$

with repeated roots $r_1 = r_2 = -4$. In the notation of Section 8.7, $c^2 - 4km = 64 - 64 = 0$, so the motion is critically damped. The general solution is

$$x(t) = (A + Bt)e^{-4t},$$

and the initial conditions yield $A = 5$ and $B = 10$. Hence

$$x(t) = (10t + 5)e^{-4t}.$$

8.7.9: The differential equation has characteristic equation $r^2 + 8r + 20 = 0$, with roots

$$r_1,\, r_2 = \frac{-8 \pm \sqrt{64 - 80}}{2} = \frac{-8 \pm 4i}{2} = -4 \pm 2i.$$

Thus the differential equation has general solution

$$x(t) = e^{-4t}(A\cos 2t + B\sin 2t).$$

Because (in the notation of Section 8.7) $c^2 = 256 < 320 = 4km$, the motion is underdamped. The initial conditions yield $A = 5$ and $B = 12$, so one form of the solution is

$$x(t) = e^{-4t}(5\cos 2t + 12\sin 2t).$$

Continuing the notation of Section 8.7, we have

$$C = \sqrt{A^2 + B^2} = 13, \quad \cos\alpha = \frac{5}{13}, \quad \text{and} \quad \sin\alpha = \frac{12}{13},$$

and hence $\alpha = \tan^{-1}\frac{12}{5}$. Thus the solution may also be written in the form

$$x(t) = 13e^{-4t}\cos\left(2t - \tan^{-1}\frac{12}{5}\right) \approx 13e^{-4t}\cos(2t - 1.17600521).$$

8.7.11: The associated homogeneous equation is $x'' + 9x = 0$, which has the complementary solution $x_c(t) = c_1\cos 3t + c_2\sin 3t$. A particular solution has the form

$$x_p(t) = A\cos 2t + B\sin 2t,$$

and substitution into the original nonhomogeneous equation yields

$$-4A\cos 2t - 4B\sin 2t + 9A\cos 2t + 9B\sin 2t = 10\cos 2t,$$

so that $A = 2$ and $B = 0$. Hence the general solution of the nonhomogeneous equation, and its derivative, are

$$x(t) = c_1\cos 3t + c_2\sin 3t - 2\cos 2t \quad \text{and}$$

$$x'(t) = 3c_2\cos 3t - 3c_1\sin 3t + 4\sin 2t.$$

The initial conditions $x(0) = x'(0) = 0$ then yield $c_1 = 2$ and $c_2 = 0$. Therefore

$$x(t) = 2\cos 2t - 2\cos 3t.$$

8.7.13: The associated homogeneous equation has characteristic equation $r^2 + 100 = 0$ and thus complementary solution

$$x_c(t) = c_1\cos 10t + c_2\sin 10t.$$

The given nonhomogeneous equation has particular solution of the form $x_p(t) = A\cos 5t + B\sin 5t$, and substitution yields

$$x_p''(t) + 100x_p(t) = 75A\cos 5t + 75B\sin 5t = 300\sin 5t,$$

so that $A = 0$ and $B = 4$. Therefore the nonhomogeneous equation has general solution

$$x(t) = c_1\cos 10t + c_2\sin 10t + 4\sin 5t.$$

465

The initial conditions $x(0) = x'(0) = 0$ yield $c_1 = 0$ and $c_2 = -2$, so the solution of the original initial value problem is

$$x(t) = 4 \sin 5t - 2 \sin 10t.$$

8.7.15: The steady periodic solution has the form

$$x_{sp}(t) = A \cos 3t + B \sin 3t$$

and satisfies the given differential equation; therefore

$$-9A \cos 3t - 9B \sin 3t - 12A \sin 3t + 12B \cos 3t + 4A \cos 3t + 4B \sin 3t = 130 \cos 3t,$$

and it follows that A and B are solutions of the simultaneous equations

$$-5A + 12B = 130,$$

$$-5B - 12A = 0, \qquad \text{so that}$$

$$A = -\frac{50}{13} \qquad \text{and} \qquad B = \frac{120}{13}.$$

Therefore

$$x_{sp}(t) = -\frac{50}{13} \cos 3t + \frac{120}{13} \sin 3t.$$

In the notation of Section 8.7, we have

$$C = \sqrt{A^2 + B^2} = 10, \qquad \cos \alpha = -\frac{5}{13}, \qquad \text{and} \qquad \sin \alpha = \frac{12}{13}.$$

Consequently $C = 1$, $\alpha = \pi - \tan^{-1}\left(\frac{12}{5}\right)$, and

$$x_{sp}(t) = 10 \cos\left(3t - \pi + \tan^{-1} \frac{12}{5}\right) \approx 10 \cos(3t - 1.96558745).$$

8.7.17: The associated homogeneous equation has characteristic equation $r^2 + 4r + 5 , = 0$, with roots

$$r_1, \ r_2 = \frac{-4 \pm \sqrt{4 - 20}}{2} = -2 \pm i,$$

so the complementary solution has the form $x_c(t) = e^{-2t}(c_1 \cos t + c_2 \sin t)$. The steady periodic solution hs the form $x_{sp}(t) = A \cos 3t + B \sin 3t$, and substitution in the original differential equation yields

$$-9A \cos 3t - 9B \sin 3t - 12A \sin 3t + 12B \cos 3t + 5A \cos 3t + 5B \sin 3t = 40 \cos 3t;$$

$$-4A + 12B = 40,$$

$$-4B - 12A = 0.$$

Therefore $A = -1$ and $B = 3$, so that $x_{sp}(t) = 3\sin 3t - \cos 3t$. The general solution $x(t)$ of the original differential equation is thus

$$x(t) = e^{-2t}(c_1 \cos t + c_2 \sin t) + 3\sin 3t - \cos 3t; \qquad \text{moreover}$$

$$x'(t) = e^{-2t}(c_2 \cos t - c_1 \sin t - 2c_1 \cos t - 2c_2 \sin t) + 9\cos 3t + 3\sin 3t.$$

The initial conditions $x(0) = x'(0) = 0$ next yield

$$c_1 - 1 = 0,$$

$$c_2 - 2c_1 + 9 = 0,$$

and thus $c_1 = 1$ and $c_2 = -7$. Hence the general solution of the given initial value problem is

$$x(t) = e^{-2t}(\cos t - 7\sin t) + 3\sin 3t - \cos 3t.$$

The transient solution is $x_{tr}(t) = e^{-2t}(\cos t - 7\sin t)$. In the notation of Section 8.7, we have

$$C = \sqrt{1 + 49} = 5\sqrt{2}, \qquad \cos\alpha = \frac{1}{5\sqrt{2}}, \qquad \text{and} \qquad \sin\alpha = -\frac{7}{5\sqrt{2}}.$$

Hence $\alpha = 2\pi - \tan^{-1}(7)$. Therefore the transient solution may be expressed in the form

$$x_{tr}(t) = 5e^{-2t}\sqrt{2}\,\cos(t - 2\pi + \tan^{-1} 7) = 5e^{-2t}\sqrt{2}\,\cos(t + \tan^{-1} 7) \approx 5e^{-2t}\sqrt{2}\,\cos(t + 1.42889927).$$

The steady periodic solution is $x_{sp}(t) = 3\sin 3t - \cos 3t$. In the notation of Section 8.7,

$$C = \sqrt{10}, \qquad \cos\alpha = -\frac{1}{\sqrt{10}}, \qquad \text{and} \qquad \sin\alpha = \frac{3}{\sqrt{10}}.$$

Therefore $\alpha = \pi + \tan^{-1}(-3) = \pi - \tan^{-1}(3)$. Hence

$$x_{sp}(t) = \sqrt{10}\,\cos(3t - \pi + \tan^{-1} 3) \approx \sqrt{10}\,\cos(3t - 1.89254688).$$

8.7.19: Equation (8) yields frequency 2 rad/s; that is, $\dfrac{1}{\pi}$ Hz. The period is π s.

8.7.21: The spring constant is $k = 15/0.2 = 75$ N/m. The solution of $3x'' + 75x = 0$ with $x(0) = 0$ and $x'(0) = -10$ is $x(t) = -2\sin 5t$. Thus the amplitude is 2 m, the frequency is 5 rad/s, and the period is $2\pi/5$ s.

8.7.23: Following the suggestion in the statement of the problem, we have

$$mx'' + cx' + kx = F(t) + mg; \qquad x(0) = x_0, \quad x'(0) = v_0. \tag{1}$$

Note that $kx_0 = mg$. Hence if we let $y(t) = x(t) - x_0$, then $y(0) = x_0 - x_0 = 0$ and $y'(0) = x'(0) = v_0$. Thus substitution in the equations in (1) yields

$$my'' + cy' + ky + kx_0 = F(t) + kx_0; \quad \text{that is,}$$

$$my'' + cy' + ky = F(t); \quad y(0) = 0, \quad y'(0) = v_0.$$

8.7.25: The fact that the buoy weighs 100 lb means that $mg = 100$, so that $m = 100/32 = 3.125$ slugs. The weight of water is 62.4 lb/ft^3, so the equation $F = ma$ of Newton's second law of motion takes the form

$$\frac{100}{32} x'' = 100 - (62.4)\pi r^2 x.$$

It follows that the circular frequency ω of the buoy is given by

$$\omega^2 = \frac{32 \cdot (62.4) \cdot \pi r^2}{100}.$$

But the fact that the period of the buoy is $\rho = 2.5$ s means that $\omega = 2\pi/(2.5)$. Equating these two results yields $r \approx 0.3173201415$ ft, approximately 3.8078 in.

8.7.27: Part (a):

$$x(t) = 50\left(e^{-2t/5} - e^{-t/2}\right).$$

Part (b): $x'(t) = 0$ when

$$25e^{-t/2} - 20e^{-2t/5} = 5e^{-2t/5}\left(5e^{-t/10} - 4\right) = 0;$$

$$t = 10\ln\frac{5}{4} \approx 2.23143551.$$

Hence the greatest distance that the mass travels to the right is

$$x\left(10\ln\frac{5}{4}\right) = \frac{512}{125} = 4.096$$

8.7.29: Part (a): With $m = \frac{12}{32} = \frac{3}{8}$ slug, $c = 3$ lb-s/ft, and $k = 24$ lb/ft, the differential equation takes the form

$$3x'' + 24x' + 129x = 0.$$

The solution satisfying $x(0) = 1$ and $x'(0) = 0$ is

$$x(t) = e^{-4t}\left(\cos 4t\sqrt{3} + \frac{1}{\sqrt{3}}\sin 4t\sqrt{3}\right)$$

$$= \frac{2}{\sqrt{3}}e^{-4t}\left(\frac{\sqrt{3}}{2}\cos 4t\sqrt{3} + \frac{1}{2}\sin 4t\sqrt{3}\right) = \frac{2}{\sqrt{3}}e^{-4t}\cos\left(4t\sqrt{3} - \frac{\pi}{6}\right).$$

Part (b): The time-varying amplitude is $2/\sqrt{3} \approx 1.1547$ ft, the frequency is $4\sqrt{3} \approx 6.9282$ rad/s, and the phase angle is $\pi/6$.

8.7.31: In the case of critical damping, we have

$$r = \frac{-c \pm \sqrt{c^2 - 4km}}{2m} = -\frac{c}{2m} = -p.$$

The general solution of the differential equation and its derivative are

$$x(t) = (c_1 + c_2 t)e^{-pt} \quad \text{and}$$

$$x'(t) = (c_1 - pc_1 t - pc_2)e^{-pt}.$$

The initial conditions yield $x_0 = c_2$ and $v_0 = c_1 - pc_2$, and it follows that $c_1 = px_0 + v_0$. Therefore

$$x(t) = (px_0 t + v_0 t + x_0)e^{-pt}.$$

8.7.33: See Problem 31. If $x(t)$ has a local extremum for $t > 0$, then $x'(t) = 0$ for some $t > 0$. Thus

$$x'(t) = (px_0 + v_0 - p^2 x_0 t - pv_0 t - px_0)e^{-pt} = 0;$$

$$v_0 - p^2 x_0 t - pv_0 t = 0;$$

$$t = \frac{v_0}{p(px_0 + v_0)}.$$

Because $p > 0$, a positive solution t of $x'(t) = 0$ exists if and only if v_0 and $px_0 + v_0$ have the same sign.

8.7.35: The motion is overdamped, so we know that $c^2 > 4km$. Substitute $x_0 = 0$ in the solution in Problem 34 to find that

$$x(t) = \frac{1}{2\gamma}\left(v_0 e^{r_1 t} - v_0 e^{r_2 t}\right) = \frac{v_0}{\gamma} \cdot \frac{e^{r_1 t} - e^{r_2 t}}{2}$$

where

$$r_1 = -p + \sqrt{p^2 - \omega_0^2}, \qquad r_2 = -p - \sqrt{p^2 - \omega_0^2}, \qquad \text{and} \qquad p = \frac{c}{2m}.$$

Thus $r_1 = -p + \gamma$ and $r_2 = -p - \gamma$. Hence

$$e^{r_1 t} - e^{r_2 t} = \exp(-pt)\left[c^{\gamma t} - e^{-\gamma t}\right] = 2\exp(-pt)\sinh \gamma t.$$

Therefore

$$x(t) = \frac{v_0}{\gamma}e^{-pt}\sinh \gamma t.$$

8.7.37: With $m = 1$, $c = 0$, $k = 9$, $F_0 = 60$, and $\omega = 3$, we have

$$x'' + 9x = 60\cos 3t;$$

$$x'(0) = x'(0) = 0.$$

Part (a): If there is a solution of the form $x(t) = A\cos 3t + B\sin 3t$, then

$$x''(t) = -9A\cos 3t - 9B\sin 3t.$$

Hence $x'' + 9x = 0$ for all t. So there can be no such solution. Part (b): If $x_p(t) = 10t\sin 3t$, then

$$x_p'(t) = 10\sin 3t + 30t\cos 3t;$$

$$x_p''(t) = 60\cos 3t - 90t\sin 3t;$$

$$x_p''(t) + 9x_p(t) = 60\cos 3t - 90t\sin 3t + 90t\sin 3t = 60\cos 3t.$$

Chapter 8 Miscellaneous Problems

8.M.1: Given $\dfrac{dy}{dx} = 2x + \cos x$, $y(0) = 0$:

$$y(x) = x^2 + \sin x + C;$$

$$0 = y(0) = C;$$

$$y(x) = x^2 + \sin x.$$

A *Mathematica* command for solving this initial value problem is

```
DSolve[ { y'[x] == 2*x + Cos[x], y[0] == 0 }, y[x], x ]
```

and other computer algebra systems, such as *Maple*, *Derive*, and MATLAB, use similar commands.

8.M.3: Given: $\dfrac{dy}{dx} = (y+1)^2$.

$$\frac{dy}{(y+1)^2} = 1 \, dx;$$

$$-\frac{1}{y+1} = x + C;$$

$$y + 1 = -\frac{1}{x+C};$$

$$y(x) = -1 - \frac{1}{x+C}.$$

470

8.M.5: Given: $\dfrac{dy}{dx} = 3x^2y^2$, $y(0) = 1$.

$$y^{-2}\,dy = 3x^2\,dx; \quad -(y^{-1}) = x^3 + C; \quad y(x) = -\dfrac{1}{x^3 + C}.$$

But $1 = y(0) = -\dfrac{1}{C}$, and therefore $y(x) = \dfrac{1}{1 - x^3}$.

8.M.7: Given: $x^2 y^2 \dfrac{dy}{dx} = 1$.

$$3y^2\,dy = 3x^{-2}dx; \quad y^3 = -3x^{-1} + C; \quad y(x) = \left(C - \dfrac{3}{x}\right)^{1/3}.$$

8.M.9: Given: $\dfrac{dy}{dx} = y^2 \cos x$, $y(0) = 1$.

$$y^{-2}\,dy = (\cos x)\,dx; \quad -(y^{-1}) = C + \sin x; \quad y(x) = -\dfrac{1}{C + \sin x}.$$

But $1 = y(0) = -\dfrac{1}{C}$, so $C = -1$. Therefore $y(x) = \dfrac{1}{1 - \sin x}$.

8.M.11: Given: $\dfrac{dy}{dx} = \dfrac{y^2\left(1 - \sqrt{x}\right)}{x^2\left(1 - \sqrt{y}\right)}$.

$$\dfrac{1 - y^{1/2}}{y^2}\,dy = \dfrac{1 - x^{1/2}}{x^2}\,dx;$$

$$(y^{-2} - y^{-3/2})\,dy = (x^{-2} - x^{-3/2})\,dx;$$

$$\dfrac{1}{y} - \dfrac{2}{\sqrt{y}} = \dfrac{1}{x} - \dfrac{2}{\sqrt{x}} + C.$$

You should leave the solution in this (implicitly defined) form because it's troublesome to solve for y explicitly as a function of x. *Mathematica* finds two solutions:

$$y(x) = \dfrac{x - 2x^{3/2} + 2x^2 + Cx^2 \pm 2\sqrt{x^3 - 2x^{7/2} + x^4 + Cx^4}}{1 - 4x^{1/2} + 4x + 2Cx - 4Cx^{3/2} + C^2x^2}.$$

8.M.13: The equation is linear, with solution $y(x) = Cx^3 + x^3 \ln x$.

8.M.15: The equation is separable, with solution $y(x) = C\exp\left(\dfrac{1 - x}{x^3}\right)$.

8.M.17: The equation is linear, with solution $y(x) = \dfrac{C + \ln x}{x^2}$.

8.M.19: The equation is linear, with solution $y(x) = (x^3 + C)e^{-3x}$.

8.M.21: The equation is linear, with solution $y(x) = 2x^{-3/2} + Cx^{-3}$.

471

8.M.23: The equation is separable, with solution $y(x) = \dfrac{x^{1/2}}{6x^2 + Cx^{1/2} + 2}$.

8.M.25: Given:

$$\frac{dy}{dx} = e^x + y; \quad \text{that is,} \quad \frac{dy}{dx} - y = e^x. \tag{1}$$

This is a linear differential equation with integrating factor $\rho(x) = e^{-x}$. Multiplication of both sides of the second equation in (1) by $\rho(x)$ yields

$$e^{-x}\frac{dy}{dx} - e^{-x}y = 1;$$

$$e^{-x}y = x + C;$$

$$y(x) = (x + C)e^x.$$

8.M.27: As a separable equation:

$$\frac{1}{y+7}\, dy = 3x^2\, dx; \qquad \ln(y+7) = x^3 + C_1;$$

$$y + 7 = \exp\left(x^3 + C_1\right) = C\exp\left(x^3\right); \qquad y(x) = -7 + C\exp\left(x^3\right).$$

As the linear equation $\dfrac{dy}{dx} - 3x^2 y = 21x^2$:

$$\text{Integrating factor:} \quad \rho(x) = \exp\left(\int (-3x^2)\, dx\right) = \exp\left(-x^3\right).$$

Thus

$$y(x) \cdot \exp\left(-x^3\right) = \int 21x^2 \exp\left(-x^3\right)\, dx = -7\exp\left(-x^3\right) + C.$$

Therefore $y(x) = -7 + C\exp\left(x^3\right)$.

8.M.29: First note that $\dfrac{1}{x^2 + 5x + 6} = \dfrac{1}{x+2} - \dfrac{1}{x+3}$. So

$$\int \frac{1}{x^2 + 5x + 6}\, dx = \int 1\, dt; \qquad \int\left(\frac{1}{x+2} - \frac{1}{x+3}\right) dx = t + C_1;$$

$$\ln\frac{x+2}{x+3} = t + C_1; \qquad \frac{x+2}{x+3} = Ce^t;$$

$$\frac{7}{8} = Ce^0 = C; \qquad 8(x+2) = 7(x+3)e^t;$$

$$(8 - 7e^t)x = 21e^t - 16; \qquad x(t) = \frac{21e^t - 16}{8 - 7e^t}.$$

8.M.31: Let τ denote the half-life of potassium, so that τ is approximately 1.28×10^9. Measure time t also in years, with $t = 0$ corresponding to the time when the rock contained only potassium, and with $t = T$ corresponding to the present. Then at time $t = 0$, the amount of potassium was $Q(0)$ and no argon was present. At present, the amount of potassium is $Q(T)$ and the amount of argon is $A(T)$, where $A(t)$ is the amount of argon in the rock at time t. Now

$$Q(t) = Q_0 e^{-(t \ln 2)/\tau}, \quad \text{so that} \quad A(t) = \tfrac{1}{9}\left(Q_0 - Q(t)\right).$$

We also are given the observation that $A(T) = Q(T)$. Thus

$$Q_0 - Q(t) = 9Q(T), \quad \text{and so} \quad Q(T) = (0.1)Q_0 = Q_0 e^{-(T \ln 2)/\tau}.$$

Therefore

$$\ln 10 = \frac{T}{\tau} \ln 2 \quad \text{and thus} \quad T = \frac{\ln 10}{\ln 2}(1.28 \times 10^9) \approx 4.2521 \times 10^9.$$

Thus the rock is approximately 4.2521×10^9 years old.

8.M.33: First, $\dfrac{dA}{dt} = -kA$, so $A(t) = A_0 e^{-kt}$.

$$\frac{3}{4} A_0 = A_0 e^{-k}, \quad \text{so} \quad k = \ln \frac{4}{3}.$$

Also

$$\frac{1}{2} A_0 = A_0 e^{-kT}$$

where T is the time required for half the sugar to dissolve. So

$$\frac{\ln 2}{k} = T = \frac{\ln 2}{\ln(\frac{4}{3})} \approx 2.40942 \quad \text{(minutes)},$$

so half of the sugar is dissolved in about 2 minutes and 25 seconds.

8.M.35: We begin with the equation $p(x) = (29.92)e^{-x/5}$.

(a) $p\left(\frac{10000}{5280}\right) \approx 20.486$ (inches); $p\left(\frac{30000}{5280}\right) \approx 9.604$ (inches).

(b) If x is the altitude in question, then we must solve

$$15 = (29.92)e^{-x/5}, \quad \text{and thus} \quad x = 5 \ln\left(\frac{29.92}{15}\right) \approx 3.4524 \quad \text{(miles)},$$

approximately 18230 feet.

(c) According to *Trails Illustrated* Topo Map 322 "Denali National Park and Preserve" (Trails Illustrated, Evergreen, CO, 1990, 1993) and other sources, the summit of Mt. McKinley is 20320 ft above sea level. We evaluate

$$p\left(\frac{20320}{5280}\right) = (29.92)\exp\left(\frac{20320}{5 \cdot 5280}\right) \approx 13.8575$$

to find that the atmospheric pressure at the summit is about 13.86 inches of mercury. For an engrossing story of an ascent of this peak, see Ruth Anne Kocour's *Facing the Extreme* (with Michael Hodgson, New York: St. Martin's Press, 1998).

8.M.37: The decay constant k satisfies the equation $140k = \ln 2$, and so $k = (\ln 2)/140$. Measuring radioactivity as a multiple of the "safe level" 1, it is then $P(t) = 5e^{-kt}$ with t measured in days. When we solve $P(t) = 1$, we find that $t \approx 325.07$, so the room should be safe to enter in a little over 325 days.

8.M.39: The characteristic equation

$$6r^2 - 19r + 15 = (2r - 3)(3r - 5) = 0$$

has the real distinct roots $r_1 = \frac{3}{2}$ and $r_2 = \frac{5}{3}$. Hence the general solution of the differential equation is

$$y(x) = Ae^{3x/2} + Be^{5x/3}, \qquad \text{for which}$$

$$y'(x) = \frac{3}{2}Ae^{3x/2} + \frac{5}{3}Be^{5x/3}.$$

The initial conditions yield

$$13 = y(0) = A + B \qquad \text{and}$$

$$21 = y'(0) = \frac{3}{2}A + \frac{5}{3}B,$$

and it quickly follows that $A = 4$ and $B = 9$. Hence $y(x) = 4e^{3x/2} + 9e^{5x/3}$.

8.M.41: The characteristic equation

$$121r^2 + 154r + 49 = (11r + 7)^2 = 0$$

has the repeated root $r_1 = r_2 = -\frac{7}{11}$, and thus the given equation has general solution

$$y(x) = (Ax + B)e^{-7x/11}, \qquad \text{for which}$$

$$y'(x) = \left(A - \frac{7}{11}Ax - \frac{7}{11}B\right)e^{-7x/11}.$$

The initial conditions then yield $A = 17$ and $B = 11$, and hence the solution of the given initial value problem is $y(x) = (17x + 11)e^{-7x/11}$.

8.M.43: First we solve the characteristic equation:

$$100r^2 + 20r + 10001 = 0; \qquad 100r^2 + 20r + 1 = -10000;$$

$$(10r + 1)^2 = (100i)^2; \qquad 10r + 1 = \pm 100i;$$

$$r = -\frac{1}{10} \pm 10i.$$

So the general solution of the differential equation may be expressed in the form

$$y(x) = e^{-x/10}(A\cos 10x + B\sin 10x), \qquad \text{for which}$$

$$y'(x) = e^{-x/10}\left(10B\cos 10x - 10A\sin 10x - \frac{1}{10}A\cos 10x - \frac{1}{10}B\sin 10x\right).$$

Then the initial conditions yield $A = 10$ and $B = 1$, so the solution of the given initial value problem is

$$y(x) = e^{-x/10}(10\cos 10x + \sin 10x)$$

.

8.M.45: Part (a): With N in thousands (of transistors) and t in years, we have $N(t) = 29e^{rt}$.

Part (b): In 1993 we have $t = 14$. So

$$3100 = N(14) = 29e^{14r}; \qquad 14r = \ln\frac{3100}{29}; \qquad r = \frac{1}{14}\ln\frac{3100}{29} \approx 0.3337.$$

Expressed as a percentage, the annual growth rate is about 33.4%. Part (c): Let τ denote the "doubling time" and let $N_0 = N(0)$. Then from the equation $N(\tau) = 2N_0 = N_0 e^{r\tau}$ we find that

$$\tau = \frac{\ln 2}{r} = \frac{14\ln 2}{\ln\frac{3100}{29}} \approx 2.0771$$

years. Thus the doubling time is 12τ, about 24.9 months. Part (d): In the year 2001 we have $t = 22$, so in that year the typical microcomputer would be expected to contain

$$N(22) = 29e^{22r} \approx 44{,}747.576$$

thousand transistors; that is, about 44.7 million transistors.

8.M.47: The differential equation leads to

$$P^{-1/2}\, dP = -k\, dt; \qquad 2P^{1/2} = C - kt; \qquad P^{1/2} = \frac{C - kt}{2};$$

$$30 = [P(0)]^{1/2} = \frac{C}{2}; \qquad C = 60; \qquad P^{1/2} = \frac{60 - kt}{2}.$$

Then the information that $P(6) = 441$ yields

$$21 = [P(6)]^{1/2} = 30 - 3k; \qquad k = 3; \qquad P(t) = \left(\frac{60 - 3t}{2}\right)^2.$$

Because we have measured time t in weeks, the answer is that all the fish will die at the end of 20 weeks.

8.M.49: Given (in effect): $P(t) = \left(\frac{1}{2}kt + \sqrt{P_0}\right)^2$, $P_0 = 100$ (we take $t = 0$ [years] in 1970 and measure population in thousands), and $P(10) = 121$. Thus

$$P(t) = \left(\tfrac{1}{2}kt + 10\right)^2,$$

and therefore $121 = P(10) = (5k + 10)^2$, so that $5k + 10 = \pm 11$. Because $k > 0$, we see that $k = \frac{1}{5}$, and hence

$$P(t) = \left(\tfrac{1}{10}t + 10\right)^2.$$

Thus in the year 2000 the population will be $P(30) = 169$ (thousand). The population will be 200 (thousand) when $P(T) = 200$:

$$\left(\tfrac{1}{10}T + 10\right) = 200; \qquad \tfrac{1}{10}T + 10 = 10\sqrt{2}; \qquad T = 100\left(\sqrt{2} - 1\right) \approx 41.42.$$

Thus the population will reach 200000 in the "year" $1970 + 41.42 = 2011.42$; that is, about June 1, 2011, assuming that $t = 0$ was January 1, 1970.

8.M.51: If $P_0 = 2$ and $P(3) = 4$ (time t is measured in months), then

$$P(t) = \frac{2}{1 - 2kt}; \qquad 4 = P(3) = \frac{2}{1 - 6k}; \qquad 1 - 6k = \frac{1}{2};$$

$$k = \frac{1}{12}; \qquad P(t) = \frac{2}{1 - \frac{1}{6}t} = \frac{12}{6 - t}.$$

Answer: $\lim\limits_{t \to 6^-} P(t) = +\infty$.

8.M.53: First we solve the initial value problem $\dfrac{dP}{dt} = -3P^{1/2}$, $P(0) = 900$, with t measured in weeks:

$$\int P^{-1/2} \, dP = -3 \, dt; \qquad 2P^{1/2} = C - 3t; \qquad 2 \cdot 30 = C - 3 \cdot 0;$$

$$2P^{1/2} = 60 - 3t; \qquad P(t) = \frac{9}{4}(20 - t)^2.$$

So all the fish will die after 20 weeks.

8.M.55: Problem 13 in Section 7.6 is to derive the initial value problem

$$\frac{dP}{dt} = rP - c, \qquad P(0) = P_0$$

where $P(t)$ is the balance owed at time t (in months), where r is the monthly interest rate (compounded continuously) and c is the monthly payment (assumed made continuously). First we need to solve this initial value problem:

$$\int \frac{r \, dP}{rP - c} = \int r \, dt; \qquad \ln(rP - c) = C + rt; \qquad rP - c = Ae^{rt};$$

$$P(t) = \frac{1}{r}\left(c + Ae^{rt}\right); \qquad P_0 = P(0) = \frac{1}{r}(c + A); \qquad P(t) = \frac{c + (rP_0 - c)e^{rt}}{r}.$$

476

In this problem, the loan is to be paid off in $25 \cdot 12 = 300$ months, and thus $P(300) = 0$. We use this information to solve for the monthly payment c:

$$\frac{c + (rP_0 - c)e^{300r}}{r} = 0;$$

$$c\left(1 - e^{300r}\right) + rP_0 e^{300r} = 0;$$

$$c = \frac{rP_0 e^{300r}}{e^{300r} - 1}.$$

With $P_0 = 120000$ and $r = 0.08/12$, we find that $c = \$925.21$. With $r = 0.12/12$ we find that $c = \$1262.87$. In the latter case the total of all 300 monthly payments is \$378862.45.

8.M.57: Let $h(t)$ denote the temperature within the freezer (in degrees Celsius) at time t (in hours), with $t = 0$ corresponding to the time the power goes off. By Newton's law of cooling, there is a positive constant k such that

$$\frac{dh}{dt} = k(20 - h); \qquad \int \frac{dh}{20 - h} = \int k\, dt; \qquad -\ln(20 - h) = kt - C;$$

$$\ln(20 - h) = C - kt; \qquad 20 - h = Ae^{-kt}; \qquad h(t) = 20 - Ae^{-kt};$$

$$-16 = h(0) = 20 - A; \qquad h(t) = 20 - 36e^{-kt}; \qquad h(7) = -10;$$

$$20 - 36e^{-7k} = -10; \qquad 36e^{-7k} = 30; \qquad k = \frac{1}{7}\ln\frac{6}{5}.$$

Finally we solve $h(T) = 0$ for

$$T = \frac{7\ln\frac{9}{5}}{\ln\frac{6}{5}} \approx 22.5673076.$$

So the critical temperature will be reached about 22 hours and 34 minutes after the power goes off; that is, at 9:34 P.M. on the following day. The data given here were drawn from a real incident.

8.M.59: The differential equation we need to solve is $\dfrac{dv}{dt} = a - \rho v$ with initial condition $v(0) = 0$:

$$\int \frac{-\rho}{a - \rho v}\, dv = \int (-\rho)\, dt; \qquad \ln(a - \rho v) = C - \rho t; \qquad a - \rho v = Ae^{-\rho t};$$

$$a - \rho \cdot 0 = A; \qquad a - \rho v = ae^{-\rho t}; \qquad \rho v = a\left(1 - e^{-\rho t}\right);$$

$$v(t) = \frac{a}{\rho}\left(1 - e^{-\rho t}\right).$$

If $a = 17.6$ and $\rho = 0.1$, then

$$v(10) = 176\left(1 - \frac{1}{e}\right) \approx 111.2532183538$$

feet per second, about 75.854 mi/h. Also $v(t) \to 176$ as $t \to +\infty$, so the limiting velocity is 176 ft/s, exactly 120 mi/h.

8.M.61: Given: $S(t) = 30e^{(0.05)t}$ (t is in years; $t = 0$ corresponds to age 30).

(a) $\Delta A = A(t + \Delta t) - A(t) \approx (0.06)A(t)\,\Delta t + (0.12)S(t)\,\Delta t.$

$$\frac{dA}{dt} = \lim_{\Delta t \to 0} \frac{\Delta A}{\Delta t} = (0.06)A(t) + (0.12)S(t);$$

$$\frac{dA}{dt} + (-0.06)A(t) = (3.6)e^{(0.05)t}.$$

(b) Now use the result in Problem 22: Take $x(t) = A(t)$, $x_0 = 0$; $a = -0.06$, $b = 3.6$, and $c = 0.05$. Then

$$A(t) = -\frac{3.6}{0.01}\left(e^{(0.05)t} - e^{(0.06)t}\right), \quad \text{so that}$$

$$A(t) = 360\left(e^{(0.06)t} - e^{(0.05)t}\right).$$

Now $A(40) = 360\left(e^{2.4} - e^2\right) \approx 1308.28330$. Because the units in this problem are in thousands of dollars, the answer is that the retirement money available will be \$1,308,283.30.

8.M.63: If we substitute $P(0) = 10^6$ and $P'(0) = 3 \times 10^5$ into the differential equation

$$P'(t) = \beta_0 e^{-\alpha t} P(t),$$

we find that $\beta_0 = 0.3$. Hence the solution given in Problem 62 is

$$P(t) = P_0 \exp\left(\frac{0.3}{\alpha}\left[1 - e^{-\alpha t}\right]\right).$$

The fact that $P(6) = 2P_0$ now yields the equation

$$f(\alpha) = (0.3)\left(1 - e^{-6\alpha}\right) - \alpha \ln 2 = 0$$

for α. We apply the iterative formula of Newton's method,

$$\alpha_{n+1} = \alpha_n - \frac{f(\alpha_n)}{f'(\alpha_n)},$$

with $f'(\alpha) = (1.8)e^{-6\alpha} - \ln 2$ and initial guess $\alpha_0 = 1$. Thereby we find that $\alpha_1 \approx 0.5381$, $\alpha_2 \approx 0.3926$, ..., and $\alpha \approx \alpha_6 \approx 0.39148754$. Therefore the limiting cell population as $t \to +\infty$ is

$$P_0 \exp\left(\frac{\beta_0}{\alpha}\right) \approx (10^6) \exp\left(\frac{0.3}{0.39148754}\right) \approx 2.152 \times 10^6.$$

Therefore the tumor does not grow much further after six months.

8.M.65: If $k = 1 - 10^{-2n}$ where (without loss of generality) n is a positive integer, then the characteristic equation has distinct real roots

$$r_1 = -1 - 10^{-n} \quad \text{and} \quad r_2 = -1 + 10^{-n}.$$

Thus the differential equation has general solution

$$x_2(t) = Ae^{r_1 t} + Be^{r_2 t}, \quad \text{so that}$$

$$x_2'(t) = r_1 Ae^{r_1 t} + r_2 Be^{r_2 t}.$$

The initial conditions yield the simultaneous equations

$$A + B = 0,$$

$$r_1 A + r_2 B = 1,$$

and it follows that

$$x_2(t) = \frac{1}{r_2 - r_1}\left(e^{r_2 t} - e^{r_1 t}\right)$$

$$= \frac{e^{-t}}{2 \cdot 10^{-n}}\left[\exp\left(10^{-n}t\right) - \exp\left(-10^{-n}t\right)\right] = 10^n e^{-t}\sinh\left(10^{-n}t\right).$$

8.M.67: If $t > 0$ is fixed, then—by l'Hôpital's rule (with $w = 10^{-n}$ as the variable)—

$$\lim_{n\to\infty} x_2(t) = \lim_{n\to\infty} 10^n e^{-t}\sinh\left(10^{-n}t\right)$$

$$= \lim_{w\to 0^+} \frac{e^{-t}\sinh wt}{w} = \lim_{w\to 0^+} te^{-t}\cosh wt = te^{-t} = x_1(t).$$

Similarly,

$$\lim_{n\to\infty} x_3(t) = \lim_{n\to\infty} 10^n e^{-t}\sin\left(10^{-n}t\right)$$

$$= \lim_{w\to 0^+} \frac{e^{-t}\sin wt}{w} = \lim_{w\to 0^+} te^{-t}\cos wt = te^{-t} = x_1(t).$$

Section 9.1

9.1.1: The given line with equation $y = -\frac{1}{2}x + \frac{5}{2}$ has slope $-\frac{1}{2}$, so the parallel line through $(1, -2)$ has equation $y + 2 = -\frac{1}{2}(x - 1)$; that is, $x + 2y + 3 = 0$.

9.1.3: The radius of the circle terminating at $(3, -4)$ has slope $-\frac{4}{3}$. Hence the line L tangent to the circle at that point (because L is perpendicular to that radius) has equation $y + 4 = \frac{3}{4}(x - 3)$; that is, $3x - 4y = 25$.

9.1.5: Given $x^2 + 2y^2 = 6$, we have by implicit differentiation

$$2x + 4y\frac{dy}{dx} = 0, \quad \text{so that} \quad \frac{dy}{dx} = -\frac{x}{2y}.$$

Therefore the tangent to the given curve at $(2, -1)$ has slope 1, so the normal to the curve there has slope -1 and thus equation $y + 1 = (-1)(x - 2) = -x + 2$; that is, $x + y = 1$.

9.1.7: Given $x^2 + 2x + y^2 = 4$, complete the square in each variable to find that $x^2 + 2x + 1 + y^2 = 5$; that is, $(x + 1)^2 + (y - 0)^2 = 5$. Hence the circle has center $(-1, 0)$ and radius $\sqrt{5}$.

9.1.9: Given $x^2 + y^2 - 4x + 6y = 3$, complete the square in each variable to find that

$$x^2 - 4x + 4 + y^2 + 6y + 9 = 16; \quad \text{that is,} \quad (x - 2)^2 + (y + 3)^2 = 16.$$

Therefore the circle has center $(2, -3)$ and radius 4.

9.1.11: Given $4x^2 + 4y^2 - 4x = 3$, complete the square in each variable as follows:

$$x^2 + y^2 - x = \frac{3}{4}; \quad x^2 - x + \frac{1}{4} + y^2 = 1; \quad \left(x - \frac{1}{2}\right)^2 + (y - 0)^2 = 1.$$

Consequently the circle has center $\left(\frac{1}{2}, 0\right)$ and radius 1.

9.1.13: Given $2x^2 + 2y^2 - 2x + 6y = 13$, complete the square in each variable as follows:

$$x^2 + y^2 - x + 3y = \frac{13}{2}; \quad x^2 - x + \frac{1}{4} + y^2 + 3y + \frac{9}{4} = 9; \quad \left(x - \frac{1}{2}\right)^2 + \left(y + \frac{3}{2}\right)^2 = 9.$$

Thus the circle has center $\left(\frac{1}{2}, -\frac{3}{2}\right)$ and radius 3.

9.1.15: Given $9x^2 + 9y^2 + 6x - 24y = 19$, complete the square in each variable:

$$x^2 + y^2 + \frac{2}{3}x - \frac{8}{3}y = \frac{19}{9}; \quad x^2 + \frac{2}{3}x + \frac{1}{9} + y^2 - \frac{8}{3}y + \frac{16}{9} = 4; \quad \left(x + \frac{1}{3}\right)^2 + \left(y - \frac{4}{3}\right)^2 = 4.$$

Therefore the circle has center $\left(-\frac{1}{3}, \frac{4}{3}\right)$ and radius 2.

9.1.17: Given $x^2 + y^2 - 6x - 4y = -13$, complete the square in each variable:

$$x^2 - 6x + 9 + y^2 - 4y + 4 = -13 + 13 = 0; \qquad (x-3)^2 + (y-2)^2 = 0.$$

Therefore the graph of the given equation consists of the single point $(3, 2)$.

9.1.19: Given $x^2 + y^2 - 6x - 10y = -84$, complete the square in each variable:

$$x^2 - 6x + 9 + y^2 - 10y + 25 = -50; \qquad (x-3)^2 + (y-5)^2 = -50.$$

The graph of the given equation has no points on it.

9.1.21: First use the distance formula to find that the radius of the circle is

$$\sqrt{(2-(-1))^2 + (3-(-2))^2} = \sqrt{3^2 + 5^2} = \sqrt{34}.$$

Thus an equation of the circle is $(x+1)^2 + (y+2)^2 = 34$.

9.1.23: Suppose that $P(a, 2a-4)$ is the point at which the circle is tangent to the line $y = 2x - 4$. This line has slope 2, so the radius terminating at P has slope $-\frac{1}{2}$. Thus

$$\frac{2a-4-6}{a-6} = -\frac{1}{2}; \qquad 4a - 20 = 6 - a; \qquad a = \frac{26}{5}; \qquad 2a - 4 = \frac{32}{5}.$$

Then the distance formula yields the fact that the circle has radius

$$\sqrt{\left(6 - \frac{26}{5}\right)^2 + \left(6 - \frac{32}{5}\right)^2} = \sqrt{\frac{16}{25} + \frac{4}{25}} = \frac{2}{5}\sqrt{5}.$$

Hence an equation of the circle is $(x-6)^2 + (y-6)^2 = \frac{4}{5}$.

9.5.25: The distance formula implies that $(x-3)^2 + (y-2)^2 = (x-7)^2 + (y-4)^2$. Therefore

$$-6x + 9 - 4y + 4 = -14x + 49 - 8y + 16;$$

$$8x + 4y = 52.$$

Hence $P(x, y)$ satisfies the equation $2x + y = 13$. The locus is a straight line; it is in fact the perpendicular bisector of the line segment joining the two given points. We omit its graph because it would occupy space unnecessarily.

9.1.27: The distance formula implies that the coordinates of $P(x, y)$ satisfy

$$3\sqrt{(x-5)^2 + (y-10)^2} = \sqrt{(x+3)^2 + (y-2)^2};$$

481

$$9x^2 - 90x + 9y^2 - 180y + 1125 = x^2 + 6x + y^2 - 4y + 13;$$

$$8x^2 - 96x + 8y^2 - 176y + 1112 = 0;$$

$$x^2 - 12x + y^2 - 22y + 139 = 0;$$

$$x^2 - 12x + 36 + y^2 - 22y + 121 = 157 - 139 = 18;$$

$$(x - 6)^2 + (y - 11)^2 = 18.$$

Therefore the locus of $P(x, y)$ is the circle with center $(6, 11)$ and radius $3\sqrt{2}$. We omit a sketch as it would occupy space unnecessarily.

9.1.29: The distance formula tells us that the point $P(x, y)$ satisfies the equations

$$\sqrt{(x - 4)^2 + y^2} + \sqrt{(x + 4)^2 + y^2} = 10;$$

$$\sqrt{(x - 4)^2 + y^2} = 10 - \sqrt{(x + 4)^2 + y^2};$$

$$(x - 4)^2 + y^2 = 100 - 20\sqrt{(x + 4)^2 + y^2} + (x + 4)^2 + y^2;$$

$$20\sqrt{(x + 4)^2 + y^2} = 100 + 16x;$$

$$5\sqrt{(x + 4)^2 + y^2} = 25 + 4x;$$

$$25x^2 + 200x + 400 + 25y^2 = 625 + 200x + 16x^2;$$

$$9x^2 + 25y^2 = 225.$$

This is an equation of the ellipse with center $(0, 0)$, horizontal major axis of length 10, vertical minor axis of length 6, and intercepts $(\pm 5, 0)$ and $(0, \pm 3)$. Its graph is shown next.

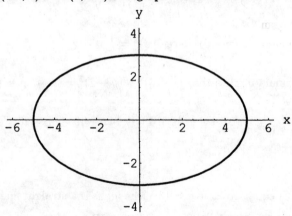

9.1.31: Let $P(a, a^2)$ be a point at which such a line is tangent to the parabola. Then

$$2a = \frac{a^2 - 1}{a - 2}; \qquad 2a^2 - 4a = a^2 - 1;$$

$$a^2 - 4a + 1 = 0; \qquad a = \frac{4 \pm \sqrt{16 - 4}}{2} = 2 \pm \sqrt{3}.$$

482

Thus there are two such lines, one with slope $4 + 2\sqrt{3}$ and one with slope $4 - 2\sqrt{3}$. Their equations are

$$y - 1 = \left(4 + 2\sqrt{3}\right)(x - 2) \quad \text{and} \quad y - 1 = \left(4 - 2\sqrt{3}\right)(x - 2).$$

9.1.33: Every line parallel to the line with equation $y = 4x$ also has slope 4. Suppose that such a line is also normal to the graph of $y = 4/x$ at the point P. The tangent at $P(x, y)$ has slope $-4/(x^2)$, so the normal there has slope $x^2/4$, which must also equal 4. It follows that $x = \pm 4$, so there are two such lines, one containing the point $(4, 1)$ and the other containing the point $(-4, -1)$. Equations of these lines are $y - 1 = 4(x - 4)$ and $y + 1 = 4(x + 4)$.

9.1.35: Equation (11) is

$$x^2(1 - e^2) - 2p(1 + e^2)x + y^2 = -p^2(1 - e^2).$$

If $e > 1$, then this equation may be written in the form

$$x^2 + 2p\frac{e^2 + 1}{e^2 - 1}x - \frac{y^2}{e^2 - 1} = -p^2;$$

$$x^2 + 2p\frac{e^2 + 1}{e - 1}x + p^2\left(\frac{e^2 + 1}{e^2 - 1}\right)^2 - \frac{y^2}{e^2 - 1} = -p^2 + p^2\left(\frac{e^2 + 1}{e^2 - 1}\right)^2;$$

$$= p^2\left[\left(\frac{e^2 + 1}{e^2 - 1}\right)^2 - 1\right] = \frac{e^4 + 2e^2 + 1 - e^4 + 2e^2 - 1}{(e^2 - 1)^2}p^2 = \frac{4e^2}{(e^2 - 1)^2}p^2.$$

Let

$$h = -p\frac{e^2 + 1}{e^2 - 1}.$$

Then Eq. (11) takes the form

$$x^2 - 2hx + h^2 - \frac{y^2}{e^2 - 1} = \frac{4e^2p^2}{(e^2 - 1)^2}.$$

Now let

$$a = \frac{2pe}{e^2 - 1}.$$

Then Eq. (11) simplifies to

$$x^2 - 2hx + h^2 - \frac{y^2}{e^2 - 1} = a^2;$$

$$\frac{(x - h)^2}{a^2} - \frac{y^2}{a^2(e^2 - 1)} = 1.$$

Finally, let $b = a\sqrt{e^2 - 1}$. Then Eq. (11) further simplifies to

$$\frac{(x - h)^2}{a^2} - \frac{y^2}{b^2} = 1$$

483

where

$$h = -p \cdot \frac{e^2+1}{e^2-1}, \quad a = \frac{2pe}{e^2-1}, \quad \text{and} \quad b = a\sqrt{e^2-1} = \frac{2pe}{\sqrt{e^2-1}}.$$

Section 9.2

9.2.1: (a): $\left(\frac{1}{2}\sqrt{2}, \frac{1}{2}\sqrt{2}\right)$; (b): $\left(1, -\sqrt{3}\right)$; (c): $\left(\frac{1}{2}, -\frac{1}{2}\sqrt{3}\right)$; (d): $(0, -3)$; (e): $\left(\sqrt{2}, -\sqrt{2}\right)$; (f): $\left(\sqrt{3}, -1\right)$; (g): $\left(-\sqrt{3}, 1\right)$.

9.2.3: $r\cos\theta = 4$: $r = 4\sec\theta$.

9.2.5: $r\cos\theta = 3r\sin\theta$: $\tan\theta = \frac{1}{3}$; $\theta = \tan^{-1}\left(\frac{1}{3}\right)$.

9.2.7: $r^2\sin\theta\cos\theta = 1$: $r^2 = \csc\theta\sec\theta$.

9.2.9: $r\sin\theta = r^2\cos^2\theta$: $r = \sec\theta\tan\theta$. Note that no points on the graph are lost when r is cancelled.

9.2.11: $r^2 = 9$: $x^2 + y^2 = 9$.

9.2.13: $r^2 = -5r\cos\theta$: $x^2 + y^2 + 5x = 0$.

9.2.15: $r = 1 - \cos 2\theta$: $r = 2 \cdot \dfrac{1-\cos 2\theta}{2} = 2\sin^2\theta$; $r^3 = 2(r\sin\theta)^2$; $(x^2+y^2)^{3/2} = 2y^2$; $(x^2+y^2)^3 = 4y^4$.

9.2.17: $r = 3\sec\theta$: $r\cos\theta = 3$; $x = 3$.

9.2.19: $x = 2$; $r = 2\sec\theta$.

9.2.21: $y + 1 = (-1) \cdot (x - 2)$; thus $x + y = 1$. $r(\cos\theta + \sin\theta) = 1$: $r = \dfrac{1}{\cos\theta + \sin\theta}$.

9.2.23: $y - 3 = x - 1$: $x - y + 2 = 0$; $r = \dfrac{2}{\sin\theta - \cos\theta}$.

9.2.25: $x^2 + (y+4)^2 = 16$: $x^2 + y^2 + 8y = 0$. $r^2 + 8r\sin\theta = 0$: $r + 8\sin\theta = 0$.

9.2.27: $(x-1)^2 + (y-1)^2 = 2$: $x^2 - 2x + y^2 - 2y = 0$. $r^2 = 2r(\cos\theta + \sin\theta)$: $r = 2\cos\theta + 2\sin\theta$.

9.2.29: Given: $r^2 = -4r\cos\theta$. Thus $x^2 + 4x + y^2 = 0$; that is, $(x+2)^2 + (y-0)^2 = 4$. This is an equation of the circle with center $(-2, 0)$ and radius 2, and thus it's the one shown in Fig. 9.2.23.

9.2.31: Given: $r = -4\cos\theta + 3\sin\theta$. Thus $r^2 = -4r\cos\theta + 3r\sin\theta$;

$$x^2 + 4x + y^2 - 3y = 0;$$

$$(x+2)^2 + \left(y - \frac{3}{2}\right)^2 = 4 + \frac{9}{4} = \frac{25}{4}.$$

Thus we are given the equation of a circle with center $\left(-2, \frac{3}{2}\right)$, radius $\frac{5}{2}$, and passing through the origin. This is the circle shown in Fig. 9.2.24.

9.2.33: First note that $r = 0$ when $6\cos\theta = 8$; that is, never. Also r is maximal when $\theta = 0$, for which $r = 14$; r is minimal when $\theta = \pi$, for which $r = 2$. So the graph of this limaçon is the one shown in Fig. 9.2.26.

9.2.35: The maximum value of r is 14, which occurs when $\theta = 0$; the minimum value of r is -4, which occurs when $\theta = \pi$. So the graph of this limaçon is the one shown in Fig. 9.2.25.

9.2.37: Assume that $a^2 + b^2 \neq 0$; that is, that neither a nor b is zero. Suppose that $r = a\cos\theta + b\sin\theta$. Then

$$r^2 = ar\cos\theta + br\sin\theta;$$

$$x^2 - ax + y^2 - by = 0;$$

$$x^2 - ax + \frac{1}{4}a^2 + y^2 - by + \frac{1}{4}b^2 = \frac{a^2 + b^2}{4};$$

$$\left(x - \frac{a}{2}\right)^2 + \left(y - \frac{b}{2}\right)^2 = \frac{a^2 + b^2}{4}.$$

Therefore the graph of $r = a\cos\theta + b\sin\theta$ is a circle with center $\left(\frac{1}{2}a, \frac{1}{2}b\right)$ and radius $\frac{1}{2}\sqrt{a^2 + b^2}$.

9.2.39: The graph of the circle with polar equation $r = 2\cos\theta$ is shown next. This graph is symmetric around the x-axis.

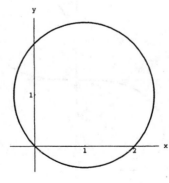

9.2.41: The graph of the cardioid with polar equation $r = 1 + \cos\theta$ is shown next. This graph is symmetric around the x-axis.

485

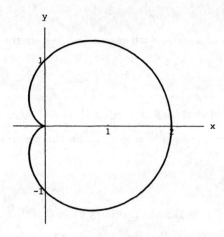

9.2.43: The graph of the limaçon with polar equation $r = 2 + 4\sin\theta$ is shown next. This graph is symmetric around the y-axis.

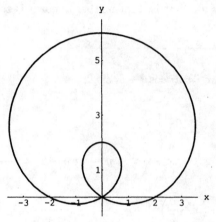

9.2.45: The graph of the lemniscate with polar equation $r^2 = 4\sin 2\theta$ is shown next. This graph is symmetric around the line $y = x$, around the line $y = -x$, and around the pole (meaning that (x, y) is on the graph if and only if $(-x, -y)$ is on the graph).

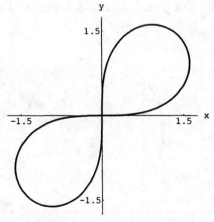

9.2.47: The graph of the four-leaved rose with polar equation $r = 2\sin 2\theta$ is shown next. This graph is symmetric around both coordinate axes, around both lines $y = x$ and $y = -x$, and around the pole.

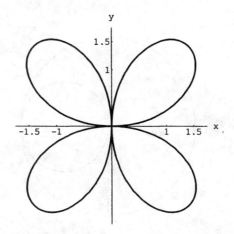

9.2.49: The graph of the three-leaved rose with polar equation $r = 3\cos 3\theta$ is shown next. This graph is symmetric around the x-axis. It is also unchanged if it is rotated any integral multiple of 120° around the origin.

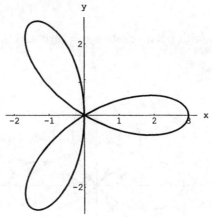

9.2.51: The graph of the five-leaved rose with polar equation $r = 2\sin 5\theta$ is shown next. This graph is symmetric around the y-axis and is also unchanged if rotated any integral multiple of 72° around the origin.

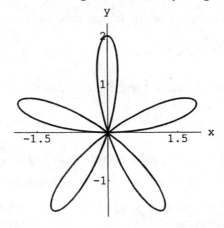

9.2.53: The graphs of the polar equations $r = 1$ and $r = \cos\theta$ are shown next.

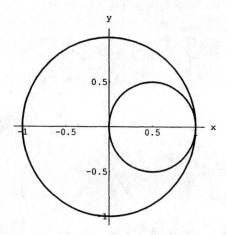

Solving $\cos\theta = 1$ yields $\theta = 0$. This corresponds to the point with polar coordinates $(1, 0)$ on both the large circle and the small circle. The graph makes it clear that there are almost certainly no other solutions. A rigorous demonstration that there are no other solutions is possible but would take this discussion too far afield.

9.2.55: The graphs of the circle with polar equation $r = \sin\theta$ and the four-leaved rose with polar equation $r = \cos 2\theta$ are shown next.

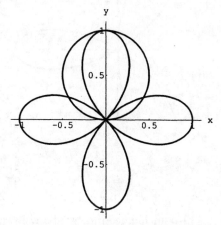

We find some of the points where they meet by solving their equations simultaneously as follows:

$$\sin\theta = \cos 2\theta; \qquad \sin\theta = 1 - 2\sin^2\theta;$$

$$2\sin^2\theta + \sin\theta - 1 = 0; \qquad (2\sin\theta - 1)(\sin\theta + 1) = 0;$$

$$\sin\theta = \frac{1}{2} \quad \text{or} \quad \sin\theta = -1; \qquad \theta = \frac{\pi}{6}, \frac{5\pi}{6}, \text{ or } \frac{3\pi}{2}.$$

Thus the curves meet at the points with polar coordinates $\left(\frac{1}{2}, \frac{1}{6}\pi\right)$ and $\left(\frac{1}{2}, \frac{5}{6}\pi\right)$. They also meet at the point with polar coordinates $\left(-1, \frac{3}{2}\pi\right)$ because it also has polar coordinates $\left(1, \frac{1}{2}\pi\right)$. Finally, they also meet at the pole because it has polar coordinates $(0, 0)$ as well as polar coordinates $\left(0, \frac{1}{4}\pi\right)$.

9.2.57: The graphs of the cardioid with polar equation $r = 1 - \cos\theta$ and the double oval with polar equation $r^2 = 4\cos\theta$ are shown next.

488

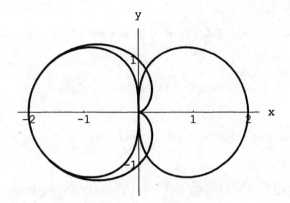

We solve their equations simultaneously as follows:

$$(1 - \cos\theta)^2 = 4\cos\theta; \qquad 1 - 2\cos\theta + \cos^2\theta = 4\cos\theta;$$

$$\cos^2\theta - 6\cos\theta + 1 = 0; \qquad \cos\theta = \frac{6 \pm \sqrt{32}}{2} = 3 \pm 2\sqrt{2};$$

$$\cos\theta = 3 - 2\sqrt{2}; \qquad \theta = \pm\cos^{-1}\left(3 - 2\sqrt{2}\right).$$

Thus we obtain two of the four points of intersection: the points with polar coordinates

$$\left(2\sqrt{2} - 2, \ \cos^{-1}\left(3 - 2\sqrt{2}\right)\right) \qquad \text{and} \qquad \left(2\sqrt{2} - 2, \ -\cos^{-1}\left(3 - 2\sqrt{2}\right)\right).$$

The curves also meet at $(0, 0)$ because it also has polar coordinates $(0, \frac{1}{2}\pi)$. Moreover, they meet at the point $(2, \pi)$ because it also has polar coordinates $(-2, 0)$. There are no other points of intersection.

9.2.59: The following figure shows a typical situation of the sort described in this problem: The point $A(p, \alpha)$ lies in the plane, the line L passes through A and is perpendicular to the line segment OA from the pole O to A. Let $B(r, \theta)$ be the polar coordinates of a typical point of L. Then OAB is a right triangle.

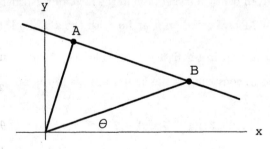

Moreover, the acute angle of this triangle at O is $\alpha - \theta$, so when we project the hypotenuse of length r onto the side OA of length p, we find that

$$r\cos(\alpha - \theta) = p;$$

$$r(\cos\alpha \cos\theta + \sin\alpha \sin\theta) = p;$$

$$r = \frac{p}{\cos\alpha \cos\theta + \sin\alpha \sin\theta}.$$

489

A rectangular coordinates equation of L can now be obtained from the second displayed equation here as follows:

$$(\cos\alpha)(r\cos\theta) + (\sin\alpha)(r\sin\theta) = p,$$

and therefore L has rectangular equation $x\cos\alpha + y\sin\alpha = p$.

9.2.61: Beginning with $a^2(x^2 + y^2) = (x^2 + y^2 - by)^2$, we convert to polar coordinates:

$$a^2 r^2 = (r^2 - br\sin\theta)^2; \qquad ar = \pm(r^2 - br\sin\theta);$$

$$a = \pm(r - b\sin\theta); \qquad r - b\sin\theta = \pm a.$$

Therefore $r = \pm a + b\sin\theta$. If $|a| = |b|$ and neither is zero, then the graph is a cardioid. If $|a| \neq |b|$ and neither a nor b is zero, then the graph is a limaçon. If either a or b is zero and the other is not, then the graph is a circle. If $a = b = 0$ then the graph consists of the pole alone.

9.2.63: The behavior of the graph of the polar equation $r = \cos(p\theta/q)$ (where p and q are positive integers) depends strongly on the values of p and q. Without loss of generality we may suppose that p and q have no integral factor in common larger than 1. First you should determine what is meant by a "loop" and what is meant by "overlapping loops." The minimum value of the positive integer k required to show the entire graph on the interval $[0, k\pi]$ appears in the following table (for the values of p and q we found practical to use). The table is followed by *Mathematica*-generated graphs of $r = \cos(p\theta/q)$ for various values of p and q. The values of p and q are given beneath each graph in the form of the ordered pair (p, q). You can probably deduce the way in which the loops depend on p and q with the aid of a little patience and imagination.

In the table, the values of q appear in the first row, in boldface; those of p appear in the first column. The values of k that we deduced in constructing the figures appear in the body of the table.

	1	**2**	**3**	**4**	**5**	**6**	**7**	**8**	**9**	**10**	**11**	**13**	**15**
1	1	4	3	8	5	12	7	16					
2	2		6		10		14		18		22	26	30
3	1	4		8	5		7	16		20	11		
4	2		6		10		14		18		22	26	30
5	1	4	3	8		12	7	16	9				
6	2			10		14					22	26	
7	1	4	3	8	5	12		16	9				

And next follow the graphs for these specified (p, q) pairs.

(1,1) and (1,2):

(1,3) and (1,4):

(1,5) and (1,6):

(1,7) and (1,8):

(2,1) and (2,3):

(2,5) and (2,7):

(3, 1) and (3, 2):

(3, 4) and (3, 5):

(3, 7) and (3, 8):

(4, 1) and (4, 3):

(4, 7) and (4, 9):

(5, 1) and (5, 2):

494

(5,3) and (5,4):

(5,6) and (5,7):

(6,1) and (6,5):

(6,7) and (6,11):

(7,1) and (7,2):

(7,3) and (7,4):

(7, 5) and (7, 6):

(7,5)

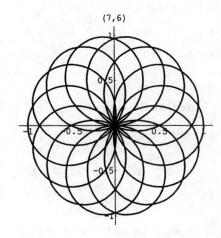
(7,6)

Section 9.3

9.3.1: The graph of $r = \theta$, $0 \leqq \theta \leqq \pi$, is shown next.

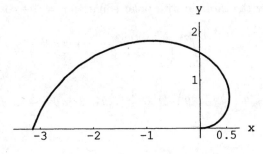

9.3.3: The graph of $r = 1/\theta$, $\pi \leqq \theta \leqq 3\pi$, is shown next.

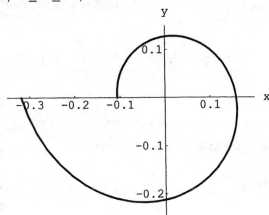

9.3.5: The graph of $r = e^{-\theta}$, $0 \leqq \theta \leqq \pi$, is next.

497

9.3.7: Note that the entire curve $r = 2\cos\theta$ is swept out as θ runs through the interval $0 \leqq \theta \leqq \pi$. Thus the area enclosed by this circle is

$$A = \frac{1}{2}\int_0^\pi 4\cos^2\theta \, d\theta = \int_0^\pi (1 + \cos 2\theta) \, d\theta = \left[\theta + \sin\theta \, \cos\theta\right]_0^\pi = \pi.$$

It might be argued that the area integral should be evaluated over the interval $-\pi/2 \leqq \theta \leqq \pi/2$. These limits will, of course, give the correct answer and are certainly more natural in this problem. On the other hand, the solution shown is correct and has the advantage that trigonometric functions are generally easier to evaluate at integral multiples of π.

9.3.9: The area enclosed by the cardioid with polar equation $r = 1 + \cos\theta$ is

$$A = \frac{1}{2}\int_0^{2\pi} (1 + \cos\theta)^2 \, d\theta = \frac{1}{2}\int_0^{2\pi} \left(1 + 2\cos\theta + \frac{1 + \cos 2\theta}{2}\right) d\theta$$

$$= \frac{1}{2}\int_0^{2\pi} \left(\frac{3}{2} + 2\cos\theta + \frac{1}{2}\cos 2\theta\right) d\theta = \frac{1}{2}\left[\frac{3}{2}\theta + 2\sin\theta + \frac{1}{2}\sin\theta \, \cos\theta\right]_0^{2\pi} = \frac{3}{4} \cdot 2\pi = \frac{3}{2}\pi.$$

9.3.11: Because $r > 0$ for all θ, the area enclosed by the limaçon with polar equation $r = 2 - \cos\theta$ is

$$A = \frac{1}{2}\int_0^{2\pi} (2 - \cos\theta)^2 \, d\theta = \frac{1}{2}\int_0^{2\pi} (4 - 4\cos\theta + \cos^2\theta) \, d\theta = \frac{1}{2}\int_0^{2\pi} \left(4 - 4\cos\theta + \frac{1 + \cos 2\theta}{2}\right) d\theta$$

$$= \frac{1}{2}\left[\frac{9}{2}\theta - 4\sin\theta + \frac{1}{4}\sin 2\theta\right]_0^{2\pi} = \frac{9}{4} \cdot 2\pi = \frac{9}{2}\pi \approx 14.137166941154.$$

9.3.13: Note that the entire circle with polar equation $r = -4\cos\theta$ is swept out as θ runs through any interval of length π. Therefore the area the circle encloses is

$$A = \int_0^\pi 8\cos^2\theta \, d\theta = 4\int_0^\pi (1 + \cos 2\theta) \, d\theta = 4\left[\theta + \sin\theta \, \cos\theta\right]_0^\pi = 4\pi \approx 12.566370614359.$$

9.3.15: An accurate scale drawing of the graph of the limaçon with polar equation $r = 3 - \cos\theta$ is shown next. It looks very much like a circle. Do you see an easy way to deduce that it is *not* a circle?

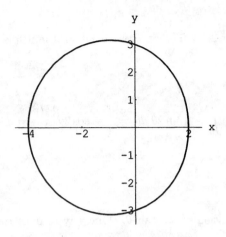

The area enclosed by this limaçon is

$$A = \frac{1}{2} \int_0^{2\pi} (3 - \cos\theta)^2 \, d\theta = \frac{1}{2} \int_0^{2\pi} \left(9 - 6\cos\theta + \frac{1 + \cos 2\theta}{2} \right) d\theta$$

$$= \frac{1}{2} \left[\frac{19}{2}\theta - 6\sin\theta + \frac{1}{4}\sin 2\theta \right]_0^{2\pi} = \frac{19}{2}\pi \approx 29.845130209103.$$

9.3.17: Given: The polar equation $r = 2\cos 2\theta$ of a four-leaved rose. The "loops," or rose petals, are formed by the curve repeatedly passing through the origin at different angles. Examine the graph of this equation in Fig. 9.2.12 (Example 6 of Section 9.2). All we need is to find when $r = 0$; that is, when $\cos 2\theta = 0$. This occurs when θ is an odd integral multiple of $\pi/4$, so we can take for the limits of integration any two consecutive such numbers. Hence the area enclosed by one loop of the rose is

$$A = \frac{1}{2} \int_{-\pi/4}^{\pi/4} 4\cos^2 2\theta \, d\theta.$$

Using the symmetry of the loop around the x-axis, we can double the area of the upper half of the loop to make the computations slightly simpler:

$$A = \int_0^{\pi/4} 2(1 + \cos 4\theta) \, d\theta = 2 \left[\theta + \frac{1}{4}\sin 4\theta \right]_0^{\pi/4} = 2 \cdot \frac{\pi}{4} = \frac{\pi}{2} \approx 1.570796326795.$$

9.3.19: Given: The polar equation $r = 2\cos 4\theta$ of an eight-leaved rose (see Fig. 9.3.13). To find the area of one loop, we need to find the limits of integration on θ, which are determined by solving the equation $r = 0$. We find that 4θ will be any odd integral multiple of $\pi/2$, and therefore that θ will be any odd integral multiple of $\pi/8$. We will also double the area of half of one loop to make the computations slightly simpler. Thus the area of one loop is

$$A = \frac{1}{2} \int_{-\pi/8}^{\pi/8} 4\cos^2 4\theta \, d\theta = \int_0^{\pi/8} 2(1 + \cos 8\theta) \, d\theta = \left[2\theta + \frac{1}{4}\sin 8\theta \right]_0^{\pi/8} = \frac{\pi}{4} \approx 0.785398163397.$$

9.3.21: The lemniscate with polar equation $r^2 = 4\sin 2\theta$ has two loops; its graph can be obtained from the one in Fig. 9.2.13 (Example 7 of Section 9.2) by a 90° rotation. To find the area of the loop in the first quadrant, find when $r = 0$: $\sin 2\theta = 0$ when θ is an integral multiple of $\pi/2$. Hence the area of the loop lying in the first quadrant is

$$A = \frac{1}{2} \int_0^{\pi/2} 4 \sin 2\theta \, d\theta = \left[-\cos 2\theta \right]_0^{\pi/2} = 1 - (-1) = 2.$$

9.3.23: Given: The polar equation $r^2 = 4 \sin \theta$. One way to construct its graph is first to construct the *Cartesian* graph $y = 4 \sin x$, which is shown next for $0 \leqq x \leqq 2\pi$.

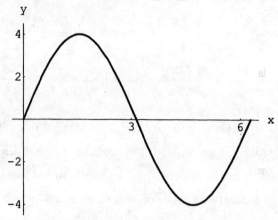

Then construct the Cartesian graph of $y = \pm\sqrt{4 \sin x}$, shown next, also for $0 \leqq x \leqq 2\pi$. Note that there is no graph for $\pi < x < 2\pi$ but *two* graphs for $0 \leqq x \leqq \pi$.

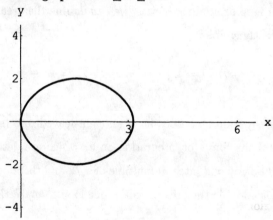

Finally, use the last graph to construct the polar graph $r^2 = 4 \sin \theta$ by sketching both $r = \sqrt{4 \sin \theta}$ and $r = -\sqrt{4 \sin \theta}$. As θ varies from 0 to π, r begins at 0, increases to a maximum $r = 2$ when $\theta = \pi/2$, then decreases to 0 as θ runs through the values from $\pi/2$ to π. Meanwhile, $-r$ sweeps out the mirror image of the previous curve, and thus we obtain the "double oval" shown in the next figure.

500

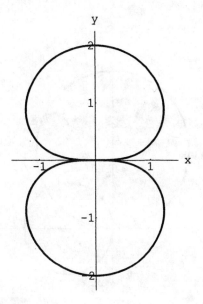

The area of the upper loop is therefore

$$A = \frac{1}{2} \int_0^\pi 4 \sin\theta \, d\theta = \left[-2\cos\theta \right]_0^\pi = 2 - (-2) = 4.$$

9.3.25: See the graph below.

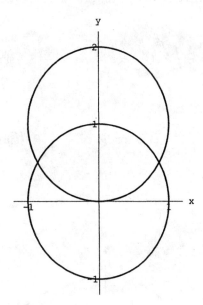

Find the area A of the region both inside the circle $r = 2\sin\theta$ and outside the circle $r = 1$. The circles cross where $2\sin\theta = 1$, thus where $\theta = \pi/6$ and where $\theta = 5\pi/6$. So

$$A = \frac{1}{2} \int_{\pi/6}^{5\pi/6} (4\sin^2\theta - 1) \, d\theta = \frac{1}{2} \int_{\pi/6}^{5\pi/6} \left[2(1 - \cos 2\theta) - 1 \right] d\theta$$

$$= \frac{1}{2} \left[\theta - \sin 2\theta \right]_{\pi/6}^{5\pi/6} = \frac{1}{2} \left(\frac{5\pi}{6} - \frac{\pi}{6} + \frac{\sqrt{3}}{2} + \frac{\sqrt{3}}{2} \right) = \frac{2\pi + 3\sqrt{3}}{6} \approx 1.913222954981.$$

9.3.27: See the figure below.

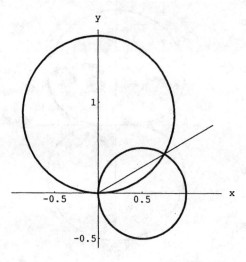

The two circles $r = \cos\theta$ and $r = \sqrt{3}\,\sin\theta$ cross where

$$\cos\theta = \sqrt{3}\,\sin\theta; \qquad \tan\theta = \frac{\sqrt{3}}{3}; \qquad \theta = \frac{\pi}{6};$$

they also cross at the pole. The region inside both is divided by the ray $\theta = \pi/6$ into a lower region of area B and an upper region of area C. Thus we find the area of the region inside both circles to be $A = B + C$. Now

$$B = \frac{1}{2}\int_0^{\pi/6} 3\sin^2\theta\,d\theta = \frac{3}{4}\int_0^{\pi/6}(1 - \cos 2\theta)\,d\theta = \frac{3}{4}\Big[\theta - \sin\theta\cos\theta\Big]_0^{\pi/6} = \frac{3}{4}\left(\frac{\pi}{6} - \frac{\sqrt{3}}{4}\right)$$

and

$$C = \frac{1}{2}\int_{\pi/6}^{\pi/2}\cos^2\theta\,d\theta = \frac{1}{4}\int_{\pi/6}^{\pi/2}(1 + \cos 2\theta)\,d\theta = \frac{1}{4}\Big[\theta + \sin\theta\cos\theta\Big]_{\pi/6}^{\pi/2} = \frac{1}{4}\left(\frac{\pi}{3} - \frac{\sqrt{3}}{4}\right).$$

Therefore $A = B + C = \dfrac{5\pi - 6\sqrt{3}}{24} \approx 0.221485767606.$

9.3.29: See the figure below.

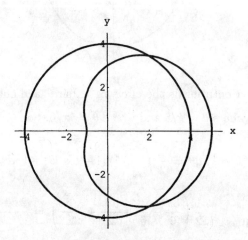

502

Let A denote the area of the region that is both inside the limaçon with polar equation $r = 3 + 2\cos\theta$ and outside the circle with equation $r = 4$. The curves cross where

$$3 + 2\cos\theta = 4; \qquad 2\cos\theta = 1; \qquad \cos\theta = \frac{1}{2}; \qquad \theta = \pm\frac{\pi}{3}.$$

Therefore

$$A = \frac{1}{2}\int_{-\pi/3}^{\pi/3} \left[(3 + 2\cos\theta)^2 - 4^2\right] d\theta = \int_0^{\pi/3} \left[9 + 12\cos\theta + 2(1 + \cos 2\theta) - 16\right] d\theta$$

$$= \left[-5\theta + 12\sin\theta + \sin 2\theta\right]_0^{\pi/3} = -\frac{5\pi}{3} + 12\cdot\frac{\sqrt{3}}{2} + \frac{\sqrt{3}}{2} = \frac{39\sqrt{3} - 10\pi}{6} \approx 6.022342493215.$$

9.3.31: See Fig. 9.3.16 of the text. The lemniscates $r^2 = \cos 2\theta$ and $r^2 = \sin 2\theta$ cross where $\cos 2\theta = \sin 2\theta$, thus where $\tan 2\theta = 1$; that is, where $\theta = \pi/8$ (and they also cross at the pole). We find the area A within both curves by doubling the area of the half to the right of the y-axis:

$$A = 2\cdot\frac{1}{2}\int_0^{\pi/8} \sin 2\theta\, d\theta + 2\cdot\frac{1}{2}\int_{\pi/8}^{\pi/4} \cos 2\theta\, d\theta$$

$$= 2\int_0^{\pi/8} \sin 2\theta\, d\theta = \left[-\cos 2\theta\right]_0^{\pi/8} = \frac{2 - \sqrt{2}}{2} \approx 0.292893218813.$$

9.3.33: See the figure below.

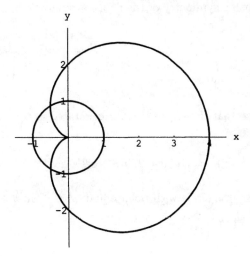

We are to find the area A of the region that is both inside the cardioid with polar equation $r = 2(1 + \cos\theta)$ and outside the circle with equation $r = 1$. The curves cross where $2 + 2\cos\theta = 1$; it follows that $\cos\theta = -\frac{1}{2}$, so that $\theta = 2\pi/3$ or $\theta = 4\pi/3$. Therefore

$$A = \frac{1}{2}\int_{4\pi/3}^{8\pi/3}\left[4(1+\cos\theta)^2 - 1\right]d\theta = \frac{1}{2}\int_{4\pi/3}^{8\pi/3}(4 + 8\cos\theta + 4\cos^2\theta - 1)\,d\theta$$

$$= \frac{1}{2}\int_{4\pi/3}^{8\pi/3}(3 + 8\cos\theta + 2 + 2\cos 2\theta)\,d\theta = \frac{1}{2}\left[5\theta + 8\sin\theta + \sin 2\theta\right]_{4\pi/3}^{8\pi/3}$$

$$= \frac{1}{2}\left(\frac{40\pi}{3} + 8\cdot\frac{\sqrt{3}}{2} - \frac{\sqrt{3}}{2}\right) - \frac{1}{2}\left(\frac{20\pi}{3} - 8\cdot\frac{\sqrt{3}}{2} + \frac{\sqrt{3}}{2}\right)$$

$$= \frac{1}{2}\left(\frac{20\pi}{3} + 8\sqrt{3} - \sqrt{3}\right) = \frac{1}{2}\left(\frac{20\pi}{3} + 7\sqrt{3}\right) = \frac{20\pi + 21\sqrt{3}}{6} \approx 16.534153338457.$$

9.3.35: See the figure below.

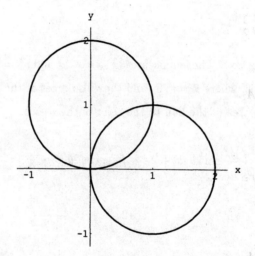

The two circles with polar equations $r = 2\cos\theta$ and $r = 2\sin\theta$ meet at the pole and where $2\cos\theta = 2\sin\theta$; that is, where $\theta = \pi/4$. So, using symmetry of the figure around the line $y = x$, the area of the region that lies within both circles is

$$A = 2\cdot\frac{1}{2}\int_0^{\pi/4}4\sin^2\theta\,d\theta = \int_0^{\pi/4}2(1-\cos 2\theta)\,d\theta = \left[2\theta - \sin 2\theta\right]_0^{\pi/4} = \frac{\pi - 2}{2} \approx 0.570796326795.$$

9.3.37: See Fig. 9.3.18. Note that the entire circle is generated as θ runs through any interval of length π and that

$$(\sin\theta + \cos\theta)^2 = \sin^2\theta + 2\sin\theta\cos\theta + \cos^2\theta = 1 + 2\sin\theta\cos\theta.$$

Therefore the area enclosed by the circle with polar equation $r = \sin\theta + \cos\theta$ is

$$A = \frac{1}{2}\int_0^{\pi}(1 + 2\sin\theta\cos\theta)\,d\theta = \left[\frac{1}{2}\theta + \frac{1}{2}\sin^2\theta\right]_0^{\pi} = \frac{\pi}{2}.$$

To write the equation of this circle in Cartesian form, proceed as follows:

$$r^2 = 4\sin\theta + r\cos\theta;$$

504

$$x^2 + y^2 = x + y;$$

$$x^2 - x + \frac{1}{4} + y^2 - y + \frac{1}{4} = \frac{1}{2};$$

$$\left(x - \frac{1}{2}\right)^2 + \left(y - \frac{1}{2}\right)^2 = \frac{1}{2}.$$

So the figure is, indeed, a circle, and the square of its radius is $\frac{1}{2}$. Therefore the area of this circle is $\pi/2$.

9.3.39: Part (a):

$$A_1 = \frac{1}{2}\int_0^{2\pi} a^2\theta^2\, d\theta = \left[\frac{1}{6}a^2\theta^3\right]_0^{2\pi} = \frac{4}{3}\pi^3 a^2 = \frac{1}{3}\pi(2\pi a)^2.$$

Part (b):

$$A_2 = \frac{1}{2}\int_{2\pi}^{4\pi} a^2\theta^2\, d\theta = \left[\frac{1}{6}a^2\theta^3\right]_{2\pi}^{4\pi} = \frac{28}{3}\pi^3 a^2 = \frac{7}{12}\pi(4\pi a)^2.$$

Part (c):

$$R_2 = A_2 - A_1 = \frac{28}{3}\pi^3 a^2 - \frac{4}{3}\pi^3 a^2 = \frac{24}{3}\pi^3 a^2 = 8\pi^3 a^2 = 6\cdot\frac{4}{3}\pi^3 a^2 = 6A_1.$$

Part (d): If $n \geqq 2$, then

$$A_n = \frac{1}{2}\int_{2(n-1)\pi}^{2n\pi} a^2\theta^2\, d\theta = \left[\frac{1}{6}a^2\theta^3\right]_{2(n-1)\pi}^{2n\pi} = \frac{1}{6}a^2\left[8n^3\pi^3 - 8(n-1)^3\pi^3\right] = \frac{4}{3}\pi^3 a^2(3n^2 - 3n + 1),$$

and therefore

$$R_{n+1} = A_{n+1} - A_n = \frac{4}{3}\pi^3 a^2(3n^2 + 6n + 3 - 3n - 3 - 3n^2 + 3n) = \frac{4}{3}\pi^3 a^2 \cdot 6n = 8\pi^3 a^2 n = nR_2.$$

9.3.41: Part (a): The area is

$$A_1 = \int_0^{2\pi} \frac{1}{2}a^2 e^{-2k\theta}\, d\theta - \int_{2\pi}^{4\pi} \frac{1}{2}a^2 e^{-2k\theta}\, d\theta = \frac{1}{2}a^2\left[-\frac{e^{-2k\theta}}{2k}\right]_0^{2\pi} - \frac{1}{2}a^2\left[-\frac{e^{-2k\theta}}{2k}\right]_{2\pi}^{4\pi}$$

$$= \frac{a^2}{4k}\left(1 - e^{-4k\pi}\right) + \frac{a^2}{4k}\left(e^{-8k\pi} - e^{-4k\pi}\right) = \frac{a^2}{4k}\left(1 - e^{-4k\pi}\right)^2.$$

With $k = \frac{1}{10}$ and $a = 1$, we obtain

$$A = \frac{5}{2}\left(1 - e^{-2\pi/5}\right)^2 \approx 1.27945876.$$

Part (b): The area is

$$A_n = \int_{2(n-1)\pi}^{2n\pi} \frac{1}{2}a^2\exp\left(-2k\theta\right)\, d\theta - \int_{2n\pi}^{2(n+1)\pi} \frac{1}{2}a^2\exp\left(-2k\theta\right)\, d\theta$$

505

$$= \frac{1}{2}a^2 \left[-\frac{\exp(-2k\theta)}{2k} \right]_{2(n-1)\pi}^{2n\pi} - \frac{1}{2}a^2 \left[-\frac{\exp(-2k\theta)}{2k} \right]_{2n\pi}^{2(n+1)\pi}$$

$$= \frac{a^2}{4k} \left[\exp(-4(n-1)k\pi) - \exp(-4nk\pi) + \exp(-4(n+1)k\pi) - \exp(-4nk\pi) \right]$$

$$= \frac{a^2}{4k} \exp(-4(n-1)k\pi) \left[1 - \exp(-4k\pi) \right]^2.$$

With $a = 1$ and $k = \frac{1}{10}$, we find that

$$A = \frac{5}{2} e^{-2(n-1)\pi/5} \left(1 - e^{-2\pi/5} \right)^2.$$

9.3.43: The point of intersection in the second quadrant is located where $\theta = \alpha \approx 2.326839$. Using symmetry, the total area of the shaded region R is approximately

$$2 \int_0^\alpha \frac{1}{2} \left(e^{-\theta/5} \right)^2 \, d\theta + 2 \int_\alpha^\pi \frac{1}{2} [2(1 + \cos\theta)]^2 \, d\theta \approx 1.58069.$$

Section 9.4

9.4.1: If $x = t + 1$, then $t = x - 1$, so that $y = 2t - 1 = 2(x - 1) - 1 = 2x - 3$. The graph is next.

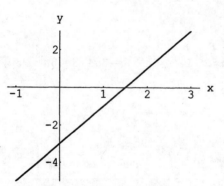

9.4.3: If $x = t^2$ and $y = t^3$, then $y = \left(t^2 \right)^{3/2} = x^{3/2}$; alternatively, $y^2 = x^3$. The graph is shown next.

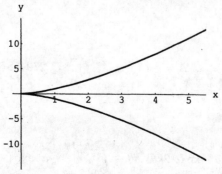

9.4.5: If $x = t + 1$, then $t = x - 1$, so that $y = 2t^2 - t - 1 = 2(x-1)^2 - (x-1) - 1 = 2x^2 - 5x + 2$. The graph is next.

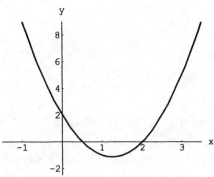

9.4.7: If $x = e^t$, then $y = 4e^{2t} = 4(e^t)^2 = 4x^2$ with the restriction that $x > 0$. The graph is next.

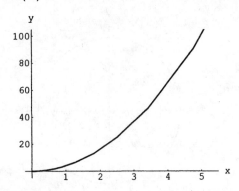

9.4.9: If $x = 5\cos t$ and $y = 3\sin t$, then

$$\left(\frac{x}{5}\right)^2 + \left(\frac{y}{3}\right)^2 = 1; \qquad \text{that is,} \qquad 9x^2 + 25y^2 = 225.$$

The graph of this ellipse is next.

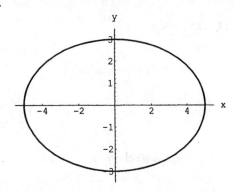

9.4.11: If $x = 2\cosh t$ and $y = 3\sinh t$, then

$$\left(\frac{x}{2}\right)^2 - \left(\frac{y}{3}\right)^2 = \cosh^2 t - \sinh^2 t = 1; \qquad \text{that is,} \quad 9x^2 - 4y^2 = 16.$$

But not all points that satisfy the last equation are on the graph, because $x = 2\cosh t \geqq 2$ for all t. Thus only points on the right half of this hyperbola form the graph of the parametric equations, shown next.

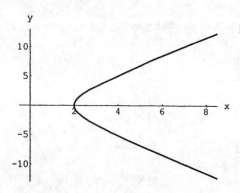

9.4.13: Given $x = \sin 2\pi t$ and $y = \cos 2\pi t$, $0 \leq t \leq 1$, it follows that $x^2 + y^2 = 1$. Thus the graph is a circle of radius 1 centered at the origin. As t runs from 0 to 1, the point (x, y) begins at $(0, 1)$ and moves once clockwise around the circle. The graph is next.

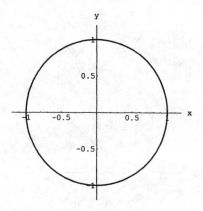

9.4.15: Given $x = \sin^2 \pi t$ and $y = \cos^2 \pi t$, $0 \leq t \leq 2$, it's clear that $x + y = 1$ and that $0 \leq x \leq 1$. So the graph is the straight line segment joining $(0, 1)$ and $(1, 0)$. As t varies from 0 to 2, the point (x, y) begins at $(0, 1)$, moves southeast until it reaches $(1, 0)$ when $t = 1$, then moves northwest until it returns to $(0, 1)$ when $t = 2$. We omit the graph to save space.

9.4.17: Given $x = 2t^2 + 1$, $y = 3t^3 + 2$, we first calculate

$$\frac{dy}{dx} = \frac{dy/dt}{dx/dt} = \frac{9t^2}{4t} = \frac{9}{4}t, \quad \text{so that} \quad \left.\frac{dy}{dx}\right|_{t=1} = \frac{9}{4}.$$

When $t = 1$, $(x, y) = (3, 5)$, so the tangent line there has equation

$$y - 5 = \frac{9}{4}(x - 3); \qquad 4y - 20 = 9x - 27; \qquad 9x = 4y + 7.$$

Next,

$$\frac{d^2y}{dx^2} = \frac{d}{dx}\left(\frac{dy}{dx}\right) = \frac{1}{dx/dt} \cdot \frac{d}{dt}\left(\frac{dy}{dx}\right) = \frac{1}{4t} \cdot \frac{9}{4} = \frac{9}{16t}.$$

The second derivative is positive when $t = 1$, so the graph is concave upward at and near the point $(3, 5)$.

508

9.4.19: Given $x = t \sin t$ and $y = t \cos t$, we first calculate

$$\frac{dy}{dx} = \frac{dy/dt}{dx/dt} = \frac{\cos t - t \sin t}{\sin t + t \cos t}; \qquad \frac{dy}{dx}\bigg|_{t=\pi/2} = -\frac{\pi}{2}.$$

Hence an equation of the line tangent to the graph at $(x, y) = (\pi/2, 0)$ is

$$y = -\frac{\pi}{2}\left(x - \frac{\pi}{2}\right); \qquad 4y = -\pi(2x - \pi); \qquad 2\pi x + 4y = \pi^2.$$

Next,

$$\frac{d^2y}{dx^2} = \frac{1}{dx/dt} \cdot \frac{d}{dt}\left(\frac{dy}{dx}\right)$$

$$= \frac{1}{\sin t + t \cos t} \cdot \frac{(\sin t + t \cos t)(-2\sin t - t \cos t) - (\cos t - t \sin t)(2 \cos t - t \sin t)}{(\sin t + t \cos t)^2}$$

$$= -\frac{(\sin t + t \cos t)(2 \sin t + t \cos t) + (\cos t - t \sin t)(2 \cos t - t \sin t)}{(\sin t + t \cos t)^3}$$

$$= -\frac{2\sin^2 t + 3t \sin t \, \cos t + t^2 \cos^2 t + 2\cos^2 t - 3t \sin t \, \cos t + t^2 \sin^2 t}{(\sin t + t \cos t)^3} = -\frac{t^2 + 2}{(\sin t + t \cos t)^2}.$$

Thus

$$\frac{d^2y}{dx^2}\bigg|_{t=\pi/2} = -\left(\frac{\pi^2}{4} + 2\right) < 0,$$

and therefore the graph is concave downward at and near the point of tangency.

9.4.21: Equation (10) tells us that

$$\cot\psi = \frac{1}{r} \cdot \frac{dr}{d\theta}$$

where $0 \leqq \psi \leqq \pi$. Thus, given $r = \exp\left(\theta\sqrt{3}\,\right)$ and the angle $\theta = \pi/2$, we find that

$$\cot\psi = \frac{1}{\exp\left(\theta\sqrt{3}\,\right)} \cdot \left(\sqrt{3}\,\right)\exp\left(\theta\sqrt{3}\,\right) = \sqrt{3}.$$

Therefore $\psi = \dfrac{\pi}{6}$.

9.4.23: Given $r = \sin 3\theta$ and the angle $\theta = \pi/6$. By Eq. (10) of the text,

$$\cot\psi = \frac{1}{\sin 3\theta} \cdot 3 \cos 3\theta = 3 \cot 3\theta.$$

Thus when $\theta = \pi/6$, we have $\cot\psi = 3\cot(\pi/2) = 0$, and thus $\psi = \dfrac{\pi}{2}$.

9.4.25: Given $x = t^2$ and $y = t^3 - 3t$,

$$\frac{dy}{dx} = \frac{3t^2 - 3}{2t}; \qquad \frac{dy}{dx} = 0 \quad \text{when} \quad t = \pm 1.$$

So the graph has horizontal tangents at the point $(1, -2)$ and $(1, 2)$. The graph crosses the x-axis when $t^3 - 3t = 0$: $t = 0$, $t = \pm\sqrt{3}$. When $t = 0$ we get a vertical tangent line at $(0, 0)$. When $t = -\sqrt{3}$ the graph passes through the point $(3, 0)$ with slope $-\sqrt{3}$; when $t = \sqrt{3}$, the graph passes through the *same* point $(3, 0)$ with slope $\sqrt{3}$. Therefore there is no line tangent to the graph of the parametric equations at the point $(3, 0)$.

9.4.27: Given the polar equation $r = 1 + \cos\theta$, we can use θ itself as parameter to obtain

$$x = r\cos\theta = \cos\theta + \cos^2\theta \quad \text{and} \quad y = r\sin\theta = \sin\theta + \sin\theta\cos\theta.$$

Thus

$$\frac{dy}{dx} = \frac{dy/d\theta}{dx/d\theta} = \frac{\cos\theta + \cos^2\theta - \sin^2\theta}{-\sin\theta - 2\sin\theta\cos\theta}.$$

Next we solve $dy/dx = 0$:

$$\cos\theta + \cos^2\theta + \cos^2\theta - 1 = 0;$$

$$2\cos^2\theta + \cos\theta - 1 = 0;$$

$$(2\cos\theta - 1)(\cos\theta + 1) = 0;$$

$$\cos\theta = \frac{1}{2} \quad \text{or} \quad \cos\theta = -1.$$

Thus $\theta = \pi/3$, π, $5\pi/3$. But we must rule out $\theta = \pi$ because the denominator in dy/dx is zero for that value of θ.

When $\theta = \dfrac{\pi}{3}$, $\quad x = \dfrac{1}{2} + \dfrac{1}{4} = \dfrac{3}{4}$ and $y = \dfrac{\sqrt{3}}{2} + \dfrac{\sqrt{3}}{4} = \dfrac{3\sqrt{3}}{4}$. There is a horizontal tangent.

When $\theta = \dfrac{5\pi}{3}$, $\quad x = \dfrac{1}{2} + \dfrac{1}{4} = \dfrac{3}{4}$ and $y = -\dfrac{\sqrt{3}}{2} - \dfrac{\sqrt{3}}{4} = -\dfrac{3\sqrt{3}}{4}$. There is a horizontal tangent.

The graph crosses the x-axis when $\cos\theta = -1$ and when $\sin\theta = 0$, so that $\theta = 0$ and $\theta = \pi$. When $\theta = 0$ the tangent line is vertical at the point with Cartesian coordinates $(2, 0)$. What happens if $\theta = \pi$? The derivative is undefined. Nevertheless, something can be done. We use l'Hôpital's rule:

$$\lim_{\theta \to \pi} \frac{dy}{dx} = \lim_{\theta \to \pi} \frac{\sin^2\theta - \cos^2\theta - \cos\theta}{2\sin\theta\cos\theta + \sin\theta} = \lim_{\theta \to \pi} \frac{4\sin\theta\cos\theta + \sin\theta}{2\cos^2\theta - 2\sin^2\theta + \cos\theta} = \frac{0 + 0}{2 - 0 - 1} = 0.$$

Thus we are justified in stating that the x-axis is tangent to the graph of this cardioid at the point $(0, 0)$.

9.4.29: Given $x = e^{-t}$ and $y = e^{2t}$, we find that

$$\frac{dy}{dx} = \frac{dy/dt}{dx/dt} = \frac{2e^{2t}}{-e^{-t}} = -2e^{3t}$$

and

$$\frac{d^2y}{dx^2} = \frac{1}{dx/dt} \cdot \frac{d}{dt}\left(\frac{dy}{dx}\right) = \frac{1}{-e^{-t}} \cdot \left(-6e^{3t}\right) = 6e^{4t},$$

510

so the second derivative is positive for all t. Thus the graph of C is concave upward for all t. The graph is shown next; note that there is no graph for $x \leqq 0$ or for $y \leqq 0$.

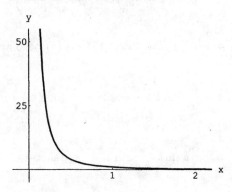

9.4.31: If the slope of the curve at $P(x, y)$ is m, then implicit differentiation yields

$$2y\frac{dy}{dx} = 4p; \qquad \frac{dy}{dx} = \frac{2p}{y}; \qquad y = \frac{2p}{m},$$

and thus

$$x = \frac{y^2}{4p} = \frac{4p^2}{4m^2p} = \frac{p}{m^2}, \qquad -\infty < m < +\infty.$$

9.4.33: The high point on the circle is $P_0(a\theta, 2a)$ and P has Cartesian coordinates $x = a(\theta - \sin\theta)$, $y = a(1 - \cos\theta)$. Therefore the slope of the line containing P_0 and P is

$$\frac{2a - a(1 - \cos\theta)}{a\theta - a(\theta - \sin\theta)} = \frac{2 - 1 + \cos\theta}{\theta - \theta + \sin\theta} = \frac{1 + \cos\theta}{\sin\theta}.$$

But the slope of the cycloid at the point P is

$$\frac{dy}{dx} = \frac{a\sin\theta}{a(1 - \cos\theta)} = \frac{(\sin\theta)(1 + \cos\theta)}{1 - \cos^2\theta} = \frac{1 + \cos\theta}{\sin\theta}.$$

We may conclude that the line containing P_0 and P is tangent to the cycloid at the point P.

9.4.35: We will need two trigonometric identities before we begin. They are

$$\cos 3t = \cos 2t \cos t - \sin 2t \sin t = \cos^3 t - \sin^2 t \cos t - 2\sin^2 t \cos t = \cos^3 t - 3\sin^2 t \cos t \qquad (1)$$

and

$$\sin 3t = \sin 2t \cos t + \cos 2t \sin t = 2\sin t \cos^2 t + \cos^2 t \sin t - \sin^3 t = 3\sin t \cos^2 t - \sin^3 t. \qquad (2)$$

We begin with the parametric equations

$$x = (a - b)\cos t + b\cos\left(\frac{a - b}{b}t\right) \qquad (3)$$

511

and

$$y = (a-b)\sin t - b\sin\left(\frac{a-b}{b}t\right). \tag{4}$$

If $b = \dfrac{a}{4}$, then $\dfrac{a-b}{b} = \dfrac{a - \frac{1}{4}a}{\frac{1}{4}a} = \dfrac{3a}{a} = 3$. Thus Eqs. (3) and (4) become

$$x = \frac{3}{4}a\cos t + \frac{a}{4}\cos 3t = \frac{a}{4}\left(3\cos t + \cos 3t\right) \tag{5}$$

and

$$y = \frac{3}{4}a\sin t - \frac{a}{4}\sin 3t = \frac{a}{4}\left(3\sin t - \sin 3t\right). \tag{6}$$

Then Eqs. (1) and (2) yield

$$x = \frac{a}{4}\left(3\cos t + \cos^3 t - 3\sin^2 t\cos t\right) = \frac{a}{4}\left(3\cos t + \cos^3 t - 3\cos t + 3\cos^3 t\right) = a\cos^3 t$$

and

$$y = \frac{a}{4}\left(3\sin t - 3\sin t\cos^2 t + \sin^3 t\right) = \frac{a}{4}\left(3\sin t - 3\sin t + 3\sin^3 t + \sin^3 t\right) = a\sin^3 t.$$

9.4.37: Extend OP the distance a to the point R at the "northeast" corner of Archimedes' rectangle. Because P has Cartesian coordinates

$$x = a\theta\cos\theta, \qquad y = a\theta\sin\theta,$$

it follows that R has coordinates

$$x = a\theta\cos\theta + a\cos\theta, \qquad y = a\theta\sin\theta + a\sin\theta.$$

Next, Q has coordinates

$$a\theta\cos\theta + a\cos\theta - a\theta\sin\theta, \qquad y = a\theta\sin\theta + a\sin\theta + a\theta\cos\theta.$$

Therefore the slope of PQ is

$$\frac{\sin\theta + \theta\cos\theta}{\cos\theta - \theta\sin\theta}.$$

The spiral has polar equation $r = a\theta$, thus parametric equations

$$x = a\theta\cos\theta, \qquad y = a\theta\sin\theta.$$

Therefore

$$\frac{dy}{dx} = \frac{a\sin\theta + a\theta\cos\theta}{a\cos\theta - a\theta\sin\theta} = \frac{\sin\theta + \theta\cos\theta}{\cos\theta - \theta\sin\theta}.$$

Hence the line containing P and Q is tangent to the spiral at the point P.

9.4.39: If ψ is constant, then by Eq. (10) of the text

$$\frac{1}{r} \cdot \frac{dr}{d\theta} = k \quad \text{(a constant)};$$

$$\frac{1}{r}\, dr = k\, d\theta;$$

$$\ln r = C + k\theta \quad \text{(where } C \text{ is constant)};$$

$$r = Ae^{k\theta} \quad \text{(where } A = e^C\text{)}.$$

9.4.41: Let $y = tx$ where $t \geqq 0$. Then this line meets the loop at exactly one point in the first quadrant. For such points on the loop, we then have

$$x^5 + t^5 x^5 = 5t^2 x^4; \quad x = \frac{5t^2}{1+t^5}, \quad y = \frac{5t^3}{1+t^5}, \quad 0 \leqq t < +\infty.$$

9.4.43: Let $f(x) = x^3 - 3x^2 + 1$. Then $f'(x) = 0$ when $x = 0$ and when $x = 2$; $f''(x) = 0$ when $x = 1$. Hence the graph of the parametric equations

$$x = t^3 - 3t^2 + 1, \quad y = t \tag{1}$$

has vertical tangents at $(-3, 2)$ and $(1, 0)$ and an inflection point at $(-1, 1)$. There are no horizontal tangents and the only critical points occur at the points where the tangent line is vertical. The graph of the equations in (1) is next.

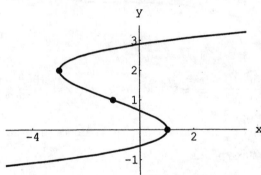

We generated this graph by executing the *Mathematica* command

```
ParametricPlot[ { t∧3 − 3*t∧2 + 1, t }, { t, −1.2, 3.2 },
            AspectRatio → Automatic ];
```

9.4.45: To use *Mathematica* to help solve this problem, we let

```
g[t_] := t∧5 − 5*t∧3 + 4
```

513

```
h[t_] := (Sign[g[t]])*(Abs[g[t]])^(1/3)
```

and

```
f[t_] := (g[t])^(1/3)
```

(note that $h(t) = f(t)$; we define h to avoid certain problems with cube roots of negative numbers). To see the graph of the parametric equations

$$x = (t^5 - 5t^3 + 4)^{1/3}, \qquad y = t,$$

we executed the *Mathematica* command

```
ParametricPlot[ { h[t], t }, { t, -2.7, 2.7 } ];
```

with the result shown next.

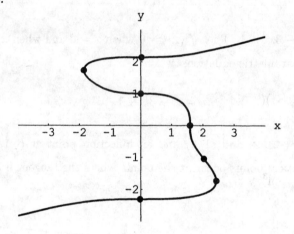

Next we found that

$$f'(t) = \frac{5t^2(t^2 - 3)}{3(t^5 - 5t^3 + 4)^{2/3}} \quad \text{and that} \quad f''(t) = \frac{10(t^8 - 9t^6 + 24t^3 - 36t)}{9(t^5 - 5t^3 + 4)^{5/3}}.$$

If follows that $f'(t) = 0$ when $t = 0$ and when $t = \pm\sqrt{3}$. Hence the graph of the equation $x^3 = y^5 - 5y^3 + 4$ has *vertical* tangents at the three points

$$\left(\left[4 - 6\sqrt{3} \right]^{1/3}, \sqrt{3} \right) \approx (-1.855891115, 1.732050808),$$

$$\left(\left[4 + 6\sqrt{3} \right]^{1/3}, -\sqrt{3} \right) \approx (2.432447355, -1.732050808), \qquad \text{and}$$

$$\left(4^{1/3}, 0 \right) \approx (1.587401052, 0).$$

The *horizontal* tangents will occur when the denominator in $f'(t) = 0$; that is, at

$$(0, -2.307699789), \qquad (0, 1), \qquad \text{and} \qquad (0, 2.143299604)$$

514

(numbers with decimal points are approximations). Finally, Newton's method yields the zeros of the numerator of $f''(t)$, and—also checking the zeros of its denominator—we find that the graph has inflection points at

$$(-5.150545372, -3.110298772), \qquad (0, -2.307699789),$$

$$(2.037032912, -1.044330352), \qquad (1.587401052, 0),$$

$$(0, 1), \qquad\qquad\qquad (0, 2.143299604), \qquad \text{and}$$

$$(4.266140637, 2.856500901).$$

The first and last of these aren't shown on the preceding graph, but the graph appears to be a straight line in their vicinity, so showing more of the graph is not useful.

Section 9.5

9.5.1: The area is

$$A = \int_{-1}^{1} (2t^2 + 1)(3t^2)\, dt = \int_{-1}^{1} (6t^4 + 3t^2)\, dt = \left[\frac{6}{5}t^5 + t^3\right]_{-1}^{1} = 2 \cdot \left(\frac{6}{5} + 1\right) = \frac{22}{5}.$$

9.5.3: The area is

$$A = \int_{0}^{\pi} \sin^3 t\, dt = \int_{0}^{\pi} (\sin t - \cos^2 t \sin t)\, dt = \left[\frac{1}{3}\cos^3 t - \cos t\right]_{0}^{\pi} = \frac{4}{3}.$$

9.5.5: The area is

$$A = \int_{0}^{\pi} e^t \sin t\, dt = \left[\frac{1}{2}e^t(\sin t - \cos t)\right]_{0}^{\pi} = \frac{1}{2}(e^\pi + 1) \approx 12.0703463164.$$

See Example 5 of Section 7.3 for the evaluation of the antiderivative using integration by parts.

9.5.7: The volume is

$$V = \int_{-1}^{1} \pi(2t^2 + 1)^2 \cdot 3t^2\, dt = \pi \int_{-1}^{1} (12t^6 + 12t^4 + 3t^2)\, dt$$

$$= \pi\left[\frac{12}{7}t^7 + \frac{12}{5}t^5 + t^3\right]_{-1}^{1} = \left(\frac{179}{35} + \frac{179}{35}\right)\pi = \frac{358}{35}\pi \approx 32.1340048567.$$

9.5.9: The volume is

$$V = \int_{0}^{\pi} \pi(\sin t)^5\, dt = \pi \int_{0}^{\pi} (1 - 2\cos^2 t + \cos^4 t)\sin t\, dt$$

$$= \pi\left[-\frac{1}{5}\cos^5 t + \frac{2}{3}\cos^3 t - \cos t\right]_{0}^{\pi} = \pi\left(\frac{8}{15} + \frac{8}{15}\right) = \frac{16}{15}\pi \approx 3.351032163829.$$

9.5.11: The arc-length element is $ds = (t+4)^{1/2}\, dt$. Hence the length of the curve is

$$L = \int_5^{12} (t+4)^{1/2}\, dt = \left[\frac{2}{3}(t+4)^{3/2} \right]_5^{12} = \frac{128}{3} - 18 = \frac{74}{3} \approx 24.6666666667.$$

9.5.13: The arc-length element is $ds = \sqrt{(\cos t + \sin t)^2 + (\cos t - \sin t)^2}\, dt = \sqrt{2}\, dt$. Therefore the length of the curve is

$$L = \int_{\pi/4}^{\pi/2} \sqrt{2}\, dt = \left[t\sqrt{2} \right]_{\pi/4}^{\pi/2} = \frac{\pi\sqrt{2}}{4} \approx 1.1107207345.$$

9.5.15: Equation (10) of the text tells us that the arc-length element in polar coordinates is

$$ds = \sqrt{r^2 + \left(\frac{dr}{d\theta} \right)^2}\, d\theta = \left(e^\theta + \frac{1}{4} e^\theta \right)^{1/2} d\theta = \frac{\sqrt{5}}{2} e^{\theta/2}\, d\theta.$$

Therefore the length of the curve is

$$L = \int_0^{4\pi} \frac{\sqrt{5}}{2} e^{\theta/2}\, d\theta = \left[e^{\theta/2}\sqrt{5} \right]_0^{4\pi} = \left(e^{2\pi} - 1 \right)\sqrt{5} \approx 1195.159675159775.$$

9.5.17: The arc-length element is $ds = \left(1 + \frac{1}{t} \right)^{1/2} dt$, so the surface area is

$$A = \int_1^4 2\pi \cdot 2t^{1/2} \cdot \left(1 + \frac{1}{t} \right)^{1/2} dt = \int_1^4 4\pi(t+1)^{1/2}\, dt = \frac{8\pi}{3}\left[(t+1)^{3/2} \right]_1^4$$

$$= \frac{40\pi\sqrt{5}}{3} - \frac{16\pi\sqrt{2}}{3} = \frac{8\pi}{3}\left(5\sqrt{5} - 2\sqrt{2} \right) \approx 69.968820743698.$$

9.5.19: The arc-length element is $ds = (9t^4 + 4)^{1/2}\, dt$, but the surface area of revolution is *not*

$$\int_{-1}^1 2\pi t^3 (9t^4 + 4)^{1/2}\, dt.$$

The reason is that the radius of the circle of revolution is t^3, which is negative for $-1 \leqq t < 0$. But symmetry of the graph allows us to double the integral over $[0,1]$ to find the area:

$$A = 2\int_0^1 2\pi t^3 (9t^4 + 4)^{1/2}\, dt = \frac{2\pi}{27}\left[(9t^4 + 4)^{3/2} \right]_0^1 = \frac{2\pi}{27}\left(13\sqrt{13} - 8 \right) \approx 9.045963922970.$$

9.5.21: The circle with polar equation $r = 4\sin\theta$, $0 \leqq \theta \leqq \pi$, is to be rotated around the x-axis. The arc-length element is

$$ds = \sqrt{(4\sin\theta)^2 + (4\cos\theta)^2}\, d\theta = 4\, d\theta$$

516

and the radius of the circle of revolution is $y = 4\sin^2\theta$, so the surface area of revolution is

$$A = 2\pi \int_0^\pi 16 \sin^2\theta \; d\theta = 16\pi \int_0^\pi (1 - \cos 2\theta) \; d\theta = 8\pi \left[2\theta - \sin 2\theta \right]_0^\pi = 16\pi^2 \approx 157.9136704174.$$

9.5.23: The cycloidal arch is the graph of the parametric equations $x = a(t - \sin t)$, $y = a(1 - \cos t)$, $0 \leqq t \leqq 2\pi$, $a > 0$. When the region between the arch and the x-axis is rotated around the x-axis, the volume swept out is

$$V = \int_{t=0}^{2\pi} \pi y^2 \; dx = \pi a^3 \int_0^{2\pi} (1 - \cos t)^3 \; dt = \pi a^3 \int_0^{2\pi} \left[1 - 3\cos t + \frac{3}{2}(1 + \cos 2t) - (1 - \sin^2 t)\cos t \right] dt$$

$$= \pi a^3 \left[t - 3\sin t + \frac{3}{2}t + \frac{3}{4}\sin 2t - \sin t + \frac{1}{3}\sin^3 t \right]_0^{2\pi} = \frac{5}{2}\pi a^3 \cdot 2\pi = 5\pi^2 a^3.$$

9.5.25: Part (a): The area of the ellipse is

$$A = 4 \int_0^{\pi/2} ab \sin^2 t \; dt = 2ab \int_0^{\pi/2} (1 - \cos 2t) \; dt = 2ab \left[t - \frac{1}{2}\sin 2t \right]_0^{\pi/2} = \pi ab.$$

Part (b): The volume generated when [the upper half of] the ellipse is rotated around the x-axis is

$$V = 2 \int_0^{\pi/2} \pi(b^2 \sin^2 t)(a \sin t) \; dt = 2\pi a b^2 \int_0^{\pi/2} (1 - \cos^2 t) \sin t \; dt = 2\pi a b^2 \left[\frac{1}{3}\cos^3 t - \cos t \right]_0^{\pi/2} = \frac{4}{3}\pi a b^2.$$

Compare this with the solution of Problem 36 in Section 6.2.

9.5.27: Using the given Cartesian parametrization, the arc-length element for the spiral is

$$ds = \sqrt{[x'(t)]^2 + [y'(t)]^2} \; dt = \sqrt{(\cos t - t\sin t)^2 + (\sin t + t\cos t)^2} \; dt = \sqrt{t^2 + 1} \; dt.$$

Therefore the arc length is

$$L = \int_0^{2\pi} \sqrt{t^2 + 1} \; dt = \frac{1}{2} \left[t\sqrt{t^2 + 1} + \ln\left(t + \sqrt{t^2 + 1} \right) \right]_0^{2\pi}$$

$$= \frac{1}{2} \left[2\pi\sqrt{1 + 4\pi^2} + \ln\left(2\pi + \sqrt{1 + 4\pi^2} \right) \right] \approx 21.2562941482.$$

The antiderivative can be obtained with the trigonometric substitution $t = \tan\theta$ or by use of integral formula 44 of the endpapers of the textbook.

9.5.29: The area bounded by the astroid is

$$A = -4 \int_{t=0}^{\pi/2} y \; dx = 4 \int_0^{\pi/2} (a\sin^3 t)(3a\cos^2 t \sin t) \; dt = 12a^2 \int_0^{\pi/2} (\sin^4 t - \sin^6 t) \; dt$$

$$= 12a^2 \left(\frac{1}{2} \cdot \frac{3}{4} \cdot \frac{\pi}{2} - \frac{1}{2} \cdot \frac{3}{4} \cdot \frac{5}{6} \cdot \frac{\pi}{2} \right) = 12a^2 \cdot \frac{3}{16}\pi \left(1 - \frac{5}{6} \right) = \frac{3}{8}\pi a^2.$$

517

The first minus sign is needed because $dx < 0$. The integral was computed using integral formula 113 of the endpapers of the text.

9.5.31: The arc-length element is

$$ds = (9a^2 \cos^4 t \sin^2 t + 9a^2 \cos^2 t \sin^4 t)^{1/2} \, dt$$

$$= 3a \left[(\sin^2 t \cos^2 t)(\cos^2 t + \sin^2 t) \right]^{1/2} \, dt = 3a \left(\frac{1}{4} \sin^2 2t \right)^{1/2} \, dt = \frac{3}{2} a \sin 2t \, dt.$$

The radius of the circle of revolution is $y = a \sin^3 t$. So the surface area of revolution around the x-axis is

$$A = 2 \int_{t=0}^{\pi/2} 2\pi y \, ds = 2 \int_0^{\pi/2} (2\pi a \sin^3 t) \left(\frac{3}{2} a \sin 2t \right) dt$$

$$= 6\pi a^2 \int_0^{\pi/2} 2 \sin^4 t \cos t \, dt = \frac{12\pi a^2}{5} \left[\sin^5 t \right]_0^{\pi/2} = \frac{12}{5} \pi a^2.$$

Compare this with the solution of Problem 42 in Section 6.4.

9.5.33: The area is

$$A = 2 \int_{t=0}^3 y \, dx = 2 \int_0^3 \left(3t - \tfrac{1}{3} t^3 \right) \left(2t\sqrt{3} \right) dt = \sqrt{3} \int_0^3 \left(12t^2 - \frac{4}{3} t^4 \right) dt$$

$$= \sqrt{3} \left[4t^3 - \frac{4}{15} t^5 \right]_0^3 = \frac{216\sqrt{3}}{5} \approx 74.8245948870.$$

9.5.35: The volume of revolution around the x-axis is

$$V = \int_{t=0}^3 \pi y^2 \, dx = \pi \int_0^3 \left(3t - \frac{1}{3} t^3 \right)^2 \cdot 2t\sqrt{3} \, dt = 2\pi\sqrt{3} \int_0^3 \left(\frac{1}{9} t^7 - 2t^5 + 9t^3 \right) dt$$

$$= 2\pi\sqrt{3} \left[\frac{1}{72} t^8 - \frac{1}{3} t^6 + \frac{9}{4} t^4 \right]_0^3 = \frac{243\pi\sqrt{3}}{4} \approx 330.5649341317.$$

9.5.37: Part (a): The parametrization found for the first-quadrant loop of the folium in Section 9.4 was

$$x = \frac{3t}{1+t^3}, \quad y = \frac{3t^2}{1+t^3}, \quad 0 \le t < +\infty.$$

We first need to compute the arc-length element.

$$[x'(t)]^2 = \frac{9(2t^3 - 1)^2}{(1+t^3)^4} \quad \text{and} \quad [y'(t)]^2 = \frac{9(t^4 - 2t)^2}{(1+t^3)^4};$$

$$ds = \frac{3\sqrt{t^8 + 4t^6 - 4t^5 - 4t^3 + 4t^2 + 1}}{(t^3 + 1)^2} \, dt.$$

518

Part (b): We will find the length of the loop by integrating ds from $t = 0$ to $t = 1$ (to avoid an improper integral) and doubling the result. The length is thus

$$L = 2 \int_0^1 \frac{3\sqrt{t^8 + 4t^6 - 4t^5 - 4t^3 + 4t^2 + 1}}{(t^3 + 1)^2}\, dt.$$

We used *Mathematica* and the command

```
6*NIntegrate[ (Sqrt[ t^8 + 4*t^6 - 4*t^5 - 4*t^3 + 4*t^2 + 1 ])/(t^3 + 1)^2,
        { t, 0, 1 }, MaxRecursion -> 18, WorkingPrecision -> 28 ]
```

to find that $L \approx 4.917488721682$.

9.5.39: We use the parametrization $x = a(t - \sin t)$, $y = a(1 - \cos t)$, $0 \leq t \leq 2\pi$, $a > 0$. By the method of nested cylindrical shells, the volume of revolution around the y-axis is

$$V = \int_{t=0}^{2\pi} 2\pi x y\, dx = \int_0^{2\pi} 2\pi a(t - \sin t) \cdot a(1 - \cos t) \cdot a(1 - \cos t)\, dt$$

$$= 2\pi a^3 \int_0^{2\pi} (t - 2t\cos t + t\cos^2 t - \sin t + 2\sin t \cos t - \sin t \cos^2 t)\, dt$$

$$= 2\pi a^3 \int_0^{2\pi} \left(\frac{3}{2}t - 2t\cos t + \frac{1}{2}t\cos 2t - \sin t + 2\sin t \cos t - \sin t \cos^2 t \right) dt$$

$$= 2\pi a^3 \left[\frac{3}{4}t^2 - 2\cos t - 2t\sin t + \frac{1}{8}\cos 2t + \frac{1}{4}t\sin 2t + \cos t + \sin^2 t + \frac{1}{3}\cos^3 t \right]_0^{2\pi} = 6\pi^3 a^3.$$

See the solution of Problem 5 in Section 7.3 for the way integration by parts can be used to find the two more troublesome antiderivatives here.

9.5.41: We will compute the area of the part of the region above the x-axis, then double the result. On the left we see a quarter-circle of radius πa, with area

$$A_1 = \frac{1}{4}\pi(\pi a)^2 = \frac{1}{4}\pi^3 a^2.$$

On the right, the area between the involute and the x-axis can be found with an integral, using the equations for $x(t)$ and $y(t)$ given in Problem 40(a):

$$A_2 = \int_0^\pi [-y(t) \cdot x'(t)]\, dt = \frac{a^2}{12}\left[3t^2 \sin 2t - 3\sin 2t + 6t\cos 2t + 2t^3 \right]_0^\pi = \frac{\pi a^2}{6}(\pi^2 + 3).$$

But we must subtract the area of the part of the water tank above the x-axis, the area of a semicircle of radius a: $A_3 = \frac{1}{2}\pi a^2$. So the total area of the region that the cow can graze is

$$A = 2(A_1 + A_2 - A_3) = \frac{5}{6}\pi^3 a^2.$$

519

9.5.43: Given $r(\theta) = 3\sin 3\theta$, remember that roses with *odd* coefficients are swept out *twice* in the interval $0 \le \theta \le 2\pi$. Therefore we should integrate

$$ds = \sqrt{[r(\theta)]^2 + [r'(\theta)]^2}\ d\theta = \sqrt{45 + 36\cos 6\theta}\ d\theta$$

from 0 to π to obtain the total length of the rose:

$$\int_{\theta=0}^{\pi} 1\ ds = \int_0^{\pi} \sqrt{45 + 36\cos 6\theta}\ d\theta \approx 20.047339830833.$$

The *Mathematica* command we used in Problem 43—we used appropriately modified versions of it for Problems 44 through 55—was

```
NIntegrate[ Sqrt[ 45 + 36*Cos[6*t] ], { t, 0, Pi },
           MaxRecursion -> 18, WorkingPrecision -> 28 ]
```

9.4.45: Given $r(\theta) = 2\cos 2\theta$, remember than a rose with an even coefficient n of θ has $2n$ "petals," and is swept out as θ ranges from 0 to 2π. The arc length element in this case is $ds = \sqrt{10 - 6\cos 4\theta}\ d\theta$, and the length of the graph is

$$\int_0^{2\pi} \sqrt{10 - 6\cos 4\theta}\ d\theta \approx 19.376896441095$$

9.5.47: Given: $r(\theta) = 5 + 9\cos\theta$, the arc length element is $ds = \sqrt{106 + 90\cos\theta}\ d\theta$, and so the total length of the limaçon is

$$\int_0^{2\pi} \sqrt{106 + 90\cos\theta}\ d\theta \approx 61.003581373850.$$

9.5.49: Given: $r(\theta) = \cos(7\theta/3)$. To sweep out all seven "petals" of this quasi-rose, you need to let θ vary from 0 to 3π. The length of the graph is

$$\int_0^{3\pi} \sqrt{\tfrac{1}{9}\left(29 - 20\cos(14\theta/3)\right)}\ d\theta \approx 16.342833373939.$$

9.5.51: Part (a): When the curve of Problem 50 is rotated around the x-axis, the surface generated is swept out twice. We will rotate the part of the curve in the first quadrant around the x-axis and double the result to get the total surface area

$$2\int_{t=0}^{\pi/2} 2\pi y\ ds \approx 16.057027566602.$$

Part (b): To find the volume of revolution around the x-axis, we evaluate

$$2\int_0^{\pi/2} \pi[y(t)]^2 \cdot x'(t)\ dt = 2\int_0^{\pi/2} 4\pi(\sin^2 t\,\cos t - \sin^4 t\,\cos t)\ dt$$

$$= 2\left[4\pi\left(\frac{1}{3}\sin^3 t - \frac{1}{5}\sin^5 t\right)\right]_0^{\pi/2} = \frac{16}{15}\pi \approx 3.351032163829.$$

520

9.5.53: The arc-length element is $ds = \sqrt{25\cos^2 5t + 9\sin^2 3t}\ dt$, and the entire Lissajous curve is obtained by letting t range from 0 to 2π. Hence the length of the graph is

$$\int_0^{2\pi} \sqrt{25\cos^2 5t + 9\sin^2 3t}\ dt \approx 24.602961618540.$$

9.5.55: The length of the graph is

$$\int_0^{2\pi} \sqrt{[x'(t)]^2 + [y'(t)]^2}\ dt \approx 39.403578712896.$$

The graph is next.

Section 9.6

9.6.1: If the vertex is at $V(0,\,0)$ and the focus is at $F(3,\,0)$, then the directrix must be the vertical line with equation $x = -3$. If $(x,\,y)$ is a point on the parabola, then by the definition of parabola, $y^2 + (x-3)^2 = (x+3)^2$. It's easy to simplify this equation to $y^2 = 12x$. The graph is next.

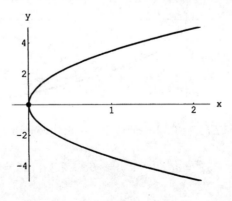

9.6.3: If the vertex of the parabola is $V(2, 3)$ and the focus is $F(2, 1)$, then the directrix must be the horizontal line $y = 5$. Then it follows from the definition of a parabola that if (x, y) is a point of the parabola, then

$$(x - 2)^2 + (y - 1)^2 = (y - 5)^2;$$
$$(x - 2)^2 + y^2 - 2y + 1 = y^2 - 10y + 25;$$
$$(x - 2)^2 = -8y + 24;$$
$$(x - 2)^2 = -8(y - 3).$$

The graph of this parabola is next.

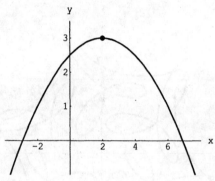

9.6.5: If the vertex is $V(2, 3)$ and the focus is $F(0, 3)$, then the directrix of this parabola must be the vertical line $x = 4$. If (x, y) is a point of the parabola, then—by definition—

$$x^2 + (y - 3)^2 = (x - 4)^2;$$
$$(y - 3)^2 - 8x + 16;$$
$$(y - 3)^2 = -8(x - 2).$$

The graph of this parabola is next.

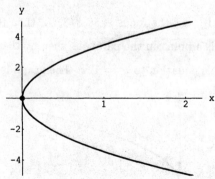

9.6.7: If a parabola has focus $F(0, -3)$, directrix $y = 0$, and contains the point (x, y), then by definition

$$x^2 + (y + 3)^2 = y^2; \qquad x^2 = -6y - 9; \qquad x^2 = -6\left(y + \frac{3}{2}\right).$$

The graph of this parabola is next.

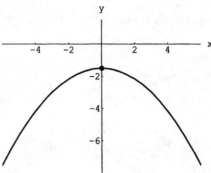

9.6.9: With focus $F(0, 0)$ and directrix $y = -2$, the definition of parabola implies that if (x, y) lies on this parabola, then

$$x^2 + y^2 = (y + 2)^2; \qquad x^2 = 4y + 4; \qquad x^2 = 4(y + 1).$$

Its graph is next.

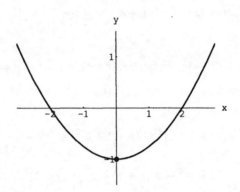

9.6.11: The parabola with equation $y^2 = 12x$ has vertex $V(0, 0)$ and horizontal axis $y = 0$, so its focus must be at $F(c, 0)$ and its directrix must be the vertical line $x = -c$ where $c > 0$. So its equation also has the form $(x - c)^2 + y^2 = (x + c)^2$; that is, $y^2 = 4cx$. Therefore $c = 3$, the focus is $F(3, 0)$, and the directrix is the vertical line $x = -3$. The graph of this parabola is next.

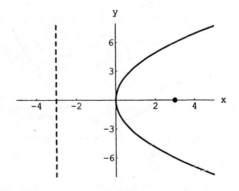

9.6.13: The parabola with equation $y^2 = -6x$ has vertex $V(0, 0)$, its axis is the x-axis, and it opens to the left. Hence it has focus $F(-c, 0)$ and directrix $x = c$ where $c > 0$. So its equation has the form

523

$(x+c)^2 + y^2 = (x-c)^2$; that is, $y^2 = -4cx$. Therefore $c = \frac{3}{2}$, the focus is $F\left(-\frac{3}{2}, 0\right)$, and the directrix is the line $x = \frac{3}{2}$. The graph of this parabola is next.

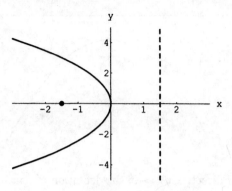

9.6.15: Given: $x^2 - 4x - 4y = 0$. We complete the square in x:

$$x^2 - 4x + 4 = 4y + 4; \qquad (x-2)^2 = 4(y+1). \tag{1}$$

Thus this parabola has vertex $V(2, -1)$, vertical axis with equation $x = 2$, and opens upward. So its focus is at $F(2, -1+c)$ and its directrix is $y = -1 - c$ where $c > 0$. Thus its equation has the form

$$(x-2)^2 + (y+1-c)^2 = (y+1+c)^2;$$

$$(x-2)^2 + (y+1)^2 - 2c(y+1) + c^2 = (y+1)^2 + 2c(y+1) + c^2;$$

$$(x-2)^2 = 4c(y+1).$$

By Eq. (1), $c = 1$. Therefore the focus is at $F(2, 0)$ and the directrix has equation $y = -2$. The graph of this parabola is next.

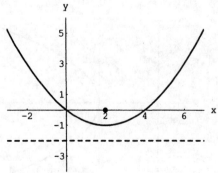

9.6.17: Given $4x^2 + 4x + 4y + 13 = 0$, we complete the square in x to find that $(2x+1)^2 = -4(y+3)$. Thus this parabola has vertex $V\left(-\frac{1}{2}, -3\right)$, vertical axis $x = -\frac{1}{2}$, and opens downward. So its focus is

524

$F\left(-\frac{1}{2}, -3-c\right)$ and its directrix has equation $y = -3+c$ where $c > 0$. So its equation has the form

$$\left(x + \tfrac{1}{2}\right)^2 + (y+3+c)^2 = (y+3-c)^2;$$

$$\left(x + \tfrac{1}{2}\right)^2 + (y+3)^2 + 2c(y+3) + c^2 = (y+3)^2 - 2c(y+3) + c^2;$$

$$\left(x + \tfrac{1}{2}\right)^2 = -4c(y+3);$$

$$(2x+1)^2 = -16c(y+3).$$

Therefore $c = \frac{1}{4}$, the focus is $F\left(-\frac{1}{2}, -\frac{13}{4}\right)$, and the directrix has equation $y = -\frac{11}{4}$. This parabola is shown next.

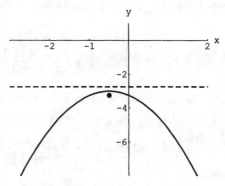

9.6.19: The location of the vertices makes it clear that the center of the ellipse is at $(0, 0)$. Therefore its equation may be written in the standard form

$$\left(\frac{x}{4}\right)^2 + \left(\frac{y}{5}\right)^2 = 1.$$

9.6.21: We use the equation $a^2 = b^2 + c^2$ with $a = 17$ and $c = 8$ to find that $b = 15$. The major axis is vertical, so an equation of this ellipse is

$$\left(\frac{x}{15}\right)^2 + \left(\frac{y}{17}\right)^2 = 1.$$

9.6.23: Because $c = 3$ and

$$a = \frac{c}{e} = 3 \cdot \frac{4}{3} = 4,$$

we use the equation $a^2 = b^2 + c^2$ to find that $b^2 = 7$. Therefore an equation of this ellipse is

$$\frac{x^2}{16} + \frac{y^2}{7} = 1.$$

9.6.25: Because $a = 10$ and $c = ea = \frac{1}{2} \cdot 10 = 5$, it follows from the equation $a^2 = b^2 + c^2$ that $b = \sqrt{75}$. Therefore an equation of this ellipse is

$$\frac{x^2}{100} + \frac{y^2}{75} = 1.$$

9.6.27: From the information given in the problem, we see that $8 = a/e$ and $a = 2/e$. It follows that $e = \frac{1}{2}$, and so $a = 4$ and $c = 2$. Consequently $b^2 = 12$, and therefore an equation of this ellipse is

$$\frac{x^2}{16} + \frac{y^2}{12} = 1.$$

9.6.29: Were the center at the origin, the equation would be $(x/4)^2 + (y/2)^2 = 1$. Because the center is at $C(2, 3)$, the translation principle implies that the equation is instead

$$\left(\frac{x-2}{4}\right)^2 + \left(\frac{y-3}{2}\right)^2 = 1.$$

9.6.31: The center of this ellipse is at $(1, 1)$, $c = 3$, and $a = 5$. Thus $b = 4$, and so an equation of this ellipse is

$$\left(\frac{x-1}{5}\right)^2 + \left(\frac{y-1}{4}\right)^2 = 1.$$

9.6.33: The center is at $C(1, 2)$, the major axis is horizontal, and $c = 3$. Next, $a = c/e$ and $e = 1/3$, so $a = 3c = 9$. Because $b^2 = a^2 - c^2$, we see that $b = \sqrt{72}$. Thus an equation of this ellipse is

$$\frac{(x-1)^2}{81} + \frac{(y-2)^2}{72} = 1.$$

9.6.35: In standard form, the equation of this ellipse is

$$\left(\frac{x}{6}\right)^2 + \left(\frac{y}{4}\right)^2 = 1,$$

so its center is at $C(0, 0)$, $a = 6$, and $b = 4$; thus $c = 2\sqrt{5}$. The foci are $(\pm 2\sqrt{5}, 0)$. The major axis is horizontal, of length 12; the minor axis has length 8. The graph of this ellipse is next.

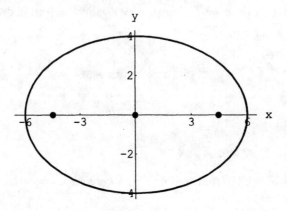

9.6.37: We complete the square in y as follows:

$$9x^2 + 4y^2 - 32y + 28 = 0; \qquad \frac{9}{4}x^2 + y^2 - 8y + 7 = 0;$$

$$\frac{9}{4}x^2 + y^2 - 8y + 16 = 9; \qquad \left(\frac{x}{2}\right)^2 + \left(\frac{y-4}{3}\right)^2 = 1.$$

526

Thus this ellipse has center $C(0, 4)$, $a = 3$, and $b = 2$; thus $c = \sqrt{5}$. The foci are $\left(0, 4 \pm \sqrt{5}\right)$, the major axis is vertical, of length 6; the minor axis has length 4. The graph of this ellipse is next.

9.6.39: The given information implies that the transverse axis is horizontal, $c = 4$, and $a = 1$. Hence $b = \sqrt{c^2 - a^2} = \sqrt{15}$. Therefore an equation of this hyperbola is

$$\frac{x^2}{1} - \frac{y^2}{15} = 1.$$

9.4.41: The given information implies that the transverse axis is horizontal, $c = 5$, and $b/a = 3/4$, so that $b = \frac{3}{4}a$. Then the equation $a^2 + b^2 = c^2$ implies that $a = 4$ and thus that $b = 3$. So an equation of this hyperbola is

$$\left(\frac{x}{4}\right)^2 - \left(\frac{y}{3}\right)^2 = 1.$$

9.6.43: The information given in the problem implies that the transverse axis is vertical, $a = 5$, and $a/b = 1$, so that $b = 5$ as well. Hence an equation of this hyperbola is

$$\left(\frac{y}{5}\right)^2 - \left(\frac{x}{5}\right)^2 = 1.$$

9.6.45: The transverse axis is vertical and $c = 6$. Hence $a = c/e = 3$, and so $b = \sqrt{27}$. Therefore an equation of this hyperbola is

$$\frac{y^2}{9} - \frac{x^2}{27} = 1.$$

9.6.47: The transverse axis is horizontal and $c = 4$. One directrix is $x = 1$, so $1 = a/e = c/e^2$. Thus $e = 2$, and so $a = 2$ and $b^2 = 12$. Thus an equation of this hyperbola is

$$\frac{x^2}{4} - \frac{y^2}{12} = 1.$$

527

9.6.49: Given: The hyperbola has center $(2, 2)$, the transverse axis is horizontal of length 6, and $e = 2$. Translate the hyperbola so that its center is at $(0, 0)$. The vertices are therefore $(-3, 0)$ and $(3, 0)$, so that $a = 3$. Then $c = ae = 6$, so that $b^2 = 27$. So the translated hyperbola has equation

$$\frac{x^2}{9} - \frac{y^2}{27} = 1.$$

Therefore an equation of the original hyperbola is $\dfrac{(x-2)^2}{9} - \dfrac{(y-2)^2}{27} = 1.$

9.6.51: Given: The hyperbola has center $C(1, -2)$, vertices $V_1(1, 1)$, and $V_2(1, -5)$, and asymptotes $3x - 2y = 7$ and $3x + 2y = -1$. Translate the hyperbola so that its center is at $(0, 0)$. The new vertices are $(0, \pm 3)$ and the new asymptotes have equations

$$3(x+1) - 2(y-2) = 7 \quad \text{and} \quad 3(x+1) + 2(y-2) = -1;$$

$$3x - 2y = 0 \quad \text{and} \quad 3x + 2y = 0.$$

Thus their equations are $y = \pm\frac{3}{2}x$. Therefore the translated parabola has $a = 3$ and $a/b = 3/2$, so that $b = 2$. Thus—because its transverse axis is vertical—it has equation

$$\left(\frac{y}{3}\right)^2 - \left(\frac{x}{2}\right)^2 = 1.$$

Therefore the original hyperbola has equation $\left(\dfrac{y+2}{3}\right)^2 - \left(\dfrac{x-1}{2}\right)^2 = 1.$

9.6.53: Given $x^2 - y^2 - 2x + 4y = 4$, we first complete the square in the two variables:

$$x^2 - 2x - (y^2 - 4y) = 4;$$

$$x^2 - 2x + 1 - (y^2 - 4y + 4) + 4 - 1 = 4;$$

$$(x - 1)^2 - (y - 2)^2 = 1.$$

Thus this hyperbola has center $C(1, 2)$. Also $a = b = 1$, so $c = \sqrt{2}$. So its foci are $\left(1 \pm \sqrt{2}, 2\right)$. If its center were $(0, 0)$, its asymptotes would be $y = \pm x$. Therefore its actual asymptotes have equations $y - 2 = \pm(x - 1)$; that is, $y = x + 1$ and $y = -x + 3$. Its graph is next.

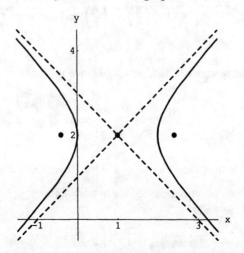

528

9.6.55: Given the equation $y^2 - 3x^2 - 6y = 0$, complete the square:

$$y^2 - 6y + 9 - 3x^2 = 9;$$

$$(y - 3)^2 - 3x^2 = 9;$$

$$\frac{(y - 3)^2}{9} - \frac{x^2}{3} = 1.$$

This hyperbola has center $(0, 3)$, $a = 3$, $b = \sqrt{3}$, and $c = 2\sqrt{3}$. If the center were at the origin, the hyperbola would have asymptotes with equations $y = \pm x\sqrt{3}$ and its foci would be $\left(0, \pm 2\sqrt{3}\right)$. So the given hyperbola has asymptotes $y = 3 \pm x\sqrt{3}$ and foci $\left(0, 3 \pm 2\sqrt{3}\right)$. Its graph is next.

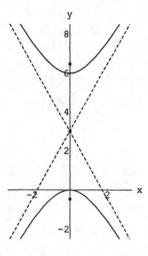

9.6.57: First complete the square in both variables:

$$9x^2 - 4y^2 + 18x + 8y = 31;$$

$$9(x^2 + 2x) - 4(y^2 - 2y) = 31;$$

$$9(x^2 + 2x + 1) - 4(y^2 - 2y + 1) = 31 + 9 - 4 = 36;$$

$$\left(\frac{x + 1}{2}\right)^2 - \left(\frac{y - 1}{3}\right)^2 = 1.$$

Thus this hyperbola has center $C(-1, 1)$. From its equation we also see that $a = 2$ and $b = 3$, so that $c = \sqrt{13}$. If its center were at the origin, its foci would be $(\pm\sqrt{13}, 0)$ and its asymptotes would be $y = \pm 3x/2$. Thus its foci are at $\left(-1 \pm \sqrt{13}, 1\right)$. Its asymptotes have equations $y - 1 = \pm 3(x + 1)/2$; that is, $2y = 3x + 5$ and $2y = -3x - 1$. The graph of this hyperbola is next.

529

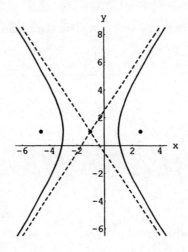

9.6.59: We read from the given equation

$$r = \frac{6}{1 + \cos\theta}$$

the information that (in the terminology of Section 9.6) $pe = 6$ and $e = 1$. Therefore the conic section is a parabola with directrix the vertical line $x = 6$; its focus is at $(0, 0)$. Conversion to Cartesian coordinates yields

$$r + r\cos\theta = 6; \qquad x^2 + y^2 = (6 - x)^2;$$

$$y^2 = 36 - 12x; \qquad x = 3 - \frac{1}{12}y^2.$$

The parabola opens to the left with vertex at $(3, 0)$; its axis is the x-axis (or the part of the x-axis for which $x \leqq 3$). To see the graph of this conic, we executed the *Mathematica* commands

```
r = 6/(1 + Cos[t]); x = r Cos[t]; y = r Sin[t];

ParametricPlot[ { x, y }, { t, 0, 2 Pi } ];
```

The result is next.

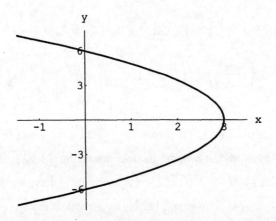

530

9.6.61: From the given polar equation

$$r = \frac{3}{1 - \cos\theta}$$

we read the information that $pe = 3$ and that $e = 1$, so that the conic section is a parabola with focus $(0, 0)$ and directrix the vertical line $x = -3$. Conversion to Cartesian coordinates yields

$$r - r\cos\theta = 3; \qquad\qquad r^2 = (x + 3)^2;$$

$$x^2 + y^2 = x^2 + 6x + 9; \qquad\qquad x = \frac{1}{6}(y^2 - 9).$$

Hence the parabola opens to the right with vertex at $\left(-\frac{3}{2}, 0\right)$ and its axis is the x-axis. To see its graph, we executed the *Mathematica* commands

```
r = 3/(1 - Cos[t]); x = r Cos[t]; y = r Sin[t];
ParametricPlot[ { x, y }, { t, 0, 2 Pi } ];
```

The result is next.

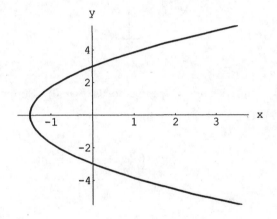

9.6.63: From the given polar equation

$$r = \frac{6}{2 - \sin\theta} = \frac{3}{1 - \frac{1}{2}\sin\theta}$$

we read the information that $pe = 3$ and that $e = \frac{1}{2}$. Hence $p = 6$, and therefore the conic section is an ellipse with one *horizontal* directrix $y = -6$ and one focus at $(0, 0)$. Conversion to Cartesian coordinates yields

$$2r - y = 6; \qquad\qquad 4(x^2 + y^2) = (y + 6)^2;$$

$$4x^2 + 3y^2 - 12y = 36; \qquad\qquad \frac{4}{3}x^2 + y^2 - 4y = 12;$$

$$\frac{4}{3}x^2 + y^2 - 4y + 4 = 16; \qquad\qquad \frac{4}{3}x^2 + (y - 2)^2 = 16.$$

531

Hence this ellipse has center at $(0, 2)$ and its other focus at $(0, 4)$. When we evaluate r for $\theta = \pm\pi/2$, we find that the vertices of this ellipse are at $(0, 6)$ and $(0, -2)$. To see its graph, we executed the *Mathematica* commands

```
r = 6/(2 - Sin[t]); x = r Cos[t]; y = r Sin[t];

ParametricPlot[ { x, y }, { t, 0, 2 Pi } ];
```

and the result is shown next.

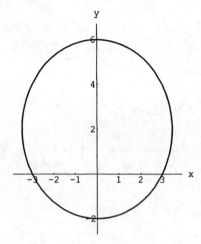

9.6.65: The parabola with equation $y^2 = 4px$ has focus $F(p, 0)$. Suppose that (x, y) is a point on the parabola. Then $x = y^2/(4p)$. Our goal is to minimize $(x - p)^2 + y^2$; that is,

$$f(y) = \left(\frac{y^2}{4p} - p\right)^2 + y^2.$$

Now

$$f'(y) = 2\left(\frac{y^2}{4p} - p\right) \cdot \frac{2y}{4p} + 2y = \frac{4y^3}{16p^2} - \frac{4py}{4p} + 2y$$

$$= \frac{y^3}{4p^2} - y + 2y = \frac{y^3}{4p^2} + y = \frac{y^3 + 4p^2y}{4p^2} = \frac{y}{4p^2}\left(y^2 + 4p^2\right).$$

Therefore $f'(y) = 0$ if and only if $y = 0$; $f'(y) < 0$ if $y < 0$ and $f'(y) > 0$ if $y > 0$. So by the first derivative test, $f(y)$ has a global minimum value and it occurs where $y = 0$, so that $x = 0$ as well. Therefore the vertex $V(0, 0)$ of this parabola is the point of the parabola closest to its focus $F(p, 0)$.

9.6.67: Given: The point $Q(x_0, y_0)$ on the graph of the parabola with equation $y^2 = 4px$ ($p \neq 0$). Using implicit differentiation,

$$2y\frac{dy}{dx} = 4p,$$

so the slope of the line tangent to the graph at Q is $4p/(2y_0)$. Thus it has equation

$$y - y_0 = \frac{4p}{2y_0}(x - x_0);$$

$$2y_0 y - 2y_0^2 = 4px - 4px_0;$$

$$y_0 y - y_0^2 = 2px - 2px_0;$$

$$y_0 y - 4px_0 = 2px - 2px_0;$$

$$2px - y_0 y + 2px_0 = 0.$$

In particular, when $y = 0$ we see that $x = -x_0$, so the tangent line meets the x-axis at the point $(-x_0, 0)$.

9.6.69: Set up coordinates so that the parabola has vertex $V(-p, 0)$. Then the equation of the comet's orbit is $y^2 = 4p(x + p)$. The line $y = x$ meets the orbit of the comet at the point (a, b), which is $100\sqrt{2}$ million miles from the origin (which is also where both the sun and the focus of the parabola are located). Therefore

$$a^2 = 4p(a + p) \quad \text{and} \quad \sqrt{a^2 + a^2} = \left(100\sqrt{2}\right)(10^6) = 10^8\sqrt{2}.$$

It follows that $a = 10^8$. Next, $a^2 = 4p(a+p)$. We apply the quadratic formula to find without difficulty that $p = \frac{1}{2}\left(\sqrt{2} - 1\right)(10^8)$. Now solve the equation of the orbit for x:

$$x = \frac{1}{4p}y^2 - p.$$

The area A_3 swept out by the line from the sun to the comet in three days is then

$$A_3 = \frac{1}{2}100^2 - \int_{2p}^{100} \left(\frac{1}{4p}y^2 - p\right) dy.$$

It now follows that

$$A_3 = 5000 - \frac{1}{12p}(10^6 - 8p^3) + 100p - 2p^2 \approx 2475.469.$$

The area of the "quarter-parabola" is

$$A_Q = \int_0^{2p} \left(p - \frac{1}{4p}y^2\right) dy = \frac{4}{3}p^2 \approx 571.9096.$$

So the comet will reach its point of closest approach in roughly 0.693 more days; that is, in about 16 h 38 min.

9.6.71: With v_0 held constant, the range

$$R = \frac{v_0^2 \sin 2\alpha}{g}$$

of the projectile will be maximized when $\sin 2\alpha = 1$; that is, when $\alpha = 45°$. Thus the maximum range will be $R_{\max} = v_0^2/g$.

9.6.73: The range will be 125 meters when

$$\frac{2500 \sin 2\alpha}{9.8} = 125;$$

$$\sin 2\alpha = \frac{(125)(9.8)}{2500} = 0.49;$$

$$2\alpha = \arcsin(0.49).$$

Therefore $\alpha \approx 14°40'13''$ and $\alpha \approx 75°19'47''$ will both produce a range of 125 meters.

9.6.75: Given $\sqrt{x} + \sqrt{y} = \sqrt{a}$, square twice to eliminate the radicals:

$$x + 2\sqrt{xy} + y = a;$$

$$2\sqrt{xy} = a - x - y;$$

$$4xy = (a - x - y)^2 = a^2 - 2a(x + y) + x^2 + 2xy + y^2;$$

$$2xy + 2a(x + y) = x^2 + y^2 + a^2.$$

Now convert to polar coordinates:

$$r^2 + a^2 = 2r^2 \sin\theta \cos\theta + 2ar(\sin\theta + \cos\theta).$$

Now rotate the graph 45° (the reason is that if you graph the original equation, it resembles a parabola with axis the line $y = x$):

$$r^2 + a^2 = 2r^2 \sin\left(\theta + \frac{\pi}{4}\right) \cos\left(\theta + \frac{\pi}{4}\right) + 2ar\left[\sin\left(\theta + \frac{\pi}{4}\right) + \cos\left(\theta + \frac{\pi}{4}\right)\right];$$

$$r^2 + a^2 = 2r^2 \cdot \frac{1}{2}(\cos^2\theta - \sin^2\theta) + 2ar\sqrt{2}\,\cos\theta.$$

Finally, return to Cartesian coordinates:

$$x^2 + y^2 + a^2 = x^2 - y^2 + 2ax\sqrt{2};$$

$$2ax\sqrt{2} = 2y^2 + a^2;$$

$$x = \frac{\sqrt{2}}{2a}y^2 + \frac{a\sqrt{2}}{4};$$

$$x - \frac{a\sqrt{2}}{4} = \frac{\sqrt{2}}{2a}y^2.$$

534

Therefore the graph of $\sqrt{x} + \sqrt{y} = \sqrt{a}$ is a parabola.

9.6.77: Part (a): In the usual notation, we have $e = 0.999925$ and $a - c = 0.13$ (AU). Now

$$b^2 = a^2 - c^2 = (a+c)(a-c) \quad \text{and} \quad a = \frac{c}{e}.$$

It follows that

$$\frac{c}{e} - c = 0.13, \quad \text{and thus} \quad c = (0.13)\frac{999925}{75} \approx 1733.203333.$$

Thus $a = c/e \approx 1733.246664$ and so $b \approx 12.25577415$. The maximum distance between Kahoutek and the sun is therefore $2a - 0.13 \approx 3466.363328$ (AU)—about 322 *billion* miles, about 20 light-days.

Part (b): In the case of Comet Hyakutake, we have $e = 0.999643856$ and $a - c = 0.2300232$. Thus

$$\frac{c}{e} - c = 0.2300232; \quad \text{hence} \quad c\left(\frac{1}{e} - 1\right) = 0.2300232.$$

Thus $c \approx 645.64130974$. But $a = c/e$, so $a \approx 645.87133294$. So the greatest distance between Hyakutake and the sun is $2a - 0.2300232 \approx 1291.51264269$ (AU). This is about 120 billion miles, about 7.45 light-days.

9.6.79: Assume that the focus on the positive y-axis is $F(0, c)$ and that the directrix is the line L with equation $y = c/e^2$ where $0 < e < 1$. Suppose that $P(x, y)$ is a point of the ellipse. Then the equation $|PF| = e \cdot |PL|$ yields

$$\sqrt{x^2 + (y-c)^2} = e \cdot \left(y - \frac{c}{e^2}\right);$$

$$x^2 + y^2 - 2cy + c^2 = e^2\left(y - \frac{c}{e^2}\right)^2;$$

$$x^2 + y^2 - 2cy + c^2 = e^2 y^2 - 2cy + \frac{c^2}{e^2};$$

$$x^2 + (1 - e^2)y^2 = \frac{c^2}{e^2} - c^2 = c^2\left(\frac{1}{e^2} - 1\right) = \frac{c^2}{e^2}(1 - e^2).$$

Now substitute $a = c/e$:

$$x^2 + (1 - e^2)y^2 = a^2(1 - e^2);$$

$$\frac{x^2}{a^2(1 - e^2)} + \frac{y^2}{a^2} = 1.$$

Let $b^2 = a^2(1 - e^2)$ where $b > 0$. This is possible because $0 < e < 1$. Then

$$b^2 = a^2 - a^2 e^2 = a^2 - c^2, \quad \text{so that} \quad a^2 + b^2 = c^2.$$

The equation of the ellipse is therefore

$$\left(\frac{x}{b}\right)^2 + \left(\frac{y}{a}\right)^2 = 1;$$

note also that $0 < b < a$ and that the directrix has equation $y = \dfrac{c}{e^2} = \dfrac{a}{e}$.

9.6.81: Let $P(x_0,\, y_0)$ be the point of tangency. Suppose that the ellipse has equation $(x/a)^2 + (y/b)^2 = 1$ where $0 < b < a$. We saw in the solution of Problem 24 that the slope of the tangent line is

$$m = -\frac{b^2 x_0}{a^2 y_0}. \tag{1}$$

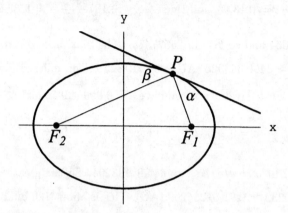

Let $F_1(c,\, 0)$ and $F_2(-c,\, 0)$ be the foci of the ellipse, where $c = \sqrt{a^2 - b^2}$. Then $m_1 = y_0/(x_0 - c)$ is the slope of the line F_1P, let $m_2 = y_0/(x_0 + c)$ is the slope of F_2P. By the trigonometric identity for $\tan(A - B)$, the tangent of the angle α between the tangent line at P and the line F_1P is

$$\tan\alpha = \frac{m - m_1}{1 + m_1 m} \tag{2}$$

while the tangent of the angle β between the tangent line at P and the line F_2P is

$$\tan\beta = \frac{m_2 - m}{1 + m_2 m} \tag{3}$$

In order to show that $\alpha = \beta$ we want to show that $\tan\alpha = \tan\beta$, and we will let *Mathematica* help us with the tedious algebra. We start by entering the definitions

```
m = - b^2 x0/(a^2 y0); c = Sqrt[a^2 - b^2];
m1 = y0/(x0 - c); m2 = y0/(x0 + c);
K = (m - m1)/(1 + m1 m); L = (m2 - m)/(1 + m2 m);
```

We want to show that $K = L$, or $K - L = 0$. Now $K - L$ is a difference of complicated fractions, so we start by finding a common denominator, and then simplifying the numerator (in hopes of showing eventually that it equals 0):

```
R = Expand[Numerator[Simplify[Together[K - L]]]]
```

$$2b^2 x_0^3 - 2a^2 b^2 x_0 + 2a^2 y_0^2 x_0$$

536

Finally, we divide R by x_0, substitute $y_0^2 = b^2(1 - x_0^2/a^2)$, and simplify:

```
S = Expand[Simplify[R/x0]];

Simplify[R /.  y0^2 -> b^2(1 - x0^2/a^2)]

   0
```

So this does it!

9.6.83: Solution (a): It's clear that the center of this ellipse is at $(1, 0)$. So the ellipse has an equation of the form

$$\frac{(x-1)^2}{a^2} + \frac{y^2}{b^2} = 1.$$

Substitution of $(x, y) = (3, 0)$ in this equation yields

$$\frac{4}{a^2} = 1, \quad \text{so that} \quad a^2 = 4.$$

Thus we may assume that $a = 2$. Substitution of $(x, y) = (0, 2)$ then yields

$$\frac{1}{4} + \frac{4}{b^2} = 1, \quad \text{so that} \quad b^2 = \frac{16}{3}.$$

Thus an equation of the ellipse through the four given points is

$$\frac{(x-1)^2}{4} + \frac{3y^2}{16} = 1.$$

Solution (b) (in case it is *not* clear where the center of the ellipse is): The *Mathematica* command

```
Solve[ { ((-1 - u)/a)^2 + ((0 - v)/b)^2 == 1,

         ((3 - u)/a)^2 + ((0 - v)/b)^2 == 1,

         ((0 - u)/a)^2 + ((2 - v)/b)^2 == 1,

         ((0 - u)/a)^2 + ((-2 - v)/b)^2 == 1 }, { a, b, u, v } ]
```

returns the solutions $u = 1$, $v = 0$, $a = \pm 2$, $b = \pm 4/\sqrt{3}$ and *no others*.

9.6.85: Given (with a change in notation):

$$\frac{x^2}{15 - q} - \frac{y^2}{q - 6} = 1. \tag{1}$$

Part (a): If $6 < q < 15$, then $15 - q > 0$ and $q - 6 > 0$. So the graph of Eq. (1) is a hyperbola with horizontal transverse axis and center $C(0, 0)$. Also $a^2 = 15 - q$ and $b^2 = q - 6$, so that $a^2 + b^2 = 9 = c^2$. Thus $c = 3$ and so the hyperbola has foci at $(\pm 3, 0)$.

Part (b): $q < 6$. Then $15 - q > 0$ and $q - 6 < 0$, so the graph of Eq. (1) is an ellipse.

Part (c): $q > 15$. In this case $15 - q < 0$ and $q - 6 > 0$, so Eq. (1) takes the form

$$\frac{x^2}{q-15} + \frac{y^2}{q-6} = -1.$$

Both denominators are positive, so there are no points on the graph.

9.6.87: The figure below shows the right branch of a hyperbola with equation

$$\frac{x^2}{a^2} - \frac{y^2}{b^2} = 1$$

(where $a > 0$ and $b > 0$), with foci $F_1(-c, 0)$ and $F_2(c, 0)$ (where $c > 0$). Let L be the line tangent to the hyperbola at the point $P(p, q)$ where $p > 0$ and $q \neq 0$. Let α be the angle between L and F_1P and let β be the angle between L and F_2P. Let θ_1 be the angle of inclination of F_1P, θ_2 the angle of inclination of F_2P, m the slope of L, and ϕ the angle of inclination of L. The goal is to prove that $\alpha = \beta$.

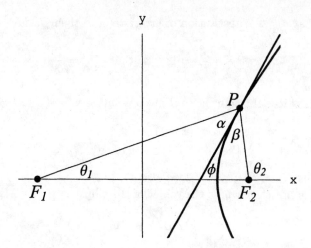

Here's what we have to work with. First, $m = \tan \phi$; we also found in the solution of Problem 86 that

$$m = \frac{b^2 p}{a^2 q} = \tan \phi.$$

We know also (Section 9.6) that $a^2 + b^2 = c^2$. Because (p, q) lies on the hyperbola, it also follows that $b^2 p^2 - a^2 q^2 = a^2 b^2$. Let m_1 be the slope of F_1P and let m_2 be the slope of F_2P. Then

$$m_1 = \tan \theta_1 = \frac{q}{p+c} \quad \text{and} \quad m_2 = \tan \theta_2 = \frac{q}{p-c}.$$

Also,

$$\theta_1 + \alpha + \pi - \phi = \pi \quad \text{and} \quad \phi + \beta + \pi - \theta_2 = \pi,$$

so that $\alpha = \phi - \theta_1$ and $\beta = \theta_2 - \phi$.

We are ready to begin. The following proof was developed interactively with *Mathematica*. We can show that $\alpha = \beta$ if we can show that $\tan \alpha = \tan \beta$. This would follow if $\tan(\phi - \theta_1) = \tan(\theta_2 - \phi)$, which follows from

$$\frac{\tan \phi - \tan \theta_1}{1 + \tan \phi \tan \theta_1} = \frac{\tan \theta_2 - \tan \phi}{1 + \tan \theta_2 \tan \phi}.$$

This equation follows from

$$\frac{m - m_1}{1 + mm_1} = \frac{m_2 - m}{1 + mm_2}; \quad \text{that is,} \quad \frac{m - m_1}{1 + mm_1} - \frac{m_2 - m}{1 + mm_2} = 0.$$

We entered the left-hand side of the last equation and applied various *Mathematica* commands to it with the following results. First we used **Together**:

$$\frac{2m - m_1 + m^2 m_1 - m_2 + m^2 m_2 - 2mm_1 m_2}{(1 + mm_1)(1 + mm_2)}.$$

Then **Numerator**:

$$2m - m_1 + m^2 m_1 - m_2 + m^2 m_2 - 2mm_1 m_2.$$

Then we entered `% /. m -> b*b*p/(a*a*q)`. This asks *Mathematica* to evaluate the previous expression (%) "subject to" the replacement of m with $b^2 p/(a^2 q)$, and we obtained

$$-m_1 - m_2 + \frac{b^4 m_1 p^2}{a^4 q^2} + \frac{b^4 m_2 p^2}{a^4 q^2} + \frac{2b^2 p}{a^2 q} - \frac{2b^2 m_1 m_2 p}{a^2 q}.$$

Similarly, we replaced m_1 with $q/(p + c)$ and m_2 with $q/(p - c)$ and thereby obtained

$$\frac{2b^2 p}{a^2 q} + \frac{b^4 p^2}{a^4(p - c)q} + \frac{b^4 p^2}{a^4(p + c)q} - \frac{q}{p - c} - \frac{q}{p + c} - \frac{2b^2 pq}{a^2(p - c)(p + c)}.$$

Another application of **Together** followed by **Numerator** yielded

$$-2(a^2 b^2 c^2 p - a^2 b^2 p^3 - b^4 p^3 + a^4 pq^2 + a^2 b^2 pq^2).$$

The command `%/(-2*p) // Cancel` produced

$$a^2 b^2 c^2 - a^2 b^2 p^2 - b^4 p^2 + a^4 q^2 + a^2 b^2 q^2,$$

and then `% /. c∧2 -> a∧2 + b∧2` yielded

$$a^2 b^2(a^2 + b^2) - a^2 b^2 p^2 - b^4 p^2 + a^4 q^2 + a^2 b^2 q^2.$$

We then asked for replacement of $b^2 p^2$ with $a^2 b^2 + a^2 q^2$ and obtained

$$a^2 b^2(a^2 + b^2) - b^4 p^2 + a^4 q^2 + a^2 b^2 q^2 - a^2(a^2 b^2 + a^2 q^2).$$

The command `Factor[%]` returned

$$b^2(a^2b^2 - b^2p^2 + a^2q^2).$$

We then cancelled b^2 to obtain $a^2b^2 - b^2p^2 + a^2q^2$, which we have already seen is zero. This establishes the desired conclusion: $\alpha = \beta$. ◀

9.6.89: First, $a^2 = \frac{9}{2} + \frac{9}{2} = 9$, so that $a = 3$ and thus $2a = 6$. Therefore Problem 24 implies that this hyperbola has equation

$$\sqrt{(x-5)^2 + (y-5)^2} + 6 = \sqrt{(x+5)^2 + (y+5)^2};$$

$$x^2 - 10x + y^2 - 10y + 50 + 12\sqrt{(x-5)^2 + (y-5)^2} + 36 = x^2 + 10x + y^2 + 10y + 50;$$

$$12\sqrt{(x-5)^2 + (y-5)^2} = 20x + 20y - 36;$$

$$3\sqrt{(x-5)^2 + (y-5)^2} = 5x + 5y - 9;$$

$$9(x^2 + y^2 - 10x - 10y + 50) = 25(x^2 + 2xy + y^2) - 90x - 90y + 81;$$

$$16x^2 + 50xy + 16y^2 = 369.$$

9.6.91: Suppose that the plane is at $P(x, y)$, that A is at $(-50, 0)$, and that B is at $(50, 0)$. Let $D = |AP|$ and $E = |BP|$, in feet. Then

$$\frac{D}{980} + \frac{E}{980} = 600 \quad \text{and} \quad \frac{D}{980} = \frac{E}{980} + 400.$$

Find D and E, observe in the process that $D = 5E$, and note that $P(x, y)$ satisfies both the equations $D = |AP|$ and $E = |BP|$. You should find that (in feet) $x \approx 218272.73$.

9.6.93: The following figure indicates the earth (as the small circle) with north pole marked N and south pole marked S; the larger curve indicates the elliptical orbit of the satellite.

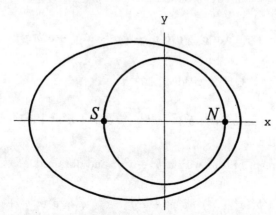

From the figure we read the information

$$\frac{pe}{1+e} = 4500 \qquad \text{and} \qquad \frac{pe}{1-e} = 9000.$$

Therefore $pe = 4500(1+e) = 9000(1-e)$, and it follows that $e = \frac{1}{3}$ and that $p = 18000$. The polar equation of the orbit of the satellite is then

$$r = \frac{6000}{1 + \frac{1}{3}\cos\theta}.$$

The satellite crosses the equatorial plane when $\theta = \pi/2$, which yields $r = 6000$. So the height of the satellite above the surface of the earth then is $h = 6000 - 4000 = 2000$ (mi).

9.6.95: Here is one solution; it may not be the simplest, and it probably isn't the most elegant—but it works. Assume that $a > b > 0$ and locate coordinate axes so that the Cartesian equation of the ellipse is

$$\frac{x^2}{a^2} + \frac{y^2}{b^2} = 1.$$

Then substitute $x = r\cos\theta$ and $y = r\sin\theta$ to convert this equation to polar coordinates. It turns out that

$$\frac{1}{2}r^2 = \frac{a^2 b^2}{a^2 + b^2 + (b^2 - a^2)\cos 2\theta}.$$

The area of the ellipse is

$$A = 4\int_0^{\pi/2} \frac{1}{2}r^2 \, d\theta.$$

The substitution $\phi = 2\theta$ then yields

$$A = 2\int_{\phi=0}^{\pi} \frac{a^2 b^2}{(a^2 + b^2) + (b^2 - a^2)\cos\phi} \, d\phi.$$

Then the substitution (see the discussion following Miscellaneous Problem 134 of Chapter 7)

$$u = \tan\frac{\phi}{2}: \qquad \phi = 2\arctan u,$$

$$d\phi = \frac{2}{1+u^2} \, du, \qquad \cos\phi = \frac{1 - u^2}{1 + u^2}$$

and the observation that $u = 0$ when $\phi = 0$ and that $u \to +\infty$ as $\phi \to \pi^-$ leads to the improper (but convergent!) integral

$$A = 2\int_0^{\infty} \frac{a^2 b^2}{(a^2 + b^2) + (b^2 - a^2) \cdot \dfrac{1 - u^2}{1 + u^2}} \cdot \frac{2}{1+u^2} \, du$$

$$= 2\int_0^{\infty} \frac{2a^2 b^2}{(a^2 + b^2) + (b^2 - a^2)(1 - u^2)} \, du = 4a^2 b^2 \int_0^{\infty} \frac{1}{2b^2 + 2a^2 u^2} \, du$$

$$= 2a^2 \int_0^\infty \frac{1}{1 + \left(\frac{au}{b}\right)^2} \, du = 2a^2 \cdot \left[\frac{b}{a} \arctan\left(\frac{au}{b}\right) \right]_0^\infty = 2ab \cdot \frac{\pi}{2} = \pi ab.$$

Here's an alternative solution that uses the standard polar form of the equation of an ellipse as presented in the final subsection of Section 9.6.

We begin with the ellipse with semiaxes a and $b < a$, eccentricity $e = c/a$, and polar equation

$$r = \frac{pe}{1 - e \cos \theta}.$$

Then Eq. (29) gives $pe = a(1 - e^2)$, so the area of the ellipse is

$$A = 2 \int_{\theta=0}^\pi \frac{1}{2} r^2 \, d\theta = a^2 (1 - e^2)^2 I \quad \text{where} \quad I = \int_0^\pi \frac{1}{(1 - e \cos \theta)^2} \, d\theta.$$

The substitution $u = \tan(\theta/2)$ mentioned in the previous solution now gives (after a bit of simplification)

$$I = \int_0^\infty \frac{2(1 + u^2)}{[(1 - e) + (1 + e)u^2]^2} \, du = \int_0^\infty \frac{2(1 + u^2)}{(B + Cu^2)^2} \, du$$

where $B = 1 - e$ and $C = 1 + e$. A simple partial-fractions expansion yields

$$I = \frac{2(C - B)}{C} I_2 + \frac{2}{C} I_1 \quad \text{where} \quad I_n = \int_0^\infty \frac{1}{(B + Cu^2)^n} \, du.$$

Then integration of I_1 by parts with

$$p = \frac{1}{B + Cu^2} \quad \text{and} \quad dq = du$$

gives

$$I_1 = \left[\frac{u}{B + Cu^2} \right]_0^\infty + \int_0^\infty \frac{2Cu^2}{(B + Cu^2)^2} \, du \quad \text{(the first term vanishes)};$$

$$I_1 = 2 \int_0^\infty \frac{B + Cu^2}{(B + Cu^2)^2} \, du - \int_0^\infty \frac{2B}{(B + Cu^2)^2} \, du = 2I_1 - 2BI_2.$$

It follows that $I_2 = \dfrac{I_1}{2B}$, so

$$I = \frac{2(C - B)}{C} \cdot \frac{I}{2B} + \frac{2}{C} I_1 = \left(\frac{1}{B} + \frac{1}{C} \right) I_1 = \left(\frac{1}{B} + \frac{1}{C} \right) \int_0^\infty \frac{1}{B + Cu^2} \, du.$$

With $k = \sqrt{\dfrac{B}{C}}$ we get

542

$$I = \frac{1}{C}\left(\frac{1}{B} + \frac{1}{C}\right)\int_0^\infty \frac{1}{k^2 + u^2}\,du = \frac{1}{C}\left(\frac{1}{B} + \frac{1}{C}\right)\cdot\left[\frac{1}{k}\tan^{-1}\frac{u}{k}\right]_0^\infty$$

$$= \frac{1}{C}\left(\frac{1}{B} + \frac{1}{C}\right)\cdot\frac{\pi}{2}\cdot\sqrt{\frac{C}{B}} = \frac{1}{\sqrt{BC}}\cdot\left(\frac{1}{B} + \frac{1}{C}\right)\cdot\frac{\pi}{2}$$

$$= \frac{1}{\sqrt{1 - e^2}}\cdot\left(\frac{1}{1 - e} + \frac{1}{1 + e}\right)\cdot\frac{\pi}{2} = \frac{\pi}{(1 - e^2)^{3/2}}.$$

Finally, we find that the area of the ellipse is

$$A = a^2(1 - e^2)^2 I = a^2(1 - e^2)^2\cdot\frac{\pi}{(1 - e^2)^{3/2}}$$

$$= \pi a^2\sqrt{1 - e^2} = \pi a^2\sqrt{1 - \left(\frac{c}{a}\right)^2} = \pi a\cdot\sqrt{a^2 - c^2} = \pi ab,$$

as desired.

Chapter 9 Miscellaneous Problems

9.M.1: Completing the square yields $(x - 1)^2 + (y - 1)^2 = 4$, so this conic section is a circle with center $C(1, 1)$ and radius 2. Its graph is next.

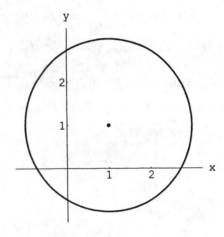

9.M.3: Completing the square in both variables yields

$$x^2 - 6x + 9 + y^2 + 2y + 1 = 1; \quad \text{that is,} \quad (x - 3)^2 + (y + 1)^2 = 1.$$

Therefore this conic section is the circle with center $C(3, -1)$ and radius 1. Its graph is next.

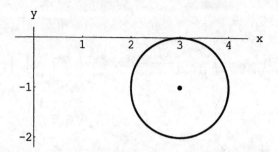

9.M.5: Completing the square in x yields

$$x^2 - 8x + 16 = -2y - 4; \quad \text{that is,} \quad (x - 4)^2 = -2(y + 2).$$

Thus this conic section is a parabola with vertex $V(4, -2)$, vertical axis with equation $x = 4$, focus at $\left(4, -\frac{5}{2}\right)$, and opening downward. Its graph is next.

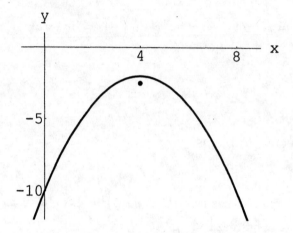

9.M.7: Given: $9x^2 + 4y^2 = 36x$. Complete the square in x as follows:

$$9(x^2 - 4x) + 4y^2 = 0; \quad 9(x^2 - 4x + 4) + 4y^2 = 36; \quad \left(\frac{x-2}{2}\right)^2 + \left(\frac{y}{3}\right)^2 = 1.$$

Hence this conic section is an ellipse with center $C(2, 0)$, vertical major axis of length 6, minor axis of length 4, foci at $\left(2, \pm\sqrt{5}\right)$, and vertices at $(0, 0)$, $(4, 0)$, $(2, 3)$, and $(2, -3)$. Its graph is next.

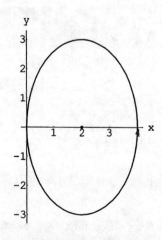

9.M.9: Given $y^2 - 2x^2 = 4x + 2y + 3$, complete the square in both variables as follows:

$$y^2 - 2y - 2x^2 - 4x = 3; \qquad y^2 - 2y - 2(x^2 + 2x) = 3;$$

$$y^2 - 2y + 1 - 2(x^2 + 2x + 1) = 2; \qquad \frac{(y-1)^2}{2} - (x+1)^2 = 1.$$

This conic section is a hyperbola with center $C(-1, 1)$, vertical transverse axis of length $2\sqrt{2}$, $a = \sqrt{2}$, $b = 1$, $c = \sqrt{3}$, foci $F\left(-1, 1 \pm \sqrt{3}\right)$, and vertices $V\left(-1, 1 \pm \sqrt{2}\right)$. Its graph is next.

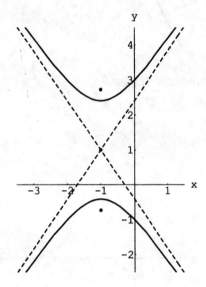

9.M.11: Complete the square in each variable:

$$x^2 + 2y^2 = 4x + 4y - 12; \qquad x^2 - 4x + 2y^2 - 4y = -12;$$

$$x^2 - 4x + 4 + 2(y^2 - 2y + 1) = -6; \qquad (x-2)^2 + 2(y-1)^2 < 0.$$

Thus there are no points on the graph of the given equation.

9.M.13: The given equation can be written in the form

$$\frac{(x-1)^2}{4} - \frac{y^2}{9} = 1,$$

so this conic section is a hyperbola with center at $(1, 0)$, horizontal transverse axis of length 4, foci at $\left(1 \pm \sqrt{13}, 0\right)$, and vertices at $(3, 0)$ and $(-1, 0)$. Its graph is next.

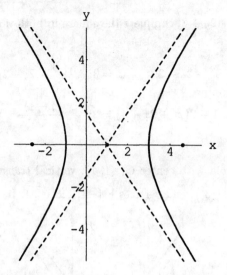

9.M.15: The equation can be written in the form $(x - 4)^2 + (y - 1)^2 = 1$; this conic section is the circle with center $(4, 1)$ and radius 1. Its graph is next.

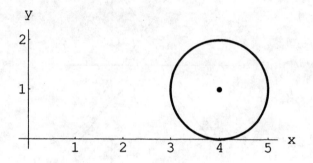

9.M.17: The given equation can be written in the form

$$\left[(x - 2)^2 + (y - 2)^2\right] \cdot (x + y)^2 = 0,$$

and therefore either $(x - 2)^2 + (y - 2)^2 = 0$ or $(x + y)^2 = 0$. In the former case the only way for the sum of two squares to be zero is if each is zero, so only $(x, y) = (2, 2)$ satisfies the equation. In the latter case $(x + y)^2 = 0$ implies that $y = -x$. So the graph consists of the line $y = -x$ together with the isolated point $(2, 2)$. It is not the graph of a conic section.

9.M.19: Convert to Cartesian coordinates, then complete the square:

$$r = -2\cos\theta; \qquad r^2 = -2r\cos\theta;$$

$$x^2 + y^2 = -2x; \qquad x^2 + 2x + 1 + y^2 = 1;$$

$$(x + 1)^2 + y^2 = 1.$$

This conic section is the circle with center $C(-1, 0)$ and radius 1. Its graph is next.

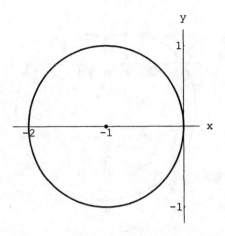

9.M.21: Given

$$r = \frac{1}{\sin\theta + \cos\theta},$$

multiply each side of this equation by the denominator to obtain $r\sin\theta - r\cos\theta = 1$. In Cartesian coordinates, that's $y = x + 1$. Hence the graph is the straight line through the point $(0, 1)$ with slope 1. One does obtain the entire graph because the denominator can take both positive and negative values arbitrarily close to zero. We omit the graph to save space.

9.M.23: Given $r = 3\csc\theta$, rewrite this as $r\sin\theta = 3$, then convert to Cartesian coordinates: $y = 3$. The graph is the horizontal line passing through the point $(0, 3)$. All of the line is the graph because $\csc\theta$ takes on arbitrarily large values. We omit the graph to save space.

9.M.25: The graph of the polar equation $r^2 = 4\cos\theta$ is a pair of tangent ovals (not a conic section). It's shown next.

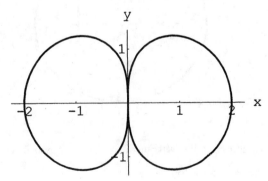

9.M.27: The graph of the polar equation $r = 3 - 2\sin\theta$ is a limaçon (from the French word for shell-snail); it is not a conic section. The graph is next.

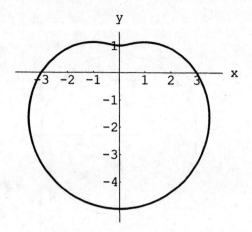

9.M.29: Given $r = \dfrac{4}{2 + \cos\theta}$, convert to Cartesian coordinates:

$$2r + r\cos\theta = 4; \qquad 2r = 4 - x;$$

$$4(x^2 + y^2) = x^2 - 8x + 16; \qquad 3x^2 + 4y^2 + 8x = 16;$$

$$3\left(x^2 + \frac{8}{3}x\right) + 4y^2 = 16; \qquad 3\left(x^2 + \frac{8}{3}x + \frac{16}{9}\right) + 4y^2 = 16 + \frac{16}{3} = \frac{64}{3};$$

$$3\left(x + \frac{4}{3}\right)^2 + 4y^2 = \frac{64}{3}; \qquad \frac{9}{64}\left(x + \frac{4}{3}\right)^2 + \frac{3}{16}y^2 = 1.$$

Thus the graph is a conic section; it is the ellipse with center $C\left(-\frac{4}{3}, 0\right)$, $a = \frac{8}{3}$, and $b = \frac{4}{3}\sqrt{3}$. Its major axis is horizontal and its eccentricity is $e = \frac{1}{2}$. Its vertices are located at $(-4, 0)$, $\left(0, \pm\frac{4}{3}\sqrt{3}\right)$, and $\left(\frac{4}{3}, 0\right)$; its foci are at $\left(-\frac{8}{3}, 0\right)$ and $(0, 0)$. Its graph is next.

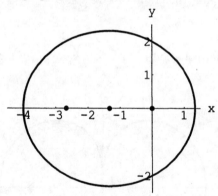

9.M.31: We consider the region that lies within both circles shown in the next figure. To find its area A, we integrate from $\theta = 0$ to $\theta = \pi/4$ and double the result:

$$A = \int_0^{\pi/4} (2\sin\theta)^2 \, d\theta = \Big[2\theta - \sin 2\theta\Big]_0^{\pi/4} = \frac{\pi}{2} - 1 = \frac{\pi - 2}{2} \approx 0.570796326795.$$

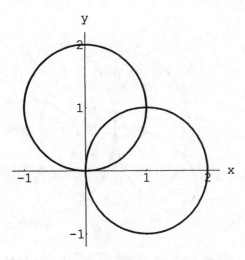

9.M.33: The circle and the limaçon are shown next. They cross where $\theta = \alpha = 7\pi/6$ and where $\theta = \beta = 11\pi/6$. To find the area A within the limaçon but outside the circle, we evaluate

$$A = \frac{1}{2}\int_\alpha^\beta \left[(3 - 2\sin\theta)^2 - 16\right] d\theta = \frac{1}{2}\int_\alpha^\beta (4\sin^2\theta - 12\sin\theta - 7) \, d\theta = \frac{1}{2}\left[12\cos\theta - \sin 2\theta - 5\theta\right]_\alpha^\beta$$

$$= \frac{1}{2}\left(\frac{13\sqrt{3}}{2} - \frac{55\pi}{6}\right) - \frac{1}{2}\left(-\frac{13\sqrt{3}}{2} - \frac{35\pi}{6}\right) = \frac{39\sqrt{3} - 10\pi}{6} \approx 6.022342493215.$$

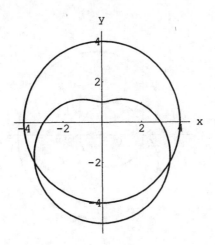

9.M.35: The circle and the four-leaved rose cross where θ is an odd integral multiple of $\pi/8$. To find the area A within the rose and outside the circle, we multiple the area in the first quadrant by 4. Thus with $\alpha = \pi/8$ and $\beta = 3\pi/8$, we have

$$A = 2\int_\alpha^\beta (-2 + 4\sin^2 2\theta) \, d\theta = \int_\alpha^\beta (-4\cos 4\theta) \, d\theta = \left[-\sin 4\theta\right]_\alpha^\beta = 1 - (-1) = 2.$$

The circle and the rose are shown next.

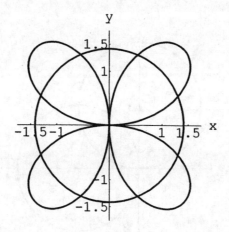

9.M.37: The circle and the cardioid are shown next. They meet only at the pole. To find the area A inside the cardioid but outside the circle, we simply find the area within the cardioid and subtract the area of a circle of radius $\frac{1}{2}$.

$$A = \frac{1}{2} \int_0^{2\pi} (1 + \cos\theta)^2 \, d\theta - \frac{\pi}{4} = -\frac{\pi}{4} + \int_0^{2\pi} 2\cos^4 \frac{\theta}{2} \, d\theta$$

$$= -\frac{\pi}{4} + \left[\frac{1}{8}(6\theta + 8\sin\theta + \sin 2\theta) \right]_0^{2\pi} = -\frac{\pi}{4} + \frac{3\pi}{2} = \frac{5\pi}{4} \approx 3.926990816987.$$

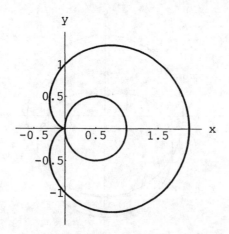

9.M.39: Elimination of the parameter yields the equation $y = x + 2$ of the straight line through $(0, 2)$ with slope 1. The graph consists of all points of this line because both $2t^3 - 1$ and $2t^3 + 1$ take on arbitrarily large positive and negative values.

9.M.41: Because $(x - 2)^2 + (y - 1)^2 = \cos^2 t + \sin^2 t = 1$, the graph is the circle with center $C(2, 1)$ and radius 1. The entire circle is obtained as the graph of the parametric equations because $\cos t$ and $\sin t$ take on all values between -1 and $+1$ as t ranges from 0 to 2π. The graph is next.

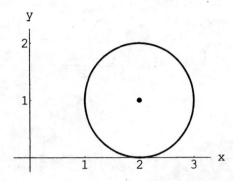

9.M.43: Because $x - 1 = t^2 = y^{2/3}$, we can write the equation in the form $y^2 = (x - 1)^3$. This curve is called by some a "semicubical parabola" even though it's not a parabola. Its graph is next.

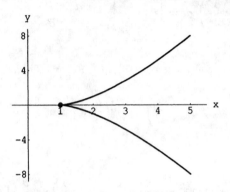

9.M.45: By the chain rule,

$$\frac{dy}{dx} = \frac{dy/dt}{dx/dt} = \frac{3\cos t}{-4\sin t} = -\frac{4}{3}\tan t.$$

When $t = \pi/4$, we have $x = \frac{3}{2}\sqrt{2}$, $y = 2\sqrt{2}$, and $dy/dx = -\frac{4}{3}$. Hence an equation of the tangent line is

$$y - 2\sqrt{2} = -\frac{4}{3}\left(x - \frac{3}{2}\sqrt{2}\right); \quad \text{that is,} \quad y = -\frac{4}{3}\left(x - 3\sqrt{2}\right).$$

9.M.47: Given the polar equation $r = \theta$, we have $x(\theta) = \theta\cos\theta$ and $y(\theta) = \theta\sin\theta$. Therefore

$$\frac{dy}{dx} = \frac{dy/d\theta}{dx/d\theta} = \frac{\theta\cos\theta + \sin\theta}{\cos\theta - \theta\sin\theta}.$$

Thus when $\theta = \pi/2$, we have $x = 0$, $y = \pi/2$, and $dy/dx = -2/\pi$. So an equation of the tangent line is

$$y = -\frac{2}{\pi}x + \frac{\pi}{2}; \quad \text{that is,} \quad 4x + 2\pi y = \pi^2.$$

9.M.49: The area is

$$\int_{t=-1}^{2} y \, dx = \left[\frac{2}{3}(9t + t^3)\right]_{-1}^{2} = \frac{52}{3} - \left(-\frac{20}{3}\right) = 24.$$

9.M.51: The area is $A = \int_{t=0}^{\pi/2} y \, dx = \int_{0}^{\pi/2} 12\cos^2 t \, dt = \left[6t + 3\sin 2t\right]_{0}^{\pi/2} = 3\pi \approx 9.424777960769.$

9.M.53: The arc length is

$$L = \int_0^1 t(4 + 9t^2)^{1/2}\, dt = \left[\frac{1}{27}(4 + 9t^2)^{3/2} \right]_0^1 = \frac{13\sqrt{13}}{27} - \frac{8}{27} = \frac{13\sqrt{13} - 8}{27} \approx 1.439709873372.$$

9.M.55: In this problem we have

$$\left(\frac{dx}{dt} \right)^2 + \left(\frac{dy}{dt} \right)^2 = 4 + \left(3t^2 - \frac{1}{3t^2} \right)^2 = 9t^4 + 2 + \frac{1}{9t^4} = \left(3t^2 + \frac{1}{3t^2} \right)^2.$$

Therefore the arc length is

$$\int_1^2 \left(3t^2 + \frac{1}{3t^2} \right) dt = \left[t^3 - \frac{1}{3t} \right]_1^2 = \frac{47}{6} - \frac{2}{3} = \frac{43}{6} \approx 7.166666666667.$$

9.M.57: Given: the polar equation $r = \sin^2(\theta/3)$, $0 \leq \theta \leq \pi$. Then

$$r^2 + \left(\frac{dr}{d\theta} \right)^2 = \sin^4 \frac{\theta}{3} + \frac{4}{9} \sin^2 \frac{\theta}{3} \cos^2 \frac{\theta}{3},$$

and therefore

$$\sqrt{r^2 + (dr/d\theta)^2} = \left(\sin \frac{\theta}{3} \right) \left(\sin^2 \frac{\theta}{3} + \frac{4}{9} \cos^2 \frac{\theta}{3} \right)^{1/2} = \frac{1}{3} \left(\sin \frac{\theta}{3} \right) \left(9 \sin^2 \frac{\theta}{3} + 4 \cos^2 \frac{\theta}{3} \right)^{1/2}$$

$$= \frac{1}{3} \left(\sin \frac{\theta}{3} \right) \left(9 \sin^2 \frac{\theta}{3} + 9 \cos^2 \frac{\theta}{3} - 5 \cos^2 \frac{\theta}{3} \right)^{1/2} = \frac{1}{3} \left(\sin \frac{\theta}{3} \right) \left(9 - 5 \cos^2 \frac{\theta}{3} \right)^{1/2}.$$

So the length of the graph is

$$L = \int_0^\pi \frac{1}{3} \left(\sin \frac{\theta}{3} \right) \left(9 - 5 \cos^2 \frac{\theta}{3} \right)^{1/2} d\theta.$$

Let

$$u = \cos \frac{\theta}{3}; \qquad \text{then} \qquad du = -\frac{1}{3} \sin \frac{\theta}{3}\, d\theta.$$

This substitution yields

$$L = \int_{u=1}^{1/2} -(9 - 5u^2)^{1/2}\, du = \int_{1/2}^1 \sqrt{9 - 5u^2}\, du = \sqrt{5} \int_{1/2}^1 \left(\frac{9}{5} - u^2 \right)^{1/2} du.$$

Then integral formula 54 of the endpapers, with $a = 3/\sqrt{5} = \frac{3}{5}\sqrt{5}$, yields

$$L = \sqrt{5} \left[\frac{u}{2} \left(\frac{9}{5} - u^2 \right)^{1/2} + \frac{9}{10} \arcsin \frac{u\sqrt{5}}{3} \right]_{1/2}^1$$

$$= \sqrt{5} \left[\frac{1}{2} \left(\frac{9}{5} - 1 \right)^{1/2} + \frac{9}{10} \arcsin \frac{\sqrt{5}}{3} - \frac{1}{4} \left(\frac{9}{5} - \frac{1}{4} \right)^{1/2} - \frac{9}{10} \arcsin \frac{\sqrt{5}}{6} \right]$$

$$= \frac{1}{2}\sqrt{9-5} + \frac{9\sqrt{5}}{10}\arcsin\frac{\sqrt{5}}{3} - \frac{1}{4}\sqrt{9 - \frac{5}{4}} - \frac{9\sqrt{5}}{10}\arcsin\frac{\sqrt{5}}{6}$$

$$= 1 + \frac{9\sqrt{5}}{10}\arcsin\frac{\sqrt{5}}{3} - \frac{\sqrt{31}}{8} - \frac{9\sqrt{5}}{10}\arcsin\frac{\sqrt{5}}{6} \approx 1.2281021668591117.$$

9.M.59: First,

$$\left[\left(\frac{dx}{dt}\right)^2 + \left(\frac{dy}{dt}\right)^2\right]^{1/2} = \sqrt{\frac{(1+t^5)^2}{t^6}} = \frac{1+t^5}{t^3}.$$

Therefore the surface area of revolution is

$$A = \int_{t=1}^{4} 2\pi y\, ds = \frac{\pi}{3}\int_{1}^{4}\left(2t^5 + 5 + 3t^{-5}\right)dt = \pi\left[\frac{1}{9}t^6 + \frac{5}{3}t - \frac{1}{4}t^{-4}\right]_{1}^{4}$$

$$= \frac{4255735\pi}{9216} - \frac{55\pi}{36} = \frac{471295\pi}{1024} \approx 1445.914950853127.$$

9.M.61: Given the polar equation $r = \exp(\theta/2)$, we have

$$x(\theta) = \exp(\theta/2)\cos\theta \quad \text{and} \quad y(\theta) = \exp(\theta/2)\sin\theta.$$

Therefore

$$[x'(\theta)]^2 + [y'(\theta)]^2 = \left(\frac{1}{2}\exp(\theta/2)\cos\theta - \exp(\theta/2)\sin\theta\right)^2 + \left(\frac{1}{2}\exp(\theta/2)\sin\theta + \exp(\theta/2)\cos\theta\right)^2$$

$$= \frac{1}{4}\exp(\theta)\cos^2\theta - \exp(\theta)\cos\theta\sin\theta + \exp(\theta)\sin^2\theta + \exp(\theta)\cos^2\theta + \exp(\theta)\cos\theta\sin\theta + \frac{1}{4}\exp(\theta)\sin^2\theta$$

$$= \frac{5}{4}\exp(\theta)\cos^2\theta + \frac{5}{4}\exp(\theta)\sin^2\theta = \frac{5}{4}\exp(\theta).$$

Therefore the surface area of revolution is

$$A = \int_{0}^{\pi} 2\pi y(\theta) \cdot \frac{\sqrt{5}}{2}\exp(\theta/2)\, d\theta = \pi\sqrt{5}\int_{0}^{\pi} e^\theta \sin\theta\, d\theta = \frac{\pi\sqrt{5}}{2}\left[(\sin\theta - \cos\theta)e^\theta\right]_{0}^{\pi}$$

$$= \frac{\pi e^\pi\sqrt{5}}{2} - \left(-\frac{\pi\sqrt{5}}{2}\right) = \frac{\pi(1+e^\pi)\sqrt{5}}{2} \approx 84.791946612137.$$

See Example 5 in Section 7.3 for how to find the antiderivative using integration by parts.

9.M.63: Suppose that the circle rolls to the right through a central angle θ. Then its center is at the point $O(a\theta, a)$. So the point $P(x, y)$ that generates the trochoid is located where $x = a\theta - b\sin\theta$ and $y = a - b\cos\theta$. This is easy to see from the small right triangle in the rolling circle shown next.

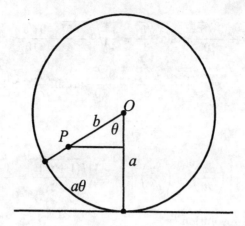

One cycle of a trochoid with $a = 3$ and $b = 1$ is shown next.

9.M.65: If $b = a$ in the parametric equations of the epicycloid of Problem 64, then its equations take the form

$$x = 2a\cos\theta - a\cos 2\theta,$$

$$y = 2a\sin\theta - a\sin 2\theta.$$

Shift this epicycloid a units to the left. Its equations will then be

$$x = 2a\cos\theta - a\cos 2\theta - a,$$

$$y = 2a\sin\theta - a\sin 2\theta.$$

Then substitution yields

$$r^2 = x^2 + y^2 = a^2(2\cos\theta - \cos^2\theta + \sin^2\theta - 1)^2 + a^2(2\sin\theta - 2\sin\theta\,\cos\theta)^2$$

$$= a^2(2\cos\theta - 2\cos^2\theta)^2 + a^2(2\sin\theta - 2\sin\theta\,\cos\theta)^2$$

$$= a^2(4\cos^2\theta - 8\cos^3\theta + 4\cos^4\theta + 4\sin^2\theta - 8\sin^2\theta\,\cos\theta + 4\sin^2\theta\,\cos^2\theta)$$

$$= a^2\big[(4\cos^2\theta)(1 - 2\cos\theta + cos^2\theta) + (4\sin^2\theta)(1 - 2\cos\theta + \cos^2\theta)\big]$$

$$= 4a^2(\cos^2\theta + \sin^2\theta)(1 - \cos\theta)^2 = 4a^2(1 - \cos\theta)^2.$$

Thus the translated epicycloid has polar equation $r = 2a(1 - \cos\theta)$, and therefore it is a cycloid.

9.M.67: Using the method of cylindrical shells, we find the volume generated to be

$$V = \int_{\theta=0}^{2\pi} 2\pi x(\theta) y(\theta)\, dx = 2\pi a^3 \int_0^{2\pi} (1 - \cos\theta)^2 (\theta - \sin\theta)\, d\theta$$

$$= \frac{1}{12}\pi a^3 \left[18\theta^2 - 18\cos\theta - 9\cos 2\theta + 2\cos 3\theta - 48\theta\sin\theta + 6\theta\sin 2\theta \right]_0^{2\pi}$$

$$= \frac{1}{12}\pi a^3 (72\pi^2 - 25 + 25) = 6\pi^3 a^3 \approx (186.037660081799)a^3.$$

The solution of Problem 3 in Section 7.3 illustrates how to use integration by parts for the antiderivatives of $\theta\cos\theta$ and $\theta\cos 2\theta$.

9.M.69: If the point C were on the x-axis, then a Cartesian equation for the circle would be

$$(x - p)^2 + y^2 = p^2; \quad \text{that is,} \quad x^2 - 2px + y^2 = 0.$$

Its polar equation is therefore $r^2 = 2px = 2pr\cos\theta$; that is, $r = 2p\cos\theta$. Because the radius to C makes the angle α with the x-axis, the actual equation we seek is therefore $r = 2p\cos(\theta - \alpha)$.

9.M.71: Assume that $a > b > 0$. Parametrize the ellipse thus:

$$x = a\cos t, \quad y = b\sin t, \quad 0 \leq t \leq 2\pi.$$

The diameter with endpoints (x, y) and $(-x, -y)$ has length

$$2\sqrt{x^2 + y^2} = 2\sqrt{a^2\cos^2 t + b^2\sin^2 t}$$

where, without loss of generality, $0 \leq t \leq \pi/2$. Thus our goal is to maximize and minimize the function $f(t) = a^2\cos^2 t + b^2\sin^2 t$ on that domain. Now

$$f'(t) = -2a^2\sin t\,\cos t + 2b^2\sin t\,\cos t = 2(b^2 - a^2)\sin t\,\cos t.$$

Because $f'(t) < 0$ if $0 < t < \pi/2$ and f is continuous on its domain, the maximum value of f occurs when $t = 0$ and its minimum when $t = \pi/2$. Therefore the diameter of this ellipse of maximum length is its major axis, of length $2a$, and its diameter of minimum length is its minor axis, of length $2b$.

9.M.73: The parabola passes through $(0, 0)$, $(b/2, h)$ (its vertex), and $(b, 0)$. Hence its equation has the form

$$y - h = c\left(x - \frac{b}{2}\right)^2.$$

Because $(0, 0)$ satisfies this equation, we find that

$$-h = c\left(-\frac{b}{2}\right)^2, \quad \text{so that} \quad c = -\frac{4h}{b^2}.$$

Therefore the desired equation of the parabola is $y = \dfrac{4hx(b-x)}{b^2}$.

9.M.75: Let θ be the angle that the segment QR makes with the x-axis (see the figure that follows this solution). Then the coordinates of the point $P(x, y)$ satisfy the equations

$$x = -a\cos\theta \quad \text{and} \quad y = b\sin\theta$$

(the figure shows only the case in which P is in the second quadrant; you should check the other three cases for yourself). It now follows that

$$\left(\frac{x}{a}\right)^2 + \left(\frac{y}{b}\right)^2 = \cos^2\theta + \sin^2\theta = 1,$$

and therefore the locus of P is an ellipse. All points of the this ellipse are obtained as θ varies from 0 to 2π.

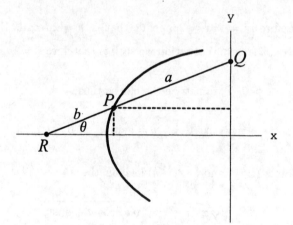

9.M.77: Please refer to the figure that follows this solution. Suppose that

$$Q_1 = Q_1\left(\frac{a^2}{4p}, a\right) \quad \text{and that} \quad Q_2 = Q_2\left(\frac{b^2}{4p}, b\right).$$

Then the slope of Q_1Q_2 will be

$$m = \frac{4p(b-a)}{b^2-a^2} = \frac{4p}{b+a}.$$

Implicit differentiation of the parabola's equation with respect to x yields

$$2y\frac{dy}{dx} = 4p, \quad \text{so that} \quad \frac{dy}{dx} = \frac{2p}{y}.$$

So we can find the y-coordinate of P (and thus of R) by solving

$$\frac{2p}{y} = \frac{4p}{b+a} : \quad y = \frac{a+b}{2}.$$

It now follows that the two right triangles with sides Q_1R and RQ_2 are congruent, and therefore R is the midpoint of the segment Q_1Q_2.

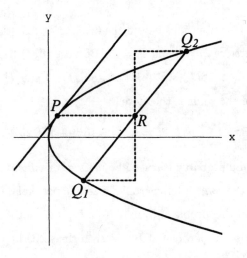

9.M.79: First we put the given equation in standard form:

$$3x^2 - y^2 + 12x + 9 = 0; \qquad 3x^2 + 12x - y^2 = -9;$$

$$3(x^2 + 4x + 4) - y^2 = 3; \qquad 3(x+2)^2 - y^2 = 3;$$

$$(x+2)^2 - \frac{y^2}{3} = 1.$$

With the usual meaning of the notation, this is a hyperbola with $a^2 = 1$ and $b^2 = 3$. By Eq. (20) of Section 9.6, $b^2 = a^2(e^2 - 1)$, we see that $3 = e^2 - 1$, so that $e^2 = 4$. Therefore this hyperbola has eccentricity $e = 2$.

9.M.81: First we convert the equation of the folium to polar coordinates:

$$x^3 + y^3 = 3xy; \qquad r^3 \cos^3\theta + r^3 \sin^3\theta = 3r^2 \sin\theta \, \cos\theta;$$

$$r\cos^3\theta + r\sin^3\theta = 3\sin\theta \, \cos\theta; \qquad r = \frac{3\sin\theta \, \cos\theta}{\sin^3\theta + \cos^3\theta};$$

$$r = \frac{3\sec\theta \, \tan\theta}{1 + \tan^3\theta}.$$

To obtain the area of the loop of the folium, we evaluate

$$A = \frac{1}{2}\int_0^{\pi/2} r^2 \, d\theta = 9\int_0^{\pi/4} \frac{\sec^2\theta \, \tan^2\theta}{(1 + \tan^3\theta)^2} \, d\theta.$$

The substitution $u = \tan\theta$, $du = \sec^2\theta \, d\theta$ transforms this integral into

$$A = 3\int_0^1 \frac{3u^2}{(1 + u^3)^2} \, du = 3\left[-\frac{1}{1+u^3}\right]_0^1 = 3\left(1 - \frac{1}{2}\right) = \frac{3}{2}.$$

9.M.83: The equation of the conic can be written in the form

$$Ax^2 + Bxy + Cy^2 + Dx + Ey + F = 0. \tag{1}$$

We may assume that $A = 1$. Because $(5, 0)$ lies on the conic, $25 + 5D + F = 0$. Because $(-5, 0)$ lies on the conic, $25 - 5D + F = 0$. Therefore $D = 0$ and $F = 25$. Because $(0, 4)$ lies on the conic, $16C + 4E + F = 0$. Because $(0, -4)$ lies on the conic, $16C - 4E + F = 0$. So $E = 0$ and $F = -16C$. Therefore $16C = 25$, so that $C = \frac{25}{16}$. The equation of the conic is therefore

$$x^2 + Bxy + \frac{25}{16}y^2 - 25 = 0: \qquad 16x^2 + 16Bxy + 25y^2 = 400. \tag{2}$$

If you have studied Eq. (1), you probably learned about its *discriminant* $B^2 - 4AC$. It is known that if the discriminant is positive, then the conic is a hyperbola; if zero, a parabola; if negative, a parabola. Degenerate cases may occur, as they do in this problem. We see from Eq. (2) that if $B < \frac{5}{2}$, then the conic is an ellipse and if $B > \frac{5}{2}$, then the conic is a hyperbola. If $B = \frac{5}{2}$ then the second equation in (2) becomes

$$16x^2 + 40xy + 25y^2 = 400; \qquad (4x + 5y)^2 = 400; \qquad 4x + 5y = \pm 20.$$

This is a *degenerate* parabola: two parallel lines.

If you have not studied Eq. (1) and its discriminant, you can proceed as follows. First, no [nondegenerate] parabola can contain the four given points. They are the vertices of a rhombus, and if three lay on a parabola, the fourth would be "within" the parabola. It is also clear that the four points can lie on an ellipse: Take $B = 0$ in Eq. (2). It is also clear that the four given points satisfy the equation $16x^2 + 100xy + 25y^2 = 400$. We claim that this is an equation of a hyperbola. To show this, we will set up a rotated rectangular uv-coordinate system in which the "mixed" term $100xy$ disappears. Please refer to the next figure.

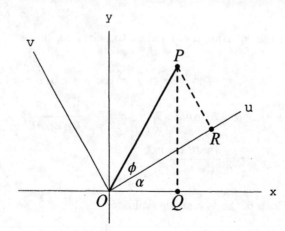

First consider the point P, with xy-coordinates (x, y) and uv-coordinates (u, v). The rectangular uv-coordinate system is obtained from the xy-coordinate system by a rotation through the angle α shown in the figure. Note that

$$x = OQ = OP\cos(\alpha + \phi) \quad \text{and} \quad y = PQ = OP\sin(\alpha + \phi). \tag{3}$$

Moreover,
$$u = OR = OP \cos \phi \quad \text{and} \quad v = PR = OP \sin \phi. \tag{4}$$

Substitution of the equations in (4) into those in (3) yields

$$x = OP(\cos \alpha \cos \phi - \sin \alpha \sin \phi) = u \cos \alpha - v \sin \alpha \tag{5}$$

and

$$y = OP(\sin \alpha \cos \phi + \cos \alpha \sin \phi) = u \sin \alpha + v \cos \alpha. \tag{6}$$

We then entered the expression

```
expr = (16*x*x + 100*x*y + 25*y*y - 400)
```

in *Mathematica*. Then we made the substitutions for x and y given in Eqs. (5) and (6), expanded it with the command Expand, then asked for the coefficient of uv with the command Coefficient[expr, u*v]. *Mathematica* returned

$$100 \cos^2 \alpha + 18 \cos \alpha \sin \alpha - 100 \sin^2 \alpha$$

We simplified this and entered the result in the form

```
expr2 = (100*Cos[2*a] + 9*Sin[2*a]),
```

using a in place of α. Then we entered Solve[expr2 == 0, a], and *Mathematica* returned two angles:

$$\alpha = \frac{1}{2} \arccos \left(\frac{-9}{\sqrt{10081}} \right), \quad \alpha = -\frac{1}{2} \arccos \left(\frac{9}{\sqrt{10081}} \right).$$

We set α equal to the second of these and entered the command expr = Expand[expr]. This caused *Mathematica* to replace α with its numerical value throughout expr. The result is too long to reproduce here. We then asked for

```
Coefficient[expr, u*v]
```

and *Mathematica* returned another long expression; when we asked *Mathematica* to Simplify the result, the answer was 0. So we have successfully eliminated the coefficient of uv. Next we asked for

```
Coefficient[expr, u*u]
```

and the result, after Simplify, became
$$\frac{1}{2} \left(41 - \sqrt{10081} \right).$$

Similarly, the coefficient of v^2 turned out to be

$$\frac{1}{2} \left(41 + \sqrt{10081} \right).$$

Because the coefficient of u^2 is approximately -29.702091589893 and the coefficient of v^2 is approximately 70.702091589893, and because the equation $16x^2 + 100xy + 15y^2 = 400$ in the uv-coordinate system is

$$\frac{1}{2}\left(41 - \sqrt{10081}\right)u^2 + \frac{1}{2}\left(41 + \sqrt{10081}\right)v^2 = 400,$$

this conic section is a hyperbola. Its graph is next.

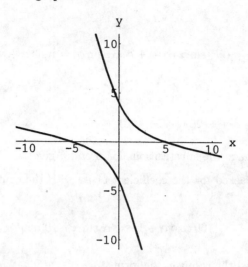

If the graph of $16x^2 + 16Bxy + 25y^2 = 400$ is normal to the y-axis at the point $(0, 4)$, then $dy/dx = 0$ there. By implicit differentiation,

$$32x + 16By + 16Bx\,\frac{dy}{dx} + 50y\,\frac{dy}{dx} = 0,$$

and when we substitute the data $x = 0$, $y = 4$, $dy/dx = 0$, we find that $64B = 0$, so that $B = 0$. In this case the graph is the ellipse with equation

$$\left(\frac{x}{5}\right)^2 + \left(\frac{y}{4}\right)^2 = 1.$$

Section 10.2

10.2.1: The most obvious pattern is that $a_n = n^2$ for $n \geq 1$.

10.2.3: The most obvious pattern is that $a_n = \dfrac{1}{3^n}$ for $n \geq 1$.

10.2.5: The most obvious pattern is that $a_n = \dfrac{1}{3n-1}$ for $n \geq 1$.

10.2.7: Perhaps the most obvious pattern is that $a_n = 1 + (-1)^n$ for $n \geq 1$.

10.2.9: $\displaystyle \lim_{n \to \infty} \frac{2n}{5n-3} = \lim_{n \to \infty} \frac{2}{5 - \dfrac{3}{n}} = \frac{2}{5-0} = \frac{2}{5}$.

10.2.11: $\displaystyle \lim_{n \to \infty} \frac{n^2 - n + 7}{2n^3 + n^2} = \lim_{n \to \infty} \frac{\dfrac{1}{n} - \dfrac{1}{n^2} + \dfrac{7}{n^3}}{2 + \dfrac{1}{n}} = \frac{0 + 0 + 0}{2 + 0} = 0$.

10.2.13: Example 9 tells us that if $|r| < 1$, then $r^n \to 0$ as $n \to +\infty$. Take $r = \frac{9}{10}$ to deduce that

$$\lim_{n \to \infty} \left[1 + \left(\tfrac{9}{10} \right)^n \right] = 1 + 0 = 1.$$

10.2.15: Given: $a_n = 1 + (-1)^n$ for $n \geq 1$. If n is odd then $a_n = 1 + (-1) = 0$; if n is even then $a_n = 1 + 1 = 2$. Therefore the sequence $\{a_n\}$ diverges. To prove this, we appeal to the definition of limit of a sequence given in Section 10.2. Suppose that $\{a_n\}$ converges to the number L. Let $\epsilon = \frac{1}{2}$ and suppose that N is a positive integer.

Case 1: $L \geq 1$. Then choose $n \geq N$ such that n is odd. Then $a_n = 0$, so

$$|a_n - L| = |0 - L| = L \geq 1 > \epsilon.$$

Case 2: $L < 1$. Then choose $n \geq N$ such that n is even. Then $a_n = 2$, so

$$|a_n - L| = |2 - L| = 2 - L > 1 > \epsilon.$$

No matter what the value of L, it cannot be made to fit the definition of the limit of the sequence $\{a_n\}$. Therefore the sequence $\{a_n\} = \{1 + (-1)^n\}$ has no limit. (We can't even say that it approaches $+\infty$ or $-\infty$; it does not.)

10.2.17: We use l'Hôpital's rule for sequences (Eq. (9) of Section 10.2):

$$\lim_{n \to \infty} a_n = \lim_{n \to \infty} \frac{1 + (-1)^n \sqrt{n}}{\left(\frac{3}{2} \right)^n} = \lim_{x \to \infty} \frac{1 \pm x^{1/2}}{\left(\frac{3}{2} \right)^x} = \pm \lim_{x \to \infty} \frac{1}{2x^{1/2} \left(\frac{3}{2} \right)^x \ln \left(\frac{3}{2} \right)} = 0.$$

10.2.19: First we need a lemma.

Lemma: If $r > 0$, then $\lim\limits_{n\to\infty} \dfrac{1}{n^r} = 0$.

Proof: Suppose that $r > 0$. Given $\epsilon > 0$, let $N = 1 + [\![1/\epsilon^{1/r}]\!]$. Then N is a positive integer, and if $n > N$, then $n > 1/\epsilon^{1/r}$, so that $n^r > 1/\epsilon$. Therefore

$$\left| \frac{1}{n^r} - 0 \right| < \epsilon.$$

Thus, by definition, $\lim\limits_{n\to\infty} \dfrac{1}{n^r} = 0$. ◀

Next, we use the squeeze law for sequences (Theorem 3 of Section 10.2): $-1 \leq \sin n \leq 1$ for all integers $n \geq 1$, and therefore

$$0 \leq \frac{\sin^2 n}{\sqrt{n}} \leq \frac{1}{\sqrt{n}}$$

for all integers $n \geq 1$. But by the preceding lemma, $1/\sqrt{n} \to 0$ as $n \to +\infty$. Therefore

$$\lim_{n\to\infty} \frac{\sin^2 n}{\sqrt{n}} = 0.$$

10.2.21: If n is a positive integer, then $\sin \pi n = 0$. Therefore $a_n = 0$ for every integer $n \geq 0$. So $\{a_n\} \to 0$ as $n \to +\infty$.

10.2.23: Suppose that $a > 0$. Then $f(x) = a^x$ is continuous, so $a^x \to 1 = a^0$ as $x \to 0$. But $-\dfrac{\sin n}{n} \to 0$ as $n \to +\infty$. Therefore $\lim\limits_{n\to\infty} \pi^{-(\sin n)/n} = 1$.

10.2.25: We use l'Hôpital's rule for sequences (Eq. (9)):

$$\lim_{n\to\infty} \frac{\ln n}{\sqrt{n}} = \lim_{x\to\infty} \frac{\ln x}{x^{1/2}} = \lim_{x\to\infty} \frac{2x^{1/2}}{x} = \lim_{x\to\infty} \frac{2}{x^{1/2}} = 0.$$

10.2.27: We use l'Hôpital's rule for sequences (Eq. (9)):

$$\lim_{n\to\infty} \frac{(\ln n)^2}{n} = \lim_{x\to\infty} \frac{(\ln x)^2}{x} = \lim_{x\to\infty} \frac{2\ln x}{x} = \lim_{x\to\infty} \frac{2}{x} = 0.$$

10.2.29: Because $-\pi/2 < \tan^{-1} x < \pi/2$ for all x,

$$-\frac{\pi}{2n} < \frac{\tan^{-1} n}{n} < \frac{\pi}{2n}$$

for all integers $n \geq 1$. Therefore, by the squeeze law for limits, $\lim\limits_{n\to\infty} \dfrac{\tan^{-1} n}{n} = 0$.

10.2.31: We use the squeeze law for limits of sequences:

$$0 < \frac{2^n + 1}{e^n} < \frac{2^n + 2^n}{e^n} = 2 \left(\frac{2}{e} \right)^n.$$

By the result in Example 9, $(2/e)^n \to 0$ as $n \to +\infty$. Therefore $\lim\limits_{n\to\infty} \dfrac{2^n+1}{e^n} = 0$.

10.2.33: By Eq. (3) in Section 4.9,

$$\lim_{n\to\infty}\left(1+\frac{1}{n}\right)^n = e.$$

10.2.35: We use l'Hôpital's rule for sequences:

$$\lim_{n\to\infty}\left(\frac{n-1}{n+1}\right)^n = \lim_{n\to\infty}\exp\left(x\ln\frac{x-1}{x+1}\right) = \exp\left(\lim_{x\to\infty}\frac{\ln(x-1)-\ln(x+1)}{\frac{1}{x}}\right)$$

$$= \exp\left(\lim_{x\to\infty}\frac{\dfrac{1}{x-1}-\dfrac{1}{x+1}}{-\dfrac{1}{x^2}}\right) = \exp\left(\lim_{x\to\infty}\frac{-2x^2}{x^2-1}\right) = e^{-2}.$$

10.2.37: Let $f(x) = 2^x$ and note that f is continuous on the set of all real numbers. Thus

$$\lim_{n\to\infty} a_n = \lim_{x\to\infty} f\left(\frac{x+1}{x}\right) = f\left(\lim_{x\to\infty}\frac{x+1}{x}\right) = f(1) = 2^1 = 2.$$

It is the continuity of f at $x = 1$ that makes the second equality valid.

10.2.39: By Example 11, $\lim\limits_{n\to\infty} n^{1/n} = 1$. Thus by Example 7,

$$\lim_{n\to\infty}\left(\frac{2}{n}\right)^{3/n} = \lim_{n\to\infty}\frac{8^{1/n}}{\left(n^{1/n}\right)^3} = \frac{1}{1^3} = 1.$$

10.2.41: Given: $a_n = \left(\dfrac{2-n^2}{3+n^2}\right)^n$. First note that

$$\lim_{n\to\infty}\frac{2-n^2}{3+n^2} = \lim_{n\to\infty}\frac{\dfrac{2}{n^2}-1}{\dfrac{3}{n^2}+1} = \frac{0-1}{0+1} = -1.$$

Moreover,

$$\lim_{n\to\infty}\left(\frac{n^2-2}{n^2+3}\right)^n = \lim_{x\to\infty}\exp\left(\frac{\ln(x^2-2)-\ln(x^2+3)}{\frac{1}{x}}\right) = \exp\left(\lim_{x\to\infty}\frac{\dfrac{2x}{x^2-2}-\dfrac{2x}{x^2+3}}{-\dfrac{1}{x^2}}\right)$$

$$= \exp\left(\lim_{x\to\infty}\frac{-x^2(2x^3+6x-2x^3+4x)}{(x^2-2)(x^2+3)}\right) = \exp\left(\lim_{x\to\infty}\frac{-10x^3}{(x^2-2)(x^2+3)}\right) = e^0 = 1.$$

Therefore, as $n \to +\infty$ through *even* values, $a_n \to 1$, whereas as $n \to +\infty$ through *odd* values, $a_n \to -1$. So we can now show that the sequence $\{a_n\}$ has no limit as $n \to \infty$.

Let $\epsilon = \frac{1}{10}$. Choose N_1 so large that if $n > N_1$ and n is even, then $|a_n - 1| < \epsilon$. Choose N_2 so large that if $n > N_2$ and n is odd, then $|a_n - (-1)| < \epsilon$. Let N be the maximum of N_1 and N_2. Then if $n > N$,

$$|a_n - 1| < \epsilon \quad \text{if } n \text{ is even};$$

$$|a_n - (-1)| < \epsilon \quad \text{if } n \text{ is odd}.$$

Put another way, if $n > N$ and n is even, then a_n lies in the interval $(0.9, 1.1)$. If $n > N$ and n is odd, then a_n lies in the interval $(-1.1, -0.9)$. It follows that no interval of length 0.2 can contain a_n for all $n > K$, no matter how large K might be. Because every real number L is the midpoint of such an interval, this means that no real number L can be the limit of the sequence $\{a_n\}$. Therefore

$$\lim_{n \to \infty} a_n \quad \text{does not exist.}$$

10.2.43: Let $f(n) = \dfrac{n-2}{n+13}$. Then the *Mathematica* command

```
Table[ { n, N[f[10^n]] }, { n, 1, 7 } ]
```

yielded the response

$$\{\{1, 0.347826\}, \{2, 0.867257\}, \{3, 0.985192\}, \{4, 0.998502\}, \{5, 0.99985\}, \{6, 0.999985\}, \{7, 0.999999\}\}.$$

Indeed,

$$\lim_{n \to \infty} \frac{n-2}{n+13} = \lim_{n \to \infty} \frac{1 - \dfrac{2}{n}}{1 + \dfrac{13}{n}} = \frac{1-0}{1+0} = 1.$$

10.2.45: Let $f(n) = a_n = \sqrt{\dfrac{4n^2 + 7}{n^2 + 3n}}$. Results of our experiment:

n	$f(n)$
10	1.76940
100	1.97083
1000	1.99701
10000	1.99970
100000	1.99997
1000000	2.00000

By Theorem 4 and l'Hôpital's rule (used twice),

$$\lim_{n \to \infty} \frac{4n^2 + 7}{n^2 + 3n} = \lim_{x \to \infty} \frac{4x^2 + 7}{x^2 + 3x} = \lim_{x \to \infty} \frac{8x}{2x + 3} = \lim_{x \to \infty} \frac{8}{2} = 4.$$

Therefore, by Theorem 2, $\lim_{n \to \infty} a_n = \sqrt{4} = 2$.

10.2.47: Let $f(n) = a_n = \exp\left(-1/\sqrt{n}\right)$. Results of an experiment:

n	$f(n)$
10	0.728893
100	0.904837
1000	0.968872
10000	0.990050
100000	0.996843
1000000	0.999000
10000000	0.999684
100000000	0.999900
1000000000	0.999968

Because $-1/\sqrt{n} \to 0$ as $n \to +\infty$ and $g(x) = e^x$ is continuous at $x = 0$, Theorem 2 implies that

$$\lim_{n \to \infty} a_n = g(0) = 1.$$

10.2.49: Let $f(n) = a_n = 4\tan^{-1} \dfrac{n-1}{n+1}$. The results of an experiment:

n	$f(n)$
10	2.74292
100	3.10159
1000	3.13759
10000	3.14119
100000	3.14155
1000000	3.14159
10000000	3.14159

The limit appears to be π. Because

$$\lim_{n \to \infty} \frac{n-1}{n+1} = \lim_{n \to \infty} \left(1 - \frac{2}{n+1}\right) = 1 - 0 = 1$$

and $g(x) = 4 \tan^{-1} x$ is continuous at $x = 1$, Theorem 2 implies that

$$\lim_{n \to \infty} a_n = g(1) = 4 \tan^{-1}(1) = 4 \cdot \frac{\pi}{4} = \pi.$$

10.2.51: Proof: Suppose that

$$\lim_{n \to \infty} a_n = A \neq 0.$$

Without loss of generality we may suppose that $A > 0$. Let $\epsilon = A/3$ and choose N so large that if $n > N$, then $|a_n - A| < \epsilon$. Then if n is even, $(-1)^n a_n = a_n$; in this case $|(-1)^n a_n - A| < \epsilon$ if $n > N$. If n is odd, then $(-1)^n a_n = -a_n$; in this case $|(-1)^n a_n - (-A)| < \epsilon$ if $n > N$. In other words, $(-1)^n a_n$ lies in the interval $I = (A - \epsilon, A + \epsilon)$ if n is even, whereas $(-1)^n a_n$ lies in the interval $J = (-A - \epsilon, -A + \epsilon)$ if n is odd. This means that no open interval of length 2ϵ can contain every number $(-1)^n a_n$ for which $n > K$, no matter how large the value of K. (Note that no such interval can contains points of both I and J because the distance between their closest endpoints is 4ϵ.) Because every real number is the midpoint of an open interval of length 2ϵ, it now follows that no real number can be the limit of the sequence $\{(-1)^n a_n\}$. This concludes the proof. ◀

10.2.53: Given: $A > 0$, $x_1 \neq 0$,

$$x_{n+1} = \frac{1}{2}\left(x_n + \frac{A}{x_n}\right) \quad \text{if} \quad n \geq 1, \quad \text{and} \quad L = \lim_{n \to \infty} x_n.$$

Then

$$\lim_{n \to \infty} x_{n+1} = L \quad \text{and} \quad \lim_{n \to \infty} \frac{1}{2}\left(x_n + \frac{A}{x_n}\right) = \frac{1}{2}\left(L + \frac{A}{L}\right).$$

It follows that

$$L = \frac{1}{2}\left(L + \frac{A}{L}\right); \qquad 2L = \frac{L^2 + A}{L};$$

$$2L^2 = L^2 + A; \qquad L^2 = A.$$

Therefore $L = \pm\sqrt{A}$.

10.2.55: Part (a): Note first that $F_1 = 1$ and $F_2 = 1$. If $n \geq 3$, then F_{n-1} is the total number of pairs present in the preceding month and F_{n-2} is the total number of productive pairs. Therefore $F_n = F_{n-1} + F_{n-2}$; that is, $F_{n+1} = F_n + F_{n-1}$ for $n \geq 2$. So $\{F_n\}$ is the Fibonacci sequence of Example 2.

Part (b): Note first that $G_1 = G_2 = G_3 = 1$. If $n \geq 4$, then G_{n-1} is the total number of pairs present in the preceding month and G_{n-3} is the total number of productive pairs. Therefore $G_n = G_{n-1} + G_{n-3}$; that is, $G_{n+1} = G_n + G_{n-2}$. The *Mathematica* commands

```
g[1] = 1; g[2] = 1; g[3] = 1;
g[n_] := g[n] = g[n - 1] + g[n - 3]
```

serve as one way to enter the formula for the recursively defined function g. Then the command

```
Table[ {n, g[n]}, {n, 4, 25} ]
```

produces the output

$\{\{4,2\}, \{5,3\}, \{6,4\}, \{7,6\}, \{8,9\}, \{9,13\}, \{10,19\}, \{11,28\}, \{12,41\},$

$\{13,60\}, \{14,88\}, \{15,129\}, \{16,189\}, \{17,277\}, \{18,406\}, \{19,595\},$

$\{20,872\}, \{21,1278\}, \{22,1873\}, \{23,2745\}, \{24,4023\}, \{25,5896\}\}$

10.2.57: **Part (a):** Clearly $a_1 < 4$. Suppose that $a_k < 4$ for some integer $k \geq 1$. Then

$$a_{k+1} = \frac{1}{2}(a_k + 4) < \frac{1}{2}(4 + 4) = 4.$$

Therefore, by induction, $a_n < 4$ for every integer $n \geq 1$. Next, $a_2 = 3$, so that $a_1 < a_2$. Suppose that $a_k < a_{k+1}$ for some integer $k \geq 1$. Then

$$a_k + 4 < a_{k+1} + 4; \qquad \frac{1}{2}(a_k + 4) < \frac{1}{2}(a_{k+1} + 4); \qquad a_{k+1} < a_{k+2}.$$

Therefore, by induction, $a_n < a_{n+1}$ for every integer $n \geq 1$.

Part (b): Part (a) establishes that $\{a_n\}$ is a bounded increasing sequence. Therefore the bounded monotonic sequence property of Section 11.2 implies that the sequence $\{a_n\}$ converges. Let L denote its limit. Then

$$L = \lim_{n \to \infty} a_{n+1} = \lim_{n \to \infty} \frac{1}{2}(a_n + 4) = \frac{1}{2}(L + 4).$$

It now follows immediately that $L = 4$.

10.2.59: Given the positive real number r, let

$$L = \frac{1 + \sqrt{1 + 4r}}{2}.$$

Define the sequence $\{a_n\}$ recursively as follows: $a_1 = \sqrt{r}$, and for each integer $n \geq 1$, $a_{n+1} = \sqrt{r + a_n}$. We plan to show first that $\{a_n\}$ is a bounded sequence, then that $\{a_n\}$ is an increasing sequence. Only then will we attempt to evaluate its limit, because our method for doing so depends on knowing that the sequence $\{a_n\}$ converges.

First,

$$a_1 = \sqrt{r} = \frac{\sqrt{4r}}{2} < \frac{1 + \sqrt{1+4r}}{2} = L.$$

Suppose that $a_k < L$ for some integer $k \geq 1$. Then

$$a_k < \frac{1 + \sqrt{1+4r}}{2}; \qquad a_k + r < L + r;$$

$$4(a_k + r) < 2 + 2\sqrt{1+4r} + 4r; \qquad 4(a_k + r) < \left(1 + \sqrt{1+4r}\right)^2;$$

$$a_k + r < L^2; \qquad a_{k+1} = \sqrt{a_k + r} < L.$$

Therefore, by induction, $a_n < L$ for all $n \geq 1$.

Next, $0 < \sqrt{r}$, so that $r < r + \sqrt{r}$. Thus $\sqrt{r} < \sqrt{r + \sqrt{r}}$. That is, $a_1 < a_2$. Suppose that $a_k < a_{k+1}$ for some integer $k \geq 1$. Then

$$r + a_k < r + a_{k+1}; \qquad \sqrt{r + a_k} < \sqrt{r + a_{k+1}}; \qquad a_{k+1} < a_{k+2}.$$

Therefore, by induction, $a_n < a_{n+1}$ for all $n \geq 1$.

Now that we know that the sequence $\{a_n\}$ converges, we may denote its limit by M. Then

$$M = \lim_{n \to \infty} a_{n+1} = \lim_{n \to \infty} \sqrt{r + a_n} = \sqrt{r + M}.$$

It now follows that $M^2 - M - r = 0$, and thus that

$$M = \frac{1 \pm \sqrt{1+4r}}{2}.$$

Because $M > 0$, we conclude that $\displaystyle\lim_{n \to \infty} a_n = \frac{1 + \sqrt{1+4r}}{2} = L$. And in conclusion, if we take $r = 20$ we then find that

$$\sqrt{20 + \sqrt{20 + \sqrt{20 + \sqrt{20 + \cdots}}}} = L = 5.$$

10.2.61: Suppose that $\{a_n\}$ is a bounded monotonic sequence; without loss of generality, suppose that it is an increasing sequence. Because the set of values of a_n is a nonempty set of real numbers with an upper bound, it has a least upper bound λ. We claim that λ is the limit of the sequence $\{a_n\}$.

Let $\epsilon > 0$ be given. Then there exists a positive integer N such that

$$\lambda - \epsilon < a_N \leq \lambda.$$

For if not, then $\lambda - \epsilon$ would be an upper bound for $\{a_n\}$ smaller than λ, its least upper bound. But because $\{a_n\}$ is an increasing sequence with upper bound λ, it now follows that if $n > N$, then $a_N \leq a_n \leq \lambda$. Thus if $n > N$, then $|a_n - \lambda| < \epsilon$. Therefore, by definition, λ is the limit of the sequence $\{a_n\}$.

10.2.63: For each integer $n \geqq$, let a_n be the largest integral multiple of $1/10^n$ such that $a_n^2 \leqq 2$. (For example, $a_1 = 1.4$, $a_2 = 1.41$, and $a_3 = 1.414$.)

Part (a): First note that the numbers 1 and $\frac{3}{2}$ are multiples of $1/10^n$ (for each $n \geqq 1$) with $1^2 < 2$ and $\left(\frac{3}{2}\right)^2 > 2$. It follows that $1 \leqq a_n \leqq \frac{3}{2}$ for each integer $n \geqq 1$, and therefore the sequence $\{a_n\}$ is bounded. Next, a_n as an integral multiple of $1/10^n$ is also an integral multiple of $1/10^{n+1}$ whose square does not exceed 2. But a_{n+1} is the *largest* multiple of $1/10^{n+1}$ whose square does not exceed 2; it follows that $a_n \leqq a_{n+1}$, and thus the sequence $\{a_n\}$ is also an increasing sequence.

Part (b): Because $\{a_n\}$ is a bounded increasing sequence, it has a limit A. Then the limit laws give

$$A^2 = \left(\lim_{n\to\infty} a_n\right)^2 = \lim_{n\to\infty} (a_n)^2 \leqq \lim_{n\to\infty} 2 = 2,$$

so we see that $A^2 \leqq 2$.

Part (c): Assume that $A^2 < 2$. Then $2 - A^2 > 0$. Choose the integer k so large that $4/10^k \leqq 2 - A^2$. Then

$$\left(a_k + \frac{1}{10^k}\right)^2 = a_k^2 + \frac{2a_k}{10^k} + \frac{1}{10^{2k}}$$

$$< a_k^2 + \frac{4}{10^k} \quad \left(\text{because } a_k < \frac{3}{2} \text{ and } \frac{1}{10^{2k}} < \frac{1}{10^k}\right)$$

$$\leqq A^2 + (2 - A^2) = 2.$$

Thus the assumption that $A^2 < 2$ implies that $(a_k + 1/10^k)^2 < 2$, which contradicts the fact that a_k is, by definition, the largest integral multiple of $1/10^k$ whose square does not exceed 2. It therefore follows that A^2 is *not* less than 2; that is, that $A^2 \geqq 2$.

Part (d): It follows immediately from the results in parts (c) and (d) that $A^2 = 2$.

Section 10.3

10.3.1: The series is geometric with first term 1 and ratio $\frac{1}{3}$. Therefore it converges to

$$\frac{1}{1 - \dfrac{1}{3}} = \frac{3}{2}.$$

10.3.3: This series diverges by the nth-term test. Alternatively, you can show by induction that

$$S_k = \sum_{n=1}^{k} (2n - 1) = k^2,$$

569

so this series diverges because $\lim\limits_{k \to \infty} S_k = +\infty$.

10.3.5: This series is geometric but its ratio is -2 and $|-2| > 1$. Therefore the given series diverges. Alternatively, it diverges by the nth-term test for divergence.

10.3.7: The given series is geometric with first term 4 and ratio $\frac{1}{3}$. Therefore it converges to

$$\frac{4}{1 - \dfrac{1}{3}} = 6.$$

10.3.9: This series is geometric with first term 1 and ratio $r = 1.01$. Because $|r| > 1$, the series diverges. Alternatively, you can apply the nth-term test for divergence.

10.3.11: The given series diverges by the nth-term test:

$$\lim_{n \to \infty} \frac{n}{n+1} = 1, \quad \text{and therefore} \quad \lim_{n \to \infty} \frac{(-1)^n n}{n+1} \neq 0.$$

10.3.13: The given series is geometric with first term 1 and ratio $r = -3/e$. It diverges because $|r| > 1$.

10.3.15: The given series is geometric with first term 1 and ratio $1/\sqrt{2}$. Therefore its sum is

$$\frac{1}{1 - \dfrac{1}{\sqrt{2}}} = \frac{\sqrt{2}}{\sqrt{2} - 1} = \frac{\sqrt{2}\left(\sqrt{2} + 1\right)}{2 - 1} = 2 + \sqrt{2} \approx 3.414213562373.$$

10.3.17: Because the limit of the nth term is

$$\lim_{n \to \infty} \frac{n}{10n + 17} = \lim_{n \to \infty} \frac{1}{10 + \dfrac{17}{n}} = \frac{1}{10} \neq 0,$$

the given series diverges.

10.3.19: The given series is the difference of two convergent geometric series, so its sum is

$$\sum_{n=1}^{\infty} \left(5^{-n} - 7^{-n}\right) = \sum_{n=1}^{\infty} \frac{1}{5^n} - \sum_{n=1}^{\infty} \frac{1}{7^n} = \frac{\dfrac{1}{5}}{1 - \dfrac{1}{5}} - \frac{\dfrac{1}{7}}{1 - \dfrac{1}{7}} = \frac{1}{4} - \frac{1}{6} = \frac{1}{12}.$$

10.3.21: The series is geometric with first term e/π and ratio $r = e/\pi$. Because $|r| < 1$, the series converges to

$$\frac{\dfrac{e}{\pi}}{1 - \dfrac{e}{\pi}} = \frac{e}{\pi - e} \approx 6.421479600999.$$

10.3.23: The given series is geometric with ratio $r = \frac{100}{99}$. But its first term is nonzero and $|r| > 1$, so the series diverges.

10.3.25: The given series is the sum of three convergent geometric series, and its sum is

$$\sum_{n=0}^{\infty} \frac{1 + 2^n + 3^n}{5^n} = \sum_{n=0}^{\infty} \frac{1}{5^n} + \sum_{n=0}^{\infty} \left(\frac{2}{5}\right)^n + \sum_{n=0}^{\infty} \left(\frac{3}{5}\right)^n$$

$$= \frac{1}{1 - \frac{1}{5}} + \frac{1}{1 - \frac{2}{5}} + \frac{1}{1 - \frac{3}{5}} = \frac{5}{4} + \frac{5}{3} + \frac{5}{2} = \frac{65}{12} \approx 5.41666666667.$$

10.3.27: We use both parts of Theorem 2:

$$\sum_{n=0}^{\infty} \frac{7 \cdot 5^n + 3 \cdot 11^n}{13^n} = 7 \cdot \left[\sum_{n=0}^{\infty} \left(\frac{5}{13}\right)^n\right] + 3 \cdot \left[\sum_{n=0}^{\infty} \left(\frac{11}{13}\right)^n\right]$$

$$= 7 \cdot \frac{1}{1 - \frac{5}{13}} + 3 \cdot \frac{1}{1 - \frac{11}{13}} = \frac{91}{8} + \frac{39}{2} = \frac{247}{8} = 30.875.$$

10.3.29: The given series converges because it is the difference of two convergent geometric series. Its sum is

$$\sum_{n=1}^{\infty} \left[\left(\frac{7}{11}\right)^n - \left(\frac{3}{5}\right)^n\right] = \sum_{n=1}^{\infty} \left(\frac{7}{11}\right)^n - \sum_{n=1}^{\infty} \left(\frac{3}{5}\right)^n = \frac{\frac{7}{11}}{1 - \frac{7}{11}} - \frac{\frac{3}{5}}{1 - \frac{3}{5}} = \frac{7}{4} - \frac{3}{2} = \frac{1}{4}.$$

10.3.31: The given series diverges by the nth-term test for divergence, because

$$\lim_{n \to \infty} \frac{n^2 - 1}{3n^2 - 1} = \lim_{n \to \infty} \frac{1 - \frac{1}{n^2}}{3 - \frac{1}{n^2}} = \frac{1 - 0}{3 - 0} = \frac{1}{3} \neq 0.$$

10.3.33: The given series is geometric with nonzero first term and ratio $r = \tan 1 \approx 1.557407724655$. Because $|r| > 1$, this series diverges.

10.3.35: This is a geometric series with first term $\pi/4$ and ratio $r = \pi/4 \approx 0.785398163397$. Because $|r| < 1$, it converges; its sum is

$$\frac{\frac{\pi}{4}}{1 - \frac{\pi}{4}} = \frac{\pi}{4 - \pi} \approx 3.659792366325.$$

10.3.37: A figure similar to Fig. 10.3.4 of the text shows that if n is an integer and $n \geq 2$, then

$$\frac{1}{n \ln n} \geq \int_n^{n+1} \frac{1}{x \ln x} \, dx.$$

Therefore the kth partial sum S_k of the given series satisfies the equalities and inequalities

$$S_k = \sum_{n=2}^{k} \frac{1}{x \ln x} \geqq \int_{2}^{k+1} \frac{1}{x \ln x}\, dx = \left[\ln(\ln x) \right]_{2}^{k+1} = \ln(\ln(k+1)) - \ln(\ln 2) = \ln\left(\frac{\ln(k+1)}{\ln 2} \right).$$

Therefore the given series diverges because $\{ S_k \} \to +\infty$ as $k \to +\infty$.

10.3.39: Here we have

$$0.47\,47\,47\,47 \cdots = \frac{47}{100} + \frac{47}{10000} + \frac{47}{1000000} + \cdots = \frac{\dfrac{47}{100}}{1 - \dfrac{1}{100}} = \frac{47}{99}.$$

10.3.41: $0.123\,123\,123 \cdots = \dfrac{123}{1000} + \dfrac{123}{1000000} + \dfrac{123}{1000000000} + \cdots = \dfrac{\dfrac{123}{1000}}{1 - \dfrac{1}{1000}} = \dfrac{123}{999} = \dfrac{41}{333}.$

10.3.43: As in Example 7,

$$3.14159\,14159\,14159 \cdots = 3 + \frac{14159}{10^5} + \frac{14150}{10^{10}} + \frac{14159}{10^{15}} + \cdots = 3 + \frac{\dfrac{14159}{100000}}{1 - \dfrac{1}{100000}}$$

$$= 3 + \frac{14159}{99999} = \frac{299997 + 14159}{99999} = \frac{314156}{99999}.$$

10.3.45: The series is geometric with ratio $x/3$. Thus it will converge when

$$\left| \frac{x}{3} \right| < 1; \quad \text{that is, when} \quad -3 < x < 3.$$

For such x, we have

$$\sum_{n=1}^{\infty} \left(\frac{x}{3} \right)^n = \frac{\dfrac{x}{3}}{1 - \dfrac{x}{3}} = \frac{x}{3 - x}.$$

10.3.47: The given series is geometric with ratio $(x-2)/3$. Hence it will converge when

$$\left| \frac{x-2}{3} \right| < 1; \quad \text{that is, when} \quad -1 < x < 5.$$

For such x, we have

$$\sum_{n=1}^{\infty} \left(\frac{x-2}{3} \right)^n = \frac{\dfrac{x-2}{3}}{1 - \dfrac{x-2}{3}} = \frac{x-2}{3 - (x-2)} = \frac{x-2}{5 - x}.$$

10.3.49: This series is geometric with ratio $5x^2/(x^2 + 16)$. Hence it will converge when

$$\frac{5x^2}{x^2 + 16} < 1 : \quad 5x^2 < x^2 + 16;$$

$$4x^2 < 16;$$

$$x^2 < 4;$$

$$-2 < x < 2.$$

For such x we have

$$\sum_{n=1}^{\infty} \left(\frac{5x^2}{x^2+16} \right)^n = \frac{\dfrac{5x^2}{x^2+16}}{1 - \dfrac{5x^2}{x^2+16}} = \frac{5x^2}{x^2+16-5x^2} = \frac{5x^2}{16-4x^2}.$$

10.3.51: The method of partial fractions yields

$$\frac{1}{9n^2+3n-2} = \frac{1}{3}\left(\frac{1}{3n-1} + \frac{-1}{3n+2} \right).$$

Therefore the kth partial sum of the given series is

$$S_k = \sum_{n=1}^{k} \frac{1}{9n^2+3n-2} = \frac{1}{3}\left(\frac{1}{2} - \frac{1}{5} + \frac{1}{5} - \frac{1}{8} + \frac{1}{8} - \frac{1}{11} + \cdots - \frac{1}{3k+2} \right) = \frac{1}{3}\left(\frac{1}{2} - \frac{1}{3k+2} \right).$$

Thus

$$\sum_{n=1}^{\infty} \frac{1}{9n^2+3n-2} = \lim_{k\to\infty} S_k = \frac{1}{6}.$$

10.3.53: The method of partial fractions yields

$$\frac{1}{16n^2-8n-3} = \frac{1}{4}\left(\frac{1}{4n-3} - \frac{1}{4n+1} \right).$$

Thus the kth partial sum of the given series is

$$S_k = \sum_{n=1}^{k} \frac{1}{16n^2-8n-3} = \frac{1}{4}\left(1 - \frac{1}{5} + \frac{1}{5} - \frac{1}{9} + \frac{1}{9} - \frac{1}{13} + \cdots - \frac{1}{4k+1} \right) = \frac{1}{4}\left(1 - \frac{1}{4k+1} \right).$$

Therefore

$$\sum_{n=1}^{\infty} \frac{1}{16n^2-8n-3} = \lim_{k\to\infty} S_k = \frac{1}{4}.$$

10.3.55: The method of partial fractions yields

$$\frac{1}{n^2-1} = \frac{1}{2}\left(\frac{1}{n-1} - \frac{1}{n+1} \right).$$

So the kth partial sum of the given series is

$$S_k = \sum_{n=2}^{k} \frac{1}{n^2 - 1}$$

$$= \frac{1}{2}\left(1 - \frac{1}{3} + \frac{1}{2} - \frac{1}{4} + \frac{1}{3} - \frac{1}{5} + \frac{1}{4} - \frac{1}{6} + \frac{1}{5} - \frac{1}{7} + \cdots + \frac{1}{k-2} - \frac{1}{k} + \frac{1}{k-1} - \frac{1}{k+1}\right)$$

$$= \frac{1}{2}\left(1 + \frac{1}{2} - \frac{1}{k} - \frac{1}{k+1}\right).$$

Thus the sum of the given series is $\displaystyle\lim_{k\to\infty} S_k = \frac{3}{4}$.

10.3.57: In *Derive* application of the command **Expand** to the nth term of the given series yields the partial fraction decomposition

$$\frac{6n^2 + 2n - 1}{n(n+1)(4n^2-1)} = \frac{1}{n} - \frac{1}{n+1} + \frac{1}{2n-1} - \frac{1}{2n+1}.$$

So the kth partial sum of the given series is

$$S_k = \sum_{n=1}^{k} \frac{6n^2 + 2n - 1}{n(n+1)(4n^2-1)}$$

$$= \left(1 - \frac{1}{2} + 1 - \frac{1}{3}\right) + \left(\frac{1}{2} - \frac{1}{3} + \frac{1}{3} - \frac{1}{5}\right) + \left(\frac{1}{3} - \frac{1}{4} + \frac{1}{5} - \frac{1}{7}\right) + \left(\frac{1}{4} - \frac{1}{5} + \frac{1}{7} - \frac{1}{9}\right)$$

$$= \left(\frac{1}{5} - \frac{1}{6} + \frac{1}{9} - \frac{1}{11}\right) + \left(\frac{1}{6} - \frac{1}{7} + \frac{1}{11} - \frac{1}{13}\right) + \cdots + \left(\frac{1}{k} - \frac{1}{k+1} + \frac{1}{2k-1} - \frac{1}{2k+1}\right)$$

$$= 1 - \frac{1}{k+1} + 1 - \frac{1}{2k+1}.$$

Therefore the sum of the given series is $\displaystyle\lim_{k\to\infty} S_k = 2$.

10.3.59: In *Mathematica* the command

```
Apart[ 6/(n*(n + 1)*(n + 2)*(n + 3)) ]
```

yields the partial fraction decomposition

$$\frac{6}{n(n+1)(n+2)(n+3)} = \frac{1}{n} - \frac{3}{n+1} + \frac{3}{n+2} - \frac{1}{n+3}.$$

Therefore the kth partial sum of the given series is

$$S_k = \sum_{n=1}^{k} \frac{6}{n(n+1)(n+2)(n+3)} = 1 - \frac{3}{2} + \frac{3}{3} - \frac{1}{4}$$

$$+ \frac{1}{2} - \frac{3}{3} + \frac{3}{4} - \frac{1}{5}$$

$$+ \frac{1}{3} - \frac{3}{4} + \frac{3}{5} - \frac{1}{6}$$

$$+ \frac{1}{4} - \frac{3}{5} + \frac{3}{6} - \frac{1}{7}$$

$$+ \frac{1}{5} - \frac{3}{6} + \frac{3}{7} - \frac{1}{8}$$

$$+ \frac{1}{6} - \frac{3}{7} + \frac{3}{8} - \frac{1}{9}$$

$$+ \cdots$$

$$+ \frac{1}{k-4} - \frac{3}{k-3} + \frac{3}{k-2} - \frac{1}{k-1}$$

$$+ \frac{1}{k-3} - \frac{3}{k-2} + \frac{3}{k-1} - \frac{1}{k}$$

$$+ \frac{1}{k-2} - \frac{3}{k-1} + \frac{3}{k} - \frac{1}{k+1}$$

$$+ \frac{1}{k-1} - \frac{3}{k} + \frac{3}{k+1} - \frac{1}{k+2}$$

$$+ \frac{1}{k} - \frac{3}{k+1} + \frac{3}{k+2} - \frac{1}{k+3}$$

Examine the diagonals that run from southwest to northeast. The four fractions with denominator 4 all cancel one another, as do those with denominators 5, 6, ..., $k-1$, and k. Thus

$$S_k = 1 - \frac{2}{2} + \frac{1}{3} - \frac{1}{k+1} + \frac{2}{k+2} - \frac{1}{k+3} = \frac{1}{3} - \frac{1}{k+1} + \frac{2}{k+2} - \frac{1}{k+3}.$$

Therefore the sum of the given series is $\displaystyle\lim_{k\to\infty} S_k = \frac{1}{3}$.

10.3.61: By part 2 of Theorem 2, if $c \neq 0$ and $\sum ca_n$ converges, then

$$\sum \frac{1}{c} \cdot ca_n = \sum a_n$$

converges. Therefore if $c \neq 0$ and $\sum a_n$ diverges, then $\sum ca_n$ diverges.

10.3.63: Let

$$S_n = \sum_{i=1}^{n} a_i \quad \text{and} \quad T_n = \sum_{i=1}^{n} b_i,$$

let k be a fixed positive integer, and suppose that $a_j = b_j$ for every integer $j \geq k$. If $n = k+1$, then

$$S_n - T_n = (S_k + a_n) - (T_k + b_n) = (S_k + a_n) - (T_k + a_n) = S_k - T_k.$$

Assume that $S_n - T_n = S_k - T_k$ for some integer $n \geq k+1$. Then

$$S_{n+1} - T_{n+1} = (S_n + a_{n+1}) - (T_n + b_{n+1}) = (S_n + a_{n+1}) - (T_n + a_{n+1}) = S_n - T_n = S_k - T_k.$$

Therefore, by induction, $S_n - T_n = S_k - T_k$ for every integer $n \geq k + 1$.

10.3.65: The total time the ball spends bouncing is

$$T = \sqrt{2a/g} + 2\sqrt{2ar/g} + 2\sqrt{2ar^2/g} + 2\sqrt{2ar^3/g} + \cdots$$

$$= -\sqrt{2a/g} + 2\sqrt{2a/g}\left(1 + r^{1/2} + r + r^{3/2} + \cdots\right) = -\sqrt{2a/g} + 2\sqrt{2a/g}\left(\frac{1}{1 - r^{1/2}}\right)$$

$$= \sqrt{2a/g}\left(-1 + \frac{2}{1 - r^{1/2}}\right) = \sqrt{2a/g}\left(\frac{-1 + r^{1/2} + 2}{1 - r^{1/2}}\right) = \sqrt{2a/g}\left(\frac{1 + r^{1/2}}{1 - r^{1/2}}\right).$$

If we take $r = 0.64$, $a = 4$, and $g = 32$, we find the total bounce time to be

$$T = \sqrt{8/32}\left(\frac{1 + 0.8}{1 - 0.8}\right) = \frac{1}{2} \cdot \frac{1.8}{0.2} = 4.5 \quad \text{(seconds)}.$$

10.3.67: Let $r = 0.95$. Then $M_1 = r M_0$, $M_2 = r M_1 = r^2 M_0$, and so on; in the general case, $M_n = r^n M_0$. Because $-1 < r < 1$, it now follows that

$$\lim_{n \to \infty} M_n = \lim_{n \to \infty} r^n M_0 = 0.$$

10.3.69: Peter's probability of winning is the sum of:

The probability that he wins in the first round;

The probability that everyone tosses tails in the first round and Peter
  wins in the second round;

The probability that everyone tosses tails in the first two rounds and
  Peter wins in the third round;

Et cetera, et cetera, et cetera.

Thus his probability of winning is

$$\frac{1}{2} + \frac{1}{2^4} + \frac{1}{2^7} + \frac{1}{2^{10}} + \cdots = \frac{\dfrac{1}{2}}{1 - \dfrac{1}{8}} = \frac{4}{7}.$$

Similarly, the probability that Paul wins is

$$\frac{1}{2^2} + \frac{1}{2^5} + \frac{1}{2^8} + \frac{1}{2^{11}} + \cdots = \frac{\dfrac{1}{4}}{1 - \dfrac{1}{8}} = \frac{2}{7}$$

and the probability that Mary wins is

$$\frac{1}{2^3} + \frac{1}{2^6} + \frac{1}{2^9} + \frac{1}{2^{12}} + \cdots = \frac{\dfrac{1}{8}}{1 - \dfrac{1}{8}} = \frac{1}{7}$$

Note that the three probabilities have sum 1, as they should.

10.3.71: The amount of light transmitted is

$$\frac{I}{2^4} + \frac{I}{2^6} + \frac{I}{2^7} + \frac{I}{2^{10}} + \cdots = I \cdot \frac{\frac{1}{16}}{1 - \frac{1}{4}} = \frac{I}{12},$$

$\frac{1}{12}$ of the incident light.

Section 10.4

10.4.1: Because $f^{(n)}(x) = (-1)^n e^{-x}$, we see that $f^{(n)}(0) = (-1)^n$ if $n \geq 0$. Thus

$$P_5(x) = 1 - x + \frac{x^2}{2!} - \frac{x^3}{3!} + \frac{x^4}{4!} - \frac{x^5}{5!} \quad \text{and}$$

$$R_5(x) = \frac{x^6}{6!} e^{-z} \quad \text{for some } z \text{ between } 0 \text{ and } x.$$

10.4.3: Given $f(x) = \cos x$ and $n = 4$, we have

$$f'(x) = -\sin x \qquad f'(0) = 0$$

$$f''(x) = -\cos x \qquad f''(0) = -1$$

$$f^{(3)}(x) = \sin x \qquad f^{(3)}(0) = 0$$

$$f^{(4)}(x) = \cos x \qquad f^{(4)}(0) = 1$$

$$f^{(5)}(x) = -\sin x$$

Therefore

$$P_4(x) = 1 - \frac{x^2}{2!} + \frac{x^4}{4!} \quad \text{and} \quad R_4(x) = -\frac{x^5}{5!} \sin z$$

for some number z between 0 and x.

10.4.5: Given $f(x) = (1+x)^{1/2}$ and $n = 3$, we compute

$$f'(x) = \frac{1}{2(1+x)^{1/2}} \qquad f'(0) = \frac{1}{2}$$

$$f''(x) = -\frac{1}{4(1+x)^{3/2}} \qquad f''(0) = -\frac{1}{4}$$

577

$$f^{(3)}(x) = \frac{3}{8(1+x)^{5/2}} \qquad f^{(3)}(0) = \frac{3}{8}$$

$$f^{(4)}(x) = -\frac{15}{16(1+x)^{7/2}}$$

Therefore

$$P_3(x) = 1 + \frac{x}{2} - \frac{x^2}{8} + \frac{x^3}{16} \quad \text{and} \quad R_3(x) = -\frac{5x^4}{128(1+z)^{7/2}}$$

for some number z between 0 and x.

10.4.7: Given $f(x) = \tan x$ and $n = 3$, we find that

$$f'(x) = \sec^2 x \qquad\qquad f'(0) = 1$$

$$f''(x) = 2\sec^2 x \tan x \qquad\qquad f''(0) = 0$$

$$f^{(3)}(x) = 2\sec^4 x + 4\sec^2 x \tan^2 x \qquad\qquad f^{(3)}(0) = 2$$

$$f^{(4)}(x) = 16\sec^4 x \tan x + 8\sec^2 x \tan^3 x$$

Therefore

$$P_3(x) = x + \frac{x^3}{3} \quad \text{and} \quad R_3(x) = \frac{x^4}{4!}(16\sec^4 z \tan z + 8\sec^2 z \tan^3 z)$$

for some number z between 0 and x.

10.4.9: Given $f(x) = \arcsin x$ and $n = 2$, we compute

$$f'(x) = \frac{1}{\sqrt{1-x^2}} \qquad\qquad f'(0) = 1$$

$$f''(x) = \frac{x}{(1-x^2)^{3/2}} \qquad\qquad f''(0) = 0$$

$$f^{(3)}(x) = \frac{1+2x^2}{(1-x^2)^{5/2}}$$

Therefore

$$P_2(x) = x \quad \text{and} \quad R_2(x) = \frac{x^3(1+2z^2)}{3!(1-z^2)^{5/2}}$$

for some number z between 0 and x.

10.4.11: Because $f^{(n)}(x) = e^x$ for all $n \geq 0$, we have $f^{(n)}(1) = e$ for such n. Therefore

$$e^x = e + e(x-1) + \frac{e}{2}(x-1)^2 + \frac{e}{6}(x-1)^3 + \frac{e}{24}(x-1)^4 + \frac{e^z}{120}(x-1)^5$$

for some z between 1 and x.

578

10.4.13: Given: $f(x) = \sin x$, $a = \pi/6$, and $n = 3$. We compute

$$f'(x) = \cos x \qquad f'(a) = \frac{\sqrt{3}}{2}$$

$$f''(x) = -\sin x \qquad f''(a) = -\frac{1}{2}$$

$$f^{(3)}(x) = -\cos x \qquad f^{(3)}(a) = -\frac{\sqrt{3}}{2}$$

$$f^{(4)}(x) = \sin x$$

Therefore

$$\sin x = \frac{1}{2} + \frac{\sqrt{3}}{2}\left(x - \frac{\pi}{6}\right) - \frac{1}{4}\left(x - \frac{\pi}{6}\right)^2 - \frac{\sqrt{3}}{12}\left(x - \frac{\pi}{6}\right)^3 + \frac{\sin z}{24}\left(x - \frac{\pi}{6}\right)^4$$

for some number z between $\pi/6$ and x.

10.4.15: Given $f(x) = (x-4)^{-2}$, $a = 5$, and $n = 5$, we compute

$$f'(x) = -2(x-4)^{-3} \qquad f'(a) = -2$$

$$f''(x) = 6(x-4)^{-4} \qquad f''(a) = 6$$

$$f^{(3)}(x) = -24(x-4)^{-5} \qquad f^{(3)}(a) = -24$$

$$f^{(4)}(x) = 120(x-4)^{-6} \qquad f^{(4)}(a) = 120$$

$$f^{(5)}(x) = -720(x-4)^{-7} \qquad f^{(5)}(a) = -720$$

$$f^{(6)}(x) = 5040(x-4)^{-8}$$

Therefore

$$\frac{1}{(x-4)^2} = 1 - 2(x-5) + 3(x-5)^2 - 4(x-5)^3 + 5(x-4)^4 - 6(x-5)^5 + \frac{(x-5)^6}{720} \cdot \frac{5040}{(z-4)^8}$$

for some number z between 5 and x.

10.4.17: Given $f(x) = \cos x$, $a = \pi$, and $n = 4$, we compute

$$f(x) = \cos x \qquad f(a) = -1$$

$$f'(x) = -\sin x \qquad f'(a) = 0$$

$$f''(x) = -\cos x \qquad f''(a) = 1$$

$$f^{(3)}(x) = \sin x \qquad f^{(3)}(a) = 0$$

$$f^{(4)}(x) = \cos x \qquad f^{(4)}(a) = -1$$

$$f^{(5)}(x) = -\sin x$$

Therefore

$$\cos x = -1 + \frac{(x-\pi)^2}{2} - \frac{(x-\pi)^4}{24} - \frac{\sin z}{120}(x-\pi)^5$$

for some number z between π and x.

10.4.19: Given $f(x) = x^{3/2}$, $a = 1$, and $n = 4$, we compute

$$f(x) = x^{3/2} \qquad f(a) = 1$$

$$f'(x) = \frac{3}{2}x^{1/2} \qquad f'(a) = \frac{3}{2}$$

$$f''(x) = \frac{3}{4}x^{-1/2} \qquad f''(a) = \frac{3}{4}$$

$$f^{(3)}(x) = -\frac{3}{8}x^{-3/2} \qquad f^{(3)}(a) = -\frac{3}{8}$$

$$f^{(4)}(x) = \frac{9}{16}x^{-5/2} \qquad f^{(4)}(a) = \frac{9}{16}$$

$$f^{(5)}(x) = -\frac{45}{32}x^{-7/2}$$

Therefore

$$x^{3/2} = 1 + \frac{3}{2}(x-1) + \frac{3}{8}(x-1)^2 - \frac{1}{16}(x-1)^3 + \frac{3}{128}(x-1)^4 - \frac{(x-1)^5}{120} \cdot \frac{45}{32z^{7/2}}$$

for some number z between 1 and x.

10.4.21: Substitution of $-x$ for x in the series in Eq. (19) yields

$$e^{-x} = 1 - x + \frac{x^2}{2!} - \frac{x^3}{3!} + \frac{x^4}{4!} - \frac{x^5}{5!} + \cdots = \sum_{n=0}^{\infty} \frac{(-1)^n x^n}{n!}.$$

This representation is valid for all x.

10.4.23: Substitution of $-3x$ for x in the series in Eq. (19) yields

$$e^{-3x} = 1 - 3x + \frac{9x^2}{2!} - \frac{27x^3}{3!} + \frac{81x^4}{4!} - \frac{243x^5}{5!} + \cdots = \sum_{n=0}^{\infty} \frac{(-1)^n 3^n x^n}{n!}.$$

This representation is valid for all x.

10.4.25: Substitution of $2x$ for x in the series in Eq. (22) yields

$$\sin 2x = 2x - \frac{8x^3}{3!} + \frac{32x^5}{5!} - \frac{128x^7}{7!} + \frac{512x^9}{9!} - \cdots = \sum_{n=0}^{\infty} \frac{(-1)^n (2x)^{2n+1}}{(2n+1)!}.$$

This representation is valid for all x.

10.4.27: Substitution of x^2 for x in the series in Eq. (22) yields

$$\sin\left(x^2\right) = x^2 - \frac{x^6}{3!} + \frac{x^{10}}{5!} - \frac{x^{14}}{7!} + \frac{x^{18}}{9!} - \cdots = \sum_{n=0}^{\infty} \frac{(-1)^n x^{4n+2}}{(2n+1)!}.$$

This representation is valid for all x.

10.4.29: Given $f(x) = \ln(1+x)$ and $a = 0$, we compute:

$$f(x) = \ln(1+x) \qquad\qquad f(a) = 0$$

$$f'(x) = \frac{1}{1+x} \qquad\qquad f'(a) = 1$$

$$f''(x) = -\frac{1}{(1+x)^2} \qquad\qquad f''(a) = -1$$

$$f^{(3)}(x) = \frac{2}{(1+x)^3} \qquad\qquad f^{(3)}(a) = 2$$

$$f^{(4)}(x) = -\frac{6}{(1+x)^4} \qquad\qquad f^{(4)}(a) = -6$$

$$f^{(5)}(x) = \frac{24}{(1+x)^5} \qquad\qquad f^{(5)}(a) = 24$$

$$f^{(6)}(x) = -\frac{120}{(1+x)^6} \qquad\qquad f^{(6)}(a) = -120$$

Evidently $f^{(n)}(a) = (-1)^{n+1}(n-1)!$ if $n \geqq 1$. (For a proof, use proof by induction. We omit the proof to save space.) Therefore the Taylor series for $f(x)$ at $a = 0$ is

$$\sum_{n=1}^{\infty} \frac{(-1)^{n+1}(n-1)! x^n}{n!} = \sum_{n=1}^{\infty} \frac{(-1)^{n+1} x^n}{n} = x - \frac{x^2}{2} + \frac{x^3}{3} - \frac{x^4}{4} + \frac{x^5}{5} - \cdots .$$

This representation of $f(x) = \ln(1+x)$ is valid if $-1 < x \leqq 1$.

10.4.31: If $f(x) = e^{-x}$, then $f^{(n)}(x) = (-1)^n e^{-x}$ for all $n \geqq 0$. With $a = 0$, this implies that $f^{(n)}(a) = (-1)^n$ for all $n \geqq 0$. Therefore the Taylor series for $f(x)$ at a is

$$\sum_{n=0}^{\infty} \frac{(-1)^n x^n}{n!} = 1 - x + \frac{x^2}{2!} - \frac{x^3}{3!} + \frac{x^4}{4!} - \frac{x^5}{5!} + \frac{x^6}{6!} - \cdots .$$

This representation of $f(x) = e^{-x}$ is valid for all x.

10.4.33: Given $f(x) = \ln x$ and $a = 1$, we compute:

$$f(x) = \ln x \qquad\qquad f(a) = 0$$

$$f'(x) = \frac{1}{x} \qquad\qquad f'(a) = 1$$

$$f''(x) = -\frac{1}{x^2} \qquad\qquad f''(a) = -1$$

$$f^{(3)}(x) = \frac{2}{x^3} \qquad\qquad f^{(3)}(a) = 2$$

$$f^{(4)}(x) = -\frac{6}{x^4} \qquad\qquad f^{(4)}(a) = -6$$

$$f^{(5)}(x) = \frac{24}{x^5} \qquad\qquad f^{(5)}(a) = 24$$

$$f^{(6)}(x) = -\frac{120}{x^6} \qquad\qquad f^{(6)}(a) = -120$$

We have here convincing evidence that if $n \geqq 1$, then $f^{(n)}(a) = (-1)^{n+1}(n-1)!$. (To prove this rigorously, use proof by induction; we omit any proof to save space.) Therefore the Taylor series for $f(x) = \ln x$ at $a = 1$ is

$$\sum_{n=1}^{\infty} \frac{(-1)^{n+1}(n-1)!(x-1)^n}{n!} = \sum_{n=1}^{\infty} \frac{(-1)^{n+1}(x-1)^n}{n}$$

$$= (x-1) - \frac{1}{2}(x-1)^2 + \frac{1}{3}(x-1)^3 - \frac{1}{4}(x-1)^4 + \frac{1}{5}(x-1)^5 - \frac{1}{6}(x-1)^6 + \cdots.$$

This representation of $f(x) = \ln x$ is valid if $0 < x \leqq 2$.

10.4.35: Given $f(x) = \cos x$ and $a = \pi/4$, we compute:

$$f(x) = \cos x \qquad\qquad f(a) = \frac{\sqrt{2}}{2}$$

$$f'(x) = -\sin x \qquad\qquad f'(a) = -\frac{\sqrt{2}}{2}$$

$$f''(x) = -\cos x \qquad\qquad f''(a) = -\frac{\sqrt{2}}{2}$$

$$f^{(3)}(x) = \sin x \qquad\qquad f^{(3)}(a) = \frac{\sqrt{2}}{2}$$

$$f^{(4)}(x) = \cos x \qquad\qquad f^{(4)}(a) = \frac{\sqrt{2}}{2}$$

$$f^{(5)}(x) = -\sin x \qquad\qquad f^{(5)}(a) = -\frac{\sqrt{2}}{2}$$

$$f^{(6)}(x) = -\cos x \qquad\qquad f^{(6)}(a) = -\frac{\sqrt{2}}{2}$$

It should be clear that

$$f^{(n)}(a) = \frac{\sqrt{2}}{2} \quad \text{if } n \text{ is of the form } 4k \text{ or } 4k+3, \text{ whereas}$$

$$f^{(n)}(a) = -\frac{\sqrt{2}}{2} \quad \text{if } n \text{ is of the form } 4k+1 \text{ or } 4k+2.$$

Therefore the Taylor series for $f(x) = \cos x$ at $a = \pi/4$ is

$$\frac{\sqrt{2}}{2} - \frac{\sqrt{2}}{2}\left(x - \frac{\pi}{4}\right) - \frac{\sqrt{2}}{2! \cdot 2}\left(x - \frac{\pi}{4}\right)^2 + \frac{\sqrt{2}}{3! \cdot 2}\left(x - \frac{\pi}{4}\right)^3 + \frac{\sqrt{2}}{4! \cdot 2}\left(x - \frac{\pi}{4}\right)^4 - \cdots .$$

This representation of $f(x) = \cos x$ is valid for all x.

10.4.37: Given $f(x) = \dfrac{1}{x}$ and $a = 1$, we compute:

$$f(x) = \frac{1}{x} \qquad f(a) = 1$$

$$f'(x) = -\frac{1}{x^2} \qquad f'(a) = -1$$

$$f''(x) = \frac{2}{x^3} \qquad f''(a) = 2$$

$$f^{(3)}(x) = -\frac{6}{x^4} \qquad f^{(3)}(a) = -6$$

$$f^{(4)}(x) = \frac{24}{x^5} \qquad f^{(4)}(a) = 24$$

$$f^{(5)}(x) = -\frac{120}{x^6} \qquad f^{(5)}(a) = -120$$

$$f^{(6)}(x) = \frac{720}{x^7} \qquad f^{(6)}(a) = 720$$

Clearly $f^{(n)}(a) = (-1)^n \cdot n!$ for $n \geqq 0$. Therefore the Taylor series for $f(x)$ at $a = 1$ is

$$\sum_{n=0}^{\infty} \frac{(-1)^n n! (x-1)^n}{n!} = \sum_{n=0}^{\infty} (-1)^n (x-1)^n$$

$$= 1 - (x-1) + (x-1)^2 - (x-1)^3 + (x-4)^4 - (x-1)^5 + (x-1)^6 - (x-1)^7 + \cdots .$$

This representation of $f(x)$ is valid for $0 < x < 2$.

10.4.39: Given $f(x) = \sin x$ and $a = \pi/4$, we compute:

$$f(x) = \sin x \qquad f(a) = \frac{\sqrt{2}}{2}$$

583

$$f'(x) = \cos x \qquad f'(a) = \frac{\sqrt{2}}{2}$$

$$f''(x) = -\sin x \qquad f''(a) = -\frac{\sqrt{2}}{2}$$

$$f^{(3)}(x) = -\cos x \qquad f^{(3)}(a) = -\frac{\sqrt{2}}{2}$$

$$f^{(4)}(x) = \sin x \qquad f^{(4)}(a) = \frac{\sqrt{2}}{2}$$

$$f^{(5)}(x) = \cos x \qquad f^{(5)}(a) = \frac{\sqrt{2}}{2}$$

$$f^{(6)}(x) = -\sin x \qquad f^{(6)}(a) = -\frac{\sqrt{2}}{2}$$

Therefore the Taylor series for $f(x) = \sin x$ at $a = \pi/4$ is

$$\frac{\sqrt{2}}{2} + \frac{\sqrt{2}}{2}\left(x - \frac{\pi}{4}\right) - \frac{\sqrt{2}}{2! \cdot 2}\left(x - \frac{\pi}{4}\right)^2 - \frac{\sqrt{2}}{3! \cdot 2}\left(x - \frac{\pi}{4}\right)^3 + \frac{\sqrt{2}}{4! \cdot 2}\left(x - \frac{\pi}{4}\right)^4 + \frac{\sqrt{2}}{5! \cdot 2}\left(x - \frac{\pi}{4}\right)^5 - \cdots .$$

This representation of $f(x) = \sin x$ is valid for all x.

10.4.41: Given $f(x) = \sin x$ and $a = 0$, we compute:

$$f(x) = \sin x \qquad f(a) = 0$$

$$f'(x) = \cos x \qquad f'(a) = 1$$

$$f''(x) = -\sin x \qquad f''(a) = 0$$

$$f^{(3)}(x) = -\cos x \qquad f^{(3)}(a) = -1$$

$$f^{(4)}(x) = \sin x \qquad f^{(4)}(a) = 0$$

$$f^{(5)}(x) = \cos x \qquad f^{(5)}(a) = 1$$

$$f^{(6)}(x) = -\sin x \qquad f^{(6)}(a) = 0$$

Therefore Taylor's formula for $f(x)$ at $a = 0$ is

$$f(x) = x - \frac{x^3}{3!} + \frac{x^5}{5!} - \cdots + (-1)^n \frac{x^{2n+1}}{(2n+1)!} + (-1)^{n+1} \frac{\sin z}{(2n+3)!} x^{2n+3} \qquad (1)$$

for some number z between 0 and x. Because $|\cos z| \leq 1$ for all z, it follows from Eq. (18) of the text that the remainder term in Eq. (1) approaches zero as $n \to \infty$. Therefore the Taylor series of $f(x) = \sin x$ at $a = 0$ is

$$\sin x = \sum_{n=0}^{\infty} \frac{(-1)^n x^{2n+1}}{(2n+1)!} = x - \frac{x^3}{3!} + \frac{x^5}{5!} - \frac{x^7}{7!} + \cdots ,$$

and this representation is valid for all x.

10.4.43: Given $f(x) = \cosh x$, $g(x) = \sinh x$, and $a = 0$, we compute:

$$f(x) = \cosh x \qquad\qquad f(a) = 1$$

$$f'(x) = \sinh x \qquad\qquad f'(a) = 0$$

$$f''(x) = \cosh x \qquad\qquad f''(a) = 1$$

$$f^{(3)}(x) = \sinh x \qquad\qquad f^{(3)}(a) = 0$$

$$f^{(4)}(x) = \cosh x \qquad\qquad f^{(4)}(a) = 1$$

$$f^{(5)}(x) = \sinh x \qquad\qquad f^{(5)}(a) = 0$$

$$f^{(6)}(x) = \cosh x \qquad\qquad f^{(6)}(a) = 1$$

Evidently $f^{(n)}(a) = 1$ if n is even and $f^{(n)}(a) = 0$ if n is odd. Therefore the Maclaurin series for $f(x) = \cosh x$ is

$$\sum_{n=0}^{\infty} \frac{x^{2n}}{(2n)!} = 1 + \frac{x^2}{2!} + \frac{x^4}{4!} + \frac{x^6}{6!} + \frac{x^8}{8!} + \cdots . \qquad (1)$$

The remainder term in Taylor's formula is

$$\frac{\sinh z}{(2n+1)!} x^{2n+1}$$

where z is between 0 and x. The remainder term approaches zero as $n \to +\infty$ by Eq. (18) of the text. Therefore the series in Eq. (1) converges to $f(x) = \cosh x$ for all x. Similarly,

$$g(x) = \sinh x \qquad\qquad g(a) = 0$$

$$g'(x) = \cosh x \qquad\qquad g'(a) = 1$$

$$g''(x) = \sinh x \qquad\qquad g''(a) = 0$$

$$g^{(3)}(x) = \cosh x \qquad\qquad g^{(3)}(a) = 1$$

$$g^{(4)}(x) = \sinh x \qquad\qquad g^{(4)}(a) = 0$$

$$g^{(5)}(x) = \cosh x \qquad\qquad g^{(5)}(a) = 1$$

$$g^{(6)}(x) = \sinh x \qquad\qquad g^{(6)}(a) = 0$$

It is clear that $g^{(n)}(a) = 0$ if n is even, whereas $g^{(n)} = 1$ if n is odd. Therefore the Maclaurin series for $g(x) = \sinh x$ is

$$\sum_{n=0}^{\infty} \frac{x^{2n+1}}{(2n+1)!} = x + \frac{x^3}{3!} + \frac{x^5}{5!} + \frac{x^7}{7!} + \frac{x^9}{9!} + \cdots . \tag{2}$$

This series converges to $g(x) = \sinh x$ for all x by an argument very similar to that given for the hyperbolic cosine series.

Next, substitution of ix for x yields

$$\cosh ix = 1 + \frac{(ix)^2}{2!} + \frac{(ix)^4}{4!} + \frac{(ix)^6}{6!} + \cdots = 1 - \frac{x^2}{2!} + \frac{x^4}{4!} - \frac{x^6}{6!} + \cdots = \cos x.$$

Similarly, $\sinh ix = \sin x$. This is one way to describe the relationship of the hyperbolic functions to the circular functions. A more prosaic response to the concluding question in Problem 43 would be that if the signs in the Maclaurin series for the cosine function are changed so that they are all plus signs, you get the Maclaurin series for the hyperbolic cosine function; the same relation hold for the sine and hyperbolic sine series.

10.4.45: Given $f(x) = e^{-x}$, its plot together with that of

$$P_3(x) = 1 - x + \frac{x^2}{2!} - \frac{x^3}{3!}$$

are shown next.

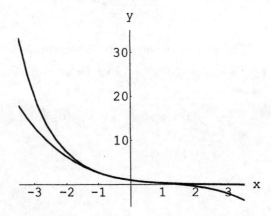

The graphs of $f(x) = e^{-x}$ and

$$P_6(x) = 1 - x + \frac{x^2}{2!} - \frac{x^3}{3!} + \frac{x^4}{4!} - \frac{x^5}{5!} + \frac{x^6}{6!}$$

are shown together next.

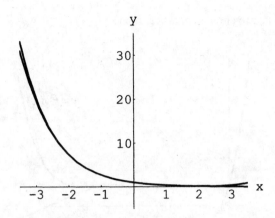

10.4.47: Given $f(x) = \cos x$, two of its Taylor polynomials are

$$P_4(x) = 1 - \frac{x^2}{2!} + \frac{x^4}{4!} \quad \text{and} \quad P_8(x) = 1 - \frac{x^2}{2!} + \frac{x^4}{4!} - \frac{x^6}{6!} + \frac{x^8}{8!}.$$

The graphs of f and P_4 are shown next, on the left; the graph of f and P_8 are on the right.

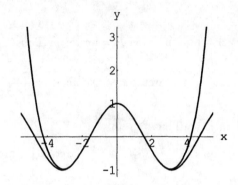

10.4.49: Given $f(x) = \dfrac{1}{1+x}$, two of its Taylor polynomials are

$$P_3(x) = 1 - x + x^2 - x^3 \quad \text{and} \quad P_4(x) = 1 - x + x^2 - x^3 + x^4.$$

The graphs of f and P_3 are shown together next, on the left; the graphs of f and P_4 are on the right.

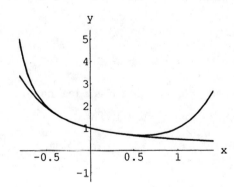

10.4.51: The graph of the Taylor polynomial

$$P_8(x) = 1 - \frac{x}{2!} + \frac{x^2}{4!} - \cdots - \frac{x^7}{14!} + \frac{x^8}{16!}$$

of $f(x)$ and the graph of $g(x)$ are shown together, next.

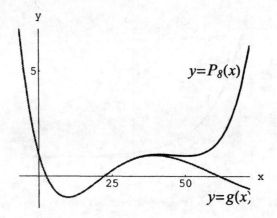

10.4.53: We begin with the formula

$$\tan(A + B) = \frac{\tan A + \tan B}{1 - \tan A \tan B}.$$

Let $A = \arctan x$ and $B = \arctan y$. Thus

$$\tan(\arctan x + \arctan y) = \frac{x + y}{1 - xy}, \quad \text{so that} \quad \arctan x + \arctan y = \arctan \frac{x + y}{1 - xy}$$

(if $xy < 1$). Thus

$$\arctan \frac{1}{2} + \arctan \frac{1}{5} = \arctan \frac{\frac{7}{10}}{\frac{9}{10}} = \arctan \frac{7}{9}.$$

Therefore

$$\arctan \frac{1}{2} + \arctan \frac{1}{5} + \arctan \frac{1}{8} = \arctan \frac{7}{9} + \arctan \frac{1}{8} = \arctan \frac{\frac{65}{72}}{\frac{65}{72}} = \arctan 1 = \frac{\pi}{4}.$$

10.4.55: Prove that

$$\lim_{n \to \infty} \frac{x^n}{n!} = 0$$

for every real number x.

Proof: Suppose that x is a real number. Choose the integer k so large that $k > |2x|$. Let $L = |x|^k / k!$. Suppose that $n = k + 1$. Then

$$\frac{|x|^n}{n!} = \frac{|x|^{k+1}}{(k+1)!} = \frac{|x|^k}{k!} \cdot \frac{|x|}{k+1} < \frac{L}{2} = \frac{L}{2^{n-k}}$$

because $|2x| < k < k + 1$ and $n - k = 1$. Next, assume that

$$\frac{|x|^m}{m!} < \frac{L}{2^{m-k}}$$

588

for some integer $m > k$. Then

$$\frac{|x|^{m+1}}{(m+1)!} = \frac{|x|^m}{m!} \cdot \frac{|x|}{m+1} < \frac{L}{2^{m-k}} \cdot \frac{1}{2} = \frac{L}{2^{m+1-k}}$$

because $|2x| < k < m$. Therefore, by induction,

$$\frac{|x|^n}{n!} < \frac{L}{2^{n-k}}$$

for every integer $n > k$. Now let $n \to +\infty$ to conclude that

$$\lim_{n \to \infty} \frac{x^n}{n!} = 0.$$

10.4.57: By Theorem 4 of Section 10.3, S is not a number. Thus attempts to do "arithmetic" with S are meaningless and may lead to all sorts of absurd results.

10.4.59: Results: With $x = 1$ in the Maclaurin series in Problem 56, we find that

$$a = \sum_{n=1}^{50} \frac{(-1)^{n+1}}{n} \approx 0.68324716057591818842565811649.$$

With $x = \frac{1}{3}$ in the second series in Problem 58, we find that

$$b = \sum_{\substack{n=1 \\ n \text{ odd}}}^{49} \frac{2}{n \cdot 3^n} \approx 0.69314718055994530941723210107.$$

Because $|a - \ln 2| \approx 0.009900019984$, whereas $|b - \ln 2| \approx 2.039 \times 10^{-26}$, it is clear that the second series of Problem 58 is far superior to the series of Problem 56 for the accurate approximation of $\ln 2$.

Section 10.5

10.5.1: $\int_1^\infty \frac{x}{x^2+1}\, dx = \left[\frac{1}{2}\ln(x^2+1) \right]_1^\infty = +\infty$. Therefore $\sum_{n=1}^\infty \frac{n}{n^2+1}$ diverges.

10.5.3: $\int_1^\infty (x+1)^{-1/2}\, dx = \left[2(x+1)^{1/2} \right]_1^\infty = +\infty$. Therefore $\sum_{n=1}^\infty \frac{1}{\sqrt{n+1}}$ diverges.

10.5.5: $\int_1^\infty \frac{1}{x^2+1}\, dx = \left[\arctan x \right]_1^\infty = \frac{\pi}{4} < +\infty$. Therefore $\sum_{n=1}^\infty \frac{1}{n^2+1}$ converges.

This is a special case of series 6.1.32 in Eldon R. Hansen, *A Table of Series and Products*, Prentice-Hall Inc. (Englewood Cliffs, N.J.), 1975. According to Hansen, its sum is

$$-\frac{1}{2} + \frac{\pi}{2}\coth \pi \approx 1.07667404746858117413405079475.$$

589

Mathematica reports that the sum of the first 1,000,000 terms of this series is approximately 1.07667.

10.5.7: $\displaystyle\int_2^\infty \frac{1}{x\ln x}\,dx = \Big[\ln(\ln x)\Big]_2^\infty = +\infty.$ Therefore $\displaystyle\sum_{n=2}^\infty \frac{1}{n\ln n}$ diverges.

10.5.9: $\displaystyle\int_1^\infty 2^{-x}\,dx = \left[-\frac{1}{2^x\ln 2}\right]_1^\infty = \frac{1}{2\ln 2} < +\infty.$ Therefore $\displaystyle\sum_{n=1}^\infty \frac{1}{2^n}$ converges (to 1).

10.5.11: For each positive integer n, let

$$I_n = \int x^n e^{-x}\,dx.$$

Let $u = x^n$ and $dv = e^{-x}\,dx$. Then $du = nx^{n-1}\,dx$; choose $v = -e^{-x}$. Then

$$I_n = -x^n e^{-x} + n\int x^{n-1}e^{-x}\,dx = -x^n e^{-x} + nI_{n-1}.$$

Therefore

$$\int x^2 e^{-x}\,dx = I_2 = -x^2 e^{-x} + 2I_1 = -x^2 e^{-x} + 2\left(-xe^{-x} + \int e^{-x}\,dx\right) = -(x^2 + 2x + 2)e^{-x} + C.$$

Hence

$$\int_1^\infty x^2 e^{-x}\,dx = -\Big[(x^2 + 2x + 2)e^{-x}\Big]_1^\infty = 5e^{-1} < +\infty,$$

and so $\displaystyle\sum_{n=1}^\infty \frac{n^2}{e^n}$ converges. It can be shown that its sum is

$$\frac{e(e+1)}{(e-1)^3} \approx 1.992294767125.$$

10.5.13: Choose $u = \ln x$ and $dv = \dfrac{1}{x^2}\,dx$. Then $du = \dfrac{1}{x}\,dx$; choose $v = -\dfrac{1}{x}$. Then

$$\int \frac{\ln x}{x^2}\,dx = -\frac{\ln x}{x} + \int \frac{1}{x^2}\,dx = -\frac{\ln x}{x} - \frac{1}{x} + C.$$

Thus

$$\int_1^\infty \frac{\ln x}{x^2}\,dx = \left[-\frac{1+\ln x}{x}\right]_1^\infty = \frac{1+0}{1} - \lim_{x\to\infty}\frac{1+\ln x}{x} = 1 - \lim_{x\to\infty}\frac{1}{x} = 1 - 0 = 1 < +\infty$$

(we used l'Hôpital's rule to find the limit). Therefore $\displaystyle\sum_{n=1}^\infty \frac{\ln n}{n^2}$ converges.

10.5.15: Because

$$\int_1^\infty \frac{x}{x^4+1}\,dx = \left[\frac{1}{2}\arctan\left(x^2\right)\right]_1^\infty = \frac{\pi}{8} < +\infty,$$

the series $\displaystyle\sum_{n=1}^{\infty} \frac{n}{n^4 + 1}$ converges.

With the aid of *Mathematica* and Theorem 2, we find that the sum of this series is approximately 0.694173022150715 (only the last digit shown here is in doubt; it may round to 6 instead of 5).

10.5.17: Because

$$\int_1^\infty \frac{2x+5}{x^2+5x+17}\,dx = \left[\ln(x^2+5x+17)\right]_1^\infty = +\infty,$$

the series $\displaystyle\sum_{n=1}^{\infty} \frac{2n+5}{n^2+5n+17}$ diverges.

10.5.19: Choose $u = \ln(1 + x^{-2})$ and $dv = dx$. Then

$$du = \frac{-2x^{-3}}{1+x^{-2}}\,dx = -\frac{2}{x^3+x}\,dx;$$

choose $v = x$. Thus

$$\int_1^\infty \ln\left(1 + \frac{1}{x^2}\right)dx = \left[x\ln(1+x^{-2})\right]_1^\infty + 2\int_1^\infty \frac{1}{1+x^2}\,dx$$

$$= \lim_{x\to\infty} \frac{\ln(1+x^{-2})}{1/x} - \ln 2 + \left[2\arctan x\right]_1^\infty$$

$$= \left[\lim_{z\to 0^+} \frac{\ln(1+z^2)}{z}\right]_1^\infty - \ln 2 + \pi - \frac{\pi}{2} = \frac{\pi}{2} - \ln 2 < +\infty$$

(use l'Hôpital's rule to evaluate the last limit). Therefore $\displaystyle\sum_{n=1}^{\infty} \ln\left(1 + \frac{1}{n^2}\right)$ converges.

10.5.21: Because

$$\int_1^\infty \frac{x}{4x^2+5}\,dx = \left[\frac{1}{8}\ln(4x^2+5)\right]_1^\infty = +\infty,$$

the series $\displaystyle\sum_{n=1}^{\infty} \frac{n}{4n^2+5}$ diverges.

10.5.23: Because

$$\int_2^\infty \frac{1}{x\sqrt{\ln x}}\,dx = \int_2^\infty \frac{(\ln x)^{-1/2}}{x}\,dx = \left[2(\ln x)^{1/2}\right]_2^\infty = +\infty,$$

the series $\displaystyle\sum_{n=2}^{\infty} \frac{1}{n\sqrt{\ln n}}$ diverges.

10.5.25: The substitution $u = 2x$ and integral formula 17 of the endpapers of the text yields

$$\int \frac{1}{4x^2+9}\,dx = \frac{1}{2}\int \frac{1}{u^2+9}\,du = \frac{1}{2}\cdot\frac{1}{3}\arctan\left(\frac{u}{3}\right) + C = \frac{1}{6}\arctan\left(\frac{2x}{3}\right) + C.$$

Therefore

$$\int_1^\infty \frac{1}{4x^2+9}\,dx = \left[\frac{1}{6}\arctan\left(\frac{2x}{3}\right)\right]_1^\infty = \frac{1}{6}\cdot\frac{\pi}{2} - \frac{1}{6}\arctan\left(\frac{2}{3}\right) < +\infty,$$

and thus $\displaystyle\sum_{n=1}^\infty \frac{1}{4n^2+9}$ converges.

This series is a special case of Eq. (6.1.32) of Eldon R. Hansen's *A Table of Series and Products*, Prentice-Hall, Inc. (Englewood Cliffs, N.J.), 1975. *Mathematica* summed this series in a fraction of a second and obtained the same answer as Hansen, viz.,

$$-\frac{1}{18} + \frac{\pi}{12}\coth\left(\frac{3\pi}{2}\right) \approx 0.20628608982235128529.$$

10.5.27: Because

$$\int_1^\infty \frac{x}{(x^2+1)^2}\,dx = \left[-\frac{1}{2(x^2+1)}\right]_1^\infty = \frac{1}{4} < +\infty,$$

the series $\displaystyle\sum_{n=1}^\infty \frac{n}{n^4+2n^2+1}$ converges. The *Mathematica* command

    NSum[n/(n∧4 + 2*n∧2 + 1), { n, 1, Infinity }, WorkingPrecision -> 28] // Timing

when executed on an aging Macintosh computer yielded the approximate sum 0.39711677137965943 in 3.45 seconds.

10.5.29: Because

$$\int_1^\infty \frac{\arctan x}{x^2+1}\,dx = \left[\frac{1}{2}(\arctan x)^2\right]_1^\infty = \frac{\pi^2}{8} - \frac{\pi^2}{32} = \frac{3}{32}\pi^2 < +\infty,$$

the series $\displaystyle\sum_{n=1}^\infty \frac{\arctan n}{n^2+1}$ converges.

10.5.31: The integral test cannot be applied because this is not a positive-term series. In Section 10.7 you will see how to prove that it converges, and in Problem 61 there you will see that its sum is $-\ln 2$.

10.5.33: The terms of this series are not monotonically decreasing. For specific examples, if we let $a_n = (2+\sin n)/(n^2)$, then

$$a_5 \approx 0.0416430290 < 0.0477940139 \approx a_6, \qquad a_{11} \approx 0.0082645437 < 0.0101626881 \approx a_{12},$$

$$a_{18} \approx 0.0038549776 < 0.0059553385 \approx a_{19}, \qquad a_{24} \approx 0.0019000376 < 0.0029882372 \approx a_{25},$$

$$a_{30} \approx 0.0011244093 < 0.0016607309 \approx a_{31}, \qquad a_{37} \approx 0.0009908414 < 0.0015902829 \approx a_{38}.$$

In Section 10.6 you will see how to use the *comparison test* to prove that this series converges.

10.5.35: There is some implication that we are to use the integral test to solve this problem. Hence we consider only the case in which $p > 0$. And if $p \neq 1$, then

$$\int_1^\infty p^{-x}\, dx = \left[-\frac{p^{-x}}{\ln p}\right]_{x=1}^\infty = \frac{1}{p \ln p} - \lim_{x \to \infty} \frac{1}{p^x \ln p}. \tag{1}$$

If $0 < p < 1$, then the limit in (1) is $+\infty$. If $p > 1$ then the limit in (1) is 0. If $p = 1$ then the series diverges because

$$\int_1^\infty \frac{1}{p^x}\, dx = \int_1^\infty 1\, dx = +\infty.$$

Answer: The series diverges if $0 < p \leqq 1$ and converges if $p > 1$.

10.5.37: If $p = 1$ then

$$\int_2^\infty \frac{1}{x(\ln x)^p}\, dx = \int_2^\infty \frac{1}{x \ln x}\, dx = \Big[\ln(\ln x) \Big]_2^\infty = +\infty.$$

Otherwise,

$$\int_2^\infty \frac{1}{x(\ln x)^p}\, dx = \int_2^\infty \frac{(\ln x)^{-p}}{x}\, dx = \left[\frac{(\ln x)^{1-p}}{1-p}\right]_2^\infty = \left[-\frac{1}{(p-1)(\ln x)^{p-1}}\right]_2^\infty.$$

So this improper integral diverges if $p < 1$. If $p > 1$ then it converges to

$$\frac{1}{(p-1)(\ln 2)^{p-1}} < +\infty.$$

Therefore the series $\displaystyle\sum_{n=2}^\infty \frac{1}{n(\ln n)^p}$ diverges if $p \leqq 1$ and converges if $p > 1$.

10.5.39: We require $R_n < 0.0001$. This will hold provided that

$$\int_n^\infty \frac{1}{x^2}\, dx < 0.0001$$

because R_n cannot exceed the integral. So we require

$$\left[-\frac{1}{x}\right]_n^\infty < 0.0001;$$

that is, that $n > 10000$.

10.5.41: We require $R_n < 0.00005$. This will hold provided that

$$\int_n^\infty \frac{1}{x^3}\, dx < 0.00005$$

because R_n cannot exceed the integral. So we require

$$\left[-\frac{1}{2x^2}\right]_n^\infty < 0.00005;$$

$$\frac{1}{2n^2} < 0.00005;$$

$$2n^2 > 20000;$$

thus we require that $n > 100$.

10.5.43: We require $R_n < 0.005$. This will hold provided that

$$\int_n^\infty \frac{1}{x^{3/2}}\, dx < 0.005;$$

that is, provided that

$$\left[-\frac{2}{x^{1/2}}\right]_n^\infty < 0.005,$$

so that $n^{1/2} > 400$, and thus $n > N = 160000$. *Mathematica* 3.0 reports that

$$S_N = \sum_{n=1}^N \frac{1}{n^{3/2}} \approx 2.607375356498 \quad \text{and that} \quad S = \sum_{n=1}^\infty \frac{1}{n^{3/2}} \approx 2.612375348685.$$

Note that $S - S_N \approx 0.004999992187 < 0.005$.

10.5.45: We require $R_n < 0.000005$. This will hold provided that

$$\int_n^\infty \frac{1}{x^5}\, dx = \frac{1}{4n^4} < 0.000005,$$

so that $n > 14.9535$. Choose $N = 15$. Then *Mathematica* 3.0 reports that

$$S_N = \sum_{n=1}^N \frac{1}{n^5} \approx 1.036923438841 \quad \text{and that} \quad S = \sum_{n=1}^\infty \frac{1}{n^5} \approx 1.036927755143.$$

Note that $S - S_N \approx 0.000004316302 < 0.000005$.

10.5.47: If $p = 1$, then

$$\int_1^\infty \frac{\ln x}{x^p}\, dx = \int_1^\infty \frac{\ln x}{x}\, dx = \left[\frac{1}{2}(\ln x)^2\right]_1^\infty = +\infty,$$

so in this case the given series $\displaystyle\sum_{n=1}^\infty \frac{\ln n}{n^p}$ diverges. Otherwise (with the aid of *Mathematica* 3.0 for the antiderivative)

$$\int_1^\infty \frac{\ln x}{x^p}\, dx = \left[\frac{x^{1-p}\ln x - x^{1-p}}{(1-p)^2} \right]_1^\infty = \left[\frac{-1 + \ln x}{(p-1)^2 x^{p-1}} \right]_1^\infty .$$

Thus if $p < 1$ the given series diverges, whereas if $p > 1$ it converges. Answer: $p > 1$.

10.5.49: From the proof of Theorem 1 (the integral test), we see that if

$$a_n = \frac{1}{n}, \quad f(x) = \frac{1}{x}, \quad \text{and} \quad S_n = \sum_{k=1}^{n} a_n$$

for each integer $n \geqq 1$, then

$$S_n \geqq \int_1^{n+1} \frac{1}{x}\, dx = \left[\ln x \right]_1^{n+1} = \ln(n+1)$$

and

$$S_n - a_1 \leqq \int_1^n \frac{1}{x}\, dx = \left[\ln x \right]_1^n = \ln n.$$

Therefore

$$\ln n < \ln(n+1) \leqq S_n \leqq 1 + \ln n;$$

put another way,

$$\ln n \leqq 1 + \frac{1}{2} + \frac{1}{3} + \frac{1}{4} + \cdots + \frac{1}{n} \leqq 1 + \ln n$$

for every integer $n \geqq 1$. So if a computer adds a million terms of the harmonic series every second, the partial sum S_n will first reach 50 when $n \leqq e^{50} \leqq e \cdot n$. This means that n must satisfy the inequalities

$$1 + \left[\!\left[e^{49} \right]\!\right] \leqq n \leqq \left[\!\left[e^{50} \right]\!\right];$$

that is,

$$1907346572495099690526 \leqq n \leqq 5184705528587072464087.$$

Divide the smaller of these bounds by one million (additions the computer carries out each second), then by 3600 to convert to hours, by 24 and then by 365.242199 to convert to years, and finally by 100 to convert to an answer: It will require over 604414 centuries. For a more precise answer, if $N = 2911002088526872100231$, then *Mathematica* reports that

$$\sum_{n=1}^{N-1} \frac{1}{n} \approx 49.999999999999999999999713 \quad \text{and}$$

$$\sum_{n=1}^{N} \frac{1}{n} \approx 50.000000000000000000000057.$$

After converting to centuries as before, we finally get the "right" answer: It will require a little over 922460 centuries.

10.5.51: Suppose that f is continuous and $f(x) > 0$ for all $x \geqq 1$. For each positive integer n, let

$$b_n = \int_1^n f(x)\, dx.$$

Part (a): Note that the sequence $\{b_n\}$ is increasing. Suppose that it is bounded, so that

$$B = \lim_{n \to \infty} b_n$$

exists. The definition of the value of an improper integral then implies that

$$\int_1^\infty f(x)\, dx = \lim_{\alpha \to \infty} \int_1^\alpha f(x)\, dx. \tag{1}$$

Therefore, by Theorem 4 in Section 10.2,

$$\int_1^\infty f(x)\, dx = \lim_{n \to \infty} \int_1^n f(x)\, dx = \lim_{n \to \infty} b_n = B.$$

Part (b): If the increasing sequence $\{b_n\}$ is not bounded, then by Problem 52 of Section 10.2,

$$\lim_{n \to \infty} b_n = +\infty.$$

Then Eq. (1) implies that

$$\int_1^\infty f(x)\, dx = +\infty$$

because $\displaystyle\int_1^\alpha f(x)\, dx$ is an increasing function of α.

Section 10.6

10.6.1: The series

$$\sum_{n-1}^\infty \frac{1}{n^2 + n + 1} \quad \text{is dominated by} \quad \sum_{n=1}^\infty \frac{1}{n^2},$$

which converges because it is a p-series with $p = 2 > 1$. Therefore the dominated series also converges.

10.6.3: The series

$$\sum_{n=1}^\infty \frac{1}{n + n^{1/2}}$$

diverges by limit-comparison with the harmonic series, demonstrated by the computation

$$\frac{\dfrac{1}{n + n^{1/2}}}{\dfrac{1}{n}} = \frac{n}{n + n^{1/2}} = \frac{1}{1 + \dfrac{1}{n^{1/2}}} \to \frac{1}{1 + 0} = 1.$$

as $n \to +\infty$.

10.6.5: The series

$$\sum_{n=1}^{\infty} \frac{1}{1+3^n} \quad \text{is dominated by} \quad \sum_{n=1}^{\infty} \frac{1}{3^n},$$

and the latter converges because it is a geometric series with ratio $\frac{1}{3} < 1$. Therefore the dominated series also converges.

10.6.7: The series

$$\sum_{n=2}^{\infty} \frac{10n^2}{n^3 - 1}$$

diverges by limit-comparison with the harmonic series, demonstrated by the computation

$$\frac{\dfrac{10n^2}{n^3 - 1}}{\dfrac{1}{n}} = \frac{10n^3}{n^3 - 1} = \frac{10}{1 - \dfrac{1}{n^3}} \to \frac{10}{1 - 0} = 10$$

as $n \to +\infty$.

10.6.9: First note that if $n \geq 1$, then $\dfrac{1}{\sqrt{37n^3 + 3}} \leq \dfrac{1}{\sqrt{n^3}} = \dfrac{1}{n^{3/2}}$. Therefore the series

$$\sum_{n=1}^{\infty} \frac{1}{\sqrt{37n^3 + 3}} \quad \text{is dominated by} \quad \sum_{n=1}^{\infty} \frac{1}{n^{3/2}},$$

and the latter series converges because it is a p-series with $p = \frac{3}{2} > 1$. Therefore the dominated series also converges.

10.6.11: Because $\dfrac{\sqrt{n}}{n^2 + n} \leq \dfrac{n^{1/2}}{n^2} = \dfrac{1}{n^{3/2}}$, the series

$$\sum_{n=1}^{\infty} \frac{\sqrt{n}}{n^2 + n} \quad \text{is dominated by} \quad \sum_{n=1}^{\infty} \frac{1}{n^{3/2}},$$

which converges because it is a p-series with $p = \frac{3}{2} > 1$. Therefore the dominated series also converges.

10.6.13: First we need a lemma: $\ln x < x$ if $x > 0$.

Proof: Let $f(x) = x - \ln x$. Then

$$f'(x) = 1 - \frac{1}{x}.$$

Because $f'(x) < 0$ if $0 < x < 1$, $f'(1) = 0$, and $f'(x) > 0$ if $1 < x$, the graph of $y = f(x)$ has a global minimum value at $x = 1$. Its minimum is $f(1) = 1 - \ln 1 = 1 > 0$, and $f(x) \geq f(1)$ if $x > 0$. Therefore $f(x) > 0$ for all $x > 0$; that is, $\ln x < x$ if $x > 0$. ◀

Therefore the series

$$\sum_{n=2}^{\infty} \frac{1}{\ln n} \quad \text{dominates} \quad \sum_{n=2}^{\infty} \frac{1}{n}.$$

The latter series diverges because it is "eventually the same" as the harmonic series, and therefore the dominating series also diverges.

10.6.15: Because $0 \leqq \sin^2 n \leqq 1$ for every integer $n \geqq 1$, the series

$$\sum_{n=1}^{\infty} \frac{\sin^2 n}{n^2 + 1} \quad \text{is dominated by} \quad \sum_{n=1}^{\infty} \frac{1}{n^2}.$$

The latter series converges because it is a p-series with $p = 2 > 1$. Therefore the dominated series also converges.

10.6.17: First we need a lemma: If n is a positive integer, then $n < 2^n$.

Proof: The lemma is true for $n = 1$ because $1 < 2$, so that $1 < 2^1$. Suppose that $k < 2^k$ for some integer $k \geqq 1$. Then

$$2^{k+1} = 2 \cdot 2^k \geqq 2 \cdot k = k + k \geqq k + 1.$$

Thus whenever the lemma holds for the integer $k \geqq 1$, it also hold for $k + 1$. Therefore, by induction, $n < 2^n$ for every integer $n \geqq 1$. ◄

Next, note that as a consequence of the lemma,

$$\frac{n + 2^n}{n + 3^n} \leqq \frac{2^n + 2^n}{3^n} = \frac{2 \cdot 2^n}{3^n} = 2 \cdot \left(\frac{2}{3}\right)^n.$$

Therefore the series

$$\sum_{n=1}^{\infty} \frac{n + 2^n}{n + 3^n} \quad \text{is dominated by} \quad \sum_{n=1}^{\infty} 2 \cdot \left(\frac{2}{3}\right)^n.$$

The latter series converges because it is geometric with ratio $\frac{2}{3} < 1$. Therefore the dominated series also converges by the comparison test.

10.6.19: Because $\dfrac{1}{n^2 \ln n} \leqq \dfrac{1}{n^2}$ if $n \geqq 3$, the given series

$$\sum_{n=2}^{\infty} \frac{1}{n^2 \ln n} \quad \text{is eventually dominated by} \quad \sum_{n=2}^{\infty} \frac{1}{n^2}.$$

The latter series converges because it is eventually the same as the p-series with $p = 2 > 1$. Therefore the dominated series converges by the comparison test. See the discussion of "eventual domination" following the proof of Theorem 1 (the comparison test) in Section 10.6 of the text.

10.6.21: First, a lemma: There is a positive integer K such that $\ln n \leqq \sqrt{n}$ if n is a positive integer and $n \geqq K$.

Proof: We use l'Hôpital's rule:

$$\lim_{n \to \infty} \frac{\ln n}{\sqrt{n}} = \lim_{x \to \infty} \frac{\ln x}{x^{1/2}} = \lim_{x \to \infty} \frac{2x^{1/2}}{x} = \lim_{x \to \infty} \frac{2}{x^{1/2}} = 0.$$

Therefore there exists a positive integer K such that $\ln n \leqq \sqrt{n}$ if $n \geqq K$. ◄

Alternatively, let $f(x) = x^{1/2} - \ln x$. Apply methods of calculus to show that $f'(x) < 0$ if $0 < x < 4$, $f'(4) = 0$, and $f'(x) > 0$ if $x > 4$. It follows that $f(x) \geqq f(4) = 2 - \ln 2 > 0$ for all $x > 0$, and hence $x^{1/2} > \ln x$ for all $x > 0$. Thus the integer K of the preceding proof may be chosen to be 1. But relying only on the lemma, we now conclude that

$$\sum_{n=1}^{\infty} \frac{\ln n}{n^2} \quad \text{is eventually dominated by} \quad \sum_{n=1}^{\infty} \frac{n^{1/2}}{n^2} = \sum_{n=1}^{\infty} \frac{1}{n^{3/2}}.$$

The last series converges because it is the p-series with $p = \frac{3}{2} > 1$. Therefore the dominated series converges by the comparison test.

10.6.23: Because $0 \leqq \sin^2(1/n) \leqq 1$ for every positive integer n, the given series

$$\sum_{n=1}^{\infty} \frac{\sin^2(1/n)}{n^2} \quad \text{is dominated by} \quad \sum_{n=1}^{\infty} \frac{1}{n^2}.$$

The latter series converges because it is the p-series with $p = 2 > 1$. Therefore the dominated series also converges by the comparison test.

10.5.25: We showed in the solution of Problem 13 that $\ln n \leqq n$ for every positive integer n. We showed in the solution of Problem 17 that $n \leqq 2^n$ for every positive integer n. Therefore

$$\sum_{n=1}^{\infty} \frac{\ln n}{e^n} \quad \text{is dominated by} \quad \sum_{n=1}^{\infty} \frac{2^n}{e^n} = \sum_{n=1}^{\infty} \left(\frac{2}{e}\right)^n.$$

The last series converges because it is geometric with ratio $2/e < 1$. Therefore the dominated series converges by the comparison test.

10.6.27: The given series

$$\sum_{n=1}^{\infty} \frac{n^{3/2}}{n^2 + 4} \quad \text{diverges by limit-comparison with} \quad \sum_{n=1}^{\infty} \frac{1}{n^{1/2}},$$

shown by the computation

$$\lim_{n \to \infty} \frac{\dfrac{n^{3/2}}{n^2 + 4}}{\dfrac{1}{n^{1/2}}} = \lim_{n \to \infty} \frac{n^2}{n^2 + 4} = \lim_{n \to \infty} \frac{1}{1 + \dfrac{4}{n^2}} = \frac{1}{1 + 0} = 1;$$

note that $\displaystyle\sum_{n=1}^{\infty} \frac{1}{n^{1/2}}$ diverges because it is a p-series with $p = \frac{1}{2} \leqq 1$.

10.6.29: First note that the series

$$\sum_{n=1}^{\infty} \frac{1}{n^{1/2}} \tag{1}$$

diverges because it is a p-series with $p = \frac{1}{2} \leqq 1$. Therefore the given series

$$\sum_{n=1}^{\infty} \frac{3}{4 + \sqrt{n}}$$

diverges by limit-comparison with the series in (1), as shown by the computation

$$\lim_{n\to\infty} \frac{\dfrac{3}{4 + n^{1/2}}}{\dfrac{1}{n^{1/2}}} = \lim_{n\to\infty} \frac{3n^{1/2}}{4 + n^{1/2}} = \lim_{n\to\infty} \frac{3}{\dfrac{4}{n^{1/2}} + 1} = \frac{3}{0 + 1} = 3.$$

10.6.31: First note that

$$\frac{2n^2 - 1}{n^2 \cdot 3^n} \leqq \frac{2n^2}{n^2 \cdot 3^n} = \frac{2}{3^n}$$

for each positive integer n. Therefore the given series

$$\sum_{n=1}^{\infty} \frac{2n^2 - 1}{n^2 \cdot 3^n} \quad \text{is dominated by} \quad \sum_{n=1}^{\infty} \frac{2}{3^n}.$$

The latter series converges because it is geometric with ratio $\frac{1}{3} < 1$. Therefore the series of Problem 31 converges by the comparison test.

10.6.33: Because $1 \leqq 2 + \sin n \leqq 3$ for each integer $n \geqq 1$, the given series

$$\sum_{n=1}^{\infty} \frac{2 + \sin n}{n^2} \quad \text{is dominated by} \quad \sum_{n=1}^{\infty} \frac{3}{n^2}.$$

The latter series converges because it is a constant multiple of the p-series with $p = 2 > 1$. Therefore the dominated series converges as well by the comparison test.

10.6.35: The given series

$$\sum_{n=1}^{\infty} \frac{(n + 1)^n}{n^{n+1}}$$

diverges by limit-comparison with the harmonic series, demonstrated by the following computation:

$$\lim_{n\to\infty} \frac{\dfrac{(n + 1)^n}{n^{n+1}}}{\dfrac{1}{n}} = \lim_{n\to\infty} \frac{(n + 1)^n}{n^n} = \lim_{n\to\infty} \left(1 + \frac{1}{n}\right)^n = e.$$

600

10.6.37: The sum of the first ten terms of the given series is

$$S_{10} = \sum_{n=1}^{10} \frac{1}{n^2 + 1} = \frac{1662222227}{1693047850} \approx 0.981792822335.$$

The error in using S_{10} to approximate the sum S of the infinite series is

$$S - S_{10} = R_{10} \leqq \int_{10}^{\infty} \frac{1}{x^2 + 1}\, dx = \left[\, \arctan x \,\right]_{10}^{\infty} = \frac{\pi}{2} - \arctan 10 \approx 0.099668652491.$$

Because $S \approx 1.076674047469$, the true value of the error is approximately 0.09488123.

10.9.39: The sum of the first ten terms of the given series is

$$\sum_{n=1}^{10} \frac{\cos^2 n}{n^2} \approx 0.528869678057.$$

The error in using S_{10} to approximate the sum S of the infinite series is

$$S - S_{10} = R_{10} \leqq \int_{10}^{\infty} \frac{1}{x^2}\, dx = \left[\, -\frac{1}{x} \,\right]_{10}^{\infty} = \frac{1}{10} = 0.1.$$

Because $S \approx 0.574137740053$, the true value of the error is approximately 0.04526806.

10.6.41: The sum of the series is

$$S = \sum_{n=1}^{\infty} \frac{1}{n^3 + 1} \approx 0.686503342339,$$

and $S - 0.005 \approx 0.681503342339$. Because

$$\sum_{n=1}^{9} \frac{1}{n^3 + 1} \approx 0.680981 < S - 0.005 < 0.681980 = \sum_{n=1}^{10} \frac{1}{n^3 + 1},$$

the smallest positive integer n such that $R_n < 0.005$ is $n = 10$. Without advance knowledge of the sum of the given series, you can obtain a conservative overestimate of n in the following way. We know that

$$R_n \leqq \int_{n}^{\infty} \frac{1}{x^3}\, dx = \left[\, -\frac{1}{2x^2} \,\right]_{n}^{\infty} = \frac{1}{2n^2}.$$

So it will be sufficient if

$$\frac{1}{2n^2} < 0.005; \qquad 2n^2 > 200; \qquad n > 10;$$

that is, if $n = 11$. More accuracy, and a smaller value of n, might be obtained had we used instead the better estimate

$$R_n \leqq \int_{n}^{\infty} \frac{1}{x^3 + 1}\, dx,$$

601

and if $n = 10$ the value of this integral is approximately 0.004998001249, but one must question whether the extra work in evaluating the antiderivative and solving the resulting inequality would be worth the trouble.

10.6.43: The sum of the series is

$$S = \sum_{n=1}^{\infty} \frac{\cos^4 n}{n^4} \approx 0.100714442927,$$

and $S - 0.005 \approx 0.095714442927$. Because

$$\sum_{n=1}^{2} \frac{\cos^4 n}{n^4} \approx 0.087095 < S - 0.005 < 0.098954 \approx \sum_{n=1}^{3} \frac{\cos^4 n}{n^4},$$

the smallest positive integer n such that $R_n < 0.005$ is $n = 3$. Without advance knowledge of the sum of the given series, you can obtain a conservative overestimate of n in the following way. We know that

$$R_n \leqq \int_{n}^{\infty} \frac{1}{x^4}\, dx = \left[-\frac{1}{3x^3} \right]_{n}^{\infty} = \frac{1}{3n^3}.$$

So it will be sufficient if

$$\frac{1}{3n^3} < 0.005; \qquad 3n^3 > 200; \qquad n^3 > 67;$$

that is, $n = 5$. There are ways to lower this estimate but they are highly technical and probably not worth the extra trouble.

10.6.45: We suppose that $\sum a_n$ is a convergent positive-term series. By Example 9 in Section 4.3, $\sin x < x$ for all $x > 0$. Moreover, the converse of Theorem 3 in Section 10.3 implies that $a_n \to 0$ as $n \to +\infty$. Thus there exists a positive integer K such that if $n \geq K$, then $a_n < \pi$. Therefore

$$0 < \sin(a_n) < a_n \qquad \text{if} \qquad n \geq K.$$

Therefore $\sum a_n$ eventually dominates the eventually positive-term series $\sum \sin(a_n)$. Therefore the latter series converges because the values of its terms for $1 \leqq n < K$ cannot affect its convergence or divergence.

10.6.47: If $\sum a_n$ is a convergent positive-term series, then we may assume that $n \geq 1$, and hence

$$0 < \frac{a_n}{n} \leqq a_n \qquad \text{for all} \qquad n.$$

Therefore $\sum a_n$ dominates the positive-term series $\sum (a_n/n)$, so by the comparison test the latter series converges as well.

10.6.49: Convergence of $\sum b_n$ implies that $\{b_n\} \to 0$ (the converse of Theorem 3 of Section 10.3). Therefore $\sum a_n b_n$ converges by Problem 48.

10.6.51: By Problem 50 in Section 10.5, if n is a positive integer then

$$0 \leq 1 + \frac{1}{2} + \frac{1}{3} + \frac{1}{4} + \cdots + \frac{1}{n} - \ln n \leq 1.$$

Therefore

$$1 + \frac{1}{2} + \frac{1}{3} + \frac{1}{4} + \cdots + \frac{1}{n} \leq 1 + \ln n$$

for every positive integer n. Hence

$$\sum_{n=1}^{\infty} \frac{1}{1 + \frac{1}{2} + \frac{1}{3} + \frac{1}{4} + \cdots + \frac{1}{n}} \quad \text{dominates} \quad \sum_{n=1}^{\infty} \frac{1}{1 + \ln n}.$$

But we showed in the solution of Problem 13 of this section that $\ln n < n$ for all $n \geq 1$. So the last series dominates

$$\sum_{n=1}^{\infty} \frac{1}{1 + n},$$

which diverges because it is eventually the same as the harmonic series. Therefore the series of Problem 51 diverges.

Section 10.7

10.7.1: The sequence $\{1/n^2\}$ is monotonically decreasing with limit zero. So the given series meets both criteria of the alternating series test and therefore converges. It is known that

$$\sum_{n=1}^{\infty} \frac{(-1)^{n+1}}{n^2} = \frac{\pi^2}{12} \approx 0.822467033424;$$

this can be derived from results in Problem 68 of Section 10.8.

10.7.3: Because

$$\lim_{n \to \infty} \frac{n}{3n + 2} = \frac{1}{3} \neq 0, \qquad \lim_{n \to \infty} \frac{(-1)^n n}{3n + 2} \quad \text{does not exist.}$$

Therefore the given series diverges by the nth-term test for divergence.

10.7.5: Because

$$\lim_{n \to \infty} \frac{n}{\sqrt{n^2 + 2}} = \lim_{n \to \infty} \frac{1}{\left(1 + \frac{2}{n^2}\right)^{1/2}} = 1 \neq 0, \qquad \lim_{n \to \infty} \frac{(-1)^{n+1} n}{\sqrt{n^2 + 2}} \quad \text{does not exist.}$$

Therefore the given series diverges by the nth-term test for divergence.

10.7.7: We showed in the solution of Problem 13 of Section 10.6 that $n > \ln n$ for every integer $n \geqq 1$. Also, by l'Hôpital's rule,

$$\lim_{n \to \infty} \frac{\ln n}{n} = 0.$$

Therefore

$$\lim_{n \to \infty} \frac{n}{\ln n} = +\infty,$$

so the given series diverges by the nth-term test for divergence.

10.7.9: First we claim that if n is a positive integer, then

$$\frac{n}{2^n} \geqq \frac{n+1}{2^{n+1}}. \tag{1}$$

This assertion is true if $n = 1$ because

$$\frac{1}{2} \geqq \frac{2}{4}, \quad \text{and thus} \quad \frac{1}{2^1} \geqq \frac{2}{2^2}.$$

Suppose that the inequality in (1) holds for some integer $k \geqq 1$. Then

$$\frac{k}{2^k} \geqq \frac{k+1}{2^{k+1}}; \qquad \frac{k}{2^k} + \frac{1}{2^k} \geqq \frac{k+1}{2^{k+1}} + \frac{2}{2^{k+1}};$$

$$\frac{k+1}{2^k} \geqq \frac{k+3}{2^{k+1}}; \qquad \frac{k+1}{2^{k+1}} > \frac{k+2}{2^{k+2}}.$$

Therefore, by induction, the inequality in (1) holds for every integer $n \geqq 1$; indeed, strict inequality holds if $n \geqq 2$. Therefore if $a_n = n/2^n$ for $n \geqq 1$, then the sequence $\{a_n\}$ is monotonically decreasing. Its limit is zero by l'Hôpital's rule:

$$\lim_{n \to \infty} a_n = \lim_{x \to \infty} \frac{x}{2^x} = \lim_{x \to \infty} \frac{1}{2^x \ln 2} = 0.$$

Therefore the given series satisfies both criteria of the alternating series test and thus it converges. To find its sum, note that

$$\sum_{n=1}^{\infty} \frac{(-1)^n n}{2^n} = f\left(\frac{1}{2}\right) \quad \text{where} \quad f(x) = \sum_{n=1}^{\infty} (-1)^{n+1} n x^n = x \sum_{n=1}^{\infty} (-1)^n n x^{n-1} = x g(x)$$

where $g(x) = h'(x)$ if we let

$$h(x) = \sum_{n=1}^{\infty} (-1)^n x^n = -x + x^2 - x^3 + x^4 - x^5 + \cdots = -\frac{x}{1+x}.$$

Thus

$$g(x) = h'(x) = -\frac{1}{(1+x)^2}, \quad \text{so that} \quad f(x) = -\frac{x}{(1+x)^2}.$$

It can be shown that all these computations are valid if $-1 < x < 1$, and therefore

$$\sum_{n=1}^{\infty} \frac{(-1)^n n}{2^n} = f\left(\frac{1}{2}\right) = -\frac{2}{9} \approx -0.222222222222.$$

10.7.11: Given the series

$$\sum_{n=1}^{\infty} \frac{(-1)^n n}{\sqrt{2^n + 1}}, \tag{1}$$

first observe that, by l'Hôpital's rule,

$$\lim_{x \to \infty} \frac{x}{(2^x + 1)^{1/2}} = \lim_{x \to \infty} \frac{2(2^x + 1)^{1/2}}{2^x \ln 2} = \lim_{x \to \infty} \frac{2(2^x + 1)^{1/2}}{(2^{2x})^{1/2} \ln 2}$$

$$= \lim_{x \to \infty} \frac{2}{\ln 2} \left(\frac{2^x + 1}{2^{2x}}\right)^{1/2} = \lim_{x \to \infty} \frac{2}{\ln 2} \left(\frac{1}{2^x} + \frac{1}{2^{2x}}\right)^{1/2} = 0.$$

Next,

$$\lim_{n \to \infty} \frac{2^{n+1} + 1}{2^n + 1} = \lim_{n \to \infty} \frac{2 + 2^{-n}}{1 + 2^{-n}} = \frac{2 + 0}{1 + 0} = 2$$

and

$$\lim_{n \to \infty} \frac{(n+1)^2}{n^2} = \lim_{n \to \infty} \left(\frac{n+1}{n}\right)^2 = 1^2 = 1.$$

Therefore there exists a positive integer K such that, if $n \geq K$, then

$$\frac{2^{n+1} + 1}{2^n + 1} > \frac{3}{2} > \frac{(n+1)^2}{n^2}.$$

For such n, it follows that

$$\frac{n^2}{2^n + 1} > \frac{(n+1)^2}{2^{n+1} + 1}; \quad \text{thus}$$

$$\frac{n}{\sqrt{2^n + 1}} > \frac{n+1}{\sqrt{2^{n+1} + 1}}.$$

This shows that the terms of the series in (1) are monotonically decreasing for $n \geq K$, and so both criteria of the alternating series test are met for $n \geq K$. Altering the terms for $n < K$ cannot change the convergence or divergence of a series, so the series in (1) converges. Its sum is approximately -0.178243455603. (By the way, the least value of K that "works" in this proof is $K = 5$, although the terms of the series begin to decrease in magnitude after $n = 3$.)

10.7.13: The values of $\sin(n\pi/2)$ for $n = 1, 2, 3, \ldots$ are $1, 0, -1, 0, 1, 0, -1, 0, 1 \cdots$. So we rewrite the given series in the form

$$\sum_{n=1}^{\infty} \frac{(-1)^{n+1}}{(2n - 1)^{2/3}}$$

to present it as an alternating series in the strict sense of the definition. Because the sequence $\{1/(2n-1)^{2/3}\}$ clearly meets the criteria of the alternating series test, this series converges. Its sum is approximately 0.711944418056.

10.7.15: Because

$$\lim_{n \to \infty} \sin\left(\frac{1}{n}\right) = \lim_{u \to 0^+} \sin u = 0$$

and because $\sin u$ decreases monotonically through positive values as $u \to 0^+$, this series converges by the alternating series test. Its sum is approximately -0.550796848134.

10.7.17: By Example 7 of Section 10.2, $2^{1/n} \to 1$ as $n \to +\infty$. So the given series diverges by the nth-term test for divergence.

10.7.19: By the result in Example 11 of Section 10.2 of the text, $n^{1/n} \to 1$ as $n \to +\infty$. Therefore the given series diverges by the nth-term test for divergence.

10.7.21: The ratio test yields

$$\rho = \lim_{n \to \infty} \frac{2^n}{2^{n+1}} = \frac{1}{2} < 1,$$

so the series

$$\sum_{n=1}^{\infty} \frac{(-1)^{n+1}}{2^n}$$

converges absolutely. Because it is geometric with ratio $r = -\frac{1}{2}$ and first term $\frac{1}{2}$, its sum is $\frac{1}{3}$.

10.7.23: If

$$f(x) = \frac{\ln x}{x}, \quad \text{then} \quad f'(x) = \frac{1 - \ln x}{x^2},$$

so the sequence $\{(\ln n)/n\}$ is monotonically decreasing if $n \geq 3$. By l'Hôpital's rule,

$$\lim_{n \to \infty} \frac{\ln n}{n} = \lim_{x \to \infty} \frac{\ln x}{x} = \lim_{x \to \infty} \frac{1}{x} = 0.$$

Therefore the series

$$\sum_{n=1}^{\infty} \frac{(-1)^n \ln n}{n}$$

converges by the alternating series test. Because

$$\int_1^{\infty} \frac{\ln x}{x} \, dx = \left[\frac{1}{2}(\ln x)^2\right]_1^{\infty} = +\infty,$$

the given series converges conditionally rather than absolutely. Its sum is approximately 0.159868903742.

10.7.25: The series

$$\sum_{n=1}^{\infty} \left(\frac{10}{n}\right)^n$$

converges absolutely by the root test, because

$$\rho = \lim_{n \to \infty} \left[\left(\frac{10}{n} \right)^n \right]^{1/n} = \lim_{n \to \infty} \frac{10}{n} = 0 < 1.$$

Its sum is approximately 186.724948614024. In spite of the relatively large sum, this series converges extremely rapidly; for example, the sum of its first 25 terms is approximately 186.724948614005. The sum is large because the first ten terms of the series are each at least 10; the largest is the fourth term, 39.0625. But the 25th term is less than 1.126×10^{-10}.

10.7.27: Given the infinite series $\displaystyle\sum_{n=0}^{\infty} \frac{(-10)^n}{n!}$, the ratio test yields

$$\rho = \lim_{n \to \infty} \frac{n! \, 10^{n+1}}{(n+1)! \, 10^n} = \lim_{n \to \infty} \frac{10}{n+1} = 0.$$

Therefore this series converges absolutely. It is the result of substitution of -10 for x in the Maclaurin series for $f(x) = e^x$ (see Eq. (19) in Section 10.4); therefore its sum is $e^{-10} \approx 0.0000453999297625$.

10.7.29: The series $\displaystyle\sum_{n=1}^{\infty} (-1)^{n+1} \left(\frac{n}{n+1} \right)^n$ diverges by the nth-term test for divergence because

$$\lim_{n \to \infty} \left(\frac{n}{n+1} \right)^n = \frac{1}{e} \neq 0.$$

10.7.31: Given the series $\displaystyle\sum_{n=1}^{\infty} \left(\frac{\ln n}{n} \right)^n$, the root test yields

$$\rho = \lim_{n \to \infty} \left[\left(\frac{\ln n}{n} \right)^n \right]^{1/n} = \lim_{n \to \infty} \frac{\ln n}{n} = \lim_{x \to \infty} \frac{\ln x}{x} = \lim_{x \to \infty} \frac{1}{x} = 0$$

(with the aid of l'Hôpital's rule). Because $\rho < 1$, the given series converges absolutely. Its sum is approximately 0.187967875056.

10.7.33: First note that

$$\sqrt{n+1} - \sqrt{n} = \frac{n+1-n}{\sqrt{n+1} + \sqrt{n}} = \frac{1}{\sqrt{n+1} + \sqrt{n}} \to 0$$

as $n \to +\infty$. Moreover, the sequence $\{ \sqrt{n+1} - \sqrt{n} \}$ is monotonically decreasing. Proof: Suppose that n is a positive integer. Then

$$n^2 + 2n + 1 > n^2 + 2n; \qquad n + 1 > \sqrt{n^2 + 2n}$$

$$2n + 2 > 2\sqrt{n^2 + 2n}; \qquad 4(n+1) > n + 2\sqrt{n^2 + 2n} + n + 2;$$

607

$$2\sqrt{n+1} > \sqrt{n} + \sqrt{n+2}; \qquad \sqrt{n+1} - \sqrt{n} > \sqrt{n+2} - \sqrt{n+1}.$$

Therefore the series $\displaystyle\sum_{n=0}^{\infty} (-1)^n \left(\sqrt{n+1} - \sqrt{n}\right)$ converges by the alternating series test.

Its sum is approximately 0.760209625219. It converges conditionally, not absolutely. The reason is that

$$\sum_{n=1}^{\infty} \frac{1}{\sqrt{n+1} + \sqrt{n}} \quad \text{dominates} \quad \sum_{n=1}^{\infty} \frac{1}{2\sqrt{n+1}},$$

which diverges because it is a constant multiple of a series eventually the same as the p-series with $p = \frac{1}{2} \leqq 1$.

10.7.35: Because

$$\lim_{n\to\infty} \left(\ln\frac{1}{n}\right)^n = \lim_{n\to\infty} (-\ln n)^n$$

does not exist, the series $\displaystyle\sum_{n=1}^{\infty} \left(\ln\frac{1}{n}\right)^n$ diverges by the nth-term test for divergence.

10.7.37: First, for each positive integer n,

$$\frac{3^n}{n(2^n + 1)} \geqq \frac{3^n}{n(2^n + 2^n)} = \frac{3^n}{2n \cdot 2^n} = \frac{1}{2n} \cdot \left(\frac{3}{2}\right)^n.$$

Next, using l'Hôpital's rule,

$$\lim_{x\to\infty} \frac{\left(\frac{3}{2}\right)^x}{2x} = \lim_{x\to\infty} \frac{\left(\frac{3}{2}\right)^x \ln\left(\frac{3}{2}\right)}{2} = +\infty$$

because $\ln\left(\frac{3}{2}\right) > 0$ and because, if $a > 1$, then $a^x \to +\infty$ as $x \to +\infty$. Therefore the series

$$\sum_{n=1}^{\infty} \frac{(-1)^{n+1} 3^n}{n(2^n + 1)}$$

diverges by the nth-term test for divergence.

10.7.39 Given the series $\displaystyle\sum_{n=1}^{\infty} \frac{(-1)^{n+1} n!}{1 \cdot 3 \cdot 5 \cdots (2n-1)}$, the ratio test yields

$$\rho = \lim_{n\to\infty} \frac{(n+1)! \cdot 1 \cdot 3 \cdot 5 \cdots (2n-1)}{n! \cdot 1 \cdot 3 \cdot 5 \cdots (2n-1) \cdot (2n+1)} = \lim_{n\to\infty} \frac{n+1}{2n+1} = \frac{1}{2}.$$

Because $\rho < 1$, the series in question converges absolutely. Its sum is approximately 0.586781998767.

10.7.41: Given the series $\displaystyle\sum_{n=1}^{\infty} \frac{(n+2)!}{3^n (n!)^2}$, the ratio test yields

$$\rho = \lim_{n\to\infty} \frac{(n+3)! 3^n (n!)^2}{(n+2)! 3^{n+1} \left[(n+1)!\right]^2} = \lim_{n\to\infty} \frac{n+3}{3(n+1)^2} = \lim_{n\to\infty} \frac{\dfrac{1}{n} + \dfrac{3}{n^2}}{3\left(1 + \dfrac{1}{n}\right)^2} = \frac{0+0}{3 \cdot 1} = 0.$$

608

Because $\rho < 1$, the given series converges absolutely. Its sum can be computed exactly, as follows. Note first that

$$\sum_{n=1}^{\infty} \frac{(n+2)!}{3^n (n!)^2} = \sum_{n=1}^{\infty} \frac{(n+2)(n+1)}{3^n \cdot (n!)} = f\left(\frac{1}{3}\right)$$

where

$$f(x) = \sum_{n=1}^{\infty} \frac{(n+2)(n+1)x^n}{n!}.$$

But $f(x) = g'(x)$ where

$$g(x) = \sum_{n=1}^{\infty} \frac{(n+2)x^{n+1}}{n!},$$

and $g(x) = h'(x)$ where

$$h(x) = \sum_{n=1}^{\infty} \frac{x^{n+2}}{n!} = x^2 \sum_{n=1}^{\infty} \frac{x^n}{n!} = x^2(e^x - 1).$$

But then, $f(x) = h''(x) = (x^2 + 4x + 2)e^x - 2$, so the sum of the series in this problem is

$$f\left(\frac{1}{3}\right) = \frac{31e^{1/3} - 18}{9} \approx 2.807109464185.$$

This is confirmed by *Mathematica*, which in response to the command

    NSum[((n + 2)*(n + 1))/((3^n)*(n!)), { n, 1, Infinity }]

returns the approximate sum 2.80711.

10.7.43: The sum of the first five terms of the given series is

$$S_5 = \sum_{n=1}^{5} \frac{(-1)^{n+1}}{n^3} = \frac{195353}{216000} \approx 0.904412037037.$$

The sixth term of the series is

$$-\frac{1}{216} \approx -0.004629629629.$$

Thus S_5 approximates the sum S of the series with error less than 0.005. Indeed, we can conclude that $S_6 \approx 0.899782 < S < 0.904412 \approx S_5$. To two decimal places, $S \approx 0.90$. *Mathematica* reports that $S \approx 0.901542677370$.

10.7.45: The sum of the first six terms of the given series is

$$S_6 = \sum_{n=1}^{6} \frac{(-1)^{n+1}}{n!} - \frac{91}{144} \approx 0.631944444444.$$

The seventh term of the series is

$$\frac{1}{5040} \approx 0.000198412698.$$

Thus S_6 approximates the sum S of the series with error less than 0.0002. Indeed, we can conclude that $S_6 \approx 0.631945 < S < 0.632142 \approx S_7$ (here we round *down* lower bounds and round *up* upper bounds). To three places, $S \approx 0.632$. *Mathematica* reports that $S \approx 0.632120558829$. Using Eq. (19) in Section 10.4, we see that the exact value of the sum is

$$S = 1 - \frac{1}{e}.$$

We have in this problem an example of a series that *Mathematica* can sum exactly (using the command Sum instead of NSum); the command

```
Sum[ ((-1)∧(n+1))/(n!), { n, 1, Infinity } ]
```

produces the exact answer in the form $-\dfrac{1-e}{e}$.

10.7.47: The sum of the first 12 terms of the series is

$$S_{12} = \sum_{n=1}^{12} \frac{(-1)^{n+1}}{n} = \frac{18107}{27720} \approx 0.653210678211.$$

The 13th term of the series is

$$\frac{1}{13} \approx 0.076923076923.$$

Thus S_{12} approximates the sum S of the series with error less than 0.08. Indeed, we may conclude that $S_{12} \approx 0.653211 < S < 0.730133 \approx S_{13}$ (we round *down* lower bounds and round *up* upper bounds). Thus to one decimal place, $S \approx 0.7$. This is a series that *Mathematica* can sum exactly; the command

```
Sum[ ((-1)∧(n+1))/n, { n, 1, Infinity } ]
```

produces the response $\ln 2$.

10.7.49: The condition

$$\frac{1}{n^4} < 0.0005 \quad \text{leads to} \quad n > 6.69,$$

so the sum of the terms through $n = 6$ will provide three-place accuracy. The sum of the first six terms of the series is

$$\sum_{n=1}^{6} \frac{(-1)^{n+1}}{n^4} = \frac{4090037}{4320000} \approx 0.946767824074,$$

so to three places, the sum of the infinite series is 0.947. The exact value of the sum of this series is

$$\frac{7\pi^4}{720} \approx 0.947032829497.$$

10.7.51: The condition

$$\frac{1}{n! \cdot 2^n} < 0.00005 \quad \text{leads to} \quad 5 < n < 6,$$

610

so the sum of the terms through $n = 5$ will provide four-place accuracy. The sum of the first six terms of the series is

$$\sum_{n=0}^{5} \frac{(-1)^n}{n! \cdot 2^n} = \frac{2329}{3840} \approx 0.606510416667,$$

so to four places, the sum of the infinite series is 0.6065. The exact value of the sum of this series is

$$\sum_{n=0}^{\infty} \frac{(-1)^n}{n! \cdot 2^n} = e^{-1/2} \approx 0.606530659713.$$

10.7.53: The condition

$$\frac{1}{(2n+1)!} \cdot \left(\frac{\pi}{3}\right)^{2n+1} < 0.000005 \quad \text{leads to} \quad 4 < n < 5,$$

so the sum of the terms through $n = 4$ will provide five-place accuracy. The sum of the first five terms of the series is

$$\sum_{n=0}^{4} \frac{(-1)^n}{(2n+1)!} \cdot \left(\frac{\pi}{3}\right)^{2n+1} \approx 0.866025445100,$$

so to five places, the sum of the infinite series is 0.86603. The exact value of the sum of the infinite series is

$$\sum_{n=0}^{\infty} \frac{(-1)^n}{(2n+1)!} \cdot \left(\frac{\pi}{3}\right)^{2n+1} = \sin\left(\frac{\pi}{3}\right) = \frac{\sqrt{3}}{2} \approx 0.866025403784.$$

10.7.55: Because

$$0 < a_n \leqq \frac{1}{n} \quad \text{for all} \quad n \geqq 1,$$

$a_n \to 0$ as $n \to +\infty$ by the squeeze law for limits (Section 2.3 of the text). The alternating series test does not apply because the sequence $\{a_n\}$ is not monotonically decreasing. The series $\sum a_n$ diverges because its $2n$th partial sum S_{2n} satisfies the inequality

$$S_{2n} > 1 + \frac{1}{3} + \frac{1}{5} + \frac{1}{7} + \cdots + \frac{1}{2n-1} > \frac{1}{2} + \frac{1}{4} + \frac{1}{6} + \frac{1}{8} + \cdots + \frac{1}{2n},$$

and the last expression is half the nth partial sum of the harmonic series. Similar remarks hold for S_{2n+1}, and hence $S_n \to +\infty$ as $n \to +\infty$. Therefore $\sum a_n$ diverges.

10.7.57: Let

$$a_n = b_n = \frac{(-1)^n}{\sqrt{n}} \quad \text{for} \quad n \geqq 1.$$

Then $\sum a_n$ and $\sum b_n$ converge by the alternating series test. But

$$\sum_{n=1}^{\infty} a_n b_n = \sum_{n=1}^{\infty} \frac{(-1)^{2n}}{n} = \sum_{n=1}^{\infty} \frac{1}{n}$$

611

diverges because it is the harmonic series.

10.7.59: Let $b = |a|$. Then the ratio test applied to $\sum (a^n/n!)$ yields

$$\rho = \lim_{n \to \infty} \frac{n! b^{n+1}}{(n+1)! b^n} = \lim_{n \to \infty} \frac{b}{n+1} = 0 < 1.$$

Therefore the series

$$\sum_{n=0}^{\infty} \frac{a^n}{n!} \tag{1}$$

converges for every real number a. Thus by the nth-term test for divergence, it follows that

$$\lim_{n \to \infty} \frac{a^n}{n!} = 0$$

for every real number a. The sum of the series in (1) is e^a.

10.7.61: We are given

$$H_n = \sum_{k=1}^{n} \frac{1}{k} \quad \text{and} \quad S_n = \sum_{k=1}^{n} \frac{(-1)^{k+1}}{k}.$$

Part (a): Note first that

$$S_2 = 1 - \frac{1}{2} \quad \text{and} \quad H_2 - H_1 = 1 + \frac{1}{2} - 1,$$

so $S_{2n} = H_{2n} - H_n$ if $n = 1$. Assume that $S_{2m} = H_{2m} - H_m$ for some integer $m \geq 1$. Then

$$S_{2(m+1)} = S_{2m} + \frac{1}{2m+1} - \frac{1}{2m+2} = H_{2m} - H_m + \frac{1}{2m+1} - \frac{1}{2m+2}$$

$$= H_{2m} + \frac{1}{2m+1} + \frac{1}{2m+2} - H_m - \frac{2}{2m+2} = H_{2(m+1)} - H_{m+1}.$$

Therefore, by induction, $S_{2n} = H_{2n} - H_n$ for every positive integer n.

Part (b): Let $m = 2n$. Then

$$\lim_{n \to \infty} (H_{2n} - \ln 2n) = \lim_{m \to \infty} (H_m - \ln m) = \gamma$$

by Problem 50 in Section 10.5.

Part (c): By the results in parts (a) and (b),

$$\lim_{n \to \infty} (H_{2n} - \ln 2n - H_n + \ln n) = 0;$$

$$\lim_{n \to \infty} (S_{2n} - \ln 2 - \ln n + \ln n) = 0;$$

$$\lim_{n \to \infty} S_{2n} = \ln 2.$$

Thus the "even" partial sums of the alternating harmonic series converge to $\ln 2$. But the alternating harmonic series converges by the alternating series test. Therefore the sequence of *all* of its partial sums converges to $\ln 2$; that is,

$$\sum_{k=1}^{\infty} \frac{1}{k} = 1 - \frac{1}{2} + \frac{1}{3} - \frac{1}{4} + \frac{1}{5} - \frac{1}{6} + \cdots = \ln 2.$$

But see the solution of Problem 28 in Section 10.6 for a better way to approximate $\ln 2$.

10.7.63: The answer consists of the first twelve terms of the following series:

$$1 + \frac{1}{3} - \frac{1}{2} + \frac{1}{5} - \frac{1}{4} + \frac{1}{7} + \frac{1}{9} - \frac{1}{6} + \frac{1}{11} + \frac{1}{13} - \frac{1}{8} + \frac{1}{15} + \frac{1}{17} - \frac{1}{10} + \frac{1}{19} + \frac{1}{21} - \frac{1}{12} + \frac{1}{23} + \frac{1}{25} - \frac{1}{14}$$

$$+ \frac{1}{27} - \frac{1}{16} + \frac{1}{29} + \frac{1}{31} - \frac{1}{18} + \frac{1}{33} + \frac{1}{35} - \frac{1}{20} + \frac{1}{37} + \frac{1}{39} - \frac{1}{22} + \frac{1}{41} + \frac{1}{43} - \frac{1}{24} + \frac{1}{45} + \frac{1}{47} - \frac{1}{26} + \cdots.$$

The 12th partial sum of the series shown here is

$$\frac{353201}{360360} \approx 0.9801337551337551$$

and the 13th is

$$\frac{6364777}{6126120} \approx 1.0389572845455198,$$

so the convergence to the sum 1 is quite slow (as might be expected when dealing with variations of the harmonic series). To generate and view many more partial sums, enter the following commands in *Mathematica* (or modify them to use in another computer algebra program):

```
u = Table[ 1/(2*n - 1), { n, 1, 2 + 1000 } ]
```
$$\left\{ 1, \frac{1}{3}, \frac{1}{5}, \frac{1}{7}, \cdots, \frac{1}{2003} \right\}$$

(Of course, the ellipsis is ours, not *Mathematica's*. And you may replace 1000 in the first two commands with as large a positive integer as you and your computer will tolerate.)

```
v = Table[ 1/(2*n), { n, 1, 2 + 1000 } ]
```
$$\left\{ \frac{1}{2}, \frac{1}{4}, \frac{1}{6}, \frac{1}{8}, \cdots, \frac{1}{2004} \right\}$$

```
x = 0;    i = 0;    j = 0;
```

(Here x denotes the running sum of the first k terms of the series; i and j are merely subscripts to be used in the arrays u and v, respectively.)

```
While[ i < 1000, {

    While[ x <= 1, { i = i + 1, x = x + u[[i]], Print[ { i, u[[i]], N[x,40] } ] } ],

    While[ x >= 1, { j = j + 1, x = x - v[[j]], Print[ { j, v[[j]], N[x,40] } ] } ] } ]
```

If you execute these commands, be prepared for 1543 lines of output, concluding with

$$\left\{ 1001, \ \frac{1}{2001}, \ 1.0003529867391675223067581695773251 55187 \right\}$$

$$\left\{ 542, \ \frac{1}{1084}, \ 0.99943047751407521384248944485074232678 \right\}$$

There is evidence that the series is converging to 1 but still stronger evidence that the convergence is painfully slow.

10.7.65: The sum of the first 50 terms of the rearrangement is $S_{50} \approx -0.00601599$. Also,

$$S_{500} \approx -0.000622656, \qquad S_{5000} \approx -0.0000624766,$$

$$S_{50000} \approx -0.0000062497656, \qquad S_{500000} \approx -0.00000062499766,$$

$$S_{5000000} \approx -0.000000062499977, \quad \text{and} \quad S_{50000000} \approx -0.0000000062499998.$$

We have strong circumstantial evidence here that the sum of the series is 0. (It is.)

Section 10.8

10.8.1: Given the series $\displaystyle\sum_{n=1}^{\infty} nx^n$, the ratio test yields

$$\lim_{n \to \infty} \frac{(n+1)|x|^{n+1}}{n|x|^n} = |x|,$$

so the series converges if $-1 < x < 1$. It clearly diverges at both endpoints of this interval, so its interval of convergence is $(-1, 1)$. To find its sum, note that

$$\sum_{n=1}^{\infty} nx^n = x \sum_{n=1}^{\infty} nx^{n-1} = xf'(x)$$

where

$$f(x) = \sum_{n=1}^{\infty} x^n = \frac{x}{1-x}, \quad \text{so that} \quad f'(x) = \frac{1}{(1-x)^2}.$$

Therefore $\displaystyle\sum_{n=1}^{\infty} nx^n = \frac{x}{(1-x)^2}$ if $-1 < x < 1$.

10.8.3: Given the series $\displaystyle\sum_{n=1}^{\infty} \frac{nx^n}{2^n}$, the ratio test yields

$$\lim_{n\to\infty} \frac{(n+1)2^n|x|^{n+1}}{n2^{n+1}|x|^n} = \frac{|x|}{2},$$

so the series converges if $-2 < x < 2$. It diverges at each endpoint of this interval by the nth-term test for divergence, so its interval of convergence is $(-2, 2)$.

10.8.5: Given the series $\displaystyle\sum_{n=1}^{\infty} n!x^n$, the ratio test yields

$$\lim_{n\to\infty} \frac{(n+1)!|x|^{n+1}}{n!|x|^n} = \lim_{n\to\infty} n|x|.$$

This limit is zero if $x = 0$ but is $+\infty$ otherwise. Therefore the series converges only at the real number $x = 0$. Thus its interval of convergence is $[0, 0]$. If you prefer the strict interpretation of the word "interval," the interval $[a, b]$ is defined only if $a < b$ according to Appendix A. If so, we must say that this series has no interval of convergence and that it converges only if $x = 0$.

10.8.7: Given the series $\displaystyle\sum_{n=1}^{\infty} \frac{3^n x^n}{n^3}$, the ratio test yields

$$\lim_{n\to\infty} \frac{n^3 3^{n+1}|x|^{n+1}}{(n+1)^3 3^n |x|^n} = 3|x|,$$

so the series converges if $-\frac{1}{3} < x < \frac{1}{3}$. When $x = \frac{1}{3}$ it is the p-series with $p = 3 > 1$, and thus it converges. When $x = -\frac{1}{3}$ the series converges by the alternating series test. Therefore its interval of convergence is $\left[-\frac{1}{3}, \frac{1}{3}\right]$.

10.8.9: Given the series $\displaystyle\sum_{n=1}^{\infty} (-1)^n n^{1/2}(2x)^n$, the ratio test yields

$$\lim_{n\to\infty} \frac{(n+1)^{1/2}2^{n+1}|x|^{n+1}}{n^{1/2}2^n |x|^n} = 2|x|,$$

so the series converges if $-\frac{1}{2} < x < \frac{1}{2}$. It diverges at each endpoint of this interval by the nth-term test for divergence, and therefore its interval of convergence is $(-\frac{1}{2}, \frac{1}{2})$.

10.8.11: Given the series $\displaystyle\sum_{n=1}^{\infty} \frac{(-1)^n nx^n}{2^n(n+1)^3}$, the ratio test yields

$$\lim_{n\to\infty} \frac{(n+1)^4 2^n |x|^{n+1}}{n(n+2)^3 2^{n+1}|x|^n} = \frac{|x|}{2},$$

so this series converges if $-2 < x < 2$. If $x = 2$ it becomes

$$\sum_{n=1}^{\infty} \frac{(-1)^n n}{(n+1)^3},$$

615

which converges by the alternating series test. If $x = -2$ it becomes

$$\sum_{n=1}^{\infty} \frac{n}{(n+1)^3},$$

which converges because it is dominated by the p-series with $p = 2 > 1$. Therefore its interval of convergence is $[-2, 2]$.

10.8.13: Given the series $\displaystyle\sum_{n=1}^{\infty} \frac{(\ln n)x^n}{3^n}$, the ratio test yields

$$\lim_{n \to \infty} \frac{[\ln(n+1)]\, 3^n |x|^{n+1}}{[\ln n]\, 3^{n+1} |x|^n} = \frac{|x|}{3},$$

so the series converges if $-3 < x < 3$. At the two endpoints of this interval it diverges by the nth-term test for divergence. Hence its interval of convergence is $(-3, 3)$. Note:

$$\lim_{n \to \infty} \frac{\ln(n+1)}{\ln n} = \lim_{x \to \infty} \frac{\ln(x+1)}{\ln x} = \lim_{x \to \infty} \frac{x}{x+1} = 1$$

by l'Hôpital's rule.

10.8.15: Given the series $\displaystyle\sum_{n=0}^{\infty} (5x - 3)^n$, the ratio test yields

$$\lim_{n \to \infty} \frac{|5x - 3|^{n+1}}{|5x - 3|^n} = |5x - 3|,$$

and we solve $|5x - 3| < 1$ as follows:

$$-1 < 5x - 3 < 1; \qquad 2 < 5x < 4; \qquad \frac{2}{5} < x < \frac{4}{5}.$$

So this series converges on the interval $I = \left(\frac{2}{5}, \frac{4}{5}\right)$. It diverges at each endpoint by the nth-term test for divergence, so I is its interval of convergence. On this interval it is a geometric series with ratio in $(-1, 1)$, so its sum is

$$\sum_{n=0}^{\infty} (5x - 3)^n = \frac{1}{1 - (5x - 3)} = \frac{1}{4 - 5x}$$

provided that x is in I.

10.8.17: Given the series $\displaystyle\sum_{n=1}^{\infty} \frac{2^n (x - 3)^n}{n^2}$, the ratio test yields

$$\lim_{n \to \infty} \frac{n^2 2^{n+1} |x - 3|^{n+1}}{(n+1)^2 2^n |x - 3|^n} = 2|x - 3|,$$

so this series converges if $2|x - 3| < 1$:

$$|x - 3| < \frac{1}{2}; \qquad -\frac{1}{2} < x - 3 < \frac{1}{2}; \qquad \frac{5}{2} < x < \frac{7}{2}.$$

616

If $x = \frac{5}{2}$, the series converges by the alternating series test. If $x = \frac{7}{2}$, it converges because it is the p-series with $p = 2 > 1$. Thus its interval of convergence is $\left[\frac{5}{2}, \frac{7}{2}\right]$.

10.8.19: Given the series $\displaystyle\sum_{n=1}^{\infty} \frac{(2n)!}{n!} x^n$, the ratio test yields

$$\lim_{n\to\infty} \frac{n!(2n+2)!|x|^{n+1}}{(n+1)!(2n)!|x|^n} = \lim_{n\to\infty} \frac{(2n+2)(2n+1)|x|}{n+1} = \lim_{n\to\infty} 2(2n+1)|x|.$$

The last limit is $+\infty$ if $x \neq 0$ but zero if $x = 0$. Therefore this series converges only at the single point $x = 0$.

10.8.21: Given the series $\displaystyle\sum_{n=1}^{\infty} \frac{n^3(x+1)^n}{3^n}$, the ratio test yields

$$\lim_{n\to\infty} \frac{(n+1)^3 3^n |x+1|^{n+1}}{n^3 3^{n+1} |x+1|^n} = \frac{|x+1|}{3},$$

so the given series converges if

$$-1 < \frac{x+1}{3} < 1; \qquad -3 < x+1 < 3; \qquad -4 < x < 2.$$

At the endpoints of this interval, the series diverges by the nth-term test for divergence. Thus its interval of convergence is $(-4, 2)$. To find its sum in closed form, let

$$f(x) = \sum_{n=1}^{\infty} \frac{n^3(x+1)^n}{3^n}.$$

Then

$$f(x) = (x+1) \sum_{n=1}^{\infty} \frac{n^3(x+1)^{n-1}}{3^n} = (x+1)g'(x)$$

where

$$g(x) = \sum_{n=1}^{\infty} \frac{n^2(x+1)^n}{3^n} = (x+1) \sum_{n=1}^{\infty} \frac{n^2(x+1)^{n-1}}{3^n}.$$

But $g(x) = (x+1)h'(x)$ where

$$h(x) = \sum_{n=1}^{\infty} \frac{n(x+1)^n}{3^n} = (x+1) \sum_{n=1}^{\infty} \frac{n(x+1)^{n-1}}{3^n} = (x+1)k'(x)$$

where

$$k(x) = \sum_{n=1}^{\infty} \frac{(x+1)^n}{3^n} = \frac{x+1}{2-x}$$

if $-4 < x < 2$ because the last series is geometric and convergent for such x. It now follows that

$$k'(x) = \frac{3}{(x-2)^2}; \qquad h(x) = \frac{3(x+1)}{(x-2)^2};$$

$$h'(x) = -\frac{3(x+4)}{(x-2)^3}; \qquad g(x) = -\frac{3(x+1)(x+4)}{(x-2)^3};$$

$$g'(x) = \frac{3(x^2+14x+22)}{(x-2)^4}; \qquad f(x) = \frac{3(x+1)(x^2+14x+22)}{(x-2)^4}.$$

10.8.23: Given the series $\displaystyle\sum_{n=1}^{\infty} \frac{(3-x)^n}{n^3}$, the ratio test yields

$$\lim_{n\to\infty} \frac{n^3|3-x|^{n+1}}{(n+1)^3|3-x|^n} = |3-x| = |x-3|,$$

so this series converges if $-1 < x - 3 < 1$; that is, if $2 < x < 4$. It also converges at $x = 2$ because it is the p-series with $p = 3 > 1$ and converges at $x = 4$ by the alternating series test. Therefore its interval of convergence is $[2, 4]$.

10.8.25: Given the series $\displaystyle\sum_{n=1}^{\infty} \frac{n!}{2^n}(x-5)^n$, the ratio test yields

$$\lim_{n\to\infty} \frac{(n+1)!\,2^n|x-5|^{n+1}}{n!\,2^{n+1}|x-5|^n} = \lim_{n\to\infty} \frac{n+1}{2}|x-5| = +\infty$$

unless $x = 5$, in which case the limit is zero. So this series converges only at the single point $x = 5$; its radius of convergence is zero.

10.8.27: Given the series $\displaystyle\sum_{n=0}^{\infty} x^{(2^n)}$, the ratio test yields

$$\lim_{n\to\infty} \frac{|x|^{(2^{n+1})}}{|x|^{(2^n)}} = \lim_{n\to\infty} |x|^{(2^n)}.$$

This limit is zero if $-1 < x < 1$, is 1 if $x = \pm 1$, and is $+\infty$ if $|x| > 1$. The series diverges if $x = \pm 1$ by the nth-term test for divergence, and hence its interval of convergence is $(-1, 1)$.

10.8.29: Given the series $\displaystyle\sum_{n=1}^{\infty} \frac{(-1)^n x^n}{1\cdot 3\cdot 5\cdots(2n-1)}$, the ratio test yields

$$\lim_{n\to\infty} \frac{1\cdot 3\cdot 5\cdots(2n-1)\cdot |x|^{n+1}}{1\cdot 3\cdot 5\cdots(2n-1)\cdot(2n+1)\cdot |x|^n} = \lim_{n\to\infty} \frac{|x|}{2n+1} = 0$$

for all x. Hence the interval of convergence of this series is $(-\infty, +\infty)$.

10.8.31: The function is the sum of a geometric series with first term and ratio x, and hence

$$f(x) = \frac{x}{1-x} = x + x^2 + x^3 + x^4 + x^5 + \cdots.$$

This series has radius of convergence 1 and interval of convergence $(-1, 1)$.

10.8.33: Substitute $-3x$ for x in the Maclaurin series for e^x in Eq. (2), then multiply by x^2 to obtain

$$f(x) = x^2 e^{-3x} = x^2\left(1 - \frac{3x}{1!} + \frac{9x^2}{2!} - \frac{27x^3}{3!} + \frac{81x^4}{4!} - \frac{243x^5}{5!} + \cdots\right)$$

$$= x^2 - \frac{3x^3}{1!} + \frac{3^2 x^4}{2!} - \frac{3^3 x^5}{3!} + \frac{3^4 x^6}{4!} - \frac{3^5 x^7}{5!} + \cdots.$$

618

The ratio test gives radius of convergence $+\infty$, so the interval of convergence of this series is $(-\infty, +\infty)$.

10.8.35: Substitute x^2 for x in the Maclaurin series in (4) to obtain

$$f(x) = \sin x^2 = x^2 - \frac{x^6}{3!} + \frac{x^{10}}{5!} - \frac{x^{14}}{7!} + \frac{x^{18}}{9!} - \cdots.$$

The ratio test yields radius of convergence $+\infty$, so the interval of convergence of this series is $(-\infty, +\infty)$.

10.8.37: Substitution of $\alpha = \frac{1}{3}$ in the binomial series in Eq. (14) yields

$$(1+x)^{1/3} = 1 + \frac{1}{3}x - \frac{1}{3} \cdot \frac{2}{3} \cdot \frac{x^2}{2!} + \frac{1}{3} \cdot \frac{2}{3} \cdot \frac{5}{3} \cdot \frac{x^3}{3!} - \frac{1}{3} \cdot \frac{2}{3} \cdot \frac{5}{3} \cdot \frac{8}{3} \cdot \frac{x^4}{4!} + \cdots$$

$$= 1 + \frac{1}{3}x - \frac{2}{3^2} \cdot \frac{x^2}{2!} + \frac{2 \cdot 5}{3^3} \cdot \frac{x^3}{3!} - \frac{2 \cdot 5 \cdot 8}{3^4} \cdot \frac{x^4}{4!} + \cdots.$$

Next, replacement of x with $-x$ yields

$$f(x) = (1-x)^{1/3} = 1 - \frac{1}{3}x - \frac{2}{3^2} \cdot \frac{x^2}{2!} - \frac{2 \cdot 5}{3^3} \cdot \frac{x^3}{3!} - \frac{2 \cdot 5 \cdot 8}{3^4} \cdot \frac{x^4}{4!} - \frac{2 \cdot 5 \cdot 8 \cdot 11}{3^5} \cdot \frac{x^5}{5!} - \cdots.$$

The radius of convergence of this series is 1.

10.8.39: Substitution of $\alpha = -3$ in the binomial series in Eq. (14) yields

$$f(x) = (1+x)^{-3} = 1 - 3x + 3 \cdot 4 \cdot \frac{x^2}{2!} - 3 \cdot 4 \cdot 5 \cdot \frac{x^3}{3!} + 3 \cdot 4 \cdot 5 \cdot 6 \cdot \frac{x^4}{4!} - \cdots.$$

The radius of convergence of this series is 1.

10.8.41: Let $g(x) = \ln(1+x)$. Then

$$g'(x) = \frac{1}{1+x} = 1 - x + x^2 - x^3 + x^4 - x^5 + \cdots, \quad -1 < x < 1.$$

So

$$g(x) = C + x - \frac{x^2}{2} + \frac{x^3}{3} - \frac{x^4}{4} + \frac{x^5}{5} - \frac{x^6}{6} + \cdots$$

by Theorem 3. Also $0 = g(0) = \ln 1 = C$, so that

$$g(x) = x - \frac{x^2}{2} + \frac{x^3}{3} - \frac{x^4}{4} + \frac{x^5}{5} - \frac{x^6}{6} + \cdots.$$

Therefore

$$f(x) = \frac{g(x)}{x} = 1 - \frac{x}{2} + \frac{x^2}{3} - \frac{x^3}{4} + \frac{x^4}{5} - \frac{x^5}{6} + \cdots.$$

619

The ratio test tells us that the radius of convergence is 1; the interval of convergence is $(-1, 1]$.

10.8.43: Termwise integration yields

$$f(x) = \int_0^x \sin t^3 \, dt = \int_0^x \left(t^3 - \frac{t^9}{3!} + \frac{t^{15}}{5!} - \frac{t^{21}}{7!} + \cdots \right) dt$$

$$= \left[\frac{t^4}{4} - \frac{t^{10}}{3! 10} + \frac{t^{16}}{5! 16} - \frac{t^{22}}{7! 22} + \cdots \right]_0^x = \frac{x^4}{4} - \frac{x^{10}}{3! 10} + \frac{x^{16}}{5! 16} - \frac{x^{22}}{7! 22} + \cdots .$$

This representation is valid for all real x.

10.8.45: Termwise integration yields

$$f(x) = \int_0^x \exp\left(-t^3\right) dt = \int_0^x \left(1 - t^3 + \frac{t^6}{2!} - \frac{t^9}{3!} + \frac{t^{12}}{4!} - \cdots \right) dt$$

$$= \left[t - \frac{t^4}{4} + \frac{t^7}{2! 7} - \frac{t^{10}}{3! 10} + \frac{t^{13}}{4! 13} - \cdots \right]_0^x = x - \frac{x^4}{4} + \frac{x^7}{2! 7} - \frac{x^{10}}{3! 10} + \frac{x^{13}}{4! 13} - \cdots .$$

This representation is valid for all real x.

10.8.47: First,

$$1 - \exp\left(-t^2\right) = 1 - \left(1 - t^2 + \frac{t^4}{2!} - \frac{t^6}{3!} + \frac{t^8}{4!} - \cdots \right) = t^2 - \frac{t^4}{2!} + \frac{t^6}{3!} - \frac{t^8}{4!} + \frac{t^{10}}{5!} - \cdots .$$

Then termwise integration yields

$$f(x) = \int_0^x \frac{1 - \exp\left(-t^2\right)}{t^2} \, dt = \int_0^x \left(1 - \frac{t^2}{2!} + \frac{t^4}{3!} - \frac{t^6}{4!} + \frac{t^8}{5!} - \cdots \right) dt$$

$$= \left[t - \frac{t^3}{2! \cdot 3} + \frac{t^5}{3! \cdot 5} - \frac{t^7}{4! \cdot 7} + \frac{t^9}{5! \cdot 9} - \cdots \right]_0^x = x - \frac{x^3}{2! \cdot 3} + \frac{x^5}{3! \cdot 5} - \frac{x^7}{4! \cdot 7} + \frac{x^9}{5! \cdot 9} - \cdots .$$

This representation is valid for all real x.

10.8.49: We begin with

$$f(x) = \sum_{n=0}^{\infty} x^n = 1 + x + x^2 + x^3 + x^4 + x^5 + \cdots = \frac{1}{1-x}, \quad -1 < x < 1.$$

Then termwise differentiation yields

$$f'(x) = \sum_{n=1}^{\infty} n x^{n-1} = \frac{1}{(1-x)^2}; \quad \text{thus}$$

$$x f'(x) = \sum_{n=1}^{\infty} n x^n = \frac{x}{(1-x)^2}, \quad -1 < x < 1.$$

10.8.51: We found in the solution of Problem 49 that if

$$f(x) = \sum_{n=0}^{\infty} x^n = \frac{1}{1-x}, \quad -1 < x < 1,$$

then

$$xf'(x) = \sum_{n=1}^{\infty} nx^n = \frac{x}{(1-x)^2}, \quad -1 < x < 1.$$

Therefore

$$D_x\left[xf'(x)\right] = \sum_{n=1}^{\infty} n^2 x^{n-1} = \frac{1+x}{(1-x)^3},$$

and hence

$$\sum_{n=1}^{\infty} n^2 x^n = \frac{x + x^2}{(1-x)^3}, \quad -1 < x < 1.$$

10.8.53: If

$$y = e^x = 1 + x + \frac{x^2}{2!} + \frac{x^3}{3!} + \cdots + \frac{x^n}{n!} + \frac{x^{n+1}}{(n+1)!} + \cdots, \quad \text{then}$$

$$\frac{dy}{dx} = 0 + 1 + \frac{2x}{2!} + \frac{3x^2}{3!} + \cdots + \frac{nx^{n-1}}{n!} + \frac{(n+1)x^n}{(n+1)!} + \cdots$$

$$= 1 + x + \frac{x^2}{2!} + \frac{x^3}{3!} + \cdots + \frac{x^{n-1}}{(n-1)!} + \frac{x^n}{n!} + \cdots = e^x = y.$$

10.5.55: If

$$y = \sinh x = x + \frac{x^3}{3!} + \frac{x^5}{5!} + \frac{x^7}{7!} + \frac{x^9}{9!} + \frac{x^{11}}{11!} + \cdots, \quad \text{then}$$

$$\frac{dy}{dx} = 1 + \frac{x^2}{2!} + \frac{x^4}{4!} + \frac{x^6}{6!} + \frac{x^8}{8!} + \frac{x^{10}}{10!} + \cdots = \cosh x \quad \text{and}$$

$$\frac{d^2y}{dx^2} = x + \frac{x^3}{3!} + \frac{x^5}{5!} + \frac{x^7}{7!} + \frac{x^9}{9!} + \cdots = \sinh x.$$

Therefore both the hyperbolic sine and hyperbolic cosine functions satisfy the differential equation

$$\frac{d^2y}{dx^2} - y = 0.$$

10.8.57: From Example 7 we have

$$J_0(x) = \sum_{n=0}^{\infty} \frac{(-1)^n x^{2n}}{2^{2n}(n!)^2}; \quad \text{we are also given} \quad J_1(x) = \sum_{n=0}^{\infty} \frac{(-1)^n x^{2n+1}}{2^{2n+1} n!(n+1)!}.$$

We apply the ratio test to the series for $J_1(x)$ with the following result:

$$\lim_{n\to\infty} \frac{2^{2n+1} n!(n+1)! |x|^{2n+3}}{2^{2n+3}(n+1)!(n+2)! |x|^{2n+1}} = \lim_{n\to\infty} \frac{x^2}{4(n+1)(n+2)} = 0$$

for all x. Therefore this series converges for all x. Next,

$$J_0'(x) = \sum_{n=1}^{\infty} \frac{(-1)^n 2n x^{2n-1}}{2^{2n}(n!)^2} = \sum_{n=1}^{\infty} \frac{(-1)^n x^{2n-1}}{2^{2n-1} n!(n-1)!}$$

$$= \sum_{n=0}^{\infty} \frac{(-1)^{n+1} x^{2n+1}}{2^{2n+1}(n+1)! n!} = -\sum_{n=0}^{\infty} \frac{(-1)^n x^{2n+1}}{2^{2n+1}(n+1)! n!} = -J_1(x).$$

10.8.59: We begin with

$$y(x) = J_0(x) = \sum_{n=0}^{\infty} \frac{(-1)^n x^{2n}}{2^{2n}(n!)^2}. \quad \text{Then:}$$

$$y'(x) = \sum_{n=1}^{\infty} \frac{(-1)^n 2n x^{2n-1}}{2^{2n}(n!)^2} = \sum_{n=0}^{\infty} \frac{(-1)^{n+1}(2n+2) x^{2n+1}}{2^{2n+2}[(n+1)!]^2} = \sum_{n=0}^{\infty} \frac{(-1)^{n+1} x^{2n+1}}{2^{2n+1} n!(n+1)!};$$

$$y''(x) = \sum_{n=0}^{\infty} \frac{(-1)^{n+1}(2n+1) x^{2n}}{2^{2n+1} n!(n+1)!};$$

$$x^2 y''(x) = \sum_{n=0}^{\infty} \frac{(-1)^{n+1}(2n+1) x^{2n+2}}{2^{2n+1} n!(n+1)!};$$

$$xy'(x) = \sum_{n=0}^{\infty} \frac{(-1)^{n+1} x^{2n+2}}{2^{2n+1} n!(n+1)!};$$

$$x^2 y(x) = \sum_{n=0}^{\infty} \frac{(-1)^n x^{2n+2}}{2^{2n}(n!)^2}.$$

Note that the coefficient n in Bessel's equation is zero. Therefore

$$x^2 y''(x) + xy'(x) + x^2 y(x) = \sum_{n=0}^{\infty} \frac{(-1)^n x^{2n+2}}{2^{2n}(n!)^2} \left[-\frac{2n+1}{2(n+1)} - \frac{1}{2(n+1)} + 1 \right]$$

$$= \sum_{n=0}^{\infty} \frac{(-1)^n x^{2n+2}}{2^{2n}(n!)^2} \left[-\frac{2n+2}{2n+2} + 1 \right] \equiv 0.$$

10.8.61: The Taylor series of f centered at $a = 0$ is

$$f(x) = \frac{\sin x}{x} = 1 - \frac{x^2}{3!} + \frac{x^4}{5!} - \frac{x^6}{7!} + \frac{x^8}{9!} - \cdots.$$

This representation is valid for all real x. We plotted the Taylor polynomials for $f(x)$ with center $a = 0$ of degrees $k = 4$, 6, and 8 and the graph of $y = f(x)$ simultaneously, with the following result.

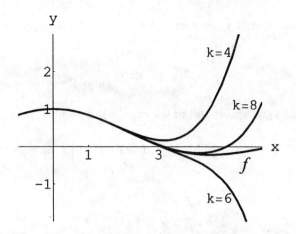

10.8.63: Equation (20) in the text is

$$\arctan x = x - \frac{x^3}{3} + \frac{x^5}{5} - \frac{x^7}{7} + \frac{x^9}{9} - \cdots, \quad -1 < x < 1.$$

Thus

$$\arctan x = \sum_{n=0}^{\infty} \frac{(-1)^n x^{2n+1}}{2n+1} = x \sum_{n=0}^{\infty} \frac{(-1)^n x^{2n}}{2n+1}, \quad -1 < x < 1.$$

Consequently,

$$\pi = 6 \cdot \frac{\pi}{6} = 6 \cdot \arctan \frac{1}{\sqrt{3}} = \frac{6}{\sqrt{3}} \sum_{n=0}^{\infty} \frac{(-1)^n}{2n+1} \cdot \left(\frac{1}{\sqrt{3}} \right)^{2n} = \frac{6}{\sqrt{3}} \sum_{n=0}^{\infty} \frac{(-1)^n}{2n+1} \cdot \frac{1}{3^n}.$$

Mathematica reports that

$$\frac{6}{\sqrt{3}} \sum_{n=0}^{10} \frac{(-1)^n}{2n+1} \cdot \frac{1}{3^n} \approx 3.1415933045030815,$$

$$\frac{6}{\sqrt{3}} \sum_{n=0}^{20} \frac{(-1)^n}{2n+1} \cdot \frac{1}{3^n} \approx 3.1415926535956350,$$

$$\frac{6}{\sqrt{3}} \sum_{n=0}^{30} \frac{(-1)^n}{2n+1} \cdot \frac{1}{3^n} \approx 3.1415926535897932, \quad \text{and}$$

$$\frac{6}{\sqrt{3}} \sum_{n=0}^{40} \frac{(-1)^n}{2n+1} \cdot \frac{1}{3^n} \approx 3.1415926535897932.$$

10.8.65: Part (a): From the Maclaurin series for the natural exponential function in Eq. (2) of this section, we derive the fact that

$$e^{-t} = \frac{1}{e^t} = \sum_{n=0}^{\infty} \frac{(-1)^n}{n!} t^n$$

for all real t. Substitute $t = x \ln x = \ln(x^x)$ to obtain

$$\frac{1}{x^x} = \sum_{n=0}^{\infty} \frac{(-1)^n}{n!} (x \ln x)^n.$$

if $x > 0$. Part (b): The formula in Problem 53 of Section 7.8 states that if m and n are fixed positive integers, then

$$\int_0^1 x^m (\ln x)^n \, dx = \frac{(-1)^n n!}{(m+1)^{n+1}}. \tag{1}$$

Moreover, term-by-term integration is valid in the case of the result in part (a), so

$$\int_0^1 \frac{1}{x^x} \, dx = \sum_{n=0}^{\infty} \frac{(-1)^n}{n!} \int_0^1 (x \ln x)^n \, dx.$$

Thus Eq. (1) here yields

$$\int_0^1 \frac{1}{x^x} \, dx = \sum_{n=0}^{\infty} \frac{(-1)^n}{n!} \cdot \frac{(-1)^n n!}{(n+1)^{n+1}} = \sum_{n=0}^{\infty} \frac{1}{(n+1)^{n+1}} = \sum_{n=1}^{\infty} \frac{1}{n^n}.$$

10.8.67: Part (a):

$$f(x) = \sum_{n=0}^{\infty} \frac{\alpha(\alpha-1)(\alpha-2)\cdots(\alpha-n+1)}{n!} x^n$$

$$= 1 + \alpha x + \frac{\alpha(\alpha-1)}{2!} x^2 + \frac{\alpha(\alpha-1)(\alpha-2)}{3!} x^3 + \cdots + \frac{\alpha(\alpha-1)(\alpha-2)\cdots(\alpha-n+1)}{n!} x^n + \cdots.$$

$$f'(x) = \alpha + \frac{\alpha(\alpha-1)}{1!} x + \frac{\alpha(\alpha-1)(\alpha-2)}{2!} x^2 + \frac{\alpha(\alpha-1)(\alpha-2)(\alpha-3)}{3!} x^3 + \cdots$$

$$+ \frac{\alpha(\alpha-1)(\alpha-2)\cdots(\alpha-n+1)}{(n-1)!} x^{n-1} + \frac{\alpha(\alpha-1)(\alpha-2)\cdots(\alpha-n)}{n!} x^n + \cdots.$$

$$xf'(x) = \alpha x + \frac{\alpha(\alpha-1)}{1!} x^2 + \frac{\alpha(\alpha-1)(\alpha-2)}{2!} x^3 + \frac{\alpha(\alpha-1)(\alpha-2)(\alpha-3)}{3!} x^4 + \cdots$$

$$+ \frac{\alpha(\alpha-1)(\alpha-2)\cdots(\alpha-n+1)}{(n-1)!} x^n + \frac{\alpha(\alpha-1)(\alpha-2)\cdots(\alpha-n)}{n!} x^{n+1} + \cdots.$$

Therefore

$$(1+x)f'(x) = f'(x) + xf'(x)$$

$$= \alpha + (\alpha^2 - \alpha + \alpha)x + \frac{1}{2!}\left[\alpha(\alpha-1)(\alpha-2+2)\right]x^2 + \frac{1}{3!}\left[\alpha(\alpha-1)(\alpha-2)(\alpha-3+3)\right]x^3 + \cdots$$

$$+ \frac{1}{n!}\left[\alpha(\alpha-1)(\alpha-2)\cdots(\alpha-n+1)(\alpha-n+n)\right]x^n + \cdots$$

$$= \alpha + \alpha^2 x + \frac{\alpha^2(\alpha-1)}{2!} x^2 + \frac{\alpha^2(\alpha-1)(\alpha-2)}{3!} x^3 + \cdots$$

$$+ \frac{\alpha^2(\alpha-1)(\alpha-2)\cdots(\alpha-n+1)}{n!} x^n + \cdots = \alpha f(x).$$

624

Part (b): From the result $(1+x)f'(x) = \alpha f(x)$ in part (a), we derive

$$\frac{f'(x)}{f(x)} = \frac{\alpha}{1+x}; \qquad \ln(f(x)) = C + \alpha \ln(1+x);$$

$$f(x) = K(1+x)^{\alpha}; \qquad 1 = f(0) = K \cdot 1^{\alpha}: \quad K = 1.$$

Therefore $f(x) = (1+x)^{\alpha}$, $-1 < x < 1$.

10.8.69: Assume that the power series

$$\sum_{n=0}^{\infty} a_n x^n$$

converges for some $x = x_0 \neq 0$. Then $\{a_n x_0^n\} \to 0$ as $n \to +\infty$. Thus there exists a positive integer K such that, if $n \geq K$ and $|x| < |x_0|$, then

$$|a_n x_0^n| \leq 1; \qquad |a_n x^n| \leq \left|\frac{x^n}{x_0^n}\right|; \qquad |a_n x^n| \leq \left|\frac{x}{x_0}\right|^n.$$

This implies that the series

$$\sum_{n=0}^{\infty} |a_n x^n| \quad \text{is eventually dominated by} \quad \sum_{n=0}^{\infty} \left|\frac{x}{x_0}\right|^n,$$

a convergent geometric series. Therefore $\sum a_n x^n$ converges absolutely if $|x| < |x_0|$.

Section 10.9

10.9.1: To estimate $65^{1/3}$ using the binomial series, first write

$$65^{1/3} = (4^3 + 1)^{1/3} = 4\left(1 + \frac{1}{64}\right)^{1/3}.$$

According to Eq. (14) in Section 10.8, the binomial series is

$$(1+x)^{\alpha} = 1 + \alpha x + \frac{\alpha(\alpha-1)}{2!}x^2 + \frac{\alpha(\alpha-1)(\alpha-2)}{3!}x^3 + \cdots;$$

it has radius of convergence $R = 1$. Therefore

$$4(1+x)^{1/3} = 4 + \frac{4}{3}x - \frac{4 \cdot 2}{3^2} \cdot \frac{x^2}{2!} + \frac{4 \cdot 2 \cdot 5}{3^3} \cdot \frac{x^3}{3!} - \frac{4 \cdot 2 \cdot 5 \cdot 8}{3^4} \cdot \frac{x^4}{4!} + \cdots.$$

With $x = \frac{1}{64}$, this series is alternating after the first two terms. For three digits correct to the right of the decimal in our approximation, we note that if $a = \frac{1}{64}$, then

$$\frac{4 \cdot 2}{3^2} \cdot \frac{a^2}{2!} < 0.00011.$$

Therefore

$$65^{1/3} \approx 4 + \frac{4}{3} \cdot \frac{1}{64} \approx 4.02083333 \approx 4.021.$$

For more accuracy, the sum of the first seven terms of the series is approximately 4.02072575858904; compare this with $65^{1/3} \approx 4.02072575858906$.

10.9.3: The Maclaurin series for the sine function yields

$$\sin(0.5) = \frac{1}{2} - \frac{1}{3! \cdot 2^3} + \frac{1}{5! \cdot 2^5} - \frac{1}{7! \cdot 2^7} - \cdots.$$

The alternating series remainder estimate (Theorem 2 in Section 10.7) tells us that because

$$\frac{1}{5! \cdot 2^5} \approx 0.000260417 < 0.0003,$$

the error in approximating $\sin(0.5)$ using only the first two terms of this series will be no greater than 0.0003. Thus

$$\sin(0.5) \approx \frac{1}{2} - \frac{1}{3! \cdot 2^3} \approx 0.47916667.$$

Therefore, to three places, $\sin(0.5) \approx 0.479$. As a separate check, the sum of the first three terms of the series is approximately 0.47942708, the sum of its first six terms is approximately 0.479425538604, and this agrees with the true value of $\sin(0.5)$ to the number of digits shown.

10.9.5: The Maclaurin series of the inverse tangent function is given in Eq. (20) of Section 10.8; it is

$$\arctan x = x - \frac{1}{3}x^3 + \frac{1}{5}x^5 - \frac{1}{7}x^7 + \frac{1}{9}x^9 - \cdots.$$

When $x = \frac{1}{2}$, the sum of the first four terms of this series is approximately 0.4634672619. With five terms we get 0.4636842758 and with six terms we get 0.4636398868. To three places, $\arctan(0.5) \approx 0.464$.

10.9.7: When $x = \pi/10$ is substituted in the Maclaurin series for the sine function (Eq. (4) in Section 10.8), we obtain

$$\sin\left(\frac{\pi}{10}\right) = \frac{\pi}{10} - \frac{\pi^3}{3! \cdot 10^3} + \frac{\pi^5}{5! \cdot 10^5} - \frac{\pi^7}{7! \cdot 10^7} + \cdots.$$

Because

$$\frac{\pi^5}{5! \cdot 10^5} \approx 0.000025501641 < 0.00003,$$

the sum of the first two terms of the series should yield three-place accuracy. Thus because

$$\sin\left(\frac{\pi}{10}\right) \approx \frac{\pi}{10} - \frac{\pi^3}{3! \cdot 10^3} \approx 0.30899155,$$

we may conclude that $\sin(\pi/10) \approx 0.309$ to three places. In fact, the sum of the first three terms of the series is approximately 0.30901699 and $\sin(\pi/10) \approx 0.30901699$ to the number of digits shown.

10.9.9: First we convert $10°$ into $\pi/18$ radians, then use the Maclaurin series for the sine function (Eq. (4) of Section 10.8):

$$\sin 10° = \sin\left(\frac{\pi}{18}\right) = \frac{\pi}{18} - \frac{\pi^3}{3! \cdot 18^3} + \frac{\pi^5}{5! \cdot 18^5} - \frac{\pi^7}{7! \cdot 18^7} + \cdots.$$

Because

$$\frac{\pi^3}{3! \cdot 18^3} \approx 0.000886096,$$

the first term of the Maclaurin series alone may not give three-place accuracy (it doesn't). But

$$\frac{\pi^5}{5! \cdot 18^5} \approx 0.0000013496016,$$

so we will certainly obtain three-place accuracy by adding the first two terms of the series:

$$\sin\left(\frac{\pi}{18}\right) \approx \frac{\pi}{18} - \frac{\pi^3}{3! \cdot 18^3} \approx 0.17364683.$$

Thus, to three places, $\sin 10° \approx 0.174$. To check, the sum of the first three terms of the series is approximately 0.1736481786 and the true value of $\sin 10°$ is approximately 0.1736481777 (to the number of digits shown).

10.9.11: Four-place accuracy demands that the error not exceed 0.00005. Here we have

$$I = \int_0^1 \frac{\sin x}{x}\,dx = \int_0^1 \left(1 - \frac{x^2}{3!} + \frac{x^4}{5!} - \frac{x^6}{7!} + \cdots\right)dx$$

$$= 1 - \frac{1}{3!3} + \frac{1}{5!5} - \frac{1}{7!7} + \frac{1}{9!9} - \frac{1}{11!11} + \cdots.$$

The alternating series remainder estimate (Theorem 2 of Section 10.7), when applied, yields

$$\frac{1}{7!7} \approx 0.000028345 \quad \text{(not good enough) and} \quad \frac{1}{9!9} \approx 0.000000008748 \quad \text{(great accuracy)},$$

so that

$$I \approx 1 - \frac{1}{3!3} + \frac{1}{5!5} - \frac{1}{7!7} \approx 0.946082766;$$

to four places, $I \approx 0.9461$. To check, the sum of the first five terms of the series is approximately 0.9460830726 and the true value of the integral is 0.9460830703671830, correct to the number of digits shown here.

10.9.13: The Maclaurin series for the inverse tangent function (Eq. (20) in Section 10.8) leads to

$$K = \int_0^{1/2} \frac{\arctan x}{x}\,dx = \int_0^{1/2} \left(1 - \frac{x^2}{3} + \frac{x^4}{5} - \frac{x^6}{7} + \cdots\right)dx$$

$$= \frac{1}{2} - \frac{1}{2^3 \cdot 3^2} + \frac{1}{2^5 \cdot 5^2} - \frac{1}{2^7 \cdot 7^2} + \frac{1}{2^9 \cdot 9^2} - \cdots.$$

Now

$$\frac{1}{2^9 \cdot 9^2} \approx 0.000024112654 < 0.00003,$$

so by the alternating series error estimate, the sum of the first four terms of the numerical series should yield four-place accuracy. Thus

$$K \approx \frac{1}{2} - \frac{1}{2^3 \cdot 3^2} + \frac{1}{2^5 \cdot 5^2} - \frac{1}{2^7 \cdot 7^2} \approx 0.4872016723;$$

that is, to four places $K \approx 0.4872$. To check this result, the sum of the first five terms of the series is approximately 0.487225784990 and the approximate value of the integral is 0.487222358295 (to the number of digits shown).

10.9.15: The Maclaurin series for $\ln(1+x)$, in Eq. (19) of Section 10.8, yields

$$I = \int_0^{1/10} \frac{\ln(1+x)}{x}\,dx = \int_0^{1/10} \left(1 - \frac{x}{2} + \frac{x^2}{3} - \frac{x^3}{4} + \frac{x^4}{5} - \frac{x^5}{6} + \cdots \right)\,dx$$

$$= \left[x - \frac{x^2}{4} + \frac{x^3}{9} - \frac{x^4}{16} + \frac{x^5}{25} - \frac{x^6}{36} + \cdots \right]_0^{1/10} = \frac{1}{10} - \frac{1}{4 \cdot 10^2} + \frac{1}{9 \cdot 10^3} - \frac{1}{16 \cdot 10^4} + \frac{1}{25 \cdot 10^5} - \cdots.$$

Because

$$\frac{1}{16 \cdot 10^4} = 0.00000625,$$

the alternating series remainder estimate tells us that the sum of the first three terms of the numerical series should give four-place accuracy. Because

$$I \approx \frac{1}{10} - \frac{1}{4 \cdot 10^2} + \frac{1}{9 \cdot 10^3} \approx 0.097611111,$$

the four-place approximation we seek is $I \approx 0.0976$. To check this result, the sum of the first four terms of the series is approximately 0.097604861111 and the value of the integral, to the number of digits shown here, is 0.0976052352293216.

10.9.17: The Maclaurin series for the natural exponential function—Eq. (2) in Section 10.8—yields

$$J = \int_0^{1/2} \frac{1 - e^{-x}}{x}\,dx = \int_0^{1/2} \left(1 - \frac{x}{2!} + \frac{x^2}{3!} - \frac{x^3}{4!} + \frac{x^4}{5!} - \cdots \right)\,dx$$

$$= \left[x - \frac{x^2}{2!2} + \frac{x^3}{3!3} - \frac{x^4}{4!4} + \frac{x^5}{5!5} - \cdots \right]_0^{1/2} = \frac{1}{2} - \frac{1}{2! \cdot 2 \cdot 2^2} + \frac{1}{3! \cdot 3 \cdot 2^3} - \frac{1}{4! \cdot 4 \cdot 2^4} + \cdots.$$

Because

$$\frac{1}{6! \cdot 6 \cdot 2^6} \approx 0.00000361690,$$

628

the alternating series remainder estimate assures us that the sum of the first five terms of the series, which is approximately 0.443845486111, will give us four-place accuracy: $J \approx 0.4438$. The value of the integral, accurate to the number of digits shown here, is 0.4438420791177484.

10.9.19: The Maclaurin series for the natural exponential function—Eq. (2) in Section 10.8—yields

$$\exp(-x^2) = 1 - x^2 + \frac{x^4}{2!} - \frac{x^6}{3!} + \frac{x^8}{4!} - \frac{x^{10}}{5!} + \cdots .$$

Then termwise integration gives

$$\int_0^1 \exp(-x^2)\, dx = 1 - \frac{1}{3} + \frac{1}{2! \cdot 5} - \frac{1}{3! \cdot 7} + \frac{1}{4! \cdot 9} - \frac{1}{5! \cdot 11} + \cdots .$$

Next, the sum of the first six terms of the numerical series is approximately 0.746729196729, the sum of the first seven terms is approximately 0.746836034336, and the sum of the first eight terms is approximately 0.746822806823. To four places, the value of the integral is 0.7468. The actual value of the integral, to the number of digits shown here, is 0.7468241328124270.

10.9.21: The binomial series (Example 8 in Section 10.8) yields

$$(1 + x^2)^{1/3} = 1 + \frac{x^2}{3} - \frac{2x^4}{2! \cdot 3^2} + \frac{2 \cdot 5 x^6}{3! \cdot 3^3} - \frac{2 \cdot 5 \cdot 8 x^8}{4! \cdot 3^4} + \frac{2 \cdot 5 \cdot 8 \cdot 11 x^{10}}{5! \cdot 3^5} - \cdots .$$

Then term-by-term integration yields

$$I = \int_0^{1/2} (1 + x^2)^{1/3}\, dx = \left[x + \frac{x^3}{3 \cdot 3} - \frac{2x^5}{2! \cdot 3^2 \cdot 5} + \frac{2 \cdot 5 x^7}{3! \cdot 3^3 \cdot 7} - \frac{2 \cdot 5 \cdot 8 x^9}{4! \cdot 3^4 \cdot 9} + \frac{2 \cdot 5 \cdot 8 \cdot 11 x^{11}}{5! \cdot 3^5 \cdot 11} - \cdots \right]_0^{1/2}$$

$$= \frac{1}{2} + \frac{1}{3 \cdot 3 \cdot 2^3} - \frac{2}{2! \cdot 3^2 \cdot 5 \cdot 2^5} + \frac{2 \cdot 5}{3! \cdot 3^3 \cdot 7 \cdot 2^7} - \frac{2 \cdot 5 \cdot 8}{4! \cdot 3^4 \cdot 9 \cdot 2^9} + \frac{2 \cdot 5 \cdot 8 \cdot 11}{5! \cdot 3^5 \cdot 11 \cdot 2^{11}} - \cdots .$$

The sum of the first five terms of the numerical series is approximately 0.513254407130 and the sum of its first six terms is approximately 0.513255746722. So to four places, $I \approx 0.5133$. The actual value of the integral, accurate to the number of digits shown here, is 0.5132555590033423.

10.9.23: The Maclaurin series for the natural exponential function (Eq. (2) in Section 10.8) yields

$$\lim_{x \to 0} \frac{1 + x - e^x}{x^2} = \lim_{x \to 0} \frac{1}{x^2} \left(-\frac{x^2}{2!} - \frac{x^3}{3!} - \frac{x^4}{4!} - \cdots \right) = \lim_{x \to 0} \left(-\frac{1}{2} - \frac{x}{6} - \frac{x^2}{24} - \cdots \right) = -\frac{1}{2}.$$

10.9.25: The series in Eqs. (2) and (3) of Section 10.8 yield

$$\lim_{x \to 0} \frac{1 - \cos x}{x(e^x - 1)} = \lim_{x \to 0} \frac{\dfrac{x^2}{2!} - \dfrac{x^4}{4!} + \dfrac{x^6}{6!} - \cdots}{x^2 + \dfrac{x^3}{2!} + \dfrac{x^4}{3!} + \cdots} = \lim_{x \to 0} \frac{\dfrac{1}{2!} - \dfrac{x^2}{4!} + \dfrac{x^4}{6!} - \cdots}{1 + \dfrac{x}{2!} + \dfrac{x^2}{3!} + \cdots} = \frac{1}{2}.$$

10.9.27: The Maclaurin series for the sine function in Eq. (4) of Section 10.8 yields

$$\lim_{x \to 0} \left(\frac{1}{x} - \frac{1}{\sin x} \right) = \lim_{x \to 0} \frac{(\sin x) - x}{x \sin x}$$

$$= \lim_{x \to 0} \frac{-\dfrac{x^3}{3!} + \dfrac{x^5}{5!} - \dfrac{x^7}{7!} + \cdots}{x^2 - \dfrac{x^4}{3!} + \dfrac{x^6}{5!} - \cdots} = \lim_{x \to 0} \frac{-\dfrac{x}{3!} + \dfrac{x^3}{5!} - \dfrac{x^5}{7!} + \cdots}{1 - \dfrac{x^2}{3!} + \dfrac{x^4}{5!} - \cdots} = \frac{0}{1} = 0.$$

10.9.29: The Taylor series with center $\pi/2$ for the sine function is

$$\sin x = 1 - \frac{1}{2!}\left(x - \frac{\pi}{2}\right)^2 + \frac{1}{4!}\left(x - \frac{\pi}{2}\right)^4 - \frac{1}{6!}\left(x - \frac{\pi}{4}\right)^6 + \cdots .$$

We convert $80°$ into $x = 4\pi/9$ and substitute:

$$\sin 80° = 1 - \frac{1}{2!}\left(\frac{\pi}{18}\right)^2 + \frac{1}{4!}\left(\frac{\pi}{18}\right)^4 - \frac{1}{6!}\left(\frac{\pi}{18}\right)^6 + \cdots . \tag{1}$$

For four-place accuracy, we need

$$\frac{1}{n!}\left(\frac{\pi}{18}\right)^n < 0.00005,$$

and the smallest even positive integer for which this inequality holds is $n = 4$. Thus only the first two terms of the series in (1) are needed to show that, to four places, $\sin 80° \approx 0.9848$. The sum of the first two terms is approximately 0.984769129011, the sum of the first three terms is approximately 0.984807792249, and $\sin 80° \approx 0.984807753012$ (all digits given here are correct).

10.9.31: The Taylor series with center $\pi/4$ for the cosine function is

$$\cos x = \frac{\sqrt{2}}{2}\left[1 - \left(x - \frac{\pi}{4}\right) - \frac{1}{2!}\left(x - \frac{\pi}{4}\right)^2 + \frac{1}{3!}\left(x - \frac{\pi}{4}\right)^3 + \frac{1}{4!}\left(x - \frac{\pi}{4}\right)^4 - \frac{1}{5!}\left(x - \frac{\pi}{4}\right)^5 - \cdots\right].$$

We convert $47°$ to radians and substitute to find that

$$\cos 47° = \frac{\sqrt{2}}{2}\left[1 - \frac{\pi}{90} - \frac{1}{2!}\left(\frac{\pi}{90}\right)^2 + \frac{1}{3!}\left(\frac{\pi}{90}\right)^3 + \frac{1}{4!}\left(\frac{\pi}{90}\right)^4 - \frac{1}{5!}\left(\frac{\pi}{90}\right)^5 - \cdots\right].$$

This series is absolutely convergent, so rearrangement will not change its convergence or its sum. We make it into an alternating series by grouping terms 2 and 3, terms 4 and 5, and so on. We seek six-place accuracy here. If $x = \pi/90$, then

$$\frac{x^3}{3!} + \frac{x^4}{4!} \approx 0.000007 \quad \text{and that} \quad \frac{x^5}{5!} + \frac{x^6}{6!} \approx 0.00000000043,$$

so the first five terms of the [ungrouped] series—those through exponent 4—yield the required six-place accuracy: $\cos 47° \approx 0.681998$. The actual value of $\cos 47°$ is approximately 0.6819983600624985 (the digits shown here are all correct or correctly rounded).

10.9.33: Note that $e^{0.1} < 1.2 = \frac{6}{5}$, and if $|x| \leqq 0.1$, then the Taylor series remainder estimate yields

$$\left| \frac{e^z}{120} x^5 \right| \leqq \frac{6}{600} \left(\frac{1}{10} \right)^5 = 10^{-7} < 0.5 \times 10^{-6},$$

so six-place accuracy is assured.

10.9.35: The Taylor series remainder estimate is difficult to work with; we use instead the cruder alternating series remainder estimate:

$$\frac{(0.1)^5}{5} < 0.5 \times 10^{-5},$$

so five-place accuracy is assured.

10.9.37: Clearly $|e^z| < \frac{5}{3}$ if $|z| < 0.5$. Hence the Taylor series remainder estimate yields

$$\left| \frac{e^z}{120} x^5 \right| \leqq \frac{5}{3 \cdot 120} \left(\frac{1}{2} \right)^5 \approx 0.434 \times 10^{-3},$$

so two-place accuracy will be obtained if $|x| \leqq 0.5$. In particular,

$$e^{1/3} \approx 1 + \frac{1}{3} + \frac{1}{18} + \frac{1}{486} + \frac{1}{1944} \approx 1.39.$$

In fact, to the number of digits shown here, $e^{1/3} \approx 1.395612425086$.

10.9.39: The Taylor series remainder estimate is

$$|R_3(x)| = \frac{\sqrt{2}}{2} \cdot \frac{\cos z}{4!} \left(x - \frac{\pi}{4} \right)^4.$$

Part (a): If $40° \leqq x° \leqq 50°$, then

$$\frac{2\pi}{9} \leqq x \leqq \frac{5\pi}{18} \quad \text{and} \quad \cos z \leqq \cos \left(\frac{\pi}{6} \right) = \frac{\sqrt{3}}{2},$$

so

$$|R_3(x)| \leqq \frac{\sqrt{2}}{2} \cdot \frac{1}{24} \cdot \left(\frac{\pi}{36} \right)^4 \cdot \frac{\sqrt{3}}{2} \approx 0.0000014797688 < 0.000002,$$

thereby giving five-place accuracy. Part (b): If $44° \leqq x° \leqq 46°$, then

$$\frac{44\pi}{180} \leqq x \leqq \frac{46\pi}{180},$$

so that

$$|R_3(x)| \leqq \frac{\sqrt{2}}{2} \cdot \frac{1}{24} \cdot \left(\frac{\pi}{180} \right)^4 \cdot \frac{\sqrt{3}}{2} \approx 0.0000000023676302 < 0.000000003,$$

thereby giving eight-place accuracy.

10.9.41: The volume of revolution around the x-axis is

$$V = 2 \int_0^\pi \pi \frac{\sin^2 x}{x^2} \, dx = 2\pi \int_0^\pi \frac{1 - \cos 2x}{2x^2} \, dx$$

$$= \pi \int_0^\pi \frac{1}{x^2} \left(\frac{(2x)^2}{2!} - \frac{(2x)^4}{4!} + \frac{(2x)^6}{6!} - \frac{(2x)^8}{8!} + \cdots \right) dx = \pi \int_0^\pi \left(\frac{2^2}{2!} - \frac{2^4 x^2}{4!} + \frac{2^6 x^4}{6!} - \frac{2^8 x^6}{8!} + \cdots \right) dx$$

$$= \pi \left[\frac{2^2 x}{2!} - \frac{2^4 x^3}{4! \cdot 3} + \frac{2^6 x^5}{6! \cdot 5} - \frac{2^8 x^7}{8! \cdot 7} + \cdots \right]_0^\pi = \frac{(2\pi)^2}{2!} - \frac{(2\pi)^4}{4! \cdot 3} + \frac{(2\pi)^6}{6! \cdot 5} - \frac{(2\pi)^8}{8! \cdot 7} + \cdots .$$

This series converges rapidly after the first 10 or 15 terms. For example, the sum of the first seven terms is about 8.927353886225. The *Mathematica* command

```
NSum[ ((-1)∧(k+1))*((2*Pi)∧(2*k))/(((2*k)!)*(2*k-1)),

    { k, 1, Infinity }, WorkingPrecision → 28 ]
```

returns the approximate sum 8.910509146510103807178167 8. The *Mathematica* 3.0 command

```
2*Integrate[ Pi*(Sin[x]/x)∧2, {x, 0, Pi} ]
```

produces the exact value of the volume:

$$V = -1 + \text{HypergeometricPFQ} \left[\left\{ -\frac{1}{2} \right\}, \left\{ \frac{1}{2}, \frac{1}{2} \right\}, -\pi^2 \right]$$

$$\approx 8.9105091465101038071781677928811594135107930070735323609643.$$

The *Mathematica* function `HypergeometricPFQ` is the generalized hypergeometric function $_pF_q$. Space prohibits further discussion; we've mentioned this only to give you a reference in case you're interested in further details.

10.9.43: The volume is

$$V = \int_0^{2\pi} 2\pi x \frac{1 - \cos x}{x^2} \, dx = 2\pi \int_0^{2\pi} \left(\frac{x}{2!} - \frac{x^3}{4!} + \frac{x^5}{6!} - \frac{x^7}{8!} + \frac{x^9}{10!} - \cdots \right) dx$$

$$= 2\pi \left[\frac{x^2}{2! \cdot 2} - \frac{x^4}{4! \cdot 4} + \frac{x^6}{6! \cdot 6} - \frac{x^8}{8! \cdot 8} + \frac{x^{10}}{10! \cdot 10} - \cdots \right]_0^{2\pi}$$

$$= \frac{(2\pi)^3}{2! \cdot 2} - \frac{(2\pi)^5}{4! \cdot 4} + \frac{(2\pi)^7}{6! \cdot 6} - \frac{(2\pi)^9}{8! \cdot 8} + \frac{(2\pi)^{11}}{10! \cdot 10} - \cdots = \sum_{n=1}^\infty \frac{(-1)^{n+1} (2\pi)^{2n+1}}{(2n)! \cdot 2n} .$$

This alternating series converges slowly at first—its ninth term is approximately 0.0127—but its 21st term is less than 4×10^{-19}, so the sum of its first 20 terms is a very accurate estimate of its value. That partial sum is approximately 15.3162279832536178, and all the digits shown here are accurate.

The *Mathematica* command

```
Integrate[ 2*Pi*x*(1 - Cos[x])/(x*x), {x, 0, 2*Pi} ]
```

produces the exact value of the volume:

$$V = 2\pi \big[\text{EulerGamma} - \text{CosIntegral}(2\pi) + \ln(2\pi) \big]$$

$$\approx 15.31622798325361781931489070705969367325235855560299990575827.$$

Here, `EulerGamma` is Euler's constant $\gamma \approx 0.577216$, which first appears in the textbook in Problem 50 of Section 10.5; `CosIntegral` is defined to be

$$\text{CosIntegral}(x) = \gamma + \ln x + \int_0^x \frac{(\cos t) - 1}{t} \, dt.$$

Again, the reference is provided only for your convenience if you care to pursue further study of this topic.

10.9.45: The long division is shown next.

$$
\begin{array}{r}
1 + x + x^2 + x^3 + \cdots \\
\hline
1 - x \,) \; x \\
1 - x \\
\hline
x \\
x - x^2 \\
\hline
x^2 - x^3 \\
\hline
x^3 \\
\cdots
\end{array}
$$

10.9.47: The equation (actually, the *identity*)

$$(1 - x)(a_0 + a_1 x + a_2 x^2 + a_3 x^3 + \cdots + a_n x^n + \cdots) = 1$$

leads to

$$a_0 + (a_1 - a_0)x + (a_2 - a_1)x^2 + (a_3 - a_2)x^3 + (a_4 - a_3)x^4 + \cdots = 1.$$

It now follows that $a_0 = 1$ and that $a_{n+1} = a_n$ if $n \geq 0$, and therefore $a_n = 1$ for every integer $n \geq 0$. Consequently

$$\frac{1}{1 - x} = 1 + x + x^2 + x^3 + x^4 + \cdots + x^n + \cdots, \quad -1 < x < 1.$$

10.9.49: The method of Example 3 uses the identity

$$\sec x \cos x = 1$$

633

and begins by assuming the existence of coefficients $\{a_i\}$ such that

$$\sec x = a_0 + a_1 x + a_2 x^2 + a_3 x^3 + a_4 x^4 + \cdots .$$

Thus

$$(a_0 + a_1 x + a_2 x^2 + a_3 x^3 + a_4 x^4 + \cdots)\left(1 - \frac{x^2}{2} + \frac{x^4}{24} - \frac{x^6}{720} + \cdots\right) = 1,$$

so that

$$a_0 + a_1 x + \left(a_2 - \frac{1}{2}a_0\right)x^2 + \left(a_3 - \frac{1}{2}a_1\right)x^3 + \left(a_4 - \frac{1}{2}a_2 + \frac{1}{24}a_0\right)x^4 + \cdots = 1.$$

It now follows that

$$a_0 = 1; \qquad a_1 = 0;$$

$$a_2 = \frac{1}{2}a_0 = \frac{1}{2}; \qquad a_3 = \frac{1}{2}a_1 = 0;$$

$$a_4 = \frac{1}{2}a_2 - \frac{1}{24}a_0 = \frac{5}{24}.$$

Therefore

$$\sec x = 1 + \frac{1}{2}x^2 + \frac{5}{24}x^4 + \frac{61}{720}x^6 + \frac{277}{8064}x^8 + \frac{50521}{3628800}x^{10} + \frac{540553}{95800320}x^{12} + \frac{199360981}{87178291200}x^{14} + \cdots .$$

We had *Mathematica* compute a few extra terms in case you did as well and want to check your work. (We used the command

```
Series[ Sec[x], { x, 0, 20 } ] // Normal
```

but to save space we show here only the first eight terms of the resulting Taylor polynomial.)

10.9.51: Example 10 in Section 10.8 shows how to derive the series

$$\ln(1+x) = x - \frac{x^2}{2} + \frac{x^3}{3} - \frac{x^4}{4} + \frac{x^5}{5} - \cdots, \qquad -1 < x < 1.$$

Hence

$$1 + x = \exp(\ln(1+x)) = \sum_{n=0}^{\infty} a_n \left(x - \frac{x^2}{2} + \frac{x^3}{3} - \frac{x^4}{4} + \frac{x^5}{5} - \cdots\right)^n$$

$$= a_0 + a_1\left(x - \frac{x^2}{2} + \frac{x^3}{3} - \frac{x^4}{4} + \frac{x^5}{5} - \cdots\right) + a_2\left(x^2 - x^3 + \left[\frac{1}{4} + \frac{2}{3}\right]x^4 - \cdots\right)$$

$$+ a_3\left(x^3 + \left[-\frac{1}{2} - 1\right]x^4 + \cdots\right) + \cdots$$

$$= a_0 + a_1 x + \left(a_2 - \frac{1}{2}a_1\right)x^2 + \left(a_3 - a_2 + \frac{1}{3}a_1\right)x^3 + \cdots .$$

Therefore

$$a_0 = 1, \qquad a_1 = 1, \qquad a_2 = \frac{1}{2}a_1 = \frac{1}{2}, \qquad \text{and} \qquad a_3 = a_2 - \frac{1}{3}a_1 = \frac{1}{6}.$$

10.9.53: Long division of the finite power series $1 + x + x^2$ into the finite power series $2 + x$ proceeds as shown next.

$$
\begin{array}{r}
2 \;-\; x \;-\; x^2 + \cdots \\[2pt]
\hline
1 + x + x^2 \;)\; 2 \;+\; x \phantom{{}+ x^2} \\
\underline{2 \;+\; 2x \;+\; 2x^2} \\
-\,x \;-\; 2x^2 \\
\underline{-\,x \;-\; x^2 \;-\; x^3} \\
-\,x^2 \;+\; x^3 \\
\underline{-\,x^2 \;-\; x^3 \;-\; x^4} \\
2x^3 \;+\; x^4 \\[4pt]
\cdots
\end{array}
$$

The next dividend is the original dividend with exponents increased by 3, so the next three terms in the numerator can be obtained by multiplying the first three by x^3. Thus we obtain the series representation

$$\frac{2+x}{1+x+x^2} = 2 - x - x^2 + 2x^3 - x^4 - x^5 + 2x^6 - x^7 - x^8 + \cdots .$$

This series is the sum of three geometric series each with ratio x^3, so it converges on the interval $(-1, 1)$. Summing the three geometric series separately, we obtain

$$\frac{2}{1-x^3} - \frac{x}{1-x^3} - \frac{x^2}{1-x^3} = \frac{2-x-x^2}{1-x^3} = \frac{(1-x)(2+x)}{(1-x)(1+x+x^2)},$$

thus verifying our computations.

10.9.55: The first two steps in the long division of $1 + x^2 + x^4$ into 1 give quotient $1 - x^2$ and remainder (and new dividend) x^6. So the process will repeat with exponents increased by 6, and thus

$$\frac{1}{1+x^2+x^4} = 1 - x^2 + x^6 - x^8 + x^{12} - x^{14} + x^{18} - x^{20} + \cdots .$$

Therefore

$$\int_0^{1/2} \frac{1}{1+x^2+x^4}\, dx = \left[x - \frac{x^3}{3} + \frac{x^7}{7} - \frac{x^9}{9} + \frac{x^{13}}{13} - \frac{x^{15}}{15} \cdots \right]_0^{1/2}$$

$$= \frac{1}{2} - \frac{1}{2^3 \cdot 3} + \frac{1}{2^7 \cdot 7} - \frac{1}{2^9 \cdot 9} + \frac{1}{2^{13} \cdot 13} - \frac{1}{2^{15} \cdot 15} + \cdots .$$

635

The sum of the first five terms of the last series is approximately 0.459239824988 and the sum of the first six terms is approximately 0.459239825000. The *Mathematica* command

```
NIntegrate[ 1/(1 + x^2 + x^4), { x, 0, 1/2 }, WorkingPrecision → 28 ]
```

returns 0.4592398249998759. The computer algebra system *Derive* yields

$$\int_0^{1/2} \frac{1}{1+x^2+x^4} \, dx$$

$$= \left[\frac{\sqrt{3}}{6} \left\{ \arctan\left(\frac{\sqrt{3}}{3}(2x+1) \right) + \arctan\left(\frac{\sqrt{3}}{3}(2x-1) \right) \right\} + \frac{1}{4} \ln\left(\frac{x^2+x+1}{x^2-x+1} \right) \right]_0^{1/2}$$

$$= \frac{\sqrt{3}}{6} \arctan\left(\frac{2\sqrt{3}}{3} \right) + \frac{1}{4} \ln\left(\frac{7}{3} \right) \approx 0.45923982499987591403.$$

10.9.57: See the solution of Problem 61 in Section 10.8. We plotted the Taylor polynomials with center zero for $f(x)$,

$$P_n(x) = \sum_{k=1}^{n} \frac{(-1)^{k+1} x^{2k-2}}{(2k-1)!},$$

for $n = 3$, 6, and 9. Their graphs, together with the graph of f, are shown next.

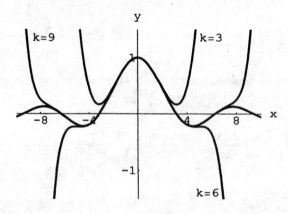

10.9.59: The four Maclaurin series we need are in Eq. (4) of Section 10.8, Eq. (7) of Section 10.9, Eq. (21) of Section 10.8, and Eq. (20) of Section 10.8. They are

$$\sin x = x - \frac{x^3}{6} + \frac{x^5}{120} - \frac{x^7}{5040} + \cdots ,$$

$$\tan x = x + \frac{x^3}{3} + \frac{2x^5}{15} + \frac{17x^7}{315} + \cdots ,$$

$$\arcsin x = x + \frac{x^3}{6} + \frac{3x^5}{40} + \frac{5x^7}{112} + \cdots , \quad \text{and}$$

$$\arctan x = x - \frac{x^3}{3} + \frac{x^5}{5} - \frac{x^7}{7} + \cdots .$$

Therefore

$$\frac{\sin x - \tan x}{\arcsin x - \arctan x} = \frac{-\frac{1}{2}x^3 - \frac{1}{8}x^5 - \cdots}{\frac{1}{2}x^3 + \frac{1}{8}x^5 + \cdots} = \frac{-\frac{1}{2} - \frac{1}{8}x^2 - \cdots}{\frac{1}{2} + \frac{1}{8}x^2 + \cdots} \rightarrow \frac{-\frac{1}{2} - 0 - 0 - \cdots}{\frac{1}{2} + 0 + 0 + \cdots} = -1$$

as $x \to 0$.

Use of l'Hôpital's rule to solve this problem—even with the aid of a computer algebra program—is troublesome. Using *Mathematica*, we first defined

$$f(x) = \sin x - \tan x \quad \text{and} \quad g(x) = \arcsin x - \arctan x.$$

Then the command

```
Limit[ f[x]/g[x], x → 0 ]
```

elicited the disappointing response "Indeterminate." But when we computed the quotient of the derivatives using

```
f'[x]/g'[x]
```

we obtained the expected fraction

$$\frac{\cos x - \sec^2 x}{\dfrac{1}{\sqrt{1-x^2}} - \dfrac{1}{1+x^2}},$$

and *Mathematica* reported that the limit of this fraction, as $x \to 0$, was -1. To avoid *Mathematica's* invocation of l'Hôpital's rule (with the intent of obtaining a fraction that was *not* indeterminate), we had *Mathematica* find the quotient of the derivatives of the last numerator and denominator; the resulting numerator was

$$-2x\sqrt{1-x^2}\left(\cos x - \sec^2 x\right) + \frac{x(1+x^2)(\cos x - \sec^2 x)}{\sqrt{1-x^2}} + (1+x^2)\sqrt{1-x^2}\left(\sin x + 2\sec^2 x \tan x\right)$$

and the corresponding denominator was

$$-2x - \frac{x}{\sqrt{1-x^2}}.$$

The quotient is still indeterminate, but repeating the process—you really don't want to see the results—next (and finally) led to a form not indeterminate, whose value at $x = 0$ was (still) -1.

10.9.61: Part (a): Assume that $a \geqq b > 0$. The parametrization $x = a\cos t$, $y = b\sin t$ yields arc length element

$$ds = (a^2 \sin^2 t + b^2 \cos^2 t)^{1/2}\, dt = [a^2 \sin^2 t + a^2 \cos^2 t + (b^2 - a^2)\cos^2 t]^{1/2}\, dt$$

$$= [a^2 + (b^2 - a^2)\cos^2 t]^{1/2}\, dt = a\left[1 - \frac{a^2 - b^2}{a^2}\cos^2 t\right]^{1/2} dt = a(1 - \epsilon^2 \cos^2 t)^{1/2}\, dt$$

637

where

$$\epsilon = \sqrt{\frac{a^2 - b^2}{a^2}} = \sqrt{1 - (b/a)^2}$$

is the eccentricity of the ellipse; recall that $0 \leq \epsilon < 1$ for the case of an ellipse. We multiply the length of the part of the ellipse in the first quadrant by 4 to find its total arc length is

$$p = 4a \int_0^{\pi/2} \sqrt{1 - \epsilon^2 \cos^2 t} \, dt.$$

Part (b): The binomial formula yields

$$(1 - x)^{1/2} = 1 - \frac{1}{2}x - \frac{1}{2^2} \cdot \frac{x^2}{2!} - \frac{3}{2^3} \cdot \frac{x^3}{3!} - \frac{3 \cdot 5}{2^4} \cdot \frac{x^4}{4!} - \frac{3 \cdot 5 \cdot 7}{2^5} \cdot \frac{x^5}{5!} - \cdots.$$

Consequently,

$$\sqrt{1 - \epsilon^2 \cos^2 t} = 1 - \frac{1}{2}\epsilon^2 \cos^2 t - \frac{1}{2^2 \cdot 2!}\epsilon^4 \cos^4 t - \frac{3}{2^3 \cdot 3!}\epsilon^6 \cos^6 t$$

$$- \frac{3 \cdot 5}{2^4 \cdot 4!}\epsilon^8 \cos^8 t - \frac{3 \cdot 5 \cdot 7}{2^5 \cdot 5!}\epsilon^{10} \cos^{10} t - \cdots.$$

Then formula 113 in the Table of Integrals in the text yields

$$p = 4a \int_0^{\pi/2} \sqrt{1 - \epsilon^2 \cos^2 t} \, dt$$

$$= 4a \left(\Big[t \Big]_0^{\pi/2} - \frac{1}{2}\epsilon^2 \cdot \frac{1}{2} \cdot \frac{\pi}{2} - \frac{1}{2^2 \cdot 2!}\epsilon^4 \cdot \frac{1}{2} \cdot \frac{3}{4} \cdot \frac{\pi}{2} \right.$$

$$\left. - \frac{3}{2^3 \cdot 3!}\epsilon^6 \cdot \frac{1}{2} \cdot \frac{3}{4} \cdot \frac{5}{6} \cdot \frac{\pi}{2} - \frac{3 \cdot 5}{2^4 \cdot 4!}\epsilon^8 \cdot \frac{1}{2} \cdot \frac{3}{4} \cdot \frac{5}{6} \cdot \frac{7}{8} \cdot \frac{\pi}{2} - \cdots \right)$$

$$= 2\pi a \left(1 - \frac{1}{4}\epsilon^2 - \frac{3}{64}\epsilon^4 - \frac{5}{256}\epsilon^6 - \frac{175}{16384}\epsilon^8 - \cdots \right).$$

Section 10.10

10.10.1: We use series methods to solve $\dfrac{dy}{dx} = y$. Assume that

$$y(x) = \sum_{n=0}^{\infty} a_n x^n,$$

so that

$$y'(x) = \sum_{n=1}^{\infty} n a_n x^{n-1} = \sum_{n=0}^{\infty} (n+1) a_{n+1} x^n.$$

638

Substitution in the given differential equation yields

$$\sum_{n=0}^{\infty} [(n+1)a_{n+1} - a_n] x^n = 0,$$

and hence

$$a_{n+1} = \frac{a_n}{n+1} \quad \text{for} \quad n \geq 0.$$

Thus

$$a_1 = a_0, \qquad a_2 = \frac{a_1}{2} = \frac{a_0}{2},$$

$$a_3 = \frac{a_2}{3} = \frac{a_0}{3 \cdot 2}, \qquad a_4 = \frac{a_3}{4} = \frac{a_0}{4!},$$

$$a_5 = \frac{a_4}{5} = \frac{a_0}{5!}, \qquad \ldots .$$

In general, $a_n = \dfrac{a_0}{n!}$ if $n \geq 0$. Hence

$$y(x) = \sum_{n=0}^{\infty} \frac{a_0}{n!} x^n = a_0 \sum_{n=0}^{\infty} \frac{x^n}{n!} = a_0 e^x.$$

Finally,

$$\lim_{n \to \infty} \left| \frac{n! x^{n+1}}{(n+1)! x^n} \right| = |x| \cdot \left(\lim_{n \to \infty} \frac{1}{n+1} \right) = 0,$$

so the radius of convergence of the series we found is $+\infty$.

10.10.3: We use series methods to solve $2 \dfrac{dy}{dx} + 3y = 0$. Assume that

$$y(x) = \sum_{n=0}^{\infty} a_n x^n,$$

so that

$$y'(x) = \sum_{n=1}^{\infty} n a_n x^{n-1} = \sum_{n=0}^{\infty} (n+1) a_{n+1} x^n.$$

Substitution in the given differential equation yields

$$\sum_{n=0}^{\infty} [2(n+1)a_{n+1} + 3a_n] x^n = 0,$$

and thus

$$a_{n+1} = -\frac{3a_n}{2(n+1)} \quad \text{if} \quad n \geq 0.$$

Therefore

$$a_1 = -\frac{3}{2}a_0, \qquad a_2 = -\frac{3}{2} \cdot \frac{a_1}{2} = \left(\frac{3}{2}\right)^2 \cdot \frac{a_0}{2},$$

$$a_3 = -\frac{3}{2} \cdot \frac{a_2}{3} = -\left(\frac{3}{2}\right)^3 \cdot \frac{a_0}{3!}, \qquad \ldots .$$

In general,

$$a_n = (-1)^n \left(\frac{3}{2}\right)^n \cdot \frac{a_0}{n!} \quad \text{for} \quad n \geq 1.$$

Therefore

$$y(x) = a_0 \sum_{n=0}^{\infty} (-1)^n \left(\frac{3}{2}\right)^n \cdot \frac{x_n}{n!} = a_0 \sum_{n=0}^{\infty} \frac{(-1)^n}{n!} \left(\frac{3x}{2}\right)^n = a_0 e^{-3x/2}.$$

By computations quite similar to those in the solution of Problem 1, this series has radius of convergence $+\infty$.

10.10.5: We use series methods to solve $\dfrac{dy}{dx} = x^2 y$. Assume that

$$y(x) = \sum_{n=0}^{\infty} a_n x^n,$$

so that

$$y'(x) = \sum_{n=1}^{\infty} n a_n x^{n-1} = \sum_{n=0}^{\infty} (n+1) a_{n+1} x^n.$$

Substitution in the given differential equation yields

$$\sum_{n=0}^{\infty} (n+1) a_{n+1} x^n = \sum_{n=0}^{\infty} a_n x^{n+2} = \sum_{n=2}^{\infty} a_{n-2} x^n.$$

Therefore $a_1 = 0$, $a_2 = 0$, and $(n+1)a_{n+1} = a_{n-2}$ if $n \geq 2$; that is,

$$a_1 = a_2 = a_4 = a_5 = a_7 = a_8 = \cdots 0 \quad \text{and} \quad a_{n+3} = \frac{a_n}{n+3}$$

if $n \geq 0$. Hence

$$a_3 = \frac{a_0}{3} = \frac{a_0}{1! \cdot 3}, \qquad a_6 = \frac{a_3}{6} = \frac{a_0}{6 \cdot 3} = \frac{a_0}{2! \cdot 3^2},$$

$$a_9 = \frac{a_6}{9} = \frac{a_0}{9 \cdot 6 \cdot 3} = \frac{a_0}{3! \cdot 3^3}, \qquad \cdots ;$$

in general,

$$a_{3n} = \frac{a_0}{n! \cdot 3^n} \quad \text{if } n \geq 1.$$

Therefore

$$y(x) = a_0 \left[1 + \frac{1}{1!} \cdot \frac{x^3}{3} + \frac{1}{2!} \left(\frac{x^3}{3}\right)^2 + \frac{1}{3!} \left(\frac{x^3}{3}\right)^3 + \cdots \right] = a_0 \exp\left(\frac{x^3}{3}\right).$$

As in previous solutions, the radius of convergence of this series is $+\infty$.

10.10.7: We use series methods to solve $(2x - 1)\dfrac{dy}{dx} + 2y = 0$. Assume that

640

$$y(x) = \sum_{n=0}^{\infty} a_n x^n,$$

so that

$$y'(x) = \sum_{n=1}^{\infty} n a_n x^{n-1} = \sum_{n=0}^{\infty} (n+1) a_{n+1} x^n.$$

Substitution in the given differential equation yields

$$\sum_{n=1}^{\infty} 2 n a_n x^n - \sum_{n=0}^{\infty} (n+1) a_{n+1} x^n + \sum_{n=0}^{\infty} 2 a_n x^n = 0.$$

If $n = 0$, we find that $-a_1 + 2 a_0 = 0$, and thus that $a_1 = 2 a_0$. If $n \geq 1$, then

$$2 n a_n - (n+1) a_{n+1} + 2 a_n = 0; \qquad (n+1) a_{n+1} = 2(n+1) a_n; \qquad a_{n+1} = 2 a_n.$$

Hence $a_1 = 2 a_0$, $a_2 = 2^2 a_0$, $a_3 = 2^3 a_0$, etc.; in general, $a_n = 2^n a_0$ if $n \geq 1$. Therefore

$$y(x) = a_0 \sum_{n=0}^{\infty} 2^n x^n = a_0 \sum_{n=0}^{\infty} (2x)^n = \frac{a_0}{1 - 2x}$$

because the series is geometric; for the same reason, its radius of convergence is $R = \frac{1}{2}$.

10.10.9: We use series methods to solve $(x - 1)\dfrac{dy}{dx} + 2y = 0$. Assume that

$$y(x) = \sum_{n=0}^{\infty} a_n x^n,$$

so that

$$y'(x) = \sum_{n=1}^{\infty} n a_n x^{n-1} = \sum_{n=0}^{\infty} (n+1) a_{n+1} x^n.$$

Substitution in the given differential equation yields

$$\sum_{n=1}^{\infty} n a_n x^n - \sum_{n=0}^{\infty} (n+1) a_{n+1} x^n + \sum_{n=0}^{\infty} 2 a_n x^n = 0.$$

When $n = 0$, we have $-a_1 + 2 a_0 = 0$, so that $a_1 = 2 a_0$. If $n \geq 1$, then

$$n a_n - (n+1) a_{n+1} + 2 a_n = 0 : \quad (n+1) a_{n+1} = (n+2) a_n,$$

and hence

$$a_{n+1} = \frac{n+2}{n+1} a_n \quad \text{if} \quad n \geq 0.$$

Therefore

$$a_2 = \frac{3}{2}a_1 = 3a_0, \qquad a_3 = \frac{4}{3}a_2 = 4a_0,$$

$$a_4 = \frac{5}{4}a_3 = 5a_0, \qquad \ldots ;$$

in general, $a_n = (n+1)a_0$ if $n \geq 1$. Therefore

$$y(x) = a_0 \sum_{n=0}^{\infty}(n+1)x^n.$$

Now $y(x) = F'(x)$ where

$$F(x) = a_0 \sum_{n=0}^{\infty} x^{n+1} = \frac{a_0 x}{1-x}.$$

Consequently,

$$y(x) = F'(x) = \frac{a_0(1-x+x)}{(1-x)^2} = \frac{a_0}{(1-x)^2}.$$

The radius of convergence of the series for $y(x)$ is $R = 1$.

10.10.11: We use series methods to solve the differential equation $y'' = y$. Assume the existence of a solution of the form

$$y(x) = \sum_{n=0}^{\infty} a_n x^n.$$

Then

$$y'(x) = \sum_{n=1}^{\infty} n a_n x^{n-1} = \sum_{n=0}^{\infty}(n+1)a_{n+1}x^n \qquad \text{and}$$

$$y''(x) = \sum_{n=2}^{\infty} n(n-1)a_n x^{n-2} = \sum_{n=0}^{\infty}(n+2)(n+1)a_{n+2}x^n.$$

Then substitution in the given differential equation yields

$$a_{n+2} = \frac{a_n}{(n+2)(n+2)} \quad \text{for} \quad n \geq 0.$$

Therefore

$$a_2 = \frac{a_0}{2 \cdot 1}, \qquad a_3 = \frac{a_1}{3 \cdot 2},$$

$$a_4 = \frac{a_0}{4!}, \qquad a_5 = \frac{a_1}{5!},$$

and so on. Hence

$$y(x) = a_0 \left(1 + \frac{x^2}{2!} + \frac{x^4}{4!} + \frac{x^6}{6!} + \cdots\right) + a_1 \left(x + \frac{x^3}{3!} + \frac{x^5}{5!} + \frac{x^7}{7!} + \cdots\right)$$

$$= a_0 \cosh x + a_1 \sinh x.$$

The radius of convergence of all series here is $R = +\infty$. The solution may also be expressed in the form $y(x) = c_1 e^x + c_2 e^{-x}$.

10.10.13: We use series methods to solve the differential equation $y'' + 9y = 0$. Assume the existence of a solution of the form

$$y(x) = \sum_{n=0}^{\infty} a_n x^n.$$

Then

$$y'(x) = \sum_{n=1}^{\infty} n a_n x^{n-1} = \sum_{n=0}^{\infty} (n+1) a_{n+1} x^n \qquad \text{and}$$

$$y''(x) = \sum_{n=2}^{\infty} n(n-1) a_n x^{n-2} = \sum_{n=0}^{\infty} (n+2)(n+1) a_{n+2} x^n.$$

Then substitution in the given differential equation leads—as in the solution of Problem 11—to the recursion formula $(n+2)(n+1)a_{n+2} + 9a_n = 0$, and thus

$$a_{n+2} = -\frac{9}{(n+2)(n+1)} a_n \qquad \text{for} \quad n \geqq 0.$$

Hence

$$a_2 = -\frac{9}{2!} a_0, \qquad a_3 = -\frac{9}{3!} a_1,$$

$$a_4 = \frac{9^2}{4!} a_0, \qquad a_5 = \frac{9^2}{5!} a_1,$$

$$a_6 = -\frac{9^3}{6!} a_0, \qquad a_7 = -\frac{9^3}{7!} a_1,$$

and so on. Hence

$$y(x) = a_0 \left(1 - \frac{9x^2}{2!} + \frac{9^2 x^4}{4!} - \frac{9^3 x^6}{6!} + \cdots \right) + a_1 \left(x - \frac{9x^3}{3!} + \frac{9^2 x^5}{5!} - \frac{9^3 x^7}{7!} + \cdots \right)$$

$$= a_0 \left(1 - \frac{(3x)^2}{2!} + \frac{(3x)^4}{4!} - \frac{(3x)^6}{6!} + \cdots \right) + \frac{a_1}{3} \left(3x - \frac{(3x)^3}{3!} + \frac{(3x)^5}{5!} - \frac{(3x)^7}{7!} + \cdots \right)$$

$$= a_0 \cos 3x + \frac{a_1}{3} \sin 3x = c_1 \cos 3x + c_2 \sin 3x.$$

The radius of convergence of each series here is $R = +\infty$.

643

10.10.15: Given the differential equation $x\dfrac{dy}{dx} + y = 0$, substitution of the series

$$y(x) = \sum_{n=0}^{\infty} a_n x^n \tag{1}$$

as in earlier solutions in this section yields

$$\sum_{n=1}^{\infty} n a_n x^n + \sum_{n=0}^{\infty} a_n x^n = 0.$$

It then follows that $a_0 = 0$ and that $n a_n + a_n = 0$ if $n \geq 1$. The latter equation implies that $a_n = 0$ if $n \geq 1$. Thus we obtain only the trivial solution $y(x) \equiv 0$, which is not part of the general solution because it contains no arbitrary constant and is not independent of any other solution. Part of the reason that the series has no solution of the form in (1) is that a general solution is

$$y(x) = \frac{C}{x}.$$

This solution is undefined at $x = 0$ and, of course, has no power series expansion with center $c = 0$. Here's an experiment for you: Assume a solution of the form

$$\sum_{n=0}^{\infty} b_n (x-1)^n$$

and see what happens. Then assume a solution of the form

$$\sum_{n=-1}^{\infty} c_n x^n$$

and see what happens. You can learn more about these ideas, and their consequences, in a standard course in differential equations (make sure that the syllabus includes the topic of series solution of ordinary differential equations).

10.10.17: Given the differential equation $x\dfrac{dy}{dx} + y = 0$, substitution of the series

$$y(x) = \sum_{n=0}^{\infty} a_n x^n \tag{1}$$

as in earlier solutions in this section yields

$$\sum_{n=1}^{\infty} n a_n x^{n+1} + \sum_{n=0}^{\infty} a_n x^n = 0;$$

$$\sum_{n=2}^{\infty} (n-1) a_{n-1} x^n + \sum_{n=0}^{\infty} a_n x^n = 0.$$

Examination of the cases $n = 0$ and $n = 1$ yields $a_0 = a_1 = 0$. If $n \geq 2$ we see that $(n-1)a_{n-1} + a_n = 0$, and hence that $a_n = 0$ for all $n \geq 0$. Thus the series method using the form in (1) uncovers only the trivial solution $y(x) \equiv 0$, not a general solution of the given differential equation. Part of the reason is that a general solution of the differential equation is

$$y(x) = C \exp\left(\frac{1}{x}\right).$$

10.10.19: Given the initial value problem

$$y'' + 4y = 0; \qquad y(0) = 0, \quad y'(0) = 3,$$

we assume the existence of a series solution of the form

$$y(x) = \sum_{n=0}^{\infty} a_n x^n.$$

Then

$$y'(x) = \sum_{n=1}^{\infty} n a_n x^{n-1} = \sum_{n=0}^{\infty} (n+1) a_{n+1} x^n \qquad \text{and}$$

$$y''(x) = \sum_{n=2}^{\infty} n(n-1) a_n x^{n-2} = \sum_{n=0}^{\infty} (n+2)(n+1) a_{n+2} x^n.$$

Substitution in the given differential equation yields $(n+2)(n+1)a_{n+2} + 4a_n = 0$, from which we obtain the recurrence relation

$$a_{n+2} = -\frac{4}{(n+2)(n+1)} a_n \qquad \text{for} \quad n \geq 0.$$

Thus we may choose a_0 and a_1 to be arbitrary constants, and find that

$$a_2 = -\frac{4}{2!} a_0, \qquad a_3 = -\frac{4}{3!} a_1,$$

$$a_4 = -\frac{4}{4 \cdot 3} a_2 = \frac{4^2}{4!} a_0, \qquad a_5 = \frac{4^2}{5!} a_1,$$

$$a_6 = -\frac{4^3}{6!} a_0, \qquad a_7 = -\frac{4^3}{7!} a_1,$$

and so on. Therefore the general solution of the given differential equation may be written in the form

$$y(x) = a_0 \left(1 - \frac{4x^2}{2!} + \frac{4^2 x^4}{4!} - \frac{4^3 x^6}{6!} + \frac{4^4 x^8}{8!} - \cdots \right) + a_1 \left(x - \frac{4x^3}{3!} + \frac{4^2 x^5}{5!} - \frac{4^3 x^7}{7!} + \frac{4^4 x^9}{9!} - \cdots \right)$$

$$= a_0 \cos 2x + \frac{a_1}{2} \sin 2x = A \cos 2x + B \sin 2x.$$

Substitution of the initial conditions yields $a_0 = y(0) = 0$ and $a_1 = y'(0) = 3$, so the particular solution of the differential equation is

$$y(x) = \frac{3}{2}\sin 2x.$$

10.10.21: Given the initial value problem

$$y'' - 2y' + y = 0; \qquad y(0) = 0, \quad y'(0) = 1,$$

we assume the existence of a series solution of the form

$$y(x) = \sum_{n=0}^{\infty} a_n x^n.$$

Then

$$y'(x) = \sum_{n=1}^{\infty} na_n x^{n-1} = \sum_{n=0}^{\infty} (n+1)a_{n+1}x^n \qquad \text{and}$$

$$y''(x) = \sum_{n=2}^{\infty} n(n-1)a_n x^{n-2} = \sum_{n=0}^{\infty} (n+2)(n+1)a_{n+2}x^n.$$

Substitution in the given differential equation then yields

$$(n+2)(n+1)a_{n+2} - 2(n+1)a_{n+1} + a_n = 0 \qquad \text{for} \quad n \geqq 0,$$

so that

$$a_{n+2} = \frac{2(n+1)a_{n+1} - a_n}{(n+2)(n+1)}, \qquad n \geqq 0.$$

At this point it would be easier to use the information that $a_0 = 0$ and $a_1 = 1$ to help find the general form of the coefficient a_n, but we choose to demonstrate that it is not necessary and, instead, find the general solution of the differential equation in terms of a_0 and a_1 as yet unspecified. Using the recursion formula just derived, we find that

$$a_2 = \frac{2a_1 - a_0}{2 \cdot 1} = \frac{2a_1 - a_0}{2!},$$

$$a_3 = \frac{4a_2 - a_1}{3 \cdot 2} = \frac{4a_1 - 2a_0 - a_1}{3 \cdot 2} = \frac{3a_1 - 2a_0}{3!},$$

$$a_4 = \frac{6a_3 - a_2}{4 \cdot 3} = \frac{3a_1 - 2a_0 - a_1 + \frac{1}{2}a_0}{4 \cdot 3} = \frac{4a_1 - 3a_0}{4!}, \qquad \text{and}$$

$$a_5 = \frac{8a_4 - a_3}{5 \cdot 4} = \frac{\frac{4}{3}a_1 - a_0 - \frac{1}{2}a_1 + \frac{1}{3}a_0}{5 \cdot 4} = \frac{8a_1 - 6a_0 - 3a_1 + 2a_0}{5!} = \frac{5a_1 - 4a_0}{5!}.$$

At this point one might conjecture that

$$a_n = \frac{na_1 - (n-1)a_0}{n!} \qquad \text{if} \quad n \geqq 2,$$

646

and this can be established using a proof by induction on n. That granted, it follows that

$$y(x) = a_0 + a_1 x + \frac{2a_1 - a_0}{2!} x^2 + \frac{3a_1 - 2a_0}{3!} x^3 + \frac{4a_1 - 3a_0}{4!} x^4 + \frac{5a_1 - 4a_0}{5!} x^5 + \cdots$$

$$= a_0 \left(1 - \frac{x^2}{2!} - \frac{2x^3}{3!} - \frac{3x^4}{4!} - \frac{4x^5}{5!} - \cdots \right) + a_1 \left(x + x^2 + \frac{x^3}{2!} + \frac{x^4}{3!} + \frac{x^5}{4!} + \cdots \right)$$

$$= a_0 \left(1 - \frac{x^2}{2!} - \frac{2x^3}{3!} - \frac{3x^4}{4!} - \frac{4x^5}{5!} - \cdots \right) + a_1 x \left(1 + x + \frac{x^2}{2!} + \frac{x^3}{3!} + \frac{x^4}{4!} + \cdots \right).$$

Let

$$F(x) = 1 - \frac{x^2}{2!} - \frac{2x^3}{3!} - \frac{3x^4}{4!} - \frac{4x^5}{5!} - \cdots .$$

Then

$$F'(x) = -x - x^2 - \frac{x^3}{2!} - \frac{x^4}{3!} - \frac{x^5}{4!} - \cdots$$

$$= -x \left(1 + x + \frac{x^2}{2!} + \frac{x^3}{3!} + \frac{x^4}{4!} + \cdots \right) = -xe^x.$$

Therefore $F(x) = (1 - x)e^x + C$. Moreover, $F(0) = 1$, so that $C = 0$. Consequently,

$$y(x) = a_0(1 - x)e^x + a_1 xe^x = a_0 e^x + (a_1 - a_0)xe^x = Ae^x + Bxe^x.$$

Finally, the given initial conditions imply that $A = y(0) = 0$ and that $B = y'(0) = 1$. Therefore the particular solution of the original initial value problem is $y(x) = xe^x$.

10.10.23: Suppose that the differential equation

$$x^2 y'' + x^2 y' + y = 0$$

has a series solution of the form

$$y(x) = \sum_{n=0}^{\infty} c_n x^n.$$

Then

$$y'(x) = \sum_{n=1}^{\infty} n c_n x^{n-1} = \sum_{n=0}^{\infty} (n+1) c_{n+1} x^n \qquad \text{and}$$

$$y''(x) = \sum_{n=2}^{\infty} n(n-1) c_n x^{n-2} = \sum_{n=0}^{\infty} (n+2)(n+1) c_{n+2} x^n.$$

Substitution in the given differential equation then yields

$$\sum_{n=2}^{\infty} n(n-1) c_n x^n + \sum_{n=1}^{\infty} n c_n x^{n+1} + \sum_{n=0}^{\infty} c_n x^n = 0;$$

647

$$\sum_{n=2}^{\infty} n(n-1)c_n x^n + \sum_{n=2}^{\infty} (n-1)c_{n-1} x^n + \sum_{n=0}^{\infty} c_n x^n = 0.$$

It then follows that $c_0 = c_1 = 0$ and that, if $n \geq 2$,

$$n(n-1)c_n + (n-1)c_{n-1} + c_n = 0; \quad \text{that is,} \quad c_n = -\frac{n-1}{n^2-n+1}c_{n-1}.$$

Thus

$$c_2 = -\frac{1}{3}c_1 = 0, \quad c_3 = -\frac{2}{7}c_2 = 0, \quad c_4 = -\frac{3}{13}c_3 = 0,$$

and so on: $c_n = 0$ for all $n \geq 0$. Therefore the only solution discovered by the series method used here is the trivial solution $y(x) \equiv 0$. Not only do we not find two linearly independent solutions, there is not even one because the trivial solution is neither independent of any solution nor has it the form of a general solution.

10.10.25: Part (a): The method of separation of variables yields

$$\frac{1}{1+y^2}\, dy = 1\, dx; \qquad \arctan y = x + C;$$

$$y(x) = \tan(x+C). \qquad 0 = y(0) = \tan C:$$

$$C = n\pi \quad (n \text{ is an integer}); \qquad y(x) = \tan(x+n\pi) = \tan x.$$

Part (b): If

$$y(x) = x + c_3 x^3 + c_5 x^5 + c_7 x^7 + c_9 x^9 + \cdots, \qquad \text{then}$$

$$y'(x) = 1 + 3c_3 x^2 + 5c_5 x^4 + 7c_7 x^6 + 9c_9 x^8 + \cdots. \qquad \text{Hence}$$

$$1 + [y(x)]^2 = 1 + x^2 + 2c_3 x^4 + (c_3^2 + 2c_5)x^6 + (2c_3 c_5 + 2c_7)x^8$$

$$+ (2c_3 c_7 + c_5^2 + 2c_9)x^{10} + (2c_3 c_9 + 2c_5 c_7 + 2c_{11})x^{12} + (2c_3 c_{11} + 2c_5 c_9 + c_7^2 + 2c_{13})x^{14} + \cdots$$

$$= y'(x) = 1 + 3c_3 x^2 + 5c_5 x^4 + 7c_7 x^6 + 9c_9 x^8 + 11c_{11}x^{10} + \cdots.$$

It follows that

$$3c_3 = 1: \qquad c_3 = \frac{1}{3}.$$

$$5c_5 = 2c_3 = \frac{2}{3}: \qquad c_5 = \frac{2}{15}.$$

$$7c_7 = c_3^2 + 2c_5 = \frac{1}{9} + \frac{4}{15} = \frac{17}{45}: \qquad c_7 = \frac{17}{315}.$$

$$9c_9 = 2c_3c_5 + 2c_7 = \frac{62}{315}: \qquad c_9 = \frac{62}{2835}.$$

$$11c_{11} = 2c_3c_7 + c_5^2 + 2c_9 = \frac{1382}{14175}: \qquad c_{11} = \frac{1382}{155925}.$$

Part (c): Continuing in this manner, we find that

$$\tan x = x + \frac{1}{3}x^3 + \frac{2}{15}x^5 + \frac{17}{315}x^7 + \frac{62}{2835}x^9 + \frac{1382}{155925}x^{11} + \frac{21844}{6081075}x^{13} + \frac{929569}{638512875}x^{15}$$

$$+ \frac{6404582}{10854518875}x^{17} + \frac{443861162}{1856156927825}x^{19} + \frac{18888466084}{194896477400625}x^{21}$$

$$+ \frac{113927491862}{2900518163668125}x^{23} + \frac{58870668456604}{3698160658676859375}x^{25} + \cdots .$$

Chapter 10 Miscellaneous Problems

10.M.1: Divide each term in numerator and denominator by n^2:

$$\lim_{n \to \infty} \frac{n^2 + 1}{n^2 + 4} = \lim_{n \to \infty} \frac{1 + \dfrac{1}{n^2}}{1 + \dfrac{4}{n^2}} = \frac{1 + 0}{1 + 0} = 1.$$

10.M.3: In Example 9 of Section 10.2, it is shown that if $|r| < 1$, then $r^n \to 0$ as $n \to +\infty$. Therefore

$$\lim_{n \to \infty} \left[10 - (0.99)^n \right] = 10 - 0 = 10.$$

10.M.5: Because

$$0 \leq |a_n| = \frac{|1 + (-1)^n\sqrt{n}|}{n + 1} \leq \frac{2\sqrt{n}}{n} = \frac{2}{\sqrt{n}} \to 0$$

as $n \to +\infty$, the sequence with the given general term converges to 0 by the squeeze law for limits. This problem is the result of a typograhical error; it was originally intended to be the somewhat more challenging problem in which

$$a_n = \frac{1 + (-1)^n n^{1/n}}{n + 1}$$

for $n \geq 1$.

10.M.7: Because $-1 \leq \sin 2n \leq 1$ for every positive integer n,

$$-\frac{1}{n} \leq \frac{\sin 2n}{n} \leq \frac{1}{n}$$

for every positive integer n. Therefore, by the squeeze law for limits (Theorem 3 of Section 10.2),

$$\lim_{n \to \infty} \frac{\sin 2n}{n} = 0.$$

10.M.9: If n is an even positive integer, then $\sin(n\pi/2) = 0$. Therefore $a_n = (-1)^0 = 1$ for arbitrarily large values of n. Hence if the sequence $\{a_n\}$ has a limit, it must be 1. But if n is an odd positive integer, then $\sin(n\pi/2) = \pm 1$. Therefore $a_n = -1$ for arbitrarily large values of n. So if the sequence $\{a_n\}$ has a limit, it must be -1. Because $1 \neq -1$, the sequence $\{a_n\}$ has no limit as $n \to +\infty$.

10.M.11: Because $-1 \leqq \sin x \leqq 1$ for all x,

$$-\frac{1}{n} \leqq \frac{1}{n} \sin \frac{1}{n} \leqq \frac{1}{n}$$

for every positive integer n. Therefore by the squeeze law for sequences, $\lim\limits_{n \to \infty} \dfrac{1}{n} \sin \dfrac{1}{n} = 0$.

10.M.13: Here we have

$$\lim_{n \to \infty} \frac{\sinh n}{n} = \lim_{n \to \infty} \frac{e^n - e^{-n}}{2n} = \lim_{n \to \infty} \frac{1 - e^{-2n}}{2ne^{-n}}.$$

The numerator in the last fraction is approaching 1 as $n \to +\infty$, but the denominator is approaching zero through positive values (by Eq. (8) of Section 4.8). Therefore $a_n \to +\infty$ as $n \to +\infty$. Alternatively, you may say that the limit in question does not exist.

10.M.15: Example 7 in Section 10.2 implies that $3^{1/n} \to 1$ as $n \to +\infty$. Example 11 in Section 10.2 shows that $n^{1/n} \to 1$ as $n \to +\infty$. Moreover, if n is a positive integer, then

$$n^{1/n} \leqq (2n^2 + 1)^{1/n} \leqq (3n^2)^{1/n} = 3^{1/n} \cdot \left(n^{1/n}\right)^2,$$

and

$$\lim_{n \to \infty} 3^{1/n} \cdot \left(n^{1/n}\right)^2 = \left(\lim_{n \to \infty} 3^{1/n}\right) \cdot \left(\lim_{n \to \infty} n^{1/n}\right)^2 = 1 \cdot 1 = 1.$$

Therefore, by the squeeze law for limits,

$$\lim_{n \to \infty} (2n^2 + 1)^{1/n} = 1.$$

10.M.17: By l'Hôpital's rule,

$$\lim_{x \to \infty} \frac{\ln x}{x} = \lim_{x \to \infty} \frac{1}{x} = 0,$$

so $(\ln n)/n \to 0$ as $n \to +\infty$ by Theorem 4 of Section 10.2. Also, if

$$f(x) = \frac{\ln x}{x}, \quad \text{then} \quad f'(x) = \frac{1 - \ln x}{x^2},$$

which is negative for $x > e$, and so the sequence $\{(\ln n)/n\}$ is monotonically decreasing if $n \geqq 3$. Therefore the given series converges by the alternating series test. Its sum is approximately 0.080357603217. The *Mathematica* command

```
Sum[ ((-1)^(n+1))*(Log[n])/n, {n, 2, Infinity} ]
```

almost immediately produces the exact value of the sum of the series; it is

$$(-\gamma + \ln 2)\ln 2 \approx 0.0803576032166697405766033928384159153690544520408140508762608$$

(Euler's constant γ is first discussed in Problem 50 of Section 10.5 of the text).

10.M.19: This series converges because the ratio test yields

$$\rho = \lim_{n\to\infty} \frac{(n+1)!\exp(n^2)}{n!\exp\left([n+1]^2\right)} = \lim_{n\to\infty} (n+1)\exp\left(n^2 - [n+1]^2\right) = \lim_{n\to\infty} \frac{n+1}{\exp(2n+1)} = 0$$

by Eq. (8) in Section 4.8 and the squeeze law for limits:

$$0 \leq \frac{n+1}{e^{2n+1}} \leq \frac{2n+1}{e^{2n+1}}$$

for every positive integer n. The sum of the series is approximately 1.405253880284.

10.M.21: For every positive integer n,

$$\left|\frac{(-2)^n}{3^n+1}\right| \leq \frac{2^n}{3^n} = \left(\frac{2}{3}\right)^n.$$

Therefore the series

$$\sum_{n=0}^{\infty}\left|\frac{(-2)^n}{3^n+1}\right| \quad \text{is dominated by} \quad \sum_{n=0}^{\infty}\left(\frac{2}{3}\right)^n.$$

The latter series converges because it is geometric with ratio $\frac{2}{3}$, and therefore the dominated series converges. Hence the original series of Problem 21 converges absolutely, and therefore it converges by Theorem 3 of Section 10.7. Its sum is approximately 0.230836643803.

10.M.23: Three applications of l'Hôpital's rule yield

$$\lim_{x\to\infty}\frac{x}{(\ln x)^3} = \lim_{x\to\infty}\frac{x}{3(\ln x)^2} = \lim_{x\to\infty}\frac{x}{6\ln x} = \lim_{x\to\infty}\frac{x}{6} = +\infty.$$

Therefore $\displaystyle\sum_{n=2}^{\infty}\frac{(-1)^n \cdot n}{(\ln n)^3}$ diverges by the nth-term test for divergence (Theorem 3 of Section 10.3).

10.M.25: For every positive integer n,

$$0 \leq \frac{n^{1/2}+n^{1/3}}{n^2+n^3} \leq \frac{2n^{1/2}}{n^3} = \frac{2}{n^{5/2}}.$$

Therefore

$$\sum_{n=1}^{\infty}\frac{n^{1/2}+n^{1/3}}{n^2+n^3} \quad \text{is dominated by} \quad \sum_{n=1}^{\infty}\frac{2}{n^{5/2}}.$$

The latter series converges because it is a constant multiple of the p-series with $p = \frac{5}{2} > 1$. Therefore the dominated series converges by the comparison test (Theorem 1 of Section 10.6). The sum of the given series is approximately 1.459973884376.

10.M.27: Given: The alternating series

$$\sum_{n=1}^{\infty} \frac{(-1)^{n+1} \arctan n}{\sqrt{n}}. \tag{1}$$

We plan to show that this series meets the criterion for convergence stated in the alternating series test (Theorem 1 of of Section 10.7). First,

$$0 \leqq \arctan n \leqq \frac{\pi}{2}$$

for every positive integer n. Thus for such n,

$$0 \leq \frac{\arctan n}{\sqrt{n}} \leqq \frac{\pi}{2\sqrt{n}}.$$

Therefore, by the squeeze law for limits,

$$\lim_{n \to \infty} \frac{\arctan n}{\sqrt{n}} = 0.$$

Now let

$$f(x) = \frac{\arctan x}{\sqrt{x}} \quad \text{for} \quad x \geqq 1. \quad \text{Then:}$$

$$f'(x) = \frac{1}{x} \cdot \left(\frac{x^{1/2}}{1+x^2} - \frac{\arctan x}{2x^{1/2}} \right) = \frac{1}{x^{3/2}} \left(\frac{x}{1+x^2} - \frac{\arctan x}{2} \right)$$

$$= \frac{1}{2x^{3/2}} \left(\frac{2x}{1+x^2} - \arctan x \right) = \frac{2x - (1+x^2)\arctan x}{2x^{3/2}(1+x^2)}.$$

Now if $x \geqq 2$, then $1 \leqq \arctan x$. Therefore

$$1 + x^2 \leqq (1+x^2)\arctan x; \qquad -(1+x^2)\arctan x \leqq -(1+x^2);$$

$$2x - (1+x^2)\arctan x \leqq 2x - (1+x^2); \qquad 2x - (1+x^2)\arctan x \leqq -(x-1)^2;$$

$$2x - (1+x^2)\arctan x < 0 \quad \text{if} \quad x > 1; \qquad f'(x) < 0 \quad \text{if} \quad x > 1.$$

Consequently the sequence of terms of the series in (1) is (after the first term) monotonically decreasing in magnitude. Because they alternate in sign, the series in (1) converges by the alternating series test. Its sum is approximately 0.378868816198.

10.M.29: We use the integral test (Theorem 1 of Section 10.5):

$$\int_3^\infty \frac{1}{x(\ln x)(\ln \ln x)}\, dx = \left[\ln(\ln \ln x)\right]_3^\infty = +\infty,$$

and therefore $\displaystyle\sum_{n=3}^\infty \frac{1}{n(\ln n)(\ln \ln n)}$ diverges.

10.M.31: The ratio test yields

$$\rho = \lim_{n\to\infty} \frac{2^{n+1} \cdot n! \cdot |x|^{n+1}}{2^n \cdot (n+1)! \cdot |x|^n} = \lim_{n\to\infty} \frac{2|x|}{n+1} = 0$$

for every real number x. Therefore the given series converges for all x; its interval of convergence is $(-\infty, +\infty)$. Its sum is e^{2x}.

10.M.33: The ratio test yields

$$\rho = \lim_{n\to\infty} \frac{n \cdot 3^n \cdot |x-1|^{n+1}}{(n+1) \cdot 3^{n+1} \cdot |x-1|^n} = \lim_{n\to\infty} \frac{n \cdot |x-1|}{3(n+1)} = \frac{|x-1|}{3}.$$

So the given series converges if $-3 < x - 1 < 3$; that is, if $-2 < x < 4$. It diverges if $x = 4$ (because it becomes the harmonic series). It converges if $x = -2$ by the alternating series test. Thus its interval of convergence is $[-2, 4)$.

10.M.35: The ratio test yields

$$\rho = \lim_{n\to\infty} \frac{(4n^2 - 1) \cdot |x|^{n+1}}{[4(n+1)^2 - 1] \cdot |x|^n} = \lim_{n\to\infty} \frac{(4n^2 - 1) \cdot |x|}{4n^2 + 8n + 3} = \lim_{n\to\infty} \frac{\left(4 - \dfrac{1}{n^2}\right) \cdot |x|}{4 + \dfrac{8}{4} + \dfrac{3}{n^2}} = |x|.$$

Thus the series converges if $-1 < x < 1$. If $x = \pm 1$ then the given series converges absolutely because it is dominated by the p-series with $p = 2 > 1$. Hence the interval of convergence of the given series is $[-1, 1]$. The *Mathematica* command

```
Sum[ ((-1)∧n)*(x∧n)/(4*n*n - 1), {n, 1, Infinity} ]
```

quickly returns the value of the sum of this series on *part* of its interval of convergence; the response is

$$\frac{\sqrt{x} - \arctan\left(\sqrt{x}\right) - x\arctan\left(\sqrt{x}\right)}{2\sqrt{x}}.$$

This result raises some intriguing new questions concerning the behavior of the series, and particularly of its sum, for $-1 \leqq x < 0$.

10.M.37: The ratio test yields

$$\rho = \lim_{n \to \infty} \frac{(n+1)! \cdot 10^n \cdot |x|^{2n+2}}{n! \cdot 10^{n+1} \cdot |x|^{2n}} = \lim_{n \to \infty} \frac{(n+1)x^2}{10} = +\infty$$

if $x \neq 0$. Therefore the given series converges only if $x = 0$.

10.M.39: Note that

$$\sum_{n=0}^{\infty} \frac{1 + (-1)^n}{n! \cdot 2} x^n = 1 + \frac{x^2}{2!} + \frac{x^4}{4!} + \frac{x^6}{6!} + \cdots = 1 + \sum_{n=1}^{\infty} \frac{x^{2n}}{(2n)!}. \tag{1}$$

So the ratio test yields

$$\rho = \lim_{n \to \infty} \frac{(2n)! \cdot x^{2n+2}}{(2n+2)! \cdot x^{2n}} = \lim_{n \to \infty} \frac{x^2}{(2n+2)(2n+1)} = 0$$

for all real x. Hence the interval of convergence of the series in (1) is $(-\infty, +\infty)$. This series converges to $f(x) = \cosh x$ (see Example 6 in Section 10.8).

10.M.41: The given series diverges for every real number x by the nth-term test for divergence.

10.M.43: The ratio test yields

$$\rho = \lim_{n \to \infty} \frac{n! \cdot e^{(n+1)x}}{(n+1)! \cdot e^{nx}} = \lim_{n \to \infty} \frac{e^x}{n+1} = 0$$

for every real number x, so the given series converges for all x. Its sum is $\exp(e^x)$.

10.M.45: Let

$$a_n = b_n = \frac{(-1)^{n+1}}{\sqrt{n}} \quad \text{for} \quad n \geq 1.$$

Then $\sum a_n$ and $\sum b_n$ converge by the alternating series test, but $\sum a_n b_n$ diverges because it is the harmonic series.

10.M.47: Assuming that A exists, we have

$$A = \lim_{n \to \infty} a_n = \lim_{n \to \infty} \left(1 + \frac{1}{1 + a_n}\right) = 1 + \frac{1}{1 + A}$$

because $A \neq -1$. Therefore $A + A^2 = 2 + A$, and it follows that $A^2 = 2$. Because $A \geq 0$ (the limit of a sequence of positive numbers cannot be negative), $A = \sqrt{2}$.

10.M.49: If $a_n = \frac{1}{n}$, then the series

$$\sum_{n=1}^{\infty} \ln(1 + a_n) = \sum_{n=1}^{\infty} \ln \left(1 + \frac{1}{n}\right)$$

654

diverges because

$$S_k = \sum_{n=1}^{k} \ln\left(1 + \frac{1}{n}\right) = \sum_{n=1}^{k} [\ln(n+1) - \ln n]$$

$$= \ln 2 - \ln 1 + \ln 3 - \ln 2 + \ln 4 - \ln 3 + \cdots + \ln(k+1) - \ln k = \ln(k+1),$$

and therefore $S_k \to +\infty$ as $k \to +\infty$. Alternatively, using the integral test,

$$J = \int_1^\infty \ln\left(1 + \frac{1}{x}\right) dx = \int_1^\infty [\ln(x+1) - \ln x] \, dx = \left[(x+1)\ln(x+1) - x\ln x\right]_1^\infty = +\infty$$

because

$$\lim_{x\to\infty} [(x+1)\ln(x+1) - x\ln x] \geq \lim_{x\to\infty} [(x+1)\ln x - x\ln x] = \lim_{x\to\infty} \ln x = +\infty$$

and, at the lower limit $x = 1$ of integration, we have $(x+1)\ln(x+1) - x\ln x = \ln 4$. Therefore, because

$$J = \int_1^\infty \ln\left(1 + \frac{1}{x}\right) dx = +\infty,$$

the infinite product $\prod_{n=1}^{\infty}\left(1 + \frac{1}{n}\right)$ diverges.

10.M.51: The binomial series is

$$(1+x)^{1/5} = 1 + \frac{x}{5} - \frac{4}{5^2}\cdot\frac{x^2}{2!} + \frac{4\cdot 9}{5^3}\cdot\frac{x^4}{3!} - \frac{4\cdot 9\cdot 14}{5^4}\cdot\frac{x^4}{4!} + \cdots .$$

Substitution of $x = \frac{1}{2}$ and summing the first five terms of this series yields 1.0839 (exactly); summing the first six terms yields 1.0843788 (exactly). So to three places, $\left(1 + \frac{1}{2}\right)^{1/5} \approx 1.084$. The true value of the expression is closer to 1.084471771198.

10.M.53: We substitute $-x^2$ for x in the Maclaurin series for the natural exponential function. Thus we find that

$$\int_0^{1/2} \exp(-x^2) \, dx = \int_0^{1/2} \left(1 - x^2 + \frac{x^4}{2!} - \frac{x^6}{3!} + \frac{x^8}{4!} - \cdots\right) dx$$

$$= \left[x - \frac{x^3}{3} + \frac{x^5}{2!\cdot 5} - \frac{x^7}{3!\cdot 7} + \frac{x^9}{4!\cdot 9} - \cdots\right]_0^{1/2}$$

$$= \frac{1}{2} - \frac{1}{3\cdot 2^3} + \frac{1}{2!\cdot 5\cdot 2^5} - \frac{1}{3!\cdot 7\cdot 2^7} + \frac{1}{4!\cdot 9\cdot 2^9} - \cdots = \sum_{n=0}^{\infty} \frac{(-1)^n}{n!\cdot(2n+1)\cdot 2^{2n+1}}.$$

The sum of the first two terms and the sum of the first three terms of this series are

$$\frac{443}{960} \approx 0.461458333333 \quad \text{and} \quad \frac{4133}{8960} \approx 0.461272321429,$$

respectively. Thus the value of the integral to three places is approximately 0.461. A closer approximation is 0.4612810064127924 (to the number of digits shown).

10.M.55: The Maclaurin series for the natural exponential function yields

$$\frac{1}{x}(1 - e^{-x}) = \frac{1}{x}\left(x - \frac{x^2}{2!} + \frac{x^3}{3!} - \frac{x^4}{4!} + \frac{x^5}{5!} - \cdots\right) = 1 - \frac{x}{2!} + \frac{x^2}{3!} - \frac{x^3}{4!} + \frac{x^4}{5!} - \frac{x^5}{6!} + \cdots.$$

Then termwise integration produces

$$\int_0^1 \frac{1 - e^{-x}}{x}\, dx = \left[x - \frac{x^2}{2! \cdot 2} + \frac{x^3}{3! \cdot 3} - \frac{x^4}{4! \cdot 4} + \frac{x^5}{5! \cdot 5} - \frac{x^6}{6! \cdot 6} + \cdots\right]_0^1$$

$$= 1 - \frac{1}{2! \cdot 2} + \frac{1}{3! \cdot 3} - \frac{1}{4! \cdot 4} + \frac{1}{5! \cdot 5} - \frac{1}{6! \cdot 6} + \cdots.$$

The sum of the first five terms of the last series and the sum of its first six terms are

$$\frac{5737}{7200} \approx 0.796805555556 \quad \text{and} \quad \frac{8603}{10800} \approx 0.796574074074,$$

respectively. So to three places, the value of the integral is 0.797. A more accurate approximation is 0.7965995992970531.

10.M.57: We will need both the recursion formula

$$\int_0^\infty t^{2n} \exp(-t^2)\, dt = \frac{2n-1}{2} \int_0^\infty t^{2n-2} \exp(-t^2)\, dt \quad (n \geq 1),$$

which follows from the formula in Problem 50 of Section 7.3, and the famous formula

$$\int_0^\infty \exp(-t^2)\, dt = \frac{\sqrt{\pi}}{2},$$

which is derived in Example 5 of Section 14.4 (it is Eq. (9) there). We begin with the Maclaurin series of the cosine function.

$$\int_0^\infty \exp(-t^2) \cos 2xt\, dt = \int_0^\infty \exp(-t^2)\left(1 - \frac{2^2 x^2 t^2}{2!} + \frac{2^4 x^4 t^4}{4!} - \frac{2^6 x^6 t^6}{6!} + \frac{2^8 x^8 t^8}{8!} - \cdots\right) dt$$

$$= \int_0^\infty \left(\exp(-t^2) - \frac{2^2 x^2}{2!} t^2 \exp(-t^2) + \frac{2^4 x^4}{4!} t^4 \exp(-t^2) - \frac{2^6 x^6}{6!} t^6 \exp(-t^2) + \cdots\right) dt$$

$$= \int_0^\infty e^{-t^2}\, dt - \frac{2^2 x^2}{2!} \int_0^\infty t^2 e^{-t^2}\, dt + \frac{2^4 x^4}{4!} \int_0^\infty t^4 e^{-t^2}\, dt - \frac{2^6 x^6}{6!} \int_0^\infty t^6 e^{-t^2}\, dt + \cdots$$

$$= \int_0^\infty e^{-t^2}\, dt - \frac{2^2 x^2}{2!} \cdot \frac{1}{2} \int_0^\infty e^{-t^2}\, dt + \frac{2^4 x^4}{4!} \cdot \frac{3}{2} \cdot \frac{1}{2} \int_0^\infty e^{-t^2}\, dt - \frac{2^6 x^6}{6!} \cdot \frac{5}{2} \cdot \frac{3}{2} \cdot \frac{1}{2} \int_0^\infty e^{-t^2}\, dt + \cdots$$

$$= \left(\int_0^\infty e^{-t^2}\, dt\right)\left(1 - \frac{2^2 x^2}{2!} \cdot \frac{1}{2} + \frac{2^4 x^4}{4!} \cdot \frac{3 \cdot 1}{2^2} - \frac{2^6 x^6}{6!} \cdot \frac{5 \cdot 3 \cdot 1}{2^3} + \frac{2^8 x^8}{8!} \cdot \frac{7 \cdot 5 \cdot 3 \cdot 1}{2^4} - \cdots\right).$$

656

The typical term in the last infinite series is

$$\frac{2^{2n}x^{2n}}{(2n)!}\cdot\frac{(2n-1)(2n-3)\cdots 5\cdot 3\cdot 1}{2^n}=\frac{2^{2n}x^{2n}}{(2n)!}\cdot\frac{(2n)!}{2^n\cdot(2n)(2n-2)\cdots 6\cdot 4\cdot 2}=\frac{2^{2n}x^{2n}}{n!\cdot 2^n\cdot 2^n}=\frac{(x^2)^n}{n!}.$$

Consequently,

$$\int_0^\infty e^{-t^2}\cos 2xt\,dt=\left(\int_0^\infty e^{-t^2}\,dt\right)\left(1-\frac{x^2}{1!}+\frac{(x^2)^2}{2!}-\frac{(x^2)^3}{3!}+\frac{(x^2)^4}{4!}-\cdots\right)$$

$$=\left(\int_0^\infty e^{-t^2}\,dt\right)e^{-x^2}=\frac{\sqrt{\pi}}{2}e^{-x^2}.$$

10.M.59: The binomial series takes the form

$$(1+t^2)^{-1/2}=1-\frac{1}{2}t^2+\frac{1\cdot 3}{2^2}\cdot\frac{t^4}{2!}-\frac{1\cdot 3\cdot 5}{2^3}\cdot\frac{t^6}{3!}+\frac{1\cdot 3\cdot 5\cdot 7}{2^4}\cdot\frac{t^8}{4!}-\cdots.$$

Thus

$$\sinh^{-1}x=\int_0^x(1+t^2)^{-1/2}\,dt=x-\frac{x^3}{3}+\frac{1\cdot 3}{2^2\cdot 5}\cdot\frac{x^5}{2!}-\frac{1\cdot 3\cdot 5}{2^3\cdot 7}\cdot\frac{x^7}{3!}+\frac{1\cdot 3\cdot 5\cdot 7}{2^4\cdot 9}\cdot\frac{x^9}{4!}-\cdots$$

$$=\sum_{n=0}^\infty\frac{1\cdot 3\cdot 5\cdots(2n-1)}{2^n\cdot n!}\cdot\frac{x^{2n+1}}{2n+1}=\sum_{n=0}^\infty\frac{1\cdot 3\cdot 5\cdots(2n-1)}{2\cdot 4\cdot 6\cdots(2n)}\cdot\frac{x^{2n+1}}{2n+1}$$

$$=\sum_{n=0}^\infty\frac{(2n)!}{2^n\cdot n!\cdot 2\cdot 4\cdot 6\cdots(2n)}\cdot\frac{x^{2n+1}}{2n+1}=\sum_{n=0}^\infty\frac{(2n)!}{2^{2n}\cdot(n!)^2}\cdot\frac{x^{2n+1}}{2n+1}$$

provided that $|x|<1$.

10.M.61: We let *Mathematica* do this problem. First we defined

```
mu = 1/(12*n) - 1/(360*n^3) + 1/(1260*n^5)
```

Then the command

```
Series[ Exp[x], { x, 0, 10 } ] // Normal
```

produced the 10th-degree Taylor polynomial with center zero for e^x:

$$1+x+\frac{x^2}{2}+\frac{x^3}{6}+\frac{x^4}{24}+\frac{x^5}{120}+\frac{x^6}{720}+\frac{x^7}{5040}+\frac{x^8}{40320}+\frac{x^9}{362880}+\frac{x^{10}}{3628800}.$$

Recall that % means "last output" to *Mathematica*. Thus the next command

```
% /.  x → mu
```

tells *Mathematica* to substitute mu for x in the Taylor polynomial, producing the response

$$1 + \left(\frac{1}{1260n^5} - \frac{1}{360n^3} + \frac{1}{12n}\right) + \frac{1}{2}\left(\frac{1}{1260n^5} - \frac{1}{360n^3} + \frac{1}{12n}\right)^2 + \frac{1}{6}\left(\frac{1}{1260n^5} - \frac{1}{360n^3} + \frac{1}{12n}\right)^3$$

$$+ \frac{1}{24}\left(\frac{1}{1260n^5} - \frac{1}{360n^3} + \frac{1}{12n}\right)^4 + \cdots + \frac{1}{3628800}\left(\frac{1}{1260n^5} - \frac{1}{360n^3} + \frac{1}{12n}\right)^{10}.$$

Finally, the Expand command resulted in almost a full page of output, including the answer:

$$\exp(\mu(n)) = 1 + \frac{1}{12n} + \frac{1}{288n^2} - \frac{139}{51840n^3} - \frac{571}{2488320n^4} + \frac{163879}{209018880n^5} + \cdots.$$

10.M.63: Proof: Assume that e is a rational number. Then $e = p/q$ where p and q are positive integers and $q > 1$ (because e is not an integer). Thus

$$\frac{p}{q} = e = 1 + \frac{1}{1!} + \frac{1}{2!} + \frac{1}{3!} + \cdots + \frac{1}{q!} + R_q$$

where

$$R_q = \frac{1}{(q+1)!} + \frac{1}{(q+2)!} + \frac{1}{(q+3)!} + \frac{1}{(q+4)!} + \cdots$$

$$= \frac{1}{q!}\cdot\left(\frac{1}{q+1} + \frac{1}{(q+1)(q+2)} + \frac{1}{(q+1)(q+2)(q+3)} + \cdots\right)$$

$$< \frac{1}{q!}\cdot\left(\frac{1}{q+1} + \frac{1}{(q+1)^2} + \frac{1}{(q+1)^3} + \cdots\right) = \frac{1}{q!}\cdot\frac{\frac{1}{q+1}}{1 - \frac{1}{q+1}} = \frac{1}{q!\cdot q}.$$

Thus

$$1 + \frac{1}{1!} + \frac{1}{2!} + \frac{1}{3!} + \cdots + \frac{1}{q!} < \frac{p}{q} < 1 + \frac{1}{1!} + \frac{1}{2!} + \frac{1}{3!} + \cdots + \frac{1}{q!} + \frac{1}{q!\cdot q}.$$

If the left member of the last inequality is multiplied by $q!$, the product is an integer; call it M. Thus when all three members of the last inequality are multiplied by $q!$, the result is

$$M < (q-1)!\cdot p < M + \frac{1}{q} < M + 1.$$

This is a contradiction because it asserts that the *integer* $(q-1)!\cdot p$ lies strictly between the *consecutive integers* M and $M + 1$. Therefore e is irrational. ◄

10.M.65: Suppose that $x^2 = 5$. Then

$$x^2 - 4 = 1; \quad x - 2 = \frac{1}{2+x}; \quad x = 2 + \frac{1}{2+x}.$$

658

Now substitute the last expression for the last x. The result is

$$x = 2 + \cfrac{1}{4 + \cfrac{1}{2 + x}}.$$

Repeat: Substitute the right-hand side of the last equation for the last x. Thus

$$x = 2 + \cfrac{1}{4 + \cfrac{1}{4 + \cfrac{1}{2 + x}}}.$$

Continue this process. It follows that $a_0 = 2$ and that $a_n = 4$ for all $n \geqq 1$.

Section 11.1

11.1.1: $\mathbf{v} = \overrightarrow{RS} = \langle 3 - 1, 5 - 2 \rangle = \langle 2, 3 \rangle$. The position vector of the point $P(2, 3)$ and $\overrightarrow{RS}$ are shown next.

11.3.3: $\mathbf{v} = \overrightarrow{RS} = \langle -5 - 5, -10 - 10 \rangle = \langle -10, -20 \rangle$. The position vector of the point P and $\overrightarrow{RS}$ are shown next.

11.1.5: $\mathbf{w} = \mathbf{u} + \mathbf{v} = \langle 1, -2 \rangle + \langle 3, 4 \rangle = \langle 1 + 3, -2 + 4 \rangle = \langle 4, 2 \rangle$. The next figure illustrates this computation in the form of the triangle law for vector addition (see Fig. 11.1.6 of the text).

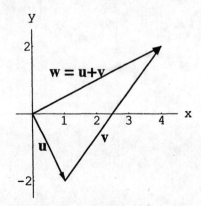

11.1.7: Given: $\mathbf{u} = 3\mathbf{i} + 5\mathbf{j}$, $\mathbf{v} = 2\mathbf{i} - 7\mathbf{j}$:

$$\mathbf{u} + \mathbf{v} = 3\mathbf{i} + 5\mathbf{j} + 2\mathbf{i} - 7\mathbf{j} = (3+2)\mathbf{i} + (5-7)\mathbf{j} = 5\mathbf{i} - 2\mathbf{j}.$$

The next figure illustrates the triangle law for vector addition using $\mathbf{u}$ and $\mathbf{v}$.

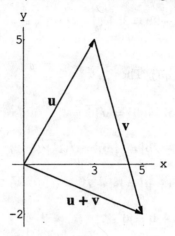

11.1.9: Given: $\mathbf{a} = \langle 1, -2 \rangle$ and $\mathbf{b} = \langle -3, 2 \rangle$. Then:

$$|\mathbf{a}| = \sqrt{(1)^2 + (-2)^2} = \sqrt{5},$$

$$|-2\mathbf{b}| = |\langle -6, 4 \rangle| = \sqrt{36 + 16} = 2\sqrt{13},$$

$$|\mathbf{a} - \mathbf{b}| = |\langle 1 - (-3), -2 - 2 \rangle| = \sqrt{16 + 16} = 4\sqrt{2},$$

$$\mathbf{a} + \mathbf{b} = \langle 1 - 3, -2 + 2 \rangle = \langle -2, 0 \rangle,$$

$$3\mathbf{a} - 2\mathbf{b} = \langle 3, -6 \rangle - \langle -6, 4 \rangle = \langle 3 - (-6), -6 - 4 \rangle = \langle 9, -10 \rangle.$$

11.1.11: Given: $\mathbf{a} = \langle -2, -2 \rangle$ and $\mathbf{b} = \langle -3, -4 \rangle$. Then:

$$|\mathbf{a}| = \sqrt{4 + 4} = \sqrt{8} = 2\sqrt{2},$$

$$|-2\mathbf{b}| = |\langle 6, 8 \rangle| = \sqrt{36 + 64} = 10,$$

$$|\mathbf{a} - \mathbf{b}| = |\langle -2 - (-3), -2 - (-4) \rangle| = |\langle 1, 2 \rangle| = \sqrt{1 + 4} = \sqrt{5},$$

$$\mathbf{a} + \mathbf{b} = \langle -2 - 3, -2 - 4 \rangle = \langle -5, -6 \rangle,$$

$$3\mathbf{a} - 2\mathbf{b} = \langle -6, -6 \rangle - \langle -6, -8 \rangle = \langle -6 - (-6), -6 - (-8) \rangle = \langle 0, 2 \rangle.$$

11.1.13: Given: $\mathbf{a} = \mathbf{i} + 3\mathbf{j}$ and $\mathbf{b} = 2\mathbf{i} - 5\mathbf{j}$. Then:

$$|\mathbf{a}| = |\mathbf{i} + 3\mathbf{j}| = \sqrt{1 + 9} = \sqrt{10},$$

661

$$|-2\mathbf{b}| = |4\mathbf{i} - 10\mathbf{j}| = \sqrt{16 + 100} = 2\sqrt{29},$$

$$|\mathbf{a} - \mathbf{b}| = |-\mathbf{i} + 8\mathbf{j}| = \sqrt{1 + 64} = \sqrt{65},$$

$$\mathbf{a} + \mathbf{b} = \mathbf{i} + 3\mathbf{j} + 2\mathbf{i} - 5\mathbf{j} = 3\mathbf{i} - 2\mathbf{j},$$

$$3\mathbf{a} - 2\mathbf{b} = 3\mathbf{i} + 9\mathbf{j} - 4\mathbf{i} + 10\mathbf{j} = -\mathbf{i} + 19\mathbf{j}.$$

11.1.15: Given: $\mathbf{a} = 4\mathbf{i}$ and $\mathbf{b} = -7\mathbf{j}$. Then:

$$|\mathbf{a}| = |4\mathbf{i}| = \sqrt{16} = 4,$$

$$|-2\mathbf{b}| = |14\mathbf{j}| = \sqrt{(14)^2} = 14,$$

$$|\mathbf{a} - \mathbf{b}| = |4\mathbf{i} + 7\mathbf{j}| = \sqrt{16 + 49} = \sqrt{65},$$

$$\mathbf{a} + \mathbf{b} = 4\mathbf{i} - 7\mathbf{j},$$

$$3\mathbf{a} - 2\mathbf{b} = 12\mathbf{i} + 14\mathbf{j}.$$

11.1.17: Because $|\mathbf{a}| = \sqrt{9 + 16} = 5,$

$$\mathbf{u} = \frac{1}{5}\mathbf{a} = -\frac{3}{5}\mathbf{i} - \frac{4}{5}\mathbf{j} \quad \text{and} \quad \mathbf{v} = -\frac{1}{5}\mathbf{a} = \frac{3}{5}\mathbf{i} + \frac{4}{5}\mathbf{j}.$$

11.1.19: Because $|\mathbf{a}| = \sqrt{64 + 225} = \sqrt{289} = 17,$

$$\mathbf{u} = \frac{1}{17}\mathbf{a} = \frac{8}{17}\mathbf{i} + \frac{15}{17}\mathbf{j} \quad \text{and} \quad \mathbf{v} = -\frac{1}{17}\mathbf{a} = -\frac{8}{17}\mathbf{i} - \frac{15}{17}\mathbf{j}.$$

11.1.21: $\mathbf{a} = \overrightarrow{PQ} = \langle 3 - 3, -2 - 2 \rangle = \langle 0, -4 \rangle = -4\mathbf{j}.$

11.1.23: $\mathbf{a} = \overrightarrow{PQ} = \langle 4 - (-4), -7 - 7 \rangle = \langle 8, -14 \rangle = 8\mathbf{i} - 14\mathbf{j}.$

11.1.25: Given $\mathbf{a} = \langle 6, 0 \rangle$ and $\mathbf{b} = \langle 0, -7 \rangle$, $\mathbf{c} = \mathbf{b} - \mathbf{a} = \langle -6, -7 \rangle$. Then

$$|\mathbf{c}|^2 = 36 + 49 = |\mathbf{a}|^2 + |\mathbf{b}|^2.$$

Therefore the vectors $\mathbf{a}$ and $\mathbf{b}$ are perpendicular because of the (true) converse of the Pythagorean theorem. See Example 5.

11.1.27: Given $\mathbf{a} = \langle 2, -1 \rangle$ and $\mathbf{b} = \langle 4, 8 \rangle$, $\mathbf{c} = \mathbf{b} - \mathbf{a} = \langle 2, 9 \rangle$. Then

$$|\mathbf{c}|^2 = 4 + 81 = 85 = 5 + 80 = |\mathbf{a}|^2 + |\mathbf{b}|^2.$$

Therefore the vectors **a** and **b** are perpendicular by the (true) converse of the Pythagorean theorem. See Example 5.

11.1.29: Given $\mathbf{a} = 2\mathbf{i} + 3\mathbf{j}$ and $\mathbf{b} = 3\mathbf{i} + 4\mathbf{j}$, we have

$$3\mathbf{a} - 2\mathbf{b} = 6\mathbf{i} + 9\mathbf{j} - 6\mathbf{i} - 8\mathbf{j} = \mathbf{j} \quad \text{and}$$

$$4\mathbf{a} - 3\mathbf{b} = 8\mathbf{i} + 12\mathbf{j} - 9\mathbf{i} - 12\mathbf{j} = -\mathbf{i}.$$

Therefore $\mathbf{i} = -4\mathbf{a} + 3\mathbf{b}$ and $\mathbf{j} = 3\mathbf{a} - 2\mathbf{b}$.

11.1.31: Given $\mathbf{a} = \mathbf{i} + \mathbf{j}$ and $\mathbf{b} = \mathbf{i} - \mathbf{j}$, we have

$$\mathbf{c} = 2\mathbf{i} - 3\mathbf{j} = r\mathbf{a} + s\mathbf{b} = r\mathbf{i} + r\mathbf{j} + s\mathbf{i} - s\mathbf{j} = (r+s)\mathbf{i} + (r-s)\mathbf{j}.$$

It follows that $r + s = 2$ and that $r - s = -3$. These simultaneous equations are easily solved, and thereby we find that $\mathbf{c} = -\frac{1}{2}\mathbf{a} + \frac{5}{2}\mathbf{b}$.

11.1.33: Part (a): $3(5\mathbf{i} - 7\mathbf{j}) = 15\mathbf{i} - 21\mathbf{j}$. Part (b): $\dfrac{1}{3}(5\mathbf{i} - 7\mathbf{j}) = \dfrac{5}{3}\mathbf{i} - \dfrac{7}{3}\mathbf{j}$.

11.1.35: Part (a): If $\mathbf{a} = 7\mathbf{i} - 3\mathbf{j}$, then $|\mathbf{a}| = \sqrt{58}$. Thus a unit vector with the same direction as **a** is

$$\mathbf{u} = \frac{\mathbf{a}}{\sqrt{58}}. \quad \text{Answer:} \quad 5\mathbf{u} = \frac{5\sqrt{58}}{58}(7\mathbf{i} - 3\mathbf{j}).$$

Part (b): If $\mathbf{b} = 8\mathbf{i} + 5\mathbf{j}$, then $|\mathbf{b}| = \sqrt{89}$. Thus a unit vector with the same direction as **b** is

$$\mathbf{v} = \frac{\mathbf{b}}{\sqrt{89}}. \quad \text{Answer:} \quad -5\mathbf{v} = -\frac{5\sqrt{89}}{89}(8\mathbf{i} + 5\mathbf{j}).$$

11.1.37: Given $\mathbf{a} = \langle 2c, -4 \rangle$ and $\mathbf{b} = \langle 3, c \rangle$, let $\mathbf{w} = \mathbf{b} - \mathbf{a} = \langle 3 - 2c, c + 4 \rangle$. Perpendicularity of **a** and **b** is equivalent to the Pythagorean relation

$$|\mathbf{a}|^2 + |\mathbf{b}|^2 = |\mathbf{w}|^2;$$

that is, $4c^2 + 16 + c^2 + 9 = 5c^2 - 4c + 25$, and thus the unique solution is $c = 0$.

11.1.39: With $\mathbf{a} = \langle a_1, a_2 \rangle$, $\mathbf{b} = \langle b_1, b_2 \rangle$, and $\mathbf{c} = \langle c_1, c_2 \rangle$, we have

$$\mathbf{a} + (\mathbf{b} + \mathbf{c}) = \langle a_1, a_2 \rangle + (\langle b_1, b_2 \rangle + \langle c_1, c_2 \rangle) = \langle a_1, a_2 \rangle + \langle b_1 + c_1, b_2 + c_2 \rangle$$

$$= \langle a_1 + (b_1 + c_1), a_2 + (b_2 + c_2) \rangle = \langle (a_1 + b_1) + c_1, (a_2 + b_2) + c_2 \rangle$$

$$= \langle a_1 + b_1, a_2 + b_2 \rangle + \langle c_1, c_2 \rangle = (\langle a_1, a_2 \rangle + \langle b_1, b_2 \rangle) + \langle c_1, c_2 \rangle = (\mathbf{a} + \mathbf{b}) + \mathbf{c}.$$

11.1.41: With $\mathbf{a} = \langle a_1, a_2 \rangle$ and scalars r and s, we have

$$(rs)\mathbf{a} = (rs)\langle a_1, a_2 \rangle = \langle (rs)a_1, (rs)a_2 \rangle = \langle r(sa_1), r(sa_2) \rangle = r\langle sa_1, sa_2 \rangle = r(s\langle a_1, a_2 \rangle) = r(s\mathbf{a}).$$

11.1.43: See Fig. 11.1.13. The angle each cable makes with the horizontal is $30°$, so

$$T_1 \cos 30° = T_2 \cos 30°. \quad \text{Therefore} \quad T_1 = T_2.$$

Also

$$T_1 \sin 30° + T_2 \sin 30° = 100; \qquad 2 \cdot \frac{1}{2} \cdot T_1 = 100; \qquad T_1 = T_2 = 100.$$

11.1.45: See Fig. 11.1.15; let T_1 be the tension in the right-hand cable and T_2 the tension in the left-hand cable. The equations that balance the horizontal and vertical forces, respectively, are

$$T_1 \cos 40° = T_2 \cos 55° \quad \text{and} \quad T_1 \sin 40° + T_2 \sin 55° = 125.$$

A *Mathematica* command for solving these equations simultaneously is

```
Solve[ { t1*Cos[40*Pi/180] == t2*Cos[55*Pi/180],

        t1*Sin[40*Pi/180] + t2*Sin[55*Pi/180] == 125 }, {t1,t2} ]
```

and it yields the solution

$$T_1 = \frac{125 \cos(11\pi/15)}{\cos(11\pi/36)\sin(2\pi/9) + \cos(2\pi/9)\sin(11\pi/36)} \approx 71.970925644575 \quad \text{(lb)},$$

$$T_2 = \frac{125 \cos(2\pi/9)}{\cos(11\pi/36)\sin(2\pi/9) + \cos(2\pi/9)\sin(11\pi/36)} \approx 96.121326055335 \quad \text{(lb)}.$$

11.1.47: $\mathbf{v}_a = \mathbf{v}_g - \mathbf{w} = \langle 500, 0 \rangle - \langle -25\sqrt{2}, -25\sqrt{2} \rangle = \langle 500 + 25\sqrt{2}, 25\sqrt{2} \rangle$. This corresponds to a compass bearing of about $86° 13'$ (about $3.778°$ north of east) with airspeed approximately 536.52 mi/h (about 467 knots).

11.1.49: $\mathbf{v}_a = \mathbf{v}_g - \mathbf{w} = \langle -250\sqrt{2}, 250\sqrt{2} \rangle - \langle -25\sqrt{2}, -25\sqrt{2} \rangle = \langle -225\sqrt{2}, 275\sqrt{2} \rangle$. This corresponds to a compass bearing of about $320° 43'$ (about $50.71°$ north of west) with airspeed approximately 502 mi/h (about 437 knots).

11.1.51: Denote the origin by O. Then $\mathbf{a} = \overrightarrow{OP}$ and $\mathbf{b} = \overrightarrow{OQ}$. Let M denote the midpoint of the line segment PQ. Then $\overrightarrow{PQ} = \mathbf{b} - \mathbf{a}$, so that

$$\overrightarrow{PM} = \frac{1}{2}(\mathbf{b} - \mathbf{a}).$$

Hence the position vector of M is

$$\overrightarrow{OP} + \overrightarrow{MP} = \mathbf{a} + \frac{1}{2}(\mathbf{b} - \mathbf{a}) = \frac{1}{2}(\mathbf{a} + \mathbf{b}).$$

11.1.53: Let A, B, C, and D be the vertices of a parallelogram, in order, counterclockwise. Then its diagonals are AC and BD. Let M be the point where the diagonals cross. Let $\mathbf{u} = \overrightarrow{AB}$ and let $\mathbf{v} = \overrightarrow{BC}$. Then

$$\overrightarrow{AC} = \mathbf{u} + \mathbf{v} \quad \text{and} \quad \overrightarrow{BD} = \mathbf{v} - \mathbf{u}.$$

Then $\overrightarrow{AM} = r(\mathbf{u} + \mathbf{v})$ and $\overrightarrow{MC} = (1 - r)(\mathbf{u} + \mathbf{v})$ for some scalar r. Similarly, $\overrightarrow{BM} = s(\mathbf{v} - \mathbf{u})$ and $\overrightarrow{MD} = (1 - s)(\mathbf{v} - \mathbf{u})$ for some scalar s. Therefore

$$\overrightarrow{AB} + \overrightarrow{BM} = \overrightarrow{AM}; \qquad \mathbf{u} + s(\mathbf{v} - \mathbf{u}) = r(\mathbf{u} + \mathbf{v});$$

$$(1 - r - s)\mathbf{u} = (r - s)\mathbf{v}; \qquad 1 - r - s = 0 = r - s$$

because $\mathbf{u}$ and $\mathbf{v}$ are not parallel. It follows that $r = s = \frac{1}{2}$, so that $|AM| = |MC|$ and $|BM| = |MD|$. Therefore the diagonals of the arbitrary parallelogram $ABCD$ bisect each other.

11.1.55: See Fig. 11.1.18, which makes it easy to see that $a_1 = r \cos \theta$ and $a_2 = r \sin \theta$. Then

$$\mathbf{a}_\perp = \left[r \cos \left(\theta + \frac{\pi}{2} \right) \right] \mathbf{i} + \left[r \sin \left(\theta + \frac{\pi}{2} \right) \right] \mathbf{j} = (-r \sin \theta)\mathbf{i} + (r \cos \theta)\mathbf{j} = -a_2 \mathbf{i} + a_1 \mathbf{j}.$$

Section 11.2

11.2.1: Given $\mathbf{a} = \langle 2, 5, -4 \rangle$ and $\mathbf{b} = \langle 1, -2, -3, \rangle$, we find that

(a): $2\mathbf{a} + \mathbf{b} = \langle 5, 8, -11 \rangle$,

(b): $3\mathbf{a} - 4\mathbf{b} = \langle 2, 23, 0 \rangle$,

(c): $\mathbf{a} \cdot \mathbf{b} = 2 \cdot 1 - 5 \cdot 2 + 4 \cdot 3 = 4$,

(d): $|\mathbf{a} - \mathbf{b}| = |\langle 1, 7, -1 \rangle| = \sqrt{51}$, and

(e): $\dfrac{\mathbf{a}}{|\mathbf{a}|} = \dfrac{1}{\sqrt{4 + 25 + 16}} \langle 2, 5, -4 \rangle = \dfrac{\sqrt{5}}{15} \langle 2, 5, -4 \rangle = \left\langle \dfrac{2\sqrt{5}}{15}, \dfrac{\sqrt{5}}{3}, -\dfrac{4\sqrt{5}}{15} \right\rangle.$

11.2.3: Given $\mathbf{a} = \langle 1, 1, 1 \rangle$ and $\mathbf{b} = \langle 0, 1, -1 \rangle$, we find that

(a): $2\mathbf{a} + \mathbf{b} = \langle 2, 3, 1 \rangle$,

(b): $3\mathbf{a} - 4\mathbf{b} = \langle 3, -1, 7 \rangle$,

(c): $\mathbf{a} \cdot \mathbf{b} = 1 \cdot 0 + 1 \cdot 1 - 1 \cdot 1 = 0$, (d): $|\mathbf{a} - \mathbf{b}| = |\langle 1, 0, 2 \rangle| = \sqrt{5}$, and

(e): $\dfrac{\mathbf{a}}{|\mathbf{a}|} = \dfrac{1}{\sqrt{3}} \langle 1, 1, 1 \rangle = \left\langle \dfrac{\sqrt{3}}{3}, \dfrac{\sqrt{3}}{3}, \dfrac{\sqrt{3}}{3} \right\rangle.$

11.2.5: Given $\mathbf{a} = \langle 2, -1, 0 \rangle$ and $\mathbf{b} = \langle 0, 1, -3 \rangle$, we find that

(a): $2\mathbf{a} + \mathbf{b} = \langle 4, -1, -3 \rangle,$

(b): $3\mathbf{a} - 4\mathbf{b} = \langle 6, -7, 12 \rangle,$

(c): $\mathbf{a} \cdot \mathbf{b} = 2 \cdot 0 - 1 \cdot 1 + 0 \cdot (-3) = -1,$

(d): $|\mathbf{a} - \mathbf{b}| = |\langle 2, -2, 3 \rangle| = \sqrt{17}$, and

(e): $\dfrac{\mathbf{a}}{|\mathbf{a}|} = \dfrac{1}{\sqrt{5}} \langle 2, -1, 0 \rangle = \left\langle \dfrac{2\sqrt{5}}{5}, -\dfrac{\sqrt{5}}{5}, 0 \right\rangle.$

11.2.7: If θ is the angle between $\mathbf{a}$ and $\mathbf{b}$, then

$$\cos\theta = \frac{\mathbf{a} \cdot \mathbf{b}}{|\mathbf{a}| \cdot |\mathbf{b}|} = \frac{4}{\sqrt{45}\,\sqrt{14}} = \frac{2\sqrt{70}}{105} \approx 0.15936371,$$

and therefore $\theta \approx 80.830028° \approx 81°$.

11.2.9: If θ is the angle between $\mathbf{a}$ and $\mathbf{b}$, then

$$\cos\theta = \frac{\mathbf{a} \cdot \mathbf{b}}{|\mathbf{a}| \cdot |\mathbf{b}|} = 0 \quad \text{(exactly)},$$

and therefore $\theta = 90°$ (exactly).

11.2.11: If θ is the angle between $\mathbf{a}$ and $\mathbf{b}$, then

$$\cos\theta = \frac{\mathbf{a} \cdot \mathbf{b}}{|\mathbf{a}| \cdot |\mathbf{b}|} = \frac{-1}{5\sqrt{2}} = -\frac{\sqrt{2}}{10} \approx -0.14142136,$$

and therefore $\theta \approx 98.130102° \approx 98°$.

11.2.13: Refer to Problem 1;

$$\text{comp}_{\mathbf{a}}\mathbf{b} = \frac{\mathbf{a} \cdot \mathbf{b}}{|\mathbf{a}|} = \frac{4\sqrt{5}}{15} \quad \text{and} \quad \text{comp}_{\mathbf{b}}\mathbf{a} = \frac{\mathbf{a} \cdot \mathbf{b}}{|\mathbf{b}|} = \frac{2\sqrt{14}}{7}.$$

11.2.15: Refer to Problem 3;

$$\text{comp}_{\mathbf{a}}\mathbf{b} = \frac{\mathbf{a} \cdot \mathbf{b}}{|\mathbf{a}|} = 0 = \text{comp}_{\mathbf{b}}\mathbf{a} = \frac{\mathbf{a} \cdot \mathbf{b}}{|\mathbf{b}|}.$$

11.2.17: Refer to Problem 5;

$$\text{comp}_{\mathbf{a}}\mathbf{b} = \frac{\mathbf{a} \cdot \mathbf{b}}{|\mathbf{a}|} = -\frac{\sqrt{5}}{5} \quad \text{and} \quad \text{comp}_{\mathbf{b}}\mathbf{a} = \frac{\mathbf{a} \cdot \mathbf{b}}{|\mathbf{b}|} = -\frac{\sqrt{10}}{10}.$$

11.2.19: An equation of the sphere is $(x-3)^2 + (y-1)^2 + (z-2)^2 = 25$; that is,

$$x^2 - 6x + y^2 - 2y + z^2 - 4z = 11.$$

11.2.21: The center of the sphere is at

$$\left(\frac{3+7}{2}, \frac{5+3}{2}, \frac{-3+1}{2} \right) = (5, 4, -1)$$

and its radius is

$$\frac{1}{2}\sqrt{(7-3)^2 + (3-5)^2 + (1+3)^2} = \frac{1}{2}\sqrt{16 + 4 + 16} = 3.$$

Therefore an equation of this sphere is $(x-5)^2 + (y-4)^2 + (z+1)^2 = 9$; that is,

$$x^2 - 10x + y^2 - 8y + z^2 + 2z + 33 = 0.$$

11.2.23: The sphere has radius 2, thus equation $x^2 + y^2 + (z-2)^2 = 4$; that is, $x^2 + y^2 + z^2 - 4z = 0$.

11.2.25: We complete the square in x and y:

$$x^2 + 4x + 4 + y^2 - 6y + 9 + z^2 = 13;$$

$$(x+2)^2 + (y-3)^2 + (z-0)^2 = 13.$$

This sphere has center at $(-2, 3, 0)$ and radius $\sqrt{13}$.

11.2.27: We complete the square in z:

$$x^2 + y^2 + z^2 - 6z + 9 = 16 + 9 = 25;$$

$$(x-0)^2 + (y-0)^2 + (z-3)^2 = 5^2.$$

This sphere has center at $(0, 0, 3)$ and radius 5.

11.2.29: The equation $z = 0$ is an equation of the xy-plane.

11.2.31: The equation $z = 10$ is an equation of the plane parallel to the xy-plane and passing through the point $(0, 0, 10)$.

11.2.33: The equation $xyz = 0$ holds when any one (or more) of the three variables is zero. Hence its graph is the union of the three coordinate planes.

11.2.35: Given the equation $x^2 + y^2 + z^2 = 0$, the sum of the three nonnegative numbers x^2, y^2, and z^2 can be zero only if none of x, y, and z is nonzero. Therefore the graph of the given equation consists of the single point $(0, 0, 0)$.

11.2.37: Complete the square in x and y:

$$x^2 + y^2 + z^2 - 6x + 8y + 25 = 0;$$

$$x^2 - 6x + 9 + y^2 + 8y + 16 + z^2 + 25 = 25;$$

$$(x - 3)^2 + (y + 4)^2 + z^2 = 0.$$

The sum of three nonnegative real numbers can be zero only if none is positive. So the graph of the given equation consists of the single point $(3, -4, 0)$.

11.2.39: If $\mathbf{a} = \langle 4, -2, 6 \rangle$ and $\mathbf{b} = \langle 6, -3, 9 \rangle$, then $\mathbf{b} = \frac{3}{2}\mathbf{a}$. Therefore $\mathbf{a}$ and $\mathbf{b}$ are parallel. Because they are nonzero and parallel, they are not perpendicular; alternatively, they are not perpendicular because $\mathbf{a} \cdot \mathbf{b} = 84 \neq 0$.

11.2.41: Because

$$\mathbf{b} = -9\mathbf{i} + 15\mathbf{j} - 12\mathbf{k} = -\frac{3}{4}(12\mathbf{i} - 20\mathbf{j} + 16\mathbf{k}) = -\frac{3}{4}\mathbf{a},$$

the vectors $\mathbf{a}$ and $\mathbf{b}$ are parallel. Because they are nonzero as well, they are not perpendicular; alternatively, $\mathbf{a} \cdot \mathbf{b} = -600 \neq 0$, and hence they are not perpendicular.

11.2.43: The distinct points P, Q, and R lie on a single straight line if and only if the vectors representing $\overrightarrow{PQ}$ and $\overrightarrow{QR}$ are parallel. Here we have

$$3\overrightarrow{PQ} = 3(\langle 1, -3, 5 \rangle - \langle 0, -2, 4 \rangle) = 3\langle 1, -1, 1 \rangle = \langle 3, -3, 3 \rangle = \langle 4, -6, 8 \rangle - \langle 1, -3, 5 \rangle = \overrightarrow{QR}.$$

Therefore $\overrightarrow{PQ}$ and $\overrightarrow{QR}$ are parallel, and so the three points P, Q, and R do lie on a single straight line.

11.2.45: The sides of triangle ABC are represented by the three vectors $\mathbf{u} = \overrightarrow{AB} = \langle -1, 1, 0 \rangle$, $\mathbf{v} = \overrightarrow{BC} = \langle 0, -1, 1 \rangle$, and $\mathbf{w} = \overrightarrow{AC} = \langle -1, 0, 1 \rangle$. Let A denote the angle of the triangle at vertex A. Then

$$\cos A = \frac{\mathbf{u} \cdot \mathbf{w}}{|\mathbf{u}| \cdot |\mathbf{w}|} = \frac{(-1)(-1)}{\sqrt{2}\,\sqrt{2}} = \frac{1}{2},$$

and therefore $A = 60°$ (exactly). Similarly,

$$\cos B = \frac{-\mathbf{u} \cdot \mathbf{v}}{|-\mathbf{u}| \cdot |\mathbf{v}|} = \frac{(-1)(-1)}{\sqrt{2}\,\sqrt{2}} = \frac{1}{2},$$

and so $B = 60°$ as well. Finally,

$$\cos C = \frac{-\mathbf{v}\cdot(-\mathbf{w})}{|-\mathbf{v}|\cdot|-\mathbf{w}|} = \frac{(-1)(-1)}{\sqrt{2}\,\sqrt{2}} = \frac{1}{2},$$

and it follows that $C = 60°$.

11.2.47: The sides of triangle ABC are represented by the three vectors $\mathbf{u} = \overrightarrow{AB} = \langle 2, -3, 2\rangle$, $\mathbf{v} = \overrightarrow{BC} = \langle 0, 6, 3\rangle$, and $\mathbf{w} = \overrightarrow{AC} = \langle 2, 3, 5\rangle$. Let A denote the angle of the triangle at vertex A. Then

$$\cos A = \frac{\mathbf{u}\cdot\mathbf{w}}{|\mathbf{u}|\cdot|\mathbf{w}|} = \frac{4-9+10}{\sqrt{17}\,\sqrt{38}} = \frac{5\sqrt{646}}{646} \approx 0.19672237,$$

and therefore $A \approx 78.654643° \approx 79°$. Similarly,

$$\cos B = \frac{-\mathbf{u}\cdot\mathbf{v}}{|-\mathbf{u}|\cdot|\mathbf{v}|} = \frac{18-6}{\sqrt{17}\,\sqrt{45}} = \frac{4\sqrt{85}}{85} \approx 0.43386082,$$

and so $B \approx 64.287166° \approx 64°$. Finally,

$$\cos C = \frac{-\mathbf{v}\cdot(-\mathbf{w})}{|-\mathbf{v}|\cdot|-\mathbf{w}|} = \frac{18+15}{\sqrt{38}\,\sqrt{45}} = \frac{11\sqrt{190}}{190} \approx 0.79802388,$$

and it follows that $C \approx 37.058191° \approx 37°$.

11.2.49: If $\mathbf{a} = \overrightarrow{PQ} = \langle 2, 5, 5\rangle$ and its direction angles are α, β, and γ (as in Fig. 11.2.14), then

$$\cos\alpha = \frac{2}{|\mathbf{a}|} = \frac{2}{\sqrt{54}} = \frac{\sqrt{6}}{9} \approx 0.27216553,$$

so that $\alpha \approx 74.206830951736°$. Also

$$\cos\beta = \cos\gamma = \frac{5}{|\mathbf{a}|} = \frac{5}{\sqrt{54}} = \frac{5\sqrt{6}}{18} \approx 0.68041382,$$

and therefore $\beta = \gamma \approx 47.124011333364°$.

11.2.51: If $\mathbf{a} = \overrightarrow{PQ} = \langle 6, 8, 10\rangle$ and its direction angles are α, β, and γ (as in Fig. 11.2.14), then

$$\cos\alpha = \frac{6}{|\mathbf{a}|} = \frac{6}{\sqrt{200}} = \frac{3\sqrt{2}}{10} \approx 0.42426407,$$

so that $\alpha \approx 64.895909749779°$. Next,

$$\cos\beta = \frac{8}{|\mathbf{a}|} = \frac{8}{\sqrt{200}} = \frac{2\sqrt{2}}{5} \approx 0.56568543,$$

and therefore $\beta \approx 55.550098012047°$. Finally,

$$\cos\gamma = \frac{10}{|\mathbf{a}|} = \frac{10}{\sqrt{200}} = \frac{\sqrt{2}}{2} \approx 0.70710678,$$

and so $\gamma = 45°$ (exactly).

11.2.53: The displacement vector is $\mathbf{D} = \overrightarrow{PQ} = 3\mathbf{i} + \mathbf{j}$, and consequently the work done is simply $\mathbf{F} \cdot \mathbf{D} = (\mathbf{i} - \mathbf{k}) \cdot (3\mathbf{i} + \mathbf{j}) = 3$.

11.2.55: $40 \cdot (\cos 40°) \cdot 1000 \cdot (0.239) \approx 7323.385$ (cal; less than 8 Cal).

11.2.57: Set up a coordinate system in which the lowest point of the inclined plane is at the origin in the xy-plane and its highest point is at (x, h). Then

$$\frac{h}{x} = \tan \alpha, \quad \text{so that} \quad x = \frac{h}{\tan \alpha} = h \cot \alpha.$$

Therefore the displacement vector in this problem is $\mathbf{D} = \langle h \cot \alpha, h \rangle$. A unit vector parallel to the inclined plane is $\langle \cos \alpha, \sin \alpha \rangle$, so a unit vector in the direction of $\mathbf{N}$ is $\langle -\sin \alpha, \cos \alpha \rangle$ (by Problem 55 in Section 11.1). Therefore $\mathbf{N} = \lambda \langle -\sin \alpha, \cos \alpha \rangle$ where λ is a positive scalar. If we denote by F the magnitude $|\mathbf{F}|$ of $\mathbf{F}$, then

$$\mathbf{F} = \langle F \cos \alpha, F \sin \alpha \rangle.$$

The horizontal forces acting on the weight must balance, as must the vertical forces, yielding (respectively) the scalar equations

$$\lambda \sin \alpha = F \cos \alpha \quad \text{and} \quad \lambda \cot \alpha + F \sin \alpha = mg.$$

From the first of these equations we see that $\lambda = F \cos \alpha$, and substitution of this in the second equation yields

$$F \cdot \left(\frac{\cos^2 \alpha}{\sin \alpha} \right) + F \cdot \left(\frac{\sin^2 \alpha}{\sin \alpha} \right) = mg,$$

and it follows that $F = mg \sin \alpha$. Therefore $\mathbf{F} = \langle mg \sin \alpha \cos \alpha, mg \sin^2 \alpha \rangle$. So the work done by $\mathbf{F}$ in moving the weight from the bottom to the top of the inclined plane is (if there's no friction)

$$W = \mathbf{F} \cdot \mathbf{D} = mgh \cos^2 \alpha + mgh \sin^2 \alpha = mgh.$$

11.2.59: By the first equation in (9), we have

$$|\mathbf{a} + \mathbf{b}|^2 = (\mathbf{a} + \mathbf{b}) \cdot (\mathbf{a} + \mathbf{b}).$$

Then we use the fact that $x \leqq |x|$ for each real number x (in the first line) and the Cauchy-Schwarz inequality of Problem 58 (in the second line):

$$|\mathbf{a} + \mathbf{b}|^2 = \mathbf{a} \cdot \mathbf{a} + 2\mathbf{a} \cdot \mathbf{b} + \mathbf{b} \cdot \mathbf{b} = |\mathbf{a}|^2 + 2\mathbf{a} \cdot \mathbf{b} + |\mathbf{b}|^2 \leqq |\mathbf{a}|^2 + 2|\mathbf{a} \cdot \mathbf{b}| + |\mathbf{b}|^2$$

$$\leqq |\mathbf{a}|^2 + 2|\mathbf{a}| \cdot |\mathbf{b}| + |\mathbf{b}|^2 = (|\mathbf{a}| + |\mathbf{b}|)^2.$$

All these expression are nonnegative. Therefore, for any two vectors $\mathbf{a}$ and $\mathbf{b}$, $|\mathbf{a} + \mathbf{b}| \leq |\mathbf{a}| + |\mathbf{b}|$.

11.2.61: Suppose that $\mathbf{u} = \langle u_1, u_2, u_3 \rangle$ and that $\mathbf{v} = \langle v_1, v_2, v_3 \rangle$. If both are the zero vector, let $\mathbf{w} = \mathbf{i}$. Otherwise we may suppose without loss of genrality that $\mathbf{u} \neq \mathbf{0}$. Indeed, we may further suppose without loss of generality that $u_1 \neq 0$. We need to choose $\mathbf{w} = \langle w_1, w_2, w_3 \rangle$ such that

$$\mathbf{u} \cdot \mathbf{w} = u_1 w_1 + u_2 w_2 + u_3 w_3 = 0 \quad \text{and}$$

$$\mathbf{v} \cdot \mathbf{w} = v_1 w_1 + v_2 w_2 + v_3 w_3 = 0.$$

This will obtain provided that

$$u_1 v_1 w_1 + u_2 v_1 w_2 + u_3 v_1 w_3 = 0 \quad \text{and}$$

$$u_1 v_1 w_1 + u_1 v_2 w_2 + u_1 v_3 w_3 = 0.$$

Subtraction of the first of these equations from the second yields the sufficient condition

$$(u_1 v_2 - u_2 v_1) w_2 + (u_1 v_3 - u_3 v_1) w_3 = 0.$$

Thus our goal will be accomplished if we let

$$w_2 = u_1 v_3 - u_3 v_1 \quad \text{and}$$

$$w_3 = -(u_1 v_2 - u_2 v_1) = u_2 v_1 - u_1 v_2.$$

Now let

$$w_1 = -\frac{u_2 w_2 + u_3 w_3}{u_1},$$

and we have reached our goal. In particular, if $\mathbf{u} = \langle 1, 2, -3 \rangle$ and $\mathbf{v} = \langle 2, 0, 1 \rangle$, then we obtain

$$w_2 = 1 \cdot 1 + 3 \cdot 2 = 7,$$

$$w_3 = 2 \cdot 2 - 1 \cdot 0 = 4, \quad \text{and}$$

$$w_1 = -(2 \cdot 7 - 3 \cdot 4) = -2,$$

and thus any nonzero multiple of $\mathbf{w} = \langle -2, 7, 4 \rangle$ is a correct answer to this problem. We will see a more efficient construction of a nonzero vector perpendicular to two given nonparallel vectors in Section 11.3.

11.2.63: Given: The points $P_1(x_1, y_1, z_1)$, $P_2(x_2, y_2, z_2)$, and

$$M\left(\frac{x_1 + x_2}{2}, \frac{y_1 + y_2}{2}, \frac{z_1 + z_2}{2}\right).$$

671

Let $\mathbf{a} = \overrightarrow{P_1 P_2} = \langle x_2 - x_1, y_2 - y_1, z_2 - z_1 \rangle$. Then $\mathbf{a}$ coincides with the line segment joining P_1 with P_2. Hence if the initial point of $\frac{1}{2}\mathbf{a}$ is placed at P_1, then its terminal point will coincide with the midpoint of the segment $P_1 P_2$. That is, $\overrightarrow{OP_1} + \frac{1}{2}\mathbf{a}$ will be the position vector of the midpoint of $P_1 P_2$. But

$$\overrightarrow{OP_1} + \frac{1}{2}\mathbf{a} = \langle x_1, y_1, z_1 \rangle + \frac{1}{2}\langle x_2 - x_1, y_2 - y_1, z_2 - z_1 \rangle = \left\langle \frac{x_1 + x_2}{2}, \frac{y_1 + y_2}{2}, \frac{z_1 + z_2}{2} \right\rangle = \overrightarrow{OM}.$$

This is enough to establish that M is the midpoint of the segment $P_1 P_2$.

11.2.65: Suppose that the vectors $\mathbf{a}$ and $\mathbf{b}$ in the plane are nonzero and nonparallel. We will provide a proof in the case that the first components of $\mathbf{a}$ and $\mathbf{b}$ are nonzero. Let $\mathbf{a} = \langle a_1, a_2, \rangle$, $\mathbf{b} = \langle b_1, b_2 \rangle$, and $\mathbf{c} = \langle c_1, c_2 \rangle$. To find α and β such that $\mathbf{c} = \alpha \mathbf{a} + \beta \mathbf{b}$, solve

$$c_1 = \alpha a_1 + \beta b_1,$$

$$c_2 = \alpha a_2 + \beta b_2;$$

$$a_2 c_1 = \alpha a_1 a_2 + \beta a_2 b_1,$$

$$a_1 c_2 = \alpha a_1 a_2 + \beta a_1 b_2.$$

Subtract the next-to-last equation from the last:

$$a_1 c_2 - a_2 c_1 = \beta(a_1 b_2 - a_2 b_1);$$

$$\beta = \frac{a_1 c_2 - a_2 c_1}{a_1 b_2 - a_2 b_1}.$$

But what if the last denominator is zero? Then $a_1 b_2 = a_2 b_1$, so that

$$\frac{a_2}{a_1} = \frac{b_2}{b_1} = k,$$

and thus $a_2 = k a_1$ and $b_2 = k b_1$. Therefore

$$\mathbf{a} = \langle a_1, k a_1 \rangle = a_1 \langle 1, k \rangle \quad \text{and}$$

$$\mathbf{b} = \langle b_1, k b_1 \rangle = b_1 \langle 1, k \rangle,$$

so that $\mathbf{a}$ and $\mathbf{b}$ are parallel. Hence $a_1 b_2 - a_2 b_1 \neq 0$.

In a similar fashion we find that

$$\alpha = \frac{b_2 c_1 - b_1 c_2}{a_1 b_2 - a_2 b_1}.$$

672

All that remains is the verification that α and β perform as advertised:

$$\alpha\mathbf{a} + \beta\mathbf{b} = \frac{b_2 c_1 - b_1 c_2}{a_1 b_2 - a_2 b_1}\langle a_1, a_2 \rangle + \frac{a_1 c_2 - a_2 c_1}{a_1 b_2 - a_2 b_1}\langle b_1, b_2 \rangle$$

$$= \frac{1}{a_1 b_2 - a_2 b_1}\left(\langle a_1 b_2 c_1 - a_1 b_1 c_2, \; a_2 b_2 c_1 - a_2 b_1 c_2 \rangle + \langle a_1 b_1 c_2 - a_2 b_1 c_1, \; a_1 b_2 c_2 - a_2 b_2 c_1 \rangle\right)$$

$$= \frac{1}{a_1 b_2 - a_2 b_1}\langle (a_1 b_2 - a_2 b_1) c_1, \; (a_1 b_2 - a_2 b_1) c_2 \rangle = \langle c_1, c_2 \rangle = \mathbf{c}.$$

11.2.67: The equation

$$(x-3)^2 + (y+2)^2 + (z-4)^2 = (x-5)^2 + (y-7)^2 + (z+1)^2$$

can be simplified to $2x + 9y - 5z = 23$. This is an equation of the plane that bisects the segment AB and is perpendicular to it.

11.2.69: The distance between the last two points listed in Problem 69 is

$$\sqrt{(1-0)^2 + (0-1)^2 + (1-1)^2} = \sqrt{2},$$

and the other five distance computations also yield the same result. Hence the given points are the vertices of a regular tetrahedron. Let $\mathbf{a}$ be the position vector of $(1, 1, 0)$ and let $\mathbf{b}$ be the position vector of $(1, 0, 1)$. Then the angle θ between $\mathbf{a}$ and $\mathbf{b}$ satisfies the equation

$$\cos\theta = \frac{\mathbf{a}\cdot\mathbf{b}}{|\mathbf{a}|\cdot|\mathbf{b}|} = \frac{1\cdot 1 + 1\cdot 0 + 0\cdot 1}{\sqrt{2}\,\sqrt{2}} = \frac{1}{2}.$$

Therefore $\theta = \dfrac{\pi}{3}$.

Section 11.3

11.3.1: If $\mathbf{a} = \langle 5, -1, -2 \rangle$ and $\mathbf{b} = \langle -3, 2, 4 \rangle$, then

$$\mathbf{a}\times\mathbf{b} = \begin{vmatrix} \mathbf{i} & \mathbf{j} & \mathbf{k} \\ 5 & -1 & -2 \\ -3 & 2 & 4 \end{vmatrix} = \langle -4+4, \; 6-20, \; 10-3 \rangle = \langle 0, -14, 7 \rangle.$$

11.3.3: If $\mathbf{a} = \mathbf{i} - \mathbf{j} + 3\mathbf{k}$ and $\mathbf{b} = -2\mathbf{i} + 3\mathbf{j} + \mathbf{k}$, then

$$\mathbf{a}\times\mathbf{b} = \begin{vmatrix} \mathbf{i} & \mathbf{j} & \mathbf{k} \\ 1 & -1 & 3 \\ -2 & 3 & 1 \end{vmatrix} = (-1-9)\mathbf{i} + (-6-1)\mathbf{j} + (3-2)\mathbf{k} = -10\mathbf{i} - 7\mathbf{j} + \mathbf{k}.$$

11.3.5: If $\mathbf{a} = \langle 2, -3, 0 \rangle$ and $\mathbf{b} = \langle 4, 5, 0 \rangle$, then

$$\mathbf{a} \times \mathbf{b} = \begin{vmatrix} \mathbf{i} & \mathbf{j} & \mathbf{k} \\ 2 & -3 & 0 \\ 4 & 5 & 0 \end{vmatrix} = \langle 0 - 0, \, 0 - 0, \, 10 + 12 \rangle = \langle 0, 0, 22 \rangle.$$

11.3.7: If $\mathbf{a} = \langle 3, 12, 0 \rangle$ and $\mathbf{b} = \langle 0, 4, 3 \rangle$, then

$$\mathbf{a} \times \mathbf{b} = \begin{vmatrix} \mathbf{i} & \mathbf{j} & \mathbf{k} \\ 3 & 12 & 0 \\ 0 & 4 & 3 \end{vmatrix} = \langle 36 - 0, \, 0 - 9, \, 12 - 0 \rangle = \langle 36, -9, 12 \rangle.$$

The magnitude of $\mathbf{a} \times \mathbf{b}$ is

$$|\mathbf{a} \times \mathbf{b}| = \sqrt{36^2 + 9^2 + 12^2} = \sqrt{1521} = 39,$$

so the two unit vectors perpendicular to both $\mathbf{a}$ and $\mathbf{b}$ are

$$\mathbf{u} = \frac{1}{39} \langle 36, -9, 12 \rangle = \left\langle \frac{12}{13}, -\frac{3}{13}, \frac{4}{13} \right\rangle \quad \text{and} \quad \mathbf{v} = -\frac{1}{39} \langle 36, -9, 12 \rangle = \left\langle -\frac{12}{13}, \frac{3}{13}, -\frac{4}{13} \right\rangle.$$

11.3.9: The definition of the cross product in Eq. (5) yields

$$\mathbf{i} \times \mathbf{j} = \begin{vmatrix} \mathbf{i} & \mathbf{j} & \mathbf{k} \\ 1 & 0 & 0 \\ 0 & 1 & 0 \end{vmatrix} = (0 - 0)\mathbf{i} + (0 - 0)\mathbf{j} + (1 - 0)\mathbf{k} = \mathbf{k},$$

$$\mathbf{j} \times \mathbf{k} = \begin{vmatrix} \mathbf{i} & \mathbf{j} & \mathbf{k} \\ 0 & 1 & 0 \\ 0 & 0 & 1 \end{vmatrix} = (1 - 0)\mathbf{i} + (0 - 0)\mathbf{j} + (0 - 0)\mathbf{k} = \mathbf{i}, \quad \text{and}$$

$$\mathbf{k} \times \mathbf{i} = \begin{vmatrix} \mathbf{i} & \mathbf{j} & \mathbf{k} \\ 0 & 0 & 1 \\ 1 & 0 & 0 \end{vmatrix} = (0 - 0)\mathbf{i} + (1 - 0)\mathbf{j} + (0 - 0)\mathbf{k} = \mathbf{j}.$$

11.3.11: If $\mathbf{a} = \mathbf{i}$, $\mathbf{b} = \mathbf{i} + \mathbf{j}$, and $\mathbf{c} = \mathbf{i} + \mathbf{j} + \mathbf{k}$, then

$$\mathbf{a} \times (\mathbf{b} \times \mathbf{c}) = \mathbf{a} \times \begin{vmatrix} \mathbf{i} & \mathbf{j} & \mathbf{k} \\ 1 & 1 & 0 \\ 1 & 1 & 1 \end{vmatrix} = \mathbf{a} \times (\mathbf{i} - \mathbf{j}) = \begin{vmatrix} \mathbf{i} & \mathbf{j} & \mathbf{k} \\ 1 & 0 & 0 \\ 1 & -1 & 0 \end{vmatrix} = -\mathbf{k},$$

whereas

$$(\mathbf{a} \times \mathbf{b}) \times \mathbf{c} = \begin{vmatrix} \mathbf{i} & \mathbf{j} & \mathbf{k} \\ 1 & 0 & 0 \\ 1 & 1 & 0 \end{vmatrix} \times \mathbf{c} = \mathbf{k} \times \mathbf{c} = \begin{vmatrix} \mathbf{i} & \mathbf{j} & \mathbf{k} \\ 0 & 0 & 1 \\ 1 & 1 & 1 \end{vmatrix} = -\mathbf{i} + \mathbf{j}.$$

Thus the vector product is not associative.

11.3.13: If $\mathbf{a}$, $\mathbf{b}$, and $\mathbf{c}$ are mutually perpendicular, note that $\mathbf{b} \times \mathbf{c}$ is perpendicular to both $\mathbf{b}$ and $\mathbf{c}$, hence $\mathbf{b} \times \mathbf{c}$ is parallel to $\mathbf{a}$. Therefore $\mathbf{a} \times (\mathbf{b} \times \mathbf{c}) = \mathbf{0}$.

11.3.15: By Eq. (10) (and the discussion that precedes and follows it), the area of triangle PQR is $A = \frac{1}{2}|\overrightarrow{PQ} + \overrightarrow{PR}|$. But

$$\overrightarrow{PQ} \times \overrightarrow{PR} = \begin{vmatrix} \mathbf{i} & \mathbf{j} & \mathbf{k} \\ 1 & 1 & 7 \\ -4 & -5 & 4 \end{vmatrix} = 39\mathbf{i} - 32\mathbf{j} - \mathbf{k}.$$

The magnitude of $39\mathbf{i} - 32\mathbf{j} - \mathbf{k}$ is

$$|39\mathbf{i} - 32\mathbf{j} - \mathbf{k}| = \sqrt{1521 + 1024 + 1} = \sqrt{2546}.$$

Therefore the area of triangle PQR is $A = \frac{1}{2}\sqrt{2546} \approx 25.2289516231$.

11.3.17: Let $\mathbf{a} = \overrightarrow{OP} = \langle 1, 3, -2 \rangle$, $\mathbf{b} = \overrightarrow{OQ} = \langle 2, 4, 5 \rangle$, and $\mathbf{c} = \overrightarrow{OR} = \langle -3, -2, 2 \rangle$. Part (a): By Theorem 4 and part (4) of Theorem 3, the volume of the parallelepiped with adjacent edges these three vectors is given by $V = |(\mathbf{a} \times \mathbf{b}) \cdot \mathbf{c}|$. Here we have

$$\mathbf{a} \times \mathbf{b} = \begin{vmatrix} \mathbf{i} & \mathbf{j} & \mathbf{k} \\ 1 & 3 & -2 \\ 2 & 4 & 5 \end{vmatrix} = \langle 23, -9, -2 \rangle,$$

and therefore $V = |\langle 23, -9, -2 \rangle \cdot \langle -3, -2, 2 \rangle| = |-55| = 55$.

Part (b): By Example 7 of Section 11.3, the volume of the pyramid is $\dfrac{55}{6}$.

11.3.19: Here we have $\overrightarrow{AB} = \langle 2, 1, 3 \rangle$, $\overrightarrow{AC} = \langle 1, -3, 0 \rangle$, and $\overrightarrow{AD} = \langle 3, 5, 6 \rangle$. By Theorem 4 and Eq. (17), the volume of the parallelepiped having these vectors as adjacent sides is the absolute value of

$$\begin{vmatrix} 2 & 1 & 3 \\ 1 & -3 & 0 \\ 3 & 5 & 6 \end{vmatrix} = (-1) \cdot (6 - 15) - 3 \cdot (12 - 9) = 9 - 9 = 0$$

(we expanded the determinant using its second row; see the discussion following Example 1 of the text). By the reasoning given in Example 8, the four given points are coplanar.

11.3.21: Here we have $\overrightarrow{AB} = \langle 1, 2, 3 \rangle$, $\overrightarrow{AC} = \langle 2, 3, 4 \rangle$, and $\overrightarrow{AD} = \langle 9, 12, 21 \rangle$. By Theorem 4 and Eq. (17), the volume of the parallelepiped having these vectors as adjacent sides is the absolute value of

$$\begin{vmatrix} 1 & 2 & 3 \\ 2 & 3 & 4 \\ 9 & 12 & 21 \end{vmatrix} = 1 \cdot (63 - 48) - 2 \cdot (42 - 36) + 3 \cdot (24 - 27) = 15 - 12 - 9 = -6$$

(we expanded the determinant using its first row; see the discussion following Example 1 of the text). Therefore the four given points are not coplanar. The parallelepiped having the three vectors as adjacent sides has volume 6 and the tetrahedron (pyramid) with the four given points as its vertices has volume 1.

11.3.23: Name the vertices of the polygon counterclockwise, beginning at the origin: O, A, and B. Then

$$\mathbf{a} = \overrightarrow{OA} = \langle 176 \cos 15°, \, 176 \sin 15°, \, 0 \rangle \approx \langle 170.002945, \, 45.552152, \, 0 \rangle \quad \text{and}$$

$$\mathbf{b} = \overrightarrow{OB} = \langle 83 \cos 52°, \, 83 \sin 52°, \, 0 \rangle \approx \langle 51.099902, \, 65.404893, \, 0 \rangle.$$

Using the exact rather than the approximate values, we entered the *Mathematica* commands

```
c = Cross[a,b]    (* to compute the vector product of a and b *)
area = (1/2)*Sqrt[c.c]    (* to compute half the magnitude of c *)
N[area,20]    (* to print a 20-place approximation to the area *)
```

The response to the last command rounds to the answer, 4395.6569291026 (m^2).

11.3.25: Name the vertices of the polygon counterclockwise, beginning at the origin: O, A, B, and C. Then

$$\mathbf{a} = \overrightarrow{OA} = \langle 220 \cos 25°, \, 220 \sin 25°, \, 0 \rangle \approx \langle 227.206664, \, 115.767577, \, 0 \rangle,$$

$$\mathbf{b} = \overrightarrow{OB} = \mathbf{a} + \langle -210 \cos 40°, \, 210 \sin 40°, \, 0 \rangle \approx \langle 38.518380, \, 227.961416, \, 0 \rangle, \quad \text{and}$$

$$\mathbf{c} = \overrightarrow{OC} = \langle -150 \cos 63°, \, 150 \sin 63°, \, 0 \rangle \approx \langle -68.098575, \, 133.650979, \, 0 \rangle.$$

Using the exact rather than the approximate values, we entered the *Mathematica* commands

```
p = Cross[a,b]    (* to compute the vector product of a and b *)
area1 = (1/2)*Sqrt[p.p]    (* to compute half the magnitude of p *)
```

```
q = Cross[b,c]

area2 = (1/2)*Sqrt[q.q]

N[area1 + area2 ,20]    (* to print a 20-place approx.  to the total area *)
```

The response to the last command rounds to the answer, 31271.643253 (ft^2).

11.3.27: Let $\mathbf{a} = \langle a_1, a_2, a_3 \rangle$ and $\mathbf{b} = \langle b_1, b_2, b_3 \rangle$. Then

$$\mathbf{a} \times \mathbf{b} = \begin{vmatrix} \mathbf{i} & \mathbf{j} & \mathbf{k} \\ a_1 & a_2 & a_3 \\ b_1 & b_2 & b_3 \end{vmatrix} = \langle a_2 b_3 - a_3 b_2, \ a_3 b_1 - a_1 b_3 \ a_1 b_2 - a_2 b_1 \rangle \quad \text{and}$$

$$\mathbf{b} \times \mathbf{a} = \begin{vmatrix} \mathbf{i} & \mathbf{j} & \mathbf{k} \\ b_1 & b_2 & b_3 \\ a_1 & a_2 & a_3 \end{vmatrix} = \langle b_2 a_3 - b_3 a_2, \ b_3 a_1 - b_1 a_3 \ b_1 a_2 - b_2 a_1 \rangle$$

$$= -\langle a_2 b_3 - a_3 b_2, \ a_3 b_1 - a_1 b_3, \ a_1 b_2 - a_2 b_1 \rangle = -(\mathbf{a} \times \mathbf{b}).$$

11.3.29: Part (a): Please refer to Fig. 11.3.13 of the text. Because the segment of length d is perpendicular to the segment PQ, we know from elementary geometry that the area of the triangle is $\frac{1}{2} |\overrightarrow{PQ}| \cdot d$. Therefore

$$\frac{1}{2} |\overrightarrow{AP} \times \overrightarrow{AQ}| = \frac{1}{2} |\overrightarrow{PQ}| \cdot d, \quad \text{and thus} \quad d = \frac{|\overrightarrow{AP} \times \overrightarrow{AQ}|}{|\overrightarrow{PQ}|}. \tag{1}$$

Part (b): With $A = (1, 0, 1)$, $P = (2, 3, 1)$, and $Q = (-3, 1, 4)$, we have

$$\overrightarrow{AP} = \langle 1, 3, 0 \rangle, \quad \overrightarrow{AQ} = \langle -4, 1, 3 \rangle, \quad \text{and} \quad \overrightarrow{PQ} = \langle -5, -2, 3 \rangle.$$

To use the formula in (1), we first compute

$$\overrightarrow{AP} \times \overrightarrow{AQ} = \begin{vmatrix} \mathbf{i} & \mathbf{j} & \mathbf{k} \\ 1 & 3 & 0 \\ -4 & 1 & 3 \end{vmatrix} = \langle 9 - 0, \ 0 - 3, \ 1 - (-12) \rangle = \langle 9, -3, 13 \rangle.$$

Then the formula in (1) implies that the distance from the given point to the given line is

$$d = \frac{|\overrightarrow{AP} \times \overrightarrow{AQ}|}{|\overrightarrow{PQ}|} = \frac{|\langle 9, -3, 13 \rangle|}{\sqrt{25 + 4 + 9}} = \frac{\sqrt{259}}{\sqrt{38}} = \frac{\sqrt{9842}}{38} \approx 2.61070670.$$

11.3.31: Let $\mathbf{u} = \overrightarrow{P_1 Q_1}$, $\mathbf{v} = \overrightarrow{P_2 Q_2}$, and $\mathbf{w} = \overrightarrow{P_1 P_2}$. A vector $\mathbf{n}$ that is perpendicular to both lines is one perpendicular to both $\mathbf{u}$ and $\mathbf{v}$, and an obvious choice is $\mathbf{n} = \mathbf{u} \times \mathbf{v}$ ($\mathbf{n}$ will be nonzero because the two

677

lines are not parallel). The projection of $\mathbf{w}$ in the direction of $\mathbf{n}$ will be $\pm d$ where d is the (perpendicular) distance between the two lines. Thus

$$d = \frac{|\mathbf{w} \cdot \mathbf{n}|}{|\mathbf{n}|} = \frac{|\overrightarrow{P_1P_2} \cdot (\overrightarrow{P_1Q_1} \times \overrightarrow{P_2Q_2})|}{|\overrightarrow{P_1Q_1} \times \overrightarrow{P_2Q_2}|}.$$

Comment: Rather than memorizing formulas such as those in Problems 29, 30, and 31, one should learn the general technique for finding the distance between two objects (points, lines, or planes) in space. First find a vector $\mathbf{n}$ normal to both objects. Then construct another vector $\mathbf{c}$ "connecting" the two objects; that is, with initial point on one and terminal point on the other. Then the projection of $\mathbf{c}$ in the direction of $\mathbf{n}$ is the (perpendicular) distance between the two objects (or its negative); that is, the distance d is given by the simple formula

$$d = \frac{|\mathbf{c} \cdot \mathbf{n}|}{|\mathbf{n}|}.$$

11.3.33: Given the vectors $\mathbf{a}$, $\mathbf{b}$, and $\mathbf{c}$ in space, we have

$$\mathbf{a} \times (\mathbf{b} \times \mathbf{c}) = -(\mathbf{b} \times \mathbf{c}) \times \mathbf{a} \quad \text{(by Eq. (12))}$$

$$= -[(\mathbf{b} \cdot \mathbf{a})\mathbf{c} - (\mathbf{c} \cdot \mathbf{a})\mathbf{b}] \quad \text{(by the result in Problem 32)}$$

$$= (\mathbf{a} \cdot \mathbf{c})\mathbf{b} - (\mathbf{a} \cdot \mathbf{b})\mathbf{c} \quad \text{(this establishes Eq. (16)).}$$

11.3.35: Given: The triangle in the plane with vertices at $P(x_1, y_1, 0)$, $Q(x_2, y_2, 0)$, and $R(x_3, y_3, 0)$. Let

$$\mathbf{u} = \overrightarrow{PQ} = \langle x_2 - x_1, \, y_2 - y_1, \, 0 \rangle \quad \text{and} \quad \mathbf{v} = \overrightarrow{PR} = \langle x_3 - x_1, \, y_3 - y_1, \, 0 \rangle.$$

The area of the triangle is $A = \frac{1}{2}|\mathbf{u} \times \mathbf{v}|$. Now

$$\mathbf{u} \times \mathbf{v} = \begin{vmatrix} \mathbf{i} & \mathbf{j} & \mathbf{k} \\ x_2 - x_1 & y_2 - y_1 & 0 \\ x_3 - x_1 & y_3 - y_1 & 0 \end{vmatrix} = \langle 0, \, 0, \, (x_2 - x_1)(y_3 - y_1) - (y_2 - y_1)(x_3 - x_1) \rangle,$$

so that

$$|\mathbf{u} \times \mathbf{v}| = |x_2y_3 - x_2y_1 - x_1y_3 + x_1y_1 - x_3y_2 + x_3y_1 + x_1y_2 - x_1y_1|$$

$$= |x_2y_3 - x_2y_1 - x_1y_3 - x_3y_2 + x_3y_1 + x_1y_2|$$

$$= |(x_2y_3 - x_3y_2) - (x_1y_3 - x_3y_1) + (x_1y_2 - x_2y_1)|,$$

678

which is the *absolute value* of the determinant

$$\begin{vmatrix} 1 & x_1 & y_1 \\ 1 & x_2 & y_2 \\ 1 & x_3 & y_3 \end{vmatrix} \tag{1}$$

(expanded using its first column—see the discussion following Eq. (3) in the text). Thus the area of triangle PQR is half the absolute value of the determinant in (1).

Section 11.4

11.4.1: Solving the vector equation $\langle x, y, z \rangle = t\mathbf{v} + \overrightarrow{OP}$ yields

$$\langle x, y, z \rangle = t\langle 1, 2, 3 \rangle + \langle 0, 0, 0 \rangle = \langle t, 2t, 3t \rangle,$$

so that the parametric equations of the line are $x = t$, $y = 2t$, $z = 3t$.

11.4.3: Solving the vector equation $\langle x, y, z \rangle = t\mathbf{v} + \overrightarrow{OP}$ yields

$$\langle x, y, z \rangle = t\langle 2, 0, -3 \rangle + \langle 4, 13, -3 \rangle = \langle 2t + 4, 13, -3t - 3 \rangle,$$

and thus the parametric equations of the line are $x = 2t + 4$, $y = 13$, $z = -3t - 3$.

11.4.5: Solving the vector equation $\langle x, y, z \rangle = \overrightarrow{OP_1} + t\overrightarrow{P_1 P_2}$ yields

$$\langle x, y, z \rangle = \langle 0, 0, 0 \rangle + t\langle -6, 3, 5 \rangle,$$

so the line has parametric equations $x = -6t$, $y = 3t$, $z = 5t$.

11.4.7: Solving the vector equation $\langle x, y, z \rangle = \overrightarrow{OP_1} + t\overrightarrow{P_1 P_2}$ yields

$$\langle x, y, z \rangle = \langle 3, 5, 7 \rangle + t\langle 3, 0, -3 \rangle = \langle 3 + 3t, 5, 7 - 3t \rangle$$

so the line has parametric equations $x = 3t + 3$, $y = 5$, $z = -3t + 7$.

11.4.9: Solving the vector equation $\langle x, y, z \rangle = \overrightarrow{OP} + t\mathbf{v}$ yields

$$\langle x, y, z \rangle = \langle 2, 3, -4 \rangle + t\langle 1, -1, -2 \rangle = \langle 2 + t, 3 - t, -4 - 2t \rangle$$

so the line has parametric equations $x = t + 2$, $y = -t + 3$, $z = -2t - 4$ and symmetric equations

$$x - 2 = -y + 3 = -\frac{z + 4}{2}.$$

11.4.11: Solving the vector equation $\langle x, y, z \rangle = \overrightarrow{OP} + t\mathbf{k}$ yields

$$\langle x, y, z \rangle = \langle 1, 1, 1 \rangle + t\langle 0, 0, 1 \rangle = \langle 1, 1, 1+t \rangle$$

so the line has parametric equations $x = 1$, $y = 1$, $z = t + 1$. In a strict sense, the line doesn't have symmetric equations, but its Cartesian equations are

$$x = 1, \quad y = 1.$$

The fact that z is not mentioned means that z is *arbitrary*.

11.4.13: A line normal to the plane with Cartesian equation $2x - y + 3z = 4$ is parallel to its normal vector $\mathbf{n} = \langle 2, -1, 3 \rangle$. The line of this problem also passes through the point $P(2, -3, 4)$ and thus has vector equation $\langle x, y, z \rangle = t\langle 2, -1, 3 \rangle + \langle 2, -3, 4 \rangle$, parametric equations $x = 2t + 2$, $y = -t - 3$, $z = 3t + 4$, and symmetric equations

$$\frac{x-2}{2} = -(y+3) = \frac{z-4}{3}.$$

11.4.15: Given: The lines L_1 and L_2 with symmetric equations

$$x - 2 = \frac{y+1}{2} = \frac{z-3}{3} \quad \text{and} \quad \frac{x-5}{3} = \frac{y-1}{2} = z - 4, \tag{1}$$

respectively. Points on L_1 include $P_1(2, -1, 3)$ and $Q_1(3, 1, 6)$, and thus L_1 is parallel to the vector $\overrightarrow{P_1Q_1} = \langle 1, 2, 3 \rangle$. Points on L_2 include $P_2(5, 1, 4)$ and $Q_2(8, 3, 5)$, and thus L_2 is parallel to the vector $\overrightarrow{P_2Q_2} = \langle 3, 2, 1 \rangle$. Clearly L_1 and L_2 are not parallel. To determine whether they intersect, the *Mathematica* command for solving the equations in (1) simultaneously is

```
Solve[ {x - 2 == (y + 1)/2, x - 2 == (z - 3)/3,
        (x - 5)/3 == (y - 1)/2, (x - 5)/3 == z - 4 }, { x, y, z } ]
```

and this command yields the response $x = 2$, $y = -1$, $z = 3$. Hence the two lines meet at the single point $(2, -1, 3)$.

11.4.17: When we try to solve the equations of L_1 and L_2 simultaneously, we find that there is no solution. Thus the lines do not intersect. Two points on L_1 are $P_1(6, 5, 7)$ and $Q_1(8, 7, 10)$, so L_1 is parallel to the vector $\mathbf{v}_1 = \overrightarrow{P_1Q_1} = \langle 2, 2, 3 \rangle$. Two points on L_2 are $P_2(7, 5, 10)$ and $Q_2(10, 8, 15)$, so L_2 is parallel to the vector $\mathbf{v}_2 = \overrightarrow{P_2Q_2} = \langle 3, 3, 5 \rangle$. The equation $\lambda \mathbf{v}_1 = \mathbf{v}_2$ has no solution, so $\mathbf{v}_1$ and $\mathbf{v}_2$ are not parallel. Therefore L_1 and L_2 are skew lines.

11.4.19: We solve the equations

$$\frac{x-7}{6} = \frac{y+5}{4}, \quad \frac{x-7}{6} = \frac{9-z}{8}, \quad \frac{11-x}{9} = \frac{7-y}{6}, \quad \text{and} \quad \frac{7-y}{6} = \frac{z-13}{12}$$

simultaneously and find that there is no solution. So L_1 and L_2 do not intersect. Two points on L_1 are $P_1(7, -5, 9)$ and $Q_1(13, -1, 1)$, so L_1 is parallel to the vector $\mathbf{v}_1 = \overrightarrow{P_1Q_1} = 6\mathbf{i} + 4\mathbf{j} - 8\mathbf{k}$. Two points of L_2 and $P_2(11, 7, 13)$ and $Q_2(2, 1, 25)$, so L_2 is parallel to the vector $\mathbf{v}_2 = \overrightarrow{P_2Q_2} = -9\mathbf{i} - 6\mathbf{j} + 12\mathbf{k}$. Because $-\frac{3}{2}\mathbf{v}_1 = \mathbf{v}_2$, the vectors $\mathbf{v}_1$ and $\mathbf{v}_2$ are parallel, and therefore L_1 and L_2 are parallel as well.

11.4.21: The vector equation of the plane is $\mathbf{n} \cdot \langle x, y, z \rangle = \mathbf{n} \cdot \overrightarrow{OP}$, and because $\overrightarrow{OP} = \mathbf{0}$ it follows that a Cartesian equation of the plane is $x + 2y + 3z = 0$.

11.4.23: The vector equation of the plane is $\mathbf{n} \cdot \langle x, y, z \rangle = \mathbf{n} \cdot \overrightarrow{OP}$; that is,

$$\langle 1, 0, -1 \rangle \cdot \langle x, y, z \rangle = \langle 1, 0, -1 \rangle \cdot \langle 5, 12, 13 \rangle,$$

and thus a Cartesian equation of the plane is $x - z + 8 = 0$.

11.4.25: A plane parallel to the xz-plane has an equation of the form $y = c$ where c is a constant. The plane of Problem 25 also passes through the point $(5, 7, -6)$, so $c = 7$. Thus a Cartesian equation of this plane is $y = 7$.

11.4.27: The vector equation of the plane is $\mathbf{n} \cdot \langle x, y, z \rangle = \mathbf{n} \cdot \overrightarrow{OP}$; that is,

$$\langle 7, 11, 0 \rangle \cdot \langle x, y, z \rangle = \langle 7, 11, 0 \rangle \cdot \langle 10, 4, -3 \rangle.$$

Thus a Cartesian equation of this plane is $7x + 11y = 114$.

11.4.29: A plane parallel to the plane with Cartesian equation $3x + 4y - z = 10$ has normal vector $\mathbf{n} = 3\mathbf{i} + 4\mathbf{j} - \mathbf{k}$, thus a Cartesian equation of the form $3x + 4y - z = c$ for some constant c. The plane of Problem 29 also passes through the origin, so that $c = 0$. Hence it has Cartesian equation $3x + 4y - z = 0$.

11.4.31: The plane through the origin $O(0, 0, 0)$ and the two points $P(1, 1, 1)$ and $Q(1, -1, 3)$ has normal vector

$$\mathbf{n} = \overrightarrow{OP} \times \overrightarrow{OQ} = \begin{vmatrix} \mathbf{i} & \mathbf{j} & \mathbf{k} \\ 1 & 1 & 1 \\ 1 & -1 & 3 \end{vmatrix} = \langle 4, -2, -2 \rangle.$$

Therefore this plane has a Cartesian equation of the form $4x - 2y - 2z = c$ where c is a constant. Because the plane of Problem 31 passes through the origin, $c = 0$. Therefore a Cartesian equation of this plane is $2x - y - z = 0$.

11.4.33: The plane $\mathcal{P}$ that contains $P(2, 4, 6)$ and the line L with parametric equations $x = 7 - 3t$, $y = 3 + 4t$, $z = 5 + 2t$ also contains the two points $Q(7, 3, 5)$ (set $t = 0$) and $R(4, 7, 4)$ (set $t = 1$) of L. So the vectors $\overrightarrow{PQ} = 5\mathbf{i} - \mathbf{j} - \mathbf{k}$ amd $\overrightarrow{PR} = 2\mathbf{i} + 3\mathbf{j} + \mathbf{k}$ are parallel to $\mathcal{P}$, and thus $\mathcal{P}$ has normal vector

$$\mathbf{n} = \vec{PQ} \times \vec{PR} = \begin{vmatrix} \mathbf{i} & \mathbf{j} & \mathbf{k} \\ 5 & -1 & -1 \\ 2 & 3 & 1 \end{vmatrix} = 2\mathbf{i} - 7\mathbf{j} + 17\mathbf{k}.$$

So $\mathcal{P}$ has vector equation $\mathbf{n} \cdot (x\mathbf{i} + y\mathbf{j} + z\mathbf{k}) = \mathbf{n} \cdot \vec{OP} = 78$ and thus Cartesian equation $2x - 7y + 17z = 78$.

11.4.35: There are several ways to solve this problem.

First solution: Two points of L are $P(7, 3, 9)$ and $Q(3, 9, 14)$, so the vector $\vec{PQ} = \langle -4, 6, 5 \rangle$ is parallel to L. The vector $\mathbf{n} = \langle 4, 1, 2 \rangle$ is normal to the plane $\mathcal{P}$. But

$$\mathbf{n} \cdot \vec{PQ} = \langle 4, 1, 2 \rangle \cdot \langle -4, 6, 5 \rangle = -16 + 6 + 10 = 0,$$

so that $\mathbf{n}$ and $\vec{PQ}$ are perpendicular. Therefore L is parallel to $\mathcal{P}$. Moreover, P does not satisfy the equation of $\mathcal{P}$ because

$$4 \cdot 7 + 1 \cdot 3 + 2 \cdot 9 = 28 + 3 + 18 = 49 \neq 17.$$

Because L is parallel to $\mathcal{P}$ and contains a point not on $\mathcal{P}$, the line and the plane cannot coincide.

Second solution: If the line and the plane both contain the point (x, y, z), then there exists a scalar t such that all four equations in Problem 35 are simultaneously true. The *Mathematica* command

```
Solve[ { x == 7 - 4*t, y == 3 + 6*t, z == 9 + 5*t,
       4*x + y + 2*z == 17 }, { x, y, z, t } ]
```

returns { }, the "empty set," telling us that these four equations have no simultaneous solution. So no point of L lies in the plane $\mathcal{P}$; the line and the plane are parallel and do not coincide.

Third solution: If $x = 7 - 4t$, $y = 3 + 6t$, and $z = 9 + 5t$ are substituted in the equation $4x + y + 2z = 17$ of the plane, the result is $49 = 17$. This is impossible, so no point of L lies in the plane $\mathcal{P}$. We reach the same conclusion as in the previous two solutions.

11.4.37: Simultaneous solution of the four equations given in the statement of Problem 37 yields the unique solution $x = \frac{9}{2}$, $y = \frac{9}{4}$, $z = \frac{17}{4}$, and $t = \frac{3}{4}$. So the line and the plane are not parallel and meet at the single point $\left(\frac{9}{2}, \frac{9}{4}, \frac{17}{4} \right)$. The easiest way to solve these equations by hand is to substitute the three parametric equations into the equation of the plane and solve for t:

$$3(3 + 2t) + 2(6 - 5t) - 4(2 + 3t) = 1; \qquad 9 + 6t + 12 - 10t - 8 - 12t = 1;$$

$$-16t = -12; \qquad t = \frac{3}{4}.$$

Then substitution in the parametric equations yields the values of x, y, and z given here.

11.4.39: The vector $\mathbf{n_1} = \langle 1, 0, 0 \rangle$ is normal to the first plane; the vector $\mathbf{n_2} = \langle 1, 1, 1 \rangle$ is normal to the second. If θ is the angle between the normals (this is, by definition, the angle between the planes), then

$$\cos\theta = \frac{\mathbf{n_1} \cdot \mathbf{n_2}}{|\mathbf{n_1}| \cdot |\mathbf{n_2}|} = \frac{1}{\sqrt{3}} = \frac{\sqrt{3}}{3},$$

so $\theta \approx 54.735610317245°$.

11.4.41: The vector $\mathbf{n_1} = \langle 1, -1, -2 \rangle$ is normal to the first plane; the vector $\mathbf{n_2} = \langle 1, -1, -2 \rangle$ is normal to the second. If θ is the angle between the normals (this is, by definition, the angle between the planes), then $\theta = 0$ because $\mathbf{n_1}$ and $\mathbf{n_2}$ are parallel.

11.4.43: By inspection, two points that lie on both planes are $P(10, 0, -10)$ and $Q(10, 1, -11)$. Hence a vector parallel to their line of intersection L is $\mathbf{v} = \overrightarrow{PQ} = \langle 0, 1, -1 \rangle$. So a vector equation of L is

$$\langle x, y, z \rangle = \overrightarrow{OP} + t\mathbf{v} = \langle 10, 0, -10 \rangle + t\langle 0, 1, -1 \rangle = \langle 10, t, -10-t \rangle,$$

its parametric equations are $x = 10$, $y = t$, $z = -10-t$, and its Cartesian equations are $x = 10$, $y = -10-z$.

11.4.45: The planes of Problem 41 are parallel, so there is no line of intersection.

11.4.47: Substitute $z = 0$, then solve the equations of the planes simultaneously to find that one point on their line of intersection L is $Q\left(\frac{7}{5}, \frac{6}{5}, 0\right)$. Repeat with $z = 1$ to find that another such point is $R\left(\frac{7}{5}, \frac{1}{5}, 1\right)$. Hence the vector $\mathbf{v} = \overrightarrow{QR} = \langle 0, -1, 1 \rangle$ is parallel to L. Therefore the parallel line through the point $P(3, 3, 1)$ has vector equation

$$\langle x, y, z \rangle = \overrightarrow{OP} + t\mathbf{v} = \langle 3, 3, 1 \rangle + t\langle 0, -1, 1 \rangle.$$

Thus L has parametric equations $x = 3$, $y = 3-t$, $z = 1+t$ and Cartesian equations $x = 3$, $-y+3 = z-1$.

11.4.49: Because the xy-plane has equation $z = 0$, the plane with equation $3x + 2y - z = 6$ intersects the xy-plane in the line with equation $3x + 2y = 6$. So three points on the plane $\mathcal{P}$ we seek are $P(1, 1, 1)$, $Q(2, 0, 0)$, and $R(0, 3, 0)$. Thus a normal to $\mathcal{P}$ is

$$\mathbf{n} = \overrightarrow{PQ} \times \overrightarrow{PR} = \begin{vmatrix} \mathbf{i} & \mathbf{j} & \mathbf{k} \\ 1 & -1 & -1 \\ -1 & 2 & -1 \end{vmatrix} = \langle 3, 2, 1 \rangle.$$

Therefore $\mathcal{P}$ has vector equation $\mathbf{n} \cdot \langle x, y, z \rangle = \mathbf{n} \cdot \overrightarrow{OP}$; that is,

$$\langle 3, 2, 1 \rangle \cdot \langle x, y, z \rangle = \langle 3, 2, 1 \rangle \cdot \langle 1, 1, 1 \rangle;$$

$$3x + 2y + z = 6.$$

11.4.51: The plane $\mathcal{P}$ whose equation we seek passes through $P(1, 0, -1)$ and $Q(2, 1, 0)$, and is thus parallel to $\overrightarrow{PQ} = \langle 1, 1, 1 \rangle$. To find two points in the line of intersection of the other two planes, set $z = 1$ and solve their equations simultaneously to find that one such point is $S(2, 2, 1)$. Repeat with $z = 5$ to find that another such point is $R(1, -1, 5)$. Hence another vector parallel to $\mathcal{P}$ is $\overrightarrow{RS} = \langle 1, 3, -4 \rangle$. If $\overrightarrow{RS}$ had turned out to be parallel to $\overrightarrow{PQ}$, then there would have been insufficient information to solve the problem, but they are not parallel. Hence a normal to $\mathcal{P}$ is

$$\mathbf{n} = \overrightarrow{PQ} \times \overrightarrow{RS} = \begin{vmatrix} \mathbf{i} & \mathbf{j} & \mathbf{k} \\ 1 & 1 & 1 \\ 1 & 3 & -4 \end{vmatrix} = \langle -7, 5, 2 \rangle.$$

Hence $\mathcal{P}$ has vector equation $\mathbf{n} \cdot \langle x, y, z \rangle = \mathbf{n} \cdot \overrightarrow{OP}$; that is,

$$\langle -7, 5, 2 \rangle \cdot \langle x, y, z \rangle = \langle -7, 5, 2 \rangle \cdot \langle 1, 0, -1 \rangle; \qquad -7x + 5y + 2z = -9; \qquad 7x - 5y - 2z = 9.$$

11.4.53: Set $z = 1$ and solve the equations of the two planes simultaneously to find that one point on their line of intersection is $P(1, 1, 1)$. Repeat with $z = 5$ to find that another such point is $Q(-5, 6, 5)$. Substitute $t = 0$ to find that a point on the given line is $R(1, 3, 2)$; substitute $t = 1$ to find that another such point is $S(7, -2, -2)$. Then $\overrightarrow{PQ} = \langle -6, 5, 4 \rangle$ and $\overrightarrow{RS} = \langle 6, -5, -4 \rangle$, so it's clear that the two lines are parallel. Note that R does not lie on the first given plane, so the given line and the line of intersection of the two given planes do not coincide. To obtain two nonparallel vectors in the plane $\mathcal{P}$ that contains both lines, use $\overrightarrow{PQ}$ and $\overrightarrow{PR} = \langle 0, 2, 1 \rangle$. Then a normal to $\mathcal{P}$ will be

$$\mathbf{n} = \overrightarrow{PQ} \times \overrightarrow{PR} = \begin{vmatrix} \mathbf{i} & \mathbf{j} & \mathbf{k} \\ -6 & 5 & 4 \\ 0 & 2 & 1 \end{vmatrix} = \langle -3, 6, -12 \rangle.$$

Replace $\mathbf{n}$ with the simpler $\langle 1, -2, 4 \rangle$ and write a vector equation of $\mathcal{P}$ in the form

$$\langle 1, -2, 4 \rangle \cdot \langle x, y, z \rangle = \langle 1, -2, 4 \rangle \cdot \langle 1, 1, 1 \rangle,$$

from which we read the Cartesian equation $x - 2y + 4z = 3$.

11.4.55: The formula in Problem 54 with $x_0 = y_0 = z_0 = 0$, $a = b = c = 1$, and $d = 10$ yields distance

$$D = \frac{10}{\sqrt{3}} = \frac{10\sqrt{3}}{3} \approx 5.773502691896.$$

11.4.57: If L_1 and L_2 are skew lines, choose two points P_1 and Q_1 in L_1 and two points P_2 and Q_2 in L_2. Let $\mathbf{n} = \overrightarrow{P_1Q_1} \times \overrightarrow{P_2Q_2}$. Let $\mathcal{P}_1$ be the plane through P_1 with normal vector $\mathbf{n}$ and let $\mathcal{P}_2$ be the plane through P_2 with normal vector $\mathbf{n}$. Clearly $\mathcal{P}_1$ contains L_1, $\mathcal{P}_2$ contains L_2, and $\mathcal{P}_1$ and $\mathcal{P}_2$ are parallel.

11.4.59: See the *Comment* in the solution of Problem 31 of Section 11.3 for our strategy in the following solution. By inspection, $P_1(1, -1, 4)$ and $Q_1(3, 0, 4)$ lie in L_1 and we are given the two points $P_2(2, 1, -3)$ and $Q_2(0, 8, 4)$ in L_2. Hence $\mathbf{v}_1 = \overrightarrow{P_1Q_1} = \langle 2, 1, 0 \rangle$ is parallel to L_1 and $\mathbf{v}_2 = \overrightarrow{P_2Q_2} = \langle -2, 7, 7 \rangle$ is parallel to L_2.

Part (a): The line L_2 has vector equation

$$\langle x, y, z \rangle = t\mathbf{v}_2 + \langle 2, 1, -3 \rangle = \langle 2 - 2t, 1 + 7t, -3 + 7t \rangle,$$

and thus symmetric equations

$$\frac{-x+2}{2} = \frac{y-1}{7} = \frac{z+3}{7}.$$

It is clear that the two lines are not parallel because $\mathbf{v}_2$ is not a scalar multiple of $\mathbf{v}_1$. The *Mathematica* command

```
Solve[ { (2 - x)/2 == (y - 1)/7, (2 - x)/2 == (z + 3)/7,
         x - 1 == 2*y + 2, z == 4 }, { x, y, z } ]
```

for solving the equations of the two lines simultaneously returns the information that there is no solution. Therefore the two lines are skew lines.

Part (b): Let $\mathbf{c} = \overrightarrow{P_1P_2}$ be a "connector" from L_1 to L_2 and note that $\mathbf{n} = \mathbf{v}_1 \times \mathbf{v}_2 = \langle 7, -14, 16 \rangle$ is normal to both lines. Therefore the distance between the lines is

$$D = \frac{|\mathbf{n} \cdot \mathbf{c}|}{|\mathbf{n}|} = \frac{133}{\sqrt{501}} = \frac{133\sqrt{501}}{501} \approx 5.942001786397.$$

Alternative Part (b): If we follow the method required in the statement of Problem 59, we find that the first line lies in the plane with equation $7x - 14y + 16z = 85$ and the second line lies in the plane with equation $7x - 14y + 16z = -48$. Therefore, by the formula in Problem 58, the distance between them is

$$D = \frac{|85 - (-48)|}{\sqrt{49 + 196 + 256}} = \frac{133}{\sqrt{501}} = \frac{133\sqrt{501}}{501}.$$

Section 11.5

11.5.1: Because $y^2 + z^2 = 1$ while x is arbitrary, the graph lies on the cylinder of radius 1 with axis the x-axis. A small part of the graph is shown in Fig. 11.5.17.

11.5.3: Because $x^2 + y^2 = t^2 = z^2$, the graph lies on the cone with axis the z-axis and equation $z^2 = x^2 + y^2$. A small part of the graph is shown in Fig. 11.5.16.

11.5.5: If $\mathbf{r}(t) = 3\mathbf{i} - 2\mathbf{j}$, then $\mathbf{r}'(t) = \mathbf{0} = \mathbf{r}''(t)$, and hence $\mathbf{r}'(1) = \mathbf{0} = \mathbf{r}''(1)$.

11.5.7: If $\mathbf{r}(t) = e^{2t}\mathbf{i} + e^{-t}\mathbf{j}$, then $\mathbf{r}'(t) = 2e^{2t}\mathbf{i} - e^{-t}\mathbf{j}$ and $\mathbf{r}''(t) = 4e^{2t}\mathbf{i} + e^{-t}\mathbf{j}$. Therefore $\mathbf{r}'(0) = 2\mathbf{i} - \mathbf{j}$ and $\mathbf{r}''(0) = 4\mathbf{i} + \mathbf{j}$.

11.5.9: If $\mathbf{r}(t) = 3\mathbf{i}\cos 2\pi t + 3\mathbf{j}\sin 2\pi t$, then

$$\mathbf{r}'(t) = -6\pi\mathbf{i}\sin 2\pi t + 6\pi\mathbf{j}\cos 2\pi t \quad \text{and} \quad \mathbf{r}''(t) = -12\pi^2\mathbf{i}\cos 2\pi t - 12\pi^2\mathbf{j}\sin 2\pi t.$$

Therefore

$$\mathbf{r}'\left(\frac{3}{4}\right) = 6\pi\mathbf{i} \quad \text{and} \quad \mathbf{r}''\left(\frac{3}{4}\right) = 12\pi^2\mathbf{j}.$$

11.5.11: If $\mathbf{r}(t) = t\mathbf{i} + t^2\mathbf{j} + t^3\mathbf{k}$, then

$$\mathbf{v}(t) = \langle 1, \, 2t, \, 3t^2 \rangle,$$

$$v(t) = \sqrt{1 + 4t^2 + 9t^4}, \quad \text{and}$$

$$\mathbf{a}(t) = \langle 0, \, 2, \, 6t \rangle.$$

11.5.13: If $\mathbf{r}(t) = \langle t, \, 3e^t, \, 4e^t \rangle$, then

$$\mathbf{v}(t) = \langle 1, \, 3e^t, \, 4e^t \rangle, \quad v(t) = \sqrt{1 + 25e^{2t}}, \quad \text{and} \quad \mathbf{a}(t) = \langle 0, \, 3e^t, \, 4e^t \rangle.$$

11.5.15: If $\mathbf{r}(t) = \langle 3\cos t, \, 3\sin t, \, -4t \rangle$, then

$$\mathbf{v}(t) = \langle -3\sin t, \, 3\cos t, \, -4 \rangle, \quad v(t) = \sqrt{9\sin^2 t + 9\cos^2 t + 16} = 5, \quad \text{and} \quad \mathbf{a}(t) = \langle -3\cos t, \, -3\sin t, \, 0 \rangle.$$

11.5.17: By Eq. (16), we have

$$\int_0^{\pi/4} \langle \sin t, \, 2\cos t \rangle \, dt = \Big[\langle -\cos t, \, 2\sin t \rangle\Big]_0^{\pi/4} = \left\langle -\frac{\sqrt{2}}{2}, \, \sqrt{2} \right\rangle - \langle -1, 0 \rangle = \left\langle \frac{2 - \sqrt{2}}{2}, \, \sqrt{2} \right\rangle.$$

686

11.5.19: By Eq. (16), we have

$$\int_0^2 \langle t^2(1+t^3)^{3/2},\, 0 \rangle \, dt = \left[\left\langle \frac{2}{15}(1+t^3)^{5/2},\, 0 \right\rangle \right]_0^2 = \left(\frac{162}{5} - \frac{2}{15} \right)\mathbf{i} = \frac{484}{15}\,\mathbf{i}.$$

11.5.21: Given $\mathbf{u}(t) = \langle 3t,\, -1 \rangle$ and $\mathbf{v}(t) = \langle 2,\, -5t \rangle$, Theorem 2 yields

$$D_t\left[\mathbf{u}(t) \cdot \mathbf{v}(t)\right] = \mathbf{u}(t) \cdot \mathbf{v}'(t) + \mathbf{u}'(t) \cdot \mathbf{v}(t) = \langle 3t,\, -1 \rangle \cdot \langle 0,\, -5 \rangle + \langle 3,\, 0 \rangle \cdot \langle 2,\, -5t \rangle = 5 + 6 = 11.$$

11.5.23: Given $\mathbf{u}(t) = \langle \cos t,\, \sin t \rangle$ and $\mathbf{v}(t) = \langle \sin t,\, -\cos t \rangle$, Theorem 2 yields

$$D_t\left[\mathbf{u}(t) \cdot \mathbf{v}(t)\right] = \mathbf{u}(t) \cdot \mathbf{v}'(t) + \mathbf{u}'(t) \cdot \mathbf{v}(t) = \langle \cos t,\, \sin t \rangle \cdot \langle \cos t,\, \sin t \rangle + \langle -\sin t,\, \cos t \rangle \cdot \langle \sin t,\, -\cos t \rangle$$

$$= \cos^2 t + \sin^2 t - \sin^2 t - \cos^2 t = 0.$$

11.5.25: Given $\mathbf{a} = \mathbf{0} = \langle 0,\, 0,\, 0 \rangle$, it follows that $\mathbf{v}(t) = \langle c_1,\, c_2,\, c_3 \rangle$ where c_1, c_2, and c_3 are constants. Then

$$\mathbf{k} = \langle 0,\, 0,\, 1 \rangle = \mathbf{v}(0) = \mathbf{v}_0 = \langle c_1,\, c_2,\, c_3 \rangle$$

implies that $c_1 = c_2 = 0$ and $c_3 = 1$. Hence $\mathbf{v}(t) = \langle 0,\, 0,\, 1 \rangle$, and therefore

$$\mathbf{r}(t) = \langle k_1,\, k_2,\, t + k_3 \rangle$$

where k_1, k_2, and k_3 are constants. Then

$$\mathbf{i} = \langle 1,\, 0,\, 0 \rangle = \mathbf{r}(0) = \mathbf{r}_0 = \langle k_1,\, k_2,\, k_3 \rangle$$

leads to $k_1 = 1$ and $k_2 = k_3 = 0$. Therefore $\mathbf{r}(t) = \langle 1,\, 0,\, t \rangle$.

11.5.27: Given $\mathbf{a} = \langle 2,\, 0,\, -4 \rangle$, it follows that $\mathbf{v}(t) = \langle 2t + c_1,\, c_2,\, -4t + c_3 \rangle$ where c_1, c_2, and c_3 are constants. Then

$$10\mathbf{j} = \langle 0,\, 10,\, 0 \rangle = \mathbf{v}(0) = \mathbf{v}_0 = \langle c_1,\, c_2,\, c_3 \rangle$$

implies that $c_1 = c_3 = 0$ and $c_2 = 10$. Hence $\mathbf{v}(t) = \langle 2t,\, 10,\, -4t \rangle$, and therefore

$$\mathbf{r}(t) = \langle t^2 + k_1,\, 10t + k_2,\, -2t^2 + k_3 \rangle$$

where k_1, k_2, and k_3 are constants. Then

$$\mathbf{0} = \langle 0,\, 0,\, 0 \rangle = \mathbf{r}(0) = \mathbf{r}_0 = \langle k_1,\, k_2,\, k_3 \rangle$$

leads to $k_1 = k_2 = k_3 = 0$. Therefore $\mathbf{r}(t) = \langle t^2,\, 10t,\, -2t^2 \rangle$.

11.5.29: Given $\mathbf{a} = \langle 0, 2, -6t \rangle$, it follows that $\mathbf{v}(t) = \langle c_1, 2t + c_2, -3t^2 + c_3 \rangle$ where c_1, c_2, and c_3 are constants. Then

$$5\mathbf{k} = \langle 0, 0, 5 \rangle = \mathbf{v}(0) = \mathbf{v}_0 = \langle c_1, c_2, c_3 \rangle$$

implies that $c_1 = c_2 = 0$ and $c_3 = 5$. Hence $\mathbf{v}(t) = \langle 0, 2t, -3t^2 + 5 \rangle$, and therefore

$$\mathbf{r}(t) = \langle k_1, t^2 + k_2, -t^3 + 5t + k_3 \rangle$$

where k_1, k_2, and k_3 are constants. Then

$$2\mathbf{i} = \langle 2, 0, 0 \rangle = \mathbf{r}(0) = \mathbf{r}_0 = \langle k_1, k_2, k_3 \rangle$$

leads to $k_1 = 2$ and $k_2 = k_3 = 0$. Therefore

$$\mathbf{r}(t) = \langle 2, t^2, -t^3 + 5t \rangle.$$

11.5.31: Given $\mathbf{a} = \langle t, t^2, t^3 \rangle$, it follows that

$$\mathbf{v}(t) = \left\langle \frac{1}{2}t^2 + c_1, \frac{1}{3}t^3 + c_2, \frac{1}{4}t^4 + c_3 \right\rangle$$

where c_1, c_2, and c_3 are constants. Then

$$10\mathbf{j} = \langle 0, 10, 0 \rangle = \mathbf{v}(0) = \mathbf{v}_0 = \langle c_1, c_2, c_3 \rangle$$

implies that $c_1 = c_3 = 0$, and $c_2 = 10$. Hence

$$\mathbf{v}(t) = \left\langle \frac{1}{2}t^2, \frac{1}{3}t^3 + 10, \frac{1}{4}t^4 \right\rangle,$$

and therefore

$$\mathbf{r}(t) = \left\langle \frac{1}{6}t^3 + k_1, \frac{1}{12}t^4 + 10t + k_2, \frac{1}{20}t^5 + k_3 \right\rangle$$

where k_1, k_2, and k_3 are constants. Then

$$10\mathbf{i} = \langle 10, 0, 0 \rangle = \mathbf{r}(0) = \mathbf{r}_0 = \langle k_1, k_2, k_3 \rangle$$

leads to $k_1 = 10$ and $k_2 = k_3 = 0$. Therefore

$$\mathbf{r}(t) = \left\langle \frac{1}{6}t^3 + 10, \frac{1}{12}t^4 + 10t, \frac{1}{20}t^5 \right\rangle.$$

11.5.33: Given $\mathbf{a} = \langle \cos t, \sin t, 0 \rangle$, it follows that

$$\mathbf{v}(t) = \langle c_1 + \sin t, c_2 - \cos t, c_3 \rangle$$

where c_1, c_2, and c_3 are constants. Then

$$-\mathbf{i} + 5\mathbf{k} = \langle -1, 0, 5 \rangle = \mathbf{v}(0) = \mathbf{v}_0 = \langle c_1, -1 + c_2, c_3 \rangle$$

implies that $c_1 = -1$, $c_2 = 1$, and $c_3 = 5$. Hence

$$\mathbf{v}(t) = \langle -1 + \sin t, 1 - \cos t, 5 \rangle,$$

and therefore

$$\mathbf{r}(t) = \langle -t - \cos t + k_1, t - \sin t + k_2, 5t + k_3 \rangle$$

where k_1, k_2, and k_3 are constants. Then

$$\mathbf{j} = \langle 0, 1, 0 \rangle = \mathbf{r}(0) = \mathbf{r}_0 = \langle -1 + k_1, k_2, k_3 \rangle$$

leads to $k_1 = 1$, $k_2 = 1$, and $k_3 = 0$. Therefore

$$\mathbf{r}(t) = \langle 1 - t - \cos t, 1 + t - \sin t, 5t \rangle.$$

11.5.35: The position vector of the moving point is $\mathbf{r}(t) = \langle 3\cos 2t, 3\sin 2t, 8t \rangle$. Hence its velocity, speed, and acceleration are

$$\mathbf{v}(t) = \langle -6\sin 2t, 6\cos 2t, 8 \rangle,$$

$$v(t) = |\mathbf{v}(t)| = \sqrt{36(\sin^2 2t + \cos^2 2t) + 64} = 10, \quad \text{and}$$

$$\mathbf{a}(t) = \langle -12\cos 2t, -12\sin 2t, 0 \rangle,$$

respectively. Therefore

$$\mathbf{v}\left(\tfrac{7}{8}\pi\right) = \langle 3\sqrt{2}, 3\sqrt{2}, 8 \rangle, \quad v\left(\tfrac{7}{8}\pi\right) = 10, \quad \text{and} \quad \mathbf{a}\left(\tfrac{7}{8}\pi\right) = \langle -6\sqrt{2}, 6\sqrt{2}, 0 \rangle.$$

11.5.37: Given $\mathbf{u}(t) = \langle 0, 3, 4t \rangle$ and $\mathbf{v}(t) = \langle 5t, 0, -4 \rangle$, we first compute

$$\mathbf{u}(t) \times \mathbf{v}(t) = \begin{vmatrix} \mathbf{i} & \mathbf{j} & \mathbf{k} \\ 0 & 3 & 4t \\ 5t & 0 & -4 \end{vmatrix} = \langle -12, 20t^2, -15t \rangle.$$

Therefore $D_t\left[\mathbf{u}(t) \times \mathbf{v}(t)\right] = \langle 0, 40t, -15 \rangle$. Next,

$$\mathbf{u}(t) \times \mathbf{v}'(t) + \mathbf{u}'(t) \times \mathbf{v}(t) = \begin{vmatrix} \mathbf{i} & \mathbf{j} & \mathbf{k} \\ 0 & 3 & 4t \\ 5 & 0 & 0 \end{vmatrix} + \begin{vmatrix} \mathbf{i} & \mathbf{j} & \mathbf{k} \\ 0 & 0 & 4 \\ 5t & 0 & -4 \end{vmatrix}$$

$$= \langle 0, 20t, -15 \rangle + \langle 0, 20t, 0 \rangle = \langle 0, 40t, -15 \rangle = D_t\left[\mathbf{u}(t) \times \mathbf{v}(t)\right].$$

11.5.39: Given: $|\mathbf{r}(t)| = R$, a constant. Let $\mathbf{v}(t) = \mathbf{r}'(t)$. Then $\mathbf{r}(t) \cdot \mathbf{r}(t) = R^2$, also a constant. Hence

$$0 = D_t\left[\mathbf{r}(t) \cdot \mathbf{r}(t)\right] = 2\mathbf{r}(t) \cdot \mathbf{v}(t),$$

so that $\mathbf{r}(t) \cdot \mathbf{v}(t) = 0$. Therefore $\mathbf{v}(t)$ is perpendicular to the radius of the sphere, so the velocity vector is everywhere tangent to the sphere.

11.5.41: The ball of Example 10 has position vector $\mathbf{r}(t) = \langle t^2, 80t, 80t - 16t^2 \rangle$ and velocity vector $\mathbf{v}(t) = \langle 2t, 80, 80 - 32t \rangle$. Its maximum height occurs when the z-component of its velocity vector is zero: $80 - 32t = 0$, so that $t = \frac{5}{2}$. The speed of the ball at time t is

$$v(t) = \sqrt{4t^2 + 6400 + 6400 - 5120t + 1024t^2},$$

so that

$$v\left(\tfrac{5}{2}\right) = \sqrt{25 + 12800 - 12800 + 6400} = \sqrt{6425} = 5\sqrt{257} \approx 80.156097709407$$

(ft/s). Its position then is

$$\mathbf{r}\left(\tfrac{5}{2}\right) = \left\langle \tfrac{25}{4}, 200, 100 \right\rangle,$$

and the z-component of this vector is its maximum height, 100 ft.

11.5.43: Because $x_0 = y_0 = 0$, the equations in (22) and (23) take the form

$$x(t) = (v_0 \cos\alpha)t, \qquad y(t) = -\frac{1}{2}gt^2 + (v_0 \sin\alpha)t.$$

To find the range, find the positive value of t for which $y(t) = 0$:

$$gt = 2v_0 \sin\alpha; \qquad t = \frac{2v_0 \sin\alpha}{g};$$

thus the range is the value of $x(t)$ then; it is

$$R = x\left(\frac{2v_0 \sin\alpha}{g}\right) = \frac{2v_0^2 \sin\alpha \cos\alpha}{g} = \frac{v_0^2 \sin 2\alpha}{g}.$$

If $\alpha = \frac{1}{4}\pi$ and $R = 5280$ (there are 5280 feet in one mile), then

$$\frac{v_0^2}{32} = 5280, \quad \text{so that} \quad v_0 = \sqrt{32 \cdot 5280} = 32\sqrt{165} \approx 411.047442517284$$

(feet per second).

11.5.45: The formula for the range is derived in the solution of Problem 43.

11.5.47: We saw in the solution of Problem 43 that the range of the projectile is

$$R = \frac{v_0^2 \sin 2\alpha}{g}.$$

To find its maximum height, first find when $y'(t) = 0$:

$$-gt = v_0 \sin \alpha, \quad \text{so that} \quad t = \frac{v_0 \sin \alpha}{g}.$$

Then to find the maximum height, evaluate $y(t)$ at that value of t:

$$y\left(\frac{v_0 \sin \alpha}{g}\right) = \frac{v_0^2 \sin^2 \alpha}{2g}.$$

Part (a): If $v_0 = 160$ and $\alpha = \frac{1}{6}\pi$, then the range is

$$R = \frac{(160)^2 \cdot \sqrt{3}}{2 \cdot 32} = 400\sqrt{3} \approx 692.820323027551$$

(feet) and the maximum height is

$$\frac{v_0 \sin^2 \alpha}{2g} = \frac{(160)^2 \cdot 1}{4 \cdot 2 \cdot 32} = 100$$

(feet). Part (b): If $v_0 = 160$ and $\alpha = \frac{1}{4}\pi$, then the range is

$$R = \frac{(160)^2 \cdot 1}{32} = 800$$

(feet) and the maximum altitude is

$$\frac{v_0^2 \sin^2 \alpha}{2g} = \frac{(160)^2 \cdot 1}{2 \cdot 2 \cdot 32} = 200$$

(feet). Part (c): If $\alpha = \frac{1}{3}\pi$ and $v_0 = 160$, then the range is

$$R = \frac{(160)^2 \cdot \sqrt{3}}{2 \cdot 32} = 400\sqrt{3} \approx 692.820323027551$$

(feet) and the maximum altitude (also in feet) is

$$y_{\max} = \frac{(160)^2 \cdot 3}{4 \cdot 2 \cdot 32} = 300.$$

11.5.49: With $x_0 = 0$ and $y_0 = 100$, the equations in (22) and (23) of the text take the form

$$x(t) = (v_0 \cos \alpha)t, \quad y(t) = -\frac{1}{2}gt^2 + (v_0 \sin \alpha)t + 100.$$

With $g = 9.8$ and $\alpha = 0$, these equations become

$$x(t) = v_0 t, \quad y(t) = 100 - \frac{1}{2}gt^2.$$

We require $x(t) = 1000$ when $y(t) = 0$. But $y(t) = 0$ when

$$t^2 = \frac{200}{9.8} = \frac{1000}{49}, \quad \text{so that} \quad t = \frac{10\sqrt{10}}{7}.$$

Thus

$$1000 = x\left(\frac{10\sqrt{10}}{7}\right) = \frac{10\sqrt{10}}{7}\,v_0,$$

and it follows that $v_0 = 70\sqrt{10} \approx 221.359436211787$ (meters per second, approximately 726.244869461242 feet per second).

11.5.51: First we analyze the behavior of the bomb. Suppose that it is dropped at time $t = 0$. If the projectile is fired from the origin, then the equations of motion of the bomb are

$$x(t) \equiv 800, \quad y(t) = 800 - \frac{1}{2}gt^2$$

where $g = 9.8$ (m/s^2). Now $y(t) = 400$ when

$$800 - \frac{1}{2}gt^2 = 400; \quad \frac{1}{2}gt^2 = 400;$$

$$t^2 = \frac{800}{9.8} = \frac{4000}{49}; \quad t = T = \frac{20\sqrt{10}}{7}.$$

Now we turn our attention to the projectile, fired at time $t = 0$ from the origin. (We will adjust for the one-second delay later in this solution.) By Eqs. (22) and (23) of the text, its equations of motion are

$$x(t) = (v_0 \cos\alpha)t, \quad y(t) = -\frac{1}{2}gt^2 + (v_0 \sin\alpha)t.$$

We require $x(T - 1) = 800$ and $y(T - 1) = 400$. (This is how we take care of the one-second delay). Thus

$$T - 1 = \frac{20\sqrt{10} - 7}{7};$$

$$(v_0 \cos\alpha)(T - 1) = 800;$$

$$-\frac{1}{2}g(T - 1)^2 + (v_0 \sin\alpha)(T - 1) = 400;$$

$$v_0 \cos\alpha = \frac{800}{T - 1}. \tag{1}$$

Also

$$-\frac{1}{2}g(T - 1) + v_0 \sin\alpha = \frac{400}{T - 1};$$

692

$$v_0 \sin \alpha = \frac{400}{T-1} + \frac{1}{2}g(T-1). \tag{2}$$

Division of Eq. (2) by Eq. (1) then yields

$$\tan \alpha = \frac{T-1}{800} \cdot \left[\frac{400}{T-1} + \frac{1}{2}g(T-1) \right] = \frac{1}{2} + \frac{9.8}{1600}(T-1)^2$$

$$= \frac{1}{2} + \frac{98}{16000} \cdot \left(\frac{20\sqrt{10} - 7}{7} \right)^2 = \frac{8049 - 280\sqrt{10}}{8000} \approx 0.895445,$$

so that $\alpha \approx 41.842705345876°$. Then, by Eq. (1),

$$v_0 = \frac{800}{(T-1)\cos \alpha} \approx 133.645951548503$$

meters per second, approximately 438.470969647319 feet per second.

11.5.53: We give proofs for vectors with two components; these proofs generalize readily to vectors with three or more components. Let $\mathbf{u}(t) = \langle u_1(t), u_2(t) \rangle$ and $\mathbf{v}(t) = \langle v_1(t), v_2(t) \rangle$. The assumptions that

$$\lim_{t \to a} \mathbf{u}(t) \quad \text{and} \quad \lim_{t \to a} \mathbf{v}(t)$$

both exist means that there exist vectors $\langle p_1, p_2 \rangle$ and $\langle q_1, q_2 \rangle$ such that

$$\lim_{t \to a} u_1(t) = p_1, \quad \lim_{t \to a} u_2(t) = p_2, \quad \lim_{t \to a} v_1(t) = q_1, \quad \text{and} \quad \lim_{t \to a} v_2(t) = q_2.$$

Part (a):

$$\lim_{t \to a} [\mathbf{u}(t) + \mathbf{v}(t)] = \lim_{t \to a} \langle u_1(t) + v_1(t), u_2(t) + v_2(t) \rangle = \langle p_1 + q_1, p_2 + q_2 \rangle$$

$$= \langle p_1, p_2 \rangle + \langle q_1, q_2 \rangle = \left[\lim_{t \to a} \mathbf{u}(t) \right] + \left[\lim_{t \to a} \mathbf{v}(t) \right].$$

Part (b):

$$\lim_{t \to a} [\mathbf{u}(t) \cdot \mathbf{v}(t)] = \lim_{t \to a} [u_1(t)v_1(t) + u_2(t)v_2(t)] = p_1 q_1 + p_2 q_2$$

$$= \langle p_1, p_2 \rangle \cdot \langle q_1, q_2 \rangle = \left(\lim_{t \to a} \mathbf{u}(t) \right) \cdot \left(\lim_{t \to a} \mathbf{v}(t) \right).$$

11.5.55: If $\mathbf{v}(t)$ is the velocity vector of the moving particle, then we are given $|\mathbf{v}(t)| = C$, a constant. Then $\mathbf{v}(t) \cdot \mathbf{v}(t) = C^2$, also a constant. Hence $0 = D_t [\mathbf{v}(t) \cdot \mathbf{v}(t)] = 2\mathbf{v}(t) \cdot \mathbf{a}(t)$ where $\mathbf{a}(t)$ is the acceleration vector of the particle. But because $\mathbf{v}(t) \cdot \mathbf{a}(t) = 0$, it follows that $\mathbf{v}$ and $\mathbf{a}$ are always perpendicular.

11.5.57: If $\mathbf{r}(t) = \langle \cosh \omega t, \sinh \omega t \rangle$, then

$$\mathbf{v}(t) = \mathbf{r}'(t) = \langle \omega \sinh \omega t, \omega \cosh \omega t \rangle \quad \text{and}$$

$$\mathbf{a}(t) = \mathbf{v}'(t) = \langle \omega^2 \cosh \omega t, \omega^2 \sinh \omega t \rangle = \omega^2 \mathbf{r}(t) = c\mathbf{r}(t)$$

where $c = \omega^2 > 0$. An external force that would produce this sort of motion would be a central repulsive force proportional to distance from the origin.

11.5.59: Given the acceleration vector $\mathbf{a} = \langle 0, a \rangle$, we first find the velocity and position vectors:

$$\mathbf{v}(t) = \langle c_1, at + c_2 \rangle \quad \text{and}$$

$$\mathbf{r}(t) = \langle c_1 t + k_1, \tfrac{1}{2} at^2 + c_2 t + k_2 \rangle$$

where c_1, c_2, k_1, and k_2 are constants. Thus the position $(x(t), y(t))$ of the moving point is given by

$$x(t) = c_1 t + k_1 \quad \text{and} \quad y(t) = \tfrac{1}{2} at^2 + c_2 t + k_2.$$

If $c_1 = 0$ then the point moves in a straight line. Otherwise, we solve the first of these equations for t and substitute in the second:

$$t = \frac{x - k_1}{c_1};$$

$$y = \frac{1}{2} a \left(\frac{x - k_1}{c_1} \right)^2 + c_2 \cdot \frac{x - k_1}{c_1} + k_2$$

$$= \frac{a}{2c_1^2} (x^2 - 2k_1 x + k_1^2) + \frac{c_2}{c_1} (x - k_1) + k_2$$

$$= \frac{a}{2c_1^2} x^2 + \left(\frac{c_2}{c_1} - \frac{ak_1}{c_1^2} \right) x + \left(\frac{ak_1^2}{2c_1^2} - \frac{c_2 k_1}{c_1} + k_2 \right)$$

$$= Ax^2 + Bx + C$$

where A, B, and C are constants. If $A \neq 0$ then the trajectory of the particle is a parabola. If $A = 0$ it is a straight line.

11.5.61: Given: $\mathbf{r}(t) = \langle r \cos \omega t, r \sin \omega t \rangle$. Part (a):

$$\mathbf{v}(t) = \langle -r\omega \sin \omega t, r\omega \cos \omega t \rangle = r\omega \langle -\sin \omega t, \cos \omega t \rangle. \quad \text{So}$$

$$\mathbf{r}(t) \cdot \mathbf{v}(t) = r^2 \omega (-\sin \omega t \cos \omega t + \sin \omega t \cos \omega t) = 0,$$

and therefore $\mathbf{r}$ and $\mathbf{v}$ are always perpendicular. Therefore $\mathbf{v}$ is always tangent to the circle. Moreover, the speed of motion is

$$v(t) = r\omega\sqrt{\sin^2 \omega t + \cos^2 \omega t} = r\omega.$$

Part (b): $\quad \mathbf{a}(t) = r\omega^2\langle -\cos \omega t, \ -\sin \omega t \rangle = -\omega^2 \mathbf{r}(t).$ Therefore $\mathbf{a}$ and $\mathbf{r}$ are always parallel and have opposite directions (because $-\omega^2 < 0$). Finally, the scalar acceleration is

$$a(t) = |\mathbf{a}(t)| = |-\omega^2| \cdot |\mathbf{r}(t)| = r\omega^2.$$

11.5.63: With north the direction of the positive x-axis, west the direction of the positive y-axis, and upward the direction of the positive z-axis, the baseball has acceleration $\mathbf{a}(t) = \langle 0.1, 0, -32 \rangle$, initial velocity $\mathbf{v}_0 = \langle 0, 0, 160 \rangle$, and initial position $\mathbf{r}_0 = \langle 0, 0, 0 \rangle$. It follows that its position vector is

$$\mathbf{r}(t) = \left\langle \frac{1}{20}t^2, \ 0, \ 160t - 16t^2 \right\rangle.$$

The ball returns to the ground at that positive value of t for which the z-component of $\mathbf{r}$ is zero; that is, $t = 10$. At that time the x-component of $\mathbf{r}$ is 5, so the ball lands 5 feet north of the point from which it was thrown.

11.5.65: In the "obvious" coordinate system, the acceleration of the projectile is $\mathbf{a}(t) = \langle 2, 0, -32 \rangle$; its initial velocity is $\mathbf{v}_0 = \langle 0, 200, 160 \rangle$ and its initial position is $\mathbf{r}_0 = \langle 0, 0, 384 \rangle$. Hence its velocity and position vectors are

$$\mathbf{v}(t) = \langle 2t, 200, 160 - 32t \rangle \quad \text{and} \quad \mathbf{r}(t) = \langle t^2, 200t, 384 - 160t - 16t^2 \rangle.$$

The projectile strikes the ground at that positive value of t from which the z-component of $\mathbf{r}$ is zero:

$$16t^2 - 160t - 384 = 0; \qquad t^2 - 10t - 24 = 0;$$

$$(t - 12)(t + 2) = 0; \qquad t = 12 \quad (not \ t = -2).$$

When $t = 12$, the position of the projectile is $\mathbf{r}(12) = \langle 144, 2400, 0 \rangle$, so it lands 2400 ft north and 144 ft east of the firing position. The projectile reaches its maximum altitude when the z-component of $\mathbf{v}(t)$ is zero; that is, when $t = 5$. Its position then is $\mathbf{r}(5) = \langle 25, 1000, 784 \rangle$, so its maximum altitude is 784 ft.

Section 11.6

11.6.1: We first compute

$$v(t) = \sqrt{(x'(t))^2 + (y'(t))^2 + (z'(t))^2} = \sqrt{64 + 36\cos^2 2t + 36\sin^2 2t} = \sqrt{100} = 10.$$

Therefore the length of the graph is

$$s = \int_0^\pi 10\, dt = \left[\, 10t \,\right]_0^\pi = 10\pi.$$

11.6.3: First,

$$v(t) = \sqrt{[x'(t)]^2 + [y'(t)]^2 + [z'(t)]^2} \; = \sqrt{(6e^t \cos t - 6e^t \sin t)^2 + (6e^t \cos t + 6e^t \sin t)^2 + (17e^t)^2}$$

$$= \sqrt{289e^{2t} + 36e^{2t} \cos^2 t - 72e^{2t} \sin t \cos t + 36e^{2t} \sin t + 36e^{2t} \cos^2 t + 72e^{2t} \sin t \cos t + 36e^{2t} \sin^2 t}$$

$$= \sqrt{289e^{2t} + 36e^{2t} + 36e^{2t}} \; = \sqrt{361e^{2t}} \; = 19e^t.$$

Therefore the arc length is

$$s = \int_0^1 19e^t \, dt = \left[\, 19e^t \,\right]_0^1 = 19e - 19 = 19(e-1) \approx 32.647354740722.$$

11.6.5: First,

$$v(t) = \sqrt{(3t \cos t + 3 \sin t)^2 + (3 \cos t - 3t \sin t)^2 + (4t)^2}$$

$$= \sqrt{16t^2 + 9 \cos^2 t + 9 \sin^2 t + 9t^2 \cos^2 t + 9t^2 \sin^2 t} = \sqrt{9 + 25t^2}.$$

Then the substitutions $u = 5t$, $a = 3$, and formula 44 of the endpapers of the text yields arc length

$$s = \int_0^{4/5} \sqrt{9 + 25t^2} \, dt = \left[\, \frac{t}{2}\sqrt{9 + 25t^2} + \frac{9}{10}\left(\ln 5t + \sqrt{9 + 25t^2}\,\right) \right]_0^{4/5}$$

$$= 2 + \frac{9}{10}\ln 9 - \frac{9}{10}\ln 3 = \frac{20 + 9\ln 3}{10} \approx 2.988751059801.$$

By comparison, *Mathematica* yields

$$s = \left[\, \frac{t}{2}\sqrt{9 + 25t^2} + \frac{9}{10}\sinh^{-1}\left(\frac{5t}{3}\right) \right]_0^{4/5},$$

which can be simplified to the same answer using identities found in Section 6.9.

11.6.7: By Eq. (13) of the text,

$$\kappa(x) = \frac{|y''(x)|}{[1 + (y'(x))^2]^{3/2}} = \frac{|6x|}{(1 + 9x^4)^{3/2}},$$

and therefore $\kappa(0) = 0$.

11.6.9: By Eq. (13) of the text,

$$\kappa(x) = \frac{|\cos x|}{(1 + \sin^2 x)^{3/2}},$$

696

and thus $\kappa(0) = 1$.

11.6.11: By Eq. (12) of the text,

$$\kappa(t) = \frac{|x'(t)y''(t) - x''(t)y'(t)|}{[(x'(t))^2 + (y'(t))^2]^{3/2}} = \frac{|-20\sin^2 t - 20\cos^2 t|}{[25\sin^2 t + 16\cos^2 t]^{3/2}} = \frac{20}{(16 + 9\sin^2 t)^{3/2}},$$

so that

$$\kappa\left(\frac{\pi}{4}\right) = \frac{20}{\left(16 + \frac{9}{2}\right)^{3/2}} = \frac{40\sqrt{82}}{1681} \approx 0.2154761484.$$

11.6.13: Given: $y = e^x$. By Eq. (13) of the text, the curvature at x is

$$\kappa(x) = \frac{|y''(x)|}{[1 + (y'(x))^2]^{3/2}} = \frac{e^x}{(1 + e^{2x})^{3/2}}.$$

Because $\kappa(x) > 0$ for all x, $\kappa(x) \to 0$ as $x \to \pm\infty$, and κ is continuous on the set of all real numbers, there is a maximum value. Next,

$$\kappa'(x) = \frac{e^x(1 - 2e^{2x})}{(1 + e^{2x})^{5/2}}; \quad \kappa'(x) = 0 \quad \text{when} \quad e^{2x} = \frac{1}{2}: \quad x = -\frac{1}{2}\ln 2.$$

Answer: The maximum curvature of the graph of $y = e^x$ occurs at the point

$$\left(-\frac{1}{2}\ln 2, \ \frac{1}{2}\sqrt{2}\right).$$

The curvature there is

$$\frac{\frac{1}{2}\sqrt{2}}{\left(1 + \frac{1}{2}\right)^{3/2}} = \frac{\sqrt{2}}{2} \cdot \frac{2^{3/2}}{3^{3/2}} = \frac{2\sqrt{3}}{9} \approx 0.3849001794597505.$$

11.6.15: Given: $x = 5\cos t$, $y = 3\sin t$. By Eq. (12) of the text, the curvature at $(x(t), y(t))$ is given by

$$\kappa(t) = \frac{|x'(t)y''(t) - x''(t)y'(t)|}{[(x'(t))^2 + (y'(t))^2]^{3/2}} = \frac{|15\sin^2 t + 15\cos^2 t|}{(25\sin^2 t + 9\cos^2 t)^{3/2}} = \frac{15}{(9 + 16\sin^2 t)^{3/2}}.$$

Nothing is lost by restriction of t to the interval $[0, 2\pi]$, and $\kappa(t)$ is continuous there, so κ has both a global maximum value and a global minimum value in that interval. To find them,

$$\kappa'(t) = \frac{-15 \cdot \frac{3}{2}(9 + 16\sin^2 t)^{1/2} \cdot 32\sin t \cos t}{(9 + 16\sin^2 t)^3} = -\frac{720\sin t \cos t}{(9 + 16\sin^2 t)^{5/2}}.$$

Because $\kappa'(t) = 0$ at every integral multiple of $\pi/2$, we check these critical points (and only these):

$$\kappa(0) = \frac{5}{9} = \kappa(\pi) = \kappa(2\pi);$$

$$\kappa(\pi/2) = \frac{3}{25} = \kappa(3\pi/2).$$

Therefore the maximum curvature of the graph of the given parametric equations is $\frac{5}{9}$, which occurs at $(5, 0)$ and at $(-5, 0)$ (corresponding to $t = 0$ and $t = \pi$); the minimum curvature is $\frac{3}{25}$ and occurs at $(0, 3)$ and at $(0, -3)$.

11.6.17: Let $\mathbf{r}(t) = \langle t, t^3 \rangle$. For the purpose of determining the direction of $\mathbf{N}$, note that the graph of $y = x^3$ is concave downward at and near the given point $(-1, -1)$. Then

$$\mathbf{v}(t) = \langle 1, 3t^2 \rangle;$$

$$\mathbf{v}(-1) = \langle 1, 3 \rangle;$$

$$\mathbf{T}(-1) = \left\langle \frac{\sqrt{10}}{10}, \frac{3\sqrt{10}}{10} \right\rangle;$$

$$\mathbf{N}(-1) = \left\langle \frac{3\sqrt{10}}{10}, -\frac{\sqrt{10}}{10} \right\rangle.$$

11.6.19: Let $\mathbf{r}(t) = \langle 3\sin 2t, 4\cos 2t \rangle$. For the purpose of determining the direction of $\mathbf{N}$, note that the graph of the given parametric equations is concave downward at and near the given point for which $t = \pi/6$. Then

$$\mathbf{v}(t) = \langle 6\cos 2t, -8\sin 2t \rangle; \qquad \mathbf{v}(\pi/6) = \langle 3, -4\sqrt{3} \rangle;$$

$$\mathbf{T}(\pi/6) = \left\langle \frac{\sqrt{57}}{19}, -\frac{4\sqrt{19}}{19} \right\rangle; \qquad \mathbf{N}(\pi/6) = \left\langle -\frac{4\sqrt{19}}{19}, -\frac{\sqrt{57}}{19} \right\rangle.$$

11.6.21: Let $\mathbf{r}(t) = \langle \cos^3 t, \sin^3 t \rangle$. For the purpose of determining the direction of $\mathbf{N}$, note that the graph of the given parametric equations is concave upward at and near the given point for which $t = 3\pi/4$. Then

$$\mathbf{v}(t) = \langle -3\cos^2 t \sin t, 3\sin^2 t \cos t \rangle; \quad \mathbf{v}(3\pi/4) = \left\langle -\frac{3\sqrt{2}}{4}, -\frac{3\sqrt{2}}{4} \right\rangle; \quad v(3\pi/4) = \frac{3}{2};$$

$$\mathbf{T}(3\pi/4) = \left\langle -\frac{\sqrt{2}}{2}, -\frac{\sqrt{2}}{2} \right\rangle; \quad \mathbf{N}(3\pi/4) = \left\langle -\frac{\sqrt{2}}{2}, \frac{\sqrt{2}}{2} \right\rangle.$$

11.6.23: We adjoin third component zero to form $\mathbf{r}(t) = \langle 2t + 1, 3t^2 - 1, 0 \rangle$. Then $\mathbf{v}(t) = \langle 2, 6t, 0 \rangle$, $\mathbf{a}(t) = \langle 0, 6, 0 \rangle$, and $v(t) = (36t^2 + 4)^{1/2}$. By Eq. (26) of the text,

$$a_T = \frac{\mathbf{v} \cdot \mathbf{a}}{v} = \frac{36t}{(36t^2 + 4)^{1/2}} = \frac{18t}{\sqrt{9t^2 + 1}},$$

and by Eq. (28),

$$a_N = \frac{|\mathbf{v} \times \mathbf{a}|}{v} = \frac{1}{v} \cdot |\langle 0, 0, 12 \rangle| = \frac{6}{\sqrt{9t^2 + 1}}.$$

698

11.6.25: We adjoin third component zero to form $\mathbf{r}(t) = \langle t\cos t,\, t\sin t,\, 0 \rangle$. Then

$$\mathbf{v}(t) = \langle \cos t - t\sin t,\, \sin t + t\cos t,\, 0 \rangle \qquad \mathbf{a}(t) = \langle -2\sin t - t\cos t,\, 2\cos t - t\sin t,\, 0 \rangle,$$

and $v(t) = \sqrt{\cos^2 t + t^2\sin^2 t + \sin^2 t + t^2\cos^2 t} = \sqrt{t^2 + 1}$. Next,

$$\mathbf{v}(t) \cdot \mathbf{a}(t) = -2\sin t \cos t + 2t\sin^2 t - t\cos^2 t + t^2\sin t\cos t + 2\sin t\cos t - t\sin^2 t + 2t\cos^2 t - t^2\sin t\cos t$$

$$= 2t - t = t.$$

So by Eq. (26) of the text,

$$a_T = \frac{\mathbf{v}(t)\cdot\mathbf{a}(t)}{v(t)} = \frac{t}{\sqrt{t^2+1}},$$

Next,

$$\mathbf{v}(t) \times \mathbf{a}(t) = \begin{vmatrix} \mathbf{i} & \mathbf{j} & \mathbf{k} \\ \cos t - t\sin t & \sin t + t\cos t & 0 \\ -2\sin t - t\cos t & 2\cos t - t\sin t & 0 \end{vmatrix}$$

$$= \langle 0,\, 0,\, 2\cos^2 t - t\sin t\cos t - 2t\sin t\cos t + t^2\sin^2 t + 2\sin^2 t + 2t\sin t\cos t + t\sin t\cos t + t^2\cos^2 t \rangle$$

$$= \langle 0,\, 0,\, t^2 + 2 \rangle.$$

Therefore by Eq. (28),

$$a_N = \frac{|\mathbf{v}(t)\times\mathbf{a}(t)|}{v(t)} = \frac{t^2+2}{\sqrt{t^2+1}}$$

11.6.27: Given $x^2 + y^2 = a^2$, implicit differentiation yields

$$2x + 2y\frac{dy}{dx} = 0, \quad \text{so that} \quad \frac{dy}{dx} = -\frac{x}{y}.$$

Differentiation of both sides of the last equation with respect to x then yields

$$\frac{d^2y}{dx^2} = \frac{x\dfrac{dy}{dx} - y}{y^2} = \frac{xy\dfrac{dy}{dx} - y^2}{y^3} = \frac{-x^2 - y^2}{y^3} = -\frac{a^2}{y^3}.$$

Next, Eq. (13) yields

$$\kappa(x) = \frac{|y''(x)|}{[1 + (y'(x))^2]^{3/2}} = \frac{a^2}{|y|^3(1 + x^2y^{-2})^{3/2}}.$$

If $y \neq 0$, then

$$\kappa(x) = \frac{a^2}{(y^2 + x^2)^{3/2}} = \frac{a^2}{a^3} = \frac{1}{a}.$$

699

If $y = 0$, then $x \neq 0$; interchange the roles of x and y and work with dx/dy to obtain the same result. In this case, instead of Eq. (13) use:

$$\kappa(y) = \frac{|x''(y)|}{[1 + (x'(y))^2]^{3/2}}.$$

11.6.29: Let $x(t) = t$, $y(t) = 1 - t^2$, and $\mathbf{r}(t) = \langle x(t), y(t) \rangle$. Then $\mathbf{r}'(t) = \langle 1, -2t \rangle$, so that $\mathbf{r}'(0) = \langle 1, 0 \rangle = \mathbf{T}(1)$. The graph of the given equation is concave down everywhere, so that the unit normal vector corresponding to $t = 0$ is $\mathbf{N} = \langle 0, -1 \rangle$. Next, $\mathbf{a}(t) = \mathbf{r}''(t) = \langle 0, -2 \rangle$, and so $\mathbf{a}(0)$ is the same. By Eq. (12) the curvature is

$$\kappa(t) = \frac{|x'(t)y''(t) - x''(t)y'(t)|}{[v(t)]^3} = \frac{2}{(1 + 4t^2)^{3/2}}.$$

Because $\kappa(0) = 2$, the radius of the osculating circle at $(0, 1)$ is $\frac{1}{2}$, and by Eq. (16) the position vector of the center of that circle is

$$\mathbf{r}(0) + \frac{1}{2}\mathbf{N}(0) = \left\langle 0, \frac{1}{2} \right\rangle.$$

Therefore an equation of the osculating circle is

$$x^2 + \left(y - \frac{1}{2}\right)^2 = \frac{1}{4}.$$

11.6.31: Let $x(t) = t$ and $y(t) = t^{-1}$; let $\mathbf{r}(t) = \langle x(t), y(t) \rangle$. Then

$$\mathbf{v}(t) = \mathbf{r}'(t) = \left\langle 1, -\frac{1}{t^2} \right\rangle \quad \text{and} \quad v(t) = |\mathbf{v}(t)| = \frac{\sqrt{t^4 + 1}}{t^2}.$$

Thus the unit tangent and unit normal vectors at $(1, 1)$ are

$$\mathbf{T}(1) = \frac{\mathbf{v}(1)}{v(1)} = \left\langle \frac{\sqrt{2}}{2}, -\frac{\sqrt{2}}{2} \right\rangle \quad \text{and} \quad \mathbf{N}(1) = \left\langle \frac{\sqrt{2}}{2}, \frac{\sqrt{2}}{2} \right\rangle,$$

respectively. By Eq. (12) of the text, the curvature at $(x(t), y(t))$ is

$$\kappa(t) = \frac{|x'(t)y''(t) - x''(t)y'(t)|}{[(x'(t))^2 + (y'(t))^2]^{3/2}} = \frac{2t^3}{(t^4 + 1)^{3/2}},$$

so the curvature at $(1, 1)$ is $\kappa(1) = \frac{1}{2}\sqrt{2}$. Therefore the osculating circle at $(1, 1)$ has radius $\rho = \sqrt{2}$. By Eq. (16) of the text, the position vector of the center of that circle is

$$\mathbf{r}(1) + \left(\sqrt{2}\right) \cdot \mathbf{N}(1) = \langle 2, 2 \rangle.$$

Therefore an equation of the osculating circle is $(x - 2)^2 + (y - 2)^2 = 2$.

11.6.33: Given $\mathbf{r}(t) = \langle t, \sin t, \cos t \rangle$, we first compute

$$\mathbf{v}(t) = \langle 1, \cos t, -\sin t \rangle, \quad v(t) = \sqrt{1 + \cos^2 t + \sin^2 t} \equiv \sqrt{2}, \quad \text{and} \quad \mathbf{a}(t) = \langle 0, -\sin t, -\cos t \rangle.$$

Then

$$\mathbf{v}(t) \times \mathbf{a}(t) = \begin{vmatrix} \mathbf{i} & \mathbf{j} & \mathbf{k} \\ 1 & \cos t & -\sin t \\ 0 & -\sin t & -\cos t \end{vmatrix} = \langle -\cos^2 t - \sin^2 t,\ \cos t,\ -\sin t \rangle = \langle -1,\ \cos t\ -\sin t \rangle.$$

Therefore, by Eq. (27), the curvature is

$$\kappa(t) = \frac{|\mathbf{v}(t) \times \mathbf{a}(t)|}{[v(t)]^3} = \frac{\sqrt{2}}{2\sqrt{2}} \equiv \frac{1}{2}.$$

11.6.35: Given: $\mathbf{r}(t) = \langle e^t \cos t,\ e^t \sin t,\ e^t \rangle$. Then

$$\mathbf{v}(t) = \langle e^t(\cos t - \sin t),\ e^t(\sin t + \cos t),\ e^t \rangle,$$

$$v(t) = e^t(\cos^2 t + \sin^2 t + \sin^2 t + \cos^2 t + 1)^{1/2} = e^t \sqrt{3}, \quad \text{and}$$

$$\mathbf{a}(t) = \langle e^t(\cos t - \sin t - \sin t - \cos t),\ e^t(\sin t + \cos t + \cos t - \sin t),\ e^t \rangle = e^t \langle -2\sin t,\ 2\cos t,\ 1 \rangle.$$

Therefore

$$\mathbf{v}(t) \times \mathbf{a}(t) = e^{2t} \begin{vmatrix} \mathbf{i} & \mathbf{j} & \mathbf{k} \\ \cos t - \sin t & \sin t + \cos t & 1 \\ -2\sin t & 2\cos t & 0 \end{vmatrix} = e^{2t} \langle \sin t - \cos t,\ -\sin t - \cos t,\ 2 \rangle.$$

Thus

$$|\mathbf{v}(t) \times \mathbf{a}(t)| = e^{2t} \sqrt{\sin^2 t + \cos^2 t + \sin^2 t + \cos^2 t + 4} = e^{2t} \sqrt{6},$$

and therefore

$$\kappa(t) = \frac{e^{2t}\sqrt{6}}{3e^{2t}\sqrt{3}} = \frac{\sqrt{2}}{3} e^{-t}.$$

11.6.37: Let $x(t) = t$, $y(t) = 2t - 1$, $z(t) = 3t + 5$, and $\mathbf{r}(t) = \langle x(t),\ y(t),\ z(t) \rangle$. It follows that $\mathbf{v}(t) = \langle 1, 2, 3 \rangle$ and $\mathbf{a}(t) = \mathbf{0}$. Hence $\mathbf{v}(t) \cdot \mathbf{a}(t) = 0$ and $\mathbf{v}(t) \times \mathbf{a}(t) = \mathbf{0}$, and therefore $a_T = 0 = a_N$.

11.6.39: Using *Mathematica*, we let

```
r[t_] := { t, t∧2, t∧3 }; v[t_] := r'[t]
```

so that $\mathbf{v}(t) = \langle 1, 2t, 3t^2 \rangle$, and we let

```
speed[t_] := Sqrt[v[t].v[t]]
```

(using the fact that $|\mathbf{u}| = \sqrt{\mathbf{u} \cdot \mathbf{u}}$), so that $v(t) = (1 + 4t^2 + 9t^4)^{1/2}$. We also defined

```
a[t_] := r''[t]
```

so that $\mathbf{a}(t) = \langle 0, 2, 6t \rangle$. Next we computed

```
v[t].a[t]
```

$$4t + 18t^3$$

and then, by Eq. (26),

```
asubT = v[t].a[t]/speed[t]
```

so that $a_T = \dfrac{4t + 18t^3}{\sqrt{1 + 4t^2 + 9t^4}}$. Next,

```
p[t_] := Cross[v[t], a[t]]
```

yielded $\mathbf{v}(t) \times \mathbf{a}(t) = \langle 6t^2, -6t, 2 \rangle$. Then

```
magp[t_] := Sqrt[p[t].p[t]]
```

revealed that $|\mathbf{v}(t) \times \mathbf{a}(t)| = (4 + 36t^2 + 36t^4)^{1/2}$. Hence, by Eq. (28),

```
asubN = magp[t]/speed[t]
```

—that is, $a_N = \dfrac{\sqrt{4 + 36t^2 + 36t^4}}{\sqrt{1 + 4t^2 + 9t^4}}$.

11.6.41: Beginning with $\mathbf{r}(t) = \langle t\sin t,\ t\cos t,\ t \rangle$, we found:

$$\mathbf{v}(t) = \mathbf{r}'(t) = \langle t\cos t + \sin t,\ \cos t - t\sin t,\ 1 \rangle,$$

$$v(t) = |\mathbf{v}(t)| = \sqrt{1 + (t\cos t + \sin t)^2 + (\cos t - t\sin t)^2} = \sqrt{t^2 + 2},$$

$$\mathbf{a}(t) = \mathbf{v}'(t) = \langle 2\cos t - t\sin t,\ -t\cos t - 2\sin t,\ 0 \rangle,$$

$$\mathbf{v}(t) \cdot \mathbf{a}(t) = (-t\cos t - 2\sin t)(\cos t - t\sin t) + (t\cos t + \sin t)(2\cos t - t\sin t) = t,$$

$$a_T = \frac{\mathbf{v}(t) \cdot \mathbf{a}(t)}{v(t)} = \frac{t}{\sqrt{t^2 + 2}},$$

$$\mathbf{v}(t) \times \mathbf{a}(t) = \langle t\cos t + 2\sin t,\ 2\cos t - t\sin t,\ -(t^2 + 2) \rangle,$$

$$|\mathbf{v}(t) \times \mathbf{a}(t)| = \sqrt{t^4 + 5t^2 + 8}, \quad \text{and}$$

$$a_N = \frac{|\mathbf{v}(t) \times \mathbf{a}(t)|}{v(t)} = \frac{\sqrt{t^4 + 5t^2 + 8}}{\sqrt{t^2 + 2}}.$$

11.6.43: Given $\mathbf{r}(t) = \langle t,\ \sin t,\ \cos t \rangle$, we will compute the unit tangent vector $\mathbf{T}$ using its definition in Eq. (17), then the principal unit normal vector $\mathbf{N}$ by means of Eq. (29). We find:

$$\mathbf{v}(t) = \mathbf{r}'(t) = \langle 1,\ \cos t,\ -\sin t \rangle,$$

$$\mathbf{v}(0) = \langle 1,\ 1,\ 0 \rangle,$$

$$v(0) = \sqrt{2},$$

$$\mathbf{T}(0) = \left\langle \frac{\sqrt{2}}{2},\ \frac{\sqrt{2}}{2},\ 0 \right\rangle,$$

$$\mathbf{a}(t) = \langle 0,\ -\sin t,\ -\cos t \rangle,$$

$$\mathbf{a}(0) = \langle 0,\ 0,\ -1 \rangle,$$

$$a_T(0) = \frac{\mathbf{v}(0) \cdot \mathbf{a}(0)}{v(0)} = 0,$$

$$a_N(0) = \frac{|\mathbf{v}(0) \times \mathbf{a}(0)|}{v(t)} = 1,$$

$$\kappa(0) = \frac{|\mathbf{v}(0) \times \mathbf{a}(0)|}{[v(t)]^3} = \frac{1}{2}, \quad \text{and}$$

$$\mathbf{N}(0) = \frac{\mathbf{a} - a_T\mathbf{T}}{a_N} = \langle 0,\ 0,\ -1 \rangle$$

11.6.45: The process of computing $\mathbf{T}$ and $\mathbf{N}$ can be carried out almost automatically using *Mathematica*. Given $\mathbf{r}(t) = \langle e^t \cos t,\ e^t \sin t,\ e^t \rangle$, we entered the following commands:

```
v0 = r'[0]

        {1, 1, 1}

a0 = r''[0]

        {0, 2, 1}

speed = Sqrt[v0.v0]

        √3

asubT = v0.a0/speed

        √3

vcrossa = Cross[ v0, a0 ]

        {-1, -1, 2}

asubN = (Sqrt[ vcrossa.vcrossa ])/speed
```

$$\sqrt{2}$$

```
kappa = asubN/(speed*speed)
```
$$\frac{\sqrt{2}}{3}$$
```
utan = v0/speed
```
$$\left\{ \frac{1}{\sqrt{3}}, \frac{1}{\sqrt{3}}, \frac{1}{\sqrt{3}} \right\}$$
```
unorm = (a0 - asubT*utan)/asubN
```
$$\left\{ -\frac{1}{\sqrt{2}}, \frac{1}{\sqrt{2}}, 0 \right\}$$

Thus we see that

$$\mathbf{T}(0) = \left\langle \frac{\sqrt{3}}{3}, \frac{\sqrt{3}}{3}, \frac{\sqrt{3}}{3} \right\rangle \quad \text{and} \quad \mathbf{N}(0) = \left\langle -\frac{\sqrt{2}}{2}, \frac{\sqrt{2}}{2}, 0 \right\rangle.$$

11.6.47: Because

$$\frac{ds}{dt} = \sqrt{[x'(t)]^2 + [y'(t)]^2 + [z'(t)]^2} = \sqrt{16 + 144 + 9} = 13,$$

we see that the arc length is given by $s = 13t$, and therefore the arc-length parametrization of the given curve is

$$x(s) = 2 + \frac{4s}{13}, \quad y(s) = 1 - \frac{12s}{13}, \quad z(s) = 3 + \frac{3s}{13}.$$

11.6.49: Because

$$\frac{ds}{dt} = \sqrt{[x'(t)]^2 + [y'(t)]^2 + [z'(t)]^2} = \sqrt{9\cos^2 t + 9\sin^2 t + 16} = \sqrt{25} = 5,$$

we see that the arc length is given by $s = 5t$, and therefore the arc-length parametrization of the given curve is

$$x(s) = 3\cos\frac{s}{5}, \quad y(s) = 3\sin\frac{s}{5}, \quad z(s) = \frac{4s}{5}.$$

11.6.51: By Newton's second law of motion, the acceleration of the particle is a scalar multiple of the force acting on the particle, so the force and acceleration vectors are parallel. Therefore the acceleration vector $\mathbf{a}$ is normal to the velocity vector $\mathbf{v}$. But then,

$$D_t(\mathbf{v} \cdot \mathbf{v}) = \mathbf{v} \cdot \mathbf{a} + \mathbf{a} \cdot \mathbf{v} = 0 + 0 = 0,$$

and therefore $\mathbf{v} \cdot \mathbf{v} = K$, a constant. Hence the speed $v(t) = \sqrt{\mathbf{v} \cdot \mathbf{v}}$ is also constant.

11.6.53: Given $x(t) = \cos t + t\sin t$ and $y(t) = \sin t - t\cos t$, we first compute

$$[v(t)]^2 = [x'(t)]^2 + [y'(t)]^2$$

$$= (t\cos t + \sin t - \sin t)^2 + (\cos t - \cos t + t\sin t)^2 = t^2\cos^2 t + t^2\sin^2 t = t^2,$$

$$v'(t) = 1,$$

$$[x''(t)]^2 = (\cos t - t\sin t)^2,$$

$$[y''(t)]^2 = (\sin t + t\cos t)^2, \quad \text{and}$$

$$[x''(t)]^2 + [y''(t)]^2 = \cos^2 t + \sin^2 t + t^2\sin^2 t + t^2\cos^2 t = t^2 + 1.$$

Then the formula in Problem 52 yields

$$\kappa(t) = \frac{\sqrt{[x''(t)]^2 + [y''(t)]^2 - [v'(t)]^2}}{[x'(t)]^2 + [y'(t)]^2} = \frac{\sqrt{t^2 + 1} - 1}{t^2} = \frac{1}{|t|}.$$

11.6.55: The six conditions listed in the statement of the problem imply, in order, that

$$0 = F,$$

$$1 = A + B + C + D + E + F,$$

$$0 = E,$$

$$1 = 5A + 4B + 3C + 2D + E,$$

$$0 = 2D, \quad \text{and}$$

$$0 = 20A + 12B + 6C + 2D.$$

The last two equations were obtained by observing—via Eq. (13)—that curvature zero is equivalent to $y''(x) = 0$. Simultaneous solution of these equations yields $A = 3$, $B = -8$, $C = 6$, and $D = E = F = 0$. Answers: $y = 3x^5 - 8x^4 + 6x^3$; because $\kappa(x)$ is zero where the curved track meets the straight tracks, because their derivatives agree at the junctions, and because

$$\kappa(x) = \frac{|60x^3 - 96x^2 + 36x|}{[1 + (15x^4 - 32x^3 + 18x^2)^2]^{3/2}}$$

is continuous on $[0, 1]$, the normal forces on the train negotiating this transitional section of track will change continuously from zero to larger values and then back to zero. Thus there will be no abrupt change in the lateral forces on the train.

11.6.57: Conversion into miles yields the semimajor axis of the orbit of Mercury to be $a = 35973972$. With eccentricity $e = 0.206$ and period $T = 87.97$ days, we use

$$b^2 = a^2(1 - e^2)$$

to find that $b \approx 35202402$. Then the formula $c^2 = a^2 - b^2$ yields $c \approx 7410638$. So at perihelion the speed of Mercury is

$$\frac{2\pi ab}{(a-c)T} \approx 3166628$$

miles per day. We divide by $24 \cdot 3600$ to convert this answer to 36.650789 miles per second. Replace $a - c$ with $a + c$ to find that its speed at aphelion is approximately 24.129956 miles per second.

11.6.59: We are given that the semimajor axis of the Moon's orbit is $a = 238900$ (miles). With eccentricity $e = 0.055$ and period $T = 27.32$ days, we use

$$b^2 = a^2(1 - e^2)$$

to find that $b \approx 238538$. Then the formula $c^2 = a^2 - b^2$ yields $c \approx 13139$. So at perigee the speed of the Moon is

$$\frac{2\pi ab}{(a-c)T} \approx 58053$$

miles per day. We divide by $24 \cdot 3600$ to convert this answer to 0.671911 miles per second. Replace $a - c$ with $a + c$ to find that the speed at apogee is approximately 0.601854 miles per second.

11.6.61: Equation (44), applied to the Earth-Moon system with units of miles and days, yields $(27.32)^2 = \gamma \cdot (238900)^3$. For a satellite with period $T = \frac{1}{12}$ (of a day—two hours), it yields $T^2 = \gamma \cdot r^3$ where r is the radius of the orbit of the satellite. Divide the second of these equations by the first to eliminate γ:

$$\frac{T^2}{(27.32)^2} = \frac{r^3}{(238900)^3},$$

so that

$$r^3 = \frac{(238900)^3}{(12)^2 \cdot (27.32)^2}, \quad \text{and thus} \quad r \approx 5024.68$$

miles, about 1065 miles above the surface of the Earth (where $r \approx 3960$ miles).

11.6.63: With the usual meaning of the symbols, the data given in the problem tell us that $a + c = 4960$ and $a - c = 4060$, which we solve for $a = 4510$ (units are in days and miles). Let T be the period of the satellite in its orbit. Equation (44), applied to the Earth-Moon system, then to the Earth-satellite system, yields

$$(27.32)^2 = \gamma \cdot (238900)^3 \quad \text{and} \quad T^2 = \gamma \cdot (4510)^2,$$

which we solve for $T \approx 0.0708632854$. Multiply by 24 to convert this answer to approximately 1.7007188486 hours—about 1 h 42 min 2.588 s.

11.6.65: We will use Eqs. (37) and (41), which are—respectively—

$$r^2\frac{d\theta}{dt} = h \quad \text{(constant)} \quad \text{and} \quad \frac{d^2r}{dt^2} = \frac{h^2}{r^2}\left(\frac{1}{r} - \frac{1}{pe}\right).$$

706

But we begin with Eq. (42),

$$\mathbf{a} = \left[\frac{d^2r}{dt^2} - r \left(\frac{d\theta}{dt} \right)^2 \right] \mathbf{u}_r.$$

Substitution of Eqs. (37) and (41) then yields

$$\mathbf{a} = \left[\frac{h^2}{r^2} \left(\frac{1}{r} - \frac{1}{pe} \right) - \frac{1}{r^3} \left(r^2 \frac{d\theta}{dt} \right)^2 \right] \mathbf{u}_r$$

$$= \left[\frac{h^2}{r^2} \left(\frac{1}{r} - \frac{1}{pe} \right) - \frac{1}{r^3} h^2 \right] \mathbf{u}_r = \left[\frac{h^2}{r^3} - \frac{h^2}{per^2} - \frac{h^2}{r^3} \right] \mathbf{u}_r = -\frac{h^2}{per^2} \mathbf{u}_r.$$

11.6.67: We begin with Eq. (33),

$$\mathbf{v} = \frac{dr}{dt} \mathbf{u}_r + r \frac{d\theta}{dt} \mathbf{u}_\theta,$$

and differentiate both sides with respect to t:

$$\mathbf{a} = \frac{d\mathbf{v}}{dt} = \left(\frac{d^2r}{dt^2} \mathbf{u}_r + \frac{dr}{dt} \cdot \frac{d\mathbf{u}_r}{dt} \right) + \left(\frac{dr}{dt} \cdot \frac{d\theta}{dt} \mathbf{u}_\theta + r \frac{d^2\theta}{dt^2} \mathbf{u}_\theta + r \frac{d\theta}{dt} \cdot \frac{d\mathbf{u}_\theta}{dt} \right)$$

$$= \frac{d^2r}{dt^2} \mathbf{u}_r + \frac{dr}{dt} \cdot \frac{d\theta}{dt} \mathbf{u}_\theta + \frac{dr}{dt} \cdot \frac{d\theta}{dt} \mathbf{u}_\theta + r \frac{d^2\theta}{dt^2} \mathbf{u}_\theta - r \left(\frac{d\theta}{dt} \right)^2 \mathbf{u}_r$$

$$= \left[\frac{d^2r}{dt^2} - r \left(\frac{d\theta}{dt} \right)^2 \right] \mathbf{u}_r + \left[2 \frac{dr}{dt} \cdot \frac{d\theta}{dt} + r \frac{d^2\theta}{dt^2} \right] \mathbf{u}_\theta$$

$$= \left[\frac{d^2r}{dt^2} - r \left(\frac{d\theta}{dt} \right)^2 \right] \mathbf{u}_r + \left[\frac{1}{r} \cdot \frac{d}{dt} \left(r^2 \frac{d\theta}{dt} \right) \right] \mathbf{u}_\theta.$$

Section 11.7

11.7.1: The graph of $3x + 2y + 10z = 20$ is a plane with intercepts $x = \frac{20}{3}$, $y = 10$, and $z = 2$. The graph produced by the *Mathematica* command

```
ParametricPlot3D[ { x, y, (20 - 3*x - 2*y)/10 }, { x, -7, 7 }, { y, -7, 7 } ];
```

is shown next. (Here and in subsequent plots we have also inserted the option Shading $\rightarrow$ False to suppress shading and therefore display a "mesh plot" of the surface.)

11.7.3: The graph of $x^2 + y^2 = 9$ is a circular cylinder of radius 3 with axis the z-axis. The graph produced by the *Mathematica* command

```
ParametricPlot3D[ { 3*Cos[t], 3*Sin[t], z }, { t, 0, 2*Pi }, { z, -4, 4 } ];
```

is shown next.

11.7.5: The graph of $xy = 4$ is a cylinder whose rulings are parallel to the z-axis. It meets the xy-plane in the hyperbola with equation $xy = 4$. The graph produced by the *Mathematica* command

```
ParametricPlot3D[ {{ x, 4/x, z }, { -x, -4/x, z }}, { x, 1/5, 10 }, { z, -7, 7 } ];
```

is shown next.

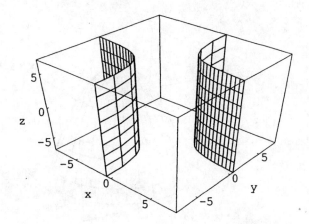

11.7.7: The graph of the equation $z = 4x^2 + y^2$ is an elliptic paraboloid with axis the z-axis and vertex at the origin. It is elliptic because, if $z = a^2$, then the horizontal cross section there has equation $4x^2 + y^2 = a^2$. It is a paraboloid because it meets every vertical plane containing the z-axis in a parabola; for example, it meets the xz-plane in the parabola with equation $z = 4x^2$. The graph produced by the *Mathematica* command

```
ParametricPlot3D[ { r*Cos[t], 2*r*Sin[t], 4*r^2 }, { r, 0, 1 }, { t, 0, 2*Pi } ];
```

is shown next.

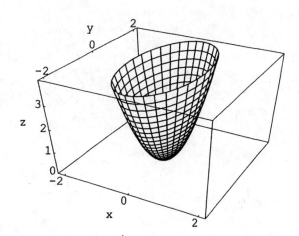

11.7.9: The given equation $z = 4 - x^2 - y^2$ has the polar coordinates form $z = 4 - r^2$, so the graph is a surface of revolution around the z-axis. The graph meets the xz-plane in the parabola $z = 4 - x^2$, so the graph is a circular paraboloid, opening downward, with axis the z-axis, and vertex at $(0, 0, 4)$. The graph produced by the *Mathematica* command

```
ParametricPlot3D[ { r*Cos[t], r*Sin[t], 4 - r^2 }, { r, 0, 3 }, { t, 0, 2*Pi } ];
```

is shown next.

709

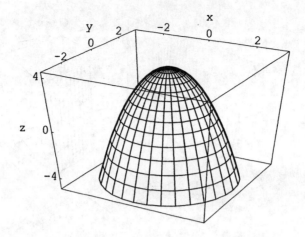

11.7.11: The given equation $2z = x^2 + y^2$ has the polar form $z = \frac{1}{2}r^2$, so the graph is a surface of revolution around the z-axis. It meets the yz-plane in the parabola with equation $z = \frac{1}{2}x^2$, so the surface is a circular paraboloid; its axis is the z-axis and its vertex is at the origin. The graph produced by the *Mathematica* command

```
ParametricPlot3D[ { r*Cos[t], r*Sin[t], (r∧2)/2 }, { t, 0, 2*Pi }, { r, 0, 2 } ];
```

is shown next.

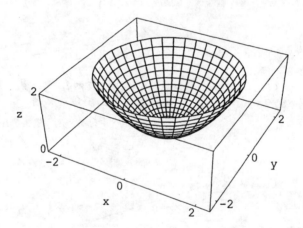

11.7.13: The polar form of the given equation $z^2 = 4(x^2 + y^2)$ is $z = \pm 2r$, so the graph is a surface of revolution around the z-axis. Because z is proportional to r, the graph consists of both nappes of a circular cone with axis the z-axis and vertex at the origin. To produce the graph of the upper nappe, we used the *Mathematica* command

```
ParametricPlot3D[ { r*Cos[t], r*Sin[t], 2*r }, { t, 0, 2*Pi }, { r, 0, 2 } ];
```

and the result is shown next.

710

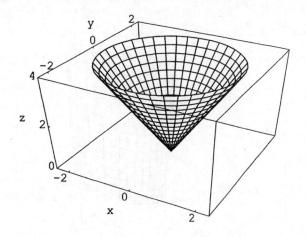

11.7.15: The graph of the equation $x^2 = 4z + 8$ is a cylinder parallel to the y-axis. It is a parabolic cylinder because its trace in the xz-plane is the parabola with equation $z = (x^2 - 8)/4$. It opens upward and its lowest points consist of line $z = -2$, $x = 0$. The graph produced by the *Mathematica* command

```
ParametricPlot3D[ { x, y, (x∧2 - 8)/4 }, { x, -4, 4 }, { y, -4, 4 } ];
```

is shown next.

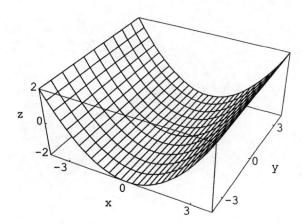

11.7.17: The graph of the equation $4x^2 + y^2 = 4$ is a cylinder parallel to the z-axis. It is an elliptical cylinder because its trace in any horizontal plane is the ellipse with equation $4x^2 + y^2 = 4$. The graph of this surface produced by the *Mathematica* command

```
ParametricPlot3D[ { Cos[t], 2*Sin[t], z }, { t, 0, 2*Pi }, { z, -3, 3 } ];
```

is shown next.

711

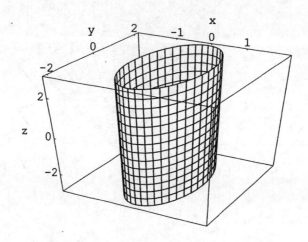

11.7.19: The graph of the equation $x^2 = 4y^2 + 9z^2$ is an elliptical cone with axis the x-axis. The graph generated by the *Mathematica* command

```
ParametricPlot3D[ {{ 6*r, 3*r*Cos[t], 2*r*Sin[t] }, { -6*r, 3*r*Cos[t], 2*r*Sin[t] }},

                  { r, 0, 1.4 }, { t, 0, 2*Pi } ];
```

is shown next.

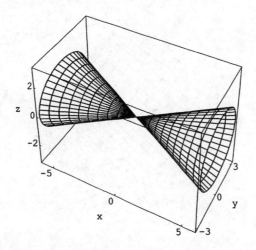

11.7.21: The polar form of the given equation $x^2 + y^2 + 4z = 0$ is $z = -\frac{1}{4}r^2$, so its graph is a paraboloid opening downward, with axis the negative z-axis and vertex at the origin. The graph generated by the *Mathematica* command

```
ParametricPlot3D[ { r*Cos[t], r*Sin[t], -r*r/4 }, { t, 0, 2*Pi }, { r, 0, 3 } ];
```

is shown next.

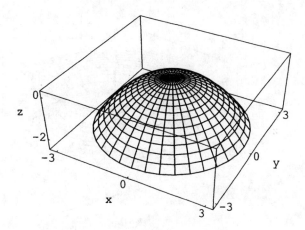

11.7.23: The graph of the equation $x = 2y^2 - z^2$ is a hyperbolic paraboloid with saddle point at the origin. It meets the xz-plane in the parabola $x = -z^2$ with vertex at the origin and opening to the left; it meets the xy-plane in the parabola $x = 2y^2$ with vertex at the origin and opening to the right. The surface meets each plane parallel to the yz-plane in both branches of a hyperbola (except for the yz-plane itself, which it meets in a pair of straight lines that meet at the origin—a degenerate hyperbola). The graph generated by the *Mathematica* command

```
ParametricPlot3D[ { 2*y∧2 - z∧2, y, z }, { y, -1, 1 }, { z, -1, 1 } ];
```

is shown next.

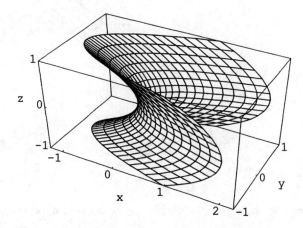

11.7.25: The graph of the equation $x^2 + y^2 - 9z^2 = 9$ is a hyperboloid of one sheet with axis the z-axis. It meets the xy-plane in a circle of radius 3 and meets parallel planes in larger circles. It meets every plane containing the z-axis in both branches of a hyperbola. The graph generated by the *Mathematica* command

```
ParametricPlot3D[ {{ r*Cos[t], r*Sin[t], Sqrt[(r∧2 - 9)/9] },

                   { r*Cos[t], r*Sin[t], -Sqrt[(r∧2 - 9)/9] }},

                 { r, 3, 5 }, { t, 0, 2*Pi } ];
```

is shown next.

713

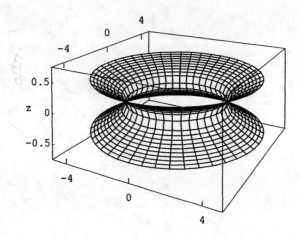

11.7.27: The graph of the equation $y = 4x^2 + 9z^2$ is an elliptic paraboloid opening in the positive y-direction, with axis the nonnegative y-axis and vertex at the origin. The graph generated by the *Mathematica* command

```
ParametricPlot3D[ { 3*r*Cos[t], 36*r^2, 2*r*Sin[t] }, { r,0,1/2 }, { t,0,2*Pi } ];
```

is shown next.

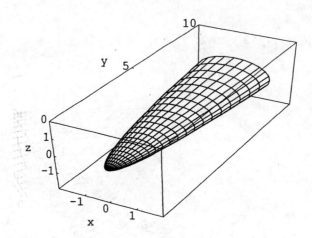

11.7.29: The graph of the equation $y^2 - 9x^2 - 4z^2 = 36$ is a hyperboloid of two sheets with axis the y-axis, center the origin, and intercepts $(0, \pm 6, 0)$. Planes containing the y-axis meet it in both branches of a hyperbola. Planes normal to the y-axis and outside the intercepts meet it in ellipses. The graph generated by the *Mathematica* command

```
ParametricPlot3D[ {{ 2*r*Cos[t], 6*Sqrt[r^2 + 1], 3*r*Sin[t] },

                   { 2*r*Cos[t], -6*Sqrt[r^2 + 1], 3*r*Sin[t] }},

                 { t, 0, 2*Pi }, { r, 0, 1 } ];
```

is shown next.

714

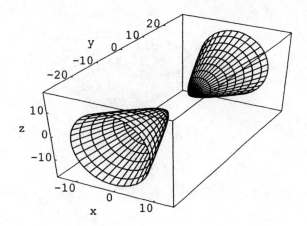

11.7.31: The graph of the curve $x = 2z^2$ (in the xz-plane) is to be rotated around the x-axis. To obtain an equation of the resulting surface, replace z with $(y^2 + z^2)^{1/2}$ to obtain $x = 2(y^2 + z^2)$. The surface is a circular paraboloid opening along the positive x-axis. The graph generated by the *Mathematica* command

```
ParametricPlot3D[ { 2*r^2, r*Cos[t], r*Sin[t] }, { t, 0, 2*Pi }, { r, 0, 2 } ];
```

is shown next.

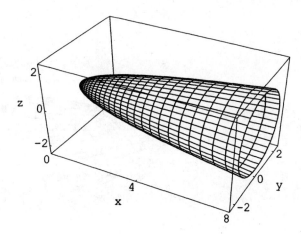

11.7.33: The curve $y^2 - z^2 = 1$ (in the yz-plane) is to be rotated around the z-axis. To obtain an equation of the resulting surface, replace y^2 with $x^2 + y^2$ to obtain $x^2 + y^2 - z^2 = 1$. The surface is a circular hyperboloid of one sheet with axis the z-axis. The graph generated by the *Mathematica* command

```
ParametricPlot3D[ {{ r*Cos[t], r*Sin[t], Sqrt[r^2 - 1] },

                   { r*Cos[t], r*Sin[t], -Sqrt[r^2 - 1] }},

                 { r, 1, 3 }, { t, 0, 2*Pi } ];
```

is shown next.

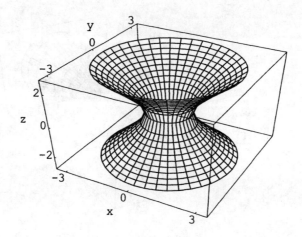

11.7.35: The curve $y^2 = 4x$ (in the xy-plane) is to be rotated around the x-axis. Replace y^2 with $y^2 + z^2$ to obtain an equation of the resulting surface: $y^2 + z^2 = 4x$. The surface is a circular paraboloid opening along the positive x-axis, with axis that axis, and with vertex at the origin. The graph generated by the *Mathematica* command

```
ParametricPlot3D[ { (r∧2)/4, r*Cos[t], r*Sin[t] }, { t, 0, 2*Pi }, { r, 0, 4 } ];
```

is shown next.

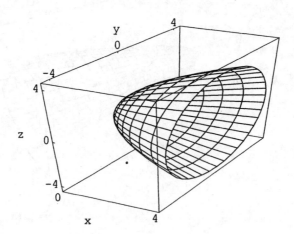

11.7.37: Given: the curve with equation $z = \exp(-x^2)$ in the xz-plane, to be rotated around the z-axis. To obtain an equation of the resulting surface, replace x^2 with $x^2 + y^2$ to obtain $z = \exp\left(-x^2 - y^2\right)$. The graph of this surface generated by the *Mathematica* 3.0 command

```
ParametricPlot3D[ { r*Cos[t], r*Sin[t], Exp[-r∧2] }, { t, 0, 2*Pi }, { r, 0, 2.5 },

                AspectRatio → Automatic, ViewPoint → { 1.3, -2.2, 0.6 } ];
```

is shown next.

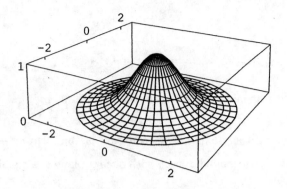

11.7.39: The curve with equation $z = 2x$ (in the xz-plane) is to be rotated around the z-axis. To obtain an equation of the surface thereby generated, replace x with $\pm\sqrt{x^2 + y^2}$. The resulting equation is $z^2 = 4(x^2 + y^2)$ and its graph consists of both nappes of a right circular cone with vertices at the origin and axis the z-axis. The graph generated by the *Mathematica* command

```
ParametricPlot3D[ {{ r*Cos[t], r*Sin[t], 2*r }, { r*Cos[t], r*Sin[t], -2*r }},
                 { t, 0, 2*Pi }, { r, 0, 3 } ];
```

is shown next.

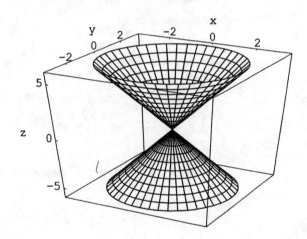

11.7.41: The graph of $x^2 + 4y^2 = 4$ is an elliptical cylinder with centerline the z-axis. Thus its traces in horizontal planes are ellipses with semiaxes 2 and 1.

11.7.43: The graph of $x^2 + 4y^2 + 4z^2 = 4$ is an ellipsoid. A plane parallel to the yz-plane is perpendicular to the x-axis and thus has an equation of the form $x = a$. So the trace of the ellipsoid in such a plane has equations

$$x = a, \quad 4y^2 + 4z^2 = 4 - a^2.$$

Thus the trace is a circle if $|a| < 2$, a single point if $|a| = 2$, and the empty set if $|a| > 2$.

11.7.45: The graph of $z = 4x^2 + 9y^2$ is an elliptical paraboloid opening upward with axis the z-axis and vertex at the origin. A plane parallel to the yz-plane has equation $x = a$ and thus the trace in such a plane

717

has equations

$$x = a, \quad z = 4a^2 + 9y^2.$$

Thus the trace is a parabola opening upward with vertex at $(a, 0, 4a^2)$.

11.7.47: A plane containing the z-axis has an equation of the form $ax + by = 0$. So if $b \neq 0$ then the intersection of the hyperbolic paraboloid $z = xy$ with such a plane has equations

$$y = -\frac{a}{b}x, \quad z = -\frac{a}{b}x^2.$$

Hence the trace is a parabola opening downward if a and b have the same sign, opening upward if they have opposite sign, and is a horizontal line if $a = 0$ (or if $b = 0$). The surface itself resembles the one shown in Fig. 11.7.22 rotated 45° around the z-axis.

11.7.49: The triangles OAC and OBC in Fig. 11.7.1 are congruent because the angles OCA and OCB are right angles, $|OA| = |OB|$ because they are both radii of the same sphere, and $|OC| = |OC|$. So the triangles have matching side-side-angle in the same order.

11.7.51: The intersection I of the plane $z = y$ and the paraboloid $z = x^2 + y^2$ satisfies both equations, and thus lies on the surface with equation

$$y = x^2 + y^2; \quad \text{that is,} \quad x^2 + \left(y - \frac{1}{2}\right)^2 = \frac{1}{4}.$$

This surface is a circular cylinder perpendicular to the xy-plane. Hence the projection of I into that plane lies on the circle C with equations

$$z = 0, \quad x^2 + \left(y - \frac{1}{2}\right)^2 = \frac{1}{4}.$$

It remains to show that all of C is obtained by projecting I into the xy-plane. Suppose that $(x, y, 0)$ lies on C. Then $y = x^2 + y^2$. Let $z = y$. Then

$$x^2 + y^2 = y = z,$$

and therefore (x, y, z) satisfies the equations of both the plane and the paraboloid. Therefore (x, y, z) lies on I. This proves that the projection of I into the xy-plane is *all* of C. Therefore the projection of I into the xy-plane is a circle.

11.7.53: The plane with equation $x + y + z = 1$ and the ellipsoid with equation $x^2 + 4y^2 + 4z^2 = 4$ have intersection that satisfies both equations, so it lies on the surface with equation

$$x^2 + 4y^2 + 4(1 - x - y)^2 = 4; \quad \text{that is,} \quad 5x^2 + 8xy + 8y^2 - 8x - 8y = 0.$$

This surface is a cylinder normal to the xy-plane, so its projection into the xy-plane—which is the projection of the intersection into the xy-plane—has the same equation. By earlier discussions of such equations, the projection must be a conic section. But it cannot be a hyperbola or a parabola because it is clearly a closed curve. (It is not any of the degenerates cases of a conic section because it contains the two points $(0, 0)$ and $(0, 1)$.) Therefore it is an ellipse. To be absolutely certain of this, the *Mathematica* command

```
ContourPlot[ 5*x∧2 + 8*x*y + 8*y∧2 - 8*x - 8*y, { x, -1, 2 }, { y, -1.5, 1.5 },

    Axes → True, AxesOrigin → (0,0), AxesLabel → { x, y }, Contours → 3,

    ContourShading → False, PlotPoints → 47, PlotRange → { -0.01, 0.01 } ];
```

generates a plot of this curve, and it's shown next.

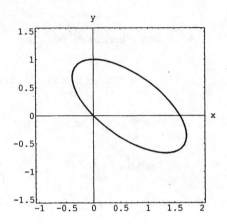

Section 11.8

11.8.1: The formulas in (3) immediately yield

$$x = r\cos\theta = 1 \cdot \cos\frac{\pi}{2} = 0, \qquad y = r\sin\theta = 1 \cdot \sin\frac{\pi}{2} = 1, \qquad z = 2.$$

Answer: $(0, 1, 2)$.

11.8.3: If $(r, \theta, z) = (2, 3\pi/4, 3)$, then $(x, y, z) = (r\cos\theta, r\sin\theta, z) = \left(-\sqrt{2}, \sqrt{2}, 3\right)$.

11.8.5: If $(r, \theta, z) = (2, \pi/3, -5)$, then $(x, y, z) = (r\cos\theta, r\sin\theta, z) = \left(1, \sqrt{3}, -5\right)$.

11.8.7: Given $(\rho, \phi, \theta) = (2, 0, \pi)$, the equations in (6) immediately yield

$$x = \rho\sin\phi\cos\theta = 0, \qquad y = \rho\sin\phi\sin\theta = 0, \qquad z = \rho\cos\phi = 2.$$

Answer: $(0, 0, 2)$.

11.8.9: If $(\rho, \phi, \theta) = (3, \pi/2, \pi)$, then

$$(x, y, z) = (\rho \sin\phi \cos\theta, \; \rho \sin\phi \sin\theta, \; \rho \cos\phi) = (-3, 0, 0).$$

11.8.11: If $(\rho, \phi, \theta) = (2, \pi/3, 3\pi/2)$, then

$$(x, y, z) = (\rho \sin\phi \cos\theta, \; \rho \sin\phi \sin\theta, \; \rho \cos\phi) = \left(0, -\sqrt{3}, 1\right).$$

11.8.13: If $P(x, y, z) = P(0, 0, 5)$, then P has cylindrical coordinates

$$r = \sqrt{x^2 + y^2} = 0, \qquad \theta \text{ (therefore) arbitrary, and} \qquad z = 5.$$

So one correct answer is $(0, 0, 5)$. Also P has spherical coordinates

$$\rho = \sqrt{x^2 + y^2 + z^2} = \pm 5, \qquad \phi = 0 \text{ (if } \rho > 0), \qquad \theta \text{ (therefore) arbitrary.}$$

Thus one correct answer is $(5, 0, 0)$. Another is $(-5, \pi, \pi/2)$.

11.8.15: Cylindrical $\left(\sqrt{2}, \pi/4, 0\right)$, spherical $\left(\sqrt{2}, \pi/2, \pi/4\right)$.

11.8.17: Given: the point with Cartesian coordinates $P(1, 1, 1)$. Its cylindrical coordinates are $\left(\sqrt{2}, \pi/4, 1\right)$. To find the spherical coordinates of P, we compute

$$\rho = \sqrt{1^2 + 1^2 + 1^2} = \sqrt{3} \qquad \text{and} \qquad \theta = \frac{\pi}{4}.$$

To find ϕ, Fig. 11.8.10 makes it clear that

$$\cos\phi = \frac{z}{\rho} = \frac{1}{\sqrt{3}} = \frac{\sqrt{3}}{3},$$

and hence

$$\phi = \arccos\left(\frac{\sqrt{3}}{3}\right) \approx 0.9553166181245093;$$

that's approximately $54.7356103162°$, about $54°\,44'\,8.197''$. Thus the spherical coordinates of P are

$$\left(\sqrt{3}, \; \cos^{-1}\frac{\sqrt{3}}{3}, \; \frac{\pi}{4}\right).$$

Other ways to express ϕ are

$$\phi = \sin^{-1}\frac{\sqrt{6}}{3} \qquad \text{and} \qquad \phi = \tan^{-1}\left(\sqrt{2}\right).$$

11.8.19: The given point P having the Cartesian coordinates $(2, 1, -2)$ has cylindrical coordinates $\left(\sqrt{5}, \tan^{-1}\left(\frac{1}{2}\right), -2\right)$. To find its spherical coordinates, it's clear that $\rho = 3$. Imagine the vertical plane that contains the z-axis and $P(2, 1, -2)$. If you mark the points $O(0, 0, 0)$ and $Q(2, 1, 0)$, draw the z-axis and the triangle OAP, then the angle ϕ is the obtuse angle reaching from the positive z-axis to the hypotenuse of triangle OAP. Let α be the acute angle of that triangle at O. Note that OA has length $r = \sqrt{5}$, that OP has length $\rho = 3$, and that AP has length $-z = 2$. Because

$$\phi = \alpha + \frac{\pi}{2}, \quad \sin\alpha = \frac{2}{3}, \quad \text{and} \quad \cos\alpha = \frac{\sqrt{5}}{3},$$

it now follows that

$$\cos\phi = \cos\left(\frac{\pi}{2} + \alpha\right) = -\sin\alpha = -\frac{2}{3}.$$

Therefore the spherical coordinates of P are

$$\left(3, \ \cos^{-1}\left(-\frac{2}{3}\right), \ \tan^{-1}\left(\frac{1}{2}\right)\right) \approx (3, \ 2.300523983822, \ 0.463647609001).$$

11.8.21: The cylindrical coordinates of $P(3, 4, 12)$ are $\left(5, \arctan\left(\frac{4}{3}\right), 12\right)$ and its spherical coordinates are $\left(13, \arcsin\left(\frac{5}{13}\right), \arctan\left(\frac{4}{3}\right)\right)$.

11.8.23: The graph of the cylindrical equation $r = 5$ is the cylinder of radius 5 with axis the z-axis.

11.8.25: The graph of the cylindrical *or* spherical equation $\theta = \pi/4$ is the vertical plane with Cartesian equation $y = x$.

11.8.27: The graph of the spherical equation $\phi = \pi/6$ consists of both nappes of the cone with Cartesian equation $z^2 = 3x^2 + 3y^2$ if ρ may be negative. If $\rho \geq 0$, then the graph consists of the upper nappe alone.

11.8.29: The graph of the spherical equation $\phi = \pi/2$ is the xy-plane.

11.8.31: The cylindrical equation $z^2 + 2r^2 = 4$ has the same graph as does the Cartesian equation $2x^2 + 2y^2 + z^2 = 4$. It is an ellipsoid (actually, a prolate spheroid) with intercepts $(\pm\sqrt{2}, 0, 0)$, $(0, \pm\sqrt{2}, 0)$, and $(0, 0, \pm 2)$.

11.8.33: The graph of the cylindrical equation $r = 4\cos\theta$ is the same as the graph of the Cartesian equation

$$x^2 + y^2 = 4x; \quad \text{that is,} \quad (x - 2)^2 + y^2 = 4.$$

Therefore its graph is a circular cylinder parallel to the z-axis, of radius 2, and centerline the line with Cartesian equations $x = 2$, $y = 0$.

11.8.35: The cylindrical equation $r^2 - 4r + 3 = 0$ can be written in the form

$$(r - 1)(r - 3) = 0,$$

so that $r = 1$ or $r = 3$. The graph consists of all points that satisfy either equation. Hence the graph consists of two concentric circular cylinders, each with axis the z-axis; their radii are 1 and 3.

11.8.37: The cylindrical equation $z^2 = r^4$ can be written in Cartesian form as

$$z = \pm(x^2 + y^2), \quad \text{so that} \quad z = x^2 + y^2 \quad \text{or} \quad z = -(x^2 + y^2).$$

The graph consists of all points that satisfy either of the last two equations, hence it consists of two circular paraboloids, each with axis the z-axis, vertex at the origin; one opens upward and the other opens downward.

11.8.39: The Cartesian equation $x^2 + y^2 + z^2 = 25$ has cylindrical form $r^2 + z^2 = 25$ and spherical form $\rho^2 = 25$; that is, $\rho = \pm 5$ (but the graph of $\rho = -5$ coincides with the graph of $\rho = 5$, so $\rho = 5$ is also a correct answer).

11.8.41: The Cartesian equation $x + y + z = 1$ takes the cylindrical form

$$r\cos\theta + r\sin\theta + z = 1$$

and the spherical form

$$\rho\sin\phi\cos\theta + \rho\sin\phi\sin\theta + \rho\cos\phi = 1.$$

11.8.43: The Cartesian equation $x^2 + y^2 + z^2 = x + y + z$ takes the cylindrical form

$$r^2 + z^2 = r\cos\theta + r\sin\theta + z$$

and the spherical form

$$\rho^2 = \rho\sin\phi\cos\theta + \rho\sin\phi\sin\theta + \rho\cos\phi.$$

The common factor ρ may be cancelled from both sides of the last equation without loss of any points on the graph.

11.8.45: The surface is the part of the cylinder of radius 3 and centerline the z-axis that lies between the planes $z = -1$ and $z = 1$. The *Mathematica* command

```
ParametricPlot3D[ { 3*Cos[t], 3*Sin[t], z }, { t, 0, 2*Pi }, { z, -1, 1 } ];
```

generates the graph, shown next.

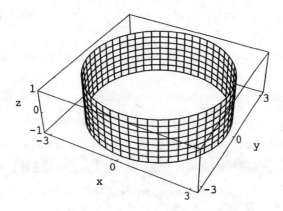

11.8.47: The surface is the part of the sphere of radius 2 and center the origin that lies between the two horizontal planes $z = -1$ and $z = 1$. The *Mathematica* command

```
ParametricPlot3D[ { 2*Sin[phi]*Cos[t], 2*Sin[phi]*Sin[t], 2*Cos[phi] },

              { t, 0, 2*Pi }, { phi, Pi/3, 2*Pi/3 } ];
```

generates the graph, shown next.

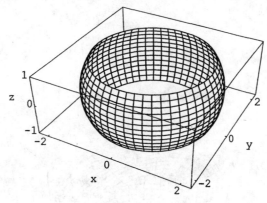

11.8.49: The solid lies between the horizontal planes $z = -2$ and $z = 2$, is bounded on the outside by the cylinder of radius 3 with centerline the z-axis, and is bounded on the inside by the cylinder of radius 1 with centerline the z-axis. To generate an image, we will create the inner and outer bounding surfaces, the top, and display them simultaneously. For the outer cylinder, we use the *Mathematica* command

```
f1 = ParametricPlot3D[ { 3*Cos[t], 3*Sin[t], z }, { t, 0, 2*Pi }, { z, -2, 2 } ];
```

For the inner cylinder, we use

```
f2 = ParametricPlot3D[ { 1*Cos[t], 1*Sin[t], z }, { t, 0, 2*Pi }, { z, -2, 2 } ];
```

For the top, we use

```
f3 = ParametricPlot3D[ { r*Cos[t], r*Sin[t], 2 }, { t, 0, 2*Pi }, { r, 1, 3 } ];
```

Finally, the command `Show[ f1, f2, f3 ]` produces an image of the solid.

11.8.51: The solid described by the spherical inequality $3 \leq \rho \leq 5$ is the region between two concentric spheres centered at the origin, one of radius 3 and the other of radius 5. We will display the half of this solid for which $y \geq 0$ (otherwise, you wouldn't be able to tell that the solid has a hollow core). We use *Mathematica* to draw the outer hemisphere, then the inner hemisphere, then make it appear "solid" by covering the gap between the two with an annulus. Here are the commands:

```
f1 = ParametricPlot3D[ { 5*Sin[phi]*Cos[t], 5*Sin[phi]*Sin[t], 5*Cos[phi] },
            { phi, 0, Pi }, { t, 0, Pi } ];

f2 = ParametricPlot3D[ { 3*Sin[phi]*Cos[t], 3*Sin[phi]*Sin[t], 3*Cos[phi] },
            { phi, 0, Pi }, { t, 0, Pi } ];

f3 = ParametricPlot3D[ { r*Cos[t], 0, r*Sin[t] }, { r, 3, 5 }, { t, 0, 2*Pi } ];

Show[ f1, f2, f3 ];
```

The figure is next.

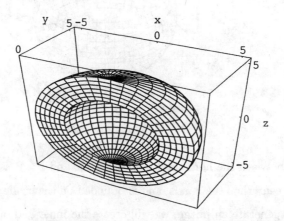

11.8.53: Given $z = x^2$, a curve in the xz-plane to be rotated around the z-axis, replace x^2 with $x^2 + y^2$ to obtain $z = x^2 + y^2$. In cylindrical coordinates, the equation is $z = r^2$.

11.8.55: A central cross section of the sphere-with-hole, shown after this solution, makes it clear that the figure is described in cylindrical coordinates by $1 \leq r \leq 2$ and $0 \leq \theta \leq 2\pi$; the third cylindrical coordinate z is determined by the cylindrical equation of the sphere, $r^2 + z^2 = 4$, so that

$$-\sqrt{4 - r^2} \leq z \leq \sqrt{4 - r^2}.$$

The figure also makes it clear that the spherical coordinates of the figure satisfy the inequalities $0 \leq \theta \leq 2\pi$ and $\pi/6 \leq \phi \leq 5\pi/6$. The value of the third spherical coordinate ρ is $\csc\phi$ on the surface of the cylinder and 2 on the surface of the sphere, and hence $\csc\phi \leq \rho \leq 2$.

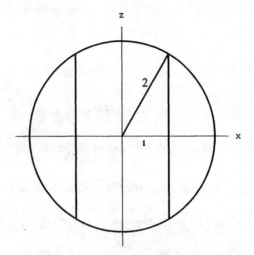

11.8.57: Take $\rho = 3960$ for the radius of the earth throughout this solution. The spherical coordinates of Fairbanks are then (ρ, ϕ, θ) where

$$\phi = \frac{7\pi}{50} \quad \text{and} \quad \theta = \frac{4243\pi}{3600}.$$

Then the formulas in (6) yield its Cartesian coordinates:

$$x_1 \approx -1427.537970311664, \quad y_1 \approx -897.229805845151, \quad z_1 \approx 3583.115127765437.$$

Similarly, the angular spherical coordinates of St. Petersburg are

$$\phi = \frac{1003\pi}{6000} \quad \text{and} \quad \theta = \frac{3043\pi}{18000},$$

and hence its Cartesian coordinates are

$$x_2 \approx 1711.895024130100, \quad y_2 \approx 1005.568095522826, \quad z_2 \approx 3426.346192611774.$$

Thus the straight-line distance between the two cities is

$$d = \sqrt{(x_1 - x_2)^2 + (y_1 - y_2)^2 + (z_1 - z_2)^2} \approx 3674.405513694586.$$

Now we use the triangle with vertices at O (the center of the earth), F (Fairbanks), and S (St. Petersburg) and the law of cosines to find the angle α between OF and OS:

$$d^2 = \rho^2 + \rho^2 - 2\rho^2 \cos\alpha;$$

$$\cos\alpha = \frac{2\rho^2 - d^2}{2\rho^2} \approx 0.569519185572;$$

$$\alpha \approx 0.964875538222.$$

Therefore the great circle distance from Fairbanks to St. Petersburg is $\rho\alpha \approx 3820.907131357640$ miles, approximately 6149.153966407630 kilometers.

11.8.59: Take $r = 3960$ as the radius of the earth. Recall from the solution of Problem 57 that the Cartesian coordinates of Fairbanks are $F(x_1, y_1, z_1)$ where

$$x_1 \approx -1427.537970311664, \quad y_1 \approx -897.229805845151, \quad z_1 \approx 3583.115127765437$$

and that those of St. Petersburg are $S(x_2, y_2, z_2)$ where

$$x_2 \approx 1711.895024130100, \quad y_2 \approx 1005.568095522826, \quad z_2 \approx 3426.346192611774.$$

Thus a normal to the plane $\mathcal{P}$ containing the center of the earth O and the two points F and P is

$$\mathbf{n} = \langle n_1, n_2, n_3 \rangle = \overrightarrow{OF} \times \overrightarrow{OS} = \langle -6677286.184221, \ 11025156.247493, \ 100476.602035 \rangle.$$

The intersection of this plane with the surface of the earth is the great-circle route between F and S. The point on this route closest to the north pole will be the point where the plane $\mathcal{Q}$ containing the z-axis and normal to $\mathcal{P}$ intersects the route. (We are in effect constructing the geodesics from F and P to the north pole and bisecting the angle between them with $\mathcal{Q}$.) Observe that $\mathcal{Q}$ has an equation of the form $Ax + By = 0$, thus a normal of the form $\mathbf{m} = \langle A, B, 0 \rangle$ having the property that $\mathbf{m} \cdot \mathbf{n} = 0$. Indeed, we choose $B = 1$ and solve for A to find that $A \approx 1.651143285358$. Then we asked *Mathematica* to solve for the point $C(x, y, z)$ on the plane's great circle route closest to the north pole:

```
Solve[ { A*x + B*y == 0, n1*x + n2*y + n3*z == 0, x*x + y*y + z*z == r*r },

    { x, y, z } ]
```

The positive value obtained for z was $z \approx 3959.879690341724$. Let $N(0, 0, r)$ denote the north pole of the earth. Then the angle γ between between OC and ON is given by

$$\gamma = \arccos\left(\frac{z}{r}\right) \approx 0.999969618773.$$

Finally, the great circle distance from C to N is $d = r\gamma \approx 30.86838704$, a bit less than 31 miles.

11.8.61: The cone has the following description in spherical coordinates:

$$0 \leqq \rho \leqq \sqrt{R^2 + H^2}, \qquad 0 \leqq \theta \leqq 2\pi, \qquad \phi = \arctan\left(\frac{R}{H}\right).$$

11.8.63: Begin with the equation of the given circle,

$$(y - a)^2 + z^2 = b^2 \qquad (x = 0),$$

726

and replace y with r to get the cylindrical-coordinates equation

$$(r - a)^2 + z^2 = b^2$$

of the torus. Then substitute $\sqrt{x^2 + y^2}$ for r and rationalize to get the rectangular-coordinates equation

$$4a^2(x^2 + y^2) = (x^2 + y^2 + z^2 + a^2 - b^2)^2$$

of the torus. We recognize the combination $x^2 + y^2 = r^2$ and $x^2 + y^2 + z^2 = \rho^2$ and substitute to obtain

$$4a^2 r^2 = (\rho^2 + a^2 - b^2)^2.$$

Finally, we take positive square roots and substitute $\rho \sin \phi$ for r to obtain the spherical-coordinates equation

$$2a\rho \sin \phi = \rho^2 + a^2 - b^2$$

of the torus.

Chapter 11 Miscellaneous Problems

11.M.1: Note that $\overrightarrow{PM} = \overrightarrow{MQ}$. Hence

$$\frac{1}{2}\left(\overrightarrow{AP} + \overrightarrow{AQ}\right) = \frac{1}{2}\left(\overrightarrow{AM} - \overrightarrow{PM} + \overrightarrow{AM} + \overrightarrow{MQ}\right) = \overrightarrow{AM}.$$

11.M.3: Given the distinct points P and Q in space, let L be the straight line through both. Then a vector equation of L is $\mathbf{r}(t) = \overrightarrow{OP} + t\overrightarrow{PQ}$. First suppose that R is a point on L. Then for some scalar t,

$$\overrightarrow{OR} = \overrightarrow{OP} + t\overrightarrow{PQ} = \overrightarrow{OP} + t\left(\overrightarrow{OQ} - \overrightarrow{OP}\right) = (1 - t)\overrightarrow{OP} + t\overrightarrow{OQ} = a\overrightarrow{OP} + b\overrightarrow{OQ}$$

where $a = 1 - t$ and $b = t$; moreover, $a + b = 1$, as desired. Next suppose that there exist scalars a and b such that $a + b = 1$ and

$$\overrightarrow{OR} = a\overrightarrow{OP} + b\overrightarrow{OQ}.$$

Let $t = b$, so that $1 - t = a$. Then

$$\overrightarrow{OR} = (1 - t)\overrightarrow{OP} + t\overrightarrow{OQ} = \overrightarrow{OP} + t\left(\overrightarrow{OQ} - \overrightarrow{OP}\right) = \overrightarrow{OP} + t\overrightarrow{PQ}.$$

Therefore R lies on the line L. Consequently an alternative form of the vector equation of L is

$$\mathbf{r}(t) = t\overrightarrow{OP} + (1 - t)\overrightarrow{OQ}.$$

11.M.5: Given $P(x_0, y_0)$, $Q(x_1, y_1)$, and $R(x_2, y_2)$, we think of $\overrightarrow{PQ}$ and $\overrightarrow{PR}$ as vectors in space, so that

$$\overrightarrow{PQ} = \langle x_1 - x_0, \, y_1 - y_0, \, 0 \rangle \quad \text{and} \quad \overrightarrow{PR} = \langle x_2 - x_0, \, y_2 - y_0, \, 0 \rangle.$$

By Eq. (10) in Section 11.3, the area A of the triangle PQR is half the magnitude of $\overrightarrow{PQ} \times \overrightarrow{OR}$. But

$$\overrightarrow{PQ} \times \overrightarrow{PR} = \begin{vmatrix} \mathbf{i} & \mathbf{j} & \mathbf{k} \\ x_1 - x_0 & y_1 - y_0 & 0 \\ x_2 - x_0 & y_2 - y_0 & 0 \end{vmatrix} = \left[(x_1 - x_0)(y_2 - y_0) - (x_2 - x_0)(y_1 - y_0) \right] \mathbf{k}.$$

Therefore

$$A = \frac{1}{2} \left| (x_1 - x_0)(y_2 - y_0) - (x_2 - x_0)(y_1 - y_0) \right|.$$

11.M.7: A vector equation of the line is

$$\mathbf{r}(t) = \overrightarrow{OP_1} + t\overrightarrow{P_1 P_2} = \langle 1, -1, 2 \rangle + t\langle 2, 3, -3 \rangle.$$

Therefore the line has parametric equations

$$x = 1 + 2t, \quad y = -1 + 3t, \quad z = 2 - 3t, \quad -\infty < t < +\infty$$

and symmetric equations

$$\frac{x-1}{2} = \frac{y+1}{3} = \frac{z-2}{-3}.$$

11.M.9: The line L_1 with symmetric equations $x - 1 = 2(y+1) = 3(z-2)$ contains the points $P_1(1, -1, 2)$ and $Q_1(7, 2, 4)$, and thus L_1 is parallel to the vector $\mathbf{u}_1 = \langle 6, 3, 2 \rangle$. The line L_2 with symmetric equations $x - 3 = 2(y - 1) = 3(z + 1)$ contains the points $P_2(3, 1, -1)$ and $Q_2(9, 4, 1)$, and thus L_2 is parallel to the vector $\mathbf{u}_2 = \langle 6, 3, 2 \rangle = \mathbf{u}_1$. Therefore L_1 and L_2 are parallel. The vector $\mathbf{v} = \overrightarrow{P_1 P_2} = \langle 2, 2, -3 \rangle$ lies in the plane $\mathcal{P}$ containing L_1 and L_2, as does $\mathbf{u}_1$, and therefore a normal to $\mathcal{P}$ is

$$\mathbf{n} = \mathbf{u}_1 \times \mathbf{v} = \begin{vmatrix} \mathbf{i} & \mathbf{j} & \mathbf{k} \\ 6 & 3 & 2 \\ 2 & 2 & -3 \end{vmatrix} = \langle -13, 22, 6 \rangle.$$

Therefore an equation of $\mathcal{P}$ has the form $13x - 22y - 6z = D$. The point $P_1(1, -1, 2)$ lies in the plane $\mathcal{P}$, and therefore $D = 13 \cdot 1 - 22 \cdot (-1) - 6 \cdot 2 = 23$. Therefore an equation of $\mathcal{P}$ is $13x - 22y - 6z = 23$.

11.M.11: Given the four points $A(2, 3, 2)$, $B(4, 1, 0)$, $C(-1, 2, 0)$, and $D(5, 4, -2)$, we are to find an equation of the plane containing A and B and parallel to $\overrightarrow{CD} = \langle 6, 2, -2 \rangle$. Because $\overrightarrow{AB} = \langle 2, -2, -2 \rangle$ is not parallel to $\overrightarrow{CD}$, a vertor normal to the plane is

$$\mathbf{m} = \overrightarrow{AB} \times \overrightarrow{CD} = \begin{vmatrix} \mathbf{i} & \mathbf{j} & \mathbf{k} \\ 2 & -2 & -2 \\ 6 & 2 & -2 \end{vmatrix} = \langle 8, -8, 16 \rangle;$$

to simplify matters slightly we will use the parallel vector $\mathbf{n} = \langle 1, -1, 2 \rangle$. Therefore the plane has a Cartesian equation of the form $x - y + 2z = K$. Because $A(2, 3, 2)$ lies in the plane, $K = 2 - 3 + 2 \cdot 2 = 3$. Answer: $x - y + 2z = 3$.

11.M.13: A vector normal to the plane $\mathcal{P}$ is $\mathbf{n} = \langle a, b, c \rangle$ and $\overrightarrow{OP_0} = \langle x_0, y_0, z_0 \rangle$ connects the origin to $\mathcal{P}$. Hence the distance D from the origin to $\mathcal{P}$ is the length of the projection of $\overrightarrow{OP_0}$ in the direction of $\mathbf{n}$. But the projection of $\overrightarrow{OP_0}$ in the direction of $\mathbf{n}$ is

$$\frac{\overrightarrow{OP_0} \cdot \mathbf{n}}{|\mathbf{n}|} = \frac{ax_0 + by_0 + cz_0}{\sqrt{a^2 + b^2 + c^2}} = \frac{-d}{\sqrt{a^2 + b^2 + c^2}}.$$

Therefore $D = \dfrac{|d|}{\sqrt{a^2 + b^2 + c^2}}$.

11.M.15: The planes have the common normal $\mathbf{n} = \langle 2, -1, 2 \rangle$, a point on the first plane $\mathcal{P}_1$ is $P_1(2, 0, 0)$, and a point on the second plane $\mathcal{P}_2$ is $P_2(7, 1, 0)$. So a vector that connects the two planes is $\mathbf{c} = \overrightarrow{P_1 P_2} = \langle 5, 1, 0 \rangle$. So the distance D between $\mathcal{P}_1$ and $\mathcal{P}_2$ is the absolute value of the projection of $\mathbf{c}$ in the direction of $\mathbf{n}$; that is,

$$D = \frac{|\mathbf{n} \cdot \mathbf{c}|}{|\mathbf{n}|} = \frac{9}{3} = 3.$$

11.M.17: Given the isosceles triangle with sides AB and AC of equal length and M the midpoint of the third side BC, let $\mathbf{u} = \overrightarrow{BM} = \overrightarrow{MC}$ and let $\mathbf{v} = \overrightarrow{AM}$. Then $\overrightarrow{AB} = \mathbf{v} - \mathbf{u}$ and $\overrightarrow{AC} = \mathbf{v} + \mathbf{u}$. Now

$$|\mathbf{v} - \mathbf{u}| = |\mathbf{v} + \mathbf{u}|, \quad \text{so}$$

$$(\mathbf{v} - \mathbf{u}) \cdot (\mathbf{v} - \mathbf{u}) = (\mathbf{v} + \mathbf{u}) \cdot (\mathbf{v} + \mathbf{u});$$

$$\mathbf{v} \cdot \mathbf{v} - 2\mathbf{v} \cdot \mathbf{u} + \mathbf{u} \cdot \mathbf{u} = \mathbf{v} \cdot \mathbf{v} + 2\mathbf{v} \cdot \mathbf{u} + \mathbf{u} \cdot \mathbf{u}.$$

Therefore $\mathbf{v} \cdot 2\mathbf{u} = 0$, and therefore the segment AM is perpendicular to the segment BC.

11.M.19: Given $\mathbf{a}(t) = \langle \sin t, -\cos t \rangle$, the velocity vector is

$$\mathbf{v}(t) = \langle c_1 - \cos t, c_2 - \sin t \rangle.$$

But $\mathbf{v}(0) = \langle -1, 0 \rangle = \langle c_1 - 1, c_2 \rangle$, and it follows that $c_1 = c_2 = 0$. So $\mathbf{v}(t) = \langle -\cos t, -\sin t \rangle$. Hence the position vector of the particle is

$$\mathbf{r}(t) = \langle k_1 - \sin t, \ k_2 + \cos t \rangle.$$

But $\langle 0, 1 \rangle = \mathbf{r}(0) = \langle k_1, k_2 + 1 \rangle$, so that $k_1 = k_2 = 0$. Thus $\mathbf{r}(t) = \langle -\sin t, \cos t \rangle$. The parametric equations of the motion of the particle are $x(t) = -\sin t$, $y(t) = \cos t$. Because $x^2 + y^2 = 1$ for all t, the particle moves in a circle of radius 1 centered at the origin.

11.M.21: The trajectory of the projectile fired by the gun is given by

$$x_1(t) = (320 \cos \alpha)t, \quad y_1(t) = -16t^2 + (320 \sin \alpha)t.$$

The trajectory of the moving target is given by

$$x_2(t) = 160 + 80t, \quad y_2(t) \equiv 0.$$

We require the angle of elevation α so that, at some time $T > 0$,

$$y_1(T) = 0 \quad \text{and} \quad x_1(T) = x_2(T); \quad \text{that is,}$$

$$-16T^2 + (320 \sin \alpha)T = 0 \quad \text{and} \quad (320 \cos \alpha)T = 160 + 80T.$$

It follows that $T = 20 \sin \alpha$ and, consequently, that

$$6400 \sin \alpha \cos \alpha = 160 + 1600 \sin \alpha;$$

$$40 \sin \alpha \cos \alpha = 1 + 10 \sin \alpha.$$

Let $u = \sin \alpha$. Then the last equation yields

$$40u\sqrt{1 - u^2} = 1 + 10u;$$

$$1600u^2(1 - u^2) = 1 + 20u + 100u^2;$$

$$1600u^4 - 1500u^2 + 20u + 1 = 0.$$

The graph of the last equation shows solutions between 0 and $\pi/2$ near 0.03 and 0.96. Five iterations of Newton's method with these starting values yields the two solutions

$$u_1 \approx 0.0333580866275847 \quad \text{and} \quad u_2 \approx 0.9611546539773131.$$

These correspond to the two angles

$$\alpha_1 \approx 0.033364276320 \quad (\text{about } 1°\, 54'\, 41.876'') \quad \text{and}$$

$$\alpha_2 \approx 1.291155565633 \quad (\text{about } 73°\, 58'\, 39.953'').$$

11.M.23: Let $\mathbf{r}(t) = \langle t, t^2, \frac{4}{3}t^{3/2} \rangle$. Then

$$\mathbf{v}(t) = \mathbf{r}'(t) = \langle 1, 2t, 2t^{1/2} \rangle,$$

so that $\mathbf{v}(1) = \langle 1, 2, 2 \rangle$. Also

$$\mathbf{a}(t) = \mathbf{r}''(t) = \langle 0, 2, t^{-1/2} \rangle,$$

and thus $\mathbf{a}(1) = \langle 0, 2, 1 \rangle$. Moreover,

$$\mathbf{v}(1) \times \mathbf{a}(1) = \begin{vmatrix} \mathbf{i} & \mathbf{j} & \mathbf{k} \\ 1 & 2 & 2 \\ 0 & 2 & 1 \end{vmatrix} = \langle -2, -1, 2 \rangle,$$

and thus $|\mathbf{v}(1) \times \mathbf{a}(1)| = 3$. Thus

$$a_T(1) = \frac{\mathbf{v}(1) \cdot \mathbf{a}(1)}{v(1)} = \frac{6}{3} = 2,$$

$$\kappa(1) = \frac{|\mathbf{v}(1) \times \mathbf{a}(1)|}{[v(1)]^3} = \frac{3}{27} = \frac{1}{9}, \quad \text{and}$$

$$a_N(1) = \frac{|\mathbf{v}(1) \times \mathbf{a}(1)|}{v(1)} = \frac{3}{3} = 1.$$

11.M.25: A vector normal to the plane is

$$\mathbf{n} = \mathbf{v}_1 \times \mathbf{v}_2 = \begin{vmatrix} \mathbf{i} & \mathbf{j} & \mathbf{k} \\ a_1 & b_1 & c_1 \\ a_2 & b_2 & c_2 \end{vmatrix} = \langle b_1 c_2 - b_2 c_1, a_2 c_1 - a_1 c_2, a_1 b_2 - a_2 b_1 \rangle.$$

Hence the plane has equation

$$(b_1 c_2 - b_2 c_1)(x - x_0) + (a_2 c_1 - a_1 c_2)(y - y_0) + (a_1 b_2 - a_2 b_1)(z - z_0) = 0.$$

But this is exactly the equation you obtain when the matrix in the equation

$$\begin{vmatrix} x - x_0 & y - y_0 & z - z_0 \\ a_1 & b_1 & c_1 \\ a_2 & b_2 & c_2 \end{vmatrix} = 0$$

given in Problem 25 is expanded along its first row.

731

11.M.27: Let $\mathbf{r}(t) = \langle\, t,\, t^2,\, t^3\,\rangle$. Then

$$\mathbf{r}'(t) = \langle\, 1,\, 2t,\, 3t^2\,\rangle, \quad \text{so that} \quad \mathbf{r}'(1) = \langle\, 1,\, 2,\, 3\,\rangle.$$

Also

$$\mathbf{r}''(t) = \langle\, 0,\, 2,\, 6t\,\rangle, \quad \text{and so} \quad \mathbf{r}''(1) = \langle\, 0,\, 2,\, 6\,\rangle.$$

Thus by the results in Problems 25 and 26, an equation of the osculating plane to the twisted cubic $\mathbf{r}(t)$ at the point $\mathbf{r}(1)$ is

$$\begin{vmatrix} x-1 & y-1 & z-1 \\ 1 & 2 & 3 \\ 0 & 2 & 6 \end{vmatrix} = 0.$$

Expansion of this determinant along its first row yields the equation of the osculating plane in more conventional form:

$$6(x-1) - 6(y-1) + 2(z-1) = 0; \quad \text{that is,} \quad 3x - 3y + z = 1.$$

11.M.29: Equation (6) in Section 11.8, with $\rho = 1$, yields

$$x = \sin\phi\,\cos\theta, \quad y = \sin\phi\,\sin\theta, \quad z = \cos\phi.$$

Mathematica can solve this problem for us.

```
x[t_] := Sin[phi[t]]*Cos[theta[t]]

y[t_] := Sin[phi[t]]*Sin[theta[t]]

z[t_] := Cos[phi[t]]

term1 = (x'[t])∧2

    (Cos[phi[t]] Cos[theta[t]] phi'[t] - Sin[phi[t]] Sin[theta[t]] theta'[t])²

term2 = (y'[t])∧2

    (Cos[phi[t]] Sin[theta[t]] phi'[t] + Cos[theta[t]] Sin[phi[t]] theta'[t])²

term3 = (z'[t])∧2

    Sin[phi[t]]² phi't]²

Simplify[ term1 + term2 + term3 ]

    phi'[t]² + Sin[phi[t]]² theta'[t]²
```

That is,

$$\left(\frac{dx}{dt}\right)^2 + \left(\frac{dy}{dt}\right)^2 + \left(\frac{dz}{dt}\right)^2 = \left(\frac{d\phi}{dt}\right)^2 + (\sin^2\phi)\left(\frac{d\theta}{dt}\right)^2.$$

Then the result in Problem 29 follows from Eq. (2) in Section 11.6.

11.M.31: The helix of Example 7 of Section 11.1 has position vector

$$\mathbf{r}(t) = \langle a\cos\omega t, \; a\sin\omega t, \; bt \rangle.$$

(Assume that a, b, and ω are all positive.) In that example we found that the velocity and acceleration vectors are

$$\mathbf{v}(t) = \langle -a\omega\sin\omega t, \; a\omega\cos\omega t, \; b \rangle \quad \text{and} \quad \mathbf{a}(t) = \langle -a\omega^2\cos\omega t, \; -a\omega^2\sin\omega t, \; 0 \rangle.$$

We also found that the speed is given by $v(t) = \sqrt{a^2\omega^2 + b^2}$. Note that $\mathbf{v}(t) \cdot \mathbf{a}(t) = 0$ and that

$$\mathbf{v}(t) \times \mathbf{a}(t) = \begin{vmatrix} \mathbf{i} & \mathbf{j} & \mathbf{k} \\ -a\omega\sin\omega t & a\omega\cos\omega t & b \\ -a\omega^2\cos\omega t & -a\omega^2\sin\omega t & 0 \end{vmatrix} = \langle ab\omega^2\sin\omega t, \; -ab\omega^2\cos\omega t, \; a^2\omega^3 \rangle.$$

By Eq. (26) of Section 11.6, the tangential component of acceleration is zero: $a_T = 0$. Next,

$$|\mathbf{v}(t) \times \mathbf{a}(t)| = \sqrt{a^2 b^2 \omega^4 + a^4 \omega^6} = a\omega^2\sqrt{b^2 + a^2\omega^2}.$$

So by Eq. (28) in Section 11.6, the normal component of acceleration is

$$a_N = \frac{|\mathbf{v}(t) \times \mathbf{a}(t)|}{v(t)} = a\omega^2.$$

Then the unit tangent vector is

$$\mathbf{T}(t) = \frac{\mathbf{v}(t)}{v(t)} = \frac{1}{\sqrt{a^2\omega^2 + b^2}} \langle -a\omega\sin\omega t, \; a\omega\cos\omega t, \; b \rangle.$$

Thus by Eq. (29) of Section 11.6, the principal unit normal vector is

$$\mathbf{N}(t) = \frac{\mathbf{a} - a_T\mathbf{T}}{a_N} = \frac{1}{a\omega^2}\left(\langle -a\omega^2\cos\omega t, \; -a\omega^2\sin\omega t, \; 0 \rangle - 0 \cdot \mathbf{T}\right) = \langle -\cos\omega t, \; -\sin\omega t, \; 0 \rangle.$$

Therefore the unit binormal vector is

$$\mathbf{B} = \mathbf{T} \times \mathbf{N} = \frac{1}{\sqrt{a^2\omega^2 + b^2}} \begin{vmatrix} \mathbf{i} & \mathbf{j} & \mathbf{k} \\ -a\omega\sin\omega t & a\omega\cos\omega t & b \\ -\cos\omega t & -\sin\omega t & 0 \end{vmatrix} = \frac{1}{\sqrt{a^2\omega^2 + b^2}} \langle b\sin\omega t, \; -b\cos\omega t, \; a\omega \rangle.$$

733

Then

$$\frac{d\mathbf{B}}{ds} = \frac{d\mathbf{B}}{dt} \cdot \frac{dt}{ds} = \frac{1}{v(t)} \cdot \frac{d\mathbf{B}}{dt} = \frac{b\omega}{a^2\omega^2 + b^2} \langle \cos\omega t, \sin\omega t, 0 \rangle = -\frac{b\omega}{a^2\omega^2 + b^2}\mathbf{N}.$$

Therefore, by definition, the torsion is

$$\tau = \frac{b\omega}{a^2\omega^2 + b^2}.$$

11.M.33: The spherical surface with center $(0, 0, 1)$ (in Cartesian coordinates) and radius 1 has Cartesian equation $x^2 + y^2 + (z-1)^2 = 1$; that is,

$$x^2 + y^2 + z^2 = 2z.$$

In spherical coordinates, this equation takes the form $\rho^2 = 2\rho\cos\phi$. No points on the surface are lost by cancellation of ρ from the last equation (take $\phi = \pi/2$), so a slightly simpler spherical equation is $\rho = 2\cos\phi$.

11.M.35: Replace y^2 with $x^2 + y^2$ to obtain a Cartesian equation of the surface of revolution:

$$(x^2 + y^2 + z^2)^2 = 2(z^2 - x^2 - y^2).$$

In cylindrical coordinates this equation takes the form

$$(r^2 + z^2)^2 = 2(z^2 - r^2),$$

and thus its spherical form is

$$\rho^4 = 2\rho^2(\cos^2\phi - \sin^2\phi); \quad \text{that is,} \quad \rho^2 = 2\cos 2\phi.$$

(No points on the graph are lost by cancellation of ρ^2: Take $\phi = \pi/2$.) The graph of this equation is next.

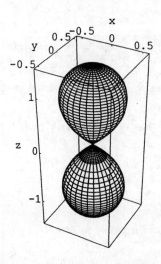

11.M.37: If $\mathbf{a} = \langle a_1, a_2, a_3 \rangle$ and $\mathbf{b} = \langle b_1, b_2, b_3 \rangle$, then by Eq. (17) in Section 11.3,

$$\mathbf{i} \cdot (\mathbf{a} \times \mathbf{b}) = \begin{vmatrix} 1 & 0 & 0 \\ a_1 & a_2 & a_3 \\ b_1 & b_2 & b_3 \end{vmatrix} = \begin{vmatrix} a_2 & a_3 \\ b_2 & b_3 \end{vmatrix};$$

similar results hold for $\mathbf{j} \cdot (\mathbf{a} \times \mathbf{b})$ and $\mathbf{k} \cdot (\mathbf{a} \times \mathbf{b})$. Hence by the result in Problem 36,

$$A^2 = \begin{vmatrix} a_2 & a_3 \\ b_2 & b_3 \end{vmatrix}^2 + \begin{vmatrix} a_3 & a_1 \\ b_3 & b_1 \end{vmatrix}^2 + \begin{vmatrix} a_1 & a_2 \\ b_1 & b_2 \end{vmatrix}^2.$$

11.M.39: By Eq. (13) in Section 11.6, the curvature of $y = \sin x$ at the point $(x, \sin x)$ is

$$\kappa(x) = \frac{|\sin x|}{(1 + \cos^2 x)^{3/2}}.$$

The curvature is minimal when it is zero, and this occurs at every integral multiple of π. To maximize the curvature, note that the numbers that maximize the numerator in the curvature formula—the odd integral multiples of $\pi/2$—also minimize the denominator and thereby maximize the curvature itself.

11.M.41: With $\mathbf{r}(t) = \langle t \cos t, \, t \sin t \rangle$, we have

$$\mathbf{v}(t) = \mathbf{r}'(t) = \langle \cos t - t \sin t, \, t \cos t + \sin t \rangle;$$

$$\mathbf{v}(\pi/2) = \langle -\pi/2, \, 1 \rangle;$$

$$v(\pi/2) = \frac{1}{2}\sqrt{\pi^2 + 4} \, ;$$

$$\mathbf{T}(\pi/2) = \frac{1}{\sqrt{\pi^2 + 4}} \langle -\pi, \, 2 \rangle.$$

Because the curve turns left as t increases,

$$\mathbf{N}\left(\frac{\pi}{2}\right) = \frac{1}{\sqrt{\pi^2 + 4}} \langle -2, \, -\pi \rangle.$$

11.M.43: We will ask *Mathematica* to solve this problem (but we will include intermediate computations so you can check your work).

```
x[t_] := r[t]*Cos[t]

y[t_] := r[t]*Sin[t]

x'[t]

    -r[t] Sin[t] + Cos[t] r'[t]
```

```
y'[t]

    Cos[t] r[t] + Sin[t] r'[t]

x''[t]

    -Cos[t] r[t] - 2 Sin[t] r'[t] + Cos[t] r''[t]

y''[t]

    -r[t] Sin[t] + 2 Cos[t] r'[t] + Sin[t] r''[t]

num = Simplify[ x'[t]*y''[t] - x''[t]*y'[t] ]

    r[t]^2 + 2 r'[t]^2 - r[t] r''[t]

den1 = Expand[ (x'[t])^2 ]

    r[t]^2 r[t]^2 - 2 Cos[t] r[t] Sin[t] r'[t] + Cos[t]^2 r'[t]^2

den2 = Expand[ (y'[t])^2 ]

    Cos[t]^2 r[t]^2 + 2 Cos[t] r[t] Sin[t] r'[t] + Sin[t]^2 r'[t]^2

den = Simplify[ den1 + den2 ]

    r[t]^2 + r'[t]^2
```

Finally, when we asked for `Abs[num]/den^(3/2)`, the response was

$$\frac{|(r(t))^2 + 2(r'(t))^2 - r(t)r''(t)|}{[(r(t))^2 + (r'(t))^2]^{3/2}}.$$

11.M.45: The function f is said to be *odd* if $f(-x) = -f(x)$ for all x; f is said to be *even* if $f(-x) = f(x)$ for all x. Because $y(x) = Ax + Bx^3 + Cx^5$ is an odd function, the condition $y(1) = 1$ will imply that $y(-1) = -1$. Because $y'(x) = A + 3Bx^2 + 5Cx^4$ is an even function, the condition $y'(1) = 0$ will imply that $y'(-1) = 0$ as well. Because the graph of y is symmetric around the origin (every odd function has this property), the condition that the curvature is zero at $(1, 1)$ will imply that it is also zero at $(-1, -1)$. By Eq. (13) of Section 11.6, the curvature at x is

$$\kappa(x) = \frac{|6Bx + 20Cx^3|}{\left[1 + (A + 3Bx^2 + 5Cx^4)^2\right]^{3/2}},$$

so the curvature at $(1, 1)$ will be zero when $6B + 20C = 0$. Thus we obtain the simultaneous equations

$$A + B + C = 1,$$

$$A + 3B + 5C = 0,$$

$$3B + 10C = 0.$$

736

These equations are easy to solve for $A = \frac{15}{8}$, $B = -\frac{5}{4}$, and $C = \frac{3}{8}$. Thus an equation of the connecting curve is

$$y(x) = \frac{15}{8}x - \frac{5}{4}x^3 + \frac{3}{8}x^5.$$

11.M.47: Without loss of generality we may suppose that the tetrahedron lies in the first octant with the solid right angle at the origin. Let the coordinates of its other three vertices be $(a, 0, 0)$, $(0, b, 0)$, and $(0, 0, c)$. Then the vectors $\mathbf{a} = \langle a, 0, 0 \rangle$, $\mathbf{b} = \langle 0, b, 0 \rangle$, and $\mathbf{c} = \langle 0, 0, c \rangle$ form three of the edges of the tetrahedron and the other three edges are $\mathbf{b} - \mathbf{a}$, $\mathbf{c} - \mathbf{b}$, and $\mathbf{a} - \mathbf{c}$. The area A of the triangle with these three edges is half the magnitude of the cross product of $\mathbf{b} - \mathbf{a}$ and $\mathbf{c} - \mathbf{a}$, and

$$(\mathbf{b} - \mathbf{a}) \times (\mathbf{c} - \mathbf{a}) = \begin{vmatrix} \mathbf{i} & \mathbf{j} & \mathbf{k} \\ -a & b & 0 \\ -a & 0 & c \end{vmatrix} = \langle bc,\ ac,\ ab \rangle.$$

Therefore

$$A^2 = \frac{1}{4}(a^2 b^2 + a^2 c^2 + b^2 c^2).$$

Part (a): The triangular face of the tetrahedron that lies in the xz plane has area $B = \frac{1}{2}ac$, the triangle in the xy plane has area $C = \frac{1}{2}ab$, and the triangle in the yz-plane has area $D = \frac{1}{2}bc$. It follows immediately that

$$A^2 = B^2 + C^2 + D^2.$$

Part (b): This is a generalization of the Pythagorean theorem to three dimensions. How would you generalize it to higher dimensions?

Section 12.2

12.2.1: Because $f(x, y) = 4 - 3x - 2y$ is defined for all x and y, the domain of f is the entire two-dimensional plane.

12.2.3: If either x or y is nonzero, then $x^2 + y^2 > 0$, and so $f(x, y)$ is defined—but not if $x = y = 0$. Hence the domain of f consists of all points (x, y) in the plane other than the origin.

12.2.5: The real number z has a unique cube root $z^{1/3}$ regardless of the value of z. Hence the domain of $f(x, y) = (y - x^2)^{1/3}$ consists of all points in the xy-plane.

12.2.7: Because $\arcsin z$ is a real number if and only if $-1 \leq z \leq 1$, the domain of the given function $f(x, y) = \sin^{-1}(x^2 + y^2)$ consists of those points (x, y) in the xy-plane for which $x^2 + y^2 \leq 1$; that is, the set of all points on and within the unit circle.

12.2.9: For every real number z, $\exp(z)$ is defined and unique. Therefore the domain of the given function $f(x, y) = \exp(-x^2 - y^2)$ consists of all points (x, y) in the entire xy-plane.

12.2.11: Because $\ln z$ is a unique real number if and only if $z > 0$, then domain of $f(x, y) = \ln(y - x)$ consists of those points (x, y) for which $y > x$. This is the region *above* the graph of the straight line with equation $y = x$ (the line itself is *not* part of the domain of f).

12.2.13: If x and y are real numbers, then so are xy, $\sin xy$, and $1 + \sin xy$. Hence the only obstruction to computation of

$$f(x, y) = \frac{1 + \sin xy}{xy}$$

is the possibility of division by zero. So the domain of f consists of those points (x, y) for which $xy \neq 0$; that is, all points in the xy-plane other than those on the coordinate axes.

12.2.15: If x and y are real numbers, then so are xy and $x^2 - y^2$. So the only obstruction to the computation of

$$f(x, y) = \frac{xy}{x^2 - y^2}$$

is the possibility that $x^2 - y^2 = 0$. If so, then $f(x, y)$ is undefined, and therefore the domain of f consists of those points (x, y) for which $x^2 \neq y^2$. That is, the domain of f consists of those points in the xy-plane other than the two straight lines with equations $y = x$ and $y = -x$.

12.2.17: If w is any real number, then $\exp(w)$ is a unique real number. So the only obstruction to the computation of

$$f(x, y, z) = \exp\left(\frac{1}{x^2 + y^2 + z^2}\right)$$

738

is the possibility that $x^2 + y^2 + z^2 = 0$. If so, then f is undefined, and therefore its domain consists of all those point (x, y, z) in space other than the origin $(0, 0, 0)$

12.2.19: If w is a positive real number, then $\ln w$ is a unique real number, but not otherwise. So the domain of $f(x, y, z) = \ln(z - x^2 - y^2)$ consists of those points (x, y, z) in space for which $x^2 + y^2 < z$. These are the points strictly above the circular paraboloid with equation $z = x^2 + y^2$; such a paraboloid is shown in Fig. 11.7.15.

12.2.21: The graph is the horizontal (parallel to the xy-plane) plane passing through the point $(0, 0, 10)$.

12.2.23: The graph of $f(x, y) = x + y$ is the plane with equation $z = x + y$. The trace of this plane in the xy-plane is the straight line with equations $z = 0$, $y = -x$. Its trace in the xz-plane is the straight line with equations $y = 0$, $z = x$ and its trace in the yz-plane is the straight line with equations $x = 0$, $z = y$.

12.2.25: The graph of $z = x^2 + y^2$ is a circular paraboloid with axis the nonnegative z-axis, opening upward, with its vertex at the origin. See Fig. 12.2.4.

12.2.27: If $z = \sqrt{4 - x^2 - y^2}$, then $z^2 = 4 - x^2 - y^2$ and $z \geq 0$; that is, $x^2 + y^2 + z^2 = 4$ and $z \geq 0$. Consequently the graph of $f(x, y) = \sqrt{4 - x^2 - y^2}$ is a hemisphere—the upper half of the sphere with radius 2 and center $(0, 0, 0)$.

12.2.29: In cylindrical coordinates, the graph of $f(x, y) = 10 - \sqrt{x^2 + y^2}$ has the equation $z = 10 - r$ where $r \geq 0$. Therefore the graph of f is the lower nappe of a circular cone with axis the z-axis and vertex at $(0, 0, 10)$.

12.2.31: The level curves of $f(x, y) = x - y$ are the straight lines $x - y = c$ where c is a constant. Some of these are shown next.

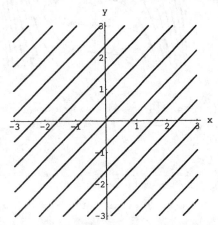

12.2.33: The level curves of $f(x, y) = x^2 + 4y^2$ are ellipses centered at the origin, with major axes on the x-axis and minor axes on the y-axis. The *Mathematica* command

```
ContourPlot[ x∧w + 4*y∧2, { x, -3, 3 }, { y, -3, 3 }, Axes -> True,
          AxesLabel -> { x, y }, AxesOrigin -> { 0, 0 }, Contours -> 15,
          ContourShading -> False, Frame -> False, PlotPoints -> 47 ];
```

generates some of these level curves; the output resulting from the preceding command is shown next.

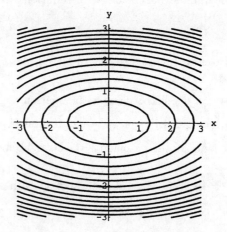

12.2.35: The level curves of $f(x, y) = y - x^3$ are curves with the equation $y = x^3 + C$ for various values of C. Their inflection points all lie on the y-axis. A few of these level curves are shown next.

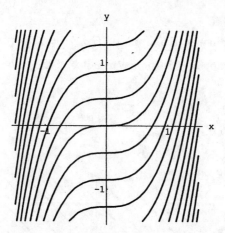

12.2.37: The level curves of $f(x, y) = x^2 + y^2 - 4x$ are circles centered at the point $(2, 0)$. The *Mathematica* command

```
ContourPlot[ x∧2 + y∧2 - 4*x, { x, -6, 6 }, { y, -6, 6 }, Axes -> True,
          AxesLabel -> { x, y }, AxesOrigin -> { 0, 0 }, Contours -> 15,
          ContourShading -> False, Frame -> False, PlotPoints -> 47 ];
```

produces several of these level curves, shown next.

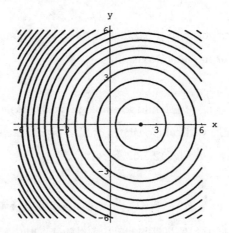

12.2.39: Because $\exp(-x^2 - y^2)$ is constant exactly when $x^2 + y^2$ is constant, the level curves of $f(x, y) = \exp(-x^2 - y^2)$ are circles centered at the origin. When they are close the graph is steep; when they are far apart the slope of the graph is more moderate. Some level curves of f are shown next.

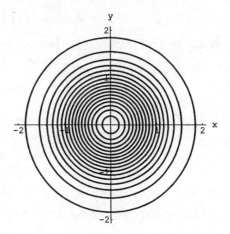

12.2.41: Because $f(x, y, z) = x^2 + y^2 - z$ is constant when $z = x^2 + y^2 + C$ (where C is a constant), the level surfaces of f are circular paraboloids with axis the z-axis, all opening upward and all having the same shape.

12.2.43: The function $f(x, y, z) = x^2 + y^2 + z^2 - 4x - 2y - 6z$ is constant when

$$(x - 2)^2 + (y - 1)^2 + (z - 3)^2 = C$$

for some nonnegative constant C. Therefore the level surfaces of f are spherical surfaces all centered at the point $(2, 1, 3)$.

12.2.45: The function $f(x, y, z) = x^2 + 4y^2 - 4x - 8y + 17$ is constant exactly when $(x - 2)^2 + 4(y - 1)^2$ is a nonnegative constant. Hence the level surfaces of f are elliptical cylinders parallel to the z-axis and centered on the vertical line that meets the xy-plane at the point $(2, 1, 0)$. The ellipse in which each such

741

cylinder meets the xy-plane has major axis parallel to the x-axis, minor axis parallel to the y-axis, and the major axis is twice the length of the minor axis.

12.2.47: The graph of $f(x, y) = x^3 + y^2$ should resemble vertical translates of $z = x^3$ in planes perpendicular to the y-axis and should resemble vertical translates of $z = y^2$ in planes perpendicular to the x-axis. Hence the graph must be the one shown in Fig. 12.2.32.

12.2.49: The graph of $f(x, y) = x^3 - 3x + \frac{1}{2}y$ will resemble vertical translates of the cubic equation $z = x^3 - 3x = x(x^2 - 3)$ in planes perpendicular to the y-axis and will resemble vertical translates of the straight line $z = \frac{1}{2}y$ in planes perpendicular to the x-axis. Hence the graph of f must be the one shown in Fig. 12.2.30.

12.2.51: The graph of $f(x, y) = x^2 + y^4 - 4y^2$ will show vertical translates of the parabola $z = x^2$ in planes perpendicular to the y-axis and vertical translates of the quartic $z = y^4 - 4y^2$ in planes perpendicular to the x-axis. Thus this graph is the one shown in Fig. 12.2.28.

12.2.53: Figure 12.2.33 has the level curves shown in Fig. 12.2.41. Note how the level curves are close together where the graph is steep.

12.2.55: Figure 12.2.35 has the level curves shown in Fig. 12.2.42. Moving outward from the center, Fig. 12.2.34 is alternately steep and almost flat; this is reflected in the behavior of the level curves, alternately close and far apart.

12.2.57: Figure 12.2.57 has the level curves shown in Fig. 12.2.44. The latter indicates three peaks in a row, or two peaks with a deep hole between them, or some variation of this idea.

12.2.59: The *Mathematica* commands

```
a = 2; b = 1;

Plot3D[ (a*x + b*y)*Exp[ -x∧2 - y∧2 ], { x, -3, 3 }, { y, -3, 3 },

    PlotRange → { -1, 1 }, PlotPoints → { 25, 25 },

    AxesLabel → { "x", "y", "z" } ];
```

produce the graph shown next.

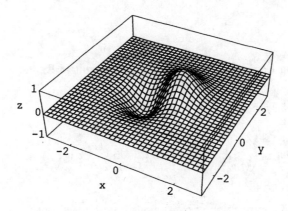

Plots with various values of a and b (not both zero) indicate that the surface always has one pit and one peak, both lying on the same straight line in the xy-plane. The values of a and b determine the orientation of this line and the distances of the pit and the peak from the origin.

12.2.61: The *Mathematica* commands

```
n = 3;

x = r*Cos[t]; y = r*Sin[t];

z = (r^2)*(Sin[ n*t ])*Exp[ -r^2 ];

ParametricPlot3D[ { x, y, z }, { r, 0, 3 }, { t, 0, 2*Pi },

    PlotPoints -> { 70, 30 }, PlotRange -> { -0.35, 0.35 } ];
```

produce the following graph.

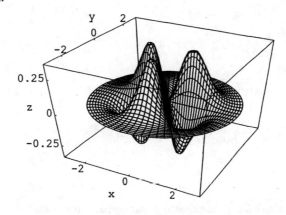

Apparently n peaks and n pits alternately surround the origin.

Section 12.3

12.3.1: The limit laws in Eqs. (5), (6), and (7) and the result in Example 4 yield

$$\lim_{(x,y)\to(0,0)} \left(7 - x^2 + 5xy\right)$$

$$= \left(\lim_{(x,y)\to(0,0)} 7\right) - \left(\lim_{(x,y)\to(0,0)} x\right)^2 + 5\left(\lim_{(x,y)\to(0,0)} x\right)\cdot\left(\lim_{(x,y)\to(0,0)} y\right) = 7 - 0^2 + 5\cdot 0\cdot 0 = 7.$$

12.3.3: The product and composition of continuous functions is continuous where defined, hence $f(x, y) = e^{-xy}$ is continuous everywhere. Then, by definition of continuity,

$$\lim_{(x,y)\to(1,-1)} f(x, y) = f(1, -1) = e^1 = e.$$

12.3.5: The sum, product, and quotient of continuous functions is continuous where defined, so

$$f(x, y) = \frac{5 - x^2}{3 + x + y}$$

is continuous where $x + y \neq -3$. Therefore

$$\lim_{(x,y)\to(0,0)} f(x, y) = \frac{5 - 0^2}{3 + 0 + 0} = \frac{5}{3}.$$

12.3.7: The sum, product, and composition of continuous functions is continuous where defined, so

$$f(x, y) = \ln\sqrt{1 - x^2 - y^2}$$

is continuous if $x^2 + y^2 < 1$. Therefore

$$\lim_{(x,y)\to(0,0)} f(x, y) = f(0, 0) = \ln\sqrt{1 - 0^2 - 0^2} = \ln 1 = 0.$$

12.3.9: Let $z = x + 2y$ and $w = 3x + 4y$. Then $z \to 0$ and $w \to 0$ as $(x, y) \to (0, 0)$ by Example 4. Hence, by continuity of the natural exponential and cosine functions and the product law for limits,

$$\lim_{(x,y)\to(0,0)} e^{x+2y}\cos(3x + 4y) = e^0\cos 0 = 1\cdot 1 = 1.$$

12.3.11: Every polynomial is continuous everywhere, and hence every rational function (even of three variables) is continuous where its denominator is not zero. Therefore

$$f(x, y, z) = \frac{x^2 + y^2 + z^2}{1 - x - y - z}$$

744

is continuous where $x + y + z \neq 1$. Hence

$$\lim_{(x,y,z)\to(1,1,1)} f(x,\,y,\,z) = f(1,1,1) = \frac{1^2 + 1^2 + 1^2}{1 - 1 - 1 - 1} = -\frac{3}{2}.$$

12.3.13: The sum, product, quotient, and composition of continuous functions is continuous where defined. Hence

$$f(x,\,y,\,z) = \frac{xy - z}{\cos xyz}$$

is continuous provided that xyz is not an odd integral multiple of $\pi/2$. Therefore

$$\lim_{(x,y,z)\to(1,1,0)} f(x,\,y,\,z) = f(1,1,0) = \frac{1 \cdot 1 - 0}{\cos 0} = \frac{1}{1} = 1.$$

12.3.15: First,

$$f(z) = \tan \frac{3\pi z}{4}$$

is continuous provided that $3\pi z/4$ is not an odd integral multiple of $\pi/2$; that is, provided that z is not two-thirds of an odd integer. Hence $f(z)$ is continuous at $z = 1$. Next,

$$g(x,\,y) = \sqrt{xy}$$

is the composition of continuous functions, so $g(x,\,y)$ is continuous provided that $xy > 0$. So g is continuous at $(2, 8)$. Finally, the product of continuous functions is continuous, so $h(x,\,y,\,z) = g(x,\,y) \cdot f(z)$ is continuous at $(2, 8, 1)$. Therefore

$$\lim_{(x,y,z)\to(2,8,1)} h(x,\,y,\,z) = h(2,8,1) = 16^{1/2} \cdot \tan \frac{3\pi}{4} = 4 \cdot (-1) = -4.$$

12.3.17: If $f(x,\,y) = xy$, then

$$\lim_{h\to 0} \frac{f(x+h,\,y) - f(x,\,y)}{h} = \lim_{h\to 0} \frac{xy + hy - xy}{h} = \lim_{h\to 0} \frac{hy}{h} = y \quad \text{and}$$

$$\lim_{k\to 0} \frac{f(x,\,y+k) - f(x,\,y)}{k} = \lim_{k\to 0} \frac{xy + kx - xy}{k} = \lim_{k\to 0} \frac{kx}{k} = x.$$

12.3.19: If $f(x,\,y) = xy^2 - 2$, then

$$\lim_{h\to 0} \frac{f(x+h,\,y) - f(x,\,y)}{h} = \lim_{h\to 0} \frac{(x+h)y^2 - 2 - xy^2 + 2}{h} = \lim_{h\to 0} \frac{hy^2}{h} = y^2 \quad \text{and}$$

$$\lim_{k\to 0} \frac{f(x,\,y+k) - f(x,\,y)}{k} = \lim_{k\to 0} \frac{x(y+k)^2 - 2 - xy^2 + 2}{k} = \lim_{k\to 0} \frac{2kxy + k^2 x}{k} = 2xy.$$

12.3.21: $\displaystyle \lim_{(x,y)\to(1,1)} \frac{1 - xy}{1 + xy} = \frac{1 - 1 \cdot 1}{1 + 1 \cdot 1} = \frac{0}{2} = 0.$

12.3.23: $\displaystyle\lim_{(x,y,z)\to(1,1,1)} \frac{xyz}{yz+xz+xy} = \frac{1\cdot1\cdot1}{1\cdot1+1\cdot1+1\cdot1} = \frac{1}{3}.$

12.3.25: $\displaystyle\lim_{(x,y)\to(0,0)} \ln(1+x^2+y^2) = \ln(1+0+0) = \ln 1 = 0.$

12.3.27: Convert to polar coordinates, then replace r^2 with z. Thus

$$\lim_{(x,y)\to(0,0)} \frac{\cot(x^2+y^2)}{x^2+y^2} = \lim_{r\to 0} \frac{\cot r^2}{r^2} = \lim_{z\to 0^+} \frac{\cos z}{z\sin z},$$

which does not exist because $\cos z \to 1$ as $z \to 0^+$, but $z\sin z \to 0$. Because the last numerator is approaching $+\infty$ as $z \to 0^+$ but the last denominator is approaching zero through *positive* values, it is also correct to write

$$\lim_{(x,y)\to(0,0)} \frac{\cot(x^2+y^2)}{x^2+y^2} = +\infty.$$

12.3.29: Convert to polar coordinates. Then

$$\lim_{(x,y)\to(0,0)} \exp\left(-\frac{1}{x^2+y^2}\right) = \lim_{r\to 0} \exp\left(-\frac{1}{r^2}\right) = \lim_{z\to\infty} e^{-z} = 0.$$

12.3.31: For continuity of $f(x,y) = \sqrt{x+y}$, we require $x+y > 0$. Thus f is continuous on the set of all points (x,y) that lie strictly above the graph of $y = -x$.

12.3.33: For continuity of $f(x,y) = \ln(x^2+y^2-1)$, we require that $x^2+y^2-1 > 0$; that is, that $x^2+y^2 > 1$. So f will be continuous on the points strictly outside the unit circle centered at $(0, 0)$; that is, strictly outside the circle with equation $x^2 + y^2 = 1$.

12.3.35: Because the inverse tangent function is continuous on the set of all real numbers, the only discontinuity of

$$f(x,y) = \arctan\left(\frac{1}{x^2+y^2}\right)$$

wlll occur when the denominator in the fraction is zero. Hence f is continuous on the set of all points in the xy-plane other than $(0, 0)$. This discontinuity is removable because conversion to polar coordinates shows that

$$\lim_{(x,y)\to(0,0)} \arctan\left(\frac{1}{x^2+y^2}\right) = \lim_{r\to 0^+} \arctan\left(\frac{1}{r}\right) = \frac{\pi}{2}.$$

12.3.37: Using polar coordinates yields

$$\lim_{(x,y)\to(0,0)} \frac{x^2-y^2}{\sqrt{x^2+y^2}} = \lim_{r\to 0} \frac{r^2\cos^2\theta - r^2\sin^2\theta}{r} = \lim_{r\to 0} \frac{r^2\cos 2\theta}{r} = \lim_{r\to 0} r\cos 2\theta = 0.$$

12.3.39: Use of polar coordinates yields

$$\lim_{(x,y)\to(0,0)} \frac{x^4+y^4}{(x^2+y^2)^{3/2}} = \lim_{r\to 0} \frac{r^4\cos^4\theta + r^4\sin^4\theta}{r^3} = \lim_{r\to 0} r(\cos^4\theta + \sin^4\theta) = 0.$$

12.3.41: Using spherical coordinates yields

$$\lim_{(x,y,z)\to(0,0,0)} \frac{xyz}{x^2+y^2+z^2} = \lim_{\rho\to 0} \frac{\rho^3(\sin^2\phi\,\cos\phi\,\sin\theta\,\cos\theta)}{\rho^2} = \lim_{\rho\to 0} \rho(\sin^2\phi\,\cos\phi\,\sin\theta\,\cos\theta) = 0.$$

12.3.43: The substitution $y = mx$ yields

$$\lim_{(x,y)\to(0,0)} \frac{x^2-y^2}{x^2+y^2} = \lim_{x\to 0} \frac{x^2(1-m^2)}{x^2(1+m^2)} = \frac{1-m^2}{1+m^2}.$$

This is the limit as $(x,\,y) \to (0,\,0)$ along the straight line of slope m. Different values of m (such as 0 and 1) give different values for the limit (such as 1 and 0), and therefore the original limit does not exist (see the Remark following Example 9).

12.3.45: If $(x,\,y,\,z) \to (0,\,0,\,0)$ along the positive x-axis—where $y = z = 0$—we obtain

$$\lim_{x\to 0^+} \frac{x+y+z}{x^2+y^2+z^2} = \lim_{x\to 0^+} \frac{1}{x} = +\infty.$$

Therefore the original limit does not exist.

12.3.47: The graph that follows this solution suggests that as $(x,\,y) \to (0,\,0)$ along straight lines of various slopes, the values of

$$\lim_{(x,y)\to(0,0)} \frac{x^2-2y^2}{x^2+y^2}$$

range from -2 to 1, so that the limit in question does not exist. To be certain that it does not, let $(x,\,y)$ approach $(0,\,0)$ along the x-axis to get limit 1, then along the y-axis to get limit -2. The figure was generated by executing the *Mathematica* command

```
ParametricPlot3D[ { r*Cos[t], r*Sin[t], (Cos[t])∧2 - 2*(Sin[t])∧2 },
    { t, 0, 2*Pi }, { r, 0.01, 2 } ];
```

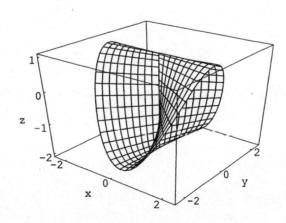

12.3.49: The graph that follows this solution suggests that as $(x, y) \to (0, 0)$ along straight lines of various slopes, the values of

$$\lim_{(x,y)\to(0,0)} \frac{xy}{2x^2 + 3y^2}$$

range from -0.2 to 0.2, so that the limit in question does not exist. To be sure that it does not, let (x, y) approach $(0, 0)$ along the line $y = x$ to get limit 0.2, then long the line $y = -x$ to get limit -0.2.

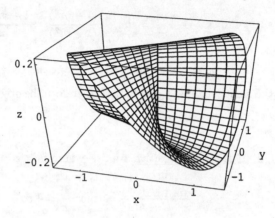

12.3.51: Given:

$$f(x, y) = \frac{2x^2 y}{x^4 + y^2}.$$

Suppose that (x, y) approaches $(0, 0)$ along the nonvertical straight line with equation $y = mx$. If $m \neq 0$, then—on that line—

$$\lim_{(x,y)\to(0,0)} f(x, y) = \lim_{x\to 0} \frac{2mx^3}{x^4 + m^2 x^2} = \lim_{x\to 0} \frac{2mx}{x^2 + m^2} = \frac{0}{m^2} = 0.$$

Clearly if $m = 0$ the result is the same. And if (x, y) approaches $(0, 0)$ along the y-axis, then—on that line—

$$\lim_{(x,y)\to(0,0)} f(x, y) = \lim_{y\to 0} \frac{2 \cdot 0 \cdot y}{0 + y^2} = 0.$$

Therefore as (x, y) approaches $(0, 0)$ along any straight line, the limit of $f(x, y)$ is zero. But on the curve $y = x^2$ we have

$$\lim_{(x,y)\to(0,0)} f(x, y) = \lim_{x\to 0} \frac{2x^4}{x^4 + x^4} = 1,$$

and therefore the limit of $f(x, y)$ does not exist at $(0, 0)$ (see the Remark following Example 9). For related paradoxical results involving functions of two variables, see Problem 60 of Section 12.5 and the miscellaneous problems for Chapter 12.

12.3.53: Given:

$$f(x, y) = \frac{xy}{x^2 + y^2}.$$

After we convert to polar coordinates, we have

$$f(r,\,\theta) = \frac{r^2 \cos\theta \sin\theta}{r^2} = \cos\theta \sin\theta = \frac{1}{2}\sin 2\theta.$$

On the hyperbolic spiral $r\theta = 1$, we have $\theta \to +\infty$ as r approaches zero through positive values. Hence $f(r,\,\theta)$ takes on all values between $-\frac{1}{2}$ and $\frac{1}{2}$ infinitely often as $r \to 0^+$. Therefore, as we discovered in Example 9, $f(x,\,y)$ has no limit as $(x,\,y) \to (0,\,0)$.

12.3.55: Let $z = x^2 - y^2$. By the discussions following Examples 4 and 5 and the "basic trigonometric limit" (Theorem 1 in Section 2.3),

$$\lim_{(x,y)\to(0,0)} \frac{\sin(x^2 - y^2)}{x^2 - y^2} = \lim_{z\to 0} \frac{\sin z}{z} = 1 = f(0,\,0).$$

Therefore f is continuous at $(0,\,0)$. By the discussion following Example 5, f is continuous at every other point of the xy-plane. Therefore f is continuous everywhere.

Section 12.4

12.4.1: If $f(x,\,y) = x^4 - x^3 y + x^2 y^2 - xy^3 + y^4$, then

$$\frac{\partial f}{\partial x} = 4x^3 - 3x^2 y + 2xy^2 - y^3 \quad \text{and}$$

$$\frac{\partial f}{\partial y} = -x^3 + 2x^2 y - 3xy^2 + 4y^3.$$

12.4.3: If $f(x,\,y) = e^x(\cos y - \sin y)$, then

$$\frac{\partial f}{\partial x} = e^x(\cos y - \sin y) \quad \text{and} \quad \frac{\partial f}{\partial y} = -e^x(\cos y + \sin y).$$

12.4.5: If $f(x,\,y) = \dfrac{x+y}{x-y}$, then

$$\frac{\partial f}{\partial x} = -\frac{2y}{(x-y)^2} \quad \text{and} \quad \frac{\partial f}{\partial y} = \frac{2x}{(x-y)^2}.$$

12.4.7: If $f(x,\,y) = \ln(x^2 + y^2)$, then

$$\frac{\partial f}{\partial x} = \frac{2x}{x^2 + y^2} \quad \text{and} \quad \frac{\partial f}{\partial y} = \frac{2y}{x^2 + y^2}.$$

12.4.9: If $f(x,\,y) = x^y$, then

$$\frac{\partial f}{\partial x} = yx^{y-1} \quad \text{and} \quad \frac{\partial f}{\partial y} = x^y \ln x.$$

749

12.4.11: If $f(x, y, z) = x^2 y^3 z^4$, then

$$\frac{\partial f}{\partial x} = 2xy^3 z^4, \quad \frac{\partial f}{\partial y} = 3x^2 y^2 z^4, \quad \text{and} \quad \frac{\partial f}{\partial z} = 4x^2 y^3 z^3.$$

12.4.13: If $f(x, y, z) = e^{xyz}$, then

$$\frac{\partial f}{\partial x} = yz e^{xyz}, \quad \frac{\partial f}{\partial y} = xz e^{xyz}, \quad \text{and} \quad \frac{\partial f}{\partial z} = xy e^{xyz}.$$

12.4.15: If $f(x, y, z) = x^2 e^y \ln z$, then

$$\frac{\partial f}{\partial x} = 2x e^y \ln z, \quad \frac{\partial f}{\partial y} = x^2 e^y \ln z, \quad \text{and} \quad \frac{\partial f}{\partial z} = \frac{x^2 e^y}{z}.$$

12.4.17: If $f(r, s) = \dfrac{r^2 - s^2}{r^2 + s^2}$. then

$$\frac{\partial f}{\partial r} = \frac{4rs^2}{(r^2 + s^2)^2} \quad \text{and} \quad \frac{\partial f}{\partial s} = -\frac{4r^2 s}{(r^2 + s^2)^2}.$$

12.4.19: If $f(u, v, w) = u e^v + v e^w + w e^u$, then

$$\frac{\partial f}{\partial u} = w e^u + e^v, \quad \frac{\partial f}{\partial v} = u e^v + e^w, \quad \text{and} \quad \frac{\partial f}{\partial w} = e^u + v e^w.$$

12.4.21: If $z(x, y) = x^2 - 4xy + 3y^2$, then

$$z_x(x, y) = 2x - 4y, \qquad z_y(x, y) = -4x + 6y,$$

$$z_{xy}(x, y) = -4, \qquad z_{yx}(x, y) = -4.$$

12.4.23: If $z(x, y) = x^2 \exp(-y^2)$, then

$$z_x(x, y) = 2x \exp(-y^2), \qquad z_y(x, y) = -2x^2 y \exp(-y^2),$$

$$z_{xy}(x, y) = -4xy \exp(-y^2), \qquad z_{yx}(x, y) = -4xy \exp(-y^2).$$

12.4.25: If $z(x, y) = \ln(x + y)$, then

$$z_x(x, y) = \frac{1}{x + y} = z_y(x, y) \quad \text{and} \quad z_{xy}(x, y) = -\frac{1}{(x + y)^2} = z_{yx}(x, y).$$

12.4.27: If $z(x, y) = e^{-3x} \cos y$, then

$$z_x(x, y) = -3e^{-3x} \cos y, \qquad z_y(x, y) = -e^{-3x} \sin y,$$

$$z_{xy}(x, y) = 3e^{-3x} \sin y, \qquad z_{yx}(x, y) = 3e^{-3x} \sin y.$$

12.4.29: If $z(x, y) = x^2 \cosh\left(\dfrac{1}{y^2}\right)$, then

$$z_x(x, y) = 2x \cosh\left(\frac{1}{y^2}\right), \qquad z_y(x, y) = -\frac{2x^2}{y^3} \sinh\left(\frac{1}{y^2}\right),$$

$$z_{xy}(x, y) = -\frac{4x}{y^3} \sinh\left(\frac{1}{y^2}\right), \qquad z_{yx}(x, y) = -\frac{4x}{y^3} \sinh\left(\frac{1}{y^2}\right).$$

12.4.31: Given: $f(x, y) = x^2 + y^2$ and the point $P(3, 4, 25)$ on its graph. Then

$$f_x(x, y) = 2x; \quad f_x(3, 4) = 6;$$

$$f_y(x, y) = 2y; \quad f_y(3, 4) = 8.$$

Hence by Eq. (11) of Section 12.4, an equation of the plane tangent to the graph of $z = f(x, y)$ at the point P is

$$z - 25 = 6(x - 3) + 8(y - 4); \quad \text{that is,} \quad 6x + 8y - z = 25.$$

12.4.33: Given: $f(x, y) = \sin\dfrac{\pi x y}{2}$ and the point $P(3, 5, -1)$ on its graph. Then

$$f_x(x, y) = \frac{\pi y}{2} \cos\frac{\pi x y}{2}; \qquad f_y(x, y) = \frac{\pi x}{2} \cos\frac{\pi x y}{2};$$

$$f_x(3, 5) = 0; \qquad f_y(3, 5) = 0.$$

The plane tangent to the graph of $z = f(x, y)$ at the point P is horizontal, so its has equation $z = -1$.

12.4.35: Given: $f(x, y) = x^3 - y^3$ and the point $P(3, 2, 19)$ on its graph. Then $f_x(x, y) = 3x^2$ and $f_y(x\,y) = -3y^2$, so $f_x(3, 2) = 27$ and $f_y(3, 2) = -12$. So by Eq. (11) an equation of the plane tangent to the graph of $z = f(x, y)$ at the point P is

$$z - 19 = 27(x - 3) - 12(y - 2); \quad \text{that is,} \quad 27x - 12y - z = 38.$$

12.4.37: Given: $f(x, y) = xy$ and the point $P(1, -1, -1)$ on its graph. Then $f_x(x, y) = y$ and $f_y(x, y) = x$, so $f_x(1, -1) = -1$ and $f_y(1, -1) = 1$. By Eq. (11) an equation of the plane tangent to the graph of $z = f(x, y)$ at the point P is $z + 1 = -(x - 1) + (y - 1)$; that is, $x - y + z = 1$.

12.4.39: Given: $f(x, y) = x^2 - 4y^2$ and the point $P(5, 2, 9)$ on its graph. Then $f_x(x, y) = 2x$ and $f_y(x, y) = -8y$, so $f_x(5, 2) = 10$ and $f_y(5, 2) = -16$. By Eq. (11) an equation of the plane tangent to the graph of $z = f(x, y)$ at the point P is

$$z - 9 = 10(x - 5) - 16(y - 2); \quad \text{that is,} \quad 10x - 16y - z = 9.$$

12.4.41: If $f_x(x, y) = 2xy^3$ and $f_y(x, y) = 3x^2y^2$, then

$$f_{xy}(x, y) = 6xy^2 = f_{yx}(x, y).$$

In Section 14.3 we will find that there must exist a function f having the given partial derivatives. Here we can find one by inspection; it is $f(x, y) = x^2y^3$.

12.4.43: If $f_x(x, y) = \cos^2 xy$ and $f_y(x, y) = \sin^2 xy$, then

$$f_{xy}(x, y) = -2x \sin xy \cos xy \neq -2y \sin xy \cos xy = f_{yx}(x, y).$$

By the *Note* preceding and following Eq. (14), there can be no function $f(x, y)$ having the given first-order partial derivatives.

12.4.45: The graph of $z = f(x, y)$ is shown in Fig. 12.4.14. The key to solving this group of six problems is first to locate f. Do this by sketching cross sections of each graph parallel to the xz- and yz-planes. This will make it easy to eliminate all but one candidate for the graph of f. Then the same sketches—the ones you created for the graph that turned out to be the graph of f—will quickly identify the other five graphs.

12.4.47: The graph of $z = f_y(x, y)$ is shown in Fig. 12.4.13.

12.4.49: The graph of $z = f_{xy}(x, y)$ is shown in Fig. 12.4.15.

12.4.51: If m and n are positive integers and $f(x, y) = x^my^n$, then $f_x(x, y) = mx^{m-1}y^n$ and $f_y(x, y) = nx^my^{n-1}$. Hence $f_{xy}(x, y) = mnx^{m-1}y^{n-1} = f_{yx}(x, y)$.

12.4.53: If $f(x, y, z) = e^{xyz}$, then

$$f_x(x, y, z) = yze^{xyz}, \quad f_y(x, y, z) = xze^{xyz}, \quad \text{and} \quad f_z(x, y, z) = xye^{xyz}.$$

Therefore

$$f_{xx}(x, y, z) = y^2z^2e^{xyz}, \quad f_{xy}(x, y, z) = f_{yx}(x, y, z) = (xyz^2 + z)e^{xyz},$$

$$f_{xz}(x, y, z) = f_{zx}(x, y, z) = (y + xy^2z)e^{xyz}, \quad f_{yz}(x, y, z) = f_{zy}(x, y, z) = (x + x^2yz)e^{xyz},$$

$$f_{yy}(x, y, z) = x^2z^2e^{xyz}, \quad f_{zz}(x, y, z) = x^2y^2e^{xyz}, \quad \text{and}$$

$$f_{xyz}(x, y, z) = (1 + 3xyz + x^2y^2z^2)e^{xyz}.$$

12.4.55: If $u(x, t) = \exp(-n^2 kt) \sin nx$ where n and k are constants, then

$$u_t(x, t) = -n^2 k \exp(-n^2 kt) \sin nx, \quad u_x(x, t) = n \exp(-n^2 kt) \cos nx,$$

$$\text{and} \quad u_{xx}(x, t) = -n^2 \exp(-n^2 kt) \sin nx.$$

Therefore $u_t = k u_{xx}$ for any choice of the constants k and n.

12.4.57: Part (a): If $y(x, t) = \sin(x + at)$ (where a is a constant), then

$$y_t(x, t) = a \cos(x + at), \qquad y_x(x, t) = \cos(x + at),$$

$$y_{tt}(x, t) = -a^2 \sin(x + at), \qquad y_{xx}(x, t) = -\sin(x + at).$$

Therefore $y_{tt} = a^2 y_{xx}$.

Part (b): If $y(x, t) = \cosh\big(3(x - at)\big)$, then

$$y_t(x, t) = -3a \sinh\big(3(x - at)\big), \qquad y_x(x, t) = 3 \sinh\big(3(x - at)\big),$$

$$y_{tt}(x, t) = 9a^2 \cosh\big(3(x - at)\big), \qquad y_{xx}(x, t) = 9 \cosh\big(3(x - at)\big).$$

Therefore $y_{tt} = a^2 y_{xx}$.

Part (c): If $y(x, t) = \sin kx \cos kat$ (where k is a constant), then

$$y_t(x, t) = -ka \sin kx \sin kat, \qquad y_x(x, t) = k \cos kx \cos kat,$$

$$y_{tt}(x, t) = -k^2 a^2 \sin kx \cos kat, \qquad y_{xx}(x, t) = -k^2 \sin kx \cos kat.$$

Therefore $y_{tt} = a^2 y_{xx}$.

12.4.59: Given: f and g are twice-differentiable functions of a single variable, a is a constant, and $y(x, t) = f(x + at) + g(x - at)$. Then

$$y_t(x, t) = af'(x + at) - ag'(x - at), \qquad y_{tt}(x, t) = a^2 f''(x + at) + a^2 g''(x - at),$$

$$y_x(x, t) = f'(x + at) + g'(x - at), \qquad y_{xx}(x, t) = f''(x + at) + g''(x - at).$$

It's now clear that $y(x, t)$ satisfies the wave equation $y_{tt} = a^2 y_{xx}$.

12.4.61: Given:

$$u(x, t) = T_0 + a_0 \exp\left(-x\sqrt{\omega/2k}\,\right) \cos\left(\omega t - x\sqrt{\omega/2k}\,\right)$$

where T_0, a_0, ω, and k are constants. First note that

$$u(0, t) = T_0 + a_0 e^0 \cos(\omega t - 0) = T_0 + a_0 \cos \omega t.$$

Next,

$$u_t(x, t) = -a_0 \omega \exp\left(-x\sqrt{\omega/2k}\right) \sin\left(\omega t - x\sqrt{\omega/2k}\right),$$

$$u_x(x, t) = -a_0 \left(\sqrt{\omega/2k}\right) \exp\left(-x\sqrt{\omega/2k}\right) \left[\cos\left(\omega t - x\sqrt{\omega/2k}\right) - \sin\left(\omega t - x\sqrt{\omega/2k}\right)\right], \quad \text{and}$$

$$u_{xx}(x, t) = -\frac{a_0 \omega}{k} \exp\left(-x\sqrt{\omega/2k}\right) \sin\left(\omega t - x\sqrt{\omega/2k}\right).$$

Therefore $y(x, t)$ satisfies the one-dimensional heat equation $u_t = k u_{xx}$.

12.4.63: Given $pV = nRT$ where n and R are constants, solve for p, V, and T in turn; then compute

$$\frac{\partial p}{\partial V} = -\frac{nRT}{V^2},$$

$$\frac{\partial V}{\partial T} = \frac{nR}{p}, \quad \text{and}$$

$$\frac{\partial T}{\partial p} = \frac{V}{nR}.$$

Then

$$\frac{\partial p}{\partial V} \cdot \frac{\partial V}{\partial T} \cdot \frac{\partial T}{\partial p} = -\frac{nRT}{pV} = -1.$$

12.4.65: If $f(x, y) = x^2 + 2xy + 2y^2 - 6x + 8y$, then the equations

$$f_x(x, y) = 0, \quad f_y(x, y) = 0$$

are

$$2x + 2y - 6 = 0, \quad 2x + 4y = -8,$$

which have the unique solution $x = 10$, $y = -7$. So the surface that is the graph of $z = f(x, y)$ contains exactly one point at which the tangent plane is horizontal, and that point is $(10, -7, -58)$.

12.4.67: Given: van der Waals' equation

$$\left(p + \frac{a}{V^2}\right)(V - b) = (82.06)T$$

where p denotes pressure (in atm), V volume (in cm^3), and T temperature (in K). For CO_2, the empirical constants are $a = 3.59 \times 10^6$ and $b = 42.7$. We are also given $V = 25600$ when $p = 1$ and $T = 313$. Part (a): We differentiate with respect to p while holding T constant. We obtain

$$\left(1 - \frac{2a}{V^3} V_p\right)(V - b) + \left(p + \frac{a}{V^2}\right) \cdot V_p = 0;$$

754

$$V - b - \frac{2a}{V^2} V_p + \frac{2ab}{V^3} V_p + p V_p + \frac{a}{V^2} V_p = 0;$$

$$\left(\frac{a}{V^2} - \frac{2ab}{V^3} - p \right) V_p = V - b;$$

$$\frac{aV - 2ab - pV^3}{V^3} V_p = V - b;$$

$$V_p = \frac{V^3(V - b)}{aV - 2ab - pV^3}.$$

If p is changed from 1 to 1.1, then the resulting change in V will be predicted by

$$\Delta V \approx V_p \, \Delta p = \frac{1}{10} V_p,$$

and substitution of the data $V = 25600$, $p = 1$ yields $\Delta V \approx -2569.76$ (cm^3).

Part (b): We hold p constant and differentiate both sides of van der Waals' equation with respect to T. The result:

$$\left(p + \frac{a}{V^2} \right) V_T - \frac{2a}{V^3} (V - b) V_T = 82.06;$$

$$\left(p V^3 + aV \right) V_T - 2a(V - b) V_T = (82.06) V^3;$$

$$V_T = \frac{(82.06) V^3}{p V^3 - aV + 2ab}.$$

If T is changed from 313 to 314 while the pressure is maintained at the constant value $p = 1$, the resulting change in V is predicted by

$$\Delta V \approx V_T \, \Delta T = 1 \cdot V_T.$$

Substitution of the data $V = 25600$, $p = 1$, $T = 313$ yields $\Delta V \approx 82.5105$.

The exact values of the two answers are closer to -2347.60 and 69.9202. Part of the discrepancy is caused by the fact that $V = 25600$ is itself only an approximation; the true value is closer to 25669.920232. Reworking parts (a) and (b) with this value of V yields the different approximations -2557.17 and 82.5095.

12.4.69: Part (a): If $f(x, y) = \sin x \, \sinh(\pi - y)$, then

$$f_x(x, y) = \cos x \, \sinh(\pi - y), \qquad f_{xx}(x, y) = -\sin x \, \sinh(\pi - y),$$

$$f_y(x, y) = -\sin x \, \cosh(\pi - y), \qquad f_{yy}(x, y) = \sin x \, \sinh(\pi - y).$$

Clearly f is harmonic.

Part (b): If $f(x, y) = \sinh 2x \sin 2y$, then

$$f_x(x, y) = 2 \cosh 2x \sin 2y, \qquad f_{xx}(x, y) = 4 \sinh 2x \sin 2y,$$

$$f_y(x, y) = 2 \sinh 2x \cos 2y, \qquad f_{yy}(x, y) = -4 \sinh 2x \sin 2y.$$

Clearly f is harmonic.

Part (c): If $f(x, y) = \sin 3x \sinh 3y$, then

$$f_x(x, y) = 3 \cos 3x \sinh 3y, \qquad f_{xx}(x, y) = -9 \sin 3x \sinh 3y,$$

$$f_y(x, y) = 3 \sin 3x \cosh 3y, \qquad f_{yy}(x, y) = 9 \sin 3x \sinh 3y.$$

Again it is clear that f is harmonic.

Part (d): If $f(x, y) = \sinh 4(\pi - x) \sin 4y$, then

$$f_x(x, y) = -4 \cosh 4(\pi - x) \sin 4y, \qquad f_{xx}(x, y) = 16 \sinh 4(\pi - x) \sin 4y,$$

$$f_y(x, y) = 4 \sinh 4(\pi - x) \cos 4y, \qquad f_{yy}(x, y) = -16 \sinh 4(\pi - y) \sin 4y.$$

Therefore f is harmonic.

12.4.71: If $f(x, y) = 100 + \dfrac{1}{100}(x^2 - 3xy + 2y^2)$, then

$$f_x(x, y) = \frac{1}{100}(2x - 3y), \qquad f_y(x, y) = \frac{1}{100}(4y - 3x),$$

$$f_x(100, 100) = -\frac{100}{100} = -1, \qquad f_y(100, 100) = \frac{100}{100} = 1.$$

Part (a): You will initially be descending at a 45° angle. Part (b): You will initially be ascending at a 45° angle.

12.4.73: Given:

$$f(x, y) = \begin{cases} \dfrac{xy}{x^2 + y^2} & \text{if } (x, y) \neq (0, 0), \\[2mm] 0 & \text{if } (x, y) = (0, 0). \end{cases}$$

Part (a): To find the first-order partial derivatives of f at a point other than $(0, 0)$, we proceed normally:

$$f_x(x, y) = \frac{y^3 - x^2 y}{(x^2 + y^2)^2} \quad \text{and} \quad f_y(x, y) = \frac{x^3 - xy^2}{(x^2 + y^2)^2}.$$

Clearly both are defined and continuous everywhere except possibly at the origin. Next,

$$f_x(0, 0) = \lim_{h \to 0} \frac{f(0 + h, 0) - f(0, 0)}{h} = \lim_{h \to 0} \frac{0}{h} = 0$$

756

and

$$f_y(0, 0) = \lim_{k \to 0} \frac{f(0, 0 + k) - f(0, 0)}{k} = \lim_{k \to 0} \frac{0}{k} = 0.$$

Therefore both f_x and f_y are defined everywhere.

Part (b): At points of the line $y = mx$ other than $(0, 0)$, we have

$$f_x(x, y) = f_x(x, mx) = \frac{m^3 - m}{(m^2 + 1)^2 x},$$

and hence (taking $m = 0$) we have $f_x(x, y) = f_x(x, 0) \equiv 0$, whereas (taking $m = 2$) we have

$$f_x(x, y) = f_x(x, 2x) = \frac{6}{25x},$$

so that f_x is not continuous at $(0, 0)$. Similarly, at points of the line $y = mx$ other than $(0, 0)$, we have

$$f_y(x, y) = f_y(x, mx) = \frac{(1 - m^2)}{(1 + m^2)^2 x},$$

and hence (taking $m = 1$) we have $f_y(x, y) = f_y(x, x) \equiv 0$, whereas (taking $m = 0$) we have

$$f_y(x, y) = f_y(x, 0) = \frac{1}{x}.$$

Hence f_y is also not continuous at $(0, 0)$.

Part (c): If $(x, y) \neq (0, 0)$, then differentiation of f_x with respect to x yields

$$f_{xx}(x, y) = \frac{2xy(x^2 - 3y^2)}{(x^2 + y^2)^3}.$$

Similarly,

$$f_{xy}(x, y) = f_{yx}(x, y) = \frac{6x^2y^2 - x^4 - y^4}{(x^2 + y^2)^3} \quad \text{and} \quad f_{yy}(x, y) = \frac{2xy(y^2 - 3x^2)}{(x^2 + y^2)^3}.$$

Therefore the second-order partial derivatives of f are all defined and continuous except possibly at the origin.

Part (d): Here we have

$$f_{xx}(0, 0) = \lim_{h \to 0} \frac{f_x(0 + h, 0) - f_x(0, 0)}{h} = \lim_{h \to 0} \frac{0}{h} = 0$$

and, similarly, $f_{yy}(0, 0) = 0$. Hence both second-order partial derivatives f_{xx} and f_{yy} exist at the origin (but you can use polar coordinates to show that neither is continuous there). On the other hand,

$$f_{xy}(0, 0) = \lim_{k \to 0} \frac{f_x(0, 0 + k) - f_x(0, 0)}{k} = \lim_{k \to 0} \frac{1}{k^2},$$

which does not exist; $f_{yx}(0, 0)$ also does not exist by a similar computation.

Section 12.5

12.5.1: If $z = x - 3y + 5$, then $z_x \equiv 1$ and $z_y \equiv -3$. Hence there is no point on the graph at which the tangent plane is horizontal. Indeed, the graph of $z = x - 3y + 5$ is itself a plane with nonvertical normal vector $\langle 1, -3, -1 \rangle$, and that's another reason why no tangent plane is horizontal.

12.5.3: If $z = xy + 5$, then $z_x = y$ and $z_y = x$. Both vanish at $(0, 0)$, so there is exactly one point on the graph of $z = xy + 5$ at which the tangent plane is horizontal—$(0, 0, 5)$.

12.5.5: If $z = f(x, y) = x^2 + y^2 - 6x + 2y + 5$, then $z_x = 2x - 6$ and $z_y = 2y + 2$. Both are zero only at the point $(3, -1)$, so the graph of $z = f(x, y)$ has a horizontal tangent plane at the point $(3, -1, -5)$.

12.5.7: If $z = f(x, y) = x^2 + 4x + y^3$, then $z_x = 2x + 4$ and $z_y = 3y^2$. Both are zero at the point $(-2, 0)$, so the graph of $z = f(x, y)$ has exactly one horizontal tangent plane—the one that is tangent at the point $(-2, 0, -4)$.

12.5.9: If $z = f(x, y) = 3x^2 + 12x + 4y^3 - 6y^2 + 5$, then $z_x = 6x + 12$ and $z_y = 12y(y - 1)$. Both partial derivatives are zero at $(-2, 0)$ and $(-2, 1)$, so the graph of $z = f(x, y)$ has two horizontal tangent planes. One is tangent at the point $(-2, 0, -7)$ and the other is tangent at the point $(-2, 1, -9)$.

12.5.11: If $f(x, y) = (2x^2 + 3y^2)\exp(-x^2 - y^2)$, then

$$\frac{\partial f}{\partial x} = 4x\exp(-x^2 - y^2) - 2x(2x^2 + 3y^2)\exp(-x^2 - y^2) = -2x(2x^2 + 3y^2 - 2)\exp(-x^2 - y^2) \quad \text{and}$$

$$\frac{\partial f}{\partial y} = 6y\exp(-x^2 - y^2) - 2y(2x^2 + 3y^2)\exp(-x^2 - y^2) = -2y(2x^2 + 3y^2 - 3)\exp(-x^2 - y^2).$$

We note that $\exp(-x^2 - y^2)$ is never zero, so to find when both partial derivatives are zero, it is enough to solve simultaneously

$$x(2x^2 + 3y^2 - 2) = 0,$$

$$y(2x^2 + 3y^2 - 3) = 0.$$

One solution is obvious: $x = y = 0$. Next, if $x \neq 0$ but $y = 0$, then the first of these equations implies that $2x^2 = 2$, so we get the two critical points $(1, 0)$ and $(-1, 0)$. If $x = 0$ but $y \neq 0$, then the second of these equations implies that $3y^2 = 3$, so we obtain two more critical points, $(0, -1)$ and $(0, 1)$. There is no solution if both x and y are nonzero, for that would imply that $2x^2 + 3y^2 = 2$ and $2x^2 + 3y^2 = 3$. But are

758

there five horizontal tangent planes? No, because two of them are tangent at two critical points. One plane is tangent to the graph of $z = f(x, y)$ at $(-1, 0, 2e^{-1})$ and $(1, 0, 2e^{-1})$, a second is tangent at $(0, -1, 3e^{-1})$ and $(0, 1, 3e^{-1})$, and the third is tangent at $(0, 0, 0)$.

12.5.13: Given: $z = f(x, y) = x^2 - 2x + y^2 - 2y + 3$. Then

$$f_x(x, y) = 2x - 2 \quad \text{and} \quad f_y(x, y) = 2y - 2,$$

so both partials are zero at only one point: $(1, 1)$. So the graph of $z = f(x, y)$ has only one horizontal tangent plane; it is tangent at the point $(1, 1, 1)$. This is clearly the lowest point on the graph of f.

12.5.15: If $z = f(x, y) = 2x - x^2 + 2y^2 - y^4$, then

$$f_x(x, y) = -2x + 2 \quad \text{and} \quad f_y(x, y) = -4y^3 + 4y = -4y(y + 1)(y - 1).$$

Thus there are three critical points: $(1, -1)$, $(1, 0)$, and $(1, 1)$. But there are only two horizontal planes tangent to the graph of f because one is tangent at two points—namely, at $(1, -1, 2)$ and at $(1, 1, 2)$; the other horizontal tangent plane is tangent to the graph at $(1, 0, 1)$. The first two of these points are the equally high highest points on the graph of f.

12.5.17: Given: $z = f(x, y) = 3x^4 - 4x^3 - 12x^2 + 2y^2 - 12y$, the following sequence of *Mathematica* commands will find the points of tangency of all horizontal tangent planes. (Recall that % refers to the "last output.")

```
f[x_, y_ ] := 3*x^4 - 4*x^3 - 12*x^2 + 2*y^2 - 12*y

d1 = D[ f[x,y], x ]
```

$$-24x - 12x^2 + 12x^3$$

```
d2 = D[ f[x,y], y ]
```

$$-12 + 4y$$

```
Solve[ { d1 == 0, d2 == 0 }, { x, y } ];
```

$$\{\{x \to -1, y \to 3\}, \{x \to 0, y \to 3\}, \{x \to 2, y \to 3\}\}$$

```
f[x,y] /. %
```

$$\{-23, -18, -50\}$$

Thus there are three horizontal planes tangent to the graph of $z = f(x, y)$; the points of tangency are $(-1, 3, -23)$, $(0, 3, -18)$, and $(2, 3, -50)$. The last of these is the lowest point on the graph of f.

12.5.19: If $f(x, y) = 2x^2 + 8xy + y^4$, then

$$f_x(x, y) = 4x + 8y = 4(x + 2y) \quad \text{and} \quad f_y(x, y) = 8x + 4y^3 = 4(2x + y^3).$$

Thus both partial derivatives are zero at the three points $(-4, 2)$, $(0, 0)$, and $(4, -2)$. But there are only two horizontal tangent planes because one is tangent to the graph of $z = f(x, y)$ at two points: $(-4, 2, -16)$ and $(4, -2, -16)$. The other plane is tangent to the graph at the origin. The two equally low lowest points on the graph of f are $(-4, 2, -16)$ and $(4, -2, -16)$.

Detail: Solving simultaneous nonlinear equations is an *ad hoc* procedure. One method that frequently works is to solve one equation for one of the variables, then substitute in the others. Here we begin with

$$4(x + 2y) = 0 \quad \text{and} \quad 4(2x + y^3) = 0.$$

We solve the first for $x = -2y$ and substitute in the second to obtain

$$-4y + y^3 = 0; \quad \text{that is,} \quad y(y + 2)(y - 2) = 0.$$

This yields the three solutions $y = -2$, $y = 0$, and $y = 2$, and the corresponding values of x are 4, 0, and -4.

12.5.21: If $z = f(x, y) = \exp(2x - 4y - x^2 - y^2)$, then

$$f_x(x, y) = (2 - 2x)\exp(2x - 4y - x^2 - y^2) \quad \text{and} \quad f_y(x, y) = -(2y + 4)\exp(2x - 4y - x^2 - y^2).$$

Hence both partial derivatives are zero when

$$2x - 2 = 0 \quad \text{and} \quad 2y + 4 = 0;$$

that is, at the single point $(1, -2)$. Hence there is exactly one horizontal plane tangent to the graph of $z = f(x, y)$; it is tangent at the point $(1, -2, e^5)$. This is the highest point on the graph of f.

12.3.23: The graph of $f(x, y) = x + 2y$ is a plane, so its maximum and minimum values on a polygonal region must occur at the vertices of the polygon. Here we have

$$f(-1, -1) = -3, \quad f(-1, 1) = 1, \quad f(1, -1) = -1, \quad \text{and} \quad f(1, 1) = 3.$$

At this point it is clear what are the maximum and minimum values of $f(x, y)$ on R.

12.5.25: Given: $f(x, y) = x^2 + y^2 - 2x$ on the triangular region R with vertices at $(0, 0)$, $(2, 0)$, and $(0, 2)$. Then

$$f_x(x, y) = 2x - 2 \quad \text{and} \quad f_y(x, y) = 2y,$$

760

so the only critical point of f is $(1, 0)$, which is a point of R. Because

$$f(x, y) = (x-1)^2 + y^2 - 1,$$

the global minimum value of f on R is $f(1, 0) = -1$. The maximum value of f must occur on the boundary of R, which we explore next.

- On the lower edge of R, we have $f(x, 0) = x^2 - 2x$, which must attain its maximum at one endpoint of that edge. Hence the maximum value of f there is $f(0, 0) = f(2, 0) = 0$.

- On the left-hand edge of R, we have $f(0, y) = y^2$, with maximum value $f(0, 2) = 4$.

- On the diagonal edge of R, which has equation $y = 2 - x$, we have

$$f(x, 2-x) = g(x) = x^2 + (2-x)^2 - 2x = 2x^2 - 6x + 4,$$

and $g'(x) = 4x - 6$, so the minimum value of f there is $g\left(\frac{3}{2}\right) = -\frac{1}{2}$ and the maximum value there (because it must occur at an endpoint of the diagonal edge) is $f(0, 2) = 4$.

In summary, the minimum value of f on R is $f(1, 0) = -1$ and its maximum value is $f(0, 2) = 4$.

12.5.27: Given: $f(x, y) = 2xy$ on the circular disk R described by the inequality $x^2 + y^2 \leq 1$. Then $f_x(x, y) = 2y$ and $f_x(x, y) = 2x$, so the only critical point of f is $(0, 0)$. On the line $y = x$ we have $f(x, x) = 2x^2$, but on the line $y = -x$ we have $f(x, -x) = -2x^2$. Therefore f does not have an extremum at $(0, 0)$. (The graph of $z = 2xy$ is a hyperbolic paraboloid with a *saddle point* at $(0, 0)$; to see its graph, rotate the graph shown in Fig. 12.10.1 45° around the z-axis.) Because f must have a global maximum and a global minimum on R, both must occur on its boundary.

- We describe the boundary $x^2 + y^2 = 1$ of R in polar coordinates: $r = 1$, $0 \leq \theta \leq 2\pi$. Thus on the boundary, we have

$$f(x, y) = 2xy = g(\theta) = 2\sin\theta\cos\theta.$$

Then

$$g'(\theta) = 2\cos^2\theta - 2\sin^2\theta = 2\cos 2\theta,$$

and $g'(\theta) = 0$ when 2θ is an odd integral multiple of $\pi/2$; that is, when θ is an odd integral multiple of $\pi/4$. Now

$$g(\pi/4) = g(5\pi/4) = 1 \quad \text{and} \quad g(3\pi/4) = g(7\pi/4) = -1,$$

so we have discovered the global extrema of f on R.

Summary: The global maximum value of f is 1 and occurs at each of the two points $\left(\frac{1}{2}\sqrt{2}, \frac{1}{2}\sqrt{2}\right)$ and $\left(-\frac{1}{2}\sqrt{2}, -\frac{1}{2}\sqrt{2}\right)$. The global minimum value of f is -1 and occurs at each of the two points $\left(-\frac{1}{2}\sqrt{2}, \frac{1}{2}\sqrt{2}\right)$ and $\left(\frac{1}{2}\sqrt{2}, -\frac{1}{2}\sqrt{2}\right)$.

12.5.29: The square of the distance between the plane and the origin is

$$f(x, y) = x^2 + y^2 + \left(\frac{169 - 12x - 4y}{3}\right)^2.$$

Setting both partials of $f(x, y)$ equal to zero yields the equations

$$2x - \frac{8}{3}(169 - 12x - 4y) = 0, \quad 2y - \frac{8}{9}(169 - 12x - 4y) = 0.$$

These equations are easiest to solve if you note first that they imply immediately that $6x = 18y$, because each is equal to $8(169 - 12x - 4y)$. It follows that their solution is $x = 12$, $y = 4$. The corresponding value of z on the given plane is 3, so the point of the plane closest to the origin is $(12, 4, 3)$. (The formula for f makes it plain that f has a global minimum and no maximum.) The distance between the plane and the origin is therefore $\sqrt{f(12, 4)} = 13$.

12.5.31: The square of the distance between the plane and the point Q is

$$f(x, y) = (x - 7)^2 + (y + 7)^2 + (49 - 2x - 3y)^2.$$

When we set both partials of f equal to zero, we get the simultaneous equations

$$2(x - 7) - 4(49 - 2x - 3y) = 0, \quad 2(y + 7) - 6(49 - 2x - 3y) = 0,$$

with solution $x = 15$, $y = 5$; the corresponding z-coordinate on the plane is 4. So the point on the plane closest to Q is $(15, 5, 4)$. The distance between the two is $4\sqrt{14}$.

12.5.33: The square of the distance from the origin to the point (x, y, z) of the surface is

$$f(x, y) = x^2 + y^2 + \frac{16}{x^4y^4}.$$

The equations $f_x(x, y) = 0 = f_y(x, y)$ are

$$2x - \frac{64}{x^5y^4} = 0, \quad 2y - \frac{64}{x^4y^5} = 0.$$

They are easiest to solve if you begin with the observation that they imply that $2x^2 = 2y^2$. There are four solutions—all possible combinations of $x = \pm\sqrt{2}$, $y = \pm\sqrt{2}$. It follows that the point on the surface in the first octant closest to the origin is $(\sqrt{2}, \sqrt{2}, 1)$; its distance from the origin is $\sqrt{5}$.

12.5.35: We will find the maximum possible product of three *nonnegative* real numbers with sum 120—the reason in a moment. If x, y, and z are the three numbers, then we are to maximize xyz given $x + y + z = 120$. So we solve for z, substitute, and maximize

$$f(x, y) = xy(120 - x - y), \quad 0 \leq x, \quad 0 \leq y, \quad x + y \leq 120.$$

762

Thus by allowing one or two of the numbers to be zero, the domain of f is now a closed and bounded subset of the plane—the triangle with two sides on the nonnegative coordinate axes and the third side part of the graph of $y = 120 - x$. Write $f(x, y) = 120xy - x^2 y - xy^2$ and set both partial derivatives equal to zero to obtain

$$120y - 2xy - y^2 = 0, \quad 120x - 2xy - x^2 = 0.$$

Because neither x nor y is zero—that would minimize the product, not maximize it—we may cancel to obtain

$$120 - 2x - y = 0, \quad 120 - 2y - x = 0,$$

and it follows that $2x + y = x + 2y$, so that $y = x$, and then the equation $3x = 120$ yields $x = 40$ and $y = 40$. It follows that $z = 40$ as well. The maximum of $f(x, y)$ does not occur on the boundary of its domain, for $f(x, y) = 0$ there. Hence this lone interior critical point must yield the global maximum, which is $40 \cdot 40 \cdot 40 = 64000$.

12.5.37: Let the dimensions of the box be x by y by z. We are to minimize total surface area $A = 2xy + 2xz + 2yz$ given $xyz = 1000$. Solve the latter equation for z and substitute in A to obtain the function to be minimized:

$$A(x, y) = \frac{2000}{x} + \frac{2000}{y} + 2xy, \quad 0 < x, \quad 0 < y.$$

Although A is continuous on its domain, the domain is neither closed nor bounded. But an argument similar to the one given in Example 7 makes it clear that A has a global minimum, so it must occur at a critical point of the domain. When we set both partial derivatives of A equal to zero, we obtain the equations

$$2y - \frac{2000}{x^2} = 0, \quad 2x - \frac{2000}{y^2} = 0.$$

These equations imply that $2x^2 y = 2xy^2$, so that $y = x$. Then either of the two preceding equations implies that $x = 10 = y$. Finally, $xyz = 1000$ implies that $z = 10$ as well. This is the only critical point, so we have found the global minimum of A. To minimize the total surface area, make a cube of edge length 10.

12.5.39: Suppose that the dimensions of the base of the box are x by y, the front and back have dimensions x by z, and the sides have dimensions y by z. The cost of the base is then $6xy$ and the total cost of the other four sides is $2 \cdot 5xz + 2 \cdot 5yz$. So we are to minimize total cost $C = 6xy + 10xz + 10yz$ given $xyz = 600$. Solve the last equation for z and substitute in the cost expression to obtain the function to be minimized:

$$C(x, y) = \frac{6000}{x} + \frac{6000}{y} + 6xy, \quad 0 < x, \quad 0 < y.$$

By an argument similar to the one used in the solution of Example 7, $C(x, y)$ has a global minimum and it occurs at a critical point. When we set both partial derivatives of C equal to zero, we get the equations

$$6y - \frac{6000}{x^2} = 0, \quad 6x - \frac{6000}{y^2} = 0.$$

It follows immediately that $x^2 y = xy^2$, so that $x = y$. Then either of the displayed equations yields $x = y = 10$. The corresponding value of z is 6, so the dimensions of the least expensive such box are these: base 10 inches by 10 inches, height 6 inches. It will cost \$18.00.

12.5.41: Let the base and top of the box have dimensions x by y, the front and back dimensions x by z, and the sides dimensions y by z. Then the top and base cost $3xy$ cents each, the front and back cost $6xz$ cents each, and the two sides cost $9yz$ cents each. Hence the total cost of the box will be $C = 6xy + 12xz + 18yz$. But $xyz = 750$, so that $z = 750/(xy)$. To substitute this into the formula for C and simplify, we use the *Mathematica* command

```
6*x*y + 12*x*z + 18*y*z /. z -> 750/(x*y)
```

(Recall that /. translates roughly as "evaluate subject to.") The response is

$$\frac{13500}{x} + \frac{9000}{y} + 6xy.$$

Next we construct the total surface area function, the quantity to be minimized:

```
f[x_, y_] := 13500/x + 9000/y + 6*x*y
```

Then we compute both partial derivatives:

```
d1 = D[ f[x,y], x]
```

$$6y - \frac{13500}{x^2}$$

```
d2 = D[ f[x,y], y]
```

$$6x - \frac{9000}{y^2}$$

Then we set both partial derivatives equal to zero and solve simultaneously:

```
Solve[ { d1 == 0, d2 == 0 }, { x, y } ]
```

The response is

$$\{\{x \to 15,\ y \to 10\},\ \{x \to -15\,(-1)^{1/3},\ y \to -10\,(-1)^{1/3}\},\ \{x \to 15\,(-1)^{2/3},\ y \to 10\,(-1)^{2/3}\}\}$$

We ignore the two pairs of non-real roots and evaluate z:

```
750/(x*y) /. {x -> 15, y -> 10}
```

5

Finally, we evaluate f at $(15, 10)$ to find the minimum cost:

`f[15,10]`

> 2700

The domain of f is not a closed and bounded set, but instead the interior of the entire first quadrant. Nevertheless, $f(x, y)$ has a global minimum at a critical point by an argument similar to the one used in Example 7. Because we have found only one critical point, we have found the global minimum as well. The box should have dimensions $x = 15$, $y = 10$, and $z = 5$ inches. It will cost $27.00.

12.5.43: Suppose that the base of the building measures x feet by y feet, that its front and back measure x by z, and that its two sides measure y by z. Then the total heating and cooling costs will be $C = 2xy + 4xz + 8yz$. Solve $xyz = 8000$ for z and substitute in the expression for cost to obtain the quantity to be minimized:

$$C(x,\, y) = \frac{64000}{x} + \frac{32000}{y} + 2xy, \quad 0 < x, \quad 0 < y.$$

Although the domain of C is not a closed and bounded subset of the plane, nevertheless $C(x,\, y)$ has a global minimum at a critical point by an argument similar to the one used in the solution of Example 7. When we set both partial derivatives of $C(x, y)$ equal to zero, we obtain

$$2y - \frac{64000}{x^2} = 0, \quad 2x - \frac{32000}{y^2} = 0,$$

having the only real solutions $x = 40$, $y = 20$. The corresponding value of z is 10, so the building should be 40 feet wide (in front), 20 feet deep, and 10 feet high. The annual heating and cooling costs will thereby have their minimum possible value, $C(40,\, 20) = 4800$ dollars per year.

12.5.45: Suppose that $(x,\, y,\, z)$ is the vertex of the box that lies on the given plane with equation $x + 3y + 7z = 11$. We are to maximize the volume $V = xyz$ of the box. Solve the equation of the plane for z and subsitute to obtain

$$V(x,\, y) = \frac{xy\,(11 - x - 3y)}{7}, \quad 0 \leqq x, \quad 0 \leqq y, \quad x + 3y \leqq 11.$$

The domain of V is a closed and bounded subset of the xy-plane—it consists of the sides and interior of the triangle with vertices at $(0, 0)$, $(11, 0)$, and $\left(0,\, \frac{11}{3}\right)$. Therefore the continuous function (it's a polynomial) $V(x, y)$ has a global maximum on its domain. The maximum does not occur on the boundary because $V(x, y)$ is identically zero there. Hence the maximum occurs at an interior critical point. When we set the partial derivatives of V equal to zero, we get the simultaneous equations

$$\frac{(11 - x - 3y)y - xy}{7} = 0, \quad \frac{(11 - x - 3y)x - 3xy}{7} = 0.$$

To solve these equations, multiply through by 7 and factor to obtain

$$(11 - 2x - 3y)y = 0, \quad (11 - x - 6y)x = 0.$$

- If $x = 0$ and $y = 0$, we have one solution.

- If $x = 0$ and $y \neq 0$, then $y = \frac{11}{3}$.

- If $x \neq 0$ and $y = 0$, then $x = 11$.

- If $x \neq 0$ and $y \neq 0$, then $2x + 3y = 11$ and $x + 6y = 11$. It follows that $x = \frac{11}{3}$ and $y = \frac{11}{9}$.

Only the last of these solutions will produce a box of positive volume. The corresponding value of z is $\frac{11}{21}$. Thus we have found the maximizing values of x, y, and z. The maximum possible volume is

$$\frac{11}{3} \cdot \frac{11}{9} \cdot \frac{11}{21} = \frac{1331}{567} \approx 2.347442680776.$$

12.5.47: Suppose that the dimensions of the rectangular box are x by y by z (in inches). Without loss of generality we may suppose that $x \leqq y \leqq z$. Then the length of the box is z, so its girth is $2x + 2y$. We are to maximize box volume $V = xyz$ given the side condition $2x + 2y + z \leqq 108$; of course, the maximum occurs when $2x + 2y + z = 108$, so that is the side condition we use. Solve for z in the last equation and substitute in the expression for volume to obtain the function to be maximized:

$$V(x, y) = xy(108 - 2x - 2y), \quad 0 \leqq x, \quad 0 \leqq y, \quad x + y \leqq 54.$$

Now V is continuous ($V(x, y)$ is a polynomial) on its domain, a closed and bounded region in the xy-plane, so V has a global maximum there. The maximum does not occur on the boundary because $V(x, y)$ is identically zero on the boundary of its domain. So the global maximum occurs at an interior critical point where both partial derivatives are zero; that is,

$$(108 - 2x - 2y)y - 2xy = 0, \quad (108 - 2x - 2y)x - 2xy = 0.$$

We may cancel y from the first of these equations and x from the second because neither is zero at the maximum. Thus we are to solve

$$108 - 4x - 2y = 0, \quad 108 - 2x - 4y = 0.$$

It follows that $4x + 2y = 2x + 4y$, and thus that $x = y$. This implies in turn that $x = y = 18$ and $z = 36$. So the maximum volume of such a box is $18 \cdot 18 \cdot 36 = 11664$ cubic inches, exactly 6.25 cubic feet. If it were filled with osmium (the heaviest element known) it would weigh over 4338 kg, about 4.78 tons.

12.5.49: Suppose that the upper corner of the box in the first octant meets the paraboloid at the point (x, y, z), so that $z = 1 - x^2 - y^2$. We are to maximize box volume $V = 2x \cdot 2y \cdot z$; that is,

$$V(x, y) = 4xy(1 - x^2 - y^2) = 4xy - 4x^3y - 4xy^2, \quad 0 \leq x, \quad 0 \leq y, \quad x^2 + y^2 \leq 1.$$

Because V is continuous ($V(x, y)$ is a polynomial) and its domain is a closed and bounded subset of the xy-plane, V has a global maximum—which does not occur on the boundary of its domain because $V(x, y)$ is identically zero there. Hence the maximum we seek occurs at an interior critical point. When we set the partial derivatives of V simultaneously equal to zero, we obtain the equations

$$4y(1 - x^2 - y^2) - 8x^2y = 0, \quad 4x(1 - x^2 - y^2) - 8xy^2 = 0;$$

that is, because neither x nor y is zero at maximum box volume,

$$1 - 3x^2 - y^2 = 0 \quad \text{and} \quad 1 - x^2 - 3y^2 = 0.$$

It follows in the usual way that $y = x$, and thus that $x = y = \frac{1}{2}$. The corresponding value of z is also $\frac{1}{2}$, so the dimensions of the box of maximum volume are 1 by 1 by $\frac{1}{2}$ and its volume is $\frac{1}{2}$.

12.5.51: Let r be the common radius of the two cones and the cylinder, h the height of the cylinder, and z the height of each cone. Note that the slant height of each cone is $(r^2 + z^2)^{1/2}$, so each has curved surface area

$$2\pi \cdot \frac{r}{2} \cdot (r^2 + z^2)^{1/2} = \pi r(r^2 + z^2)^{1/2}.$$

We are to minimize the total surface area

$$A = 2\pi r(r^2 + z^2)^{1/2} + 2\pi rh \tag{1}$$

of the buoy given fixed volume $V = \frac{2}{3}\pi r^2 z + \pi r^2 h$. We first solve this last equation for

$$h = \frac{3V - 2\pi r^2 z}{3\pi r^2},$$

then substitute in (1) to express A as a function of r and z:

$$A(r, z) = \frac{2V}{r} - \frac{4\pi rz}{3} + 2\pi r(r^2 + z^2)^{1/2}.$$

The domain of A is described by the inequalities

$$0 < r, \quad 0 \leq z, \quad r^2 z \leq \frac{3V}{2\pi},$$

and though it is neither closed nor bounded, it can be shown by an argument similar to the one in Example 7 that $A(r, z)$ has a global minimum that does not occur on the boundary of its domain (unless it occurs where

767

$h = 0$ or where $z = 0$; we will attend to those possibilities later). Moreover, intuition and experience suggest that the minimal surface area will occur when the figure can be inscribed in a nearly spherical ellipsoid.

Next,

$$A_z(r, z) = \frac{2[3\pi r z - 2\pi r (r^2 + z^2)^{1/2}]}{3(r^2 + z^2)^{1/2}},$$

so $A_z(r, z) = 0$ when $3z = 2(r^2 + z^2)^{1/2}$. Therefore, when $A_z(r, z) = 0$, we have both $2r = z\sqrt{5}$ and $(r^2 + z^2)^{1/2} = \frac{3}{2}z$.

Also,

$$A_r(r, z) = \frac{2[6\pi r^4 + 3\pi r^2 z^2 - 3V(r^2 + z^2)^{1/2} - 2\pi r^2 z (r^2 + z^2)^{1/2}]}{3r^2(r^2 + z^2)^{1/2}},$$

so $A_r(r, z) = 0$ when

$$6\pi r^4 + 3\pi r^2 z^2 - 3V(r^2 + z^2)^{1/2} - 2\pi r^2 z (r^2 + z^2)^{1/2} = 0.$$

We substitute $\frac{3}{2}z$ for $(r^2 + z^2)^{1/2}$ in this last equation to find that when both partials vanish, also

$$\frac{9Vz}{2} = 6\pi r^4,$$

then we replace r with $\frac{1}{2}z\sqrt{5}$ to find that when both partials vanish,

$$25\pi z^3 = 12V.$$

Thus the minimum surface area seems to occur when

$$z = \left(\frac{12V}{25\pi}\right)^{1/3} \approx (0.534601847029)V^{1/3},$$

for which

$$r = \left(\frac{9V^2}{20\pi^2}\right)^{1/6} \approx (0.597703035427)V^{1/3}$$

and

$$h = \left(\frac{12V}{25\pi}\right)^{1/3} \approx (0.534601847029)V^{1/3}.$$

At these values of the variables, the surface area is

$$A = 5^{1/6}(18\pi V^2)^{1/3} \approx (5.019214931473)V^{2/3}.$$

The symmetry of the solution—that $h = z$—suggests that we have found the minimum, but we have yet to check the cases $h = 0$ and $z = 0$.

- If $z = 0$, then the buoy is a cylinder with radius r, height h, total surface area $A = 2\pi r^2 + 2\pi r h$, and fixed volume $V = \pi r^2 h$; we are to minimize its total surface area. We substitute

$$h = \frac{V}{\pi r^2}$$

in the surface area formula to obtain the function to be minimized:

$$A(r) = 2\pi r^2 + \frac{2V}{r}, \quad 0 < r < \infty.$$

Then

$$A'(r) = 4\pi r - \frac{2V}{r^2} = \frac{4\pi r^3 - 2V}{r^2}.$$

Thus $A'(r) = 0$ when

$$r = r_0 = \left(\frac{V}{2\pi}\right)^{1/3} \approx (0.541926070139)V^{1/3}.$$

The corresponding value of $A(r)$ is a global minimum by the first derivative test, and it is

$$A(r_0) = (54\pi V^2)^{1/3} \approx (5.535810445932)V^{2/3}.$$

This is somewhat larger than the minimum we found in the case $z > 0$.

• If $h = 0$, then the buoy consists of two congruent right circular cones with their bases, circles of radii r, coinciding, and each cone of height z. The total volume of the two cones is

$$V = \frac{2}{3}\pi r^2 z,$$

which we solve for z and substitute in the surface area formula $A = 2\pi r(r^2 + z^2)^{1/2}$ to obtain the function to be minimized,

$$A(r) = 2\pi r\left(r^2 + \frac{9V^2}{4\pi^2 r^4}\right)^{1/2} = \frac{(4\pi^2 r^6 + 9V^2)^{1/2}}{r}, \quad 0 < r.$$

Now

$$A'(r) = \frac{8\pi^2 r^6 - 9V^2}{r^2(4\pi^2 r^6 + 9V^2)^{1/2}};$$

$A'(r) = 0$ when $8\pi^2 r^6 = 9V^2$, so that

$$r = r_0 = \frac{\sqrt{2}}{2}\left(\frac{3V}{\pi}\right)^{1/3} \approx (0.696319882685)V^{1/3}.$$

Then

$$A(r_0) = 3^{7/6} \cdot \pi^{1/3} \cdot V^{2/3} \approx (5.276647566071)V^{2/3},$$

which is larger than the minimum found in the first part of this solution. Therefore the minimum possible surface area of the buoy is $5^{1/6}(18\pi V^2)^{1/3} \approx (5.019214931473)V^{2/3}$.

A final observation: The buoy of minimal possible surface area cannot be inscribed in a sphere. If $V = 1$ (say), then it can be inscribed in an ellipsoid (actually, a prolate spheroid) with approximate equation

$$\frac{x^2}{(0.69016802)^2} + \frac{y^2}{(0.69016802)^2} + \frac{z^2}{(1.06923700)^2} = 1$$

769

with the axis of symmetry of the buoy on the z-axis. A figure showing the elliptical cross-section of this ellipsoid in the xz-plane and the cross-section of the buoy in the xz plane (for the case $V = 1$) is next.

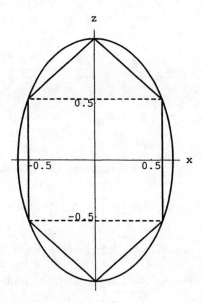

12.5.53: We want to minimize

$$f(x, y) = x^2 + (y - 1)^2 + x^2 + y^2 + (x - 2)^2 + y^2 = (x - 2)^2 + 2x^2 + (y - 1)^2 + 2y^2$$

with domain the entire xy-plane. An argument similar to the one used in Example 7 establishes that $f(x, y)$ has a global minimum value, which must be at a point where both partial derivatives are zero. These equations are $6x - 4 = 0$, $6y - 2 = 0$, with solution $x = \frac{2}{3}$, $y = \frac{1}{3}$. Because this is the only critical point, we have located the point that minimizes $f(x, y)$. It is $\left(\frac{2}{3}, \frac{1}{3}\right)$, and the value of $f(x, y)$ there is $\frac{10}{3}$.

It would seem somewhat more practical to find the point (x, y) such that the sum of the distances (not their squares) from (x, y) to the three points $(0, 1)$, $(0, 0)$, and $(2, 0)$ is a minimum. For example, where should a power company be located to minimize the total length of its cables to industries located at the three points $(0, 1)$, $(0, 0)$, and $(2, 0)$? This is a much more difficult problem. The function to be minimized is

$$h(x, y) = \sqrt{x^2 + (y - 1)^2} + \sqrt{x^2 + y^2} + \sqrt{(x - 2)^2 + y^2},$$

and its partial derivatives are

$$f(x, y) = h_x(x, y) = \frac{x}{\sqrt{x^2 + (y - 1)^2}} + \frac{x - 2}{\sqrt{(x - 2)^2 + y^2}} + \frac{x}{\sqrt{x^2 + y^2}} \quad \text{and}$$

$$g(x, y) = h_y(x, y) = \frac{y - 1}{\sqrt{x^2 + (y - 1)^2}} + \frac{y}{\sqrt{(x - 2)^2 + y^2}} + \frac{y}{\sqrt{x^2 + y^2}}.$$

It seems doubtful that any present-day computer algebra system could solve the simultaneous equations

$$f(x, y) = 0, \qquad g(x, y) = 0 \tag{1}$$

exactly with a reasonable amount of memory and in a reasonable time. But recall Newton's method from the last section of Chapter 3. Techniques of this chapter can be used to extend Newton's method to two simultaneous equations in two unknowns, as in Eq. (1). Here is the extension.

The graph of $f(x, y) = 0$ is, generally, a curve in the xy-plane, as is the graph of $g(x, y) = 0$. The point where these curves meet is the simultaneous solution that we seek. If (x_0, y_0) is an initial "guess" for the simultaneous solution, then $(x_0, y_0, f(x_0, y_0))$ is a point on the surface $z = f(x, y)$ and $(x_0, y_0, g(x_0, y_0))$ is a point on the surface $z = g(x, y)$. The tangent planes to these respective surfaces at these respective points generally meet the xy-plane in a pair of lines, whose intersection should be close to the desired solution $(x_\star, y_\star)$ of the simultaneous equations in (1). Let (x_1, y_1) denote the intersection of these lines, and repeat the process with (x_0, y_0) replaced with (x_1, y_1). This leads to the pair of iterative formulas

$$x_{k+1} = x_k - \frac{f(x_k, y_k)g_y(x_k, y_k) - g(x_k, y_k)f_y(x_k, y_k)}{f_x(x_k, y_k)g_y(x_k, y_k) - g_x(x_k, y_k)f_y(x_k, y_k)},$$

$$y_{k+1} = y_k - \frac{g(x_k, y_k)f_x(x_k, y_k) - f(x_k, y_k)g_x(x_k, y_k)}{f_x(x_k, y_k)g_y(x_k, y_k) - g_x(x_k, y_k)f_y(x_k, y_k)}$$

for $k = 0, 1, 2, \ldots$.

We wrote a simple *Mathematica* program to implement this procedure with the initial guess $(x_0, y_0) = (0.5, 0.5)$. Here are the results (rounded, of course):

$(x_1, y_1) = (0.201722209269, 0.369098300563),$ $(x_2, y_2) = (0.250232659645, 0.283132873843),$

$(x_3, y_3) = (0.254890752792, 0.304117462143),$ $(x_4, y_4) = (0.254568996440, 0.304503089648),$

$(x_5, y_5) = (0.254569313597, 0.304503701206),$ $(x_6, y_6) = (0.254569313597, 0.304503701206),$

and—to the number of digits shown—$(x_7, y_7) = (x_6, y_6)$. The sum of the distances from (x_6, y_6) to the three points $(0, 1)$, $(0, 0)$, and $(2, 0)$ is approximately 2.909312911180.

Warning: A much better "initial guess" is required than in the case of a function of a single variable. For example, with the initial guess $(x_0, y_0) = (2/3, 1/3)$ (the solution to the original version of Problem 53, which one would think would be a fairly good initial guess), $(x_6, y_6) \approx (1422.779, 732.866)$. Thus it is clear that Newton's method is not converging to the correct solution.

12.5.55: Let $2x$ denote the length of the base of each isosceles triangle, z the height of each triangle, and y the distance between the triangles. We are to minimize the area $A = 2xz + 2y\sqrt{x^2 + z^2}$ given the house has fixed volume $V = xyz$. Solve the latter equation for y and subsitute in the area formula to obtain the function

$$A(x, z) = 2xz + \frac{2V\sqrt{x^2 + z^2}}{xz}, \quad 0 < x, \quad 0 < z.$$

An argument similar to the one used in the solution of Example 7 shows that A must have a global minimum value even though its domain is neither closed nor bounded. Therefore, if there is a unique critical point of A in its domain, that will be the location of its global minimum.

To find the critical point or points, we use *Mathematica* and first define

```
a[x_, z_] := 2*x*z + (2*v*Sqrt[x*x + z*z])/(x*z)
```

Then we compute and simplify both partial derivatives.

```
d1 = D[ a[x,z], x ]
```

$$2z + \frac{2v}{z\sqrt{x^2 + z^2}} - \frac{2v\sqrt{x^2 + z^2}}{x^2 z}$$

```
d1 = Together[d1]
```

$$\frac{2z\left(-v + x^2\sqrt{x^2 + z^2}\right)}{x^2\sqrt{x^2 + z^2}}$$

```
n1 = Numerator[ d1 ]
```

$$2z\left(-v + x^2\sqrt{x^2 + z^2}\right)$$

```
d2 = D[ a[x,z], z ]
```

$$2x + \frac{2v}{x\sqrt{x^2 + z^2}} - \frac{2v\sqrt{x^2 + z^2}}{xz^2}$$

```
d2 = Together[ d2 ]
```

$$\frac{2x\left(-v + z^2\sqrt{x^2 + z^2}\right)}{z^2\sqrt{x^2 + z^2}}$$

```
n2 = Numerator[ d2 ]
```

$$2x\left(-v + z^2\sqrt{x^2 + z^2}\right)$$

Because neither $x = 0$ nor $z = 0$ (because $V > 0$), we may cancel:

```
n1 = n1/(2*z) // Cancel
```

$$-v + x^2\sqrt{x^2 + z^2}$$

```
n2 = n2/(2*x) // Cancel
```

$$-v + z^2\sqrt{x^2 + z^2}$$

Both of the last expressions must be zero when both partial derivatives are set equal to zero, and it follows that $z = x$. We substitute this information into the numerator $n1$ of $A_x(x, z)$ and set the result equal to zero.

```
n1 /.  z -> x
```

$$-v + \sqrt{2}\ (x^2)^{3/2}$$

```
Solve[ % == 0, x ]
```

And *Mathematica* returns six solutions, only two of which are real, and only one of the two is positive: We find that

$$x = z = \frac{V^{1/3}}{2^{1/6}}, \quad \text{and} \quad y = 2^{1/3} \cdot V^{1/3},$$

so that the minimum possible surface area is $3 \cdot 2^{2/3} \cdot V^{2/3}$.

12.5.57: Let x, y, and z be the lengths of the edges of the three squares. We are to maximize and minimize their total area $A = x^2 + y^2 + z^2$ given the condition $4x + 4y + 4z = 120$; that is, $x + y + z = 30$. Using this side condition to eliminate z in the expression for A, we obtain the function

$$A(x, y) = x^2 + y^2 + (30 - x - y)^2, \quad 0 \le x, \quad 0 \le y, \quad x + y \le 30.$$

The domain D of A is the triangular region in the xy-plane with vertices at $(0, 0)$, $(30, 0)$, and $(0, 30)$, including the boundary segments. Hence D is a closed and bounded subset of the plane and A is continuous there (because $A(x, y)$, when expanded, is a polynomial). Therefore there is both a global maximum and a global minimum value of $A(x, y)$. We proceed in the usual way, first setting both partial derivatives equal to zero:

$$2x - 2(30 - x - y) = 0 \quad \text{and} \quad 2y - 2(30 - x - y) = 0.$$

It follows that $y = x$. By symmetry (rework the problem eliminating y rather than z) $z = x$, so that $x = y = z$. Therefore $x = y = z = 10$ may yield an extremum of A.

- On the boundary segment of D on which $y = 0$,

$$A(x, 0) = f(x) = x^2 + (30 - x)^2, \quad 0 \le x \le 30.$$

Methods of single-variable calculus yield the critical point $x = 15$; we have also the two endpoints of the domain of f to check.

- On the boundary segment of D on which $x = 0$,

$$A(0, y) = g(y) = y^2 + (30 - y)^2, \quad 0 \le y \le 30.$$

773

Methods of single-variable calculus yield the critical point $y = 15$; we have also the two endpoints of the domain of g to check.

- On the boundary segment of D with equation $x + y = 30$, we have $y = 30 - x$, so that

$$A(x, 30 - x) = h(x) = x^2 + (30 - x)^2, \quad 0 \leqq x \leqq 30.$$

Again, methods of single-variable calculus yield the critical point $x = 15$; the endpoints of this boundary segment will be checked in the other two cases.

Results:

$$A(10, 10) = 300, \qquad A(15, 0) = 450,$$

$$A(0, 15) = 450, \qquad A(0, 0) = 900,$$

$$A(30, 0) = 900, \qquad A(0, 30) = 900, \quad \text{and}$$

$$A(15, 15) = 450.$$

Answer: For maximum total area, make only one square, measuring 30 cm on each side, with area 900 cm^2. For minimum total area, make three equal squares, each measuring 10 cm on each side, with total area 300 cm^2.

12.5.59: Using the notation in Fig. 13.5.16, we have cross-sectional area

$$A(x, \theta) = (L - 2x)x \sin\theta + x^2 \sin\theta \cos\theta, \quad 0 \leqq \theta \leqq \frac{\pi}{2}, \quad 0 \leqq x \leqq \frac{L}{2}.$$

to be maximized. When we set the partial derivatives of A equal to zero, we obtain the equations

$$(L - 2x)\sin\theta - 2x\sin\theta + 2x\sin\theta\cos\theta = 0, \quad (L - 2x)x\cos\theta + x^2\cos^2\theta - x^2\sin^2\theta = 0. \tag{1}$$

The first equation in (1) yields

$$(L - 2x) - 2x + 2x\cos\theta = 0 \quad \text{or} \quad \sin\theta = 0,$$

but the second of these can be rejected as $\theta = 0$ minimizes the cross-sectional area. The second equation in (1) yields

$$(L - 2x)\cos\theta + x\cos^2\theta - x\sin^2\theta = 0 \quad \text{or} \quad x = 0,$$

and the second of these can also be rejected because $x = 0$ minimizes the area. Thus we have the simultaneous equations

$$L - 2x - 2x + 2x\cos\theta = 0 \quad \text{and} \quad (L - 2x)\cos\theta + x\cos^2\theta - x\sin^2\theta = 0. \tag{2}$$

We solve the first of these for $L - 2x = 2x - 2x\cos\theta$ and substitute in the second to obtain

$$2x(1 - \cos\theta)\cos\theta + x\cos^2\theta - x\sin^2\theta = 0;$$

$$2(1 - \cos\theta)\cos\theta + \cos^2\theta - \sin^2\theta = 0;$$

$$2\cos\theta - 2\cos^2\theta + \cos^2\theta - 1 + \cos^2\theta = 0;$$

$$\cos\theta = \frac{1}{2};$$

$$\theta = \frac{\pi}{3}.$$

In the second step we used the fact that $x = 0$ minimizes A to cancel x with impunity; in the last step we used the domain of A to determine θ. In any case, the first equation in (2) lets us determine the maximizing value of x as well:

$$L - 4x + x = 0, \quad \text{and thus} \quad x = \frac{L}{3}.$$

We have found only one critical point, and the only endpoint that might produce a larger cross-sectional area occurs when $\theta = \pi/2$, in which case the first equation in (2) implies that

$$L - 4x = 0, \quad \text{so that} \quad x = \frac{L}{4}.$$

In the latter case the cross-sectional area of the gutter is $L^2/8$, but in the previous case, when $\theta = \pi/3$, we find that $A = \left(L^2\sqrt{3}\right)/12 \approx (0.1443)L^2$, so this is the maximum possible cross-sectional area of the gutter.

12.5.61: Given:

$$P(x) = -2x^2 + 12x + xy - y - 10,$$

$$Q(y) = -3y^2 + 18y + 2xy - 2x - 15.$$

Part (a): This is known as a *game of perfect information*—each player (manager) knows every strategy available to his opponent. Each manager computes

$$P'(x) = 12 - 4x + y, \quad Q'(y) = 18 + 2x - 6y,$$

sets both equal to zero (knowing that the other manager is doing the same), and solves for $x = 45/11$, $y = 48/11$. Thus each maximizes his profit knowing that the other manager is doing the same; indeed, if either player (manager) deviates from his *optimal strategy*, his profit will decrease and that of his opponent (the other manager) is likely to increase. With these values of x and y, the profits will be

$$P = \frac{2312}{121} \approx 19.107 \quad \text{and} \quad Q = \frac{4107}{121} \approx 33.942.$$

Part (b): After the merger and the agreement to maximize total profit, the new partners plan to maximize

$$R(x, y) = P(x) + Q(y) = -2x^2 - 3y^2 + 3xy + 10x + 17y - 25.$$

The junior partner computes both partial derivatives and sets both equal to zero:

$$10 - 4x + 3y = 0, \quad 17 + 3x - 6y = 0.$$

The simultaneous solution is

$$x = \frac{37}{5}, \quad y = \frac{98}{15}$$

for a combined profit of

$$\frac{1013}{15} \approx 67.533 > 53.050 \approx \frac{6419}{121} = \frac{2312}{121} + \frac{4107}{121}.$$

Thus the merger increases total profit.

12.5.63: Let x be the number of sheep, y the number of hogs, and z the number of head of cattle. Suppose that 60 cattle use 1 unit of land. Then each head of cattle uses $\frac{1}{60}$ units of land. By similar reasoning, each hog uses $\frac{1}{120}$ units of land and each sheep uses $\frac{1}{80}$ units of land. This leads to the side condition

$$\frac{x}{80} + \frac{y}{120} + \frac{z}{60} = 1$$

for each unit of land available. Let's write this in the simpler form $3x + 2y + 4z = 240$. An additional condition in the problem is that $y \geq x + z$. We are now to maximize the profit P per unit of land, given by $P(x, y, z) = 10x + 8y + 20z$.

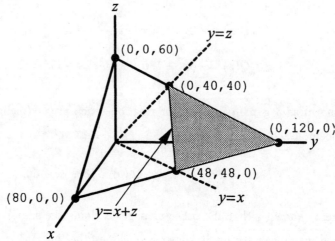

The domain of P is the triangular region shown shaded in the preceding figure. It is obtained as follows: Draw the part of the plane $3x + 2y + 4z = 240$ that lies in the first octant (because none of x, y, or z can be negative). The intersections of that plane with the positive coordinate planes are shown as solid lines. The intersection of the plane $y = x + z$ with the positive coordinate planes is shown as a pair of dashed lines. The condition $y \geqq x + z$ implies that only the part of the first plane to the *right* of the second may be used as the domain of P. Therefore we arrive at the shaded triangle as the domain of the profit function. Finally, because P is a linear function of x, y, and z, its maximum and minimum values occur at the vertices of the shaded triangle. Here, then, are the results:

Vertex: (48, 48, 0);   Profit: \$864   per unit of land.

Vertex: (0, 120, 0);   Profit: \$960   per unit of land.

Vertex: (0, 40, 40);   Profit: \$1120   per unit of land.

Were it not for the restriction of the state law mentioned in the problem, the farmer could maximize her profit per unit of land by raising only cattle: $P(0, 0, 60) = 1200$. *Answer:* Raise 40 hogs and 40 cattle per unit of land, but no sheep.

12.5.65: With $x = r\cos(t)$, $y = r\sin(t)$, $z = ax^2 + 2bxy + cz^2$ and $a = 3$, $b = 1$, $c = 2$, the *Mathematica* commands

```
x = r Cos[t]; y = r Sin[t];

z = a x^2 + 2b x y + c y^2;

ParametricPlot3D[ { x, y, z }, { r, 0, 1 }, { t, 0, 2*Pi } ];
```

(with additional plot parameters specified as necessary) produce the following graph.

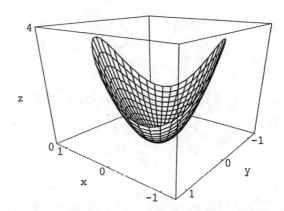

With $a = 3$, $b = 1$, $c = 2$ we get the graph
.

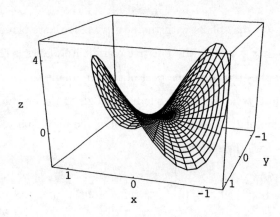

These two examples support the conclusions given in the statement of Problem 65 (why?), and you should try additional combinations of the parameters a, b, c.

12.5.67: Upon changing to polar coordinates we get $f(x, y) = x^4 + 2bx^2y^2 + y^4 = r^4g(\theta)$ where

$$g(\theta) = \cos^4\theta + 2b\cos^2\theta\sin^2\theta + \sin^4\theta.$$

Upon differentiating and simplifying we find that

$$g'(\theta) = 4(b-1)(\cos^2\theta - \sin^2\theta)\cos\theta\sin\theta.$$

Hence the critical points of g are

- multiples of $\pi/2$, where $g(\theta) = 1$, and
- odd multiples of $\pi/4$, where $\cos^2\theta = \sin^2\theta = \frac{1}{2}$ so $g(\theta) = \frac{1}{4} + \frac{1}{2}b + \frac{1}{4} = \frac{1}{2}(b+1)$.

If $b > -1$ it follows that $g(\theta)$ is always positive, so $f(x, y) = r^4g(\theta)$ is positive except at the origin. But if $b < -1$ it follows that $g(\theta)$ attains both positive and negative values, so $z = f(x, y) = r^4g(\theta)$ exhibits a saddle point at the origin.

12.5.69: When we set all first-order partial derivative of g equal to zero, we get the simultaneous equations

$$g_x(x, y, z) = 4x^3 - 16xy^2 = 0,$$

$$g_y(x, y, z) = 4y^3 - 16x^2y = 0,$$

$$g_z(x, y, z) = z = 0.$$

The only solution of these equations is $x = y = z = 0$, at which $g(0, 0, 0) = 12$. But

$$g(1, 1, 0) = 6 < 12 \quad \text{and} \quad g(0, 0, 2) = 28 > 12.$$

778

Therefore g has no global maximum or minimum at the origin. Examination of the behavior of $g(x, y, z)$ on the two lines

$$y = x, \quad z = 0 \qquad \text{and} \qquad x = 0, \quad y = 0$$

is enough to establish that g has no extrema, global or local. Thus one cannot conclude that g has an extremum at a point where all its partial derivatives are zero.

Section 12.6

12.6.1: If $w = 3x^2 + 4xy - 2y^3$, then $dw = (6x + 4y)\,dx + (4x - 6y^2)\,dy$.

12.6.3: If $w = \sqrt{1 + x^2 + y^2}$, then

$$dw = \frac{x}{\sqrt{1 + x^2 + y^2}}\,dx + \frac{y}{\sqrt{1 + x^2 + y^2}}\,dy = \frac{x\,dx + y\,dy}{\sqrt{1 + x^2 + y^2}}.$$

12.6.5: If $w(x, y) = \arctan\left(\dfrac{x}{y}\right)$, then $dw = \dfrac{y\,dx - x\,dy}{x^2 + y^2}$.

12.6.7: If $w = \ln(x^2 + y^2 + z^2)$, then

$$dw = \frac{2x\,dx}{x^2 + y^2 + z^2} + \frac{2y\,dy}{x^2 + y^2 + z^2} + \frac{2z\,dz}{x^2 + y^2 + z^2} = \frac{2x\,dx + 2y\,dy + 2z\,dz}{x^2 + y^2 + z^2}.$$

12.6.9: If $w = x\tan yz$, then $dw = \tan yz\,dx + xz\sec^2 yz\,dy + xy\sec^2 yz\,dz$.

12.6.11: If $w = e^{-xyz}$, then $dw = -yze^{-xyz}\,dx - xze^{-xyz}\,dy - xye^{-xyz}\,dz$.

12.6.13: If $w = u^2\exp(-v^2)$, then $dw = 2u\exp(-v^2)\,du - 2u^2 v\exp(-v^2)\,dv$.

12.6.15: If $w = \sqrt{x^2 + y^2 + z^2}$, then

$$dw = \frac{x\,dx}{\sqrt{x^2 + y^2 + z^2}} + \frac{y\,dy}{\sqrt{x^2 + y^2 + z^2}} + \frac{z\,dz}{\sqrt{x^2 + y^2 + z^2}} = \frac{x\,dx + y\,dy + z\,dz}{\sqrt{x^2 + y^2 + z^2}}.$$

12.6.17: If $w = f(x, y) = \sqrt{x^2 + y^2}$, then

$$dw = \frac{x\,dx + y\,dy}{\sqrt{x^2 + y^2}}.$$

Choose $x = 3$, $y = 4$, $dx = -0.03$, and $dy = 0.04$. Then

$$f(2.97, 4.04) \approx f(3, 4) + \frac{3\cdot(-0.03) + 4\cdot(0.04)}{\sqrt{3^2 + 4^2}} = \frac{2507}{500} = 5.014.$$

779

Compare with the true value of

$$f(2.97, 4.04) = \frac{\sqrt{10057}}{20} \approx 5.014229751417.$$

12.6.19: If $w = f(x, y) = \dfrac{1}{1 + x + y}$, then

$$dw = -\frac{dx + dy}{(1 + x + y)^2}.$$

Choose $x = 3$, $y = 6$, $dx = 0.02$, and $dy = 0.05$. Then

$$f(3.02, 6.05) \approx f(3, 6) - \frac{0.02 + 0.05}{(1 + 3 + 6)^2} = \frac{1}{10} - \frac{7}{10000} = \frac{993}{10000} = 0.0993.$$

Compare with the true value of

$$f(3.02, 6.05) = \frac{100}{1007} \approx 0.0993048659384310.$$

12.6.21: If $w = f(x, y, z) = \sqrt{x^2 + y^2 + z^2}$, then

$$dw = \frac{x\,dx + y\,dy + z\,dz}{\sqrt{x^2 + y^2 + z^2}}.$$

Choose $x = 3$, $y = 4$, $z = 12$, $dx = 0.03$, $dy = -0.04$, and $dz = 0.05$. Then

$$f(3.03, 3.96, 12.05) \approx f(3, 4, 12) + \frac{3 \cdot (0.03) - 4 \cdot (0.04) + 12 \cdot (0.05)}{\sqrt{3^2 + 4^2 + 12^2}}$$

$$= 13 + \frac{53}{1300} = \frac{16953}{1300} \approx 13.040769230769.$$

Compare with the true value of

$$f(3.03, 3.96, 12.05) = \frac{\sqrt{68026}}{20} \approx 13.040897208398.$$

12.6.23: If $w = f(x, y, z) = e^{-xyz}$, then

$$dw = -e^{-xyz}(yz\,dx + xz\,dy + xy\,dz).$$

Take $x = 1$, $y = 0$, $z = -2$, $dx = 0.02$, $dy = 0.03$, and $dz = -0.02$. Then

$$f(1.02, 0.03, -2.02) \approx f(1, 0, -2) - e^0(0 - 2 \cdot (0.03) + 0) = 1 + \frac{3}{50} = 1.06.$$

Compare with the exact value, which is

$$f(1.02, 0.03, -2.02) = \exp\left(\frac{15453}{250000}\right) \approx 1.0637623386083891.$$

12.6.25: If $w = f(x, y) = \left(\sqrt{x} + \sqrt{y}\right)^2$, then

$$dw = \frac{\sqrt{x} + \sqrt{y}}{\sqrt{x}}\, dx + \frac{\sqrt{x} + \sqrt{y}}{\sqrt{y}}\, dy.$$

Take $x = 16$, $y = 100$, $dx = -1$, and $dy = -1$. Then

$$f(15, 99) \approx f(16, 100) + \frac{4 + 10}{4} \cdot (-1) + \frac{4 + 10}{10} \cdot (-1) = 196 - \frac{49}{10} = \frac{1911}{10} = 191.1.$$

By comparison, the exact value is

$$f(15, 99) = \left(\sqrt{15} + \sqrt{99}\right)^2 \approx 191.0713954719907741.$$

12.6.27: If $w = f(x, y) = \exp(x^2 - y^2)$, then

$$dw = 2x \exp(x^2 - y^2)\, dx - 2y \exp(x^2 - y^2)\, dy.$$

Take $x = 1$, $y = 1$, $dx = 0.1$, and $dy = -0.1$. Then

$$f(1.1, 0.9) \approx f(1, 1) + 2 \cdot (0.1) - 2 \cdot (-0.1) = 1 + \frac{2}{5} = \frac{7}{5} = 1.4.$$

Compare with the true value, which is

$$f(1.1, 0.9) = e^{2/5} \approx 1.4918246976412703.$$

12.6.29: If $w = f(x, y, z) = \sqrt{x^2 + y^2 + z^2}$, then

$$dw = \frac{x\, dx + y\, dy + z\, dz}{\sqrt{x^2 + y^2 + z^2}}.$$

Take $x = 3$, $y = 4$, $z = 12$, $dx = 0.1$, $dy = 0.2$, and $dz = -0.3$. Then

$$f(3.1, 4.2, 11.7) \approx f(3, 4, 12) + \frac{0.3 + 0.8 - 3.6}{13} = 13 - \frac{5}{26} = \frac{333}{26} \approx 12.8076923076923077.$$

The true value is

$$f(3.1, 4.2, 11.7) = \frac{\sqrt{16414}}{10} \approx 12.8117133904876283.$$

12.6.31: Given: The point $Q(1, 2)$ on the curve $f(x, y) = 0$, where $f(x, y) = 2x^3 + 2y^3 - 9xy$. Then

$$df = (6x^2 - 9y)\, dx + (6y^2 - 9x)\, dy = 0.$$

Choose $x = 1$, $y = 2$, and $dx = 0.1$. Then

$$(6 - 18) \cdot \frac{1}{10} + (24 - 9)\, dy = 0;$$

$$dy = \frac{1}{15} \cdot \frac{12}{10} = \frac{12}{150} = \frac{2}{25} = 0.08.$$

So the point P on the curve $f(x, y) = 0$ near Q and with x-coordinate 1.1 has y-coordinate

$$y \approx 2 + \frac{2}{25} = \frac{52}{25} = 2.08.$$

The true value of the y-coordinate is approximately 2.0757642703016864.

12.6.33: Suppose that the base of the rectangle has length x and that its height is y. Then its area is $w = f(x, y) = xy$, and $dw = y\, dx + x\, dy$. Choose $x = 10$, $y = 15$, $dx = 0.1$, and $dy = 0.1$. Then $dw = 2.5$; this is the estimate of the maximum error in computing the area of the rectangle. The actual maximum error possible is $f(10.1, 15.1) - f(10, 15) = 2.51$.

12.6.35: The volume of the cone is given by

$$w = f(r, h) = \frac{\pi}{3} r^2 h, \quad \text{so that} \quad dw = \frac{2\pi}{3} rh\, dr + \frac{\pi}{3} r^2\, dh.$$

Choose $r = 5$, $h = 10$, and $dr = dh = 0.1$. Then

$$dw = \frac{2\pi}{3} \cdot 5 \cdot 10 \cdot (0.1) + \frac{\pi}{3} \cdot 5^2 \cdot (0.1) = \frac{25\pi}{6} \approx 13.0899693899574718$$

is an estimate of the maximum error in measuring the volume of the cylinder. The true value of the maximum error is

$$f(5.1, 10.1) - f(5, 10) = \frac{12701\pi}{3000} \approx 13.3004560977479880.$$

12.6.37: If the sides of the field are x and y and the angle between them is θ, then the area of the field is given by

$$w = f(x, y, \theta) = \frac{1}{2} xy \sin\theta,$$

so that

$$dw = \frac{1}{2} y \sin\theta\, dx + \frac{1}{2} x \sin\theta\, dy + \frac{1}{2} xy \cos\theta\, d\theta.$$

If $x = 500$, $y = 700$, $\theta = \pi/6$, $dx = dy = 1$, and $d\theta = \pi/720$, then

$$dw = 350 \cdot \frac{1}{2} \cdot 1 + 250 \cdot \frac{1}{2} \cdot 1 + \frac{1}{2} \cdot 500 \cdot 700 \cdot \frac{\sqrt{3}}{2} \cdot \frac{\pi}{720} = 300 + \frac{4375\pi\sqrt{3}}{36} \approx 961.2810182103919247$$

(in square feet) is an estimate of the maximum error in computing the area of the field. The true value of the maximum error is

$$f(501, 701, (\pi/6) + (\pi/720)) - f(500, 700, \pi/6) \approx 962.9622561829376760$$

(in square feet). The former amounts to approximately 0.0220679756246646 acres (there are 43560 square feet in one acre).

12.6.39: The period T of a pendulum of length L is given (approximately) by

$$T = 2\pi \left(\frac{L}{g}\right)^{1/2}, \quad \text{for which} \quad dT = \left(\frac{g}{L}\right)^{1/2} \cdot \frac{\pi g\, dL - \pi L\, dg}{g^2}.$$

If $L = 2$, $dL = 1/12$, $g = 32$, and $dg = 0.2$, then

$$dT = \frac{17\pi}{1920} \approx 0.0278161849536596.$$

The true value of the increase in the period is

$$T(2 + 1/12, 32.2) - T(2, 32) \approx 0.0274043631738259.$$

12.6.41: Given: $R(v_0, \alpha) = \dfrac{1}{32}(v_0)^2 \sin 2\alpha$, we first compute

$$dR = \frac{1}{16}\left(v_0 \sin 2\alpha\, dv_0 + (v_0)^2 \cos 2\alpha\, d\alpha\right).$$

Substitution of $v_0 = 400$, $dv_0 = 10$, $\alpha = \pi/6$, and $d\alpha = \pi/180$ yields

$$dR = 125\sqrt{3} + \frac{250\pi}{9} \approx 303.7728135458261405$$

as an estimate of the increase in the range. The true value of the increase is

$$R(410, (\pi/6) + (\pi/180)) - R(400, \pi/6) = -2500\sqrt{3} + \frac{42025}{8} \sin\left(\frac{31\pi}{90}\right) \approx 308.1070548148573585.$$

12.6.43: Part (a): If $(x, y) \to (0, 0)$ along the line $y = x$, then

$$\lim_{(x,y)\to(0,0)} f(x, y) = \lim_{x\to 0} f(x, x) = \lim_{x\to 0} 1 = 1.$$

But if $(x, y) \to (0, 0)$ along the line $y = 0$, then

$$\lim_{(x,y)\to(0,0)} f(x, y) = \lim_{x\to 0} f(x, 0) = \lim_{x\to 0} 0 = 0.$$

Therefore f is not continuous at $(0, 0)$.

Part (b): We compute the partial derivatives of f at $(0, 0)$ by the definition:

$$f_x(0, 0) = \lim_{h \to 0} \frac{f(0 + h, 0) - f(0, 0)}{h} = \lim_{h \to 0} \frac{f(h, 0)}{h} = \lim_{h \to 0} \frac{0}{h} = 0;$$

$$f_y(0, 0) = \lim_{k \to 0} \frac{f(0, 0 + k) - f(0, 0)}{k} = \lim_{k \to 0} \frac{f(0, k)}{k} = \lim_{k \to 0} \frac{0}{k} = 0.$$

Therefore both f_x and f_y exist at $(0, 0)$ but f is not continuous at $(0, 0)$.

12.6.45: Given:

$$f(x, y) = y^2 + x^2 \sin \frac{1}{x}$$

if $x \neq 0$; $f(0, y) = y^2$. Then

$$f_x(0, 0) = \lim_{h \to 0} \frac{f(h, 0) - f(0, 0)}{h} = \lim_{h \to 0} \frac{1}{h} \cdot h^2 \sin \frac{1}{h} = \lim_{h \to 0} h \sin \frac{1}{h} = 0$$

and

$$f_y(0, 0) = \lim_{k \to 0} \frac{f(0, k) - f(0, 0)}{k} = \lim_{k \to 0} \frac{k^2}{k} = \lim_{k \to 0} k = 0.$$

Therefore the linear approximation to f at $(0, 0)$ can only be $z = 0$. Moreover,

$$0 \leqq f(x, y) \leqq x^2 + y^2 = g(x, y)$$

for all (x, y), and $z = g(x, y)$ has the tangent plane $z = 0$ at $(0, 0)$. Therefore $z = 0$ does approximate $f(x, y)$ accurately at and near $(0, 0)$. That is, f is differentiable at $(0, 0)$. But if $x \neq 0$, then

$$f_x(x, y) = 2x \sin \frac{1}{x} - \cos \frac{1}{x},$$

so $f_x(x, y)$ has no limit as $(x, y) \to (0, 0)$ along the x-axis. Therefore $f_x(x, y)$ is not continuous at $(0, 0)$, and thus f is not continuously differentiable at $(0, 0)$ even though it is differentiable there.

12.6.47: Suppose that the function f of $n \geqq 2$ variables is differentiable at $\mathbf{a}$. Then there exists a constant vector $\mathbf{c} = \langle c_1, c_2, \ldots, c_n \rangle$ such that

$$\lim_{\mathbf{h} \to 0} \frac{f(\mathbf{a} + \mathbf{h}) - f(\mathbf{a}) - \mathbf{c} \cdot \mathbf{h}}{|\mathbf{h}|} = 0.$$

Therefore

$$\left(\lim_{\mathbf{h} \to 0} |\mathbf{h}| \right) \cdot \left(\lim_{\mathbf{h} \to 0} \frac{f(\mathbf{a} + \mathbf{h}) - f(\mathbf{a}) - \mathbf{c} \cdot \mathbf{h}}{|\mathbf{h}|} \right) = 0 \cdot 0 = 0,$$

and therefore

$$\lim_{\mathbf{h} \to 0} \left[f(\mathbf{a} + \mathbf{h}) - f(\mathbf{a}) - \mathbf{c} \cdot \mathbf{h} \right] = 0.$$

But $\mathbf{c} \cdot \mathbf{h} \to 0$ as $\mathbf{h} \to 0$ because $\mathbf{c}$ is a constant vector. Therefore

$$\left(\lim_{\mathbf{h} \to 0} [f(\mathbf{a} + \mathbf{h}) - f(\mathbf{a}) - \mathbf{c} \cdot \mathbf{h}] \right) + \left(\lim_{\mathbf{h} \to 0} [\mathbf{c} \cdot \mathbf{h} + f(\mathbf{a})] \right) = 0 + 0 + f(\mathbf{a}) = f(\mathbf{a}),$$

and thus we see that

$$\lim_{\mathbf{h} \to 0} f(\mathbf{a} + \mathbf{h}) = f(\mathbf{a}).$$

Therefore f is continuous at $\mathbf{x} = \mathbf{a}$. That is, the function f is continuous wherever it is differentiable.

Section 12.7

12.7.1: If $w = \exp(-x^2 - y^2)$, $x = t$, and $y = t^{1/2}$, then

$$\frac{dw}{dt} = \frac{\partial w}{\partial x} \cdot \frac{dx}{dt} + \frac{\partial w}{\partial y} \cdot \frac{dy}{dt}$$

$$= -2x \exp(-x^2 - y^2) - y t^{-1/2} \exp(-x^2 - y^2) = -2t \exp(-t^2 - t) - \exp(-t^2 - t).$$

Alternatively, $w = \exp(-t^2 - t)$, and hence

$$\frac{dw}{dt} = -(2t + 1) \exp(-t^2 - t).$$

12.7.3: If $w(x, y, z) = \sin xyz$, $x = t$, $y = t^2$, and $z = t^3$, then

$$\frac{dw}{dt} = \frac{\partial w}{\partial x} \cdot \frac{dx}{dt} + \frac{\partial w}{\partial y} \cdot \frac{dy}{dt} + \frac{\partial w}{\partial z} \cdot \frac{dz}{dt}$$

$$= yz \cos xyz + 2txz \cos xyz + 3t^2 xy \cos xyz = t^5 \cos t^6 + 2t^5 \cos t^6 + 3t^5 \cos t^6 = 6t^5 \cos t^6.$$

Alternatively, $w = \sin t^6$, and thus $\dfrac{dw}{dt} = 6t^5 \cos t^6$.

12.7.5: If $w(x, y, z) = \ln(x^2 + y^2 + z^2)$, $x = s - t$, $y = s + t$, and $z = 2(st)^{1/2}$, then

$$\frac{\partial w}{\partial s} = \frac{\partial w}{\partial x} \cdot \frac{\partial x}{\partial s} + \frac{\partial w}{\partial y} \cdot \frac{\partial y}{\partial s} + \frac{\partial w}{\partial z} \cdot \frac{\partial z}{\partial s}$$

$$= \frac{2x}{x^2 + y^2 + z^2} + \frac{2y}{x^2 + y^2 + z^2} + \frac{2tz}{(st)^{1/2}(x^2 + y^2 + z^2)} = \frac{2(st)^{1/2}x + 2(st)^{1/2}y + 2tz}{(st)^{1/2}(x^2 + y^2 + z^2)}$$

$$= \frac{2(s - t)(st)^{1/2} + 4t(st)^{1/2} + 2(s + t)(st)^{1/2}}{(st)^{1/2}[(s - t)^2 + 4st + (s + t)^2]} = \frac{2(2s + 2t)}{2s^2 + 4st + 2t^2} = \frac{2}{s + t}$$

and

$$\frac{\partial w}{\partial t} = \frac{\partial w}{\partial x} \cdot \frac{\partial x}{\partial t} + \frac{\partial w}{\partial y} \cdot \frac{\partial y}{\partial t} + \frac{\partial w}{\partial z} \cdot \frac{\partial z}{\partial t}$$

$$= -\frac{2x}{x^2 + y^2 + z^2} + \frac{2y}{x^2 + y^2 + z^2} + \frac{2sz}{(st)^{1/2}(x^2 + y^2 + z^2)} = \frac{-2(st)^{1/2}x + 2(st)^{1/2}y + 2sz}{(st)^{1/2}(x^2 + y^2 + z^2)}$$

$$= \frac{4s(st)^{1/2} - 2(s - t)(st)^{1/2} + 2(s + t)(st)^{1/2}}{(st)^{1/2}[(s - t)^2 + 4st + (s + t)^2]} = \frac{2(2s + 2t)}{2s^2 + 4st + 2t^2} = \frac{2}{s + t}.$$

Alternatively,

$$w(s,\, t) = \ln\left((s-t)^2 + 4st + (s+t)^2\right) = \ln(2s^2 + 4st + 2t^2),$$

and therefore

$$\frac{\partial w}{\partial s} = \frac{2}{s+t} \quad \text{and} \quad \frac{\partial w}{\partial t} = \frac{2}{s+t}.$$

12.7.7: If $w(u,\, v,\, z) = \sqrt{u^2 + v^2 + z^2}$, $u = 3e^t \sin s$, $v = 3e^t \cos s$, and $z = 4e^t$, then

$$\frac{\partial w}{\partial s} = \frac{3ue^t \cos s}{\sqrt{u^2 + v^2 + z^2}} - \frac{3ve^t \sin s}{\sqrt{u^2 + v^2 + z^2}} + 0 = \frac{3e^t(u \cos s - v \sin s)}{\sqrt{u^2 + v^2 + z^2}} = 0$$

because $u \cos s - v \sin s = 3e^t \sin s \cos s - 3e^t \cos s \sin s = 0$. But

$$\frac{\partial w}{\partial t} = \frac{3ue^t \sin s}{\sqrt{u^2 + v^2 + z^2}} + \frac{3ve^t \cos s}{\sqrt{u^2 + v^2 + z^2}} + \frac{4ze^t}{\sqrt{u^2 + v^2 + z^2}}$$

$$= \frac{e^t(3u \sin s + 3v \cos s + 4z)}{\sqrt{u^2 + v^2 + z^2}} = \frac{e^t(16e^t + 9e^t \cos^2 s + 9e^t \sin^2 s)}{\sqrt{16e^{2t} + 9e^{2t} \cos^2 s + 9e^{2t} \sin^2 s}} = 5e^t.$$

Alternatively,

$$w(s,\, t) = \sqrt{16e^{2t} + 9e^{2t} \cos^2 s + 9e^{2t} \sin^2 s} = 5e^t,$$

and therefore

$$\frac{\partial w}{\partial s} = 0 \quad \text{and} \quad \frac{\partial w}{\partial t} = 5e^t.$$

12.7.9: Because $r(x,\, y,\, z) = \exp(yz + xz + xy)$, we find that

$$\frac{\partial r}{\partial x} = (y + z)\exp(yz + xz + xy), \qquad \frac{\partial r}{\partial y} = (x + z)\exp(yz + xz + xy),$$

$$\text{and} \qquad \frac{\partial r}{\partial z} = (x + y)\exp(yz + xz + xy).$$

12.7.11: Because

$$r(x,\, y,\, z) = \sin\left(\frac{\sqrt{xy^2z^3}}{\sqrt{x + 2y + 3z}}\right),$$

we find that

$$\frac{\partial r}{\partial x} = \frac{(2y + 3z)\sqrt{xy^2z^3}}{2x(x + 2y + 3z)^{3/2}} \cos\left(\frac{\sqrt{xy^2z^3}}{\sqrt{x + 2y + 3z}}\right),$$

$$\frac{\partial r}{\partial y} = \frac{(x + y + 3z)\sqrt{xy^2z^3}}{y(x + 2y + 3z)^{3/2}} \cos\left(\frac{\sqrt{xy^2z^3}}{\sqrt{x + 2y + 3z}}\right), \quad \text{and}$$

$$\frac{\partial r}{\partial z} = \frac{3(x + 2y + 2z)\sqrt{xy^2z^3}}{2z(x + 2y + 3z)^{3/2}} \cos\left(\frac{\sqrt{xy^2z^3}}{\sqrt{x + 2y + 3z}}\right).$$

12.7.13: If $p = f(x, y)$, $x = x(u, v, w)$, and $y = y(u, v, w)$, then

$$\frac{\partial p}{\partial u} = \frac{\partial f}{\partial x} \cdot \frac{\partial x}{\partial u} + \frac{\partial f}{\partial y} \cdot \frac{\partial y}{\partial u},$$

$$\frac{\partial p}{\partial v} = \frac{\partial f}{\partial x} \cdot \frac{\partial x}{\partial v} + \frac{\partial f}{\partial y} \cdot \frac{\partial y}{\partial v}, \quad \text{and}$$

$$\frac{\partial p}{\partial w} = \frac{\partial f}{\partial x} \cdot \frac{\partial x}{\partial w} + \frac{\partial f}{\partial y} \cdot \frac{\partial y}{\partial w}.$$

12.7.15: If $p = f(u, v, w)$, $u = u(x, y, z)$, $v = v(x, y, z)$, and $w = w(x, y, z)$, then

$$\frac{\partial p}{\partial x} = \frac{\partial f}{\partial u} \cdot \frac{\partial u}{\partial x} + \frac{\partial f}{\partial v} \cdot \frac{\partial v}{\partial x} + \frac{\partial f}{\partial w} \cdot \frac{\partial w}{\partial x},$$

$$\frac{\partial p}{\partial y} = \frac{\partial f}{\partial u} \cdot \frac{\partial u}{\partial y} + \frac{\partial f}{\partial v} \cdot \frac{\partial v}{\partial y} + \frac{\partial f}{\partial w} \cdot \frac{\partial w}{\partial y}, \quad \text{and}$$

$$\frac{\partial p}{\partial z} = \frac{\partial f}{\partial u} \cdot \frac{\partial u}{\partial z} + \frac{\partial f}{\partial v} \cdot \frac{\partial v}{\partial z} + \frac{\partial f}{\partial w} \cdot \frac{\partial w}{\partial z}.$$

12.7.17: If $p = f(w)$ and $w = w(x, y, z, u, v)$, then

$$\frac{\partial p}{\partial x} = f'(w) \cdot \frac{\partial w}{\partial x}, \quad \frac{\partial p}{\partial y} = f'(w) \cdot \frac{\partial w}{\partial y}, \quad \frac{\partial p}{\partial z} = f'(w) \cdot \frac{\partial w}{\partial z}, \quad \frac{\partial p}{\partial u} = f'(w) \cdot \frac{\partial w}{\partial u}, \quad \text{and} \quad \frac{\partial p}{\partial v} = f'(w) \cdot \frac{\partial w}{\partial v}.$$

12.7.19: Let $F(x, y, z) = x^{2/3} + y^{2/3} + z^{2/3} - 1$. Then

$$\frac{\partial z}{\partial x} = -\frac{F_x}{F_z} = -\frac{\frac{2}{3}x^{-1/3}}{\frac{2}{3}z^{-1/3}} = -\frac{z^{1/3}}{x^{1/3}} \quad \text{and}$$

$$\frac{\partial z}{\partial y} = -\frac{F_y}{F_z} = -\frac{\frac{2}{3}y^{-1/3}}{\frac{2}{3}z^{-1/3}} = -\frac{z^{1/3}}{y^{1/3}}.$$

12.7.21: Let $F(x, y, z) = xe^{xy} + ye^{zx} + ze^{xy} - 3$. Then

$$\frac{\partial z}{\partial x} = -\frac{F_x}{F_z} = -\frac{e^{xy} + xye^{xy} + yze^{zx} + yze^{xy}}{xye^{zx} + e^{xy}} \quad \text{and} \quad \frac{\partial z}{\partial y} = -\frac{F_y}{F_z} = -\frac{x^2e^{xy} + e^{zx} + xze^{xy}}{xye^{zx} + e^{xy}}.$$

12.7.23: Let

$$F(x, y, z) = \frac{x^2}{a^2} + \frac{y^2}{b^2} + \frac{z^2}{c^2} - 1.$$

Then

$$\frac{\partial z}{\partial x} = -\frac{F_x}{F_z} = -\frac{c^2 x}{a^2 z} \quad \text{and} \quad \frac{\partial z}{\partial y} = -\frac{F_y}{F_z} = -\frac{c^2 y}{b^2 z}.$$

12.7.25: If $w = u^2 + v^2 + x^2 + y^2$, $u = x - y$, and $v - x + y$, then

$$\frac{\partial w}{\partial x} = 2u \cdot u_x + 2v \cdot v_x + 2x = 2(x - y) + 2(x + y) + 2x = 6x \quad \text{and}$$

$$\frac{\partial w}{\partial y} = 2u \cdot u_y + 2v \cdot v_y + 2y = -2(x - y) + 2(x + y) + 2y = 6y.$$

12.7.27: If $w(u, v, x, y) = xy \ln(u + v)$, $u = (x^2 + y^2)^{1/3}$, and $v = (x^3 + y^3)^{1/2}$, then

$$w_x = w_u u_x + w_v v_x + w_x \cdot 1 + w_y \cdot 0$$

$$= \frac{2x^2 y}{3(u + v)(x^2 + y^2)^{2/3}} + \frac{3x^3 y}{2(u + v)(x^3 + y^3)^{1/2}} + y \ln(u + v).$$

Then substitution of the formulas for u and v yields

$$\frac{\partial w}{\partial x} = \frac{2x^2 y}{3(x^2 + y^2)^{2/3} \left[(x^2 + y^2)^{1/3} + (x^3 + y^3)^{1/2} \right]}$$

$$+ \frac{3x^3 y}{2(x^3 + y^3)^{1/2} \left[(x^2 + y^2)^{1/3} + (x^3 + y^3)^{1/2} \right]} + y \ln \left((x^2 + y^2)^{1/3} + (x^3 + y^3)^{1/2} \right).$$

Similarly,

$$\frac{\partial w}{\partial y} = \frac{2xy^2}{3(x^2 + y^2)^{2/3} \left[(x^2 + y^2)^{1/3} + (x^3 + y^3)^{1/2} \right]}$$

$$+ \frac{3xy^3}{2(x^3 + y^3)^{1/2} \left[(x^2 + y^2)^{1/3} + (x^3 + y^3)^{1/2} \right]} + x \ln \left((x^2 + y^2)^{1/3} + (x^3 + y^3)^{1/2} \right).$$

12.7.29: We differentiate the equation $x^2 + y^2 + z^2 = 9$ implicitly, first with respect to x, then with respect to y, and obtain

$$2x + 2z \cdot z_x = 0, \qquad 2y + 2z \cdot z_y = 0.$$

We substitute the coordinates of the point of tangency $P(1, 2, 2)$ and find that

$$2 + 4z_x = 0 \quad \text{and} \quad 4 + 4z_y = 0,$$

and it follows that at P, $z_x = -\frac{1}{2}$ and $z_y = -1$. Hence an equation of the plane tangent to the given surface at the point P is

$$z - 2 = -\frac{1}{2}(x - 1) - (y - 2);$$

that is, $x + 2y + 2z = 9$.

12.7.31: Given: The surface with equation $x^3 + y^3 + z^3 = 5xyz$ and the point $P(2, 1, 1)$ on it. We differentiate the equation implicitly, first with respect to x, then with respect to y, and thereby obtain

$$3x^2 + 3z^2 \cdot z_x = 5yz + 5xy \cdot z_x, \qquad 3y^2 + 3z^2 \cdot z_y = 5xz + 5xy \cdot z_y.$$

Then we substitute the coordinates of P and find that

$$12 + 3z_x = 5 + 10z_x \quad \text{and} \quad 3 + 3z_y = 10 + 10z_y,$$

788

and it follows that $z_x = 1$ and $z_y = -1$ at the point P. Hence an equation of the plane tangent to the surface at P is

$$z - 1 = (x - 2) - (y - 1);$$

that is, $x - y - z = 0$.

12.3.33: Suppose that the square base of the box measures x (inches) on each side and that its height is z. Suppose also that time t is measured in hours. Then the volume of the box is $V = x^2 z$, and by the chain rule

$$\frac{dV}{dt} = \frac{\partial V}{\partial x} \cdot \frac{dx}{dt} + \frac{\partial V}{\partial z} \cdot \frac{dz}{dt} = 2xz \cdot (-3) + x^2 \cdot (-2).$$

Thus when $x = 24$ and $z = 12$, we have

$$\frac{dV}{dt} = 2 \cdot 24 \cdot 12 \cdot (-3) + 24^2 \cdot (-2) = -2880$$

cubic inches per hour; that is, $-\frac{5}{3}$ cubic feet per hour.

12.7.35: Let r denote the radius of the conical sandpile and h its height. Units: feet and minutes. We are given that, at the time t when when $h = 5$ and $r = 2$,

$$\frac{dh}{dt} = 0.4 \quad \text{and} \quad \frac{dr}{dt} = 0.7.$$

The volume of the sandpile is given by $V = \frac{1}{3}\pi r^2 h$, and thus

$$\frac{dV}{dt} = \frac{\partial V}{\partial r} \cdot \frac{dr}{dt} + \frac{\partial V}{\partial h} \cdot \frac{dh}{dt} = \left(\frac{2}{3}\pi r h\right) \cdot \frac{7}{10} + \left(\frac{1}{3}\pi r^2\right) \cdot \frac{2}{5}.$$

Thus when $h = 5$ and $r = 2$,

$$\frac{dV}{dt} = \frac{20\pi}{3} \cdot \frac{7}{10} + \frac{4\pi}{3} \cdot \frac{2}{5} = \frac{26\pi}{5} \approx 16.336282$$

(cubic feet per minute).

12.7.37: For this gas sample, we have $V = 10$ when $p = 2$ and $T = 300$. Substitution in the equation $pV = nRT$ yields $nR = \frac{1}{15}$. Moreover, with time t measured in minutes, we have

$$V = \frac{nRT}{p}, \quad \text{so that} \quad \frac{dV}{dt} = nR\left(\frac{1}{p} \cdot \frac{dT}{dt} - \frac{T}{p^2} \cdot \frac{dp}{dt}\right).$$

Finally, substitution of the data $nR = \frac{1}{15}$, $V = 10$, $p = 2$, $T = 300$, $dT/dt = 10$, and $dp/dt = 1$ yields the conclusion that the volume of the gas sample is decreasing at the rate of $\frac{14}{3}$ liters per minute at the time in question.

12.7.39: Given: $x = h(y, z)$ satisfies the equation $F(x, y, z) = 0$. Thus $F(h(y, z), y, z) \equiv 0$, and so implicit differentiation with respect to y yields

$$\frac{\partial F}{\partial x} \cdot \frac{\partial x}{\partial y} + \frac{\partial F}{\partial y} \cdot \frac{\partial y}{\partial y} + \frac{\partial F}{\partial z} \cdot 0 = 0.$$

Thus if $F_x \neq 0$, we find that $\dfrac{\partial x}{\partial y} = -\dfrac{F_y}{F_x}.$

12.7.41: If $w = f(u)$ and $u = x + y$, then

$$\frac{\partial w}{\partial x} = f'(u) \cdot \frac{\partial u}{\partial x} = f'(u) = f'(u) \cdot \frac{\partial u}{\partial y} = \frac{\partial w}{\partial y}.$$

12.7.43: If $w = f(x, y)$, $x = u + v$, and $y = u - v$, then

$$\frac{\partial w}{\partial v} = \frac{\partial w}{\partial x} \cdot \frac{\partial x}{\partial v} + \frac{\partial w}{\partial y} \cdot \frac{\partial y}{\partial v} = \frac{\partial w}{\partial x} - \frac{\partial w}{\partial y}.$$

Therefore

$$\frac{\partial^2 w}{\partial u \partial v} = \frac{\partial}{\partial u} \left(\frac{\partial w}{\partial x} - \frac{\partial w}{\partial y} \right)$$

$$= \frac{\partial^2 w}{\partial x^2} \cdot \frac{\partial x}{\partial u} + \frac{\partial^2 w}{\partial y \partial x} \cdot \frac{\partial y}{\partial u} - \frac{\partial^2 w}{\partial x \partial y} \cdot \frac{\partial x}{\partial u} - \frac{\partial^2 w}{\partial y^2} \cdot \frac{\partial y}{\partial u}$$

$$= \frac{\partial^2 w}{\partial x^2} + \frac{\partial^2 w}{\partial y \partial x} - \frac{\partial^2 w}{\partial x \partial y} - \frac{\partial^2 w}{\partial y^2} = \frac{\partial^2 w}{\partial x^2} - \frac{\partial^2 w}{\partial y^2}.$$

12.7.45: If $w = f(x, y)$, $x = r \cos \theta$, and $y = r \sin \theta$, then

$$\frac{\partial w}{\partial r} = \frac{\partial w}{\partial x} \cos \theta + \frac{\partial w}{\partial y} \sin \theta;$$

$$\frac{\partial^2 w}{\partial r^2} = \frac{\partial}{\partial r} \left(\frac{\partial w}{\partial x} \cos \theta + \frac{\partial w}{\partial y} \sin \theta \right)$$

$$= \frac{\partial^2 w}{\partial x^2} \cdot \frac{\partial x}{\partial r} \cos \theta + \frac{\partial^2 w}{\partial y \partial x} \cdot \frac{\partial y}{\partial r} \cos \theta + \frac{\partial^2 w}{\partial x \partial y} \cdot \frac{\partial x}{\partial r} \sin \theta + \frac{\partial^2 w}{\partial y^2} \cdot \frac{\partial y}{\partial r} \sin \theta$$

$$= \frac{\partial^2 w}{\partial x^2} \cos^2 \theta + \frac{\partial^2 w}{\partial y \partial x} \sin \theta \cos \theta + \frac{\partial^2 w}{\partial x \partial y} \sin \theta \cos \theta + \frac{\partial^2 w}{\partial y^2} \sin^2 \theta;$$

$$\frac{1}{r} \cdot \frac{\partial w}{\partial r} = \frac{1}{r} \cdot \frac{\partial w}{\partial x} \cos \theta + \frac{1}{r} \cdot \frac{\partial w}{\partial y} \sin \theta;$$

$$\frac{\partial w}{\partial \theta} = \frac{\partial w}{\partial x} \cdot \frac{\partial x}{\partial \theta} + \frac{\partial w}{\partial y} \cdot \frac{\partial y}{\partial \theta} = \frac{\partial w}{\partial x} (-r \sin \theta) + \frac{\partial w}{\partial y} (r \cos \theta);$$

$$\frac{\partial^2 w}{\partial \theta^2} = \frac{\partial}{\partial \theta} \left(-r \cdot \frac{\partial w}{\partial x} \sin \theta + r \cdot \frac{\partial w}{\partial y} \cos \theta \right)$$

$$= -r\left(\frac{\partial w}{\partial x}\cos\theta + \frac{\partial^2 w}{\partial x^2}\cdot\frac{\partial x}{\partial\theta}\sin\theta + \frac{\partial^2 w}{\partial y\,\partial x}\cdot\frac{\partial y}{\partial\theta}\sin\theta\right)$$

$$+ r\left(-\frac{\partial w}{\partial y}\sin\theta + \frac{\partial^2 w}{\partial x\,\partial y}\cdot\frac{\partial x}{\partial\theta}\cos\theta + \frac{\partial^2 w}{\partial y^2}\cdot\frac{\partial y}{\partial\theta}\cos\theta\right)$$

$$= -r\cdot\frac{\partial w}{\partial x}\cos\theta + r^2\cdot\frac{\partial^2 w}{\partial x^2}\sin^2\theta - r^2\cdot\frac{\partial^2 w}{\partial y\,\partial x}\cos\theta\sin\theta$$

$$- r\cdot\frac{\partial w}{\partial y}\sin\theta - r^2\cdot\frac{\partial^2 w}{\partial x\,\partial y}\sin\theta\cos\theta + r^2\cdot\frac{\partial^2 w}{\partial y^2}\cos^2\theta.$$

Therefore

$$\frac{\partial^2 w}{\partial r^2} + \frac{1}{r}\cdot\frac{\partial w}{\partial r} + \frac{1}{r^2}\cdot\frac{\partial^2 w}{\partial\theta^2} = \frac{\partial^2 w}{\partial x^2}\cos^2\theta + 2\frac{\partial^2 w}{\partial y\,\partial x}\sin\theta\cos\theta + \frac{\partial^2 w}{\partial y^2}\sin^2\theta$$

$$+ \frac{1}{r}\cdot\frac{\partial w}{\partial x}\cos\theta + \frac{1}{r}\cdot\frac{\partial w}{\partial y}\sin\theta - \frac{1}{r}\cdot\frac{\partial w}{\partial x}\cos\theta - \frac{1}{r}\cdot\frac{\partial w}{\partial y}\sin\theta$$

$$+ \frac{\partial^2 w}{\partial x^2}\sin^2\theta + \frac{\partial^2 w}{\partial y^2}\cos^2\theta - 2\frac{\partial^2 w}{\partial y\,\partial x}\sin\theta\cos\theta = \frac{\partial^2 w}{\partial x^2} + \frac{\partial^2 w}{\partial y^2}.$$

12.7.47: Suppose that $w = f(r)$ where $r = (x^2 + y^2 + z^2)^{1/2}$. Then

$$\frac{\partial w}{\partial x} = f'(r)\frac{\partial r}{\partial x} = \frac{x}{(x^2 + y^2 + z^2)^{1/2}}f'(r),$$

and thus

$$\frac{\partial^2 w}{\partial x^2} = \frac{\partial}{\partial x}\left(\frac{x}{(x^2 + y^2 + z^2)^{1/2}}f'(r)\right)$$

$$= \frac{x}{(x^2 + y^2 + z^2)^{1/2}}\cdot\frac{\partial}{\partial x}\big(f'(r)\big) + f'(r)\cdot\frac{\partial}{\partial x}\left(\frac{x}{(x^2 + y^2 + z^2)^{1/2}}\right)$$

$$= -\frac{x^2 f'(r)}{(x^2 + y^2 + z^2)^{3/2}} + \frac{f'(r)}{(x^2 + y^2 + z^2)^{1/2}} + \frac{x^2 f''(r)}{x^2 + y^2 + z^2}$$

$$= -\frac{x^2 f'(r)}{r^3} + \frac{f'(r)}{r} + \frac{x^2 f''(r)}{r^2}.$$

Similarly,

$$\frac{\partial^2 w}{\partial y^2} = -\frac{y^2 f'(r)}{r^3} + \frac{f'(r)}{r} + \frac{y^2 f''(r)}{r^2} \quad\text{and}\quad \frac{\partial^2 w}{\partial z^2} = -\frac{z^2 f'(r)}{r^3} + \frac{f'(r)}{r} + \frac{z^2 f''(r)}{r^2}.$$

Hence

$$\frac{\partial^2 w}{\partial x^2} + \frac{\partial^2 w}{\partial y^2} + \frac{\partial^2 w}{\partial z^2} = -\frac{r^2 f'(r)}{r^3} + \frac{3f'(r)}{r} + \frac{r^2 f''(r)}{r^2}$$

$$= -\frac{f'(r)}{r} + \frac{3f'(r)}{r} + f''(r) = \frac{d^2 w}{dr^2} + \frac{2}{r}\cdot\frac{dw}{dr}.$$

791

12.7.49: If $w = f(u, v)$, $u = x + y$, and $v = x - y$, then

$$\frac{\partial w}{\partial x} = \frac{\partial w}{\partial u} \cdot \frac{\partial u}{\partial x} + \frac{\partial w}{\partial v} \cdot \frac{\partial v}{\partial x} = \frac{\partial w}{\partial u} + \frac{\partial w}{\partial v} \quad \text{and}$$

$$\frac{\partial w}{\partial y} = \frac{\partial w}{\partial u} \cdot \frac{\partial u}{\partial y} + \frac{\partial w}{\partial v} \cdot \frac{\partial v}{\partial y} = \frac{\partial w}{\partial u} - \frac{\partial w}{\partial v}.$$

Therefore

$$\frac{\partial w}{\partial x} \cdot \frac{\partial w}{\partial y} = \left(\frac{\partial w}{\partial u}\right)^2 - \left(\frac{\partial w}{\partial v}\right)^2.$$

12.7.51: We are given $w = f(x, y)$ and the existence of a constant α such that

$$x = u \cos \alpha - v \sin \alpha \quad \text{and} \quad y = u \sin \alpha + v \cos \alpha.$$

Then

$$w_u = w_x x_u + w_y y_u = w_x \cos \alpha + w_y \sin \alpha;$$

$$w_v = w_x x_v + w_y y_v = -w_x \sin \alpha + w_y \cos \alpha;$$

$$(w_u)^2 + (w_v)^2 = (w_x)^2 \cos^2 \alpha + 2w_x w_y \sin \alpha \cos \alpha + (w_y)^2 \sin^2 \alpha$$

$$+ (w_x)^2 \sin^2 \alpha - 2w_x w_y \sin \alpha \cos \alpha + (w_y)^2 \cos^2 \alpha$$

$$= (w_x)^2 + (w_y)^2.$$

12.7.53: Using the *Suggestion* and the notation in the equations in (17), we have

$$\left(\frac{\partial x}{\partial y}\right)_z \cdot \left(\frac{\partial y}{\partial z}\right)_x \cdot \left(\frac{\partial z}{\partial x}\right)_y = \left(-\frac{F_y}{F_x}\right) \cdot \left(-\frac{F_z}{F_y}\right) \cdot \left(-\frac{F_x}{F_z}\right) = -1.$$

12.7.55: If the equation $pV - nRT = 0$ implicitly defines the functions $T = f(p, V)$, $V = g(p, T)$, and $p = h(V, T)$, then

$$p + V\frac{\partial h}{\partial V} = 0, \quad \text{so} \quad -\frac{p}{V} = \frac{\partial h}{\partial V} = \left(\frac{\partial p}{\partial V}\right)_T,$$

$$p\frac{\partial g}{\partial T} - nR = 0, \quad \text{so} \quad \frac{nR}{p} = \frac{\partial g}{\partial T} = \left(\frac{\partial V}{\partial T}\right)_p, \quad \text{and}$$

$$V - nR\frac{\partial f}{\partial p} = 0, \quad \text{so} \quad \frac{V}{nR} = \frac{\partial f}{\partial p} = \left(\frac{\partial T}{\partial p}\right)_V.$$

Therefore

$$\left(\frac{\partial p}{\partial V}\right)_T \left(\frac{\partial V}{\partial T}\right)_p \left(\frac{\partial T}{\partial p}\right)_V = \left(-\frac{p}{V}\right)\left(\frac{nR}{p}\right)\left(\frac{V}{nR}\right) = -1.$$

12.7.57: First note that

$$\frac{\partial p}{\partial T} = \frac{\alpha}{\beta} = \frac{1.8 \times 10^6}{3.9 \times 10^4} = \frac{600}{13}.$$

Hence an increase of 5° in the Celsius temperature multiplies the initial pressure (1 atm) by $\frac{3000}{13} \approx 230.77$, so the bulb will burst.

12.7.60: Here we have

$$T'(\rho, \phi, \theta) = \begin{bmatrix} \sin\phi\cos\theta & \rho\cos\phi\cos\theta & -\rho\sin\phi\sin\theta \\ \sin\phi\sin\theta & \rho\cos\phi\sin\theta & \rho\sin\phi\cos\theta \\ \cos\phi & -\rho\sin\phi & 0 \end{bmatrix}.$$

Therefore

$$|T'(\rho, \phi, \theta)| = \rho^2 \sin^3\phi \sin^2\theta + \rho^2 \sin\phi \cos^2\phi \cos^2\theta + \rho^2 \cos^2\phi \sin\phi \sin^2\theta + \rho^2 \sin^3\phi \cos^2\theta$$

$$= \rho^2(\sin^3\phi + \sin\phi \cos^2\phi) = \rho^2 \sin\phi.$$

12.7.61: Here we compute

$$\begin{bmatrix} F_x & F_y & F_z \end{bmatrix} \begin{bmatrix} \sin\phi\cos\theta & \rho\cos\phi\cos\theta & -\rho\sin\phi\sin\theta \\ \sin\phi\sin\theta & \rho\cos\phi\sin\theta & \rho\sin\phi\cos\theta \\ \cos\phi & -\rho\sin\phi & 0 \end{bmatrix} = \begin{bmatrix} \dfrac{\partial w}{\partial \rho} & \dfrac{\partial w}{\partial \phi} & \dfrac{\partial w}{\partial \theta} \end{bmatrix},$$

which has first component

$$\frac{\partial w}{\partial \rho} = F_x \sin\phi \cos\theta + F_y \sin\phi \sin\theta + F_z \cos\phi,$$

second component

$$\frac{\partial w}{\partial \phi} = F_x \rho \cos\phi \cos\theta + F_y \rho \cos\phi \sin\theta - F_z \rho \sin\phi,$$

and third component

$$\frac{\partial w}{\partial \theta} = -F_x \rho \sin\phi \sin\theta + F_y \rho \sin\phi \cos\theta.$$

Section 12.8

12.8.1: If $f(x, y) = 3x - 7y$ and $P(17, 39)$ are given, then

$$\nabla f(x, y) = \langle 3, -7 \rangle, \quad \text{and thus} \quad \nabla f(17, 39) = \langle 3, -7 \rangle.$$

12.8.3: If $f(x, y) = \exp(-x^2 - y^2)$ and $P(0, 0)$ are given, then

$$\nabla f(x, y) = \langle -2x \exp(-x^2 - y^2), -2y \exp(-x^2 - y^2) \rangle, \quad \text{and thus} \quad \nabla f(0, 0) = \langle 0, 0 \rangle = \mathbf{0}.$$

12.8.5: Given $f(x, y, z) = y^2 - z^2$ and $P(17, 3, 2)$, then

$$\nabla f(x, y, z) = \langle 0, 2y, -2z \rangle, \quad \text{and therefore} \quad \nabla f(17, 3, 2) = \langle 0, 6, -4 \rangle.$$

12.8.7: Given $f(x, y, z) = e^x \sin y + e^y \sin z + e^z \sin x$ and $P(0, 0, 0)$, then

$$\nabla f(x, y, z) = \langle e^z \cos x + e^x \sin y, \; e^x \cos y + e^y \sin z, \; e^y \cos z + e^z \sin x \rangle,$$

and therefore $\nabla f(0, 0, 0) = \langle 1, 1, 1 \rangle$.

12.8.9: Given $f(x, y, z) = 2\sqrt{xyz}$ and $P(3, -4, -3)$, then

$$\nabla f(x, y, z) = \left\langle \frac{yz}{\sqrt{xyz}}, \frac{xz}{\sqrt{xyz}}, \frac{xy}{\sqrt{xyz}} \right\rangle, \quad \text{and so} \quad \nabla f(3, -4, -3) = \left\langle 2, -\frac{3}{2}, -2 \right\rangle.$$

12.8.11: Given $f(x, y) = x^2 + 2xy + 3y^2$, $P(2, 1)$, and $\mathbf{v} = \langle 1, 1 \rangle$, we first compute a unit vector with the same direction as $\mathbf{v}$:

$$\mathbf{u} = \frac{\mathbf{v}}{|\mathbf{v}|} = \left\langle \tfrac{1}{2}\sqrt{2}, \tfrac{1}{2}\sqrt{2} \right\rangle.$$

Also $\nabla f(x, y) = \langle 2x + 2y, 2x + 6y \rangle$, so $\nabla f(P) = \langle 6, 10 \rangle$. Therefore

$$D_{\mathbf{u}} f(P) = (\nabla f(P)) \cdot \mathbf{u} = \langle 6, 10 \rangle \cdot \left\langle \tfrac{1}{2}\sqrt{2}, \tfrac{1}{2}\sqrt{2} \right\rangle = 8\sqrt{2}.$$

12.8.13: Given $f(x, y) = x^3 - x^2 y + xy^2 + y^3$, $P(1, -1)$, and $\mathbf{v} = \langle 2, 3 \rangle$, we first compute a unit vector with the same direction as $\mathbf{v}$:

$$\mathbf{u} = \frac{\mathbf{v}}{|\mathbf{v}|} = \left\langle \tfrac{2}{13}\sqrt{13}, \tfrac{3}{13}\sqrt{13} \right\rangle.$$

Also $\nabla f(x, y) = \langle 3x^2 - 2xy + y^2, 3y^2 + 2xy - x^2 \rangle$, so $\nabla f(P) = \langle 6, 0 \rangle$. Therefore

$$D_{\mathbf{u}} f(P) = (\nabla f(P)) \cdot \mathbf{u} = \langle 6, 0 \rangle \cdot \left\langle \tfrac{2}{13}\sqrt{13}, \tfrac{3}{13}\sqrt{13} \right\rangle = \tfrac{12}{13}\sqrt{13}.$$

12.8.15: Given: $f(x, y) = \sin x \cos y$, the point $P\left(\tfrac{1}{3}\pi, -\tfrac{2}{3}\pi\right)$, and the vector $\mathbf{v} = \langle 4, -3 \rangle$, we first construct a unit vector with the same direction as $\mathbf{v}$:

$$\mathbf{u} = \frac{\mathbf{v}}{|\mathbf{v}|} = \left\langle \frac{4}{5}, -\frac{3}{5} \right\rangle.$$

Next, $\nabla f(x, y) = \langle \cos x \cos y, -\sin x \sin y \rangle$, and hence $\nabla f(P) = \langle -\frac{1}{4}, \frac{3}{4} \rangle$. Therefore

$$D_{\mathbf{u}}f(P) = \big(\nabla f(P)\big) \cdot \mathbf{u} = -\frac{13}{20}.$$

12.8.17: Given $f(x, y, z) = \sqrt{xyz}$, the point $P(2, -1, -2)$, and the vector $\mathbf{v} = \langle 1, 2, -2 \rangle$, we first construct a unit vector $\mathbf{u}$ with the same direction as $\mathbf{v}$:

$$\mathbf{u} = \frac{\mathbf{v}}{|\mathbf{v}|} = \left\langle \frac{1}{3}, \frac{2}{3}, -\frac{2}{3} \right\rangle.$$

Next,

$$\nabla f(x, y, z) = \left\langle \frac{yz}{2\sqrt{xyz}}, \frac{xz}{2\sqrt{xyz}}, \frac{xy}{2\sqrt{xyz}} \right\rangle,$$

and hence $D_{\mathbf{u}}f(P) = \big(\nabla f(P)\big) \cdot \mathbf{u} = \left\langle \frac{1}{2}, -1, -\frac{1}{2} \right\rangle \cdot \left\langle \frac{1}{3}, \frac{2}{3}, -\frac{2}{3} \right\rangle = -\frac{1}{6}$.

12.8.19: Given $f(x, y, z) = \exp(xyz)$, the point $P(4, 0, -3)$, and the vector $\mathbf{v} = \langle 0, 1, -1 \rangle$, we first construct a unit vector with the same direction as $\mathbf{v}$:

$$\mathbf{u} = \frac{\mathbf{v}}{|\mathbf{v}|} = \left\langle 0, \frac{\sqrt{2}}{2}, -\frac{\sqrt{2}}{2} \right\rangle.$$

Next, $\nabla f(x, y, z) = \exp(xyz)\langle yz, xz, xy \rangle$, and so $\nabla f(P) = \langle 0, -12, 0 \rangle$. Therefore

$$D_{\mathbf{u}}f(P) = \big(\nabla f(P)\big) \cdot \mathbf{u} = -6\sqrt{2}.$$

12.8.21: Given $f(x, y) = 2x^2 + 3xy + 4y^2$ and the point $P(1, 1)$, we first compute

$$\nabla f(x, y) = \langle 4x + 3y, \, 3x + 8y \rangle.$$

So the direction in which f is increasing the most rapidly at P is $\nabla f(P) = \langle 7, 11 \rangle$ and its rate of increase in that direction is $|\langle 7, 11 \rangle| = \sqrt{170}$.

12.8.23: Given $f(x, y) = \ln(x^2 + y^2)$ and the point $P(3, 4)$, we first compute

$$\nabla f(x, y) = \left\langle \frac{2x}{x^2 + y^2}, \frac{2y}{x^2 + y^2} \right\rangle.$$

Therefore the direction in which f is increasing the most rapidly at P is

$$\mathbf{v} = \nabla f(P) = \left\langle \frac{6}{25}, \frac{8}{25} \right\rangle,$$

and its rate of increase in that direction is $|\mathbf{v}| = \dfrac{2}{5}$.

12.8.25: Given $f(x, y, z) = 3x^2 + y^2 + 4z^2$ and the point $P(1, 5, -2)$, we first compute

$$\nabla f(x, y, z) = \langle 6x, 2y, 8z \rangle.$$

Therefore the direction in which f is increasing the most rapidly at P is $\mathbf{v} = \nabla f(P) = \langle 6, 10, -16 \rangle$ and its rate of increase in that direction is $|\mathbf{v}| = 14\sqrt{2}$.

12.8.27: We are given $f(x, y, z) = \sqrt{xy^2z^3}$ and the point $P(2, 2, 2)$. We first compute the gradient of f:

$$\nabla f(x, y, z) = \left\langle \frac{y^2z^3}{2\sqrt{xy^2z^3}}, \frac{xyz^3}{\sqrt{xy^2z^3}}, \frac{3xy^2z^2}{2\sqrt{xy^2z^3}} \right\rangle.$$

Thus the direction in which f is increasing the most rapidly at P is $\nabla f(P) = \langle 2, 4, 6 \rangle$ and its rate of increase in that direction is $|\nabla f(P)| = 2\sqrt{14}$.

12.8.29: Let $f(x, y) = \exp(25 - x^2 - y^2) - 1$. Then

$$\nabla f(x, y) = \langle -2x \exp(25 - x^2 - y^2), -2y \exp(25 - x^2 - y^2) \rangle,$$

so at $P(3, 4)$ we have $\nabla f(P) = \langle -6, -8 \rangle$, a vector normal to the graph of $f(x, y) = 0$ at the point P. Hence, as in Example 7, an equation of the line tangent to the graph at P is $-6(x - 3) - 8(y - 4) = 0$; simplified, this is $3x + 4y = 25$.

12.8.31: Let $f(x, y) = x^4 + xy + y^2 - 19$. Then $\nabla f(x, y) = \langle 4x^3 + y, x + 2y \rangle$, so a vector normal to the graph of $f(x, y) = 0$ at the point $P(2, -3)$ is $\nabla f(P) = \langle 29, -4 \rangle$. So the tangent line at P has equation $29(x - 2) - 4(y + 3) = 0$; that is, $29x - 4y = 70$.

12.8.33: Let $f(x, y, z) = x^{1/3} + y^{1/3} + z^{1/3} - 1$. Then

$$\nabla f(x, y, z) = \left\langle \frac{1}{3x^{2/3}}, \frac{1}{3y^{2/3}}, \frac{1}{3z^{2/3}} \right\rangle,$$

and thus a vector normal to the surface $f(x, y, z) = 0$ at the point $P(1, -1, 1)$ is

$$\nabla f(P) = \left\langle \frac{1}{3}, \frac{1}{3}, \frac{1}{3} \right\rangle.$$

Therefore an equation of the plane tangent to the surface at P is

$$\frac{1}{3}(x - 1) + \frac{1}{3}(y + 1) + \frac{1}{3}(z - 1) = 0; \quad \text{that is,} \quad x + y + z = 1.$$

12.8.35: If u and v are differentiable functions of x and y and a and b are constants, then

$$\nabla\big(au(x, y) + bv(x, y)\big) = \left\langle \frac{\partial}{\partial x}\big(au(x, y) + bv(x, y)\big), \frac{\partial}{\partial y}\big(au(x, y) + bv(x, y)\big) \right\rangle$$

$$= \langle au_x + bv_x, au_y + bv_y \rangle = \langle au_x, au_y \rangle + \langle bv_x, bv_y \rangle$$

$$= a\langle u_x, u_y \rangle + b\langle v_x, v_y \rangle = a\nabla u(x, y) + b\nabla v(x, y).$$

12.8.37: If u and v are differentiable functions of x and y and $v(x, y) \neq 0$, then

$$\nabla \left(\frac{u(x, y)}{v(x, y)} \right) = \left\langle \frac{\partial}{\partial x} \left(\frac{u}{v} \right), \frac{\partial}{\partial y} \left(\frac{u}{v} \right) \right\rangle = \left\langle \frac{vu_x - uv_x}{v^2}, \frac{vu_y - uv_y}{v^2} \right\rangle$$

$$= \left\langle \frac{vu_x}{v^2}, \frac{vu_y}{v^2} \right\rangle - \left\langle \frac{uv_x}{v^2}, \frac{uv_y}{v^2} \right\rangle = \frac{v\nabla u}{v^2} - \frac{u\nabla v}{v^2} = \frac{v\nabla u - u\nabla v}{v^2}.$$

12.8.39: We know that $\mathbf{v} = \nabla f(P)$ gives the direction in which f is increasing the most rapidly at P. Then $\mathbf{v}$ is the direction in which $-f$ is decreasing the most rapidly at P. But $\nabla(-f(P)) = -\nabla f(P) = -\mathbf{v}$, so that $-\mathbf{v}$ is the direction in which $-f$ is increasing the most rapidly at P and, therefore, is the direction in which f is decreasing the most rapidly at P.

12.8.41: Let $f(x, y) = Ax^2 + Bxy + Cy^2 - D$. Then

$$\nabla f(x, y) = \langle 2Ax + By, 2Cy + Bx \rangle,$$

so a vector normal to the graph of $f(x, y) = 0$ at the point $P(x_0, y_0)$ is

$$\nabla f(P) = \langle 2Ax_0 + By_0, 2Cy_0 + Bx_0 \rangle.$$

Hence, as in Example 7, an equation of the line tangent to the graph at P is

$$(2Ax_0 + By_0)(x - x_0) + (2Cy_0 + Bx_0)(y - y_0) = 0;$$

$$2Ax_0x + By_0x - 2A(x_0)^2 - Bx_0y_0 + 2Cy_0y + Bx_0y - 2C(y_0)^2 - Bx_0y_0 = 0;$$

$$2Ax_0x + By_0x + Bx_0y + 2Cy_0y = 2A(x_0)^2 + 2Bx_0y_0 + 2C(y_0)^2;$$

$$2(Ax_0)x + B(y_0x + x_0y) + 2(Cy_0)y = 2D;$$

$$(Ax_0)x + \frac{1}{2}B(y_0x + x_0y) + (Cy_0)y = D.$$

12.8.43: The equation of the paraboloid can be written in the form

$$H(x, y, z) = Ax^2 + By^2 - z = 0, \quad \text{and} \quad \nabla H(x, y, z) = \langle 2Ax, 2By, -1 \rangle.$$

A vector normal to the paraboloid at the point $P(x_0, y_0, z_0)$ is $\mathbf{n} = \langle 2Ax_0, 2By_0, -1 \rangle$, and hence the plane tangent to the paraboloid at P has an equation of the form

$$2Ax_0x + 2By_0y - z = d.$$

But the point P also lies on the plane, and hence

$$d = 2A(x_0)^2 + 2B(y_0)^2 - z_0 = 2(Ax_0^2 + By_0^2) - z_0 \doteq 2z_0 - z_0 = z_0.$$

Hence an equation of the tangent plane is $2Ax_0x + 2By_0y - z = z_0$, and the result in Problem 43 follows immediately.

12.8.45: In the solution of Problem 44 we calculated $\nabla f(P) = \langle 5, 4, 3 \rangle$, and the unit vector in the direction from P to Q is

$$\mathbf{u} = \frac{\overrightarrow{PQ}}{|\overrightarrow{PQ}|} = \left\langle \frac{2}{3}, \frac{2}{3}, \frac{1}{3} \right\rangle.$$

Then

$$D_{\mathbf{u}}f(P) = \nabla f(P) \cdot \mathbf{u} = \langle 5, 4, 3 \rangle \cdot \langle \tfrac{2}{3}, \tfrac{2}{3}, \tfrac{1}{3} \rangle = 7$$

(degrees per kilometer). Hence

$$\frac{dw}{dt} = \frac{dw}{ds} \cdot \frac{ds}{dt} = \left(7 \, \frac{\deg}{\mathrm{km}} \right) \left(2 \, \frac{\mathrm{km}}{\min} \right) = 14 \, \frac{\deg}{\min}$$

as the hawk's rate of change of temperature at P.

12.8.47: Part (a): If $W(x, y, z) = 50 + xyz$, then $\nabla W = \langle yz, xz, xy \rangle$, so at the point $P(3, 4, 1)$ we have $\nabla W(P) = \langle 4, 3, 12 \rangle$. The unit vector with the same direction as $\mathbf{v} = \langle 1, 2, 2 \rangle$ is

$$\mathbf{u} = \frac{\mathbf{v}}{|\mathbf{v}|} = \left\langle \frac{1}{3}, \frac{2}{3}, \frac{2}{3} \right\rangle,$$

and so the rate of change of temperature at P in the direction of $\mathbf{v}$ is

$$\left(\nabla W(P) \right) \cdot \mathbf{u} = \langle 4, 3, 12 \rangle \cdot \left\langle \frac{1}{3}, \frac{2}{3}, \frac{2}{3} \right\rangle = \frac{34}{3}.$$

Because distance is measured in feet, the units for this rate of change are degrees Celsius per foot.

Part (b): The maximal directional derivative of W at P is $|\nabla W(P)| = |\langle 4, 3, 12 \rangle| = 13$ and the direction in which it occurs is $\nabla W(P) = \langle 4, 3, 12 \rangle$.

12.8.49: Part (a): Given $f(x, y) = \frac{1}{10}(x^2 - xy + 2y^2)$, let

$$g(x, y, z) = z - f(x, y); \quad \text{then} \quad \nabla g(x, y, z) = \frac{1}{10} \langle y - 2x, x - 4y, 10 \rangle.$$

Thus a normal to the surface $z = f(x, y)$ at the point $P\left(2, 1, \frac{2}{5}\right)$ is $\frac{1}{10} \langle -3, -2, 10 \rangle$. Hence an equation of the plane tangent to this surface at P is

$$3x + 2y - 10z = 4; \quad \text{that is,} \quad z = \frac{3}{10}x + \frac{1}{5}y - \frac{2}{5}.$$

Part (b): Let Q denote the point $(2, 1)$ and R the point $(2.2, 0.9)$. Then $\mathbf{v} = \overrightarrow{QR} = \langle 0.2, -0.1 \rangle$. Thus an approximation to $f(2.2, 0.9)$ is

$$f(2, 1) + \left(\nabla f(Q) \right) \cdot \mathbf{v} = 0.4 + \langle 0.3, 0.2 \rangle \cdot \langle 0.2, -0.1 \rangle = \frac{11}{25} = 0.44.$$

The true value is $f(2.2, 0.9) = \frac{56}{125} = 0.448$.

12.8.51: Let $F(x, y, z) = z^2 - x^2 - y^2$ and $G(x, y, z) = 2x + 3y + 4z + 2$. Then the cone is the graph of $F(x, y, z) = 0$ and the plane is the graph of $G(x, y, z) = 0$. At the given point $P(3, 4, -5)$ we have

$$\nabla F(3, 4, -5) = \langle -6, -8, -10 \rangle \quad \text{and} \quad \nabla G(3, 4, -5) = \langle 2, 3, 4 \rangle.$$

Let $\mathcal{P}$ denote the plane normal to the ellipse (the intersection of the cone and the first plane) at the point P. Then a normal to $\mathcal{P}$ is

$$\mathbf{n} = \langle -6, -8, -10 \rangle \times \langle 2, 3, 4 \rangle = \begin{vmatrix} \mathbf{i} & \mathbf{j} & \mathbf{k} \\ -6 & -8 & -10 \\ 2 & 3 & 4 \end{vmatrix} = \langle -2, 4, -2 \rangle.$$

We will use instead the parallel vector $\langle 1, -2, 1 \rangle$. In the usual way we find that $\mathcal{P}$ has Cartesian equation $x - 2y + z + 10 = 0$.

12.8.53: Let $F(x, y, z) = x^2 + y^2 + z^2 - r^2$ and $G(x, y, z) = z^2 - a^2x^2 - b^2y^2$. Then the sphere is the graph of $F(x, y, z) = 0$ and the cone is the graph of $G(x, y, z) = 0$. At a point where the sphere and the cone meet, these vectors are the normals to their tangent planes. To show that the tangent planes are perpendicular, it is sufficient to show that their normals are perpendicular. But

$$(\nabla F) \cdot (\nabla G) = \langle 2x, 2y, 2z \rangle \cdot \langle -2a^2x, -2b^2y, 2z \rangle = -4a^2x^2 - 4b^2y^2 + 4z^2 = 4(z^2 - a^2x^2 - b^2y^2) = 0$$

because (x, y, z) lies on the cone. Therefore the tangent planes are perpendicular at every point of the intersection of the sphere and the cone.

12.8.55: The surface is the graph of the equation $G(x, y, z) = 0$ where

$$G(x, y, z) = xyz - 1, \quad \text{so that} \quad \nabla G(x, y, z) = \langle yz, xz, xy \rangle.$$

Suppose that $P(a, b, c)$ is a point strictly within the first octant (so that a, b, and c are all positive). Note that $abc = 1$. A vector normal to the surface at P is $\mathbf{n} = \langle bc, ac, ab \rangle$, and hence the plane tangent to the surface at P has equation

$$bcx + acy + abz = d$$

for some constant d. Moreover, because P is a point of the surface,

$$bca + acb + abc = d; \quad \text{that is,} \quad d = 3abc.$$

Hence an equation of the tangent plane is $bcx + acy + abz = 3abc$. The intercepts of the pyramid therefore occur at $(3a, 0, 0)$, $(0, 3b, 0)$, and $(0, 0, 3c)$. Therefore, because of the right angle at the origin, the pyramid has volume

$$V = \frac{1}{6}(3a)(3b)(3c) = \frac{27}{6} abc = \frac{9}{2},$$

independent of the choice of P, as we were to show.

12.8.57: The hill is steepest in the direction of $\nabla z(-100, -100) = \langle 0.6, 0.8 \rangle$. The slope of the hill in that direction is $|\langle 0.6, 0.8 \rangle| = 1$, so that your initial angle of climb would be $45°$. The compass heading in the direction you are climbing is

$$\frac{\pi}{2} - \arctan \frac{4}{3}$$

radians, approximately $36° \ 52' \ 11.632''$.

12.8.59: Given

$$z = f(x, y) = \frac{1000}{1 + (0.00003)x^2 + (0.00007)y^2},$$

we first compute

$$\nabla f(x, y) = \left\langle -\frac{600\,000\,000x}{(100000 + 3x^2 + 7y^2)^2}, -\frac{1\,400\,000\,000y}{(100000 + 3x^2 + 7y^2)^2} \right\rangle.$$

Hence to climb the most steeply, the initial direction should be

$$\nabla f(100, 100) = \left\langle -\frac{3}{2}, -\frac{7}{2} \right\rangle,$$

and the initial rate of ascent will be $|\nabla f(100, 100)| = \frac{2}{2}\sqrt{58} \approx 3.807886553$ feet per foot. The initial angle of climb will be $\arctan\left(\frac{1}{2}\sqrt{58}\right) \approx 1.313982409$ radians, approximately $75° \ 17' \ 8.327''$. The compass heading is

$$270° - \arctan\left(\frac{7}{3}\right)° \approx 203° \ 11' \ 54.926''.$$

12.8.61: Let

$$f(x, y) = \frac{1}{1000}(3x^2 - 5xy + y^2).$$

Then

$$\nabla f(x, y) = \frac{1}{1000}\langle 6x - 5y, 2y - 5x \rangle,$$

and therefore $\mathbf{v} = \nabla f(100, 100) = \left\langle \frac{1}{10}, -\frac{3}{10} \right\rangle$.

Part (a): A unit vector in the northeast direction is $\mathbf{u} = \frac{1}{2}\langle \sqrt{2}, \sqrt{2} \rangle$, so the directional derivative of f at $(100, 100)$ in the northeast direction is

$$\mathbf{v} \cdot \mathbf{u} = \left\langle \tfrac{1}{10}, -\tfrac{3}{10} \right\rangle \cdot \tfrac{1}{2}\left\langle \sqrt{2}, \sqrt{2} \right\rangle = -\tfrac{1}{10}\sqrt{2}.$$

Hence you will initially be descending the hill, and at an angle of $\arctan\left(\frac{1}{10}\sqrt{2}\right)$ below the horizontal, approximately $8°$ $2'$ $58.081''$.

Part (b): A unit vector in the direction $30°$ north of east is $\mathbf{u} = \frac{1}{2}\left\langle\sqrt{3}, 1\right\rangle$, so the directional derivative of f at $(100, 100)$ in the direction of $\mathbf{u}$ is

$$\left\langle\tfrac{1}{10}, -\tfrac{3}{10}\right\rangle \cdot \tfrac{1}{2}\left\langle\sqrt{3}, 1\right\rangle = -\frac{3-\sqrt{3}}{20} \approx -0.06339746.$$

Hence you will initially be descending the hill, and at an angle of approximately $3°$ $37'$ $39.187''$.

12.8.63: Because $\mathbf{u} = \langle a, b\rangle$ and $\mathbf{v} = \langle c, d\rangle$ are not collinear, neither is zero and neither is a scalar multiple of the other. Hence, as vectors, they are linearly independent, and this implies that the simultaneous equations

$$a f_x(P) + b f_y(P) = D_{\mathbf{u}}f(P),$$

$$c f_x(P) + d f_y(P) = D_{\mathbf{v}}f(P)$$

have a unique solution for the values of $f_x(P)$ and $f_y(P)$. Thus $\nabla f(P) = \langle f_x(P), f_y(P)\rangle$ is uniquely determined, and therefore so is the directional derivative

$$D_{\mathbf{w}}f(P) = \nabla f(P) \cdot \mathbf{w}$$

in the direction of the arbitrary unit vector $\mathbf{w}$.

Section 12.9

12.9.1: Given $f(x, y) = 2x + y$ and the constraint $g(x, y) = x^2 + y^2 - 1 = 0$, the equation $\nabla f = \lambda \nabla g$ yields

$$2 = 2\lambda x \quad\text{and}\quad 1 = 2\lambda y,$$

so that $\lambda \neq 0$. Hence

$$\frac{1}{\lambda} = x = 2y, \quad\text{and thus}\quad 4y^2 + y^2 = 1.$$

Thus we have two solutions:

$$(x, y) = \left(\frac{2}{5}\sqrt{5}, \frac{1}{5}\sqrt{5}\right) \quad\text{and}\quad (x, y) = \left(-\frac{2}{5}\sqrt{5}, -\frac{1}{5}\sqrt{5}\right).$$

Clearly the first maximizes $f(x, y)$ and the second minimizes $f(x, y)$. So the global maximum value of $f(x, y)$ is $\sqrt{5}$ and its global minimum value is $-\sqrt{5}$.

Note: The function f is continuous on the circle $x^2 + y^2 = 1$ and a continuous function defined on a closed and bounded subset of euclidean space (such as that circle) must have both a global maximum value and a global minimum value. We will use this argument to identify extrema when possible and without stating the argument explicitly in various solutions in this section.

12.9.3: The Lagrange multiplier equation $\langle 2x, -2y \rangle = \lambda \langle 2x, 2y \rangle$ yields the scalar equations $x = \lambda x$ and $-y = \lambda y$. To solve them with a minimum number of cases, note that multiplication of the first by y and the second by x yields

$$xy = \lambda xy = -xy, \quad \text{so that} \quad xy = 0.$$

If $x = 0$ then $y = \pm 2$; if $y = 0$ then $x = \pm 2$. So there are four critical points:

$$f(2, 0) = 4: \quad \text{global maximum}; \quad f(-2, 0) = 4: \quad \text{global maximum};$$

$$f(0, 2) = -4: \quad \text{global minimum}; \quad f(0, -2) = -4: \quad \text{global minimum}.$$

12.9.5: The Lagrange multiplier equation $\langle y, x \rangle = \lambda \langle 8x, 18y \rangle$ yields the scalar equations $y = 8\lambda x$ and $x = 18\lambda y$. Multiply the first by $9y$ and the second by $4x$ to obtain

$$9y^2 = 72\lambda xy = 4x^2, \quad \text{so that} \quad 3y = \pm 2x.$$

The constraint equation takes the form $18y^2 = 36$, so that $y^2 = 2$. Thus there are four critical points. But $f(x, y)$ must have both a global maximum and a global minimum value on the ellipse $4x^2 + 9y^2 = 36$, so we may identify all of the critical points:

$$f\left(\tfrac{3}{2}\sqrt{2}, \sqrt{2}\right) = 3: \quad \text{global maximum}; \quad f\left(-\tfrac{3}{2}\sqrt{2}, \sqrt{2}\right) = -3: \quad \text{global minimum};$$

$$f\left(\tfrac{3}{2}\sqrt{2}, -\sqrt{2}\right) = -3: \quad \text{global minimum}; \quad f\left(-\tfrac{3}{2}\sqrt{2}, -\sqrt{2}\right) = 3: \quad \text{global maximum}.$$

12.9.7: It is clear that $f(x, y, z) = x^2 + y^2 + z^2$ can have no global maximum value on the plane $g(x, y, z) = 3x + 2y + z - 6 = 0$, and almost as clear that there is a unique global minimum. The Lagrange multiplier equation

$$\langle 2x, 2y, 2z \rangle = \lambda \langle 3, 2, 1 \rangle$$

then yields $4x = 6y = 12z = 6\lambda$, so that $x = 3z$ and $y = 2z$. Then the constraint takes the form $9z + 4z + z = 6$, and therefore the global minimum value of f on the plane $g(x, y, z) = 0$ is $f\left(\tfrac{9}{7}, \tfrac{6}{7}, \tfrac{3}{7}\right) = \tfrac{18}{7}$. We have also discovered three distinct positive rational numbers whose sum is equal to the sum of their squares.

12.9.9: The continuous function $f(x, y, z) = x + y + z$ must have both a global maximum value and a global minimum value on the ellipsoidal surface with equation $g(x, y, z) = x^2 + 4y^2 + 9z^2 - 36 = 0$. The Lagrange multiplier method yields $\langle 1, 1, 1 \rangle = \lambda \langle 2x, 8y, 18z \rangle$, and thus $\lambda \neq 0$. Therefore

$$\frac{1}{\lambda} = 2x = 8y = 18z, \quad \text{so that} \quad x = 4y = 9z.$$

Substitution of $x = 9z$ and $y = \frac{9}{4}z$ in the equation $g(x, y, z) = 0$ then yields $z = \pm\frac{4}{7}$. Results:

$$\text{Global maximum:} \quad f\left(\tfrac{36}{7}, \tfrac{9}{7}, \tfrac{4}{7}\right) = 7;$$

$$\text{Global minimum:} \quad f\left(-\tfrac{36}{7}, -\tfrac{9}{7}, -\tfrac{4}{7}\right) = -7.$$

12.9.11: Clearly $f(x, y, z) = xy + 2z$ is continuous on the closed and bounded spherical surface $g(x, y, z) = x^2 + y^2 + z^2 - 36 = 0$, and hence f has both a global maximum and a global minimum there. The Lagrange multiplier equation is

$$\langle y, x, 2 \rangle = \lambda \langle 2x, 2y, 2z \rangle, \quad \text{and hence} \quad y = 2\lambda x, \quad x = 2\lambda y, \quad \text{and} \quad 2 = 2\lambda z.$$

Now $\lambda z = 1$, so $\lambda \neq 0$. If neither x nor y is zero, then

$$\frac{1}{\lambda} = \frac{2x}{y} = \frac{2y}{x} = z.$$

So $y^2 = x^2$ in this case. If $y = x$, then $z = 2$, so $2x^2 + 4 = 36$ and we obtain two critical points and the values $f(4, 4, 2) = 20$ and $f(-4, -4, 2) = 20$. If $y = -x$, then $z = -2$, and we obtain two critical points and the values $f(4, -4, -2) = -20$ and $f(-4, 4, -2) = -20$. Finally, if either of x and y is zero, then so is the other; $z^2 = 36$, and we obtain two more critical points and the values $f(0, 0, 6) = 12$ and $f(0, 0, -6) = -12$. Hence the global maximum value of $f(x, y, z)$ is 20 and its global minimum value is -20.

12.9.13: The continuous function $f(x, y, z) = x^2 y^2 z^2$ clearly has a global maximum value and a global minimum value on the ellipsoidal surface $g(x, y, z) = x^2 + 4y^2 + 9z^2 - 27 = 0$. The Lagrange multiplier method yields the equations

$$xy^2 z^2 = \lambda x, \quad x^2 y z^2 = 4\lambda y, \quad \text{and} \quad x^2 y^2 z = 9\lambda z.$$

If any one of x, y, or z is zero, then $f(x, y, z) = 0$, and this is clearly the global minimum value of $f(x, y, z)$. Otherwise, $\lambda \neq 0$, and hence

$$x^2 y^2 z^2 = \lambda x^2 = 4\lambda y^2 = 9\lambda z^2,$$

and thus $x^2 = 4y^2 = 9z^2 = 9$; that is, $x^2 = 9$, $y^2 = \frac{9}{4}$, and $z^2 = 1$. Thus the global maximum value of $f(x, y, z)$ occurs at eight different critical points, and that maximum value is $9 \cdot \frac{9}{4} \cdot 1 = \frac{81}{4}$.

12.9.15: We are to find the extrema of $f(x, y, z) = x^2 + y^2 + z^2$ subject to the two constraints $g(x, y, z) = x + y + z - 1 = 0$ and $h(x, y, z) = x + 2y + 3z - 6 = 0$. The Lagrange multiplier equation is

$$\langle 2x, 2y, 2z \rangle = \lambda \langle 1, 1, 1 \rangle + \mu \langle 1, 2, 3 \rangle,$$

from which we obtain the scalar equations $\lambda = 2x - \mu = 2y - 2\mu = 2z - 3\mu$, so that

$$2x + \mu = 2y \quad \text{and} \quad 2y + \mu = 2z, \quad \text{and thus}$$

$$\mu = 2(y - x) = 2(z - y) : \quad 2y = x + z.$$

Thus we are to solve the following three *linear* equations in three unknowns:

$$x + y + z = 1,$$

$$x - 2y + z = 0,$$

$$x + 2y + 3z = 6.$$

We find that $x = -\frac{5}{3}$, $y = \frac{1}{3}$, and $z = \frac{7}{3}$. A geometric interpretation of this problem is to find the point on the intersection of two planes—a line—closest to and farthest from the origin. Hence there is no maximum but surely a minimum, and we have found it: $f\left(-\frac{5}{3}, \frac{1}{3}, \frac{7}{3}\right) = \frac{25}{3}$.

12.9.17: We are to find the extrema of the function $f(x, y, z) = z$ subject to the *two* constraints $g(x, y, z) = x + y + z - 1 = 0$ and $h(x, y, z) = x^2 + y^2 - 1 = 0$. Geometrically, this is the problem of finding the highest (maximum z-coordinate) and lowest points on the ellipse formed by the intersection of a plane and a vertical cylinder. Hence there will be a unique global maximum and a unique global minimum unless the plane is horizontal or vertical—which it is not. The Lagrange multiplier equation is

$$\langle 0, 0, 1 \rangle = \lambda \langle 1, 1, 1 \rangle + \mu \langle 2x, 2y, 0 \rangle.$$

We thus obtain the three (simultaneous) scalar equations

$$2\mu x + \lambda = 0, \quad 2\mu y + \lambda = 0, \quad \lambda = 1.$$

Thus $2\mu x = 2\mu y = -1$ and so, because $\mu \neq 0$, we find that $y = x$. Then the second constraint implies that $2x^2 = 1$, and because $y = x$ there are only two cases:

Case 1: $x = \frac{1}{2}\sqrt{2} = y$, $z = 1 - \sqrt{2}$. This is the lowest point on the ellipse.

Case 2: $x = -\frac{1}{2}\sqrt{2} = y$, $z = 1 + \sqrt{2}$. This is the highest point on the ellipse.

12.9.19: We should find only one possible extremum because there is no point on the line farthest from the origin. We minimize $f(x, y) = x^2 + y^2$ given the constraint $g(x, y) = 3x + 4y - 100 = 0$. The Lagrange multiplier method yields

$$\langle 2x, 2y \rangle = \lambda \langle 3, 4 \rangle, \quad \text{so that} \quad 2x = 3\lambda \quad \text{and} \quad 2y = 4\lambda.$$

Thus $8x = 12\lambda = 6y$, so that $y = \frac{4}{3}x$. Substitution in the constraint yields $x = 12$ and $y = 16$. Answer: The point on the line $g(x, y) = 0$ closest to the origin is $(12, 16)$.

12.9.21: Please refer to Problem 29 of Section 12.5. We are to minimize

$$f(x, y, z) = x^2 + y^2 + z^2 \quad \text{given} \quad g(x, y, z) = 12x + 4y + 3z - 169 = 0.$$

Geometrically, we are to find the point (x, y, z) on the plane with equation $g(x, y, z) = 0$ closest to the origin, hence we can be sure that a unique solution exists. The Lagrange multiplier method yields the scalar equations

$$2x = 12\lambda, \quad 2y = 4\lambda, \quad \text{and} \quad 2z = 3\lambda,$$

so that $2x = 6y = 8z = 12\lambda$, and thus $y = \frac{1}{3}x$ and $z = \frac{1}{4}x$. Then the constraint yields

$$12x + \frac{4}{3}x + \frac{3}{4}x = 169;$$

$$144x + 16x + 9x = 169 \cdot 12;$$

$$x = 12, \quad y = 4, \quad z = 3.$$

Answer: The point on the plane $g(x, y, z) = 0$ closest to the origin is $(12, 4, 3)$.

12.9.23: Please refer to Problem 31 of Section 12.5. We are to minimize

$$f(x, y, z) = (x - 7)^2 + (y + 7)^2 + z^2 \quad \text{given} \quad g(x, y, z) = 2x + 3y + z - 49 = 0.$$

This is the geometric problem of finding the point (x, y, z) on the plane $g(x, y, z) = 0$ closest to the point $Q(7, -7, 0)$. Hence we can be sure that a unique minimum exists and, because there can be no maximum of $f(x, y, z)$, we anticipate a single critical point. The Lagrange multiplier method yields the scalar equations

$$2(x - 7) = 2\lambda, \quad 2(y + 7) = 3\lambda, \quad 2z = \lambda.$$

It follows that $6\lambda = 6x - 42 = 4y + 28 = 12z$, and thus

$$x = \frac{12z + 42}{6} = 2z + 7 \quad \text{and}$$

$$y = \frac{12z - 28}{4} = 3z - 7.$$

Substitution in the constraint gives the equation

$$4z + 14 + 9z - 21 + z = 49, \quad \text{so that} \quad 14z = 56.$$

Therefore $z = 4$, $y = 5$, and $x = 15$. The point on the plane $g(x, y, z) = 0$ closest to Q is thus $(15, 5, 4)$.

12.9.25: Please refer to Problem 33 of Section 12.5. We are to find the point (x, y, z) in the first octant and on the surface $g(x, y, z) = x^2 y^2 z - 4 = 0$ closest to the origin, so we minimize

$$f(x, y, z) = x^2 + y^2 + z^2 \quad \text{subject to the constraint} \quad g(x, y, z) = 0.$$

The surface is unbounded, so we anticipate no maximum and only one minimum in the first octant. The Lagrange multiplier method yields the scalar equations

$$2x = 2\lambda xy^2 z, \quad 2y = 2\lambda x^2 yz, \quad 2z = \lambda x^2 y^2.$$

It follows that $2\lambda x^2 y^2 z = 2x^2 = 2y^2 = 4z^2$, and hence that $x^2 = y^2 = 2z^2$. The restriction to the first octant then implies that $x = y = z\sqrt{2}$, and then the constraint $x^2 y^2 z = 4$ implies that

$$(2z^2)(2z^2)(z) = 4; \quad 4z^5 = 4; \quad z = 1.$$

So $x = y = \sqrt{2}$. The point on $g(x, y, z) = 0$ in the first octant closest to the origin is $\left(\sqrt{2}, \sqrt{2}, 1\right)$. Its distance from the origin is $\sqrt{5}$.

12.9.27: Please see Problem 35 of Section 12.5. We are to maximize $f(x, y, z) = xyz$ given $g(x, y, z) = x + y + z - 120 = 0$ and the additional condition that x, y, and z are all positive. The Lagrange multiplier equations are

$$\lambda = yz = xz = xy, \quad \text{and hence} \quad x = y = z = 40.$$

Therefore the maximum value of f is $f(40, 40, 40) = 40^3 = 64000$.

To establish that $f(x, y, z)$ actually has a global maximum value on its domain (the set of all triples of positive real numbers (x, y, z) such that $x + y + z = 120$), extend the domain slightly to include all triples of nonnegative triples of real numbers (x, y, z) such that $x + y + z = 120$. Then f is continuous on this domain, which is a closed and bounded subset of three-dimensional space (it is the part of the plane $x + y + z = 120$ that lies in the first octant or on the adjacent coordinate planes). Hence f has a global maximum there, and the maximum does not occur on the boundary because $xyz = 0$ if any of x, y, or z is zero. Hence this maximum must occur at an interior critical point, and in the previous paragraph we found only one such critical point. This is enough to establish that 64000 is the correct answer.

12.9.29: Please see Problem 37 of Section 12.5. Suppose that the box has dimensions x by y by z (units are in inches). Then we are to minimize total surface area $A(x, y, z) = 2xy + 2xz + 2yz$ given $V(x, y, z) = xyz - 1000 = 0$. The Lagrange multipliers equations are

$$2(y + z) = \lambda yz, \quad 2(x + z) = \lambda xz, \quad \text{and} \quad 2(x + y) = \lambda xy;$$

note also that in the solution, all three of x, y, and z must be positive. Hence we can eliminate λ from all three of the previous equations:

$$\frac{\lambda}{2} = \frac{y + z}{yz} = \frac{x + z}{xz} = \frac{x + y}{xy};$$

$$\frac{1}{z} + \frac{1}{y} = \frac{1}{z} + \frac{1}{x} = \frac{1}{y} + \frac{1}{x};$$

$$x = y = z \quad \text{and} \quad xyz = 1000;$$

$$x = y = z = 10.$$

Answer: The minimum possible surface area occurs when the box is a cube measuring 10 in. along each edge. The minimum possible surface area is 600 in.[2]

12.9.31: Please see Problem 39 of Section 12.5. Units in this problem will be cents and inches. Suppose that the bottom of the box has dimensions x by y and that its height is z. We are to minimize total cost $C(x, y, z) = 6xy + 10xz + 10yz$ given the constraint $V(x, y, z) = xyz - 600 = 0$. The Lagrange multiplier equations are

$$6y + 10z = \lambda yz, \quad 6x + 10z = \lambda xz, \quad \text{and} \quad 10x + 10y = \lambda xy;$$

also note that x, y, and y are all positive. Multiply each of the three previous equations by the "missing" variable to find that

$$\lambda xyz = 6xy + 10xz = 6xy + 10yz = 10xz + 10yz,$$

and thus $x = y$ and $6y = 10z$; that is, $x = y = \frac{5}{3}z$. Substitution in the constraint equation yields

$$600 = xyz = \frac{25}{9}z^3: \quad z^3 = \frac{5400}{25} = 216.$$

Therefore $z = 6$ and $x = y = 10$. Answer: The dimensions that minimize the total cost are these: base 10 in. by 10 in., height 6 in.; its total cost will be \$18.00.

12.9.33: Please refer to Problem 41 of Section 12.5. Suppose that the front of the box has width x and height z (units are in inches and cents) and that the bottom of the box measures x by y. We are to minimize total cost $C(x, y, z) = 6xy + 12xz + 18yz$ given the constraint $V(x, y, z) = xyz - 750 = 0$. The Lagrange multiplier equations are

$$6y + 12z = \lambda yz, \quad 6x + 18z = \lambda xz, \quad \text{and} \quad 12x + 18y = \lambda xy.$$

Note that x, y, and z are positive. Multiply each of the three preceding equations by the "missing" variable to obtain

$$\lambda xyz = 6xy + 12xz = 6xy + 18yz = 12xz + 18yz.$$

It follows that $12xz = 18yz$, so that $2x = 3y$; also, $6xy = 12xz$, so that $y = 2z$. Thus $x = \frac{3}{2}y = 3z$. Then substitution in the constraint equation yields

$$750 = xyz = 6z^3: \quad z^3 = 125,$$

and so $z = 5$, $y = 10$, and $x = 15$. The front of the cheapest box should be 15 in. wide and 5 in. high; its depth (from front to back) should be 10 in. (This box will cost \$27.00.)

12.9.35: We are to minimize $f(x, y, z) = x^2 + y^2 + z^2$ given (x, y, z) lies on the surface with equation $g(x, y, z) = xy + 5 - z = 0$. The Lagrange multiplier equations are

$$2x = \lambda y, \quad 2y = \lambda x, \quad \text{and} \quad 2z = -\lambda.$$

Therefore $\lambda = -2z$, and thus $2x = -2yz$ and $2y = -2xz$. So

$$2x^2 = 2y^2 = -2xyz: \quad x^2 = y^2 = -xyz.$$

Case 1: $y = x$. Then $x^2 = -x^2z$, so $x = 0$ or $z = -1$. In the latter case, the constraint equation yields $x^2 = -6$, so that case is rejected. We obtain only the critical point $(0, 0, 5)$.

Case 2: $y = -x$. Then $x^2 = x^2z$, so $x = 0$ or $z = 1$. If $x = 0$, then we obtain only the critical point of Case 1. If $z = 1$ then the constraint equation yields $x^2 = 4$, and we obtain two more critical points: $(2, -2, 1)$ and $(-2, 2, 1)$.

Now $f(0, 0, 5) = 25$, $f(2, -2, 1) = 9$, and $f(-2, 2, 1) = 9$. Hence there are two points of the surface $g(x, y, z) = 0$ closest to the origin. They are $(2, -2, 1)$ and $(-2, 2, 1)$; each is at distance $\sqrt{9} = 3$ from the origin.

12.9.37: See Fig. 12.9.9 of the text. There we see three small isosceles triangles, each with two equal sides of length 1 meeting at the center of the circle. Their total area

$$A = \frac{1}{2}\sin\alpha + \frac{1}{2}\sin\beta + \frac{1}{2}\sin\gamma$$

is the area of the large triangle, the quantity to be maximized given the constraint

$$g(\alpha, \beta, \gamma) = \alpha + \beta + \gamma - 2\pi = 0.$$

It is clear that, at maximum area, the center of the circle is *within* the large triangle or, at worst, on its boundary. Thus we have the additional restrictions

$$0 \leqq \alpha \leqq \pi, \quad 0 \leqq \beta \leqq \pi, \quad \text{and} \quad 0 \leqq \gamma \leqq \pi. \tag{1}$$

The Lagrange multiplier equations are

$$\lambda = \frac{1}{2}\cos\alpha = \frac{1}{2}\cos\beta = \frac{1}{2}\cos\gamma,$$

and the restrictions in (1) imply that $\alpha = \beta = \gamma = \frac{2}{3}\pi$. Hence the triangle of maximal area is equilateral (because it is equiangular). If x denotes the length of each side, then by the law of cosines

$$x^2 = 1^2 + 1^2 - 2 \cdot 1 \cdot 1 \cdot \cos\frac{2\pi}{3} = 3,$$

and so each side of the maximal-area triangle has length $\sqrt{3}$. The area of the maximal-area triangle is

$$A = \frac{3}{2}\sin\frac{2\pi}{3} = \frac{3}{2} \cdot \frac{\sqrt{3}}{2} = \frac{3\sqrt{3}}{4} \approx 1.299038106,$$

and the ratio of this area to that of the circumscribed circle is

$$\frac{A}{\pi} \approx 0.413496672.$$

So the inscribed triangle of maximal area occupies about 41% of the area of the circle; this result meets the test of plausibility.

12.9.39: We are to use the *Suggestion* in Problem 38 to find the point or points on the hyperbola $g(x, y) = x^2 + 12xy + 6y^2 - 130 = 0$ that are closest to the origin. To do so we minimize $f(x, y) = x^2 + y^2$ subject to the constraint $g(x, y) = 0$. The Lagrange multiplier equations are

$$2x = \lambda(2x + 12y) \quad \text{and} \quad 2y = \lambda(12y + 12x). \tag{1}$$

When put into the form suggested in Problem 38, they become

$$(1 - \lambda)x - 6\lambda y = 0,$$

$$6\lambda x + (6\lambda - 1)y = 0.$$

Because $(x, y) = (0, 0)$ is not a solution of this system and because it must therefore have a nontrivial solution, the determinant of this system must be zero:

$$(1 - \lambda)(6\lambda - 1) + 36\lambda^2 = 0;$$

$$30\lambda^2 + 7\lambda - 1 = 0;$$

$$(3\lambda + 1)(10\lambda - 1) = 0.$$

Case 1: $\lambda = -\frac{1}{3}$. Then (1) implies that

$$\frac{4}{3}x + 2y = 0: \quad y = -\frac{2}{3}x.$$

Substitution of the last equation for y into the constraint equation $g(x, y) = 0$ yields $x^2 + 30 = 0$, so there is no solution in Case 1.

Case 2: $\lambda = \frac{1}{10}$. Then (1) implies that

$$\frac{9}{10}x - \frac{6}{10}y = 0: \quad y = \frac{3}{2}x.$$

Substitution of the last equation for y into the constraint equation yields $x^2 = 4$, so we obtain two solutions in this case: $(-2, -3)$ and $(2, 3)$. These two points of the hyperbola are its points closest to $(0, 0)$, each at distance $\sqrt{13}$.

12.9.41: The highest point has the largest z-coordinate; the lowest point, the smallest. Consequently we are to maximize and minimize

$$f(x, y\ z) = z \quad \text{given} \quad g(x, y, z) = x^2 + y^2 - 1 = 0 \quad \text{and} \quad h(x, y, z) = 2x + y - z - 4 = 0.$$

Because the ellipse is formed by the intersection of a plane with a vertical cylinder, there will be a unique highest point and a unique lowest point unless the plane is horizontal or vertical—and it is not; one of its normal vectors is $\mathbf{n} = \langle 2, 1, -1 \rangle$. So we expect to find exactly two critical points. The Lagrange multiplier equations are

$$2\lambda x + 2\mu = 0, \quad 2\lambda y + \mu = 0, \quad -\mu = 1,$$

and when the third is substituted into the first two, we find that

$$2\lambda x = -2 = 4\lambda y.$$

But then $\lambda \neq 0$, and therefore $x = 2y$. Substitution of this information into the first constraint yields $5y^2 = 1$ and thus leads to the following two cases.

Case 1: $y = \frac{1}{5}\sqrt{5}$. Then $x = \frac{2}{5}\sqrt{5}$, and the second constraint implies that $z = \sqrt{5} - 4$.

Case 2: $y = -\frac{1}{5}\sqrt{5}$. Then $x = -\frac{2}{5}\sqrt{5}$, and the second constraint implies that $z = -\sqrt{5} - 4$.

Therefore the lowest point on the ellipse is $\left(-\frac{2}{5}\sqrt{5}, -\frac{1}{5}\sqrt{5}, -\sqrt{5} - 4\right)$ and the highest point on the ellipse is $\left(\frac{2}{5}\sqrt{5}, \frac{1}{5}\sqrt{5}, \sqrt{5} - 4\right)$.

12.9.43: We are to maximize and minimize $f(x, y, z) = x^2 + y^2 + z^2$ given the two constraints $g(x, y, z) = x^2 + y^2 - z^2 = 0$ and $h(x, y, z) = x + 2y + 3z - 3 = 0$. From the geometry of the problem we see that there will be a unique maximum distance and a unique minimum distance, but there may be more than two critical points. The Lagrange multiplier vector equation is

$$\langle\, 2x,\, 2y,\, 2z \,\rangle = \lambda\langle\, 2x,\, 2y,\, -2z \,\rangle + \mu\langle\, 1,\, 2,\, 3 \,\rangle,$$

and the corresponding scalar equations are

$$2x = 2\lambda x + \mu, \qquad 2y = 2\lambda y + 2\mu, \qquad \text{and} \qquad 2z = -2\lambda z + 3\mu.$$

It follows that

$$6\mu = 12x - 12\lambda x = 6y - 6\lambda y = 4z + 2\lambda z,$$

and hence $12x(1-\lambda) = 6y(1-\lambda)$.

Case 1: $\lambda = 1$. Then $6z = 0$, and so $z = 0$. Then the first constraint implies that $x = y = 0$, which contradicts the second constraint. This leaves only the second case.

Case 2: $\lambda \neq 1$. Then $y = 2x$. Then the first constraint takes the form $5x^2 = z^2$, and it is convenient to consider separately two subcases.

Case 2a: $z = x\sqrt{5}$. Then the second constraint yields

$$x = \frac{3}{20}\left(-5 + 3\sqrt{5}\right), \quad y = \frac{3}{10}\left(-5 + 3\sqrt{5}\right), \quad \text{and} \quad z = \frac{3}{4}\left(3 - \sqrt{5}\right).$$

Case 2b: $z = -x\sqrt{5}$. Then the second constraint yields

$$x = -\frac{3}{20}\left(5 + 3\sqrt{5}\right), \quad y = -\frac{3}{10}\left(5 + 3\sqrt{5}\right), \quad \text{and} \quad z = \frac{3}{4}\left(3 + \sqrt{5}\right).$$

The coordinates in Case 2a are those of the point closest to the origin; its distance from the origin is approximately 0.81027227. The coordinates of the point farthest from the origin are those given in Case 2b; the distance from this point to the origin is approximately 5.55368876.

12.9.45: Suppose that $f(x, y, z)$ and $g(x, y, z)$ have continuous first-order partial derivatives. Suppose also that the maximum (or minimum) of $f(x, y, z)$ subject to the constraint

$$g(x, y, z) = 0$$

occurs at a point P at which $\nabla g(P) \neq \mathbf{0}$. Prove that

$$\nabla f(P) = \lambda \nabla g(P)$$

for some number λ.

Proof: Parametrize the surface $g(x, y, z) = 0$ at and near the point P with a smooth function $\mathbf{r}(u, v)$ and in such a way that $\mathbf{r}_u(u, v)$ and $\mathbf{r}_v(u, v)$ are nonzero at and near P. Suppose that $\mathbf{r}(u_0, v_0) = \overrightarrow{OP}$. Then $f(\mathbf{r}(u, v))$ has a maximum (or minimum) at (u_0, v_0). Hence

$$D_u(f(\mathbf{r}(u, v))) = 0 \quad \text{and} \quad D_v(f(\mathbf{r}(u, v))) = 0.$$

Hence

$$\nabla f(P) \cdot \mathbf{r}_u(u_0, v_0) = 0 \quad \text{and} \quad \nabla f(P) \cdot \mathbf{r}_v(u_0, v_0) = 0.$$

But $g(\mathbf{r}(u, v)) \equiv 0$, and hence

$$\nabla g(P) \cdot \mathbf{r}_u(u_0, v_0) = 0 \quad \text{and} \quad \nabla g(P) \cdot \mathbf{r}_v(u_0, v_0) = 0.$$

Therefore $\nabla f(P)$ and $\nabla g(P)$ are parallel to $\mathbf{r}_u(u_0, v_0) \times \mathbf{r}_v(u_0, v_0)$. Consequently, because $\nabla g(P) \neq \mathbf{0}$, $\nabla f(P) = \lambda \nabla g(P)$ for some scalar λ. ◄

12.9.47: Given the constant P, we are to maximize

$$A(x, y, z) = \frac{1}{2} xy$$

given the constraints $g(x, y, z) = x + y + z - P = 0$ and $h(x, y, z) = x^2 + y^2 - z^2 = 0$. The Lagrange multiplier equations are

$$\frac{1}{2} y = \lambda + 2\mu x, \quad \frac{1}{2} x = \lambda + 2\mu y, \quad \text{and} \quad 0 = \lambda - 2\mu z.$$

We solve the last equation for $\lambda = 2\mu z$ and substitute in the other two to obtain

$$y = 4\mu z + 4\mu x,$$

$$x = 4\mu z + 4\mu y.$$

We subtract the second of these from the first to find that $y - x = -4\mu(y - x)$. There are two cases to consider.

Case 1: $y = x$. Then the constraint equations become $2x + z = P$ and $z^2 = 2x^2$. We solve the first for z and substitute in the second to obtain

$$(P - 2x)^2 = 2x^2;$$

$$P - 2x = \pm x\sqrt{2};$$

$$2x \pm x\sqrt{2} = P;$$

$$x = \frac{P}{2 \pm \sqrt{2}}.$$

We must take the plus sign in the last denominator because $x \leqq P$. Thus we obtain our first critical point:

$$x = \frac{P}{2+\sqrt{2}} = \frac{2-\sqrt{2}}{2}P, \quad y = x, \quad z = P - 2x = \left(\sqrt{2}-1\right)P.$$

In this case the area of the triangle is

$$A = \frac{1}{2}xy = \frac{\left(2-\sqrt{2}\right)^2}{8}P^2 = \frac{3-2\sqrt{2}}{4}P^2 \approx (0.042893218813)P^2.$$

Case 2: $y \neq x$. Then $4\mu = -1$, so that $\mu = -\frac{1}{4}$. Our earlier equations

$$y = 4\mu(z+x) \quad \text{and} \quad x = 4\mu(z+y)$$

now become $y = -z - x$ and $x = -z - y$, each of which is impossible as, at maximum, x, y, and z are all positive.

Thus Case 1 is the only case that produces a critical point. We may conclude that $y = x$ and that the right triangle with fixed perimeter and maximum area is isosceles.

12.9.49: The hexagon in Fig. 12.9.13 is the union of four congruent trapezoids, one in each quadrant. The area of the trapezoid in the first quadrant is $\frac{1}{2}(1+y)x$, so the total area of the hexagon is

$$A(x, y) = 2x(1+y),$$

which we are to maximize subject to the constraint $x^2 + y^2 - 1 = 0$. Note also that $0 \leqq x \leqq 1$ and $0 \leqq y \leqq 1$. The Lagrange multiplier equations are

$$2 + 2y = 2\lambda x \quad \text{and} \quad 2x = 2\lambda y,$$

which yield (multiply the first by y, the second by x)

$$2y + 2y^2 = 2\lambda xy = 2x^2, \quad \text{so that} \quad y^2 + y = x^2.$$

Substitute for x^2 in the constraint equation to obtain $2y^2 + y - 1 = 0$, so that $(2y-1)(y+1) = 0$. Thus $y = \frac{1}{2}$ because $y \neq -1$. So the only critical point is $\left(\frac{1}{2}\sqrt{3}, \frac{1}{2}\right)$. To verify that the resulting hexagon is regular, it is sufficient (by the various symmetries in the figure) to verify that the distances from $\left(\frac{1}{2}\sqrt{3}, \frac{1}{2}\right)$ to $(0, 1)$ and from $\left(\frac{1}{2}\sqrt{3}, \frac{1}{2}\right)$ to $\left(\frac{1}{2}\sqrt{3}, -\frac{1}{2}\right)$ are equal. They are; each distance is 1.

12.9.51: We are to minimize $f(x, y) = x^2 + y^2$ given the constraint $(x-1)^2 - y = 0$. The Lagrange multiplier equations are

$$2x = 2\lambda(x-1) \quad \text{and} \quad 2y = -\lambda,$$

813

and elimination of λ leads to the equation $x = -2y(x - 1)$. We combine this with the constraint equation and ask *Mathematica* to solve the resulting two equations:

```
Solve[ { x == -2*y*(x - 1), y == (x - 1)^2 }, { x, y } ]
```

The only real solution is

$$x = 1 + \frac{\left(-9 + \sqrt{87}\right)^{1/3}}{6^{2/3}} - \frac{1}{\left[6 \cdot \left(-9 + \sqrt{87}\right)\right]^{1/3}} \approx 0.4102454876985416,$$

$$y = \frac{1}{36}\left\{ -12 + \frac{6^{4/3}}{\left(-9 + \sqrt{87}\right)^{2/3}} + \left[6 \cdot \left(-9 + \sqrt{87}\right)\right]^{2/3} \right\} \approx 0.3478103847799310.$$

12.9.53: We are to find the first-quadrant point of the hyperbola $xy = 24$ that is closest to the point $P(1, 4)$, so we minimize $f(x, y) = (x - 1)^2 + (y - 4)^2$ subject to the constraint $xy = 24$ and the observation that $x > 0$ and $y > 0$. The Lagrange multiplier equations are

$$2(x - 1) = \lambda y \quad \text{and} \quad 2(y - 4) = \lambda x.$$

To eliminate λ, multiply the first equation by x and the second by y to obtain

$$2x(x - 1) = \lambda xy = 2y(y - 4).$$

We simplified this equation and asked *Mathematica* to solve it simultaneously with the constraint equation:

```
Solve[ { x*x - x == y*y - y, x*y == 24 }, { x, y } ]
```

The resulting output is too long to reproduce here, but to evaluate the results numerically to 20 places we entered

```
N[ %, 20 ]
```

and obtained four pairs of solutions, only two of which were real:

$$x_1 = 4, \ y_1 = 6 \quad \text{and} \quad x_2 \approx -5.5338281384822297, \ y_2 \approx -4.3369615751352589.$$

Hence the point in the first quadrant on the hyperbola $xy = 24$ closest to P is $(4, 6)$. It seems very likely that (x_2, y_2) is the point on the hyperbola in the third quadrant closest to P.

12.9.55: To find the point on the spherical surface with equation

$$(x - 1)^2 + (y - 2)^2 + (z - 3)^2 - 36 = 0 \tag{1}$$

that are closest to and farthest from the origin, note that there do exist such points, and probably only one of each type, so we expect to find only two critical points. We maximize and minimize $f(x, y, z) = x^2 + y^2 + z^2$ subject to the constraint in Eq. (1). The Lagrange multiplier equations are

$$2x = 2\lambda(x - 1), \quad 2y = 2\lambda(y - 2), \quad \text{and} \quad 2z = 2\lambda(z - 3).$$

To solve these, we entered the *Mathematica* command

```
Solve[ { x == lambda*(x - 1), y == lambda*(y - 2), z == lambda*(z - 3),
    (x - 1)∧2 + (y - 2)∧2 + (z - 3)∧2 == 36 }, { x, y, z, lambda } ]
```

As expected, there are two solutions. Omitting the values of λ, they are

$$x_1 = \frac{7 - 3\sqrt{14}}{7}, \quad y_1 = \frac{14 - 6\sqrt{14}}{7}, \quad z_1 = \frac{21 - 9\sqrt{14}}{7} \quad \text{and}$$

$$x_2 = \frac{7 + 3\sqrt{14}}{7}, \quad y_2 = \frac{14 + 6\sqrt{14}}{7}, \quad z_2 = \frac{21 + 9\sqrt{14}}{7}.$$

Their numerical values are

$$x_1 \approx -0.6035674514745463, \quad y_1 \approx -1.2071349029490926, \quad z_1 \approx -1.8107023544236389 \quad \text{and}$$

$$x_2 \approx 2.6035674514745463, \quad y_2 \approx 5.2071349029490926, \quad z_2 \approx 7.8107023544236389.$$

The first of these is obviously much closer to the origin than the second, so (x_1, y_1, z_1) is the point of the spherical surface closest to the origin and (x_2, y_2, z_2) is the point of the spherical surface farthest from the origin.

12.9.57: We are to find the points of the ellipse with equation $4x^2 + 9y^2 = 36$ closest to, and farthest from, the line with equation $x + y = 10$. What if the ellipse and the line intersect? If they do, then the equation $4x^2 + 9(10 - x)^2 = 36$ will have one or two real solutions. But this equation reduces to $13x^2 - 180x + 864 = 0$, which has discriminant

$$\Delta = 180^2 - 4 \cdot 13 \cdot 864 = -12528 < 0.$$

Because the quadratic has no real solutions, the ellipse and the line do not meet. To find the answers, we could assume that (x, y) is a point of the ellipse, that (u, v) is a point of the line, and maximize and minimize

$$f(x, y, u, v) = (x - u)^2 + (y - v)^2$$

subject to the constraints $4x^2 + 9y^2 - 36 = 0$ and $u + v - 10 = 0$. The Lagrange multiplier equations are

$$2(x - u) = 8\lambda x, \qquad 2(y - v) = 18\lambda y,$$

$$-2(x - u) = \mu, \qquad -2(y - v) = \mu.$$

Mathematica can solve the system of six simultaneous equations (the four Lagrange multiplier equations and the two constraint equations) exactly in a few tenths of a second, but this problem can be solved by hand. Observe that if the line is moved without rotation toward the ellipse, when it first touches the ellipse it will be touching the point of the ellipse closest to the original line and the moving line will be tangent to the ellipse at that point. As the line continues to move across the ellipse, it will last touch the ellipse at the point of the ellipse farthest from the original line and the moving line will be tangent to the ellipse at that point. Because the line has slope -1, all we need do is find the points of the ellipse where the tangent line has slope -1. Implicit differentiation of the equation of the ellipse yields

$$8x + 18y\frac{dy}{dx} = 0, \qquad \text{so that} \qquad \frac{dy}{dx} = -\frac{8x}{18y} = -\frac{4x}{9y}.$$

The extrema occur when

$$\frac{dy}{dx} = -1; \qquad 4x = 9y; \qquad y = \frac{4}{9}x.$$

Substitution of the last equation for y in the equation of the ellipse yields the two solutions $x = \pm\frac{9}{13}\sqrt{13}$. So the points of the ellipse closest to, and farthest from, the line are (respectively)

$$\left(\frac{9}{13}\sqrt{13}, \frac{4}{13}\sqrt{13}\right) \qquad \text{and} \qquad \left(-\frac{9}{13}\sqrt{13}, -\frac{4}{13}\sqrt{13}\right).$$

(You can tell the maximum from the minimum by observing that the line is to the "northeast" of the ellipse.)

12.9.59: The sides of the box must be parallel to the coordinate planes. Let (x, y, z) be the upper vertex of the box that lies in the first octant, so that at maximum volume we have x, y, and z all positive. The box has width $2x$, depth $2y$, and height $z = 9 - x^2 - 2y^2$, so we are to maximize box volume $V(x, y, z) = 4xyz$ given the constraint $x^2 + 2y^2 + z - 9 = 0$. The Lagrange multiplier equations are

$$4yz = 2\lambda x, \quad 4xz = 4\lambda y, \quad \text{and} \quad 4xy = \lambda.$$

These equations (together with the constraint equation) are relatively easy to solve by hand. To solve them using *Mathematica*, we entered the command

```
Solve[ { 4*y*z == 2*lambda*x, 4*x*z == 4*lambda*y, 4*x*y == lambda,

        x*x + 2*y*y + z == 9 }, { x, y, z, lambda } ]
```

The computer returned nine solutions, but eight involve negative or zero values of x, y, or z; the only viable solution is

$$x = \frac{3}{2}, \quad y = \frac{3}{4}\sqrt{2}, \quad z = \frac{9}{2},$$

and therefore the box of maximum volume has volume $V = \frac{81}{4}\sqrt{2}$.

12.9.61: With $n = 3$ and $k = 2$ (for instance), the equations in (15) are

$$g_1(x_1, x_2, x_3, x_4) = 0,$$

$$g_2(x_1, x_2, x_3, x_4) = 0,$$

$$g_3(x_1, x_2, x_3, x_4) = 0$$

and the scalar component equations of the vector equation

$$\nabla f(x_1, x_2, x_3, x_4) = \lambda_1 \nabla g_1(x_1, x_2, x_3, x_4) + \lambda_2 \nabla g_2(x_1, x_2, x_3, x_4) + \lambda_3 \nabla g_3(x_1, x_2, x_3, x_4)$$

are

$$D_1 f(x_1, x_2, x_3, x_4) = \lambda_1 D_1 g_1(x_1, x_2, x_3, x_4) + \lambda_2 D_1 g_2(x_1, x_2, x_3, x_4) + \lambda_3 D_1 g_3(x_1, x_2, x_3, x_4),$$

$$D_2 f(x_1, x_2, x_3, x_4) = \lambda_1 D_2 g_1(x_1, x_2, x_3, x_4) + \lambda_2 D_2 g_2(x_1, x_2, x_3, x_4) + \lambda_3 D_2 g_3(x_1, x_2, x_3, x_4),$$

$$D_3 f(x_1, x_2, x_3, x_4) = \lambda_1 D_3 g_1(x_1, x_2, x_3, x_4) + \lambda_2 D_3 g_2(x_1, x_2, x_3, x_4) + \lambda_3 D_3 g_3(x_1, x_2, x_3, x_4),$$

$$D_4 f(x_1, x_2, x_3, x_4) = \lambda_1 D_4 g_1(x_1, x_2, x_3, x_4) + \lambda_2 D_4 g_2(x_1, x_2, x_3, x_4) + \lambda_3 D_4 g_3(x_1, x_2, x_3, x_4).$$

Consequently we have altogether $3 + 4 = 7$ equations in the seven unknowns λ_1, λ_2, λ_3, x_1, x_2, x_3, and x_4.

12.9.63: With

$$f(x, y) = x^2 + y^2 \quad \text{and} \quad g(x, y) = \frac{a}{x} + \frac{b}{y} - 1,$$

the Lagrange multiplier equations are

$$2x = -\frac{\lambda}{x^2} \quad \text{and} \quad 2y = -\frac{\lambda}{y^2},$$

and it follows that $2x^2 = -\lambda a$ and $2y^2 = -\lambda b$. Then division of the last equation by the one before it yields $y = xb^{1/3}a^{-1/3}$, and then substitution of this value in the constraint equation readily gives $x = a^{1/3}(a^{2/3} + b^{2/3})$. Thus $y = b^{1/3}(a^{2/3} + b^{2/3})$, and substitution of these values of x and y gives

$$L_{\min} = \sqrt{x^2 + y^2} = \left(a^{2/3} + b^{2/3}\right)^{3/2}.$$

12.9.65: We write $P_1(x, y, z)$ and $P_2(u, v, w)$ because the two points are independent. *Mathematica* then yields the solution, as follows.

```
f = (x - u)∧2 + (y - v)∧2 + (z - w)∧2;

eq1 = 2*x + y + 2*z == 15;

eq2 = x + 2*y + 3*z == 30;

eq3 = u - v - 2 *w == 15;

eq4 = 3*u - 2*v - 3*w == 20;

eq5 = 2*(x -u) == 2*λ₁ + λ₂;

eq6 = 2*(y -v) == λ₁ + 2*λ₂;

eq7 = 2*(z -w) == 2*λ₁ + 3*λ₂;

eq8 = -2*(x - u) == λ₃ + 3*λ₄;

eq9 = -2*(y -v) == -λ₃ - 2*λ₄;

eq10 = -2*(z - w) == -2*λ₃ - 3*λ₄;

eqs = { eq1, eq2, eq3, eq4, eq5, eq6, eq7, eq8, eq9, eq10 };

vars = { x, y, z, u, v, w, λ₁, λ₂, λ₃, λ₄ };

soln = Solve[ eqs, vars ]
```

$$x = 7, \qquad y = 43, \qquad z = -21, \qquad u = 12, \qquad v = 41, \qquad w = -22,$$

$$\lambda_1 = -8, \qquad \lambda_2 = 6, \qquad \lambda_3 = -8, \qquad \lambda_4 = 6.$$

Thus the closest points are $P_1(7, 43, -21)$ on line L_1 and $P_2(12, 41, -22)$ on line L_2.

Section 12.10

12.10.1: Given $f(x, y) = 2x^2 + y^2 + 4x - 4y + 5$, in the notation of Section 12.10 we have $A = 4$, $B = 0$, and $C = 2$. Hence $\Delta = 8 > 0$ and $A > 0$ for all points (x, y), and so—by Theorem 1—any critical point where both partials vanish is a local minimum. Here we have

$$f_x(x, y) = 4x + 4 \quad \text{and} \quad f_y(x, y) = 2y - 4,$$

so the only such critical point is $(-1, 2)$. The *Mathematica* command

```
ContourPlot[ 2*x∧2 + y∧2 + 4*x - 4*y + 5, {x, -4, 2}, {y, -2, 4},
```

AspectRatio → 1.0, Axes → True, Contours → 15, ContourShading → False,

Frame → False, PlotPoints → 31, AxesOrigin → {0,0}];

generates level curves for f in the vicinity of the critical point $(-1, 2)$, shown next. Because the graph of f is an elliptic paraboloid opening upward, the local minimum at $(-1, 2)$ is actually a global minimum.

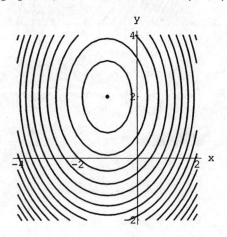

12.10.3: Given $f(x, y) = 2x^2 - 3y^2 + 2x - 3y + 7$, both partial derivatives are zero at the single point $P\left(-\frac{1}{2}, -\frac{1}{2}\right)$. In the notation of Section 13.10 we have $A = 4$, $B = 0$, and $C = -6$. Hence $\Delta = -24 < 0$, and so—by Theorem 2—there is a saddle point at P. The *Mathematica* command `ContourPlot` on the rectangle $-3 \leq x \leq 2$, $-3 \leq y \leq 2$ produced the level curve diagram shown next.

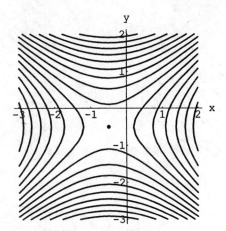

12.10.5: Given $f(x, y) = 2x^2 + 2xy + y^2 + 4x - 2y + 1$, both partial derivatives are zero at the single point $P(-3, 4)$. In the notation of Section 13.10, $A = 4$, $B = 2$, and $C = 2$, so that $\Delta = 4 > 0$. Because $A > 0$, there is a local minimum at P. A diagram of some level curves of f near P generated by *Mathematica* is next.

12.10.7: Given $f(x, y) = x^3 + y^3 + 3xy + 3$, both partials vanish at the two points $P(-1, -1)$ and $Q(0, 0)$. Also $f_{xx}(x, y) = 6x$, $f_{xy}(x, y) = 3$, and $f_{yy}(x, y) = 6y$. At P we have $A = -6$, $B = 3$, and $C = -6$, so $\Delta = 27 > 0$. Thus there is a local maximum at P. At Q we have $A = 0$, $B = 3$, and $C = 0$, so $\Delta = -9 < 0$. So there is a saddle point at Q. Level curves of f near P are shown next, on the left; level curves near Q are shown on the right. The figure on the left was generated by the *Mathematica* command

```
ContourPlot[ f[x,y], {x, -1.5, -0.5}, {y, -1.5, -0.5}, AspectRatio → 1.0

        Contours → 15, ContourShading → False, PlotPoints → 31 ];
```

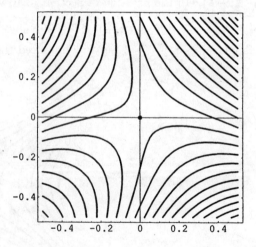

12.10.9: Given $f(x, y) = 6x - x^3 - y^3$, both partial derivatives are zero at the two points $P\left(-\sqrt{2}, 0\right)$ and $Q\left(\sqrt{2}, 0\right)$. Because $f_{xx}(x, y) = -6x$, $f_{xy}(x, y) = 0$, and $f_{yy}(x, y) = -6y$, at P we have $A = 6\sqrt{2}$ and $B = C = 0$, so $\Delta = 0$; $\Delta = 0$ at Q as well. Hence Theorem 2 provides us with no information. But on the vertical line $x = -\sqrt{2}$ through P, $f(x, y)$ takes the form $-4\sqrt{2} - y^3$, so there is no extremum at P; similarly, there is none at Q. Hence f has no extrema. We generated level curves of f near P and Q using *Mathematica* commands similar to those in the solution of Problem 7. The results are shown next.

820

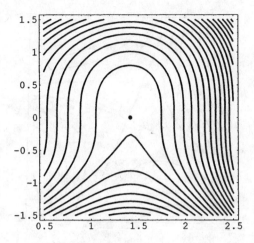

12.10.11: Given $f(x, y) = x^4 + y^4 - 4xy$, both partial derivatives are zero at $P(-1, -1)$, $Q(0, 0)$, and $R(1, 1)$. Also $f_{xx}(x, y) = 12x^2$, $f_{xy}(x, y) = -4$, and $f_{yy}(x, y) = 12y^2$. Thus at both P and R, we have $A = 12$, $B = -4$, and $C = 12$, so that $\Delta = 128 > 0$; thus there are local minima at P and R. At Q we have $A = C = 0$ and $B = -4$, so that $\Delta = -16 < 0$: There is a saddle point at Q. The minima at P and R are global because $x^4 + y^4$ greatly exceeds $4xy$ if x and y are large in magnitude. Level curves of f near the two critical points P and Q, generated by *Mathematica*, are shown next.

 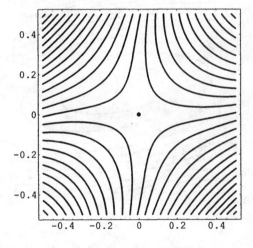

12.10.13: Given $f(x, y) = x^3 + 6xy + 3y^2 - 9x$, both partial derivatives are zero at the two points $P(-1, 1)$ and $Q(3, -3)$. Next, $f_{xx}(x, y) = 6x$ and $f_{xy}(x, y) = f_{yy}(x, y) = 6$. Hence at P we have $A = -6$ and $B = C = 6$, so that $\Delta = -72 < 0$: There is a saddle point at P. At Q we have $A = 18$ and $B = C = 6$, so $\Delta = 72 > 0$. Because $A > 0$ there is a local minimum at Q. It cannot be global because the dominant term in $f(x, y)$ is x^3. We used *Mathematica* with commands similar to those in previous solutions to show level curves of f near these two critical points; the diagrams are next.

821

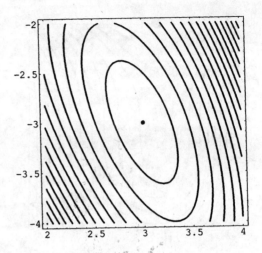

12.10.15: If $f(x, y) = 3x^2 + 6xy + 2y^3 + 12x - 24y$, then both partial derivatives of f are zero at the two points $P(-5, 3)$ and $Q(0, -2)$. Next, $f_{xx}(x, y) = f_{xy}(x, y) = 6$ and $f_{yy}(x, y) = 12y$. Thus at P we have $A = B = 6$ and $C = 36$, so $\Delta = 180 > 0$. Because $A > 0$ there is a local minimum at P (it cannot be global because the dominant term in $f(x, y)$ is $2y^3$.) At Q we have $A = B = 6$ and $C = -24$, so $\Delta = -180 < 0$; there is a saddle point at Q. We used *Mathematica* commands similar to those in previous solutions to generate level curves of f near its two critical points; these diagrams are next.

 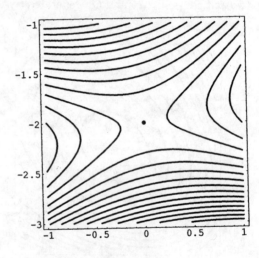

12.10.17: If $f(x, y) = 4xy - 2x^4 - y^2$, then both partial derivatives are zero when

$$8x^3 = 4y \quad \text{and} \quad 4x = 2y;$$

$$y = 2x^3 \quad \text{and} \quad y = 2x;$$

$$x^3 - x = 0; \quad \text{that is,} \quad x(x - 1)(x + 1) = 0.$$

Thus we obtain three critical points: $P(-1, -2)$, $Q(0, 0)$, and $R(1, 2)$. For the next step, we compute $f_{xx}(x, y) = -24x^2$, $f_{xy}(x, y) = 4$, and $f_{yy}(x, y) = -2$. Thus an application of Theorem 2 yields this

822

information.

$$\text{At } P: \quad A = -24, \quad B = 4, \quad C = -2, \quad \Delta = 32: \qquad \text{Local maximum.}$$

$$\text{At } Q: \quad A = 0, \qquad B = 4, \quad C = -2, \quad \Delta = -16: \qquad \text{Saddle point.}$$

$$\text{At } R: \quad A = -24, \quad B = 4, \quad C = -2, \quad \Delta = 32: \qquad \text{Local maximum.}$$

The extrema are global maxima with the same value, $f(-1, -2) = 2 = f(1, 2)$. To see that they are global, examine the behavior of $f(x, y)$ on straight lines through the origin. Diagrams of level curves near P and Q were generated using *Mathematica* and appear next.

 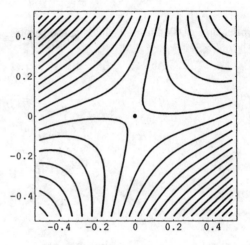

12.10.19: Given $f(x, y) = 2x^3 - 3x^2 + y^2 - 12x + 10$, we set both partial derivatives equal to zero and solve to find the only two critical points: $P(-1, 0)$ and $Q(2, 0)$. At P we find that $A = -18$, $B = 0$, and $C = 2$, so there is a saddle point at P. At Q we find that $A = 18$, $B = 0$, and $C = 2$, so there is a local minimum at Q. (There are no global extrema because the dominant term in $f(x, y)$ is $2x^3$.) We used *Mathematica* in the usual way to generate level curves of f near these two critical points; the resulting diagram is next.

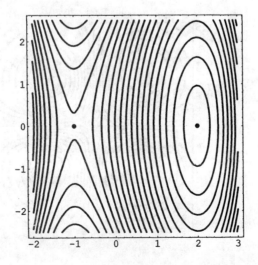

12.10.21: If $f(x, y) = xy \exp(-x^2 - y^2)$, then

$$f_x(x, y) = (1 - 2x^2)y \exp(-x^2 - y^2) \quad \text{and} \quad f_y(x, y) = x(1 - 2y^2) \exp(-x^2 - y^2).$$

Thus there are a total of five critical points. They are $P(0, 0)$, $Q\left(-\frac{1}{2}\sqrt{2}, -\frac{1}{2}\sqrt{2}\right)$, $R\left(-\frac{1}{2}\sqrt{2}, \frac{1}{2}\sqrt{2}\right)$, $S\left(\frac{1}{2}\sqrt{2}, -\frac{1}{2}\sqrt{2}\right)$, and $T\left(\frac{1}{2}\sqrt{2}, \frac{1}{2}\sqrt{2}\right)$. The values of f at these points are (in the same order) 0, $1/(2e)$, $-1/(2e)$, $-1/(2e)$, and $1/(2e)$. The second-order partial derivatives of f are

$$f_{xx}(x, y) = (4x^3y - 6xy) \exp(-x^2 - y^2) = 2xy(2x^2 - 3) \exp(-x^2 - y^2),$$

$$f_{xy}(x, y) = (4x^2y^2 - 2x^2 - 2y^2 + 1) \exp(-x^2 - y^2), \quad \text{and}$$

$$f_{yy}(x, y) = (4xy^3 - 6xy) \exp(-x^2 - y^2) = 2xy(2y^2 - 3) \exp(-x^2 - y^2).$$

Application of Theorem 2 yields these results:

At P: $A = 0$, $B = 1$, $C = 0$, $\Delta = -1$: Saddle point.

At Q: $A = -\dfrac{2}{e}$, $B = 0$, $C = -\dfrac{2}{e}$, $\Delta = \dfrac{4}{e^2}$: Local maximum.

At R: $A = \dfrac{2}{e}$, $B = 0$, $C = \dfrac{2}{e}$, $\Delta = \dfrac{4}{e^2}$: Local minimum.

At S: $A = \dfrac{2}{e}$, $B = 0$, $C = \dfrac{2}{e}$, $\Delta = \dfrac{4}{e^2}$: Local minimum.

At T: $A = -\dfrac{2}{e}$, $B = 0$, $C = -\dfrac{2}{e}$, $\Delta = \dfrac{4}{e^2}$: Local maximum.

The extrema are all global because $f(x, y) \to 0$ as $|x|$ and/or $|y|$ increase without bound. To generate a diagram of these critical points and nearby level curves of f, we executed the *Mathematica* command

```
ContourPlot[ f[x,y], {x, -1.5, 1.5}, {y, -1.5, 1.5}, AspectRatio → 1.0,
        Contours → 15, ContourShading → False, PlotPoints → 48 ];
```

and the result is shown next.

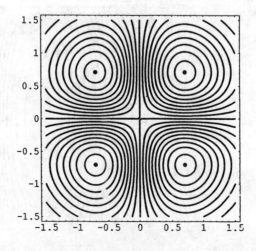

12.10.23: If $f(x, y) = x^4 + y^4$, then

$$\Delta = AC - B^2 = (12x^2)(12y^2) - 0^2 = 144x^2y^2,$$

and so $\Delta(0, 0) = 0$. But $f(x, y)$ has the global minimum value 0 at $(0, 0)$ because $f(x, y) > 0$ for all $(x, y) \neq (0, 0)$.

12.10.25: If $f(x, y) = \exp(-x^4 - y^4)$, then

$$\Delta = AC - B^2$$

$$= \left[4x^2(4x^4 - 3) \exp(-x^4 - y^4) \right] \cdot \left[4y^2(4y^4 - 3) \exp(-x^4 - y^4) \right] - \left[16x^3y^3 \exp(-x^4 - y^4) \right]^2,$$

and so $\Delta(0, 0) = 0$. To maximize $f(x, y)$, make $x^4 + y^4$ as small as possible, and hence $1 = f(0, 0)$ is the global maximum value of $f(x, y)$.

12.10.27: Let $f(x, y) = x^2 + y^2 + (xy - 2)^2$. Then

$$f_x(x, y) = 2x + 2y(xy - 2) = 2x + 2xy^2 - 4y \quad \text{and}$$

$$f_y(x, y) = 2y + 2x(xy - 2) = 2y + 2x^2y - 4x.$$

Both partial derivatives are zero when

$$x + xy^2 - 2y = 0 \quad \text{and} \quad y + x^2y - 2x = 0;$$

$$x^2 + x^2y^2 - 2xy = 0 \quad \text{and} \quad y^2 + x^2y^2 - 2xy = 0;$$

$$x^2 = y^2, \quad \text{so that} \quad y = x \quad \text{or} \quad y = -x.$$

If $y = x$, then $x^3 - x = 0$, and we obtain the three critical points $P(-1, -1)$, $Q(0, 0)$, and $R(1, 1)$. If $y = -x$ then $x^3 + 3x = 0$ and we obtain the critical point Q again but no others. Next

$$f_{xx}(x, y) = 2(y^2 + 1), \quad f_{xy}(x, y) = 4(xy - 1), \quad \text{and} \quad f_{yy}(x, y) = 2(x^2 + 1).$$

At P and at R, we find that $A = 4$, $B = 0$, and $C = 4$, so $f(x, y)$ has a local minimum at those two points. Its value at each is 3. But at Q we find that $A = 2$, $B = -4$, and $C = 2$, so $\Delta = -12$ and hence there is no extremum at Q. It is geometrically evident that $f(x, y)$ has a global minimum, so we have found it: Its global minimum value is 3, which occurs at both the points P and R.

12.10.29: The sum of the areas of the squares is

$$f(x, y) = \left(\frac{x}{4}\right)^2 + \left(\frac{y}{4}\right)^2 + \left(\frac{120 - x - y}{4}\right)^2, \quad 0 \leq x, \ 0 \leq y, \ x + y \leq 120.$$

Thus

$$\frac{\partial f}{\partial x} = \frac{1}{8}(2x + y - 120) \quad \text{and} \quad \frac{\partial f}{\partial y} = \frac{1}{8}(x + 2y - 120),$$

and both partial derivatives are zero at the single point $(x, y) = (40, 40)$. Because

$$\frac{\partial^2 f}{\partial x^2} = \frac{1}{4}, \quad \frac{\partial^2 f}{\partial y\, \partial x} = \frac{1}{8}, \quad \text{and} \quad \frac{\partial^2 f}{\partial y^2} = \frac{1}{4},$$

we see that $\Delta = \frac{3}{64} > 0$, so that $f(x, y)$ has a local minimum at $(40, 40)$. Because there must be a global maximum, it must occur on the boundary of the domain of f. If $y = 0$ then it is easy to show that $x = 60$ is a critical point of $f(x, 0)$, and $f(60, 0) = 450$. Similar results hold if $x = 0$ and if $x + y = 120$. At the corner points of the domain we find the global maximum value of $f(x, y)$ to be $f(0, 120) = 900$. It also follows that the local minimum at $(40, 40)$ is global.

12.10.31: Given

$$f(x, y) = \sin\left(\frac{\pi x}{2}\right) \sin\left(\frac{\pi y}{2}\right),$$

we first compute

$$f_x(x, y) = \frac{\pi}{2} \cos\left(\frac{\pi x}{2}\right) \sin\left(\frac{\pi y}{2}\right) \quad \text{and} \quad f_y(x, y) = \frac{\pi}{2} \sin\left(\frac{\pi x}{2}\right) \cos\left(\frac{\pi y}{2}\right).$$

Now $\cos(\pi x/2) = 0$ when x is an odd integer and $\sin(\pi x/2) = 0$ when x is an even integer. In order that both partial derivatives are zero, it is necessary and sufficient that x and y are both odd integers or both even integers. Next,

$$f_{xx}(x, y) = -\frac{\pi^2}{4} \sin\left(\frac{\pi x}{2}\right) \sin\left(\frac{\pi y}{2}\right),$$

$$f_{xy}(x, y) = \frac{\pi^2}{4} \cos\left(\frac{\pi x}{2}\right) \cos\left(\frac{\pi y}{2}\right), \quad \text{and}$$

$$f_{yy}(x, y) = -\frac{\pi^2}{4} \sin\left(\frac{\pi x}{2}\right) \sin\left(\frac{\pi y}{2}\right).$$

Note that every integer has exactly one of the following forms: $4k$, $4k + 1$, $4k + 2$, $4k + 3$ (where k is an integer). The reason is that division of an integer by 4 produces eactly one of the remainders 0, 1, 2, or 3.

Case 1: x and y are both of the form $4k$ or both of the form $4k + 2$. Then (in the notation of Theorem 1) $A = 0$, $B = \frac{1}{4}\pi^2$, and $C = 0$. Hence $\Delta < 0$, and so (x, y) is a saddle point.

Case 2: One of x and y is of the form $4k$ and the other is of the form $4k + 2$. Then $A = 0$, $B = -\frac{1}{4}\pi^2$, and $C = 0$. Hence $\Delta < 0$, and so (x, y) is a saddle point.

Case 3: x and y are both of the form $4k + 1$ or both of the form $4k + 3$. Then $A = C = -\frac{1}{4}\pi^2$ and $B = 0$. Thus $\Delta > 0$ and $A < 0$, so (x, y) is a local maximum (actually, a global maximum; the value of f there is 1).

Case 4: One of x and y is of the form $4k+1$ and the other is of the form $4k+3$. Then $A = C = \frac{1}{4}\pi^2$ and $B = 0$. Thus $\Delta > 0$ and $A > 0$, so (x, y) is a local minimum (actually, a global minimum; the value of f there is -1).

Execute the *Mathematica* command

```
ContourPlot[ f[x,y], {x, -3.5, 3.5}, {y, -3.5, 3.5}, AspectRatio → 1.0,
             Contours → 21, ContourShading → False, PlotPoints → 61 ];
```

to see the maxima and minima in a checkerboard pattern with saddle points arranged neatly among them.

12.10.33: If $f(x, y) = 4xy(x^2 - y^2)$, then

$$f_x(x, y) = 4y(3x^2 - y^2) \quad \text{and} \quad f_y(x, y) = 4x(x^2 - 3y^2).$$

Hence $(0, 0)$ is the only critical point of f. On the line $y = mx$ we have

$$f(x, y) = f(x, mx) = -4m(m+1)(m-1)x^4.$$

In the first-quadrant region between the lines $y = 0$ and $y = x$, where $0 < m < 1$, $f(x, y) > 0$. In the first-quadrant region between the lines $y = x$ and $x = 0$, where $1 < m$, $f(x, y) < 0$. Continuing in this way counterclockwise around the origin, we find eight regions on which $f(x, y)$ is alternately positive and negative. Hence the origin qualifies as a "dog saddle."

12.10.35: Given $f(x, y) = 2x^4 - 12x^2 + y^2 + 8x$, note first that $f(x, y)$ is dominated by the term $2x^4$. Therefore it can have no global maximum and must have a global minimum. We continue with the aid of *Mathematica*, but as usual, we will rewrite its responses for a little more clarity.

```
{D[f[x,y],x], D[f[x,y],y]}
```

$$f_x(x,y) = 8x^3 - 24x + 8, \qquad f_y(x, y) = 2y.$$

```
N[Solve[ % == 0, {x,y} ], 20]
```

(We let *Mathematica* find the exact solutions, then approximate them to 20 places.)

$$x_1 \approx 1.532088886238, \quad y_1 = 0;$$
$$x_2 \approx 0.347296245334, \quad y_2 = 0;$$
$$x_3 \approx -1.879385241572, \quad y_3 = 0.$$

```
{D[f[x,y], {x,2}], D[f[x,y], x, y], D[f[x,y], {y,2}]}
```

$$f_{xx}(x, y) = 24x^2 - 24, \quad f_{xy}(x, y) = 0, \quad f_{yy}(x, y) = 2.$$

Then we asked *Mathematica* to evaluate the second-order partial derivatives and $f(x, y)$ itself at the three critical points. The response was

$$(x_1, y_1): \quad A \approx 32.335113, \quad B = 0, \quad C = 2, \quad f(x_1, y_1) \approx -4.891245,$$

$$(x_2, y_2): \quad A \approx -21.105246, \quad B = 0, \quad C = 2, \quad f(x_2, y_2) \approx 1.360090,$$

$$(x_3, y_3): \quad A \approx 60.770133, \quad B = 0, \quad C = 2, \quad f(x_3, y_3) \approx -32.468845.$$

Hence there is a local minimum at (x_1, y_1), a saddle point at (x_2, y_2), and the global minimum at (x_3, y_3).

12.10.37: Given $f(x, y) = x^4 + 12xy + 6y^2 + 4x + 10$, note first that $f(x, y)$ is dominated by the term x^4. Therefore it can have no global maximum. We continue with the aid of *Mathematica*, but as usual, we will rewrite its responses for a little more clarity.

```
{D[f[x,y],x], D[f[x,y],y]}
```

$$f_x(x, y) = 4x^3 + 12y + 4, \qquad f_y(x, y) = 12x + 12y.$$

```
N[Solve[ % == 0, {x,y} ], 20]
```

(We let *Mathematica* find the exact solutions, then approximate them to 20 places.)

$$x_1 \approx -1.879385241572, \quad y_1 \approx 1.879385241572;$$

$$x_2 \approx 0.347296355334, \quad y_2 \approx -0.347296355334;$$

$$x_3 \approx 1.532088886238, \quad y_3 \approx -1.532088886238.$$

```
{D[f[x,y], {x,2}], D[f[x,y], x, y], D[f[x,y], {y,2}]}
```

$$f_{xx}(x, y) = 12x^2, \quad f_{xy}(x, y) = 12, \quad f_{yy}(x, y) = 12.$$

Then we asked *Mathematica* to evaluate the second-order partial derivatives and $f(x, y)$ itself at the three critical points. The response was

$$(x_1, y_1): \quad A \approx 42.385067, \quad B = 12, \quad C = 12, \quad f(x_1, y_1) \approx -6.234422,$$

$$(x_2, y_2): \quad A \approx 1.447377, \quad B = 12, \quad C = 12, \quad f(x_2, y_2) \approx 10.680045,$$

$$(x_3, y_3): \quad A \approx 28.167556, \quad B = 12, \quad C = 12, \quad f(x_3, y_3) \approx 7.554378.$$

Hence there is a local (indeed, global) minimum at (x_1, y_1), a saddle point at (x_2, y_2), and a local minimum at (x_3, y_3).

12.10.39: Given $f(x, y) = x^4 + 2y^4 - 12xy^2 - 20y^2$, note first that $f(x, y)$ is dominated by the terms x^4 and $2y^4$. Therefore it can have no global maximum and must have a global minimum. We continue with the aid of *Mathematica*, but as usual, we will rewrite its responses for a little more clarity.

```
{D[f[x,y],x], D[f[x,y],y]}
```

$$f_x(x, y) = 4x^3 - 12y^2, \qquad f_y(x, y) = 8y^3 - 24xy - 40y.$$

```
N[Solve[ % == 0, {x,y} ], 20]
```

(We let *Mathematica* find the exact solutions, then approximate them to 20 places.)

$$x_1 = 0, \quad y_1 = 0;$$
$$x_2 \approx 3.624678956480, \quad y_2 \approx -3.984223496422;$$
$$x_3 \approx 3.624678956480, \quad y_3 \approx 3.984223496422.$$

```
{D[f[x,y], {x,2}], D[f[x,y], x, y], D[f[x,y], {y,2}]}
```

$$f_{xx}(x, y) = 12x^2, \quad f_{xy}(x, y) = -24y, \quad f_{yy}(x, y) = 24y^2 - 24x - 40.$$

Then we asked *Mathematica* to evaluate the second-order partial derivatives and $f(x, y)$ itself at the three critical points. The response was

$$(x_1, y_1): \quad A = 0, \quad B = 0, \quad C = -40, \quad f(x_1, y_1) = 0,$$
$$(x_2, y_2): \quad A \approx 157.659570, \quad B \approx 95.621364, \quad C \approx 253.984590, \quad f(x_2, y_2) \approx -331.355231,$$
$$(x_3, y_3): \quad A \approx 157.659570, \quad B \approx -95.621364, \quad C \approx 253.984590, \quad f(x_3, y_3) \approx -331.355231.$$

Hence there is a global minimum at (x_2, y_2), and at (x_3, y_3). A plot of the level curves of $f(x, y)$ using the ContourPlot command in *Mathematica* indicates that the origin is an (ordinary) saddle point. You can verify that there is no extremum there by showing that $f(x, 0)$ has a local minimum at the origin whereas $f(0, y)$ has a local maximum there.

Chapter 12 Miscellaneous Problems

12.M.1: Using polar coordinates, we get

$$\lim_{(x,y)\to(0,0)} \frac{x^2y^2}{x^2 + y^2} = \lim_{r\to 0} \frac{r^4 \sin^2\theta \cos^2\theta}{r^2} = \lim_{r\to 0} (r^2 \sin^2\theta \cos^2\theta) = 0$$

because $0 \leq \sin^2\theta \leq 1$ and $0 \leq \cos^2\theta \leq 1$ for all θ.

12.M.3: First we note that $\lim\limits_{(x,y)\to(0,0)} g(x, y) \neq 0$ because

$$\lim_{x\to 0} g(x, x) = \lim_{x\to 0} \frac{x^2}{2x^2} = \frac{1}{2}.$$

Therefore g is not continuous at $(0, 0)$.

12.M.5: If $f_x(x, y) = 2xy^3 + e^x \sin y$, then

$$f(x, y) = x^2 y^3 + e^x \sin y + g(y).$$

Hence

$$f_y(x, y) = 3x^2 y^2 + e^x \cos y + g'(y) = 3x^2 y^2 + e^x \cos y + 1.$$

Therefore $g'(y) = 1$, and thus $g(y) = y + C$. Thus every solution of this problem has the form

$$f(x, y) = x^2 y^3 + e^x \sin y + y + C$$

where C is a constant.

12.M.7: The paraboloid is a level surface of $f(x, y, z) = x^2 + y^2 - z$ with gradient $\langle 2a, 2b, -1 \rangle$ at the point $(a, b, a^2 + b^2)$. The normal line L through that point has vector equation

$$\langle x, y, z \rangle = \langle a, b, a^2 + b^2 \rangle + t \langle 2a, 2b, -1 \rangle$$

and thus parametric equations

$$x = 2at + a, \quad y = 2bt + b, \quad z = a^2 + b^2 - t.$$

Suppose that this line passes through the point $(0, 0, 1)$. Set $x = 0$, $y = 0$, and $z = 1$ in the parametric equations of L, solve the third equation for t, then substitute the result in the other two equations to find that

$$2a(a^2 + b^2) - a = 0 = 2b(a^2 + b^2) - b.$$

It now follows that $a^2 + b^2 = \frac{1}{2}$ or $a = 0 = b$. Therefore the points on the paraboloid where the normal vector points at $(0, 0, 1)$ are the origin $(0, 0, 0)$ and the points on the circle formed by the intersection of the paraboloid and the horizontal plane $z = \frac{1}{2}$.

12.M.9: Let $f(x, y, z) = x^2 + y^2 - z^2$. Then $\nabla f(x, y, z) = \langle 2x, 2y, -2z \rangle$ is normal to the cone at (x, y, z). So a normal at $P(a, b, c)$ is $\mathbf{n} = \langle a, b, -c \rangle$. An equation of the line through P with direction $\mathbf{n}$ is

$$\langle x, y, z \rangle = \langle a, b, c \rangle + t \langle a, b, -c \rangle;$$

that is,

$$x = a + ta, \quad y = b + tb, \quad z = c - tc.$$

When $x = y = 0$, we have $t = -1$, and thus $z = 2c$. Hence $(0, 0, 2c)$ is the point where the line meets the z-axis. Thus the normal vector $\mathbf{n}$ (extended in length, if necessary) intersects the z-axis.

12.M.11: If

$$u(x, y, t) = \frac{1}{4\pi kt} \exp\left(-\frac{x^2 + y^2}{4kt}\right),$$

then

$$u_t(x, y, t) = \frac{x^2 + y^2 - 4kt}{16k^2\pi t^3} \exp\left(-\frac{x^2 + y^2}{4kt}\right),$$

$$u_x(x, y, t) = -\frac{x}{8k^2\pi t^2} \exp\left(-\frac{x^2 + y^2}{4kt}\right),$$

$$u_{xx}(x, y, t) = \frac{x^2 - 2kt}{16k^3\pi t^3} \exp\left(-\frac{x^2 + y^2}{4kt}\right),$$

$$u_y(x, y, t) = -\frac{y}{8k^2\pi t^2} \exp\left(-\frac{x^2 + y^2}{4kt}\right),$$

$$u_{yy}(x, y, t) = \frac{y^2 - 2kt}{16k^3\pi t^3} \exp\left(-\frac{x^2 + y^2}{4kt}\right), \quad \text{and}$$

$$u_{xx}(x, y, t) + u_{yy}(x, y, t) = \frac{x^2 + y^2 - 4kt}{16k^3\pi t^3} \exp\left(-\frac{x^2 + y^2}{4kt}\right).$$

Therefore $u_t = k(u_{xx} + u_{yy})$.

12.M.13: If $\mathbf{r}(x, y) = \langle x, y, f(x, y) \rangle$, then

$$\mathbf{r}_x(x, y) = \langle 1, 0, f_x(x, y) \rangle \quad \text{and} \quad \mathbf{r}_y(x, y) = \langle 0, 1, f_y(x, y) \rangle,$$

and hence

$$\mathbf{r}_x \times \mathbf{r}_y = \begin{vmatrix} \mathbf{i} & \mathbf{j} & \mathbf{k} \\ 1 & 0 & f_x(x, y) \\ 0 & 1 & f_y(x, y) \end{vmatrix} = \langle -f_x(x, y), -f_y(x, y), 1 \rangle.$$

Let $g(x, y, z) = z - f(x, y)$. Then

$$\nabla g(x, y, z) = \langle -f_x(x, y), -f_y(x, y), 1 \rangle = \mathbf{r}_x \times \mathbf{r}_y,$$

and therefore $\mathbf{r}_x \times \mathbf{r}_y$ is normal to the surface $z = f(x, y)$.

12.M.15: Suppose that the bottom of the crate measures x by y (units are in feet and dollars) and that its height is z. We are to minimize its total cost $C(x, y, z) = 5xy + 2xz + 2yz$ given the constraint $V(x, y, z) = xyz - 60 = 0$. The Lagrange multiplier equations are

$$5y + 2z = \lambda yz, \quad 5x + 2z = \lambda xz, \quad \text{and} \quad 2x + 2y = \lambda xy.$$

Because x, y, and z are all positive (because of the constraint) and $\lambda \neq 0$, it follows that

$$\lambda xyz = 5xy + 2xz = 5xy + 2yz = 2xz + 2yz,$$

and thus $5x = 5y = 2z$, so that $x = y = \frac{2}{5}z$. Substitution in the constraint yields $z = 5 \cdot 3^{1/3}$, then $x = y = 2 \cdot 3^{1/3}$. The base of the shipping crate will be a square $2 \cdot 3^{1/3} \approx 2.884449914$ feet on each side and the height of the crate will be $5 \cdot 3^{1/3} \approx 7.21124785$ feet. (Its cost will be \$124.81!)

12.M.17: Given

$$\frac{1}{R} = \frac{1}{R_1} + \frac{1}{R_2},$$

we differentiate implicitly with respect to R_1 and thereby find that

$$-\frac{1}{R^2} \cdot \frac{\partial R}{\partial R_1} = -\frac{1}{R_1^2}, \quad \text{and thus} \quad \frac{\partial R}{\partial R_1} = \left(\frac{R}{R_1}\right)^2.$$

Similarly,

$$\frac{\partial R}{\partial R_2} = \left(\frac{R}{R_2}\right)^2.$$

Hence if the errors in measuring R_1 and R_2 are $\Delta R_1 = 3$ and $\Delta R_2 = 6$, we estimate the resulting error in computation of R as follows:

$$\Delta R \approx \frac{\partial R}{\partial R_1} \cdot \Delta R_1 + \frac{\partial R}{\partial R_2} \cdot \Delta R_2 = \left(\frac{R}{R_1}\right)^2 \Delta R_1 + \left(\frac{R}{R_2}\right)^2 \Delta R_2$$

$$= \left(\frac{200}{300}\right)^2 \cdot 3 + \left(\frac{200}{600}\right)^2 \cdot 6 = \frac{4}{9} \cdot 3 + \frac{1}{9} \cdot 6 = 2.$$

In fact, when $R_1 = 303$ and $R_2 = 606$, the value of R is exactly 202, so there is no error in this approximation.

12.M.19: Suppose that a, b, and c are positive. The ellipsoid with equation

$$\left(\frac{x}{a}\right)^2 + \left(\frac{y}{b}\right)^2 + \left(\frac{z}{c}\right)^2 = 1$$

and, therefore, semiaxes of lengths a, b, and c, has volume $V = \frac{4}{3}\pi abc$. Let

$$V(x, y, z) = \frac{4}{3}\pi xyz.$$

832

Then

$$V_x(x, y, z) = \frac{4}{3}\pi yz, \quad V_y(x, y, z) = \frac{4}{3}\pi xz, \quad \text{and} \quad V_z(x, y, z) = \frac{4}{3}\pi xy.$$

Assume that errors of at most 1% are made in measuring x, y, and z. Then the error in computing V will be at most

$$\Delta V \approx dV = V_x\, dx + V_y\, dy + V_z\, dz = \frac{4}{3}\pi yz \cdot \frac{x}{100} + \frac{4}{3}\pi xz \cdot \frac{y}{100} + \frac{4}{3}\pi xy \cdot \frac{z}{100} = \frac{1}{25}\pi xyz,$$

and hence the percentage error in computing V will be at most

$$100 \cdot \frac{\Delta V}{V} \approx 100 \cdot \frac{1}{25}\pi xyz \cdot \frac{3}{4\pi xyz} = 3;$$

that is, the maximum error is approximately 3%.

12.M.21: The ellipsoidal surface S is the level surface

$$f(x, y, z) = x^2 + 4y^2 + 9z^2 - 16 = 0,$$

and hence a vector normal to S at the point (x, y, z) is $\mathbf{n} = \nabla f = \langle 2x, 8y, 18z \rangle$. If $\mathbf{n}$ is parallel to the position vector $\langle x, y, z \rangle$, then $\mathbf{n}$ (extended if necessary) will pass through $(0, 0, 0)$. This leads to the equation

$$\langle 2x, 8y, 18z \rangle = \lambda \langle x, y, z \rangle;$$

that is,

$$2x = \lambda x, \quad 8y = \lambda y, \quad 18z = \lambda z. \tag{1}$$

Multiply both sides of the first equation by yz, the second by xz, and the third by xy to eliminate λ:

$$\lambda xyz = 2xyz = 8xyz = 18xyz.$$

We obtain a contradiction if $xyz \neq 0$, and thus at least one of x, y, and z is zero. If (say) $x \neq 0$, then multiply the second equation in (1) by z and the third by y to obtain $\lambda yz = 8yz = 18yz$. If $yz \neq 0$ we obtain a contradiction, so at least one of y and z is zero. Repeating with the other two cases, we conclude that at least two of x, y, and z are zero. Of course the third cannot be zero because of the condition $f(x, y, z) = 0$. Therefore there are six points on S at which the normal vector points toward the origin; they are $(\pm 4, 0, 0)$, $(0, \pm 2, 0)$, and $\left(0, 0, \pm\frac{4}{3}\right)$.

12.M.23: Given: $\mathbf{a}$, $\mathbf{b}$, and $\mathbf{c}$ are mutually perpendicular unit vectors in space and f is a function of the three variables x, y, and z. Rename the unit vectors if necessary so that $\mathbf{a}$, $\mathbf{b}$, $\mathbf{c}$ forms a right-handed triple. Then

$$\mathbf{a} \times \mathbf{b} = \mathbf{c}, \quad \mathbf{b} \times \mathbf{c} = \mathbf{a}, \quad \text{and} \quad \mathbf{c} \times \mathbf{a} = \mathbf{b}. \tag{1}$$

833

Write $\mathbf{a} = \langle a_1, a_2, a_3 \rangle$, $\mathbf{b} = \langle b_1, b_2, b_3 \rangle$, and $\mathbf{c} = \langle c_1, c_2, c_3 \rangle$. Then

$$\mathbf{a}D_\mathbf{a}f + \mathbf{b}D_\mathbf{b}f + \mathbf{c}D_\mathbf{c}f = \mathbf{a}(\nabla f) \cdot \mathbf{a} + \mathbf{b}(\nabla f) \cdot \mathbf{b} + \mathbf{c}(\nabla f) \cdot \mathbf{c}$$

$$= (a_1 f_x + a_2 f_y + a_3 f_z)\langle a_1, a_2, a_3 \rangle + (b_1 f_x + b_2 f_y + b_3 f_z)\langle b_1, b_2, b_3 \rangle + (c_1 f_x + c_2 f_y + c_3 f_z)\langle c_1, c_2, c_3 \rangle.$$

If the scalar multiplications and vector additions in the last line are carried out, the resulting vector will have first component

$$(a_1^2 + b_1^2 + c_1^2)f_x + (a_1 a_2 + b_1 b_2 + c_1 c_2)f_y + (a_1 a_3 + b_1 b_3 + c_1 c_3)f_z. \tag{2}$$

By Eq. (1),

$$c_3 = a_1 b_2 - a_2 b_1, \quad a_3 = b_1 c_2 - b_2 c_1, \quad \text{and} \quad b_3 = c_1 a_2 - c_2 a_1.$$

Hence substitution for a_3, b_3, and c_3 in Eq. (2) yields

$$(a_1^2 + b_1^2 + c_1^2)f_x + (a_1 a_2 + b_1 b_2 + c_1 c_2)f_y + (a_1 b_1 c_2 - a_1 b_2 c_1 + b_1 c_1 a_2 - b_1 c_2 a_1 + c_1 a_1 b_2 - c_1 a_2 b_1)f_z$$

$$= (a_1^2 + b_1^2 + c_1^2)f_x + (a_1 a_2 + b_1 b_2 + c_1 c_2)f_y + 0 \cdot f_z.$$

Also by Eq. (1),

$$a_2 = b_3 c_1 - b_1 c_3, \quad b_2 = c_3 a_1 - c_1 a_3, \quad \text{and} \quad c_2 = a_3 b_1 - a_1 b_3.$$

Hence substitution for a_2, b_2, and c_2 in the last expression yields

$$(a_1^2 + b_1^2 + c_1^2)f_x + (a_1 a_2 + b_1 b_2 + c_1 c_2)f_y$$

$$= (a_1^2 + b_1^2 + c_1^2)f_x + (a_1 b_3 c_1 - a_1 b_1 c_3 + b_1 c_3 a_1 - b_1 c_1 a_3 + c_1 a_3 b_1 - c_1 a_1 b_3)f_y$$

$$= (a_1^2 + b_1^2 + c_1^2)f_x + 0 \cdot f_y = (a_1^2 + b_1^2 + c_1^2)f_x.$$

Moreover, Eq. (1) implies that

$$a_1 = b_2 c_3 - b_3 c_2, \quad b_1 = c_2 a_3 - c_3 a_2, \quad \text{and} \quad c_1 = a_2 b_3 - a_3 b_2.$$

Hence substitution for *only one* of each of a_1, b_1, and c_1 in the last expression yields

$$(a_1 b_2 c_3 - a_1 b_3 c_2 + b_1 c_2 a_3 - b_1 c_3 a_2 + c_1 a_2 b_3 - c_1 a_3 b_2)f_x = \mathbf{a} \cdot (\mathbf{b} \times \mathbf{c})f_x = f_x$$

by Eq. (17) of Section 11.3. Similarly, the second component of $\mathbf{a}D_\mathbf{a}f + \mathbf{b}D_\mathbf{b}f + \mathbf{c}D_\mathbf{c}f$ is f_y and its third component is f_z. Therefore $\nabla f = \mathbf{a}D_\mathbf{a}f + \mathbf{b}D_\mathbf{b}f + \mathbf{c}D_\mathbf{c}f$. ◄

12.M.25: If $f(x, y) = 500 - (0.003)x^2 - (0.004)y^2$, then

$$\nabla f(x, y) = \left\langle -\frac{3x}{500}, -\frac{4y}{500} \right\rangle, \quad \text{so} \quad \nabla f(-100, -100) = \left\langle \frac{3}{5}, \frac{4}{5} \right\rangle.$$

To maintain a constant altitude, you should move in a direction normal to the gradient vector; that is, you should initially move in the direction of either $\langle -4, 3 \rangle$ or $\langle 4, -3 \rangle$.

12.M.27: The given surface S is the level surface

$$f(x, y, z) = x^{2/3} + y^{2/3} + z^{2/3} - 1 = 0,$$

and

$$\nabla f(x, y, z) = \left\langle \frac{2}{3x^{1/3}}, \frac{2}{3y^{1/3}}, \frac{2}{3z^{1/3}} \right\rangle.$$

Hence the plane $\mathcal{P}$ tangent to S at the point $P(a, b, c)$ has normal vector

$$\mathbf{n} = \nabla f(a, b, c) = \left\langle \frac{2}{3a^{1/3}}, \frac{2}{3b^{1/3}}, \frac{2}{3c^{1/3}} \right\rangle,$$

and thus an equation of $\mathcal{P}$ is

$$\frac{x - a}{a^{1/3}} + \frac{y - b}{b^{1/3}} + \frac{z - c}{c^{1/3}} = 0.$$

Set x and y equal to zero to find the z-intercept of $\mathcal{P}$:

$$\frac{z - c}{c^{1/3}} = a^{2/3} + b^{2/3}, \quad \text{so that} \quad z = (a^{2/3} + b^{2/3} + c^{2/3})c^{1/3} = c^{1/3}.$$

Similarly, the x-intercept of $\mathcal{P}$ is $a^{1/3}$ and its y-intercept is $b^{1/3}$. So the sum of the squares of its intercepts is $a^{2/3} + b^{2/3} + c^{2/3} = 1$.

12.M.29: Given:

$$f(x, y) = \frac{x^2 y^2}{x^2 + y^2}$$

if $(x, y) \neq (0, 0)$; $f(0, 0) = 0$. Then

$$f_x(0, 0) = \lim_{h \to 0} \frac{f(h, 0) - f(0, 0)}{h} = \lim_{h \to 0} \frac{0}{h^3} = 0.$$

Also $f_y(0, 0) = 0$ by a very similar computation. If $(x, y) \neq (0, 0)$, then

$$f_x(x, y) = \frac{2xy^4}{(x^2 + y^2)^2} \quad \text{and} \quad f_y(x, y) = \frac{2x^4 y}{(x^2 + y^2)^2},$$

and (using polar coordinates)

$$\lim_{(x,y) \to (0,0)} f_x(x, y) = \lim_{r \to 0} \frac{2r^5 \cos\theta \sin^4\theta}{r^4} = 0,$$

835

so that f_x is continuous at $(0, 0)$ (as is f_y, by a similar argument). Hence both first-order partial derivatives are continuous everywhere, and therefore f is differentiable at the origin. Finally, $f(x, y) \geqq 0 = f(0, 0)$, so $0 = f(0, 0)$ is a local (indeed, the global) minimum value of $f(x, y)$.

12.M.31: The absence of first-degree terms in its equation implies that the center of the given rotated ellipse $73x^2 + 72xy + 52y^2 = 100$ is the origin. Thus to find its semiaxes, we maximize and minimize $f(x, y) = x^2 + y^2$ subject to the constraint

$$g(x, y) = 73x^2 + 72xy + 52y^2 - 100 = 0.$$

The Lagrange multiplier equations are

$$2x = \lambda(146x + 72y) \quad \text{and} \quad 2y = \lambda(72x + 104y).$$

The methods of Problem 38 of Section 12.9 lead to the simultaneous equations

$$(1 - 73\lambda)x - 36\lambda y = 0,$$

$$36\lambda x + (52\lambda - 1)y = 0.$$

Because this system has a nontrivial solution, the determinant of its coefficient matrix must be zero:

$$\begin{vmatrix} 1 - 73\lambda & -36\lambda \\ 36\lambda & 52\lambda - 1 \end{vmatrix} = -2500\lambda^2 + 125\lambda - 1 = 0,$$

and hence $(100\lambda - 1)(25\lambda - 1) = 0$.

Case 1: $\lambda = \frac{1}{100}$. Then the simultaneous equations shown here lead to $3x = 4y$; then the constraint equation yields

$$(x_1, y_1) = \left(-\frac{4}{5}, -\frac{3}{5}\right) \quad \text{and} \quad (x_2, y_2) = \left(\frac{4}{5}, \frac{3}{5}\right).$$

Case 2: $\lambda = \frac{1}{25}$. Then the simultaneous equations shown here lead to $4x + 3y = 0$; then the constraint equation yields

$$(x_3, y_3) = \left(-\frac{6}{5}, \frac{8}{5}\right) \quad \text{and} \quad (x_4, y_4) = \left(\frac{6}{5}, -\frac{8}{5}\right).$$

Thus these four points are the endpoints of the semiaxes of the ellipse; the minor semiaxis (Case 1) has length 1 and the major semiaxis (Case 2) has length 2.

12.M.33: Let us find both the maximum and the minimum perimeter of a triangle inscribed in the unit circle. The center of the circle will be within the triangle of maximum perimeter or on its boundary by the following argument. The following figure on the left shows a triangle with the center of the circle outside it. By moving the chord AB to the new position CD shown next on the right, you increase the perimeter of the circle and now the center of the circle is within the triangle. (Note: CD and AB are parallel and have the same length.) Similarly, the center of the circle is outside the triangle of minimum perimeter, for if inside—as in the figure on the right—move the chord CD to position AB to obtain a triangle of smaller perimeter with the center of the circle now outside the triangle.

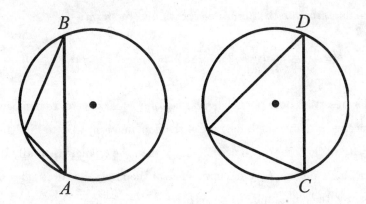

Thus to find the triangle of maximum perimeter, use the notation in Fig. 13.9.9 and maximize the perimeter

$$p(\alpha, \beta, \gamma) = \sqrt{2 - 2\cos\alpha} + \sqrt{2 - 2\cos\beta} + \sqrt{2 - 2\cos\gamma} = 2\sin\frac{\alpha}{2} + 2\sin\frac{\beta}{2} + 2\sin\frac{\gamma}{2}$$

given the constraint $g(\alpha, \beta, \gamma) = \alpha + \beta + \gamma - 2\pi = 0$. The Lagrange multiplier equations are

$$\cos\frac{\alpha}{2} = \lambda, \qquad \cos\frac{\beta}{2} = \lambda, \qquad \cos\frac{\gamma}{2} = \lambda,$$

and therefore, because $f(x) = \cos\left(\frac{1}{2}x\right)$ is single-valued on the interval $0 \le x \le 2\pi$, we may conclude that $\alpha = \beta = \gamma$. Then the constraint equation implies that $\alpha = \beta = \gamma = \frac{2}{3}\pi$, and thus we find that the triangle of maximum perimeter inscribed in the unit circle is equilateral and its perimeter is

$$p\left(\frac{2}{3}\pi, \frac{2}{3}\pi, \frac{2}{3}\pi\right) = 3\sqrt{3} \approx 5.1961524227066319.$$

Next we seek the triangle of minimum perimeter. The figure resembles the one on the left in the preceding illustration. Let C denote the third vertex of the triangle. Let α be the central angle that subtends the arc BC, let β be the central angle that subtends the arc CA, and let γ be the central angle that subtends the short arc BA. Then we are to minimize the perimeter of the triangle, given by

$$p(\alpha, \beta, \gamma) = 2\sin\frac{\alpha}{2} + 2\sin\frac{\beta}{2} + 2\sin\frac{\gamma}{2}$$

(exactly as in the maximum-perimeter investigation); the constraint is now $\alpha + \beta - \gamma = 0$, and we note that $0 \le \gamma \le \pi$ as well. The Lagrange multiplier equations are

$$\cos\frac{\alpha}{2} = \lambda, \qquad \cos\frac{\beta}{2} = \lambda, \qquad \cos\frac{\gamma}{2} = -\lambda.$$

But all three of $\frac{1}{2}\alpha$, $\frac{1}{2}\beta$, and $\frac{1}{2}\gamma$ lie in the first quadrant, so these equations are impossible unless $\lambda = 0$, in which case $\alpha = \beta = \pi$ and $\gamma = 2\pi$ or, equivalently in this case, $\gamma = 0$. This implies that AC and BC are the same diameter of the circle and that AB is one endpoint of that diameter. Thus we obtain a degenerate triangle of perimeter $|AC| + |BC| = 4$. But this is not the minimum perimeter; the minimum occurs at a boundary point extremum missed by the Lagrange multiplier method.

To find the minimum, rewrite the perimeter function in the form

$$p(\alpha, \beta) = 2\sin\frac{\alpha}{2} + 2\sin\frac{\beta}{2} + 2\sin\frac{\alpha+\beta}{2}, \quad 0 \le \alpha, \ 0 \le \beta, \ \alpha + \beta \le \pi.$$

There is no point in setting both partial derivatives of p equal to zero; remember, we are seeking a boundary extremum. Hence we examine the boundary of the domain of $p(\alpha, \beta)$ to find the minimum. One possibility is that $\alpha = \beta = 0$. In this case $\gamma = 0$ as well. If so, then the triangle is totally degenerate; its three vertices all coincide at a single point on the circumference of the circle, its three sides all have length zero, and its perimeter is (finally, the global minimum) zero.

12.M.35: First we make sure that the line and the ellipse do not intersect. Substitution of $y = 2 - x$ in the equation of the ellipse yields

$$x^2 + 2(2 - x)^2 = 1; \quad 3x^2 - 8x + 7 = 0.$$

The discriminiant of the quadratic is $\Delta = 64 - 84 < 0$, so the last equation has no real solutions. Therefore the ellipse and the line do not meet.

Next, following the *Suggestion,* we let

$$f(x, y, u, v) = (x - u)^2 + (y - v)^2$$

and maximize and minimize $f(x, y, u, v)$ subject to the two constraints

$$g(x, y, u, v) = x^2 + 2y^2 - 1 = 0 \quad \text{and} \quad h(x, y, u, v) = u + v - 2 = 0.$$

The Lagrange multiplier equations are

$$2(x - u) = 2\lambda x, \quad 2(y - v) = 4\lambda y, \quad 2(u - x) = \mu, \quad \text{and} \quad 2(v - y) = \mu.$$

Elimination of μ yields $u - x = v - y$. Then elimination of λ yields

$$2y(x - u) = x(y - v) \quad \text{and} \quad u + y = v + x.$$

Finally, solving the last two equations simultaneously with the two constraint equations yields the expected number of solutions—two. They are

$$x = -\frac{1}{3}\sqrt{6}, \quad y = -\frac{1}{6}\sqrt{6}, \quad u = \frac{12 - \sqrt{6}}{6}, \quad v = \frac{12 + \sqrt{6}}{6} \quad \text{and}$$

$$x = \frac{1}{3}\sqrt{6}, \quad y = \frac{1}{6}\sqrt{6}, \quad u = \frac{12 + \sqrt{6}}{6}, \quad v = \frac{12 - \sqrt{6}}{6}.$$

Because the ellipse is "southwest" of the line, the first solution listed here gives the point (x, y) of the ellipse farthest from the line and the second gives the point closest to the line.

12.M.37: Let $f(x_1, x_2, \ldots, x_n) = x_1 + x_2 + \cdots + x_n$ where n is a fixed positive integer. Suppose that a is a nonnegative real number. To maximize $f(x_1, x_2, \ldots, x_n)$ subject to the constraint

$$g(x_1, x_2, \ldots, x_n) = x_1^2 + x_2^2 + \cdots + x_n^2 - a^2 = 0, \tag{1}$$

the Lagrange multiplier equations are

$$2\lambda x_i = 1, \quad 1 \leq i \leq n.$$

Hence $\lambda \neq 0$, and thus $x_1 = x_2 = \cdots = x_n$. Then the constraint equation implies that

$$nx_1^2 = a^2, \quad \text{so that} \quad x_1 = \pm\frac{a}{\sqrt{n}}.$$

Therefore the maximum value of f subject to the constraint in (1) is $nx_1 = a\sqrt{n}$.

It now follows that $|x_1 + x_2 + \ldots + x_n| \leq a\sqrt{n}$ for every point $(x_1, x_2, \ldots, x_n)$ satisfying Eq. (1). Therefore

$$(x_1 + x_2 + \cdots + x_n)^2 \leq na^2 = n(x_1^2 + x_2^2 + \cdots + x_n^2)$$

for all numbers $x_1, x_2, \ldots, x_n$ satisfying Eq. (1). But every such set of numbers satisfies Eq. (1) for some nonnegative value of a, and a is arbitrary. Therefore

$$\left(\sum_{i=1}^{n} x_i\right)^2 \leq n \sum_{i=1}^{n} (x_i)^2 \tag{2}$$

for all n-tuples of real numbers $\{x_1, x_2, \ldots, x_n\}$. Note that equality holds *only* if $x_1 = x_2 = \cdots = x_n$; this will be important in the solution of Problem 51. Finally, divide both sides in Eq. (2) by n^2 and take square roots to obtain the result

$$\frac{x_1 + x_2 + \cdots + x_n}{n} \leqq \sqrt{\frac{x_1^2 + x_2^2 + \cdots + x_n^2}{n}}$$

in Problem 37.

12.M.39: We are to find the extrema of $f(x, y, z) = x^2 - yz$ subject to the constraint or side condition $g(x, y, z) = x^2 + y^2 + z^2 - 1 = 0$. The Lagrange multiplier equations are

$$2x = 2\lambda x, \quad -z = 2\lambda y, \quad \text{and} \quad -y = 2\lambda z. \tag{1}$$

Thus

$$2xyz = 2\lambda xyz, \quad -xz^2 = 2\lambda xyz, \quad \text{and} \quad -xy^2 = 2\lambda xyz,$$

and hence $-xy^2 = -xz^2 = 2xyz$.

Case 1: $x = 0$. Multiply the second equation in (1) by z and the third by y to obtain $-z^2 = 2\lambda yz = -y^2$, so that $y^2 = z^2$. If $y = z$, then the constraint equation yields $y^2 = \frac{1}{2}$, and we obtain the two critical points

$$A\left(0, \frac{1}{2}\sqrt{2}, \frac{1}{2}\sqrt{2}\right) \quad \text{and} \quad B\left(0, -\frac{1}{2}\sqrt{2}, -\frac{1}{2}\sqrt{2}\right).$$

If $y = -z$, then the constraint equation yields $y^2 = \frac{1}{2}$, and we obtain two more critical points,

$$C\left(0, \frac{1}{2}\sqrt{2}, -\frac{1}{2}\sqrt{2}\right) \quad \text{and} \quad D\left(0, -\frac{1}{2}\sqrt{2}, \frac{1}{2}\sqrt{2}\right).$$

Case 2: $x \neq 0$. Then $y^2 = z^2 = -2yz$. If $z = y$ then $y^2 = -2y^2$, so that $y = z = 0$. If $z = -y$ then $y^2 = 2y^2$, and again $y = z = 0$. Then the constraint equation yields $x = \pm 1$, so in this case we obtain the two critical points

$$E(1, 0, 0) \quad \text{and} \quad F(-1, 0, 0).$$

The values of $f(x, y, z)$ at these six points are

$$f(A) = -\frac{1}{2} = f(B), \quad f(C) = \frac{1}{2} = f(D), \quad \text{and} \quad f(E) = f(F) = 1.$$

Therefore the maximum value of $f(x, y, z)$ is 1 and occurs at the two critical points E and F. Its minimum value is $-\frac{1}{2}$ and occurs at the two critical points A and B. Numerical experimentation with

$$h(y, z) = 1 - y^2 - z^2 + yz$$

for y and z near the critical points C and D shows that there is not even a local extremum at either of these points.

12.M.41: If $f(x, y) = x^3 y - 3xy + y^2$, then when we equate both partial derivatives to zero we obtain the equations

$$3y(x^2 - 1) = 0 \quad \text{and} \quad x^3 - 3x + 2y = 0.$$

840

The first equation holds when $x = \pm 1$ and when $y = 0$. Then the second equation yields the critical points $P(-1, -1)$, $Q(0, 0)$, $R\left(-\sqrt{3}, 0\right)$, $S\left(\sqrt{3}, 0\right)$, and $T(1, 1)$. In the notation of Theorem 2 of Section 13.10, we find that

$$A = 6xy, \quad B = 3x^2 - 3, \quad \text{and} \quad C = 2,$$

and application of Theorem 2 yields the following results:

At $P(-1, -1)$: $A = 6$, $B = 0$, $C = 2$, $\Delta = 12$, $f(P) = -1$:   Local minimum;

At $Q(0, 0)$: $A = 0$, $B = -3$, $C = 2$, $\Delta = -9$, $f(Q) = 0$:   Saddle point;

At $R\left(-\sqrt{3}, 0\right)$: $A = 0$, $B = 6$, $C = 2$, $\Delta = -36$, $f(R) = 0$:   Saddle point;

At $S\left(\sqrt{3}, 0\right)$: $A = 0$, $B = 6$, $C = 2$, $\Delta = -36$, $f(S) = 0$:   Saddle point;

At $T(1, 1)$: $A = 6$, $B = 0$, $C = 2$, $\Delta = 12$, $f(T) = -1$:   Local minimum.

There are no global extrema; consider the behavior of $f(x, y)$ on the two lines $y = x$ and $y = -x$.

12.M.43: If $f(x, y) = x^3 - 6xy + y^3$, then the equations $f_x(x, y) = 0 = f_y(x, y)$ take the form

$$3x^2 - 6y = 0 \quad \text{and} \quad 3y^2 - 6x = 0.$$

Thus $2y = x^2$ and $y^2 = 2x$, so that $8x = x^4$. So the only two critical points are $(0, 0)$ and $(2, 2)$. In the notation of Section 13.10 we have $A = 6x$, $B = -6$, and $C = 6y$. Hence $\Delta = -36$ at $(0, 0)$; $A = 12$ and $\Delta = 108$ at $(2, 2)$. Therefore the graph of $z = f(x, y)$ has a saddle point at the origin and a local minimum at $(2, 2)$. There are no global extrema; examine the behavior of $f(x, 0)$.

12.M.45: Given $f(x, y) = x^3 y^2 (1 - x - y)$, the equations $f_x(x, y) = 0 = f_y(x, y)$ become

$$3x^2 y^2 (1 - x - y) - x^3 y^2 = 0 \quad \text{and} \quad 2x^3 y (1 - x - y) - x^3 y^2 = 0. \tag{1}$$

Clearly both partial derivatives are zero at every point where $x = 0$ (the y-axis), at every point where $y = 0$ (the x-axis). Note that $f(x, y) = 0$ at all such points. Moreover, there is one additional critical point; if $x \neq 0$ and $y \neq 0$, then the equations in (1) may be simplified to

$$4x + 3y = 3 \quad \text{and} \quad 2x + 3y = 2,$$

with the unique solution $(x, y) = \left(\frac{1}{2}, \frac{1}{3}\right)$. At this critical point the value of $f(x, y)$ is $\frac{1}{432}$. Because we have infinitely many critical points, we will have to deal with them by *ad hoc* methods. The next diagram shows the three lines where $f(x, y) = 0$. They divide the xy-plane into seven regions, and the sign of $f(x, y)$ is indicated on each.

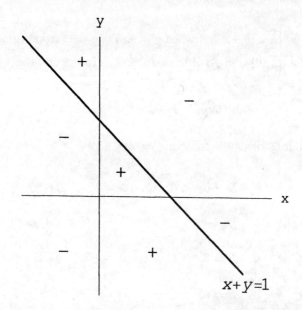

It is clear from this diagram that no point on the line $x + y = 1$ is an extremum, but that there is a saddle point at $(0, 1)$. (Theorem 2 of Section 12.10 fails at this point because $\Delta = 0$ there.) It is also clear from the diagram that there is a local maximum at every point of the negative x-axis and at every point of the x-axis for which $x > 1$; there is a local minimum at every point of the x-axis for which $0 < x < 1$. Finally, no point of the y-axis is an extremum. The fact that $f\left(\frac{1}{2}, \frac{1}{3}\right)$ is positive shows that there is a local maximum at $\left(\frac{1}{2}, \frac{1}{3}\right)$, and this conclusion is supported by Theorem 2 of Section 13.10; at that point we have $A = -\frac{1}{9}, B = -\frac{1}{12}$, and $C = -\frac{1}{8}$. Finally, because $f(x, x) = x^5(1 - 2x)$, there are no global minima; because $f(x, -2x) = 4x^5(1 + x)$, there are no global maxima.

12.M.47: If $f(x, y) = e^{xy} - 2xy$, then the equations $f_x(x, y) = 0 = f_y(x, y)$ take the form

$$(e^{xy} - 2)y = 0 \quad \text{and} \quad (e^{xy} - 2)x = 0.$$

Hence $(0, 0)$ is an isolated critical point and the points for which $e^{xy} = 2$ (the points on both branches of the hyperbola $xy = \ln 2$) are all critical points. Note that $f(0, 0) = 1$ and that $f(x, y) \equiv 2 - 2\ln 2 \approx 0.61370564$ for all points on the hyperbola $xy = \ln 2$. Next,

$$f_{xx}(x, y) = y^2 e^{xy}, \quad f_{xy}(x, y) = (xy + 1)e^{xy} - 2, \quad \text{and} \quad f_{yy}(x, y) = x^2 e^{xy}.$$

Hence Theorem 2 of Section 12.10 yields $A = 0$, $B = -1$, and $C = 0$ at the critical point $(0, 0)$, and therefore it is a saddle point. It is easy to show that the global minimum value of $g(x) = e^x - 2x$ is $2 - 2\ln 2$, and therefore every point of the hyperbola $xy = \ln 2$ is a location of the global minimum value of $f(x, y)$. There are no other extrema.

12.M.49: If $f(x, y) = (x - y)(xy - 1)$, then when we equate both partial derivatives to zero we obtain

$$2xy - y^2 - 1 = 0 \quad \text{and} \quad 2xy - x^2 - 1 = 0. \tag{1}$$

842

Therefore $y^2 = x^2$. If $y = x$ then either equation in (1) yields $x^2 = 1$, and thus we obtain the two critical points $(-1, -1)$ and $(1, 1)$. If $y = -x$ then each equation in (1) yields $3x^2 + 1 = 0$, and so there are no other critical points. When Theorem 2 of Section 12.10 is applied to the first critical point, we find that $A = -2$, $B = 0$, and $C = 2$; at the second critical point we find that $A = 2$, $B = 0$, and $C = -2$. Therefore both these points are saddle points and there are no extrema.

12.M.51: Given the data $\{(x_i, y_i)\}$ for $i = 1, 2, 3, \ldots, n$ (where n is a positive integer), we are to minimize

$$f(m, b) = \sum_{i=1}^{n} \left[y_i - (mx_i + b) \right]^2.$$

When we equate both partial derivatives of f to zero, we obtain the equations

$$\sum_{i=1}^{n} \left(y_i - mx_i - b \right) x_i = 0 \quad \text{and} \quad \sum_{i=1}^{n} \left(y_i - mx_i - b \right) = 0.$$

Rewrite the first of these in the form

$$b \sum_{i=1}^{n} x_i + m \sum_{i=1}^{n} (x_i)^2 = \sum_{i=1}^{n} x_i y_i \tag{1}$$

and the second in the form

$$b \sum_{i=1}^{n} 1 + m \sum_{i=1}^{n} x_i = \sum_{i=1}^{n} y_i. \tag{2}$$

Let

$$P = \sum_{i=1}^{n} x_i, \quad Q = \sum_{i=1}^{n} y_i, \quad R = \sum_{i=1}^{n} (x_i)^2, \quad \text{and} \quad S = \sum_{i=1}^{n} x_i y_i.$$

Then Eqs. (1) and (2) take the form

$$Pb + Rm = S, \qquad nb + Pm = Q. \tag{3}$$

Because P, Q, R, and S are the results of experimental observations, it is highly unlikely that the determinant of coefficients in (3) is zero, and hence these equations will normally have a unique solution (although the experimenter should be wary if the determinant is near zero). Next,

$$f_{mm}(m, b) = 2 \sum_{i=1}^{n} (x_i)^2 = 2R, \qquad f_{mb}(m, b) = 2 \sum_{i=1}^{n} x_i = 2P, \qquad \text{and} \qquad f_{bb}(m, b) = 2n.$$

Application of Theorem 2 of Section 12.10 yields

$$A = 2R, \quad B = 2P, \quad C = 2n, \quad \text{and} \quad \Delta = 4Rn - 4P^2 = 4(Rn - P^2).$$

Hence our sole critical point will be a local minimum (and, by the geometry of the problem, a global minimum) provided that

$$P^2 < nR; \quad \text{that is,} \quad \left(\sum_{i=1}^{n} x_i\right)^2 < n \sum_{i=1}^{n} (x_i)^2.$$

The inequality in Miscellaneous Problem 37 is strict unless all the x_i are equal, and in an experimental problem in which the least squares method is used, they will not be equal. This establishes that the critical point is a global minimum.

Section 13.1

13.1.1: Part (a):

$$f(1, -2) \cdot 1 + f(2, -2) \cdot 1 + f(1, -1) \cdot 1 + f(2, -1) \cdot 1 + f(1, 0) \cdot 1 + f(2, 0) \cdot 1 = 198.$$

Part (b):

$$f(2, -1) \cdot 1 + f(3, -1) \cdot 1 + f(2, 0) \cdot 1 + f(3, 0) \cdot 1 + f(2, 1) \cdot 1 + f(3, 1) \cdot 1 = 480.$$

The average of the two answers is 335, fairly close to the exact value 312 of the integral.

13.1.3: We omit Δx and Δy from the computation because each is equal to 1.

$$f\left(\tfrac{1}{2}, \tfrac{1}{2}\right) + f\left(\tfrac{3}{2}, \tfrac{1}{2}\right) + f\left(\tfrac{1}{2}, \tfrac{3}{2}\right) + f\left(\tfrac{3}{2}, \tfrac{3}{2}\right) = 8.$$

This is also the exact value of the iterated integral.

13.1.5: The Riemann sum is

$$f(2, -1) \cdot 2 + f(4, -1) \cdot 2 + f(2, 0) \cdot 2 + f(4, 0) \cdot 2 = 88.$$

The true value of the integral is $\dfrac{416}{3} \approx 138.666666666667$.

13.1.7: We factor out of each term in the sum the product $\Delta x \cdot \Delta y = \tfrac{1}{4}\pi^2$. The Riemann sum then takes the form

$$\frac{1}{4}\pi^2 \cdot \left[f\left(\tfrac{1}{4}\pi, \tfrac{1}{4}\pi\right) + f\left(\tfrac{3}{4}\pi, \tfrac{1}{4}\pi\right) + f\left(\tfrac{1}{4}\pi, \tfrac{3}{4}\pi\right) + f\left(\tfrac{3}{4}\pi, \tfrac{3}{4}\pi\right) \right] = \tfrac{1}{2}\pi^2 \approx 4.935.$$

The true value of the integral is 4.

13.1.9: Because $f(x, y) = x^2 y^2$ is increasing in both the positive x-direction and the positive y-direction on $[1, 3] \times [2, 5]$, $L \leqq M \leqq U$.

13.1.11: We integrate first with respect to x, then with respect to y:

$$\int_0^2 \int_0^4 (3x + 4y)\, dx\, dy = \int_0^2 \left[\frac{3}{2}x^2 + 4xy \right]_0^4 dy = \int_0^2 (24 + 16y)\, dy = \left[24y + 8y^2 \right]_0^2 = 80.$$

13.1.13: We integrate first with respect to y, then with respect to x:

$$\int_{-1}^2 \int_1^3 (2x - 7y)\, dy\, dx = \int_{-1}^2 \left[2xy - \frac{7}{2}y^2 \right]_1^3 dx = \int_{-1}^2 (4x - 28)\, dx = \left[2x^2 - 28x \right]_{-1}^2 = -48 - 30 = -78.$$

13.1.15: We integrate first with respect to x, then with respect to y:

$$\int_0^3 \int_0^3 (xy + 7x + y)\, dx\, dy = \int_0^3 \left[xy + \frac{7}{2}x^2 + \frac{1}{2}x^2 y \right]_0^3 dy$$

$$= \int_0^3 \left(\frac{3}{2}(5y + 21) \right) dy = \left[\frac{1}{4}(15y^2 + 126y) \right]_0^3 = \frac{513}{4} = 128.25.$$

13.1.17: We integrate first with respect to y, then with respect to x:

$$\int_{-1}^2 \int_{-1}^2 (2xy^2 - 3x^2 y)\, dy\, dx = \int_{-1}^2 \left[\frac{2}{3}xy^3 - \frac{3}{2}x^2 y^2 \right]_{-1}^2 dx$$

$$= \int_{-1}^2 \frac{3}{2}(4x - 3x^2)\, dx = \left[3x^2 - \frac{3}{2}x^3 \right]_{-1}^2 = 0 - \frac{9}{2} = -\frac{9}{2} = -4.5.$$

13.1.19: We integrate first with respect to x, then with respect to y:

$$\int_0^{\pi/2} \int_0^{\pi/2} \sin x \cos y\, dx\, dy = \int_0^{\pi/2} \left[-\cos x \cos y \right]_0^{\pi/2} dy$$

$$= \int_0^{\pi/2} \cos y\, dy = \left[\sin y \right]_0^{\pi/2} = 1 - 0 = 1.$$

13.1.21: We integrate first with respect to y, then with respect to x:

$$\int_0^1 \int_0^1 xe^y\, dy\, dx = \int_0^1 \left[xe^y \right]_0^1 dx$$

$$= \int_0^1 (ex - x)\, dx = \left[\frac{1}{2}(e-1)x^2 \right]_0^1 = \frac{1}{2}(e-1) \approx 0.8591409142295226.$$

13.1.23: We integrate first with respect to y, then with respect to x:

$$\int_0^1 \int_0^\pi e^x \sin y\, dy\, dx = \int_0^1 \left[-e^x \cos y \right]_0^\pi dx$$

$$= \int_0^1 2e^x\, dx = \left[2e^x \right]_0^1 = 2e - 2 \approx 3.436563656918.$$

13.1.25: We integrate first with respect to x, then with respect to y:

$$\int_0^\pi \int_0^\pi (xy + \sin x)\, dx\, dy = \int_0^\pi \left[\frac{1}{2}x^2 y - \cos x \right]_0^\pi dy = \int_0^\pi \left(2 + \frac{1}{2}\pi^2 y \right) dy$$

$$= \left[2y + \frac{1}{4}\pi^2 y^2 \right]_0^\pi = \frac{1}{4}\left(\pi^4 + 8\pi \right) \approx 30.635458065680.$$

13.1.27: We integrate first with respect to x, then with respect to y:

$$\int_0^{\pi/2} \int_1^e \frac{\sin y}{x} \, dx \, dy = \int_0^{\pi/2} \left[(\ln x) \sin y \right]_1^e \, dy$$

$$= \int_0^{\pi/2} \sin y \, dy = \left[- \cos y \right]_0^{\pi/2} = 0 - (-1) = 1.$$

13.1.29: We integrate first with respect to x, then with respect to y:

$$\int_0^1 \int_0^1 \left(\frac{1}{x+1} + \frac{1}{y+1} \right) dx \, dy = \int_0^1 \left[\frac{x}{y+1} + \ln(x+1) \right]_0^1 dy = \int_0^1 \left(\frac{1}{y+1} + \ln 2 \right) dy$$

$$= \left[\ln(y+1) + y \ln 2 \right]_0^1 = 2 \ln 2 - 0 = 2 \ln 2 \approx 1.3862943611198906.$$

13.1.31: The first evaluation yields

$$\int_{-2}^2 \int_{-1}^1 (2xy - 3y^2) \, dx \, dy = \int_{-2}^2 \left[x^2 y - 3xy^2 \right]_{-1}^1 dy$$

$$= \int_{-2}^2 (-6y^2) \, dy = \left[- 2y^3 \right]_{-2}^2 = -16 - 16 = -32.$$

The second yields

$$\int_{-1}^1 \int_{-2}^2 (2xy - 3y^2) \, dy \, dx = \int_{-1}^1 \left[xy^2 - y^3 \right]_{-2}^2 dx$$

$$= \int_{-1}^1 (-16) \, dx = \left[- 16x \right]_{-1}^1 = -16 - 16 = -32.$$

13.1.33: The first evaluation yields

$$\int_1^2 \int_0^1 (x+y)^{1/2} \, dx \, dy = \int_1^2 \left[\frac{2}{3}(x+y)^{3/2} \right]_0^1 dy = \int_1^2 \left(\frac{2}{3}(y+1)^{3/2} - \frac{2}{3} y^{3/2} \right) dy$$

$$= \left[\frac{4}{15}(y+1)^{5/2} - \frac{4}{15} y^{5/2} \right]_1^2 = \frac{4}{15} \left(9\sqrt{3} - 8\sqrt{2} + 1 \right) \approx 1.406599671769.$$

The second yields

$$\int_0^1 \int_1^2 (x+y)^{1/2} \, dy \, dx = \int_0^1 \left[\frac{2}{3}(x+y)^{3/2} \right]_1^2 dx = \int_0^1 \left(\frac{2}{3}(x+2)^{3/2} - \frac{2}{3}(x+1)^{3/2} \right) dx$$

$$= \left[\frac{4}{15}(x+2)^{5/2} - \frac{4}{15}(x+1)^{5/2} \right]_0^1 = \frac{4}{15} \left(9\sqrt{3} - 8\sqrt{2} + 1 \right).$$

13.1.35: We may assume that $n \geq 1$ and, if you wish, even that n is a positive integer. Then

$$\int_0^1 \int_0^1 x^n y^n \, dx \, dy = \int_0^1 \left[\frac{x^{n+1} y^n}{n+1} \right]_0^1 dy = \int_0^1 \frac{y^n}{n+1} \, dy = \left[\frac{y^{n+1}}{(n+1)^2} \right]_0^1 = \frac{1}{(n+1)^2}.$$

Therefore

$$\lim_{n \to \infty} \int_0^1 \int_0^1 x^n y^n \, dx \, dy = \lim_{n \to \infty} \frac{1}{(n+1)^2} = 0.$$

13.1.37: Let $a(R)$ denote the area of R. If $0 \leq x \leq \pi$ and $0 \leq y \leq \pi$, then $0 \leq f(x, y) \leq \sin \frac{1}{2}\pi = 1$. Hence every Riemann sum lies between $0 \cdot a(R)$ and $1 \cdot a(R)$. Therefore

$$0 \leq \int_0^\pi \int_0^\pi \sin \sqrt{xy} \, dx \, dy \leq a(R) = \pi^2 \approx 9.869604401.$$

The exact value of the integral is

$$\int_0^\pi \int_0^\pi \sin \sqrt{xy} \, dx \, dy = 4 \int_0^\pi \frac{\sin t}{t} \, dt \approx 7.4077482079298646814442134806319654533832.$$

13.1.39: The corresponding relation among Riemann sums is

$$\sum_{i=1}^n [f(x_i^\star, y_i^\star) + g(x_i^\star, y_i^\star)] \cdot \Delta A_i = \left[\sum_{i=1}^n f(x_i^\star, y_i^\star) \cdot \Delta A_i \right] + \left[\sum_{i=1}^n g(x_i^\star, y_i^\star) \cdot \Delta A_i \right].$$

Section 13.2

13.2.1: $\displaystyle \int_0^1 \int_0^x (1+x) \, dy \, dx = \int_0^1 \left[y + xy \right]_{y=0}^x \, dx = \int_0^1 (x + x^2) \, dx = \left[\frac{1}{2}x^2 + \frac{1}{3}x^3 \right]_0^1 = \frac{5}{6}.$

13.2.3: $\displaystyle \int_0^1 \int_y^1 (x+y) \, dx \, dy = \int_0^1 \left[\frac{1}{2}x^2 + xy \right]_y^1 \, dy$

$$= \int_0^1 \left(\frac{1}{2} + y - \frac{3}{2}y^2 \right) dy = \left[\frac{1}{2}(y + y^2 - y^3) \right]_0^1 = \frac{1}{2}.$$

13.2.5: $\displaystyle \int_0^1 \int_0^{x^2} xy \, dy \, dx = \int_0^1 \left[\frac{1}{2}xy^2 \right]_0^{x^2} \, dx = \int_0^1 \frac{1}{2}x^5 \, dx = \left[\frac{1}{12}x^6 \right]_0^1 = \frac{1}{12}.$

13.2.7: $\displaystyle \int_0^1 \int_x^{\sqrt{x}} (2x - y) \, dy \, dx = \int_0^1 \left[2xy - \frac{1}{2}y^2 \right]_x^{\sqrt{x}} \, dx$

$$= \int_0^1 \left(-\frac{1}{2}x + 2x^{3/2} - \frac{3}{2}x^2 \right) dx = \left[-\frac{1}{4}x^2 + \frac{4}{5}x^{5/2} - \frac{1}{2}x^3 \right]_0^1 = \frac{1}{20}.$

13.2.9: $\displaystyle \int_0^1 \int_{x^4}^x (y - x) \, dy \, dx = \int_0^1 \left[\frac{1}{2}y^2 - xy \right]_{x^4}^x \, dx$

$$= \int_0^1 \left(-\frac{1}{2}x^2 + x^5 - \frac{1}{2}x^8 \right) dx = \left[-\frac{1}{6}x^3 + \frac{1}{6}x^6 - \frac{1}{18}x^9 \right]_0^1 = -\frac{1}{18}.$$

848

13.2.11: $\displaystyle\int_0^1 \int_0^{x^3} e^{y/x} \, dy \, dx = \int_0^1 \left[x e^{y/x} \right]_0^{x^3} dx = \int_0^1 \left(x \exp(x^2) - x \right) dx$

$$= \left[\frac{1}{2} \left(\exp(x^2) - x^2 \right) \right]_0^1 = \frac{e-2}{2} \approx 0.3591409142295226.$$

13.2.13: $\displaystyle\int_0^3 \int_0^y (y^2 + 16)^{1/2} \, dx \, dy = \int_0^3 \left[x(y^2 + 16)^{1/2} \right]_0^y dy$

$$= \int_0^3 y(y^2 + 16)^{1/2} \, dy = \left[\frac{1}{3} (y^2 + 16)^{3/2} \right]_0^3 = \frac{125}{3} - \frac{64}{3} = \frac{61}{3}.$$

13.2.15: The following sketch of the graphs of $y = x^2$ and $y \equiv 4$ is extremely helpful in finding the limits of integration.

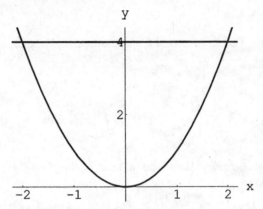

Answer:

$$\int_{-2}^2 \int_{x^2}^4 xy \, dy \, dx = \int_{-2}^2 \left[\frac{1}{2} xy^2 \right]_{x^2}^4 dx = \int_{-2}^2 \left(8x - \frac{1}{2} x^5 \right) dx = \left[4x^2 - \frac{1}{12} x^6 \right]_{-2}^2 = \frac{32}{3} - \frac{32}{3} = 0.$$

13.1.17: The following diagram, showing the graphs of $y = x^2$ and $y = 8 - x^2$, is useful in finding the limits of integration. (Solve $y = x^2$ and $y = 8 - x^2$ simultaneously to find where the two curves cross.)

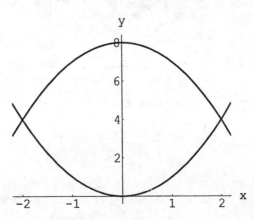

Answer:

$$\int_{-2}^{2}\int_{x^2}^{8-x^2} x\, dy\, dx = \int_{-2}^{2}\left[\, xy\,\right]_{x^2}^{8-x^2} dx = \int_{-2}^{2}(8x - 2x^3)\, dx = \left[4x^2 - \frac{1}{2}x^4\right]_{-2}^{2} = 8 - 8 = 0.$$

13.1.19: The following graph of $y = \sin x$ is helpful in determining the limits of integration.

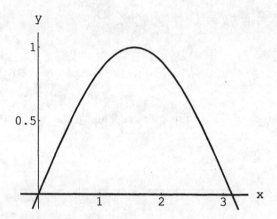

Answer:

$$\int_{0}^{\pi}\int_{0}^{\sin x} x\, dy\, dx = \int_{0}^{\pi}\left[\, xy\,\right]_{0}^{\sin x} dx = \int_{0}^{\pi} x\sin x\, dx = \left[\, \sin x - x\cos x\,\right]_{0}^{\pi} = \pi.$$

13.2.21: A *Mathematica* command to draw the domain of the double integral is

```
ParametricPlot[ {{t,1}, {E,t}, {t,t}}, {t,1,E}, AxesOrigin → {0,0},
        PlotRange → {{-0.3,3.3}, {-0.3,3.3}} ];
```

The graph produced by this command is shown next.

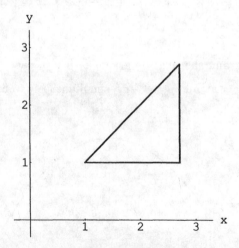

Answer:

$$\int_{1}^{e}\int_{1}^{x} \frac{1}{y}\, dy\, dx = \int_{1}^{e}\left[\, \ln y\,\right]_{1}^{x} dx = \int_{1}^{e} \ln x\, dx = \left[\, -x + x\ln x\,\right]_{1}^{e} = 0 - (-1) = 1.$$

13.2.23: A *Mathematica* command to draw the domain of the double integral is

```
Plot[ {x, -x/2, 1}, {x, -2, 1}, AspectRatio → Automatic,
    PlotRange → {{-2,1}, {0,1}} ];
```

the resulting figure is next.

The value of the double integral is

$$\int_0^1 \int_{-2y}^y (1-x) \, dx \, dy = \int_0^1 \left[x - \frac{1}{2}x^2 \right]_{-2y}^y \, dy = \int_0^1 \left(3y + \frac{3}{2}y^2 \right) \, dy = \left[\frac{1}{2}(3y^2 + y^3) \right]_0^1 = 2 - 0 = 2.$$

13.2.25: The domain of the double integral can be plotted by executing the following *Mathematica* command:

```
Plot[ {x∧2, 4}, {x, -2, 2} ];
```

and the resulting figure is next.

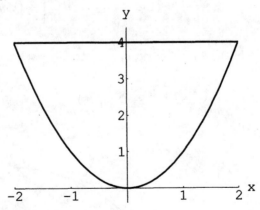

When the order of integration is reversed, we obtain

$$\int_0^4 \int_{-\sqrt{y}}^{\sqrt{y}} x^2 y \, dx \, dy = \int_0^4 \left[\frac{1}{3}x^3 y \right]_{-\sqrt{y}}^{\sqrt{y}} \, dy = \int_0^4 \frac{2}{3} y^{5/2} \, dy = \left[\frac{4}{21} y^{7/2} \right]_0^4 = \frac{512}{21} \approx 24.380952380952.$$

13.2.27: The *Mathematica* command

```
Plot[ { x^2, 2*x + 3 }, { x, -1, 3 } ];
```

produces a figure showing the domain of the double integral; it appears next.

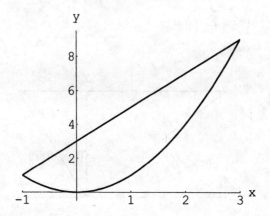

When the order of integration is reversed, two integrals are required—one for the part of the region for which $0 \leq y \leq 1$, the other for the part when $1 \leq y \leq 9$. The reason is that the lower limit of integration for x changes where $y = 1$. The first integral is

$$I_1 = \int_0^1 \int_{-\sqrt{y}}^{\sqrt{y}} x \, dx \, dy = \int_0^1 \left[\frac{1}{2} x^2 \right]_{-\sqrt{y}}^{\sqrt{y}} dy = \int_0^1 0 \, dy = 0.$$

The second integral is

$$I_2 = \int_1^9 \int_{(y-3)/2}^{\sqrt{y}} x \, dx \, dy = \int_1^9 \left[\frac{1}{2} x^2 \right]_{(y-3)/2}^{\sqrt{y}} dy$$

$$= \int_1^9 \frac{1}{8} (10y - 9 - y^2) \, dy = \left[\frac{1}{24} (15y^2 - 27y - y^3) \right]_1^9 = \frac{32}{3}.$$

13.2.29: To see the domain of the double integral, execute the *Mathematica* command

```
Plot[ { 2*x, 4*x - x^2 }, { x, 0, 2 } ];
```

the result is shown next.

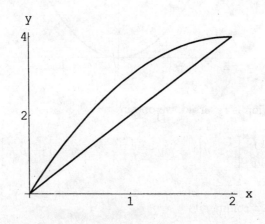

852

When the order of integration is reversed, the given integral becomes

$$\int_0^4 \int_{2-\sqrt{4-y}}^{y/2} 1 \; dx \; dy = \int_0^4 \left[x \right]_{2-\sqrt{4-y}}^{y/2} dy$$

$$= \int_0^4 \left(\frac{1}{2} y + (4-y)^{1/2} - 2 \right) dy = \left[\frac{1}{4} y^2 - 2y - \frac{2}{3}(4-y)^{3/2} \right]_0^4 = \frac{4}{3}.$$

13.2.31: The domain of the given integral is bounded above by the line $y = \pi$, on the left by the y-axis, and on the right by the line $y = x$. When the order of integration is reversed, we obtain

$$\int_0^\pi \int_0^y \frac{\sin y}{y} \; dx \; dy = \int_0^\pi \left[\frac{x \sin y}{y} \right]_0^y dy = \int_0^\pi \sin y \; dy = \left[-\cos y \right]_0^\pi = 2.$$

If the improper integral is disturbing, merely define the integrand to have the value 1 at $y = 0$. Then it will be continuous and the integral will no longer be improper. Because the antiderivative

$$F(t) = \int_0^t \frac{\sin y}{y} \; dy$$

is known to be nonelementary, the only way to evaluate the given integral by hand is first to reverse the order of integration.

13.2.33: To generate a figure showing the domain of the given integral, use the *Mathematica* command

```
ParametricPlot[ {{t,t}, {1,t}, {t,0}}, {t,0,1}, AspectRatio → Automatic,
        PlotRange → {{-0.1, 1.1}, {-0.1, 1.1}} ];
```

The result is shown next.

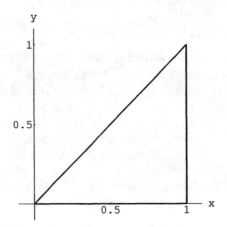

When the order of integration is reversed, we obtain

$$\int_0^1 \int_0^x \frac{1}{1+x^4} \; dy \; dx = \int_0^1 \left[\frac{y}{1+x^4} \right]_0^x dx = \int_0^1 \frac{x}{1+x^4} \; dx = \left[\frac{1}{2} \arctan x^2 \right]_0^1 = \frac{\pi}{8}.$$

853

The integration can be carried out in the order given in the textbook, but finding the partial fraction decomposition of the integrand is long and complex if machine aid is not available.

13.2.35: We used *Mathematica* in this problem. First we entered

```
Solve[ x∧3 + 1 == 3*x∧2, x ]
```

and the computer returned the exact answers. Then we asked for the numerical values to 40 places, and we found that the curves intersect at the three points with approximate coordinates

$$(a, 3a^2) \approx (-0.5320888862379561, 0.8493557485738457),$$

$$(b, 3b^2) \approx (0.6527036446661393, 1.270661432813855), \quad \text{and}$$

$$(c, 3c^2) \approx (2.8793852415718168, 24.8725781081447688).$$

We then entered the command

```
Plot[ {x∧3 + 1, 3*x∧2}, {x, a, b}, PlotRange → {-1, 25} ];
```

and thereby discovered that the cubic graph is above the quadratic on (a, b) but below it on (b, c). Thus we needed to compute two integrals:

```
i1 = Integrate[ Integrate[ x, {y, 3*x∧2, x∧3 + 1} ], {x, a, b} ]
```

and

```
i2 = Integrate[ Integrate[ x, {y, x∧3 + 1, 3*x∧2} ], {x, b, c} ]
```

Results:

$$I_1 \approx 0.0276702414879754 \quad \text{and} \quad I_2 \approx 7.9240769481663325.$$

All the digits are correct or correctly rounded because we used at least 32 decimal digits in every computation. The answer, therefore, is

$$I_1 + I_2 \approx 7.9517471896543079.$$

13.2.37: We began with the *Mathematica* command

```
Solve[ x∧2 - 1 == 1/(1 + x∧2), x ]
```

and were rewarded with the exact solutions—two real, two complex non-real. The two real solutions are $a = -2^{1/4}$ and $b = 2^{1/4}$, so we used these exact values in the following computations. The double integral has the value

```
Integrate[ Integrate[ x, {y, x∧2 - 1, 1/(x∧2 + 1)}, {x, a, b} ]
```

$$\frac{1}{2}\left[1 - \sqrt{2} - \ln\left(1 + \sqrt{2}\right)\right] + \frac{1}{2}\left[-1 + \sqrt{2} + \ln\left(1 + \sqrt{2}\right)\right] = 0.$$

13.2.39: We used *Mathematica* to automate Newton's method for solving the equation $x^2 = \cos x$:

```
f[x_] := x∧2 - Cos[x]

g[x_] := N[x - f[x]/f'[x], 60]
```

The function g carries out the iteration of Newton's method, carrying 60 digits in its computations. A graph indicated that the positive solution of $f(x) = 0$ is close to 4/5, hence we entered the successive commands

```
g[4/5]

g[%]
```

(Recall that % refers to the "last output.")

```
g[%]
```

After six iterations the results agreed to over 50 decimal degits, and thus we find that the graphs cross at the two points

$$(b, b^2) \approx (0.8241323123025224, 0.6791940681811024) \quad \text{and} \quad (a, a^2)$$

where $a = -b$. Hence the approximate value of the double integral is

```
Integrate[ Integrate[ x, {y, x∧2, Cos[x]}], {x, a, b} ]
```

$$0.0 \times 10^{-58}$$

A moment's thought about Riemann sums, symmetry, and the sign of the integrand reveals that the exact value of the integral is zero.

13.2.41: The integral is zero. Each term $f(x_i^\star, y_i^\star)\,\Delta A_i = x_i^\star\,\Delta A_i$ in every Riemann sum is cancelled by a similar term in which $x_i^\star$ has the opposite sign.

13.2.43: Every term of the form $f(x_i^\star, y_i^\star)\,\Delta A_i$ in every Riemann sum is cancelled by a similar term in which $x_i^\star$ has the opposite sign (but $y_i^\star$ is the same). Therefore the value of the integral is zero.

13.2.45: Suppose that the rectangle R consists of those points (x, y) for which both $a \leq x \leq b$ and $c \leq y \leq d$. Suppose that k is a positive constant and that f is a function continuous on R. Then

$$I = \iint_R f(x, y)\,dA$$

exists. Suppose that $\epsilon > 0$ is given. Then there exists a number $\delta_1 > 0$ such that, for every partition $\mathcal{P} = \{R_1, R_2, \ldots R_n\}$ of R such that $|\mathcal{P}| < \delta_1$ and for every selection $(x_i^\star, y_i^\star)$ in R_i $(i = 1, 2, \ldots, n)$,

$$\left| \sum_{i=1}^{n} f(x_i^\star, y_i^\star)\, \Delta A_i - I \right| < \frac{\epsilon}{2k}$$

(where ΔA_i is the area $a(R_i)$ of R_i). Consequently,

$$\left| k \sum_{i=1}^{n} f(x_i^\star, y_i^\star)\, \Delta A_i - kI \right| < \frac{\epsilon}{2}. \tag{1}$$

Moreover, kf is continuous on R, and hence

$$J = \iint_R kf(x, y)\, dA$$

exists. So there exists a number $\delta_2 > 0$ such that, for every partition $\mathcal{P} = \{R_1, R_2, \ldots R_n\}$ of R such that $|\mathcal{P}| < \delta_2$ and every selection $(x_i^\star, y_i^\star)$ in R_i $(i = 1, 2, \ldots n)$,

$$\left| \sum_{i=1}^{n} kf(x_i^\star, y_i^\star)\, \Delta A_i - J \right| < \frac{\epsilon}{2}; \quad \text{that is,}$$

$$\left| k \sum_{i=1}^{n} f(x_i^\star, y_i^\star)\, \Delta A_i - J \right| < \frac{\epsilon}{2}.$$

Let δ be the minimum of δ_1 and δ_2. Then, for every partition $\mathcal{P}$ of R with $|\mathcal{P}| < \delta$ and every selection $(x_i^\star, y_i^\star)$ in R_i $(i = 1, 2, \ldots n)$, we have both

$$\left| kI - k \sum_{i=1}^{n} f(x_i^\star, y_i^\star)\, \Delta A_i \right| < \frac{\epsilon}{2} \quad \text{(by (1)) and}$$

$$\left| k \sum_{i=1}^{n} f(x_i^\star, y_i^\star)\, \Delta A_i - J \right| < \frac{\epsilon}{2}.$$

Add the last two inequalities. Then, by the triangle inequality (Theorem 1 of Appendix A, page A-2),

$$|kI - J| < \frac{\epsilon}{2} + \frac{\epsilon}{2} = \epsilon.$$

Because ϵ is an *arbitrary* positive number, this proves that $J = kI$; that is, we have shown that

$$\iint_R kf(x, y)\, dA = k \iint_R f(x, y)\, dA.$$

The proof is similar in the case $k < 0$, and if $k = 0$ there is nothing to prove. ◄

856

For a shorter proof, one that exploits both the continuity of f and the fact that R is a rectangle with sides parallel to the coordinate axes, choose a continuous function F such that $F_x = f$. Then choose a continuous function P such that $P_y = F$. Then

$$\iint_R kf(x, y)\, dA = \int_c^d \int_a^b kf(x, y)\, dx\, dy = \int_c^d \left[kF(x, y) \right]_a^b dy$$

$$= \int_c^d \left[kF(b, y) - kF(a, y) \right] dy = \left[kP(b, y) - kP(a, y) \right]_c^d$$

$$= kP(b, d) - kP(a, d) - kP(b, c) + kP(a, c) = k\left[P(b, d) - P(a, d) - P(b, c) + P(a, c) \right]$$

$$= k\left[P(b, y) - P(a, y) \right]_c^d = k\int_c^d \left[F(b, y) - F(a, y) \right] dy$$

$$= k\int_c^d \left[F(x, y) \right]_a^b dy = k\int_c^d \int_a^b f(x, y)\, dx\, dy = k\iint_R f(x, y)\, dA.$$

13.2.47: Suppose that R is a plane rectangle with sides parallel to the coordinate axes, so that R consists of those points (x, y) for which both $a \leq x \leq b$ and $c \leq y \leq d$ for some numbers a, b, c, and d. Suppose that f is continuous on R and that $m \leq f(x, y) \leq M$ for all (x, y) in R.

Let $g(x, y) \equiv m$ and $h(x, y) \equiv M$ for (x, y) in R. Then $g(x, y) \leq f(x, y) \leq h(x, y)$ for all points (x, y) in R. Let $\mathcal{P} = \{R_1, R_2, \ldots, R_n\}$ be a partition of R and let $(x_i^\star, y_i^\star)$ be a selected point in R_i for $1 \leq i \leq n$. As usual, let $\Delta A_i = a(R_i)$ for $1 \leq i \leq n$. Then for each integer i, $1 \leq i \leq n$, we have

$$g(x_i^\star, y_i^\star) \leq f(x_i^\star, y_i^\star) \leq h(x_i^\star, y_i^\star);$$

thus

$$g(x_i^\star, y_i^\star)\, \Delta A_i \leq f(x_i^\star, y_i^\star)\, \Delta A_i \leq h(x_i^\star, y_i^\star)\, \Delta A_i$$

for $1 \leq i \leq n$. Add these inequalities to find that

$$\sum_{i=1}^n m \cdot \Delta A_i \leq \sum_{i=1}^n f(x_i^\star, y_i^\star)\, \Delta A_i \leq \sum_{i=1}^n M \cdot \Delta A_i.$$

Therefore

$$m \cdot \sum_{i=1}^n a(R_i) \leq \sum_{i=1}^n f(x_i^\star, y_i^\star)\, \Delta A_i \leq M \cdot \sum_{i=1}^n a(R_i);$$

that is,

$$m \cdot a(R) \leq \sum_{i=1}^n f(x_i^\star, y_i^\star)\, \Delta A_i \leq M \cdot a(R) \tag{1}$$

for every partition $\mathcal{P}$ of R and every selection $(x_i^\star, y_i^\star)$ for $\mathcal{P}$. That is, the inequalities in (1) hold for every Riemann sum for f on R. Because the double integral of f on R is the limit of such sums, we may conclude that

$$m \cdot a(R) \leq \iint_R f(x, y)\, dA \leq M \cdot a(R).$$

13.2.49: Suppose that R is a rectangle with sides parallel to the coordinate axes, that $f(x, y) \leqq g(x, y)$ for all (x, y) in R, and that both

$$I = \iint_R f(x, y)\, dA \quad \text{and} \quad J = \iint_R g(x, y)\, dA$$

exist. Suppose by way of contradiction that $J < I$. Let $\epsilon = I - J$. Note that $\epsilon/3 > 0$. Choose $\delta > 0$ so small that if $\mathcal{P} = \{R_1, R_2, \ldots, R_n\}$ is a partition of R with $|\mathcal{P}| < \delta$ and $(x_i^\star, y_i^\star)$ is a selection for $\mathcal{P}$ with $(x_i^\star, y_i^\star)$ in R_i for $1 \leqq i \leqq n$, then

$$\left| \sum_{i=1}^{n} f(x_i^\star, y_i^\star)\, \Delta A_i - I \right| < \frac{\epsilon}{3} \quad \text{and} \quad \left| \sum_{i=1}^{n} g(x_i^\star, y_i^\star)\, \Delta A_i - J \right| < \frac{\epsilon}{3}.$$

With such a partition and such a selection, note that

$$\sum_{i=1}^{n} g(x_i^\star, y_i^\star)\, \Delta A_i < J + \frac{\epsilon}{3} < I - \frac{\epsilon}{3} < \sum_{i=1}^{n} f(x_i^\star, y_i^\star)\, \Delta A_i,$$

and thus

$$\sum_{i=1}^{n} \left[f(x_i^\star, y_i^\star) - g(x_i^\star, y_i^\star) \right] \Delta A_i > 0.$$

Because $\Delta A_i > 0$ for $1 \leqq i \leqq n$, it follows that

$$f(x_j^\star, y_j^\star) > g(x_j^\star, y_j^\star)$$

for some j, $1 \leqq j \leqq n$, contrary to hypothesis. Therefore $I \leqq J$. ◀

13.2.51: Recall that R is the region in the first quadrant bounded by the circle $x^2 + y^2 = 1$ and the coordinate axes. Hence

$$\iint_R (x + y)\, dA = \int_{x=0}^{1} \int_{y=0}^{\sqrt{1-x^2}} (x + y)\, dy\, dx$$

$$= \int_0^1 \left[xy + \frac{1}{2}y^2 \right]_0^{\sqrt{1-x^2}} dx = \int_0^1 \left(x\sqrt{1 - x^2} + \frac{1}{2}(1 - x^2) \right) dx$$

$$= \left[-\frac{1}{3}(1 - x^2)^{3/2} + \frac{1}{2}x - \frac{1}{6}x^3 \right]_0^1 = \frac{1}{2} - \frac{1}{6} + \frac{1}{3} = \frac{2}{3}.$$

13.2.53: The domain of the integral and the partition using $n = 5$ subintervals of equal length in each direction is shown next.

858

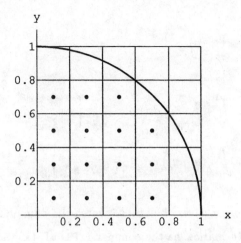

The midpoints of the subrectangles of the inner partition are indicated with "bullets" in the figure. Let $f(x, y) = xy \exp\left(y^2\right)$. Then the corresponding midpoint sum for the given integral is

$$S = \frac{1}{n^2}\Big[f(0.1, 0.1) + f(0.3, 0.1) + f(0.5, 0.1) + f(0.7, 0.1) + f(0.3, 0.1) + f(0.3, 0.3)$$

$$+ f(0.3, 0.5) + f(0.3, 0.7) + f(0.5, 0.1) + f(0.5, 0.3) + f(0.5, 0.5) + f(0.5, 0.7)$$

$$+ f(0.7, 0.1) + f(0.7, 0.3) + f(0.7, 0.5)\Big]$$

$$= \frac{1}{25}\left[\frac{4}{25}e^{0.01} + \frac{12}{25}e^{0.09} + \frac{4}{5}e^{0.25} + \frac{63}{100}e^{0.49}\right] \approx 0.109696.$$

The exact value of the integral is

$$\iint_R f(x, y)\, dA = \int_{y=0}^1 \int_{x=0}^{\sqrt{1-y^2}} xy \exp\left(y^2\right)\, dx\, dy$$

$$= \int_0^1 \left[\frac{1}{2}x^2 y \exp\left(y^2\right)\right]_0^{\sqrt{1-y^2}}\, dy = \int_0^1 \frac{1}{2}(y - y^3) \exp\left(y^2\right)\, dy$$

$$= \left[\frac{2 - y^2}{4}\exp\left(y^2\right)\right]_0^1 = \frac{e-2}{4} \approx 0.17957045711476130884000718678.$$

If you prefer the other order of integration—which avoids the integration by parts—it is

$$\iint_R f(x, y)\, dA = \int_{x=0}^1 \int_{y=0}^{\sqrt{1-x^2}} xy \exp\left(y^2\right)\, dy\, dx$$

$$= \int_0^1 \left[\frac{1}{2}x \exp\left(y^2\right)\right]_0^{\sqrt{1-x^2}}\, dx = \int_0^1 \frac{1}{2}\left[x \exp\left(1 - x^2\right) - x\right]\, dx$$

$$= \left[-\frac{1}{4}x^2 - \frac{1}{4}\exp\left(1 - x^2\right)\right]_0^1 = \frac{e-2}{4}.$$

859

Section 13.3

13.3.1: The area is

$$A = \int_{y=0}^{1} \int_{x=y^2}^{y} 1 \, dx \, dy = \int_{y=0}^{1} \left[x \right]_{x=y^2}^{y} dy$$

$$= \int_{y=0}^{1} (y - y^2) \, dy = \left[\frac{1}{2} y^2 - \frac{1}{3} y^3 \right]_{y=0}^{1} = \frac{1}{2} - \frac{1}{3} = \frac{1}{6}.$$

To find the limits of integration, it is very helpful to sketch the domain of the double integral. The figure is next; it was produced by *Mathematica* via the command `Plot[ {x, Sqrt[x]}, {x, 0, 1} ]`.

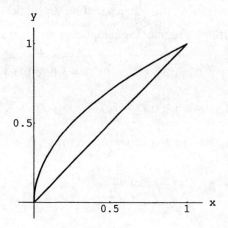

13.3.3: The graphs cross where $x^2 = 2x + 3$; that is, where $x = -1$ and where $x = 3$. A sketch of the domain of the integral is next; it was produced by *Mathematica* via the command

`Plot[ {x∧2, 2*x + 3}, {x, -1, 3} ]`.

The area of the region is

$$A = \int_{-1}^{3} \int_{y=x^2}^{2x+3} 1 \, dy \, dx = \int_{-1}^{3} (3 + 2x - x^2) \, dx = \left[3x + x^2 - \frac{1}{3}x^3 \right]_{-1}^{3} = \frac{32}{3}.$$

13.3.5: The graphs $y = x^2$ and $y = 2 - x$ cross where $x^2 = 2 - x$; that is, where $x = -2$ and where $x = 1$. And the x-axis is also part of the boundary of the region, so the figure below indicates the specified region whose area is sought. It was produced by the *Mathematica* command

```
Plot[ {x^2, 2 - x, 0}, {x, -0.4, 2.2} ]
```

The area of the roughly triangular-shaped region bounded by all three graphs is

$$A = \int_{y=0}^{1} \int_{x=\sqrt{y}}^{2-y} 1 \, dx \, dy = \int_{0}^{1} (2 - \sqrt{y} - y) \, dy = \left[2y - \frac{2}{3}y^{3/2} - \frac{1}{2}y^2 \right]_{0}^{1} = \frac{5}{6}.$$

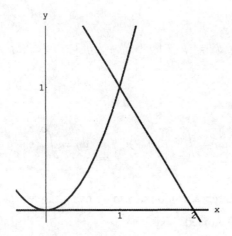

13.3.7: The graphs cross where $x^2 + 1 = 2x^2 - 3$; that is, where $x = -2$ and where $x = 2$. The area between them is

$$A = \int_{x=-2}^{2} \int_{y=2x^2-3}^{x^2+1} 1 \, dy \, dx = \int_{-2}^{2} (4 - x^2) \, dx = \left[4x - \frac{1}{3}x^3 \right]_{-2}^{2} = \frac{32}{3}.$$

13.3.9: The first-quadrant region bounded by these three curves is shown next; the figure was generated using the *Mathematica* command `Plot[ {x, 2x, 2/x}, {x, -0.2, 2.2} ]`.

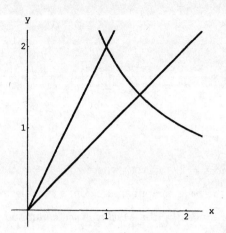

The bounding curves cross at the origin and at the points $(1, 2)$, and $(\sqrt{2}, \sqrt{2})$. The area they bound (in the first quadrant) is

$$A = \int_{x=0}^{1} \int_{y=x}^{2x} 1 \, dy \, dx + \int_{x=1}^{\sqrt{2}} \int_{y=x}^{2/x} 1 \, dy \, dx = \int_{0}^{1} x \, dx + \int_{1}^{\sqrt{2}} \left(\frac{2}{x} - x \right) \, dx$$

$$= \left[\frac{1}{2} x^2 \right]_{0}^{1} + \left[2 \ln x - \frac{1}{2} x^2 \right]_{1}^{\sqrt{2}} = \frac{1}{2} + \ln 2 - \frac{1}{2} = \ln 2 \approx 0.6931471805599453309417232.$$

13.3.11: The volume is

$$V = \int_{x=0}^{1} \int_{y=0}^{1} (1 + x + y) \, dy \, dx = \int_{x=0}^{1} \left[y + xy + \frac{1}{2} y^2 \right]_{y=0}^{1} dx = \int_{0}^{1} \left(x + \frac{3}{2} \right) dx = \left[\frac{3}{2} x + \frac{1}{2} x^2 \right]_{0}^{1} = 2.$$

13.3.13: The volume is

$$V = \int_{y=0}^{2} \int_{x=0}^{1} (y + e^x) \, dx \, dy = \int_{y=0}^{2} \left[xy + e^x \right]_{x=0}^{1} dy = \int_{0}^{2} (e + y - 1) \, dy = \left[ey + \frac{1}{2} y^2 - y \right]_{0}^{2} = 2e.$$

13.3.15: The domain of the integral can be drawn by using the *Mathematica* command Plot[1 - x, {x, 0, 1}] and the result is shown next.

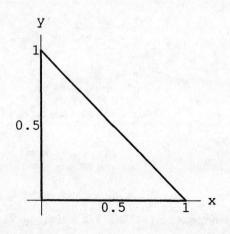

862

The volume is

$$V = \int_{x=0}^{1} \int_{y=0}^{1-x} (x+y)\, dy\, dx = \int_{x=0}^{1} \left[xy + \frac{1}{2}y^2 \right]_{y=0}^{1-x} dx = \int_{0}^{1} \frac{1}{2}(1 - x^2)\, dx = \left[\frac{1}{2}x - \frac{1}{6}x^3 \right]_{0}^{1} = \frac{1}{3}.$$

13.3.17: The domain of the integral can be drawn using the *Mathematica* command

```
ParametricPlot[ {{1,t}, {t,0}, {t,t^2}}, {t,0,1}, AspectRatio → 1 ];
```

and the result is shown next.

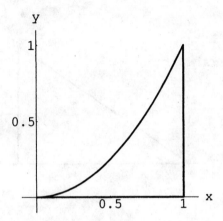

The volume is

$$V = \int_{x=0}^{1} \int_{y=0}^{x^2} (1 + x + y)\, dy\, dx = \int_{x=0}^{1} \left[y + xy + \frac{1}{2}y^2 \right]_{y=0}^{x^2} dx = \int_{0}^{1} \left(x^2 + x^3 + \frac{1}{2}x^4 \right) dx$$

$$= \left[\frac{1}{3}x^3 + \frac{1}{4}x^4 + \frac{1}{10}x^5 \right]_{0}^{1} = \frac{41}{60} \approx 0.683333333333.$$

13.3.19: The domain of the integral is shown next.

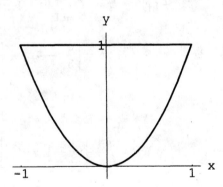

The volume of the solid is

$$V = \int_{x=-1}^{1} \int_{y=x^2}^{1} x^2\, dy\, dx = \int_{x=-1}^{1} \left[x^2 y \right]_{y=x^2}^{1} dx = \int_{-1}^{1} (x^2 - x^4)\, dx = \left[\frac{1}{3}x^3 - \frac{1}{5}x^5 \right]_{-1}^{1} = \frac{4}{15}.$$

863

13.3.21: The volume of the solid is

$$V = \int_{y=0}^{2} \int_{x=0}^{1} (x^2 + y^2)\, dx\, dy = \int_{y=0}^{2} \left[\frac{1}{3}x^3 + xy^2 \right]_{x=0}^{1} dy = \int_{0}^{2} \left(\frac{1}{3} + y^2 \right) dy = \left[\frac{1}{3}(y + y^3) \right]_{0}^{2} = \frac{10}{3}.$$

13.3.23: The domain of the integral can be drawn by executing the *Mathematica* command

```
ParametricPlot[ {{3, 2*t/3}, {t, 0}, {t, 2*t/3}}, {t, 0, 3} ];
```

and the result is shown next.

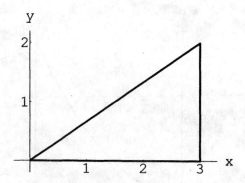

The volume of the solid is

$$V = \int_{x=0}^{3} \int_{y=0}^{2x/3} (9 - x - y)\, dy\, dx = \int_{x=0}^{3} \left[9y - xy - \frac{1}{2}y^2 \right]_{y=0}^{2x/3} dx$$

$$= \int_{0}^{3} \left(6x - \frac{8}{9}x^2 \right) dx = \left[3x^2 - \frac{8}{27}x^3 \right]_{0}^{3} = 19.$$

13.3.25: The volume is

$$V = \int_{x=0}^{1} \int_{y=0}^{2-2x} (4x^2 + y^2)\, dy\, dx = \int_{x=0}^{1} \left[4x^2 y + \frac{1}{3}y^3 \right]_{y=0}^{2-2x} dx$$

$$= \int_{0}^{1} \frac{8}{3}(1 - 3x + 6x^2 - 4x^3)\, dx = \left[\frac{8}{3}x - 4x^2 + \frac{16}{3}x^3 - \frac{8}{3}x^4 \right]_{0}^{1} = \frac{4}{3}.$$

13.3.27: The volume is

$$V = \int_{x=0}^{2} \int_{y=0}^{(6-3x)/2} (6 - 3x - 2y)\, dy\, dx = \int_{x=0}^{2} \left[6y - 3xy - y^2 \right]_{y=0}^{(6-3x)/2} dx$$

$$= \int_{0}^{2} \left(9 - 9x + \frac{9}{4}x^2 \right) dx = \left[9x - \frac{9}{2}x^2 + \frac{3}{4}x^3 \right]_{0}^{2} = 6.$$

13.3.29: The triangular domain of the integral can be drawn by executing the *Mathematica* command

```
ParametricPlot[ {{1, 2 + 2*t}, {1 + 4*t, 2}, {1 + 4*t, 4 - 2*t}},

                {t, 0, 1}, PlotRange → {{-0.5, 5.5}, {-0.5, 4.5}},

                AspectRatio → Automatic, AxesOrigin → {0, 0} ];
```

and the result is shown next.

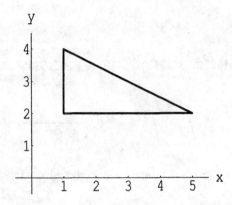

The volume of the solid is

$$V = \int_{x=1}^{5} \int_{y=2}^{(9-x)/2} xy \, dy \, dx = \int_{x=1}^{5} \left[\frac{1}{2} xy^2 \right]_{y=2}^{(9-x)/2} dx$$

$$= \int_{1}^{5} \left(\frac{65}{8} x - \frac{9}{4} x^2 + \frac{1}{8} x^3 \right) dx = \left[\frac{65}{16} x^2 - \frac{3}{4} x^3 + \frac{1}{32} x^4 \right]_{1}^{5} = 24.$$

13.3.31: The volume is

$$V = \int_{y=-1}^{1} \int_{x=-\sqrt{1-y^2}}^{\sqrt{1-y^2}} (x + 1) \, dx \, dy.$$

This integral can be evaluated exactly with a single command in *Mathematica*, but we will evaluate it one step at a time. As usual, the *Mathematica* output is rewritten slightly for more clarity.

```
Integrate[ x + 1, x ]
```

$$x + \frac{1}{2} x^2$$

```
(% /. x → Sqrt[1 - y*y]) - (% /. x → -Sqrt[1 - y*y])
```

$$2\sqrt{1 - y^2} + \frac{1}{2}(1 - y^2) + \frac{1}{2}(y^2 - 1)$$

```
Simplify[ % ]
```

$$2\sqrt{1 - y^2}$$

```
Integrate[ %, y ]
```

865

$$y\sqrt{1-y^2} + \arcsin y$$

```
(% /.  y → 1) - (% /.  y → -1)
```

$$\pi$$

```
N[ %, 60 ]
```

3.14159265358979323846264338327950288419716939937510582097494

13.3.33: The volume is

$$V = \int_{x=-1}^{1} \int_{y=-\sqrt{1-x^2}}^{\sqrt{1-x^2}} 2\sqrt{4 - x^2 - y^2} \, dy \, dx.$$

We used *Derive* to evaluate this integral in a step-by-step fashion much as in the solution of Problem 31. The results:

$$V = \int_{x=-1}^{1} \left[(4 - x^2) \arctan \frac{y}{\sqrt{4 - x^2 - y^2}} + y\sqrt{4 - x^2 - y^2} \right]_{y=-\sqrt{1-x^2}}^{\sqrt{1-x^2}} dx$$

$$= \int_{-1}^{1} \left[2(4 - x^2) \arctan \frac{\sqrt{3}\sqrt{1-x^2}}{3} + 2\sqrt{3}\sqrt{1-x^2} \right] dx$$

$$= \left[\frac{2}{3} x(12 - x^2) \arctan \frac{\sqrt{3}\sqrt{1-x^2}}{3} + \frac{16}{3} \arctan \frac{\sqrt{3}(2x+1)}{3\sqrt{1-x^2}} \right.$$

$$\left. + \frac{16}{3} \arctan \frac{\sqrt{3}(2x-1)}{3\sqrt{1-x^2}} - 4\sqrt{3} \arcsin x + \frac{2\sqrt{3}}{3} x\sqrt{1-x^2} \right]_{-1}^{1}$$

$$= \frac{\pi}{3} \left(32 - 12\sqrt{3} \right) \approx 11.7447292674805137.$$

In Section 13.4 we will find that this integral is quite easy to evaluate if we first convert to polar coordinates. If so, the integral takes the form

$$\int_{\theta=0}^{2\pi} \int_{r=0}^{1} 2r\sqrt{4 - r^2} \, dr \, d\theta = \int_{\theta=0}^{2\pi} \left[-\frac{2}{3}(4 - r^2)^{3/2} \right]_{r=0}^{1} d\theta = \int_{0}^{2\pi} \left(\frac{16}{3} - 2\sqrt{3} \right) d\theta = 2\pi \left(\frac{16}{3} - 2\sqrt{3} \right).$$

13.3.35: Given: the plane with Cartesian equation

$$\frac{x}{a} + \frac{y}{b} + \frac{z}{c} = 1$$

cutting off a tetrahedron in the first octant. We set $z = 0$ and solve for

$$y = \frac{b}{a}(a - x);$$

866

the triangular region in the first quadrant bounded by this line and the coordinate axes is the domain of the volume integral. Hence the volume of the tetrahedron is

$$V = \int_{x=0}^{a} \int_{y=0}^{b(a-x)/a} c\left(1 - \frac{x}{a} - \frac{y}{b}\right)\, dy\, dx = \int_{x=0}^{a} \left[\frac{2abcy - 2bcxy - acy^2}{2ab}\right]_{y=0}^{b(a-x)/a} dx$$

$$= \int_{0}^{a} \frac{bc(a-x)^2}{2a^2}\, dx = \left[\frac{3a^2bcx - 3abcx^2 + bcx^3}{6a^2}\right]_{0}^{a} = \frac{abc}{6}.$$

13.3.37: The volume is

$$V = \int_{y=0}^{1} \int_{x=0}^{\sqrt{1-y^2}} \sqrt{1-y^2}\, dx\, dy = \int_{y=0}^{1} \left[x\sqrt{1-y^2}\right]_{x=0}^{\sqrt{1-y^2}} dy = \int_{0}^{1}(1-y^2)\, dy = \left[y - \frac{1}{3}y^3\right]_{0}^{1} = \frac{2}{3}.$$

As the text indicates, the other order of integration provides more difficulties. You obtain

$$V = \int_{x=0}^{1} \int_{y=0}^{\sqrt{1-x^2}} \sqrt{1-y^2}\, dy\, dx = \int_{x=0}^{1} \left[\frac{1}{2}y\sqrt{1-y^2} + \frac{1}{2}\arcsin y\right]_{y=0}^{\sqrt{1-x^2}} dx$$

$$= \int_{0}^{1} \left[\frac{1}{2}x\sqrt{1-x^2} + \frac{1}{2}\arcsin\left(\sqrt{1-x^2}\right)\right] dx$$

$$= \left[\left(\frac{1}{6}x^2 - \frac{2}{3}\right)\sqrt{1-x^2} + \frac{1}{2}x\arcsin\left(\sqrt{1-x^2}\right)\right]_{0}^{1} = \frac{2}{3}.$$

13.3.39: We integrate to find the volume of an eighth of the sphere, then multiply by 8. Thus the volume of a sphere of radius a is

$$V = 8 \int_{x=0}^{a} \int_{y=0}^{\sqrt{a^2-x^2}} \sqrt{a^2 - x^2 - y^2}\, dy\, dx.$$

Let $y = (a^2 - x^2)^{1/2}\sin\theta$. Then $dy = (a^2 - x^2)^{1/2}\cos\theta\, d\theta$. This substitution yields

$$V = 8 \int_{x=0}^{a} \int_{\theta=0}^{\pi/2} \left[(a^2 - x^2) - (a^2 - x^2)\sin^2\theta\right]^{1/2}(a^2 - x^2)^{1/2}\cos\theta\, d\theta\, dx$$

$$= 8 \int_{x=0}^{a} \int_{\theta=0}^{\pi/2} (a^2 - x^2)\cos^2\theta\, d\theta\, dx$$

$$= 8 \int_{x=0}^{a/2} \int_{\theta=0}^{\pi/2} (a^2 - x^2) \cdot \frac{1 + \cos 2\theta}{2}\, d\theta\, dx = 8 \int_{x=0}^{a} (a^2 - x^2)\left[\frac{1}{2}\theta + \frac{1}{4}\sin 2\theta\right]_{0}^{\pi/2} dx$$

$$= 8 \int_{0}^{a} \frac{\pi}{4}(a^2 - x^2)\, dx = 2\pi\left[a^2 x - \frac{1}{3}x^3\right]_{0}^{a} = 2\pi \cdot \frac{2}{3}a^3 = \frac{4}{3}\pi a^3.$$

13.3.41: We integrate over the quarter-circle of radius 5 and center $(0,0)$ in the first quadrant, then multiply by 4. Hence the volume is

$$V = 4 \int_{x=0}^{5} \int_{y=0}^{\sqrt{25-x^2}} (25 - x^2 - y^2)\, dy\, dx = 4 \int_{x=0}^{5} \left[25y - x^2 y - \frac{1}{3}y^3\right]_{y=0}^{\sqrt{25-x^2}} dx$$

$$= \int_{0}^{5} \frac{8}{3}(25 - x^2)^{3/2}\, dx = \left[\frac{8}{3}\sqrt{25-x^2}\left(\frac{125}{8}x - \frac{1}{4}x^3\right) + 625\arcsin\left(\frac{x}{5}\right)\right]_{0}^{5}$$

$$= \frac{625}{2}\pi \approx 981.74770424681038701957605.$$

867

The techniques of Section 14.4 will transform this problem into one that is remarkably simple.

13.3.43: Suppose that the cylinder is the one with equation $x^2 + z^2 = R^2$ and that the square hole is centered on the z-axis and its sides are parallel to the coordinate planes. Thus the hole meets the xy-plane in the square with vertices at $(\pm\frac{1}{2}R, \pm\frac{1}{2}R)$. We will integrate over the quarter of that square that lies in the first quadrant, then multiply by 4. Hence the volume of material removed by the drill is

$$V = 4 \int_{x=0}^{R/2} \int_{y=0}^{R/2} 2\sqrt{R^2 - x^2} \, dy \, dx = 4 \int_{x=0}^{R/2} \left[2y\sqrt{R^2 - x^2} \right]_{y=0}^{R/2} dx$$

$$= 4 \int_0^{R/2} R(R^2 - x^2)^{1/2} \, dx = 4 \left[\frac{1}{2} Rx(R^2 - x^2)^{1/2} - \frac{1}{2} R^3 \arctan\left(\frac{x}{(R^2 - x^2)^{1/2}} \right) \right]_0^{R/2}$$

$$= \left[\frac{1}{2}\sqrt{3} + 2\arctan\left(\frac{1}{3}\sqrt{3} \right) \right] \cdot R^3 = \frac{3\sqrt{3} + 2\pi}{6} \cdot R^3 \approx (1.913222954981)R^3.$$

13.3.45: The region bounded by the parabolas $y = x^2$ and $y = 8 - x^2$ in the xy-plane is a suitable domain for a double integral that gives the volume of the solid. Hence the volume of the solid is

$$V = \int_{x=-2}^{2} \int_{y=x^2}^{8-x^2} (2x^2 - x^2) \, dy \, dx = \int_{x=-2}^{2} \left[x^2 y \right]_{y=x^2}^{8-x^2} dx = \int_{-2}^{2} (8x^2 - 2x^4) \, dx$$

$$= \left[\frac{8}{3}x^3 - \frac{2}{5}x^5 \right]_{-2}^{2} = \frac{256}{15} \approx 17.066666666667.$$

13.3.47: We used *Mathematica* in the usual way; the volume of the solid is

$$V = \int_{x=-\pi/2}^{\pi/2} \int_{y=-\cos x}^{\cos x} \cos y \, dy \, dx = \int_{x=-\pi/2}^{\pi/2} \left[\sin y \right]_{y=-\cos x}^{\cos x} dx$$

$$= \int_{-\pi/2}^{\pi/2} 2\sin(\cos x) \, dx \approx 3.572974963900010467337.$$

Mathematica reports that the exact value of the integral is

$$4*\text{HypergeometricPFQ}\left[\{1\}, \left\{ \frac{3}{2}, \frac{3}{2} \right\}, -\frac{1}{4} \right].$$

13.3.49: A *Mathematica* solution:

```
eq1 = z == 2*x + 3;

eq2 = z == x∧2 + y∧2;

Eliminate[ { eq1, eq2 }, z ]
```

$$y^2 = -x^2 + 2x + 3$$

This is the circle $(x-1)^2 + y^2 = 4$ with center $(1, 0)$ and radius 2. Therefore the volume of the solid is

```
Integrate[ 3 + 2*x - x∧2 - y∧2, { x, -1, 3 },
              { y, -Sqrt[ 3 + 2*x - x{∧2 ], Sqrt[ 3 + 2*x - x∧2 ] } ]
```

8π

13.3.51: A *Mathematica* solution:

```
eq1 = z == -16*x - 18*y;

eq2 = z == 11 - 4*x∧2 - 9*y∧2;

Eliminate[ { eq1, eq2 }, z ]
```

$$-9y^2 + 18y + 11 = 4x^2 - 16x$$

This is the ellipse $4(x-2)^2 + 9(y-1)^2 = 36$ with center $(2, 1)$ and semiaxes $a = 3$ and $b = 2$. Hence the volume of the solid is

```
Integrate[ 11 - 4*x∧2 - 9*y∧2 + 16*x + 18*y, { x, -1, 5 },
              { y, 1 - 1/3*Sqrt[ 36 - 4*(x - 2)∧2 ], 1 + 1/3*Sqrt[ 36 - 4*(x - 2)∧2 ] } ]
```

108π

13.3.53: First let's put the center of the sphere at the point $(-2, 0, 0)$. A *Mathematica* solution:

```
V = FullSimplify[ 2*Integrate[ Integrate[
              Sqrt[ 16 - (x + 2)∧2 - y∧2 ], { y, -1, 1 } ], { x, -1, 1 } ] ]
```

$$\frac{2}{3}\left[6\sqrt{6} - 2\sqrt{14} + 29\cot^{-1}\left(\sqrt{6}\right) + 41\cot^{-1}\left(\sqrt{14}\right) - 47\csc^{-1}\left(\sqrt{15}\right)\right.$$
$$+ 47\sin^{-1}\left(\sqrt{3/5}\right) + 20\tan^{-1}\left(\sqrt{3/2}\right) - 108\tan^{-1}\left(9\sqrt{3/2}\right)$$
$$\left. - 20\tan^{-1}\left(11/\sqrt{14}\right) + 108\tan^{-1}\left(19/\sqrt{14}\right)\right]$$

```
N[V]
```

26.7782

```
100*%/(4*Pi*64/3)*percent
```

9.98878 percent

Section 13.4

13.4.1: The circle with center $(0, 0)$ and radius $a > 0$ has polar description $r = a$, $0 \leq \theta \leq 2\pi$. Therefore its area is

$$A = \int_{\theta=0}^{2\pi} \int_{r=0}^{a} r \, dr \, d\theta = \int_{\theta=0}^{2\pi} \left[\frac{1}{2} r^2 \right]_{r=0}^{a} d\theta = \int_{0}^{2\pi} \frac{1}{2} a^2 \, d\theta = \left[\frac{1}{2} a^2 \theta \right]_{0}^{2\pi} = \pi a^2.$$

13.4.3: The area bounded by the cardioid with polar description $r = 1 + \cos\theta$, $0 \leq \theta \leq 2\pi$, is

$$A = \int_{\theta=0}^{2\pi} \int_{r=0}^{1+\cos\theta} r \, dr \, d\theta = \int_{\theta=0}^{2\pi} \left[\frac{1}{2} r^2 \right]_{r=0}^{1+\cos\theta} d\theta = \int_{0}^{2\pi} \left(\frac{1}{2} + \cos\theta + \frac{1}{2} \cos^2\theta \right) d\theta$$

$$= \frac{1}{8} \left[6\theta + 8\sin\theta + \sin 2\theta \right]_{0}^{2\pi} = \frac{3}{2} \pi.$$

13.4.5: To see the two circles, execute the *Mathematica* command

```
ParametricPlot[ {{Cos[t], Sin[t]}, {2*Sin[t]*Cos[t], 2*Sin[t]*Sin[t]}},
        {t, 0, 2*Pi}, AspectRatio -> Automatic ];
```

the result is shown next.

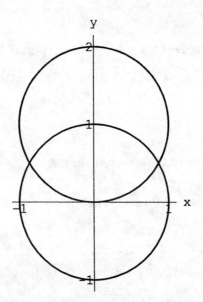

To find where the two circles meet, solve their equations simultaneously:

$$2\sin\theta = 1; \qquad \sin\theta = \frac{1}{2}; \qquad \theta = \frac{1}{6}\pi, \quad \theta = \frac{5}{6}\pi.$$

To find the area between them, two integrals are required. Each is doubled because we are actually finding the area of the right half of the intersection of the circles.

$$A_1 = 2 \int_{\theta=0}^{\pi/6} \int_{r=0}^{2\sin\theta} r \, dr \, d\theta = 2 \int_{0}^{\pi/6} 2\sin^2\theta \, d\theta = \left[2\theta - \sin 2\theta \right]_{0}^{\pi/6} = \frac{2\pi - 3\sqrt{3}}{6};$$

$$A_2 = 2 \int_{\theta=\pi/6}^{\pi/2} \int_{r=0}^{1} r \, dr \, d\theta = \int_{\theta=\pi/6}^{\pi/2} 1 \, d\theta = \frac{\pi}{3}.$$

Therefore the total area enclosed by both circles is

$$A = A_1 + A_2 = \frac{4\pi - 3\sqrt{3}}{6} \approx 1.2283696986087568.$$

13.4.7: To find where the limaçon $r = 1 - 2\sin\theta$ passes through the origin, solve $1 - 2\sin\theta = 0$ for $\theta = \frac{1}{6}\pi$, $\theta = \frac{5}{6}\pi$. The small loop of the limaçon is generated by the values of θ between these two angles, and is thus given by

$$A = \int_{\theta=\pi/6}^{5\pi/6} \int_{r=0}^{1-2\sin\theta} r\, dr\, d\theta = \int_{\theta=\pi/6}^{5\pi/6} \left[\frac{1}{2}r^2\right]_{r=0}^{1-2\sin\theta} d\theta = \int_{\pi/6}^{5\pi/6} \left[\frac{1}{2}(1 - 2\sin\theta)^2\right] d\theta$$

$$= \left[\frac{3}{2}\theta + 2\cos\theta - \frac{1}{2}\sin 2\theta\right]_{\pi/6}^{5\pi/6} = \frac{2\pi - 3\sqrt{3}}{2} \approx 0.5435164422364773.$$

13.4.9: In polar coordinates the integral takes the form

$$I = \int_{\theta=0}^{2\pi} \int_{r=0}^{2} r^2\, dr\, d\theta = 2\pi \left[\frac{1}{3}r^3\right]_0^2 = \frac{16}{3}\pi \approx 16.7551608191455639.$$

Because the inner integral does not involve θ (either in the integrand or in the limits of integration), it is constant with respect to θ. Therefore to integrate it with respect to θ over the interval $0 \leq \theta \leq 2\pi$, simply multiply the inner integral by 2π. We will use this time-saving technique frequently and without further comment.

13.4.11: Note that the domain of the integral is generated as θ varies from 0 to π. In polar coordinates the integral takes the form

$$I = \int_{\theta=0}^{\pi} \int_{r=0}^{\sin\theta} (10 + 2r\cos\theta + 3r\sin\theta) \cdot r\, dr\, d\theta = \int_{\theta=0}^{\pi} \left[5r^2 + \frac{1}{3}r^3(2\cos\theta + 3\sin\theta)\right]_{r=0}^{\sin\theta} d\theta$$

$$= \int_0^{\pi} \left(5\sin^2\theta + \frac{2}{3}\sin^3\theta\cos\theta + \sin^4\theta\right) d\theta$$

$$= \frac{1}{96} \left[276\theta - 8\cos 2\theta + 2\cos 4\theta - 144\sin 2\theta + 3\sin 4\theta\right]_0^{\pi} = \frac{23}{8}\pi \approx 9.03207887907065556058.$$

The reduction formulas in Problems 53 and 54 of Section 7.3, and the formulas of Problem 58 there and Formula (113) of the endpapers, may be used if you prefer to avoid trigonometric identities.

13.4.13: In polar form the given integral becomes

$$J = \int_{\theta=0}^{\pi/2} \int_{r=0}^{1} \frac{r}{1 + r^2}\, dr\, d\theta = \frac{\pi}{2} \cdot \left[\frac{1}{2}\ln(1 + r^2)\right]_0^1 = \frac{1}{4}\pi\ln 2 \approx 0.5443965225759005.$$

To evaluate the integral in Cartesian coordinates, you would first need to evaluate

$$\int_{x=0}^{\sqrt{1-y^2}} \frac{1}{1 + x^2 + y^2}\, dx = \frac{1}{\sqrt{1 + y^2}} \arctan\left(\frac{\sqrt{1 - y^2}}{\sqrt{1 + y^2}}\right). \tag{1}$$

Then you would need to antidifferentiate the last expression in (1). Neither *Mathematica* nor *Derive* could do so.

13.4.15: In polar coordinates the integral becomes

$$I = \int_{\theta=0}^{\pi/2} \int_{r=0}^{2} r^4 \, dr \, d\theta = \frac{\pi}{2} \cdot \left[\frac{1}{5} r^5 \right]_0^2 = \frac{\pi}{2} \cdot \frac{32}{5} = \frac{16}{5} \pi \approx 10.0530964914873384.$$

This integral can be evaluated in Cartesian coordinates. You should obtain

$$I = \int_0^2 \int_0^{\sqrt{4-x^2}} (x^2 + y^2)^{3/2} \, dy \, dx$$

$$= \int_0^2 \left[\left(\frac{5}{8} x^2 y + \frac{1}{4} y^3 \right) \sqrt{x^2 + y^2} + \frac{3}{8} x^4 \ln \left(y + \sqrt{x^2 + y^2} \right) \right]_0^{\sqrt{4-x^2}} dx$$

$$= \int_0^2 \left[\frac{5}{4} x^2 \sqrt{4-x^2} + \frac{1}{2} (4-x^2)^{3/2} - \frac{3}{4} x^4 \ln x + \frac{3}{8} x^4 \ln \left(2 + \sqrt{4-x^2} \right) \right] dx$$

$$= \left[\left(\frac{2}{5} x + \frac{3}{20} x^3 \right) \sqrt{4-x^2} + \frac{32}{5} \arcsin \left(\frac{x}{2} \right) - \frac{3}{20} x^5 \ln x + \frac{3}{40} x^5 \ln \left(2 + \sqrt{4-x^2} \right) \right]_0^2 = \frac{16}{5} \pi.$$

The final equality requires l'Hôpital's rule.

13.4.17: In polar coordinates the integral becomes

$$K = \int_{\theta=0}^{\pi/2} \int_{r=0}^{1} r \sin r^2 \, dr \, d\theta = \frac{\pi}{2} \cdot \left[-\frac{1}{2} \cos r^2 \right]_0^1 d\theta = \frac{\pi}{4} (1 - \cos 1) \approx 0.3610457246892050.$$

Exact evaluation of the given integral in Cartesian coordinates may be impossible. *Mathematica* reports that

$$\int \sin(x^2 + y^2) \, dx = \sqrt{\frac{\pi}{2}} \left[(\cos y^2) \cdot \text{FresnelS} \left(x \sqrt{\frac{2}{\pi}} \right) + (\sin y^2) \cdot \text{FresnelC} \left(x \sqrt{\frac{2}{\pi}} \right) \right]$$

where

$$\text{FresnelS}(x) = \int_0^x \sin \left(\frac{\pi t^2}{2} \right) dt \quad \text{and} \quad \text{FresnelC}(x) = \int_0^x \cos \left(\frac{\pi t^2}{2} \right) dt.$$

13.4.19: The volume is

$$V = \int_{\theta=0}^{2\pi} \int_{r=0}^{1} (2 + r \cos \theta + r \sin \theta) \cdot r \, dr \, d\theta = \int_{\theta=0}^{2\pi} \left[r^2 + \frac{1}{3} r^3 (\cos \theta + \sin \theta) \right]_{r=0}^{1} d\theta$$

$$= \int_0^{2\pi} \left(1 + \frac{1}{3} \cos \theta + \frac{1}{3} \sin \theta \right) d\theta = \left[\theta - \frac{1}{3} \cos \theta + \frac{1}{3} \sin \theta \right]_0^{2\pi} = 2\pi.$$

13.4.21: The volume is

$$V = \int_{\theta=0}^{\pi} \int_{r=0}^{2\sin\theta} (3 + r\cos\theta + r\sin\theta) \cdot r \, dr \, d\theta = \int_{\theta=0}^{\pi} \left[\frac{3}{2}r^2 + \frac{1}{3}r^3(\cos\theta + \sin\theta) \right]_{r=0}^{2\sin\theta} d\theta$$

$$= \int_0^{\pi} \left(6\sin^2\theta + \frac{8}{3}\sin^3\theta\cos\theta + \frac{8}{3}\sin^4\theta \right) d\theta = \frac{1}{12}\left[48\theta - 4\cos 2\theta + \cos 4\theta - 26\sin 2\theta + \sin 4\theta \right]_0^{\pi}$$

$$= \frac{1}{4} + \frac{48\pi - 3}{12} = 4\pi \approx 12.56637061435917295385.$$

13.4.23: We will find the volume of the sphere of radius a centered at the origin:

$$V = \int_{\theta=0}^{2\pi} \int_{r=0}^{a} 2r\sqrt{a^2 - r^2} \, dr \, d\theta = 2\pi \cdot \left[-\frac{2}{3}(a^2 - r^2)^{3/2} \right]_{r=0}^{a} = 2\pi \cdot \frac{2}{3}a^3 = \frac{4}{3}\pi a^3.$$

13.4.25: The volume of the solid is

$$V = \int_{\theta=0}^{2\pi} \int_{r=0}^{a} (h + r\cos\theta) \cdot r \, dr \, d\theta = \int_{\theta=0}^{2\pi} \left[\frac{1}{2}r^2 h + \frac{1}{3}r^2\cos\theta \right]_{r=0}^{a} d\theta = \int_0^{2\pi} \left(\frac{1}{2}a^2 h + \frac{1}{3}a^3\cos\theta \right) d\theta$$

$$= \frac{1}{6}\left[3a^2 h\theta + 2a^3\sin\theta \right]_0^{2\pi} = \pi a^2 h.$$

13.4.27: When we solve the equations of the paraboloids simultaneously, we find that one consequence is that $x^2 + y^2 = 1$. Thus the curve in which the paraboloids meet lies on that cylinder, and hence the circular disk $x^2 + y^2 \leq 1$ in the xy-plane is a suitable domain for the double integral that yields the volume between the paraboloids. That volume is therefore

$$V = \int_{\theta=0}^{2\pi} \int_{r=0}^{1} (4 - 4r^2) \cdot r \, dr \, d\theta = 2\pi \cdot \left[2r^2 - r^4 \right]_{r=0}^{1} d\theta = 2\pi \approx 6.2831853071795865.$$

13.4.29: When the equations of the sphere and the cone are solved simultaneously, one consequence is that $x^2 + y^2 = \frac{1}{2}a^2$. Therefore the circle in which the sphere and cone meet lies on the cylinder with that equation. Thus a suitable domain for the double integral that gives the volume in question is the circle in the xy-plane with equation $x^2 + y^2 = \frac{1}{2}a^2$. Hence the volume of the "ice-cream cone" is

$$V = \int_{\theta=0}^{2\pi} \int_{r=0}^{a/\sqrt{2}} \left(\sqrt{a^2 - r^2} - r \right) \cdot r \, dr \, d\theta = 2\pi \cdot \left[-\frac{1}{3}\left((a^2 - r^2)^{3/2} + r^3 \right) \right]_{r=0}^{a/\sqrt{2}}$$

$$= 2\pi \cdot \frac{1}{12}\left(4 - 2\sqrt{2} \right)a^3 = \frac{1}{3}\pi\left(2 - \sqrt{2} \right)a^3 \approx (0.6134341230070734)a^3.$$

13.4.31: The lemniscate defined by $r^2 = 2\sin 2\theta$ is shown below, as generated by the *Mathematica* command

```
ParametricPlot[ {{(Sqrt[2*Sin[2*t]])*Cos[t], (Sqrt[2*Sin[2*t]])*Sin[t]},
            {-(Sqrt[2*Sin[2*t]])*Cos[t], -(Sqrt[2*Sin[2*t]])*Sin[t]}},
            {t, 0, Pi/2}, AspectRatio → Automatic ];
```

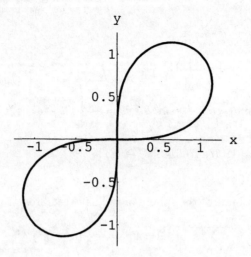

We find the volume over the first-quadrant loop and under the paraboloid $z = r^2$ to be

$$V = \int_{\theta=0}^{\pi/2} \int_{r=0}^{\sqrt{2\sin 2\theta}} r^3 \, dr \, d\theta = \int_{\theta=0}^{\pi/2} \left[\frac{1}{4} r^4 \right]_{r=0}^{\sqrt{2\sin 2\theta}} d\theta = \int_0^{\pi/2} \sin^2 2\theta \, d\theta$$

$$= \frac{1}{8} \left[4\theta - \sin 4\theta \right]_0^{\pi/2} = \frac{1}{4}\pi \approx 0.7853981633974483.$$

13.4.33: Part(a): A cross section of Fig. 13.4.23 in the xz-plane reveals a right triangle with legs b and $a - h$ and hypotenuse a, and it follows immediately that $b^2 = 2ah - h^2$. Part (b): The volume of the spherical segment is

$$V = \int_0^{2\pi} \int_0^b r\left[\sqrt{a^2 - r^2} - (a - h) \right] dr \, d\theta = 2\pi \left[-\frac{1}{3}(a^2 - r^2)^{3/2} - \frac{1}{2}ar^2 + \frac{1}{2}hr^2 \right]_0^b$$

$$= \frac{\pi}{3}\left[3r^2 h - 3r^2 a - 2(a^2 - r^2)^{3/2} \right]_0^b = \frac{\pi}{3}\left[3b^2 h - 3b^2 a - 2(a^2 - b^2)^{3/2} + 2a^3 \right]$$

$$= \frac{\pi}{3}\left[3b^2 h - 3b^2 a - 2(a^2 - 2ah + h^2)^{3/2} + 2a^3 \right] = \frac{\pi}{3}\left[3b^2 h - 3b^2 a - 2(a - h)^3 + 2a^3 \right]$$

$$= \frac{\pi}{3}(3b^2 h - 3b^2 a + 6a^2 h - 6ah^2 + 2h^3) = \frac{\pi}{3}\left[3b^2 h - 3a(2ah - h^2) + 6a^2 h - 6ah^2 + 2h^3 \right]$$

$$= \frac{\pi}{3}(3b^2 h - 6a^2 h + 3ah^2 + 6a^2 h - 6ah^2 + 2h^3) = \frac{\pi}{3}(3b^2 h - 3ah^2 + 2h^3) = \frac{\pi}{3}h(3b^2 - 3ah + 2h^2).$$

Recall that $2ah = b^2 + h^2$, so we may substitute $\frac{3}{2}b^2 + \frac{3}{2}h^2$ for $3ah$ in the last expression. Therefore the volume of the spherical cap is

$$V = \frac{1}{3}\pi h\left(3b^2 - \frac{3}{2}b^2 - \frac{3}{2}h^2 + 2h^2 \right) = \frac{1}{6}\pi h(6b^2 - 3b^2 - 3h^2 + 4h^2) = \frac{1}{6}\pi h(3b^2 + h^2).$$

13.4.35: Using the *Suggestion*, we have

$$2\pi(b - x) \, dA = 2\pi(b - r\cos\theta)r \, dr \, d\theta,$$

and hence the volume of the torus is

$$V = \int_0^{2\pi} \int_0^a 2\pi r(b - r\cos\theta) \, dr \, d\theta = \int_0^{2\pi} \left[b\pi r^2 - \frac{2}{3}\pi r^3 \cos\theta \right]_0^a d\theta$$

$$= \left[\frac{1}{3}(3\pi a^2 b\theta - 2\pi a^3 \sin\theta) \right]_0^{2\pi} = 2\pi^2 a^2 b.$$

Read the *First Theorem of Pappus* in Section 13.5 and apply it to this circular disk of radius a rotated around a circle of radius b to obtain the same answer in a tenth of the time.

13.4.37: The plane and the paraboloid meet in a curve that lies on the cylinder $x^2 + y^2 = 4$, so the circular disk $x^2 + y^2 \leq 4$ in the xy-plane is a suitable domain for the volume integral. The volume of the solid is thus

$$V = \int_0^{2\pi} \int_0^2 (4 - r^2) \cdot r \, dr \, d\theta = 2\pi \cdot \left[2r^2 - \frac{1}{4}r^4 \right]_0^2 = 8\pi \approx 25.1327412287183459.$$

13.4.39: When we solve the equations of the paraboloids simultaneously, we find that $x^2 + y^2 = 4$. Hence the curve in which the surfaces intersect lies on the cylinder $x^2 + y^2 = 4$. Thus the disk in the xy-plane bounded by the circle $x^2 + y^2 = 4$ is appropriate for the domain of the volume integral. So the volume of the solid bounded by the paraboloids is

$$V = \int_{\theta=0}^{2\pi} \int_{r=0}^2 (12 - 3r^2) \cdot r \, dr \, d\theta = 2\pi \cdot \left[6r^2 - \frac{3}{4}r^4 \right]_{r=0}^2 = 24\pi \approx 75.3982236861550377.$$

13.4.41: A *Mathematica* solution:

```
V = Simplify[ .2*Integrate[ Integrate[
             r*Sqrt[ b∧2 - r∧2 ], { r, 0, a } ], { θ, 0, 2*Pi } ] ]
```

$$\frac{3}{2}\left(b^3 - (b^2 - a^2)^{3/2} \right) \pi$$

```
V1 = V /. { a → 2, b → 4 }
```

$$\frac{4}{3}\left(64 - 24\sqrt{3} \right) \pi$$

Percent of material removed:

```
N[ 100*V1/(4*Pi*64/3) ]
```

    35.0481

13.4.43: A *Mathematica* solution for the case of a square hole:

875

```
n = 4;

V = 4*n*Integrate[ Integrate[ r*Sqrt[ 4 - r^2 ],

             { r, 0, Cos[ Pi/n ]*Sec[ θ ] } ], { θ, 0, Pi/n } ];
```

(We suppress the output; it's not attractive.)

```
V = Chop[ N[V] ]
```

7.6562

Percent:

```
100*V/(4*Pi*8/3)
```

22.8473

Now for the pentagonal, hexagonal, heptagonal, and 17-sided holes. First, the pentagonal hole:

```
n = 5;

V = 4*n*Integrate[ Integrate[ r*Sqrt[ 4 - r^2 ],

             { r, 0, Cos[ Pi/n ]*Sec[ θ ] } ], { θ, 0, Pi/n } ];

V = Chop[ N[V] ]
```

9.03688

Percent:

```
100*V/(4*Pi*8/3)
```

26.9675

Next, the hexagonal hole:

```
n = 6;

V = 4*n*Integrate[ Integrate[ r*Sqrt[ 4 - r^2 ],

             { r, 0, Cos[ Pi/n ]*Sec[ θ ] } ], { θ, 0, Pi/n } ];

V = Chop[ N[V] ]
```

9.83041

Percent:

```
100*V/(4*Pi*8/3)
```

  29.3355

For the heptagonal and 17-sided holes, we use *Maple*.

```
n:=7;

V:=4*n*int(int(r*sqrt(4-r^2), r=0..cos(Pi/n)*sec(t)), t=0..Pi/n);

V:=evalf(V);
```

  $V := 10.32346688$

Percent:

```
evalf(100*V/(4*Pi*8/3));
```

  30.80682719

```
n:=17;

V:=4*n*int(int(r*sqrt(4-r^2), r=0..cos(Pi/n)*sec(t)), t=0..Pi/n);

V:=evalf(V);
```

  $V := 11.49809060$

Percent:

```
evalf(100*V/(4*Pi*8/3));
```

  34.31208666

Section 13.5

Note: To integrate positive integral powers of sines and cosines (and their products), there are effective techniques not illustrated in the text. They use the following identities, which are consequences of the Euler-DeMoivre formula

$$(\cos\theta + i\sin\theta)^n = \left(e^{i\theta}\right)^n = e^{in\theta} = \cos n\theta + i\sin n\theta.$$

---

1. $\sin^2\theta = \dfrac{1}{2}(1 - \cos 2\theta)$.

2. $\sin^3 \theta = \dfrac{1}{4}(3\sin\theta - \sin 3\theta).$

3. $\sin^4 \theta = \dfrac{1}{8}(3 - 4\cos 2\theta + \cos 4\theta).$

4. $\sin^5 \theta = \dfrac{1}{16}(10\sin\theta - 5\sin 3\theta + \sin 5\theta).$

5. $\sin^6 \theta = \dfrac{1}{32}(10 - 15\cos 2\theta + 6\cos 4\theta - \cos 6\theta).$

6. $\sin^7 \theta = \dfrac{1}{64}(35\sin\theta - 21\sin 3\theta + 7\sin 5\theta - \sin 7\theta).$

7. $\sin^8 \theta = \dfrac{1}{128}(35 - 56\cos 2\theta + 28\cos 4\theta - 8\cos 6\theta + \cos 8\theta).$

8. $\sin^9 \theta = \dfrac{1}{256}(126\sin\theta - 84\sin 3\theta + 36\sin 5\theta - 9\sin 7\theta + \sin 9\theta).$

9. $\sin^{10} \theta = \dfrac{1}{512}(126 - 210\cos 2\theta + 120\cos 4\theta - 45\cos 6\theta + 10\cos 8\theta - \cos 10\theta).$

10. $\cos^2 \theta = \dfrac{1}{2}(1 + \cos 2\theta).$

11. $\cos^3 \theta = \dfrac{1}{4}(3\cos\theta + \cos 3\theta).$

12. $\cos^4 \theta = \dfrac{1}{8}(3 + 4\cos 2\theta + \cos 4\theta).$

13. $\cos^5 \theta = \dfrac{1}{16}(10\cos\theta + 5\cos 3\theta + \cos 5\theta).$

14. $\cos^6 \theta = \dfrac{1}{32}(10 + 15\cos 2\theta + 6\cos 4\theta + \cos 6\theta).$

15. $\cos^7 \theta = \dfrac{1}{64}(35\cos\theta + 21\cos 3\theta + 7\cos 5\theta + \cos 7\theta).$

16. $\cos^8 \theta = \dfrac{1}{128}(35 + 56\cos 2\theta + 28\cos 4\theta + 8\cos 6\theta + \cos 8\theta).$

17. $\cos^9 \theta = \dfrac{1}{256}(126\cos\theta + 84\cos 3\theta + 36\cos 5\theta + 9\cos 7\theta + \cos 9\theta).$

18. $\cos^{10} \theta = \dfrac{1}{512}(126 + 210\cos 2\theta + 120\cos 4\theta + 45\cos 6\theta + 10\cos 8\theta + \cos 10\theta).$

19. $\sin^2 \theta \cos^2 \theta = \dfrac{1}{8}(1 - \cos 4\theta).$

20. $\sin^3 \theta \cos^2 \theta = \dfrac{1}{16}(2\sin\theta + \sin 3\theta - \sin 5\theta).$

21. $\sin^4 \theta \cos^2 \theta = \dfrac{1}{32}(2 - \cos 2\theta - 2\cos 4\theta + \cos 6\theta).$

22. $\sin^5 \theta \cos^2 \theta = \dfrac{1}{64} (5 \sin \theta + \sin 3\theta - 3 \sin 5\theta + \sin 7\theta).$

23. $\sin^6 \theta \cos^2 \theta = \dfrac{1}{128} (5 - 4 \cos 2\theta - 4 \cos 4\theta + 4 \cos 6\theta - \cos 8\theta).$

24. $\sin^2 \theta \cos^3 \theta = \dfrac{1}{16} (2 \cos \theta - \cos 3\theta - \cos 5\theta).$

25. $\sin^2 \theta \cos^4 \theta = \dfrac{1}{32} (2 + \cos 2\theta - 2 \cos 4\theta - \cos 6\theta).$

26. $\sin^2 \theta \cos^5 \theta = \dfrac{1}{64} (5 \cos \theta - \cos 3\theta - 3 \cos 5\theta - \cos 7\theta).$

27. $\sin^2 \theta \cos^6 \theta = \dfrac{1}{128} (5 + 4 \cos 2\theta - 4 \cos 4\theta - 4 \cos 6\theta - \cos 8\theta).$

28. $\sin^3 \theta \cos^3 \theta = \dfrac{1}{32} (3 \sin 2\theta - \sin 6\theta).$

29. $\sin^4 \theta \cos^3 \theta = \dfrac{1}{64} (3 \cos \theta - 3 \cos 3\theta - \cos 5\theta + \cos 7\theta).$

30. $\sin^5 \theta \cos^3 \theta = \dfrac{1}{128} (6 \sin 2\theta - 2 \sin 4\theta - 2 \sin 6\theta + \sin 8\theta).$

40. $\sin^6 \theta \cos^3 \theta = \dfrac{1}{256} (6 \cos \theta - 8 \cos 3\theta + 3 \cos 7\theta - \cos 9\theta).$

41. $\sin^4 \theta \cos^4 \theta = \dfrac{1}{128} (3 - 4 \cos 4\theta + \cos 8\theta).$

42. $\sin^5 \theta \cos^4 \theta = \dfrac{1}{256} (6 \sin \theta + 4 \sin 3\theta - 4 \sin 5\theta - \sin 7\theta + \sin 9\theta).$

43. $\sin^6 \theta \cos^4 \theta = \dfrac{1}{512} (6 - 2 \cos 2\theta - 8 \cos 4\theta + 3 \cos 6\theta + 2 \cos 8\theta - \cos 10\theta).$

44. $\sin^5 \theta \cos^5 \theta = \dfrac{1}{512} (10 \sin 2\theta - 5 \sin 6\theta + \sin 10\theta).$

45. $\sin^6 \theta \cos^5 \theta = \dfrac{1}{1024} (10 \cos \theta - 10 \cos 3\theta - 5 \cos 5\theta + 5 \cos 7\theta + \cos 9\theta - \cos 11\theta).$

46. $\sin^6 \theta \cos^6 \theta = \dfrac{1}{2048} (10 - 15 \cos 4\theta + 6 \cos 8\theta - \cos 12\theta).$

13.5.1: By the symmetry principle, the centroid is at $(\overline{x}, \overline{y}) = (2, 3)$.

13.5.3: By the symmetry principle, the centroid is at $(\overline{x}, \overline{y}) = (1, 1)$.

13.5.5: The mass and moments are

$$m = \int_0^4 \int_0^{(4-x)/2} 1 \, dy \, dx = \int_0^4 \frac{4-x}{2} \, dx = \left[2x - \frac{1}{4}x^2 \right]_0^4 = 4,$$

$$M_y = \int_0^4 \int_0^{(4-x)/2} x \, dy \, dx = \int_0^4 \left(2x - \frac{1}{2}x^2 \right) dx = \left[x^2 - \frac{1}{6}x^3 \right]_0^4 = \frac{16}{3}, \quad \text{and}$$

$$M_x = \int_0^4 \int_0^{(4-x)/2} y \, dy \, dx = \int_0^4 \frac{1}{8}(4-x)^2 \, dx = \left[\frac{1}{24}(x-4)^3 \right]_0^4 = \frac{8}{3}.$$

Therefore the centroid is located at $(\overline{x}, \overline{y}) = \left(\frac{4}{3}, \frac{2}{3} \right)$.

13.5.7: The mass and moments are

$$m = \int_0^2 \int_0^{x^2} 1 \, dy \, dx = \int_0^2 x^2 \, dx = \left[\frac{1}{3}x^3 \right]_0^2 = \frac{8}{3},$$

$$M_y = \int_0^2 \int_0^{x^2} x \, dy \, dx = \int_0^2 x^3 \, dx = \left[\frac{1}{4}x^4 \right]_0^2 = 4, \quad \text{and}$$

$$M_x = \int_0^2 \int_0^{x^2} y \, dy \, dx = \int_0^2 \frac{1}{2}x^4 \, dx = \left[\frac{1}{10}x^5 \right]_0^2 = \frac{16}{5}.$$

Therefore the centroid is $(\overline{x}, \overline{y}) = \left(\frac{3}{2}, \frac{6}{5} \right)$.

13.5.9: By symmetry, $M_y = 0$. Next,

$$m = \int_{-2}^2 \int_{x^2-4}^0 1 \, dy \, dx = \int_{-2}^2 (4-x^2) \, dx = \left[4x - \frac{1}{3}x^3 \right]_{-2}^2 = \frac{32}{3} \quad \text{and}$$

$$M_x = \int_{-2}^2 \int_{x^2-4}^0 y \, dy \, dx = \int_{-2}^2 -\frac{1}{2}(4-x^2)^2 \, dx = \left[\frac{4}{3}x^3 - 8x - \frac{1}{10}x^5 \right]_{-2}^2 = -\frac{256}{15}.$$

Therefore the centroid is at the point $\left(0, -\frac{8}{5} \right)$.

13.5.11: The mass and moments are

$$m = \int_0^1 \int_0^{1-x} xy \, dy \, dx = \int_0^1 \frac{1}{2}(x - 2x^2 + x^3) \, dx = \frac{1}{24}\left[6x^2 - 8x^3 + 3x^4 \right]_0^1 = \frac{1}{24},$$

$$M_y = \int_0^1 \int_0^{1-x} x^2 y \, dy \, dx = \int_0^1 \left[\frac{1}{2}x^2 y^2 \right]_0^{1-x} dx = \int_0^1 \frac{1}{2}(x^2 - 2x^3 + x^4) \, dx$$

$$= \frac{1}{60}\left[10x^3 - 15x^4 + 6x^5 \right]_0^1 = \frac{1}{60},$$

$$M_x = \int_0^1 \int_0^{1-x} xy^2 \, dy \, dx = \int_0^1 \left[\frac{1}{3}xy^3 \right]_0^{1-x} dx = \int_0^1 \frac{1}{3}(x - 3x^2 + 3x^3 - x^4) \, dx$$

$$= \frac{1}{60}\left[10x^2 - 20x^3 + 15x^4 - 4x^5 \right]_0^1 = \frac{1}{60}.$$

Therefore the centroid is located at the point $\left(\frac{2}{5}, \frac{2}{5}\right)$.

13.5.13: The mass and moments of the lamina are

$$m = \int_{-2}^{2} \int_{0}^{4-x^2} y \, dy \, dx = \int_{-2}^{2} \frac{1}{2}(4-x^2)^2 \, dx = \frac{1}{30}\left[240x - 40x^3 + 3x^5\right]_{-2}^{2} = \frac{256}{15},$$

$$M_y = \int_{-2}^{2} \int_{0}^{4-x^2} xy \, dy \, dx = \int_{-2}^{2} \frac{1}{2}x(4-x^2)^2 \, dx = \frac{1}{12}\left[48x^2 - 12x^4 + x^6\right]_{-2}^{2} = 0, \quad \text{and}$$

$$M_x = \int_{-2}^{2} \int_{0}^{4-x^2} y^2 \, dy \, dx = \int_{-2}^{2} \frac{1}{3}(4-x^2)^3 \, dx = \left[\frac{64}{3}x - \frac{16}{3}x^3 + \frac{4}{5}x^5 - \frac{1}{21}x^7\right]_{-2}^{2} = \frac{4096}{105}.$$

Therefore the centroid of the lamina is $(\overline{x}, \overline{y}) = \left(0, \frac{16}{7}\right)$.

13.5.15: The mass and moments of the lamina are

$$m = \int_{0}^{1} \int_{x^2}^{\sqrt{x}} xy \, dy \, dx = \int_{0}^{1} \frac{1}{2}(x^2 - x^5) \, dx = \frac{1}{12}\left[2x^3 - x^6\right]_{0}^{1} = \frac{1}{12},$$

$$M_y = \int_{0}^{1} \int_{x^2}^{\sqrt{x}} x^2 y \, dy \, dx = \int_{0}^{1} \frac{1}{2}(x^3 - x^6) \, dx = \frac{1}{56}\left[7x^4 - 4x^7\right]_{0}^{1} = \frac{3}{56},$$

$$M_x = \int_{0}^{1} \int_{x^2}^{\sqrt{x}} xy^2 \, dy \, dx = \int_{0}^{1} \frac{1}{3}(x^{5/2} - x^7) \, dx = \frac{1}{168}\left[16x^{7/2} - 7x^8\right]_{0}^{1} = \frac{3}{56}.$$

Hence the centroid of the lamina is at $(\overline{x}, \overline{y}) = \left(\frac{9}{14}, \frac{9}{14}\right)$.

13.5.17: The mass and moments of the lamina are

$$m = \int_{-1}^{1} \int_{x^2}^{2-x^2} y \, dy \, dx = \int_{-1}^{1} (2 - 2x^2) \, dx = \left[2x - \frac{2}{3}x^3\right]_{-1}^{1} = \frac{8}{3},$$

$$M_y = \int_{-1}^{1} \int_{x^2}^{2-x^2} xy \, dy \, dx = \int_{-1}^{1} (2x - 2x^3) \, dx = \left[x^2 - \frac{1}{2}x^4\right]_{-1}^{1} = 0,$$

$$M_x = \int_{-1}^{1} \int_{x^2}^{2-x^2} y^2 \, dy \, dx = \int_{-1}^{1} \left[\frac{1}{3}y^3\right]_{x^2}^{2-x^2} dx = \int_{-1}^{1} \left(\frac{1}{3}(2-x^2)^3 - \frac{1}{3}x^6\right) dx$$

$$= \left[\frac{8}{3}x - \frac{4}{3}x^3 + \frac{2}{5}x^5 - \frac{2}{21}x^7\right]_{-1}^{1} = \frac{344}{105}.$$

Hence the centroid of the lamina is at $(\overline{x}, \overline{y}) = \left(0, \frac{43}{35}\right)$.

13.5.19: The mass and moments of the lamina are

$$m = \int_{0}^{\pi} \int_{0}^{\sin x} 1 \, dy \, dx = \int_{0}^{\pi} \sin x \, dx = \left[-\cos x\right]_{0}^{\pi} = 2,$$

$$M_y = \int_{0}^{\pi} \int_{0}^{\sin x} x \, dy \, dx = \int_{0}^{\pi} x \sin x \, dx = \left[\sin x - x \cos x\right]_{0}^{\pi} = \pi,$$

$$M_x = \int_{0}^{\pi} \int_{0}^{\sin x} y \, dy \, dx = \int_{0}^{\pi} \frac{1}{2}\sin^2 x \, dx = \left[\frac{1}{8}(2x - \sin 2x)\right]_{0}^{\pi} = \frac{1}{4}\pi.$$

Therefore the centroid of the lamina is located at

$$(\overline{x}, \overline{y}) = \left(\frac{\pi}{2}, \frac{\pi}{8}\right) \approx (1.5707963267948966, 0.3926990816987242).$$

13.5.21: The mass and moments of the lamina are

$$m = \int_0^a \int_0^a (x+y)\,dy\,dx = \int_0^a \left(\frac{1}{2}a^2 + ax\right) dx = \left[\frac{1}{2}(a^2 x + ax^2)\right]_0^a = a^3,$$

$$M_y = \int_0^a \int_0^a x(x+y)\,dy\,dx = \int_0^a \left(\frac{1}{2}a^2 x + ax^2\right) dx = \left[\frac{1}{4}a^2 x^2 + \frac{1}{3}ax^3\right]_0^a = \frac{7}{12}a^4, \quad \text{and}$$

$$M_x = \int_0^a \int_0^a (x+y)y\,dy\,dx = \int_0^a \left(\frac{1}{3}a^3 + \frac{1}{2}a^2 x\right) dx = \left[\frac{1}{3}a^3 x + \frac{1}{4}a^2 x^2\right]_0^a = \frac{7}{12}a^4.$$

Therefore its centroid is located at the point $(\overline{x}, \overline{y}) = \left(\frac{7}{12}a, \frac{7}{12}a\right)$.

13.5.23: The mass and moments of the lamina are

$$m = \int_{-2}^2 \int_{x^2}^4 y\,dy\,dx = \int_{-2}^2 \left(8 - \frac{1}{2}x^4\right) dx = \left[8x - \frac{1}{10}x^5\right]_{-2}^2 = \frac{128}{5},$$

$$M_y = \int_{-2}^2 \int_{x^2}^4 xy\,dy\,dx = \int_{-2}^2 \left(8x - \frac{1}{2}x^5\right) dx = \left[4x^2 - \frac{1}{12}x^6\right]_{-2}^2 = 0, \quad \text{and}$$

$$M_x = \int_{-2}^2 \int_{x^2}^4 y^2\,dy\,dx = \int_{-2}^2 \left(\frac{64}{3} - \frac{1}{3}x^6\right) dx = \left[\frac{64}{3}x - \frac{1}{21}x^7\right]_{-2}^2 = \frac{512}{7}.$$

Thus the centroid of the lamina is located at the point $(\overline{x}, \overline{y}) = \left(0, \frac{20}{7}\right)$.

13.5.25: The mass and moments are

$$m = \int_0^\pi \int_0^{\sin x} x\,dy\,dx = \int_0^\pi x \sin x\,dx = \left[\sin x - x\cos x\right]_0^\pi = \pi,$$

$$M_y = \int_0^\pi \int_0^{\sin x} x^2\,dy\,dx = \int_0^\pi x^2 \sin x\,dx = \left[2\cos x - x^2 \cos x + 2x \sin x\right]_0^\pi = \pi^2 - 4, \quad \text{and}$$

$$M_x = \int_0^\pi \int_0^{\sin x} xy\,dy\,dx = \int_0^\pi \frac{1}{2}x \sin^2 x\,dx = \left[\frac{1}{16}\left(2x^2 - \cos 2x - 2x \sin 2x\right)\right]_0^\pi = \frac{1}{8}\pi^2.$$

Consequently the centroid of the lamina is at

$$(\overline{x}, \overline{y}) = \left(\frac{\pi^2 - 4}{\pi}, \frac{\pi}{8}\right) \approx (1.8683531088546306, 0.3926990816987242).$$

13.5.27: The mass and moments are

$$m = \int_{\theta=0}^\pi \int_{r=0}^a r^2\,dr\,d\theta = \pi \cdot \left[\frac{1}{3}r^3\right]_0^a = \frac{1}{3}\pi a^3,$$

$$M_y = \int_{\theta=0}^\pi \int_{r=0}^a r^3 \cos\theta\,dr\,d\theta = \int_0^\pi \frac{1}{4}a^4 \cos\theta\,d\theta = \left[\frac{1}{4}a^4 \sin\theta\right]_0^\pi = 0, \quad \text{and}$$

$$M_x = \int_{\theta=0}^\pi \int_{r=0}^a r^3 \sin\theta\,dr\,d\theta = \int_0^\pi \frac{1}{4}a^4 \sin\theta\,d\theta = \left[-\frac{1}{4}a^4 \cos\theta\right]_0^\pi = \frac{1}{2}a^4.$$

Therefore the centroid of the lamina is located at the point

$$(\bar{x}, \bar{y}) = \left(0, \frac{3a}{2\pi}\right) \approx \left(0, (0.47746483) \cdot a\right).$$

13.5.29: The following figure, generated by *Mathematica*, shows the two circles.

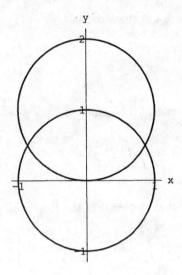

They cross where $2\sin\theta = 1$; that is, where $\theta = \frac{1}{6}\pi$ and where $\theta = \frac{5}{6}\pi$. The lamina in question is the region outside the lower circle and inside the upper circle. Its mass and moments are

$$m = \int_{\pi/6}^{5\pi/6} \int_1^{2\sin\theta} r^2 \sin\theta \, dr \, d\theta = \int_{\pi/6}^{5\pi/6} \left(\frac{8}{3}\sin^4\theta - \frac{1}{3}\sin\theta\right) d\theta$$

$$= \frac{1}{12}\left[12\theta + 4\cos\theta - 8\sin 2\theta + \sin 4\theta\right]_{\pi/6}^{5\pi/6} = \frac{8\pi + 3\sqrt{3}}{12},$$

$$M_y = \int_{\pi/6}^{5\pi/6} \int_1^{2\sin\theta} r^3 \sin\theta \cos\theta \, dr \, d\theta = \int_{\pi/6}^{5\pi/6} \left(4\sin^5\theta\cos\theta - \frac{1}{4}\sin\theta\cos\theta\right) d\theta$$

$$= \frac{1}{48}\left[6\cos 4\theta - \cos 6\theta - 12\cos 2\theta\right]_{\pi/6}^{5\pi/6} = 0,$$

$$M_x = \int_{\pi/6}^{5\pi/6} \int_1^{2\sin\theta} r^3 \sin^2\theta \, dr \, d\theta = \int_{\pi/6}^{5\pi/6} \left(4\sin^6\theta - \frac{1}{4}\sin^2\theta\right) d\theta$$

$$= \frac{1}{48}\left[54\theta - 42\sin 2\theta + 9\sin 4\theta - \sin 6\theta\right]_{\pi/6}^{5\pi/6} = \frac{12\pi + 11\sqrt{3}}{16}.$$

Thus its centroid is at the point

$$(\bar{x}, \bar{y}) = \left(0, \frac{36\pi + 33\sqrt{3}}{32\pi + 12\sqrt{3}}\right) \approx (0, 1.4034060567438982)$$

and its mass is approximately 2.5274078042854148.

13.5.31: The polar moment of inertia of the lamina is

$$I_0 = \int_0^{2\pi} \int_0^a r^{n+3} \, dr \, d\theta = 2\pi \cdot \left[\frac{r^{n+4}}{n+4} \right]_0^a = \frac{2\pi a^{n+4}}{n+4}.$$

13.5.33: The polar moment of inertia of the lamina is

$$I_0 = \int_{-\pi/2}^{\pi/2} \int_0^{2\cos\theta} kr^3 \, dr \, d\theta = \int_{-\pi/2}^{\pi/2} 4k\cos^4\theta \, d\theta = \frac{1}{8}k \left[12\theta + 8\sin 2\theta + \sin 4\theta \right]_{-\pi/2}^{\pi/2} = \frac{3}{2}\pi k.$$

13.5.35: The polar moment of inertia of the lamina is

$$I_0 = \int_{-\pi/4}^{\pi/4} \int_0^{\sqrt{\cos 2\theta}} r^5 \, dr \, d\theta = \int_{-\pi/4}^{\pi/4} \frac{1}{6}\cos^3 2\theta \, d\theta = \frac{1}{144} \left[9\sin 2\theta + \sin 6\theta \right]_{-\pi/4}^{\pi/4} = \frac{1}{9}.$$

13.5.37: In Problem 23 we found that the mass of the lamina is $m = \frac{128}{5}$. Next,

$$I_y = \int_{-2}^2 \int_{x^2}^4 x^2 y \, dy \, dx = \int_{-2}^2 \left(8x^2 - \frac{1}{2}x^6 \right) dx = \left[\frac{8}{3}x^3 - \frac{1}{14}x^7 \right]_{-2}^2 = \frac{512}{21} \quad \text{and}$$

$$I_x = \int_{-2}^2 \int_{x^2}^4 y^3 \, dy \, dx = \int_{-2}^2 \left(64 - \frac{1}{4}x^8 \right) dx = \left[64x - \frac{1}{36}x^9 \right]_{-2}^2 = \frac{2048}{9}.$$

Therefore $\hat{x} = \frac{2}{21}\sqrt{105}$ and $\hat{y} = \frac{4}{3}\sqrt{5}$.

13.5.39: In the solution of Problem 27 we found that the mass of the lamina is $m = \frac{1}{3}\pi a^3$. Next,

$$I_y = \int_0^\pi \int_0^a r^4 \cos^2 t \, dr \, d\theta = \int_0^\pi \frac{1}{5}a^5 \cos^2\theta \, d\theta = \left[\frac{1}{20}a^5(2\theta + \sin 2\theta) \right]_0^\pi = \frac{1}{10}\pi a^5 \quad \text{and}$$

$$I_x = \int_0^\pi \int_0^a r^4 \sin^2\theta \, dr \, d\theta = \int_0^\pi \frac{1}{5}a^5 \sin^2\theta \, d\theta = \left[\frac{1}{20}a^5(2\theta - \sin 2\theta) \right]_0^\pi = \frac{1}{10}\pi a^5.$$

Therefore $\hat{x} = \hat{y} = \frac{1}{10}a\sqrt{30}$.

13.5.41: The quarter of the circular disk $x^2 + y^2 \leq a^2$ that lies in the first quadrant has the polar-coordinates description $0 \leq r \leq a$, $0 \leq \theta \leq \frac{1}{2}\pi$. If we assume that it has uniform density $\delta = 1$, then its mass is $m = \frac{1}{4}\pi a^2$. Next,

$$M_y = \int_0^{\pi/2} \int_0^a r^2 \cos\theta \, dr \, d\theta = \int_0^{\pi/2} \frac{1}{3}a^3 \cos\theta \, d\theta = \left[\frac{1}{3}a^2 \sin\theta \right]_0^{\pi/2} = \frac{1}{3}a^3,$$

and $M_x = M_y$ by symmetry. Therefore the centroid is located at the point

$$(\overline{x}, \overline{y}) = \left(\frac{4a}{3\pi}, \frac{4a}{3\pi} \right) \approx ([0.4244131815783876]a, [0.4244131815783876]a).$$

13.5.43: The arc $x^2 + y^2 = r^2$ can be parametrized in this way:

$$x(t) = r\cos\theta, \quad y(t) = r\sin\theta, \quad 0 \leq \theta \leq \frac{\pi}{2}.$$

We may also assume that it has unit density. The arc length element is

$$ds = \sqrt{[x'(\theta)]^2 + [y'(\theta)]^2} \; d\theta = r \, d\theta,$$

and hence its mass is

$$m = \int_0^{\pi/2} r \, d\theta = \frac{1}{2}\pi r.$$

Its moment around the y-axis is

$$M_y = \int_0^{\pi/2} r^2 \cos\theta \; d\theta = \left[r^2 \sin\theta \right]_0^{\pi/2} = r^2.$$

Therefore, because $\bar{y} = \bar{x}$ by symmetry,

$$(\bar{x}, \bar{y}) \; = \; \left(\frac{2r}{\pi}, \frac{2r}{\pi} \right).$$

13.5.45: We may assume that the triangle has unit density. The hypotenuse of this right triangle has equation

$$y = f(x) = h\left(1 - \frac{x}{r}\right),$$

and hence the mass and moments of the triangle are

$$m = \int_0^r \int_0^{f(x)} 1 \; dy \; dx = \int_0^r h\left(1 - \frac{x}{r}\right) dx = \left[hx - \frac{h}{2r}x^2 \right]_0^r = \frac{1}{2}hr,$$

$$M_y = \int_0^r \int_0^{f(x)} x \; dy \; dx = \int_0^r hx\left(1 - \frac{x}{r}\right) dx = \left[\frac{h}{2}x^2 - \frac{h}{3r}x^3 \right]_0^r = \frac{1}{6}hr^2, \quad \text{and}$$

$$M_x = \int_0^r \int_0^{f(x)} y \; dy \; dx = \int_0^r \left(\frac{1}{2}h^2 - \frac{h^2}{r}x + \frac{h^2}{2r^2}x^2 \right) dx$$

$$= \left[\frac{h^2}{2}x - \frac{h^2}{2r}x^2 + \frac{h^2}{6r^2}x^3 \right]_0^r = \frac{1}{6}h^2 r.$$

Therefore the centroid of the triangle is

$$C = (\bar{x}, \bar{y}) \; = \; \left(\frac{1}{3}r, \frac{1}{3}h \right).$$

The midpoint of the hypotenuse is the point $M\left(\frac{1}{2}r, \frac{1}{2}h\right)$ and the line from the origin to M has equation $y = hx/r$, so it is clear that C lies on this line and is two-thirds of the way from the origin to M.

13.5.47: Here we simply observe that $2\pi\dfrac{r}{2}\cdot L = \pi r L$.

13.5.49: The diagonal side of the trapezoid has centroid at its midpoint, so that $\overline{x} = \frac{1}{2}(r_1 + r_2)$. By the second theorem of Pappus, the curved surface area of the conical frustum—which is generated by the diagonal side—is

$$A = 2\pi\overline{x}\cdot L = \pi L(r_1 + r_2).$$

13.5.51: First we compute the mass and moments of the rectangle (we assume that it has unit density):

$$m = \int_{-a}^{a}\int_{0}^{b} 1\, dy\, dx = \int_{-a}^{a} b\, dx = 2ab;$$

$$M_y = \int_{-a}^{a}\int_{0}^{b} x\, dy\, dx = \int_{-a}^{a} bx\, dx = \left[\frac{1}{2}bx^2\right]_{-a}^{a} = 0;$$

$$M_x = \int_{-a}^{a}\int_{0}^{b} y\, dy\, dx = \int_{-a}^{a} \frac{1}{2}b^2\, dx = ab^2.$$

Next we compute the mass and moments of the semicircle. Note that its curved boundary has equation $y = b + \sqrt{a^2 - x^2} = f(x)$. Because we have assumed unit density, its mass is $m = \frac{1}{2}\pi a^2$. By symmetry, $M_y = 0$. And

$$M_x = \int_{-a}^{a}\int_{b}^{f(x)} y\, dy\, dx = \int_{-a}^{a}\left[\frac{1}{2}\left(b + \sqrt{a^2 - x^2}\right)^2 - \frac{1}{2}b^2\right] dx$$

$$= \int_{-a}^{a}\left[b(a^2 - x^2)^{1/2} + \frac{1}{2}(a^2 - x^2)\right] dx = b\int_{-a}^{a}(a^2 - x^2)^{1/2}\, dx + \left[\frac{1}{2}a^2 x - \frac{1}{6}x^3\right]_{-a}^{a}$$

$$= b\cdot\frac{\pi a^2}{2} + a^3 - \frac{1}{3}a^3 = \frac{3\pi a^2 b + 4a^3}{6}.$$

Moments, like masses, are additive. Thus we find the mass and moments for the entire lamina by adding those of the rectangle and semicircle:

$$m = \frac{4ab + \pi a^2}{2}, \quad M_y = 0, \quad \text{and} \quad M_x = \frac{6ab^2 + 3\pi a^2 b + 4a^3}{6}.$$

Therefore $\overline{x} = 0$ and

$$\overline{y} = \frac{6ab^2 + 3\pi a^2 b + 4a^3}{6}\cdot\frac{2}{4ab + \pi a^2} = \frac{6b^2 + 3\pi ab + 4a^2}{12b + 3\pi a}.$$

Finally, we apply the first theorem of Pappus to find the volume of the solid generated by rotation of the lamina around the x-axis:

$$V = 2\pi\overline{y}\cdot\frac{4ab + \pi a^2}{2} = \pi\cdot\frac{6b^2 + 3\pi ab + 4a^2}{12b + 3\pi a}\cdot(4ab + \pi a^2) = \frac{\pi a}{3}\cdot(6b^2 + 3\pi ab + 4a^2).$$

886

To check the answer, note that it has the correct domensions of volume and that in the extreme case $b = 0$, it becomes $\frac{4}{3}\pi a^3$, the correct formula for the volume of a sphere of radius a.

13.5.53: The plate is a rectangle of area ab, area density δ, and mass m, and hence $m = ab\delta$. The moment of inertia with respect to the y-axis is

$$I_y = \int_{-a/2}^{a/2} \int_{-b/2}^{b/2} \delta x^2 \, dy \, dx = \int_{-a/2}^{a/2} \delta b x^2 \, dx = \left[\frac{1}{3} \delta b x^3 \right]_{-a/2}^{a/2} = \frac{1}{12} \delta a^3 b = \frac{1}{12} m a^2.$$

By symmetry or by a very similar computation $I_x = \frac{1}{12} m b^2$. So, by the comment following Eq. (6) of the text,

$$I_0 = I_x + I_y = \frac{1}{12} m(a^2 + b^2).$$

13.5.55: Suppose that the plane lamina L is the union of the two nonoverlapping laminae R and S. Let I_L, I_R, and I_S denote the polar moments of inertia of L, R, and S, respectively. Let $\delta(x, y)$ denote the density of L at the point (x, y). Then

$$I_L = \iint_L (x^2 + y^2)\delta(x, y) \, dA = \iint_R (x^2 + y^2)\delta(x, y) \, dA + \iint_S (x^2 + y^2)\delta(x, y) \, dA = I_R + I_S.$$

Next, let I_1 denote the polar moment of inertia of the lower rectangle in Fig. 14.5.25 and let I_2 denote the polar moment of inertia of the upper rectangle. Then

$$I_1 = \int_{-1}^{1} \int_{0}^{3} (x^2 + y^2) \cdot k \, dy \, dx = k \int_{-1}^{1} (3x^2 + 9) \, dx = k \left[x^3 + 9x \right]_{-1}^{1} = 20k;$$

$$I_2 = \int_{-4}^{4} \int_{3}^{4} (x^2 + y^2) \cdot k \, dy \, dx = k \int_{-4}^{4} \left[x^2 y + \frac{1}{3} y^3 \right]_{3}^{4} \, dx = k \int_{-4}^{4} \left(x^2 + \frac{37}{3} \right) \, dx$$

$$= k \left[\frac{1}{3} x^3 + \frac{37}{3} x \right]_{-4}^{4} = k \cdot \left(\frac{128}{3} + \frac{296}{3} \right) = \frac{424}{3} k.$$

Thus, by the first result in this solution, the polar moment of inertia of the T-shaped lamina is

$$I_0 = I_1 + I_2 = \frac{484}{3} k.$$

13.5.57: The mass and moments are

$$m = \int_{0}^{\pi} \int_{0}^{2\sin\theta} r^2 \sin\theta \, dr \, d\theta = \int_{0}^{\pi} \frac{8}{3} \sin^4\theta \, d\theta = \frac{1}{12} \left[12\theta - 8\sin 2\theta + \sin 4\theta \right]_{0}^{\pi} = \pi;$$

$$M_y = \int_{0}^{\pi} \int_{0}^{2\sin\theta} r^3 \sin\theta \cos\theta \, dr \, d\theta = \int_{0}^{\pi} 4\sin^5\theta \cos\theta \, d\theta = \left[\frac{2}{3} \sin^6\theta \right]_{0}^{\pi} = 0;$$

$$M_x = \int_{0}^{\pi} \int_{0}^{2\sin\theta} r^3 \sin^2\theta \, dr \, d\theta = \int_{0}^{\pi} 4\sin^6\theta \, d\theta = \frac{1}{48} \left[60\theta - 45\sin 2\theta + 9\sin 4\theta - \sin 6\theta \right]_{0}^{\pi} = \frac{5}{4}\pi.$$

887

Therefore the centroid is at $(\overline{x}, \overline{y}) = \left(0, \dfrac{5}{4}\right)$.

13.5.59: The mass and moments are

$$m = \int_0^{\pi/2} \int_0^{2\cos\theta} r^2 \cos\theta \, dr \, d\theta = \int_0^{\pi/2} \frac{8}{3} \cos^4\theta \, d\theta = \frac{1}{12}\left[12\theta + 8\sin 2\theta + \sin 4\theta\right]_0^{\pi/2} = \frac{1}{2}\pi;$$

$$M_y = \int_0^{\pi/2} \int_0^{2\cos\theta} r^3 \cos^2\theta \, dr \, d\theta = \int_0^{\pi/2} 4\cos^6\theta \, d\theta = \frac{1}{48}\left[60\theta + 45\sin 2\theta + 9\sin 4\theta + \sin 6\theta\right]_0^{\pi/2} = \frac{5}{8}\pi;$$

$$M_x = \int_0^{\pi/2} \int_0^{2\cos\theta} r^3 \sin\theta \cos\theta \, dr \, d\theta = \int_0^{\pi/2} 4\sin\theta \cos^5\theta \, d\theta = \left[-\frac{2}{3}\cos^6\theta\right]_0^{\pi/2} = \frac{2}{3}.$$

Therefore the centroid is located at $(\overline{x}, \overline{y}) = \left(\dfrac{5}{4}, \dfrac{4}{3\pi}\right)$.

Section 13.6

Note: For some triple integrals, there are six possible orders of integration of the corresponding iterated integrals; for many triple integrals, there are two. To save space, in this section we do not show all possibilities, but only the one that seems most natural.

13.6.1: The value of the triple integral is

$$I = \int_{z=0}^1 \int_{y=0}^3 \int_{x=0}^2 (x + y + z) \, dx \, dy \, dz = \int_{z=0}^1 \int_{y=0}^3 \left[\frac{1}{2}x^2 + xy + xz\right]_{x=0}^2 dy \, dz$$

$$= \int_{z=0}^1 \int_{y=0}^3 (2 + 2y + 2z) \, dy \, dz = \int_{z=0}^1 \left[2y + y^2 + 2yz\right]_{y=0}^3 dz = \int_0^1 (15 + 6z) \, dz = \left[15z + 3z^2\right]_0^1 = 18.$$

13.6.3: The value of the triple integral is

$$K = \int_{z=-2}^6 \int_{y=0}^2 \int_{x=-1}^3 xyz \, dx \, dy \, dz = \int_{z=-2}^6 \int_{y=0}^2 \left[\frac{1}{2}x^2yz\right]_{x=-1}^3 dy \, dz = \int_{z=-2}^6 \int_{y=0}^2 4yz \, dy \, dz$$

$$= \int_{z=-2}^6 \left[2y^2 z\right]_{y=0}^2 dz = \int_{z=-2}^6 8z \, dz = \left[4z^2\right]_{z=-2}^6 = 128.$$

13.6.5: The value of the triple integral is

$$J = \int_{x=0}^1 \int_{y=0}^{1-x} \int_{z=0}^{1-x-y} x^2 \, dz \, dy \, dx = \int_{x=0}^1 \int_{y=0}^{1-x} \left[x^2 z\right]_{z=0}^{1-x-y} dy \, dx = \int_{x=0}^1 \int_{y=0}^{1-x} (x^2 - x^3 - x^2 y) \, dy \, dx$$

$$= \int_{x=0}^1 \left[x^2 y - x^3 y - \frac{1}{2}x^2 y^2\right]_{y=0}^{1-x} dx = \int_0^1 \left(\frac{1}{2}x^2 - x^3 + \frac{1}{2}x^4\right) dx = \left[\frac{1}{6}x^3 - \frac{1}{4}x^4 + \frac{1}{10}x^5\right]_0^1 = \frac{1}{60}.$$

13.6.7: The value of the triple integral is

$$I = \int_{x=-1}^{0} \int_{y=0}^{2} \int_{z=0}^{1-x^2} xyz \, dz \, dy \, dx = \int_{x=-1}^{0} \int_{y=0}^{2} \left[\frac{1}{2} xyz^2 \right]_{z=0}^{1-x^2} dy \, dx$$

$$= \int_{x=-1}^{0} \int_{y=0}^{2} \left(\frac{1}{2} xy - x^3 y + \frac{1}{2} x^5 y \right) dy \, dx = \int_{x=-1}^{0} \left[\frac{1}{4} xy^2 - \frac{1}{2} x^3 y^2 + \frac{1}{4} x^5 y^2 \right]_{y=0}^{2} dx$$

$$= \int_{x=-1}^{0} (x - 2x^3 + x^5) \, dx = \frac{1}{6} \left[3x^2 - 3x^4 + x^6 \right]_{x=-1}^{0} = -\frac{1}{6}.$$

13.6.9: The value of this triple integral is

$$K = \int_{-1}^{1} \int_{0}^{3} \int_{x^2}^{2-x^2} (x + y) \, dz \, dy \, dx = \int_{-1}^{1} \int_{0}^{3} \left[xy + xz \right]_{x^2}^{2-x^2} dy \, dx$$

$$= \int_{-1}^{1} \int_{0}^{3} (2x - 2x^3 + 2y - 2x^2 y) \, dy \, dx = \int_{-1}^{1} \left[2xy - 2x^3 y + y^2 - x^2 y^2 \right]_{0}^{3} dx$$

$$= \int_{-1}^{1} (9 + 6x - 9x^2 - 6x^3) \, dx = \left[9x + 3x^2 - 3x^3 - \frac{3}{2} x^4 \right]_{-1}^{1} = 12.$$

13.6.11: The solid resembles the tetrahedron shown in Fig. 13.6.4 of the text. You will see it if you execute the *Mathematica* command

```
ParametricPlot3D[ {{x, 0, 6 - 2*x - 3*y}, {x, y, 6 - 2*x - 3*y}},
    {x, 0, 3}, {y, 0, 2}, PlotRange → {0, 3}, {0, 2}, {0, 6}},
    AspectRatio → 1.0, ViewPoint → {3.5, -1.6, 3.0},
    LightSources → {{{1., 0., 1.}, RGBColor[1., 0., 0.]},
    {{1., 1., 1.}, RGBColor[0., 1., 0.]}, {{0., 1., 1.}, RGBColor[1., 0., 1.]}} ];
```

If you change the `Viewpoint` parameters, you will see that we plotted only the diagonal face of the tetrahedron and the face that lies in the xz-plane. The volume of this tetrahedron is given by

$$V = \int_{x=0}^{3} \int_{y=0}^{(6-2x)/3} \int_{z=0}^{6-2x-3y} 1 \, dz \, dy \, dx = \int_{x=0}^{3} \int_{y=0}^{(6-2x)/3} (6 - 2x - 3y) \, dy \, dx$$

$$= \int_{x=0}^{3} \left[6y - 2xy - \frac{3}{2} y^2 \right]_{y=0}^{(6-2x)/3} dx = \int_{x=0}^{3} \left(6 - 4x + \frac{2}{3} x^2 \right) dx = \left[6x - 2x^2 + \frac{2}{9} x^3 \right]_{0}^{3} = 6.$$

13.6.13: The view of the figure on the left below was generated with the *Mathematica* command

```
ParametricPlot3D[ {{u, 0, v}, {u, 4 - u∧2, v}, {u, v, 4 - v}},
    {u, -2, 2}, {v, 0, 4}, AspectRatio → Automatic, Shading → False,
```

PlotRange → {{-2.2, 2.2}, {-0.2, 4.2}, { -0.2, 4.2}}, ViewPoint → {4, 4, 5}];

Change the `ViewPoint` command parameters to {4, -4, 7} to see the figure from another angle (this view is shown above, on the right). The volume of the solid is

$$V = \int_{x=-2}^{2} \int_{y=0}^{4-x^2} \int_{z=0}^{4-y} 1 \, dz \, dy \, dx = \int_{x=-2}^{2} \int_{y=0}^{4-x^2} (4-y) \, dy \, dx = \int_{x=-2}^{2} \left[4y - \frac{y^2}{2} \right]_{y=0}^{4-x^2} dx$$

$$= \int_{-2}^{2} \left(8 - \frac{1}{2} x^4 \right) dx = \left[8x - \frac{1}{10} x^5 \right]_{-2}^{2} = \frac{128}{5} = 25.6.$$

13.6.15: It's not easy to get a clear three-dimensional plot of this solid. We got fair results with the *Mathematica* command

```
ParametricPlot3D[ {{u, u∧2, v}, {u*u, u, v}, {u*Sqrt[10], v, 10 - u*u - v*v}},

    {u, 0, 1}, {v, 0, 10}, AspectRatio → 0.8,

    PlotRange → {{0, Sqrt[10]}, {0, Sqrt[10]}, {0, 10}}, ViewPoint → {6, -4, 12} ];
```

If you have the time and interest, experiment with changing the `AspectRatio` and `ViewPoint` parameters until you get a clear view of the solid. Its volume is

$$V = \int_{x=0}^{1} \int_{y=x^2}^{\sqrt{x}} \int_{z=0}^{10-x^2-y^2} 1 \, dz \, dy \, dx = \int_{x=0}^{1} \int_{y=x^2}^{\sqrt{x}} (10 - x^2 - y^2) \, dy \, dx$$

$$= \int_{x=0}^{1} \left[\frac{1}{3} (30y - 3x^2 y - y^3) \right]_{y=x^2}^{y=\sqrt{x}} dx = \int_{0}^{1} \left(10x^{1/2} - \frac{1}{3} x^{3/2} - 10x^2 - x^{5/2} + x^4 + \frac{1}{3} x^6 \right) dx$$

$$= \left[\frac{20}{3} x^{3/2} - \frac{2}{15} x^{5/2} - \frac{10}{3} x^3 - \frac{2}{7} x^{7/2} + \frac{1}{5} x^5 + \frac{1}{21} x^7 \right]_{0}^{1} = \frac{332}{105} \approx 3.1619047619047619.$$

13.6.17: The volume is

$$V = \int_{x=-2}^{2} \int_{z=x^2}^{4} \int_{y=0}^{4-z} 1 \, dy \, dz \, dx = \int_{x=-2}^{2} \int_{z=x^2}^{4} (4-z) \, dz \, dx = \int_{x=-2}^{2} \left[4z - \frac{1}{2} z^2 \right]_{z=x^2}^{4} dx$$

$$= \int_{x=-2}^{2} \left(8 - 4x^2 + \frac{1}{2} x^4 \right) dx = \left[8x - \frac{4}{3} x^3 + \frac{1}{10} x^5 \right]_{-2}^{2} = \frac{256}{15} \approx 17.0666666666666667.$$

13.6.19: The volume is

$$V = \int_{y=0}^{1} \int_{z=y^2}^{\sqrt{y}} \int_{x=0}^{2-y-z} 1 \, dx \, dz \, dy = \int_{y=0}^{1} \int_{z=y^2}^{\sqrt{y}} (2-y-z) \, dz \, dy = \int_{y=0}^{1} \left[2z - yz - \frac{1}{2} z^2 \right]_{z=y^2}^{\sqrt{y}} dy$$

$$= \int_{y=0}^{1} \frac{1}{2} \left(4y^{1/2} - y - 2y^{3/2} - 4y^2 + 2y^3 + y^4 \right) dy$$

$$= \frac{1}{60} \left[80y^{3/2} - 15y^2 - 24y^{5/2} - 40y^3 + 15y^4 + 6y^5 \right]_{0}^{1} = \frac{11}{30} \approx 0.3666666666666667.$$

13.6.21: Because the solid has constant density $\delta = 1$, its mass is

$$m = \int_{x=-2}^{2} \int_{y=x^2}^{4} \int_{z=0}^{y} 1 \, dz \, dy \, dx = \int_{x=-2}^{2} \int_{y=x^2}^{4} y \, dy \, dx = \int_{x=-2}^{2} \left[\frac{1}{2} y^2 \right]_{y=x^2}^{4} dx$$

$$= \int_{x=-2}^{2} \left(8 - \frac{1}{2} x^4 \right) dx = \left[8x - \frac{1}{10} x^5 \right]_{-2}^{2} = \frac{128}{5}.$$

Its moments are

$$M_{yz} = \int_{x=-2}^{2} \int_{y=x^2}^{4} \int_{z=0}^{y} x \, dz \, dy \, dx = \int_{x=-2}^{2} \int_{y=x^2}^{4} xy \, dy \, dx = \int_{x=-2}^{2} \left[\frac{1}{2} xy^2 \right]_{y=x^2}^{4} dx$$

$$= \int_{x=-2}^{2} \left(8x - \frac{1}{2} x^5 \right) dx = \left[4x^2 - \frac{1}{12} x^6 \right]_{-2}^{2} = 0,$$

$$M_{xz} = \int_{x=-2}^{2} \int_{y=x^2}^{4} \int_{z=0}^{y} y \, dz \, dy \, dz = \int_{x=-2}^{2} \int_{y=x^2}^{4} \frac{1}{2} y^2 \, dy \, dx = \int_{x=-2}^{2} \left[\frac{1}{3} y^3 \right]_{y=x^2}^{4} dx$$

$$= \int_{x=-2}^{2} \left(\frac{64}{3} - \frac{1}{3} x^6 \right) dx = \left[\frac{64}{3} x - \frac{1}{21} x^7 \right]_{-2}^{2} = \frac{512}{7},$$

$$M_{xy} = \int_{x=-2}^{2} \int_{y=x^2}^{4} \int_{z=0}^{y} z \, dz \, dy \, dx = \int_{x=-2}^{2} \int_{y=x^2}^{4} \left[\frac{1}{2} z^2 \right]_{z=0}^{y} dy \, dx$$

$$= \frac{1}{2} \int_{x=-2}^{2} \int_{y=x^2}^{4} y^2 \, dy \, dx = \frac{1}{2} \int_{x=-2}^{2} \left[\frac{1}{3} y^3 \right]_{y=x^2}^{4} dx = \int_{x=-2}^{2} \left(\frac{32}{3} - \frac{1}{6} x^6 \right) dx$$

$$= \left[\frac{32}{3} x - \frac{1}{42} x^7 \right]_{-2}^{2} = \frac{256}{7}.$$

Therefore the centroid of the solid is located at the point $(\overline{x}, \overline{y}, \overline{z}) = \left(0, \dfrac{20}{7}, \dfrac{10}{7}\right)$.

13.6.23: Because the solid has unit density $\delta = 1$, its mass and moments are

$$m = \int_{x=-2}^{2} \int_{z=x^2}^{4} \int_{y=0}^{4-z} 1 \, dy \, dz \, dx = \int_{x=-2}^{2} \int_{z=x^2}^{4} (4-z) \, dz \, dx = \int_{x=-2}^{2} \left[4z - \frac{1}{2}z^2\right]_{z=x^2}^{4} dx$$

$$= \int_{-2}^{2} \left(8 - 4x^2 + \frac{1}{2}x^4\right) dx = \left[8x - \frac{4}{3}x^3 + \frac{1}{10}x^5\right]_{-2}^{2} = \frac{256}{15};$$

$$M_{yz} = \int_{x=-2}^{2} \int_{z=x^2}^{4} \int_{y=0}^{4-z} x \, dy \, dz \, dx = \int_{x=-2}^{2} \int_{z=x^2}^{4} (4x - xz) \, dz \, dx = \int_{x=-2}^{2} \left[4xz - \frac{1}{2}xz^2\right]_{z=x^2}^{4} dx$$

$$= \int_{-2}^{2} \left(8x - 4x^3 + \frac{1}{2}x^5\right) dx = \left[4x^2 - x^4 + \frac{1}{12}x^6\right]_{-2}^{2} = 0;$$

$$M_{xz} = \int_{x=-2}^{2} \int_{z=x^2}^{4} \int_{y=0}^{4-z} y \, dy \, dz \, dx = \int_{x=-2}^{2} \int_{z=x^2}^{4} \frac{1}{2}(4-z)^2 \, dz \, dx = \int_{x=-2}^{2} \left[8z - 2z^2 + \frac{1}{6}z^4\right]_{z=x^2}^{4} dx$$

$$= \int_{-2}^{2} \left(\frac{32}{3} - 8x^2 + 2x^4 - \frac{1}{6}x^6\right) dx = \left[\frac{32}{3}x - \frac{8}{3}x^3 + \frac{2}{5}x^5 - \frac{1}{42}x^7\right]_{-2}^{2} = \frac{2048}{105};$$

$$M_{xy} = \int_{x=-2}^{2} \int_{z=x^2}^{4} \int_{y=0}^{4-z} z \, dy \, dz \, dx = \int_{x=-2}^{2} \int_{z=x^2}^{4} (4z - z^2) \, dz \, dx = \int_{x=-2}^{2} \left[2z^2 - \frac{1}{3}z^3\right]_{z=x^2}^{4} dx$$

$$= \int_{-2}^{2} \left(\frac{32}{3} - 2x^4 + \frac{1}{3}x^6\right) dx = \left[\frac{32}{3}x - \frac{2}{5}x^5 + \frac{1}{21}x^7\right]_{-2}^{2} = \frac{1024}{35}.$$

Therefore the centroid of the solid is located at the point $(\overline{x}, \overline{y}, \overline{z}) = \left(0, \dfrac{8}{7}, \dfrac{12}{7}\right)$.

13.6.25: Because the solid has unit density $\delta = 1$, its mass and moments are given by

$$m = \int_{x=-\pi/2}^{\pi/2} \int_{z=0}^{\cos x} \int_{y=0}^{1-z} 1 \, dy \, dz \, dx = \int_{x=-\pi/2}^{\pi/2} \left[z - \frac{1}{2}z^2\right]_{z=0}^{\cos x} dx$$

$$= \int_{-\pi/2}^{\pi/2} \left(\cos x - \frac{1}{2}\cos^2 x\right) dx = \int_{0}^{\pi/2} (2\cos x - \cos^2 x) \, dx = \frac{1}{4}\left[-2x + 8\sin x - \sin 2x\right]_{0}^{\pi/2} = \frac{8-\pi}{4};$$

$$M_{yz} = 0 \quad \text{(by symmetry)};$$

$$M_{xz} = \int_{x=-\pi/2}^{\pi/2} \int_{z=0}^{\cos x} \int_{y=0}^{1-z} y \, dy \, dz \, dx = \int_{x=-\pi/2}^{\pi/2} \int_{z=0}^{\cos x} \frac{1}{2}(1-z)^2 \, dz \, dx$$

$$= \int_{x=-\pi/2}^{\pi/2} \left[\frac{1}{2}z - \frac{1}{2}z^2 + \frac{1}{6}z^3\right]_{z=0}^{\cos x} dx = \int_{-\pi/2}^{\pi/2} \left(\frac{1}{2}\cos x - \frac{1}{2}\cos^2 x + \frac{1}{6}\cos^3 x\right) dx$$

$$= \frac{1}{72}\left[-18x + 45\sin x - 9\sin 2x + \sin 3x\right]_{-\pi/2}^{\pi/2} = \frac{44 - 9\pi}{36};$$

$$M_{xy} = \int_{x=-\pi/2}^{\pi/2} \int_{z=0}^{\cos x} \int_{y=0}^{1-z} z\, dy\, dz\, dx = \int_{x=-\pi/2}^{\pi/2} \int_{z=0}^{\cos x} (z - z^2)\, dz\, dx = \int_{x=-\pi/2}^{\pi/2} \left[\frac{1}{2}z^2 - \frac{1}{3}z^3 \right]_{z=0}^{\cos x} dx$$

$$= \int_{-\pi/2}^{\pi/2} \left(\frac{1}{2}\cos^2 x - \frac{1}{3}\cos^3 x \right) dx = \frac{1}{72} \left[18x - 18\sin x + 9\sin 2x - 2\sin 3x \right]_{-\pi/2}^{\pi/2} = \frac{9\pi - 16}{36}.$$

Therefore the centroid of the solid is located at the point

$$(\bar{x}, \bar{y}, \bar{z}) = \left(0, \frac{44 - 9\pi}{72 - 9\pi}, \frac{9\pi - 16}{72 - 9\pi} \right) \approx (0,\ 0.359643831963,\ 0.280712336074).$$

13.6.27: (See Problem 24.) The moment of inertia of the solid (with density $\delta = 1$) with respect to the y-axis is

$$I_y = \int_{x=-1}^{1} \int_{y=-1}^{1} \int_{z=0}^{1-x^2} (x^2 + z^2)\, dz\, dy\, dx = \int_{x=-1}^{1} \int_{y=-1}^{1} \left[x^2 z + \frac{1}{3}z^3 \right]_{z=0}^{1-x^2} dy\, dx$$

$$= \int_{x=-1}^{1} \int_{-1}^{1} \frac{1}{3}(1 - x^6)\, dy\, dx = \int_{-1}^{1} \frac{2}{3}(1 - x^6)\, dx = \left[\frac{2}{3}x - \frac{2}{21}x^7 \right]_{-1}^{1} = \frac{8}{7}.$$

13.6.29: With unit density $\delta = 1$, the moment of inertia of the given tetrahedron with respect to the z-axis is

$$I_z = \int_{z=0}^{1} \int_{y=0}^{1-z} \int_{x=0}^{1-y-z} (x^2 + y^2)\, dx\, dy\, dz = \int_{z=0}^{1} \int_{y=0}^{1-z} \left[\frac{1}{3}x^3 + xy^2 \right]_{x=0}^{1-y-z} dy\, dz$$

$$= \int_{z=0}^{1} \int_{y=0}^{1-z} \left(\frac{1}{3} - y + 2y^2 - \frac{4}{3}y^3 - z + 2yz - 2y^2 z + z^2 - yz^2 - \frac{1}{3}z^3 \right) dy\, dz$$

$$= \int_{z=0}^{1} \left[-\frac{1}{3}y^4 + \frac{2}{3}y^3(1 - z) - \frac{1}{2}y^2(1 - z)^2 + \frac{1}{3}y(1 - z)^3 \right]_{y=0}^{1-z} dz$$

$$= \int_{0}^{1} \frac{1}{6}(1 - 4z + 6z^2 - 4z^3 + z^4)\, dz = \frac{1}{30} \left[5z - 10z^2 + 10z^3 - 5z^4 + z^5 \right]_{0}^{1} = \frac{1}{30}.$$

13.6.31: It should be clear that $\bar{x} = \bar{y} = 0$ by symmetry, but there is a slight suggestion in the wording of the problem that we should *prove* this more rigorously. Hence we will compute the mass and all three moments. Assuming unit density $\delta = 1$, we have

$$m = \int_{x=-\sqrt{h}}^{\sqrt{h}} \int_{y=-\sqrt{h-x^2}}^{\sqrt{h-x^2}} \int_{z=x^2+y^2}^{h} 1\, dz\, dy\, dx = \int_{x=-\sqrt{h}}^{\sqrt{h}} \int_{y=-\sqrt{h-x^2}}^{\sqrt{h-x^2}} (h - x^2 - y^2)\, dy\, dx$$

$$= \int_{x=-\sqrt{h}}^{\sqrt{h}} \left[hy - x^2 y - \frac{1}{3}y^3 \right]_{y=-\sqrt{h-x^2}}^{\sqrt{h-x^2}} dx = \int_{x=-\sqrt{h}}^{\sqrt{h}} \frac{4}{3}(h - x^2)^{3/2}\, dx.$$

The substitution $x = \sqrt{h}\,\sin\theta$, $dx = \sqrt{h}\,\cos\theta\, d\theta$ then yields

$$m = \int_{\theta=-\pi/2}^{\pi/2} \frac{4}{3}h^2 \cos^4\theta\, d\theta = \frac{1}{24}h^2 \left[12\theta + 8\sin 2\theta + \sin 4\theta \right]_{-\pi/2}^{\pi/2} = \frac{1}{2}\pi h^2.$$

893

Next,

$$M_{yz} = \int_{x=-\sqrt{h}}^{\sqrt{h}} \int_{y=-\sqrt{h-x^2}}^{\sqrt{h-x^2}} \int_{z=x^2+y^2}^{h} x \, dz \, dy \, dx = \int_{x=-\sqrt{h}}^{\sqrt{h}} \int_{y=-\sqrt{h-x^2}}^{\sqrt{h-x^2}} (hx - x^3 - xy^2) \, dy \, dx$$

$$= \int_{x=-\sqrt{h}}^{\sqrt{h}} \left[hxy - x^3y - \frac{1}{3}xy^3 \right]_{y=-\sqrt{h-x^2}}^{\sqrt{h-x^2}} dx = \int_{-\sqrt{h}}^{\sqrt{h}} \frac{4}{3} x(h - x^2)^{3/2} \, dx$$

$$= \left[-\frac{4}{15}(h - x^2)^{5/2} \right]_{-\sqrt{h}}^{\sqrt{h}} = 0;$$

$$M_{xz} = \int_{x=-\sqrt{h}}^{\sqrt{h}} \int_{y=-\sqrt{h-x^2}}^{\sqrt{h-x^2}} \int_{z=x^2+y^2}^{h} y \, dz \, dy \, dx = \int_{x=-\sqrt{h}}^{\sqrt{h}} \int_{y=-\sqrt{h-x^2}}^{\sqrt{h-x^2}} (hy - x^2y - y^3) \, dy \, dx$$

$$= \int_{x=-\sqrt{h}}^{\sqrt{h}} \left[\frac{1}{2}hy^2 - \frac{1}{2}x^2y^2 - \frac{1}{4}y^4 \right]_{y=-\sqrt{h-x^2}}^{\sqrt{h-x^2}} dx = \int_{-\sqrt{h}}^{\sqrt{h}} 0 \, dx = 0;$$

$$M_{xy} = \int_{x=-\sqrt{h}}^{\sqrt{h}} \int_{y=-\sqrt{h-x^2}}^{\sqrt{h-x^2}} \int_{z=x^2+y^2}^{h} z \, dz \, dy \, dx = \int_{x=-\sqrt{h}}^{\sqrt{h}} \int_{y=-\sqrt{h-x^2}}^{\sqrt{h-x^2}} \left(\frac{1}{2}h^2 - \frac{1}{2}(x^2 + y^2)^2 \right) \, dy \, dx.$$

Now rewrite the last integral in polar coordinates. Then

$$M_{xy} = \int_{\theta=0}^{2\pi} \int_{r=0}^{\sqrt{h}} \frac{1}{2}(h^2 - r^4) \cdot r \, dr \, d\theta = \int_{\theta=0}^{2\pi} \left[\frac{1}{4}h^2r^2 - \frac{1}{12}r^6 \right]_{r=0}^{\sqrt{h}} d\theta = 2\pi \cdot \frac{1}{6}h^3 = \frac{1}{3}\pi h^3.$$

Thus the centroid of the parabolic segment is located at the point

$$(\overline{x}, \overline{y}, \overline{z}) = \left(0, 0, \frac{2}{3}h \right),$$

and therefore the centroid is on the axis of symmetry of the segment, two-thirds of the way from the vertex $(0, 0, 0)$ to the base.

13.6.33: Place the cube in the first octant with three of its faces in the coordinate planes, one vertex at $(0, 0, 0)$, and the opposite vertex at (a, a, a). With density $\delta = 1$, its moment of inertia with respect to the z-axis is

$$I_z = \int_{z=0}^{a} \int_{y=0}^{a} \int_{x=0}^{a} (x^2 + y^2) \, dx \, dy \, dz = \int_{0}^{a} \int_{0}^{a} \left[\frac{1}{3}x^3 + xy^2 \right]_{0}^{a} dy \, dz = \int_{0}^{a} \int_{0}^{a} \left(\frac{1}{3}a^3 + ay^2 \right) dy \, dz$$

$$= \int_{0}^{a} \left[\frac{1}{3}a^3y + \frac{1}{3}ay^3 \right]_{0}^{a} dz = \int_{0}^{a} \frac{2}{3}a^4 \, dz = a \cdot \frac{2}{3}a^4 = \frac{2}{3}a^5.$$

Because the mass of the cube is $m = a^3$, we see that $I_z = \frac{2}{3}ma^2$, which is dimensionally correct.

13.6.35: With density $\delta(x, y, z) = k(x^2 + y^2 + z^2)$ at the point (x, y, z) (k is a positive constant), the moment of inertia of the cube of Problem 34 with respect to the z-axis is

$$I_z = \int_{0}^{a} \int_{0}^{a} \int_{0}^{a} k(x^2 + y^2 + z^2)(x^2 + y^2) \, dz \, dy \, dx$$

$$= k \int_0^a \int_0^a \left[x^4 z + 2x^2 y^2 z + y^4 z + \frac{1}{3} x^2 z^3 + \frac{1}{3} y^2 z^3 \right]_0^a \, dy \, dx$$

$$= k \int_0^a \int_0^a \left(\frac{1}{3} a^3 x^2 + a x^4 + \frac{1}{3} a^3 y^2 + 2a x^2 y^2 + a y^4 \right) \, dy \, dx$$

$$= k \int_0^a \left[\frac{1}{3} a^3 x^2 y + a x^4 y + \frac{1}{9} a^3 y^3 + \frac{2}{3} a x^2 y^3 + \frac{1}{5} a y^5 \right]_0^a \, dx$$

$$= k \int_0^a \left(\frac{14}{45} a^6 + a^4 x^2 + a^2 x^4 \right) \, dx = k \left[\frac{14}{45} a^6 x + \frac{1}{3} a^4 x^3 + \frac{1}{5} a^2 x^5 \right]_0^a = \frac{38}{45} k a^7.$$

13.6.37: The moment of inertia of the cube of Problem 36 with respect to the z-axis is

$$I_z = \int_0^1 \int_0^1 \int_0^1 k(x^2 + y^2) z \, dz \, dy \, dx = \int_0^1 \int_0^1 \frac{1}{2} k(x^2 + y^2) \, dy \, dx$$

$$= \int_0^1 \left[\frac{1}{2} k x^2 y + \frac{1}{6} k y^3 \right]_0^1 \, dx = \int_0^1 \left(\frac{1}{2} k x^2 + \frac{1}{6} k \right) \, dx = k \cdot \left[\frac{1}{6} x^3 + \frac{1}{6} x \right]_0^1 = \frac{1}{3} k.$$

13.6.39: With constant density $\delta = 1$, the mass and moments are

$$m = \int_{z=0}^1 \int_{y=0}^{\sqrt{1-z^2}} \int_{x=0}^{\sqrt{1-z^2}} 1 \, dx \, dy \, dz = \int_{z=0}^1 \int_{y=0}^{\sqrt{1-z^2}} \sqrt{1-z^2} \, dy \, dz$$

$$= \int_{z=0}^1 (1 - z^2) \, dz = \left[z - \frac{1}{3} z^3 \right]_0^1 = \frac{2}{3};$$

$$M_{yz} = \int_{z=0}^1 \int_{y=0}^{\sqrt{1-z^2}} \int_{x=0}^{\sqrt{1-z^2}} x \, dx \, dy \, dz = \int_{z=0}^1 \int_{y=0}^{\sqrt{1-z^2}} \frac{1}{2}(1 - z^2) \, dy \, dz$$

$$= \int_{z=0}^1 \frac{1}{2}(1 - z^2)^{3/2} \, dz = \left[\frac{5z - 2z^3}{16} \cdot (1 - z^2)^{1/2} + \frac{3}{16} \arcsin z \right]_0^1 = \frac{3}{32} \pi;$$

$$M_{xz} = \int_{z=0}^1 \int_{y=0}^{\sqrt{1-z^2}} \int_{x=0}^{\sqrt{1-z^2}} y \, dx \, dy \, dz = \int_{z=0}^1 \int_{y=0}^{\sqrt{1-z^2}} y(1 - z^2)^{1/2} \, dy \, dz$$

$$= \int_{z=0}^1 \frac{1}{2}(1 - z^2)^{3/2} \, dz = \left[\frac{5z - 2z^3}{16} \cdot (1 - z^2)^{1/2} + \frac{3}{16} \arcsin z \right]_0^1 = \frac{3}{32} \pi;$$

$$M_{xy} = \int_{z=0}^1 \int_{y=0}^{\sqrt{1-z^2}} \int_{x=0}^{\sqrt{1-z^2}} z \, dx \, dy \, dz = \int_{z=0}^1 \int_{y=0}^{\sqrt{1-z^2}} z(1 - z^2)^{1/2} \, dy \, dz$$

$$= \int_{z=0}^1 (z - z^3) \, dz = \left[\frac{1}{2} z^2 - \frac{1}{4} z^4 \right]_0^1 = \frac{1}{4}.$$

Therefore the centroid of the solid is located at the point $\left(\dfrac{9}{64} \pi, \ \dfrac{9}{64} \pi, \ \dfrac{3}{8} \right)$.

13.6.41: The given solid projects onto the circular disk D with radius 2 and center $(0, 0)$ in the xy-plane. Hence the volume of the solid is

$$V = \iint_D \left(\int_{z=2x^2+y^2}^{12-x^2-2y^2} 1 \, dz \right) dA = \iint_D (12 - 3x^2 - 3y^2) \, dA.$$

Rewrite the last double integral as an iterated integral in polar coordinates. Thus

$$V = \int_{\theta=0}^{2\pi} \int_{r=0}^{2} (12 - 3r^2) \cdot r \, dr \, d\theta = 2\pi \cdot \left[6r^2 - \frac{3}{4}r^4 \right]_0^2 = 24\pi \approx 75.3982236861550377.$$

13.6.43: Following the *Suggestion*, the volume is

$$V = \int_{z=0}^{1} \int_{y=-z/2}^{z/2} \int_{x=-\sqrt{z^2-4y^2}}^{\sqrt{z^2-4y^2}} 1 \, dx \, dy \, dz = \int_{z=0}^{1} \int_{y=-z/2}^{z/2} 2\sqrt{z^2 - 4y^2} \, dy \, dz.$$

Let $y = \frac{1}{2} z \sin u$, $dy = \frac{1}{2} z \cos u \, du$. This substitution yields

$$V = \int_{z=0}^{1} \int_{u=-\pi/2}^{\pi/2} z^2 \cos^2 u \, du \, dz = \int_{z=0}^{1} \left[\frac{1}{4} z^2 (2u + \sin 2u) \right]_{u=-\pi/2}^{\pi/2} dz$$

$$= \int_0^1 \frac{1}{2} \pi z^2 \, dz = \left[\frac{1}{6} \pi z^3 \right]_0^1 = \frac{1}{6} \pi \approx 0.5235987755982989.$$

Methods of single-variable calculus also succeed here. A horizontal cross section of the solid at $z = h$ $(0 < h \leq 1)$ is an ellipse with equation $x^2 + 4y^2 = h^2$. This ellipse has major semiaxis of length h and minor semiaxis of length $\frac{1}{2} h$. Therefore its area is $\frac{1}{2} \pi h^2$. So, by the method of parallel cross sections (see Eq. (3) of Section 6.2), the volume of the solid is

$$\int_0^1 \frac{1}{2} \pi h^2 \, dh = \left[\frac{1}{6} \pi h^3 \right]_0^1 = \frac{1}{6} \pi.$$

13.6.45: If the pyramid (tetrahedron) of Example 2 has density z at the point (x, y, z), then its mass and moments are

$$m = \int_{x=0}^{2} \int_{y=0}^{(6-3x)/2} \int_{z=0}^{6-3x-2y} z \, dz \, dy \, dx = \int_0^2 \int_0^{(6-3x)/2} \frac{1}{2} (6 - 3x - 2y)^2 \, dy \, dx$$

$$= \int_0^2 \left[18y - 18xy + \frac{9}{2}x^2 y - 6y^2 + 3xy^2 + \frac{2}{3}y^3 \right]_0^{(6-3x)/2} dx$$

$$= \int_0^2 \left[9(6 - 3x) - \frac{3}{2}(6 - 3x)^2 + \frac{1}{12}(6 - 3x)^3 - 9x(6 - 3x) + \frac{3}{4}x(6 - 3x)^2 + \frac{9}{4}x^2(6 - 3x) \right] dx$$

$$= \int_0^2 \left(18 - 27x + \frac{27}{2}x^2 - \frac{9}{4}x^3 \right) dx = \left[18x - \frac{27}{2}x^2 + \frac{9}{2}x^3 - \frac{9}{16}x^4 \right]_0^2 = 9;$$

$$M_{yz} = \int_{x=0}^{2} \int_{y=0}^{(6-3x)/2} \int_{z=0}^{6-3x-2y} xz \, dz \, dy \, dx = \int_{0}^{2} \int_{0}^{(6-3x)/2} \frac{1}{2} x(6 - 3x - 2y)^2 \, dy \, dx$$

$$= \int_{0}^{2} \int_{0}^{(6-3x)/2} \left(18x - 18x^2 + \frac{9}{2}x^3 - 12xy + 6x^2 y + 2xy^2 \right) dy \, dx$$

$$= \int_{0}^{2} \left[\frac{9}{2}xy(x-2)^2 + 3xy^2(x-2) + \frac{2}{3}xy^3 \right]_{0}^{(6-3x)/2} dx$$

$$= \int_{0}^{2} \left[9x(6-2x) - \frac{3}{2}x(6-3x)^2 + \frac{1}{12}x(6-3x)^3 - 9x^2(6-3x) + \frac{3}{4}x^2(6-3x)^2 + \frac{9}{4}x^3(6-3x) \right] dx$$

$$= \int_{0}^{2} \left(18x - 27x^2 + \frac{27}{2}x^3 - \frac{9}{4}x^4 \right) dx = \left[9x^2 - 9x^3 + \frac{27}{8}x^4 - \frac{9}{20}x^5 \right]_{0}^{2} = \frac{18}{5};$$

$$M_{xz} = \int_{x=0}^{2} \int_{y=0}^{(6-3x)/2} \int_{z=0}^{6-3x-2y} yz \, dz \, dy \, dx = \int_{0}^{2} \int_{0}^{(6-3x)/2} \frac{1}{2} y(6 - 3x - 2y)^2 \, dy \, dx$$

$$= \int_{0}^{2} \int_{0}^{(6-3x)/2} \left(18y - 18xy + \frac{9}{2}x^2 y - 12y^2 + 6xy^2 + 2y^3 \right) dy \, dx$$

$$= \int_{0}^{2} \left[9y^2 - 9xy^2 + \frac{9}{4}x^2 y^2 - 4y^3 + 2xy^3 + \frac{1}{2}y^4 \right]_{0}^{(6-3x)/2} dx$$

$$= \int_{0}^{2} \left[\frac{9}{4}(6-3x)^2 - \frac{1}{2}(6-3x)^3 + \frac{1}{32}(6-3x)^4 - \frac{9}{4}x(6-3x)^2 + \frac{1}{4}x(6-3x)^3 + \frac{9}{16}x^2(6-3x)^2 \right] dx$$

$$= \int_{0}^{2} \left(\frac{27}{2} - 27x + \frac{81}{4}x^2 - \frac{27}{4}x^3 + \frac{27}{32}x^4 \right) dx = \left[\frac{27}{2}x - \frac{27}{2}x^2 + \frac{27}{4}x^3 - \frac{27}{16}x^4 + \frac{27}{160}x^5 \right]_{0}^{2} = \frac{27}{5};$$

$$M_{xy} = \int_{x=0}^{2} \int_{y=0}^{(6-3x)/2} \int_{z=0}^{6-3x-2y} z^2 \, dz \, dy \, dx = \int_{0}^{2} \int_{0}^{(6-3x)/2} \frac{1}{3}(6 - 3x - 2y)^3 \, dy \, dx$$

$$= \int_{0}^{2} \int_{0}^{(6-3x)/2} \left(72 - 108x + 54x^2 - 9x^3 - 72y + 72xy - 18x^2 y + 24y^2 - 12xy^2 - \frac{8}{3}y^3 \right) dy \, dx$$

$$= \int_{0}^{2} \left[72y - 108xy + 54x^2 y - 9x^3 y - 36y^2 + 36xy^2 - 9x^2 y^2 + 8y^3 - 4xy^3 - \frac{2}{3}y^4 \right]_{y=0}^{(6-3x)/2} dx$$

$$= \int_{0}^{2} \left(54 - 108x + 81x^2 - 27x^3 + \frac{27}{8}x^4 \right) dx = \left[54x - 54x^2 + 27x^3 - \frac{27}{4}x^4 + \frac{27}{40}x^5 \right]_{0}^{2} = \frac{108}{5}.$$

Therefore the centroid of the pyramid is located at the point $\left(\dfrac{2}{5}, \dfrac{3}{5}, \dfrac{12}{5} \right)$.

13.6.47: First we compute

$$\int_{x=0}^{2} \int_{y=0}^{(6-3x)/2} \int_{z=0}^{6-3x-2y} z \, dz \, dy \, dx = \int_{0}^{2} \int_{0}^{(6-3x)/2} \frac{1}{2}(6 - 3x - 2y)^2 \, dy \, dx$$

$$= \int_{0}^{2} \int_{0}^{(6-3x)/2} \left(18 - 18x + \frac{9}{2}x^2 - 12y + 6xy + 2y^2 \right) dy \, dx$$

$$= \int_0^2 \left[18y - 18xy + \frac{9}{2}x^2y - 6y^2 + 3xy^2 + \frac{2}{3}y^3 \right]_0^{(6-3x)/2} dx$$

$$= \int_0^2 \left(18 - 27x + \frac{27}{2}x^2 - \frac{9}{4}x^3 \right) dx = \left[18x - \frac{27}{2}x^2 + \frac{9}{2}x^3 - \frac{9}{16}x^4 \right]_0^2 = 9.$$

The volume of the pyramid is 6, and hence the average value of the density function $\delta(x, y, z) = z$ on the pyramid is $\bar{\delta} = \frac{9}{6} = \frac{3}{2}$.

13.6.49: The centroid of the cube of Problem 48 is, by symmetry, its midpoint $\left(\frac{1}{2}, \frac{1}{2}, \frac{1}{2} \right)$. Because the cube has volume 1, the average value of

$$g(x, y, z) = \left(x - \frac{1}{2} \right)^2 + \left(y - \frac{1}{2} \right)^2 + \left(z - \frac{1}{2} \right)^2$$

on the cube is

$$\bar{g} = \int_{x=0}^1 \int_{y=0}^1 \int_{z=0}^1 g(x, y, z)\, dz\, dy\, dx = \int_0^1 \int_0^1 \left[\frac{3}{4}z - xz + x^2z - yz + y^2z - \frac{1}{2}z^2 + \frac{1}{3}z^3 \right]_0^1 dy\, dx$$

$$= \int_0^1 \int_0^1 \left(\frac{7}{12} - x + x^2 - y + y^2 \right) dy\, dx = \int_0^1 \left[\frac{7}{12}y - xy + x^2y - \frac{1}{2}y^2 + \frac{1}{3}y^3 \right]_0^1 dx$$

$$= \int_0^1 \left(\frac{5}{12} - x + x^2 \right) dx = \left[\frac{5}{12}x - \frac{1}{2}x^2 + \frac{1}{3}x^3 \right]_0^1 = \frac{1}{4}.$$

13.6.51: First we compute

$$J = \int_{x=0}^2 \int_{y=0}^{(6-3x)/2} \int_{z=0}^{6-3x-2y} (x^2 + y^2 + z^2)\, dz\, dy\, dx = \int_0^2 \int_0^{(6-3x)/2} \left[x^2z + y^2z + \frac{1}{3}z^3 \right]_0^{6-3x-2y} dy\, dx$$

$$= \int_0^2 \int_0^{(6-3x)/2} \left(72 - 108x + 60x^2 - 12x^3 - 72y + 72xy - 20x^2y + 30y^2 - 15xy^2 - \frac{14}{3}y^3 \right) dy\, dx$$

$$= \int_0^2 \left[72y - 108xy + 60x^2y - 12x^3y - 36y^2 + 36xy^2 - 10x^2y^2 + 10y^3 - 5xy^3 - \frac{7}{6}y^4 \right]_0^{(6-3x)/2} dx$$

$$= \int_0^2 \left(\frac{135}{2} - 135x + \frac{441}{4}x^2 - \frac{171}{4}x^3 + \frac{207}{32}x^4 \right) dx = \left[\frac{135}{2}x - \frac{135}{2}x^2 + \frac{147}{4}x^3 - \frac{171}{16}x^4 + \frac{207}{160}x^5 \right]_0^2$$

$$= \frac{147}{5}.$$

Because the pyramid has volume $V = 6$, the average squared distance of its points from its centroid is

$$\bar{d} = \frac{J}{V} = \frac{147}{30} = \frac{49}{10} = 4.9.$$

898

13.6.53: Using *Mathematica* for each single-integration step, we find that the average distance of points of the cube of Problem 48 from the origin is

$$\bar{d} = \int_{x=0}^{1} \int_{y=0}^{1} \int_{z=0}^{1} \sqrt{x^2 + y^2 + z^2} \; dz \, dy \, dx$$

$$= \int_0^1 \int_0^1 \left[\frac{1}{2} z\sqrt{x^2 + y^2 + z^2} + \frac{1}{2}(x^2 + y^2) \ln\left(z + \sqrt{x^2 + y^2 + z^2} \right) \right]_0^1 dy \, dx$$

$$= \int_0^1 \int_0^1 \left[\frac{1}{2}\sqrt{x^2 + y^2 + 1} - \frac{1}{2}(x^2 + y^2) \ln\left(\sqrt{x^2 + y^2} \right) + \frac{1}{2}(x^2 + y^2)\ln\left(1 + \sqrt{x^2 + y^2 + 1} \right) \right] dy \, dx$$

$$= \int_0^1 \left[\frac{1}{3} y\sqrt{x^2 + y^2 + 1} - \frac{1}{3} x^2 \arctan\left(\frac{y}{x\sqrt{x^2 + y^2 + 1}} \right) - \frac{1}{6} y(3x^2 + y^2) \ln\left(\sqrt{x^2 + y^2} \right) \right.$$

$$\left. + \frac{1}{6} y(3x^2 + y^2)\ln\left(1 + \sqrt{x^2 + y^2 + 1} \right) + \frac{1}{6}(3x^2 + 1)\ln\left(y + \sqrt{x^2 + y^2 + 1} \right) \right]_0^1 dx$$

$$= \int_0^1 \left[\frac{1}{3}\sqrt{x^2 + 2} - \frac{1}{3} x^3 \arctan\left(\frac{1}{x\sqrt{x^2 + 2}} \right) - \frac{1}{3}(3x^2 + 1)\ln\left(\sqrt{x^2 + 1} \right) \right.$$

$$\left. + \frac{1}{3}(3x^2 + 1)\ln\left(1 + \sqrt{x^2 + 2} \right) \right] dx$$

$$= \left[\frac{1}{4} x\sqrt{x^2 + 2} + \frac{1}{3}\operatorname{arcsinh}\left(\frac{x}{\sqrt{2}} \right) - \frac{1}{12} x^4 \arctan\left(\frac{1}{x\sqrt{x^2 + 2}} \right) \right.$$

$$\left. - \frac{1}{6}\arctan\left(\frac{x}{\sqrt{x^2 + 2}} \right) - \frac{1}{3} x(x^2 + 1)\ln\left(\sqrt{x^2 + 1} \right) + \frac{1}{3} x(x^2 + 1)\ln\left(1 + \sqrt{x^2 + 2} \right) \right]_0^1$$

$$= \frac{1}{24}\left[6\sqrt{3} - \pi + 12\ln\left(2 + \sqrt{3} \right) \right] \approx 0.960591956455052959425108.$$

Section 13.7

13.7.1: The volume is

$$V = \int_{\theta=0}^{2\pi} \int_{r=0}^{2} \int_{z=r^2}^{4} r \, dz \, dr \, d\theta = 2\pi \int_{r=0}^{2} (4r - r^3) \, dr = 2\pi \left[2r^2 - \frac{1}{4} r^4 \right]_0^2 = 8\pi.$$

13.7.3: Place the center of the sphere at the origin. Then its volume is

$$V = \int_{\theta=0}^{2\pi} \int_{r=0}^{a} \int_{z=-\sqrt{a^2-r^2}}^{\sqrt{a^2-r^2}} r \, dz \, dr \, d\theta = 2\pi \int_{r=0}^{a} 2r\sqrt{a^2 - r^2} \; dr = 2\pi \cdot \left[-\frac{2}{3}(a^2 - r^2)^{3/2} \right]_0^a = \frac{4}{3}\pi a^3.$$

13.7.5: The volume is

$$V = \int_{\theta=0}^{2\pi} \int_{r=0}^{1} \int_{z=-\sqrt{4-r^2}}^{\sqrt{4-r^2}} r \, dz \, dr \, d\theta = 2\pi \int_{r=0}^{1} 2r(4 - r^2)^{1/2} \, dr = 2\pi \cdot \left[-\frac{2}{3}(4 - r^2)^{3/2} \right]_0^1$$

$$= 2\pi\left(\frac{16}{3} - 2\sqrt{3} \right) = \frac{4}{3}\pi\left(8 - 3\sqrt{3} \right) \approx 11.7447292674805137.$$

13.7.7: The mass of the cylinder is

$$m = \int_{\theta=0}^{2\pi} \int_{r=0}^{a} \int_{z=0}^{h} rz \, dz \, dr \, d\theta = 2\pi \int_{r=0}^{a} \frac{1}{2} rh^2 \, dr = 2\pi \cdot \left[\frac{1}{4} r^2 h^2 \right]_0^a = \frac{1}{2} \pi a^2 h^2.$$

13.7.9: The moment of inertia of the cylinder of Problem 7 with respect to the z-axis is

$$I_z = \int_{\theta=0}^{2\pi} \int_{r=0}^{a} \int_{z=0}^{h} r^3 z \, dz \, dr \, d\theta = 2\pi \int_{r=0}^{a} \left[\frac{1}{2} r^3 z^2 \right]_{z=0}^{h} dr$$

$$= 2\pi \int_{r=0}^{a} \frac{1}{2} r^3 h^2 \, dr = 2\pi \cdot \left[\frac{1}{8} r^4 h^2 \right]_{r=0}^{a} = 2\pi \cdot \frac{1}{8} a^4 h^2 = \frac{1}{4} \pi a^4 h^2.$$

13.7.11: The volume is

$$V = \int_{\theta=0}^{2\pi} \int_{r=0}^{3} \int_{z=0}^{9-r^2} r \, dz \, dr \, d\theta = 2\pi \int_{r=0}^{3} (9r - r^3) \, dr = 2\pi \cdot \left[\frac{9}{2} r^2 - \frac{1}{4} r^4 \right]_0^3 = \frac{81}{2} \pi.$$

By symmetry, $\bar{x} = \bar{y} = 0$. The moment of the solid with respect to the xy-plane is

$$M_{xy} = \int_{\theta=0}^{2\pi} \int_{r=0}^{3} \int_{z=0}^{9-r^2} rz \, dz \, dr \, d\theta = 2\pi \int_{r=0}^{3} \frac{1}{2} r (9 - r^2)^2 \, dr$$

$$= 2\pi \int_{r=0}^{3} \left(\frac{81}{2} r - 9r^3 + \frac{1}{2} r^5 \right) dr = 2\pi \cdot \left[\frac{81}{4} r^2 - \frac{9}{4} r^4 + \frac{1}{12} r^6 \right]_0^3 = \frac{243}{2} \pi.$$

Therefore the z-coordinate of the centroid is $\bar{z} = 3$. *Suggestion:* Compare this answer with the answer obtained by using the result in Problem 31 of Section 13.6.

13.7.13: The curve formed by the intersection of the paraboloids lies on the cylinder $x^2 + y^2 = 4$, and hence the solid projects vertically onto the disk D with boundary $x^2 + y^2 = 4$ in the xy-plane. Therefore the volume of the solid is

$$V = \int_{\theta=0}^{2\pi} \int_{r=0}^{2} \int_{z=r^2+r^2\cos^2\theta}^{12-r^2-r^2\sin^2\theta} r \, dz \, dr \, d\theta = \int_{\theta=0}^{2\pi} \int_{r=0}^{2} (12r - 3r^3) \, dr \, d\theta = \int_{\theta=0}^{2\pi} \left[6r^2 - \frac{3}{4} r^4 \right]_{r=0}^{2} d\theta$$

$$= \int_0^{2\pi} 12 \, d\theta = 2\pi \cdot 12 = 24\pi \approx 75.3982236861550377.$$

13.7.15: The spherical surface $r^2 + z^2 = 2$ and the paraboloid $z = r^2$ meet in a horizontal circle that projects vertically onto the circle $x^2 + y^2 = 1$ in the xy-plane. Hence the volume between the two surfaces is

$$V = \int_{\theta=0}^{2\pi} \int_{r=0}^{1} \int_{z=r^2}^{\sqrt{2-r^2}} r \, dz \, dr \, d\theta = \int_0^{2\pi} \int_0^1 \left[r(2 - r^2)^{1/2} - r^3 \right] dr \, d\theta = 2\pi \cdot \left[-\frac{1}{3} (2 - r^2)^{3/2} - \frac{1}{4} r^4 \right]_0^1$$

$$= 2\pi \cdot \left(\frac{2}{3} \sqrt{2} - \frac{7}{12} \right) = \frac{1}{6} \pi \left(8\sqrt{2} - 7 \right) \approx 2.2586524883563962.$$

13.7.17: Set up a coordinate system in which the points of the cylinder are described by

$$0 \le r \le a, \quad 0 \le \theta \le 2\pi, \quad 0 \le z \le h.$$

Because the cylinder has constant density δ, its mass is $m = \pi \delta a^2 h$. One diameter of its base coincides with the x-axis, so we will find the moment of inertia I of the cylinder with respect to that axis. The square of the distance of the point (x, y, z) from the x-axis is $y^2 + z^2 = z^2 + r^2 \sin^2 \theta$. Therefore

$$I = I_x = \int_{\theta=0}^{2\pi} \int_{r=0}^{a} \int_{z=0}^{h} \delta(z^2 + r^2 \sin^2 \theta) \cdot r \, dz \, dr \, d\theta = \delta \int_0^{2\pi} \int_0^a \left(\frac{1}{3} h^3 r + h r^3 \sin^2 \theta \right) dr \, d\theta$$

$$= \delta \int_0^{2\pi} \left[\frac{1}{6} h^3 r^2 + \frac{1}{4} h r^4 \sin^2 \theta \right]_0^a d\theta = \delta \int_0^{2\pi} \left(\frac{1}{6} a^2 h^3 + \frac{1}{4} a^4 h \sin^2 \theta \right) d\theta$$

$$= \delta \left[\frac{6a^4 h \theta + 8a^2 h^3 \theta - 3a^4 h \sin 2\theta}{48} \right]_0^{2\pi} = \frac{1}{12} \delta \pi a^2 h (3a^2 + 4h^2) = \frac{1}{12} m (3a^2 + 4h^2)$$

where m is the mass of the cylinder.

13.7.19: The volume is

$$V = \int_{\theta=0}^{2\pi} \int_{r=0}^{1} \int_{z=r}^{1} r \, dz \, dr \, d\theta = 2\pi \int_0^1 (r - r^2) \, dr = 2\pi \cdot \left[\frac{1}{2} r^2 - \frac{1}{3} r^3 \right]_0^1 = \frac{1}{3} \pi.$$

13.7.21: Without loss of generality we may assume that the hemispherical solid has density $\delta = 1$. Choose a coordinate system in which the solid is bounded above by the spherical surface $\rho = a$ and below by the xy-plane. Then its mass and moments are

$$m = \int_{\theta=0}^{2\pi} \int_{\phi=0}^{\pi/2} \int_{\rho=0}^{a} \rho^2 \sin \phi \, d\rho \, d\phi \, d\theta = 2\pi \int_{\phi=0}^{\pi/2} \frac{1}{3} a^3 \sin \phi \, d\phi = \frac{2}{3} \pi a^3 \left[-\cos \phi \right]_0^{\pi/2} = \frac{2}{3} \pi a^3;$$

$$M_{yz} = \int_{\theta=0}^{2\pi} \int_{\phi=0}^{\pi/2} \int_{\rho=0}^{a} \rho^3 \sin^2 \phi \cos \theta \, d\rho \, d\phi \, d\theta = \int_{\theta=0}^{2\pi} \int_{\phi=0}^{\pi/2} \frac{1}{4} a^4 \sin^2 \phi \cos \theta \, d\phi \, d\theta$$

$$= \int_{\theta=0}^{2\pi} \left(\int_{\phi=0}^{\pi/2} \frac{1}{4} a^2 \sin^2 \phi \, d\phi \right) \cos \theta \, d\theta = \left(\int_{\phi=0}^{\pi/2} \frac{1}{4} a^2 \sin^2 \phi \, d\phi \right) \cdot \left[\sin \theta \right]_{\theta=0}^{2\pi} = 0;$$

$$M_{xz} = \int_{\theta=0}^{2\pi} \int_{\phi=0}^{\pi/2} \int_{\rho=0}^{a} \rho^3 \sin^2 \phi \sin \theta \, d\rho \, d\phi \, d\theta = \int_{\theta=0}^{2\pi} \int_{\phi=0}^{\pi/2} \frac{1}{4} a^4 \sin^2 \phi \sin \theta \, d\phi \, d\theta$$

$$= \int_{\theta=0}^{2\pi} \left(\int_{\phi=0}^{\pi/2} \frac{1}{4} a^2 \sin^2 \phi \, d\phi \right) \sin \theta \, d\theta = \left(\int_{\phi=0}^{\pi/2} \frac{1}{4} a^2 \sin^2 \phi \, d\phi \right) \cdot \left[-\cos \theta \right]_{\theta=0}^{2\pi} = 0;$$

$$M_{yx} = \int_{\theta=0}^{2\pi} \int_{\phi=0}^{\pi/2} \int_{\rho=0}^{a} \rho^3 \sin \phi \cos \phi \, d\rho \, d\phi \, d\theta = 2\pi \int_{\phi=0}^{\pi/2} \frac{1}{4} a^4 \sin \phi \cos \phi \, d\phi$$

$$= \frac{1}{2} \pi a^4 \left[\frac{1}{2} \sin^2 \phi \right]_0^{\pi/2} = \frac{1}{4} \pi a^4.$$

Therefore the centroid of the hemispherical solid is located at the point $\left(0,\, 0,\, \dfrac{3}{8} a \right)$.

13.7.23: The plane $z = 1$ has the spherical-coordinates equation $\rho = \sec \phi$; the cone with cylindrical-coordinates equation $r = z$ has spherical-coordinates equation $\phi = \frac{1}{4}\pi$. Hence the volume bounded by the plane and the cone is

$$V = \int_{\theta=0}^{2\pi} \int_{\phi=0}^{\pi/4} \int_{\rho=0}^{\sec \phi} \rho^2 \sin \phi \, d\rho \, d\phi \, d\theta = 2\pi \int_{\phi=0}^{\pi/4} \frac{1}{3} \sec^2 \phi \tan \phi \, d\phi = 2\pi \cdot \left[\frac{1}{6} \sec^2 \phi \right]_{\phi=0}^{\pi/4} = \frac{1}{3} \pi.$$

13.7.25: Assume unit density. Then the mass and the volume are numerically the same; they and the moments are

$$m = V = \int_0^{2\pi} \int_0^{\pi/4} \int_0^a \rho^2 \sin \phi \, d\rho \, d\phi \, d\theta = 2\pi \cdot \frac{1}{3} a^3 \left[-\cos \phi \right]_0^{\pi/4}$$

$$= \frac{2}{3} \pi a^2 \left(1 - \frac{1}{2}\sqrt{2} \right) = \frac{1}{3} \pi \left(2 - \sqrt{2} \right) a^3;$$

$$M_{yz} = \int_0^{2\pi} \int_0^{\pi/4} \int_0^a \rho^3 \sin^2 \phi \cos \theta \, d\rho \, d\phi \, d\theta = \left(\int_0^{\pi/4} \int_0^a \rho^3 \sin^2 \phi \, d\rho \, d\phi \right) \cdot \left[\sin \theta \right]_0^{2\pi} = 0;$$

$$M_{xz} = \int_0^{2\pi} \int_0^{\pi/4} \int_0^a \rho^3 \sin^2 \phi \sin \theta \, d\rho \, d\phi \, d\theta = \left(\int_0^{\pi/4} \int_0^a \rho^3 \sin^2 \phi \, d\rho \, d\phi \right) \cdot \left[-\cos \theta \right]_0^{2\pi} = 0;$$

$$M_{xy} = \int_0^{2\pi} \int_0^{\pi/4} \int_0^a \rho^3 \sin \phi \cos \phi \, d\rho \, d\phi \, d\theta = 2\pi \cdot \frac{1}{4} a^4 \cdot \left[\frac{1}{2} \sin^2 \phi \right]_0^{\pi/4} = \frac{1}{8} \pi a^4.$$

So the z-coordinate of the centroid is

$$\overline{z} = \frac{3\pi a^4}{8\pi \left(2 - \sqrt{2} \right) a^3} = \frac{3a}{8 \left(2 - \sqrt{2} \right)} = \frac{3 \left(2 + \sqrt{2} \right) a}{16} = \frac{3}{16} \left(2 + \sqrt{2} \right) a;$$

clearly $\overline{x} = \overline{y} = 0$. For a plausibility check, note that $\overline{z} \approx (0.6401650429449553)a$.

13.7.27: Set up a coordinate system so that the center of the sphere is at the point with Cartesian coordinates $(a, 0, 0)$. Then its Cartesian equation is

$$(x - a)^2 + y^2 + z^2 = a^2; \quad x^2 - 2ax + y^2 + z^2 = 0;$$

and thus it has spherical-coordinates equation $\rho = 2a \sin \phi \cos \theta$ (but note that θ ranges from $-\frac{1}{2}\pi$ to $\frac{1}{2}\pi$). We plan to find its moment of inertia with respect to the z-axis, and the square of the distance of a point of the sphere from the z-axis is

$$x^2 + y^2 = \rho^2 \sin^2 \phi \cos^2 \theta + \rho^2 \sin^2 \phi \sin^2 \theta = \rho^2 \sin^2 \phi.$$

Moreover, if the sphere has mass m and constant density δ, then we also have $m = \frac{4}{3}\pi a^3 \delta$. Finally,

$$I_z = \int_{-\pi/2}^{\pi/2} \int_0^\pi \int_0^{2a\sin\phi\cos\theta} \delta\rho^2 \sin^2\phi \, d\rho \, d\phi \, d\theta = 2\delta \int_0^{\pi/2} \int_0^\pi \frac{1}{5}(2a\sin\phi\cos\theta)^5 \sin^3\phi \, d\phi \, d\theta$$

$$= \frac{64}{5}\delta a^5 \int_0^{\pi/2} \int_0^\pi \sin^8\phi\cos^5\theta \, d\phi \, d\theta = \frac{128}{5}\delta a^5 \int_0^{\pi/2} \int_0^{\pi/2} \sin^8\phi\cos^5\theta \, d\phi \, d\theta$$

$$= \frac{128}{5}\delta a^5 \cdot \left(\int_0^{\pi/2} \sin^8\phi \, d\phi\right) \cdot \left(\int_0^{\pi/2} \cos^5\theta \, d\theta\right).$$

Then Formula (113) from the long table of integrals (see the endpapers) yields

$$I_z = \frac{128}{5}\delta a^5 \cdot \frac{1}{2}\cdot\frac{3}{4}\cdot\frac{5}{6}\cdot\frac{7}{8}\cdot\frac{\pi}{2}\cdot\frac{2}{3}\cdot\frac{4}{5} = \frac{28}{15}\pi\delta a^5 = \frac{4}{3}\pi\delta a^3 \cdot \frac{7}{5}a^2 = \frac{7}{5}ma^2.$$

13.7.29: The surface with spherical-coordinates equation $\rho = 2a\sin\phi$ is generated as follows. Draw the circle in the xz-plane with center $(a, 0)$ and radius a. Rotate this circle around the z-axis. This generates the surface with the given equation. It is called a *pinched torus*—a doughnut with an infinitesimal hole. Its volume is

$$V = \int_{\theta=0}^{2\pi} \int_{\phi=0}^{\pi} \int_{\rho=0}^{2a\sin\phi} \rho^2 \sin\phi \, d\rho \, d\phi \, d\theta = 2\pi \int_{\phi=0}^{\pi} \frac{1}{3}(2a\sin\phi)^3 \sin\phi \, d\phi$$

$$= \frac{16}{3}\pi a^3 \cdot 2\int_{\phi=0}^{\pi/2} \sin^4\phi \, d\phi = \frac{32}{3}\pi a^3 \cdot \frac{1}{2}\cdot\frac{3}{4}\cdot\frac{\pi}{2} = 2\pi^2 a^3.$$

We evaluated the last integral with the aid of Formula (113) from the long table of integrals (see the endpapers). The volume of the pinched torus is also easy to evaluate using the first theorem of Pappus (Section 13.5).

13.7.31: Assuming constant density δ, we have

$$I_x = 2\int_0^{2\pi} \int_0^a \int_0^{\sqrt{4a^2-r^2}} \delta(r^2\sin^2\theta + z^2)\cdot r \, dz \, dr \, d\theta$$

$$= 2\delta \int_0^{2\pi} \int_0^a \left(\frac{1}{3}r(4a^2-r^2)^{3/2} + r^3(4a^2-r^2)^{1/2}\sin^2\theta\right) dr \, d\theta$$

$$= 2\delta \int_0^{2\pi} \frac{1}{30}\left[(4a^2-r^2)^{3/2}\big[(8a^2+3r^2)\cos 2\theta - 16a^2 - r^2\big]\right]_{r=0}^a d\theta$$

$$= 2\delta \int_0^{2\pi} \frac{1}{30}a^5\big[128 - 51\sqrt{3} + \left(33\sqrt{3} - 64\right)\cos 2\theta\big] d\theta$$

$$= \frac{1}{30}\delta a^5\left[\left(128 - 5\sqrt{3}\right)\cdot 2\theta + \left(33\sqrt{3} - 64\right)\sin 2\theta\right]_0^{2\pi} = \frac{2}{15}\left(128 - 51\sqrt{3}\right)\delta\pi a^5.$$

903

13.7.33: If the density at (x, y, z) of the ice-cream cone is z, then its mass and moments are

$$m = \int_0^{2\pi} \int_0^{\pi/6} \int_0^{2a\cos\phi} \rho^3 \sin\phi \cos\phi \, d\rho \, d\phi \, d\theta = 2\pi \int_0^{\pi/6} 4a^4 \sin\phi \cos^5\phi \, d\phi$$

$$= \left[-\frac{4}{3}\pi a^4 \cos^6\phi \right]_0^{\pi/6} = \frac{37}{48}\pi a^4;$$

$$M_{yz} = \int_0^{2\pi} \left(\int_0^{\pi/6} \int_0^{2a\cos\phi} \rho^4 \sin^2\phi \cos\phi \, d\rho \, d\phi \right) \cos\theta \, d\theta$$

$$= \left(\int_0^{\pi/6} \int_0^{2a\cos\phi} \rho^4 \sin^2\phi \cos\phi \, d\rho \, d\phi \right) \cdot \Big[\sin\theta \Big]_0^{2\pi} = 0;$$

$$M_{xz} = 0 \quad \text{(by a similar computation)};$$

$$M_{xy} = \int_0^{2\pi} \int_0^{\pi/6} \int_0^{2a\cos\phi} \rho^4 \sin\phi \cos^2\phi \, d\rho \, d\phi \, d\theta = 2\pi \int_0^{\pi/6} \frac{32}{5} a^5 \cos^7\phi \sin\phi \, d\phi$$

$$= 2\pi \cdot \left[-\frac{4}{5} a^5 \cos^8\phi \right]_0^{\pi/6} = \frac{35}{32}\pi a^5.$$

Hence the centroid is located at the point $\left(0, \ 0, \ \frac{105}{74} a \right)$.

13.7.35: The similar star with uniform density k has mass $m_2 = \frac{4}{3} k\pi a^3$. The other star has mass

$$m_1 = \int_0^{2\pi} \int_0^{\pi} \int_0^a k \left[1 - \left(\frac{\rho}{a} \right)^2 \right] \cdot \rho^2 \sin\phi \, d\rho \, d\phi \, d\theta = 2\pi \int_0^{\pi} \left[\frac{5ka^2\rho^3 \sin\phi - 3k\rho^5 \sin\phi}{15a^2} \right]_0^a d\phi$$

$$= 2\pi \int_0^{\pi} \frac{2}{15} ka^3 \sin\phi \, d\phi = \left[-\frac{4}{15} k\pi a^3 \cos\phi \right]_0^{\pi} = \frac{8}{15} k\pi a^3.$$

Finally, $\dfrac{m_1}{m_2} = \dfrac{2}{5}$.

13.7.37: The given triple integral takes the following form:

$$\int_0^{2\pi} \int_0^{\pi} \int_0^a \rho^2 \exp\left(-\rho^3\right) \sin\phi \, d\rho \, d\phi \, d\theta = 2\pi \int_0^{\pi} \left[-\frac{1}{3} \exp\left(-\rho^3\right) \sin\phi \right]_0^a d\phi$$

$$= 2\pi \int_0^{\pi} \frac{1}{3} \left[1 - \exp\left(-a^3\right) \right] \sin\phi \, d\phi = \frac{4}{3} \pi \left[1 - \exp\left(-a^3\right) \right].$$

Clearly the value of the integral approaches $\dfrac{4}{3}\pi$ as $a \to +\infty$.

13.7.39: Let $V = \frac{4}{3}\pi a^3$, the volume of the ball. The average distance of points of such a ball from its center is then

$$\bar{d} = \frac{1}{V} \int_0^{2\pi} \int_0^{\pi} \int_0^a \rho^3 \sin\phi \, d\rho \, d\phi \, d\theta = \frac{2\pi}{V} \int_0^{\pi} \frac{1}{4} a^4 \sin\phi \, d\phi = \frac{\pi a^4}{2V} \Big[-\cos\phi \Big]_0^{\pi} = \frac{\pi a^4}{V} = \frac{3}{4} a.$$

Note that the answer is both plausible and dimensionally correct.

13.7.41: A *Mathematica* solution:

```
m = Integrate[ Integrate[ delta*a∧2*Sin[ phi ], { phi, 0, Pi } ],
                { theta, 0, 2*Pi } ]
```

$$4\pi a^2 \delta$$

```
r = a*Sin[ phi ];

I0 = Integrate[ Integrate[ r∧2*delta*a∧2*Sin[ phi ], { phi, 0, Pi } ],
                { theta, 0, 2*Pi } ]
```

$$\frac{8}{3}\pi a^4 \delta$$

```
I0/m
```

$$\frac{2}{3}a^2$$

Second solution, by hand:. The spherical surface S of radius a is described by $\rho = a$, $0 \le \phi \le \pi$, $0 \le \theta \le 2\pi$. Hence its moment of inertia with respect to the z-axis is

$$I_z = \iint_S \delta(x^2 + y^2)\, dA = \int_{\theta=0}^{2\pi}\int_{\phi=0}^{\pi} \delta a^4 \sin^3\phi\, d\phi\, d\theta$$

$$= 2\pi\delta a^4 \int_0^\pi (1-\cos^2\phi)\sin\phi\, d\phi = 2\pi\delta a^4 \left[\frac{1}{3}\cos^3\phi - \cos\phi\right]_0^\pi = \left(\frac{2}{3}a^2\right)\cdot 4\pi\delta a^2 = \frac{2}{3}ma^2.$$

13.7.43: A *Mathematica* solution:

```
z = Sqrt[ b∧2 − a∧2 ];

m = 2*Integrate[ Integrate[ delta*r*z, { r, a, b } ], { theta, 0, 2*Pi } ]
```

$$\frac{4}{3}\pi\delta(b^2 - a^2)^{3/2}$$

```
I0 = 2*Integrate[ Integrate[ delta*r∧3*z, { r, a, b } ], { theta, 0, 2*Pi } ]
```

$$\frac{4}{15}\pi\delta(2b^4 + b^2a^2 - 3a^4)\sqrt{b^2 - a^2}$$

```
Simplify[ I0/m ]
```

$$\frac{1}{5}(3a^2 + 2b^2)$$

Second solution, by hand: Choose a coordinate system so that the z-axis is the axis of symmetry of the sphere-with-hole. The central cross section of the solid in the xz-plane is bounded by the circle with polar (or cylindrical coordinates) equation $r^2 + z^2 = b^2$. Hence the mass of the solid is

$$m = 2 \int_{\theta=0}^{2\pi} \int_{r=a}^{b} \int_{z=0}^{\sqrt{b^2-r^2}} \delta r \, dz \, dr \, d\theta = 4\pi\delta \int_{a}^{b} r(b^2 - r^2)^{1/2} \, dr$$

$$= 4\pi\delta \left[-\frac{1}{3}(b^2 - r^2)^{3/2} \right]_{a}^{b} = \frac{4}{3}\pi\delta(b^2 - a^2)^{3/2}.$$

The moment of inertia of this solid with respect to the z-axis is

$$I_z = 2 \int_{\theta=0}^{2\pi} \int_{r=a}^{b} \int_{z=0}^{\sqrt{b^2-r^2}} \delta r^3 \, dz \, dr \, d\theta = 4\pi\delta \int_{a}^{b} r^3(b^2 - r^2)^{1/2} \, dr.$$

Integration by parts with $u = r^2$, $dv = (b^2 - r^2)^{1/2} \, dr$, so that

$$du = 2r \, dr \quad \text{and} \quad v = -\frac{1}{3}(b^2 - r^2)^{3/2},$$

then yields

$$I_z = 4\pi\delta \left(\left[-\frac{1}{3}r^2(b^2 - r^2)^{3/2} \right]_{a}^{b} + \frac{2}{3} \int_{a}^{b} r(b^2 - r^2)^{3/2} \, dr \right)$$

$$= 4\pi\delta \left(\frac{1}{3}a^2(b^2 - a^2)^{3/2} + \frac{2}{3} \left[-\frac{1}{5}(b^2 - r^2)^{5/2} \right]_{a}^{b} \right)$$

$$= \frac{4}{15}\pi\delta \left[5a^2(b^2 - a^2)^{3/2} + 2(b^2 - a^2)(b^2 - a^2)^{3/2} \right] = \frac{4}{15}\pi\delta(b^2 - a^2)^{3/2}(5a^2 + 2b^2 - 2a^2)$$

$$= \frac{4}{3}\pi\delta(b^2 - a^2)^{3/2} \cdot \frac{1}{5}(3a^2 + 2b^2) = \frac{1}{5}m(3a^2 + 2b^2).$$

13.7.45: First we let $g(\phi, \theta) = 6 + 3\cos 3\theta \sin 5\phi$, so that the boundary of the bumpy sphere B has spherical equation $\rho = g(\phi, \theta)$. Then the volume V of B is simply

$$V = \int_{\theta=0}^{2\pi} \int_{\phi=0}^{\pi} \int_{\rho=0}^{g(\phi,\theta)} \rho^2 \sin\phi \, d\rho \, d\phi \, d\theta.$$

To evaluate V with the aid of *Mathematica*, we write $\mathtt{r}$ in place of ρ, $\mathtt{f}$ instead of ϕ, and $\mathtt{t}$ for θ, as usual. Then, to find the volume V of B step-by-step, we proceed as follows:

```
Integrate[ r*r*Sin[f], r ]
```

$$\frac{1}{3}\rho^3 \sin\phi$$

```
% /.  r → g[f,t]
```

906

$$\frac{1}{3}(\sin\phi)(6 + 3\cos 3\theta \, \sin 5\phi)^3$$

```
Integrate[ %, f ]
```

$$\frac{3}{78848}\left[-2247168\cos\phi - 19712\cos 9\phi + 16128\cos 11\phi - 177408\cos(\phi - 6\theta)\right.$$

$$- 9856\cos(9\phi - 6\theta) + 8064\cos(11\phi - 6\theta) - 177408\cos(\phi + 6\theta)$$

$$- 9856\cos(9\phi + 6\theta) + 8064\cos(11\phi + 6\theta) + 2772\sin(4\phi - 9\theta)$$

$$- 1848\sin(6\phi - 9\theta) - 264\sin(14\phi - 9\theta) + 231\sin(16\phi - 9\theta)$$

$$+ 185724\sin(4\phi - 3\theta) - 123816\sin(6\phi - 3\theta) - 792\sin(14\phi - 3\theta)$$

$$+ 693\sin(16\phi - 3\theta) + 185724\sin(4\phi + 3\theta) - 123816\sin(6\phi + 3\theta)$$

$$- 792\sin(14\phi + 3\theta) + 693\sin(16\phi + 3\theta) + 2772\sin(4\phi + 9\theta) - 1848\sin(6\phi + 9\theta)$$

$$\left. - 264\sin(14\phi + 9\theta) + 231\sin(16\phi + 9\theta)\right]$$

```
(% /. f → Pi) - (% /. f → 0) // FullSimplify
```

$$\frac{12\,(157 + 25\cos 6\theta)}{11}$$

```
Integrate[ %, t ]
```

$$\frac{1884\theta + 50\sin 6\theta}{11}$$

```
(% /. t → 2*Pi) - (% /. t → 0)
```

$$\frac{3768\pi}{11}$$

```
N[ %, 60 ]
```

$$1076.13828352057644750247638801792426606959031178950 3624849396$$

Thus $V = \frac{3768}{11}\pi \approx 1076.13828352$.

13.7.47: The following figure makes some of the equations we use easy to derive.

$(0, 0, c)$

α

w

c

P

ρ

ϕ

$(0, 0, 0)$

A mass element $\delta\, dV$ located at the point $P(\rho,\, \phi,\, \theta)$ of the ball exerts a force on the mass m at $(0,\, 0,\, c)$ that has vertical component

$$dF = -\frac{Gm\delta\cos\alpha}{w^2}\, dV.$$

Note the following:

$$M = \frac{4}{3}\pi\delta a^3;$$

$$w^2 = \rho^2 + c^2 - 2\rho c\cos\phi;$$

$$2w\, dw = 2\rho c\sin\phi\, d\phi;$$

$$w\cos\alpha + \rho\cos\phi = c;$$

$$\rho\cos\phi = \frac{\rho^2 + c^2 - w^2}{2c};$$

$$\rho\sin\phi\, d\phi = \frac{w}{c}\, dw.$$

Hence the total force exerted by the ball on m is

$$F = -\int_{\theta=0}^{2\pi}\int_{\rho=0}^{a}\int_{\phi=0}^{\pi}\frac{Gm\delta\cos\alpha}{w^2}\cdot\rho^2\sin\phi\, d\phi\, d\rho\, d\theta = -\int_{0}^{2\pi}\int_{0}^{a}\int_{0}^{\pi}\frac{Gm\delta w\cos\alpha}{w^3}\cdot\rho^2\sin\phi\, d\phi\, d\rho\, d\theta$$

$$= -2\pi\int_{0}^{a}\int_{0}^{\pi}\frac{Gm\delta}{w^3}(c - \rho\cos\phi)\cdot\rho^2\sin\phi\, d\phi\, d\rho = -2\pi\int_{0}^{a}\int_{c-\rho}^{c+\rho}\frac{Gm\delta}{w^3}\left(c - \frac{\rho^2 + c^2 - w^2}{2c}\right)\cdot\frac{\rho w}{c}\, dw\, d\rho$$

$$= -2\pi\int_{\rho=0}^{a}\int_{w=c-\rho}^{c+\rho}\frac{Gm\delta}{w^3}\cdot\frac{2c^2 - \rho^2 - c^2 + w^2}{2c}\cdot\rho\cdot\frac{w}{c}\, dw\, d\rho$$

$$= -\pi Gm\delta\int_{0}^{a}\int_{c-\rho}^{c+\rho}\frac{1}{w^2}\cdot\frac{c^2 - \rho^2 + w^2}{c^2}\cdot\rho\, dw\, d\rho = -\frac{\pi Gm\delta}{c^2}\int_{0}^{a}\int_{c-\rho}^{c+\rho}\left(\frac{c^2 - \rho^2}{w^2} + 1\right)\cdot\rho\, dw\, d\rho$$

$$= -\frac{\pi Gm\delta}{c^2} \int_0^a \rho \cdot \left[\frac{\rho^2 - c^2}{w} + w\right]_{w=c-\rho}^{c+\rho} d\rho = -\frac{\pi Gm\delta}{c^2} \int_0^a \rho \cdot \left(\frac{\rho^2 - c^2}{\rho + c} + c + \rho + \frac{c^2 - \rho^2}{c - \rho} - c + \rho\right) d\rho$$

$$= -\frac{\pi Gm\delta}{c^2} \int_0^a \rho(\rho - c + \rho + c + c + \rho - c + \rho)\, d\rho = -\frac{\pi Gm\delta}{c^2} \int_0^a 4\rho^2\, d\rho = -\frac{\pi Gm\delta}{c^2} \cdot \left[\frac{4}{3}\rho^3\right]_0^a$$

$$= -\frac{\pi Gm\delta}{c^2} \cdot \frac{4}{3}a^3 = -\frac{4}{3}\pi\delta a^3 \cdot \frac{Gm}{c^2} = -\frac{GMm}{c^2}.$$

Magnificent! You can even extend this result to show that if the density of the ball varies only as a function of ρ, then the same conclusion follows: The ball acts, for purposes of gravitational attraction of an external mass m, as if all its mass M were concentrated at its center. And note one additional item of interest: This is one of the rare spherical triple integrals *not* that is best evaluated in the order $d\rho\, d\phi\, d\theta$.

13.7.49: A *Mathematica* solution:

```
r = 6370*1000;    k = 0.371;

d1 = 11000;    d2 = 5000;

m1 = (4/3)*Pi*d1*x∧3;

m2 = (4/3)*Pi*d2*(r∧3 - x∧3);

m = m1 + m2;

I1 = (2/5)*m1*x∧2;
```

By Problem 42, the moment of inertia of the mantle with respect to the polar axis (through the poles of the planet) is

```
I2 = (2/5)*m2*(r∧5 - x∧5)/(r∧3 - x∧3);
```

Therefore we proceed to solve for x as follows.

```
I0 = I1 + I2;

eq1 = I0 == k*m*r∧2;

soln = NSolve[ eq1, x ];
```

We suppress the output, but there are five solutions. Only two are real and positive,

$$x_1 \approx 2.76447 \times 10^6 \quad \text{and} \quad x_2 \approx 5.87447 \times 10^6.$$

909

We are given the information that the mantle is "a few thousand kilometers thick," and x_2 does not satisfy this condition, as it implies that the mantle is less than 496 km thick. We conclude that the radius of the core is $x_1/1000$ km, and hence that the thickness of the mantle is $(r - x_1)/1000 \approx 3605.53$ km.

Section 13.8

13.8.1: The surface area element is

$$dS = \sqrt{1 + 1^2 + 3^2} \, dA = \sqrt{11} \, dA.$$

So the area in question is

$$A = 4 \int_{x=0}^{2} \int_{y=0}^{3\sqrt{1-(x/2)^2}} \sqrt{11} \, dy \, dx = \int_{0}^{2} 6\sqrt{11} \sqrt{4 - x^2} \, dx$$

$$= \left[3x\sqrt{11} \sqrt{4 - x^2} + 12\sqrt{11} \arcsin\left(\frac{x}{2}\right) \right]_{0}^{2} = 6\pi\sqrt{11} \approx 62.5169044565658738.$$

13.8.3: The paraboloid meets the plane in the circle with equation $x^2 + y^2 = 4$, $z = 5$. Let D denote the circular disk $x^2 + y^2 \leq 4$ in the xy-plane. The surface area element is

$$dS = \sqrt{1 + 4x^2 + 4y^2} \, dA,$$

so the surface area in question is

$$A = \iint_{D} \sqrt{1 + 4x^2 + 4y^2} \, dA = \int_{\theta=0}^{2\pi} \int_{r=0}^{2} r\sqrt{1 + 4r^2} \, dr \, d\theta = 2\pi \cdot \left[\frac{1}{12}(1 + 4r^2)^{3/2} \right]_{0}^{2}$$

$$= 2\pi \left(\frac{17\sqrt{17} - 1}{12} \right) = \frac{1}{6}\pi \left(17\sqrt{17} - 1 \right) \approx 36.1769031974114084.$$

13.8.5: The surface area element is

$$dS = \sqrt{2 + 4y^2} \, dA,$$

so the area is

$$A = \int_{y=0}^{2} \int_{x=0}^{1} (2 + 4y^2)^{1/2} \, dx \, dy = \int_{y=0}^{2} (2 + 4y^2)^{1/2} \, dy = \left[\frac{1}{2}y(2 + 4y^2)^{1/2} + \frac{1}{2}\text{arcsinh}\left(y\sqrt{2}\right) \right]_{0}^{2}$$

$$= 3\sqrt{2} + \frac{1}{2}\text{arcsinh}\left(2\sqrt{2}\right) = 3\sqrt{2} + \frac{1}{2}\ln\left(3 + 2\sqrt{2}\right) \approx 5.1240142741388282.$$

13.8.7: The surface area element is $dS = \sqrt{14} \, dA$, so the surface area is

$$A = \int_{x=0}^{3} \int_{y=0}^{(6-2x)/3} \sqrt{14} \, dy \, dx = \int_{0}^{3} \frac{1}{3}(6 - 2x)\sqrt{14} \, dx = \left[2x\sqrt{14} - \frac{1}{3}x^2\sqrt{14} \right]_{0}^{3} = 3\sqrt{14}.$$

Alternatively, the vectors $\mathbf{u} = \langle -3, 2, 0 \rangle$ and $\mathbf{v} = \langle -3, 0, 6 \rangle$ span two adjacent sides of the triangular surface. So its area is half the magnitude of their cross product:

$$A = \frac{1}{2}|\mathbf{u} \times \mathbf{v}| = \frac{1}{2}|\langle 12, 18, 6 \rangle| = \frac{1}{2}\sqrt{504} = 3\sqrt{14} \approx 11.224972160321824415675.$$

13.8.9: The surface area element is $dS = \sqrt{1 + x^2 + y^2}\ dA$, so the surface area is

$$A = \int_{\theta=0}^{2\pi}\int_{r=0}^{1} r(1+r^2)^{1/2}\ dr\ d\theta = 2\pi \cdot \left[\frac{1}{3}(1+r^2)^{3/2}\right]_{0}^{1} = \frac{2}{3}\pi\left(2\sqrt{2} - 1\right) \approx 3.82944881515112928.$$

13.8.11: The paraboloid meets the xy-plane in the circle with equation $x^2 + y^2 = 16$. The surface area element is

$$dS = \sqrt{1 + 4x^2 + 4y^2}\ dA,$$

so the area is

$$A = \int_{0}^{2\pi}\int_{0}^{4} r(1+4r^2)^{1/2}\ dr\ d\theta = 2\pi \cdot \left[\frac{1}{12}(1+4r^2)^{3/2}\right]_{0}^{4} = \frac{1}{6}\pi\left(65\sqrt{65} - 1\right) \approx 273.866639786258.$$

13.8.13: Let $\mathbf{r}(\theta, z) = \langle a\cos\theta,\ a\sin\theta,\ z \rangle$. Then

$$\mathbf{r}_\theta \times \mathbf{r}_z = \begin{vmatrix} \mathbf{i} & \mathbf{j} & \mathbf{k} \\ -a\sin\theta & a\cos\theta & 0 \\ 0 & 0 & 1 \end{vmatrix} = \langle a\cos\theta,\ a\sin\theta,\ 0 \rangle$$

and hence

$$|\mathbf{r}_\theta \times \mathbf{r}_z| = \sqrt{a^2\cos^2\theta + a^2\sin^2\theta} = a.$$

Therefore, by Eq. (8), the area of the zone is

$$A = \int_{\theta=0}^{2\pi}\int_{z=0}^{h} a\ dz\ d\theta = 2\pi \cdot \left[az\right]_{z=0}^{h} = 2\pi ah.$$

13.8.15: Let $z(x, y) = \sqrt{a^2 - x^2}$. Then the surface area element in Cartesian coordinates is

$$dS = \frac{a}{\sqrt{a^2 - x^2}}\ dA.$$

Let D be the disk in which the vertical cylinder meets the xy-plane. Then the area of the part of the horizontal cylinder—top and bottom—that lies within the vertical cylinder is

$$A = 2\iint_D \frac{a}{\sqrt{a^2 - x^2}}\ dA = \int_{\theta=0}^{2\pi}\int_{r=0}^{a} \frac{2a}{\sqrt{a^2 - r^2\cos^2\theta}} \cdot r\ dr\ d\theta$$

$$= \int_{\theta=0}^{2\pi}\left[\frac{-2a(a^2 - r^2\cos^2\theta)^{1/2}}{\cos^2\theta}\right]_{r=0}^{a}\ d\theta = 4\int_{\theta=0}^{\pi/2} 2a^2(\sec^2\theta - \sec\theta\tan\theta)\ d\theta = 8a^2\left[\tan\theta - \sec\theta\right]_{0}^{\pi/2}$$

$$= 8a^2 + 8a^2 \cdot \left(\lim_{\theta\to(\pi/2)^-}\left[\tan\theta - \sec\theta\right]\right) = 8a^2 + 8a^2 \cdot 0 = 8a^2.$$

The change in the limits of integration in the second line was necessary because we needed the simplification $(1 - \cos^2\theta)^{1/2} = \sin\theta$, which is not valid if $\pi < \theta < 2\pi$; moreover, we needed to avoid the discontinuity of the following improper integral at $\theta = \pi/2$.

13.8.17: The surface $y = f(x, z)$ is parametrized b

$$\mathbf{r}(x, z) = \langle x,\ f(x, z),\ z \rangle$$

for (x, z) in the region R in the xz-plane. Then

$$\mathbf{r}_x \times \mathbf{r}_z = \begin{vmatrix} \mathbf{i} & \mathbf{j} & \mathbf{k} \\ 1 & f_x & 0 \\ 0 & f_z & 1 \end{vmatrix} = \langle f_x,\ -1,\ f_z \rangle.$$

Therefore the area of the surface $y = f(x, z)$ lying "over" the region R is

$$A = \iint_R \sqrt{1 + \left(\frac{\partial f}{\partial x}\right)^2 + \left(\frac{\partial f}{\partial z}\right)^2}\ dx\ dz.$$

The surface $x = f(y, z)$ is parametrized by

$$\mathbf{r}(y, z) = \langle f(y, z),\ y,\ z \rangle$$

for (y, z) in the region R in the xz-plane. Then

$$\mathbf{r}_y \times \mathbf{r}_z = \begin{vmatrix} \mathbf{i} & \mathbf{j} & \mathbf{k} \\ f_y & 1 & 0 \\ f_z & 0 & 1 \end{vmatrix} = \langle 1,\ -f_y,\ -f_z \rangle.$$

Therefore the area of the surface $x = f(y, z)$ lying "over" the region R is

$$A = \iint_R \sqrt{1 + \left(\frac{\partial f}{\partial y}\right)^2 + \left(\frac{\partial f}{\partial z}\right)^2}\ dy\ dz.$$

13.8.19: By Problem 50 of Section 13.2,

$$A = \int_{\theta_1}^{\theta_2} \int_{\phi_1}^{\phi_2} a^2 \sin\phi\ d\phi\ d\theta = a^2 \sin\hat{\phi}\ \Delta\phi\ \Delta\theta$$

for some $\hat{\phi}$ in (ϕ_1, ϕ_2). Therefore

$$\Delta V = \int_{\rho_1}^{\rho_2} (\rho^2 \sin\hat{\phi}\ \Delta\phi\ \Delta\theta)\ d\rho = \frac{1}{3}(\rho_2^3 - \rho_1^3) \sin\hat{\phi}\ \Delta\phi\ \Delta\theta$$

$$= \frac{1}{3} \cdot \frac{\rho_2^3 - \rho_1^3}{\rho_2 - \rho_1} \sin\hat{\phi}\ \Delta\phi\ \Delta\theta\ \Delta\rho = \frac{1}{3} \cdot 3\hat{\rho}^2 \sin\hat{\phi}\ \Delta\phi\ \Delta\theta\ \Delta\rho = \hat{\rho}^2 \sin\hat{\phi}\ \Delta\rho\ \Delta\phi\ \Delta\theta$$

for some $\hat{\rho}$ in (ρ_1, ρ_2), and this is Eq. (8) of Section 13.7.

13.8.21: Given:

$$x = f(z)\cos\theta, \quad y = f(z)\sin\theta, \quad z = z$$

where $0 \leq \theta \leq 2\pi$ and $a \leq z \leq b$. The surface thereby generated is thereby parametrized by

$$\mathbf{r}(\theta, z) = \langle\, f(z)\cos\theta,\; f(z)\sin\theta,\; z\,\rangle,$$

and thus

$$\mathbf{r}_\theta = \langle\, -f(z)\sin\theta,\; f(z)\cos\theta,\; 0\,\rangle \quad \text{and}$$

$$\mathbf{r}_z = \langle\, f'(z)\cos\theta,\; f'(z)\sin\theta,\; 1\,\rangle.$$

Therefore

$$\mathbf{r}_\theta \times \mathbf{r}_z = \begin{vmatrix} \mathbf{i} & \mathbf{j} & \mathbf{k} \\ -f(z)\sin\theta & f(z)\cos\theta & 0 \\ f'(z)\cos\theta & f'(z)\sin\theta & 1 \end{vmatrix} = \langle\, f(z)\cos\theta,\; f(z)\sin\theta,\; -f(z)\cdot f'(z)\,\rangle$$

and hence

$$|\mathbf{r}_\theta \times \mathbf{r}_z| = \sqrt{[f(z)]^2 + [f(z)\cdot f'(z)]^2} = f(z)\sqrt{1 + [f'(z)]^2}\,.$$

Thus, by Eq. (8), the area of the surface of revolution is

$$A = \int_{\theta=0}^{2\pi}\int_{z=a}^{b} f(z)\sqrt{1 + [f'(z)]^2}\; dz\, d\theta = \int_{z=a}^{b} 2\pi f(z)\sqrt{1 + [f'(z)]^2}\; dz.$$

Compare this with Eq. (8) in Section 6.4.

13.8.23: In the result in Problem 21, take $f(z) = r$ (the constant radius of the cylinder). Then

$$f(z)\sqrt{1 + \left[f'(z)\right]^2} = r,$$

so the curved surface area of the cylinder is

$$A = \int_{\theta=0}^{2\pi}\int_{z=0}^{h} r\; dz\, d\theta = 2\pi \cdot \left[rz\right]_{z=0}^{h} = 2\pi rh.$$

13.8.25: Part (a): The *Mathematica* command

```
Plot3D[ x∧2 + y∧2, {x, -1, 1}, {y, -1, 1} ];
```

913

produced the view of the surface that is shown next.

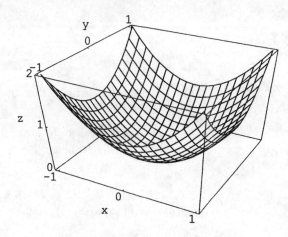

The surface area element is $dS = \sqrt{1 + 4x^2 + 4y^2}\, dA$, so the area of the surface is

$$A = \int_{x=-1}^{1} \int_{y=-1}^{1} \sqrt{1 + 4x^2 + 4y^2}\, dy\, dx$$

$$= \int_{-1}^{1} \left[\frac{1}{2} y \sqrt{1 + 4x^2 + 4y^2} + \frac{1}{4}(4x^2 + 1) \ln\left(2y + \sqrt{1 + 4x^2 + 4y^2}\right) \right]_{-1}^{1} dx$$

$$= \int_{-1}^{1} \left[\sqrt{4x^2 + 5} - \frac{1}{4}(4x^2 + 1) \ln\left(-2 + \sqrt{4x^2 + 5}\right) + \frac{1}{4}(4x^2 + 1) \ln\left(2 + \sqrt{4x^2 + 5}\right) \right] dx$$

$$= \left[\frac{2}{3} x \sqrt{4x^2 + 5} + \frac{7}{6} \operatorname{arcsinh}\left(\frac{2x}{\sqrt{5}}\right) - \frac{1}{6} \arctan\left(\frac{4x}{\sqrt{4x^2 + 5}}\right) \right.$$

$$\left. + \frac{1}{12} x(4x^2 + 3) \ln\left(-2 + \sqrt{4x^2 + 5}\right) + \frac{1}{12} x(4x^2 + 3) \ln\left(2 + \sqrt{4x^2 + 5}\right) \right]_{-1}^{1}$$

$$= 4 + \frac{7}{3} \operatorname{arcsinh}\left(\frac{2\sqrt{5}}{5}\right) - \frac{1}{3} \arctan\left(\frac{4}{3}\right) + \frac{7}{6} \ln 5 \approx 7.44625672301236346326.$$

Part (b): By symmetry, we integrate over the quarter of the square that lies in the first quadrant and multiply by 4. Thus the area is

$$A = 4 \int_{x=0}^{1} \int_{y=0}^{1-x} \sqrt{1 + 4x^2 + 4y^2}\, dy\, dx$$

$$= \int_{0}^{1} \left[2y \sqrt{1 + 4x^2 + 4y^2} + (4x^2 + 1) \ln\left(2y + \sqrt{1 + 4x^2 + 4y^2}\right) \right]_{0}^{1-x} dx$$

$$= \int_{0}^{1} \left[2(1 - x) \sqrt{1 + 4(1 - x)^2 + 4x^2} - (1 + 4x^2) \ln\left(\sqrt{1 + 4x^2}\right) \right.$$

$$\left. + (1 + 4x^2) \ln\left(2(1 - x) + \sqrt{1 + 4(1 - x)^2 + x^2}\right) \right] dx$$

$$= \left[\frac{1}{3}(4x - 2x^2 - 1) \sqrt{8x^2 - 8x + 5} + \frac{5\sqrt{2}}{6} \operatorname{arcsinh}\left(\frac{\sqrt{6}}{3}[2x - 1]\right) \right.$$

914

$$-\frac{1}{6}\arctan\left(\frac{72+288x^2-25(4x+1)\sqrt{8x^2-8x+5}}{184x^2-200x-29}\right)$$

$$+\frac{1}{6}\arctan\left(\frac{-72-288x^2+25(4x+1)\sqrt{8x^2-8x+5}}{184x^2-200x-29}\right)$$

$$-\frac{1}{3}x(4x^2+3)\ln\left(\sqrt{1+4x^2}\right)+\frac{1}{3}x(4x^2+3)\ln\left(2-2x+\sqrt{8x^2-8x+5}\right)\Bigg]_0^1$$

$$=\frac{2\sqrt{5}}{3}+\frac{5\sqrt{2}}{3}\operatorname{arcsinh}\left(\frac{\sqrt{6}}{3}\right)-\frac{1}{6}\arctan\left(\frac{72-25\sqrt{5}}{71}\right)-\frac{1}{6}\arctan\left(\frac{72-25\sqrt{5}}{29}\right)$$

$$+\frac{1}{6}\arctan\left(\frac{25\sqrt{5}-72}{29}\right)+\frac{1}{6}\arctan\left(\frac{25\sqrt{5}-72}{71}\right)\approx 3.0046254342814410.$$

Of course it was *Mathematica* that computed and evaluated the antiderivatives in this solution.

13.8.27: Part (a): The following graph of the surface was generated by the *Mathematica* command

```
Plot3D[ 1 + x*y, {x, -1, 1}, {y, -1, 1} ];
```

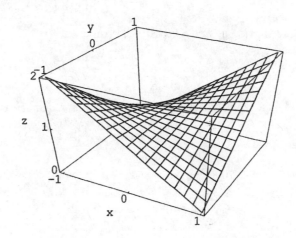

The surface area element is $dS=\sqrt{1+x^2+y^2}\;dA$. Hence the surface area is

$$\int_{x=-1}^{1}\int_{y=-1}^{1}\sqrt{1+x^2+y^2}\;dy\,dx=\int_{-1}^{1}\left[\frac{1}{2}y\sqrt{1+x^2+y^2}+\frac{1}{2}(1+x^2)\ln\left(y+\sqrt{1+x^2+y^2}\right)\right]_{-1}^{1}dx$$

$$=\int_{-1}^{1}\left[\sqrt{2+x^2}-\frac{1}{2}(1+x^2)\ln\left(-1+\sqrt{2+x^2}\right)+\frac{1}{2}(1+x^2)\ln\left(1+\sqrt{2+x^2}\right)\right]dx$$

$$=\left[\frac{2}{3}x\sqrt{2+x^2}+\frac{4}{3}\operatorname{arcsinh}\left(\frac{x\sqrt{2}}{2}\right)-\frac{2}{3}\arctan\left(\frac{x}{\sqrt{2+x^2}}\right)\right.$$

$$\left.-\frac{1}{6}x(3+x^2)\ln\left(-1+\sqrt{2+x^2}\right)+\frac{1}{6}x(3+x^2)\ln\left(1+\sqrt{2+x^2}\right)\right]_{-1}^{1}$$

915

$$= \frac{4\sqrt{3}}{3} - \frac{2}{9}\pi + \frac{8}{3}\text{arcsinh}\left(\frac{\sqrt{2}}{2}\right) - \frac{4}{3}\ln\left(-1+\sqrt{3}\right) + \frac{4}{3}\ln\left(1+\sqrt{3}\right) \approx 5.123157101094.$$

Part (b): Using symmetry, we integrate over the quarter of the square in the first quadrant and multiply the answer by 4. Thus the surface area is

$$4\int_{x=0}^{1}\int_{y=0}^{1-x}\sqrt{1+x^2+y^2}\,dy\,dx = \int_0^1\left[2y\sqrt{1+x^2+y^2} + 2(1+x^2)\ln\left(y+\sqrt{1+x^2+y^2}\right)\right]_0^1 dx$$

$$= \int_0^1\left[2(1-x)\sqrt{2x^2-2x+2} - 2(1+x^2)\ln\left(\sqrt{1+x^2}\right)\right.$$

$$\left. + 2(1+x^2)\ln\left(1-x+\sqrt{2x^2-2x+2}\right)\right]dx$$

$$= \left[\frac{1}{3}(4x-2x^2-1)\sqrt{2x^2-2x+2} + \frac{7\sqrt{2}}{6}\text{arcsinh}\left(\frac{2x-1}{\sqrt{3}}\right) + \frac{4}{3}\arctan\left(\frac{\sqrt{2x^2-2x+2}}{x+1}\right)\right.$$

$$\left. - \frac{2}{3}x(3+x^2)\ln\left(\sqrt{1+x^2}\right) + \frac{2}{3}x(3+x^2)\ln\left(1-x+\sqrt{2x^2-2x+2}\right)\right]_0^1$$

$$= \frac{2\sqrt{2}}{3} + \frac{7\sqrt{2}}{3}\text{arcsinh}\left(\frac{\sqrt{3}}{3}\right) + \frac{4}{3}\arctan\left(\frac{\sqrt{2}}{2}\right) - \frac{4}{3}\arctan\left(\sqrt{2}\right) \approx 2.302310960471.$$

13.8.29: We readily verify that x, y, and z satisfy the equation

$$\frac{x^2}{a^2} + \frac{y^2}{b^2} = \frac{z}{c}$$

of an elliptic paraboloid. For a typical graph, we executed the *Mathematica* commmands

```
a = 2; b = 1; c = 3;

x = a*u*Cos[v]; y = b*u*Sin[v]; z = c*u^2;

ParametricPlot3D[ { x, y, z }, { u, 0, 1 }, { v, 0, 2*Pi } ];
```

and the resulting graph is next.

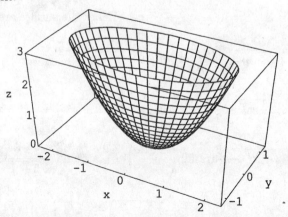

916

13.8.31: We readily verify that x, y, and z satisfy the equation

$$\frac{z^2}{c^2} - \frac{x^2}{a^2} - \frac{y^2}{b^2} = 1$$

of a hyperboloid of two sheets. To see the upper half of a typical example, we executed the following *Mathematica* commands:

```
a = 2; b = 1; c = 4;

x = a*Sinh[u]*Cos[v]; y = b*Sinh[u]*Sin[v]; z = c*Cosh[u];

ParametricPlot3D[ { x, y, z }, { u, 0, 1 }, { v, 0, 2*Pi } ];
```

and the result is next.

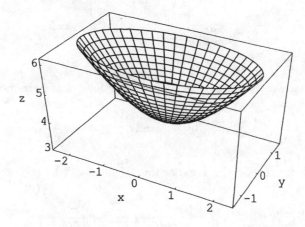

13.8.33: The ellipsoid is parametrized via

$$\mathbf{r}(\phi,\,\theta) = \langle 4\sin\phi\cos\theta,\ 3\sin\phi\sin\theta,\ 2cos\phi \rangle.$$

Then

$$\mathbf{r}_\phi = \langle 4\cos\phi\cos\theta,\ 3\cos\phi\sin\theta,\ -2\sin\phi \rangle \quad \text{and}$$

$$\mathbf{r}_\theta = \langle -4\sin\phi\sin\theta,\ 3\sin\phi\cos\theta,\ 0 \rangle.$$

It now follows that

$$\mathbf{r}_\phi \times \mathbf{r}_\theta = \langle 6\sin^2\phi\cos\theta, 8\sin^2\phi\sin\theta,\ 12\sin\phi\cos\phi \rangle,$$

so that

$$|\mathbf{r}_\phi \times \mathbf{r}_\theta| = \sqrt{36\sin^4\phi\cos^2\theta + 144\sin^2\phi\cos^2\phi + 64\sin^4\phi\sin^2\theta}.$$

917

We used the `NIntegrate` command in *Mathematica* to approximate the surface area of the ellipsoid; it is

$$\int_{\theta=0}^{2\pi} \int_{\phi=0}^{\pi} (36 \sin^4 \phi \cos^2 \theta + 144 \sin^2 \phi \cos^2 \phi + 64 \sin^4 \phi \sin^2 \theta)^{1/2} \, d\phi \, d\theta \approx 111.545774984838.$$

Section 13.9

13.9.1: It is easy to solve the given equations for

$$x = \frac{u+v}{2}, \quad y = \frac{u-v}{2}.$$

Hence

$$\frac{\partial(x, y)}{\partial(u, v)} = \begin{vmatrix} \dfrac{1}{2} & \dfrac{1}{2} \\ \dfrac{1}{2} & -\dfrac{1}{2} \end{vmatrix} = -\frac{1}{2}.$$

13.9.3: When we solve the equations $u = xy$ and $v = y/x$ for x and y, we find that there are two solutions:

$$x = \sqrt{\frac{u}{v}}, \quad y = \sqrt{uv} \qquad \text{and} \qquad x = -\sqrt{\frac{u}{v}}, \quad y = -\sqrt{uv}.$$

It doesn't matter which we choose; the value of the Jacobian will be the same. (Why?) So we choose the first solution. Then

$$\frac{\partial(x, y)}{\partial(u, v)} = \begin{vmatrix} \dfrac{1}{2u^{1/2}v^{1/2}} & -\dfrac{u^{1/2}}{2v^{3/2}} \\ \dfrac{v^{1/2}}{2u^{1/2}} & \dfrac{u^{1/2}}{2v^{1/2}} \end{vmatrix} = \frac{1}{2v}.$$

13.9.5: When we solve the equations $u = x + 2y^2$, $v = x - 2y^2$ for x and y, we get two solutions:

$$x = \frac{u+v}{2}, \qquad y = \pm \frac{\sqrt{u-v}}{2}.$$

We choose the solution for which y is nonnegative. Then

$$\frac{\partial(x, y)}{\partial(u, v)} = \begin{vmatrix} \dfrac{1}{2} & \dfrac{1}{2} \\ \dfrac{1}{4\sqrt{u-v}} & -\dfrac{1}{4\sqrt{u-v}} \end{vmatrix} = -\frac{1}{4\sqrt{u-v}}.$$

13.9.7: First we solve the equations $u = x + y$ and $v = 2x - 3y$ for

$$x = \frac{3u+v}{5}, \qquad y = \frac{2u-v}{5}.$$

918

Substitution in the equation $x + y = 1$ then yields

$$1 = \frac{3u + v}{5} + \frac{2u - v}{5} = \frac{5u}{5} = u.$$

Similarly, $x + y = 2$ yields $u = 2$, $2x - 3y = 2$ yields $v = 2$, and $2x - 3y = 5$ yields $v = 5$. Moreover,

$$\frac{\partial(x, y)}{\partial(u, v)} = \begin{vmatrix} \dfrac{3}{5} & \dfrac{1}{5} \\[2mm] \dfrac{2}{5} & -\dfrac{1}{5} \end{vmatrix} = -\frac{1}{5}.$$

Therefore

$$\iint_R 1 \; dy \; dx = \int_{v=2}^5 \int_{u=1}^2 \frac{1}{5} \; du \; dv = 3 \cdot 1 \cdot \frac{1}{5} = \frac{3}{5}.$$

Note: Because R is a parallelogram with adjacent sides represented by the two vectors $\mathbf{a} = \langle \frac{3}{5}, \frac{2}{5}, 0 \rangle$ and $\mathbf{b} = \langle \frac{3}{5}, -\frac{3}{5}, 0 \rangle$, we have the following alternative method of finding the area A of R:

$$\mathbf{a} \times \mathbf{b} = \begin{vmatrix} \mathbf{i} & \mathbf{j} & \mathbf{k} \\[2mm] \dfrac{3}{5} & \dfrac{2}{5} & 0 \\[2mm] \dfrac{3}{5} & -\dfrac{3}{5} & 0 \end{vmatrix} = \left\langle 0, 0, -\frac{3}{5} \right\rangle,$$

and therefore $A = |\mathbf{a} \times \mathbf{b}| = \dfrac{3}{5}$.

13.9.9: If $u = xy$ and $v = xy^3$, then

$$uy^2 = xy^3 = v, \quad \text{so that} \quad y^2 = \frac{v}{u}; \quad y = \frac{v^{1/2}}{u^{1/2}}.$$

Then

$$x = \frac{u}{y} = u \cdot \frac{u^{1/2}}{v^{1/2}} = \frac{u^{3/2}}{v^{1/2}}.$$

(We do not need the solution in which x and y are negative.) Then

$$\frac{\partial(x, y)}{\partial(u, v)} = \begin{vmatrix} \dfrac{3u^{1/2}}{2v^{1/2}} & -\dfrac{u^{3/2}}{2v^{3/2}} \\[3mm] -\dfrac{v^{1/2}}{2u^{3/2}} & \dfrac{1}{2u^{1/2}v^{1/2}} \end{vmatrix} = \frac{3}{4v} - \frac{1}{4v} = \frac{1}{2v}.$$

We also find by substitution that $xy = 2$ corresponds to $u = 2$, $xy = 4$ corresponds to $u = 4$, $xy^3 = 3$ corresponds to $v = 3$, and $xy^3 = 6$ corresponds to $v = 6$. Hence the area of the region shown in Fig. 13.9.8 is

$$A = \iint_D 1 \; dx \; dy = \int_{v=3}^6 \int_{u=2}^4 \frac{1}{2v} \; du \; dv = \int_3^6 \frac{1}{v} \; dv = \ln 2 \approx 0.6931471805599453.$$

13.9.11: Given: the region R bounded by the curves $y = x^3$, $y = 2x^3$, $x = y^3$, and $x = 4y^3$. Choose u and v so that $y = ux^3$ and $x = vy^3$. Then

$$y = uv^3 y^9; \qquad y^8 = \frac{1}{uv^3};$$

$$y = \frac{1}{u^{1/8}v^{3/8}}; \qquad x = vy^3 = \frac{v}{u^{3/8}v^{9/8}} = \frac{1}{u^{3/8}v^{1/8}}.$$

Then the curve $y = x^3$ can be written as

$$\frac{1}{u^{1/8}v^{3/8}} = \frac{1}{u^{9/8}v^{3/8}};$$

$$u^{1/8}v^{3/8} = u^{9/8}v^{3/8};$$

$$u = 1.$$

Similarly, the curve $y = 2x^3$ corresponds to $u = 2$, the curve $x = y^3$ corresponds to $v = 1$, and the curve $x = 4y^3$ corresponds to $v = 4$. Next,

$$\frac{\partial(x, y)}{\partial(u, v)} = \begin{vmatrix} -\dfrac{3}{8u^{11/8}v^{1/8}} & -\dfrac{1}{8u^{3/8}v^{9/8}} \\[2ex] -\dfrac{1}{8u^{9/8}v^{3/8}} & -\dfrac{3}{8u^{1/8}v^{11/8}} \end{vmatrix} = \frac{1}{8u^{3/2}v^{3/2}}.$$

Hence the area of R is

$$\iint_R 1 \, dx \, dy = \int_{v=1}^{4} \int_{u=1}^{2} \frac{1}{8u^{3/2}v^{3/2}} \, du \, dv = \int_{v=1}^{4} \left[-\frac{1}{4u^{1/2}v^{3/2}} \right]_{u=1}^{2} dv = \int_1^4 \left(\frac{1}{4v^{3/2}} - \frac{1}{4\sqrt{2}\, v^{3/2}} \right) dv$$

$$= \left[\frac{\sqrt{2} - 2}{4v^{1/2}} \right]_1^4 = \frac{2 - \sqrt{2}}{8} \approx 0.07322330470336311890.$$

13.9.13: The Jacobian of the transformation $x = 3r\cos\theta$, $y = 2r\sin\theta$ is

$$\frac{\partial(x, y)}{\partial(r, \theta)} = \begin{vmatrix} 3\cos\theta & -3r\sin\theta \\[2ex] 2\sin\theta & 2r\cos\theta \end{vmatrix} = 6r\cos^2\theta + 6r\sin^2\theta = 6r.$$

The ellipse $\dfrac{x^2}{9} + \dfrac{y^2}{4} = 1$ is transformed into

$$\frac{9r^2\cos^2\theta}{9} + \frac{4r^2\sin^2\theta}{4} = 1: \quad \text{the circle} \quad r = 1.$$

The paraboloid has equation

$$z = x^2 + y^2 = 9r^2\cos^2\theta + 4r^2\sin^2\theta.$$

920

Therefore the volume of the solid is

$$V = \int_{\theta=0}^{2\pi} \int_{r=0}^{1} (9r^2 \cos^2 \theta + 4r^2 \sin^2 \theta) \cdot 6r \, dr \, d\theta = \int_0^{2\pi} \left(\frac{27}{2} \cos^2 \theta + 6 \sin^2 \theta \right) d\theta$$

$$= \int_0^{2\pi} \left(6 + \frac{15}{2} \cdot \frac{1 + \cos 2\theta}{2} \right) d\theta = \left[\frac{39}{4} \theta + \frac{15}{8} \sin 2\theta \right]_0^{2\pi} = \frac{39}{2} \pi \approx 61.26105674500096815.$$

13.9.15: We are given the transformation $u = xy$, $v = xz$, $w = yz$. Then $uvw = x^2 y^2 z^2$. Hence

$$u^{1/2} v^{1/2} w^{1/2} = xyz = uz : \qquad z = \frac{v^{1/2} w^{1/2}}{u^{1/2}};$$

$$u^{1/2} v^{1/2} w^{1/2} = xyz = vy : \qquad y = \frac{u^{1/2} w^{1/2}}{v^{1/2}};$$

$$u^{1/2} v^{1/2} w^{1/2} = xyz = wx : \qquad x = \frac{u^{1/2} v^{1/2}}{w^{1/2}}.$$

The surface $xy = 1$ corresponds to the plane $u = 1$. Similarly, the other surfaces correspond to the planes $u = 4$, $v = 1$, $v = 9$, $w = 4$, and $w = 9$. The Jacobian of the given transformation is

$$\frac{\partial(x, y, z)}{\partial(u, v, w)} = \begin{vmatrix} \dfrac{v^{1/2}}{2u^{1/2}w^{1/2}} & \dfrac{u^{1/2}}{2v^{1/2}w^{1/2}} & -\dfrac{u^{1/2}v^{1/2}}{2w^{3/2}} \\[2mm] \dfrac{w^{1/2}}{2u^{1/2}v^{1/2}} & -\dfrac{u^{1/2}w^{1/2}}{2v^{3/2}} & \dfrac{u^{1/2}}{2v^{1/2}w^{1/2}} \\[2mm] -\dfrac{v^{1/2}w^{1/2}}{2u^{3/2}} & \dfrac{w^{1/2}}{2u^{1/2}v^{1/2}} & \dfrac{v^{1/2}}{2u^{1/2}w^{1/2}} \end{vmatrix} = -\frac{1}{2u^{1/2}v^{1/2}w^{1/2}}.$$

Therefore the volume bounded by the surfaces is

$$V = \int_{w=4}^{9} \int_{v=1}^{9} \int_{u=1}^{4} \frac{1}{2u^{1/2}v^{1/2}w^{1/2}} \, du \, dv \, dw = \int_4^9 \int_1^9 \left[\frac{u^{1/2}}{v^{1/2}w^{1/2}} \right]_{u=1}^{4} dv \, dw$$

$$= \int_4^9 \int_1^9 \frac{1}{v^{1/2}w^{1/2}} \, dv \, dw = \int_4^9 \left[\frac{2v^{1/2}}{w^{1/2}} \right]_{v=1}^{9} dw = \int_4^9 \frac{4}{w^{1/2}} \, dw = \left[8w^{1/2} \right]_4^9 = 8.$$

13.9.17: The substitution $x = u + v$, $y = u - v$ transforms the rotated ellipse $x^2 + xy + y^2 = 3$ into the ellipse S in "standard position," in which its axes lie on the coordinate axes. The resulting equation of S (in the uv-plane) is $3u^2 + v^2 = 3$. The Jacobian of this transformation is

$$\frac{\partial(x, y)}{\partial(u, v)} = \begin{vmatrix} 1 & 1 \\ 1 & -1 \end{vmatrix} = -2.$$

Therefore

$$I = \iint_R \exp(-x^2 - xy - y^2) \, dx \, dy = 2 \iint_S \exp(-3u^2 - v^2) \, du \, dv. \qquad (1)$$

921

The substitution $u = r\cos\theta$, $v = r\sqrt{3}\,\sin\theta$ has Jacobian

$$\frac{\partial(u, v)}{\partial(r, \theta)} = \begin{vmatrix} \cos\theta & -r\sin\theta \\ \sqrt{3}\,\sin\theta & r\sqrt{3}\,\cos\theta \end{vmatrix} = r\sqrt{3}\,(\cos^2\theta + \sin^2\theta) = r\sqrt{3}\,.$$

This transformation, applied to the bounding ellipse of the region S, yields

$$3 = 3u^2 + v^2 = 3r^2\cos^2\theta + 3r^2\sin^2\theta = 3r^2,$$

and thereby transforms it into the circle with polar equation $r = 1$. Then substitution in the second integral in Eq. (1) yields

$$I = 2\iint_S \exp(-3u^2 - u^2)\,du\,dv = 2\int_{\theta=0}^{2\pi}\int_{r=0}^{1} r\sqrt{3}\,\exp(-3r^2)\,dr\,d\theta$$

$$= 4\pi\sqrt{3}\left[-\frac{1}{6}\exp(-3r^2)\right]_0^1 = \frac{2}{3}\pi\sqrt{3}\,(1 - e^{-3}) \approx 3.44699122256300138528.$$

13.9.19: Suppose that k is a positive constant. First we need an integration by parts with

$$u = \rho^2 \quad\text{and}\quad dv = \rho\exp(-k\rho^2)\,d\rho :$$

$$du = 2\rho\,d\rho \quad\text{and}\quad v = -\frac{1}{2k}\exp(-k\rho^2).$$

Thus

$$\int \rho^3 \exp(-k\rho^2)\,d\rho = -\frac{1}{2k}\rho^2\exp(-k\rho^2) + \int \frac{1}{k}\rho\exp(-k\rho^2)\,d\rho$$

$$= -\frac{1}{2k}\rho^2\exp(-k\rho^2) - \frac{1}{2k^2}\exp(-k\rho^2) + C.$$

Then the improper triple integral given in Problem 19 will be the limit of I_a as $a \to +\infty$, where

$$I_a = \int_{\theta=0}^{2\pi}\int_{\phi=0}^{\pi}\int_{\rho=0}^{a} \rho^3 \exp(-k\rho^2)\sin\phi\,d\rho\,d\phi\,d\theta$$

$$= 2\pi\left[-\cos\phi\right]_{\phi=0}^{\pi}\left[-\frac{1}{2k}\rho^2\exp(-k\rho^2) - \frac{1}{2k^2}\exp(-k\rho^2)\right]_{\rho=0}^{a}$$

$$= 4\pi\left[-\frac{1}{2k}a^2\exp(-ka^2) - \frac{1}{2k^2}\exp(-ka^2) + \frac{1}{2k^2}\right].$$

Because $k > 0$, it is clear that $I_a \to \dfrac{2\pi}{k^2}$ as $a \to +\infty$.

13.9.21: Given: The solid ellipsoid R with constant density δ and boundary surface with equation

$$\frac{x^2}{a^2} + \frac{y^2}{b^2} + \frac{z^2}{c^2} = 1 \tag{1}$$

(where a, b, and c are positive constants). The transformation

$$x = a\rho \sin\phi \cos\theta, \quad y = b\rho \sin\phi \sin\theta, \quad z = c\rho \cos\phi$$

has Jacobian

$$J = \frac{\partial(x, y, z)}{\partial(\rho, \phi, \theta)} = \begin{vmatrix} a\sin\phi\cos\theta & a\rho\cos\phi\cos\theta & -a\rho\sin\phi\sin\theta \\ b\sin\phi\sin\theta & b\rho\cos\phi\sin\theta & b\rho\sin\phi\cos\theta \\ c\cos\phi & -c\rho\sin\phi & 0 \end{vmatrix}$$

$$= abc\rho^2 \cos^2\phi \sin\phi \cos^2\theta + abc\rho^2 \sin^3\phi \cos^2\theta + abc\rho^2 \cos^2\phi \sin\phi \sin^2\theta + abc\rho^2 \sin^3\phi \sin^2\theta$$

$$= abc\rho^2 \sin\phi.$$

This transformation also transforms the ellipsoidal surface of Eq. (1) into

$$\rho^2 \sin^2\phi \cos^2\theta + \rho^2 \sin^2\phi \sin^2\theta + \rho^2 \cos^2\phi = \rho^2 \sin^2\phi + \rho^2 \cos^2\phi = \rho^2 = 1,$$

and thereby transforms R into the solid ball B of radius 1 and center at the origin. Therefore the moment of inertia of the ellipsoid with respect to the z-axis is

$$I_z = \int_{\theta=0}^{2\pi} \int_{\phi=0}^{\pi} \int_{\rho=0}^{1} (\rho^2 \sin^2\phi)(a^2\cos^2\theta + b^2\sin^2\theta)\delta abc\rho^2 \sin\phi \, d\rho \, d\phi \, d\theta$$

$$= \int_0^{2\pi} \int_0^{\pi} \left[\frac{1}{5}(\delta abc\rho^5 \sin^3\phi)(a^2\cos^2\theta + b^2\sin^2\theta) \right]_{\rho=0}^{1} d\phi \, d\theta$$

$$= \int_0^{2\pi} \int_0^{\pi} \frac{1}{5}(\delta abc \sin^3\phi)(a^2\cos^2\theta + b^2\cos^2\theta) \, d\phi \, d\theta$$

$$= \frac{1}{60}\delta abc \int_0^{2\pi} \left[(\cos 3\phi - 9\cos\phi)(a^2\cos^2\theta + b^2\sin^2\theta) \right]_0^{\pi} d\theta$$

$$= \frac{4}{15}\delta abc \int_0^{2\pi} (a^2\cos^2\theta + b^2\sin^2\theta) \, d\theta = \frac{1}{15}\delta abc \left[2a^2\theta + 2b^2\theta + a^2\sin 2\theta - b^2\sin 2\theta \right]_0^{2\pi}$$

$$= \frac{1}{15}\delta abc(4\pi a^2 + 4\pi b^2).$$

Because the mass of the sphere (found in the solution of Problem 20) is $M = \frac{4}{3}\pi\delta abc$, we see that

$$\frac{I_z}{M} = \frac{1}{5}(a^2 + b^2), \quad \text{and hence that} \quad I_z = \frac{1}{5}M(a^2 + b^2).$$

923

13.9.23: If $u = xy$ and $v = xy^3$, then

$$uy^2 = xy^3 = v, \quad \text{so that} \quad y^2 = \frac{v}{u}; \quad y = \frac{v^{1/2}}{u^{1/2}}.$$

Then

$$x = \frac{u}{y} = u \cdot \frac{u^{1/2}}{v^{1/2}} = \frac{u^{3/2}}{v^{1/2}}.$$

(We do not need the solution in which x and y are negative.) Then

$$\frac{\partial(x, y)}{\partial(u, v)} = \begin{vmatrix} \dfrac{3u^{1/2}}{2v^{1/2}} & -\dfrac{u^{3/2}}{2v^{3/2}} \\[2mm] -\dfrac{v^{1/2}}{2u^{3/2}} & \dfrac{1}{2u^{1/2}v^{1/2}} \end{vmatrix} = \frac{3}{4v} - \frac{1}{4v} = \frac{1}{2v}.$$

We also find by substitution that $xy = 2$ corresponds to $u = 2$, $xy = 4$ corresponds to $u = 4$, $xy^3 = 3$ corresponds to $v = 3$, and $xy^3 = 6$ corresponds to $v = 6$. Hence the area of the region shown in Fig. 13.9.8 is

$$A = \iint_D 1 \, dx \, dy = \int_{v=3}^6 \int_{u=2}^4 \frac{1}{2v} \, du \, dv = \int_3^6 \frac{1}{v} \, dv = \ln 2 \approx 0.6931471805599453.$$

Its moments with respect to the coordinate axes are

$$M_y = \int_{v=3}^6 \int_{u=2}^4 \frac{1}{2v} \cdot \frac{u^{3/2}}{v^{1/2}} \, du \, dv = \int_{v=3}^6 \left[\frac{u^{5/2}}{5v^{3/2}} \right]_{u=2}^4 dv = \int_3^6 \frac{32 - 4\sqrt{2}}{5v^{3/2}} \, dv$$

$$= \left[\frac{8\sqrt{2} - 64}{5v^{1/2}} \right]_3^6 = \frac{72\sqrt{3} - 40\sqrt{6}}{15};$$

$$M_x = \int_{v=3}^6 \int_{u=2}^4 \frac{1}{2v} \cdot \frac{v^{1/2}}{u^{1/2}} \, du \, dv = \int_{v=3}^6 \left[\frac{u^{1/2}}{v^{1/2}} \right]_{u=2}^4 dv = \int_3^6 \frac{2 - \sqrt{2}}{v^{1/2}} \, dv$$

$$= \left[\left(4 - 2\sqrt{2} \right) v^{1/2} \right]_3^6 = 6\sqrt{6} - 8\sqrt{3}.$$

Therefore its centroid is located at the point

$$(\overline{x}, \overline{y}) = \left(\frac{72\sqrt{3} - 40\sqrt{6}}{15 \ln 2}, \frac{6\sqrt{6} - 8\sqrt{3}}{\ln 2} \right) \approx (2.570696785449, 1.212631342551).$$

13.9.25: Given: The solid ellipsoid R with constant density δ and boundary surface with equation

$$\frac{x^2}{a^2} + \frac{y^2}{b^2} + \frac{z^2}{c^2} = 1 \tag{1}$$

(where a, b, and c are positive constants). The transformation

$$x = a\rho \sin\phi \cos\theta, \quad y = b\rho \sin\phi \sin\theta, \quad z = c\rho \cos\phi$$

924

has Jacobian

$$J = \frac{\partial(x, y, z)}{\partial(\rho, \phi, \theta)} = \begin{vmatrix} a\sin\phi\cos\theta & a\rho\cos\phi\cos\theta & -a\rho\sin\phi\sin\theta \\ b\sin\phi\sin\theta & b\rho\cos\phi\sin\theta & b\rho\sin\phi\cos\theta \\ c\cos\phi & -c\rho\sin\phi & 0 \end{vmatrix}$$

$$= abc\rho^2\cos^2\phi\sin\phi\cos^2\theta + abc\rho^2\sin^3\phi\cos^2\theta + abc\rho^2\cos^2\phi\sin\phi\sin^2\theta + abc\rho^2\sin^3\phi\sin^2\theta$$

$$= abc\rho^2\sin\phi.$$

This transformation also transforms the ellipsoidal surface of Eq. (1) into

$$\rho^2\sin^2\phi\cos^2\theta + \rho^2\sin^2\phi\sin^2\theta + \rho^2\cos^2\phi = \rho^2\sin^2\phi + \rho^2\cos^2\phi = \rho^2 = 1,$$

and thereby transforms R into the solid ball B of radius 1 and center at the origin. Note also that

$$x^2 + y^2 = a^2\rho^2\sin^2\phi\cos^2\theta + b^2r^2\sin^2\phi\sin^2\theta = (\rho^2\sin^2\phi)(a^2\cos^2\theta + b^2\sin^2\theta).$$

Assume that the solid R has constant density δ. Then its moment of inertia with respect to the z-axis is

$$I_z = \int_{\theta=0}^{2\pi}\int_{\phi=0}^{\pi}\int_{\rho=0}^{1}(\rho^2\sin^2\phi)(a^2\cos^2\theta + b^2\sin^2\theta)\delta abc\rho^2\sin\phi\,d\rho\,d\phi\,d\theta$$

$$= \int_0^{2\pi}\int_0^{\pi}\frac{1}{5}(\delta abc\sin^3\phi)(a^2\cos^2\theta + b^2\sin^2\theta)\,d\phi\,d\theta$$

$$= \int_0^{2\pi}\frac{4}{15}(\delta abc)(a^2\cos^2\theta + b^2\sin^2\theta)\,d\theta = \frac{4}{15}\pi\delta abc(a^2 + b^2) = \frac{1}{5}M(a^2 + b^2)$$

where M is the mass of the ellipsoid. By symmetry,

$$I_y = \frac{4}{15}\pi\delta abc(a^2 + c^2) = \frac{1}{5}M(a^2 + c^2) \quad \text{and} \quad I_x = \frac{4}{15}\pi\delta abc(b^2 + c^2) = \frac{1}{5}M(b^2 + c^2).$$

13.9.27: The average distance of points of the ellipsoid from its center at $(0, 0, 0)$ is

$$\overline{d} = \frac{1}{V}\int_0^{2\pi}\int_0^{\pi}\int_0^1 (abc\rho^2\sin\phi)\sqrt{(a\rho\sin\phi\cos\theta)^2 + (b\rho\sin\phi\sin\theta)^2 + (c\rho\cos\phi)^2}\,\,d\rho\,d\phi\,d\theta$$

where $V = \frac{4}{3}\pi abc$ is the volume of the ellipsoid. In particular, if $a = 4$, $b = 3$, and $c = 2$, we find (using the `NIntegrate` command in *Mathematica*) that $\overline{d} \approx 2.300268522983$.

13.9.29: Part (a): First note that

$$\int_0^1\int_0^1\left(\frac{1}{1 - xy} - \frac{1}{1 + xy}\right)\,dx\,dy = \int_0^1\int_0^1\frac{2xy}{1 - x^2y^2}\,dx\,dy.$$

925

The Jacobian of the substitution $u = x^2$, $v = y^2$ is

$$\frac{\partial(u, v)}{\partial(x, y)} = \begin{vmatrix} 2x & 0 \\ 0 & 2y \end{vmatrix} = 4xy,$$

so

$$\int_0^1 \int_0^1 \frac{2xy}{1 - x^2 y^2}\, dx\, dy = \frac{1}{2} \int_0^1 \int_0^1 \frac{1}{1 - x^2 y^2} \cdot 4xy\, dx\, dy = \frac{1}{2} \int_0^1 \int_0^1 \frac{1}{1 - uv}\, du\, dv = \frac{1}{2}\zeta(2).$$

Part (b): Addition as indicated in Problem 29, and cancellation of the integrals involving $1/(1 + xy)$, yields the equation

$$2 \int_0^1 \int_0^1 \frac{1}{1 - xy}\, dx\, dy = \frac{1}{2}\zeta(2) + 2 \int_0^1 \int_0^1 \frac{1}{1 - x^2 y^2}\, dx\, dy,$$

which we readily solve for

$$\int_0^1 \int_0^1 \frac{1}{1 - x^2 y^2}\, dx\, dy = \int_0^1 \int_0^1 \frac{1}{1 - xy}\, dx\, dy - \frac{1}{4}\zeta(2) = \frac{3}{4}\zeta(2).$$

Part (c): The Jacobian of the transformation $T : \mathbf{R}^2_{uv} \to \mathbf{R}^2_{xy}$ that we define by $x = (\sin v)/(\cos u)$, $y = (\sin u)/(\cos v)$ is

$$J_T = \begin{vmatrix} \dfrac{\cos u}{\cos v} & -\dfrac{\sin u \sin v}{\cos^2 v} \\[2ex] -\dfrac{\sin u \sin v}{\cos^2 u} & \dfrac{\cos v}{\cos u} \end{vmatrix} = 1 - \frac{\sin^2 u \sin^2 v}{\cos^2 u \cos^2 v} = 1 - \tan^2 u \tan^2 v.$$

Reading the limits for the transformed integral from Fig. 14.9.10(a) in the text, we therefore find that

$$\zeta(2) = \frac{4}{3} \int_0^1 \int_0^1 \frac{1}{1 - x^2 y^2}\, dx\, dy$$

$$= \frac{4}{3} \int_0^{\pi/2} \int_0^{(\pi/2)-v} \left(1 - \frac{\sin^2 u \sin^2 v}{\cos^2 u \cos^2 v}\right)^{-1} \cdot (1 - \tan^2 u \tan^2 v)\, du\, dv = \frac{4}{3} \int_0^{\pi/2} \int_0^{(\pi/2)-v} 1\, du\, dv$$

$$= \frac{4}{3} \int_0^{\pi/2} \left(\frac{\pi}{2} - v\right) dv = \frac{4}{3} \left[\frac{\pi}{2} v - \frac{1}{2} v^2\right]_0^{\pi/2} = \frac{4}{3} \cdot \frac{\pi^2}{8} = \frac{\pi^2}{6}.$$

Chapter 13 Miscellaneous Problems

13.M.1: The domain of the given integral is bounded above by the graph of $y = x^3$, below by the x-axis, and on the right by the vertical line $x = 1$. When its order of integration is reversed, the given integral becomes

$$\int_{x=0}^1 \int_{y=0}^{x^3} \frac{1}{\sqrt{1 + x^2}}\, dy\, dx = \int_0^1 \frac{x^3}{\sqrt{1 + x^2}}\, dx$$

$$= \left[\frac{1}{3}(x^2 - 2)\sqrt{1 + x^2}\right]_0^1 = \frac{2 - \sqrt{2}}{3} \approx 0.1952621458756350.$$

Mathematica can evaluate the given integral without first reversing the order of integration. It obtains

$$\int_{y=0}^{1} \int_{x=y^{1/3}}^{1} \frac{1}{\sqrt{1+x^2}} \, dx \, dy = \int_{0}^{1} \left[\operatorname{arcsinh}(1) - \operatorname{arcsinh}(y^{1/3}) \right] dy$$

$$= \left[\frac{y^{2/3} - 2}{3} \sqrt{1 + y^{2/3}} + y \operatorname{arcsinh}(1) - y \operatorname{arcsinh}(y^{1/3}) \right]_{0}^{1} = \frac{2 - \sqrt{2}}{3}.$$

13.M.3: The domain of the given integral is bounded above by the line $y = 1$, below and on the right by the line $y = x$, and on the left by the y-axis. When its order of integration is reversed, it becomes

$$\int_{y=0}^{1} \int_{x=0}^{y} \exp(-y^2) \, dx \, dy = \int_{0}^{1} y \exp(-y^2) \, dy = \left[-\frac{1}{2} \exp(-y^2) \right]_{0}^{1}$$

$$= \frac{1}{2} - \frac{1}{2e} = \frac{e-1}{2e} \approx 0.3160602794142788.$$

Mathematica can evaluate the given integral without first reversing the order of integration:

$$\int_{x=0}^{1} \int_{y=x}^{1} \exp(-y^2) \, dy \, dx = \int_{x=0}^{1} \left[\frac{1}{2} \sqrt{\pi} \, \operatorname{erf}(y) \right]_{y=x}^{1} dx = \int_{0}^{1} \frac{1}{2} \sqrt{\pi} \left[\operatorname{erf}(1) - \operatorname{erf}(x) \right] dx$$

$$= \frac{1}{2} \left[\sqrt{\pi} \, x \left[\operatorname{erf}(1) - \operatorname{erf}(x) \right] - \exp(-x^2) \right]_{0}^{1} = \frac{e-1}{2e}.$$

Note: By definition, the *error function* is $\operatorname{erf}(x) = \dfrac{2}{\sqrt{\pi}} \displaystyle\int_{0}^{x} \exp(-t^2) \, dt$.

13.M.5: The domain of the given integral is bounded above by the graph of $y = x^2$, below by the x-axis, and on the right by the line $x = 2$. When its order of integration is reversed, it becomes

$$\int_{x=0}^{2} \int_{y=0}^{x^2} \frac{y \exp(x^2)}{x^3} \, dy \, dx = \int_{x=0}^{2} \left[\frac{y^2 \exp(x^2)}{2x^3} \right]_{y=0}^{x^2} dx$$

$$= \int_{0}^{2} \frac{1}{2} x \exp(x^2) \, dx = \left[\frac{1}{4} \exp(x^2) \right]_{0}^{2} = \frac{e^4 - 1}{4} \approx 13.3995375082860598.$$

Mathematica can evaluate the given integral without first reversing the order of integration, but the intermediate antiderivatives involve the *exponential integral function*

$$\operatorname{Ei}(z) = \int_{z}^{\infty} \frac{e^{-t}}{t} \, dt$$

and hence we omit the details.

13.M.7: The volume is

$$V = \int_{y=0}^{1} \int_{x=y}^{2-y} (x^2 + y^2) \, dx \, dy = \int_{y=0}^{1} \left[\frac{1}{3} x^3 + x y^2 \right]_{x=y}^{2-y} dy$$

$$= \int_{0}^{1} \frac{4}{3} (2 - 3y + 3y^2 - 2y^3) \, dy = \left[\frac{8}{3} y - 2y^2 + \frac{4}{3} y^3 - \frac{2}{3} y^4 \right]_{0}^{1} = \frac{4}{3}.$$

13.M.9: By symmetry, the centroid lies on the z-axis. Assume that the solid has unit density. Then its mass and volume are

$$m = V = \int_{\theta=0}^{2\pi} \int_{\phi=\pi/3}^{\pi/2} \int_{\rho=0}^{3} \rho^2 \sin\phi \, d\rho \, d\phi \, d\theta$$

$$= 2\pi \cdot \left[-\cos\phi \right]_{\pi/3}^{\pi/2} \cdot \left[\frac{1}{3}\rho^3 \right]_0^3 = 18\pi \cdot \left[0 - \left(-\frac{1}{2} \right) \right] = 9\pi \approx 28.2743338823081391.$$

The moment of the solid with respect to the xy-plane is

$$M_{xy} = \int_{\theta=0}^{2\pi} \int_{\phi=\pi/3}^{\pi/2} \int_{\rho=0}^{3} \rho^3 \sin\phi \cos\phi \, d\rho \, d\phi \, d\theta = 2\pi \int_{\phi=\pi/3}^{\pi/2} \left[\frac{1}{4}\rho^4 \sin\phi \cos\phi \right]_{\rho=0}^{3} d\phi$$

$$= 2\pi \int_{\phi=\pi/3}^{\pi/2} \frac{81}{4} \sin\phi \cos\phi \, d\phi = 2\pi \left[-\frac{81}{8} \cos^2\phi \right]_{\phi=\pi/3}^{\pi/2} = \frac{81}{16}\pi.$$

Therefore the centroid of the solid is located at the point with coordinates

$$(\overline{x}, \overline{y}, \overline{z}) = \left(0, 0, \frac{9}{16} \right).$$

13.M.11: First interchange y and z: We are to find the volume bounded by the paraboloid $z = x^2 + 3y^2$ and the cylinder $z = 4 - y^2$. These surfaces intersect in a curve that lies on the elliptic cylinder $x^2 + 4y^2 = 4$, bounding the region R in the xy-plane. Hence the volume is

$$V = \iint_R (4 - x^2 - 4y^2) \, dx \, dy.$$

Apply the transformation $x = 2r\cos\theta$, $y = r\sin\theta$. This transforms R into the region $0 \leqq r \leqq 1$, $0 \leqq \theta \leqq 2\pi$. Moreover, the Jacobian of this transformation is

$$\frac{\partial(x, y)}{\partial(r, \theta)} = \begin{vmatrix} 2\cos\theta & -2r\sin\theta \\ \sin\theta & r\cos\theta \end{vmatrix} = 2r.$$

Hence

$$V = \int_{\theta=0}^{2\pi} \int_{r=0}^{1} (4 - 4r^2) \cdot 2r \, dr \, d\theta = 2\pi \cdot \left[4r^2 - 2r^4 \right]_0^1 = 4\pi \approx 12.56637061435917295385.$$

13.M.13: First interchange x and z: We are to find the volume enclosed by the elliptical cylinder $4x^2 + y^2 = 4$ and between the planes $z = 0$ and $z = y + 2$. Let R denote the plane region in which the elliptical cylinder meets the xy-plane. Then the volume is

$$V = \iint_R (y + 2) \, dx \, dy.$$

928

The transformation $x = r\cos\theta$, $y = 2r\sin\theta$ transforms R into the rectangle $0 \le r \le 1$, $0 \le \theta \le 2\pi$. The Jacobian of this transformation is

$$\frac{\partial(x,\,y)}{\partial(r,\,\theta)} = \begin{vmatrix} \cos\theta & -r\sin\theta \\[4pt] 2\sin\theta & 2r\cos\theta \end{vmatrix} = 2r.$$

Therefore

$$V = \int_{\theta=0}^{2\pi} \int_{r=0}^{1} (2 + 2r\sin\theta) \cdot 2r \; dr \; d\theta = \int_0^{2\pi} \left[2r^2 + \frac{4}{3}r^3 \sin\theta \right]_{r=0}^{1} d\theta$$

$$= \int_0^{2\pi} \left(2 + \frac{4}{3}\sin\theta \right) d\theta = \left[2\theta - \frac{4}{3}\cos\theta \right]_0^{2\pi} = 4\pi \approx 12.56637061435917295385.$$

13.M.15: The graph of $x^4 + x^2 y^2 = y^2$ in the first quadrant is the graph of

$$y = \frac{x^2}{\sqrt{1-x^2}}, \qquad 0 \le x < 1.$$

This curve meets the line $y = x$ at the point $\left(\frac{1}{2}\sqrt{2}, \frac{1}{2}\sqrt{2} \right)$ and, of course, at the point $(x,\,y) = (0,\,0)$. Conversion of the first equation into polar form yields

$$r^4 \cos^4\theta + r^4 \cos^2\theta \sin^2\theta = r^2 \sin^2\theta;$$

$$(r^2 \cos^2\theta)(\cos^2\theta + \sin^2\theta) = \sin^2\theta;$$

$$r^2 \cos^2\theta = \sin^2\theta;$$

thus $r = \tan\theta$. Noting that the line $y = x$ has polar equation $\theta = \frac{1}{4}\pi$, we find that

$$\iint_R \frac{1}{(1 + x^2 + y^2)^2} \; dA = \int_{\theta=0}^{\pi/4} \int_{r=0}^{\tan\theta} \frac{r}{(1+r^2)^2} \; dr \; d\theta = \int_0^{\pi/4} \left[-\frac{1}{2(1+r^2)} \right]_{r=0}^{\tan\theta} d\theta$$

$$= \int_0^{\pi/4} \frac{1}{2}(1 - \cos^2\theta) \; d\theta = \frac{1}{8}\left[2\theta - \sin 2\theta \right]_0^{\pi/4} = \frac{\pi - 2}{16} \approx 0.0713495408493621.$$

13.M.17: The mass and and moments are

$$m = \int_{-2}^{2} \int_{2y^2}^{4+y^2} y^2 \; dx \; dy = \int_{-2}^{2} \left[xy^2 \right]_{2y^2}^{4+y^2} dy$$

$$= \int_{-2}^{2} (4y^2 - y^4) \; dy = \left[\frac{4}{3}y^3 - \frac{1}{5}y^5 \right]_{-2}^{2} = \frac{128}{15} \approx 8.5333333333333333;$$

$$M_x = \int_{-2}^{2} \int_{2y^2}^{4+y^2} y^3 \; dx \; dy = \int_{-2}^{2} \left[xy^3 \right]_{2y^2}^{4+y^2} dx = \int_{-2}^{2} (4y^3 - y^5) \; dy = \left[y^4 - \frac{1}{6}y^6 \right]_{-2}^{2} = 0;$$

$$M_y = \int_{-2}^{2} \int_{2y^2}^{4+y^2} xy^2 \; dx \; dy = \int_{-2}^{2} \left(8y^2 + 4y^4 - \frac{3}{2}y^6 \right) dy = \left[\frac{8}{3}y^3 + \frac{4}{5}y^5 - \frac{3}{14}y^7 \right]_{-2}^{2} = \frac{4096}{105}.$$

929

Therefore the centroid of the lamina is located at the point with coordinates

$$(\bar{x}, \bar{y}) = \left(\frac{32}{7}, 0\right) \approx (4.5714285714285714, 0).$$

13.M.19: The mass and moments are

$$m = \int_{\theta=-\pi/2}^{\pi/2} \int_{r=0}^{2\cos\theta} kr \, dr \, d\theta = \int_{\theta=-\pi/2}^{\pi/2} \left[\frac{1}{2}kr^2\right]_{r=0}^{2\cos\theta} d\theta$$

$$= \int_{\theta=-\pi/2}^{\pi/2} 2k\cos^2\theta \, d\theta = \left[\frac{1}{2}k(2\theta+\sin 2\theta)\right]_{\theta=-\pi/2}^{\pi/2} = k\pi;$$

$$M_y = \int_{\theta=-\pi/2}^{\pi/2} \int_{r=0}^{2\cos\theta} kr^2\cos\theta \, dr \, d\theta = \int_{\theta=-\pi/2}^{\pi/2} \left[\frac{1}{3}kr^3\cos\theta\right]_{r=0}^{2\cos\theta} d\theta$$

$$= \int_{\theta=-\pi/2}^{\pi/2} \frac{8}{3}k\cos^4\theta \, d\theta = \left[\frac{1}{12}k(12\theta+8\sin 2\theta+\sin 4\theta)\right]_{\theta=-\pi/2}^{\pi/2} = k\pi;$$

$$M_x = \int_{\theta=-\pi/2}^{\pi/2} \int_{r=0}^{2\cos\theta} kr^2\sin\theta \, dr \, d\theta = \int_{\theta=-\pi/2}^{\pi/2} \left[\frac{1}{3}kr^3\sin\theta\right]_{r=0}^{2\cos\theta} d\theta$$

$$= \int_{\theta=-\pi/2}^{\pi/2} \frac{8}{3}k\cos^3\theta\sin\theta \, d\theta = \left[-\frac{2}{3}k\cos^4\theta\right]_{\theta=-\pi/2}^{\pi/2} = 0.$$

Therefore the centroid of the lamina is located at the point $(1, 0)$.

13.M.21: By the first theorem of Pappus, the y-coordinate $\bar{y}$ of the centroid must satisfy the equation

$$2\pi\bar{y}\cdot\frac{1}{2}\pi ab = \frac{4}{3}\pi ab^2,$$

and it follows immediately that $\bar{y} = \dfrac{4b}{3\pi}$.

13.M.23: Assume that the lamina has constant density δ. By symmetry, $\bar{x} = 0$. The mass of the lamina and its moment with respect to the x-axis are

$$m = \int_{x=-2}^{2} \int_{y=0}^{4-x^2} \delta \, dy \, dx = \int_{x=-2}^{2} \delta(4-x^2) \, dx = \delta\left[4x - \frac{1}{3}x^3\right]_{-2}^{2} = \frac{32}{3}\delta \quad \text{and}$$

$$M_x = \int_{x=-2}^{2} \int_{y=0}^{4-x^2} \delta y \, dy \, dx = \int_{x=-2}^{2} \left[\frac{1}{2}\delta y^2\right]_{y=0}^{4-x^2} dx = \frac{1}{2}\delta\int_{x=-2}^{2}(16-8x^2+x^4) \, dx$$

$$= \frac{1}{2}\delta\left[16x - \frac{8}{3}x^3 + \frac{1}{5}x^5\right]_{-2}^{2} = \frac{256}{15}\delta.$$

Therefore the centroid of the lamina is at the point $\left(0, \frac{8}{5}\right)$.

13.M.25: The volume of the ice-cream cone is

$$V = \int_{\theta=0}^{2\pi} \int_{r=0}^{1} \int_{z=2r}^{\sqrt{5-r^2}} r \, dz \, dr \, d\theta = 2\pi \int_{r=0}^{1} \left(r\sqrt{5-r^2} - 2r^2 \right) dr$$

$$= -\frac{2}{3}\pi \cdot \left[(5-r^2)^{3/2} + 2r^3 \right]_{0}^{1} = \frac{10}{3}\pi \left(\sqrt{5} - 2 \right) \approx 2.47209807953713305410362 6.$$

13.M.27: Let δ be the [constant] density of the cone and let h denote its height. Place the cone with its vertex at the origin and with its axis lying on the nonnegative z-axis. Then its mass is $M = \frac{1}{3}\pi\delta a^2 h$. The side of the cone has cylindrical equation $z = hr/a$, so the moment of inertia of the cone with respect to the z-axis is

$$I_z = \int_{\theta=0}^{2\pi} \int_{r=0}^{a} \int_{z=hr/a}^{h} \delta r^3 \, dz \, dr \, d\theta = 2\pi\delta \int_{r=0}^{a} \left(r^3 h - \frac{r^4 h}{a} \right) dr$$

$$= 2\pi\delta \left[\frac{1}{4}r^4 h - \frac{1}{5a}r^5 h \right]_{r=0}^{a} = \frac{1}{10}\pi\delta a^4 h = \frac{3}{10}Ma^2.$$

Note that the answer is plausible and dimensionally correct. One of our physics teachers, Prof. J. J. Kyame of Tulane University (retired), always insisted that we express moment of inertia in terms of mass, as here, so that the answer can be inspected for plausibility and dimensional accuracy.

13.M.29: We are given the solid ellipsoid E with constant density $\delta = 1$ and boundary the surface with Cartesian equation

$$\frac{x^2}{a^2} + \frac{y^2}{b^2} + \frac{z^2}{c^2} = 1.$$

Its moment of inertia with respect to the x-axis is then

$$I_x = \iiint_E (y^2 + z^2) \, dV.$$

We use the transformation

$$x = a\rho \sin\phi \cos\theta, \quad y = b\rho \sin\phi \sin\theta, \quad z = c\rho \cos\phi. \tag{1}$$

Under this transformation, E is replaced with the solid B determined by

$$0 \leqq \theta \leqq 2\pi, \quad 0 \leqq \phi \leqq \pi, \quad 0 \leqq \rho \leqq 1.$$

The Jacobian of the transformation in (1) is

$$\frac{\partial(x,y,z)}{\partial(\rho,\phi,\theta)} = \begin{vmatrix} a\sin\phi\cos\theta & a\rho\cos\phi\cos\theta & -a\rho\sin\phi\sin\theta \\ b\sin\phi\sin\theta & b\rho\cos\phi\sin\theta & b\rho\sin\phi\cos\theta \\ c\cos\phi & -c\rho\sin\phi & 0 \end{vmatrix} = abc\rho^2 \sin\phi.$$

Therefore

$$I_x = \int_0^{2\pi} \int_0^\pi \int_0^1 \left[(b\rho\sin\phi\sin\theta)^2 + (c\rho\cos\phi)^2\right] \cdot abc\rho^2 \sin\phi \, d\rho \, d\phi \, d\theta$$

$$= \int_0^{2\pi} \int_0^\pi \int_0^1 (b^2\sin^2\phi\sin^2\theta + c^2\cos^2\phi) \cdot abc\rho^4 \sin\phi \, d\rho \, d\phi \, d\theta$$

$$= \frac{1}{5}abc \int_0^{2\pi} \int_0^\pi (b^2\sin^3\phi\sin^2\theta + c^2\sin\phi\cos^2\phi) \, d\phi \, d\rho$$

$$= \frac{1}{5}abc \int_0^{2\pi} \int_0^\pi \left[b^2(1 - \cos^2\phi)\sin\phi\sin^2\theta + c^2\sin\phi\cos^2\phi\right] d\phi \, d\theta$$

$$= \frac{1}{5}abc \int_0^{2\pi} \left[\frac{1}{3}b^2\cos^3\phi\sin^2\theta - b^2\cos\phi\sin^2\theta - \frac{1}{3}c^2\cos^3\phi\right]_0^\pi d\theta$$

$$= \frac{1}{5}abc \int_0^{2\pi} \left[-\frac{1}{3}b^2(1 - \cos 2\theta) + b^2(1 - \cos 2\theta) + \frac{2}{3}c^2\right] d\theta$$

$$== \frac{1}{5}abc \left[-\frac{1}{3}b^2\theta + \frac{1}{6}b^2\sin 2\theta + b^2\theta - \frac{1}{2}b^2\sin 2\theta + \frac{2}{3}c^2\theta\right]_0^{2\pi}$$

$$= \frac{1}{5}abc \left(-\frac{2}{3}\pi b^2 + 2\pi b^2 + \frac{4}{3}\pi c^2\right) = \frac{4}{15}\pi abc(b^2 + c^2) = \frac{1}{5}M(b^2 + c^2)$$

where $M = \frac{4}{3}\pi abc$ is the mass of E.

13.M.31: The cylinder $r = 2\cos\theta$ meets the xy-plane in the circle with equation $r = 2\cos\theta$, $-\frac{1}{2}\pi \leq \theta \leq \frac{1}{2}\pi$. With density $\delta = 1$, the moment of inertia of the solid region with respect to the z-axis is

$$I_z = 2\int_{-\pi/2}^{\pi/2} \int_0^{2\cos\theta} \int_0^{\sqrt{4-r^2}} r^3 \, dz \, dr \, d\theta = 4\int_0^{\pi/2} \int_0^{2\cos\theta} r^3\sqrt{4-r^2} \, dr \, d\theta$$

$$= 4\int_0^{\pi/2} \left[\frac{1}{15}(3r^4 - 4r^2 - 32)\sqrt{4-r^2}\right]_0^{2\cos\theta} d\theta$$

$$= \frac{4}{15}\int_0^{\pi/2} (64 - 64\sin\theta - 32\sin\theta\cos^2\theta + 96\sin\theta\cos^4\theta) \, d\theta$$

$$= \frac{8}{225}\left[480\theta + 450\cos\theta - 25\cos 3\theta - 9\cos 5\theta\right]_0^{\pi/2} = \frac{128}{225}(15\pi - 26) \approx 12.0171461995217912.$$

The student who obtains the incorrect answer $\frac{128}{15}\pi$ may well have overlooked the fact that

$$\sqrt{4 - 4\cos^2\theta} = |2\sin\theta|.$$

13.M.33: The area element $r \, dr \, d\theta$ moves around a circle of radius $y = r\sin\theta$, and therefore of circumference $2\pi r\sin\theta$. Hence the volume swept out is

$$V = \int_{\theta=0}^\pi \int_{r=0}^{1+\cos\theta} 2\pi r^2\sin\theta \, dr \, d\theta = \int_{\theta=0}^\pi \left[\frac{2}{3}\pi r^3\sin\theta\right]_{r=0}^{1+\cos\theta} d\theta$$

$$= \int_0^\pi \frac{2}{3}\pi(1 + \cos\theta)^3\sin\theta \, d\theta = \left[-\frac{1}{6}\pi(1 + \cos\theta)^4\right]_0^\pi = \frac{8}{3}\pi \approx 8.37758040957278196923.$$

932

13.M.35: The moment of inertia of the torus of Problem 34 with respect to the line $x = -b$ (where $b \geq 0$), its natural axis of symmetry, is

$$I = \int_{\theta=0}^{2\pi} \int_{r=0}^{a} 2\pi\delta(b + r\cos\theta)^3 \cdot r \, dr \, d\theta = \int_{\theta=0}^{2\pi} \pi\delta \left[b^3 r^2 + 2b^2 r^3 \cos\theta + \frac{3}{2} br^4 \cos^2\theta + \frac{2}{5} r^5 \cos^3\theta \right]_{r=0}^{a} d\theta$$

$$= \pi\delta \int_0^{2\pi} \left(a^2 b^3 + 2a^3 b^2 \cos\theta + \frac{3}{2} a^4 b \cos^2\theta + \frac{2}{5} a^5 \cos^3\theta \right) d\theta$$

$$= \frac{1}{120} \pi\delta a^2 \left[90a^2 b\theta + 120b^3\theta + 36a^3 \sin\theta + 240ab^2 \sin\theta + 45a^2 b \sin 2\theta + 4a^3 \sin 3\theta \right]_0^{2\pi}$$

$$= \frac{1}{120} \pi\delta a^2 (180\pi a^2 b + 240\pi b^3) = \frac{1}{2} \pi^2 \delta a^2 b (3a^2 + 4b^2) = \frac{1}{4} M(3a^2 + 4b^2)$$

where $M = 2\pi^2 \delta a^2 b$ is the mass of the torus.

13.M.37: Use the disk bounded by the circle with polar equation $r = 2a\sin\theta$, $0 \leq \theta \leq \pi$. Then the origin is a point on the boundary of the disk, and the average distance of points of this disk from the origin is

$$\bar{d} = \frac{1}{\pi a^2} \int_{\theta=0}^{\pi} \int_{r=0}^{2a\sin\theta} r^2 \, dr \, d\theta = \frac{1}{\pi a^2} \int_0^{\pi} \frac{8}{3} a^3 \sin^3\theta \, d\theta$$

$$= \frac{2a^3}{9\pi a^2} \left[\cos 3\theta - 9\cos\theta \right]_0^{\pi} = \frac{32}{9\pi} a \approx (1.131768484209)a.$$

13.M.39: We use the ball bounded by the surface with spherical-coordinates equation $\rho = a$. Then the average distance of points of this ball from its center is

$$\bar{d} = \frac{3}{4\pi a^3} \int_{\theta=0}^{2\pi} \int_{\phi=0}^{\pi} \int_{\rho=0}^{a} \rho^3 \sin\phi \, d\rho \, d\phi \, d\theta$$

$$= \frac{3}{4\pi a^3} \int_{\theta=0}^{2\pi} \int_{\phi=0}^{\pi} \frac{1}{4} a^4 \sin\phi \, d\phi \, d\theta = \frac{3}{4\pi a^3} \cdot 2\pi \cdot \left[-\frac{1}{4} a^4 \cos\phi \right]_0^{\pi} = \frac{3}{4\pi a^3} \cdot 2\pi \cdot \frac{1}{2} a^4 = \frac{3}{4} a.$$

13.M.41: We will use the spheres with spherical-coordinates equations $\rho = 2\cos\phi$ and $\rho = 4\cos\phi$, which have a mutual point of tangency at the origin. Then the average distance of points outside the smaller and inside the larger sphere from the origin is

$$\bar{d} = \frac{3}{28\pi} \int_{\theta=0}^{2\pi} \int_{\phi=0}^{\pi/2} \int_{\rho=2\cos\phi}^{4\cos\phi} \rho^3 \sin\phi \, d\rho \, d\phi \, d\theta = \frac{3}{28\pi} \cdot 2\pi \cdot \int_{\phi=0}^{\pi/2} \left[\frac{1}{4} \rho^4 \sin\phi \right]_{\rho=2\cos\phi}^{4\cos\phi} d\phi$$

$$= \frac{3}{14} \int_0^{\pi/2} 60 \sin\phi \cos^4\phi \, d\phi = \frac{3}{14} \left[-12\cos^5\phi \right]_0^{\pi/2} = \frac{18}{7} \approx 2.5714285714285714.$$

13.M.43: The part of the paraboloid that lies between the two given planes also is the part between the cylinders $r = 2$ and $r = 3$. Let R denote the part of the xy-plane between those two cylinders. Then the

933

surface area in question is

$$A = \iint_R \sqrt{r^2 + (rz_r)^2 + (z_\theta)^2} \; dr \; d\theta = \int_{\theta=0}^{2\pi} \int_{r=2}^3 \sqrt{r^2 + 4r^4} \; dr \; d\theta = 2\pi \int_2^3 r(1 + 4r^2)^{1/2} \; dr$$

$$= 2\pi \left[\frac{1}{12}(1 + 4r^2)^{3/2} \right]_2^3 = \frac{1}{6}\pi \left(37\sqrt{37} - 17\sqrt{17} \right) \approx 81.1417975124065455.$$

13.M.45: Let R be the region in the $\phi\theta$-plane determined by the inequalities $\phi_2 \leqq \phi \leqq \phi_1$ and $0 \leqq \theta \leqq 2\pi$, where

$$\cos\phi_2 = \frac{z_2}{a} \quad \text{and} \quad \cos\phi_1 = \frac{z_1}{a}.$$

Thus the part of the sphere $\rho = a$ for which the spherical coordinates ϕ and θ satisfy these inequalities is the part of the sphere between the planes $z = z_1$ and $z = z_2$. Thus the formula in Problem 18 of Section 14.8 yields the area of this surface to be

$$A = \iint_R a^2 \sin\phi \; d\phi \; d\theta = \int_{\theta=0}^{2\pi} \int_{\phi=\phi_2}^{\phi_1} a^2 \sin\phi \; d\phi \; d\theta$$

$$= 2\pi a^2 \left[+\cos\phi \right]_{\phi_1}^{\phi_2} = 2\pi a^2 \left(\frac{z_2 - z_1}{a} \right) = 2\pi a(z_2 - z_1) = 2\pi ah$$

because $h = z_2 - z_1$.

13.M.47: Position the cone with its vertex at the origin and its axis on the nonnegative z-axis. The side of the cone has Cartesian equation $z = \sqrt{x^2 + y^2}$, and hence

$$dS = \sqrt{1 + (z_x)^2 + (z_y)^2} \; dx \; dy = \sqrt{1 + \frac{x^2}{x^2 + y^2} + \frac{y^2}{x^2 + y^2}} \; dx \; dy = \sqrt{2} \; dx \; dy.$$

Let S be the square with vertices at $(\pm 1, \pm 1)$. Because the area of S is 4, we see with no additional computations that the area of the part of the cone that lies directly above S is

$$\iint_S \sqrt{2} \; dx \; dy = 4\sqrt{2} \approx 5.6568542494923802.$$

13.M.49: Given: $x = x(t)$, $y = y(t)$, $z = z$, $a \leqq t \leqq b$, $0 \leqq z \leqq h(t)$: Let

$$\mathbf{r}(t, z) = \langle x(t), y(t), z \rangle, \quad \text{so that} \quad \mathbf{r}_t = \langle x'(t), y'(t), 0 \rangle \quad \text{and} \quad \mathbf{r}_z = \langle 0, 0, 1 \rangle.$$

Then

$$\mathbf{r}_t \times \mathbf{r}_z = \begin{vmatrix} \mathbf{i} & \mathbf{j} & \mathbf{k} \\ x'(t) & y'(t) & 0 \\ 0 & 0 & 1 \end{vmatrix} = \langle y'(t), -x'(t), 0 \rangle,$$

934

and hence

$$|\mathbf{r}_t \times \mathbf{r}_z| = \sqrt{\left[x'(t)\right]^2 + \left[y'(t)\right]^2}.$$

Therefore the area of the "fence" is

$$A = \int_{t=a}^{b} \int_{z=0}^{h(t)} \left(\left[x'(t)\right]^2 + \left[y'(t)\right]^2\right)^{1/2} \, dz \, dt.$$

13.M.51: We are given the region R bounded by the curves $x^2 - y^2 = 1$, $x^2 - y^2 = 4$, $xy = 1$, and $xy = 3$, of constant density δ. Its polar moment of inertia is

$$I_0 = \iint_R (x^2 + y^2) \, \delta \, dx \, dy.$$

The hyperbolas bounding R are u-curves and v-curves if we let $u = xy$ and $v = x^2 - y^2$. If we make this substitution, then

$$4u^2 + v^2 = 4x^2 y^2 - (x^2 - y^2)^2 = (x^2 + y^2)^2,$$

and therefore we will substitute $\sqrt{4u^2 + v^2}$ for $x^2 + y^2$ in the integral for I_0. Moreover, it is not necessary to solve for x and y in terms of u and v because of a result in Section 13.9 (see the proof in Problem 18 there). Thus

$$\frac{\partial(u, v)}{\partial(x, y)} = \begin{vmatrix} y & x \\ 2x & -2y \end{vmatrix} = -2(x^2 + y^2),$$

and therefore

$$\frac{\partial(x, y)}{\partial(u, v)} = -\frac{1}{2(x^2 + y^2)} = -\frac{1}{2\sqrt{4u^2 + v^2}}.$$

Therefore

$$I_0 = \int_{v=1}^{4} \int_{u=1}^{3} \frac{\sqrt{4u^2 + v^2}}{2\sqrt{4u^2 + v^2}} \, \delta \, du \, dv = \int_{1}^{4} \int_{1}^{3} \frac{1}{2} \delta \, du \, dv = 3\delta.$$

13.M.53: We use the transformation

$$x = a\rho \sin \phi \cos \theta, \quad y = b\rho \sin \phi \sin \theta, \quad z = c\rho \cos \phi.$$

We saw in the solution of Problem 29 that the Jacobian of this transformation is

$$\frac{\partial(x, y, z)}{\partial(\rho, \phi, \theta)} = abc\rho^2 \sin \phi.$$

Moreover, it follows from work shown in the solution of Problem 29 that the density function takes the form $\delta(\rho, \phi, \theta) = 1 - \rho^2$. Finally, the ellipsoidal surface

$$\frac{x^2}{a^2} + \frac{y^2}{b^2} + \frac{z^2}{c^2} = 1$$

is transformed into the surface $\rho = 1$, and therefore the mass of the solid ellipsoid is

$$m = \int_{\theta=0}^{2\pi} \int_{\phi=0}^{\pi} \int_{\rho=0}^{1} (1 - \rho^2)abc\rho^2 \sin\phi \, d\rho \, d\phi \, d\theta = 2\pi abc \int_{\phi=0}^{\pi} \int_{\rho=0}^{1} (\rho^2 - \rho^4) \sin\phi \, d\rho \, d\phi$$

$$= 2\pi abc \int_0^{\pi} \frac{2}{15} \sin\phi \, d\phi = \frac{4}{15}\pi abc \left[-\cos\phi \right]_0^{\pi} = \frac{8}{15}\pi abc.$$

13.M.55: The spherical surface with radius $\sqrt{3}$ centered at the origin has equation $x^2 + y^2 + z^2 = 3$, and hence the upper hemisphere has equation $z = \sqrt{3 - x^2 - y^2}$. Next,

$$1 + (z_x)^2 + (z_y)^2 = 1 + \frac{x^2}{3 - x^2 - y^2} + \frac{y^2}{3 - x^2 - y^2} = \frac{3}{3 - x^2 - y^2}.$$

We integrate over the unit square in the xy-plane, quadruple the result to find the area of the part of the surface above the 2-by-2 square, then double it to account for the spherical surface *below* the xy-plane. Thus the surface area is

$$A = 8 \int_{x=0}^{1} \int_{y=0}^{1} \frac{\sqrt{3}}{\sqrt{3 - x^2 - y^2}} \, dy \, dx = 8 \int_{x=0}^{1} \left[\sqrt{3} \arctan\left(\frac{y}{\sqrt{3 - x^2 - y^2}} \right) \right]_{y=0}^{1}$$

$$= 8 \int_0^1 \sqrt{3} \arctan\left(\frac{1}{\sqrt{2 - x^2}} \right) dx = \int_{x=0}^{1} 8\sqrt{3} \arcsin\left(\frac{1}{\sqrt{3 - x^2}} \right) dx.$$

Now use integration by parts with

$$u = 8\sqrt{3} \arcsin\frac{1}{\sqrt{3 - x^2}}, \qquad dv = dx;$$

$$du = \frac{8x\sqrt{3}}{(3 - x^2)\sqrt{2 - x^2}} \, dx, \qquad v = x.$$

Thus we find that

$$A = \left[8x\sqrt{3} \arcsin\frac{1}{\sqrt{3 - x^2}} \right]_0^1 - 8\sqrt{3} \int_0^1 \frac{x^2}{(3 - x^2)\sqrt{2 - x^2}} \, dx$$

$$= 8\sqrt{3} \arcsin\frac{1}{\sqrt{2}} - 8\sqrt{3} \int_0^1 \frac{x^2}{(3 - x^2)\sqrt{2 - x^2}} \, dx = 2\pi\sqrt{3} - 8\sqrt{3} \int_0^1 \frac{x^2}{(3 - x^2)\sqrt{2 - x^2}} \, dx.$$

Now make the substitution $x = \sqrt{2} \sin\theta$, $dx = \sqrt{2} \cos\theta \, d\theta$. This yields

$$A = 2\pi\sqrt{3} - 8\sqrt{3} \int_0^{\pi/4} \frac{2\sin^2\theta}{(3 - 2\sin^2\theta)} \, d\theta$$

$$= 2\pi\sqrt{3} - 8\sqrt{3} \int_0^{\pi/4} \frac{1 - \cos 2\theta}{3 - 2\sin^2\theta} \, d\theta = 2\pi\sqrt{3} - 4\sqrt{3} \int_{\phi=0}^{\pi/2} \frac{1 - \cos\phi}{2 + \cos\phi} \, d\phi$$

where $\phi = 2\theta$. Now substitute

$$u = \tan\frac{\phi}{2}, \qquad \sin\phi = \frac{2u}{1 + u^2}, \qquad \cos\phi = \frac{1 - u^2}{1 + u^2}, \qquad d\phi = \frac{2 \, du}{1 + u^2}$$

(see the discussion immediately following Miscellaneous Problem 134 of Chapter 7). This yields

$$A = 2\pi\sqrt{3} - 4\sqrt{3} \int_0^1 \frac{4u^2}{(u^2+1)(u^2+3)}\, du = 2\pi\sqrt{3} - 8\sqrt{3} \int_0^1 \left(\frac{3}{u^2+3} - \frac{1}{u^2+1}\right) du$$

$$= 2\pi\sqrt{3} - 8\sqrt{3}\left[\sqrt{3}\arctan\left(\frac{u}{\sqrt{3}}\right) - \arctan u\right]_0^1 = 2\pi\sqrt{3} - \left(8\sqrt{3}\right)\cdot \frac{\pi}{12}\left(2\sqrt{3}-3\right)$$

$$= 4\pi\left(\sqrt{3}-1\right) \approx 9.19922175645144125328.$$

13.M.57: We will find the volume of the part of the solid that lies in the first octant, then multiply by 8. Thus the volume of the entire solid is

$$V = 8 \int_{x=0}^a \int_{y=0}^{(a^{1/3}-x^{1/3})^3} (a^{1/3} - x^{1/3} - y^{1/3})^3 \, dy \, dx.$$

The substitution $y = b\sin^6\theta$ transforms the integrand into

$$(a^{1/3} - x^{1/3} - b^{1/3}\sin^2\theta)^3,$$

and hence will be useful provided that $b^{1/3} = a^{1/3} - x^{1/3}$. Thus we choose $b = (a^{1/3} - x^{1/3})^3$, and the substitution

$$y = (a^{1/3} - x^{1/3})^3 \sin^6\theta, \quad dy = 6(a^{1/3}-x^{1/3})^3 \sin^5\theta\,\cos\theta\,d\theta$$

then yields

$$V = 8 \int_{x=0}^a \int_{\theta=0}^{\pi/2} \left[(a^{1/3}-x^{1/3})\cos^2\theta\right]^3 \cdot 6(a^{1/3}-x^{1/3})^3 \sin^5\theta\,\cos\theta\,d\theta$$

$$= 48 \int_0^a \int_0^{\pi/2} (a^{1/3}-x^{1/3})^6 \sin^5\theta\,\cos^7\theta\,d\theta$$

$$= 48 \int_0^a (a^{1/3}-x^{1/3})^6 \left[\frac{1}{122880}(-600\cos 2\theta - 75\cos 4\theta \right.$$

$$\left. + 100\cos 6\theta + 30\cos 8\theta - 12\cos 10\theta - 5\cos 12\theta)\right]_0^{\pi/2} dx$$

$$= 48 \int_0^a \frac{1}{120}(a^{1/3}-x^{1/3})^6 \, dx = \frac{2}{5}\int_0^a (a^{1/3}-x^{1/3})^6\,dx$$

$$= \frac{2}{5}\left[a^2 x - \frac{9}{2}a^{5/3}x^{4/3} + 9a^{4/3}x^{5/3} - 10ax^2 + \frac{45}{7}a^{2/3}x^{7/3} - \frac{9}{4}a^{1/3}x^{8/3} + \frac{1}{3}x^3\right]_0^a = \frac{2}{5}\cdot\frac{1}{84}a^3 = \frac{1}{210}a^3.$$

13.M.59: Locate the unit cube C as shown in the next figure, with one vertex at the origin and the opposite vertex at the point $(1, 1, 1)$ in space. Let L be the line through these two points; we will rotate C around the line L to generate the solid S. We also install a coordinate system on L; it becomes the w-axis, with $w = 0$ at the origin and $w = \sqrt{3}$ at the cube vertex with Cartesian coordinates $(1, 1, 1)$. Thus distance is measured on the w-axis in exactly the same way it is measured on the three Cartesian coordinate axes.

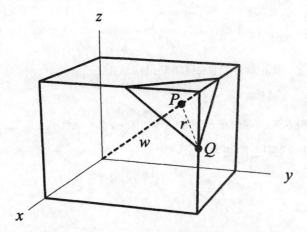

Choose a point v on the x-axis with $\frac{1}{2} \leqq v \leqq 1$. We first deal with the case $\frac{2}{3} \leqq v \leqq 1$. Then the point with Cartesian coordinates $P(v, v, v)$ is the point on the w-axis where $w = v\sqrt{3}$, and the plane normal to the w-axis at this point intersects C in a triangle, also shown in the preceding figure. It is clear that the plane has equation $x + y + z = 3v = w\sqrt{3}$ and that it meets one edge of the cube at the point $Q(1, 1, z)$ for some z between 0 and 1. In fact, because $(1, 1, z)$ satisfies the equation of the plane, it follows that $z = -2 + w\sqrt{3}$.

When the cube is rotated around L, the resulting solid S meets the plane $x + y + z = 3v$ in a circular disk centered at $P(v, v, v)$, and the radius of this disk is the distance from $P(v, v, v)$ to $Q(1, 1, z)$, which is

$$r = \sqrt{(v-1)^2 + (v-1)^2 + (w\sqrt{3} - 2 - v)^2} = \sqrt{2(v-1)^2 + (3v-2-v)^2}$$

$$= \sqrt{6(v-1)^2} = (1-v)\sqrt{6} = \left(1 - \tfrac{1}{3}w\sqrt{3}\right) \cdot \sqrt{6}.$$

Now we turn to the case $\frac{1}{2} \leqq v \leqq \frac{2}{3}$. In this case the plane through $P(v, v, v)$ meets the surface of the cube in a semi-regular hexagon, one in which each interior angle is $2\pi/3$ and whose sides are of only two different lengths a and b, alternating as one moves around the hexagon. Such a hexagon is shown in the figure below.

One of the vertices of the hexagon is located at the point $R(1, y, 0)$ on one edge of the cube. The distance from $P(v, v, v)$ to this point R is the radius r of the circular disk in which the plane normal to the w-axis at P meets the solid S. It is easy to show that $y = 3v - 1$, and it follows that the distance in question is

$$r = \sqrt{(v-1)^2 + (2v-1)^2 + v^2} = \sqrt{6v^2 - 6v + 2} = \sqrt{2w^2 - 2w\sqrt{3} + 2}.$$

938

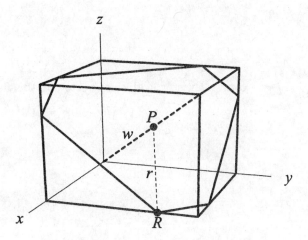

By considering only values of v for which $\frac{1}{2} \leqq v \leqq 1$, we obtain only half of the solid S, so we now can find the volume V of S as follows. We shift to coordinates on the w-axis and remember that $dw = \sqrt{3}\, dv$. The result is that

$$V = 2 \int_{v=1/2}^{2/3} \pi(2w^2 - 2w\sqrt{3} + 2)\sqrt{3}\, dw + 2 \int_{v=2/3}^{1} 6\pi \left(1 - \tfrac{1}{3}w\sqrt{3}\right)^2 \sqrt{3}\, dw.$$

The adjusted limits of integration are $w = \frac{1}{2}\sqrt{3}$ to $\frac{2}{3}\sqrt{3}$ in the first integral and $w = \frac{2}{3}\sqrt{3}$ to $\sqrt{3}$ in the second. Then *Mathematica* promptly reports that

$$V = \frac{\pi}{\sqrt{3}} \approx 1.81379936423421785059407825764215573.$$

14.1.1: $\mathbf{F}(x, y) = \langle 1, 1 \rangle$ is a constant vector field; some vectors in this field are shown next.

14.1.3: Some typical vectors in the field $\mathbf{F}(x, y) = \langle x, -y \rangle$ are shown next.

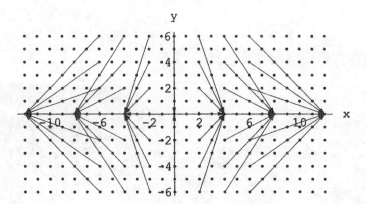

14.1.5: Some typical vectors in the field $\mathbf{F}(x, y) = \langle (x^2 + y^2)^{1/2} \langle x, y \rangle$ are shown next. Note that the length of each vector is proportional to the square of the distance from the origin to its initial point and that each vector points directly away from the origin.

940

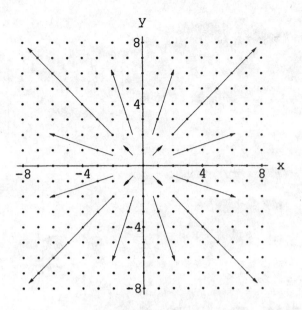

14.1.7: The vector field $\mathbf{F}(x,\, y,\, z) = \langle\, 0,\, 1,\, 1\,\rangle$ is a constant vector field. All vectors in this field are parallel translates of the one shown in the next figure.

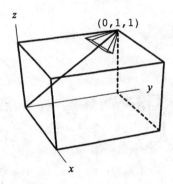

14.1.9: Each vector in the field $\mathbf{F}(x,\, y,\, z) = \langle -x,\, -y\,\rangle$ is parallel to the xy-plane and reaches from its initial point at $(x,\, y,\, z)$ to its terminal point $(0,\, 0,\, z)$ on the z-axis.

14.1.11: The vector field $\nabla(xy) = \langle\, y,\, x\,\rangle$ is shown in Fig. 14.1.8. To verify this, evaluate the gradient at $(2,\, 0)$.

14.9.13: The gradient vector field

$$\nabla\left(\sin \tfrac{1}{2}(x^2 + y^2)\right) = \left\langle\, x\cos\tfrac{1}{2}(x^2 + y^2),\, y\cos\tfrac{1}{2}(x^2 + y^2)\,\right\rangle$$

is shown in Fig. 14.1.10. To verify this, evaluate the gradient at $(1,\, 1)$ and at $(0,\, 1)$.

941

14.9.15: If $\mathbf{F}(x, y, z) = \langle x, y, z \rangle$, then

$$\nabla \cdot \mathbf{F} = 1 + 1 + 1 = 3 \quad \text{and} \quad \nabla \times \mathbf{F} = \begin{vmatrix} \mathbf{i} & \mathbf{j} & \mathbf{k} \\ \dfrac{\partial}{\partial x} & \dfrac{\partial}{\partial y} & \dfrac{\partial}{\partial z} \\ x & y & z \end{vmatrix} = \langle 0, 0, 0 \rangle = \mathbf{0}.$$

14.1.17: If $\mathbf{F}(x, y, z) = \langle yz, xz, xy \rangle$, then

$$\nabla \cdot \mathbf{F} = 0 + 0 + 0 = 0 \quad \text{and} \quad \nabla \times \mathbf{F} = \begin{vmatrix} \mathbf{i} & \mathbf{j} & \mathbf{k} \\ \dfrac{\partial}{\partial x} & \dfrac{\partial}{\partial y} & \dfrac{\partial}{\partial z} \\ yz & xz & xy \end{vmatrix} = \langle x - x, \, y - y, \, z - z \rangle = \mathbf{0}.$$

14.1.19: If $\mathbf{F}(x, y, z) = \langle xy^2, yz^2, zx^2 \rangle$, then

$$\nabla \cdot \mathbf{F} = y^2 + z^2 + x^2 \quad \text{and} \quad \nabla \times \mathbf{F} = \begin{vmatrix} \mathbf{i} & \mathbf{j} & \mathbf{k} \\ \dfrac{\partial}{\partial x} & \dfrac{\partial}{\partial y} & \dfrac{\partial}{\partial z} \\ xy^2 & yz^2 & zx^2 \end{vmatrix} = \langle -2yz, \, -2xz, \, -2xy \rangle.$$

14.1.21: If $\mathbf{F}(x, y, z) = \langle y^2 + z^2, \, x^2 + z^2, \, x^2 + y^2 \rangle$, then

$$\nabla \cdot \mathbf{F} = 0 + 0 + 0 = 0 \quad \text{and} \quad \nabla \times \mathbf{F} = \begin{vmatrix} \mathbf{i} & \mathbf{j} & \mathbf{k} \\ \dfrac{\partial}{\partial x} & \dfrac{\partial}{\partial y} & \dfrac{\partial}{\partial z} \\ y^2 + z^2 & x^2 + z^2 & x^2 + y^2 \end{vmatrix} = \langle 2y - 2z, \, 2z - 2x, \, 2x - 2y \rangle.$$

14.1.23: If $\mathbf{F}(x, y, z) = \langle x + \sin yz, \, y + \sin xz, \, z + \sin xy \rangle$, then

$$\nabla \cdot \mathbf{F} = 1 + 1 + 1 = 3 \quad \text{and}$$

$$\nabla \times \mathbf{F} = \begin{vmatrix} \mathbf{i} & \mathbf{j} & \mathbf{k} \\ \dfrac{\partial}{\partial x} & \dfrac{\partial}{\partial y} & \dfrac{\partial}{\partial z} \\ x + \sin yz & y + \sin xz & z + \sin xy \end{vmatrix}$$

$$= \langle x \cos xy - x \cos xz, \, y \cos yz - y \cos xy, \, z \cos xz - z \cos yz \rangle.$$

14.1.25: If a and b are constants and f and g are differentiable functions of two variables, then

$$\nabla(af + bg) = \left\langle \frac{\partial}{\partial x}(af + bg), \, \frac{\partial}{\partial y}(af + bg) \right\rangle = \langle af_x + bg_x, \, af_y + bg_y \rangle$$

$$= \langle af_x, \, af_y \rangle + \langle bg_x, \, bg_y \rangle = a\langle f_x, \, f_y \rangle + b\langle g_x, \, g_y \rangle = a\nabla f + b\nabla g.$$

If f and g are functions of three variables, then the proof is similar, merely longer.

14.1.27: If a and b are constants and $\mathbf{F} = \langle P, Q, R \rangle$ and $\mathbf{G} = \langle S, T, U \rangle$ where P, Q, R, S, T, and U are each differentiable functions of three variables, then

$$\nabla \times (a\mathbf{F} + b\mathbf{G}) = \nabla \times \langle aP + bS, \ aQ + bT, \ aR + bU \rangle$$

$$= \begin{vmatrix} \mathbf{i} & \mathbf{j} & \mathbf{k} \\ \dfrac{\partial}{\partial x} & \dfrac{\partial}{\partial y} & \dfrac{\partial}{\partial z} \\ aP + bS & aQ + bT & aR + bU \end{vmatrix}$$

$$= \langle aR_y + bU_y - aQ_z - bT_z, \ aP_z + bS_z - aR_x - bU_x, \ aQ_x + bT_x - aP_y - bS_y \rangle$$

$$= a \langle R_y - Q_z, \ P_z - R_x, \ Q_x - P_y \rangle + b \langle U_y - T_z, \ S_z - U_x, \ T_x - S_y \rangle$$

$$= a \, (\nabla \times \mathbf{F}) + b \, (\nabla \times \mathbf{G}).$$

14.1.29: Suppose that $\mathbf{G} = \langle S, T, U \rangle$ where S, T, U, and f are differentiable functions of x, y, and z. Then

$$\nabla \times (f\mathbf{G}) = \nabla \times \langle fS, \ fT, \ fU \rangle = \begin{vmatrix} \mathbf{i} & \mathbf{j} & \mathbf{k} \\ \dfrac{\partial}{\partial x} & \dfrac{\partial}{\partial y} & \dfrac{\partial}{\partial z} \\ fS & fT & fU \end{vmatrix}$$

$$= \langle f_y U + fU_y - f_z T - fT_z, \ f_z S + fS_z - f_x U - fU_x, \ f_x T + fT_x - f_y S - fS_y \rangle$$

$$= \langle f_y U - f_z T, \ f_z S - f_x U, \ f_x T - f_y S \rangle + \langle fU_y - fT_z, \ fS_z - fU_x, \ fT_x - fS_y \rangle$$

$$= \begin{vmatrix} \mathbf{i} & \mathbf{j} & \mathbf{k} \\ f_x & f_y & f_z \\ S & T & U \end{vmatrix} + (f) \begin{vmatrix} \mathbf{i} & \mathbf{j} & \mathbf{k} \\ \dfrac{\partial}{\partial x} & \dfrac{\partial}{\partial y} & \dfrac{\partial}{\partial z} \\ S & T & U \end{vmatrix} = (\nabla f) \times \mathbf{G} + (f)(\nabla \times \mathbf{G}).$$

14.1.31: If $\mathbf{F} = \langle P, Q, R \rangle$ and $\mathbf{G} = \langle S, T, U \rangle$ where P, Q, R, S, T, and U are each differentiable functions of three variables, then

$$\nabla \cdot (\mathbf{F} \times \mathbf{G}) = \nabla \cdot \langle QU - RT, \ RS - PU, \ PT - QS \rangle$$

$$= Q_x U + QU_x - R_x T - RT_x + R_y S + RS_y - P_y U - PU_y + P_z T + PT_z - Q_z S - QS_z$$

and

$$\mathbf{G} \cdot (\nabla \times \mathbf{F}) - \mathbf{F} \cdot (\nabla \times \mathbf{G}) = \mathbf{G} \cdot \langle R_y - Q_z, \ P_z - R_x, \ Q_x - P_y \rangle \ - \ \mathbf{F} \cdot \langle U_y - T_z, \ S_z - U_x, \ T_x - S_y \rangle$$

$$= SR_y - SQ_z + TP_z - TR_x + UQ_x - UP_y - PU_y + PT_z - QS_z + QU_x - RT_x + RS_y.$$

Then comparison of the last expressions in each of the two computations reveals that

$$\nabla \cdot (\mathbf{F} \times \mathbf{G}) \ = \ \mathbf{G} \cdot (\nabla \times \mathbf{F}) - \mathbf{F} \cdot (\nabla \times \mathbf{G}).$$

14.1.33: Suppose that f and g are twice-differentiable functions of the three variables x, y, and z. Then

$$\nabla \cdot \left[\nabla(fg) \right] = \nabla \cdot \langle f_x g + f g_x, \ f_y g + f g_y, \ f_z g + f g_z \rangle$$

$$= f_{xx} g + 2 f_x g_x + f g_{xx} + f_{yy} g + 2 f_y g_y + f g_{yy} + f_{zz} g + 2 f_z g_z + f g_{zz}$$

$$= (f)(g_{xx} + g_{yy} + g_{zz}) + (g)(f_{xx} + f_{yy} + f_{zz}) + 2(f_x g_x + f_y g_y + f_z g_z)$$

$$= f \nabla \cdot \langle g_x, g_y, g_z \rangle \ + \ g \nabla \cdot \langle f_x, f_y, f_z \rangle \ + \ 2 \langle f_x, f_y, f_z \rangle \cdot \langle g_x, g_y, g_z \rangle$$

$$= f \nabla \cdot (\nabla g) + g \nabla \cdot (\nabla f) + 2(\nabla f) \cdot (\nabla g).$$

14.1.35: If $\mathbf{r} = \langle x, y, z \rangle$, then $\nabla \cdot \mathbf{r} = 3$ and $\nabla \times \mathbf{r} = \mathbf{0}$ by the solution of Problem 15.

14.1.37: By the results in Problems 28 and 35,

$$\nabla \cdot \frac{\mathbf{r}}{r^3} = \frac{1}{r^3} \left(\nabla \cdot \mathbf{r} \right) - \frac{3}{2} (x^2 + y^2 + z^2)^{-5/2} \langle 2x, 2y, 2z \rangle \cdot \mathbf{r} = \frac{3}{r^3} - 3r^{-5}(x^2 + y^2 + z^2) = \frac{3}{r^3} - \frac{3r^2}{r^5} = 0.$$

14.1.39: If $r = |\mathbf{r}| = (x^2 + y^2 + z^2)^{1/2}$, then

$$\nabla r = \nabla (x^2 + y^2 + z^2)^{1/2} = \frac{1}{2} (x^2 + y^2 + z^2)^{-1/2} \langle 2x, 2y, 2z \rangle = \frac{\mathbf{r}}{r}.$$

14.1.41: By the results in Problems 28, 35, and 39,

$$\nabla \cdot (r\mathbf{r}) = r \nabla \cdot \mathbf{r} + (\nabla r) \cdot \mathbf{r} = 3r + \frac{\mathbf{r} \cdot \mathbf{r}}{r} = 3r + \frac{r^2}{r} = 4r.$$

14.1.43: Here we have

$$\nabla(\ln r) = \nabla \left(\ln \sqrt{x^2 + y^2 + z^2} \right) = \frac{1}{2} \left\langle \frac{2x}{x^2 + y^2 + z^2}, \ \frac{2y}{x^2 + y^2 + z^2}, \ \frac{2z}{x^2 + y^2 + z^2} \right\rangle$$

$$= \frac{1}{x^2 + y^2 + z^2} \langle x, y, z \rangle = \frac{\mathbf{r}}{r^2}.$$

Section 14.2

14.2.1: If $x(t) = 4t - 1$ and $y(t) = 3t + 1$, $-1 \le t \le 1$, then

$$\int_C (x^2 + y^2) \, ds = \int_{-1}^{1} (125t^2 - 10t + 10) \, dt = \left[\frac{125}{3} t^3 - 5t^2 + 10t \right]_{-1}^{1} = \frac{310}{3},$$

$$\int_C (x^2 + y^2) \, dx = \int_{-1}^{1} (100t^2 - 8t + 8) \, dt = \left[\frac{100}{3} t^3 - 4t^2 + 8t \right]_{-1}^{1} = \frac{248}{3}, \quad \text{and}$$

$$\int_C (x^2 + y^2) \, dy = \int_{-1}^{1} (75t^2 - 6t + 6) \, dt = \left[25t^3 - 3t^2 + 6t \right]_{-1}^{1} = 62.$$

14.2.3: If $x(t) = e^t + 1$ and $y(t) = e^t - 1$, $0 \le t \le \ln 2$, then

$$\int_C (x + y) \, ds = \int_0^{\ln 2} 2^{3/2} e^{2t} \, dt = \left[2^{1/2} e^{2t} \right]_0^{\ln 2} = 3\sqrt{2} \approx 4.2426406871192851,$$

$$\int_C (x + y) \, dx = \int_0^{\ln 2} 2e^{2t} \, dt = \left[e^{2t} \right]_0^{\ln 2} = 3, \quad \text{and} \quad \int_C (x + y) \, dy = \int_0^{\ln 2} 2e^{2t} \, dt = \left[e^{2t} \right]_0^{\ln 2} = 3.$$

14.2.5: If $x(t) = 3t$ and $y(t) = t^4$, $0 \le t \le 1$, the

$$\int_C xy \, ds = \int_0^1 3t^5 \sqrt{9 + 16t^6} \, dt = \left[\frac{1}{48} (9 + 16t^6)^{3/2} \right]_0^1 = \frac{49}{24} \approx 2.0416666666666667,$$

$$\int_C xy \, dx = \int_0^1 9t^5 \, dt = \left[\frac{3}{2} t^6 \right]_0^1 = \frac{3}{2}, \quad \text{and} \quad \int_C xy \, dy = \int_0^1 12t^8 \, dt = \left[\frac{4}{3} t^9 \right]_0^1 = \frac{4}{3}.$$

14.2.7: One parametrization of the path C is this: Let $y(t) = t$ and $x(t) = t^3$, $-1 \le t \le 1$. Then

$$\int_C P(x, y) \, dx + Q(x, y) \, dy = \int_C y^2 \, dx + x \, dy = \int_{-1}^{1} (t^3 + 3t^4) \, dt = \left[\frac{1}{4} t^4 + \frac{3}{5} t^5 \right]_{-1}^{1} = \frac{6}{5}.$$

14.2.9: Parametrize the path C in two parts. Let

$$x_1(t) = t, \ y_1(t) = 1, \ -1 \le t \le 2 \quad \text{and let} \quad x_2(t) = 2, \ y_2(t) = t, \ 1 \le t \le 5.$$

Then

$$\int_C P(x, y) \, dx + Q(x, y) \, dy = \int_C x^2 y \, dx + xy^3 \, dy = \int_{-1}^{2} t^2 \, dt + \int_{1}^{5} 2t^3 \, dt$$

$$= \left[\frac{1}{3} t^3 \right]_{-1}^{2} + \left[\frac{1}{2} t^4 \right]_{1}^{5} = 3 + 312 = 315.$$

14.2.11: Because $\mathbf{r}(t) = \langle t, t^2, t^3 \rangle$ for $0 \leqq t \leqq 1$, we have

$$d\mathbf{r} = \langle 1, 2t, 3t^2 \rangle \quad \text{and} \quad \mathbf{F}(t) = \langle t^3, t, -t^2 \rangle.$$

Therefore

$$\int_C \mathbf{F} \cdot d\mathbf{r} = \int_0^1 (2t^2 + t^3 - 3t^4) \, dt = \left[\frac{2}{3}t^3 + \frac{1}{4}t^4 - \frac{3}{5}t^5 \right]_0^1 = \frac{19}{60} \approx 0.316666666667.$$

14.2.13: The path C is parametrized by $\mathbf{r}(t) = \langle \sin t, \cos t, 2t \rangle$, $0 \leqq t \leqq \pi$. Hence

$$d\mathbf{r} = \langle \cos t, -\sin t, 2 \rangle \quad \text{and} \quad \mathbf{F}(t) = \langle \cos t, -\sin t, 2t \rangle,$$

and thus

$$\int_C \mathbf{F} \cdot d\mathbf{r} = \int_0^\pi (4t + 1) \, dt = \left[2t^2 + t \right]_0^\pi = 2\pi^2 + \pi \approx 22.8808014557685105.$$

14.2.15: Parametrize the path C in three sections, as follows:

$$x_1(t) = -1, \qquad y_1(t) = 2, \qquad z_1(t) = -2 + 4t, \quad 0 \leqq t \leqq 1;$$

$$x_2(t) = -1 + 2t, \quad y_2(t) = 2, \qquad z_2(t) = 2, \qquad 0 \leqq t \leqq 1;$$

$$x_3(t) = 1, \qquad y_3(t) = 2 + 3t, \quad z_3(t) = 2, \qquad 0 \leqq t \leqq 1.$$

Then

$$\int_C \mathbf{F} \cdot d\mathbf{r} = I_1 + I_2 + I_3$$

where

$$I_1 = \int_0^1 (32 - 64t) \, dt = \left[32t - 32t^2 \right]_0^1 = 0,$$

$$I_2 = \int_0^1 16 \, dt = \left[16t \right]_0^1 = 16, \quad \text{and}$$

$$I_3 = \int_0^1 12 \, dt = \left[12t \right]_0^1 = 12.$$

Therefore $\displaystyle\int_C \mathbf{F} \cdot d\mathbf{r} = 0 + 16 + 12 = 28.$

14.2.17: Here we have

$$\int_C (2x + 9xy) \, ds = \int_0^1 (2t + 9t^3)(1 + 4t^2 + 9t^4)^{1/2} \, dt$$

$$= \left[\frac{1}{6}(1 + 4t^2 + 9t^4)^{3/2} \right]_0^1 = \frac{14\sqrt{14} - 1}{6} \approx 8.56386723580586532.$$

946

14.2.19: Because the wire W is uniform, we may assume that its density is $\delta = 1$. Moreover, $\bar{x} = 0$ by symmetry. Parametrize the wire by

$$\mathbf{r}(t) = \langle a\cos t, \ a\sin t \rangle, \quad 0 \leq t \leq \pi.$$

Then $ds = a\,dt$. Also the mass of the wire is πa, so it remains only to compute the moment

$$M_x = \int_W ay\,dt = \int_0^\pi a^2 \sin t\,dt = \left[-a^2 \cos t\right]_0^\pi = 2a^2.$$

Therefore the centroid of the wire is located at the point $\left(0, \dfrac{2a}{\pi}\right)$.

14.2.21: First, the arc length element is

$$ds = \sqrt{9\sin^2 t + 9\cos^2 t + 16}\ dt = 5\,dt.$$

The mass and moments of the helical wire W are

$$m = \int_W 5k\,dt = \int_0^{2\pi} 5k\,dt = 10k\pi;$$

$$M_{yz} = \int_0^{2\pi} 15k\cos t\,dt = \left[15k\sin t\right]_0^{2\pi} = 0;$$

$$M_{xz} = \int_0^{2\pi} 15k\sin t\,dt = \left[-15k\cos t\right]_0^{2\pi} = 0;$$

$$M_{xy} = \int_0^{2\pi} 20kt\,dt = \left[10kt^2\right]_0^{2\pi} = 40k\pi^2.$$

Therefore the coordinates of the centroid are

$$\bar{x} = \bar{y} = 0, \quad \bar{z} = \frac{40k\pi^2}{10k\pi} = 4\pi.$$

14.2.23: Parametrize the wire W via

$$\mathbf{r}(t) = \langle a\cos t, \ a\sin t, \ 0 \rangle, \quad 0 \leq t \leq \frac{\pi}{2}.$$

Then the arc-length element is $ds = |\mathbf{r}'(t)|\,dt = a\,dt$. The mass element is $dm = a \cdot k \cdot a^2 \sin t \cos t\,dt$, and hence the mass of W is

$$m = \int_0^{\pi/2} ka^3 \sin t \cos t\,dt = \left[\frac{1}{2}ka^3 \sin^2 t\right]_0^{\pi/2} = \frac{1}{2}ka^3.$$

Clearly the z-coordinate of the centroid is $\bar{z} = 0$, and $\bar{y} = \bar{x}$ by symmetry. The moment of the wire with respect to the y-axis is

$$M_y = \int_0^{\pi/2} ka^4 \sin t \cos^2 t \, dt = -\frac{1}{3} ka^4 \left[\cos^3 t \right]_0^{\pi/2} = \frac{1}{3} ka^4.$$

Therefore the x-coordinate of the centroid is

$$\bar{x} = \frac{2ka^4}{3ka^3} = \frac{2}{3} a.$$

The moments of W with respect to the coordinate axes are

$$I_x = \int_0^{\pi/2} ka^5 \sin^2 t \cos t \, dt = \frac{1}{4} ka^5 \left[\sin^4 t \right]_0^{\pi/2} = \frac{1}{4} ka^5 = \frac{1}{2} ma^2;$$

$I_y = I_x$ by symmetry, and $I_0 = I_x + I_y = \frac{1}{2} ka^5 = ma^2$.

14.2.25: Using the given parametrization we find that the arc-length element is $ds = \frac{3}{2} |\sin 2t| \, dt$, and hence the polar moment of inertia of the wire is

$$I_0 = 4 \int_0^{\pi/2} k(\cos^6 t + \sin^6 t) \, ds = -\frac{3}{16} k \left[\cos 6t + 7\cos 2t \right]_0^{\pi/2} = 3k.$$

Because the mass of the wire is

$$m = 4 \int_0^{\pi/2} k \, ds = -3k \left[\cos 2t \right]_0^{\pi/2} = 6k,$$

we can also write $I_0 = \frac{1}{2} m$.

14.2.27: We are given the circle C of radius a centered at the origin and the point $(a, 0)$ on C. Suppose that (x, y) is a point of C. Let t be the angular polar coordinate of (x, y). Then, by the law of cosines, the distance w between $(a, 0)$ and (x, y) satisfies the equation

$$w^2 = a^2 + a^2 - 2a^2 \cos t = 2a^2(1 - \cos t) = 4a^2 \cdot \frac{1 - \cos t}{2} = 4a^2 \sin^2 \frac{t}{2}.$$

Because $0 \leq t \leq 2\pi$, it now follows that $w = 2a \sin \frac{t}{2}$, and hence the average value of w on C is

$$\bar{d} = \frac{1}{2\pi a} \int_0^{2\pi} 2a^2 \sin \frac{t}{2} \, dt = -\frac{4a^2}{2\pi a} \left[\cos \frac{t}{2} \right]_0^{2\pi} = \frac{8a^2}{2\pi a} = \frac{4}{\pi} a \approx (1.2732395447351627)a.$$

14.2.29: The parametrization $x(t) = \cos^3 t$, $y(t) = \sin^3 t$, $0 \leq t \leq 2\pi$ of the astroid yields the arc-length element

$$ds = \sqrt{9\cos^4 t \sin^2 t + 9\cos^2 t \sin^4 t} \, dt = \frac{3}{2} \sqrt{\sin^2 2t} \, dt = \frac{3}{2} \sin 2t \, dt,$$

948

although the last equality is valid only if $\sin 2t$ is nonnegative. Hence we will find the average distance of points of the astroid in the first quadrant from the origin; by symmetry, this will be the same as the average distance of all of its points from the origin. Noting that

$$[x(t)]^2 + [y(t)]^2 = \cos^6 t + \sin^6 t$$

and noting also that the length of the first-quadrant arc of the astroid is $\frac{3}{2}$ (a consequence of the solution of Problem 30 in Section 9.5), we find that the average distance of points of the astroid from the origin is

$$\bar{d} = \frac{2}{3} \int_0^{\pi/2} \frac{3}{2} \sqrt{\cos^6 t + \sin^6 t} \, \sin 2t \, dt$$

$$= \left[\frac{\sqrt{3}}{24} \operatorname{arctanh}\left(\frac{\sqrt{6} \cos 2t}{\sqrt{5 + 3 \cos 4t}} \right) - \frac{\sqrt{2}}{16} (\cos 2t) \sqrt{5 + 3 \cos 4t} \right]_0^{\pi/2}$$

$$= \frac{1}{2} + \frac{\sqrt{3}}{12} \operatorname{arctanh} \frac{\sqrt{3}}{2} \approx 0.6900864990752365868827735637 2.$$

Of course we used *Mathematica* to find and simplify both the antiderivative and the value of the definite integral. By contrast, *Derive* yields the result

$$\bar{d} = \frac{2}{3} \int_0^{\pi/2} \frac{3}{2} \sqrt{\cos^6 t + \sin^6 t} \, \sin 2t \, dt$$

$$= \left[-\frac{\sqrt{3}}{24} \ln\left(2\sqrt{3 \cos^4 t - 3 \cos^2 t + 1} + 2\sqrt{3} \, \cos^2 t - \sqrt{3} \right) - \frac{1 - 2\cos^2 t}{4} \sqrt{3 \cos^4 t - 3 \cos^2 t + 1} \right]_0^{\pi/2}$$

$$= \frac{6 + \sqrt{3} \, \ln\left(2 + \sqrt{3} \right)}{12} \approx 0.6900864990752365868827735637 2.$$

14.2.31: With the given parametrization, we find that $ds = \sqrt{2} \, e^{-t} \, dt$ and that

$$\sqrt{[x(t)]^2 + [y(t)]^2} = e^{-t}.$$

The length of the spiral is

$$\int_0^\infty \sqrt{2} \, e^{-t} \, dt = \left[-\sqrt{2} \, e^{-t} \right]_0^\infty = \sqrt{2} \,,$$

and thus the average distance of points of the spiral from the origin is

$$\bar{d} = \frac{1}{\sqrt{2}} \int_0^\infty \sqrt{2} \, e^{-2t} \, dt = \left[-\frac{1}{2} e^{-2t} \right]_0^\infty = \frac{1}{2}.$$

14.2.33: Part (a): Parametrize the path by $x(t) = 1$, $y(t) = t$, $0 \leq t \leq 1$. Then the force is

$$\mathbf{F}(t) = \left\langle \frac{k}{1 + t^2}, \frac{kt}{1 + t^2} \right\rangle,$$

and so the work is

$$W = \int_0^1 \mathbf{F} \cdot d\mathbf{r} = \int_0^1 \frac{kt}{1+t^2}\, dt = \left[\frac{1}{2}k\ln(1+t^2)\right]_0^1 = \frac{1}{2}k\ln 2.$$

Part (b): Parametrize the path by $x(t) = 1 - t$, $y(t) = 1$, $0 \le t \le 1$. Then the force is

$$\mathbf{F}(t) = \left\langle \frac{k(1-t)}{1+(1-t)^2},\ \frac{k}{1+(1-t)^2} \right\rangle,$$

and thus the work is

$$W = \int_0^1 -\frac{k(1-t)}{1+(1-t)^2}\, dt = \left[\frac{1}{2}k\ln\left(1+(1-t)^2\right)\right]_0^1 = -\frac{1}{2}k\ln 2.$$

14.2.35: Parametrize the unit circle C in the usual way: $x(t) = \cos t$, $y(t) = \sin t$, $0 \le t \le 2\pi$. Then the force function is $\mathbf{F}(t) = \langle k\cos t, k\sin t \rangle$. Hence $\mathbf{F} \cdot d\mathbf{r} = 0$, and therefore

$$\int_C \mathbf{F} \cdot d\mathbf{r} = 0.$$

14.2.37: The work done in moving along a path on the sphere is zero because $\mathbf{F}$ is normal to the sphere. Therefore $\mathbf{F} \cdot \mathbf{T}$ is identically zero on any path on the sphere.

14.2.39: The force function is $\mathbf{F}(t) = \langle 0, -150 \rangle$ and the path may be parametrized as follows: $x(t) = 100\sin t$, $y(t) = 100\cos t$, $0 \le t \le \frac{1}{2}\pi$. Hence the work done is

$$\int_C \mathbf{F} \cdot d\mathbf{r} = \int_0^{\pi/2} 15000\sin t\, dt = -15000\left[\cos t\right]_0^{\pi/2} = 15000 \quad \text{(ft·lb)}.$$

14.2.41: The force function is $\mathbf{F}(t) = \langle 0, 0, -200 \rangle$ and the path may be parametrized as follows: $x(t) = 25\cos t$, $y(t) = 25\sin t$, $z(t) = 100 - 100t/(10\pi)$, $0 \le t \le 10\pi$. Hence the work done is

$$\int_C \mathbf{F} \cdot d\mathbf{r} = \int_0^{10\pi} \frac{2000}{\pi}\, dt = \left[\frac{2000}{\pi}t\right]_0^{10\pi} = 20000 \quad \text{(ft·lb)}.$$

Section 14.3

Note: In the solutions of Problems 1 through 16, $P(x, y)$ always denotes the first component of the given vector field $\mathbf{F}(x, y)$, $Q(x, y)$ always denotes its second component, and $\phi(x, y)$ denotes a scalar potential function with gradient $\mathbf{F}(x, y)$.

14.3.1: If $P(x, y) = 2x + 3y$ and $Q(x, y) = 3x + 2y$, then

$$\frac{\partial P}{\partial y} = 3 = \frac{\partial Q}{\partial x}.$$

Hence $\mathbf{F}$ is conservative. By inspection, $\phi(x, y) = x^2 + 3xy + y^2$.

14.3.3: If $P(x, y) = 3x^2 + 2y^2$ and $Q(x, y) = 4xy + 6y^2$, then

$$\frac{\partial P}{\partial y} = 4y = \frac{\partial Q}{\partial x}.$$

Hence $\mathbf{F}$ is conservative. By inspection, $\phi(x, y) = x^3 + 2xy^2 + 2y^3$.

14.3.5: If $P(x, y) = 2y + \sin 2x$ and $Q(x, y) = 3x + \cos 3y$, then

$$\frac{\partial P}{\partial y} = 2 \neq 3 = \frac{\partial Q}{\partial x}.$$

Hence $\mathbf{F}$ is not conservative.

14.3.7: If $P(x, y) = x^3 + \dfrac{y}{x}$ and $Q(x, y) = y^2 + \ln x$, then

$$\frac{\partial P}{\partial y} = \frac{1}{x} = \frac{\partial Q}{\partial x}.$$

Hence $\mathbf{F}$ is conservative. By inspection, $\phi(x, y) = \dfrac{1}{4}x^4 + y \ln x + \dfrac{1}{3}y^3$.

14.3.9: If $P(x, y) = \cos x + \ln y$ and $Q(x, y) = \dfrac{x}{y} + e^y$, then

$$\frac{\partial P}{\partial y} = \frac{1}{y} = \frac{\partial Q}{\partial x}.$$

Hence $\mathbf{F}$ is conservative. By inspection, $\phi(x, y) = \sin x + x \ln y + e^y$.

14.3.11: If $P(x, y) = x \cos y + \sin y$ and $Q(x, y) = y \cos x + \sin x$, then

$$\frac{\partial P}{\partial y} = \cos y - x \sin y \neq \cos x - y \sin x = \frac{\partial Q}{\partial x}.$$

Hence $\mathbf{F}$ is not conservative.

14.3.13: If $P(x, y) = 3x^2y^3 + y^4$ and $Q(x, y) = 3x^3y^2 + y^4 + 4xy^3$, then

$$\frac{\partial P}{\partial y} = 9x^2y^2 + 4y^3 = \frac{\partial Q}{\partial x}.$$

Hence $\mathbf{F}$ is conservative. By inspection, $\phi(x, y) = x^3y^3 + xy^4 + \dfrac{1}{5}y^5$.

14.3.15: If $P(x, y) = \dfrac{2x}{y} - \dfrac{3y^2}{x^4}$ and $Q(x, y) = \dfrac{2y}{x^3} - \dfrac{x^2}{y^2} + \dfrac{1}{\sqrt{y}}$, then

$$\frac{\partial P}{\partial y} = -\frac{2x}{y^2} - \frac{6y}{x^4} = \frac{\partial Q}{\partial x}.$$

Hence $\mathbf{F}$ is conservative. By inspection, $\phi(x, y) = \dfrac{x^2}{y} + 2\sqrt{y} + \dfrac{y^2}{x^3}$.

14.3.17: We let $x(t) = x_1 t$ and $y(t) = y_1 t$ for $0 \le t \le 1$. Also let $\mathbf{r}(t) = \langle x(t), y(t) \rangle$ and $\mathbf{F}(x, y) = \langle 3x^2 + 2y^2, \ 4xy + 6y^2 \rangle$. Then

$$\int_{t=0}^{1} \mathbf{F}(x(t), y(t)) \cdot \mathbf{r}'(t) \, dt = \int_{0}^{1} 3(x_1^3 + 2x_1 y_1^2 + 2y_1^3)t^2 \, dt = \left[(x_1^3 + 2x_1 y_1^2 + 2y_1^3)t^3 \right]_0^1 = x_1^3 + 2x_1 y_1^2 + 2y_1^3.$$

Then, as in Example 3, a scalar potential for $\mathbf{F}(x, y)$ is $\phi(x, y) = x^3 + 2xy^2 + 2y^3$.

14.3.19: We let $x(t) = x_1 t$ and $y(t) = y_1 t$ for $0 \le t \le 1$. Also let $\mathbf{r}(t) = \langle x(t), y(t) \rangle$ and $\mathbf{F}(x, y) = \langle 3x^2 y^3 + y^4, \ 3x^3 y^2 + y^4 + 4xy^3 \rangle$. Then

$$\int_{t=0}^{1} \mathbf{F}(x(t), y(t)) \cdot \mathbf{r}'(t) \, dt = \int_{0}^{1} (6x_1^3 y_1^3 t^5 + 5x_1 y_1^4 t^4 + y_1^5 t^4) \, dt$$

$$= \left[x_1^3 y_1^3 t^6 + x_1 y_1^4 t^5 + \frac{1}{5} y_1^5 t^5 \right]_0^1 = x_1^3 y_1^3 + x_1 y_1^4 + \frac{1}{5} y_1^5.$$

Therefore a scalar potential for $\mathbf{F}$ is $\phi(x, y) = x^3 y^3 + xy^4 + \dfrac{1}{5} y^5$.

14.3.21: Let $P(x, y) = y^2 + 2xy$ and $Q(x, y) = x^2 + 2xy$. Then

$$\frac{\partial P}{\partial y} = 2x + 2y = \frac{\partial Q}{\partial x},$$

and therefore $\mathbf{F}(x, y)$ is conservative with potential function $\phi(x, y) = x^2 y + xy^2$. Therefore

$$\int_{(0,0)}^{(1,2)} P \, dx + Q \, dy = \left[x^2 y + xy^2 \right]_{(0,0)}^{(1,2)} = 6 - 0 = 6.$$

14.3.23: Let $P(x, y) = 2xe^y$ and $Q(x, y) = x^2 e^y$. Then

$$\frac{\partial P}{\partial y} = 2xe^y = \frac{\partial Q}{\partial x},$$

and therefore $\mathbf{F}(x, y)$ is conservative with potential function $\phi(x, y) = x^2 e^y$. Consequently

$$\int_{(0,0)}^{(1,-1)} P \, dx + Q \, dy = \left[x^2 e^y \right]_{(0,0)}^{(1,-1)} = \frac{1}{e} - 0 = \frac{1}{e}.$$

14.3.25: Let $P(x, y) = \sin y + y \cos x$ and $Q(x, y) = \sin x + x \cos y$. Then

$$\frac{\partial P}{\partial y} = \cos x + \cos y = \frac{\partial Q}{\partial x},$$

and therefore $\mathbf{F}(x, y)$ is conservative with potential function $\phi(x, y) = y \sin x + x \sin y$. Therefore

$$\int_{(\pi/2,\pi/2)}^{(\pi,\pi)} P \, dx \; + \; Q \, dy = \left[y \sin x + x \sin y \right]_{(\pi/2,\pi/2)}^{(\pi,\pi)} = 0 - \pi = -\pi.$$

14.3.27: By inspection, $\phi(x, y, z) = xyz$. For an analytic solution, write

$$\mathbf{F}(x, y, z) = \langle\, P(x, y, z), \, Q(x, y, z), \, R(x, y, z) \,\rangle$$

where $P(x, y, z) = yz$, $Q(x, y, z) = xz$, and $R(x, y, z) = xy$. Then let

$$g(x, y, z) = \int P(x, y, z) \, dx = \int yz \, dx = xyz + h(y, z).$$

Then

$$\frac{\partial g}{\partial y} = Q(x, y, z) = xz = xz + \frac{\partial h}{\partial y} \quad \text{and} \quad \frac{\partial g}{\partial z} = R(x, y, z) = xy = xy + \frac{\partial h}{\partial z},$$

and therefore the choice $h(y, z) \equiv 0$ yields a scalar potential for $\mathbf{F}$.

14.3.29: Write

$$\mathbf{F}(x, y, z) = \langle\, P(x, y, z), \, Q(x, y, z), \, R(x, y, z) \,\rangle$$

where $P(x, y, z) = y \cos z - yze^x$, $Q(x, y, z) = x \cos z - ze^x$, and $R(x, y, z) = -xy \sin z - ye^x$. Let

$$\phi(x, y, z) = \int P(x, y, z) \, dx = xy \cos z - yze^x + g(y, z).$$

Then

$$Q(x, y, z) = x \cos z - ze^x = \frac{\partial \phi}{\partial y} = x \cos z - ze^x + \frac{\partial g}{\partial y}.$$

Hence $g_y(x, z) = 0$, and so g is a function of z alone. Thus

$$\phi(x, y, z) = xy \cos z - yze^x + g(z),$$

and therefore

$$R(x, y, z) = -xy \sin z - ye^x = \frac{\partial \phi}{\partial z} = -xy \sin z - ye^x + g'(z).$$

Thus $g(z) = C$, a constant. So every scalar potential for $\mathbf{F}$ has the form $\phi(x, y, z) = xy \cos z - yze^x + C$. For a particular scalar potential, simply choose $C = 0$.

14.3.31: Suppose that the force field $\mathbf{F} = \langle P, Q \rangle$ is conservative in the plane region D. Then there exists a potential function $\phi(x, y)$ for $\mathbf{F}$; that is, $\nabla \phi = \mathbf{F}$, so that

$$\frac{\partial \phi}{\partial x} = P(x, y) \quad \text{and} \quad \frac{\partial \phi}{\partial y} = Q(x, y)$$

on the interior of D. But then

$$\frac{\partial P}{\partial y} = \frac{\partial^2 \phi}{\partial y\, \partial x} = \frac{\partial^2 \phi}{\partial x\, \partial y} = \frac{\partial Q}{\partial x}$$

on the interior of D, under the assumption that the second-order mixed partial derivatives of ϕ are continuous there. Of course, continuity of P_y and Q_x on D is enough to guarantee this.

14.3.33: The given integral is not independent of the path because

$$\frac{\partial x^2}{\partial z} = 0 \neq 2y = \frac{\partial y^2}{\partial y}.$$

14.3.35: Part (a): If

$$f(x, y) = \arctan\left(\frac{y}{x}\right),$$

then

$$\frac{\partial f}{\partial x} = \frac{-\dfrac{y}{x^2}}{1 + \dfrac{y^2}{x^2}} = -\frac{y}{x^2 + y^2} \quad \text{and}$$

$$\frac{\partial f}{\partial y} = \frac{\dfrac{1}{x}}{1 + \dfrac{y^2}{x^2}} = \frac{x}{x^2 + y^2}.$$

Part (b): Because $\mathbf{F} = \nabla f$ has a potential on the right half-plane $x > 0$, line integrals of $\mathbf{F}$ will be independent of the path C provided that the path always remains in the right half-plane. Hence if C is such a path from $A(x_1, y_1) = (r_1, \theta_1)$ to $B(x_2, y_2) = (r_2, \theta_2)$, then

$$\int_C \mathbf{F} \cdot \mathbf{T}\, ds = \int_A^B \mathbf{F} \cdot d\mathbf{r} = \left[f(x, y) \right]_A^B = \theta_2 - \theta_1.$$

Part (c): Parametrize the unit circle using $x(t) = \cos t$, $y(t) = \sin t$, $-\pi \leqq t \leqq \pi$. It is easy to verify that

$$\mathbf{F}(x(t), y(t)) = \langle -\sin t, \cos t \rangle$$

and that $\mathbf{F}(t) \cdot \mathbf{r}'(t) = 1$. Therefore

$$\int_{C_1} \mathbf{F} \cdot d\mathbf{r} = \int_0^\pi 1\, dt = \pi \quad \text{and} \quad \int_{C_2} \mathbf{F} \cdot d\mathbf{r} = \int_0^{-\pi} 1\, dt = -\pi.$$

This does not contradict the fundamental theorem of calculus for line integrals because $\mathbf{F}$ does not have a scalar potential defined in a region containing both C_1 and C_2 (see the solution of Problem 30).

14.3.37: The units are mks units throughout. We use $M = 5.97 \times 10^{24}$, $G = 6.67 \times 10^{-11}$, $m = 10000$, and let $k = GMm = 3.98199 \times 10^{18}$ in the formula in Problem 36. We also must convert r_1 and r_2 into meters: $r_1 = 9000 \cdot 1000$ and $r_2 = 11000 \cdot 1000$. Then substitution in the formula in Problem 36 yields $W = 8.04442 \times 10^{10}$ N·m.

Section 14.4

Note: As in the text, the notation

$$\oint_C P(x, y)\,dx + Q(x, y)\,dy$$

(and variations thereof) always denotes an integral around the *closed* path C with *counterclockwise* positive orientation.

14.4.1: By Green's theorem,

$$\oint_C (x + y^2)\,dx + (y + x^2)\,dy = \int_{y=-1}^{1} \int_{x=-1}^{1} (2x - 2y)\,dx\,dy$$

$$= \int_{y=-1}^{1} \left[x^2 - 2xy \right]_{x=-1}^{1}\,dy = \int_{-1}^{1} -4y\,dy = \left[-2y^2 \right]_{0}^{1} = 0.$$

14.4.3: By Green's theorem,

$$\oint_C (y + e^x)\,dx + (2x^2 + \cos y)\,dy = \int_0^1 \int_y^{2-y} (4x - 1)\,dx\,dy$$

$$= \int_0^1 \left[2x^2 - x \right]_y^{2-y}\,dy = \int_0^1 (6 - 6y)\,dy = \left[6y - 3y^2 \right]_0^1 = 3.$$

14.4.5: By Green's theorem,

$$\oint_C \left[-y^2 + \exp(e^x) \right] dx + (\arctan y)\,dy = \int_0^1 \int_{x^2}^{\sqrt{x}} 2y\,dy\,dx = \int_0^1 \left[y^2 \right]_{x^2}^{\sqrt{x}}\,dx$$

$$= \int_0^1 (x - x^4)\,dx = \left[\frac{1}{2}x^2 - \frac{1}{5}x^5 \right]_0^1 = \frac{3}{10}.$$

14.4.7: By Green's theorem,

$$\oint_C (x - y)\,dx + y\,dy = \int_0^\pi \int_0^{\sin x} 1\,dy\,dx = \int_0^\pi \sin x\,dx = \left[-\cos x \right]_0^\pi = 2.$$

14.4.9: Let D be the bounded region bounded by the curve C. Then by Green's theorem,

$$\oint_C y^2\,dx + xy\,dy = \iint_D (y - 2y)\,dA = -\iint_D y\,dA = 0$$

955

by symmetry.

14.4.11: Let R denote the bounded plane region bounded by the curve C. Then by Green's theorem,

$$\oint_C xy\,dx + x^2\,dy = \iint_R (2x - x)\,dA = \int_{\theta=0}^{\pi/2} \int_{r=0}^{\sin 2\theta} r^2 \cos\theta\,dr\,d\theta = \int_0^{\pi/2} \left[\frac{1}{3} r^3 \cos\theta\right]_0^{\sin 2\theta} d\theta$$

$$= \int_0^{\pi/2} \frac{8}{3} \sin^3\theta\,\cos^4\theta\,d\theta = \frac{1}{840}\left[5\cos 7\theta + 7\cos 5\theta - 35\cos 3\theta - 105\cos\theta\right]_0^{\pi/2}$$

$$= \frac{16}{105} \approx 0.15238095238095238095.$$

14.4.13: The given parametrization and the corollary to Green's theorem yield area

$$A = \oint_C x\,dy = \int_0^{2\pi} a^2 \cos^2 t\,dt = a^2\left[\frac{1}{2}t + \frac{1}{2}\sin t\,\cos t\right]_0^{2\pi} = \pi a^2.$$

14.4.15: We'll use the given parametrization, find the area of the part of the astroid in the first quadrant, then multiply by 4. The corollary to Green's theorem yields area

$$A = \oint_C x\,dy = 4\int_0^{\pi/2} 3\sin^2 t\,\cos^4 t\,dt$$

$$= \frac{1}{16}\left[12t + 3\sin 2t - 3\sin 4t - \sin 6t\right]_0^{\pi/2} = \frac{3}{8}\pi \approx 1.17809724509617246442.$$

There's no need to evaluate the line integral along the x- or y-axes because $x\,dy = 0$ there.

14.4.17: Denote by E the bounded plane region bounded by the given curve C. Then the work is

$$W = \oint_C \mathbf{F}\cdot\mathbf{T}\,ds = \oint_C -2y\,dx + 3x\,dy = \iint_E 5\,dA = 5\cdot 6\pi = 30\pi \approx 94.2477796076937972.$$

(The area of the ellipse is 6π because it has semiaxes of lengths 2 and 3.)

14.4.19: Denote by T the triangular region bounded by the given triangle C. Then the work done is

$$W = \oint_C \mathbf{F}\cdot\mathbf{T}\,ds = \oint_C 5x^2 y^3\,dx + 7x^3 y^2\,dy = \iint_T (21x^2 y^2 - 15x^2 y^2)\,dA = \int_{x=0}^3 \int_{y=0}^{6-2x} 6x^2 y^2\,dy\,dx$$

$$= \int_0^3 \left[2x^2 y^3\right]_{y=0}^{6-2x} dx = \int_0^3 (432x^2 - 432x^3 + 144x^4 - 16x^5)\,dx$$

$$= \left[144x^3 - 108x^4 + \frac{144}{5}x^5 - \frac{8}{3}x^6\right]_0^3 = \frac{972}{5} = 194.4.$$

14.4.21: Denote by R the bounded plane region bounded by the given curve C. Then the outward flux of $\mathbf{F}$ across C is

$$\phi = \oint_C \mathbf{F} \cdot \mathbf{n}\, ds = \iint_R \boldsymbol{\nabla} \cdot \mathbf{F}\, dA = \iint_R \boldsymbol{\nabla} \cdot \langle 2x, 3y \rangle\, dA = \iint_R 5\, dA = 30\pi \approx 94.2477796076937972.$$

14.4.23: Denote by T the triangular region bounded by the given path C. Then the outward flux of $\mathbf{F}$ across C is

$$\phi = \oint_C \mathbf{F} \cdot \mathbf{n}\, ds = \iint_T \boldsymbol{\nabla} \cdot \mathbf{F}\, dA$$

$$= \iint_T \boldsymbol{\nabla} \cdot \langle 3x + \sqrt{1 + y^2}\,,\ 2y - (1 + x^4)^{1/3} \rangle\, dA = \iint_T 5\, dA = 5 \cdot \frac{1}{2} \cdot 3 \cdot 6 = 45$$

because T is a triangle with base 3 and height 6.

14.4.25: Given f, a twice-differentiable function of x and y, we have

$$\nabla^2 f = \boldsymbol{\nabla} \cdot (\nabla f) = \boldsymbol{\nabla} \cdot \langle f_x, f_y \rangle = \frac{\partial^2 f}{\partial x^2} + \frac{\partial^2 f}{\partial y^2}.$$

14.4.27: If f and g are twice-differentiable functions of x and y, then

$$\nabla^2(fg) = \boldsymbol{\nabla} \cdot \left[\boldsymbol{\nabla}(fg) \right] = \boldsymbol{\nabla} \cdot \langle f_x g + f g_x,\ f_y g + f g_y \rangle$$

$$= f_{xx} g + f_x g_x + f_x g_x + f g_{xx} + f_{yy} g + f_y g_y + f_y g_y + f g_{yy}$$

$$= (f)(g_{xx} + g_{yy}) + 2(f_x g_x + f_y g_y) + (g)(f_{xx} + f_{yy})$$

$$= f \nabla^2 g + g \nabla^2 f + 2 \langle f_x, f_y \rangle \cdot \langle g_x, g_y \rangle = f \nabla^2 g + g \nabla^2 f + 2 \nabla f \cdot \nabla g.$$

Compare this with Problem 33 of Section 14.1.

14.4.29: We may assume constant density $\delta = 1$. Then

$$A = \oint_C x\, dy = \oint_C -y\, dx.$$

Now

$$M_y = \iint_R x\, dA \quad \text{and} \quad M_x = \iint_R y\, dA.$$

Hence, by Green's theorem,

$$M_y = \oint_C \frac{1}{2} x^2\, dy \quad \text{and} \quad M_x = -\oint_C \frac{1}{2} y^2\, dx.$$

Therefore

$$\bar{x} = \frac{M_y}{A} = \frac{1}{2A} \oint_C x^2 \, dy \quad \text{and} \quad \bar{y} = \frac{M_x}{A} = -\frac{1}{2A} \oint_C y^2 \, dx.$$

14.4.31: Suppose that the plane region R is bounded by the piecewise smooth simple closed curve C, oriented counterclockwise, and that R has constant density δ. Then, by Green's theorem,

$$I_x = \iint_R \delta y^2 \, dA = \delta \oint_C -\frac{1}{3} y^3 \, dx = -\frac{1}{3} \delta \oint_C y^3 \, dx \quad \text{and}$$

$$I_y = \iint_R \delta x^2 \, dA = \delta \oint_C \frac{1}{3} x^3 \, dy = \frac{1}{3} \delta \oint_C x^3 \, dy.$$

14.4.33: As in Problem 30 of Section 9.4, the substitution $y = tx$ in the equation $x^3 + y^3 = 3xy$ of the folium yields $x^3 + t^3 x^3 = 3tx^2$, and thereby the parametrization

$$x(t) = \frac{3t}{1+t^3}, \quad y(t) = \frac{3t^2}{1+t^3}, \quad 0 \leq t < +\infty$$

of the first-quadrant loop of the folium. If C is the half of its loop that stretches from $(0, 0)$ to $\left(\frac{3}{2}, \frac{3}{2}\right)$ along the lower half of the folium, then C is swept out by this parametrization as t varies from 0 to 1. Let J be the straight line segment joining $\left(\frac{3}{2}, \frac{3}{2}\right)$ with $(0, 0)$; parametrize J with $x = \frac{3}{2}(1-t)$, $y = \frac{3}{2}(1-t)$, $0 \leq t \leq 1$. Then the area of the folium is

$$A = 2 \cdot \frac{1}{2} \oint_{C \cup J} x \, dy - y \, dx = \int_C x \, dy - y \, dx + \int_J x \, dy - y \, dx.$$

The last integral is

$$\int_{t=0}^1 \left[-\frac{3}{2}(1-t) + \frac{3}{2}(1-t) \right] dt = 0,$$

and hence the area of the folium is

$$A = \int_0^1 \left[x(t)y'(t) - y(t)x'(t) \right] dt = \int_0^1 \left[\frac{9t(2t - t^4)}{(1+t^3)^3} - \frac{9t^2(2t^3 - 1)}{(1+t^3)^3} \right] dt$$

$$= \int_0^1 \frac{9t^2}{(1+t^3)^2} \, dt = \left[-\frac{3}{1+t^3} \right]_0^1 = \frac{3}{2}.$$

14.4.35: We substitute $f\nabla g$ for $\mathbf{F}$ in Eq. (9),

$$\oint_C \mathbf{F} \cdot \mathbf{n} \, ds = \iint_R \nabla \cdot \mathbf{F} \, dA.$$

With the aid of the result in Problem 28 of Section 15.1, this yields

$$\oint_C f\nabla g \cdot \mathbf{n} \, ds = \iint_R \nabla \cdot (f\nabla g) \, dA = \iint_R (f\nabla \cdot \nabla g + \nabla f \cdot \nabla g) \, dA.$$

14.4.37: It suffices to show the result in the case that $R = R_1 \cup R_1$ is the union of two regions, with C the boundary of R, $C_1 \cup C_3$ the boundary of R_1, and $-C_3 \cup C_2$ the boundary of R_2. Then $C = C_1 \cup C_2$. Perhaps the next figure will clarify all this.

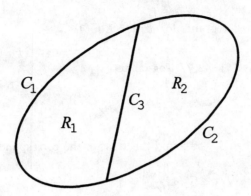

Under the assumption that Green's theorem holds for R_1 and for R_2, we have

$$\oint_{C_1 \cup C_3} P \, dx + Q \, dy = \iint_{R_1} (Q_x - P_y) \, dA \quad \text{and}$$

$$\oint_{-C_3 \cup C_2} P \, dx + Q \, dy = \iint_{R_2} (Q_x - P_y) \, dA.$$

Addition of these equations yields

$$\int_{C_1} (P \, dx + Q \, dy) + \int_{C_2} (P \, dx + Q \, dy) + \int_{C_3} (P \, dx + Q \, dy) - \int_{C_3} (P \, dx + Q \, dy) = \iint_R (Q_x - P_y) \, dA.$$

Therefore

$$\oint_C P \, dx + Q \, dy = \iint_R (Q_x - P_y) \, dA.$$

14.4.39: Part (a): With $x_1 = 1$, $y_1 = 0$, $x_2 = \cos(2\pi/3)$, $y_2 = \sin(2\pi/3)$, $x_3 = \cos(4\pi/3)$, and $y_3 = \sin(4\pi/3)$, we obtain area

$$A = \frac{1}{2}(x_1 y_2 - x_2 y_1 + x_2 y_3 - x_3 y_2 + x_3 y_1 - x_1 y_3) = \frac{1}{4}\sqrt{3} + \frac{1}{4}\sqrt{3} + \frac{1}{4}\sqrt{3} = \frac{3}{4}\sqrt{3} \approx 1.29903810567666.$$

Part (b): With the points (x_i, y_i) $(1 \leqq i \leqq 5)$ chosen in a way analogous to that in part (a), we obtain area

$$A = \frac{1}{2}(x_1 y_2 - x_2 y_1 + x_2 y_3 - x_3 y_2 + x_3 y_4 - x_4 y_3 + x_4 y_5 - x_5 y_4 + x_5 y_1 - x_1 y_5)$$

$$= \frac{5}{8}\sqrt{10 + 2\sqrt{5}} \approx 2.37764129073788393029.$$

(We used a computer algebra program to evaluate the first line, then simplified the result by hand.)

14.4.41: Using *Mathematica*, we entered the parametric functions

$$x(t) = \frac{(2n+1)t^n}{t^{2n+1} + 1} \quad \text{and} \quad y(t) = \frac{(2n+1)t^{n+1}}{t^{2n+1} + 1}.$$

Then we computed one of the integrands for area in Eq. (4):

```
(-y'[t]*x[t] + x[t]*y'[t])/2 // Together
```

$$\frac{(2n + 1)^2 t^{2n}}{2(t^{2n+1} + 1)^2}$$

Then we set $n = 1$, then 2, then 3, and found the areas of the corresponding loops as follows:

```
Integrate[ (9/2)*(t∧2)/(1 + t∧3)∧2, t ]
```

$$-\frac{3}{2(t^3 + 1)}$$

```
( % /. t → Infinity ) - ( % /. t → 0 )
```

$$\frac{3}{2}$$

```
Integrate[ (25/2)*(t∧4)/(1 + t∧5)∧2, t ]
```

$$-\frac{5}{2(t^5 + 1)}$$

```
( % /. t → Infinity ) - ( % /. t → 0 )
```

$$\frac{5}{2}$$

```
Integrate[ (49/2)*(t∧6)/(1 + t∧7)∧2, t ]
```

$$-\frac{7}{2(t^7 + 1)}$$

```
( % /. t → Infinity ) - ( % /. t → 0 )
```

$$\frac{7}{2}$$

In fact, the integral is easy to find in the general case:

```
Integrate[ (1/2)*((2*n + 1)∧2)*(t∧(2*n))/(1 + t∧(2*n + 1))∧2, t ]
```

$$-\frac{2n + 1}{2(1 + t^{2n+1})}$$

Therefore

$$A_n = \left[-\frac{2n + 1}{2(1 + t^{2n+1})} \right]_0^\infty = 0 + \frac{2n + 1}{2} = n + \frac{1}{2}.$$

To avoid the improper integral, let C denote the simple closed curve consisting of the lower half of the loop (swept out as t ranges from 0 to 1) together with the "return path" along the line $y = x$ back to the origin.

On the return path P we have the parametrization $x = y = n + \frac{1}{2} - t$ as t ranges from 0 to $n + \frac{1}{2}$; but on this path, $-y\,dx + x\,dy = 0$, and hence

$$\frac{1}{2}\int_P -y\,dx + x\,dy = 0.$$

The area enclosed by the loop is double that enclosed by C, and thus

$$A_n = 2 \cdot \left[-\frac{2n+1}{2(1+t^{2n+1})} \right]_0^1 = -\frac{2n+1}{2} + (2n+1) = n + \frac{1}{2}.$$

The only danger in this "short cut" is that you may use one of the other two formulas in Eq. (4) of the text and forget the line integral along P.

Section 14.5

14.5.1: Here S is the surface $z = h(x, y) = 1 - x - y$ over the plane triangle bounded by the nonnegative coordinate axes and the graph of $y = 1 - x$. So

$$dS = \sqrt{1 + (h_x)^2 + (h_y)^2}\ dx\,dy = \sqrt{3}\ dx\,dy.$$

Therefore

$$\iint_S (x + y)\,dS = \int_{x=0}^1 \int_{y=0}^{1-x} (x+y)\sqrt{3}\ dy\,dx = \sqrt{3}\int_0^1 \left[xy + \frac{1}{2}y^2 \right]_0^{1-x} dx$$

$$= \sqrt{3}\int_0^1 \left(\frac{1}{2} - \frac{1}{2}x^2 \right)\,dx = \sqrt{3}\left[\frac{1}{2}x - \frac{1}{6}x^3 \right]_0^1 = \frac{1}{3}\sqrt{3} \approx 0.5773502691896258.$$

14.5.3: First, S is the surface $z = h(x, y) = 2x + 3y$ lying over the circular disk D with center $(0, 0)$ and radius 3 in the xy-plane. Also

$$dS = \sqrt{1 + (h_x)^2 + (h_y)^2}\ dx\,dy = \sqrt{14}\ dx\,dy,$$

and thus

$$\iint_S (y + z + 3)\,dS = \iint_D (y + 2x + 3y + 3)\sqrt{14}\ dA = \int_{\theta=0}^{2\pi} \int_{r=0}^3 (4r\sin\theta + 2r\cos\theta + 3)\left(r\sqrt{14} \right)\,dr\,d\theta$$

$$= \sqrt{14}\int_0^{2\pi} \left[\frac{3}{2}r^2 + \frac{2}{3}r^3(\cos\theta + 2\sin\theta) \right]_0^3 d\theta = \sqrt{14}\int_0^{2\pi} \left(\frac{27}{2} + 18\cos\theta + 36\sin\theta \right)\,d\theta$$

$$= \sqrt{14}\left[\frac{27}{2}\theta + 18\sin\theta - 36\cos\theta \right]_0^{2\pi} = 27\pi\sqrt{14} \approx 317.3786106805529421.$$

14.5.5: The surface S is the part of the paraboloid $z = h(x, y) = x^2 + y^2$ that lies over the circular disk D with center $(0, 0)$ and radius 2 in the xy-plane. Also

$$dS = \sqrt{1 + (h_x)^2 + (h_y)^2}\ dx\ dy = \sqrt{1 + 4x^2 + 4y^2}\ dx\ dy,$$

and thus

$$\iint_S (xy + 1)\ dS = \iint_D (xy + 1)\sqrt{1 + 4x^2 + 4y^2}\ dx\ dy = \int_0^{2\pi} \int_0^2 (1 + r^2 \sin\theta \cos\theta) \cdot r(1 + 4r^2)^{1/2}\ dr\ d\theta$$

$$= \int_0^{2\pi} \frac{1}{240}(1 + 4r^2)^{3/2}\left[20 + (6r^2 - 1)\sin 2\theta\right]_0^2\ d\theta$$

$$= \int_0^{2\pi} \frac{1}{240}\left[340\sqrt{17} - 20 + \left(1 + 391\sqrt{17}\right)\sin 2\theta\right]\ d\theta$$

$$= \frac{1}{480}\left[40\left(17\sqrt{17} - 1\right)\theta - \left(1 + 391\sqrt{17}\right)\cos 2\theta\right]_0^{2\pi}$$

$$= \frac{1}{480}\left[1 + 391\sqrt{17} - 1 - 391\sqrt{17} + 80\pi\left(17\sqrt{17} - 1\right)\right]$$

$$= \frac{1}{6}\pi\left(-1 + 17\sqrt{17}\right) \approx 36.176903197411408364756.$$

14.5.7: The surface S is the part of the graph of $z = h(x, y) = x + y$ that lies over the circular disk D with center $(0, 0)$ and radius 3 in the xy-plane. Also $dS = \sqrt{1 + 1 + 1}\ dx\ dy = \sqrt{3}\ dx\ dy$, and hence

$$I_z = \iint_S \delta(x^2 + y^2)\ dS = \iint_D \delta r^2 \sqrt{3}\ dA = \delta\sqrt{3}\int_0^{2\pi}\int_0^3 r^3\ dr\ d\theta = 2\pi\delta\sqrt{3}\left[\frac{1}{4}r^4\right]_0^3 = \frac{81}{2}\pi\delta\sqrt{3}.$$

The mass of S is

$$m = \delta\sqrt{3}\int_0^{2\pi}\int_0^3 r\ dr\ d\theta = 2\pi\delta\sqrt{3}\left[\frac{1}{2}r^2\right]_0^3 = 9\pi\delta\sqrt{3},$$

and therefore $I_z = \dfrac{9}{2}m$.

14.5.9: Suppose that (x, z) is a point in the xz-plane. Let $\mathbf{w}$ be the radius vector from the origin in the xz-plane to (x, z) and let θ be the angle that $\mathbf{w}$ makes with the nonnegative x-axis. Then points in the cylindrical surface S are described by

$$x = \cos\theta, \quad y = y, \quad z = \sin\theta, \quad -1 \leqq y \leqq 1, \quad 0 \leqq \theta \leqq 2\pi.$$

Thus the cylindrical surface S is parametrized by $\mathbf{r}(y, \theta) = \langle \cos\theta, y, \sin\theta \rangle$, for which $\mathbf{r}_y = \langle 0, 1, 0 \rangle$ and $\mathbf{r}_\theta = \langle -\sin\theta, 0, \cos\theta \rangle$. Hence

$$\mathbf{r}_y \times \mathbf{r}_\theta = \begin{vmatrix} \mathbf{i} & \mathbf{j} & \mathbf{k} \\ 0 & 1 & 0 \\ -\sin\theta & 0 & \cos\theta \end{vmatrix} = \langle \cos\theta, 0, \sin\theta \rangle.$$

Therefore $|\mathbf{r}_y \times \mathbf{r}_\theta| = 1$, so that $dS = dy\,d\theta$. Let D denote the rectangle $-1 \leqq y \leqq 1,\ 0 \leqq \theta \leqq 2\pi$. Because S has constant density δ, its moment of inertia with respect to the z-axis is therefore

$$I_z = \iint_S (x^2 + y^2)\delta\,dS = \iint_D (y^2 + \cos^2\theta)\delta\,dA = \int_{\theta=0}^{2\pi} \int_{y=-1}^{1} (y^2 + \cos^2\theta)\delta\,dy\,d\theta$$

$$= \delta \int_0^{2\pi} \left[\frac{1}{3}y^3 + y\cos^2\theta\right]_{-1}^{1} d\theta = \delta \int_0^{2\pi} \left(\frac{2}{3} + 2\cos^2\theta\right) d\theta = \frac{1}{6}\delta\left[10\theta + 3\sin 2\theta\right]_0^{2\pi} = \frac{10}{3}\pi\delta.$$

Because the mass m of S is the product of its surface area and its density, we have $m = 4\pi\delta$, and hence we may also express I_z in the form

$$I_z = \frac{5}{6}m = m \cdot (0.9128709291752769)^2.$$

14.5.11: The surface S has the spherical-coordinates parametrization

$$\mathbf{r}(\phi,\,\theta) = \langle 5\sin\phi\cos\theta,\ 5\sin\phi\sin\theta,\ 5\cos\phi\rangle, \quad 0 \leqq \phi \leqq \arccos\left(\frac{3}{5}\right),\ 0 \leqq \theta \leqq 2\pi.$$

Therefore

$$\mathbf{r}_\phi \times \mathbf{r}_\theta = \begin{vmatrix} \mathbf{i} & \mathbf{j} & \mathbf{k} \\ 5\cos\phi\cos\theta & 5\cos\phi\sin\theta & -5\sin\phi \\ -5\sin\phi\sin\theta & 5\sin\phi\cos\theta & 0 \end{vmatrix} = \langle 25\sin^2\phi\cos\theta,\ 25\sin^2\phi\sin\theta,\ 25\sin\phi\cos\phi\rangle,$$

and thus

$$|\mathbf{r}_\phi \times \mathbf{r}_\theta| = \sqrt{625\sin^2\phi\cos^2\phi + 625\sin^4\phi\cos^2\theta + 625\sin^4\phi\sin^2\theta} = 25\sin\phi$$

(because $0 \leqq \phi \leqq \pi/2$). Hence the mass of S is

$$m = \iint_S \delta\,dS = \int_{\theta=0}^{2\pi} \int_{\phi=0}^{\arccos(3/5)} 25\delta\sin\phi\,d\phi\,d\theta = 2\pi\delta\left[-25\cos\phi\right]_0^{\arccos(3/5)}$$

$$= 20\pi\delta \approx (62.8318530717958648)\delta.$$

Next,

$$x^2 + y^2 = (5\sin\phi\cos\theta)^2 + (5\sin\phi\sin\theta)^2 = 25\sin^2\phi,$$

and hence the moment of inertia of S with respect to the z-axis is

$$I_z = \iint_S (x^2 + y^2)\delta\,dS = \int_{\theta=0}^{2\pi} \int_{\phi=0}^{\arccos(3/5)} 625\delta\sin^3\phi\,d\phi\,d\theta = 2\pi\delta\left[\frac{625}{3}\cos^3\phi - 625\cos\phi\right]_0^{\arccos(3/5)}$$

$$= \frac{520}{3}\pi\delta \approx (544.5427266222308280)\delta.$$

963

The moment of inertia may also be expressed in the form $I_z = \dfrac{26}{3} m \approx m \cdot (2.9439202887759490)^2$.

14.5.13: An upward unit vector normal to S is

$$\mathbf{n} = \left\langle \frac{x}{3}, \frac{y}{3}, \frac{z}{3} \right\rangle.$$

The surface has equation $z = h(x, y) = \sqrt{9 - x^2 - y^2}$, and therefore

$$dS = \sqrt{1 + \frac{x^2}{9 - x^2 - y^2} + \frac{y^2}{9 - x^2 - y^2}}\; dA = \frac{3}{\sqrt{9 - x^2 - y^2}}\; dA.$$

Next, $\mathbf{F} \cdot \mathbf{n} = \dfrac{1}{3}(x^2 + y^2)$, and S lies over the circular disk D in the plane with center $(0, 0)$ and radius 3. Therefore

$$\iint_S \mathbf{F} \cdot \mathbf{n}\; dS = \iint_D \frac{x^2 + y^2}{\sqrt{9 - x^2 - y^2}}\; dA = \int_{\theta=0}^{2\pi} \int_{r=0}^{3} \frac{r^3}{\sqrt{9 - r^2}}\; dr\; d\theta$$

$$= 2\pi \left[-\frac{1}{3}(r^2 + 18)\sqrt{9 - r^2} \right]_0^3 = 36\pi \approx 113.0973355292325566.$$

14.5.15: An upward unit vector normal to S is

$$\mathbf{n} = \left\langle -\frac{3}{10}\sqrt{10},\; 0,\; \frac{1}{10}\sqrt{10} \right\rangle.$$

The surface S has equation $z = h(x, y) = 3x + 2$, and therefore $dS = \sqrt{10}\; dA$. Also, S lies over the circular disk S in the xy-plane with center $(0, 0)$ and radius 2. Therefore

$$\iint_S \mathbf{F} \cdot \mathbf{n}\; dS = \iint_D 3z\; dA = \iint_D (9x + 6)\; dA = \int_{\theta=0}^{2\pi} \int_{r=0}^{2} (9r^2 \cos\theta + 6r)\; dr\; d\theta = \int_0^{2\pi} \left[3r^3 \cos\theta + 3r^2 \right]_0^2$$

$$= \int_0^{2\pi} (24\cos\theta + 12)\; d\theta = \left[24\sin\theta + 12\theta \right]_0^{2\pi} = 24\pi \approx 75.3982236861550377.$$

14.5.17: The surface S has Cartesian equation $z = h(x, y) = \sqrt{x^2 + y^2}$, and thus has normal vector

$$\mathbf{n}_1 = \langle h_x, h_y, -1 \rangle = \left\langle \frac{x}{\sqrt{x^2 + y^2}},\; \frac{y}{\sqrt{x^2 + y^2}},\; -1 \right\rangle,$$

and thus (in polar coordinates) a upward-pointing vector normal to S is $\mathbf{n}_2 = \langle -\cos\theta,\; -\sin\theta,\; 1 \rangle$. Therefore an upward-pointing unit vector normal to S is

$$\mathbf{n} = \frac{\mathbf{n}_2}{|\mathbf{n}_2|} = \frac{\sqrt{2}}{2} \langle -\cos\theta,\; -\sin\theta,\; 1 \rangle.$$

Next,

$$dS = \sqrt{1 + (h_x)^2 + (h_y)^2}\ dA = \left(1 + \frac{x^2}{x^2 + y^2} + \frac{y^2}{x^2 + y^2}\right)^{1/2} dA = \sqrt{2}\ dA,$$

and in polar coordinates we have $\mathbf{F} = \langle r\sin\theta,\ -r\cos\theta,\ 0\rangle$. But then $\mathbf{F} \cdot \mathbf{n}\ dS = 0$, so the surface integral is zero as well.

14.5.19: On the face of the cube in the xy-plane, $z = 0$, and so

$$\mathbf{F}(x, y, z) \cdot \mathbf{n} = \langle x, 2y, 0\rangle \cdot \langle 0, 0, -1\rangle = 0,$$

and hence the flux of $\mathbf{F}$ across that face is zero. Similarly, the flux across the faces in the other two coordinate planes is zero. On the top face we have $z = 1$, and hence

$$\mathbf{F}(x, y, z) \cdot \mathbf{n} = \langle x, 2y, 3\rangle \cdot \langle 0, 0, 1\rangle = 3.$$

Similarly, the flux across the face in the plane $y = 1$ is 2 and the flux across the face in the plane $x = 1$ is 1. Hence the total flux of $\mathbf{F}$ across S is $3 + 2 + 1 = 6$.

14.5.21: For the same reasons given in the solution of Problem 19, $\mathbf{F} \cdot \mathbf{n} = 0$ on the three faces of the pyramid in the coordinate planes. On the fourth face T a unit normal vector is

$$\mathbf{n} = \frac{1}{\sqrt{26}}\langle 3, 4, 1\rangle,$$

and because this face is the graph of $z = h(x, y) = 12 - 3x - 4y$, we have

$$dS = \sqrt{1 + (h_x)^2 + (h_y)^2}\ dA = \sqrt{26}\ dx\ dy.$$

Therefore $\mathbf{F} \cdot \mathbf{n}\ dS = (3x - 4y)\ dy\ dx$, and consequently

$$\iint_T \mathbf{F} \cdot \mathbf{n}\ dS = \int_0^4 \int_0^{(12-3x)/4} (3x - 4y)\ dy\ dx = \int_0^4 \left[3xy - 2y^2\right]_0^{(12-3x)/4} dx$$

$$= \int_0^4 \left(-18 + 18x - \frac{27}{8}x^2\right) dx = \left[-18x + 9x^2 - \frac{9}{8}x^3\right]_0^4 = 0.$$

Thus the total flux of $\mathbf{F}$ across S is zero. If you now turn two pages ahead in your textbook, you will see how the divergence theorem enables you to obtain the same result in less than two seconds and without need of pencil, paper, or computer.

14.5.23: The paraboloids meet in the circle $x^2 + y^2 = 9$, $z = 9$, so both the upper surface and the lower surface lie over the disk D in the xy-plane with center $(0, 0)$ and radius 3. The lower surface L is the graph of $h(x, y) = x^2 + y^2$ and the upper surface U is the graph of $j(x, y) = 18 - x^2 - y^2$ for (x, y) in D. A vector normal to L is

$$\langle h_x, h_y, -1 \rangle = \langle 2x, 2y, -1 \rangle$$

and hence the outer unit vector normal to L is

$$\mathbf{n}_1 = \frac{1}{\sqrt{1 + 4x^2 + 4y^2}} \langle 2x, 2y, -1 \rangle;$$

similarly, the outer unit vector normal to U is

$$\mathbf{n}_2 = \frac{1}{\sqrt{1 + 4x^2 + 4y^2}} \langle 2x, 2y, 1 \rangle.$$

The surface area element for L is

$$dS = \sqrt{1 + (h_x)^2 + (h_y)^2} \, dA = \sqrt{1 + 4x^2 + 4y^2} \, dA$$

and the surface area element for U is the same. Next,

$$\mathbf{F} \cdot \mathbf{n}_1 \, dS = -z^2 = -(x^2 + y^2)^2 \, dA \quad \text{and} \quad \mathbf{F} \cdot \mathbf{n}_2 \, dS = z^2 = (18 - x^2 - y^2)^2 \, dA.$$

Thus

$$\iint_L \mathbf{F} \cdot \mathbf{n}_1 \, dS = -\iint_D (x^2 + y^2)^2 \, dA = -\int_0^{2\pi} \int_0^3 r^5 \, dr \, d\theta = -2\pi \left[\frac{1}{6} r^6 \right]_0^3 = -243\pi$$

and

$$\iint_U \mathbf{F} \cdot \mathbf{n}_2 \, dS = \iint_D (18 - x^2 - y^2)^2 \, dA = \int_0^{2\pi} \int_0^3 (18 - r^2)^2 \cdot r \, dr \, d\theta = 2\pi \left[-\frac{1}{6} (18 - r^2)^3 \right]_0^3 = 1701\pi.$$

Therefore

$$\iint_S \mathbf{F} \cdot \mathbf{n} \, dS = 1701\pi - 243\pi = 1458\pi \approx 4580.4420889339185417.$$

14.5.25: The surface S may be parametrized by

$$x(\phi, \theta) = a \sin\phi \, \cos\theta, \ y(\phi, \theta) = a \sin\phi \, \sin\theta, \ z(\phi, \theta) = a \cos\phi, \quad 0 \le \phi \le \frac{1}{2}\pi, \ 0 \le \theta \le \frac{1}{2}\pi.$$

Then, if $\mathbf{r}(\phi, \theta) = \langle x(\phi, \theta), y(\phi, \theta), z(\phi, \theta) \rangle$, we find that

$$\mathbf{r}_\phi \times \mathbf{r}_\theta = \begin{vmatrix} \mathbf{i} & \mathbf{j} & \mathbf{k} \\ a\cos\phi\cos\theta & a\cos\phi\sin\theta & -a\sin\phi \\ -a\sin\phi\sin\theta & a\sin\phi\cos\theta & 0 \end{vmatrix} = \langle a^2 \sin^2\phi \cos\theta, \ a^2 \sin^2\phi \sin\theta, \ a^2 \sin\phi \cos\phi \rangle,$$

and therefore

$$|\mathbf{r}_\phi \times \mathbf{r}_\theta| = \sqrt{a^4 \cos^2 \phi \sin^2 \phi + a^4 \sin^4 \phi \cos^2 \theta + a^4 \sin^4 \phi \sin^2 \theta} = a^2 \sin \phi$$

because $0 \le \phi \le \pi$. Because the surface has unit density, its mass is

$$m = \iint_S 1 \, dS = \int_0^{\pi/2} \int_0^{\pi/2} a^2 \sin \phi \, d\phi \, d\theta = \frac{1}{2} \pi \left[-a^2 \cos \phi \right]_0^{\pi/2} = \frac{1}{2} \pi a^2.$$

The moment of S with respect to the yz-plane is

$$M_{yz} = \iint_S x(\phi, \theta) \, dS = \int_0^{\pi/2} \int_0^{\pi/2} a^3 \sin^2 \phi \cos \theta \, d\phi \, d\theta$$

$$= \int_0^{\pi/2} \left[\frac{1}{4} a^3 (\cos \theta)(2\phi - \sin 2\phi) \right]_0^{\pi/2} d\theta = \int_0^{\pi/2} \frac{1}{4} \pi a^3 \cos \theta \, d\theta = \left[\frac{1}{4} \pi a^3 \sin \theta \right]_0^{\pi/2} = \frac{1}{4} \pi a^3.$$

Therefore (by symmetry) $(\overline{x}, \overline{y}, \overline{z}) = \left(\frac{1}{2} a, \frac{1}{2} a, \frac{1}{2} a \right)$.

14.5.27: The surface $z = r^2$, $0 \le r \le a$ is described in Cartesian coordinates by $z = h(x, y) = x^2 + y^2$, and thus

$$dS = \sqrt{1 + (h_x)^2 + (h_y)^2} \, dA = \sqrt{1 + 4x^2 + 4y^2} \, dA.$$

Because the surface S lies over the circular disk D in the xy-plane with center $(0, 0)$ and radius a and because S has constant density δ, its mass is

$$m = \iint_S \delta \, dS = \iint_D \delta \sqrt{1 + 4x^2 + 4y^2} \, dA = \int_0^{2\pi} \int_0^a \delta r \sqrt{1 + 4r^2} \, dr \, d\theta$$

$$= 2\pi\delta \left[\frac{1}{12} (1 + 4r^2)^{3/2} \right]_0^a = \frac{1}{6} \pi\delta \left[(1 + 4a^2)^{3/2} - 1 \right].$$

The moment of S with respect to the xy-plane is

$$M_{xy} = \iint_S \delta z \, dS = \iint_D \delta(x^2 + y^2) \sqrt{1 + 4x^2 + 4y^2} \, dA = \int_0^{2\pi} \int_0^a \delta r^3 \sqrt{1 + 4r^2} \, dr \, d\theta$$

$$= 2\pi\delta \left[\left(\frac{1}{5} r^4 + \frac{1}{60} r^2 - \frac{1}{120} \right) \sqrt{1 + 4r^2} \right]_0^a = \frac{1}{60} \pi\delta \left[(24a^4 + 2a^2 - 1)\sqrt{1 + 4a^2} + 1 \right].$$

Hence the z-coordinate of the centroid of S is

$$\overline{z} = \frac{M_{yx}}{m} = \frac{(24a^4 + 2a^2 - 1)\sqrt{1 + 4a^2} + 1}{10 \left[(1 + 4a^2)^{3/2} - 1 \right]}.$$

For example, if $a = 1$, then

$$\overline{z} = \frac{1 + 25\sqrt{5}}{10 \left(-1 + 5\sqrt{5} \right)} \approx 0.5589371284878981.$$

By symmetry, $\bar{x} = \bar{y} = 0$. Finally, the moment of inertia of S with respect to the z-axis is

$$I_z = \iint_S \delta(x^2 + y^2)\, dS = \iint_D \delta(x^2 + y^2)\sqrt{1 + 4x^2 + 4y^2}\, dA = \frac{1}{60}\pi\delta\left[(24a^4 + 2a^2 - 1)\sqrt{1 + 4a^2} + 1\right]$$

(the computations are exactly the same as those in the evaluation of M_{xy}).

14.5.29: The surface S is described by $z = h(x, y) = \sqrt{4 - x^2 - y^2}$, and

$$1 + (h_x)^2 + (h_y)^2 = 1 + \frac{x^2}{4 - x^2 - y^2} + \frac{y^2}{4 - x^2 - y^2} = \frac{4}{4 - x^2 - y^2}.$$

Thus

$$dS = \frac{2}{\sqrt{4 - x^2 - y^2}}\, dA.$$

In cylindrical coordinates, S is described by

$$z = \sqrt{4 - r^2}, \quad 0 \leq r \leq 2\cos\theta, \quad -\frac{1}{2}\pi \leq \theta \leq \frac{1}{2}\pi.$$

We may without loss of generality assume that S has constant density $\delta = 1$. Hence its mass is

$$m = \iint_S 1\, dS = 2\int_0^{\pi/2}\int_0^{2\cos\theta} \frac{2r}{\sqrt{4 - r^2}}\, dr\, d\theta = 2\int_0^{\pi/2} \left[-2(4 - r^2)^{1/2}\right]_0^{2\cos\theta} d\theta$$

$$= 4\int_0^{\pi/2} \left(2 - \sqrt{4 - 4\cos^2\theta}\right) d\theta = 8\int_0^{\pi/2}(1 - \sin\theta)\, d\theta = 8\left[\theta + \cos\theta\right]_0^{\pi/2} = 4\pi - 8.$$

Clearly $\bar{y} = 0$ by symmetry. The moment of S with respect to the xy-plane is

$$M_{xy} = \iint_S z\, dS = 2\int_0^{\pi/2}\int_0^{2\cos\theta} \frac{2r\sqrt{4 - r^2}}{\sqrt{4 - r^2}}\, dr\, d\theta = 2\int_0^{\pi/2} \left[r^2\right]_0^{2\cos\theta} d\theta$$

$$= 2\int_0^{\pi/2} 4\cos^2\theta\, d\theta = 4\int_0^{\pi/2}(1 + \cos 2\theta)\, d\theta = 4\left[\theta + \frac{1}{2}\sin 2\theta\right]_0^{\pi/2} = 2\pi.$$

The moment of S with respect to the yz-plane is

$$M_{yz} = \iint_S x\, dS = 2\int_0^{\pi/2}\int_0^{2\cos\theta} \frac{2r^2\cos\theta}{\sqrt{4 - r^2}}\, dr\, d\theta.$$

Let $r = 2\sin\psi$. Then $dr = 2\cos\psi\, d\psi$ and $\sqrt{4 - r^2} = 2\cos\psi$. Thus

$$\int \frac{r^2}{\sqrt{4 - r^2}}\, dr = \int \frac{4\sin^2\psi}{2\cos\psi} \cdot 2\cos\psi\, d\psi = 4\int \frac{1 - \cos 2\psi}{2}\, d\psi$$

$$= 2\psi - \sin 2\psi + C = 2\psi - 2\sin\psi\cos\psi + C = 2\arcsin\left(\frac{r}{2}\right) - \frac{1}{2}r\sqrt{4 - r^2} + C.$$

968

Therefore

$$M_{yz} = 2\int_0^{\pi/2} \left[\left(4\arcsin\frac{r}{2} - r\sqrt{4-r^2}\right)\cos\theta \right]_0^{2\cos\theta} d\theta$$

$$= 2\int_0^{\pi/2} \left[4(\cos\theta)\arcsin(\cos\theta) - 4\cos^2\theta\,\sin\theta \right] d\theta.$$

To evaluate

$$J = \int (\cos\theta)\arcsin(\cos\theta)\,d\theta,$$

we use integration by parts. Let

$$u = \arcsin(\cos\theta), \qquad dv = \cos\theta\,d\theta; \quad \text{then}$$

$$du = -\frac{\sin\theta}{\sqrt{1-\cos^2\theta}}\,d\theta, \qquad v = \sin\theta.$$

Thus

$$J = (\sin\theta)\arcsin(\cos\theta) + \int \frac{\sin^2\theta}{\sin\theta}\,d\theta = (\sin\theta)\arcsin(\cos\theta) - \cos\theta + C.$$

Consequently,

$$M_{yz} = 2\left[4(\sin\theta)\arcsin(\cos\theta) - 4\cos\theta + \frac{4}{3}\cos^3\theta \right]_0^{\pi/2} = 2\left(4 - \frac{4}{3}\right) = \frac{16}{3}.$$

Therefore

$$\overline{x} = \frac{M_{yz}}{m} = \frac{4}{3\pi - 6} \approx 1.16795892925607244080260 6 \quad \text{and}$$

$$\overline{z} = \frac{M_{xy}}{m} = \frac{\pi}{2\pi - 4} \approx 1.375969196942054330601955.$$

14.5.31: The surface S is described by $h(x,y) = 4 - y^2$, and hence $dS = \sqrt{1+4y^2}\,dA$. Thus the moment of inertia of S with respect to the z-axis is

$$I_z = \int_{-2}^{2}\int_{-1}^{1} (x^2+y^2)\sqrt{1+4y^2}\,dx\,dy \int_{-2}^{2} \left[\left(\frac{1}{3}x^3 + xy^2\right)\sqrt{1+4y^2} \right]_{-1}^{1} dy$$

$$= \int_{-2}^{2} \left(\frac{2}{3} + 2y^2\right)\sqrt{1+4y^2}\,dy = \left[\frac{24y^3 + 19y}{48}\sqrt{1+4y^2} + \frac{13}{96}\operatorname{arcsinh}(2y) \right]_{-2}^{2}$$

$$= \frac{460\sqrt{17} + 13\operatorname{arcsinh}(4)}{48} \approx 40.080413560385795202979817.$$

14.5.33: By Eq. (12) in Section 14.5,

$$\cos\gamma = \frac{1}{|\mathbf{N}|}\cdot\frac{\partial(x,y)}{\partial(x,y)} = \frac{1}{|\mathbf{N}|}.$$

969

Therefore $|\mathbf{N}|\,dS = \sec\gamma\,dx\,dy$.

14.5.35: The temperature within the ball is $u(x,\,y,\,z) = 4(x^2+y^2+z^2)$. With position vector $\mathbf{r} = \langle\,x,\,y,\,z\,\rangle$ for points of B, we find that

$$\mathbf{q} = -K\nabla u = -2\cdot 4\langle\,2x,\,2y,\,2z\,\rangle = -16\langle\,x,\,y,\,z\,\rangle = -16\mathbf{r}.$$

A unit vector normal to the concentric spherical surface S of radius 3 is $\mathbf{n} = \frac{1}{3}\mathbf{r}$, so

$$\mathbf{q}\cdot\mathbf{n} = -\frac{16}{3}(x^2+y^2+z^2) = -\frac{16}{3}\cdot 9 = -48.$$

Because S is a spherical surface of radius 3, its surface area is $4\pi\cdot 9 = 36\pi$. Therefore the rate of heat flow across S is

$$\iint_S \mathbf{q}\cdot\mathbf{n}\,dS = -\iint_S 48\,dS = -48\cdot 36\pi = -1728\pi.$$

14.5.37: The given parametrization yields $\mathbf{N} = \langle\,-2bu^2\cos v,\,-2au^2\sin v,\,abu\,\rangle$, so the area of the paraboloid is

$$A = \int_0^{2\pi}\int_0^c u\sqrt{(2au\sin v)^2 + (2bu\cos v)^2 + (ab)^2}\,du\,dv$$

$$= \int_0^{2\pi} \frac{\left[(ab)^2 + 2(ac)^2 + 2(bc)^2 - 2(a^2-b^2)c^2\cos 2v\right]^{3/2} - (ab)^3}{6\left[a^2+b^2-(a^2-b^2)\cos 2v\right]}\,dv.$$

We believe the last integral to be nonelementary (because *Mathematica* uses elliptic functions to compute the antiderivative). With $a = 4$, $b = 3$, and $c = 2$ it reduces to

$$\int_0^{2\pi} \frac{-1728 + (344 - 56\cos 2v)^{3/2}}{150 - 42\cos 2v}\,dv.$$

The *Mathematica* `NIntegrate` command yields the result $A \approx 194.702812872043$. To compute the moment of inertia of the paraboloid with respect to the z-axis, we insert the factor

$$x^2 + y^2 = (au\cos v)^2 + (bu\sin v)^2 = (4u\cos v)^2 + (3u\sin v)^2$$

into the first integral, and *Mathematica* yields the result $I_z \approx 5157.168115181396$.

14.5.39: The given parametrization yields

$$|\mathbf{N}| = (\cosh u)\sqrt{(b\cosh u\cos v)^2 + (a\cosh u\sin v)^2 + (ab\sinh u)^2}\,,$$

and hence (using $a = 4$, $b = 3$, $c = 2$, and density $\delta = 1$) we find that the hyperboloid has surface area

$$A = \int_0^{2\pi}\int_{-c}^c |\mathbf{N}|\,du\,dv \approx 1057.350512779488$$

and moment of inertia with respect to the z-axis

$$I_z = \int_0^{2\pi} \int_{-c}^{c} (\cosh^2 u) \left[(a\cos v)^2 + (b\sin v)^2 \right] \cdot |\mathbf{N}| \, du \, dv \approx 98546.9348740325.$$

14.5.41: We use Fig. 13.7.17 of the text and the notation there; the only change is replacement of the variable ρ with the constant radius a of the spherical surface. The spherical shell has constant density δ and total mass $M = 4\pi a^2 \delta$. The "sum" of the vertical components of the gravitational forces exerted by mass elements $\delta \, dS$ of the spherical surface S on the mass m is

$$F = \iint_S \frac{Gm\delta \cos \alpha}{w^2} \, dS.$$

We saw in the solution of Problem 25 (among others) that $dS = a^2 \sin \phi \, dA$. Figure 13.7.17 also shows us that

$$w \cos \alpha = c - a \cos \phi \quad \text{and} \quad w^2 = a^2 + c^2 - 2ac \cos \phi \qquad (1)$$

(by the law of cosines (Appendix L)). Note that

$$F = 2\pi Gm\delta \int_{\phi=0}^{\pi} \frac{a^2 \cos \alpha \sin \phi}{w^2} \, d\phi.$$

Substitute $\cos \alpha = \dfrac{c - a\cos\phi}{w}$ to obtain

$$F = 2\pi Gm\delta \int_0^{\pi} \frac{a^2(c - a\cos\phi)\sin\phi}{w^3} \, d\phi.$$

Next note that $\phi = 0$ corresponds to $w = c - a$ and that $\phi = \pi$ corresponds to $w = c + a$. Moreover, by the second equation in (1),

$$\cos \phi = \frac{a^2 + c^2 - w^2}{2ac} \quad \text{and thus}$$

$$-\sin \phi \, d\phi = -\frac{w}{ac} \, dw.$$

These substitutions yield

$$F = 2\pi Gm\delta \int_{w=c-a}^{c+a} \frac{a^2}{w^3} \left(c - \frac{a^2 + c^2 - w^2}{2c} \right) \cdot \frac{w}{ac} \, dw$$

$$= 2\pi Gm\delta \int_{c-a}^{c+a} \frac{a}{w^2 c} \cdot \frac{1}{2c} \cdot (c^2 + w^2 - a^2) \, dw = \frac{2\pi Gm\delta a}{2c^2} \int_{c-a}^{c+a} \left(\frac{c^2 - a^2}{w^2} + 1 \right) dw$$

$$= \frac{\pi Gm\delta a}{c^2} \left[\frac{a^2 - c^2}{w} + w \right]_{c-a}^{c+a} = \frac{\pi Gm\delta a}{c^2} (a - c + a + c + c + a - c + a) = \frac{4\pi Gm\delta a^2}{c^2} = \frac{GMm}{c^2}.$$

971

Section 14.6

14.6.1: The right-hand side in the divergence theorem (Eq. (1)) is

$$\iiint_B \nabla \cdot \mathbf{F}\, dV = \iiint_B 3\, dV = 3 \cdot \frac{4}{3}\pi \cdot 1^3 = 4\pi$$

and the left-hand side in the divergence theorem is

$$\iint_S \mathbf{F} \cdot \mathbf{n}\, dS = \iint_S \langle x, y, z \rangle \cdot \langle x, y, z \rangle\, dS = \iint_S (x^2 + y^2 + z^2)\, dS = \iint_S 1\, dS = 1 \cdot 4\pi \cdot 1^2 = 4\pi.$$

Note that we integrate a constant function by multiplying its value by the size (length, area, or volume) of the domain of the integral. We will continue to do so without further comment.

14.6.3: On the face F of the cube in the plane $x = 2$, a unit vector normal to F is $\mathbf{n} = \mathbf{i}$, and $\mathbf{F} \cdot \mathbf{i} = x = 2$. Hence

$$\iint_F \mathbf{F} \cdot \mathbf{n}\, dS = 2 \cdot \text{area}(F) = 8.$$

By symmetry, the same result obtains on the faces in the planes $y = 2$ and $z = 2$. On the face G of the cube in the plane $x = 0$, a unit vector normal to G is $\mathbf{n} = -\mathbf{i}$, and $\mathbf{F} \cdot (-\mathbf{i}) = -x = 0$. Hence

$$\iint_G \mathbf{F} \cdot \mathbf{n}\, dS = 0.$$

By symmetry, the same result holds on the faces in the other two coordinate planes. Hence

$$\iint_S \mathbf{F} \cdot \mathbf{n}\, dS = 3 \cdot 8 + 3 \cdot 0 = 24.$$

Let B denote the solid cube bounded by S and let V denote the volume of B. Because $\nabla \cdot \mathbf{F} = 3$, we also have

$$\iiint_B \nabla \cdot \mathbf{F}\, dV = 3 \cdot V = 3 \cdot 8 = 24.$$

14.6.5: On the face F of the tetrahedron that lies in the plane $x = 0$, a unit vector normal to F is $\mathbf{n} = -\mathbf{i}$, and $\mathbf{F} \cdot \mathbf{n} = -x - y = -y$. Hence

$$\iint_F \mathbf{F} \cdot \mathbf{n}\, dS = \int_0^1 \int_0^{1-z} (-y)\, dy\, dz = \int_0^1 -\frac{1}{2}(1-z)^2\, dz = \left[\frac{1}{6}(1-z)^3 \right]_0^1 = -\frac{1}{6}.$$

By symmetry the same result holds on the faces in the other two coordinate planes. On the fourth face G of the tetrahedron, a unit vector normal to G is

$$\mathbf{n} = \frac{\sqrt{3}}{3}\langle 1, 1, 1 \rangle,$$

and G is part of the graph of $z = h(x, y) = 1 - x - y$, so that

$$dS = \sqrt{1 + (h_x)^2 + (h_y)^2} \; dA = \sqrt{3} \; dA.$$

Therefore

$$\iint_G \mathbf{F} \cdot \mathbf{n} \; dS = \iint_G (2x + 2y + 2z) \; dA = \iint_G 2 \; dA = \int_0^1 \int_0^{1-x} 2 \; dy \; dx = \int_0^1 \left[2y\right]_0^{1-x} dx = \left[2x - x^2\right]_0^1 = 1,$$

and therefore

$$\iint_S \mathbf{F} \cdot \mathbf{n} \; dS = 1 - 3 \cdot \frac{1}{6} = \frac{1}{2}.$$

Let B denote the solid tetrahedron itself, with volume V. Then

$$\iiint_B \boldsymbol{\nabla} \cdot \mathbf{F} \; dV = \iiint_B 3 \; dV = 3 \cdot V = 3 \cdot \frac{1}{6} \cdot 1 \cdot 1 = \frac{1}{2}.$$

14.6.7: Let B denote the solid cylinder bounded by the surface S. Then

$$\iint_S \mathbf{F} \cdot \mathbf{n} \; dS = \iiint_B \boldsymbol{\nabla} \cdot \mathbf{F} \; dV = \iiint_B 3(x^2 + y^2 + z^2) \; dV = \int_0^{2\pi} \int_0^3 \int_{-1}^4 3(r^2 + z^2) \cdot r \; dz \; dr \; d\theta$$

$$= 2\pi \int_0^3 \left[3r^3 z + rz^3\right]_{-1}^4 dr = 2\pi \int_0^3 (65r + 15r^3) \; dr = 2\pi \left[\frac{65}{2}r^2 + \frac{15}{4}r^4\right]_0^3$$

$$= 2\pi \cdot \frac{2385}{4} = \frac{2385}{2}\pi \approx 3746.3492394058284367.$$

14.6.9: Let B denote the tetrahedron bounded by the given surface S. Then

$$\iint_S \mathbf{F} \cdot \mathbf{n} \; dS = \iiint_B \boldsymbol{\nabla} \cdot \mathbf{F} \; dV = \int_0^1 \int_0^{1-x} \int_0^{1-x-y} (2x + 1) \; dz \; dy \; dx$$

$$= \int_0^1 \int_0^{1-x} (1 + x - 2x^2 - y - 2xy) \; dy \; dx = \int_0^1 \left[(1 + x - 2x^2)y - \frac{1}{2}(2x + 1)y^2\right]_0^{1-x} dx$$

$$= \int_0^1 \left(\frac{1}{2} - \frac{3}{2}x^2 + x^3\right) dx = \left[\frac{1}{2}x - \frac{1}{2}x^3 + \frac{1}{4}x^4\right]_0^1 = \frac{1}{4}.$$

14.6.11: Let B denote the solid paraboloid bounded by the surface S. Then

$$\iint_S \mathbf{F} \cdot \mathbf{n} \; dS = \iiint_B \boldsymbol{\nabla} \cdot \mathbf{F} \; dV = \iiint_B 5(x^2 + y^2 + z^2) \; dV = \int_0^{2\pi} \int_0^5 \int_0^{25-r^2} 5(r^2 + z^2) \cdot r \; dz \; dr \; d\theta$$

$$= 2\pi \int_0^5 \left[5r^3 z + \frac{5}{3}rz^3\right]_0^{25-r^2} dr = 2\pi \int_0^5 \left(\frac{78125}{3}r - 3000r^3 + 120r^5 - \frac{5}{3}r^7\right) dr$$

$$= 2\pi \left[\frac{78125}{6} r^2 - 750 r^4 + 20 r^6 - \frac{5}{24} r^8 \right]_0^5 = 2\pi \cdot \frac{703125}{8} = \frac{703125}{4} \pi \approx 552233.08363883.$$

14.6.13: Here is a step-by-step illustration of the solution using *Mathematica*.

```
f = {x, y, 3}        (* First we define the vector function F. *)
```

$\{x, y, 3\}$

```
D[%[[1]],x] + D[%[[2]],y] + D[%[[3]],z]        (* Then we compute div F. *)
```

2

```
Integrate[ 2*r, z ]        (* Begin the triple integral in cylindrical coordinates.  *)
```

$2rz$

```
(% /.  z → 4) - (% /.  z → r∧2)        (* Substitute the limits on z.  *)
```

$8r - 2r^3$

```
Integrate[ %, r]
```

$4r^2 - \frac{1}{2} r^4$

```
(% /.  r → 2) - (% /.  r → 0)        (* Substitute the limits on r.  *)
```

8

```
2*Pi*%        (* Integrate the constant by multiplying by 2π.  *)
```

16π

```
N[%, 18]        (* Approximate the answer.  *)
```

50.2654824574366918

14.6.15: Compare this problem and its solution with Problem 25 of Section 14.4. If f is a twice-differentiable scalar function, then

$$\nabla^2 f = \nabla \cdot (\nabla f) = \nabla \cdot \langle f_x, f_y, f_z \rangle = f_{xx} + f_{yy} + f_{zz}.$$

974

14.6.17: If $\nabla^2 f \equiv 0$ in the region T with boundary surface S, then by the divergence theorem and Problem 28 in Section 14.1,

$$\iint_S f \frac{\partial f}{\partial n}\, dS = \iint_S (f)(\nabla f) \cdot \mathbf{n}\, dS = \iiint_T \nabla \cdot [(f)(\nabla f)]\, dV$$

$$= \iiint_T [(f)\nabla \cdot (\nabla f) + (\nabla f) \cdot (\nabla f)]\, dV = \iiint_T [(f)\nabla^2 f + |\nabla f|^2]\, dV = \iiint_T |\nabla f|^2\, dV.$$

14.6.19: Green's first identity states that if the space region T has surface S with a piecewise smooth parametrization, if f and g are twice-differentiable scalar functions, and if $\partial f/\partial n = (\nabla f) \cdot \mathbf{n}$ where $\mathbf{n}$ is the unit vector normal to S with outer direction, then

$$\iint_S f \frac{\partial g}{\partial n}\, dS = \iiint_T (f \nabla^2 g + \nabla f \cdot \nabla g)\, dV.$$

Interchanging the roles of f and g yields the immediate consequence

$$\iint_S g \frac{\partial f}{\partial n}\, dS = \iiint_T (g \nabla^2 f + \nabla g \cdot \nabla f)\, dV.$$

Then substraction of the second of these equations from the first yields

$$\iint_S \left(f \frac{\partial g}{\partial n} - g \frac{\partial f}{\partial n} \right) dS = \iiint_B (f \nabla^2 g + \nabla f \cdot \nabla g - g \nabla^2 f - \nabla g \cdot \nabla f)\, dV = \iiint_B (f \nabla^2 g - g \nabla^2 f)\, dV,$$

and this is Green's second identity.

14.6.21: By the result in Problem 20, we have

$$\mathbf{B} = -\iint_S p\mathbf{n}\, dS = -\iiint_T \nabla(\delta g z)\, dV = -\iiint_T \langle 0, 0, \delta g \rangle\, dV = -\mathbf{k} \iiint_T \delta g\, dV = -W\mathbf{k}$$

because

$$\iiint_T \delta g\, dV = mg = W$$

is the weight of the fluid displaced by the body.

14.6.23: Let B denote the region bounded by the paraboloid and the plane. Because

$$\mathbf{F}(x, y, z) = \left\langle x\sqrt{x^2 + y^2 + z^2},\ y\sqrt{x^2 + y^2 + z^2},\ z\sqrt{x^2 + y^2 + z^2} \right\rangle,$$

we have

$$\nabla \cdot \mathbf{F} = \frac{x^2}{\sqrt{x^2 + y^2 + z^2}} + \frac{y^2}{\sqrt{x^2 + y^2 + z^2}} + \frac{z^2}{\sqrt{x^2 + y^2 + z^2}} + 3\sqrt{x^2 + y^2 + z^2} = 4\sqrt{x^2 + y^2 + z^2}.$$

Thus in cylindrical coordinates, $\nabla \cdot \mathbf{F} = 4\sqrt{r^2 + z^2}$. Then, by the divergence theorem,

$$I = \iint_S \mathbf{F} \cdot \mathbf{n}\, dS = \iiint_B \nabla \cdot \mathbf{F}\, dV$$

$$= \int_0^{2\pi} \int_0^{25} \int_0^{\sqrt{25-z}} 4r\sqrt{r^2 + z^2}\, dr\, dz\, d\theta = \frac{8}{3}\pi \int_0^{25} \left[(25 - z + z^2)^{3/2} - z^3 \right] dz.$$

Let

$$J = \int_0^c \left((z - a)^2 + b^2 \right)^{3/2} dz.$$

Later we will use the following values: $b = \frac{3}{2}\sqrt{11}$, $a = \frac{1}{2}$, and $c = 25 = a^2 + b^2$. The substitution $z = a + b\tan u$ yields

$$J = b^4 \int_{z=0}^c \sec^5 u\, du,$$

and the integral formulas in 37 and 28 of the endpapers of the text then yield

$$J = b^4 \int_{z=0}^c \sec^5 u\, du = b^4 \left(\left[\frac{1}{4} \sec^3 u \tan u \right]_{z=0}^c + \frac{3}{4} \int_{z=0}^c \sec^3 u\, du \right)$$

$$= b^4 \left(\left[\frac{1}{4} \sec^3 u \tan u \right]_{z=0}^c + \frac{3}{4} \left[\frac{1}{2} \sec u \tan u + \frac{1}{2} \ln|\sec u + \tan u| \right]_{z=0}^c \right).$$

The substitution $z = a + b\tan u$ also yields

$$\sec u = \frac{1}{b} \left[(z - a)^2 + b^2 \right]^{1/2} \quad \text{and} \quad \tan u = \frac{1}{b}(z - a).$$

It now follows that

$$J = b^4 \left[\frac{3(z-a)\left[(z-a)^2 + b^2\right]^{3/2}}{4b^4} + \frac{3(z-a)\left[(z-a)^2 + b^2\right]^{1/2}}{8b^2} + \frac{3}{8} \ln \left| \frac{z - a + \left[(z-a)^2 + b^2\right]^{1/2}}{b} \right| \right]_{z=0}^c$$

$$= b^4 \left[\frac{3a(a^2 + b^2)^{1/2}}{8b^2} + \frac{a(a^2 + b^2)^{3/2}}{4b^4} + \frac{3(c-a)\left[(c-a)^2 + b^2\right]^{1/2}}{8b^2} \right.$$

$$\left. + \frac{(c-a)\left[(c-a)^2 + b^2\right]^{3/2}}{4b^4} - \frac{3}{8} \ln \left(\frac{-a + (a^2 + b^2)^{1/2}}{b} \right) + \frac{3}{8} \ln \left(\frac{c - a + \left[(c-a)^2 + b^2\right]^{1/2}}{b} \right) \right].$$

Then substitution of the numerical values of a, b, and c yields

$$J = \frac{9801}{16} \left[\frac{1081885}{6534} - \frac{3}{8} \ln \left(\frac{3}{\sqrt{11}} \right) + \frac{3}{8} \ln \left(3\sqrt{11} \right) \right] = \frac{3}{128} \left(4327540 + 9801 \ln 11 \right).$$

And, finally,

$$I = \frac{8}{3}\pi \left(J - \int_0^{25} z^3\, dz \right) = \frac{8}{3}\pi \left[\frac{1}{128} \left(482620 + 29403 \ln 11 \right) \right]$$

$$= \frac{482620 + 29403 \ln 11}{48} \pi \approx 36201.96719156658969933477411 5.$$

976

We integrated $\sec^5 u$ using the integral formulas mentioned earlier; all the subsequent work was done by *Mathematica*.

14.6.25: We begin with Gauss's law in the form

$$\iint_S \mathbf{F} \cdot \mathbf{n} \, dS = -4\pi GM.$$

Because $\mathbf{F}$ and $\mathbf{n}$ have opposite directions and $\mathbf{n}$ is a unit vector, we may in this case deduce that

$$\iint_S |\mathbf{F}| \, dS = 4\pi GM = 4\pi G \cdot 0 = 0.$$

Therefore $4\pi r^2 |\mathbf{F}| = 0$, so that $|\mathbf{F}| = 0$. Therefore $\mathbf{F} = \mathbf{0}$.

14.6.27: Imagine a cylindrical surface of radius r and length L concentric around the wire. Because the electric field $\mathbf{E}$ is normal to the wire, there is no flux of $\mathbf{E}$ across the top and bottom of the cylinder, so the surface S of Gauss's law may be regarded as the curved side of the cylinder. It follows that

$$\iint_S \mathbf{E} \cdot \mathbf{n} \, dS = \frac{Q}{\epsilon_0} = \frac{Lq}{\epsilon_0}.$$

Because $\mathbf{E}$ and $\mathbf{n}$ are parallel and $\mathbf{n}$ is a unit vector, it now follows that

$$\iint_S |\mathbf{E}| \, dS = \frac{Lq}{\epsilon_0};$$

$$2\pi r L |\mathbf{E}| = \frac{Lq}{\epsilon_0};$$

$$|\mathbf{E}| = \frac{q}{2\pi \epsilon_0 r}.$$

Section 14.7

14.7.1: Because $\mathbf{n}$ is to be the upper unit normal vector, we have

$$\mathbf{n} = \mathbf{n}(x, y, z) = \frac{1}{2} \langle x, y, z \rangle.$$

The boundary curve C of the hemispherical surface S has the parametrization

$$x = 2\cos\theta, \quad y = 2\sin\theta, \quad z = 0, \quad 0 \leq \theta \leq 2\pi.$$

Therefore

$$\iint_S (\nabla \times \mathbf{F}) \cdot \mathbf{n} \, dS = \oint_C 3y \, dx - 2x \, dy + xyz \, dz$$

$$= \int_0^{2\pi} (-12\sin^2\theta - 8\cos^2\theta) \, d\theta = \left[-10\theta + \sin 2\theta \right]_0^{2\pi} = -20\pi.$$

14.7.3: Parametrize the boundary curve C of the surface S as follows:

$$x = 3\cos t, \quad y = 3\sin t, \quad z = 0, \quad 0 \leq t \leq 2\pi.$$

Then

$$\iint_S (\nabla \times \mathbf{F}) \cdot \mathbf{n} \, dS = \oint_C \mathbf{F} \cdot \mathbf{T} \, ds = \int_0^{2\pi} (-6\cos t - 27\sin^2 t \, \cos t) \, dt = \left[\frac{3}{4}(3\sin 3t - 17\sin t) \right]_0^{2\pi} = 0.$$

14.7.5: Parametrize the boundary curves of the surface S as follows:

$$C_1: \quad x = \cos t, \quad y = -\sin t, \quad z = 1, \quad 0 \leq t \leq 2\pi;$$

$$C_2: \quad x = 3\cos t, \quad y = 3\sin t, \quad z = 3, \quad 0 \leq t \leq 2\pi.$$

Then

$$\iint_S (\nabla \times \mathbf{F}) \cdot \mathbf{n} \, dS = \oint_{C_1} \mathbf{F} \cdot \mathbf{T} \, ds + \oint_{C_2} \mathbf{F} \cdot \mathbf{T} \, ds$$

$$= \int_0^{2\pi} (\cos^2 t + \sin^2 t) \, dt + \int_0^{2\pi} -27(\cos^2 t + \sin^2 t) \, dt = \left[-26t \right]_0^{2\pi} = -52\pi.$$

14.7.7: Parametrize S (the elliptical region bounded by C) as follows:

$$x = z = r\cos t, \quad y = r\sin t, \quad 0 \leq t \leq 2\pi.$$

Then $\mathbf{r}_r \times \mathbf{r}_t = \langle -r, 0, r \rangle$, $dS = r\sqrt{2} \, dr \, dt$, the upper unit normal for S is

$$\mathbf{n} = \frac{1}{2}\sqrt{2} \, \langle -1, 0, 1 \rangle,$$

and $\nabla \times \mathbf{F} = \langle 3, 2, 1 \rangle$. Therefore $(\nabla \times \mathbf{F}) \cdot \mathbf{n} = -\sqrt{2}$. Consequently,

$$\oint_C \mathbf{F} \cdot \mathbf{T} \, ds = \iint_S (\nabla \times \mathbf{F}) \cdot \mathbf{n} \, dS = \int_0^{2\pi} \int_0^2 (-2r) \, dr \, dt$$

$$= 2\pi \left[-r^2 \right]_0^2 = 2\pi \cdot (-4) = -8\pi \approx -25.1327412287183459.$$

14.7.9: If $\mathbf{F}(x, y, z) = \langle y - x, x - z, x - y \rangle$, then

$$\nabla \times \mathbf{F} = \begin{vmatrix} \mathbf{i} & \mathbf{j} & \mathbf{k} \\ \dfrac{\partial}{\partial x} & \dfrac{\partial}{\partial y} & \dfrac{\partial}{\partial z} \\ y - x & x - z & x - y \end{vmatrix} = \langle 0, -1, 0 \rangle.$$

The surface S bounded by C is part of the plane with equation $x + 2y + z = 2$, so an upward normal to S is $\langle 1, 2, 1 \rangle$. Hence the unit normal vector we need is

$$\mathbf{n} = \frac{1}{6}\sqrt{6}\,\langle 1, 2, 1 \rangle.$$

Then $(\nabla \times \mathbf{F}) \cdot \mathbf{n} = -\frac{1}{3}\sqrt{6}$, a constant, so it remains only to find the area of S. Its vertices are at $A(2, 0, 0)$, $B(0, 1, 0)$, and $C(0, 0, 2)$, so we compute the "edge vectors"

$$\mathbf{u} = \overrightarrow{AB} = \langle -2, 1, 0 \rangle \quad \text{and} \quad \mathbf{v} = \overrightarrow{AC} = \langle -2, 0, 2 \rangle$$

and their cross product

$$\mathbf{u} \times \mathbf{v} = \begin{vmatrix} \mathbf{i} & \mathbf{j} & \mathbf{k} \\ -2 & 1 & 0 \\ -2 & 0 & 2 \end{vmatrix} = \langle 2, 4, 2 \rangle;$$

the area of S is then $\frac{1}{2}|\mathbf{u} \times \mathbf{v}| = \sqrt{6}$. Therefore

$$\oint_C \mathbf{F} \cdot \mathbf{T}\, ds = \iint_S (\nabla \times \mathbf{F}) \cdot \mathbf{n}\, dS = \left(-\frac{1}{3}\sqrt{6}\right) \cdot \sqrt{6} = -2.$$

14.7.11: If $\mathbf{F}(x, y, z) = \langle 3y - 2z, 3x + z, y - 2x \rangle$, then

$$\nabla \times \mathbf{F} = \begin{vmatrix} \mathbf{i} & \mathbf{j} & \mathbf{k} \\ \dfrac{\partial}{\partial x} & \dfrac{\partial}{\partial y} & \dfrac{\partial}{\partial z} \\ 3y - 2z & 3x + z & y - 2x \end{vmatrix} = \langle 1 - 1,\ -2 + 2,\ 3 - 3 \rangle = \mathbf{0}.$$

Therefore $\mathbf{F}$ is irrotational. Next, let C be the straight line segment from $(0, 0, 0)$ to the point (u, v, w) of space. Parametrize C as follows: $x = tu$, $y = tv$, $z = tw$, $0 \leqq t \leqq 1$. Then

$$\int_C \mathbf{F} \cdot \mathbf{T}\, ds = \int_{t=0}^1 (6tuv - 4tuw + 2tvw)\, dt = \Big[(3uv - 2uw + vw)t^2\Big]_0^1 = 3uv - 2uw + vw.$$

Now replace u with x, v with y, and w with z (because (u, v, w) represents an arbitrary point of space). This yields the potential function

$$\phi(x, y, z) = 3xy - 2xz + yz.$$

To be absolutely certain of this, verify for yourself that $\nabla \phi = \mathbf{F}$.

14.7.13: If $\mathbf{F}(x, y, z) = \langle 3e^z - 5y \sin x,\ 5\cos x,\ 17 + 3xe^z \rangle$, then

$$\nabla \times \mathbf{F} = \begin{vmatrix} \mathbf{i} & \mathbf{j} & \mathbf{k} \\ \dfrac{\partial}{\partial x} & \dfrac{\partial}{\partial y} & \dfrac{\partial}{\partial z} \\ 3e^z z - 5y \sin x & 5\cos x & 17 + 3xe^z \end{vmatrix} = \langle 0,\ 3e^z - 3e^z,\ -5\sin x + 5\sin x \rangle = \mathbf{0}.$$

Therefore **F** is irrotational. Then the method of Example 3 yields

$$x(t) = tu, \quad y(t) = tv, \quad z(t) = tw, \quad 0 \leqq t \leqq 1,$$

$$\mathbf{r}(t) = \langle\, x(t),\, y(t),\, z(t)\,\rangle = \langle\, tu,\, tv,\, tw\,\rangle, \quad \text{and}$$

$$\mathbf{F}(t) = \langle\, 3\exp(tw) - 5tv\sin tu,\, 5\cos tu,\, 17 + 3tu\exp(tw)\,\rangle,$$

so that

$$\int_C \mathbf{F}\cdot d\mathbf{r} = \int_{t=0}^{1} \left(17w + 3tuwe^{tw} + 5v\cos tu + 3ue^{tw} - 5tuv\sin tu\right) dt$$

$$= \left[3tue^{tw} + 17tw + 5tv\cos tu\right]_0^1 = 3ue^w + 17w + 5v\cos u.$$

Therefore $\phi(x, y, z) = 3xe^z + 17z + 5y\cos x$.

14.7.15: We are given $\mathbf{r} = \langle x, y, z \rangle$; suppose that $\mathbf{a} = \langle b, c, d \rangle$ is a constant vector. Part (a):

$$\mathbf{a}\times\mathbf{r} = \begin{vmatrix} \mathbf{i} & \mathbf{j} & \mathbf{k} \\ b & c & d \\ x & y & z \end{vmatrix} = \langle\, cz - dy,\, dx - bz,\, by - cx\,\rangle,$$

and hence $\nabla\cdot(\mathbf{a}\times\mathbf{r}) = 0 + 0 + 0 = 0$. Part (b): Using some of the results in part (a), we have

$$\nabla\times(\mathbf{a}\times\mathbf{r}) = \begin{vmatrix} \mathbf{i} & \mathbf{j} & \mathbf{k} \\ \dfrac{\partial}{\partial x} & \dfrac{\partial}{\partial y} & \dfrac{\partial}{\partial z} \\ cz - dy & dx - bz & by - cx \end{vmatrix} = \langle\, 2b,\, 2c,\, 2d\,\rangle = 2\mathbf{a}.$$

Part (c): First,

$$(\mathbf{r}\cdot\mathbf{r})\mathbf{a} = \langle\, b(x^2 + y^2 + z^2),\, c(x^2 + y^2 + z^2),\, d(x^2 + y^2 + z^2)\,\rangle.$$

Thus $\nabla\cdot\big[(\mathbf{r}\cdot\mathbf{r})\mathbf{a}\big] = 2bx + 2cy + 2dz = 2\mathbf{r}\cdot\mathbf{a}$. Part (d): Using some of the results in parts (a) and (c), we have

$$\nabla\times\big[(\mathbf{r}\cdot\mathbf{r})\mathbf{a}\big] = \begin{vmatrix} \mathbf{i} & \mathbf{j} & \mathbf{k} \\ \dfrac{\partial}{\partial x} & \dfrac{\partial}{\partial y} & \dfrac{\partial}{\partial z} \\ b(x^2 + y^2 + z^2) & c(x^2 + y^2 + z^2) & d(x^2 + y^2 + z^2) \end{vmatrix}$$

$$= \langle\, 2dy - 2cz,\, 2bz - 2dx,\, 2cx - 2by\,\rangle$$

$$= -2\langle\, cz - dy,\, dx - bz,\, by - cx\,\rangle = -2(\mathbf{a}\times\mathbf{r}) = 2(\mathbf{r}\times\mathbf{a}).$$

14.7.17: Assume that S is a closed surface having a piecewise smooth parametrization, that $\mathbf{n}$ is the outer unit normal vector for S, and that $\mathbf{F}$ is continuous and has continuous first-order partial derivatives on an open region containing S and the solid T that it bounds. Part (a): By the divergence theorem,

$$\iint_S (\nabla \times \mathbf{F}) \cdot \mathbf{n} \, dS = \iiint_T \nabla \cdot (\nabla \times \mathbf{F}) \, dV = \iiint_T 0 \, dV = 0$$

by Problem 32 of Section 14.1. Part (b): Let C be a simple closed curve on S having a suitably differentiable parametrization and a given orientation. Then C is the common boundary of the two surfaces S_1 and S_2 into which it divides S; that is, S is the union of S_1 and S_2, S_1 and S_2 meet in the curve C, and C has positive orientation on (say) S_1 and the opposite orientation on S_2. Then

$$\iint_{S_1} (\nabla \times \mathbf{F}) \cdot \mathbf{n} \, dS = \int_C \mathbf{F} \cdot \mathbf{T} \, ds = -\iint_{S_2} (\nabla \times \mathbf{F}) \cdot \mathbf{n} \, dS.$$

Therefore

$$\iint_S (\nabla \times \mathbf{F}) \cdot \mathbf{n} \, dS = \iint_{S_1} (\nabla \times \mathbf{F}) \cdot \mathbf{n} \, dS + \iint_{S_2} (\nabla \times \mathbf{F}) \cdot \mathbf{n} \, dS = 0.$$

14.7.19: We are given the constant vector $\mathbf{a}$ and the vector $\mathbf{r} = \langle x, y, z \rangle$. Let $\mathbf{F} = \mathbf{a} \times \mathbf{r}$. Then by Stokes' theorem,

$$\int_C (\mathbf{a} \times \mathbf{r}) \cdot \mathbf{T} \, ds = \iint_S [\nabla \times (\mathbf{a} \times \mathbf{r})] \cdot \mathbf{n} \, dS = \iint_S 2\mathbf{a} \cdot \mathbf{n} \, dS$$

by part (b) of Problem 15. But because integration is carried out componentwise, it now follows that

$$\int_C (\mathbf{a} \times \mathbf{r}) \cdot \mathbf{T} \, ds = 2\mathbf{a} \iint_S \mathbf{n} \, dS.$$

14.7.21: Beginning with the *Suggestion* given in the statement of Problem 21, we find that

$$\phi_x = \lim_{h \to 0} \frac{\phi(x+h, y, z) - \phi(x, y, z)}{h} = \lim_{h \to 0} \frac{1}{h} \int_{(x,y,z)}^{(x+h,y,z)} P(t, y, z) \, dt = \lim_{h \to 0} \frac{h \cdot P(x^\star, y, z)}{h} = P(x, y, z)$$

where $x^\star$ is between x and $x+h$. A similar argument shows that $\phi_y = Q$ and $\phi_z = R$. Adding these results establishes that $\nabla \phi = \mathbf{F} = \langle P, Q, R \rangle$.

Chapter 14 Miscellaneous Problems

14.M.1: Parametrize C: $x = t$, $y = \frac{4}{3}t$, $0 \leq t \leq 3$. Then

$$ds = \sqrt{[x'(t)]^2 + [y'(t)]^2} \, dt = \frac{5}{3} \, dt \quad \text{and} \quad [x(t)]^2 + [y(t)]^2 = \frac{25}{9}t^2.$$

Therefore
$$\int_C (x^2 + y^2)\, ds = \int_0^3 \frac{125}{27} t^2\, dt = \left[\frac{125}{81} t^3\right]_0^3 = \frac{125}{3} \approx 41.666666666667.$$

14.M.3: We are given the curve C parametrized by $\mathbf{r}(t) = \langle e^{2t}, e^t, e^{-t}\rangle$, $0 \leqq t \leqq \ln 2$. Because $\mathbf{F}(t) = \langle x, y, z\rangle$, we can also write
$$\mathbf{F}(t) = \langle e^{2t}, e^t, e^{-t}\rangle,$$

and therefore
$$\int_C \mathbf{F} \cdot \mathbf{T}\, ds = \int_C \mathbf{F} \cdot d\mathbf{r} = \int_0^{\ln 2} (2e^{4t} + e^{2t} - e^{-2t})\, dt = \frac{1}{2}\left[e^{4t} + e^{2t} + e^{-2t}\right]_0^{\ln 2} = \frac{69}{8} = 8.625.$$

14.M.5: Given the curve C with parametrization
$$x(t) = t, \quad y(t) = t^{3/2}, \quad z(t) = t^2, \quad 0 \leqq t \leqq 4,$$

we substitute and find that
$$\int_C z^{1/2}\, dx + x^{1/2}\, dy + y^2\, dz = \int_0^4 \left(t + \frac{3}{2}t + 2t^4\right) dt = \left[\frac{5}{4}t^2 + \frac{2}{5}t^5\right]_0^4 = \frac{2148}{5} = 429.6.$$

14.M.7: Suppose that there exists a function $\phi(x, y)$ such that $\nabla\phi = \langle x^2 y, xy^2\rangle$. Then
$$\phi(x, y) = \int x^2 y\, dx = \frac{1}{3}x^3 y + g(y),$$

and hence
$$\frac{\partial \phi}{\partial y} = \frac{1}{3}x^3 + g'(y). \tag{1}$$

But there is no choice of $g(y)$ such that the last expression in Eq. (1) can equal xy^2 unless x is constant. This is not possible on any path C from $(0, 0)$ to $(1, 1)$. Thus there is no such function ϕ, and thus by Theorem 2 the given integral is not independent of the path from $(0, 0)$ to $(1, 1)$.

14.M.9: Parametrize the wire W using $x(t) = t$, $y(t) = \frac{1}{2}t^2$, $0 \leqq t \leqq 2$. Then
$$ds = \sqrt{[x'(t)]^2 + [y'(t)]^2}\, dt = \sqrt{1 + t^2}\, dt.$$

The density of the wire is $\delta(t) = x(t) = t$. Therefore the mass of the wire is
$$m = \int_W \delta\, ds = \int_0^2 t\sqrt{1 + t^2}\, dt = \left[\frac{1}{3}(1 + t^2)^{3/2}\right]_0^2 = \frac{5\sqrt{5} - 1}{3} \approx 3.3934466291663162.$$

Its moment of inertia with respect to the y-axis is

$$I_y = \int_W \delta x^2 \, ds = \int_0^2 t^3 (1+t^2)^{1/2} \, dt$$

$$= \left[\frac{1}{15}(1+t^2)^{1/2}(3t^4 + t^2 - 2) \right]_0^2 = \frac{50\sqrt{5}+2}{15} \approx 7.5868932583326323.$$

If you prefer, $I_y \approx m \cdot (1.4952419583303542)^2$.

14.M.11: Let R denote the region bounded by C. Then Green's theorem yields

$$\oint_C x^2 y \, dx + xy^2 \, dy = \iint_R (y^2 - x^2) \, dA = \int_{x=-2}^2 \int_{y=x^2}^{8-x^2} (y^2 - x^2) \, dy \, dx$$

$$= \int_{-2}^2 \left[\frac{1}{3} y^3 - x^2 y \right]_{x^2}^{8-x^2} dx = \int_{-2}^2 \left(\frac{512}{3} - 72x^2 + 10x^4 - \frac{2}{3}x^6 \right) dx$$

$$= \left[\frac{512}{3} x - 24x^3 + 2x^5 - \frac{2}{21} x^7 \right]_{-2}^2 = \frac{2816}{7} \approx 402.2857142857142857.$$

14.M.13: Suppose that C is any positively oriented piecewise smooth simple closed curve in the xy-plane. Let $\mathbf{n}$ be the outwardly directed unit vector normal to C. We will apply the vector form of Green's theorem in Eq. (9) of Section 14.4,

$$\oint_C \mathbf{F} \cdot \mathbf{n} \, ds = \iint_R \boldsymbol{\nabla} \cdot \mathbf{F} \, dA$$

where R denotes the bounded plane region with boundary C and $\mathbf{F}$ is a two-dimensional vector function of x and y with continuously differentiable component functions. If $\mathbf{F}(x, y) = \langle x^2 y, -xy^2 \rangle$, then

$$\oint_C \mathbf{F} \cdot \mathbf{n} \, ds = \iint_R (2xy - 2xy) \, dA = 0.$$

Therefore the integrals given in the statement of Problem 13 are equal because each is equal to zero. It is also possible to verify this by a direct computation. For example, using the parametrization $x = \cos t$, $y = \sin t$, $0 \leq t \leq 2\pi$ and the outer unit normal vector $\mathbf{n} = \langle \cos t, \sin t \rangle$, the first integral in Problem 13 becomes

$$\oint_C \mathbf{F} \cdot \mathbf{n} \, ds = \int_0^{2\pi} (\cos^3 t \, \sin t - \cos t \, \sin^3 t) dt = -\frac{1}{4} \left[\cos^4 t + \sin^4 t \right]_0^{2\pi} = 0 - 0 = 0.$$

14.M.15: Suppose that

$$\int_C P \, dx + Q \, dy$$

983

is independent of the path in the plane region D. By Theorem 2 of Section 15.3, $\mathbf{F} = \langle P, Q \rangle = \nabla \phi$ in D where ϕ is some differentiable scalar potential function. Suppose that C is a simple closed curve in D. Choose a point (a, b) on C. Then by Theorem 1 of Section 14.3,

$$\oint_C P\, dx + Q\, dy = \left[\phi(x, y) \right]_{(a,b)}^{(a,b)} = \phi(a, b) - \phi(a, b) = 0.$$

14.M.17: The surface S is described by $h(x, y) = 2 - x^2 - y^2$, and thus

$$dS = \sqrt{1 + (h_x)^2 + (h_y)^2}\, dA = \sqrt{1 + 4x^2 + 4y^2}\, dA.$$

Therefore

$$\iint_S (x^2 + y^2 + 2z)\, dS = \iint_S (4 - x^2 - y^2)\, dS = \int_{\theta=0}^{2\pi} \int_{r=0}^{\sqrt{2}} r(4 - r^2)\sqrt{1 + 4r^2}\, dr\, d\theta$$

$$= 2\pi \left[\frac{1}{120}(41 + 158r^2 - 24r^4)\sqrt{1 + 4r^2} \right]_0^{\sqrt{2}} = 2\pi \cdot \frac{371}{60} = \frac{371}{30}\pi \approx 38.8510291493937764.$$

14.M.19: The upper surface S_1 is described by $h(x, y) = 12 - 2x^2 - y^2$, and thus

$$dS = \sqrt{1 + (h_x)^2 + (h_y)^2}\, dA = \sqrt{1 + 16x^2 + 4y^2}\, dA.$$

An outwardly pointing vector normal to S_1 is

$$\left\langle -\frac{\partial h}{\partial x}, -\frac{\partial h}{\partial y}, 1 \right\rangle = \langle 4x, 2y, 1 \rangle,$$

and therefore a outwardly pointing unit vector normal to S_1 is

$$\mathbf{n} = \frac{1}{\sqrt{1 + 16x^2 + 4y^2}} \langle 4x, 2y, 1 \rangle.$$

Hence the outward flux of $\mathbf{F} = \langle x, y, z \rangle$ across S_1 is

$$\iint_{S_1} \mathbf{F} \cdot \mathbf{n}\, dS = \iint_{S_1} \frac{4x^2 + 2y^2 + z}{\sqrt{1 + 16x^2 + 4y^2}} \cdot \sqrt{1 + 16x^2 + 4y^2}\, dS = \iint_{S_1} (4x^2 + 2y^2 + z)\, dS$$

$$= \iint_S (12 + 2x^2 + y^2)\, dS = \int_{\theta=0}^{2\pi} \int_{r=0}^{2} (12 + 2r^2 \cos^2 \theta + r^2 \sin^2 \theta) \cdot r\, dr\, d\theta$$

$$= \int_0^{2\pi} \left[6r^2 + \frac{1}{8}r^4(3 + \cos 2\theta) \right]_0^2 d\theta = \int_0^{2\pi} (30 + 2\cos 2\theta)\, d\theta = \left[30\theta + \sin 2\theta \right]_0^{2\pi} = 60\pi.$$

The lower surface S_2 is described by $h(x, y) = x^2 + 2y^2$. By computations similar to those shown earlier in this solution, we find that

$$dS = \sqrt{1 + 4x^2 + 16y^2}\, dA$$

984

and that an outer unit vector normal to S_2 is

$$\mathbf{n} = \frac{1}{\sqrt{1 + 4x^2 + 16y^2}} \langle 2x, 4y, -1 \rangle.$$

Thus the flux of $\mathbf{F}$ across S_2 is

$$\iint_{S_2} \mathbf{F} \cdot \mathbf{n} \, dS = \iint_{S_2} (2x^2 + 4y^2 - z) \, dS = \iint_{S_2} (x^2 + 2y^2) \, dS = \int_0^{2\pi} \int_0^2 (r^2 \cos^2 \theta + 2r^2 \sin^2 \theta) \cdot r \, dr \, d\theta$$

$$= \int_0^{2\pi} \left[\frac{1}{4} r^4 \cos^2 \theta + \frac{1}{2} r^4 \sin^2 \theta \right]_0^2 d\theta = \int_0^{2\pi} (4 \cos^2 \theta + 8 \sin^2 \theta) \, d\theta = \left[6\theta - \sin 2\theta \right]_0^{2\pi} = 12\pi.$$

Therefore the total outward flux of $\mathbf{F}$ across the boundary of T is

$$\phi = 60\pi + 12\pi = 72\pi \approx 226.19467105846511316931.$$

14.M.21: We compute the three Jacobians in Eq. (17) of Section 14.5 using the parameters y and z. The result is

$$\frac{\partial(y, z)}{\partial(y, z)} = \begin{vmatrix} y_y & y_z \\ z_y & z_z \end{vmatrix} = 1,$$

$$\frac{\partial(z, x)}{\partial(y, z)} = \begin{vmatrix} z_y & z_z \\ x_y & x_z \end{vmatrix} = -\frac{\partial x}{\partial y}, \quad \text{and}$$

$$\frac{\partial(x, y)}{\partial(y, z)} = \begin{vmatrix} x_y & x_z \\ y_y & y_z \end{vmatrix} = -\frac{\partial x}{\partial z}.$$

Therefore

$$\iint_S P \, dy \, dz + Q \, dz \, dx + R \, dx \, dy = \iint_D \left(P - Q \frac{\partial x}{\partial y} - R \frac{\partial x}{\partial z} \right) dy \, dz.$$

14.M.23: Here we have

$$\bar{z} = \frac{1}{V} \iiint_T z \, dV = \frac{1}{V} \iiint_T \nabla \cdot \langle 0, 0, \tfrac{1}{2} z^2 \rangle \, dV = \frac{1}{V} \iint_S \frac{1}{2} z^2 \, dx \, dy = \frac{1}{2V} \iint_S z^2 \, dx \, dy$$

by Eq. (4) of Section 14.6.

14.M.25: By Eq. (23) of Section 14.5, the heat flow across the boundary sphere S into B is given by

$$R = \iint_S K(\nabla u) \cdot \mathbf{n} \, dS.$$

The divergence theorem then gives

$$R = \iiint_B \nabla \cdot (K \nabla u)\, dV = \iiint_B K \nabla^2 u\, dV.$$

14.M.27: Problems 25 and 26 imply that

$$\iiint_B c \cdot \frac{\partial u}{\partial t}\, dV = \iiint_B K \nabla^2 u\, dV$$

for any small ball B within the body. This can be so only if $cu_t \equiv K\nabla^2 u$; that is, $u_t = k\nabla^2 u$ (because $k = K/c$).

14.M.29: We begin with $\mathbf{r} = \langle x, y, z \rangle$ and $\phi = \phi(r)$ where $r = |\mathbf{r}| = \sqrt{x^2 + y^2 + z^2}$. Part (a):

$$\nabla \phi(r) = \nabla \phi\left(\sqrt{x^2 + y^2 + z^2}\right)$$

$$= \left\langle \phi'(r) \cdot \frac{x}{\sqrt{x^2 + y^2 + z^2}},\ \phi'(r) \cdot \frac{y}{\sqrt{x^2 + y^2 + z^2}},\ \phi'(r) \cdot \frac{z}{\sqrt{x^2 + y^2 + z^2}} \right\rangle$$

$$= \phi'(r) \left\langle \frac{x}{r}, \frac{y}{r}, \frac{z}{r} \right\rangle = \frac{\mathbf{r}}{r} \phi'(r).$$

Part (b): We use the result in part (a) and the result in Problem 28 of Section 14.1:

$$\nabla \cdot [\phi(r)\mathbf{r}] = \phi(r)(\nabla \cdot \mathbf{r}) + (\nabla \phi) \cdot \mathbf{r} = \phi(r)(1 + 1 + 1) + \frac{\mathbf{r} \cdot \mathbf{r}}{r} \cdot \frac{d\phi}{dr} = 3\phi(r) + r\frac{d\phi}{dr}.$$

Part (c): We use the result in part (a) and the results in Problems 29 and 35 of Section 15.1:

$$\nabla \times (\phi(r)\mathbf{r}) = (\phi(r))(\nabla \times \mathbf{r}) + (\nabla \phi) \times \mathbf{r} = 0 + \frac{\mathbf{r} \times \mathbf{r}}{r} \cdot \frac{d\phi}{dr} = 0.$$

14.M.31 Let us envision a Möbius strip M in space constructed from a long narrow rectangular strip of paper by matching its ends with a half-twist. Let the strip of paper be subdivided into smaller rectangles $R_1, R_2, \ldots, R_n$ as indicated in the following figure, with the boundary curve of each of these rectangles oriented in the positive fashion described in Section 15.7. Then the arrows cancel along any interior segment indicated in the figure, and the arrows on the ends of the strip are as indicated there—upward on the right and downward on the left.

986

The Möbius strip is formed by matching these two ends of the rectangular paper strip, with the two end arrows matching in direction (thus providing the necessary half-twist). If we denote by S_i the ith curvilinear rectangle on the Möbius strip (corresponding to the original R_i), then Stokes' theorem gives

$$\oint_{C_i} \mathbf{F} \cdot \mathbf{T}\, ds = \iint_{S_i} (\nabla \times \mathbf{F}) \cdot \mathbf{n}\, dS$$

for each i. Taking account of cancellation of line integrals in opposite directions along segments corresponding to interior segments in the figure, summation then yields

$$\iint_M (\nabla \times \mathbf{F}) \cdot \mathbf{n}\, dS = \sum_{i=1}^{n} \iint_{S_i} (\nabla \times \mathbf{F}) \cdot \mathbf{n}\, dS$$

$$= \sum_{i=1}^{n} \oint_{C_i} \mathbf{F} \cdot \mathbf{T}\, ds = \oint_C \mathbf{F} \cdot \mathbf{T}\, ds + 2 \int_J \mathbf{F} \cdot \mathbf{T}\, ds,$$

where C denotes the boundary curve of the Möbius strip and J denotes the single interior segment along which the arrows match, so that the line integrals along J do *not* cancel. Because of the final term in the last equation, this calculation does *not* yield Stokes' theorem in the form

$$\iint_M (\nabla \times \mathbf{F}) \cdot \mathbf{n}\, dS = \oint_C \mathbf{F} \cdot \mathbf{T}\, ds$$

for the Möbius strip. Greater generality would take us too far afield, but surely you can well imagine that a similar "failure to cancel" would occur for any subdivision of the original narrow rectangular strip.

14.M.33: Part (a): If ΔS_i is a small piece of the boundary sphere S of the small ball B, and δ_i, $\mathbf{v}_i$, and $\mathbf{n}_i$ denote (respectively) the density, fluid flow velocity vector, and outward unit normal at time t at a typical point of ΔS_i, then the rate of *outward* fluid flow across this area element ΔS_i is approximately $\delta_i \mathbf{v}_i \cdot \mathbf{n}_i \, \Delta S_i$. Hence the rate of flow of fluid *into* B at time t is given by

$$Q'(t) \approx - \sum_{i=1}^{n} \delta_i \mathbf{v}_i \cdot \mathbf{n}_i \, \Delta S_i \ \rightarrow \ - \iint_S \delta \mathbf{v} \cdot \mathbf{n}\, dS.$$

Part (b): Equating our two expressions for $Q'(t)$, we get

$$\iiint_B \frac{\partial \delta}{\partial t}\, dV = - \iint_S \delta \mathbf{v} \cdot \mathbf{n}\, dS = - \iiint_B \nabla \cdot (\delta \mathbf{v})\, dV$$

(applying the divergence theorem on the right). The fact that

$$\iiint_B \left[\frac{\partial \delta}{\partial t} + \boldsymbol{\nabla} \cdot (\delta \mathbf{v}) \right] dV = 0$$

therefore holds for *any* small ball B within the fluid flow region implies that the integrand must vanish identically; that is,

$$\frac{\partial \delta}{\partial t} + \boldsymbol{\nabla} \cdot (\delta \mathbf{v}) = 0.$$

Appendix A

A.1: $|3 - 17| = |-14| = -(-14) = 14.$

A.3: $\left|-0.25 - \frac{1}{4}\right| = \left|-\frac{1}{2}\right| = -\left(-\frac{1}{2}\right) = \frac{1}{2}.$

A.5: $|(-5)(4 - 9)| = |(-5)(-5)| = |25| = 25.$

A.7: $|(-3)^3| = |-27| = 27.$

A.9: According to *Mathematica*,

$$\pi \approx 3.14159265358979323846264338327950288419716939937510582097494, \quad \text{whereas}$$

$$\frac{22}{7} \approx 3.14285714285714285714285714285714285714285714285714285714286.$$

Therefore

$$\pi - \frac{22}{7} < 0, \quad \text{and thus} \quad \left|\pi - \frac{22}{7}\right| = \frac{22}{7} - \pi.$$

For a more elegant proof of the inequality, see Miscellaneous Problems 110 and 111 of Chapter 7.

A.11: If $x < 3$ then $x - 3 < 0$, so in this case $|x - 3| = -(x - 3) = 3 - x.$

A.13: $2x - 7 < -3$: $2x < 4$; $x < 2.$ Solution set: $(-\infty, 2).$

A.15: $3x - 4 \geqq 17$: $3x \geqq 21$; $x \geqq 7.$ Solution set: $[7, +\infty).$

A.17: $2 - 3x < 7$: $-3x < 5$; $3x > -5$; $x > -\frac{5}{3}.$ Solution set: $\left(-\frac{5}{3}, +\infty\right).$

A.19: $-3 < 2x + 5 < 7$: $-8 < 2x < 2$; $-4 < x < 1.$ Solution set: $(-4, 1).$

A.21: $-6 \leqq 5 - 2x < 2$: $-11 \leqq -2x < -3$; $3 < 2x \leqq 11$; $\frac{3}{2} < x \leqq \frac{11}{2}.$ Solution set: $\left(\frac{3}{2}, \frac{11}{2}\right].$

A.23: If $|3 - 2x| < 5$, then by the fourth property of absolute value in (3) we have $-5 < 3 - 2x < 5.$ Thus $-8 < -2x < 2$, and hence $-2 < 2x < 8.$ Therefore $-1 < x < 4.$ So the solution set of the original inequality is $(-1, 4).$

A.25: We will find the *complement* of the solution set—those real numbers *not* in the solution set—by solving instead the inequality $|1 - 3x| \leqq 2.$ By an extension of the fourth property of absolute value in (3), we then have $-2 \leqq 1 - 3x \leqq 2.$ It follows that $-3 \leqq -3x \leqq 1$, and hence $-1 \leqq 3x \leqq 3.$ So the complement of the solution set of the original equality is the closed interval $\left[-\frac{1}{3}, 1\right].$ Therefore the solution set of the inequality $|1 - 3x| > 2$ is $(-\infty, \frac{1}{3}) \cup (1, +\infty).$

A.27: Given: $2 \leqq |4-5x| \leqq 4$. Case 1: $4-5x \geqq 0$. This is equivalent to the assertion that $x \leqq \frac{4}{5}$; moreover, in this case the original inequality takes the form $2 \leqq 4 - 5x \leqq 4$. Thus $-2 \leqq -5x \leqq 0$, so that $0 \leqq 5x \leqq 2$. So the solutions we obtain in Case 1 are the real numbers in *both* of the intervals $\left[0, \frac{2}{5}\right]$ and $\left(-\infty, \frac{4}{5}\right]$. So Case 1 contributes the interval $\left[0, \frac{2}{5}\right]$ to the solution set. Case 2: $4 - 5x < 0$. This is equivalent to the assertion that $x > \frac{4}{5}$. Also, the original inequality takes the form $2 \leqq 5x - 4 \leqq 4$; that is, $6 \leqq 5x \leqq 8$, so that $\frac{6}{5} \leqq x \leqq \frac{8}{5}$. All these numbers satisfy the other condition that $x > \frac{4}{5}$, and hence Case 2 contributes the interval $\left[\frac{6}{5}, \frac{8}{5}\right]$ to the solution set. Therefore the solution set of the original inequality is $\left[0, \frac{2}{5}\right] \cup \left[\frac{6}{5}, \frac{8}{5}\right]$.

A.29: The given inequality

$$\frac{2}{7-3x} \leqq -5$$

implies that the denominator in the fraction is negative, so when we multiply both sides by that denominator we must reverse the inequality:

$$2 \geqq 15x - 35; \qquad 15x \leqq 37; \qquad x \leqq \frac{37}{15} \approx 2.466666666667.$$

In addition, we have the condition

$$7 - 3x < 0; \qquad 3x > 7; \qquad x > \frac{7}{3} \approx 2.333333333333.$$

So the original inequality has solution set $\left(\frac{7}{3}, \frac{37}{15}\right]$.

A.31: If $x \neq \frac{1}{5}$ then $|1 - 5x| > 0$, so that

$$\frac{1}{|1-5x|} > 0 > -\frac{1}{3},$$

and if $x = \frac{1}{5}$ then the left-hand side of the given inequality is undefined. Consequently the solution set we seek is $\left(-\infty, \frac{1}{5}\right) \cup \left(\frac{1}{5}, +\infty\right)$.

A.33: Given $x^2 - 2x - 8 > 0$, first factor: $(x + 2)(x - 4) > 0$. Case 1: $x > -2$ and $x > 4$, so that $x > 4$. Case 2: $x < -2$ and $x < 4$, so that $x < -2$. The cases are exclusive and exhaustive, so the solution set is $(-\infty, -2) \cup (4, +\infty)$.

A.35: Given: $4x^2 - 8x + 3 \geqq 0$, first factor: $(2x - 1)(2x - 3) \geqq 0$. Case 1: $2x - 1 \geqq 0$ and $2x - 3 \geqq 0$. Then $x \geqq \frac{1}{2}$ and $x \geqq \frac{3}{2}$, so that $x \geqq \frac{3}{2}$. Case 2: $2x - 1 \leqq 0$ and $2x - 3 \leqq 0$. Then $x \leqq \frac{1}{2}$ and $x \leqq \frac{3}{2}$, so that $x \leqq \frac{1}{2}$. Solution set: $\left(-\infty, \frac{1}{2}\right] \cup \left[\frac{3}{2}, +\infty\right)$.

A.37: Because $100 \leqq V \leqq 200$ and $pV = 800$, we have

$$100p \leqq 800 \leqq 200p; \qquad p \leqq 8 \quad \text{and} \quad 4 \leqq p; \qquad 4 \leqq p \leqq 8.$$

A.39: If $25 < R < 50$, then

$$25I < IR < 50I; \qquad 25I < 100 < 50I; \qquad I < 4 \quad \text{and} \quad 2 < I; \qquad 2 < I < 4.$$

A.41: Suppose that a and b are real numbers such that $0 < a$ and $0 < b$. By the second property in (1), $0 + b < a + b$; that is, $b < a + b$. But $0 < b$, so by the first property in (1), $0 < a + b$. Therefore the sum of two positive numbers is positive.

A.43: If a and b are both negative, then $a < 0$, so $a \cdot b > 0 \cdot b$ by the fourth property in (1). Thus ab is positive. If a is positive and b is negative, then $b < 0$, so $a \cdot b < a \cdot 0 = 0$ by the third property in (3). Thus $ab < 0$, and hence ab is negative.

A.45: If a, b, and c are real numbers, then

$$|a+b+c| = |(a+b)+c| \leqq |a+b| + |c| \leqq |a| + |b| + |c|.$$

That is, $|a+b+c| \leqq |a| + |b| + |c|$.

A.47: Suppose first that $|a| < b$. Then there are two cases. Case 1: $a \geqq 0$. Then $a < b$ and $b > 0$, so $-b < 0 < a < b$, and therefore $-b < a < b$. Case 2: $a < 0$. Then $-a < b$, so $b > 0$. Thus $-b < a < 0 < b$, so that $-b < a < b$. Therefore if $|a| < b$, then $-b < a < b$.

Next suppose that $-b < a < b$. Again there are two cases. Case 1: $a \geqq 0$. Then $a = |a|$, and thus $|a| < b$. Case 2: $a < 0$. Then $|a| = -a$. But from the hypothesis $-b < a < b$, it follows that

$$(-1) \cdot (-b) > (-1) \cdot a > (-1) \cdot b, \qquad \text{so that} \qquad -b < -a < b.$$

Hence $-b < |a| < b$, and consequently $|a| < b$.

Therefore $|a| < b$ if and only if $-b < a < b$. ◀

Appendix B

B.1: Both the segments AB and BC have the same slope 1. So the three given points do lie on a single straight line.

B.3: The segment AB has slope -2 but the segment BC has slope $-\frac{4}{3}$. Thus the three given points do not lie on a single straight line.

B.5: Both AB and CD have slope $-\frac{1}{2}$; both BC and DA have slope 2. This parallelogram is a rectangle!

B.7: AB has slope 2 and AC has slope $-\frac{1}{2}$. Hence triangle ABC has a right angle at the vertex A.

B.9: The given equation can be put in the form $y = \frac{2}{3}x$, so the line has slope $\frac{2}{3}$ and y-intercept 0.

B.11: The given equation takes the form $y = 2x + 3$, so the line has slope 2 and y-intercept 3.

B.13: The given equation can be put in the form $y = -\frac{2}{5}x + \frac{3}{5}$, and consequently the line with this equation has slope $-\frac{2}{5}$ and y-intercept $\frac{3}{5}$.

B.15: $y = -5$.

B.17: The slope of L is 2, so a point-slope equation of L is $y - 3 = 2(x - 5)$; its slope-intercept equation is $y = 2x - 7$.

B.19: The slope of L is $\tan 135° = -1$, so a point-slope equation of L is $y - 2 = -(x - 4)$; another equation of L is $x + y = 6$.

B.21: The equation $2x + y = 10$ can be written in the form $y = -2x + 10$, so the line with this equation has slope -2. So a parallel line passing through $(1, 5)$ has point-slope equation $y - 5 = -2(x - 1)$; the slope-intercept equation of this line is $y = -2x + 7$.

B.23: The given line segment has slope 2 and midpoint $(1, 6)$. Hence L has slope $-\frac{1}{2}$ and thus point-slope equation $y - 6 = -\frac{1}{2}(x - 1)$. Another equation of L is $x + 2y = 13$.

B.25: The parallel lines have slope 5, so the line $y = -\frac{1}{5}x + 1$ is perpendicular to both of them. It meets the line $y = 5x + 1$ at the point $(0, 1)$ and the line $y = 5x + 9$ at the point $\left(-\frac{20}{13}, \frac{17}{13}\right)$. The answer is the distance between the last two points, which is

$$\sqrt{\left(\frac{20}{13}\right)^2 + \left(\frac{4}{13}\right)^2} = \frac{4}{13}\sqrt{26} \approx 1.56892908110547225539.$$

B.27: With $A(-1, 6)$, $B(0, 0)$, $C(3, 1)$, and $D(2, 7)$ forming the consecutive vertices of the parallelogram, its diagonals are AC and BD. The midpoint of AC is $\left(1, \frac{7}{2}\right)$ and the midpoint of BD is the same. Therefore the diagonals of the parallelogram bisect each other.

B.29: The midpoint of AB is $P\left(\frac{5}{2}, 3\right)$ and the midpoint of BC is $Q(5, 4)$. The slope of PQ is $\frac{2}{5}$, the same as the slope of AC, and hence PQ is parallel to AC.

B.31: The slope of $P_1 M$ is

$$\frac{y_1 - \bar{y}}{x_1 - \bar{x}} = \frac{y_1 - \frac{1}{2}(y_1 + y_2)}{x_1 - \frac{1}{2}(x_1 + x_2)} = \frac{2y_1 - y_1 - y_2}{2x_1 - x_1 - x_2} = \frac{y_1 - y_2}{x_1 - x_2},$$

and the slope of MP_2 is the same.

B.33: Because F and K satisfy a linear equation, $K = mF + b$ for some constants m and b. From the data given in the statement of the problem, we find that

$$273.16 = 32m + b \quad \text{and}$$

$$373.16 = 212m + b.$$

Subtraction of the first equation from the second yields $180m = 100$, so that $m = \dfrac{5}{9}$. Then the first equation here yields

$$273.16 = \frac{160}{9} + b, \quad \text{so that} \quad b = \frac{6829}{25} - \frac{160}{9} = \frac{57461}{225} \approx 255.382222222222.$$

Therefore $K = \dfrac{1}{225}(125F + 57461)$ provided that $F > -459.688$.

B.35: If s denotes weekly sales in gallons and p the selling price per gallon, then $s = mp + b$ for some constants m and b. From the data given in the statement of the problem, we have

$$980 = (1.69)m + b \quad \text{and} \quad 1220 = (1.49)m + b. \tag{1}$$

Subtract the second of these equations from the first to find that

$$(0.20)m = -240, \quad \text{so that} \quad m = -1200.$$

Then substitution of this value of m in the first equation in (1) yields

$$980 = -2028 + b, \quad \text{and thus} \quad b = 3008.$$

Therefore $s = s(p) = -1200p + 3008$. At \$1.56 per gallon one would expect to sell $s(1.56) = 1136$ gallons per week.

B.37: Given the simultaneous equations

$$2x + 3y = 5,$$

$$2x + 5y = 12,$$

the *Mathematica* command

```
Solve[ { 2*x + 3*y == 5, 2*x + 5*y == 12 }, { x, y } ]
```

993

yields the exact solution $x = -\dfrac{11}{4}, \quad y = \dfrac{7}{2}.$

B.39: $x = \dfrac{37}{6}, \quad y = -\dfrac{1}{2}.$

B.41: $x = \dfrac{22}{5}, \quad y = -\dfrac{1}{5}.$

B.43: $x = -\dfrac{7}{4}, \quad y = \dfrac{33}{8}.$

B.45: $x = \dfrac{119}{12}, \quad y = -\dfrac{19}{4}.$

B.47: If (x_1, y_1) is a point of the plane not on the line L, then the line segment with endpoints (x_0, y_0) and (x_1, y_1) cannot have slope m. Therefore the equation $y_1 - y_0 = m(x_1 - x_0)$ cannot hold.

Appendix C

C.1: $40 \cdot \dfrac{\pi}{180} = \dfrac{2}{9}\pi \quad \text{(rad)}.$

C.3: $315 \cdot \dfrac{\pi}{180} = \dfrac{7}{4}\pi \quad \text{(rad)}.$

C.5: $-150 \cdot \dfrac{\pi}{180} = -\dfrac{5}{6}\pi \quad \text{(rad)}.$

C.7: $\dfrac{2\pi}{5} \cdot \dfrac{180}{\pi} = 72°.$

C.9: $\dfrac{15\pi}{4} \cdot \dfrac{180}{\pi} = 675°.$

C.11: If $x = -\dfrac{\pi}{3}$, then the values of the six trigonometric functions are given in the following table.

$\sin x$	$\cos x$	$\tan x$	$\sec x$	$\csc x$	$\cot x$
$-\dfrac{\sqrt{3}}{2}$	$\dfrac{1}{2}$	$-\sqrt{3}$	2	$-\dfrac{2\sqrt{3}}{3}$	$-\dfrac{\sqrt{3}}{3}$

C.13: If $x = \dfrac{7\pi}{6}$, then the values of the six trigonometric functions are given in the following table.

$\sin x$	$\cos x$	$\tan x$	$\sec x$	$\csc x$	$\cot x$
$-\dfrac{1}{2}$	$-\dfrac{\sqrt{3}}{2}$	$\dfrac{\sqrt{3}}{3}$	$-\dfrac{2\sqrt{3}}{3}$	-2	$\sqrt{3}$

C.15: Solutions: $x = n\pi$ where n is an integer.

C.17: Solutions: $x = 2n\pi - \dfrac{\pi}{2}$ where n is an integer.

C.19: Solutions: $x = 2n\pi$ where n is an integer.

C.21: Solutions: $x = n\pi$ where n is an integer.

C.23: Solutions: $x = n\pi - \dfrac{\pi}{4}$ where n is an integer.

C.25: Draw a right triangle with acute angle x and label the side opposite x as -3, the side adjacent x as 4, and the hypotenuse as 5. Then the values of the six trigonometric functions at x can be read from this triangle. The results appear in the next table.

$\sin x$	$\cos x$	$\tan x$	$\sec x$	$\csc x$	$\cot x$
$-\dfrac{3}{5}$	$\dfrac{4}{5}$	$-\dfrac{3}{4}$	$\dfrac{5}{4}$	$-\dfrac{5}{3}$	$-\dfrac{4}{3}$

C.27: Begin with the identity $\sin^2\theta + \cos^2\theta = 1$:

$$\frac{\sin^2\theta}{\sin^2\theta} + \frac{\cos^2\theta}{\sin^2\theta} = \frac{1}{\sin^2\theta};$$

$$1 + \cot^2\theta = \csc^2\theta.$$

C.29: The sine addition formula yields

$$\sin\frac{5\pi}{6} = \sin\left(\frac{\pi}{2} + \frac{\pi}{3}\right) = \left(\sin\frac{\pi}{2}\right)\cdot\left(\cos\frac{\pi}{3}\right) + \left(\cos\frac{\pi}{2}\right)\cdot\left(\sin\frac{\pi}{3}\right) = 1\cdot\frac{1}{2} + 0\cdot\frac{\sqrt{3}}{2} = \frac{1}{2}.$$

C.31: The sine addition formula yields

$$\sin\frac{11\pi}{6} = \sin\left(2\pi - \frac{\pi}{6}\right) = (\sin 2\pi)\cdot\left(\cos\frac{\pi}{6}\right) - (\cos 2\pi)\cdot\left(\sin\frac{\pi}{6}\right) = 0\cdot\frac{\sqrt{3}}{2} - 1\cdot\frac{1}{2} = -\frac{1}{2}.$$

C.33: By the sine addition formula,

$$\sin\frac{2\pi}{3} = \sin\left(\frac{\pi}{2} + \frac{\pi}{6}\right) = \left(\sin\frac{\pi}{2}\right)\cdot\left(\cos\frac{\pi}{6}\right) + \left(\cos\frac{\pi}{2}\right)\cdot\left(\sin\frac{\pi}{6}\right) = 1\cdot\frac{\sqrt{3}}{2} + 0\cdot\frac{1}{2} = \frac{\sqrt{3}}{2}.$$

C.35: The sine addition formula yields

$$\sin\frac{5\pi}{3} = \sin\left(2\pi - \frac{\pi}{3}\right) = (\sin 2\pi)\cdot\left(\cos\frac{\pi}{3}\right) - (\cos 2\pi)\cdot\left(\sin\frac{\pi}{3}\right) = 0\cdot\frac{1}{2} - 1\cdot\frac{\sqrt{3}}{2} = -\frac{\sqrt{3}}{2}.$$

C.37: Part (a):

$$\cos\left(\frac{\pi}{2} - \theta\right) = \left(\cos\frac{\pi}{2}\right)\cdot\cos\theta + \left(\sin\frac{\pi}{2}\right)\cdot\sin\theta = 0\cdot\cos\theta + 1\cdot\sin\theta = \sin\theta.$$

Part (b):

$$\sin\left(\frac{\pi}{2} - \theta\right) = \left(\sin\frac{\pi}{2}\right) \cdot \cos\theta - \left(\cos\frac{\pi}{2}\right) \cdot \sin\theta = 1 \cdot \cos\theta - 0 \cdot \sin\theta = \cos\theta.$$

Part (c): We use the results in parts (a) and (b).

$$\cot\left(\frac{\pi}{2} - \theta\right) = \frac{\cos\left(\frac{\pi}{2} - \theta\right)}{\sin\left(\frac{\pi}{2} - \theta\right)} = \frac{\sin\theta}{\cos\theta} = \tan\theta.$$

C.39: The cosine addition formula yields

$$\cos(\pi \pm \theta) = \cos\pi\cos\theta \mp \sin\pi\sin\theta = (-1)\cdot\cos\theta \mp 0\cdot\sin\theta = -\cos\theta.$$

C.41: Because AC and BD have the same length, we use the distance formula and obtain the equation

$$(\cos\phi - \cos\theta)^2 + (\sin\phi + \sin\theta)^2 = [1 - \cos(\theta + \phi)]^2 + [0 - \sin(\theta + \phi)]^2;$$

$$\cos^2\phi - 2\cos\phi\cos\theta + \cos^2\theta + \sin^2\phi + 2\sin\phi\sin\theta + \sin^2\theta$$

$$= 1 - 2\cos(\theta + \phi) + \cos^2(\theta + \phi) + \sin^2(\theta + \phi);$$

$$2 - 2\cos\phi\cos\theta + 2\sin\phi\sin\theta = 2 - 2\cos(\theta + \phi);$$

$$\cos(\theta + \phi) = \cos\phi\cos\theta - \sin\phi\sin\theta.$$

C.43: Let $u = \sin x$. Then

$$3u^2 - 1 + u^2 = 2; \qquad 4u^2 = 3; \qquad \sin x = u = \pm\frac{\sqrt{3}}{2}.$$

Thus the solutions in $[0, \pi]$ are $x = \frac{\pi}{3}$ and $x = \frac{2\pi}{3}$.

C.45: Let $u = \sin x$. Then

$$2(1 - u^2) + 3u^2 = 3; \qquad u^2 = 1; \qquad \sin x = u = \pm 1.$$

The only solution in $[0, \pi]$ is $x = \frac{\pi}{2}$.

C.47: If $8\sin^2 x \cos^2 x = 1$, then:

$$\sin^2 x \cos^2 x = \frac{1}{8}; \qquad \sin x \cos x = \pm\frac{\sqrt{2}}{4};$$

$$2\sin x \cos x = \pm\frac{\sqrt{2}}{2}; \qquad \sin 2x = \pm\frac{\sqrt{2}}{2};$$

$$2x = \frac{\pi}{4}, \frac{3\pi}{4}, \frac{5\pi}{4}, \frac{7\pi}{4}; \qquad x = \frac{\pi}{8}, \frac{3\pi}{8}, \frac{5\pi}{8}, \frac{7\pi}{8}.$$

Appendix D

D.1: Given $\epsilon > 0$, let $\delta = \epsilon$. Suppose that $0 < |x - a| < \delta$. Then $|x - a| < \epsilon$. Therefore

$$\lim_{x \to a} x = a.$$

D.3: Given $\epsilon > 0$, let $\delta = \epsilon$. Suppose that $0 < |x - 2| < \delta$. Then $|(x + 3) - 5| < \delta = \epsilon$. Therefore

$$\lim_{x \to 2} (x + 3) = 5.$$

D.5: Given $\epsilon > 0$, let δ be the minimum of 1 and $\epsilon/3$. Suppose that $0 < |x - 1| < \delta$. Then $|x - 1| < 1$; consequently,

$$-1 < x - 1 < 1;$$

$$1 < x + 1 < 3;$$

$$|x + 1| < 3.$$

Moreover, if $|x - 1| < \delta$ then $|x - 1| < \epsilon/3$. Hence

$$|x^2 - 1| = |x + 1| \cdot |x - 1| < 3 \cdot \frac{\epsilon}{3} = \epsilon.$$

Therefore, by definition, $\lim_{x \to 1} x^2 = 1$.

D.7: Given $\epsilon > 0$, let δ be the minimum of 1 and $\epsilon/6$. Suppose that $0 < |x - (-1)| < \delta$. Then $|x + 1| < \delta$, and so

$$|x + 1| < 1; \quad -1 < x + 1 < 1; \quad -3 < x - 1 < -1 < 3; \quad |x - 1| < 3.$$

Hence

$$|(2x^2 - 1) - 1| = 2|x - 1| \cdot |x + 1| < 2 \cdot 3 \cdot \frac{\epsilon}{6} = \epsilon.$$

Therefore, by definition, $\lim_{x \to -1} (2x^2 - 1) = 1$.

D.9: Case 1: $a > 0$. Let $\epsilon > 0$ be given. Choose

$$\delta = \min\left\{\frac{a}{2}, \frac{(a^2 + 1)(a^2 + 4)}{10a}\epsilon\right\}.$$

Suppose that $0 < |x - a| < \delta$. Then

$$|x - a| < \frac{a}{2}; \qquad -\frac{a}{2} < x - a < \frac{a}{2}; \qquad \frac{a}{2} < x < \frac{3a}{2}.$$

Therefore $x > 0$ and $x + a > 0$. In addition,

$$\frac{3a}{2} < x + a < \frac{5a}{2}; \qquad |x + a| < \frac{5a}{2}.$$

Also

$$\frac{a^2}{4} < x^2 < \frac{9a^2}{4}; \qquad x^2 > \frac{a^2}{4}; \qquad x^2 + 1 > \frac{a^2 + 4}{4}; \qquad (x^2 + 1)(a^2 + 1) > \frac{(a^2 + 1)(a^2 + 4)}{4}.$$

Consequently,

$$\left| \frac{1}{x^2 + 1} - \frac{1}{a^2 + 1} \right| = \frac{|x^2 - a^2|}{|(x^2 + 1)(a^2 + 1)|} = \frac{|x - a| \cdot |x + a|}{(x^2 + 1)(a^2 + 1)} < \frac{5a}{2} \cdot \frac{4}{(a^2 + 1)(a^2 + 4)} \cdot \delta$$

$$= \frac{10a}{(a^2 + 1)(a^2 + 4)} \cdot \delta \leq \frac{10a}{(a^2 + 1)(a^2 + 4)} \cdot \frac{(a^2 + 1)(a^2 + 4)}{10a} \cdot \epsilon = \epsilon.$$

Therefore $\displaystyle \lim_{x \to a} \frac{1}{x^2 + 1} = \frac{1}{a^2 + 1}$ if $a > 0$.

Case 2: $a < 0$. The proof is similar to the one given in Case 1.

Case 3: $a = 0$. In this case we are to show that

$$\lim_{x \to 0} \frac{1}{x^2 + 1} = 1. \tag{1}$$

Given $\epsilon < 0$, let $\delta = \sqrt{\epsilon}$. Suppose that $0 < |x - 0| < \delta$. Then $x^2 < \epsilon$. Therefore

$$\left| \frac{1}{x^2 + 1} - 1 \right| = \left| \frac{x^2}{x^2 + 1} \right| = \frac{x^2}{x^2 + 1} < x^2 < \epsilon.$$

This establishes Eq. (1).

D.11: Given $\epsilon > 0$, note that $\epsilon/2$ is also positive. Choose δ_1 so that

$$|f(x) - L| < \frac{\epsilon}{2} \quad \text{if} \quad 0 < |x - a| < \delta_1;$$

choose δ_2 so that

$$|f(x) - M| < \frac{\epsilon}{2} \quad \text{if} \quad 0 < |x - a| < \delta_2.$$

Then for such x we have

$$|L - f(x)| + |f(x) - M| < \frac{\epsilon}{2} + \frac{\epsilon}{2} = \epsilon,$$

so by the triangle inequality

$$|L - M| = |L - f(x) + f(x) - M| < \epsilon.$$

Thus L and M differ by an amount smaller than any positive real number. Therefore $L = M$.

D.13: We treat only the case $L > 0$; the case $L < 0$ is quite similar. Let $\epsilon > 0$ be given. Choose $\delta > 0$ so small that

$$|f(x) - L| < \frac{L}{2} \quad \text{and} \quad |f(x) - L| < \frac{L^2 \epsilon}{2}$$

if $0 < |x - a| < \delta$. Then

$$-\frac{L}{2} < f(x) - L < \frac{L}{2}, \quad \text{so that} \quad \frac{L}{2} < f(x) < \frac{3L}{2};$$

moreover, $f(x) > 0$. Also

$$\frac{L^2}{2} < L \cdot f(x), \quad \text{and hence} \quad \frac{1}{|L \cdot f(x)|} < \frac{2}{L^2}.$$

Therefore, if $0 < |x - a| < \delta$, then

$$\left| \frac{1}{f(x)} - \frac{1}{L} \right| = \frac{|f(x) - L|}{|L \cdot f(x)|} < \frac{2}{L^2} \cdot \frac{L^2 \epsilon}{2} = \epsilon.$$

Consequently $\displaystyle\lim_{x \to a} \frac{1}{f(x)} = \frac{1}{L}$.

D.15: Suppose that $a > 0$ and let $\epsilon > 0$ be given. Let

$$\delta = \min\left\{ \frac{a}{2}, \epsilon\sqrt{2a} \right\}.$$

Suppose that $0 < |x - a| < \delta$. Then

$$|x - a| < \frac{a}{2}, \quad \text{so that} \quad \frac{a}{2} < x < \frac{3a}{2};$$

moreover, $x > 0$. Hence

$$\sqrt{\frac{a}{2}} < \sqrt{x} < \sqrt{\frac{3a}{2}};$$

$$\sqrt{\frac{a}{2}} + \sqrt{a} < \sqrt{x} + \sqrt{a} < \sqrt{\frac{3a}{2}} + \sqrt{a};$$

$$\frac{1}{\sqrt{x} + \sqrt{a}} < \frac{1}{\sqrt{\frac{a}{2}} + \sqrt{a}} < \frac{1}{\sqrt{2a}}.$$

So if $0 < |x - a| < \delta$, then

$$|\sqrt{x} - \sqrt{a}| = \frac{|x - a|}{|\sqrt{x} + \sqrt{a}|} < \frac{|x - a|}{\sqrt{2a}} \leq \frac{\epsilon\sqrt{2a}}{\sqrt{2a}} = \epsilon.$$

Therefore $\displaystyle\lim_{x \to a} \sqrt{x} = \sqrt{a}$ if $a > 0$.

Appendix G

G.1: The limit is

$$\int_0^1 x^2 \, dx = \left[\frac{1}{3} x^3 \right]_0^1 = \frac{1}{3}.$$

G.3: The limit is $\displaystyle\int_0^2 x\sqrt{4-x^2} \, dx = \left[-\frac{1}{3}(4-x^2)^{3/2} \right]_0^2 = \frac{8}{3}.$

G.5: The limit is $\displaystyle\int_0^{\pi/2} \sin x \cos x \, dx = \left[\frac{1}{2} \sin^2 x \right]_0^{\pi/2} = \frac{1}{2}.$

G.7: The limit is

$$\int_0^2 \sqrt{x^4 + x^7} \, dx = \int_0^2 x^2 (1+x^3)^{1/2} \, dx = \left[\frac{2}{9}(1+x^3)^{3/2} \right]_0^2 = \frac{2}{9}(27-1) = \frac{52}{9} \approx 5.7777777777777778.$$

G.9: Take $G(y) = 2\pi y$ and $H(y) = \sqrt{1 + \left[g'(y) \right]^2}$ in Theorem 1. Then

$$\sum_{i=1}^n 2\pi y_i^{**} \sqrt{1 + \left[g'(y_i^\star) \right]^2} \, \Delta y \to A = \int_c^d 2\pi y \sqrt{1 + \left[g'(y) \right]^2} \, dy$$

by Theorem 1.